S0-AIT-596

STANDARD HANDBOOK OF PLANT ENGINEERING

STANDARD HANDBOOK OF PLANT ENGINEERING

Robert C. Rosaler, P.E. Editor-in-Chief

Third Edition

McGRAW-HILL

New York Chicago San Francisco Lisbon London Madrid
Mexico City Milan New Delhi San Juan Seoul
Singapore Sydney Toronto

Library of Congress Cataloging-in-Publication Data

Standard handbook of plant engineering / Robert C. Rosaler, editor in chief.—3rd ed.
p. cm.
Includes index.
ISBN 0-07-136192-8
1. Plant engineering—Handbooks, manuals, etc. I. Rosaler, Robert C.
TS184.S7 2002
658.2—dc21 2001044874

McGraw-Hill 

*A Division of The **McGraw-Hill** Companies*

6 7 8 9 0 QVR/QVR 0 7 6 5 4 3

ISBN 0-07-136192-8

The sponsoring editor for this book was Kenneth P. McCombs, the editing supervisor was M. R. Carey, and the production supervisor was Sherri Souffrance. It was set in Times Roman by North Market Street Graphics.

Printed and bound by R. R. Donnelley & Sons Company.

This book is printed on acid-free paper.

CONTENTS

Section 5 Basic Maintenance Technology 5.1

Section 6 In-Plant Prime Power Generation and Cogeneration 6.1

BOARD OF ADVISORS

Roger Engstrom *Protek/CA, San Jose, California*

Richard Greco *EYP Mission Critical Facilities, Inc., San Francisco, California*

William Jackson *Consulting Engineer, Dallas, Texas*

William Maggard *Consulting Engineer, Blountville, Tennessee*

Mike Williamsen *Consulting Engineer, Cobden, Illinois*

CONTRIBUTORS

John C. Andreas *Consultant, MagneTek, St. Louis, Missouri*

Frank A. Baczek *Baker Process, Salt Lake City, Utah*

Robert L. Bays *Scientist, Shared Technical Services, Ecolab Professional Products, St. Paul, Minnesota*

William N. Berryman *Engineering Consultant, Morgan Hill, California*

Hugh Blackwell *Alcoa/Mt. Holly, Goose Creek, South Carolina*

William A. Bradley III *American Gear Manufacturers Association (AGMA), Alexandria, Virginia*

Building Commissioning Association *Seattle, Washington*

Michael C. Carey *Roy F. Weston, Inc., West Chester, Pennsylvania*

William H. Conner *Elf Lubricants, Linden, New Jersey*

Paul Crawford *Zenon Environmental, Inc., Oakville, Ontario*

R. B. Curry *Senior Engineer, Rexnord Corporation, Indianapolis, Indiana*

Dresser Valve Division *Stafford, Texas*

Richard J. Eisman *Coombs-Hopkins Company, Carlsbad, California*

B. W. Elliott Manufacturing Co., L.L.C. *Binghamton, New York*

Robert Emmett *Emmett Process Consulting Company, Salt Lake City, Utah*

Envirex Products *U.S. Filter, Waukesha, Wisconsin*

Peter M. Fairbanks, P.E. *Bluestone Energy Services, Inc., Braintree, Massachusetts*

Fasco Motors Group *Eldon, Missouri*

James R. Fox, P.E. *Arcadis-Giffels, Inc., Southfield, Michigan*

Ricardo R. Gamboa *Manager, Engineering and Technical Services, Manville Mechanical Specialty Insulations, Schuller International, Inc., Denver, Colorado*

General Battery Corporation *Reading, Pennsylvania*

Ernest H. Graf, P.E. *Assistant Director, Mechanical Engineering, Giffels Associates, Inc., Southfield, Michigan*

Casey C. Grant *Assistant Vice President, Codes and Standards Administration, National Fire Protection Association, Quincy, Massachusetts*

Michael R. Harrison *Vice President and General Manager, Manville Mechanical Specialty Insulations, Schuller International, Inc., Denver, Colorado*

C. Phillip Headley *Technical Specialist, Waste Equipment Technology Association (WASTEC), Washington, D.C.*

David H. Helwig *Arcadis-Giffels, Inc., Southfield, Michigan*

Terry Hoffmann *Johnson Controls, Inc., Milwaukee, Wisconsin*

Industrial Division Engineering Staff *Torrington Company, Torrington, Connecticut*

Jim Iverson *Manager, Technical Marketing, Industrial Business Group, Onan Corporation, Minneapolis, Minnesota*

William V. Jackson *President, H. H. Felton & Associates, Dallas, Texas*

Wayne H. Lawton *Arcadis-Giffels, Inc., Southfield, Michigan*

William S. Lytle, P.E. *Project Engineer, Mechanical Engineering, Giffels Associates, Inc., Southfield, Michigan*

Mary Martis *Water 3 Engineering, Escondido, California*

John McNary *Senior Industrial Hygienist, Clayton Group Services, Inc., Santa Ana, California*

Peter Metcalf *Koch Industries, San Diego, California*

Russell N. Mosher *Assistant Executive Director, American Boiler Manufacturers Association, Arlington, Virginia*

Stanley A. Mruk *Executive Director, Plastics Pipe Institute, Division of the Society of the Plastics Industry, Washington, D.C.*

National Electrical Contractors Association, Inc. *Washington, D.C.*

William C. Newman *Arcadis-Giffels, Inc., Southfield, Michigan*

Robert W. Okey *University of Utah, Salt Lake City, Utah*

Brent Oman *Application Engineer, Power Transmission Applications Department, Gates Rubber Company, Denver, Colorado*

P. Eric Ralston *Vice President and General Manager, Environmental Equipment Division, Babcock and Wilcox, Barberton, Ohio*

Paul C. Siebert *Roy F. Weston, Inc., West Chester, Pennsylvania*

Simpson Electric Company *Elgin, Illinois*

J. Stephen Slottee *Baker Process, Salt Lake City, Utah*

Gary Smith *Director, Product Management, Venture Lighting Technologies, Inc., Solon, Ohio*

Douglas H. Sturz *Senior Acoustical Consultant, Acentech, Inc., Cambridge, Massachusetts*

Donald C. Taylor *Baker Process, Salt Lake City, Utah*

Clemens M. Thoennes *Sales Programs Development, General Electric Company, Schenectady, New York*

Tower Performance, Inc. *Florham Park, New Jersey*

K. W. Tunnell Co., Inc. *King of Prussia, Pennsylvania*

Eric E. Ungar *Chief Engineering Scientist, Acentech, Inc., Cambridge, Massachusetts*

Jose L. Villalobos *President, V & A Consulting Engineers, Oakland, California*

Robert C. Walther, P.E. *President, Industrial Power Technology, Santa Rosa, California*

Waukesha Engine Division *Dresser Industries, Inc., Waukesha, Wisconsin*

Travis West *President, Building Air Quality, The Woodlands, Texas*

Thomas E. Wiles *Golf System, Inc., Plano, Texas*

S. E. Winegardner *Senior Engineer, Rexnord Corporation, Indianapolis, Indiana*

James R. Wright, P.E. *Manager of Codes and Standards, Siemens Energy & Automation, Inc., Batavia, Illinois*

Jeffrey A. Zapfe *Senior Consultant, Acentech, Inc., Cambridge, Massachusetts*

PREFACE TO THIRD EDITION

This volume is dedicated to my three young grandsons, Jonathan, Gregory, and Maxwell, whose generation holds the keys to a safe, just, prosperous, and tolerant new world. I hope this book contributes to that lofty ideal.

I express my appreciation to the authors and their sponsoring organizations for their contributions. Also, my thanks to the Board of Advisors and McGraw-Hill editor Ken McCombs for their advice and guidance, as well as to M. R. Carey of North Market Street Graphics for an outstanding copyediting job.

PROTECTING FACILITIES PERSONNEL AGAINST TERRORIST ATTACKS

As this book is nearing publication, our country has been subjected to a series of unprovoked attacks ranging from the deliberate destruction of several buildings, with accompanying loss of life, to the deliberate spreading of toxic anthrax spores, along with threats to broaden these actions.

Our plant engineering colleagues have responded to these events by offering specific advice to minimize the impact of the spread of toxic biological substances:

- Keep the ventilation system on, so as to maintain indoor air pressure slightly higher than that outside.
- Add high-efficiency particulate air (HEPA) filters to the HVAC system immediately.

By the time you read this, other procedures will have been suggested as well.

The reader is urged to implement this advice. Also, it is important to maintain close communications with your engineering society to keep abreast of this problem and its solutions.

CONTACTS

American Society of Heating, Refrigerating and Air-Conditioning Engineers (ASHRAE): (404) 636-8400; *www.ashrae.org*

American Society of Mechanical Engineers (ASME) International: (800) 843-2763; *www.asme.org*

Association for Facilities Engineering (AFE): (513) 489-2473; *www.afe.org*

Robert C. Rosaler, P.E.

PREFACE TO SECOND EDITION

In addition to updating all technical discussions, this Second Edition adds coverage of plant engineering management, with particular stress on maintenance and workplace safety. This reflects the increasing role of the plant engineer in corporate management.

The editor expresses appreciation for the excellent cooperation shown by the authors and their organizations in updating or entirely rewriting individual chapters as appropriate. Also, a thank you to the Board of Advisors of this edition, who helped redirect the emphasis of the work. Finally, a word of gratitude to Robert W. Hauserman and Stephen M. Smith of McGraw-Hill, who willingly provided wise editorial advice.

This Second Edition is dedicated to my dear departed brother George, who embodied all that is good in the world.

Robert C. Rosaler, P.E.

PREFACE TO FIRST EDITION

Virtually every industrial activity has been affected, often in revolutionary ways, by the surge of technology. Because it is so central to virtually all manufacturing and service facilities, plant engineering is *uniquely* affected. It is "in the middle" in the sense that the plant engineer must, increasingly, have a broader knowledge of an ever-widening universe.

Events of the past decade have further served to accent the importance of the plant engineer's role in corporate operations, notably the demands for energy conservation and pollution control.

This Handbook is a response to these changing conditions and needs.

Arranging a logical structure and index to meet the needs of all engineers required considerable thought. The structure finally developed here is a reflection of the procedural sequences that occur in the plant facility itself: Planning and Construction, Plant Equipment Procurement and Operation, Maintenance. Individual equipment is covered broadly with descriptions of operational features, installation, and maintenance. Managerial aspects are included only where they interface closely with technical matter.

The objective of the book is to provide the reader with sufficient data on any specific equipment to permit judgment on choices and an insight into "how it works" and how to maintain it.

We want to express our appreciation to the authors and their organizations for their generous contributions and prompt execution of their tasks, and to the Board of Advisors for their guidance, particularly Leo Spector, Chairman of the Board and Editor of *Plant Engineering* magazine. For initial suggestions on the outline, we wish to thank Stewart Burkland. We also received excellent guidance and encouragement from Harold Crawford, Ruth Weine, and M. Joseph Dooher of McGraw-Hill. Our thanks go, too, to Dorothy Smith and Betsy Watson for helping to keep the project moving.

Saul Poliak, to whom this Handbook is dedicated, founded the National Plant Engineering Exposition and Conference which has been held both nationally and regionally since 1950. He has also pioneered similar expositions and conferences in the United Kingdom, Europe, Central America, and the Far East. It is widely recognized that he has been a major force in advancing the awareness of the critical plant engineering function in the industrial societies.

Robert C. Rosaler, P.E.
James O. Rice

STANDARD HANDBOOK OF PLANT ENGINEERING

SECTION 1

THE PLANT ENGINEER AND THE ORGANIZATION

CHAPTER 1.1
OBJECTIVES AND PHILOSOPHY

Hugh Blackwell
Alcoa/Mt. Holly
Goose Creek, South Carolina

INTRODUCTION

The degree of success of a plant engineer will be measured not by his or her ability to recite equations, balance budgets, complete capital projects, or maintain equipment, but by the ability to lead others in the face of insufficient personnel, resources, and time to do the job comfortably. In years past, internal workloads determined our pace of progress. Today, external information and customer demands drive behavior and pace. In order to successfully *manage* information and *lead* people, plant engineers must:

- Be a part of the management team
- Know the workforce culture
- Understand and implement strategic planning
- Thrive, not survive

How one goes about addressing and prioritizing these concepts will determine the success or failure of the organization.

THE MANAGEMENT TEAM

The term *management* is misleading because it implies that one is managing *people.* In fact, people don't follow people (managers), they follow *vision.* Therefore, the key to a successful management team is not in its ability to tell people what to do but in its ability to help them align their vision with that of the overall organization.

It has been said that organizations are much like people. Both have five senses: purpose, community, urgency, responsibility, and commitment. A sense of purpose refers to *mission* and *vision.* As a plant engineer, you need to ask yourself why you are there. Do your personal goals align with those of your organization? If not, one of three things is apt to happen: you will either convert your goals to those of the organization, comply with them because they allow you to remain in your "comfort zone," or you will eventually leave.

A sense of community simply means don't reinvent the wheel! Many others have gone before us. How did they do it? Cross-functional teams are a great way of accelerating the learning process. Having access to the Internet or to intranet web sites is another great way of creating a sense of community. The success of any engineering or maintenance organization hinges on its ability to *communicate* through crucial windows of opportunity. D. Edward Deming once said, "There's no such thing as instant pudding!" Developing a sense of community is absolutely crucial and essential to long-term survival and growth. It doesn't happen overnight, as a result of a promotion, or with a change in top management.

Probably the most important of the five senses is the sense of *urgency.* The leaders among our ranks must have a sense to *act,* not to *wait.* We've all heard that there are three types of people: those who *make* things happen, those who *watch* things happen, and those who *wonder* what happened. The speed at which engineering organizations advance will be measured not by the number of computer programs or software packages employed, but by the speed at which people learn and apply new technologies and concepts.

Simply stated, priorities change. Therefore, we must be flexible. We must be willing to "get out of the box," yet stay within the realm of reality. The concept of *breakthrough thinking* comes to mind. Paraphrasing what Joel Barker once said, "We will live out the remainder of our working lives in a state of change." In many cases, there is no longer time to adapt our processes to new demands; rather, we should adopt *new* processes and concepts.

One definition of insanity is "doing the same thing over and over yet expecting different results." If we expect or desire different results, we must do things differently. Said another way, if you don't like what you're getting from others, change what they are doing. Most people naturally resist change; therefore, a sense of urgency is essential to identifying the sources of resistance to change so progress can begin. Don't spend all your time trying to manage change. Instead, *plan* for change. None of us has a crystal ball, but time spent thinking about the future is better spent than thinking about the past or present. It's much easier to plan for change than to change plans.

The word *responsibility* brings to mind two words: *leadership* and *accountability.* A sense of responsibility is accepting accountability for your actions and the outcome of your work. It matters not whether you are a process engineer, project engineer, plant engineer, engineering manager, engineering team leader, or corporate vice president of engineering. We all are leaders at various times. Engineers often lead bid meetings, frequent project reviews, periodic budget reviews, safety briefs, and postproject completion reviews.

Leadership should be an *enabler* to success, not a push to get things done. Enabling leaders do two things well—they both create and sustain an environment where people can grow professionally and personally. Enabling leaders don't focus on doing just the right things, but on doing things right! Success is a shared responsibility.

Last, but not least, is creating a sense of *commitment.* Commitment is cooperation with communication. As you communicate with others, ask yourself these three questions: why are you here, what do you want, and what have you learned? We are all in the business of lifelong learning. So, if your answers to these three questions are not consistent with your personal mission statement and aligned with the organization's vision, you've got an important decision to make.

It's often been said, "You are what you do, not what you say!" Leading by example is the best measure of commitment. Vince Lombardi once said, "It's not whether you get knocked down, but whether or not you get back up." Commitment and continuous improvement go hand in glove. Not unlike encouragement, commitment is a gift we give each other.

THE WORKFORCE CULTURE

Plant engineers must know the culture of the workforce. How do things get done around here? Many hierarchical organizations of the past are gone, replaced by flatter and more flexible relational organizations. Today, many plant engineers effectively get their work done horizontally rather than vertically. Successful engineering organizations have commonly shared values (at all levels within the organization), identified key-results areas, and dynamic metrics to track performance.

Values tell us how to accomplish our mission. In short, values govern behavior. Unfortunately, all organizations have embedded cultural filters that filter ideas, information, and data. Once filtered, ideas yield action and drive results. *Proactive* plant engineers want ideas (based on sound values) governing future operations. It has been said that managing an operation from behind a desk is a dangerous thing. To be understood, your values must be seen on the shop floor by your actions and involvement in day-to-day activity, not by your title or level of education.

Values also provide a common language for aligning leadership with rest of the organization. It is the plant engineer's sole responsibility to define and document the values of the engineering organization. Typically, these values include such things as involvement and participation, continuous improvement, a focus on people, maintaining levels of quality, exceeding customer expectations, and maintaining an awareness of costs. Once understood by all, values not only govern behavior, they also define "organizational north."

Organizations that base their vision on values seldom fail. With a clear vision, values lead to ideas and results. Without a clear vision, values aren't important and outcomes are uncertain or unpredictable. Do your homework. Share your values and ideas with others. Admiral Hyman Rickover, renowned as the father of the nuclear Navy, once said, "Simple minds discuss people, average minds discuss events, great minds discuss ideas."

STRATEGIC PLANNING

Within the past 5 to 8 years, there has been a tremendous amount of activity within industry centered on the concept of strategic planning. The concept is not new, but getting the entire organization involved in the process *is* a change from the past. It's often referred to as "genius-level thinking"—that is, no one person is smarter than the collective experience and knowledge of a group of people. Collectively, we are smarter than any one of us alone. The success of strategic planning is attributable to just that—genius-level thinking at the group level.

Organizations without strategic plans are at risk. Topics typically addressed in strategic plans include such things as safety, revenue, facilities, infrastructure, information systems, competition, and customers. The key to successful strategic planning lies in the timely execution of related tactics, but each of these topics is important for the following reasons:

- *Safety.* People are still getting hurt.
- *Revenue.* Long-term price declines are prevalent.
- *Facilities.* Older plants cost more to sustain and maintain.
- *Infrastructure.* Reliable equipment is essential to profitability.
- *Competition.* It's global and getting tougher.
- *Customers.* They are demanding more for less.

The bottom-line purpose of strategic planning is to ensure long-term viability and growth, the cornerstones of which are *quality* to customers, *returns* to owners, and *opportunities* for employees. None of these happen in a vacuum and none should be a strange concept to a plant engineer. In short, plant engineers must be actively engaged in strategic planning, not stereotyped as just a technical resource when needed.

THRIVING, NOT SURVIVING

Successful plant engineers of the twenty-first century will be those who are regarded as thrivers, not survivors. *Survivors* tend to stay out of sight and do only what is asked. Although strong technically, they are not change agents and tend to do things the way they have always done them. *Thrivers,* on the other hand, typically bring energy, insightfulness, concern for the future, and recognition to individuals and groups. They work to become part of the management team that adds value to the bottom line. Their contribution to profitability is by design, not coincidence. Thrivers aren't consumed by process changes—they invent them.

In the absence of good reliable information, perception becomes reality. Perceptions are not right or wrong, but they *are* good and bad. Too often plant engineers are looked at as being comfortable, passive, and unimaginative. None of these conditions is remotely related to reality in a progressive organization. We operate in a worldwide competitive market governed by four *C*s: continuously changing, competitive climate. The plant engineering organization must keep ahead of the game or get out of the way. An engaged plant engineering group can see the direct relationship between what it does daily and the financial impact on the company's bottom line.

Plant engineers must understand the business case for action. Again, why are we here? Determine the current condition. How are things done around here? Are there opportunities for improvement? If so, define the target condition. What's possible and achievable? The key to this improvement process is developing a realistic action plan to get from the current to the target condition. Timing is everything.

Successful plant engineers know and understand the following very clearly:

- The current and desired state of the engineering function
- The bottom-line impact on plant profitability
- Their internal vision of the future
- Their mission, vision, and organizational values

Don't underestimate the power of values. Values govern behavior ("walk the talk"). Behavior defines your work ethic (what gets measured gets done). Work ethics enable profitability (continuous improvement). Profitability drives survivability (carried out by thrivers). And survivability overcomes the competition (benchmark the best).

SUMMARY

In summary, folks on the floor want leadership by example, not leadership lip service. The folks in the front office want acceptable returns on investment, not cost overruns. The folks under your charge want a caring, consistent, enabling leader who can create a sense of urgency when needed, understand and share concerns, communicate up and down the line, energize folks for broad-based action, focus on short- and long-term results, and never lose his or her sense of humor. Engineering organizations that thrive are characterized by the following six attributes:

- Work is interesting and challenging.
- People, not events, make the decisions.
- Management sets the direction, and then gets out of the way.
- Paradigms are allowed and encouraged to shift.
- Success is a shared responsibility.
- Confidence, not comfort, is sought.

Where will you be when margins are close? The challenge is real, and the choice is yours.

CHAPTER 1.2
THE PLANT ENGINEERING ORGANIZATION

William V. Jackson
President, H.H. Felton & Associates
Dallas, Texas

In 1983, when the first edition of the *Standard Handbook of Plant Engineering* was published, a discussion of the structure of the plant engineering organization would have been straightforward. Organizational design parameters would have centered on the size of the plant, the relative size of the maintenance organization in relation to the other departments, and the complexity of the equipment and processes to be maintained. Alternative designs would have been limited to variations of a traditional, functionally oriented structure.

Today, however, it seems that all organizations, large and small, are replacing traditional organizations with multiskilled teams working together. *Self-directed* work teams are taking over the responsibilities formerly given to the first-line supervisors, who, by the way, have now become *team resources. Empowerment* has been the management buzzword since the 1990s.

Plant engineering organizations are not immune to the changing roles of workers, first-line supervisors, and even upper management. Service organizations, like plant engineering, are frequently caught in the middle between the movement away from recognition of functional excellence (and the resulting organizational structure), and the functional expertise required to keep equipment and processes running at ever-increasing levels of quality and reliability.

ORGANIZATIONAL DESIGN ALTERNATIVES

Before discussing plant engineering organizations in detail, it is necessary to begin with an overview of organization design in a broader sense. The three basic ways to organize will be discussed, and the effect on each of these of the changing role of the first-line supervisor will be analyzed.

Three Types of Organizations[1]

Organizations can be structured by grouping together individuals with the same general work specialty (functional organization), collecting them by the output of the organization (product or project team organization), or a mixture of both types (the matrix organization). Each type of organization has its strengths and weaknesses.

Functional Organization. This is the traditional structure for plant and plant engineering organizations. All of the technical personnel (engineering and maintenance) are grouped together. Although within the plant engineering organization there may be some small project teams, for the most part the organization is structured functionally. Figure 1.1 shows an example of a plant functional organization.

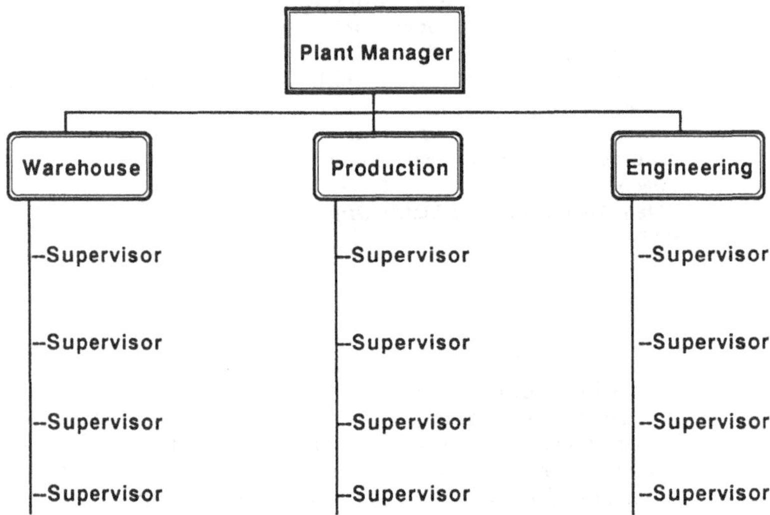

FIGURE 1.1 Functional plant organization.

Common characteristics of the functional organization are as follows:

- The division of labor, promotions and demotions, compensation system, and operating budgets are based on the functional competence of the organization and the individuals within the organization.
- Managers of the functional organizations have the most influence within the plant.
- Each function strives, and is rewarded for, maximizing its own goals; the goals of the organization as a whole are secondary.

Strengths of the functional organization are as follows:

- There is organizational support for technical competence; members all "speak the same language."
- Organization members can specialize in their technical area of competence and let others be responsible for *the big picture.*
- Individuals are secure within the walls of their own stable environment.

Weaknesses are as follows:

- Conflict between different functional organizations is unavoidable.
- The vertical hierarchy mandates decision making at the top; decisions are, therefore, slow in being made.
- Most of the organization members never see *the big picture.*
- Changing outputs of the organization take a long time to accomplish; bureaucracy is a frequent attribute of a functional organization.

Product Organization. This type organization is a popular one for companies wanting to move away from the inherent bureaucracy of a functional organization. This structure is well-suited to a rapidly changing environment. Under this form of organization, plant engineering personnel are combined into various product teams. Team members do several tasks to maximize the quality and quantity of the output of the team. Figure 1.2 shows an example of a product organization.

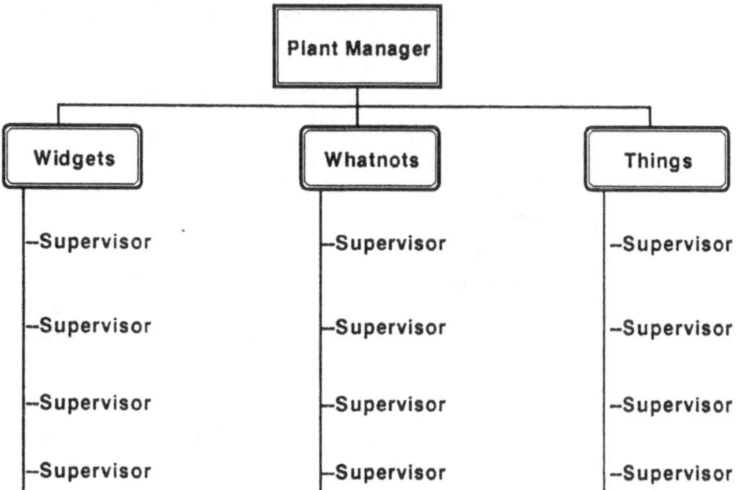

FIGURE 1.2 Product organization.

Common characteristics of the product organization are as follows:

- There is a minimal need to coordinate with other teams.
- Promotions and demotions, monetary compensation, and influence depend on the members' ability to work together as a team to produce the desired output.
- Team leaders have the most influence in the organization.

Strengths of the product organization are as follows:

- The organization is responsive to rapidly changing conditions.
- Conflicts with other teams are minimized.
- Team members all can easily see *the big picture.*
- Team members have an opportunity to develop additional skills and obtain more responsibility.

Weaknesses of this organization are as follows:

- Technical competence of individual team members decreases as individuals attempt to learn additional skills. Generalists are rewarded; specialists are not.
- It can become difficult to attract technical specialists.
- Innovation is restricted to the specific product or products of the team.
- Teams compete for pooled staff resources.

Matrix Organization. This organization is a combination of the functional and product organizations. The matrix organization attempts to combine the strengths of the other two types and eliminate, or at least minimize, the weaknesses of each. To some extent the matrix organization successfully accomplishes this, but not without some drawbacks of its own.

In a matrix organization, some parts of the plant are organized functionally and others by product. While plant engineering is typically one of the functional organizations, many members are assigned to the product teams. These people usually have dual reporting relationships; they are responsible to the product team leader for their normal day-to-day team activities, but are also responsible to the plant engineering organization for proper maintenance of their equipment and processes. Figure 1.3 shows a matrix organization.

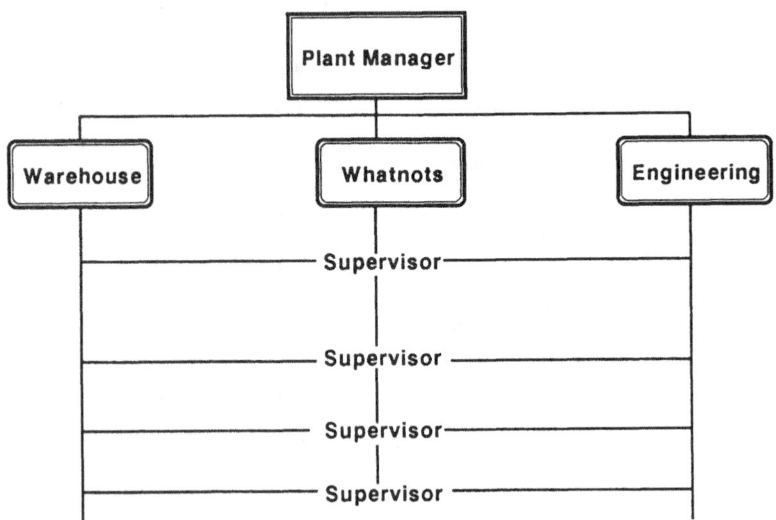

FIGURE 1.3 Matrix organization.

Strengths of the matrix organization are as follows:

- This type of organization provides maximum flexibility.
- Multiple career paths are provided; both generalists and specialists are rewarded.

Weaknesses are as follows:

- Conflict management is difficult because two bosses must be dealt with.
- Dual compensation systems are necessary to reward both generalists and specialists.
- Few people have the experience and training to work within this type of complex environment.

THE ROLE OF THE FIRST-LINE SUPERVISOR

As organizations have changed from the traditional functional structure to the product or matrix structure, the role of the first-line supervisor is changing too. Since this position has the most impact on attempts to move toward participative management and empowerment of the workers, an understanding of the supervisor's role is necessary.

The relationship of the first-line supervisor to the workers in the organization undergoes a natural transition as the organization develops and workers obtain more and higher skill levels. Some roles to be discussed will occur naturally; others must be formally introduced to encourage the transition.

Factors Affecting the Supervisor's Role

The factors that have had a major influence on the changing supervisor's role are as follows:

- The movement from specialized to generalized jobs
- The merging of line and staff positions with fewer levels of management
- Decision making being pushed to the lowest level possible
- Movement toward group instead of individual accountabilities
- Increased emphasis on team problem solving instead of individual problem solving

A Developmental Model

A developmental model of the first-line supervisor's changing role is shown in Fig. 1.4 and described as follows.[2]

The Leadperson. The leadperson has a dual role of supervisor and worker. Typically, the individual chosen for the position is the highest qualified from a technical standpoint and serves as a role model for the group. As the individual workers develop higher skill levels, the leadperson can assign specific jobs and, if the organization permits, move to the role of a one-on-one supervisor.

One-on-One Supervisor. This is the traditional role of the first-line supervisor. The supervisor is responsible for directing and controlling a group of workers. He or she is totally responsible for the group's output, but gets others to do the work. The supervisor's interpersonal skills are more important in this role than technical skills.

As workers further develop their skills they require less direct supervision. In addition, the workers tend to form their own informal subgroups. The supervisor then, often without realizing it, becomes a subgroup supervisor.

Subgroup Supervisor. In this role the supervisor manages by communicating with the subgroup leaders. The worker who does not become a part of a subgroup must still be managed individually. Some organizations tend to discourage the formation of informal subgroups, thinking that the authority of the supervisor will be challenged. This attempt to discourage subgroups usually fails and is a waste of time. More enlightened organizations recognize the process and attempt to use this role to their benefit.

As the subgroups develop, the supervisor may recognize the groups formally and create the position of group (or team) leader.

Team Leader. The team leader is responsible for the activities and output of a group of workers who share the same values, goals, and other common characteristics. The team leader

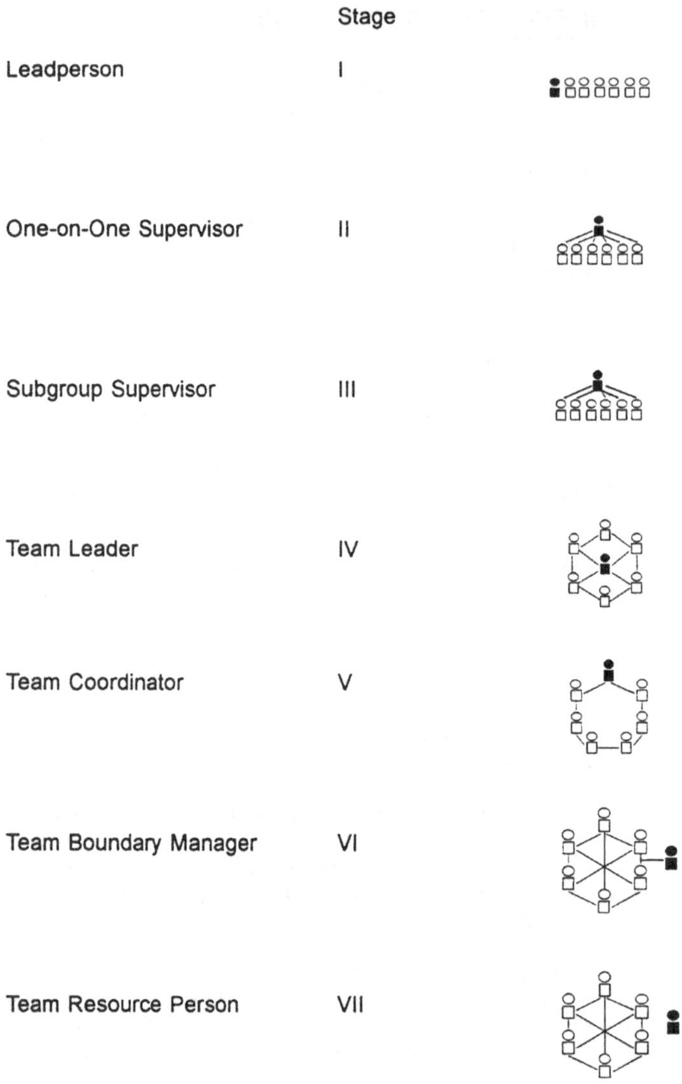

	Stage
Leadperson	I
One-on-One Supervisor	II
Subgroup Supervisor	III
Team Leader	IV
Team Coordinator	V
Team Boundary Manager	VI
Team Resource Person	VII

FIGURE 1.4 First-line supervisor's changing role.

manages the group by facilitating group interaction, problem solving, and decision making. Social skills of the team leader are much more important than technical skills. As team members develop production, troubleshooting, and problem-solving skills and become more adapt at leadership, the team leader becomes a team coordinator.

Team Coordinator. A team coordinator shares many leadership functions with other team members. Individual team members accept specific management-type activities. The team gradually develops the ability to manage its own responsibilities. When this happens, the team coordinator is free to become involved in other activities outside the team. As close contact with individual team members becomes less and less frequent, the supervisor assumes the role of team boundary manager.

Team Boundary Manager. The team boundary manager is removed from daily individual contact with team members. The manager still maintains responsibility for the team's activities and output, however, and must rejoin the team, as necessary, to ensure the quality and quantity of production. As the need to rejoin the team becomes infrequent, the boundary manager moves to the final supervisory role of team resource.

Team Resource. A team resource serves as a consultant to several work teams that are held accountable for their own work. At this point, the teams are truly *self-directed,* and the first-line supervisor's position no longer exists.

DESIGN OF THE PLANT ENGINEERING ORGANIZATION

The changing role of the first-line supervisor has many implications for the design of the plant engineering organization. No one type of organizational structure is ideal for all situations; each depends to a large extent on the role of the first-line supervisor or the organization's goal for what that role should become. Another primary factor influencing organization design is the relative maturity of the organization.

Plant Start-Ups

Plant start-ups are best managed by having the first-line supervisor function in the leadperson role. In these situations, the technical expertise of the workers is low. Supervisors should be selected, therefore, primarily for their technical abilities. Team training should be provided, however, to all workers and managers when possible to prepare them for an *eventual* transition into a team organization. Some organizations have attempted start-ups with self-directed work teams, usually with disastrous results. A *functional organization* works best for start-ups.

As the start-up is completed and workers gain in technical skills, the leadperson becomes a one-on-one supervisor. Many organizations remain at this stage of development for the duration of their existence. Since greater participation of workers usually leads to improvements in productivity and quality, however, further organization development is recommended. A one-on-one supervisor works best in a *functional organization.*

The subgroup supervisor usually functions in this role informally. As mentioned earlier, some organizations try to eliminate subgroups, usually without much success. Subgroups can exist in a functional organization and are typically the last stage of development before a formal transition into a team organization.

Transition to a Team Organization

Organizations that want to move from an authoritative to a participative type of management frequently do so by changing their structure from a functional type to a *team organization.* Unfortunately, *calling* a group a team does not make it so. As discussed, a real team exists because of the changing role of the first-line supervisor. Calling a supervisor a *team leader* accomplishes nothing. Real teams can exist in a functional organization just as well as in a team organization.

Creating a Real Team Organization

Creating effective work teams requires a high level of commitment by the organization. Both workers and managers need extensive training in team skills, social skills, technical skills, and problem-solving skills. In addition, changes in attitudes are required for individuals to effectively work in the new environment. Finally, management must be prepared to provide workers with the tools they will need to eventually become true self-directed teams.

Pseudoteams

Plant engineering organizations are affected by the movement to pseudoteams in two ways. First, the plant engineering organization is affected itself, just like any other organization. Second, since it is a service organization, plant engineering must function within the parameters set forth by the larger organization of which it is a part.

Plant Engineering in a Matrix Organization

Plant engineering organizations work best as part of a matrix organizational structure. The weaknesses of a product team organization eventually lead to major issues with effective maintenance. This is due to two primary factors. First, maintenance must be managed by using tools not normally a part of the production-oriented manager's toolbox. Second, a significant portion of the maintenance effort is more efficiently performed by a core team of specialists. Examples are major repairs and overhauls, master preventive maintenance scheduling, planning and estimating of maintenance work, and operation of a computerized maintenance management system (see Section 2).

As organizations develop and mature, work teams become truly self-directed and supervisors are replaced by team resource persons. The key elements here are *develop* and *mature.* This type of organization is not created by outside influences. It is created from within with *support* from the outside.

REFERENCES

1. Raab, A., "Three Ways to Organize," unpublished manuscript, 1986.
2. Bramlette, C. A., "Free to Change," *Training and Development Journal,* March 1984, pp. 32–39.

SECTION 2

EFFECTIVE MAINTENANCE MANAGEMENT

CHAPTER 2.1
PRINCIPLES AND PHILOSOPHY

William N. Berryman
Engineering Consultant
Morgan Hill, California

Condition-based maintenance (CBM) programs are established based on information collected, such as equipment failure and adjustment points, and determination of mean time between failure (MTBF) of equipment. This information can be gathered in many ways, through data collection processes in the program architecture, predictive technologies (e.g., vibration analysis, ferrography, and thermography), and building automated systems that provide input based on the various adjustments that take place.

This information is compiled and, through either software or statistical analysis, the condition of the equipment can be established at a point in time, or a prediction of when equipment maintenance should be performed (in terms of frequency or run time) can be made.

This is one of the most cost-effective methodologies of maintaining equipment, because only *required* maintenance is performed. Establishing *criticality* of the equipment plays a large part in these cost savings.

Reliability-centered maintenance (RCM) is a common application of time-based scheduled preventive maintenance procedures, and of predictive maintenance technologies applied to a specific application that allows for equipment life optimization.

RCM is a very effective methodology for many maintenance programs and, if the program architecture is designed appropriately, could provide cost savings and cost avoidance opportunities.

The keys to an effective RCM are the following:

1. Identifying a delivery method—that is, computerized maintenance management system (CMMS) procedures
2. Ensuring that all equipment is identified
3. Establishing criticality of all equipment
4. Deciding what equipment to target for *predictive* maintenance
5. Deciding what equipment to target for *preventive* maintenance
6. Deciding what equipment should be run to failure
7. Deciding how data should be collected in the field
8. Selecting data and determining how the data should be used

With these tools, an effective RCM program can be established. Sustaining any maintenance program can be a challenge, but can be accomplished by utilizing processes and procedures that establish the core of the program.

Time-based and task-based maintenance has been an effective methodology for many businesses, especially businesses whose budgets do not allow for implementation of costly software and hardware. Although not as effective as RCM and CBM programs, these programs do have their place in today's environment, but there *are* risks.

Time-based programs, without the use of predictive tools, will only extend the life of equipment, but *all* rotating equipment will fail in time. However, time-based programs could be improved by allowing for data collection. There is additional cost involved based on the time spent on the equipment, but some other costs are deferred (CMMS software, etc.). Using a simple spreadsheet, this program could be effective for small preventive maintenance (PM) programs.

Run-to-failure methodology is generally the most costly method of maintenance for the following reasons:

1. When rotating equipment does fail, it is usually catastrophic, causing more damage and raising the cost of repairs.
2. If the failure event is on a critical piece of equipment, bringing the equipment or system back on line will usually take more time and be more costly.
3. To reduce the downtime in a run-to-failure program, additional spare parts must be available.
4. All failures using this methodology are unplanned events and in many cases have other consequences that usually equate to some additional cost or customer impact.

Many environments still utilize this methodology, but not all in these categories: office environments, restaurants, and warehouses. In many cases, there is no business impact if equipment fails. It is just an inconvenience, but it usually equates to additional cost of repair due to its catastrophic nature.

CHAPTER 2.2

TYPES OF MAINTENANCE MANAGEMENT

William N. Berryman
Engineering Consultant
Morgan Hill, California

COMPUTER-BASED MAINTENANCE

The following list shows the steps required to implement an effective computerized maintenance management system (CMMS).

Verify File Server Functionality. This refers to testing communication between the file server and the database server.

Verify Database Functionality. This is to ensure that the database is functional and accessible by client PCs—for example, by launching the software from a client PC and testing basic transactions.

Review Organization Mission Statement. This is suggested for documenting and analyzing how the CMMS supports the maintenance management philosophy and overall maintenance program to ensure the two are in synch. Understanding this should impact the project scope and degree of CMMS utilization.

Review Current Maintenance Processes. This activity is the start of a very important phase. This can be an opportunity to modify workflow processes for optimal efficiency. Review how maintenance is done currently before modifying existing processes. Deliverables information from this step includes the following:

- Enumerating tasks performed by maintenance staff (planners, supervisors, techs, etc.)
- Defining data flow and how, when, and why it is entered into the existing CMMS
- Establishing the reports that will be used by staff, their frequency of generation, how the reports will be used, and so on.
- Defining how users interact with the current CMMS, security and access rights, and work-order distribution

Review Existing Policies and Procedures. Examples of policies to be documented include the following:

- *Work request policy.* This is for customers and maintenance personnel to request work orders. This includes call-center requests and field personnel work orders.
- *Work-order priority policy.* This establishes standard priorities for work orders based on criticality, delinquency of planning and scheduled work orders, and location.
- *Work-order status policy.* This records the status of every work order in the system. Examples include parts and materials, waiting authorization, work in progress, closed, and cancelled.
- *Work-order types policy.* This is used to define categories of maintenance work orders. Examples include preventive maintenance, predictive maintenance, projects, and call center.
- *Inventory policy.* This provides accountability for budgets, spare parts, and materials. Ordering policies (just-in-time delivery, min/max levels, auto reorder, etc.), cover critical equipment, cycle counting, and establishing and maintaining critical spare parts, among others.
- *Service contracts policy.* See "Contract Maintenance" later in this chapter.

Document All New Processes and Policies. New maintenance processes and policies that are to be adopted and implemented should be documented, and close management is recommended for all policies and procedures. Examples include the following:

- Tasks performed by maintenance staff (planners, supervisors, techs, and others)
- Definition of the dataflow process for the CMMS
- Types and distribution of reports
- Interaction with the CMMS; security and access rights
- Policy manual covering the following policies: criticality, work request, work-order priority, work-order status, work-order types, inventory, and service contracts

Review Location Hierarchy Structure. A recommended approach for multicampus facilities is geographic location, campus, building, floor, and room or grid. This allows for the exact equipment location to be identified; by using general ledger (GL) accounts, you can track cost of maintenance at a specific location or department. For single buildings, use individual building, floor, or room, as this allows for expansion if another facility is entered into the database.

Analyze Asset ID Number Scheme. This provides an opportunity for standardization on an equipment identification convention when there is more than one site and each is using a different naming convention. Using the location hierarchy as part of this equipment identification will allow crafts personnel to easily locate equipment by asset number.

Define Craft Codes. Use a consistent craft code identification scheme (e.g., electrician, mechanic, plumber) which allows for analysis of craft scheduling. This allows for balancing maintenance crews.

Define Shifts and Calendars. Most CMMSs are capable of building craft and individual-level resource calendars to keep track of craft personnel availability.

Set Preventive Maintenance (PM) Schedules and Define PM Procedures. The basis of a CMMS allows for scheduling of PM, specifying the PM tasks to be performed, the labor required by craft per task, the parts and materials required to perform the PM, tool requirements, and any special notes (e.g., safety guidelines).

Define Failure Codes and Scheme. Standardize failure codes across the organization to facilitate statistical analysis of failure data on common equipment. This allows for program improvement and root-cause analysis.

Prepare Parts ID Scheme and Equipment Bills of Materials (BOMs) for Critical Spares. By using equipment criticality, parts critical for program support can be specified, then those parts can be stocked in a convenient location. In the case of spare parts used during normal planned activities (PM), arrange to have them delivered just in time, depending on the inventory requirements.

Define Value Lists for Selected Fields. This refers to various pull-down menus or value lists in a CMMS (e.g., equipment condition: new, good, average, poor, replace). This standard pull-down menu will allow for consistency in information provided and utilized. For example, when equipment condition information is collected during PM, the information becomes immediately available to management so it may be utilized for capital planning. This allows for the identification of equipment needing replacement and, along with criticality, these items can be prioritized for replacement.

Verify any Additional Data to Be Added. Undoubtedly, there will be additional data to be loaded into the new CMMS (e.g., additional equipment, additional PM, and inventory). A procedure is required to track new equipment, relocation of equipment, or removal or abandonment.

CMMS Customization. The system architecture will be based on the business processes. This may include changing the canned field names and terminology, adding additional fields, deleting and/or hiding certain fields, or adding new screens. The system administrator should do this, and it should be based on a process requirement.

Develop Customized Reports. Most CMMSs have standard reports on specific business requirements, and customized reports may be developed. These reports are usually defined by management or engineering.

Define User Training. Now that the CMMS has been customized and set up for implementation, end-user training requirements should be based on the business process. The type of training that is required needs to be identified and then developed, based on work process considerations, usage policies, and identification of who needs training.

Functional Testing Phase. Functionally testing of the data conversion and data load is necessary to ensure that all the data are in the system correctly before implementation.

PREDICTIVE MAINTENANCE

The basic application of predictive maintenance involves taking measurements and applying the technology or processes to predict failures.

The approach should include the assessment of the equipment as it relates to the business or personnel safety, prioritizing the equipment, and targeting the critical equipment. For predictive maintenance, one other criterion must be utilized in the assessment process: the monetary value of the equipment. Even if the equipment is not critical, its value may warrant the use of predictive maintenance to reduce potential catastrophic failure of expensive equipment.

It is important to target this most critical and valuable equipment for predictive maintenance because there is an associated cost to predictive maintenance implementation. Predictive maintenance technologies are valuable tools if applied appropriately. These include the following:

- Vibration analysis
- Motor circuit analysis

- Infrared imaging
- Ultrasound
- Oil analysis
- Ferrography (wear-particle oil analysis)

These technologies, when applied properly, can reduce catastrophic failure, and thus maintenance cost. One other application is statistical process control (SPC). This predictive tool can be used to predict failures, but a plan must be in place first, for the data collection process is critical. If a CMMS is used, then the proper system architecture must be developed, along with associated processes and procedures that allow for accurate data collection.

Mean time between failures (MTBF) is determined using the technologies and processes listed previously. This will allow for proper planning of preventive maintenance based on information, not just on recommended schedules. Using these technologies and SPC should reduce the cost of equipment maintenance over time.

One of the key approaches to a good predictive maintenance program is consistency or standardization.

For example, if data on equipment failures (problem, cause, and remedy) do not include a list of standards, utilizing SPC becomes difficult and will require extensive research to identify what the data really mean. The cost savings will be lost because of the hours expended researching this information.

Developing a strategy and approach is the key to program success.

Predictive Maintenance Cost Savings and Avoidance Potential. Utilizing predictive maintenance can reduce the overall cost of facilities operation. In a facility where rotating equipment is prevalent, applying vibration analysis, ferrography, and laser alignment to coupled equipment may reduce power consumption. This is assuming that the coupled equipment is out of alignment, or has undetected wear condition.

Utilizing thermographic imaging will reduce power loss by detecting loose connections and following up with corrective action.

RELIABILITY-CENTERED MAINTENANCE (RCM)

The purpose of reliability-centered maintenance is to reduce defects, downtime, and accidents to as close to zero as possible; to maximize production capacity and product quality; and to keep maintenance cost per unit to a minimum. But, the *main purpose* of RCM is to create a systematic approach to maintenance that introduces controlled preventive maintenance while properly applying predictive maintenance technologies.

RCM is a method for establishing a progressive, scheduled, preventive maintenance program while integrating predictive technologies that will efficiently and effectively achieve safe, and inherently reliable, plants and systems.

The first step in our RCM process is to define the critical equipment (see the criticality assessment example in Table 2.1). You must first go through a systematic assessment of each major asset and its subassemblies, creating a priority list that shows the relative importance of each item as it relates to the business. Following is a list of questions to ask:

1. What would happen to plant availability if this item failed? Would it result in a forced outage? If so, what is the cost?
2. Would there be secondary damage? If so, what is the cost?
3. If failure occurs, what would be the repair cost?
4. Would product quality be affected by failure? If so, what is the cost?
5. What is the frequency of failures in this plant?

TABLE 2.1 Criticality Assessment Example

Assessment	Explanation
Critical without redundancy	Systems that would impact production, product quality, or are politically critical
Critical with redundancy	Systems that have redundant systems in place ($N + 1$)
Noncritical system	Systems that would have no impact upon production or product quality

The second step in the RCM process is to carry out the failure code analysis.

The third step in the RCM process is to decide on maintenance tasks. To do this, combine the criticality list and failure types to come up with the most appropriate maintenance program for each piece of equipment. Request input from the technicians, engineers, supervisors, and department managers.

The fourth step in the RCM process is to carry out the maintenance. All work should be issued via work orders. The technicians should complete the work orders giving full details, including time taken, materials used, and plant downtime. Use fault codes for plant failures, with full details of the work carried out. The supervisors, to insure compliance with completeness of work orders, should close out the work.

The fifth and final step is experience and analysis. Every 6 months there should be technical, financial, and organizational reporting. The technical reports should emanate from the technician work sheets, history files, and recommendations on job plans. The financial and organizational reports should be based on maintenance cost per unit, fault code analysis, plant availability, and related factors.

RCM is a continually evolving system of constant changes and improvements. It creates a total maintenance strategy that is flexible to a company's needs at all times.

INTEGRATED SYSTEMS

In today's environment, software integration is becoming commonplace. Some of the integration taking place in the maintenance environment today provides for ease of moving information back and forth accurately from the field to the CMMS. This is being accomplished using handheld devices or laptop computers and software that integrates with the CMMS, allowing the proper information to be transmitted to the field, field data to be loaded into the device, and the information to be transmitted back to the CMMS.

This methodology reduces the risk of data loss and transposition. This is very important if you are using reliability models or statistical processes in your maintenance programs. If data is lost or transposed, your models will be corrupted, thus producing inaccurate output.

Using integrated technologies will allow information to be transferred quickly and accurately. When integrated predictive reports are available, they allow for infrared, vibration analysis, oil analysis, and ferrography data to reside in one application and report, thus allowing for ease of information retrieval.

CONTRACT MAINTENANCE

There are several types of contract maintenance. Two are considered here.

Individual contracts for maintenance of systems or structures [e.g., heating, ventilation, and air-conditioning (HVAC) systems; painting; and roofing management] allows the company to manage the scheduling and cost of individual companies' activities.

However, these require extensive management time and monitoring of contractor quality, and could become difficult from a scheduling perspective if the selected contractor is not available when the work is scheduled. Rescheduling is then required and this can add to maintenance costs. The other additional cost is the contract management. If the facility is large and requires multiple contractors, each contractor will require a contract that defines the tasks, cost, and timelines. This also requires a contact administrator and multi-invoice management, as well as additional financial support.

This type can be effective if planned properly and if contractors are selected carefully. Some of the keys to using multiple contractor sources are: defining the processes and procedures for scheduling, quality control, billing, and service requirements for facilities customers, and specifying the tasks in the contract.

Outsourcing is another method of contract maintenance. In this model, the work to be outsourced, usually all or most facility operations, is defined. A single contractor is selected for maintenance of systems and structures. This contractor will usually provide the management, support, maintenance management software, and technical support personnel (craft personnel) as its own employees.

This method provides on-site support personnel; scheduling is defined by the contractor as agreed upon by customer requirements, and is usually a long-term contract (3 to 5 years). It also provides the advantage of having a single point of contact and the support to provide administration of financial, scheduling, documentation, and personnel management. This model permits the customer company's management, mutually with the contractor's management, to strategically plan maintenance of the facility and equipment, yet reduces the customer's involvement in day-to-day operations.

The key to success in outsourcing is to plan well by specifying the scope of work with attention to detail. Define the selection process, understand the scope, define the budget, and communicate this clearly in a request for proposals (RFP). Understanding the outsource contractor's ability to fully support all facets of the contract is very important, so first understand what you need to have accomplished and then identify the contractor's core competencies. Also, ask for specifics and recent successful projects.

In this model, the outsource contractor may manage subcontractors for special maintenance areas in which it may not have core competencies (e.g., elevator maintenance, roofing repairs, or crane and lifting device load testing). These will relate to the project and support capabilities.

STAND-ALONE SYSTEMS

Stand-alone systems can be effective within their own capabilities, but usually do not provide for a comprehensive maintenance program. They should be used only after an analysis to determine which system should be used and what the goals of your implementation are.

EXAMPLE 1. Vibration analysis may be used as a stand-alone system in specific situations where cost is a concern, and if the machinery involved is rotating. This may expedite failure prediction of critical equipment.

EXAMPLE 2. Infrared imaging cameras may be used effectively on electrical systems to determine if loose electrical connections or critical current situations exist. These conditions may also be evaluated after corrective actions have been taken to ensure that they were effective. Infrared cameras are also effective for qualitative inspections of mechanical systems (e.g., pump packages, boilers, and HVAC). They may also be employed for certain roof inspections to locate leaks.

It is important to have a strategy based on the expectations of your final result before deciding on a stand-alone system.

CHAPTER 2.3
PLANNING AND CONTROLS

William N. Berryman
Engineering Consultant
Morgan Hill, California

SCHEDULING

Scheduling of work depends on the nature of the business, personnel availability, equipment availability, and unexpected events. Effective scheduling of maintenance personnel in a perfect situation is simple. It is called *area maintenance.* That means to schedule the proper crafts to a specific area and complete all the maintenance tasks, rather than scheduling them to move from area to area, then back to the previous area to accomplish tasks on adjacent equipment.

Travel time is lost time in most cases, but in an effective program it can be used as an opportunity to identify other maintenance issues, as long as the area maintenance methodology is employed.

Scheduling with your internal managers ("customers") will reduce false starts. Using good scheduling tools and communicating your planned schedule will reduce scheduling conflicts with production, operations, and other customer-planned activities. Follow these simple steps:

1. Understand whom your "customers" are (stakeholders).
2. Acquire their schedules.
3. Understand their special needs.
4. Develop a plan.
5. Develop a schedule.
6. Provide customers with a horizon schedule of at least two months.

Develop a "lessons learned" policy and use this tool to improve your scheduling techniques.

COST CONTROL

Maintenance cost control is accomplished by the following means:

1. Understanding the number of equipment sets
2. Defining criticality, so that noncritical or low-cost equipment will not be maintained and only critical and high-value equipment will be maintained

3. Proper scheduling of activities (area maintenance—see "Scheduling").
4. Using predictive technologies appropriately
5. Collecting data for statistical process control (SPC) to evaluate mean time between failures (MTBF), cost of maintenance, failure rates, and so on
6. Performing root-cause analysis on failures to develop various approaches to maintaining the equipment, or providing training as required to improve skills based on new technologies, or identifying poorly designed equipment to be replaced.

RELIABILITY AND MAINTAINABILITY

Reliability is the ability of equipment to function as designed for a given period of time. The criterion for how long a period the equipment operates at its design specification without failure is mean time between failures (MTBF).

Reliability metrics for equipment can be established using reliability models. This can be accomplished either by measuring the equipment device (e.g., air handler, air compressor, or boiler) or breaking down the equipment into major components and measuring the "parent" and "children" (e.g., air handler = parent; motor, coil, valves, fan, etc. = children), then defining the expected life of the components or device and plotting these failures against equipment life expectancy and comparing the information.

You can also contact the manufacturer and request the MTBF data on that specific equipment. Alternatively, benchmarking for the required data can provide the equipment life-expectancy information needed to build a model. Once the model has been established, the MTBF data can be extracted from maintenance records, or a computerized maintenance management system (CMMS) can be set up to provide the data.

Maintainability is the ability of equipment to allow access to perform the required maintenance tasks. This is usually a function of the equipment design and location and any special tools or fixtures required for access, calibration, lubrication, or other maintenance activities.

Equipment that is inaccessible or even difficult to access is a prime cause of reduced maintainability. Generally, equipment should be designed for ease of maintenance. Maintainability issues will arise when the equipment is installed or when other equipment is installed adjacent to it. Make sure all access points are clear; if height of access points is an issue, for example, ensure that catwalks and lift devices are properly positioned for ease of access.

BENCHMARKING

Maintenance benchmarking is accomplished in two ways, internally and externally. *Internal benchmarking* is accomplished within the company, usually against other similar facilities. *External benchmarking* is accomplished by benchmarking against another company in a similar area.

Benchmarking takes many forms, but in the maintenance environment the areas benchmarked are usually similar. In most cases, benchmarking begins by defining the goal. Is it based on cost, equipment reliability, or the number of personnel required to accomplish equipment maintenance tasks?

The second phase is the data collection method. Is it based on time studies, survey results, field data collected, or CMMS data results? These are as varied as methods allow. The third phase is deciding where the data is going to be collected, externally or internally. The fourth phase is deciding how you are going to use the results.

It is extremely important to plan benchmarking activities well:

1. Keep the goals in mind.
2. Develop a standard for information collection.

3. Properly target areas for benchmarking.
4. Use standard methods for compiling information.
5. Understand what the final results will affect.
6. Develop a plan to use the information correctly.

MAINTENANCE MEASUREMENT

Measuring maintenance has many facets. It is important to define what is to be measured, how it is to be measured, why it is to be measured, and how the information is to be used.

Following are some measurements and recommendations for using the data.

1. You might measure *preventive maintenance (PM) procedures scheduled versus completed.* The purpose of this measurement is to gauge the performance of PM activities against a schedule. This information will help determine whether your support staff is large enough. If you are unable to perform 100 percent of PM every month, your staff is too small; or, if you measure by craft (e.g., electrician or mechanic) you may find an imbalance in craft staffing.
2. You might measure *corrective activities on equipment.* This will provide MTBF information and provide data to substantiate whether your PM program is effective. It may also define such issues as lubrication and incorrect power input parameters (voltage, frequency, power factor, etc.).
3. *Cost measurement* is another important type of metric. Measuring the costs of maintaining equipment is very important and can be accomplished through good recordkeeping or CMMS architecture. Knowing how long a repair activity takes, what parts are used, the cost of parts, and the actual downtime are important to good cost measurement.

These are just a few of the parameters, but the list can be as long as the tasks require.

SECTION 3

MANAGING THE FACILITY

CHAPTER 3.1
COMMISSIONING

Building Commissioning Association
Seattle, Washington

INTRODUCTION

As commercial and institutional buildings have become more complex, the traditional methods for building start-up and final acceptance have proven inadequate. The increased complexity of building systems is a response to energy conservation requirements, to the need for safer work environments, to demands for improved indoor air quality and better comfort control, and to technological advancements in teaching and research methodologies. Distributed digital control systems help to meet these needs but add sophistication and complexity. Thus, the entire process of acquiring new facilities must also become more sophisticated in order to keep pace with the complexity of new technologies and to get the required performance from them.

The process of commissioning is relatively new to commercial and institutional building construction. The model for building commissioning derives from commissioning industrial facilities and naval ships. Industrial facility controls are dry-lab tested extensively before installation in the facility. After installation, equipment and controls are further extensively tested and calibrated to ensure that production starts up on schedule and continues with maximum operational reliability.

Similarly, components and systems on naval ships are tested at dockside and then pushed to the limits during sea trials. Commissioning attempts to break whatever is prone to failure and to discover whatever design and installation errors exist. The serious consequences of failure in naval and industrial systems drive the effort to systematically wring out problems before systems are brought online. The failure of traditional building start-up, combined with the health, safety, and energy consequences of failure, is the impetus driving adaptation of the commissioning philosophy to commercial and institutional facilities.

Commissioning, as it applies to commercial and institutional construction, is in the early stages of development. This developmental process will span many years before it matures to the point that commissioning is well understood by owners, developers, design professionals, and contractors.

Definition

An all-inclusive definition for commissioning has not been universally adopted. As one listens to early entrants into the field, it becomes apparent that definitions most often reflect the specialty within a facility that is being commissioned. Electrical utilities concentrate on energy conservation measures for which they have contributed funding, in order to optimize energy consumption. Fire departments concentrate on commissioning fire- and life-safety provisions required by related codes. Other regulatory agencies insist that provisions within their jurisdictions be proven fully operational in accordance with their specific requirements.

As these narrow definitions persist, they negatively affect a proper understanding of the comprehensive requirements for commissioning an entire facility—total building commissioning. Whereas utilities and regulatory agencies focus attention on commissioning certain systems, commercial or institutional owners must be assured that *all* systems within the facility have been commissioned. For hospitals and other health care facilities, commissioning is virtually mandatory. Commissioning is now a critical requirement for high-technology laboratories and research facilities. It is also becoming important for classrooms and office buildings to meet the specific and diverse needs of the occupants. Hence, a broad definition for commissioning is imperative.

The institutional owner views the entire building as a system, a system that usually must meet a wide range of occupancies and functional needs. Therefore, the definition from the summary report of the 1993 National Conference on Building Commissioning may be the most appropriate for commercial and institutional facilities: "Commissioning is a systematic process of assuring that a *building* performs in accordance with the design intent and the owner's operational needs." (Emphasis on *building* rather than *systems.*)

Another relevant term (virtually a synonym) is *discovery.* Commissioning is a process which ensures *discovery* of flaws in the design or construction that preclude facility operation in accordance with parameters set forth by the owner or developer. Of course, discovery will inevitably occur, but it usually occurs under the most unfavorable circumstances, resulting in operating difficulties which could be critical or (in the extreme) even fatal. The least that will occur is inconvenience for building occupants and maintenance and operations staff. Commissioning forces discovery to occur under controlled conditions and at a time when dire consequences are least likely to result.

Further, if discovery occurs before the construction contract is accepted as complete, the design professionals and contractors will bear the burden of taking corrective action and, generally, all related costs. When discovery occurs later, the owner inherits these responsibilities and costs with little or no recourse back to those responsible for the failure.

Goals

The overall goal of building commissioning is to have a facility that operates as intended. However, it is important to recognize several subgoals which will be achieved as a direct result of the commissioning process.

The *primary* goal is to provide a safe and healthy facility for all occupants: office staff, the public, students, faculty, researchers, and operations and maintenance staff. Commissioning minimizes functional and operational deficiencies which have been shown to be responsible for the majority of indoor air-quality problems and comfort complaints. Commissioning also minimizes liabilities inherent in laboratory building operations.

The *second* goal is to improve energy performance. Commissioning is the tune-up that gets the most efficient performance out of the installed equipment. Commissioning tailors system operating parameters to the conditions of actual usage.

The *third* goal is to reduce operating costs. Equipment operating improperly is operating inefficiently. Maloperation usually induces more frequent maintenance activity and results in shorter life expectancy for the equipment. Annual operating costs increase and capital replacement costs occur more frequently.

The *fourth* goal is to improve orientation and training of the operations and maintenance staff. The sophisticated systems being installed will be disabled if operations and maintenance staff do not fully understand operation and maintenance requirements. No matter how well the equipment and systems are operating at the outset, systems operation will degenerate without proper care.

The *fifth* goal is improved documentation. Specifications and drawings do not provide all of the information needed for operation, troubleshooting, and renovation of the facility. Design intent and design criteria documentation, one-line diagrams, and operating descriptions help to communicate the designers' intentions to current and future operators and designers. Fully documented testing procedures and results verify the capacity and operating parameters of the facility and systems, and facilitate recommissioning as needed in the future.

The *final* and most important goal is to meet the clients' needs. When the design and construction is subjected to systematic scrutiny and verification, the design intent and customer satisfaction will be achieved.

Process

Once it is recognized that all building systems must be commissioned, the process for doing so becomes very specific and disciplined. In order for commissioning to become an effective program, all participants in the design and construction community must understand the requirements and willingly accept their responsibilities accordingly.

In brief, the owner must make clear during the programming phases of each project what the owner's commissioning expectations will be, since expectations may vary with different kinds of projects. Then the design professionals must translate that program into construction documents. Since commissioning scheduling, procedures, and activities are currently not well understood throughout the construction industry, contract documents must be much more comprehensive and specific regarding this process. New roles may develop, for example, those of commissioning authority (CA) and testing contractor (TC). As appropriate for each, their roles must be defined and the working relationships identified. The commissioning authority will be working directly for the owner, and the testing contractor will be working directly for the contractor; hence, the specifics of each role must be clearly defined to ensure program continuity and to avoid conflict and redundancy.

So, it is no longer enough to just build the building. Now, the contractors and design professionals have to *make it work*. The following sections will elaborate on the steps required to implement an effective building commissioning program.

DESIGN REQUIREMENTS

Considerably more information is required to commission a building than traditionally has been presented in the construction documents. Heretofore, the philosophy has been that the contractor does not have to know the design intent or the design considerations in order to build the building.

Upon completion of construction, the as-built drawings and operations and maintenance (O&M) manuals are turned over to the owner. The owner soon discovers that these documents may be adequate for maintenance but not for operating the building, particularly at operational extremes or when operational problems develop.

Generally, most of the information required to effectively operate the building has been developed during the design process. Unfortunately, that information is either buried in the design files or is still in the head of the design engineer. In either case, it is essentially lost to the owner's operations staff. This is a critical circumstance that must be corrected. Therefore, design requirements placed upon the design professionals must be changed to ensure that the owner receives all of the data and documentation critical to effective operation of the facility. This documentation, added to the extensive documentation developed specifically for and by the commissioning process, has resulted in the development of the systems manual, in addition to the traditional O&M manuals.

Traditionally, architectural and engineering (design professionals') agreements have been fairly generic in regard to design and document requirements. Today, these agreements should be augmented to be much more specific about what the design professional is expected to provide upon completion of each design phase. In order to maintain flexibility, some owners have a formal design professional agreement which refers to an attachment that can be conveniently tailored to suit specific project requirements. The attachment is used to specify the detailed documentation required.

To achieve the desired results, the owner must prepare a document which clearly defines the expectations for the design professionals. If the owner has a facilities design manual which is given to design professionals, all of the requirements should be covered in that publication. If there is no such manual, a commissioning document should be prepared.

These design requirements are referred to as *deliverables.* They are specifically required at various stages in the design process. Many of these deliverables are not universally understood, or accepted, by design professionals, so elaboration and negotiation is necessary. Specific to this discussion, it is important to determine when a commissioning authority will become involved with the project and the extent of involvement to be required. For an owner with minimal design experience and staff resources, the involvement of a commissioning authority should come early in the design process. In many cases, that person could be the only effective direct representative of the owner's operations and maintenance staff. In any event, the "extra services" agreement with the design professional must clearly indicate direct expectations of the design professional and the role of the commissioning authority in the process (even though the commissioning authority is hired directly by the owner).

Design Phases

Most owners are accustomed to the following basic design phases: programming, schematic, design development, construction documents, and bidding. *Programming* results in a detailed statement of owner requirements. The *schematic* process results in the first translation of written requirements into a conceptual facility design. *Design development* corrects misinterpretations and presents an organized facility that the design professionals will execute, which will meet the owner's requirements. *Construction documents* provide all details necessary for a contractor to build the facility. *Bidding* commits a contractor to construct the facility for a specific price by a specified completion date.

Two of these phases are major plateaus, or major documentation points, for the commissioning process: the end of the design development phase and the end of the construction documents phase. There are other points along the way that are also important to the overall design process and commissioning.

Programming Phase. The owner should have prepared a program prior to engaging the design professional. This is the ideal time to include a brief discussion of the owner's expectations regarding commissioning: Does the owner plan to engage a commissioning authority? If so, during the design process or only during the construction process? Will the contractor be expected to engage a testing contractor and perform a major share of the commissioning functions? These questions get the thinking started for later refinement during design. Some owners are engaging their commissioning authority during this phase to provide specific assistance, that they otherwise might be lacking, in the development of the building program.

The programming phase should produce four associated documents: a *functional program,* a *technical program,* the *design intent,* and the beginning of the *systems manual.*

Predesign Conference. It is very important to have a predesign conference with the design professionals. This conference must include appropriate representatives from the owner's staff who will be responsible for review of the construction documents, construction oversight, start-up and commissioning, and operation and maintenance of the completed facility.

The predesign conference should discuss a range of design and operations philosophies that are important for the design professionals to understand. They have a much better chance of designing a building that will function in accordance with all of the owner's needs once they understand the capabilities and limitations of the maintenance and operations organization. The discussion regarding commissioning is also important, particularly because many design professionals do not properly understand commissioning.

It is at this stage that the relative importance of commissioning should be recognized and factored into the design process. The type of facility being developed tends to dictate relative need and requirements accordingly. For example, hospitals and complex science facilities require the maximum commissioning effort. Even relatively simple office or classroom facilities should have some level of formal commissioning beyond the usual testing, adjusting, and balancing (TAB) work. Another important factor is the quality and reliability of the construction community to successfully complete the facility. And last, there is consideration of the owner's operations staff's ability to work with the contractors to wring out the systems and optimize design-intended performance.

Regardless of the decision about the level of formal commissioning to be performed, the requirements for the construction documents should not change since all of the resulting information is critical to the successful operation of the facility by the owner's staff. The following paragraphs indicate requirements which are important to the owner under any circumstance, many of which have not heretofore been provided to the owner by the design professionals.

Schematic Phase. It is important to expand commissioning discussions at this early stage in the project. If for no other reason, the more the design professionals are required to think about how to actually get the building running right, the better the design is likely to be. This is also the time for the owner to begin to seriously consider engaging a commissioning authority, if one was not engaged during the programming phase.

The commissioning deliverables at the end of the schematic phase are the following:

- A scope statement for the commissioning program
- The commissioning outline specification

These documents help to validate the design professional's understanding of the owner's design intent and operational expectations.

Design Development Phase. The design development phase could be considered to be the most important phase of the entire design process. All design criteria and operational parameters must be thoroughly considered by the design professionals and the owner, and documented. All philosophical design issues must (ideally) be resolved by the end of this phase.

The commissioning deliverables at the end of the design development phase are the following:

- Documentation of design criteria after the design professionals and owner have completely agreed upon all system operating philosophies. Documentation shall include one-line diagrams depicting operations at various design conditions, including fluid flow rates, temperatures, and pressures as necessary to comprehend the intended operation.
- Publication of functional and technical programs with amendments thereto, such as refined design intent and design criteria assumptions, in an appropriately bound document. This is a continuation of the systems manual, to be used by the owner and others for operation of the facility and as future alterations and revisions are imposed upon the original facility.

Construction Documents Phase. The construction documents phase is relatively simple. The major decisions have been made, agreed to, and documented. The commissioning deliverables at the end of the construction documents phase are the following:

- Include in the construction documents all one-line diagrams and sequence of operations discussions necessary to fully identify how all systems are intended to operate during all design conditions. (It is important that all of this be included on the drawings, not in the specifications, for convenient use by owner's operations personnel.)
- Specify contractor participation requirements for component, equipment, and system testing before and during commissioning.
- Specify prime contractor responsibilities during construction regarding commissioning coordination of prime and subcontractor responsibilities in cooperation with the commissioning authority.
- Specify contractor requirements for the O&M manual and the systems manual, the definition of *substantial completion,* the definition of *functional completion,* and requirements for final acceptance.
- Specify contractor requirements for training of O&M staff, both on-site and at special schools, as necessary.
- Publish all amendments to the functional and technical programs, design intent, and design criteria which have evolved during the construction document phase; include with the documentation published at the end of the design development phase.

One area in which owners have not imposed on the design professionals is the arrangement of construction documents for the benefit of the owner's operations staff. There are some relatively simple and basic steps that can be taken which will facilitate the operations staff's work, yet not significantly affect the work of the design professionals.

Of greatest importance to the operations staff are mechanical and electrical one-line diagrams for the many systems within the facilities; related sequence-of-operations discussions; equipment schedules indicating design requirements; and final operational set-points after testing, adjusting, balancing, and commissioning have been completed. Therefore, it is desirable to coordinate these items onto consecutively arranged sheets, rather than have them occur randomly throughout the documents as space might allow.

CONTRACT DOCUMENT REQUIREMENTS

There are three contract documents required to appropriately incorporate and coordinate the commissioning process: the *design professional agreement,* the *commissioning authority agreement,* and the *construction contract.*

Design Professional Agreement

The design professional agreement must be tailored to include commissioning-related services from the design professionals. These are primarily expansions of the design phase requirements, most of which are not yet included in the American Institute of Architects (AIA) Standard Form of Agreement. Since some of these deliverables are not universally understood by design professionals, elaboration is necessary. This can conveniently be handled by preparing an attachment to the design professional agreement.

The design professionals' scope of services should include their assistance in development of the scope of commissioning services and participation in the selection of a commissioning authority (if the commissioning authority has not already been engaged). Assisting in the selection of a commissioning authority helps to engender some sense of ownership of the commissioning program and foster improved working relations with the selected commissioning authority.

The design professional agreement must address communication with the commissioning authority during design and construction. The extent of this communication must take into account how the commissioning authority agreement will be written. Construction-phase activities of the design professionals should include review of commissioning-related submittals, though primary responsibility for such review lies with the commissioning authority.

The design professionals must participate in resolution of problems and conflicts during construction. The design professionals must retain the final authority during construction, as is current practice, unless that has been specifically delegated elsewhere. Still, the design professionals' scope should require their direction to be consistent with the recommendations of the commissioning authority on commissioning-related issues.

Commissioning Authority Agreement

The owner negotiates an agreement for commissioning services directly with the selected commissioning services provider. This agreement shall incorporate provisions relating to conflicts of interest, the scope of commissioning services, lines of communication, and authority.

The owner must have the full allegiance of the commissioning authority during the project. Accordingly, the agreement prohibits the commissioning authority from having any business affiliation with, financial interest in, or contract with the design professionals, the contractor, subcontractors, or suppliers for the duration of the agreement. Violation of such prohibitions constitutes a conflict of interest and is cause for the owner to terminate the agreement.

The scope of services includes responsibilities during the design, construction, and postoccupancy periods. In the design process, the commissioning authority should review each design submittal for commissioning-related qualities. These qualities include design consistency with design intent, design criteria, maintainability, serviceability, and physical provisions for testing. The services provided should also include commissioning specifications, with emphasis on identifying systems to be tested and the associated test criteria.

The commissioning authority participates in onboard review sessions and various other design meetings with the design professionals. The intent is to ensure that the commissioning authority has as much familiarity with the design as is feasible. This allows an understanding of the design for effective reviewing, providing input to the design professionals regarding commissioning requirements which would not be readily evident to many design professionals. Commissioning authority participation in the design process results in increased effectiveness during the construction and postoccupancy phases of the project.

During construction, the commissioning authority performs a quality-assurance role relative to the contractor's commissioning activities. The scope includes review of the qualifications of the contractor's selected testing contractor, all equipment submittals and shop drawings related to systems to be commissioned, commissioning submittals, O&M and systems manuals, and training plans. Commissioning submittals include the commissioning plan and schedule, static and component testing procedures (verification testing), and systems functional performance testing procedures.

The commissioning authority's scope also includes witnessing and verifying the results of air and hydronic balancing, static tests, component tests, and systems functional performance tests. To the extent the owner's staff is involved in witnessing the balancing, equipment testing, and systems functional performance testing, the commissioning authority's scope can be reduced to witnessing critical functional performance tests and a sample of other verification and functional tests. The owner's staff benefits from witnessing as much of the balancing and functional performance testing as possible. The commissioning authority's function, then, is to ensure that the testing contractor and test technicians properly understand and execute verification and systems functional performance testing procedures.

The commissioning authority's agreement should also include analysis of the functional performance test results, review of the contractor's proposed corrective measures when test results are not acceptable, and recommendation of alternate or additional corrective measures, as appropriate in the commissioning authority's scope.

Clear lines of communication and authority must be indicated. Communications and authority of the commissioning authority should be tailored to the level of involvement of the owner in the project.

If the owner is *intimately* involved in all aspects of design and construction, the owner should manage the commissioning authority's involvement. In this case, the commissioning authority would communicate formally with the design professionals through the owner. During construction, the commissioning authority should communicate formally with the contractor only through the established lines of communication—that is, directly through the design professional, or indirectly through the design professional via the owner. In either case, it is essential that the owner be kept informed of problems and decisions evolving during the commissioning process.

In cases where the owner is only *marginally* involved in the day-to-day business of the project, it may be desirable to allow the commissioning authority to communicate with the design professionals directly on commissioning issues. This is recommended only when the owner is very confident of the expertise and judgment of the selected commissioning authority, and only when the commissioning authority and design professionals have a good working relationship.

The authority of the commissioning authority should be limited to recommending improvements to the design or operation of the systems, solutions to problems encountered, and acceptance or rejection of test results. The commissioning authority should not directly order the contractor, or design professionals, to make changes. Only the design professionals may make changes in the design or order construction changes. The owner must speak with only one voice.

Construction Contract

The construction specifications define the scope of the contractor's participation in commissioning. A brief summary of the recommended specification sections follows.

Sections 01810 through 01819, Commissioning. The Construction Specifications Institute (CSI) has allocated Sections 01810 through 01819 for the primary commissioning specifications.

Section 01810, Commissioning, General Requirements. This section is intended to clearly indicate that the prime contractor is responsible for the overall commissioning program specified in the construction contract. Further, it requires the prime contractor to hire a testing contractor to carry out these responsibilities and work with the owner, or the owner's commissioning authority. It lists the minimum qualifications of the testing contractor and identifies the scope of the testing contractor's responsibilities. The contractor must submit the qualifications of the proposed testing contractor for approval by the design professional and owner. It also refers to Sections 01811 through 01819, and Sections 15995 and 16995 requirements, making the prime contractor responsible for all Division 15 and Division 16 commissioning scheduling and coordination, including any other Division 2 through 14 sections providing equipment and/or systems requiring commissioning.

The overall cost for commissioning should be less if the contractors are required to perform virtually all of the start-up, verification, and testing activities. Further, knowing in advance they will be required to make everything operate per the design intent, and having been provided the contractual tools to do so, is likely to result in a significant improvement in the overall quality of the submitted equipment, installation, and workmanship.

Section 01810 describes the testing contractor's functions in detail. These tasks include refining the commissioning plan and schedule to integrate commissioning activities throughout the prime contractor's construction plan; updating the initial systems manual to reflect changes made during construction; providing system operating descriptions and one-line diagrams; reviewing software documentation; providing continuous updating input to the commissioning and construction schedule; writing verification and systems functional performance test procedures; verifying installation; performing pre-start-up checks; coordinating work completion, verification tests, and system functional performance tests; assembling O&M manuals; supervising system start-up, TAB, and functional performance tests; planning and implementing owner staff training; and final assembly of the systems manual and record drawings.

Section 01810 also provides a summary of all of the generic areas wherein functional performance testing of systems is required and lists, or refers to, the systems accordingly. It also refers to examples of functional performance testing documentation which the contractor is expected to use as a reference in developing specific testing procedures and documentation appropriate to the systems included in the project.

Sections 01811–01819, Commissioning. The rest of the sections should be organized to relate to equipment and systems specified for construction in Divisions 2 through 16, to specifically identify components and equipment to be verification tested and systems to be functional performance tested, and to identify acceptable results to be expected in each and every case.

A paragraph, "Acceptance Procedures," in each section must be completed by the design professionals. The design professionals develop the functional performance testing (FPT) criteria under these sections. Each feature or function to be verified as a condition of acceptance of the subject system must be incorporated. The design professional lists the various conditions under which each system is to be tested, the acceptable results, and the allowable deviation therefrom for each test.

The subconsultant responsible for design of the particular systems writes these functional performance testing criteria in collaboration with the commissioning authority. These are not the actual test procedures. The actual test procedures must be equipment-specific and, therefore, cannot be prepared until all submittals have been approved, the actual equipment has been purchased, and the manufacturer's technical data and operations information have been received.

Section 01810 requires the contractor's testing contractor to elaborate the functional performance testing criteria into functional performance testing procedures—step-by-step instructions to field technicians on how to perform each test, including acceptable results which must be achieved. They are as follows: (1) identify the component, equipment, or system to be tested; (2) identify the conditions under which the tests are to be performed, including various modes of operation as may be appropriate; (3) identify the functions to be tested; and (4) identify the acceptable results that must be achieved in each case. Such functional performance testing procedures must be prepared for every component, piece of equipment, and system in the project.

The criteria for acceptance must be written to allow objective determination of whether the test results are acceptable. For example, rather than just requiring that a temperature sensor be "accurate," the criteria must specify the range of acceptable performance at various set-points.

Section 15995, Commissioning. This section is included in the mechanical division of the construction documents. It is intended to clearly indicate the role and responsibilities of the mechanical subcontractors and vendors regarding their participation in the commissioning program, and it refers to Section 01810–01819 requirements.

Section 16995, Commissioning. This section is included in the electrical division of the contract documents. It is virtually identical to Section 15995, except it relates to the roles and responsibilities of the electrical subcontractors and vendors regarding their participation in the commissioning program, and it refers to Section 01810–01819 requirements.

Section xx995, Commissioning. Each division preceding Divisions 15 and 16 requires a similar xx995 section identifying commissioning responsibilities for equipment provided therein, and referring to Section 01810–01819 requirements.

THE COMMISSIONING PROCESS

The commissioning process will vary from project to project. It can begin at the onset of the design phases or can come as late as when the job has been accepted by the owner and it is discovered that systems are not operating as they should. Ideally, the process will start during programming and continue throughout design and construction.

The most fundamental step in the commissioning process is to first identify what systems are to be commissioned. This should have been done during the design process. Based on what systems are to be commissioned, the commissioning authority can develop a proposal for services.

The requirements of an effective commissioning program are consistent. For the most part, the following are typical services to be provided by the commissioning authority (CA) and the testing contractor (TC).

During Design

1. CA and owner decide which systems will be commissioned.
2. CA develops initial commissioning plan and schedule.
3. CA reviews schematic, design development, and construction documents for commissionability of the design and finished facility.
4. CA and design professionals write commissioning specifications.
5. CA and design professionals refine commissioning plan and schedule.

During Bidding

6. Design professionals respond to prebid questions, with input from CA.

During Construction

7. TC and CA review design intent, design criteria, and construction documents.
8. TC refines the commissioning plan; CA reviews and approves.
9. TC and CA review control strategy and software; CA approves.
10. TC develops component, equipment, and subsystems verification testing procedures; CA reviews and approves.
11. TC develops system functional performance testing procedures; CA reviews and approves.
12. TC and CA perform on-site construction inspections.

13. TC develops testing, adjusting, and balancing schedule; CA reviews and approves.
14. TC develops commissioning schedule; CA reviews and approves.
15. TC performs testing, adjusting, and balancing; CA observes, witnesses, and approves.
16. TC performs commissioning using the functional performance testing procedures; CA observes, witnesses, and approves.
17. TC performs final job documentation and closeout; CA reviews and approves.
18. CA prepares a final commissioning report.

Deciding Which Systems Will Be Commissioned. During the programming phase, or the schematic design phase, the CA can be a major help to the owner, assisting the owner in understanding the purpose and value of a comprehensive commissioning program, including ranking the importance of commissioning various systems so that the owner can make informed judgments in developing the optimum commissioning program for the project at hand. The CA can also assist the owner in developing the functional program, technical program, and design intent documents, followed by defining the scope of the commissioning program. If this is done during programming, the CA can also assist the owner with development of the design professional agreement, to ensure that the agreement includes all requirements for the design professional's participation in the commissioning program.

Developing the Initial Commissioning Plan and Schedule. During the programming phase, or the schematic design phase, the CA can assist the owner in developing an initial commissioning plan and schedule. This initial plan and schedule is important to understanding the integration of commissioning into the overall construction schedule, something that design professionals generally do not fully appreciate unless they have had prior experience with a total building commissioning program. A properly integrated commissioning process will help to ensure that commissioning does not delay the construction schedule and, in fact, will help to ensure that the construction schedule will be maintained as planned.

Reviewing the Design Documents for Commissionability. During the several design phases there are planned design review periods. These are the owner's opportunities to ensure that the design will meet the owner's design intent requirements. The CA will complement the owner's review by concentrating experienced attention on the commissionability considerations of the design. The CA's work is not intended to be a peer review; it is only for the purpose of ensuring that the systems include all necessary components required for functional performance testing and that accessibility to all necessary components has been provided.

Writing the Commissioning Specifications. Commissioning specifications are a unique subset of the overall construction specification document. It is safe to say, at this stage in the development of the building commissioning industry, that most design professionals are not skilled in the art of building commissioning specification writing. Thus, because of the kind of detail required for the commissioning specification, it is very important that a CA assist the design professional in the development of the commissioning specifications, or that the CA actually write them for the design professional, if that is preferred.

Refining the Commissioning Plan and Schedule. At each review period the design professional, assisted by the CA, should update the original commissioning plan and schedule in recognition of the complexities of the systems being designed. During the final review period the commissioning plan and schedule should receive very thoughtful attention to ensure continuity with the construction activity and the plan and schedule incorporated into the construction documents, so that bidders will understand the owner's intentions regarding the commissioning program.

Responding to Prebid Questions. The CA must assist the design professionals as prebid questions arise to ensure that questions regarding the commissioning program receive appropriate responses.

Reviewing the Design Intent, Design Criteria, and Construction Documents. The TC, with help from the CA, must review the project's design intent, design criteria, and construction documents in order to know the basis for system functional performance testing procedures. During this review, the CA should bring to the attention of the owner any items that might be considered questionable. The primary purpose of this review is for the TC to understand the design intent and the design criteria.

Examples of design criteria are: (1) a minimum flow rate of 5 ft/s in the piping of a pure water system, (2) a 0.05 in (0.1 cm) water pressure differential between spaces in a clean room environment, and (3) an outside dry-bulb and wet-bulb design temperature.

A review of the construction documents must be made to acquaint the TC with how the systems are to be laid out in the building, how the mechanical and electrical systems are to operate in conjunction with one another, and what quality standards the installed equipment must meet.

Refining the Commissioning Plan. The final integrated commissioning plan is prepared by the TC to provide a comprehensive document that clarifies the responsibilities of the design professionals, contractor, subcontractors, vendors, CA, and owner during the commissioning process. This plan identifies the systems that are to be commissioned and the documentation processes that are to be used. The plan also includes the testing, adjusting, and balancing schedule and the commissioning schedule.

Reviewing the Control Strategy and Control Software. The CA and TC need to review the control strategy prior to development of the functional testing procedures. This includes the review of the control contractor's submittals and allows the CA and TC to identify how the contractor intends to meet the design intent and accommodate the design criteria. It allows the CA a chance to see the types of controls, the number of controls, and how they will sequence with each other to control the systems. Since the TC is writing testing procedures for these systems, he or she must know its specific component parts prior to writing the tests.

Prior to performing the commissioning of systems utilizing direct digital control (DDC), the CA needs to have adequate time to review the control systems software. This review is twofold. One purpose is to acquaint the CA with the idiosyncrasies of the software being developed, and the second is to ensure that the control strategy will meet the design intent and the owner's requirements.

Developing the Component, Equipment, and Subsystems Verification Testing Procedures. Verification testing of individual components, units of equipment, and subsystems, such as control of air-conditioning terminal units by local thermostats, is done before testing, adjusting, and balancing commences, which is then followed by system functional performance testing. In order to develop these testing procedures, the TC must have reviewed the submitted data, including the manufacturer's functional operation literature. Based on this information, written testing procedures and test record sheets will be prepared by the TC. Each procedure must be explicit and exact enough so it can be easily repeated at a later date, by a different tester, in the exact same manner as originally done, and achieve the same results. Verification testing validates performance in accordance with the construction documents—validating the contractor's equipment submittal selections and installation work.

Developing the System Functional Performance Testing Procedures. The development of system functional performance testing procedures is crucial to proper performance of the commissioning process. In order to develop these procedures, the TC must have reviewed the design criteria, the construction documents, and the approved environmental control and building automation systems. Based on this information, written testing procedures and test record sheets will be prepared by the TC. Each procedure must be explicit and exact enough that it can be easily repeated at a later date, by a different tester, in the exact same manner as originally done, and achieve the same results. The procedures should also include the design criteria. System functional performance testing validates performance in accordance with the design intent and basis of design documents—validating the design professional's design concepts and execution.

After the testing procedures have been developed they should be reviewed by the CA, design professionals, subcontractors, and owner. The review by the design professionals serves as a check to ensure that the TC and CA have correctly interpreted the intent of the design. The review by the owner serves to ensure that the design intent will meet the owner's needs as an operating entity. The review by the subcontractors lets them know how their installations are to be tested.

After the review, the TC will modify the testing procedures as required, gain approval of the CA, and issue the finalized tests.

Performing Construction Inspection. The CA and TC should have a relatively minor involvement in the construction process. The primary responsibility of performing construction inspection should remain with the project manager and design professionals. The CA's and TC's involvement is that of a second set of eyes in relation to the design professionals and project manager. They should become aware of how and where systems are going in, which allows the TC to interface the testing, balancing, and commissioning functions into the project schedule so all work can be completed before owner occupancy.

Basic CA service includes periodic site visits and scheduling the owner's staff participation in the commissioning activity. During this time the CA will relay any supplemental observations of construction activities to either the construction manager, the design professionals, or the owner for attention and remedy. Detailed punch lists, enforcement of contractor work to meet the schedule, or other activities required to bring the job to a state of completeness prior to testing, balancing, and commissioning are not intended, but may be necessary to meet the owner's occupancy date.

The CA's site inspections are primarily focused on ensuring that the systems being installed are accessible and lend themselves to being tested, commissioned, and maintained once the job is complete. Often general contractors will indicate the work is done when, in fact, perhaps the power is not connected, or the controls that make it function are not ready. False starts create delays and frustration.

It is the CA's responsibility to inform the owner of whether the contractor is meeting the schedule. Often there will be a disagreement between the contractor and the CA with regard to how well the contractor is meeting the schedule. Although this can place the CA in an uncomfortable position, it is the CA's responsibility to always give the owner an honest and unbiased analysis of the situation.

Developing the Testing, Adjusting, and Balancing Schedule. It is the responsibility of the TC to develop the testing, adjusting, and balancing (TAB) schedule. This will require the TC to have a good understanding both of how realistic the construction schedule is and of the timing of the completion of construction. After reviewing the construction schedule, the TAB schedule should be developed and integrated into the construction schedule. It is important that the schedule be reviewed and updated on a weekly basis with the contractors, construction manager, and owner.

Adequate time must be provided for the TAB work, after verification testing is completed and before systems functional performance testing begins. Often, construction completion is delayed, while the occupation date is not. The CA must keep the owner aware of construction delays and how they will impact TAB and commissioning.

Developing the Commissioning Schedule. Once the TAB schedule has been finalized, the TC must develop the commissioning schedule. This schedule will be based on the construction completion schedule and the TAB schedule.

Air and hydronic systems should be tested for functional performance after they are balanced. If the occupancy of certain spaces is critical, the schedule should reflect that these spaces are to be commissioned prior to those which are not as critical. Once the schedule is developed, it should be given to the contractors, construction manager, CA, design professionals, and owner for their review and comment. After this review the schedule should be finalized.

The schedule should be reviewed with the contractors, construction manager, CA, design professionals, and owner and be updated by the TC on a weekly basis. As commissioning is in process, construction omissions and design problems may be identified that will need to be corrected before commissioning can continue. As with testing, adjusting, and balancing, it is imperative that the CA keep the owner aware of any commissioning problems that may have an adverse effect on the commissioning schedule and subsequent occupancy by the owner.

Performing Testing, Adjusting, and Balancing. Prior to beginning the testing, adjusting, and balancing, the TC should develop the balancing logs, schematic one-line diagrams of the systems to be balanced (if not already available), and balancing agendas for each system. These should be prepared and presented to the CA for review prior to commencement of the balancing.

During the CA's on-site construction inspections, the CA looks for such items as missing balancing dampers, improper sensor location, inaccessible terminal units, and any other items that would impede balancing and testing. Deficiencies should be documented and transmitted to the owner. Prior to commencing the TAB, a final pass should be made to ensure that the systems are ready for balancing. The TC should have completed all related start-up procedures and verification tests, as approved by the CA. The TAB should then begin, following the TAB schedule. If there are items which adversely affect the balance, the CA should immediately advise the owner so resolution can be determined and TAB completed in a timely manner.

Once TAB is complete, a rough draft of the final report should be prepared by the TC. After commissioning is complete, any rebalancing that was required for commissioning should be incorporated into the report, and the final report should be submitted to the CA for approval.

Performing Commissioning Using the Verification and System Functional Performance Testing Procedures. The actual commissioning of the systems should be performed by the TC in association with technicians from the appropriate subcontractors. The CA should observe final verification and system functional performance testing in association with technicians from the owner's operations staff. Commissioning should follow the commissioning schedule.

To minimize wasted time in trying to commission systems that aren't ready for commissioning, the construction documents should require that the TC pretest all components, equipment, and systems prior to inviting the CA to witness the final tests, as follows. When the contractor has completed installation of a system or piece of equipment (designated for commissioning) per the construction documents, and the work is complete and functional, the TC should perform commissioning in accordance with the approved procedures and work out all discrepancies. Upon completion of successful verification or system functional performance tests (FPTs), the TC should notify the owner in writing (a certificate of readiness letter), through formal communications channels, that the systems are ready for witnessing and approval by the CA. The CA and the owner's operations personnel should observe the final FPTs.

During final system functional performance testing, the TC should fill out the functional test record sheets for approval by the CA. If the systems fail due to a correctable contractor error or design error, the contractors and/or design professionals should implement appropriate corrections. Upon completion, systems should be retested.

Performing Final Job Documentation and Closeout. Once commissioning is complete, a final report should be prepared by the TC that includes the final verification testing and systems functional performance testing record sheets and lists any major unresolved discrepancies. This report will be reviewed and approved by the CA.

Preparing the Final Report. Upon receipt of the final report prepared by the TC, the CA prepares a final report to the owner explaining how the commissioning program evolved, the problems encountered, the corrections made, problems not resolved, and the benefits gained as a result of the commissioning program.

THE COMMISSIONING AUTHORITY DURING DESIGN

The commissioning authority can provide valuable input to the design professionals on the owner's behalf during the design phase. From a design perspective, a seasoned commissioning authority will have firsthand experience relative to the pros and cons of most systems being designed. Most of the commissioning authority's perspective on the design is on how well the system can be operated and maintained over its lifetime.

The commissioning authority should insist that all design criteria be properly documented. This information will be necessary to develop the functional testing procedures. This documentation will be reviewed with the design professionals and the owner so that the owner knows the design limitations at the beginning of the project rather than discovering them during the commissioning process. This understanding can allow the owner to make budget decisions with better knowledge of the consequences.

The commissioning authority will typically stress simplicity in design. This can help prevent overdesign and guard the owner from purchasing additional sophistication which is not necessary or not likely to be used by the operations staff.

The commissioning authority's review of the design drawings and construction documents before bidding will provide valuable feedback as to how clear the documents are and how well they will be interpreted by contractors. This review can identify areas that, if not modified prior to bidding, may result in additional clarification and possible change orders after the project is under way.

The input of the commissioning authority during any value engineering process can help keep that process from becoming strictly a cost-cutting process. For example, eliminating such items as balancing valves and as-built drawings is inappropriate.

SELECTING A COMMISSIONING AUTHORITY

The commissioning authority is becoming a major player in the overall process of design and construction. In fact, that role is becoming every bit as important as that of a specialist doing a constructability review during design, or of a construction manager assisting with project management during construction. Therefore, selection of a fully qualified commissioning authority is critical to the success of the commissioning process, and considerable care should be taken by the owner to ensure that a good selection is made.

Criteria

The first step is to decide what activities the commissioning authority is expected to participate in. This is a function of the qualifications of the owner's staff to review design documents and to participate during construction. It is also a function of the amount of money the owner can allocate to the commissioning process.

If the owner has experienced staff capable of effectively reviewing the design, it may not be necessary to engage the commissioning authority until construction begins. However, the latest practical time to engage the commissioning authority is shortly before the construction contract is awarded so the commissioning authority can be involved with review of schedules, submittals, and testing programs to be developed by the contractor.

The criteria for a good commissioning authority are heavily weighted toward extensive field experience with the installation, testing, and operation of mechanical and electrical equipment and systems. High in the experience priorities is experience with testing and commissioning life-safety systems, environmental control systems, and HVAC systems. In fact, if funding is extremely limited, the two systems that should be commissioned, above all others, are life-safety and environmental control systems. So doing will lead to discovery of components in other systems which may not be operating correctly, even though the life-safety and environmental control systems are operating properly.

Request for Qualifications

Once the scope of the commissioning program has been developed, the owner can determine what kind of skills and experience will be required to carry it out. This leads to development of a request for qualifications (RFQ) to be placed as an advertisement in the appropriate local or regional media—for example, a daily journal of commerce or newspaper, or a trade journal such as the *Engineering News Record,* if no other appropriate medium is available. However, one should shy away from the national media; a more local or regional firm is preferable if available and qualified.

It is important that the RFQ clearly indicate the minimum qualifications which will be acceptable for the project. It is important to identify the nature of the construction project to be commissioned. It is also important to indicate the types of systems to be commissioned. This gives respondents a sense of the kinds of skills and expertise required to commission the building. A well-written RFQ will not only encourage qualified firms to apply, it will also serve to discourage unqualified firms from applying.

In addition, there are several specific items which should be required in the submittals from respondents:

1. History of the company
2. Commissioning expertise and capability
3. Commissioning experience during the past 5 years on projects of similar size and type
4. Resumes of personnel to be assigned to the project, with project responsibilities
5. Special attention to the on-site commissioning project manager
6. References for projects of similar size and type

Further, the advertisement should state that finalists to be interviewed will be required to present comprehensive evidence of a prior similar commissioning project: commissioning schedules, component and equipment verification testing procedures, system functional performance testing procedures, and final documentation of all tests. This will discourage unqualified firms from applying since they will not be able to meet the interview requirements.

Selection Process

At the same time the RFQ is being prepared, one should also prepare a selection evaluation form. This evaluation form can be used both in the original screening of submittals and during the interviews of the finalists. In order to be objective, it is best to develop this evaluation form before the submittals are received.

There are several important considerations which warrant specific comment:

1. This is a request for qualifications (RFQ) and does not ask for cost proposals. The commissioning contract should be negotiated with the selected firm, at which point the scope

of commissioning work and the commissioning services contract cost will be finalized. If the commissioning authority is engaged late in the construction documents design phase of the project, or later, a single-phased commissioning services contract can be negotiated. However, if the commissioning authority is selected and engaged during the programming or schematic design phases of the project, it will be necessary to first negotiate a commissioning services contract for the design phases only; later, after all systems have been designed and the true scope of the commissioning program can be defined, negotiate a commissioning services contract for the construction phase of the project.

2. It is important to identify the firm's experience in the commissioning field. Specifically, it is important to know of successful commissioning projects of similar type and complexity.
3. It is important that the firm identify the project manager who will be the on-site person in charge of the commissioning program for your project. Experience has shown that the person assigned to lead the project in the field is the key to the success of the work to be done. The project manager's background and references must be provided for evaluation.
4. The firm should also indicate anyone else who will be involved, on which aspects of the project, and their prior relevant experience.

One must be cautious about what is purported to be commissioning experience. Testing, adjusting, and balancing (TAB) firms may suggest they do commissioning. TAB work is not commissioning. Commissioning includes TAB work, but goes beyond the average TAB firm's experience.

Consulting engineering firms will likely apply, claiming that they do final inspections, punch out buildings, and do start-up—and, therefore, commissioning. Although most design firms follow their projects completely through the construction process to acceptance, not many have experienced, full-time, field commissioning engineers on staff. To be acceptable, commissioning must be identified in the firm's organization as a principal business enterprise of the firm, headed by a principal in the firm.

And one last critical issue that must be considered—should one of the consulting firms or contractors already associated with the project be considered to be the commissioning authority? The preferred answer is no. The commissioning process should be viewed as the major quality-control element of the project. One has to be concerned as to whether the design professionals will be completely objective about problems discovered, particularly if they are found to be design problems. Similar lack of objectivity can occur with a contractor's work. Therefore, it is best that the commissioning authority be an independent, third-party participant which has no emotional or economic tie to the project (other than the commissioning fee). Complete unfettered allegiance must be to the owner.

Interviews

The interview process is important. One should learn a great deal about each firm's commissioning experience and the different perspectives represented by the firms. One can also get a sense of the firm's culture, which can be important in relation to the project, design professionals, and contractors associated with it. The commissioning authority must be fully accepted by all other participants.

It is recommended that no more than four firms be interviewed and that all interviews take place on the same day. This tends to ensure that all interview team members will be available for all of the interviews, and that important points will not be forgotten over a period of several intervening days between interviews.

Selection of the interview team is important. Since few commercial owners or institutions have completely similar capital improvement program management structures, it is difficult to make precise suggestions. However, a general philosophy suggests that team members should include the owner's project manager, a representative of the maintenance and operations staff, a representative of the client (end user), and a representative of the design professionals.

The client should participate for two reasons: First, since it is difficult to explain to the client what all is involved with commissioning and its importance to the successful completion of the project, the client stands to gain considerable education through the interview process. Second, it is important for the client to have confidence that the best selection was made. The same philosophy would apply in regard to the maintenance and operations representative.

If the design professionals have already been engaged by the owner, they should participate in the interview and selection process because they will be working closely with the CA and must be able to establish a good working rapport and have confidence in the observations and recommendations of the CA.

Negotiation—Scope and Cost

The selected commissioning authority is then asked to prepare a commissioning program proposal. The proposal should be prepared in detail so the merit of doing or not doing various components of the program can be discussed. Ideally, the entire building, including all systems, should be commissioned. However, available funding may dictate a curtailed scope involving only the most critical systems.

THE COST OF COMMISSIONING

Experience has proven time and again that the cost of commissioning is less than the cost of not commissioning. Thus, one of the major benefits of commissioning is avoidance of the costs of not commissioning. The costs of not commissioning may exceed 20 percent of the construction costs of the uncommissioned systems when a broad view is considered. Several studies have been done that support this comment; one example follows later in this discussion.

In the early programming or design stages of a project it is not possible to estimate the costs of commissioning. However, for initial budgeting purposes, before detailed proposals can be developed, a reasonable approach is to budget between 2 and 3 percent of the mechanical system construction cost, plus between 1 and 2 percent of the electrical system construction cost. Factors affecting the magnitude of the cost include the type and complexity of the systems involved. If the building is considered to be very high-tech, then budget on the high side of the percentages. Also, take into account the amount of equipment to be specified in Divisions 2 through 14 of the construction documents. If there are likely to be commissioning requirements as a result of these divisions, again budget on the high side of the percentages.

Benefits: Avoiding the Costs of Not Commissioning

By understanding the costs of not commissioning, one can gain a measure of the benefits of commissioning. The costs of not commissioning are the expenses that owners incur to correct design and construction deficiencies in order to make new or renovated facilities work as they were intended. The costs of not commissioning include staff time to correct problems, additional or replacement equipment or materials, lost time and productivity of the occupants, health and safety impacts, excess energy costs, impacts on central plant utilities, legal fees, and the intangible "hassle factor."

At the 1993 National Conference on Building Commissioning in Sacramento, a British Columbia Buildings Corporation (BCBC) representative cited estimates of the costs of not commissioning mechanical systems from a study done in 1989. The study drew information from five BCBC buildings which were described as "not problem buildings," but which were not commissioned. Mechanical construction costs averaged $10 per square foot.

The study attributes to building commissioning a $0.05 per square foot improvement in operations, a 5 to 10 percent reduction in energy consumption, and a cost of 0.75 percent of the original mechanical construction cost to make corrections after the fact. They concluded that the operation and maintenance costs of not commissioning ranged between \$0.45 and \$0.75 per square foot for the first 3 years of operation. So the cost of not commissioning these five buildings during the first 3 years amounted to 4.5 to 7.5 percent of the mechanical costs.

The presenter went further to estimate the costs of not commissioning related to the occupants, and indicated that occupant dissatisfaction with environmental conditions in their work spaces results in decreased productivity. At an annual payroll cost of only \$200 per square foot of occupied space, if every fifth employee spent 30 min of work time per month on such activities (gripe sessions, letters of complaint, meetings, investigations, etc.), the resulting cost would be \$0.13 per square foot per year. Using the same payroll costs, increased absenteeism (every fifth employee takes an extra 2 days off per year) may cost an additional \$0.33 per square foot per year. Over 3 years, these lost productivity figures may total as much as \$1.38 per square foot, or 13.8 percent of the original mechanical contract.

When the occupant impacts are added to the operations and maintenance costs, the total is \$2.13 per square foot, or 21.3 percent of the mechanical construction cost. Even if the real costs are only one-quarter of the estimates, they still show the substantial magnitude of the costs of not commissioning.

In addition to the preceding, there are also significant intangible costs involved in not commissioning. What does one say when the president of a major client or occupant calls, exasperated with the endless series of failures in a newly occupied facility? How does one explain the departure of a leading faculty member who can no longer tolerate the interruptions to research that result from building systems failures? How does one counteract the bad press that results from lingering indoor air quality complaints in a new building?

Commissioning Approaches Comparison

The commissioning approach selected has an impact on the costs. The following are five commissioning scenarios:

1. All commissioning functions performed by an independent third party (independent commissioning authority)
2. All commissioning functions performed by the design team (designer as commissioning authority)
3. All commissioning functions performed by the contractor (contractor as commissioning authority)
4. All commissioning functions performed by the owner (owner as commissioning authority)
5. Commissioning functions split between an independent commissioning authority, the design team, the contractor, and the owner (integrated commissioning)

Independent Commissioning Authority. The cost of the independent commissioning authority approach for the mechanical systems is about 2 to 3 percent of the mechanical construction costs. Commissioning electrical systems costs about 1 to 2 percent of the electrical construction contract amount. Costs for commissioning other systems, such as elevators or laboratory equipment, are likely to be similar to those for the electrical system. Mechanical systems cost more to commission because there are more dynamic functions to commission than with other systems.

Duplication of effort and negotiated (instead of competitively bid) commissioning services agreements make the independent commissioning authority the highest-cost approach. However, this is also the method with the lowest potential for conflict of interest.

The downside of the independent commissioning authority is the potential for delay. Due to the need to keep the site activities of the commissioning authority and the contractor from interfering with each other, much of the commissioning testing that will discover problems will not occur until after the contractor has completed installation and demobilized. Corrections take longer when subcontractors are called back after they have left the site.

Designer as Commissioning Authority. The designer as commissioning authority approach may cost less than employing an independent commissioning authority. The anticipated cost saving accrues from the design professional's knowledge of the design.

However, there are two factors that make this a less desirable option. As with the independent commissioning authority, the designer as commissioning authority approach suffers the liability of high potential for delays, for the same reasons. However, the most critical issue is the potential for conflict of interest. Remember that commissioning must subject the design to the same scrutiny as the construction. The designer is not in a good position for this task. The same reason the designer is a less expensive commissioning authority is also the reason the designer is less likely to be objective in evaluating the design.

Contractor as Commissioning Authority. The contractor as commissioning authority approach minimizes project delays by integrating commissioning tasks into the construction schedule. Commissioning costs are among the lowest when the contractor is the commissioning authority because the costs are determined in a competitive bid environment. A low bid, unfortunately, does not always result in high quality.

Like the designer as commissioning authority method, the contractor as commissioning authority approach suffers a high potential for conflict of interest. For the contractor, the obvious conflict is that correction of construction problems is an added cost. These costs cut into the contractor's profit margin. Further, in a bid project, the contractor is not available to perform commissioning tasks during design, making this option inappropriate for bid work. The contractor as commissioning authority option appears desirable at first glance due to the low cost and low potential for delay. But there is little difference between this method and the status quo; therefore, it is not recommended.

Owner as Commissioning Authority. If the owner has a sufficiently sophisticated staff, the owner may perform all commissioning services. This may be the lowest-cost option because there are no markups on the staff's time. Also, the staff retains all the expertise gained during commissioning, and there is no conflict of interest. In fact, there is probably no one with a greater interest in seeing that systems are designed and operating properly and are maintainable than the people who operate and maintain them. Most owners, however, lack the staff to provide their own commissioning services. This method may also suffer from project delays for the same reason as with the independent commissioning authority approach.

Integrated Commissioning. The last approach to be considered for this discussion is the integrated approach. The combination of players and tasks developed for integrated commissioning is intended to utilize the strengths of each player while keeping costs low and minimizing delays and conflicts of interest. All players are responsible for the commissioning functions they are best qualified to perform.

The design professionals and the commissioning authority work together during design to establish the scope of commissioning and to specify the tests to be performed. The commissioning authority also provides review of the design.

The contractor (via the testing contractor) assembles documentation; conducts training; and coordinates, schedules, and performs commissioning testing. Construction corrections and delays are the contractor's responsibility. But the commissioning authority is there to verify the results of tests and help identify solutions, as needed.

When contractors begin to understand the benefits they reap from integrated building commissioning, their bids will reflect the difference. Commissioning produces cost savings for the contractor due to several factors: improved on-site efficiency, most deficiencies discovered early and corrected early, fewer deficiencies to resolve at substantial completion, and fewer hidden deficiencies discovered after demobilization.

The contractor's on-site efficiency improves because the commissioning process refines the scheduling of subcontractors' work. Previously, the general contractor had to deal with anxious subcontractors, each demanding priority for their work. The commissioning process imposes a logical flow to many aspects of the work, ensuring that the requisites for each subsequent stage of work are scheduled in a timely fashion.

The contractor's costs decrease due to a reduction of the deficiencies at substantial completion. Subcontractors tend to be more careful in the completion of their work when they know a process is in place to discover deficiencies. Of the deficiencies that do remain, most are discovered during verification testing and system functional performance testing, while the subcontractors are still on-site.

The contractor's costs also decrease due to a reduction in the number of deficiencies discovered after the contractor has left the site. Deficiencies which would previously have remained hidden after the contractor had demobilized will, instead, be corrected prior to substantial completion.

It can be seen that contractor cost savings may well exceed the additional costs incurred by participating in commissioning. Unfortunately, in the short term, owners are likely to see higher bids as contractors bid the commissioning work conservatively, with ample margins of safety built in. As the relative magnitude of the contractor's costs and benefits associated with commissioning become better understood by contractors, their bids will reflect the cost benefit.

In this scenario, the independent commissioning authority can be replaced by the owner's staff, if such expertise is available in-house. In addition to the other advantages of integrated commissioning, if the owner serves as the commissioning authority, the owner retains the systems expertise gained during commissioning. In the long run, integrated commissioning with the owner performing the tasks of the commissioning authority is likely to be the optimum approach to building commissioning.

Costs

The BCBC representative cited the following mechanical systems commissioning cost estimates at the 1993 National Conference on Building Commissioning. The figures are based on an approach in which an independent commissioning authority, the designers, and the contractor share responsibility for mechanical systems commissioning. In many respects the BCBC approach to building commissioning is similar to the integrated commissioning method.

Based on a mechanical construction cost of $10 per square foot, the study estimates the following range of costs per square foot:

Extra designer time	$0.025 to $0.010
Commissioning authority	$0.100 to $0.300
Owner's operations staff	$0.025 to $0.200

These costs total $0.15 to $0.60 per square foot, which is 1.5 to 6.0 percent of the mechanical construction costs. This compares with the study's previously cited estimates of the costs of not commissioning of 4.5 to 7.5 percent of the mechanical construction costs, not including occupant impacts.

Relative Cost Comparison

Table 3.1 summarizes the merit of the five methods of performing commissioning. The designer or contractor as commissioning authority approaches are not recommended due to the high potential for conflicts of interest. If the owner has staff expertise to carry the function of commissioning authority in-house, the owner as commissioning authority option is acceptable. The independent commissioning authority approach is also acceptable, especially if commissioning was not initiated during design.

The optimum approach is integrated commissioning. Costs are reasonable. The potential for conflicts of interest is low. The risk of project delay is low. Integrated commissioning is at its best if the owner can perform the duties of the commissioning authority in order to retain the expertise gained in the commissioning process.

TABLE 3.1 Comparison of Commissioning Methods

Commissioner	Cost	Conflict	Delay
Designer	High	High	High
Contractor	Low	Highest	Lowest
Independent agent	Highest	Low	High
Owner	Low	Lowest	High
Integrated commissioning	Low	Low	Low

Budgeting for the Commissioning Program

There is little agreement on how best to establish an initial budget for the commissioning program. This is due, in large part, to the fact that when initial construction budgets are established, the complexity of the systems, and hence the commissioning requirements, are not known. One university in the Pacific Northwest, which has done extensive commissioning of a number of new science buildings, has had good success with the following rule of thumb: Use 1 to 2 percent of the estimated construction cost for the electrical work, and 2 to 3 percent of the estimated construction cost for the mechanical work; add them together, and that becomes a reasonable order-of-magnitude amount to budget for the commissioning program. If the building will be highly sophisticated for its type of construction and utilization, move toward the higher percentages. If the sophistication is nominal, move toward the lower percentages. Although not directly taken into account via this process, don't forget that there will likely be some equipment to be commissioned that is specified in divisions other than 15 and 16. Informal consideration of the likelihood of commissioning requirements for some of this equipment may lead to staying with the higher percentages to allow for these additional commissioning requirements.

Conclusion

In conclusion, the costs of not commissioning exceed the costs of commissioning. Commissioning costs are best estimated based on the construction costs of the commissioned systems: 2 to 3 percent of mechanical construction costs and 1 to 2 percent of other systems costs. The cost of commissioning depends on the commissioning approach used. Integrated commissioning offers an optimal mix of low cost, low potential for conflicts of interest, and low potential for project delay.

There are significant intangible costs involved in *not* commissioning. Can anyone afford the embarrassment and political costs of *not* commissioning new facilities? Can anyone afford to embarrass a company president or a department chair with an endless series of failures in a new facility that he or she has helped to fund? Can anyone afford to lose a leading faculty member because he or she can no longer tolerate the interruptions to their research that result from building systems failures? Can anyone afford the bad press that results from lingering indoor air quality complaints in a new building?

By not commissioning new facilities there is risk of creating situations which call into question the stewardship of the physical plants. Ensuring that new facilities work as they were intended is a requirement of responsible physical plant construction and administration. Building commissioning is the key to delivering facilities that function as intended and meet the client's needs.

CHAPTER 3.2
ELECTRIC SYSTEMS MANAGEMENT

Robert C. Walther, P.E.
President
Industrial Power Technology
Santa Rosa, California

GLOSSARY

Demand Level of power supplied from the electric system during a specific period of time.

Demand-side management (DSM) Measures taken by a utility to influence the level or timing of a customer's energy demand. By optimizing the use of existing utility assets, DSM programs enable utilities to defer expenditures for adding new generating capacity.

Harmonic distortion Continuous distortion of the normal sine wave, occurring at frequencies between 60 Hz and 3 kHz.

Linear load A predictable nonprocess energy load that has a profile that changes with time and condition.

Noise Continuous distortion of the normal sine wave occurring at frequencies above 5 kHz, usually of constant duration.

Nonlinear load A load profile composed of process and cyclic loads that may have a broad swing in energy and demand requirements.

Off-peak Hours during the day or night when utility system loading is low.

On-peak Period during the day when the energy provider (utility) experiences the highest demand.

Power factor The fraction of power actually used by a customer's electric equipment, compared to the total apparent power supplied; usually expressed as a percentage. Applies only to ac circuits; dc circuits always exhibit a power factor of 100 percent.

Sag A decrease in voltage up to 20 percent below the normal voltage, lasting less than 2.5 s. Also called *undervoltage.* Can result in memory loss, data errors, flickering lights, and equipment shutdown.

Silicon-controlled rectifiers (SCRs) A control system utilized to control speed by modifying the sine wave profile of power to variable-speed ac motors.

Spike A sharp, sudden increase in voltage of up to several thousand volts lasting less than 0.001 s. Can cause catastrophic memory loss or equipment damage.

Surge An increase in voltage up to 20 percent above the normal voltage, lasting less than 2.5 s. Also called *overvoltage.* Can result in memory loss, data errors, flickering lights, and equipment shutdown.

Total harmonic distortion (THD) Term used to quantify distortion as a percentage of the fundamental (pure sine) of voltage and current waveforms.

Uninterruptible power supply (UPS) A system consisting of a rectifier/charger, a battery bank, a static inverter and a bypass switch, used to protect against short-term service interruptions and outside power disturbances.

INTRODUCTION

Electric system management is the process through which a facility management team ensures that a plant or building receives a sufficient, reliable supply of power at the level of quality required. This process also involves seeking to obtain that power at the lowest possible cost for the required class of service.

Until recently the main objective of electric system management was to guarantee uninterrupted operation of a facility's lighting, process, and environmental [heating, ventilating, and air conditioning (HVAC)] systems. This "lights on, motors turn" approach originated during the late nineteenth and early twentieth centuries, when energy consumption was dominated by linear loads. In recent years, however, nonlinear loads such as variable-speed motors, programmable logic controllers, and other electronic equipment have proliferated. Compared with linear loads, these nonlinear loads are much more sensitive to overvoltage, undervoltage, and other disturbances that have always existed on the utility power line. Such routine disturbances can cause problems ranging from minor equipment malfunctions to costly system shutdowns and damage to equipment. In addition, nonlinear devices can create their own power disturbances, which in turn cause problems in other parts of the facility and may feed back onto the utility distribution system.

The increased reliance on nonlinear loads has added new objectives to electric system management. While the guaranteed supply of power remains critical, issues of reliability and power quality are becoming paramount, and capacity requirements have increased. In addition, the need continues to control energy use and costs to remain competitive. In the face of these challenges, *plant managers must seek to identify their facility's risk of experiencing reliability and power quality problems and assess the economic impact of problems that might occur.* Then, working with a utility or consulting engineers, managers can design and implement a cost-effective risk management program.

This chapter gives an overview of the reliability and power quality issues involved in electric system management, and describes available remedies to power quality problems. In addition, it discusses utility rate structures, as well as demand-side management programs and other strategies for reducing or controlling energy use and/or demand.

RECENT DEVELOPMENTS*

Current upheavals in the electric utility industry have resulted in sharp price increases and unscheduled outages in many areas of the United States. The plant engineer can best deal with these problems with a thorough review of the parameters described in this chapter and by working closely with the local utility. Alternative and emergency power sources now warrant increased investigation. Inputs from established electric power consultants at this time would be particularly prudent to coordinate long-range planning.

RELIABILITY

A reliable power supply is one that delivers electricity sufficient to serve a facility's load at the grade of power quality desired, and one that provides for enough power during curtailment or other emergency conditions to ensure the safety of personnel and protection of critical processes and process equipment.

Quantity

To determine if a facility's electric service is sufficient for the load being supplied, it is necessary for plant managers to conduct a facility load profile. The profile will provide the management team with a thorough understanding of how a facility's electricity consumption varies hourly, daily, and seasonally.

One way to identify electricity consumption patterns is through analysis of a facility's demand chart. Electric utilities usually maintain records on kW (or kVA) demand in 15- or 30-min intervals to permit identification of and subsequent billing for the peak demand established during the billing period. Alternatively, the utility's metering system may incorporate electronic pulse recording that transmits and records similar information on magnetic tape. These data may be retrieved and used to compile a demand chart. If no utility records of customer demand are available, the utility should be asked to install demand recording instrumentation so load patterns can be analyzed.

Service voltages should also be examined to determine if voltage control measures are required. Table 3.2 shows national voltage standards set by the American National Standards Institute (ANSI) Guideline C84.1 for electrically operated equipment. In practice the voltage delivered to service entrances can vary, but most major utilities adhere to minimum and maximum voltage standards that are well within the ANSI standards.

* Contributed by the editor-in-chief.

TABLE 3.2 Customer Service Voltages

Nominal service voltage	ANSI C84.1 minimum utilization voltage*	PG&E's minimum service voltage	PG&E's maximum service voltage
120	108	114	126
208	187	197	218
240	216	228	252
277	249	263	291
480	432	456	504

* American National Standard Institute's C84.1 shown for customer information only. The utility has no control over voltage drop in customer wiring.

Quality

In terms of power quality, conventional utility service "to the fence" is not (nor has it ever been) 100 percent reliable. For some utility customers willing to pay a premium, the power supply may be made nearer to 100 percent reliable. Even at this higher level of service it may be necessary for some users to provide an in-house power conditioning system. (A more detailed discussion of power quality and power conditioning appears below.)

Performing routine maintenance procedures on process equipment on a regular basis will also help ensure improved reliability. See Sec. 5, "Basic Maintenance Technology."

Despite the use of diagnostic and preventive measures by plant management, unexpected power outages and other failures do occur. In such instances, a well-managed plant electric system provides emergency power—enough to permit an orderly shutdown of equipment. Backup power may be supplied by diesel generators or a device known as an uninterruptible power supply (UPS), described below in "Correcting Power Quality Problems." In a growing number of facilities it may be economically feasible to provide on-site power generation through a cogeneration system. Cogeneration systems utilize waste or purchased fuels to generate power and recover waste heat, which then can be used to produce process heat in the form of steam, hot water, or hot air. Determining the potential benefits, both tangible and intangible, of a cogeneration system requires an in-depth analysis of all aspects of a facility's energy needs.

POWER QUALITY

Sources of Power Quality Problems

Even though today's utilities use advanced hardware and software at their substations and on their distribution systems, power disturbances occur. These irregularities can result from transmission or distribution system switching faults on the utility distribution system, lightning strikes, simultaneous operation of equipment (either within the plant or by customers nearby) or other causes. In many cases the disturbances can be traced to wiring and grounding problems within the plant itself. Common disturbances are outages, under- or overvoltages, spikes, sags or surges, or noise. (See Fig. 3.1.) These disturbances can range in duration from sustained outages lasting several hours to surges lasting only a few microseconds and undiscernible to plant operators.

The introduction of harmonic distortion to a system may take many forms. Determining acceptable levels of total harmonic distortion (THD) may require input from equipment manufacturers as well as engineering expertise. Inattention to THD in a sensitive circuit can lead to chasing phantoms.

Older electrical equipment such as motors, solenoids, and electromechanical controls are largely unaffected by disturbances of short duration. However, solid-state electronic equipment is far more susceptible to a wide range of disturbances. This vulnerability stems from the way an electronic device consumes the alternating-current (ac) power supplied to it—"chopping" it into the low-voltage, high-speed power it needs for digital processing. Problems arise when the alternating-current (ac) sine wave in the power supply deviates from its normal "clean" waveshape. When the sine wave becomes distorted, or "dirty," electronic devices are unable to convert ac power to direct-current (dc) power. As a result, they can experience interruption, data errors, memory loss, and even shutdown; in some cases the device may sustain damage.

Moreover, because of the way they draw current, *electronic devices can actually create their own power disturbances* (in the form of harmonic distortion, impulses, and voltage loss), and introduce these disturbances to the power distribution system—within a facility and on the utility line.

Problem	Definition	Duration	Cause	Effect
Outage 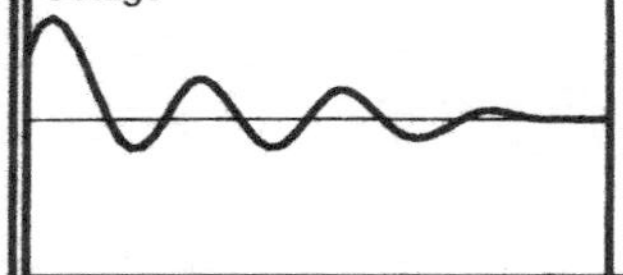	Planned or accidental total loss of power in a localized area. A blackout is a wide-ranging outage.	Minutes to a few days	Catastrophic system failure, weather, small animals, human error (auto accidents, kites, etc.).	System shutdowns.
Dropout 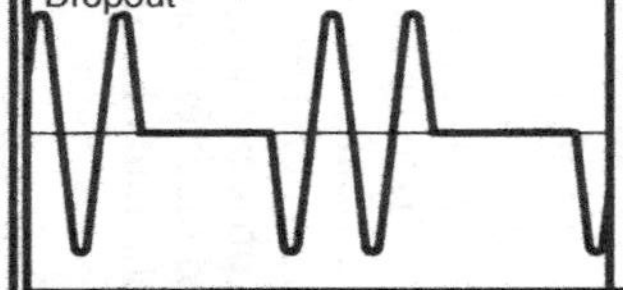	A very short planned or accidental power loss.	1 ms to 1 s	Utility switching operations attempting to maintain power to your area despite a failure somewhere on the system.	Equipment resets, data loss.
Sag / Surge 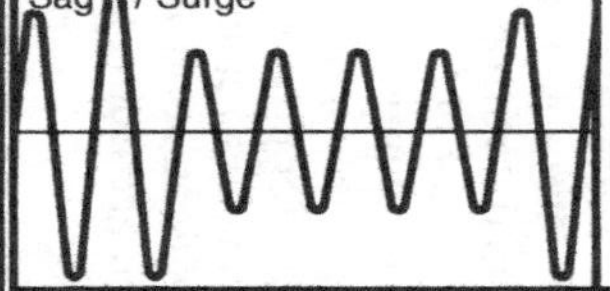	A decrease or increase in voltage that can be up to 20% above or below the normal voltage level (also called an over- or undervoltage).	Less than 2.5 s	Heavy load switching, air conditioning, disk drives, transformers, and other equipment drawing large amounts of power.	Memory loss, data errors, flickering or dimming lights, shrinking display screen, equipment shutdown.
Spike 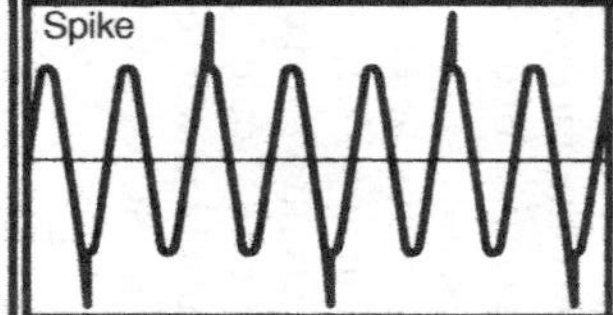	A sharp sudden increase or decrease in voltage of up to several thousand volts. Also called an impulse, transient, or notch.	1 μs to 1 ms	Utility switching operations, on-and-off switching of heavy equipment or office machinery, SCRs' firing, elevators, welding equipment, static discharges, and lightning.	Loss of data, burned circuit boards.
Noise 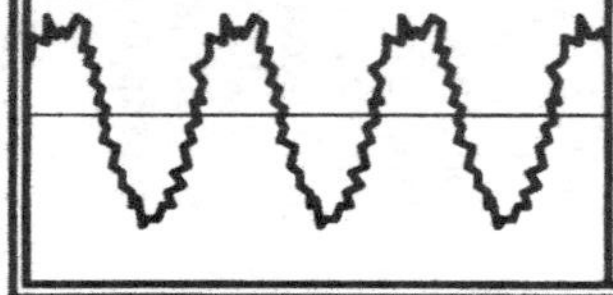	A high-frequency interference from 7000 Hz to 50 MHz.	Usually of constant duration	Electromagnetic interference, microwave and radar transmissions, radio and TV broadcasts, arc welding, heaters, laser printers, thermostats, electric typewriters, loose wiring, improper grounding.	Although generally not destructive, it can garble or wipe out stored data.

FIGURE 3.1 Power problem categories.

Monitoring Power Quality

When power quality problems are suspected, the plant manager should initiate a power quality survey of the plant to determine whether equipment troubles are attributable to utility operations or to conditions within the facility. The survey should include a thorough inspection, including an infrared scan, of the site's electric system, including wiring and connections, grounding, equipment closets and utility rooms, transformers and power conditioning equipment, and main and subbreaker panels. It is advisable to monitor at multiple locations, including the problem locations, such as transformers, service entrance, and any other suspect areas. Monitoring can be conducted by either qualified in-house staff or with the assistance of utility personnel or consulting engineers.

Before monitoring begins, plant managers should establish acceptable limits for sensitive equipment. Some standard thresholds recommended by Basic Measuring Instruments of Foster City, Calif., are shown below. These may be modified according to the sensitivity of equipment at the site.

- Frequency tolerance: 0.1 Hz
- Swell voltage: 5 to 10 percent above nominal
- Sag voltage: 10 to 15 percent below nominal
- Impulse: two times nominal voltage
- Neutral-to-ground voltage: 2 to 5 V
- Neutral-to-ground impulse: one times nominal voltage
- High-frequency noise: 5 V
- Radio-frequency interference: 3 V/m
- Temperature: high and low temperatures depend on application; rate of change should not exceed 10°F (5°C) per hour.
- Humidity: high—70 percent relative humidity; low—30 to 40 percent relative humidity; rate of change should not exceed 20 percent relative humidity per hour

Figure 3.2 shows a checklist to follow in analyzing power quality problems.

Correcting Power Quality Problems

There are several types of remedies available to protect solid-state, power-sensitive equipment from power disturbances; most are simple and inexpensive.* In addition, disturbances can be prevented altogether by conditioning the power supply to smooth out the sine waveshape. Because power conditioning equipment is costly, it is best suited only for those applications requiring the highest grade of power.

Wiring and Grounding. According to the Electric Power Research Institute, approximately 80 percent of power quality problems at commercial and industrial facilities can be traced to problems within the facilities—improper grounding, inadequate wiring, loose connections, and the accumulation of dust and dirt from poor maintenance practices. The importance of a good, low-resistance ground cannot be overemphasized, especially since solid-state systems depend on the grounding for a reference to operate by and for dissipating stray power that could cause damage if left on the circuit. Adequate wiring and proper grounding are the lowest-cost prevention and cure for power quality problems.

In addition, care should be taken to ensure proper sizing of in-house transformers and conductors that supply power to silicon controlled rectifiers (SCRs). It is imperative that the circuit supplying an SCR be sized according to the manufacturer's recommendations and those

* The following material is based on "Power Quality in Your Business," Pacific Gas & Electric Co., San Francisco.

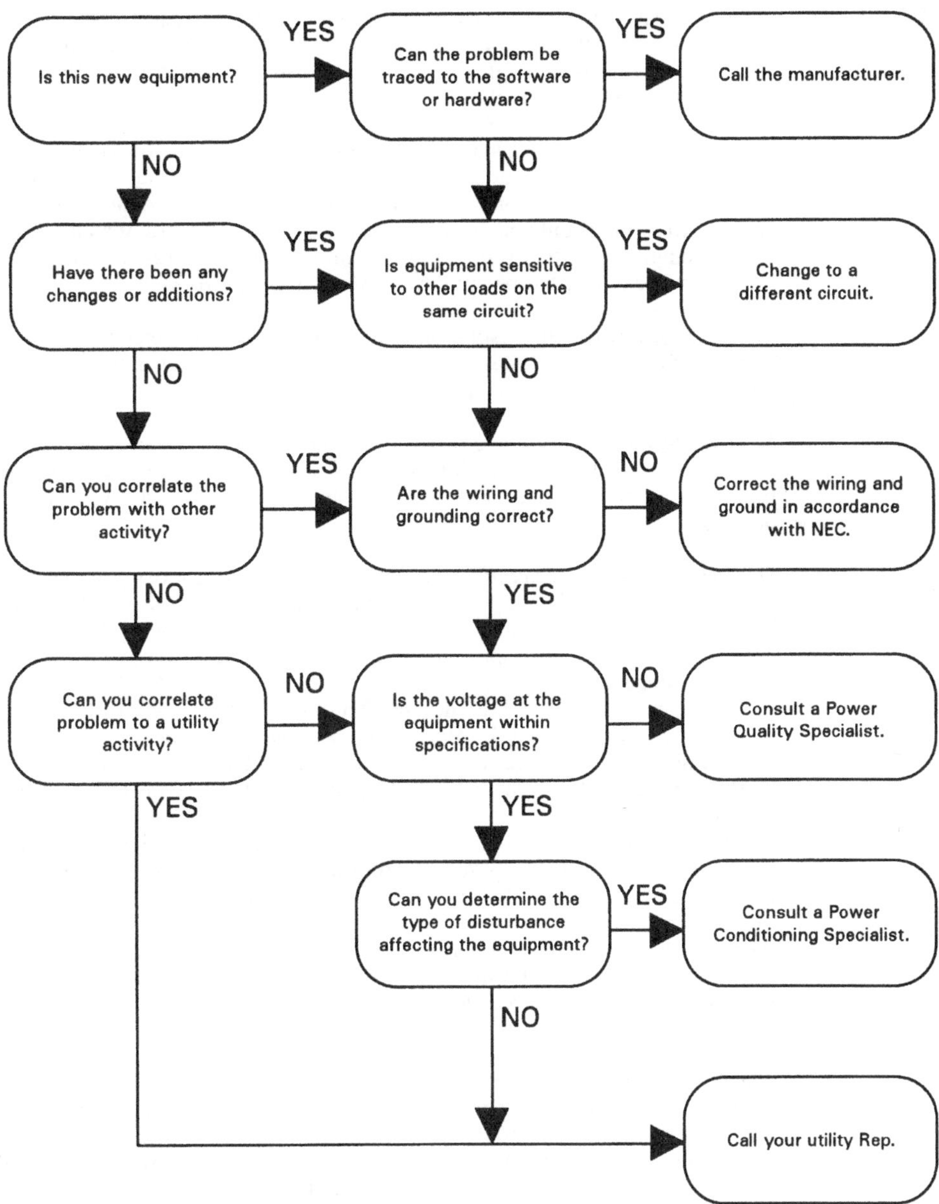

FIGURE 3.2 Power problem flowchart.

of NEC and IEEE. SCRs have a potential to severely affect the total harmonic distortion of a power system. Utilization of isolation transformers may be necessary to prevent export of distortion throughout the facility.

Dedicated Circuits. Most power disturbances in the form of *noise* (distortions at frequencies above 5 kHz) are generated within the plant itself. As a result, an effective method of protecting critical or highly sensitive equipment is to locate the equipment on its own isolated circuit to protect it from power disturbances caused by other equipment in proximity. Dedicated circuits also prevent a circuit from being overloaded by tying it directly to the power source and by restricting access to it.

Spike Suppressors. Spike suppressors reduce the amplitude of voltage spikes to safe levels and can eliminate many sudden changes in voltage. They are the simplest and least expensive protective devices; however, their capability depends on the quality of the suppressor purchased. Attention should be given to the specific nomenclature of a unit prior to installation. The attenuation ability of some units is minimal. Some suppressors feature a diagnostic light to indicate the device has been hit by a spike but is still working.

Isolation Transformers. Isolation transformers filter out electrical noise and distortion from other on-site equipment or incoming power. They cannot, however, protect against other types of disturbances such as spikes and surges.

Voltage Regulators. Voltage regulators maintain a relatively constant voltage by protecting against surges and sags through mechanical or electronic means. This option is more costly than those listed above, but is at the midpoint of the cost spectrum for power enhancement devices.

Uninterruptible Power Supply. A UPS protects against short-term power interruptions and outside power disturbances. UPS systems typically consist of a rectifier/charger, a battery bank, a static inverter and an automatic or manual bypass switch. DC power is supplied to the inverter by the rectifier or batteries and converted to ac power, which is then fed to the protected equipment.

There are two types of UPS designs. An on-line UPS is continually fed the supply power and produces clean output. It offers protection against all power quality problems, including momentary outages. Protection against sustained outages is limited to the size of the battery bank. An on-line UPS typically has the capability to switch to the ac power source if some element of the UPS fails.

In an off-line UPS, electric supply power is connected directly to the load; the UPS is utilized only during ac power source interruptions. It alerts the user that supply power failure has occurred and that the load is now on battery power, which will last from about 5 to 15 min. The off-line UPS design does not protect against transients, sags, swells, or other abnormalities, though some models have integrated a few on-line power conditioning features that operate when the unit is engaged on-line. Off-line UPSs are best suited for small loads (up to 1.5 kVA) that can tolerate the milliseconds of switching time required when the power supply is interrupted.

Electric Motor-Generators (MG Set). An MG set uses a motor driven by incoming utility power to spin an ac or dc generator that, in turn, produces electric power for the protected load. As a result, the protected load is electrically isolated from the incoming utility supply. An MG set has the capability to ride through momentaries; during sags it provides buffering to prevent transients from occurring on its output.

If equipped for both ac and dc generation, an MG set also can charge batteries for a UPS that supplies power during an outage for the duration of the battery charge—generally long enough to allow an orderly shutdown of a facility's computer or processing systems. MG sets are less energy-efficient and more expensive than voltage regulators, and require maintenance; however, they provide more complete protection.

UPS with Diesel Generator. These systems combine an on-line UPS with backup generation to deliver a full, clean power supply during extended power outages. A battery bank keeps computers and processes in operation until the generator comes on-line.

Static Automatic Transfer Switches. Static automatic transfer switches provide an alternative to full UPS protection if a facility is served by two synchronized utility power feeds. When the power supply to one feed is interrupted, the load is automatically transferred to the second power feed without affecting equipment.

Managing Risk

The majority of power quality problems can be corrected or avoided by implementing simple measures. Good risk management entails spending only as much as required to avoid catastrophic losses of data and/or equipment. In most instances, expensive power-conditioning equipment is unnecessary; basic mitigation measures (e.g., rewiring or reconfiguring grounding) will suffice.

As knowledge of power quality issues grows, educational opportunities for facility management teams are increasing. Many utilities, industry associations, and universities offer curricula designed to assist participants in diagnosing power problems and identifying practical, cost-effective solutions.

UTILITY RATE STRUCTURES

Utility rate structures form the basis for utility compensation for energy and demand capacity delivered. Knowledge of available options and their corresponding rate schedules is fundamental to developing a good managerial strategy. This section describes basic concepts and terms related to utility rate design and outlines some strategies for controlling consumption and reducing costs.

Utility rate structures are typically complex contracts individually tailored to each utility's load profile, power generating methods, fuel, and energy sources. Rates are also influenced by the distance between the utility's generating plants and the user, which affects power transmission costs. Because it is impossible to define a general rate schedule, the responsibility rests with the plant manager to become familiar with the rate structures of the electric utility serving each facility. Utility representatives will assist plant staff in interpreting rate schedules.

Energy and Demand Charges

Electric utility rates have two main components: energy charges and demand charges. Energy charges are based on the amount of energy consumed, measured in kilowatthours (kWh), during the billing period. Demand charges are based on the customer's maximum demand, measured in kilowatts (kW), during a specified period, usually monthly or during specific intervals when a utility experiences its highest demand (peak demand). Demand charges are designed to compensate a utility for capital and operating expenditures required to meet customer demand. The kW demand is defined as the average rate of energy draw from the utility system during a specified time segment—usually 15 or 30 min.

Demand charges also vary by season. Most utilities experience peak demand during the summer months, when air-conditioning use is highest. Other utilities experience peak demand in the winter months, when electric heating is the dominant load. In an effort to impose uniformity of loads throughout the year, utilities may also assess a demand penalty. For example, a summer-peaking utility may apply the demand charge established during the summer (period A) as a minimum demand charge during the winter (period B). This billing formula is commonly referred to as a *ratchet*.

Demand Reduction

Because energy costs affect a business's bottom line and its ability to remain competitive, every plant manager should develop a plan to control or reduce demand.

A facility's demand has two components: base load and variable load. The base load, which is fairly constant, represents the electric service required to maintain a facility's environment and comfort conditions (e.g., lighting and HVAC). The variable load, which is superimposed on the base load, depends on the business activity of the facility, weather, and other variables.

Base load reduction can be achieved by implementing programs to optimize the efficiency of lighting, motors, HVAC, process or other systems. (For further discussion of energy conservation options, see Chap. 4.14, "Energy Conservation.")

Variable load reduction can be obtained in several ways. In evaluating specific approaches, consideration should be given to the utility rate structure, the facility's electric system, and its relation to operations. Some potential strategies include:

- Inhibiting the simultaneous operation of large loads (e.g., large chillers or heat-treatment furnaces) during the period used by the utility to establish demand
- Rescheduling certain operations to other periods of the day or night to minimize their impact on peak demand
- Generating power on-site (using emergency equipment) to serve a particular load or group of loads during periods of high activity, thereby avoiding establishment of a new, higher peak demand
- Storing energy during inactive periods and retrieving it during high-demand periods (e.g., precooling; thermal storage, such as chilled water or ice storage; or preheating domestic water or other process fluids)

Rates

Utility rate schedules typically are designed to encourage use of electricity during off-peak hours when the utility's generating costs are lower. For example, a utility may offer off-peak rates that are considerably lower than on-peak rates, or it may provide incentives, such as reduced demand charges, to maximize load during nondaylight hours. This *time-of-day* billing strategy offers utility customers the potential to substantially reduce their energy costs by judiciously rescheduling certain plant operations to off-peak hours.

In recent years a few utilities have begun to experiment with another rate option known as *real-time pricing*. Under real-time pricing, the rate paid by customers can vary by the hour (or even smaller time intervals) to reflect the changes in the marginal cost to the utility of producing electricity throughout the day and night. Plant managers should evaluate real-time pricing carefully to determine its cost-effectiveness for particular applications; in some cases real-time pricing may actually raise energy costs.

Demand-Side Management

The term *demand-side management* (DSM) refers to measures taken by utilities to influence the level or timing of their customers' energy use in order to optimize the use of existing utility assets and defer the addition of new generating capacity. These measures take many forms, including:

- Financial incentives such as rebates and low-cost financing to encourage upgrading to higher-efficiency appliances or systems

- Informational services such as energy audits, publications, and seminars to alert customers to energy-saving opportunities
- Rate incentives such as real-time pricing or time-of-day rates to encourage customers to clip or shift loads during peak periods
- Fuel substitution programs to promote the use of one fuel over another

The main customer benefits derived from implementing DSM measures are increased efficiency and lower energy costs. It is important to note, however, that some DSM measures may also introduce power quality problems. For example, some electronic ballasts for fluorescent lighting can cause harmonic distortion. Similarly, variable-speed motors, though highly efficient, often have detrimental effects on the operation of nearby equipment requiring clean power because they introduce harmonics and/or noise to the system. For this reason, careful evaluation of the power quality characteristics of new, high-efficiency equipment is required before installation.

Power Factor Correction (Improvement)

Power factor indicates the degree to which a customer's electrical equipment causes the electric current delivered to the site to lag or lead with the voltage sine wave; in other words, it is a measure of how much reactive power the equipment requires to operate. Power factor is calculated by using metered measures of the amount of power used (in kilowatts) and the amount of reactive power used (in kilovolt-amperes—kVA). Lower power factors indicate the use of greater amounts of reactive power.

To recover the cost of supplying large amounts of reactive power, some utilities impose a financial penalty on customers with large electric loads. Even so, the economic incentive to improve power factor is minimal. When penalties are assessed, they may not be high enough to justify the cost of power factor correction.

If power factor improvement is desired, individual load correction procedures include:

- Applying high-power-factor utilization equipment such as high-power-factor lighting ballasts and distribution transformers
- Operating induction motors at near full load to improve the power factor of the motor
- Adding capacitors in stepped banks with automatic controls

Adding capacitors, however, can introduce system disturbances, including transient voltage, harmonic resonance and transient voltage magnification.

In general, it is best to avoid power factor problems altogether by purchasing or upgrading to equipment and systems that promote good power factor, and by correctly sizing all equipment, including conductors and transformers.

Customer-Owned Substations

In an effort to reduce energy costs, some industrial customers have opted to purchase distribution substations from their local utility. This option may reduce energy costs up to 10 percent or more, but it carries some risk because the utility is no longer responsible for repairing and maintaining the substation. In addition to the added repair and maintenance costs, plant staff may encounter difficulty in locating replacement transformers or other parts when problems occur. Service and maintenance expertise also may not be readily available.

BIBLIOGRAPHY

Basic Measuring Instruments, Foster City, Calif.:

Application Note 227: "How to Do a Power Quality Survey," undated.

Application Note 229: "How to Do IEEE Standard 519-1992 Compliance Testing," undated.

"8800 Powerscope: Performance, Productivity and Practicality for Power Disturbance Analysis," 1993.

"Electric Power Analyzers: Tools for Measuring Electric Power Quality, Power Flow, Disturbances, and Demand," 1993.

McEachern, Alexander: *Handbook of Power Signatures,* 1989.

Electric Power Research Institute, Palo Alto, Calif.:

"Wiring and Grounding for Power Quality," CU.2026.3.90, 1990.

"Power Quality Considerations for Adjustable Speed Drives," CU.3036.4.91, 1991.

PG&E Power Quality Enhancement, San Francisco:

"Power Notes on Power System Harmonics," January 1993.

"Power Notes on Surge Suppressors," July 1990.

"Power Notes on Uninterruptible Power Supply," January 1993.

CHAPTER 3.3
WATER PURIFICATION AND TREATMENT

J. Stephen Slottee
Donald C. Taylor
Frank A. Baczek
Baker Process
Salt Lake City, Utah

Peter Metcalf
Koch Industries
San Diego, California

Richard J. Eisman
The Coombs-Hopkins Company
Carlsbad, California

Mary Martis
WATER 3 ENGINEERING
Escondido, California

Paul Crawford
ZENON Environmental Inc.
Oakville, Ontario

Robert W. Okey
University of Utah
Salt Lake City, Utah

Robert Emmett
Emmett Process Consulting Company
Salt Lake City, Utah

GLOSSARY

Activated carbon A highly adsorptive material used to remove organic substances from water.

Aeration The process of bringing water and air into close contact to remove or modify constituents in the water.

Chemical coagulation Destabilization of colloidal and suspended matter by the reduction of electrostatic repulsive forces between particles with chemicals.

Disinfection The water treatment process that kills disease-causing organisms in water, usually by adding an oxidizing agent.

Filtration The water treatment process involving the removal of suspended matter by passing the water through a porous medium.

Flocculation The water treatment process following coagulation, which uses gentle stirring to bring suspended particles together so they will form larger, more settleable particles called *floc.*

Hardness A characteristic of water, caused primarily by the salts of calcium and magnesium.

Ion exchange A reversible process where ions of a given species are exchanged between a solid (ion-exchange resins) and a liquid for an ion of another species.

Precipitation An aqueous chemical reaction that forms an insoluble compound which is commonly removed by sedimentation or filtration.

Reactivate To remove the adsorbed materials from spent activated carbon and restore the carbon's porous structure so it can be used again.

Sedimentation The water treatment process that involves reducing the velocity of water in basins so the suspended material can settle out by gravity.

Sludge dewatering A process to remove a portion of water from sludge.

Softening The water treatment process that removes the hardness-causing constituents in water—calcium and magnesium—by precipitation or ion exchange.

Solids contact A process where coagulation, flocculation, and sedimentation rates and efficiency are enhanced by previously formed solids.

INTRODUCTION

The unit operations for water purification and treatment include the following:

Precipitation	Crystallization
Sedimentation	Evaporation
Filtration	Activated-carbon treatment
Ion exchange	Reverse osmosis
Disinfection	Other membrane separations

The following industries are major users of purified water:

Power generation	Petroleum
Beverage	Pulp and paper
Aluminum	Chemical
Iron and stee	lFood
Textile	Electronics

Figure 3.3 illustrates the unit operations most commonly used for treating water. Water is used primarily for steam generation (boiler feed), industrial processes, cooling tower makeup, and potable water supply.

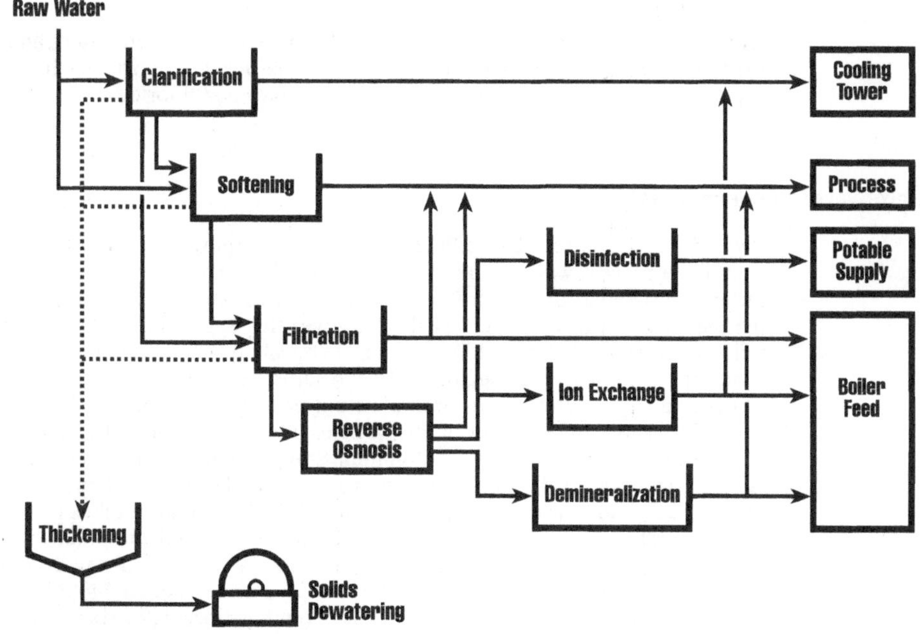

FIGURE 3.3 Water treatment diagram.

IMPURITIES IN WATER

All natural waters contain suspended or dissolved inorganic or organic chemicals to some degree. Whether they are present in high enough concentration to be considered as impurities depends on the water use.

The impurities can be classified as follows:

Inorganic	Organic	Biologically active
Suspended	Suspended	Bacteria
Colloidal	Immiscible	Viruses
Dissolved	Miscible	Algae
	Soluble	Protozoa

In turbulent streams, suspended solids range from small pebbles down to colloidal clay particles 0.1 to 0.001 μm in diameter. The water may also contain organic solids, algae, and bacteria. Dissolved inorganic solids are usually bicarbonates, sulfates, and chlorides of calcium, magnesium, and sodium, as well as compounds of silica, iron, and manganese. Nonferrous metals and organic compounds may be present in low concentrations, which nevertheless exceed the EPA's proposed limits.

The presence of dissolved calcium and magnesium compounds (termed *hardness*) leads to scale formation in boilers and fouling of evaporative cooling systems, and is the biggest cause of plant water problems. Other dissolved solids may be considered as impurities, depending on their concentrations and the intended water use.

SEDIMENTATION PROCESSES AND EQUIPMENT

Sedimentation processes are used to clarify turbid and/or colored waters, and are used in conjunction with chemical precipitation to remove dissolved impurities such as iron, manganese, calcium, and magnesium compounds, as well as silica and fluorides.

Removal of Suspended Solids

On a few occasions, unaided sedimentation will be employed to remove suspended solids, but usually the process will include the addition of chemicals to improve removal of solids. The particles suspended in water result in a turbid, or colored, appearance that is objectionable, and they have a static charge (usually negative) that causes the particles to repel each other and remain suspended. By adding certain chemicals, it is possible to neutralize these charges, permitting the particles to agglomerate and settle from the liquid more effectively. The current practice is to refer to this neutralization or destabilization step as *coagulation* and the subsequent gathering together of the particles into larger, more settleable "flocs" as *flocculation.* Inorganic chemicals such as aluminum sulfate (alum), ferrous sulfate (copperas), ferric chloride, and sodium aluminate, as well as a long list of organic polymers, are used for coagulating and flocculating suspended matter in water. Coagulation takes place very quickly—a few seconds to 1 minute—and the chemicals should be added with intense mixing in order to obtain maximum efficiency. Flocculation, on the other hand, normally requires detention periods of 20 to 45 min, and, once the chemicals have been added, should be accomplished by relatively gentle mixing. The mixing is carried out in a flocculation basin, and its purpose is to bring about the maximum collisions between the suspended particles without shearing or breaking apart the particles that have already been formed.

This will limit the particle velocities to the range of 1 to 6 ft/s (0.305 to 1.83 m/s), depending on the "toughness" of the floc that is produced. Equipment should be provided for varying the intensity of the mixing so that the optimum velocities can be achieved.

Organic polymers (polyelectrolytes) are frequently used as flocculant aids. They facilitate the gathering of the already coagulated or destabilized particles into larger and less fragile floc particles that have better settling characteristics. The type and amount of chemicals required to treat a given supply of water are best determined by laboratory jar tests. These tests should be made on fresh samples at the same temperatures and other conditions that will be present in the full-scale plant. No other reliable method has been found for predicting the best chemicals and optimum dosages for clarifying water. Typically, dosages of inorganic coagulants (alum, ferric chloride, etc.) might range from 10 to 100 mg/L, cationic organic coagulants from 1 to 5 mg/L, and anionic organic flocculants from 0.1 to 1.0 mg/L. Many chemical suppliers and equipment manufacturers will perform jar-test studies at reasonable fees and make recommendations as to the best chemical, equipment, and sizing for treating a water supply.

Removal of Dissolved Impurities

Generally, sedimentation processes are used primarily for the removal of suspended material from water, but the removal of dissolved mineral impurities such as in the lime or lime–soda ash softening process is an equally important aspect of sedimentation in the treatment of water. In the softening process, hydrated lime or hydrated lime and soda ash are added to react with the dissolved CO_2 and the calcium and magnesium salts that commonly cause the hardness of water. The following equations describe some of the reactions that take place in the formation of the calcium carbonate and magnesium hydroxide precipitates.

$$CO_2 + Ca(OH)_2 \rightarrow CaCO_3 + H_2O$$

$$Ca(HCO_3)_2 + Ca(OH)_2 \rightarrow 2CaCO_3 + 2H_2O$$

$$Mg(HCO_3)_2 + 2Ca(OH)_2 \rightarrow 2CaCO_3 + Mg(OH)_2 + 2H_2O$$

$$CaSO_4 + Na_2CO_3 \rightarrow CaCO_3 + Na_2SO_4$$

A coagulant is normally added along with the lime and soda ash to improve clarity of the product water. When the lime requirements are high, economies can usually be realized by using quicklime (CaO) in a lime slaker to convert it to the hydrated lime [$Ca(OH)_2$]. When requirements reach approximately 200 lb/h (90 kg/h), the economies or quicklime should be investigated.

Neither calcium carbonate nor magnesium hydroxide is completely insoluble, so some amount, depending upon the type of treatment, temperature, and other conditions, will remain in solution. Unlike most compounds, $Mg(OH)_2$ and $CaCO_3$ are less soluble at higher temperatures.

Temperature and changes in temperature of the liquid being treated are extremely important considerations in the design and operation of sedimentation units. The rate at which a particle settles in water is inversely proportional to the kinematic viscosity, a property that varies with the temperature. Thus, the settling rate of a given particle at 40°F (4.4°C) is only 63 percent of what it would be at 70°F (21.1°C). Rapid changes of inlet water temperatures to settling units will cause thermal currents, which at best are disruptive to the settling of particles and at worst are totally upsetting. Manufacturers typically limit changes to 2°F (1°C) per hour in their performance warranties. The rate at which chemical reactions proceed in water is higher at higher water temperatures. A common rule of thumb is a doubling of the rate for each 18°F (10°C) increase in temperature. In general, warm but constant water temperature is desirable in treating water.

Conventional Sedimentation

Sedimentation, with or without chemical treatment, is usually carried out in continuous, flow-through settling units with horizontal flow patterns (Fig. 3.4). Important exceptions to this are the solids contact clarifiers and reactors, which are discussed later. Settling units are normally rectangular, with the flow along the length of the basin, or circular, with the flow radially outward from a central inlet compartment. Mechanical scraping mechanisms are employed to move the settled sludge along the bottom to hoppers or sumps from which it is discharged. The mechanical design of the scraper mechanism should be carefully evaluated when considering such equipment, as should the inlet distribution and effluent collection system. The settling zone of the unit must provide sufficient area and volume so the bulk of the solids will settle before reaching the effluent collector. Proper design of the inlet distributor and effluent collector will result in the most effective operation of the settling zone. Typical settling-area designs should allow liquid flow to range from 0.35 to 0.75 gal/min · ft^2 (0.85 to 1.80 m^3/h · m^2) and settling-unit detention times from 2.5 to 4.0 h.

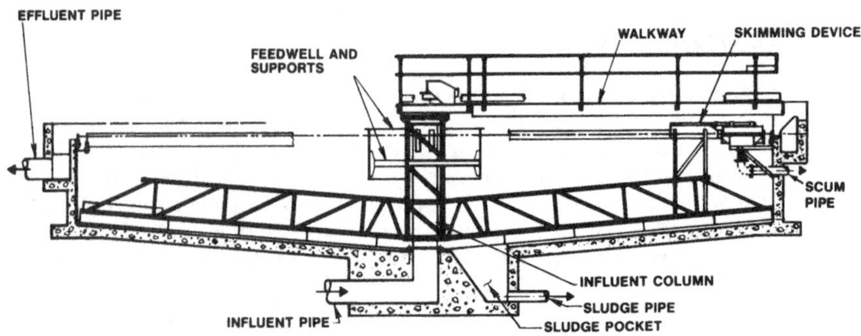

FIGURE 3.4 Illustrative section of a settling unit.

Solids Contact Sedimentation

Solids contact units combine, within the settling unit, the coagulation and flocculation functions with the ability to internally recirculate solids that have been formed by earlier reactions (Fig. 3.5). Besides the economies attainable by combining these functions within a single unit, improvements in settling characteristics and reaction rates permit higher design rates for this type of unit. Most industrial water treatment applications now use some form of solids contact or combination treatment equipment. Various units are available from manufacturers of such equipment, and a careful comparison of the process and the mechanical features of each is advisable.

Solids contact units are particularly well suited to lime–soda softening applications. Calcium carbonate and, to some degree, magnesium hydroxide have a tendency to supersaturate (remain in solution at considerably higher than theoretical concentrations), and solids contact operation reduces this tendency. By mixing the chemicals and the untreated water in the presence of recycled solids, a large surface area is provided on which the precipitate will form. This seeding not only reduces the supersaturation, but also results in the growth of larger particles that settle more rapidly. Overall, chemical usage efficiency is improved because the lower calcium and magnesium values attained through solids contact require no additional chemicals.

The hot-process softener, sometimes used for treating boiler feedwater, uses steam to heat the water to more than 200°F (93°C). Hot-process lime treatment has been used to remove silica from boiler feedwater where extremely low concentrations requiring ion exchange are not necessary. The silica is removed by adsorption on freshly precipitated magnesium hydroxide. The removal is therefore dependent on the amount of $Mg(OH)_2$ precipitated. The effective-

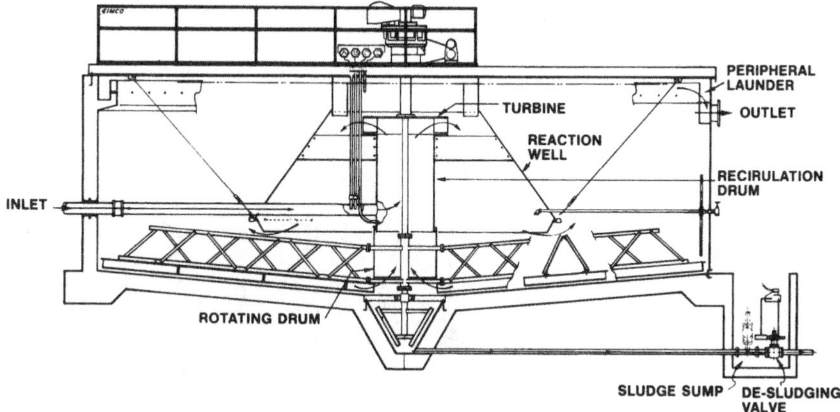

FIGURE 3.5 Illustrative section of a solids contact unit. *(Eimco)*

ness of the process is enhanced by the high temperatures and can be further improved by solids contact. Often there is insufficient natural magnesium available in the water, so it is supplemented by the addition of dolomite lime, which contains a high percentage of magnesium oxide. This is feasible in a high-temperature process where the magnesium oxide will hydrate, something it will not usually do at normal water temperatures.

Cold lime-treatment processes are now being employed to reduce silica for cooling-water systems in specially designed solids contact units. By providing intense mixing and pumpage for high solids contact concentration and long detention times, the silica can be reduced at cold-water temperatures to levels formerly attainable only in hot-process treatment. Some of these systems use a two-stage sedimentation process where lime treatment at a high pH is used in the first stage. Soda ash and CO_2 (carbonic acid) are added in the final stage for stabilization.

Tube and Plate Settlers

Tube and plate settlers have been used for many years as sedimentation units. There are a variety of designs, but basically all of them use submerged inclined surfaces with relatively close spacing that increases the separation surface area in a vessel that is smaller than conventional sedimentation units (Fig. 3.6). The water to be treated is passed between the surfaces at velocities which permit suspended solids to settle and to coalesce on the lower tube or plate surface. The angle of inclination (45° or more) enables the settled solids to slide downward into a sludge-concentration compartment located at the bottom of the treatment unit.

These units permit efficient clarification at detention times substantially less than those used in conventional clarifiers. Total settling detention times may be as low as 10 min, greatly reducing the plant size.

The units are cost-effective for many applications, even though the inclined tubes or plates add appreciably to the equipment cost. These settling units are best for discrete solids settling where detention time is not a significant factor.

GRANULAR-MEDIA FILTRATION

Granular-media filters are used in water treatment to remove relatively low concentrations of suspended matter. Typically the feed contains around 50 mg/L suspended solids, or less, and 90

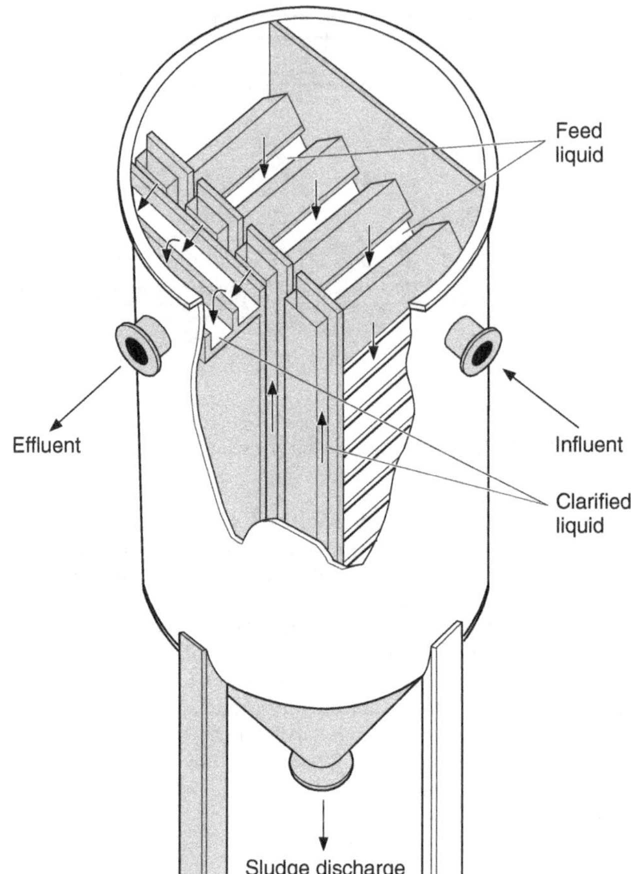

FIGURE 3.6 Stacked chevron clarifier (Delta-Stak™ clarifier). *(Eimco)*

percent removal can be expected. They are constructed in a variety of materials (concrete, steel, fiberglass, etc.), are either open-gravity or enclosed pressure type, and for downflow or upflow filtering. The filter media is typically silica sand, or a combination of sand and anthracite. The media is selected and sized for the particular application and performance desired.

The traditional filter is a batch operation where solids are collected in the filter media until the pressure drop becomes excessive or there is *breakthrough* and solids begin to pass with the effluent. The unit is then taken off-line, backwashed to remove the solids, and returned to service. The washwater volume is normally less than 5 percent of the filtered water produced. Many filters employ an air scour prior to, or simultaneously with, the water wash. If air is not used, a surface wash using high-pressure water jets should be included. A popular style filter used in many industrial applications incorporates a backwash water storage compartment above the filtering compartment (Fig. 3.7).

Filters are available in many different designs and configurations, most of which work well for their particular application. In installing new equipment, care should be taken in selecting equipment from a reputable supplier who has had experience with similar applications.

More recently, continuous-flow filters have received acceptance for many applications. They have the advantage of not requiring interruption of service for backwashing. Dirty media are continually removed from the bottom of the filter, cleaned in a separate media washer, and returned to the top of the media bed. The small amount (less than 5 percent) of dirty backwash

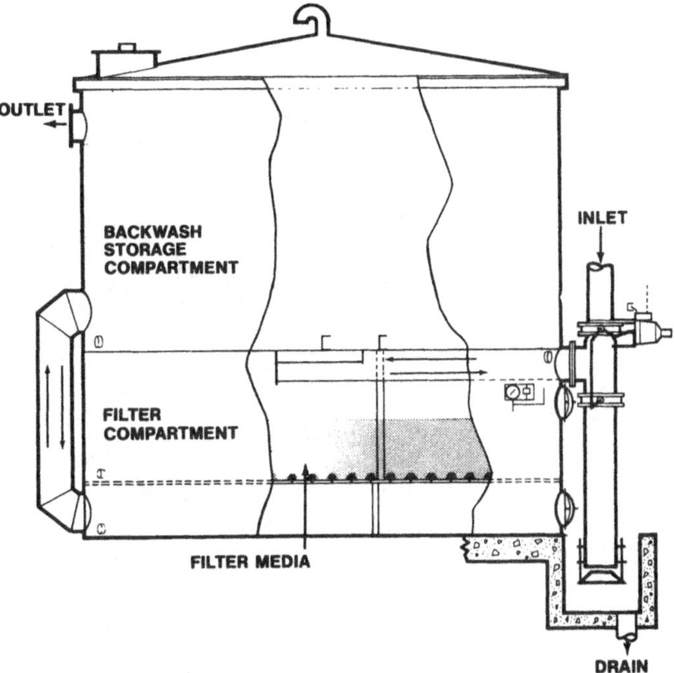

FIGURE 3.7 Granular-media filter.

water flows continually and can be more easily recycled to an upstream clarifier or otherwise disposed of than can the intermittent, high-volume backwash flows from traditional filters (Fig. 3.8).

CARTRIDGE FILTERS

Cartridge filters are broadly used in industry for filtering water and many other liquids. In water applications, they are generally the final polishing unit where high clarity is needed for the product or to protect downstream equipment or processes. Since the cost of replacing disposable cartridges can be significant, sizing the equipment and selecting the filtering material should be done carefully to optimize overall costs.

ION EXCHANGE

Applications of ion exchange for softening, dealkalizing, or demineralizing water include municipal water supply, boiler feedwater, and industrial process water.

Ion exchange is a process where ions of a given species are exchanged with another species. The most common form of ion exchange is found in domestic water softening. When water containing salts of calcium and magnesium is passed through a sodium exchanger, the sodium ions of the bed replace the calcium and magnesium to produce an effluent of close to zero hardness.

In addition to sodium ion exchange for softening water, other cation and anion exchangers are used in series and combination (mixed-bed ion exchange) to reduce the concentrations of all dissolved solids, and thus increase the water purity. The terms *demineralization* and *deion-*

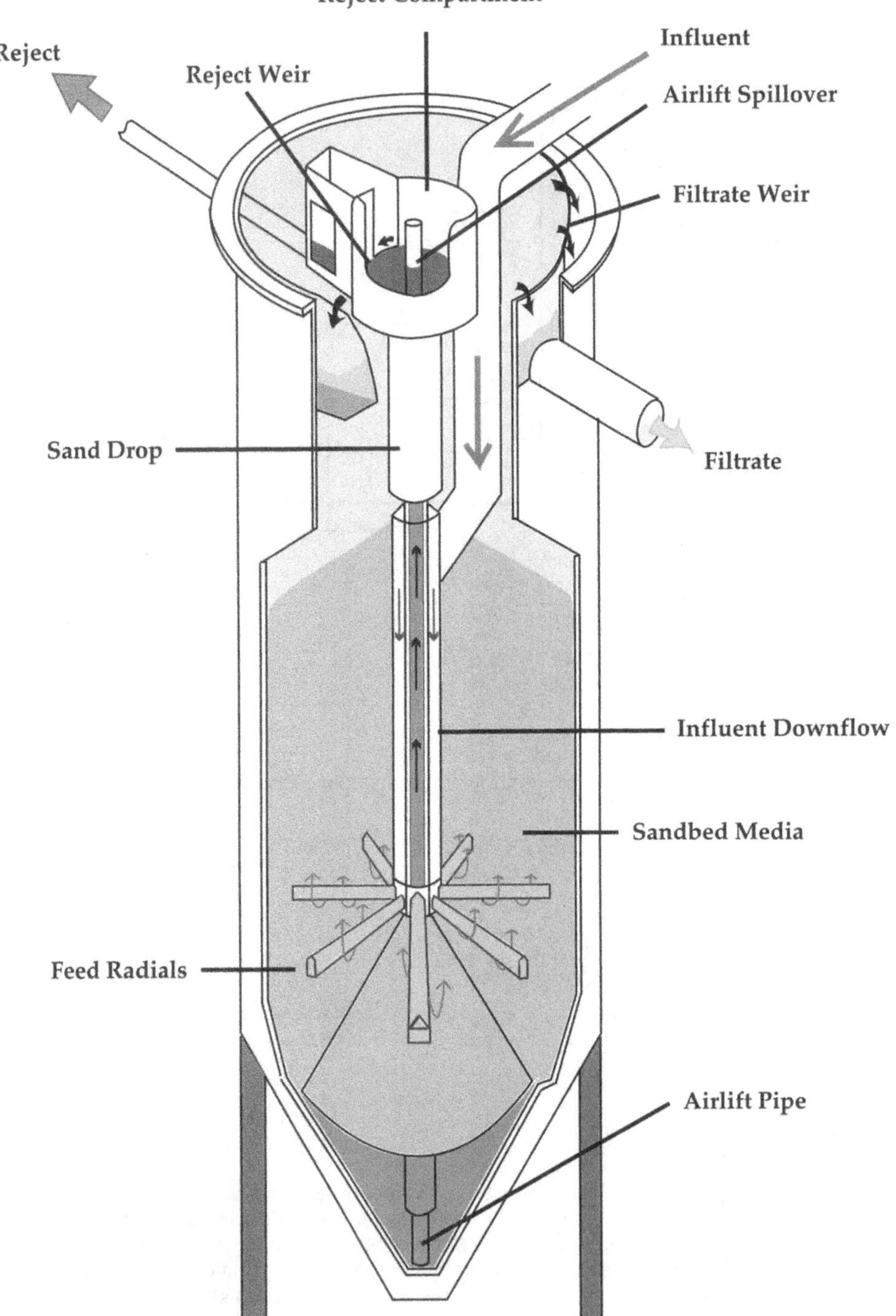

FIGURE 3.8 DynaSand® continuous-flow granular-media filter.

ization refer to the removal of all cations and anions from water. For demineralization, a cationic exchanger replaces hydrogen ions for cations (such as Ca^{2+}, Mg^{2+}, and Na^{+}), and an anionic exchanger replaces the anions (such as Cl^{-}, and SO_4^{2-}) with hydroxide, thus forming water and reducing the total dissolved salts concentration. Where high-purity water is required, mixed cation-anion resin beds are used.

When an ion-exchange resin is exhausted, the resin is regenerated. For the cationic sodium ion exchanger, a salt solution is used for regeneration. The rate and efficiency of regeneration are a function of the regenerating salt solution concentration and contact time with the resin. For example, the theoretical salt requirement for sodium ion exchange is about 0.17 lb (0.08 kg) per 1000 gr of hardness removed, and the typical efficiency is 40 to 60 percent. It is possible to obtain higher exchange capacities by increasing the amount of salt used per regeneration; for average industrial applications, the amount of salt used with the high-capacity resin cation exchanger is usually 0.4 lb/1000 gr (0.18 kg/1000 gr).

For other cationic exchangers, either sulfuric or hydrochloric acid is used. Sulfuric acid is generally used because of its low cost, but hydrochloric acid should be used when calcium-loaded resin would result in the precipitation of calcium sulfate and fouling of the resin. Anion-exchange resins are commonly regenerated with sodium hydroxide. Regeneration of mixed beds consists of separating the anion resin from the cation resin and regenerating each separately.

Cation- and anion-exchange resins may be weak or strong. Weak-acid resins are capable of removing metals that are associated with carbonate and bicarbonate alkalinity. Strong-acid resins are not as selective, and they can exchange a wide range of cations. The weakly basic resins remove the strongly ionized acids but not the weakly ionized ones. The effluent contains the same amount of silica as the influent water plus carbon dioxide equivalent to the bicarbonate alkalinity and free carbon dioxide. The concentration of carbon dioxide can be reduced to 5 to 10 mg/L in a degasifier or a vacuum deaerator. Strongly basic resins remove both strongly ionized and weakly ionized acids. At the end of each operating run, the weakly basic anion exchanger is backwashed, regenerated with a solution of sodium carbonate or caustic soda, rinsed, and returned to service. Caustic soda is used to regenerate the strongly basic resins.

Figure 3.9 illustrates the operating options for demineralizing systems. Selection of ion-exchange processes is based on the following factors:

- Raw-water characteristics
- Treated-water characteristics required
- Operating costs: chemicals, energy, waste disposal, qualified operating labor, and maintenance
- Capital costs
- Available space

In most cases, the services of a qualified consulting engineer should be used to evaluate alternatives and recommend the optimum method of treatment.

It is essential that the influent water to ion exchangers be free of suspended solids. Silt and organic matter are particularly objectionable, since these materials can deposit in the exchanger beds and reduce the capacity of the units by coating the exchange resins with films which either prevent or retard the movement of anions and cations through the resins. The influent water may require a preliminary treatment of sedimentation and/or filtration.

PRESSURE-DRIVEN MEMBRANE SEPARATION

Membrane separation processes are characterized by a thin film, a membrane, which acts as a selective barrier between two solutions. The membrane allows the passage of certain components, while restricting the passage of others, to accomplish a separation. The driving force

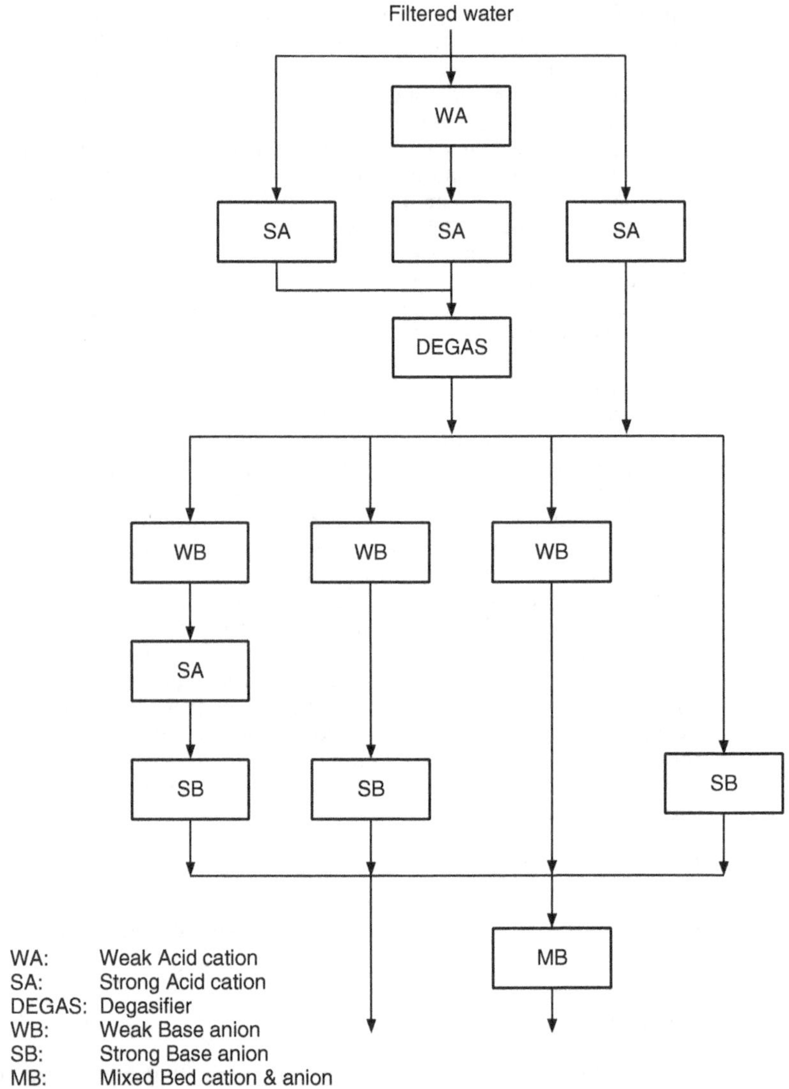

FIGURE 3.9 Alternatives for a demineralization system.

may be electrical (electrodialysis), concentration (dialysis, pervaporation), or pressure (reverse osmosis).

Pressure membrane separation is increasingly used for potable water treatment and in water reuse processes. Applications include softening, organic removal, and desalination of brackish well water, surface water, and seawater. For some applications, membrane processes offer the advantages of superior water quality, reduced chemical usage, less chemical waste production, and lower energy consumption.

Pressure-driven membrane processes include the following:

1. *Reverse osmosis* products separate dissolved inorganic salts from water. They are sometimes referred to as *hyperfiltration membranes.* Reverse osmosis membranes can be char-

acterized as being capable of passing <5 percent of feed-side chloride ions (monovalent, molecular weight = 35.5) to the permeate (or product) side of the membrane (single element basis).

2. *Nanofiltration* products also separate dissolved inorganic salts from water, but are "looser" than the reverse osmosis products. They can be described as passing 5 to 95 percent of the feed-side chloride ions to the permeate, while maintaining lower passage of divalent ions (typically 5 to 20 percent), and also low passage of larger molecules (molecular weight < 300–400).
3. *Ultrafiltration* products do not prevent the passage of dissolved inorganic salts. They do prevent the passage of particulates in the feed stream and higher-molecular-weight molecules. They are typically characterized using dextran or polyethylene glycol marker solutions of specific molecular weights to determine the molecular weight cutoff (MWCO) for the membrane, rather than defining a pore size. Products are available from 1000 through 500,000 MWCO. If pore size is quoted, it is typically <0.2 μm.
4. *Microfiltration* products typically have a pore size in the range of 0.3–0.8 μm, and prevent the passage of particulates and fine colloidal suspensions across the membrane. Passage of dissolved species may be hindered by the buildup of material on the membrane surface, but the membrane itself does not restrict passage.

Reverse osmosis, nanofiltration, and ultrafiltration are typically used in the *cross-flow* configuration, where the feed stream flows across or tangential to the membrane surface. The constant flow across the membrane surface minimizes the buildup in concentration of salts at the membrane surface for reverse osmosis and nanofiltration products, and inhibits the formation of a gel or particulate layer for ultrafiltration and microfiltration products. The *permeate* passes through the membrane, and the *concentrate* or *retentate* retains the dissolved and suspended solids rejected by the membrane.

REVERSE OSMOSIS

System Description

A reverse-osmosis (RO) system may consist of the following essential parts:

Pretreatment Section. The choice of pretreatment will depend on the source of the water to be treated, but can be broadly categorized as follows:

Ground (Well) Water. If the source is of low turbidity and silt density index (SDI), a 10-μm cartridge filter system should be sufficient. A pressure filter at moderate filtration rate [4 to 6 gpm/ft^2 (10 to 15 m/h)] may be added if necessary to handle any solids loading.

Lake, River, and Canal Water. These are typically higher solids loading cases. "Conventional" design would typically use a clarifier, followed by deep-bed media filtration and cartridge filtration. Ultrafiltration or microfiltration pretreatment is increasingly being used to replace all these steps. The advantages are the following:

- Smaller plant footprint
- Much reduced chemical consumption (and disposal) costs
- Approximately the same capital cost
- Much better quality feed to the RO, which enables the RO to be operated at higher flux

Wastewater. This can be subdivided as follows:

- *Municipal (sewage) reclamation.* This is a growing application for RO with microfiltration or ultrafiltration pretreatment. Large-scale systems are now in operation in the western

United States [5 mgd (800 m^3/h) or larger]. The reclaimed water is used for process applications.

- *Industrial waste reclamation.* A very wide field, in many cases requiring microfiltration or ultrafiltration pretreatment to condition the water prior to the RO membrane. Piloting is often required due to the one-off nature of these streams. Some thought at the design stage for industrial plant may identify some waste streams that are easily treated for recycle—they should be kept separate from the difficult streams.

Note: A reverse-osmosis product is designed to remove dissolved salts. Trying to make it remove particulates or other foulants is a misapplication of the technology. Foulants should be dealt with by the pretreatment.

Water Chemistry Issues. The RO system concentrates all dissolved salts on the feed side of the membrane from the feed inlet to the concentrate discharge. This includes sparingly soluble salts. A detailed ion-by-ion analysis of the feedwater, including the following ions, is required at the design stage to make the RO system design, and to check for problems with sparingly soluble salts.

Cations	Ca^{2+}, Mg^{2+}, Na^+, K^+, NH_4^+, Sr^{2+}, Ba^{2+}, Fe^{2+}, Mn^{2+}
Anions	HCO_3^-, SO_4^{2-}, Cl^-, F^-, NO_3^-
Others	Total dissolved solids; conductivity; pH; silica (SiO_2), both total and reactive; water temperature range

If the concentration of sparingly soluble salts becomes too high, it will result in scale formation on the surface of the membrane and loss in performance. Depending on the type of scale formed, it may be difficult or impossible to remove.

Scale control can be made possible to a limited extent by using antiscalant compounds and/or acid. The system manufacturer should have access to software from the RO membrane manufacturers and antiscalant manufacturers to determine a safe operating condition and recovery method for the system. *This step is an essential part of the design process, and should not be overlooked—mistakes can be costly!*

Reverse-Osmosis System. The RO system itself consists of the following components:

- High-pressure pump (or pumps)
- Inlet and concentrate control valves
- Pressure vessels to contain the RO membranes (typically up to six membrane elements 40 in long, or four membrane elements 60 in long)
- High-pressure feed and concentrate side manifolds and piping
- Low-pressure permeate manifolds and piping
- Control and instrumentation package [flowmeters, pressure measurement, and programmable logic controller (PLC)]
- Skid

Figure 3.10 presents a schematic drawing of a typical RO system.

Posttreatment. This may include the following:

- Degasser (if acid is used on the pretreatment, for CO_2 removal)
- Chemical addition to improve the water stability for potable water distribution
- Chlorine (or chloramine) addition for sterilization purposes

Other Equipment. The following equipment may also be included:

Cleaning Skid. Most RO membranes will need to be cleaned sooner or later, depending on the pretreatment and water source. Cleaning consists of recirculating chemical solutions

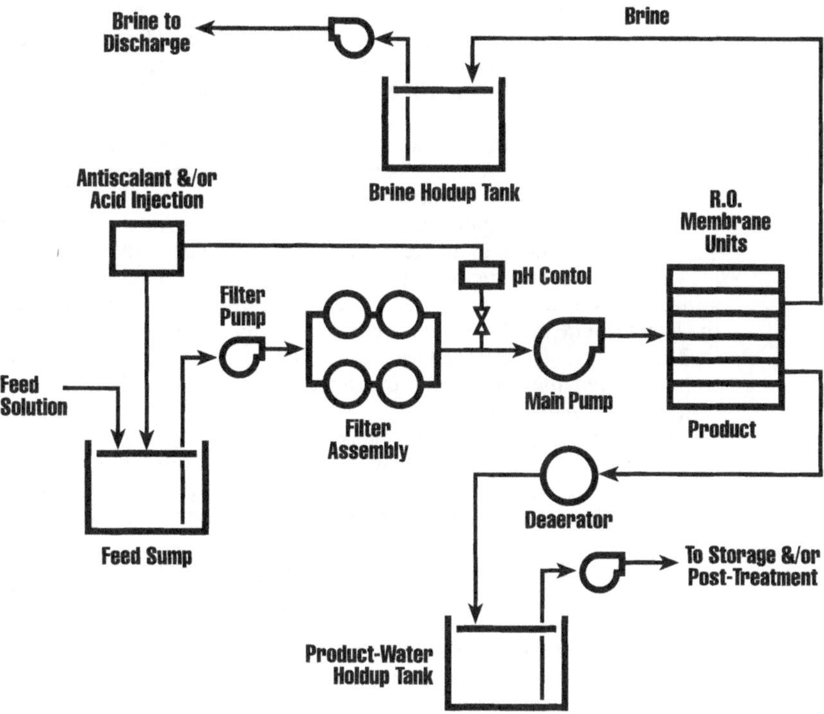

FIGURE 3.10 Basic flow diagram for a single-stage reverse-osmosis plant.

through the membrane elements while they are located in the system. The equipment consists of the following components:

- Tank, with mixing and heating capability
- Recirculation pump
- Cartridge filter

Provision should be available on the RO system to allow each bank to be cleaned independently.

Single-Element Tester. On larger systems, it is very useful to have the capability to independently test elements to determine actual performance at the site. The tester consists of a single-element vessel with suitable instrumentation to characterize the element.

Process Description

Reverse-osmosis membranes make a separation between the dissolved solids phase and the liquid phase by reversing the flow that normally occurs through a semipermeable membrane when the concentration of a given salt is different on the two sides of the membrane. Normally, the flow will proceed in such a fashion as to equalize the concentration; hence, pressure equal to the osmotic pressure plus an additional driving force is provided to permit flow from the more concentrated to the less concentrated side—thus the term *reverse osmosis.*

The flow proceeds through the membrane, passing between the molecules in the polymer lattice. Flow is accompanied by a loose association between the transported species and the membrane driven by the pressure differential. Typically, brackish water systems operate at 150 to 600 psig (1035 to 4137 kPa) and seawater systems at 800 to 1200 psig (5516 to 8275 kPa).

A key feature of the RO process is cross-flow filtration. While some water is passing through the membranes there is a constant flow of water flushing the rejected salts away from the membrane surface. The ratio of these two flows is determined by the design recovery rate, the product flow divided by the feed flow. This is an important design factor that should be determined by the membrane manufacturer. Most manufacturers have special computer programs to calculate the maximum recovery rate based on a feedwater analysis.

In general, the flow per square foot of membrane surface is quite low. The range is 8 to 20 $gal/ft^2 \cdot day$ depending on the level of suspended solids in the feedwater.

There are several methods to measure the fouling tendency of the feedwater. Most are approximations and cannot be used as a guarantee of toning control. The only way to positively determine the fouling nature of a given feed is to perform a pilot test.

The quality of the water with respect to its suitability for introduction into a reverse-osmosis system may be measured in a crude fashion through the use of a test yielding what is called the *silt density index* (SDI).

The SDI test is performed by measuring the time T_1 to collect 500 mL of filtrate through a 0.45-μm filter at exactly 30 psi. The filter is allowed to flow for 15 min, then the time T_2 for another 500 mL is measured. These times are compared in the formula given below. If the filter plugs before 15 min, the feedwater needs additional treatment before going to the RO.

$$\mathrm{SDI} = \left(1 - \frac{T_1}{T_2}\right) \frac{100}{15}$$

In general, SDI readings should be less than 5.0. Less than 3.0 is considered ideal.

Applications

Reverse-osmosis systems find their application in a number of areas. However, the most significant are the following:

1. Inclusion in systems for providing low-conductivity water for boiler makeup and related purposes
2. The production of potable water where such supplies are not conveniently available
3. Industrial waste treatment and water recovery

Of the preceding, perhaps the most significant application of reverse osmosis is in the deionization systems for the production of low-conductivity boiler feedwater.

In most instances where the raw-water supply contains more than 300 to 400 mg/L total dissolved solids (TDS), an overall system economy may be shown by utilizing reverse osmosis for removing the bulk of the dissolved solids prior to the ion-exchange polishing system. The approach is to replace the cationic exchange, degasification, and anionic-exchange subsystems with the reverse-osmosis system and to employ the mixed-bed polishing subsystem for the removal of residual dissolved solids. In some instances, it can be shown that this puts a somewhat higher load on the mixed-bed polisher; however, if this load is less than 50 mg/L TDS, it does not constitute an improper or unreasonable burden in terms of cycle time or regeneration frequency. This type of system usually produces a less expensive water at roughly the same recoveries as would be experienced with the full deionizing (DI) system and often does not produce the quantity of solids in the backwash and regenerating stream that the DI system will produce. The preceding is a generalization only, and a careful cost comparison should be made in any specific situation.

Sometimes only high-salinity waters are available for process and other purposes, and the supply of potable water is either limited or nonexistent. In these cases, a reverse osmosis facility can be installed to produce water for the plant at comparatively low cost.

Many industrial wastes are highly amenable to treatment by reverse-osmosis systems. The most commonly encountered RO application is in the area of metal contamination or the presence of excessive dissolved materials in the effluent. Often when RO systems are employed for treatment, the water is suitable or can easily be made suitable for reuse and can constitute a supply for processes or sometimes even boiler feedwater.

Cost of RO Systems

Table 3.3 contains a brief summary of capital and operating costs of RO systems as a function of salinity. The actual cost will vary as a function of the overall recovery employed in the system.

TABLE 3.3 RO System Costs, 1992 Dollars

Capacity, gal/day	Pressure, psi	Feed salinity, Mg/L NaCl	Capital cost, $	O&M cost, $/1000 gal
50,000	250	1,000	55,000	0.90
150,000	250	1,000	160,000	0.80
500,000	250	1,000	475,000	0.75
1,000,000	250	1,000	850,000	0.65
2,500,000	250	1,000	2,000,000	0.60
10,000	850	35,000	60,000	4.30
50,000	850	35,000	200,000	3.50
100,000	850	35,000	380,000	3.20
500,000	850	35,000	1,500,000	3.00

Data supplied by Fluid Systems Corporation, San Diego, California. To convert gallons to cubic meters, divide by 264 gal/m^3. Based on power at $0.10/kWh, element replacement $0.17/1000 gal brackish, $0.30/1000 gal seawater; chemicals $.04/1000 gal brackish, $0.05/1000 gal seawater; operator $0.15 to $0.50/1000 gal; maintenance $0.06/1000 gal; RO recovery rate 75% for brackish, 25 to 40% for seawater.

NANOFILTRATION

Nanofiltration (NF), also referred to as *ultra-low-pressure reverse osmosis,* was developed to fill the gap between reverse osmosis and ultrafiltration. In water purification, nanofiltration has been found to effectively remove selected salts to reduce total hardness at lower pressures than RO systems. The nanofiltration membrane employed for this application has been referred to as a *softening membrane.* The softening membrane also effectively removes color and trihalomethane (THM) precursors.

Softening membranes operate at pressures of 75 to 250 psi (517 to 1724 kPa), as compared to 200 to 600 psi (1379 to 4137 kPa) for RO and 50 to 150 psi (345 to 1034 kPa) for ultrafiltration membranes. The process flow scheme, including pretreatment, is similar to that of a typical brackish water reverse-osmosis plant.

Nanofiltration softening membranes may be an economic alternative to conventional softening. Possible advantages of membrane filtration for softening include smaller space requirements, no lime requirement, superior quality water, and less operator attendance. As for all membrane separation processes, suitability of softening membranes for a given application must be made on a case-by-case analysis.

ULTRAFILTRATION

Ultrafiltration (UF) is a pressure-driven membrane process similar in some ways to reverse osmosis. However, in this case, as opposed to the situation encountered in an RO system, the flow of water through the membrane is generally through pores and not through the space between the lattices in the polymer, so osmotic pressure is not a factor. Furthermore, there is little or no chemical interaction between the transported species and the membrane itself. UF membranes may be tailor-made to meet virtually any type of removal specification.

Although they cannot reject any dissolved salts or other low-molecular-weight soluble matter, UF systems can remove very fine particulate material and high-molecular-weight organic matter from water streams. To remove any low-molecular-weight soluble species with a UF membrane, a process must occur to convert the soluble matter to particulate form. As examples, soluble phosphorus may be precipitated with a metal salt, soluble organics may be adsorbed onto powdered activated carbon, and soluble iron may be oxidized to particulate form. All of these processes and others will allow a UF membrane to remove even soluble matter.

There are several different membrane module geometries on the market. *Spiral-wound* membrane modules are similar to RO membranes. Although they have a high membrane density, it is not possible to feed them with a high suspended solids concentration because of the narrow passages available through the module. This geometry is suitable for the separation of high-molecular-weight organics in combination with low suspended solids. A second geometry is *tubular,* in which the feed flow passes through tubes at high velocity. The tube diameters range from about 0.5 to 1 in (12 to 25 mm). Because of the large tube diameter, tubular membranes are able to effectively separate liquid from biomass slurries of relatively high concentration, about 2 to 3 percent solids by weight. The permeate flow in the tubes is from the inside to the outside. Pressure, typically on the order of 70 to 90 psig (483 to 621 kPa), is required to force the permeate through the membrane pores. A high feed-flow velocity [12 to 15 fps (3.7 to 4.6 m/s)] through the tubes is required to ensure sufficient shear at the membrane surface to keep the membrane clean and reduce concentration polarization. A third variety of geometry is *immersed hollow fiber,* in which the membranes are submerged directly in the feed fluid without the need for a pressure vessel. Hundreds of small-diameter [0.07-inch (1.9-mm)] vertically oriented hollow fibers are supported at the top and bottom. The permeate flow is from the outside to the inside so that fibers can handle very high solids concentrations, as only clean, pure water flows inside the membranes. The vacuum required to operate the hollow fiber design is very small, about −2 to −4 psig (−13.8 to 27.6 kPa). This vacuum is normally provided by a standard pump connected to the membranes via a piping manifold. To create the required shear on the membrane surface, low-pressure air is diffused intermittently under the fibers. The air rises through the fiber bundle, providing the necessary shear.

The nature of the solids–liquid separation requirement will dictate the kind of membrane module geometry that is best suited for a project. For small flows and very low suspended solids, the spiral-wound membrane can be used. Care must be taken to ensure that high solids concentrations do not develop within the membrane module by keeping the recirculation flow high. For small flows and high suspended or emulsified solids concentrations, including tramp oils, tubular membranes should be selected. Free oils are very fouling to the membrane and must be prevented from coming in contact with the membrane. For high flows and high suspended solids concentrations, immersed hollow-fiber membranes are usually the most economical because of their lower pressure operation and hence lower energy requirements.

Figures 3.11*a* and 3.11*b* show examples of ultrafiltration systems.

Applications

Spiral Wound. As discussed previously, spiral-wound UF membranes have somewhat limited applications because of their inability to handle high solids concentrations. For certain specific applications for the separation of high-molecular-weight organics in industrial processing, they are very well suited. As they are usually available in a wide range of molecular weight cutoffs, it is possible to achieve very specific separations.

Tubular. Tubular membranes are the workhorses of industrial wastewater solids–liquid separation. Their ability to handle high solids concentrations and high concentrations of emulsi-

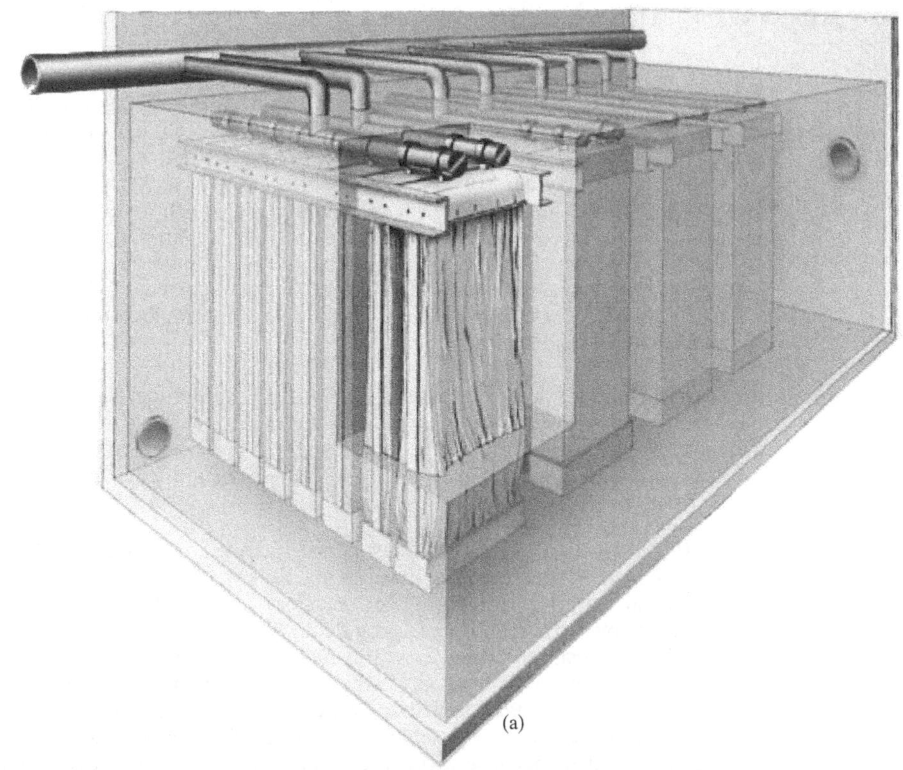
(a)

(b)

FIGURE 3.11 (*a*) and (*b*) typical ultrafiltration installations.

fied oils with little or no pretreatment leads to many applications for both waste reduction and material recovery. They have also been adapted to work as a solids–liquid separation device to replace the clarifier in conventional activated sludge treatment. Since the membranes provide a true barrier to the bacteria, very high effluent quality is achievable. In addition, because the biomass solids concentrations can be much higher than those limited by gravity separation, the sludge ages achieved are also much higher, reducing the amount of waste sludge production. In applications treating oily wastewater, not only are all of the suspended solids retained within the system by the membrane but also the higher-molecular-weight organics (mainly oils). This means that they are subjected to biological degradation for a time equal to the sludge age rather than simply the hydraulic retention time of the bioreactor as in conventional treatment.

Immersed Hollow Fiber. Immersed hollow-fiber membranes have catapulted UF technology into the high-flow regime prevalent in the municipal market. Applications of this technology have penetrated not only the wastewater side of the market, but also the water side, meeting and exceeding the new challenges for higher-quality drinking and industrial process water.

1. *Municipal water treatment.* New concerns over contamination of raw-water supplies with giardia and cryptosporidium have piqued a strong interest in the use of UF technology for municipal water treatment. With pore sizes far less than the size of these bacteria, immersed hollow-fiber membranes provide an absolute barrier against them, resulting in a typical 6-log removal rate using direct filtration only, with no chemical additions.

Colored water has long been an aesthetic problem with many raw-surface-water sources. UF technology, in combination with metallic salt addition to coagulate the color bodies to a particulate form, can be used to achieve as much as 90 percent removal of color.

High iron and manganese levels are sometimes prevalent in raw-groundwater sources. Oxidation of these constituents to produce particulates from the soluble forms present in the raw water prior to UF treatment provides economical treatment in cases in which conventional technology is inadequate because of high concentrations.

2. *Municipal wastewater treatment.* Increasingly over the last few years, there has been a demand for higher levels of nutrient removal. Conventional technology has not been able to meet these requirements because of the limitations of suspended solids removal even with tertiary multimedia filtration. Because UF technology is an absolute barrier to the biomass used in conventional activated sludge and biological nutrient removal treatment plants, much lower levels of total nitrogen and phosphorus are achievable. Also, because the concentration of mixed-liquor suspended solids can be elevated significantly compared with conventional treatment, sludge age can be increased, both reducing the quantity of waste sludge and increasing the reliability of nitrification.

3. *Water reuse.* In areas of the world that have a shortage of raw water supplies, there is increasing interest in high-volume water reuse for industrial processes such as cooling water and boiler feed water. In this context, the availability of a secondary effluent-quality municipal wastewater source can provide a feasible solution. Using UF technology, this water can be treated without chemicals to yield a solids-free product suitable for direct use as cooling water. In addition, because its silt density index is consistently below 3, the permeate can be used directly to feed an RO system to produce water suitable for boiler feedwater.

Cleaning

All UF membranes require periodic cleaning to maintain their hydraulic capacity. Modern systems are designed and built with integral PLC controls to completely automate the cleaning procedures. Selection of the proper membrane module geometry for the specific application will minimize the cleaning frequency and chemical consumption.

MICROFILTRATION

Microfiltration (MF), like ultrafiltration, is a pressure-driven membrane process. In the case of microfiltration, the pore size is typically in the 0.03- to 0.1-μm range. In this range, individual bacteria and viruses will pass through the membrane. However, colloidal suspended solids, floc particles, and parasite cysts are prevented from passing. No removal of dissolved solids is accomplished by microfiltration membranes.

Microfiltration membranes come in both in-line or cartridge filter arrangements and cross-flow arrangements. For cross-flow systems, the rejection rate is usually 5 to 10 percent, that is, 90 to 95 percent recovery. Higher recovery rates are feasible; however, the overall flux through the membrane is reduced.

Cross-flow microfiltration membranes can be further subdivided into tubular and immersed-membrane types. Tubular membranes are arranged such that the raw-water source is introduced under pressure in a tube that surrounds the membrane. Typically, the permeate water proceeds from the outside of the membrane into the lumen in the center of the membrane, where the permeate is then conducted back to a manifold for collection. The solids remain on the outside of the membrane in the pressure tube and are periodically blown down from the system.

Immersed membranes can be placed directly in a process tank where the permeate is removed through the membrane by a suction pump on the permeate collection manifold. This arrangement is possible because of the low transmembrane pressure exhibited by this type of membrane.

The flux through the membrane during operation is dependent on suspended solids in the feedwater as well as temperature (viscosity). A typical flux range for the membrane surface is 10 to 50 gal/day · ft^2. Increasing transmembrane pressure also increases the flux.

Microfiltration membranes are typically periodically cleaned by backpulsing these membranes. This can be accomplished through a variety of ways, some incorporating air and some using just product water or water containing a small amount of hypochlorite. Some immersed-membrane systems also use diffused air to agitate the membranes and prevent solids from caking on the membrane surface.

Capital costs for microfiltration membrane system capacity range from $0.50 to approximately $1.00 per gallon per day. This range is primarily a function of solids concentration in the feed stream. Also note that membrane systems must be sized on peak flow rather than average daily flow, which can significantly affect the cost of a microfiltration membrane system.

Operational costs are primarily associated with power and vary from 50 to 120 hp/mgd. Pretreatment chemicals include coagulants, and when biofouling is a problem, hypochlorite solution can be used on some membranes for cleaning. Citric acid can also be used for membranes that do not tolerate chlorine. Microfiltration membranes are typically not compatible with polymer addition, which is not required, since even small "pin flocs" cannot permeate the membranes.

EVAPORATIVE SYSTEMS

Evaporative systems are used in some applications for the production of very high quality water from saline waters or wastewater. With high energy costs, evaporative systems (except under very special circumstances) appear to offer a less satisfactory solution to many desalination problems when compared to the alternative systems such as membrane filtration. Nevertheless, the need for ultrapure water in process applications such as electronics and pharmaceutical manufacture, production of high-pressure steam for electrical power generation, and increasingly stringent discharge regulations provide applications for evaporative systems. Where waste energy is available, evaporation procedures should be considered as a possible candidate in any water recovery system analysis.

Evaporator system types such as multistage flash (MSF), multiple-effect distillation (MED), and vapor recompression, both thermal and mechanical, are applied to water recovery situations. Evaporator configurations such as horizontal tube, falling film, rising film, and forced circulation are utilized, depending on the requirements for the application. Generally speaking, economy ratios of 5 to 12 lb of water per 1000 Btu (2 to 5 kg H_2O per 10^6 J) are possible. Maintenance requirements such as cleaning of heat-transfer surfaces and pump and compressor maintenance will vary with type of evaporator system and should be examined closely along with capital and energy costs. The substantial advantage that an evaporator offers is that, even if the source has variable salinity or extremely high salinity, the quality of the product will be essentially unchanged. Furthermore, the amount of energy required to run the system is independent of the salinity.

System costs for some evaporator systems are shown in Table 3.4.

TABLE 3.4 Evaporator Capital and Operating Costs, 1992 Dollars

Evaporation capacity, gal/day	Capital costs, $	Operating costs, $/1000 gal	Principal energy source
13,000	450,000	47.85	Low-pressure steam
283,000	1,200,000	14.21	Low-pressure steam
346,000	1,800,000	10.81	Electric power

The data are based on the following assumptions:

1. Materials of construction for process contact parts are type 316 stainless steel.
2. Steam costs are $5/1000 lb.
3. Electric power cost is $0.035/kWh.
4. Operating labor cost is $25/h.
5. The annual cost of maintenance and depreciation is 10% of capital costs.
6. Capital costs are inclusive of installation and structural supports, but do not include foundations or buildings.

Prices supplied by Dedert Corporation, Olympia Fields, Illinois. To convert 1000 gal to cubic meters, multiply by 3.8 m^3/1000 gal.

CHLORINATION

Chlorine compounds (sodium hypochlorite, calcium hypochlorite, and chlorine gas) are strong oxidizing agents commonly used for disinfection, as well as taste and odor control. Chlorine will oxidize ferrous iron, manganese, and sulfide ions, as well as react with ammonia or amines to form chloramines. Chloramines are weaker disinfectants than chlorine but are useful for maintaining a residual chlorine content in water mains. The advantage of chlorine is that it is an inexpensive method of disinfection. The disadvantage of chlorination is its reaction with organic material to form chloroorganic compounds, including trihalomethanes. The maximum contaminant level (MCL) for trihalomethanes for drinking water is set at 0.10 mg/L. Chlorine dioxide is also used as an oxidant and disinfectant to a limited extent. When it is used for the treatment of potable water, trihalomethanes are not produced.

OZONIZATION

Ozone (O_3) is a powerful oxidizing agent used for color, taste, and odor removal and for organic oxidation, bacterial disinfection, and virus inactivation. Ozonization followed by activated carbon treatment appears to be effective in meeting proposed drinking water standards. The advantage of ozone is that it reacts quickly, leaving no residual or trihalomethanes. The disadvantage is high capital and energy costs and the inherent danger of using a toxic substance.

ACTIVATED-CARBON TREATMENT

Activated carbon is an adsorbent used for removing taste- and odor-causing material and chlorinated compounds (e.g., trihalomethanes). It is available in powdered (PAC) and granular (GAG) forms. High surface area, pore size distribution, and particle surface chemistry give activated carbon its adsorbent nature. A variety of materials, including bituminous coal, lignite, petroleum coke, wood, and nutshell, are used to make activated carbon. The pore size distribution and surface chemistry of the activated carbon is dependent on the original material.

When the efficiency of the activated carbon is diminished by coating the surface of the activated carbon with adsorbed material, reactivation is required. This is accomplished by oxidizing the adsorbed material in regeneration furnaces at temperatures around 1470°F (800°C), or chemical treatment with phosphoric acid, potassium hydroxide, or zinc chloride.

APPLICATIONS (FLOWSHEETS)

Figure 3.12 depicts many treatment systems that can be developed by applying the various treatment methods that have been discussed. Many options are shown, and the most suitable can be selected on the basis of raw-water characteristics and the treated-water requirements. The diagram is by no means complete; it does, however, illustrate some common treatment scenarios.

Obviously, raw-water supplies vary widely in quality and, although not indicated on the diagram, can be used without any treatment in some cases. Potable supplies will all require disinfection. When selecting any system to be used for potable purposes, the local governing health departments should be consulted so all their requirements are met.

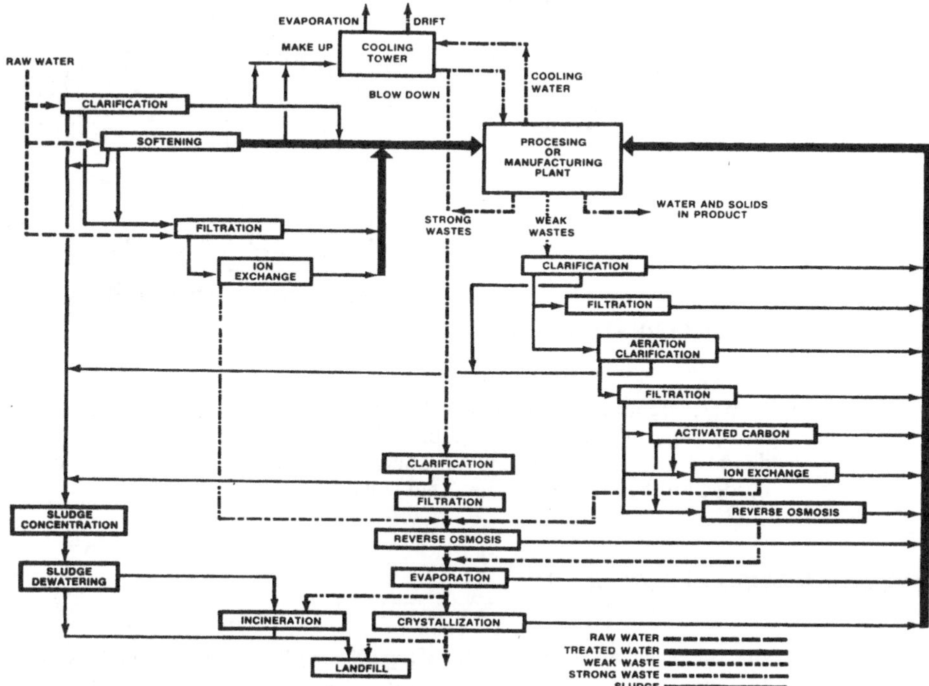

FIGURE 3.12 Diagram of water and waste treatment.

The minimum water quality required at each point of use must be established and the method or series of treatments that can conservatively meet these requirements under any and all raw-water conditions must be selected. Consideration must be given to the degree of automation and the quality and training of operators and maintenance personnel.

Chemical feeders, controllers, and instrumentation are an important part of water treatment installations. These accessories are probably the source of most of the operational and maintenance problems, and the same care should be given to their selection as to the major equipment.

Water treatment is a continuous operation that must be monitored by testing. The complexity depends on the process. Qualified personnel familiar with the laboratory techniques must be assigned to this task.

WATER RECYCLING AND REUSE

A reuse treatment plant is usually nothing more than a water treatment plant using as a source of supply the discharge from either a domestic or an industrial wastewater treatment plant. Figure 3.13 shows a flow diagram of such a plant using treated domestic sewage as a source and treating it further for use in an electrical generating station. Except for the nitrification towers which are included because of a considerable savings in operating chemicals, the balance of the plant is a typical water treatment plant.

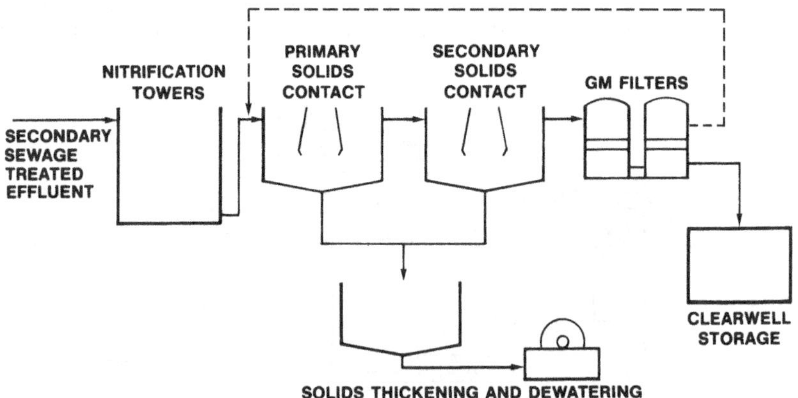

FIGURE 3.13 Flow diagram for reuse treatment.

Waste treatment plants, particularly those using biological treatment, are subject to upsets on occasion, and this can be troublesome to the operation of the reuse treatment plant. The design of a reuse treatment plant should include adequate storage, recycle, or bypass facilities needed for the inevitable upsets that will occur upstream.

RECOMMENDED REFERENCES

The Betz Handbook of Industrial Water Conditioning, 9th ed., Betz Laboratories, Trevose, Pa., 1991.

Drew Principals of Industrial Water and Other Uses, 11th ed., Ashland Chemical Inc., Boonton, N.J., 1992.

The Nalco Water Handbook, 2d ed., Nalco Chemical Co., Naperville, Ill., 1988.

Principles and Practices of Water Supply Operations, Vol. 2—Introduction to Water Treatment, American Water Works Association, 1984.

Standard Methods for the Examination of Water and Waste Water, 17th ed., American Public Health Association, Washington, D.C., 1989.

Water Treatment Plant Design, 2d ed., American Society of Civil Engineers, Denver, Colo., 1990.

Water Treatment Principles and Design, James M. Montgomery Consulting Engineers, John Wiley & Sons, New York, 1985.

Practical Principles of Ion Exchange, Dean L. Owens, Tall Oaks Publishing, Inc., Voorhees, N.J., 1985.

Solid-Liquid Separation, Ladislar Svarovsky, ed., Butterworth Heinemann, Boston, 1990.

CHAPTER 3.4
WATER COOLING SYSTEMS

Tower Performance, Inc.
Florham Park, New Jersey

GLOSSARY

Approach The difference between the cold-water temperature and the ambient or inlet wet-bulb temperature. Units: °F (°C)

Blowdown Water discharged from the system to control concentration of salts or other impurities in the circulating water.

Cell The smallest tower subdivision which can function as an independent unit with regard to air and water flow; it is bounded by walls or partitions. Each cell may have one or more fans or stacks and one or more distribution systems.

Cold-water temperature (CWT) Temperature of the water entering the cold-water basin before addition of makeup or removal of blowdown. Units: °F (°C).

Counterflow tower One in which air, drawn in through the air intakes (induced draft) or forced in at the base by the fan (forced draft), flows upward through the fill material and interfaces countercurrently with the falling hot water.

Crossflow tower One in which air, drawn or forced in through the air intakes by the fan, flows horizontally across the fill section and interfaces perpendicularly with the falling hot water.

Design conditions Defined as the hot-water temperature (HWT), cold-water temperature (CWT), gallons (liters) per minute, and wet-bulb temperature (WBT) in mechanical-draft towers. In natural-draft towers, design conditions are HWT, CWT, GPM, and WBT plus either dry-bulb temperature (DBT) or relative humidity (RH).

Distribution system Those parts of a tower, beginning with the inlet connection, which distribute the hot circulating water within the tower to the points where it contacts the air. In a *counterflow* tower, this includes the header, laterals, and distribution nozzles. In a *crossflow* tower, the system includes the header or manifold, valves, distribution box, hot water basin, and nozzles.

Drift Water lost from the tower as liquid droplets entrained in the exhaust air. It is independent of water lost by evaporation. Units may be in pounds (kilograms) per hour or percentage of circulating water flow. Drift eliminators minimize this loss from the cooling tower.

Drift eliminator An assembly constructed of wood, plastic, cement asbestos board, steel, or other material which serves to minimize entrained droplets from the discharged air.

Fan stack Cylindrical or modified cylindrical structure in which the fan operates. Fan stacks are used on both induced-draft and forced-draft axial-flow propeller fans. Also known as *fan cylinder.*

Fill The section of the cooling tower consisting of materials which are placed within the cooling tower to effect heat and mass transfer between the circulating water and the air flowing through the tower.

Hot-water temperature (HWT) Temperature of circulating water entering the distribution system. Units: °F (°C).

Makeup Water added to the circulating water system to replace water lost from the system by evaporation, drift, blowdown, and leakage.

Plenum The enclosed space between the drift eliminators and the fan stack in induced-draft towers or the enclosed space between the fan and the filling in forced-draft towers.

Pumping head Minimum pressure required to lift the water from basin curb to the top of the system. Pumping head is equal to static head plus friction loss through the distribution system. Units: ft (m).

Range Difference between the hot-water temperature and the cold-water temperature. Units: °F (°C). Also known as *cooling range.*

Recirculation This term describes a condition in which a portion of the discharge air enters the tower along with the fresh air. The amount of recirculation is determined by tower design, tower placement, and atmospheric conditions. The effect is generally evaluated on the basis of the increase in the entering wet-bulb temperature compared to the ambient.

Water loading Circulating water flow of effective horizontal wetted area of the tower. Units: gal/min $\cdot$ ft^2 ($m^3/h \cdot m^2$).

Wet-bulb temperature (WBT) Temperature measured by passing air over the bulb of a thermometer covered by a wick saturated with water. Also known as the *thermodynamic wet-bulb temperature* or the *temperature of adiabatic saturation.* Units: °F (°C).

INTRODUCTION

With the growth in the number and sizes of manufacturing plants of all types and the attendant high heat-rejection rates, cooling-tower requirements have increased dramatically. These trends are coupled with environmental aspects, including water conservation and limitations on thermal and chemical discharges. As a result, the plant engineer has witnessed an upsurge in the specification and use of cooling towers.

COOLING SYSTEM OPTIONS

Once-Through Cooling Systems

Many plants operating today are once-through cooling systems, as shown in Fig. 3.14, utilizing water from a lake or river to supply cooling water to the heat exchangers. The heated water is then returned to the body of water.

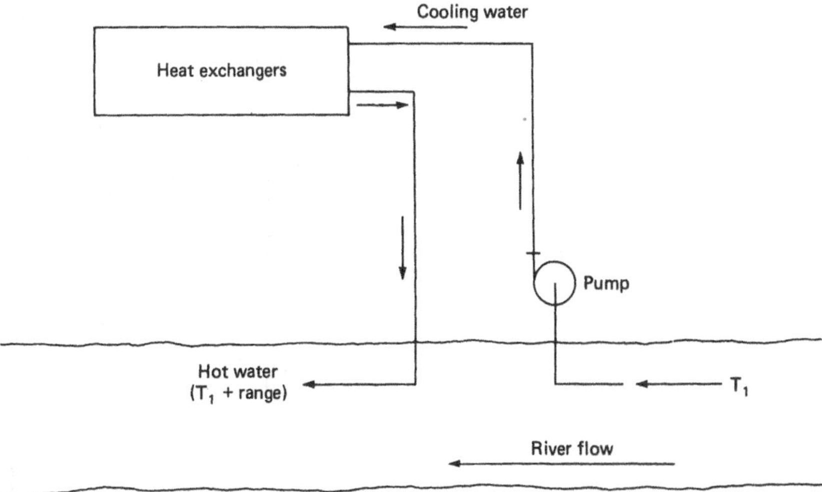

FIGURE 3.14 Once-through cooling.

As a result of all the heat being discharged to rivers, lakes, etc. by plants operating with once-through cooling, the term *thermal pollution* has assumed greater significance and has resulted in the enactment of environment-related legislation. Consequently, once-through cooling is not available as an option in many cases.

Closed-Cycle Cooling Systems

Closed-cycle cooling refers to the water side of the system and generally takes the form of a cooling tower. Figure 3.15 shows the relationship of the cooling tower to the cooling system. The cooling water is continuously recirculated through the plant. The cooling tower is used to remove the heat added to the circulating cooling water by the heat exchangers. Water withdrawn from the natural source would be used only for makeup of losses.

Cooling Towers. Cooling towers currently available to plant designers use natural-draft and mechanical-draft designs. The natural-draft design utilizes large-dimension concrete chimneys to induce air through the fill. In the mechanical-draft design, large-diameter fans driven by electric motors induce or force the air through the fill. Mechanical-draft cooling towers use either induced-draft (pulling air through cooling tower) or forced-draft (pushing air into cooling tower) designs. In all cooling towers, water is sprayed over the fill while air passes through the fill. The fill is provided to interrupt the flow of water and increase the time of contact between air and water, thereby permitting the efficient transfer of heat from the water to the air.

Spray Ponds. An alternative to cooling towers in closed-cycle cooling systems is a spray pond, where warm water is pumped through pipes from the heat exchangers and then out of the spray nozzles. The nozzles atomize the warm water into fine droplets. The basic arrangement of a spray pond is shown in Fig. 3.16. The spray nozzles are usually located about 5 ft (1.52 m) above the pond surface. Height of the spray is about 6 ft (2 m). A nominal water loading rate of 1 gal/min · ft^2 (2.44 $m^3/h \cdot m^2$) of pond area and wind speed of 5 mi/h (8.05 km/h) would be typical design parameters for such a pond. Performance is strongly dependent on

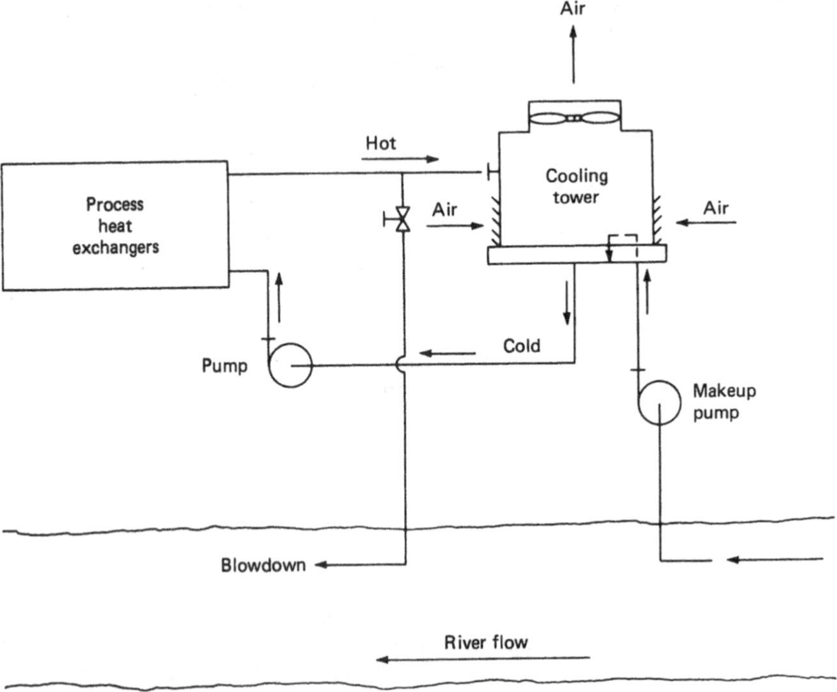

FIGURE 3.15 Closed-cycle cooling-tower system.

wind speed and direction and is limited by the relatively short contact time between the air and water spray.

Objections to spray ponds include excessive water losses due to drift, which may cause localized icing and fogging, and relatively high pumping costs. The land area required for a spray pond system compared to a cooling tower installation is about 8 to 1.

Atmospheric Cooling Towers. When there is a need for larger cooling ranges and closer approaches, the natural-draft or atmospheric towers might be considered. Fill installed in

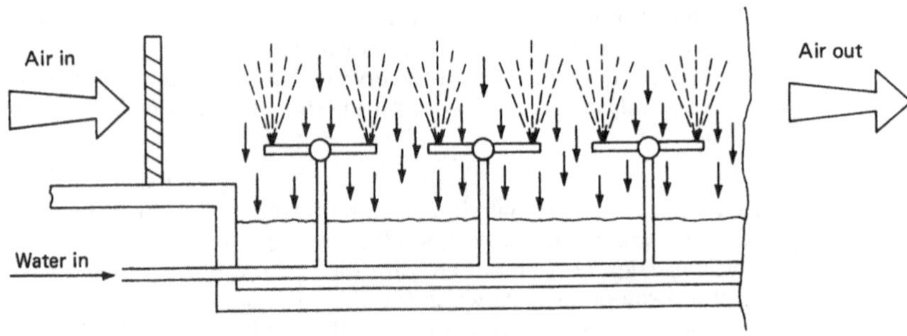

FIGURE 3.16 Spray pond.

natural-draft cooling towers increases the time of contact between water and air. Various types of fill and spacing are utilized, and tower heights vary in relation to the extent of cooling to be accomplished. The cooling is dependent on the efficiency of the fill and the air velocity through the cooling tower as the water descends through the fill.

The advantages are (1) no electric power is required except for pumping head and (2) no mechanical equipment is necessary, reducing maintenance requirements. The disadvantages are (1) atmospheric towers have limited capacities, since they are solely dependent on ambient atmospheric conditions, (2) water loss as a result of high wind velocities can be appreciable, and (3) a rather high pumping head is required to allow for maximum air–water contact time.

The large natural-draft hyperbolic cooling towers are found only in utility power station service in the United States. The economics of plant designs will favor the mechanical-draft cooling towers because of the rather short amortization period. Natural-draft towers perform better when wet-bulb temperatures are low and relative humidity is high or if demand is higher in winter. A combination of low design wet-bulb temperature and high inlet and exit water temperatures would enhance the operation of a hyperbolic tower.

Because of the size of these units, 500 ft (155 m) high and 400 ft (122 m) in diameter at the base, they are more practical when the circulating cooling water flow rate is about 200,000 gal/min and higher.

Mechanical-draft towers have a positive control of air delivery through the fill by the use of large-diameter fans. Therefore, they can be designed for close control of cold-water temperature. Counterflow and crossflow designs are indicated in Figs. 3.17 and 3.18.

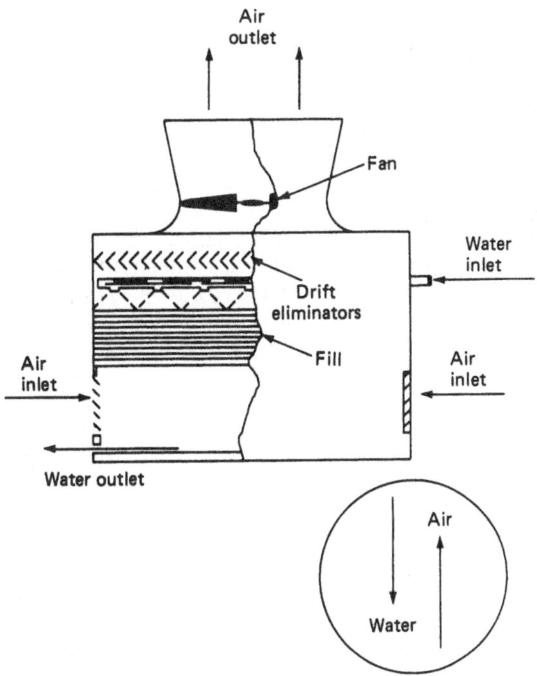

FIGURE 3.17 Mechanical-draft counterflow tower.

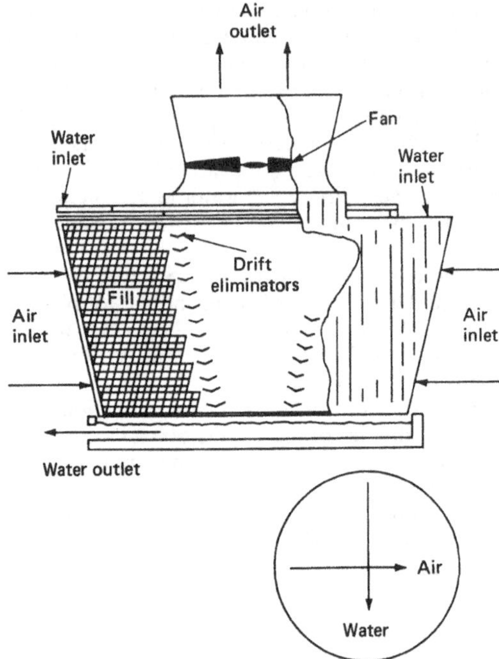

FIGURE 3.18 Mechanical-draft crossflow tower.

COOLING-TOWER OPERATION

Theoretical Concepts

The basic equations covering combined mass- and heat-transfer phenomena have been covered in the literature.[1] The analysis combines the sensible and latent heat transfer into an overall process based on enthalpy potential as the driving force.

The process is shown schematically in Fig. 3.19, where each particle of bulk water in the tower is assumed to be surrounded by an interfacial film to which heat is transferred from the water. This heat is then transferred from the interface to the main air mass by (1) a transfer of sensible heat and (2) mass heat transfer (latent) resulting from the evaporation of a portion of the bulk water. This can be represented by the equation

$$\frac{KaV}{L} = \int_{T_2}^{T_1} \frac{dT}{h_w - h_a} \tag{1}$$

where KaV/L = tower characteristic
T_1 = hot-water temperature, °F (°C)
T_2 = cold-water temperature, °F (°C)
T = bulk water temperature, °F (°C)
h_w = enthalpy of air–water vapor mixture at bulk water temperature, Btu/lb dry air (J/kg)
h_a = enthalpy of air–water vapor mixture at wet-bulb temperature, Btu/lb dry air (J/kg)

This equation is commonly referred to as the Merkel equation. The derivation can be found in Ref. 2.

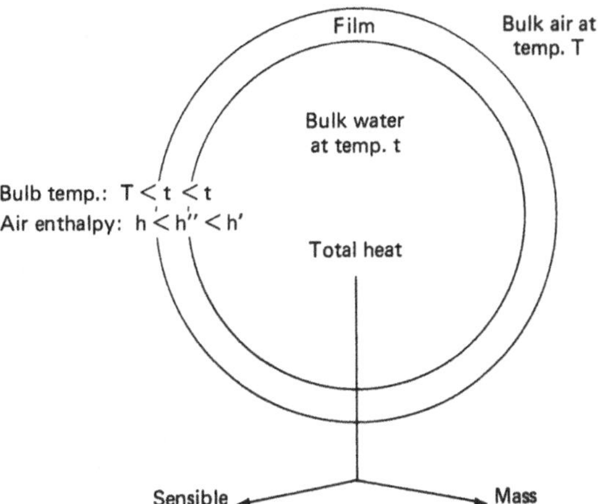

FIGURE 3.19 Schematic of a water droplet with interface film.

The left side of the equation is called the *tower characteristic.* The laws of thermodynamics demand that the heat discharged by the water descending down through the cooling tower must equal the heat absorbed by the air rising upward through the tower, or:

$$L * (T_1 - T_2) = G * (h_2 - h_1) \tag{2}$$

$$\frac{L}{G} = \frac{h_2 - h_1}{T_1 - T_2}$$

where L = mass water flow, lb/h · ft² (kg/h · m² plan area)
T_1 = hot-water temperature, °F (°C)
T_2 = cold-water temperature, °F (°C)
G = mass airflow, lb dry air/h · ft² (kg/h · m²)
h_2 = enthalpy of air–water vapor mixture at exhaust wet-bulb temperature, Btu/lb (J/kg) dry air
h_1 = enthalpy of air–water vapor mixture at inlet wet-bulb temperature, Btu/lb (J/kg) dry air
L/G = liquid-to-gas ratio, lb water/lb dry air (kg/kg)

Equations (1) and (2) or the tower characteristic can be represented graphically by the diagram in Fig. 3.20. The interfacial film is assumed to be saturated with water vapor at the bulk water temperature T_1 (*A* in Fig. 3.20). As the water is cooled to temperature T_2, the film enthalpy follows the saturation curve to *B*.

Air entering the tower at wet-bulb temperature T_{wb} has an enthalpy *C′*. The origin of the air operating line, point *C*, is vertically below *B* and is positioned to have an enthalpy corresponding to that of the entering wet-bulb temperature. The heat removed from the water is added to the air so its enthalpy increases along line *CD*, having a slope equaling the *L/G* ratio. The vertical distance *BC* represents the initial driving force.

Point *D* represents the air leaving the cooling tower. It is the point on the air operating line vertically below *A*. The projected length *CD* (or *AB*) is the cooling range.

The coordinates refer directly to the temperatures and enthalpy of the water operating line *AB*, but refer directly only to the enthalpy of a point on the air operating line *CD*. The corresponding wet-bulb temperature of any point on *CD* is found by projecting the point horizontally to the saturation curve, then vertically down to the temperature coordinate. *DEF*

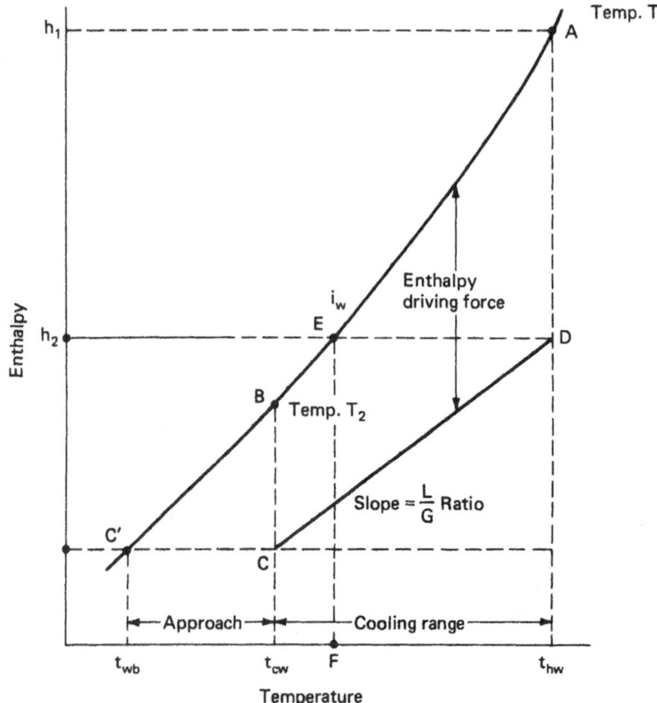

FIGURE 3.20 Graphical representation of tower characteristic.

shows this projection for the outlet air wet-bulb temperature of point *D*. Point *F* is the outlet air wet-bulb temperature.

The following integral is represented by the area *ABCD:*

$$\int_{T_2}^{T_1} \frac{dT}{h_w - h_a}$$

where T_1 = hot-water temperature, °F (°C)
T_2 = cold-water temperature, °F (°C)
T = bulk water temperature, °F (°C)
h_w = enthalpy of air–water vapor mixture at bulk water temperature, Btu/lb dry air (J/kg)
h_a = enthalpy of air–water vapor mixture at wet-bulb temperature, Btu/lb dry air (J/kg)

This value is characteristic of the tower, varying with the rates of water- and airflow. An increase in the entering air wet-bulb temperature moves the origin *C* upward and the line *CD* shifts to the right to establish equilibrium. Both the inlet and outlet water temperatures increase, while the approach decreases. The curvature of the saturation line is such that the approach decreases at a progressively slower rate as the wet-bulb temperature increases.

An increase in the heat load increases the cooling range and increases the length of line *CD*. To maintain equilibrium, the line shifts to the right, increasing hot- and cold-water temperatures and the approach.

The increase causes the hot-water temperature to increase considerably faster than does the cold-water temperature.

In both these cases, the area *ABCD* should remain constant—actually it decreases about 2 percent for every 10°F (5.6°C) increase in hot-water temperature. The cooling tower designers take this into consideration in their initial design by applying a hot-water temperature correction to design calculations when the design hot-water temperature exceeds 110°F (43.5°C). See Fig. 3.21.

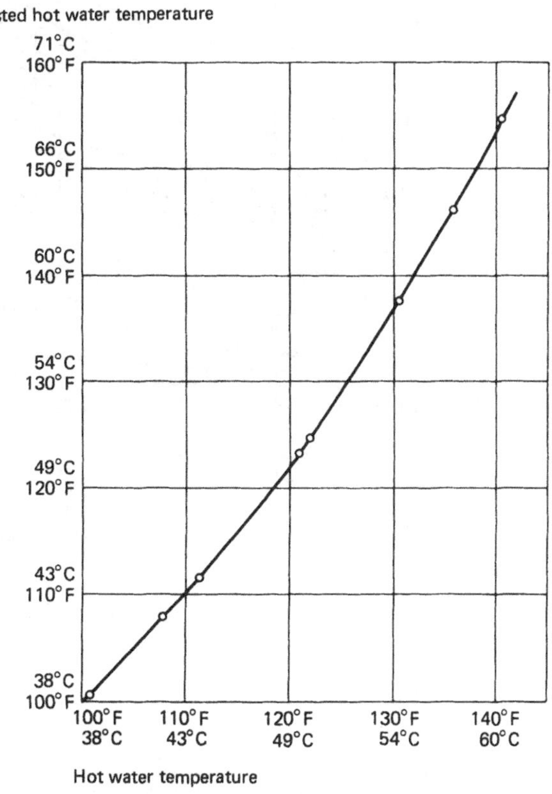

FIGURE 3.21 Plot of hot-water temperature adjustment.

However, a change in *L/G* will change this area. It has been found that a logarithmic plot of *L/G* versus *KaV/L* at a constant airflow results in a straight line. This line, when plotted on the demand curve for the design conditions, is the tower characteristic curve. The slope of the curve depends on the tower fill. In the absence of more specific data, splash-type fill will have a slope of –0.6.

Knowing the wet-bulb temperature, the range, the approach, and the *L/G* ratio, we can determine *KaV/L* by referring to the charts in the Cooling Technology Institute's *Blue Book.*[3] A typical tower characteristic curve which would be submitted by a manufacturer is shown in Fig. 3.22. The complete set of cooling tower performance curves in the CTI *Blue Book* is to the cooling tower engineer what the steam tables are to the turbine engineer. The set of curves can be used to predict the performance of a given cooling tower under widely varying conditions of service.

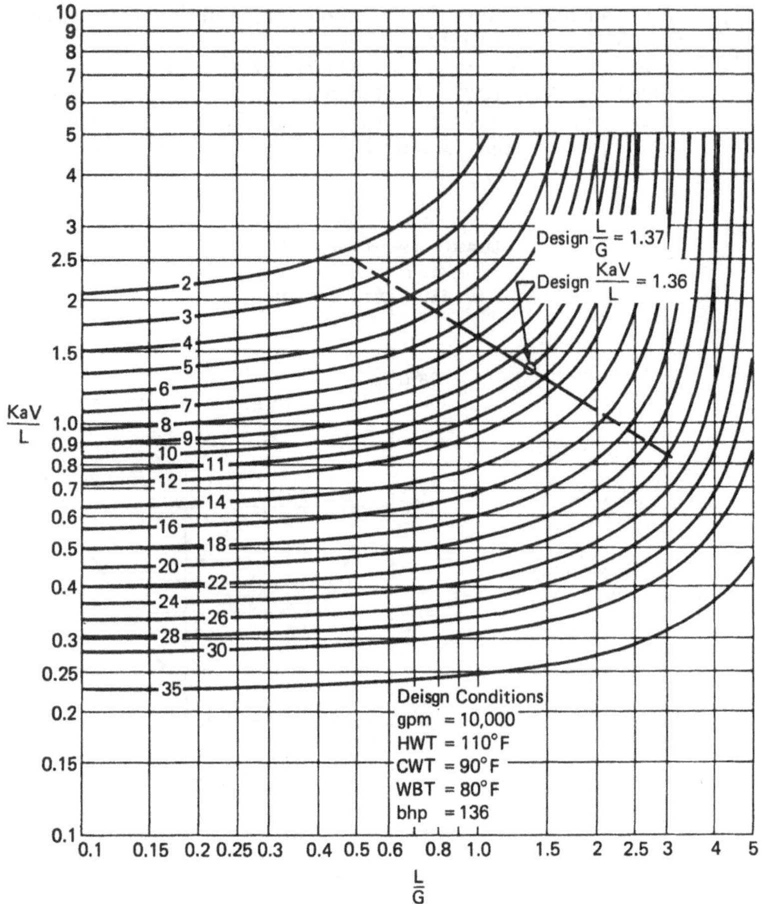

FIGURE 3.22 Tower characteristic curves.

The most important design characteristic is *L/G*. The plant engineer should have on file the design *L/G* ratio for each cooling tower in the plant. When bids are solicited for a new cooling tower, the specifications should require that the characteristic curve for the tower proposed be submitted with the bid package.

Design Parameters

Cooling towers are designed to meet a condition of operation specified by the plant engineer. This condition requires the removal of a heat load of a specified magnitude. When the cooling water flow rate is selected, the specification can be set. The cooling tower is specified to cool a water quantity [gal/min (m^3/h)] through a definite temperature gradient (range) to a final temperature which is a certain number of degrees above the design wet-bulb temperature (approach). Only infrequently will the tower operate at this point, since the plant will normally level out at a slightly different requirement and/or the wet-bulb temperature will be other than the design value. For this reason the tower characteristic curve should be supplied by the manufacturer.

The design wet-bulb temperature is usually based on National Weather Service records and is chosen as the temperature which will not be experienced more than, say 1, 2½, or 5 percent of the time during the summer months of June, July, August, and September. The design wet-bulb temperature should be selected only after some reference has been made to the economics of the plant being served, the seasonal requirements of cooling, and the tabulated Weather Service record for the locality. Choosing the 1-percent frequency level would be judicious when the cooling tower is serving a temperature-sensitive process, or in production of a high-profit product.

The three design levels shown in Table 3.5 are 1, 2½, and 5 percent of the 2928 summer hours, June through September, rounded off to the nearest whole degree. For selecting design temperatures at locations between the cities shown in the table, use data from the city or town nearest the locality of the plant installation. For major installations involving large expenditures or critical temperature balance, make further detailed studies with the assistance of a meteorologist.

STANDARDS AND SPECIFICATIONS

Over the years, the Cooling Technology Institute has developed several standards which are important to industrial cooling-tower design. When writing a specification for a new tower, if you state, "This is to be a CTI Code cooling tower," the CTI standards immediately become a part of your specifications and your contract with the manufacturer. Applicable standards include the following:

1. "Redwood Lumber Specification," Standard 103
2. "Gear Speed Reducers," Standard 111
3. "Pressure Preservative Treatment of Lumber," Standard WMS-112
4. "Douglas Fir Lumber Specifications," Standard 114
5. "Timber Fastener Specification," Standard 119
6. "Asbestos Cement Materials," Standard 127
7. "Acceptance Test Code," Bulletin ATC-105

When standards are specified, manufacturers are protected since all proposals will be on the same basis; the buyer is protected with the assurance of getting a quality product.

Suggested Cooling-Tower Specifications

I. General

This specification covers the construction of an induced-draft, counterflow water-cooling tower at (location) for (company) (hereafter referred to as Owner).

Each cell of the cooling tower is to be capable of individual operation with its own water supply and mechanical equipment. The design and construction of the cooling tower shall conform to the latest applicable provisions of the Cooling Technology Institute standards and shall be a CTI Code tower.

Bids are to be submitted on CTI bid forms with all items completed.

Attached plant safety requirements are a part of this specification and will become a part of the issued contract.

II. Facilities Furnished by Others

A. Power wiring, motor controls, and all electrical labor.
B. Materials and installation labor for external piping to and from the tower, including valves.

TABLE 3.5 Summer, June–September, Design Wet-Bulb Temperature

	1%	2½%	5%
Alabama			
Birmingham	79	78	77
Huntsville	78	79	76
Mobile	80	79	79
Alaska			
Anchorage	61	61	59
Fairbanks	64	63	61
Arizona			
Flagstaff	61	60	59
Phoenix	77	76	75
Tucson	74	73	72
Arkansas			
El Dorado	81	80	79
Fayetteville	77	76	75
Little Rock	80	79	78
California			
Arcata/Eureka	60	59	58
Bakersfield	72	71	70
Fresno	73	72	71
Los Angeles	69	68	67
Sacramento	72	70	69
San Diego	71	70	68
San Francisco	65	63	62
Colorado			
Denver	65	64	63
Grand Junction	64	63	62
Pueblo	68	67	66
Connecticut			
Hartford	77	76	74
New Haven	77	76	75
Delaware			
Wilmington	79	77	76
District of Columbia			
Washington	78	77	76
Florida			
Jacksonville	80	79	79
Miami	80	79	79
Orlando	80	79	78
Pensacola	82	81	80
Tampa	81	80	79
Georgia			
Atlanta	78	77	76
Macon	80	79	78
Valdosta	80	79	78
Hawaii			
Honolulu	75	74	73
Idaho			
Boise	68	66	65
Idaho Falls	65	64	62
Pocatello	65	63	62
Illinois			
Chicago	78	76	75
Peoria	78	77	76
Springfield	79	78	77
Indiana			
Evansville	79	78	77
Indianapolis	78	77	76
South Bend	77	76	74
Iowa			
Des Moines	79	77	76
Mason City	77	75	74
Sioux City	79	77	76
Kansas			
Dodge City	74	73	72
Goodland	71	70	69
Topeka	79	78	77
Wichita	77	76	74
Kentucky			
Lexington	78	77	76
Louisville	79	78	77
Paducah	80	79	78
Louisiana			
Baton Rouge	81	80	79
New Orleans	81	80	79
Shreveport	81	80	79
Maine			
Augusta	74	73	71
Portland	75	73	71
Maryland			
Baltimore	79	78	77
Massachusetts			
Boston	76	74	73
Worcester	75	73	71
Michigan			
Battle Creek	76	74	73
Detroit	76	75	74
Saginaw	76	75	73
Minnesota			
Alexandria	76	74	72
Duluth	73	71	69
Minneapolis/St. Paul	77	75	74
Rochester	77	75	74
Mississippi			
Greenwood	81	80	79
Meridian	80	79	78
McComb	80	79	79
Missouri			
Joplin	79	78	77
Kansas City	79	77	76
Springfield	78	77	76

TABLE 3.5 Summer, June–September, Design Wet-Bulb Temperature (*Continued*)

	1%	2½%	5%		1%	2½%	5%
Montana				Rhode Island			
Billings	68	66	65	Providence	76	75	74
Butte	60	59	57	South Carolina			
Great Falls	64	63	61	Charleston	81	80	79
Nebraska				Columbia	79	79	78
Grand Island	76	75	74	Spartanburg	77	76	75
North Platte	74	73	72	South Dakota			
Omaha	79	78	76	Pierre	76	74	73
Scotts Bluff	70	69	67	Rapid City	72	71	69
Nevada				Sioux Falls	77	75	74
Elko	64	62	61	Tennessee			
Las Vegas	72	71	70	Chattanooga	78	78	77
Reno	64	62	61	Knoxville	77	76	75
New Hampshire				Memphis	80	79	78
Concord	75	73	72	Nashville	79	78	77
Manchester	76	74	73	Texas			
New Jersey				Abilene	76	75	74
Newark	77	76	75	Amarillo	72	71	70
Trenton	78	77	76	Austin	79	78	77
New Mexico				Big Spring	75	73	72
Albuquerque	66	65	64	Corpus Christi	81	80	80
Carlsbad	72	71	70	Dallas	79	78	78
Roswell	71	70	69	El Paso	70	69	68
New York				Houston	80	80	79
Albany	76	74	73	Utah			
Binghamton	74	72	71	Richfield	66	65	64
Buffalo	75	73	72	St. George	71	70	69
New York	77	76	75	Salt Lake City	67	66	65
North Carolina				Vermont			
Asheville	75	74	73	Burlington	74	73	71
Charlotte	78	77	76	Virginia			
Raleigh	79	78	77	Norfolk	79	78	78
North Dakota				Richmond	79	78	77
Bismarck	74	72	70	Roanoke	76	75	74
Fargo	76	74	72	Washington			
Minot	72	70	68	Ellensburg	67	65	63
Ohio				Seattle/Tacoma	66	64	63
Cincinnati	78	77	76	Spokane	66	64	63
Cleveland	76	75	74	West Virginia			
Columbus	77	76	75	Charleston	76	75	74
Oklahoma				Morgantown	76	74	73
Ponca City	78	77	76	Parkersburg	77	76	75
Oklahoma City	78	77	76	Wisconsin			
Tulsa	79	78	77	Green Bay	75	73	72
Oregon				Madison	77	75	73
Medford	70	68	66	Milwaukee	77	75	73
Pendleton	66	65	63	Wyoming			
Portland	69	67	66	Casper	63	62	60
Pennsylvania				Cheyenne	63	62	61
Altoona	74	73	72	Rock Springs	58	57	56
Harrisburg	76	75	74				
Philadelphia	78	77	76				
Pittsburgh	75	74	73				

C. Necessary concrete cold-water basin.
D. 110-V, 60-Hz, ________ -kW, single-phase power to contractor at one location at the tower site. (Additional facilities which others are willing to supply to contractor.)

III. Design Data

Circulation rate ________________________ gal/min (m^3/h)
Water temperature to tower ________________ °F (°C)
Water temperature from tower ______________ °F (°C)
Inlet wet-bulb temperature ________________ °F (°C)
Range ______________________________ °F (°C)
Approach ____________________________ °F (°C)
Wind velocity, maximum _______________ mi/h (m/s)
Wind loading, maximum ________________ lb/ft^2 (kg/m^2)
Basin depth __________________________ ft (m)
Tower location (above sea level) ____________ ft (m)
Drift loss ____________________________ percent

Bidder to include tower characteristic curve with bid and state design *L/G* ratio.

IV. Evaluation

A. Fan horsepower evaluation will he added to the base price of the tower at $________ per horsepower (kilowatt) (or $________ per horsepower (kilowatt) per year for ________ years). Typical values of the cost of horsepower are $500 to $1500 per horsepower.
B. Pumping head evaluation will be added to the base price of the tower at $________ per foot (meter) of head. Typical values of cost of pumping are $5000 to $100,000 per foot.
C. Concrete cold water basin evaluation will be added to the base price of the tower at $________ per square foot (meter) of plan area at the base of the tower. Typical values of the cost of concrete basin are $500 to $1000 per square foot.

V. Materials

A. General
  1. *Lumber.* All lumber used shall be structural no. 1 redwood as graded and specified in CTI Standard 103, Grades II and III. No plywood is allowed in any structural area of the tower. (Acceptable alternative: Douglas fir.)
  2. *Preservative treatment.* All lumber shall be treated. Lumber shall be cut to dimensions, notched, and drilled prior to preservative treatment. Treatment to be with chromated copper arsenate, Type B (CCA-b), and in accordance with CTI Standard WMS-112.
  3. *Hardware.* (See CTI Standard TPR-126 for charting of materials.) All bolts, nuts, and washers shall be ________. All nails shall be ________. Other hardware, such as connectors and base anchors, shall be stainless steel, galvanized steel, or epoxy-coated cast iron.

B. Component Parts
  1. *Framework*
    a. All tower columns shall not be less than 4 × 4 in (10 × 10 cm) nominal.
    b. All connections and joints are to be carefully fitted and bolted. Nailing or notching of structural members will not be permitted. Nonframework members

such as fill, sheathing, and louvers shall not be used to furnish the structural strength of the cooling tower.

2. *Fan deck*
   a. Fan deck shall be designed for a live load of 60 lb/ft^2 and shall be reinforced for any concentrated or distributed dead loads. Fan decking shall be tongue-and-groove planking with nominal 2-in (5-cm) thickness or 1⅛-in (2.9-cm) treated Douglas fir plywood.
   b. For counterflow cooling towers, one access door per fan cell shall be furnished through the fan deck. A ladder is to be supplied for access from the fan deck to the drift eliminators. A small platform shall be installed even with the top of the drift eliminators.
3. *Fill.* Tower fill shall be pressure-treated, clear heart redwood, treated Douglas fir, or polyvinyl chloride (PVC) modules. For crossflow cooling towers, fill supports shall be PVC-coated steel or stainless steel hangers. Bidder to specify vertical and horizontal spacing of fill bars, fill bar size, and fill depth. If film fill is used, air-travel dimension shall be stated.
4. *Drift eliminators.* Shall be treated redwood, treated Douglas fir, or PVC modules.
5. *Hot-water distribution system*
   a. *Counterflow.* The tower shall be provided with a complete water distribution system fabricated from PVC or fiberglass-reinforced plastic pipe. The header and laterals shall have self-draining, nonclogging, full-pattern spray nozzles. Piping shall terminate with one flanged connection for each cell approximately 1 ft (0.3 m) outside the tower casing to permit shutdown of any cell without affecting operation of other cells.
   b. *Crossflow.* The hot-water distribution basin floor shall be constructed of tongue-and-groove redwood (or Douglas fir) or Douglas fir plywood with downtake orifice nozzles constructed of polypropylene with integral splash surface diffusers. Each basin shall have a flow control valve capable of full shutoff.
6. *Fans and drives*
   a. Fan shall be propeller type with at least six adjustable-pitch blades of fiberglass-reinforced epoxy, Air Moving and Conditioning Association (AMCA) rated, and statically and dynamically balanced prior to shipment, if possible.
   b. All motors shall be suitable for across-the-line starting and shall be designed for cooling-tower service. Motors shall be installed outside the exit air stream and nameplate rating shall not be exceeded when tower is operating within the limits of design conditions specified. Motors shall be wired for 3-phase, 60-Hz, __________-V power.
   c. The fans shall be driven through right-angle, heavy-duty, cooling-tower reduction-gear assemblies having a minimum service factor of 2.0, based on motor nameplate rating. Reduction gears shall be provided with vent lines and oil fill lines extending outside of fan stack. Oil fill lines shall include oil level sight gauges. Reducers shall conform to CTI Standard 111.
   d. The driveshaft assembly connecting the motor and gear reducer shall be the nonlubricated design. Two driveshaft guards shall be supplied, with one at each end of each shaft.
   e. Supports for motor and reducer assembly shall be of unitized galvanized steel construction. Minimum thickness of steel employed shall be ¼ in (6.4 mm).
   f. One vibration cutout switch shall be provided with each fan.
7. *Fan stack.* Fan stacks shall be eased-inlet entrance type, not less than 6 ft (1.33 m) high, constructed of fiberglass-reinforced polyester.
8. *Partitions.* Towers consisting of two or more cells shall have a solid transverse partition wall between all cells which extends from louver face to louver face and from basin-curb level to fan-deck (counterflow towers) and distribution-deck lev-

els (crossflow towers). All partition walls will be constructed of treated redwood, Douglas fir, or corrugated, fiberglass-reinforced, polyester plastic.

9. *Casing.* Tower casing shall be single-wall, 8-oz, 4.2 or 5.33 corrugated, fiberglass-reinforced, polyester plastic. Casing sheets shall have a minimum of one corrugation overlap at all seams and shall be of watertight design.
10. *Air-inlet louvers.* Louvers shall be of treated redwood, Douglas fir, or fiberglass-reinforced, polyester plastic supported on spans of 4 ft (1.22 m) or less.
11. *Access*
    - **a.** At least one ladder and one stairway at opposite ends of the tower shall be provided extending from the ground level to the top of the fan deck.
    - **b.** Stairways and ladders shall be in accordance with OSHA requirements.
    - **c.** Handrails around the top of the tower shall be provided in accordance with OSHA requirements.
    - **d.** Access to plenum chamber on crossflow towers shall be through the end-wall casing at basin-curb elevation. Access doors shall be provided through each partition wall. A walkway shall be provided at basin-curb level from end wall to end wall.

VI. Drawings

The cooling-tower manufacturer shall submit three copies of general arrangement drawings for approval. Catalog drawings will not be considered acceptable as approval drawings. Approval drawings shall clearly show exact dimensions and all construction details.

VII. Testing

Acceptance Test. The cooling-tower manufacturer shall conduct an acceptance test in accordance with the Cooling Technology Institute's "Acceptance Test Procedure for Industrial Water Cooling Towers," CTI Bulletin ATC-105, latest revision. The cooling-tower manufacturer shall quote a separate price for conducting the acceptance test.

VIII. Guarantee

The cooling tower shall be guaranteed for a period of 1 year after structural completion or 18 months after shipment, whichever occurs first. Any defective parts or workmanship shall be repaired at the cooling-tower manufacturer's expense.

OPERATION AND MAINTENANCE

Evaporation Loss

In the usual cooling-tower operation, the water evaporation rate is essentially fixed by the rate of removal of sensible heat from the water, and the evaporation loss can be roughly estimated as 0.1 percent of the circulating water flow for each degree Fahrenheit of cooling range.

Drift Loss

Cooling-tower drift loss is the entrained liquid water droplets discharged with the exit air. The function of drift eliminators is to limit the number of escaping droplets to an acceptable level. Most design specifications state the permissible drift loss as a percentage of the circulating water flow. Most modern cooling towers are designed with drift-eliminator face velocities below 650 ft/min (198.1 m/min), and entrainment losses less than 0.008 percent of the water circulation rate.

Blowdown. Cooling-tower blowdown is a portion of the circulating water that is discharged from the system to prevent excessive buildup of solids. The maximum concentration of solids that can be tolerated is usually determined by the effects on the various components of the cooling system, such as piping, pumps, heat exchangers, and the cooling tower itself. The required blowdown rate is determined from a material balance, yielding the equation:

$$b = \frac{e}{r-1} - d$$

where b = blowdown rate, gal/mm (m^3/h)
e = evaporation loss, gal/min (m^3/h)
r = ratio of solids in blowdown to solids in makeup (cycles of concentration)
d = drift loss, gal/min (m^3/h)

Makeup Water Rate. To hold a given solids concentration ratio, sufficient water must be added to the recirculating water system to make up for evaporation, blowdown, drift, and other losses. The required makeup rate may be computed by either of the equations:

$$M = b + e + d$$

or

$$M = \frac{r}{r-1} e$$

where M = makeup rate, gal/min (m^3/h)
b = blowdown rate, gal/min (m^3/h)
e = evaporation loss, gal/min (m^3/h)
r = ratio of solids in blowdown to solids in makeup (cycles of concentration)
d = drift loss, gal/min (m^3/h)

Proper maintenance of the mechanical equipment and distribution system will ensure optimum operation from the cooling tower over a long period of time. Motors must remain properly lubricated, gearbox oil should be maintained at the proper level, and driveshaft alignment should be checked on a regular basis. Uniform distribution of hot water within the tower is essential in order to maintain optimum tower performance. Refer to Table 3.6 for a suggested preventive maintenance schedule for key tower components.

COLD-WEATHER OPERATION

General

The successful operation of induced-draft cooling towers during extremely cold weather very often presents a problem to the plant operators. Ice tends to form on the air-intake louvers and filling immediately adjacent to the louvers. This is because the water in contact with the airstream will intermittently splash on the louver boards, where it freezes and will eventually build up to the point where the flow of air is restricted. Also, subcooling will be obtained in the area of the filling immediately adjacent to the louvers, where ice will build up during periods of light load or high air velocity. Particular care must be taken in starting up fans that have been shut down for an appreciable length of time because of the possibility of unequal ice loading on the blades. If a fan is started up in an unbalanced condition, the gear unit could be torn from its mounting, causing permanent damage to the mechanical equipment and surrounding tower structure.

Control of Ice on Air-Intake Louvers and Filling

Every effort should be made to maintain the design water quantity and heat load per cell. In the event of a reduction in the plant heat load, it is extremely important that the water quan-

TABLE 3.6 Master Periodic Inspection Chart

Weekly	What to look for	What to do
Fan	Noise, vibration, tower sway	Visually check for damage, check weep holes for water in fan blades.
Speed reducer	Noise, rapid vibration, oil leaks	Check oil level; check oil for water or other contamination; check breather pipe for clogging; check shaft for misalignment. Run idle units for 10 min.
Driveshaft	Vibration, broken disks	Replace broken disks. Tighten bolts.
Suction pit screen	Debris	Remove debris.
Monthly	**What to look for**	**What to do**
Speed reducers	Routine check	Check oil level and contamination. Inspect oil on high- and low-speed shaft for lubrication. Check alignment.
Driveshaft	Routine check	Clean; replace if damaged.
Nozzles	Scale, corrosion, debris	Spot-check nozzles on side of tower opposite from risers.
Headers and laterals	Scale or clogging	Clean.
Distribution decks	Algae, debris, channeling deposits of lime, scale, etc.	Clean with steam, high-pressure hose, or stiff brush.
Fill	Clogging	Clean or remove, as necessary.
Yearly	**What to look for**	**What to do**
Tower	Routine check	Shut down; clean thoroughly from top to bottom including basin.
Fill	Warping, water channeling	Replace as needed.
Structure	Decay, excessive delignification	Use ice pick. Replace structural members, if necessary.
Bolts	Looseness, corrosion	Tighten all bolts; replace those corroded.
Wall sheathing	Leaks	Caulk as necessary. Keep tower wet.
Mechanical equipment	Fan blade damage, unbalance, pitch, fan shaft looseness, speed reducer wear, shaft alignment	Make checks and corrections as necessary.
Basin	Dirt, debris, signs of oil	Clean thoroughly.

tity be reduced proportionately and that the cells be shut down to maintain the design quantity per cell; that is, riser valves should be shut on idle cells. In addition, the water temperature in the basin should be maintained at a reasonable value, such as 60° to 70°F (15.6° to 21.1°C), by reducing the volume of air entering the tower.

The greater the air volume moved by the fans, the greater the amount of ice formation. It is our recommendation that the fan motors on a multicell cooling tower be reduced from full speed to half speed as required to maintain the temperature of the water in the basin at 60°F to 70°F (15.6°C to 21.1°C). In the event that all fans are running at half speed and the basin temperature falls below 60°F (15.6°C), it will then be necessary to shut down the fans as required to maintain this temperature.

Usually the water concentration in gallons per square foot (liters per square meter) of cell area is sufficient to cause a reversal of airflow when the fans are not operating. This will tend to melt any ice that is formed on the fill and the louvers. In a multicell tower, the fans should remain turned off for approximately 12 h and then turned back to low speed. The adjoining cell should then have its fan turned off for the same period of time, and this operation should be repeated on the other cells during cold weather.

Should this procedure prove ineffective in controlling the ice formation, one of the following recommendations should be followed:

1. *Remove ice from the louvers with a steam hose.* Be careful not to allow the ice load on the fill to exceed its design value, causing it to collapse during thawing.
2. *Remove louver boards.* Some counterflow and straight-sided crossflow towers have the top louver board and every fifth louver beneath of double width. The intermediate louver boards may be removed, thereby reducing the amount of ice forming between these louver blades and choking off the air supply.
3. *Install reversing switches on low-speed fan motor terminals.* This will reverse the flow of air through the cooling tower, which will quickly melt any ice that is formed on the louvers and fill. The fan motors should not be operated in reverse for more than about 30 min at a time. During extremely cold weather, each of the fans operating at half speed should be reversed once a day to remove ice from the air-intake louvers.

NOMENCLATURE

a = area of transfer surface per unit of tower volume, ft^2/ft^3
b = blowdown rate, gal/min
d = drift loss, gal/min
e = evaporation loss rate, gal/min
g = mass airflow, lb dry air/$h \cdot ft^2$
h = enthalpy of air–water vapor mixture, Btu/lb dry air
h_a = enthalpy of air–water vapor mixture at wet-bulb temperature, Btu/lb dry air
h_w = enthalpy of air–water vapor mixture at bulk water temperature, Btu/lb dry air
h_1 = enthalpy of air–water vapor mixture at inlet wet-bulb temperature, Btu/lb dry air
h_2 = enthalpy of air–water vapor mixture at exhaust wet-bulb temperature, Btu/lb dry air
k = overall enthalpy transfer coefficient, lb/$h \cdot ft^2$ per lb water/lb dry air
l = mass water flow, lb/$h \cdot ft^2$ plan area
m = makeup rate, gal/min
r = ratio of solids in blowdown to solids in makeup (cycles of concentration)
t = bulk water temperature, °F (°C)
t_1 = hot-water temperature, °F (°C)
t_2 = cold-water temperature, °F (°C)
v = effective cooling-tower volume, ft^2/ft^3 plan area
KaV/L = tower characteristic
L/G = liquid-to-gas ratio, lb water/lb dry air

REFERENCES

1. Sherwood, T. K., and R. L. Pigford: *Absorption and Extraction,* 2d ed., McGraw-Hill, New York, 1952, pp. 102–104.
2. Kern, D. Q.: *Process Heat Transfer,* McGraw-Hill, New York, 1950.
3. "Cooling Tower Performance Curves," *Blue Book,* Cooling Technology Institute, Houston, Tex., 1970.

CHAPTER 3.5

APPLICATIONS OF HEATING, VENTILATING, AND AIR-CONDITIONING SYSTEMS*

Ernest H. Graf, P.E.
Assistant Director, Mechanical Engineering, Giffels Associates, Inc., Southfield, Michigan

William S. Lytle, P.E.
Project Engineer, Mechanical Engineering, Giffels Associates, Inc., Southfield, Michigan

GENERAL CONSIDERATIONS

As a system design develops from concept to final contract documents, the following subjects should be considered throughout the heating, ventilating, and air-conditioning (HVAC) design period. These subjects are of a general nature inasmuch as they are applicable to all HVAC designs, and they may become specific requirements inasmuch as codes are continually updated.

* Adapted from Robert C. Rosaler (ed.), *HVAC Systems and Components Handbook,* 2d ed., McGraw-Hill, New York, 1998; used with permission. Updated for this handbook by James R. Fox, P.E., William C. Newman, Wayne H. Lawton, and David H. Helwig, Arcadis-Giffels, Inc

Cooling Towers and Legionnaires' Disease

Since the 1976 outbreak of pneumonia in Philadelphia, cooling towers have frequently been linked with the *Legionella pneumophila* bacteria, or Legionnaires' disease. Several precautions should be taken:

1. Keep basins and sumps free of mud, silt, and organic debris.
2. Use chemical and/or biological inhibitors as recommended by water-treatment specialists. Do not overfeed, because high concentrations of some inhibitors are nutrients for microbes.
3. Do not permit the water to stagnate. The water should be circulated throughout the system for at least 1 h each day regardless of the water temperature at the tower. The water temperature within the indoor piping will probably be 60°F (15.6°C) or warmer, and one purpose of circulating the water is to disperse active inhibitors throughout the system.
4. Minimize leaks from processes to cooling water, especially at food plants. Again, the processes may contain nutrients for microbes.

Elevator Machine Rooms

These spaces are of primary importance to the safe and reliable operation of elevators. In the United States, all ductwork or piping in these rooms must be for the sole purpose of serving equipment in these rooms unless the designer obtains permission from the authorities in charge of administering American National Standards Institute/American Society of Mechanical Engineers (ANSI/ASME) Standard 17.1, *Safety Code for Elevators and Escalators.* If architectural or structural features tend to cause an infringement of this rule, the duct or pipe must be enclosed in an approved manner.

Energy Conservation

A consequence of the 1973 increase in world oil prices is legislation governing the design of buildings and their HVAC systems. Numerous U.S. states and municipalities include an energy code or invoke a particular issue of American Society of Heating, Refrigeration, and Air Conditioning Engineers (ASHRAE) Standard 90 as a part of their building code. Standard 90.1 establishes indoor and outdoor design conditions, limits the overall U-factor for walls and roofs, limits reheat systems, requires the economizer cycle on certain fan systems, limits fan motor power, requires minimum duct and pipe insulation, requires minimum efficiencies for heating and cooling equipment, and so on.

In the interest of freedom of design, the energy codes permit tradeoffs between specified criteria as long as the annual consumption of depletable energy does not exceed that of a system built in strict conformance with the standard. Certain municipalities require that the drawings submitted for building-permit purposes include a statement to the effect that the design complies with the municipality's energy code. Some states issue their own preprinted forms that must be completed to show compliance with the state's energy code.

Equipment Maintenance

The adage "out of sight, out of mind" applies to maintenance. Equipment that a designer knows should be periodically checked and maintained may get neither when access is difficult. Maintenance instructions are available from equipment manufacturers; the system designer should be acquainted with these instructions, and the design should include reasonable access, including walk space and headroom, for ease of maintenance. Some features for ease of maintenance will increase project costs, and the client should be included in the decision to accept or reject these features.

Penthouse and rooftop equipment should be serviceable via stairs or elevators and via roof walkways (to protect the roofing). Ship's ladders are inadequate when tools, parts, chemicals, and so on are to be carried. Rooftop air handlers, especially those used in cold climates, should have enclosed service corridors (where project budgets allow). If heavy rooftop replacement parts, filters, or equipment are expected to be skidded or rolled across a roof, the architect must be advised of the loading to permit proper roof system design.

Truss-mounted air handlers, unit heaters, valves, exhaust fans, and so on should be over aisles (for servicing from mechanized lifts and rolling platforms) when catwalks are impractical. Locate isolation valves and traps within reach of building columns and trusses to provide a degree of stability for service personnel on ladders.

It is important that access to ceiling spaces be coordinated with the architect. Lay-in ceilings provide unlimited access to the space above, except possibly at lights, speakers, sprinklers, and the like. When possible, locate valves, dampers, air boxes, coils, and the like above corridors and janitor closets so as to disturb the client's operations the least.

Piping system diagrams are important and should be provided by the construction documents. Piping should be labeled with service and flow direction arrows, and valves should be numbered as to an identification type number. The valves normally used in modern-day piping systems fall into two categories, multiturn and quarter-turn. Quarter-turn valves would include ball valves, butterfly valves, and plug valves. The multiturn variety includes the traditional gate and globe type. Quarter-turn valves are becoming more popular and usually can be provided at less cost, depending on service, but care should be used in their application. Valves in pipe sizes up to 2 in (5 cm) are normally manufactured with threaded bodies, though some types are available with flanged ends. Valves in these sizes are ball, plug, gate, or globe type in design. For pipe sizes 2½ in (6.3 cm) and larger, threaded ends are rarely used or available. The multiturn designs would normally be provided with flanged ends. Plug valves would also be flanged. The butterfly valve is furnished with either flanged ends, though these are rarely used in aboveground work, lug body, or wafer style. The wafer style would be the least expensive but is not a good choice because there is no way to disconnect the service downstream of the valve without a short flanged spool bolted up to the valve (see Chap. 5.11). The best choice for the butterfly valve is the lug body because it is the best of both worlds. It is installed like a flanged-body valve without the flanged-body cost.

Most experienced piping designers shy away from quarter-turn valves for steam service at all temperatures and pressures. This is, for the most part, out of consideration for safety, especially for the smaller pipe sizes where lever-type operators would be expected. For larger pipe sizes that carry steam, high-performance butterfly valves with gear operators are gaining popularity. The gear operator provides a measure of safety where slow operation is desired. Warm-up bypasses should always be considered for all steam isolation valves without regard to whether they are gate or butterfly type.

Pressure gauges are normally provided to monitor changes in pressure loss across pumps, heat exchangers, and strainers. Figure 3.23 suggests a piping scheme that allows one gauge, which can be used to increase the accuracy of gauge reading where the piping configuration permits. Gauge valves should be either small globe [⅜ in (10 mm)] or ball valves. These types of valves are not as likely to seize with disuse, and they allow throttling to steady a bouncing gauge needle.

The observation of steam-trap operation can be facilitated by having a ⅜-in (10-mm) test valve at the trap discharge pipe (Fig. 3.24). With valve V-1 closed, trap leakage and cycling may be observed at an open test valve. The test valve can be used to monitor reverse-flow leaks at check valves.

Equipment Noise and Vibration

Noise and vibration can reach unacceptable levels in manufacturing plants as well as in offices, auditoriums, and the like. Once an unacceptable level is built in, it is very costly to correct. The noise and vibration control recommendations in Chaps. 4.11 and 4.12 of this hand-

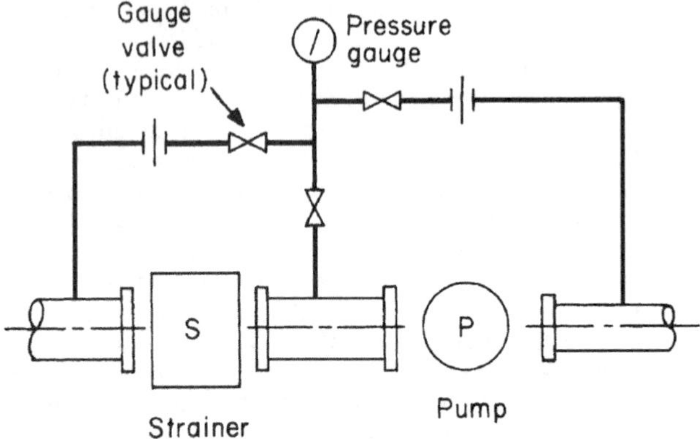

FIGURE 3.23 Multiple-point pressure gauge.

book and in the *1999 ASHRAE Handbook, HVAC Applications,* should be followed. Sound and vibration specialists should be consulted for HVAC systems serving auditoriums and other sensitive areas. Fans, variable-air-volume boxes, dampers, diffusers, pumps, valves, ducts, and pipes which have sudden size changes or interior protrusions or which are undersized can be sources of unwelcome noise.

Fans are the quietest when operating near maximum efficiency, yet even then they may require sound attenuation at the inlet and outlet. Silencers and/or a sufficient length of acoustically lined ductwork are commonly used to "protect" room air grilles nearest the fan. Noise through duct and fan sides must also be considered. In the United States, do not use acoustic duct lining in hospitals except as permitted by the U.S. Department of Health and Human Services (DHHS) Publication HRS-M-HF 84-1.

Dampers with abrupt edges and those used for balancing or throttling air flows cause turbulence in the air stream, which in turn is a potential noise source. Like dampers, diffusers (as well as registers, grilles, and slots) are potential noise sources because of their abrupt edges and integral balancing dampers. Diffuser selection, however, is more advanced in that sound criteria are readily available in the manufacturers' catalogs. Note, however, that a background noise (white noise) is preferable in office spaces because it imparts a degree of privacy to conversation. Diffusers can provide this.

Pumps are also the quietest when operating near maximum efficiency. Flexible connectors will help to dampen vibration transmission to the pipe wall but will not stop water- or liquid-borne noise.

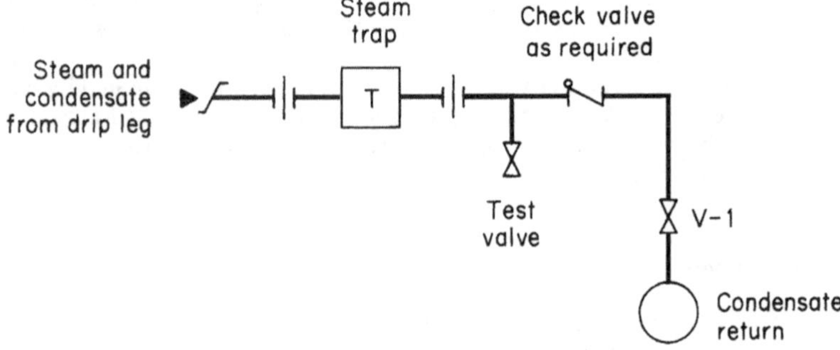

FIGURE 3.24 Test valve at steam trap.

Valves for water, steam, and compressed-air service can be a noise source or even a source of damaging vibration, depending on the valve pattern and on the degree of throttling or pressure reduction. Here again, the findings of manufacturers' research are available for the designer's use.

Equipment rooms with large fans, pumps, boilers, chillers, compressors, and cooling towers should not be located adjacent to sound- or vibration-sensitive spaces. General office, commercial, and institutional occupancies usually require that this equipment be mounted on springs or vibration isolation pads (with or without inertia bases) to mitigate the transfer of vibration to the building's structure. Spring-mounted equipment requires spring pipe hangers and flexible duct and conduit connections. Air-mixing boxes and variable-volume boxes are best located above corridors, toilet rooms, and public spaces. Roof fans, exhaust pipes from diesel-driven generators, louvers, and the like should be designed and located to minimize noise levels, especially when near residential areas.

Evaporative Cooling

An airstream will approach at its wet bulb temperature a 100-percent saturated condition after intimate contact with recirculated water. Evaporative cooling can provide considerable relief without the cost of refrigeration equipment for people working in otherwise unbearably hot commercial and industrial surroundings, such as laundries, boiler rooms, and foundries. Motors and transformers have been cooled (and their efficiency increased) by an evaporatively cooled airstream.

Figure 3.25 shows the equipment and psychrometric elements of a *direct* evaporative cooler. Its greatest application is in hot, arid climates. For example, the 100°F (38°C), 15-percent relative humidity (RH) outdoor air in Arizona could be cooled to 70°F (21°C), 82-percent RH with an 88-percent efficient unit. Efficiency is the quotient of the dry-bulb conditions shown at (2), (3), and (4) in Fig. 3.25*b*. Note that the discharge air from a direct evaporative cooler is near 100-percent humidity and that condensation will result if the air is in contact with surfaces below its dew point. The discharge dew point in the above example is 64°F (18°C).

Figure 3.26 schematically shows an *indirect* evaporative cooler. Whereas a direct evaporative cooler increases the airstream's moisture, an indirect evaporative cooler does not; that is, there is sensible cooling only at (1) to (2) in Fig. 3.26*b*. Air is expelled externally at (5). When an indirect cooler's discharge (2) is ducted to a direct cooler's inlet, the final discharge (3) will

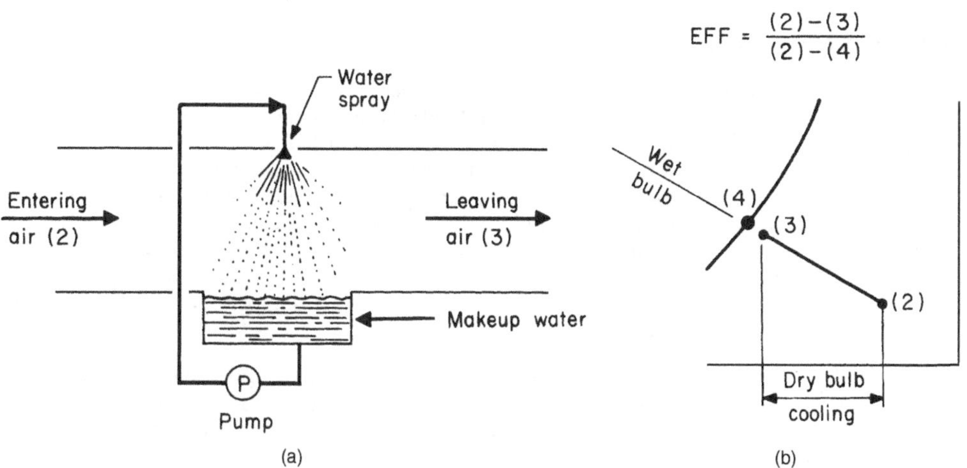

FIGURE 3.25 Direct evaporative cooling: (*a*) equipment, and (*b*) psychrometrics.

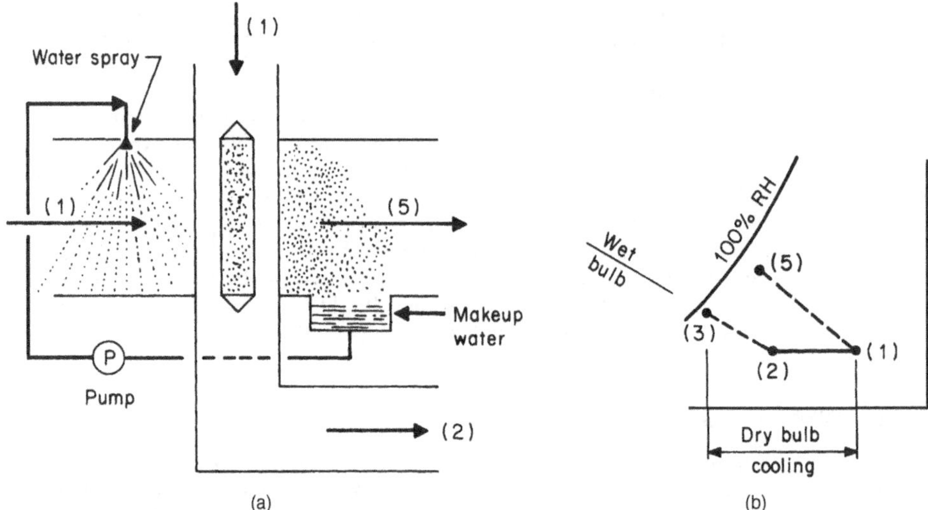

FIGURE 3.26 Indirect evaporative cooling: (*a*) equipment, and (*b*) psychrometrics.

be somewhat cooler and include less moisture than that of a direct cooler only. Various combinations of direct and indirect equipment have been used as stand-alone equipment or to augment refrigeration equipment for reduced overall operating costs. Refer to the *2000 ASHRAE Handbook, Systems and Equipment,* and the *1999 ASHRAE Handbook, HVAC Applications.*

Some evaporative cooling equipment operates with an atomizing water spray only, with any overspray going to the drain. Some additional air cooling is available when the water temperature is less than the air wet-bulb temperature. Evaporative cooling involves 100 percent outdoor air, and there must be provisions to exhaust the air. Evaporative cooling has also been applied to roof cooling; a roof is wetted by fine sprays, and the water evaporation causes cooler temperatures at the roof's upper and lower surfaces. The water supply for all applications must be analyzed for suitability and, as needed, treated to control scale, algae, and bacteria.

Fire and Smoke Control Dampers

Wherever practical and/or necessary, building walls and floors are made of fire-resistant material to hinder the spread of fire. Frequently, HVAC ducts must penetrate walls and floors. In order to maintain the fire resistance of a penetrated wall, fire dampers or equal protection must be provided whenever a fire-resistance-rated wall, floor, or ceiling is penetrated by ducts or grilles. Fire dampers are approved devices (approved by administrators of the building code, fire marshall, and/or insurance underwriter) that automatically close in the presence of higher-than-normal temperatures to restrict the passage of air and flame. Smoke dampers are approved devices that automatically close on a control signal to restrict the passage of smoke.

The following are general applications for fire or smoke dampers per the National Fire Protection Association Standard NFPA-90A, 1996 edition:

- Provide 3-h fire dampers in ducts that penetrate walls and partitions which require a 3-h or higher resistance rating, provide 1½-h dampers in ducts that penetrate those requiring a rating of 2 h or higher but less than 3 h, and provide 1½-h dampers in ducts that penetrate shaft walls requiring a rating of 1 to 2 h.
- Provide fire dampers in all nonducted air-transfer openings that penetrate partitions if they require a fire-resistance rating.

- Provide smoke dampers and associated detectors at air-handling equipment whose supply capacity exceeds 2000 ft^3/min (944 L/s). Provide supply and return smoke detectors and associated dampers for units whose return capacity exceeds 15,000 ft^3/min (7080 L/s). The dampers shall isolate the equipment (including filters) from the remainder of the system except that the smoke dampers may be omitted (subject to approval by the authority having jurisdiction) when the entire air-handling system is within the space served or when rooftop air handlers serve ducts in large open spaces directly below the air handler.

Exceptions to the preceding are allowed when the facility design includes an engineered smoke control system. Note that schools, hospitals, nursing homes, and jails may have more stringent requirements. Kitchen hood exhaust systems must be coordinated with local jurisdictions.

Dampers that snap closed have often incurred sufficient vacuum on the downstream side to collapse the duct (see Ref. 1). Smoke and other control dampers that close normally and restrict the total air flow of a rotating fan can cause pressure (or vacuum) within the duct equal to fan shutoff pressure. A fan might require a full minute after the motor is deenergized before coasting to a safe speed (pressure). Provide adequate duct construction, relief doors, or delayed damper closure (as approved by the authority having jurisdiction).

Refer to the building codes, local fire marshall rules, insurance underwriter's rules, and NFPA-90A for criteria regarding fire and smoke dampers.

Outdoor Air

This is needed to make up for air removed by exhaust fans; to pressurize buildings so as to reduce the infiltration of unwanted hot, cold, moist, or dirty outdoor air; to dilute exhaled carbon dioxide, off-gassing of plastic materials, tobacco smoke, and body odors; and to replenish oxygen.

A frequently used rule of thumb to provide building pressurization is to size the return fan's air flow for 85 percent of the supply fan's, thereby leaving 15 percent for pressurization and small toilet-exhaust makeup. This is acceptable for simple, constant-volume systems and buildings. The required outdoor air can also be established by estimating the air flow through window and door cracks, open windows and doors, curtain walls, exhaust fans, etc. Building pressurization should be less than 0.15 in (4 mm) water gauge (WG) on ground floors that have doors to the outside so that doors do not hang open from outflow of air. The building's roof and walls must be basically airtight to attain pressurization. If there are numerous cracks, poor construction joints, and other air leaks throughout the walls, it is impractical to pressurize the building—and worse, the wind will merely blow in through the leaks on one side of the building and out through the leaks on the other side. Variable-air-volume (VAV) systems require special attention regarding outdoor air because as the supply fan's air flow is reduced, the outdoor and return air entering this fan tend to reduce proportionately. Provide minimum outdoor air at the minimum unit airflow for VAV systems.

The National Fire Protection Association (NFPA) standards recommend minimum outdoor air quantities for hazardous occupancies. NFPA standards are a requirement insofar as building codes have adopted them by reference. Building codes frequently specify minimum outdoor air requirements for numerous hazardous and nonhazardous occupancies. ASHRAE Standard 62 recommends minimum quantities of outdoor air for numerous activities. In the interest of energy conservation, 5 ft^3/min (2.4 L/s) per person had been considered acceptable for sedentary non-smoking activities, but this was later determined to be inadequate. ASHRAE 62-1989 typically requires at least 20 ft^3/min (9.5 L/s) per person. Refer to the ASHRAE standards for additional requirements.

Perimeter Heating

The heat loss through outside walls, whether solid or with windows, must be analyzed for occupant comfort. The floor temperature should be no less than 65°F (18°C), especially for sedentary activities. In order to have comfortable floor temperatures, it is important that perimeter insulation be continuous from the wall through the floor slab and continue below per Refs. 2 and 3.

Walls with less than 250 Btu/h · lin ft (240 W/lin m) loss may generally be heated by ceiling diffusers that provide airflow down the window—unless the occupants would be especially sensitive to cold, such as in hospitals, nursing homes, day-care centers, and swimming pools. Walls with 250 to 450 Btu/h · lin ft (240 to 433 W/lin m) of loss can be heated by warm air flowing down from air slots in the ceiling; the air supply should be approximately 85 to 110°F (29 to 43°C). Walls with more than 450 Btu/h · lin ft (433 W/lin m) of loss should be heated by underwindow air supply or radiation. See Ref. 4 for additional discussion. The radiant effect of cold surfaces may be determined from the procedures in ANSI/ASHRAE Standard 55-1992.

Curtain-wall construction, custom-designed wall-to-roof closures, and architectural details at transitions between differing materials have, at times, been poorly constructed and sealed, with the result that cold winter air is admitted to the ceiling plenum and/or occupied spaces. Considering that the infiltration rates published by curtain-wall manufacturers are frequently exceeded because of poor construction practices, it is prudent to provide overcapacity in lieu of undercapacity in heating equipment. The design of finned radiation systems should provide for a *continuous* finned element along the wall requiring heat. Do not design short lengths of finned element connected by bare pipe all within a continuous enclosure. Cold downdrafts can occur in the area of bare pipe. Reduce the heating-water supply temperature and then the finned-element size as required to provide the needed heat output and water velocity.

The surface temperatures of glass, window frames, ceiling plenums, structural steel, vapor barriers, and so on should be analyzed for potential condensation, especially when humidifiers or wet processes are installed.

Process Loads

Heat release from manufacturing processes is frequently a major portion of an industrial air-conditioning load. Motors, transformers, hot tanks, ovens, etc. form the process load. If all motors and other equipment in large plants are assumed to be fully loaded and to be operating continuously, then invariably the air-conditioning system will be greatly oversized. The designer and client should mutually establish diversity factors that consider actual motor loads and operating periods, large equipment with motors near the roof (here the motor heat may be directly exhausted and not affect the air-conditioned zone), amount of motor input energy carried off by coolants, etc. Diversity factors could be as much as 0.5 or even 0.3 for research and development shops containing numerous machines that are used only occasionally by the few operators assigned to the shop.

Room Air Motion

Ideally, occupied portions [or the lower 6 ft (2 m)] of air-conditioned spaces for sedentary activities would have 20- to 40-ft/min (0.1- to 0.2-m/s) velocity of air movement, with the air being within 2°F (1°C) of a set point. It is impractical to expect this velocity throughout an entire area at all times, inasmuch as air would have to be supplied at approximately a 2-ft^3/min · ft^2 (10.2-L/s · m^2) rate or higher. This rate is easily incurred by the design load of perimeter offices, laboratories, and computer rooms but would only occur in an interior office when there is considerable heat-releasing equipment. The supply air temperature should be selected such that, at design conditions, a flow rate of at least 0.8 ft^3/min · ft^2 (4.1 L/s · m^2), but never less than 0.5 ft^3/min · ft^2 (2.5 L/s · m^2), is provided.

People doing moderate levels of work in non-air-conditioned industrial plants might require as much as a 250-ft/min (1.3-m/s) velocity of air movement in order to be able to continue working as the air temperature approaches 90°F (32°C). This would not necessarily provide a full-comfort condition, but it would provide acceptable relief. Loose paper, hair, and other light objects may start to be blown about at air movements of 160 ft/min (0.8 m/s); see Ref. 5. Workers influenced by high ambient temperatures and radiant heat may need as much as a 4000-ft/min (20-m/s) velocity of a 90°F (32°C) airstream to increase their convective and evaporative heat loss. These high velocities would be in the form of spot cooling or of a relief station that the worker could enter and exit at will. Air movement can only compensate for, but not stop, low levels of radiant heat. Only effective shielding will stop radiant energy. Continuous air movement of approximately 300 ft/min (1.5 m/s) and higher can be disturbing to workers.

Situations involving these higher air movements and temperatures should be analyzed by the methods in Refs. 6 to 9.

OCCUPANCIES

Clean Rooms

For some manufacturing facilities, an interior room that is conditioned by a unitary air conditioner with 2-in- (5-cm-) thick throwaway filters might be called a *clean room;* that is, it is clean relative to the atmosphere of the surrounding plant. Generally, however, clean rooms are spaces associated with the microchip, laser optics, medical, and other industries where airborne particles as small as 0.5 μm and less are removed. One micrometer equals one-millionth of a meter, or 0.000039 in (0.000001 m).

Clean rooms are identified by the maximum permissible number of 0.5-μm particles per cubic foot. For example, a class 100 clean room will have no more than 100 of these particles per cubic foot, a class 10 clean room no more than 10, and so on. This degree of cleanliness can be attained by passing the air through a high-efficiency particulate air (HEPA) filter installed in the plane of the clean-room ceiling, after which the air continues in a downward vertical laminar flow (VLF) to return grilles located in the floor or in the walls at the floor. Horizontal laminar flow (HLF) rooms are also built wherein the HEPA filters are in one wall and the return grilles are in the opposite wall. A disadvantage with an HLF room is that downstream activities may receive contaminants from upstream activities.

An alternative to an entire space being ultraclean is to provide ultraclean chambers within a clean room (e.g., class 100 chambers in a class 10,000 room). This is feasible when a product requires the class 100 conditions for only a few operations along the entire assembly line.

The air-conditioning system frequently includes a three-fan configuration (primary, secondary, and makeup) similar to that shown in Fig. 3.27. The primary fan maintains the high air change through the room and through the final HEPA filters. The secondary fan maintains a side-stream (to the primary circuit) airflow through chilled-water or -brine cooling coils, humidifiers, and heating coils. The makeup fan injects conditioned outdoor air into the secondary circuit, thereby providing clean-room pressurization and makeup for exhaust fans. Clean-room air changes are high, such that the total room air might be replaced every 7 s, and this generally results in the fan energy being the major portion of the internal heat gain. Whenever space permits, locate filters downstream of fans so as to intercept containments from the lubrication and wear of drive belts, couplings, bearings, etc.

For additional discussions, refer to the *1999 ASHRAE HVAC Handbook, Applications,* and to the latest issue of Federal Standard 209 *Clean Room and Work Station Requirements, Controlled Environment.* ISO Standard 14 644-1 is a comparable international standard for clean rooms.

Computer Rooms

These rooms are required to house computer equipment that is sensitive to swings in temperature and humidity. Equipment of this type normally requires controlled conditions 24 h per day, 7 days per week. Computer equipment can be classified as (1) data processing, (2) computer-

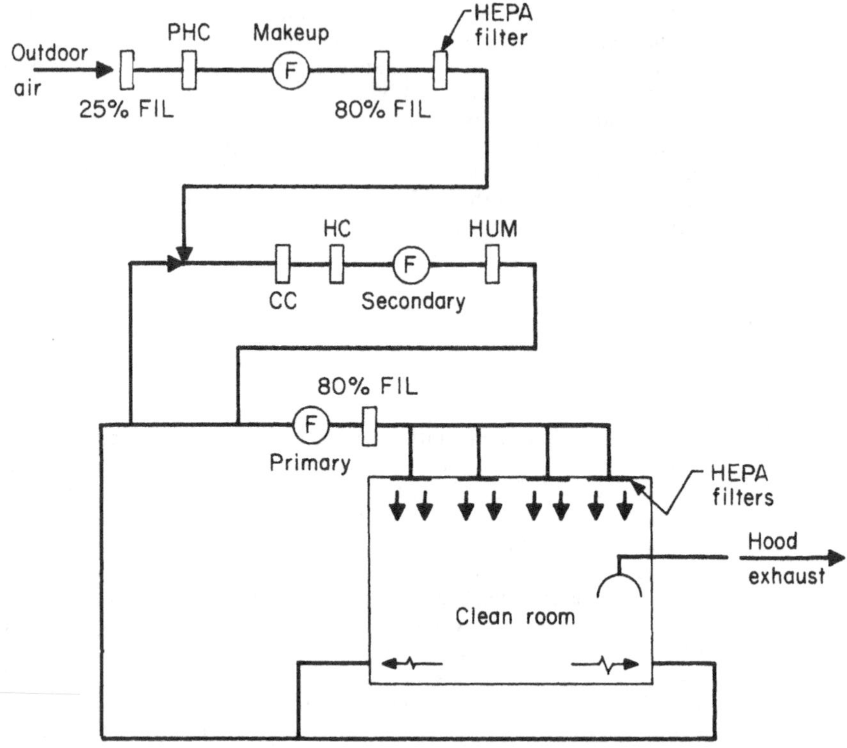

FIGURE 3.27 Three-fan clean-room air system.

aided design and drafting (CADD), and (3) microcomputer. Microcomputers are generally similar to standard office equipment and require no special treatment. Some CADD equipment is also microcomputer-based and falls into the same category. Data processing and larger CADD systems fall into the realm of specialized computer rooms, and these are discussed below.

Data processing systems operate on a multiple-shift basis, requiring air-conditioning during other than normal working hours. Humidity stability is of prime importance with data processing equipment and CADD plotters. The equipment is inherently sensitive to rapid changes in moisture content and temperature.

To provide for the air-conditioning requirements of computer equipment, two components are necessary: a space to house the equipment and a system to provide cooling and humidity control. Fundamental to space construction is a high-quality vapor barrier and complete sealing of all space penetrations, such as piping, ductwork, and cables. To control moisture penetration into the space effectively, it is necessary to extend the vapor barrier up over the ceiling in the form of a plenum enclosure. Vapor-sealing the ceiling itself is not generally adequate due to the nature of its construction and to penetration from lighting and other devices.

A straightforward approach to providing conditioning to computer spaces is to use packaged, self-contained computer-room units specifically designed for the service. Controls for these units have the necessary accuracy and response to provide the required room conditions. An added advantage to packaged computer-room units is flexibility. As the needs of the computer room change and as the equipment and heat loads move around, the air-conditioning units can be relocated to suit the new configuration. The units can be purchased either with chilled-water or direct-expansion coils, as desired. Remote condensers or liquid coolers can also be provided. Large installations lend themselves quite well to heat recovery; therefore, the designer should be aware of possible potential uses for the energy.

Centrally located air-handling units external to the computer space offer benefits on large installations. More options are available with regard to introduction of ventilation air, energy recovery, and control systems. Maintenance is also more convenient where systems are centrally located. There are obvious additional benefits with noise and vibration control. Use of a centrally located system must be carefully evaluated with regard to first cost and to potential savings, as the former will carry a heavy impact.

The load in the room will be primarily sensible. This will require a fairly high airflow rate as compared to comfort applications. High airflow rates require a high degree of care with air distribution devices in order to avoid drafts. One way to alleviate this problem is to utilize underfloor distribution where a raised floor is provided for computer cable access. A typical computer-room arrangement is shown in Fig. 3.28. Major obstructions to air flow below the floor must be minimized so as to avoid dead spots.

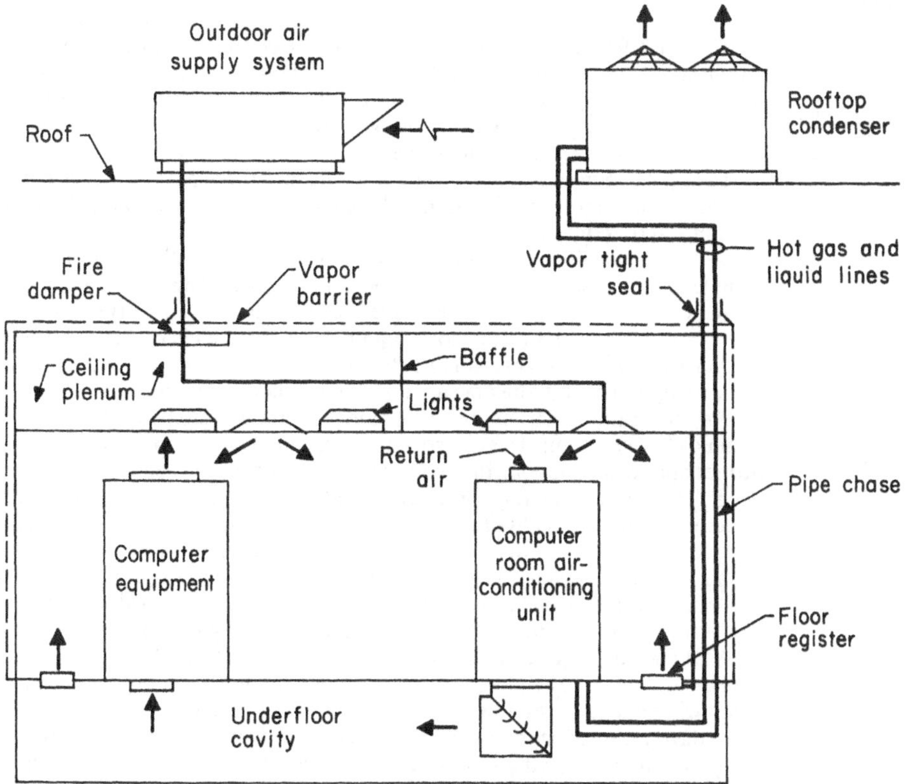

FIGURE 3.28 Typical computer-room layout. Locate floor registers so as to be in nontraffic areas and free from obstruction. Ceiling plenum baffles are located where and as directed by local codes and insurance underwriters.

In summary, important points to remember are the following:

1. Completely surround the room with an effective vapor barrier.
2. Provide well-sealed wall penetrations where ductwork and piping pass into the computer space.
3. Provide high-quality humidity and temperature controls capable of holding close tolerances: ±1°F (0.6°C) for temperature and ±5 percent for relative humidity.

4. Pay close attention to air distribution, avoiding major obstructions under floors where underfloor distribution is used.
5. Be alert to opportunities for energy recovery.
6. Make sure that the chosen control parameters and design temperatures and conditions satisfy the equipment manufacturer's specifications.
7. Be attentive to operating-noise levels within the computer space.
8. If chilled water or cooling water is piped to computer-room units within the computer-room space, provide a looped- or grid-type distribution system with extra valved outlets for flexibility.
9. Provide minimum outside air per ASHRAE Standard 62-1999.

Offices

Cooling and heating systems for office buildings and spaces are usually designed with an emphasis on the occupants' comfort and well-being. The designer should remain aware that not only the mechanical systems but also the architectural features of the space affect the comfort of the occupants. And the designer will do well to remember that the mechanical system should in all respects be invisible to the casual observer.

The application of system design is divided into three parts: the method of energy transfer, the method of energy distribution, and the method of control. Controls are discussed in Chap. 5.3 and will therefore not be discussed here.

To properly apply a mechanical system to control the office environment, it is necessary to completely understand the nature of the load involved. This load will have a different character depending on the part of the office that is being served. Perimeter zones will have relatively large load swings due to solar loading and heat loss because of thermal conduction. The loading from the occupants will be relatively minor. Core zones, on the other hand, will impart more loading from building occupants and installed equipment.

For the office environment, the more common system used today is the variable-air-volume (VAV) system. This approach was originally developed as a cooling system, but with proper application of control it will serve equally well for heating. In climates where there is need for extensive heating, perimeter treatment is required to replace the skin loss of the building structure. An old but reliable method is fin-tube radiation supplied with hot water to replace the skin loss. A system that is being seen with more regularity is in the form of perimeter air supply. Care should be taken with the application of perimeter air systems to ensure that wall U-values are at least to the level of ASHRAE Standard 90.1. If this is not done, interior surface temperatures will be too low and the occupants in the vicinity will feel cold.

Avoid striking the surface of exterior windows with conditioned air, as this will probably cool even double-pane glass to below the dew point of the outdoor air in the summer. The result will be fogged windows and a less-than-happy client.

In the interest of economy from a final cost and operating basis, it is best to return the bulk of the air circulated to the supply fan unit. Only enough outdoor air should be made up to the building space to provide ventilation air (ASHRAE 62-1999), replace toilet exhaust, and pressurize the building. For large office systems, it is generally more practical to return spent air to the central unit or units through a ceiling plenum. If the plenum volume is excessively large, a better approach would be to duct the return air directly back to the unit. The ceiling plenum will be warmer during the cooling season when the return air is ducted, and this will require a somewhat greater room air supply because more heat will be transmitted to the room space from the ceiling rather than directly back to the coil through the return air.

Terminal devices require special attention when applied to VAV systems. At low flow rates, the diffuser will tend to dump unless care is taken in the selection to maintain adequate throw. Slot-type diffusers tend to perform well in this application, but there are other diffuser designs, such as the perforated type, that are more economical and will have adequate performance.

The air-handling, refrigeration, and heating equipment could be located either within an enclosed mechanical-equipment room or on the building roof in the form of unitary self-contained equipment. For larger systems, of 200 tons (703 kW) of refrigeration or more, the mechanical-equipment room offers distinct advantages from the standpoint of maintenance; however, the impact on building cost must be evaluated carefully. An alternate approach to the enclosed equipment room is a custom-designed, factory-fabricated equipment room. These are shipped to the jobsite in preassembled, bolted-together, ready-to-run modules. For small offices and retail stores, the most appropriate approach would be roof-mounted, packaged, self-contained, unitary equipment. It will probably be found that this is the lowest in first cost, but it will not fare well in a lifecycle analysis because of increased maintenance costs after 5 to 10 yr of service.

Test Cells

The cooling and heating of test cells poses many problems.

Within the automotive industry, test cells are used for the following purposes:

- Endurance testing of transmissions and engines
- Hot and cold testing of engines
- Barometric testing and production testing
- Emissions testing

The treatment of production test cells would be very similar to the treatment of noisy areas in other parts of an industrial environment. These areas are generally a little more open in design, with localized protection to contain the scattering of loose pieces in the event of a mechanical failure of the equipment being tested. Hot and cold rooms and barometric cells are usually better left to a package purchase from a manufacturer engaged in that work as a specialty.

Endurance cells, on the other hand, are generally done as a part of the building package (Fig. 3.29). It will be found that these spaces are air conditioned for personnel comfort during setup only. The cell would be ventilated while a test is under way. Heat gains for the nontest air-conditioned mode would be from the normal sources: ambient surroundings, lights, people, and so on. Air distribution for air conditioning would be similar to any space with a nominal loading of 200 to 400 ft^2/ton (5.3 to 10.6 m^2/kW) of refrigeration. It should be remembered, however, that sufficient outdoor air will be needed to make up for trench and floor exhaust while maintaining the cell at a negative-pressure condition relative to other areas. Consult local building codes to ensure compliance with regulations concerning exhaust requirements in areas of this nature.

During testing, as stated previously, the cell would only be ventilated. Outdoor air would be provided at a rate of 100 percent in sufficient quantity to maintain reasonable conditions within the cell. Temperatures within the cell could often be in excess of 120°F (49°C) during a test. Internal-combustion engines are generally liquid-cooled, but even so, the frame losses are substantial and large amounts of outdoor air will be required in order to maintain space conditions to even these high temperature limits. In cold climates, it is necessary to temper ventilation air to something above freezing; 50°F (10°C) is usually appropriate, but each situation needs to be evaluated on its own merit. The engine losses are best obtained from the manufacturer, but in the absence of these data there is information in the *1999 ASHRAE Handbook, HVAC Applications,* that will aid in completing an adequate heat balance. The dynamometer is most often air-cooled and can be thought of as similar to an electric motor. The engine horsepower (wattage) output will be converted to electricity, which is usually fed into the building's electrical system; therefore, the dynamometer losses to the cell will be on the order of 15 to 20 percent of the engine shaft output.

The engine test cell will require a two-stage exhaust system for cooling. The first stage would be to provide low-level floor and trench exhaust to remove heavy fuel vapors and to

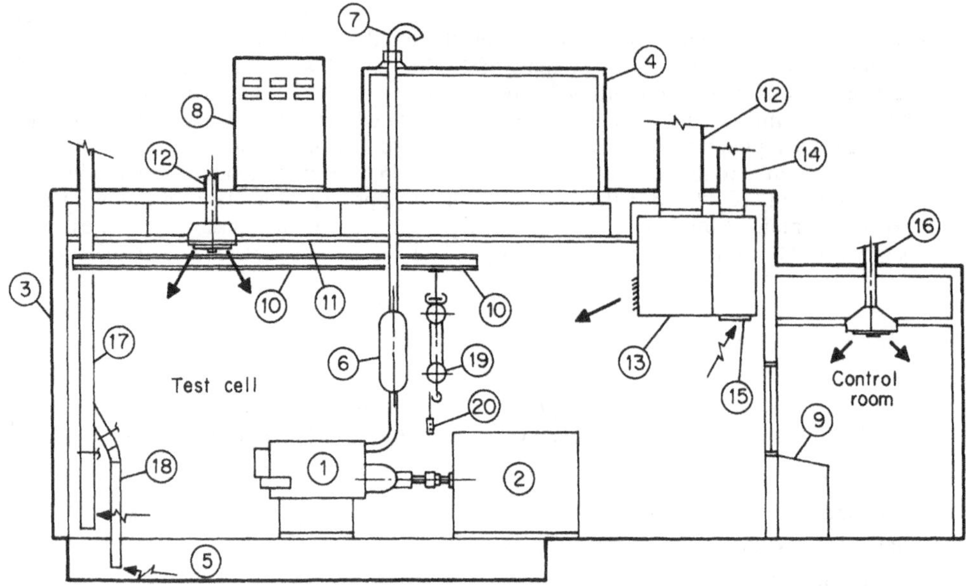

1. Engine
2. Dynamometer
3. Blast wall
4. Blast cupola
5. Fuel and service trench
6. Muffler
7. Engine exhaust
8. Dynamometer
9. Control panel
10. Crane
11. Suspended ceiling
12. Supply air (conditioned, unconditioned)
13. Supply air plenum
14. Cell exhaust
15. Exhaust plenum
16. Control room supply (conditioned)
17. Exhaust duct
18. Trench exhaust duct
19. Electric hoist
20. Hoist electric control

FIGURE 3.29 Typical test-cell layout.

maintain negative conditions in the cell at all times. The second stage would be interlocked with the ventilation system and would come on during testing and would exhaust at a rate about 5 to 10 percent greater than the supply rate to maintain negative-pressure conditions. The second stage would also be activated in the event of a fuel spill to purge the cell as quickly as possible. Activation of the purge should be by automatic control in the event that excessive fumes are detected. An emergency manual override for the automatic purge should be provided. Shutdown of the purge should be manual. Consult local codes for explicit requirements.

Depending on the extent of the engine exhaust system, a booster fan may be required to preclude excessive backpressure on the engine. Where more than one cell is involved, one fan would probably serve multiple cells. Controls would need to be provided to hold the back pressure constant at the engine (Fig. 3.30).

Air conditioning for the test cell could be via either direct-expansion or chilled-water coils. During a test, the cell conditioning would be shut down in all areas except the control room. Depending on equipment size, it usually is an advantage to have a separate system cooling the control room. One approach to heating and cooling an endurance-type test cell is shown

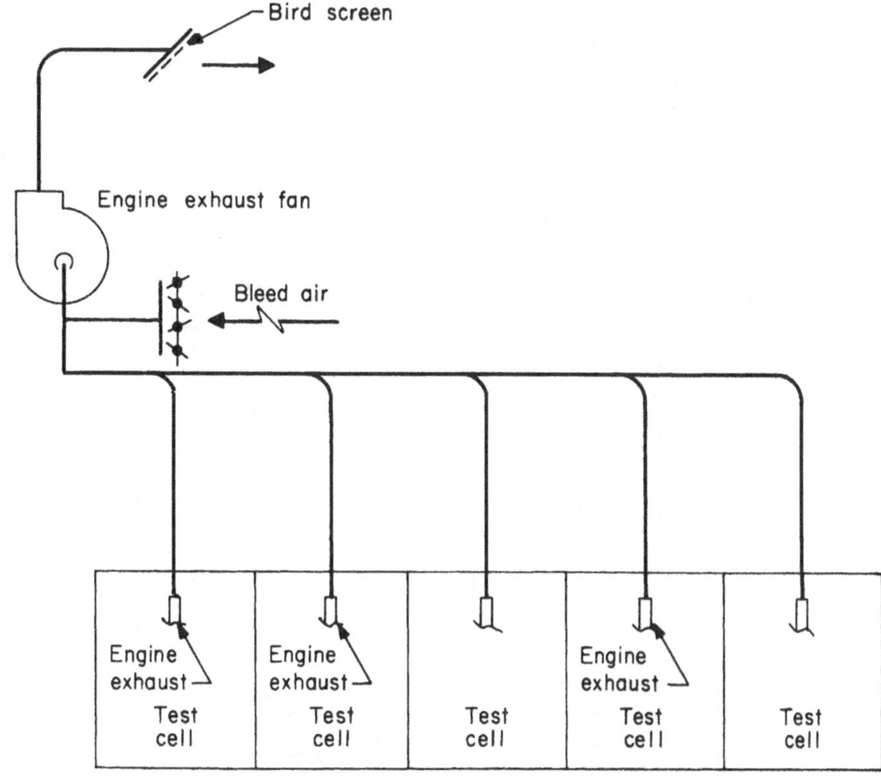

FIGURE 3.30 Engine exhaust booster fan.

schematically in Fig. 3.31. Local building codes and the latest volumes of NFPA should be reviewed to ensure that local requirements are being meet. Fuel vapors within the cell should be continually monitored. The cell should purge automatically in the event that dangerous concentrations are approached.

The following is suggested as the sequence of events for the control cycle of the test cell depicted in Fig. 3.31:

Setup Mode

1. AC-1 and RF-1 are running. Outdoor-air and relief-air dampers are modulated in an economizer arrangement.
2. EF-2 is controlled manually and runs at all times, maintaining negative conditions in the cell and the control room.
3. EF-1 is off and D-1 is shut.

Emergency Ventilation Mode

1. If vapors are detected, D-2 shuts and D-1 opens.
2. EF-1 starts and AC-1 changes to high-volume delivery with cooling coil shut down and outdoor-air damper open.
3. HV-1 starts and its outdoor-air damper opens.
4. System should be returned to normal manually.

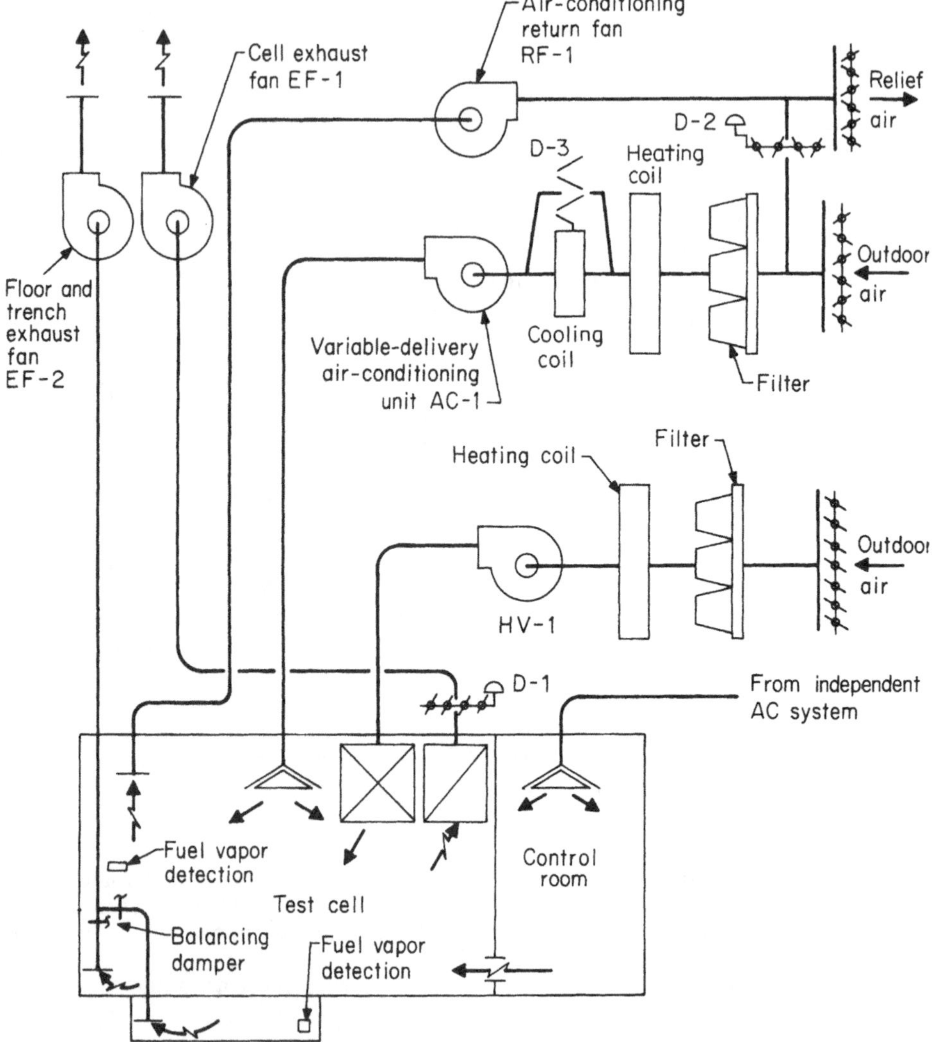

FIGURE 3.31 Test-cell heating, ventilating, and cooling.

Test Mode

1. AC-1 cooling coil shuts down.
2. AC-1 changes to high-volume delivery with outdoor-air damper fully open. D-2 closes and D-3 opens.
3. HV-1 starts and EF-1 starts.

EXHAUST SYSTEMS

One of the early considerations in the design of an exhaust (or ventilation) system should be the ultimate discharge point into the atmosphere. Most of the emissions from ventilation systems are nontoxic or inert and thus will not require a permit for installation or building operation. But should the exhaust air stream contain any of the criteria pollutants—those pollutants for which emissions and ambient concentration criteria have been established, such as CO, NO_x, SO_2, lead, particulate matter (PM), and hydrocarbons (HCs)—it is likely that a permit to install the system will be required.

Once it is determined that a permit will be necessary, an emissions estimate must be made to determine estimates of both uncontrolled (before a pollution control device) and controlled emissions. The emissions estimate may be obtained from either the supplier of the equipment being contemplated for installation or from the Environmental Protection Agency (EPA) Publication AP-42, *Compilation of Air Pollutant Emission Factors.* AP-42 contains emission factors for many common industrial processes, which, when applied to process weight figures, yield emission rates in pounds (kilograms) per hour or tons per year, depending on process operating time. The permit to install an application may be obtained from the state agency responsible for enforcing the federal Clean Air Act. In most states, the Department of Environmental Protection or Department of Natural Resources will have jurisdiction. In general, the permit-to-install application requires the information and data listed in Fig. 3.32.

When designing an area containing a process exhaust system and a control system for the exhaust, it would be well to keep in mind that federal and local air-quality regulations may govern the type of emission control equipment installed and the maximum allowable emissions. The factors dictating what regulations apply include the type of process or equipment being exhausted, the type and quantity of emissions, the maximum emission rate, and the geographic location of the exhausted process. In order to determine what specific rules and regulations apply, the requirements of the U.S. Code of Federal Regulations Title 40 (40 CFR) should be understood early in the project stages so that all applicable rules may be accommodated. Should the design office lack the necessary expertise in this area, a qualified consultant should be engaged. The federal government has issued a list entitled "Major Stationary Sources." The exhaust system's designer should be acquainted with this list, for it identifies the pollutant sources governed by special requirements. Several of the more common sources are listed in Fig. 3.33, and 40 CFR should be consulted for the complete listing. One of the major sets of rules included in 40 CFR is the Prevention of Significant Deterioration (PSD) rules, which establish the extent of pollution control necessary for the major stationary sources.

If a source is determined to be major for any pollutant, the PSD rules may require that the installation include the best available control technology (BACT). The BACT is dependent on the energy impact, environmental impact, economic impact, and other incidental costs associated with the equipment. In addition, the following items are prerequisites to the issue of a permit for pollutants from a major source:

1. Applicant's name and address.
2. Person to contact and telephone number.
3. Proposed facility location.
4. Standard Industrial Classification (SIC) Code.
5. Amount of each air contaminant from each source in pounds per hour (pph) and tons per year (tpy) at maximum and average.
6. What federal requirement will apply to the source?
   - National Emission Standards for Hazardous Air Pollutants (NESHAPs)
   - New Source Performance Standards (NSPS)
   - Prevention of Significant Deterioration (PSD)
   - Emission Offset Policy (EOP)
7. Will best available control technology (BACT) be used?
8. Will the new source cause significant degradation of air quality?
9. How will the new source affect the ambient air quality standard?
10. What monitoring will be installed to monitor the process, exhaust, or control device?
11. What is the construction schedule and the estimated cost of the pollution abatement devices?

FIGURE 3.32 Commonly requested information for air-quality permit applications.

1. Fossil-fuel-fired generating plants greater than 250-million-Btu/h (73-MW) input
2. Kraft pulp mills
3. Portland cement plants
4. Iron- and steel-mill plants
5. Municipal incinerators greater than 250-ton/day charging
6. Petroleum refineries
7. Fuel-conversion plants
8. Chemical process plants
9. Fossil-fuel boilers, or combination thereof totaling more than 250-million-Btu/h (73-MW) input
10. Petroleum storage and transfer units exceeding 300,000-barrel storage
11. Glass-fiber processing plants

FIGURE 3.33 Major stationary sources—partial list.

1. Review and compliance of control technology with the following:
   a. State Air-Quality Implementation Plan (SIP)
   b. New Source Performance Standards (NSPS) (see Fig. 3.34)
   c. National Emissions Standards for Hazardous Air Pollutants (NESHAPs)
   d. BACT
2. Evidence that the source's allowable emissions will not cause or contribute to the deterioration of the National Ambient Air-Quality Standard (NAAQS) or the increment over baseline, which is the amount the source is allowed to increase the background concentration of the particular pollutant.
3. The results of an approved computerized air-quality model that demonstrates the acceptability of emissions in terms of health-related criteria.
4. The monitoring of any existing NAAQS pollutant for up to 1 yr or for such time as is approved.

1. Fossil-fuel-fired steam generators with construction commencing after August 17, 1971
2. Electric utility steam generators with construction commencing after September 18, 1978
3. Incinerators
4. Portland cement plants
5. Sulfuric acid plants
6. Asphalt concrete plants
7. Petroleum refineries
8. Petroleum liquid storage vessels constructed after June 11, 1973, and prior to May 19, 1978
9. Petroleum liquid storage vessels constructed after May 18, 1978, and prior to July 23, 1984
10. Sewage treatment plants
11. Phosphate fertilizer industry—wet process phosphoric acid plants
12. Steel plants—electric arc furnaces
13. Steel plants—electric arc furnaces and argon decarburization vessels constructed after August 17, 1983
14. Kraft pulp mills
15. Grain elevators
16. Surface coating of metal furniture
17. Stationary gas turbines
18. Automobile and light-duty truck painting
19. Graphic arts industry—rotogravure printing
20. Pressure-sensitive tape and label surface coating operations
21. Industrial surface coating: large appliance
22. Asphalt processing and asphalt roofing manufacture
23. Bulk gasoline terminals
24. Petroleum dry cleaners

FIGURE 3.34 New Source Performance Standards (NSPS)—partial list.

5. Documentation of the existing (if any) source's impact and growth since August 7, 1977, in the affected area.
6. A report of the projected impact on visibility, soils, and vegetation.
7. A report of the projected impact on residential, industrial, commercial, and other growth associated with the area.
8. Promulgation of the proposed major source to allow for public comment. Normally, the agency processing the permit application will provide for public notice.

One of the first steps regarding potential pollutant sources is to determine the applicable regulations. For this, an emissions estimate must be made, and the *in-attainment* or *nonattainment* classification of the area in which the source is to be located must be determined. The EPA has classified all areas throughout the United States, including all U.S. possessions and territories. The area is classified as either in-attainment (air quality is better than federal standards) or nonattainment (air quality is worse than federal standards).

If the source is to be located in a nonattainment area, the PSD rules and regulations do not apply, but all sources that contribute to the violation of the NAAQS are subject to the Emissions Offset Policy (EOP). The following items must be considered when reviewing a source that is to be located in a nonattainment area:

1. The lowest achievable emission rate (LAER), which is defined as the most stringent emission limit that can be achieved in practice
2. The emission limitation compliance with the SIP, NSPS, and NESHAPs
3. The contribution of the source to the violation of the NAAQS
4. The impact on the nonattainment area of the fugitive dust sources accompanying the major source

In general, the EOP requires that for a source locating in a nonattainment area, more than equivalent offsetting emission reductions must be obtained from existing emissions prior to approval of the new major source or major modification. The *bubble concept,* wherein the total emissions from the entire facility with the new source do not exceed the emissions prior to addition of the new source, may be used to determine the emission rate. If there were emission reductions at *existing* sources, they would offset the contributions from the new source, or *offset* the new emissions. This same bubble concept may be used for sources that qualify for in-attainment or PSD review.

In the design of a polluting or pollution control facility, stack design should be considered. A stackhead rain-protection device (Figs. 3.35 and 3.39) should be used in lieu of the weather cap found on many older installations, since this cap does not allow for adequate dispersion of the exhaust gas. When specifying or designing stack heights, it should be noted that the EPA has promulgated rules governing the minimum stack height; these rules are known as *good engineering practice* (GEP). A GEP stack has sufficient height to ensure that emissions from the stack do not result in excessive concentrations of any air pollutant in the vicinity of the source as a result of atmospheric downwash, eddy currents, or wakes caused by the building itself or by nearby structures (Figs. 3.36 and 3.37). For uninfluenced stacks, the GEP height is 98 ft (30 m). For stacks on or near structures, the GEP height is (1) 1.5 times the lesser of the height or width of the structure, plus the height of the structure, or (2) such height that the owner of the building can show is necessary for proper dispersion. In addition to GEP stack height, stack exit velocity must be maintained for proper dispersion characteristics.

Figures 3.38 and 3.39 illustrate the relationship between velocity at discharge and the velocity at various distances for the weather-cap and stackhead-type rain hoods, respectively. Maintaining an adequate exit velocity ensures that the exhaust gases will not reenter the building through open windows, doors, or mechanical ventilation equipment. Depending on normal ambient atmospheric conditions, the exit velocities may range from 2700 to 5400 ft/min (14 to 28 m/s). In practice, it has been found that 3500 ft/min (18 m/s) is a good average figure for stack exit velocity, giving adequate plume rise yet maintaining an acceptable noise level within the vicinity of the stack.

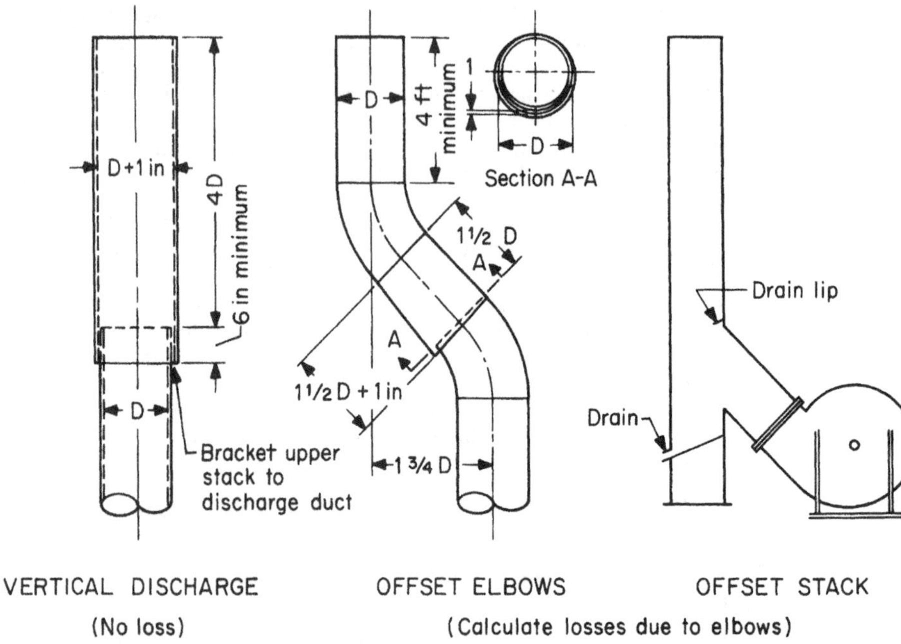

FIGURE 3.35 Typical rain-protection devices. Rain protection characteristics of these caps are superior to a deflecting cap located 0.75D from top of stack. The length of the upper stack is related to rain protection. Excessive additional distance may cause blowout of effluent at the gap between upper and lower sections. (*From* Industrial Ventilation—A Manual of Recommended Practice, *23d ed., Committee on Industrial Ventilation, American Conference of Governmental Industrial Hygienists, copyright 1998, pp. 5–67.*)

Care must be taken when designing exhaust systems handling pollutants for which no specific federal emission limit exists (noncriteria pollutants). All pollutants not included in the criteria pollutant category or the NESHAPS category are considered noncriteria pollutants. When establishing or attempting to determine acceptable concentration levels for noncriteria pollutants, the local authority responsible for regulating air pollution should be consulted since policy varies from district to district. In general, however, noncriteria pollutants' allow-

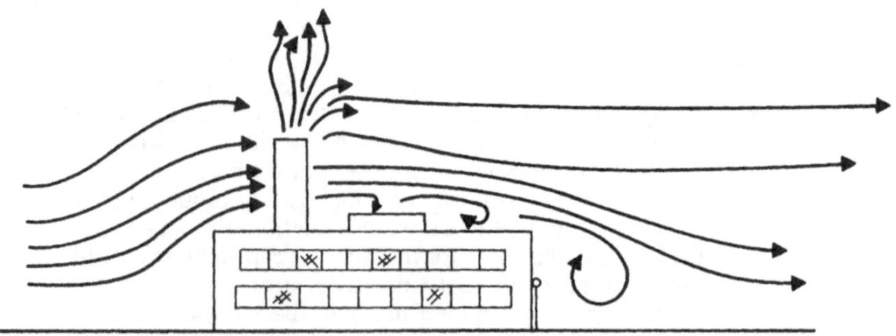

FIGURE 3.36 Good engineering practice (GEP) stack. Stack height minimizes reentrainment of exhaust gases into air which might enter building ventilation system.

FIGURE 3.37 Non-GEP stack. Stack allows exhaust gases to be entrained in building wakes and eddy currents.

able emission rates are based on the American Conference of Governmental Industrial Hygienists (ACGIH) time-weighted average acceptable exposure levels.

A hazardous air pollutant is one for which no ambient air-quality standard is applicable, but which may cause or contribute to increased mortality or illness in the general population. Emission standards for such pollutants are required to be set at levels that protect the public health. These allowable pollutants' emission levels are known as NESHAPS and include levels for radon-222, beryllium, mercury, vinyl chloride, radionuclides, benzene, asbestos, arsenic, and fugitive organic leaks from equipment.

An exhaust stream that includes numerous pollutants, with some being noncriteria pollutants, can be quickly reviewed by assuming that all the exhaust consists of the most toxic pollutant compound. If the emission levels are acceptable for that review, they will be acceptable for all other compounds.

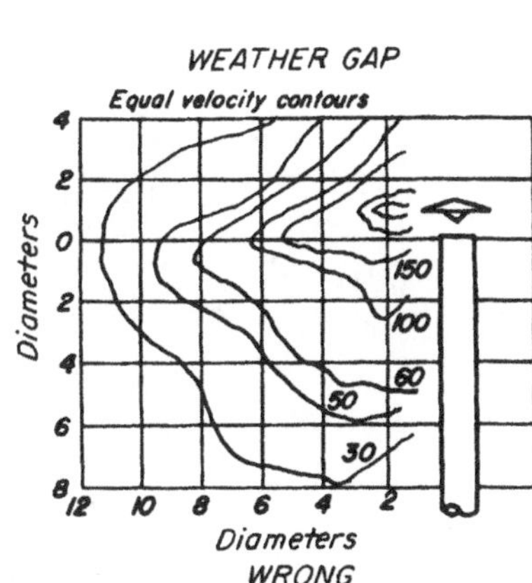

FIGURE 3.38 Weather-cap dispersion characteristics. Deflecting weather cap discharges downward. (*From* Industrial Ventilation—A Manual of Recommended Practice, *23d ed., Committee on Industrial Ventilation, American Conference of Governmental Industrial Hygienists, copyright 1998, pp. 5–67.*)

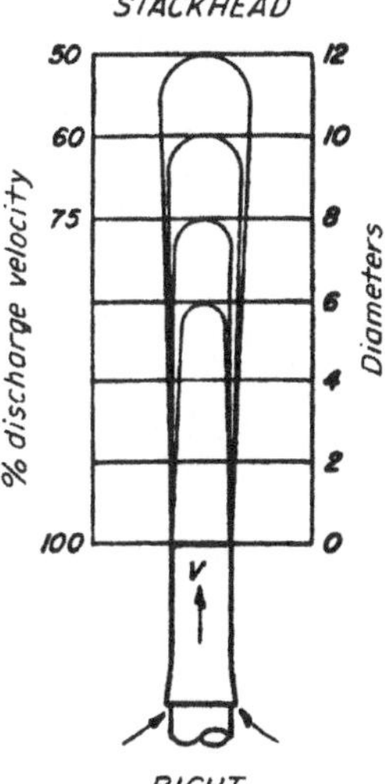

FIGURE 3.39 Stackhead dispersion characteristics. Vertical discharge cap throws discharge upward where dilution will take place. (*From* Industrial Ventilation—A Manual of Recommended Practice, *23d ed., Committee on Industrial Ventilation, American Conference of Governmental Industrial Hygienists, copyright 1998, pp. 5–67.*)

REFERENCES

1. United McGill Corporation, *Engineering Bulletin,* vol. 2, no. 9, copyright 1990.
2. *Energy Conservation in New Building Design,* ASHRAE Standard 90A-1980, American Society of Heating, Refrigerating and Air-Conditioning Engineers, Atlanta, Ga., p. 18, para. 4.4.2.4.
3. *1997 ASHRAE Handbook, Fundamentals,* American Society of Heating, Refrigerating and Air-Conditioning Engineers, Atlanta, GA, 1997.
4. Tom Zych, "Overhead Heating of Perimeter Zones in VAV Systems," *Contracting Business,* August 1985, pp. 75–78.
5. *Thermal Environmental Conditions for Human Occupancy,* ANSI/ASHRAE Standard 55-192, American Society of Heating, Refrigerating and Air-Conditioning Engineers, Atlanta, Ga., p. 4, para 5.1.4.
6. Knowlton J. Caplan, "Heat Stress Measurements," *Heating/Piping/Air Conditioning,* February 1980, pp. 55–62.
7. *Industrial Ventilation—A Manual of Recommended Practice,* 21st ed., Committee on Industrial Ventilation, American Conference of Governmental Industrial Hygienists, Lansing, Mich., 1998.
8. *1999 ASHRAE Handbook, HVAC Applications,* ASHRAE, Atlanta, Ga., chaps. 28 and 29.
9. W. C. L. Hemeon, *Plant and Process Ventilation,* 2d ed., Industrial Press, New York, 1963 chap. 13, pp. 325–334.

CHAPTER 3.6
COMMUNICATION AND COMPUTER NETWORKS

Thomas E. Wiles
The Golf System, Inc.
Plano, TX

INTRODUCTION

A computer is a tool. People use tools to facilitate the accomplishment of tasks. An expert woodcarver requires a very special knife to produce the desired output. In the hands of an expert, the woodcarver's knife produces objects of beauty; in the hands of a novice, the same tool produces cut fingers and pain. The computer, like the woodcarver's knife, is a highly specialized tool that has tremendous potential to increase productivity (quantity and quality) or inflict pain. Computers in production facilities are very sharp tools. Used well they can be invaluable.

The manager of a woodcarving facility needs to be able to do the following:

1. Evaluate and hire expert woodcarvers.
2. Evaluate and purchase appropriate knives.
3. Provide an environment that will maximize the output of the woodcarvers.

There is no requirement that the manager be able to carve wood. Touching one of those sharp pointy instruments is definitely not in the job description.

The intent of this chapter is to provide the production manager with a starting point for dealing with the modernization (i.e., computerization) of the facility. Much of this chapter will be negative in content (i.e., what *not* to do rather than what to do). Knowing which pitfalls to avoid is frequently as important as knowing how to accomplish a goal. The manager's job is to ensure that the installation, oversight, and maintenance of the computer system are done well. This chapter provides you with guidelines for accomplishing these tasks.

THE NETWORK

The backbone of your computer system is the network. The network consists of wire, hubs, routers, gateways, patch cords, punch-down blocks, and other associated components. This skeleton is the most critical component of any computer system. If the backbone is not correct, the computers on the network will not run reliably (headaches, pain, and suffering). Networks in manufacturing facilities have special problems (noise). To install a network in a plant requires excellent equipment, careful planning, and professional installation. In many corporations manufacturing gets the administration's discarded equipment (i.e., no planning), and the maintenance staff is tasked to run the wire (frequently CAT-3 telephone cable). If you have problems with your existing network, look right here for the root of the problem.

One Network or Two?

When you wire a production facility, the first (and very serious) question is whether to put in one network or two independent networks. Here are the questions:

- Is any of the machinery programmed directly from the network?
- Will production be affected if the network goes down and stays down?
- Will the company's sales, production, and competitiveness be adversely impacted if any information on the network is published on every web site in the world?

If the answer to just one of these questions is *yes,* then an isolated production network is essential.

CASE STUDY A colleague recently took a job as network administrator for a medium-sized aerospace company in Dallas. The previous administrator had set up two networks, one for production and one for administration with Internet access. He had a dedicated machine in his office to administer each network. Being brand new, my colleague got multiple requests for access to the other net. Also, it would be more convenient if he could manage both networks from a single machine. He asked the CEO if he could connect the networks and was told to go ahead, provided he could guarantee the security of the production network. My colleague considered putting a Cisco firewall between the networks and then carefully limiting access. Before implementing this he called for an opinion. Our recommendation was to go with his predecessor's judgment and *not* connect the networks. While my colleague was probably good enough to successfully make the bridge, why take the risk? There was no material advantage to the company by implementing the bridge, and the inconvenience of the current system was acceptable. If one of the trusted machines were to be infected from the Internet, the intruder would have unlimited access to the secure network. The firewall provides no security against a trusted agent. My colleague, after due consideration, left the networks separate.

Wiring the Building

The objective here is to provide enough knowledge to determine whether the job is being done correctly. *Do not attempt to wire your own facility.* Wiring a production facility is a job for professionals.

There are four types of cable that can (or may already) be strung: 10/100 Base-T, 10 Base-2, 10 Base-5, and glass fiberoptic. Glass is the most efficient, reliable, and expensive. All routers and hubs should be connected using glass. If the servers are at a single (isolated) location with their own hub, there is no need to run glass directly to the servers. If this condition is not met, then very seriously consider running glass to the servers. Where there is plastic fiberoptic cable, glass should *not* be used.

Coax cable (10 Base-2 or 10 Base-5) is no longer being supported by the industry. Coax cable has more bandwidth, better shielding, and overall is superior to twisted-pair cable. Unfortunately, coax protocols are no longer being developed so converting to twisted-pair (10/100 base T) should be done as soon as convenient. There are strong security reasons for choosing twisted-pair over coax. (Security considerations will be covered later.)

For the preceding reasons the remainder of this section assumes you are using twisted-pair wire. If possible, have the local phone company string your wire. They will charge you a fair price (maybe a little on the high side), but will do the job faster and more professionally than almost any of their competitors. They will use the correct wire, comply with fire codes, avoid the high-noise areas, properly label each segment, check each segment for continuity, check each segment for signal attenuation, use Category 5 (CAT-5) punch-down blocks, have all the correct fittings and tubing in stock, and be reasonably responsive to any question or complaint you may have. The following is a list of items to check for:

- CAT-5 plenum cable. Belden cable should be used if you have a choice. AT&T cable is excellent but usually 30 percent more expensive than Belden cable. Cable should be eight-wire (four twisted pairs). There are quality cable makers besides Belden and AT&T but not many. (Cheap cable equates to pain and suffering!)
- Plenum cable *must* be strung in all restricted areas (walls, ceilings, crawlways, etc.). Both state and federal fire codes require this. If the fire inspector finds polyvinyl chloride (PVC) in any restricted area, your doors will be closed until all the PVC is removed.
- RJ-45 connectors are wired to CAT-5 Category B specifications. For CAT-5 Category B, two of the wires in the center pair are crossed. If the four colored pairs (orange, brown, green, and blue) are ordered tip, ring, tip, ring, tip, ring, tip, ring, then the wiring is *not* CAT-5. The best order is orange stripe, orange solid, green stripe, blue solid, blue stripe, green solid, brown stripe, and brown solid. If you have any doubt that the wiring is correct, have a phone company employee check it for you. This wiring sequence is known as T568B and is referred to as the *optional* wiring sequence. The *preferred* wiring sequence is T568A. In the United States we use the optional wiring sequence because that is what AT&T uses. Functionally both wiring sequences are identical, and wire segments can be mixed without affecting the functionality of the network. In Europe the preferred wiring sequence is more common.
- Wire must be strung at least 12 in (30 cm) from any power line. If power and network wire are run through the same conduit, the network will not run [high radio-frequency (RF) noise].
- Wire cannot be strung next to fluorescent lights (high RF noise).
- Use a continuity checker that has a liquid-crystal display (LCD) and takes about 5 s to run a test on assembled twisted-pair wire. The Jensen unit costs about $400. Have the contractor demonstrate the use of a continuity checker with an LCD display to you.
- Wire must *not* be laid on top of removable ceilings. To be in compliance with most state fire codes, it must be either hung with hooks or run through conduits.
- Wire needs to be clearly marked with a tab at both ends (where it goes and its number). Unfortunately no standard wire labeling system has been adopted; however, there are sev-

eral corporate labeling systems that are commonly used. The cable labeling system provided by the Panduit Corporation is an excellent choice. No matter how you label your cables, require absolute consistency (no variations).

- Wall faceplates should screw into a box, not the wall. Do not cut costs here—the plastic insert box costs less than a dollar.
- If you buy your patch cords prebuilt, PVC is fine; otherwise, use plenum. The price difference between PVC and plenum is not sufficient to justify the risk of PVC (which you have in stock) accidentally being strung in a location where plenum is required by fire codes.

A competent contractor will avoid all the pitfalls mentioned here. If you notice violations, do not correct the contractor, *replace the contractor.* You cannot live with the errors you do not catch (most users are not experienced enough to monitor a contractor's work).

The last component of the network backbone is the hubs. Here it is best to just go on faith—put in 3Com 10/100 switching hubs. Do not price-shop, do not question it, just do it. There are other manufacturers of high-quality switching hubs, but unless you are positive of the quality of the product you are purchasing, do not experiment. The hub is just too crucial.

The hub must be a switching hub—no other type of hub is acceptable. First, a switching hub handles traffic faster and more reliably than any other type of hub. A conventional hub, sometimes called a *shared media hub,* uses what is called *Carrier Sense Multiple Access with Collision Detection* (CSMA/CD.) A switching hub uses a high-speed processor running Media Access Control (MAC). The problem with a shared media hub (besides the fact that it is many times slower at the same data transfer rate) is that the network is susceptible to packet capture. This is a huge security hole. Switching hubs almost plugs this hole. For instance, Cisco catalyst switches support SPAN, which can be exploited, providing packet capture. (If the "enemy" is good enough to do this, you are beaten anyway.)

CASE STUDY. This study is instructive. It illustrates the pitfalls of acting on assumptions without verifying the validity of those assumptions first. The production manager asked several foremen to build checklists, with color photographs, which could be attached to assembly line equipment. The purpose of these checklists was to aid new employees in the operation of this equipment and thus ensure the quality of the product. After several months, no checklists. The explanation was that the pictures never came through the printer. Investigation finally revealed that the pictures were being sent correctly from computers in the production facility, but never arrived at the printer located in the administration building 100 yd (91.4 m) away. All the machines and the printer shared the same network. The management information services (MIS) department was unable to solve this problem.

Shortly thereafter, a new MIS manager was hired. The new manager was a fan of Microsoft and assumed the printer problem was associated with the old Novell 3.12 network. Other complaints of the network being slow, and computers being unable to attach properly to their servers, were attributed to the Novell network. The solution was to replace the five Novell servers with five brand-new NT servers (all new hardware and NT 4.0 licenses). This was accomplished with the associated expense ($5000 in NT licenses plus $25 licenses for over 200 workstations), not including the cost of five brand-new servers ($5000 apiece). All the previously mentioned problems continued. One month after the new servers were put in place, it was noticed that inventory control was perfect (stock on hand was identical to that of the previous month). Another month passed, and inventory control was still perfect (stock on hand was identical to that of the previous month). The warehouse was starting to have trouble finding the parts required to keep the assembly lines running. There were still no checklists for the manufacturing stations. Government contracts were being delayed because the inventory system could not be certified as Y2K-compliant.

The production manager brought us in to look at the problem. The production facility's network was modern and efficient. The backbone between the buildings was glass. When we moved our NT server from one location to another to attempt to isolate the connectivity problem, it became apparent that the problem was associated with the administration building, not the server room itself. They were using shared media hubs, which affects network speed, but this could not be the connectivity problem. Removing one of the wall plates we discovered that instead of stringing CAT-5, they had used the extra two pairs of wires contained in the telephone cord. Two problems here: first, this cable is CAT-3 at best; second, the telephone signal interferes with the network signal. In the early 1980s an attempt was made to use existing phone line in old buildings where stringing new wire would have been very difficult. Old LANs running at 2 MB/s had some success with this technique, but it is totally unacceptable for modern LANs that run at 10 MB/s. Recommendation: rewire the entire building.

The problem with the inventory system was more serious. Their older inventory software used a Novell file-locking protocol which was not supported by NT. As a result, NT put a write lock on the inventory system's data files. The inventory software did not detect this lock and had no way of knowing that the records being written to the database were, in fact, being discarded. There are only two solutions to this problem. Solution 1 is to install a new inventory system that runs under Windows NT. Solution 2 is to reinstall the old Novell 3.12 file servers. Neither of these solutions was an option at the time. Fortunately, the production manager found a third solution that no one else had thought of. He looked up the name of the person who had been in charge of inventory 15 years before. He discovered that this person still worked for the company in a different capacity. The production manager had this individual reassigned to inventory to implement the manual system that had been in place 15 years before (journals and 5- × 7-in cards). Production was halted on a Wednesday evening, 4 days were used to take physical inventory, and a manual system was started on Monday.

There were two strange results from this implementation. First, the total number of worker-hours to run the manual system was about the same as that required by the computerized system. Second, they were able to immediately file their Y2K-compliance forms. No computers, no Y2K problems. (Pencils are Y2K-compliant.)

It took 5 months to let the contract and string new cable. At the same time, switching hubs replaced the older units. The assembly lines got beautiful checklists done in color, and all network connectivity problems went away. Bench tests indicated that effective data transfer across the net had more than doubled.

By the spring of 2001, this company's new ERP system was installed and running. We are sure this system will have many advantages over the interim manual system. The important lesson here is that there was life before computers, and that computerized industry is not an end in itself. A new MIS manager will be implementing the new inventory and ERP system.

HARDWARE (COMPUTERS)

This is a very contentious area. Heavy friction frequently occurs between the MIS and production departments over hardware needs, installation, and maintenance. We highly recommend consulting with the MIS manager prior to submitting any form of equipment requisition. Going around the MIS department and contracting directly with a vendor for equipment needs is a sure road to disaster. Remember, you do not want MIS as an active enemy (you will not win). Keeping this in mind, there are some points to look for.

- Computers and manufacturing facilities do not like each other. For a variety of reasons they just do not seem to get along well, so keep the number of computers to an absolute minimum.
- We have seen employees spend 3 h a day reading their e-mail. Providing line foremen with e-mail hurts production and costs the company money.
- Most commercial computers have undersized power supplies. This causes problems in manufacturing facilities because of noisy power (causing lockups, reboots, and reduced computer life). The solution to this problem is a *line conditioner.* A line conditioner is *not* an uninterruptible power supply (UPS). Many MIS departments consider a UPS to be a substitute for a line conditioner; it is *not.* One exception is made by SOLA Corporation. SOLA

UPSs do provide conditioned power. Inexpensive line conditioners provide low-voltage regulation only. If the line conditioner does not regulate both low and high voltage, do not purchase it.

- Computers use switching power supplies. Switching power supplies compensate for variations in line voltage. The size of the "window" within which the switcher can maintain stable output voltages depends on the percentage of rated output that is being drawn from the switcher. Your computers should not draw more than half the rated output of the power supply. If the computer actually draws over 130 W, use a 300-W power supply. The power supply in a modern computer (using an ATX-style motherboard) can be changed in less than 5 min. A good 300-W power supply costs less than $70; stocking and changing them is not a major task.

Mission-Critical Machines

A mission-critical computer is essential to production. If the computer goes down, production is interrupted. The following list applies to mission-critical machines:

- *No screen saver.* Mission-critical machines do not run screen savers or any other form of recreational software.
- *No power management.* All power management features are disabled (both in the operating system and the BIOS).
- *No virus checker.* MIS can check for viruses manually when the machine is not in operation.
- *No unneeded software.* No software is loaded on the machine that is not required to perform the machine's primary task.
- *No secondary tasks.*
- *No remote maintenance software.* Especially VNC, NetBus, PCAnywhere, and the like.
- *No dirt.* The machine is taken to a relatively clean room and opened, and all the dust, dirt, and insects are blown out with compressed air. This needs to be done once every 3 months (more or less, depending on conditions).
- *No power variations.* Consider running this machine on a ferroresonant standby power system (SPS) or an equivalent SOLA UPS. Modern microprocessors run on 2 V. A 1-V ground event can bring the machine down. Ferroresonant SPSs protect against ground events; this is not true for most UPSs and line conditioners.
- *No overheating.* Both microprocessor and motherboard temperature monitoring need to be enabled in the BIOS. MIS should be called immediately if the machine starts wailing (indication of impending overtemperature shutdown).
- *No old computers.* The machine should be replaced after 3 yr. *No mission-critical machine should ever be more than 3 yr old.* These are nice machines; send them to administration for a change.

General Rules

- No machines with integrated motherboards.
- No high-end video cards (Voodoo, Gforce, etc.).
- No very fast computers (they draw more power and run less reliably). This does not mean inexpensive; stay away from low-cost generic computers. When you purchase a computer you should have the option of selecting the speed of the microprocessor that will be

installed in the computer. Do not select the fastest microprocessor available, select a medium-speed microprocessor.

- No leading-edge technology; try to buy machines based on older technology with a proven track record of reliability. (That is, select the Microstar Athalon motherboard over the ASUS VIA K133 motherboard. The ASUS benchmarks 15 percent faster, but the Microstar has demonstrated unusual reliability and stability since its introduction.)
- Machines provided by Dell, Gateway, Hewlett-Packard, Compaq, and Packard Bell routinely have problems in manufacturing facilities. Most of these companies do have premium models that perform better in your area, but it is hard to identify them. Purchasing never seems to get this right.
- If you have a choice, require that the microprocessor in your computer be a *boxed* microprocessor. First, boxed microprocessors have an unconditional three-year warranty from the manufacturer (both Intel and AMD). Second, the microprocessor you are buying is in fact the microprocessor you are paying for. Third, a study we conducted established to my satisfaction that boxed microprocessors have fewer problems and a lower failure rate than *tray* microprocessors. The fact that boxed microprocessors are substantially more expensive than tray microprocessors implies that the manufacturers think so too. (The warranty form is included with the microprocessor; if there is no warranty form, then the microprocessor is a tray microprocessor.)

Unauthorized Software

Under no circumstance should unauthorized software ever be put on a production computer, no matter how innocent it seems. Because of the environment, production computers run more erratically, have more downtime, and require more maintenance than their counterparts in the administration building. MIS departments always seem to be of the opinion that operator misuse is the reason for this increased workload. (They have been right on occasion.) If you give MIS personnel any confirmation of this opinion, they will not even start to look for the real problem. (Remember the color pictures that would not print.)

Computers Are Tools

Remember, computers are tools, not an end in themselves. Give every machine you have a really hard look. Unless there is an obvious financial benefit to the machine's existence, get rid of it. In all too many cases, computers cause more grief than they solve. If you are unsure, try removing a machine for awhile, and see what happens. In spite of the picture we have painted of MIS, the MIS manager is your friend; consult often.

CASE STUDY. This concerns a billion-dollar company with multiple production facilities nationwide. Headquarters was at an isolated location. A directive came down from headquarters to all site MIS departments. The directive dictated that all computers would be Y2K-compliant by July 15, 1999. Any non-Y2K machines would be replaced by the July 15 date. There would be no exceptions.

Site X hired a summer college student from the local community college and gave the student the purchased test software, a log book, and green and red tags. After the survey, MIS counted the number of red tags and purchased that number of replacement machines. They then sent this student out to exchange the older machines for the new machines. (Unfortunately, they saw no reason to inform anybody that they were going to do this.) Company policy was that no data was to be kept on corporate machines; therefore, there should be no problem with the exchange. (Everything was supposed to be on the network servers.)

This site's product was shipped in lots. Each lot had to be manufactured to the client's specifications. To guarantee this, a number of units from each lot were tested, and the test results were reduced and plotted. A copy of these results was shipped with each lot. A lot could not be shipped without being certified, and the test results included with the product.

Our college student worked over the weekend, replacing the old machine with a brand-new Windows 98 Pentium machine with all the whistles and bells. All very nice. It had a CD-ROM, sound system, and a huge hard drive.

When the testing crew came in Monday morning, they had a problem. First, there was no slot in the front of the new machine to insert a 5¼-inch 360K low-density diskette. Second, the new machine would not turn on with the test interface card in it. MIS sent the student over to help, which made matters worse. The lot that had been tested on Thursday and Friday was scheduled to be shipped Tuesday, and the data could not be reduced without a running computer. MIS refused to return the old machine because it was past July 15, and not their problem. Headquarters would not budge on Y2K compliancy, so some other solution had to be found.

The old computer took 15 min to reduce the data and plot the required output. Three people working with calculators produced the same results on graph paper in 6 h. We were asked to look at the situation, but it was hopeless. The original machine had been there 18 yr. It ran Control Program Microprocessor (CPM) and was pre-Microsoft. The interface card was wire-wrapped and totally undocumented.

The company contracted for a replacement unit, but the contractor was not up to the task. Another contract was let with hopefully better results. Meanwhile, 6 months later the data was still being reduced manually.

The sad part is that it never should have happened. The problem occurred when the old computer was certified unacceptable for Y2K. They should have passed the old machine instead of failing it. They should have used the negative approach. While the old machine could not *pass* the Y2K tests, it also could not *fail* the Y2K tests. The reason for this was that the machine had *no clock* and *no calendar.* It had been procured before these became standard features on microcomputers. Think of the grief that could have been avoided if the right person had been sent out to test the machine in the first place!

BACKING UP DATA

You *do not* have a backup of your data. No matter what backup procedures you have in place, no matter what assurances you have been given by your information systems (IS) department, a magnetic backup of critical data does not exist. Unfortunately, most managers do not discover this fact until they need the data. The bad news is that there is no solution for this problem; the good news is that there are procedures you can put in place that will minimize the risk.

Causes

- Magnetic media loses its polarization with time. Most magnetic media have a shelf life of about 5 yr, depending on the storage environment. Given enough time, backup media will not be readable.
- Magnetic oxide wears (rubs) off magnetic tape each time the tape is used. Every time a tape is written to, its storage life and restore probability are reduced. Check with the media manufacturer to obtain the recommended number of backups it will deliver, then divide that number by 3. In spite of the apparent price, tape media is cheap.
- IS departments like to do differential backups. Differential backups save time and cut costs. The theory behind this type of backup is, *you back up only the data that has changed.* The problem is that any failure in the restore chain can result in older data being lost.

- Tapes do break. This problem can be reduced by retensioning the tape prior to use (almost never done). Also, tensioning the tape prior to use increases the probability of a good restore. Note that some backup systems always retension the tape prior to use. (In the Utilities menu of most tape systems there will be a selection labeled tension tape.)
- The hardware unit itself requires periodic maintenance (cleaning) and head replacement (they do wear out). This is almost never done.
- Failure to verify the backup. Most tape backup systems write to the tape but never read the tape to see if anything has actually been written. A perfectly clean error log does not guarantee that any actual data was written to the tape. All backup systems have a verification option. If this is turned on, it will take about three times longer to actually do the backup, but the data on the tape will be verified as correct.
- If the backup system has problems during the backup process it records these problems in the error log. Nobody bothers to read the error log, which is erased the next time a backup occurs (and all record of the problems is conveniently destroyed).

Solutions

We do not have a direct solution, but we can give you a list of suggestions that may reduce the risk. Since your database is probably administered by MIS, and the MIS manager does not answer to you, your input may be severely limited.

- Require that the error log for each backup be printed and initialed by the employee responsible for making the backup. This guarantees that the error log has been read.
- A full system server backup should be made at least once a month and stored off-site.
- A database restore should be done at least once a month to a designated machine (not the server). This exercises the restore process and identifies procedural and hardware problems before they become critical.
- Have your secretary's machine (or equivalent) equipped with a CD burner. Have reports run on all critical data (inventory, receivables, etc.) and saved in a special directory. Have these reports periodically burned to a CD. Good CDs cost less than $1 and hold 640 MB of data. A modern burner will burn a CD in minutes (640 MB CD in 6 min). If your production equipment is controlled or programmed by computer, key files associated with production can be archived in a similar manner. Optical storage lasts 150 yr and *cannot* be accidentally erased. All routine data that is generated as a report can be archived in this manner. It is not a bad idea to archive all submitted reports and correspondence in this manner (by date). This seldom requires more than one CD a month (one a week at the most, say on Fridays). These CDs are convenient to store, allow for instant retrieval, and provide an archival solution that you control.

CASE STUDY. A small retail establishment put in a computerized point-of-sale (POS) system consisting of five POS stations plus five administrative stations. The entire system was expensive and first-class. A complete system backup was done automatically every evening. The company appended each backup to the tape until the tape was full, then started a new tape. No data was ever erased, nor was a tape ever reused. The company that installed the hardware provided a backup checklist which was followed religiously for 18 months. Then a mechanical failure of the server's hard drive resulted in complete data loss. The repaired server and the tapes were shipped to us for restoration. No one on location had ever done a restore, and everyone was afraid to try—they did not want to be responsible for a disaster. Of the 500 perfectly done backups, 2 were restorable. The most recent good backup was 40 days old. This company did 800 line items of business a day, and every line item had to be individually reentered.

One of the administrative stations was routinely left running at night. The employee who occupied that station would minimize the inventory system, turn the monitor off, and go home. The inventory system held two critical data files open. The backup system recognized that the files were in use and did not back them up (duly noting the omission in the log file). Reading the log file was not on the checklist, and nobody on-site knew to check it! The cause of the disaster was never checking whether the backup system was functioning. Checking that procedures are being accomplished correctly is part of what managers are paid for! You do not have to know anything about the fire-prevention system to run a fire drill.

PERIPHERALS

Windows does not like to change peripherals. Install all peripherals with permanence in mind. Select the computer where the peripheral will have the most utility, install the peripheral on that machine, and do not relocate the peripheral to another machine. Let the users come to the machine, not vice versa.

Do not install inexpensive printers in the manufacturing or industrial environment. All printers should be laser printers (no ink-jet or bubble-jet printers). Do not purchase a laser jet printer that has a retail price of under $700. Small printers designed for use in the home are fragile, and historically have not held up in a production environment. Do not acquire any printer that uses bin feed rather than tray feed. Trays always feed the paper with perfect alignment. Bin-feed printers have too many internal jams to be usable. Printers do require care and preventive maintenance. Designate an individual who will be responsible for the condition of the printer.

Scanners, cameras, and all other peripherals pertaining to the facility's operation must be of good quality. Procedures and checklists need to be established for the maintenance and use of these products.

Employees who are to use computer peripherals need to be trained on proper use and handling. This instruction should include how to properly attach cables, operate the equipment, and service it. Employees should be warned against overtightening screws. No screw attaching a cable to the back of a computer should be more than lightly finger tight. Half of all peripheral downtime can be attributed to overenthusiasm with a screwdriver.

Computer peripheral cables are delicate. The pins in the cable are fragile and easily damaged. These cables should be removed as infrequently as possible. Do not attempt to attach cables when the lighting is bad or the port (plug) is not easily accessible. Attaching cables by feel results in bent pins, unreliable operation, and possibly damage to the peripheral and the computer.

Computers and peripherals are expensive, delicate tools; treat them with respect.

ENVIRONMENT

When selecting the location for a computer, consider where the computer will be the happiest, not necessarily the most convenient location for the users.

- Choose an air-conditioned environment that is relatively clean of airborne dust. Do the best you can; the location is unlikely to be ideal.
- Select a location where air circulates freely behind and in front of the computer. Air flows in the front of the case and is pushed out the back of the case. If the exhaust route is blocked, the computer will overheat, resulting in unreliable operation and shorter life.
- Select a location where cables are unlikely to be pulled, twisted, or tripped over.
- Make sure power receptacles are carefully marked. Many plants have both 115- and 220-V outlets. The computer has a slide switch on the back of the machine that allows it to be plugged into either a 115- or a 220-V receptacle. If the machine is set on 115-V and is

plugged into a 220-V line, it will most likely be destroyed. If the machine is set on 220 V and is plugged into a 115-V line, no damage will be done. If the computer's power supply is large enough, it is possible for a computer set to 220-V to run when plugged into a 115-V receptacle (not reliably, but it may run).

CASE STUDY. We were once asked to look at a client's machine that kept crashing. It failed 3 to 5 times a day. The client had returned the machine to the vendor twice for maintenance and no problems could be found. The vendor was convinced that poor power at the client's location was the cause of the problem. The client was extremely unhappy and had no idea what to do. We moved the slide switch on the computer's power supply from 220 V to 115 V. The computer has not crashed since.

SECURITY

Introduction

It is not possible to cover all security considerations, so do not consider this section to be complete. This introduction is lengthy because it is very important. The corporate executive must understand the mechanics driving network security issues. First, some definitions:

cracker, Type A Someone who breaks into a computer or network because it is a challenge. This cracker frequently has the permission of the network owner to attack the network. Embarrassing the government does not qualify as damage for our purposes. Replacing the NASA web site with a cracker web site (November 1999) qualifies as Type A cracking.

cracker, Type B Someone who breaks into a machine or network with the intent to steal or damage.

hacker Someone who writes code and does not intentionally inflict damage on the target machine.

script kiddy A cracker who is not capable of writing code. Script kiddies tend to be malicious, and frequently seem too young to know better.

While viruses have been around since the late 1980s, we consider the release of MS-DOS 6 (DOS 6) the start of the current security-sensitive environment. With DOS 6 Microsoft made a strong move to control the microcomputer platform (in our opinion). This version of DOS restricted what operating systems could run on a machine through the use of a new protocol called the Master Boot Record (MBR). In 1986, our work computer had 15 partitions on its hard drive, ran seven operating systems, had two different internal microprocessors, and was selectively bootable from the ROM monitor. DOS 6 ended this type of flexibility and strongly limited operating system (OS) options.

The hacker community was unwilling to accept these restrictions, and with the Internet coming of age, hackers were able to pool their resources in decompiling, analyzing, and documenting the OS loader in the MBR. This soon led to bootstrap loaders such as System Commander (proprietary) and LILO (open source).

The hacker community is a worldwide organization with over 250,000 (estimated) members. The community has bylaws, codes of conduct, copyrighted products, and a standard general public license (GPL) agreement. Hackers publish, maintain, and distribute LINUX, the most powerful operating system available. They wrote and copyrighted TCP/IP, the protocol that runs the Internet. They have successfully defeated every attempt by the world's most powerful corporations (and some countries) to carve up or control the Internet.

The fundamental goal of the hacker community is to acquire and distribute knowledge. Hackers operate under the belief that knowledge cannot be owned—that there is no such thing as intellectual property. Knowledge, in their minds, should be distributed without charge to all who have a desire to acquire it.

The cracker community is fragmented and breaks into two groups. One group sees operating systems and networks as puzzles. This group formed liaisons which evolved into our nation's top security firms. These crackers attack a network only after being asked to evaluate that network's security by the network owner. The tools this group produces can be purchased, and have value both as network security tools and as network penetration tools (two-edged swords).

The second group of crackers uses the tools provided by hackers and other cracker groups (sometimes modified) to break into computers illegally for profit or other, frequently political, motives.

Script kiddies use the tools provided by the rest of the community to produce as much corporate-user chaos and suffering as possible.

CASE STUDY. When Microsoft put its new web server on line in the winter of 1999, the company was so sure of its security that it challenged the entire cracker community to attempt to break into the Microsoft network. Rewards were offered for a successful penetration. Within 24 h, these crackers were turning the Microsoft servers on and off like kids playing with a light switch. (More information is available at www.l0pht.com.)

Viruses

There are over 30,000 documented viruses, belonging to over 8000 families. New viruses are being written at the rate of about 3000 a year, or roughly 10 a day. There are over 200 families of viruses that are undetectable by today's virus checkers. There have been instances of infected programs being shipped by major software providers. There are no procedures or precautions that will guarantee the integrity of your computers.

Boot-Sector Viruses. This class of viruses resides in the boot sector of a floppy disk, and is implanted in the computer's MBR if an attempt is made to boot the computer from the floppy disk. This is the largest class of viruses. These viruses exploit the security hole created when Microsoft created the MBR. A famous member of this class is the Michelangelo virus, which erases key tracks on the hard drive on Michelangelo's birthday.

All viruses of this class can be defeated by *not* booting from a floppy disk. There are two settings in the ROM BIOS of computers manufactured after about 1995 that will protect your computers from this class of viruses:

- *Disable floppy boot.*
- *Enable BIOS-level virus protection.* This is a small program embedded in the computer's BIOS that will notify you if an attempt is made by any program to write to the MBR. Be advised that this feature can be circumvented.

This class of viruses cannot be transmitted over a network (Internet).

Because of the wealth of information available on this class of viruses (theory, descriptions, and sample code) many new boot sector viruses are being written. Taking the precautions listed here, and checking your computers weekly with a virus checker, should defeat this class of viruses.

Executable Viruses. An *executable virus* is a virus that is embedded inside a legitimate program (.exe, .com, or .dll extension). An executable virus *canNOT* infect a data file! When the program containing the virus is executed, the virus infects the OS (it may or may not infect the MBR) and attempts to transfer itself to other executable programs on the infected computer. Viruses of this class may also be able to infect programs across a network.

This is the second-largest class of viruses. The most common virus being written today belongs to this class. One of the most notorious of this class of virus is the Natas virus (Satan spelled backward), sometimes referred to as the Satan virus). During a major Dallas computer show, one of the vendors distributed 15,000 infected samples of its software.

The code that allows this type of virus to infect other programs leaves a signature by which it can be detected. Virus-detection software companies (like McAfee) use analytical methodology to determine this signature and then scan for it. Virus writers who understand the methodologies being used to detect their viruses are now using cryptography and other techniques to mask or hide the signatures of their viruses (stealth viruses).

Parasite Viruses. A *parasite virus* is a virus that attaches to another virus. The host virus moves from computer to computer. When the host virus infects a machine, it infects the machine with the parasite virus instead of itself. In some cases, both viruses become active if they reside in different portions of the OS. Because they do not replicate, parasite viruses have no signature. When an infected computer is scanned by virus-detection software, the host virus is detected and removed, leaving the parasite virus intact and undetectable.

CASE STUDY. A good friend brought us a machine. The virus detection software was reporting the presence of the NYB virus. My friend was afraid to touch the machine. We used F-PROT virus-detection and -removal software from a clean-booted diskette to verify the presence of the virus and remove it. After removing the virus, we checked the machine with F-PROT and both IBM and Norton antivirus software. The computer reported clean. We then loaded F-PROT on the machine and ran the program again (after a clean boot). F-PROT reported a corrupted signature file. File compare (FC) verified that 3 bytes of the signature file had been altered. The three altered bytes were the signature of the NYB virus. We then loaded IBM antivirus software. It reported no viruses. Having come this far, we inserted a floppy disk that contained the European Community Institute Antivirus Research ECIAR.COM file. IBM reported the floppy as having no viruses. Normally the IBM software will report the ECIAR.COM file as a virus. This demonstrates that the antivirus software was disabled by the running virus. At this point we erased the hard drive and reloaded the operating system.

Had we not loaded F-PROT on my friend's computer we would never have known that a virus remained. Nobody, that we know of, has ever directly detected or identified this virus. The virus author got a little too cute in trying to mask the F-PROT signature file, which indicates the virus was probably of European origin. If it had not been for this alteration, we never would have suspected the presence of the second virus.

Remedies. While absolute protection against viruses is not possible, the following recommendations will reduce the risk of one of your computers becoming infected:

- Never leave a floppy disk in the A drive.
- Never boot from a floppy disk.
- Disable floppy boot in the ROM BIOS.
- Enable ROM BIOS virus checking.
- Do not load programs that are not required by the computer's mission.
- Always load programs from the original distribution media.
- Have the MIS department scan your computers periodically. This scan entails enabling floppy boot, booting from an antivirus diskette, and scanning the computer using the software on the diskette.
- Maintain and use two different virus-scanning programs.
- Update your virus-scanning program's signature file monthly.
- Keep a copy of the ECIAR.COM file on floppy. Scan that floppy to verify that your antivirus software sees the test file.

Antivirus software that is used while the Windows 9X operating system is running is only partially effective. To reliably scan for MBR viruses, the boot disk must be DOS 5 or earlier. The surest way to check for an MBR virus is to have MIS build a special DOS 5 bootable disk that has a utility that will write the MBR out to a file. Visually verify the integrity of this file.

Worms

Worms are programs designed to penetrate as many machines as possible. These programs exploit the communication protocols used by computers connected to networks. Worms do not intentionally do damage, but they do replicate. As a result, they can be a considerable nuisance. A worm that misbehaves can bring a machine or a network down (no data should be lost). Programmers sufficiently skilled to write a worm are normally mature enough not to do so. Very few worms have been written, and you will likely not see one. A released worm is normally detected by an Internet monitoring site very quickly. A description of the worm and software to remove it is normally published on the Internet within 48 h of the worm's release.

Trojan Horses

A *Trojan horse* is a program that attaches itself to the OS. This program then allows an unauthorized user access to the computer on which the Trojan horse resides. In the early years, Trojan horses were used almost exclusively for password interception. Today they have become a fine art, enabling many forms of subtle manipulation of your computer.

The deployment of Windows 9X created three huge security holes which were so inviting that the energy of the cracker community moved almost exclusively to exploiting these holes. The three holes are Internet Explorer (IE), Outlook (and Outlook Express), and Java. (Java is not itself a security hole, but is the mechanism by which the security hole is exploited.) Distributed processing is inherently a security disaster.

There are four mechanisms for transmitting a Trojan horse to your computer. The first, buffer overflow, generally exploits weaknesses in the mail system. To exploit this hole requires a level of expertise that excludes all but a very few crackers. Your risk of being subject to this form of attack is small. A variant of this attack does not load a Trojan horse but, instead, endeavors to crash the machine. This denial of service (DOS) attack is much less sophisticated, and sample scripts are in the hands of the script kiddies. The second mechanism is to implant a cookie in your cookie file. This can be done by any web site you connect to. The third method is to e-mail you a file that executes either a Visual Basic or a Java script. It can also be an attached Java applet.)

The fourth method is to entice you into loading the Trojan horse on your computer. There is a fairly entertaining game floating around the Internet called Wack-A-Mole. If you run this game it loads the NetBus Trojan horse on your machine. Deleting the delivery program (Wack-A-Mole) does not affect the Trojan horse. The NetBus Trojan horse allows an intruder to log the user's key strokes; take regular screen shots of whatever the user is viewing; retrieve the user's cached passwords; examine browser bookmarks; read e-mail; and view, run, edit, upload, and delete files. A cracker can scan several hundred IP addresses a minute. If the cracker hits a machine with the NetBus Trojan horse loaded, the target computer returns its current IP address and the presence of the Trojan horse. Any cracker can then take control of your computer. UltraAccess Networks (the owner of NetBus) has negotiated an agreement with McAfee by which McAfee antivirus software will no longer report the presence of NetBus to the computer's user (owner).

Some corporations wish to monitor how their employees use corporate computers. If MIS personnel wish to install NetBus on the corporate computers, they have to use McAfee's antivirus software so that the user will not be alerted to the presence of the Trojan horse.

Securing the Net

We will present one general solution. Corporate policy, security needs, and your MIS department will customize this general approach. Net security is started at the connection of your corporation's local network and the Internet. This is frequently referred to as the *border router.* This router should either double as a firewall or be connected directly to a firewall computer (We prefer the latter). Next your company should use a *proxy server.* All Internet packets are sent with the proxy's return address; the packet is then routed by the proxy server to the originating machine. This defeats many types of attacks. By *spoofing* (using a return IP address that does not belong to the sending machine) all Internet packets, all machines downstream are hidden from the Internet. The firewall (border router) will also prevent crackers from detecting individual machines on the network. One additional advantage of this setup is that your company only needs one fixed Internet IP address. The IP addresses of the individual computers on your network are *illegal* Internet IP addresses. Several million of these illegal addresses are set aside for corporations to use. You do not have to register an illegal address to use it, making it more difficult for a cracker to scan your network. Remember, firewalls provide only a relative degree of protection, they are not absolute.

Securing the Browser Holes

Both Trojan horses and cookies can be planted on your machine through your web browser. Since the browser is a trusted agent, the corporate firewall provides very little protection against this attack. The hacker community is becoming irritated by these intrusions onto private computers and is reacting to the threat. We do not know of a central depository for this class of software, but we can give you some sources. Gibson Research (www.grc.com) has a program call OptOut which looks for known browser-related Trojan horses and removes them. For cookies and banner ads, the hacker web site www.junkbusters.com provides a product called Junkbusters 2.2. This software is covered by the hacker's GPL and is free (paying for it is *not* an option; however, there are restrictions, so read the GPL).

Close-in Defense

Your last line of defense should be a personal computer firewall. The one we recommend is ZoneAlarm Pro by Zone Labs. This product costs $39 for a single machine license (one-time fee, upgrades are free). The company's web site is www.zonelabs.com. This firewall has many powerful features:

- It allows you to lock your computer down when you are not using the network. All network traffic (legitimate or otherwise) is blocked. A log file tells you the name of every computer that tries to contact your computer.
- It requires you to give access privileges to every program on your machine that has network communication capabilities. If a Trojan horse is loaded on your machine, ZoneAlarm will alert you that it has requested network access and will ask you to authorize the access. You now know that the Trojan horse is on your machine and can deny the penetrator access.
- It maintains a handy list of all programs on your machine that have been given network access. Check this list periodically just in case your MIS section has loaded a Trojan horse to spy on you.
- It detects, blocks, and highlights most e-mail–attached viruses and Trojan horses.

One last security measure is to disable Power On LAN in the BIOS. This prevents someone on the network from powering up your computer while you are gone and your office is locked. Physical access to your computer is now required.

Mail Viruses

Attaching a Visual Basic script to electronic mail is a recent fad among script kiddies. The Melissa virus, and more recently the "I Love You" and "Resume" viruses, are examples. These scripts are not true viruses; in order to become active they must be executed (opened) by the user. Because these viruses come from a legitimate trusted agent (your mail software—generally Outlook), there is no way to prevent them being written to your machine. The defense against this virus is *not to open it.* Any e-mail that has an unexpected attachment should not be opened without first checking with the sender. Once the script is executed, the damage is done. The key element here is, if you do not know what it is, do not open it. If you are in doubt, have the MIS department check the script with an editor before you open it; the code will be obvious.

Help

There are basically two types of help available:

- *Monitoring help.* There are companies that will monitor your net for security holes. They will hit your net with the latest known cracker techniques and forward the results to you. Fixing the hole is your company's responsibility. One such company is Vigilante, at www.vigilante.com.
- *Consulting help.* Many security consulting firms have sprung up lately. Unfortunately, many of these consultants lack the basic knowledge to do the job. One company that has the right people is @stack (www.astake.com).

CONCLUSION

Don't panic, it's not that bad. Remember, computers are tools. Each computer has specific tasks to perform. Screen your computers for misuse, nonmission software, and unnecessary network access. If you always remain conscious that something bad can happen, very likely nothing bad will happen. Take reasonable security precautions, and consult professionals when you need help.

SECTION 4

PLANT OPERATIONS

CHAPTER 4.1
MATERIALS HANDLING: PLANNING

K. W. Tunnell Co., Inc.
King of Prussia, Pennsylvania

MATERIALS-HANDLING DEFINITIONS

Materials-handling technology includes hardware and systems which can be categorized as follows:

- Containerization
- Fixed-path handling
- Mobile handling
- Warehousing

Containerization. This classification covers the broad spectrum of confinement methods that are used for storage through all phases of the manufacturing, or process, cycle. The materials-handling engineer employs the unit-size principle to optimize the quantity, size, and weight of the load to be handled or moved and is able to specify the best container after considering material and other production-system parameters. Pallets, skids, tote boxes, and wire-mesh containers, covering a wide range of sizes and materials, are included within the category.

Fixed-Path Handling. This classification applies to movement and storage of unit loads of material with an intermittent or a continuous flow over a fixed path from one point to another. Fixed-path-handling equipment is secured to, and is considered part of, the facility. Once installed it is more difficult to modify or replace; therefore, a considerable amount of planning and interfacing with other functions has to be considered. This equipment, if installed above the floor surface, can effectively utilize what would otherwise be dead space. Chutes, conveyors, elevators, bridge cranes, palletizing equipment, and robots are examples of fixed-path-handling equipment.

Mobile Handling. This classification includes all handling systems that move material over various paths within a manufacturing or processing cycle. Handling equipment allows a considerable degree of flexibility in moving material in an intermittent flow but requires special facility requirements, such as aisle sizes, clearances, door openings, and running and maneuvering surfaces. Equipment in this category consumes more energy per unit load moved than most other systems and generally requires trained personnel for operation. Equipment in this category ranges from simple two-wheeled hand trucks to specially designed over-the-road vehicles and also includes skid trucks, floor trucks, powered walkie lift trucks, powered lift trucks, and mobile hydraulic cranes.

Warehousing. This classification of materials handling considers the systems, equipment practices, and requirements dedicated to the following operations within the manufacturing, processing, or distribution cycle:

- Receiving
- Storage of raw, in-process, and finished materials
- Movement in and out of storage
- Order picking and accumulation
- Containerization for shipping
- Loading and shipping

This area of materials handling involves a wider range of planning and analysis. Consider some of the following factors:

1. Shipment volumes
2. Location of activity
3. Sizing and physical characteristics relating to product size, type, and volume
4. Number of stockkeeping units
5. Storage equipment
6. Selection of materials-movement methods
7. Packaging methods for shipping
8. Labeling and carrier selection

The range of solutions covers the full spectrum, from a single warehouse that shares movement equipment with other parts of the operation to a self-contained, specially equipped, fully automated warehouse.

INTRODUCTION

Materials handling deals with the movement of materials from receiving, through fabrication, to the shipment of the finished product. More broadly, handling and distribution are considered to be one overall system. This viewpoint gives consideration to all handling activities involved in movement of materials from all sources of supply through the various plant and central warehouse operations to the customer distribution network.

The activities related to the flow of the materials are either viewed as separate activities or treated as one element in an integrated handling system. Not everyone agrees that materials-handling activities should be viewed as an overall system. Progressive plant engineering personnel, however, recognize that materials handling represents the efficient integration of workers, materials-control systems, and equipment, as well as the movement of materials. Materials-handling applications must take into account operating costs and time-phased material flow.

Inefficient materials handling and storage increase product cost, delay product delivery, and consume excessive square feet of plant and warehouse space. Studies have indicated that actual materials-handling cost runs between 20 and 50 percent of the total product cost *even though it does not add any value to the finished product.* In addition, from 80 to 95 percent of the total overall time devoted to processing a customer order from fabrication to shipment is devoted to materials handling and storage. Product manufacturing time, therefore, is only a small percentage of the overall process time.

A properly designed and integrated materials-handling system provides tremendous cost-saving opportunities and customer-service improvement potential. The correct selection of a handling method can reduce handling costs per unit by as much as 200 percent. Improvement in storage, such as high-rise equipment and high-density storage applications, can reduce unit storage space cost by 20 to 40 percent. Work-in-process inventory can be reduced 30 to 50 percent through compressed cycle times. The reduced cycle times will also result in shorter customer delivery cycles.

The significant impact of materials-handling costs on total product cost has resulted in a great deal of attention and substantial resources being directed toward discovering more efficient methods to reduce handling costs. This effort is expected to receive even greater concentration in the future. The following trends are beginning:

- Most manufacturing managers now recognize that materials handling is a prime area of cost-reduction opportunities.
- Many manufacturers have become environmentally aware, which is reflected in products such as lift trucks with reduced emissions and recyclable plastic and corrugated pallets.
- Preengineered storage and handling systems with proven success records are being offered to help reduce installation and start-up times.
- Computer systems are being designed to enable the manager to electronically track materials from the point of entry to exit.
- Computer-based technology will employ more powerful and sophisticated techniques such as queueing theory, simulation facilities, and flow-planning techniques to consider and select optimum solutions from a wide range of variables and materials-handling options.
- Handling systems will become more automated by employing computer-controlled systems, robot loading and unloading, driverless vehicles, and automatic storage and retrieval systems. These automation principles will be applied in receiving, manufacturing operations, warehousing, and shipping.
- OSHA and other safety requirements and systems will increase the complexity of designs and the applications of various systems.

It should be recognized that materials handling is an extremely broad-based subject which more often deals with the application of equipment and mechanical devices than fundamental engineering principles or basic physical laws. At least in part, it requires the application of subjective and experienced judgment and has even been described (with some justification) as more of an art than a science. Based on this fact and the trends in materials handling already discussed, this chapter outlines the methodology for solving materials-handling problems as well as the classification of hardware.

However, it is suggested that plant engineers involved continually with materials-handling projects should be familiar with additional sources of current information not only from reference books but also from specialized trade periodicals, professional societies, and trade associations. (See the reference section at the end of this chapter.)

SOLVING MATERIALS-HANDLING PROBLEMS

Materials-handling activities and installations vary in complexity from operation to operation. The plant engineer can solve most materials-handling problems by keeping two points in mind: one is to thoroughly understand the materials-handling principles; the other is to recognize that a materials-handling system is composed of a series of interrelated handling activities. The plant engineer should therefore apply materials-handling principles to improve each separate materials-handling activity and then interrelate the handling activities by applying flow-planning principles.

Principles of Materials Handling

The collective experience and knowledge of many materials-handling experts has been organized into a framework of generalized principles. The principles are basic and can be used universally. They include:

1. Integrate as many handling activities, such as receiving, storage, production, and inspection, as is practical into a coordinated system.
2. Arrange operation sequence and equipment layout so as to optimize materials flow.
3. Simplify handling by reducing, eliminating, or combining unnecessary movement and/or equipment.
4. Use gravity to move material wherever practical.
5. Make optimum use of the building cube.
6. Increase the quantity, size, and weight of the load handled.
7. Use mechanized or automated handling equipment whenever it can be economically or safely justified.
8. Select handling equipment on the basis of lowest overall cost when considering the material to be handled, the move to be made, and the methods to be utilized.
9. Standardize methods as well as types and sizes of handling equipment.
10. Use methods and equipment that perform a variety of tasks and applications.
11. Plan preventive maintenance, and schedule regular repairs on all handling equipment.
12. Determine the effectiveness of handling performance in terms of expense per unit handled.
13. Move materials in as direct a path as possible, minimizing backtracking.
14. Deliver materials directly to work areas whenever practical, and plan the minimum of material in the area.
15. Move the greatest weight or bulk the shortest distance.
16. Provide alternative plans in case of a breakdown.
17. Use the appropriate level of proven technology and systems.

Steps in Solving Handling Problems

The general methods that are used for solving other operational problems are applicable in the materials-handling area. The factors that must be considered relate to how to most efficiently move certain volumes and types of materials by a particular method. The steps involved in systematically solving these problems consist of:

- Problem identification
- Problem analysis and quantification
- Selection and evaluation of alternatives
- Project justification

Problem Identification. Identification of materials-handling problems includes determining the impact of interfacing activities such as production control, manufacturing, vendors, shipping, and receiving. The buildup of material in front of a machine or a truck dock may not be a problem of too little storage but rather one of lot sizing or inefficient truck-loading systems. Most importantly, a costly handling route between two distant machines may not be caused by the handling mechanism but by the location of the equipment itself.

Problem Analysis and Quantification. Qualitative and quantitative answers are obtained through use of industrial engineering techniques such as flow diagrams, flowcharts, from-to charts, and activity-relationship charts. For the detailed application of these manual techniques see Refs. 1 to 4.

Computer-aided techniques are useful when large amounts of data are involved.[5]

Selection and Evaluation of Alternatives. In the selection of alternatives three general types of criteria are involved.

Movement. Movement involves the study of routes in terms of the combination of handling equipment and containers jointly rather than on an individual product basis. Under this criterion, distances from and to locations, outside travel, and frequencies would be minimized.

Criteria Which Cannot Be Directly Costed. These involve criteria such as:

1. *Performance.* The potentials for relocation of equipment at future time periods, as related to the handling equipment, and for material design changes, as they relate to the containers themselves.
2. *Delicacy.* The nature of the part and its requirements for special handling, dunnage, or containers—particularly to avoid damage.
3. *Interfaces.* Production control and manufacturing departments, vendors, shipping, receiving, and intersite movement requirements.
4. *Uniformity.* The need to standardize or at least unitize containers to provide uniform handling characteristics. This provides for the use of idealized container sizing and packing techniques, including proper dunnage for irregular loads. On the other hand, it requires the special handling and designs for special irregular-sized parts.

Cost-Effectiveness. This involves the analysis in concept of all standard operating costs, equipment life, maintenance and spare usage for equipment, and intermediate storage.

Generally, the analysis will be dominated by the cost-effectiveness of the alternatives involving cost components as follows.

- *Capital Investment Costs.* One-time charges incurred at the time of equipment procurement that include:
  1. Equipment cost, including freight
  2. Installation costs
  3. Special maintenance requirements
  4. Special power and fuel facilities
  5. Rearrangement and alteration of facilities to accommodate equipment
  6. Engineering support
  7. Supplies
- *Fixed Costs.* Determined or assigned to a system, a piece of equipment, or an activity on a time-period basis; include:
  1. Depreciation
  2. Taxes
  3. Insurance
  4. Supervision
- *Variable Costs.* Can be considered the cost of performing an operation or activity. In the case of equipment, it is the cost of using the equipment. The following items are included within this component.

1. Equipment-operating personnel or personnel manually performing a materials-handling task
2. Power and fuel costs
3. Lubricants
4. Maintenance labor supplies

- *Indirect Costs.* Affected in other areas of company operation as a result of changing a method, adding new equipment, or changing the materials-handling system, and may consist of:

  1. Space occupied
  2. Effect on taxes
  3. Values of repair parts
  4. Changes in production rate and quality of product
  5. Downtime

A summary of such an analysis is contained in Table 4.1.

TABLE 4.1 Annual Operating Cost

Cost component	Present manual method, $	Proposed method, $		
		Conveyor	Fork lift, electric	Fork lift, propane
1. Capital equip. investment				
Equipment cost		6,000	25,000	27,000
Freight		500	800	800
Installation		8,000		
Fuel-power facilities			4,000	
Alterations to facility		10,000		
Engineering support		1,000		
Supplies				
Total capital investment		25,500	29,800	27,800
2. Fixed cost				
Depreciation		5,100	6,560	6,160
Taxes		750	3,500	4,000
Insurance		200	1,000	1,000
Supervision			1,400	1,400
Total fixed cost		6,050	12,460	12,560
3. Variable cost				
Operators-loaders	(5)* 50,000	(2)* 20,000	(1)* 12,000	(1)* 12,000
Power-fuel		1,200	2,300	800
Lubrication		20	150	150
Maintenance			1,500	2,000
Total variable cost	50,000	21,220	15,950	14,950
4. Indirect cost				
Space occupied		1,500		
Effect on taxes		1,000		
Changes in prod. rate				
Downtime		200		
Total indirect		2,700		
Total operating cost (2 + 3 + 4)	50,000	29,970	28,410	27,510
Annual cost savings		20,030	21,590	22,490

* Number of units employed.

JUSTIFICATION OF MATERIALS-HANDLING PROJECTS

The most common methods of determining the profitability of materials-handling investment are payoff period, return on investment (ROI), and discounted cash flow.

The *payoff-period* method indicates the amount of time that new equipment or a system will take to produce the savings to recover the capital investment. The investment is divided by the annual savings to give the time (in years) needed to break even. In Table 4.1:

$$\text{Payoff period} = \frac{\text{total capital investment}}{\text{annual cost savings}}$$

This method is a good risk indicator and measure which can be useful to indicate the projects that would be worth considering for closer study, but the actual profitability of new equipment depends on how much useful life is left after the payoff period. Some caution is therefore advised if the payoff period is to be the sole determinant for equipment justification, because cheaper equipment having a low useful life will always appear to be the best investment opportunity.

Simple ROI is another gross indicator that can be used to set priorities for capital investments. Here again, the effect of useful equipment life is not considered, so this method should not be used for determining the profitability of the proposed equipment:

$$\text{ROI} = \frac{\text{annual cost savings}}{\text{total capital investment}}$$

The ROI method that considers the effect of useful equipment life is

$$\text{ROI} = \frac{\text{annual savings} - \text{capital investment/useful equipment life}}{\text{capital investment}}$$

The discounted-cash-flow method of determining ROI indicates in a more realistic manner the equipment cost and return on investment by considering:

1. Savings and cost over equipment life period.
2. Net cash flow of the savings and depreciation.
3. Present worth of each year's cash flow. A factor is used to reduce the cash flow for each year to the amount of cash that would be required today to earn a desired rate of interest.
4. The effect of taxes on the rate of return.

The ROI is calculated as follows (Table 4.2):

1. Determine the cost savings for each year of the equipment useful life.
2. Deduct the estimated percentage for taxes for each year of equipment life.
3. Add depreciation for each year of the depreciation period.
4. Determine net cash flow, which is the algebraic total of items 1, 2, and 3 above.
5. Consult present-worth value table (Table 4.3) and select an interest value for the first trial.
6. Multiply the net cash flow by the factor selected in step 5.
7. If the present-worth cash flow is higher than the capital investment, select the present-worth factor for the higher percentage; if lower, select present-worth factor for the lower percentage.
8. Continue the discounted-cash-flow trials until the total discounted cash flow equals the capital investment cost. Interpolation will generally be necessary to determine the exact percent of ROI.

TABLE 4.2 Example of Calculating ROI* by Cash-Flow Analysis

Factors	Amounts
Total capital investment from annual operating cost	$30,800
Equipment life	5 years
Depreciation straight line	5 years, 20%/year
Savings per year	$22,490

	Cash flow					Trial 1 @ 45%		Trial 2 @ 50%	
Year	Cost savings	Taxes	Savings after taxes	Depre- ciation	Net cash flow	Factor	Present worth	Factor	Present worth
1	22,490	11,245	11,245	6,160	17,405	0.690	12,009.45	0.667	11,609.14
2	22,490	11,245	11,245	6,160	17,405	0.476	8,284.78	0.444	7,727.82
3	22,490	11,245	11,245	6,160	17,405	0.328	5,708.84	0.296	5,151.88
4	22,490	11,245	11,245	6,160	17,405	0.226	3,933.53	0.198	3,446.19
5	22,490	11,245	11,245	6,160	17,405	0.156	2,715.18	0.132	2,297.46
					78,715		32,651.78		30,232.49

* Interpolating

$$\frac{32{,}651.78 - 30{,}800}{32{,}651.78 - 30{,}232.49} \times 5 = \frac{1{,}851.78}{2{,}419.29} \times 5 = 3.83$$

Add 3.83% + 45% = 48.83% return on investment

TABLE 4.3 Present-Worth Values

	Interest, %											
Years	6	8	10	12	15	20	25	30	35	40	45	50
1	0.943	0.926	0.909	0.893	0.870	0.833	0.800	0.769	0.741	0.714	0.690	0.667
2	0.890	0.857	0.826	0.797	0.756	0.694	0.640	0.592	0.549	0.510	0.476	0.444
3	0.840	0.794	0.751	0.712	0.658	0.579	0.512	0.455	0.406	0.364	0.328	0.296
4	0.792	0.735	0.683	0.636	0.572	0.482	0.410	0.350	0.301	0.260	0.226	0.198
5	0.747	0.681	0.621	0.568	0.497	0.402	0.328	0.269	0.223	0.186	0.156	0.132
6	0.705	0.630	0.564	0.507	0.432	0.335	0.262	0.207	0.165	0.133	0.108	0.088
7	0.665	0.583	0.513	0.452	0.376	0.279	0.210	0.159	0.122	0.095	0.074	0.058
8	0.627	0.540	0.466	0.404	0.327	0.323	0.168	0.123	0.091	0.068	0.051	0.039
9	0.592	0.500	0.424	0.361	0.284	0.194	0.134	0.094	0.067	0.048	0.035	0.026
10	0.558	0.463	0.386	0.322	0.247	0.162	0.107	0.072	0.050	0.035	0.024	0.017
11	0.527	0.429	0.350	0.288	0.215	0.135	0.086	0.056	0.037	0.025	0.017	0.012
12	0.497	0.397	0.319	0.257	0.187	0.112	0.069	0.043	0.027	0.018	0.012	0.008
13	0.469	0.368	0.290	0.229	0.162	0.094	0.055	0.033	0.020	0.013	0.008	0.005
14	0.442	0.340	0.263	0.205	0.141	0.078	0.044	0.025	0.015	0.009	0.006	0.003
15	0.417	0.315	0.239	0.183	0.123	0.065	0.035	0.020	0.011	0.006	0.004	0.002
16	0.394	0.292	0.218	0.163	0.107	0.054	0.028	0.015	0.008	0.005	0.003	0.002
17	0.371	0.270	0.198	0.146	0.093	0.045	0.022	0.012	0.006	0.003	0.002	0.001
18	0.350	0.250	0.180	0.130	0.081	0.038	0.018	0.009	0.004	0.002	0.001	0.001
19	0.330	0.232	0.164	0.116	0.070	0.031	0.014	0.007	0.003	0.002	0.001	0.000
20	0.312	0.214	0.149	0.104	0.061	0.026	0.012	0.005	0.002	0.001	0.001	0.000
	1.030	1.039	1.049	1.059	1.073	1.097	1.120	1.143	1.166	1.189	1.211	1.233

9. The trial present-worth calculations that equal the net cash flow total are those that are used to determine the ROI.

REFERENCES AND BIBLIOGRAPHY

1. Sims, E. Ralph, Jr.: *Planning and Managing Material Flow*, Industrial Education Institute, Boston, 1968.
2. Apple, James M.: *Material Handling Systems Design,* Ronald, New York, 1971.
3. Muther, Richard, and Knut Haganas: *Systematic Handling Analysis,* Management & Industrial Research Publications, 1969.
4. Muther, Richard: *Systematic Layout Planning,* Industrial Education Institute, Boston, 1961.
5. Tompkins, J. A.: "Computer-Aided Plant Layout," *Modern Material Handling,* 7-part series, May–September 1978.
6. Merkle, W.: "Dock Planning Guide," *Material Handling Engineering,* August 1980.
7. Bolz, Harold A., et al. (eds.): *Materials Handling Handbook,* Wiley-Interscience, New York, 1958.
8. Ann Christopher, Esq.: "OSHA's Other Standard, *Warehousing Management*, April 2001.
9. "Warehouse Terminology Guide," *APICS—The Performance Advantage*, September 1999.

CHAPTER 4.2
MATERIALS HANDLING: CONTAINERIZATION

K. W. TUNNELL CO., INC.
King of Prussia, Pennsylvania

INTRODUCTION

One of the basic principles of materials handling is that materials should be converted wherever possible to unit loads to avoid manual handling. A unit load is defined as a standard container package containing one or more items that can be handled in a standard way. The *unit-load principle* suggests that the larger the load to be handled or moved, the lower the overall handling cost. To meet this objective, materials-handling systems must be designed to handle the materials-handling volume within the constraints imposed by load size as well as the material properties involved in the production or process cycle. The decisions regarding size, shape, and configuration of the unit load should also take into account compatibility.

Some *guidelines for the specification of unit load sizes* leading to the design of containerization methods and hardware to transport and store materials include:

1. Use the same pallet or container throughout the system, or at least standardize on a limited number of containers wherever possible.
2. Plan to use raw material or parts directly out of the original container.
3. Use stackable containers to permit stacking without racks.
4. Consider collapsible containers to save space and freight costs, if they are to be used also as returnable shipping containers.
5. Use nesting.
6. Be sure that the size selected fits efficiently into standard trailers and/or railcars if containers are to be used for shipping.
7. Design or select containers suitable for mechanical handling.
8. Plan containers to accommodate a wide range of products and parts.
9. Design containers to fit into building geometry.
10. Design containers that do not require special orientation to accomplish movement.

11. Use the lightest-weight material possible.
12. Consider the use of recyclable and reusable materials to reduce waste.
13. Use containers through which contents can be identified.
14. Keep the design simple and inexpensive.
15. Design must include all required safety considerations.

CONTAINERIZATION HARDWARE

Containerization hardware can generally be grouped into five main categories:

- Pallets
- Containers
- Tote boxes and bins
- Dunnage
- Outer securement

Standard Pallets

Pallets are used mainly as supports, carrier surfaces, or storing structures for unit loads.

The most commonly used material is *wood,* and pallets are available in a number of different hardwood and softwood varieties (Table 4.4). The type of wood, like any other material that is specified, should depend on load capacity, load requirements, durability, and the handling and storage environment. In general, softwood pallets are lighter and suitable for shipping pallets, while hardwood pallets are stronger, have a longer life, and are less susceptible to the wear and tear associated with interplant movement. Local, indigenous woods should be specified wherever possible to minimize costs.

Principles of Pallet Construction

Nomenclature, Design, Style, and Size. The principal pallet parts and the most commonly used construction features are indicated in Fig. 4.1. By convention, the length of the pallet is the first-stated dimension, the dimensions are always stated in inches, and the width is the dimension that is parallel to the top of the deck boards.

Types. Wood pallets fall into three general groups:

1. *Expendable* (one-way pallets). Cost is the major factor and the design and construction must meet the requirements for this purpose.
2. *Special-purpose.* Design and construction must meet the special requirements for the product or material to be moved or stored.
3. *General-purpose.* Uses standard design and features which enable the pallet to be used in a wide range of applications and also to be replaced and exchanged easily.

Pallet Configuration. This is specified by a combination of design, style, and construction features. The National Wooden Pallet and Container Association has established the descriptions of each parameter.

Typical pallet configurations are shown in Fig. 4.2. Pallets are available in a wide range of sizes; however, the most popular size is the 48- × 40-in (1.7- × 1.4-m) pallet which accounts for over 27 percent of all pallets produced.

There is movement within some industries, particularly the food and grocery industries, to standardize pallet sizes. It has been determined that size standardization, among other obvi-

TABLE 4.4 Strength Properties of Commercial Woods Employed for Pallets
(Figures shown are for 12-percent moisture content.)

Species	Static bending fiber stress at proportional limit, lb/in²*	Compression perpendicular to grain, lb/in²*	General properties
Group IV			
Oak, red	8,400	1,260	Heaviest hardwood species; greatest nail-holding power and beam strength; best shock-resisting capacity; greatest tendency to split at nails; difficult to dry
Oak, white	7,900	1,410	
Maple, sugar	9,500	1,810	
Beech	8,700	1,250	
Birch	10,100	1,250	
Hickory, true	10,900	2,310	
Ash, white	8,900	1,510	
Pecan	9,100	2,040	
Group III			
Ash, black	7,200	940	More inclined to split when nailed; greater nail-holding and shock-resisting power, beam strength, and easier to dry than group IV
Gum, black	7,300	1,150	
Maple, silver	6,200	910	
Gum, red	8,100	860	
Sycamore	6,400	860	
Tupelo	7,200	1,070	
Elm, white	7,600	850	
Group II			
Douglas fir	7,400	950	Relatively free from splitting when nailed; moderate nail-holding power and shock-resisting capacity; lightweight, easy to work, holds shape well, and easy to dry
Hemlock (W)	6,800	680	
Larch (Tamarack)	8,000	990	
N.C. pine	7,700	1,000	
Southern yellow pine (longleaf)	9,300	1,190	
Group I			
Aspen	5,600	460	
Cottonwood	5,700	470	
Redwood	6,900	860	
Spruce	6,700	710	
Sugar pine	5,700	590	
Ponderosa pine	6,300	740	
White fir	6,300	610	
White pine (N)	6,300	550	
White pine (W)	6,200	540	
Yellow poplar	6,100	580	

* Multiply by 6900 for newtons per square meter.

ous benefits, could also increase the use of pallet pools or exchange programs, which would have cost advantages throughout the distribution cycle.

DESIGN. The most common designs of wood pallets are:

1. *Two-way pallets.* Permit the entry of forklift or hand pallet trucks from two sides only and in opposite directions.
2. *Four-way pallets.* Permit entry on all four sides.
   a. *Notched stringer design.* Has four-way entry *only* with forklift trucks, and two-way entry with hand pallet trucks
   b. *Block design.* Has four-way entry with both forklift and hand pallet trucks.

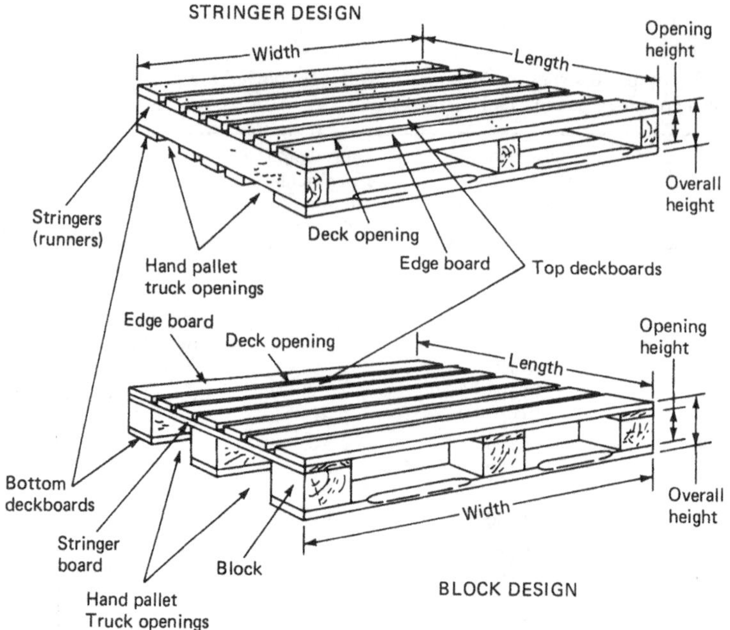

FIGURE 4.1 Principal parts of wooden pallets. (*National Wooden Pallet and Container Association.*)

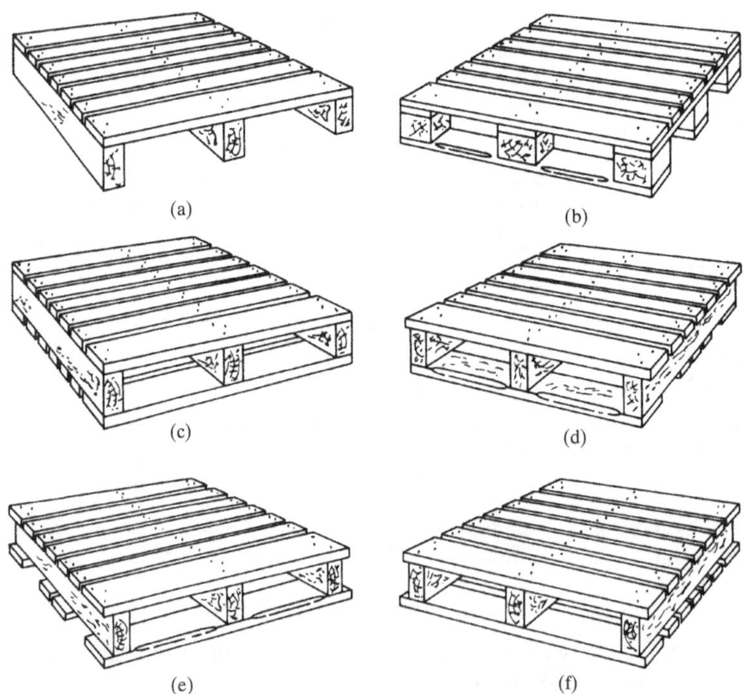

FIGURE 4.2 Typical pallet configurations. (*a*) Single-face; (*b*) double-face, reversible; (*c*) double-wing, double-face, nonreversible; (*d*) double-face, nonreversible; (*e*) single-wing, double-face, nonreversible; (*f*) double-wing, double-face, reversible. (*National Wooden Pallet Association.*)

STYLE. There are two styles of wood pallets, and they are (Fig. 4.2):

1. *Single-face pallet.* Has only one deck as the top surface.
2. *Double-face pallet.* Has both top and bottom decks and comes in two different designs, viz.:
   a. *Reversible.* Has identical top and bottom decks, and goods may be stacked on either deck.
   b. *Nonreversible.* Top and bottom decks have different configurations, and substitute goods may be stacked only on the top deck.

CONSTRUCTIONS. Wood pallet constructions are as follows:

1. *Flush stringer.* A pallet in which the outside stringers or blocks are flush with the ends of the deckboards.
2. *Single wing.* A pallet in which the outside stringers are set inboard of the top deck, while the stringers are flush with the ends of the bottom deckboards.
3. *Double wing.* A pallet in which the outside stringers are set inboard of both top and bottom deckboards to accommodate bar slings or other devices for handling pallets.

Maintenance and Repair. Procedures should also be established within the system to identify worn pallets that require repair or disposal. To accomplish this effectively, the acquisition date should be marked on the pallet and older pallets should be inspected periodically to detect wear. The following are guidelines for repair operations:

1. Never repair a pallet a second time.
2. Never repair more than three deckboards or one stringer on a given pallet. If the *average* replacement is more than 1½ boards per pallet, repair is uneconomical.
3. Productivity should average 100 repaired pallets per worker per 8-h shift for those on the repair line—forklift support and supervisory personnel excluded.
4. Cost of repair should not exceed half the price of a new, similar pallet.
5. Deckboard or stringer replacement should be made utilizing the same wood species as the original.

Pallets for Use with Forklifts

Expendable Wood Pallets. These pallets are used to support a unit load for one-way and one-time use. Pallets of this type must be specified with the capacity to carry unit load but do not require the durability of reusable types. The single-face style (Fig. 4.2) is primarily used for this purpose. Plywood deck surfaces are frequently used for expendable pallets.

Metal Pallets. These pallets can be made of corrugated steel, expanded metal, steel wire, aluminum, and combinations of metal and wood. Metal pallets are more expensive than wood pallets and are used mainly for movement of materials inside the plant where additional strength and life are required.

Corrugated-Metal Pallet Bases. These pallets (Fig. 4.3) are often integrated into the design of corrugated-steel containers with a number of other features. This permits wide versatility in parts handling and storage in the plant. The style variations available are similar to those of their wood pallet counterparts to permit movement in both two- and four-way entry bases by forklift trucks and pallet hand trucks.

All-Steel, Single-Face Pallets. Supported on three runners, this type is designed to handle heavy loads and containers. Recessed side channels bound in flanges can be incorporated in the design to permit safe movement by hand. Their double-faced, reversible design eliminates sharp edges, and thus prevents damage to bagged materials.

One-Piece, Formed-Metal Pallets. These pallets have a built-in nesting feature that permits a number of empty pallets to be stored conveniently. They are useful where pallet storage space is scarce.

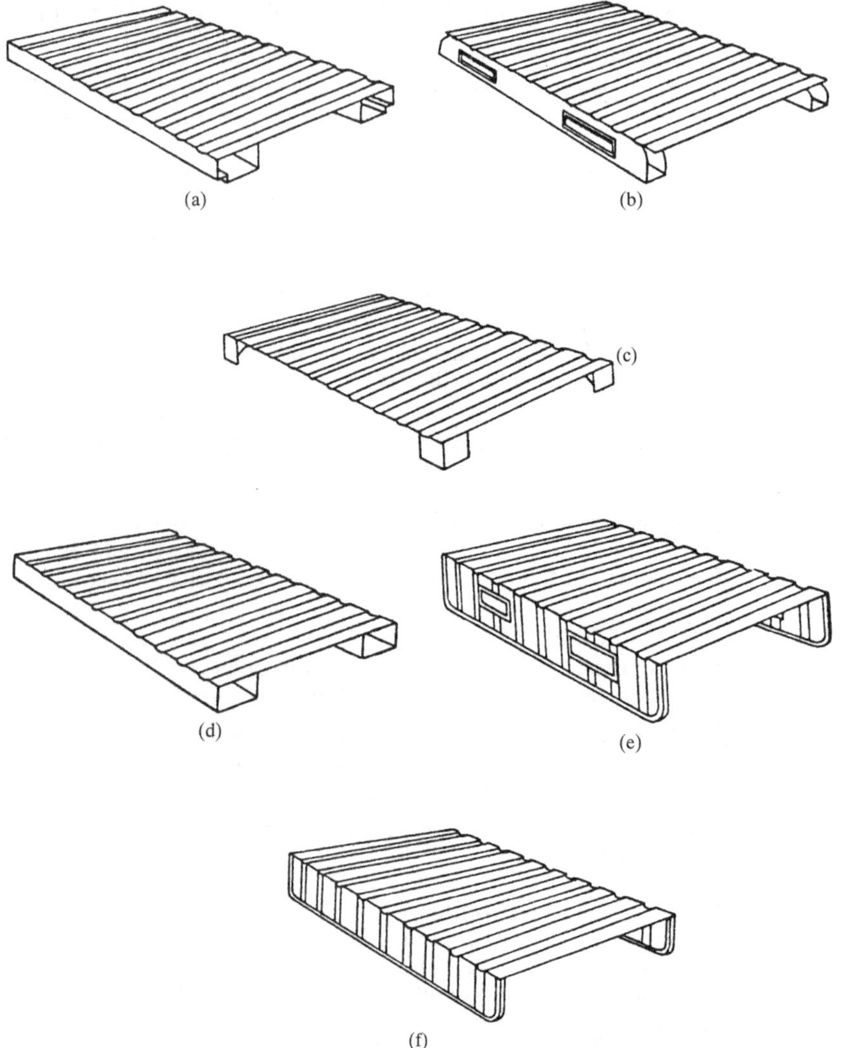

FIGURE 4.3 Corrugated-steel pallet container bases. (*a*) Two-way entry pallet style, (*b*) four-way entry pallet style, (*c*) box- or angle-style pallet, (*d*) two-way box runner, (*e*) skid or pallet truck, (*f*) standard pallet with runners.

Wire-Mesh Pallets. These pallets use galvanized or painted steel or aluminum deck sections with formed, corrugated support structures and are used where durability and light weight are required. The wire-mesh pallet, like the corrugated-metal pallets, are often incorporated as bases in wire-mesh containers.

Corrugated Pallets. These pallets (Fig. 4.4) are useful for light unit loads that are less than 1500 lb (700 kg) and for stacked loads that are less than 1000 lb (450 kg) per pallet leg. If employees handle pallets or if employee injury is a problem, these lightweight pallets should be considered. The corrugated pallets have a low cost and can be recycled. Because of their light weight, the pallets will also save money on shipping or airfreight charges.

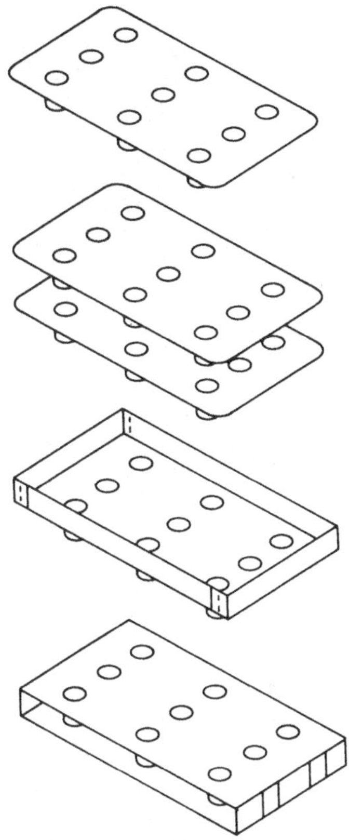

FIGURE 4.4 Cardboard expandable pallet. (*Memosha Corporation.*)

Plastic Pallets. These pallets are more expensive than wood ones and in some cases are more expensive than metal ones. However, good or broken plastic pallets can be recycled. Sometimes, a broken pallet can also be repaired. Many manufacturers guarantee life of the pallet to be 5 to 10 times longer than an ordinary wood pallet. Food and pharmaceutical industries have been the traditional users of plastic pallets, since they require a high standard of cleanliness.

Other Types of Pallets. In addition to using forklift trucks or hand-operated equipment for major movements in the plant, there are other handling requirements for movement of materials within and between manufacturing operations. Generally there is very little standardization in this area, since the carrier surfaces have to be specified to be compatible with equipment or product.

Slip-Sheet Systems. Such systems enable a unit load to be handled and moved without being supported on a pallet type of platform. Slip sheets (Fig. 4.5) are made of heavy corrugated paperboard, plastic, or kraft fiber composition and function as the base surface for the unit load. Special equipment or *push-pull* forklift truck attachments are required to move and handle loads unitized by this method. The cost benefits of slip-sheet systems are obvious: in addition to lower initial cost, storage space requirements are $\frac{1}{100}$ of the cube required for empty pallets and shipping costs are less than for comparable loads using wood pallets.

Slave Pallets. These are used for assembling unit loads before transfer to other containers, for moving odd-shaped loads on conveyors, for serving as accumulating and transfer platforms for automated computer-controlled storage and retrieval systems, and for supporting unit-load containers that are not designed for use on conveyors.

Slave pallets are normally plywood-sheet surfaces, but if interim storage is required in racks where the pallet-edge surfaces are supported by shelf angles, pallet specifications become more critical from the standpoint of supporting loads and safety. As a general rule, to achieve maximum strength and stiffness, face grain should be across supports. Design criteria include pallet size, total uniform load, permissible deflection, clear span, and uniform load in pounds per square foot (newtons per square meter). Selection of the proper grade and thickness of plywood can then be determined. Information regarding recommended maximum uniform loads and deflections is available from the American Plywood Association.

Air Pallets. This type uses a bed of air to support a unit load and enables large loads to be moved and maneuvered. Air pallets or air-film equipment can be used to convey parts, rotate work stock, and move palletized loads in and out of buildings as well as trucks, railcars, and other conveyances.

Portable Stacking Racks. These racks are used for storage of palletized loads that cannot be stacked on each other. Pallet stacking frames (Fig. 4.6) are used to confine and protect irregularly shaped, fragile, or nonuniform loads during in-process or temporary storage. The pallet itself is the base unit and rests on the frame of the pallet beneath it. The second kind of portable stacking rack (Fig. 4.7) consists of a base unit and removable post and end

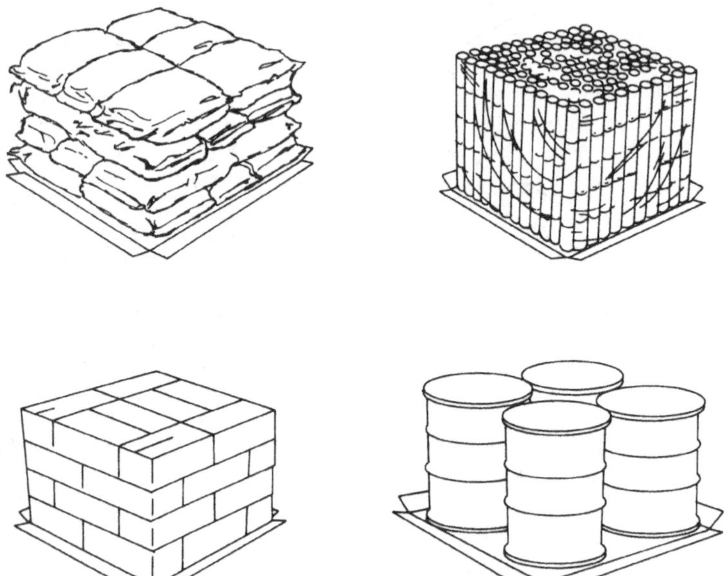

FIGURE 4.5 Typical uses of slip-sheet system. (*Little Giant Products.*)

frames. Pallets are stored on the base units and the base units, when stacked, nest in the end frame.

Pallet Loading Patterns. A pallet pattern is an arrangement of units on a pallet and ideally is the most effective way of loading a pallet with the least loss of cube. There are a number of ways to select the optimum pattern, ranging from trial and error to the use of computer models. No matter what technique is used, the following factors must be taken into consideration:

1. *Size of material.* There may be several ways, one way, or no way to place a given-size material onto a given pallet.
2. *Weight of material.* In the case of very heavy material, fewer layers will be stacked on a pallet. To a certain extent the number of layers will depend on the strength of the containers, if any are used.

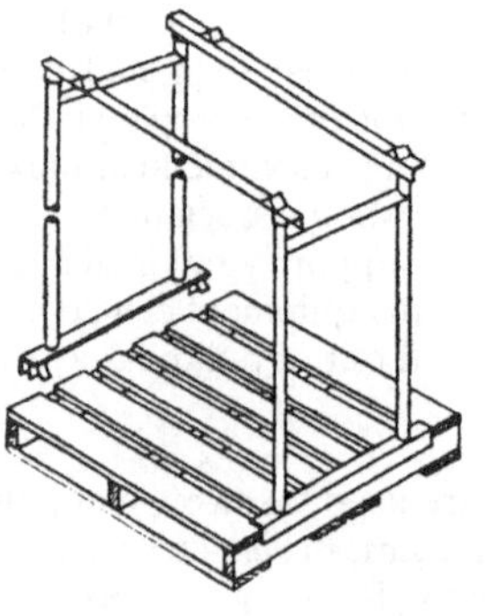

FIGURE 4.6 Pallet stacking frame using pallet as base unit.

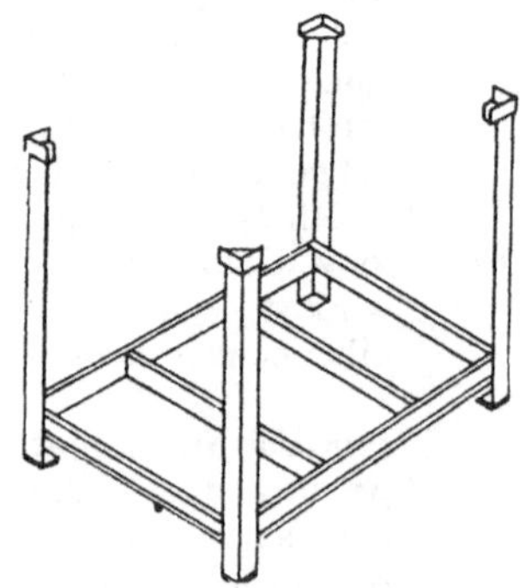

FIGURE 4.7 Pallet stacking frame with base unit.

3. *Size of unit load.* Taken as a whole, the length, width, and especially the height of the load must be considered.
4. *Loss of space within unit load.* Some patterns have too many large gaps between units. This kind of piling is particularly bad when paper pallets are used because the weight should be distributed evenly and the units should brace each other.
5. *Compactness.* Some patterns do not tie together well; they will not interlock.
6. *Methods of binding products in patterns.* If the units of a load are glued together, one kind of pattern may be ideal; with strapping, another type may be the best; and if no fastening at all is used, some combination of stacking may be the most suitable method to interlock and hold the load together.

Some general rules that should be followed in establishing pallet patterns are:

1. Interlocking unit loads should be used when possible to make the most effective use of the cube and to provide load stability.
2. Overhang, where unit loads extend beyond the edge of a pallet, should be avoided or minimized to a point where container damage or load stability is not affected. The added dimensions caused by this condition should not exceed the width or length of the shipping conveyance or reduce the utilization of the conveyance.
3. Underhang, where unit loads do not fill the deck surface and where there are large voids, should be avoided.
4. Utilize the basic pallet patterns (Fig. 4.8) effectively. Use block patterns for containers of equal width and length. This type of pattern is the least stable and may require bonding and fastening if considerable movement is involved.
5. Brick, row, and pinwheel patterns are used for containers of unequal length or width. All three patterns result in the interlocking that stabilizes a load.

Pallet patterns have been developed empirically for materials that have rectangular dimensions. The U.S. Navy Research and Development Facility has developed such a pattern.[1]

Container capacities range from 500 to 6000 lb (230 to 2700 kg), and standard base sizes cover the range of sizes of wood pallets, including 40 × 48 in (1.4 × 1.7 m). The 44- × 54- × 40-in (1.6- × 2- × 1.4-m) size is ideal for use when making shipments by rail or truck because the container dimensions are exact multiples of trailer and railcar dimensions and therefore can use the cube of these conveyances fully.

Metal Containers

Three types of metal containers are in general use: wire mesh, noncorrugated steel, and corrugated steel. Current development trends tend toward increasing versatility by incorporating features such as stacking and dumping capability and pallet-type bases to allow and facilitate movement.

Welded-Wire-Mesh Containers. These containers (Fig. 4.9) are fabricated from welded wire for containment of materials. Additional structural sections are added for additional strength, and optional features can be included for specific uses and applications. The kinds of material and product that can be handled are limited only by the size that would fall through the wire mesh and total container volume. Mesh openings from ½ × ½ in (1.3 × 1.3 cm) to 4 × 4 in (10 × 10 cm) are available to accommodate a wide range of product or material sizes.

Advantages that are generally cited in using this type of container are:

- Lightweight as compared with other metal containers
- Allows visibility of product for easy and quick identification

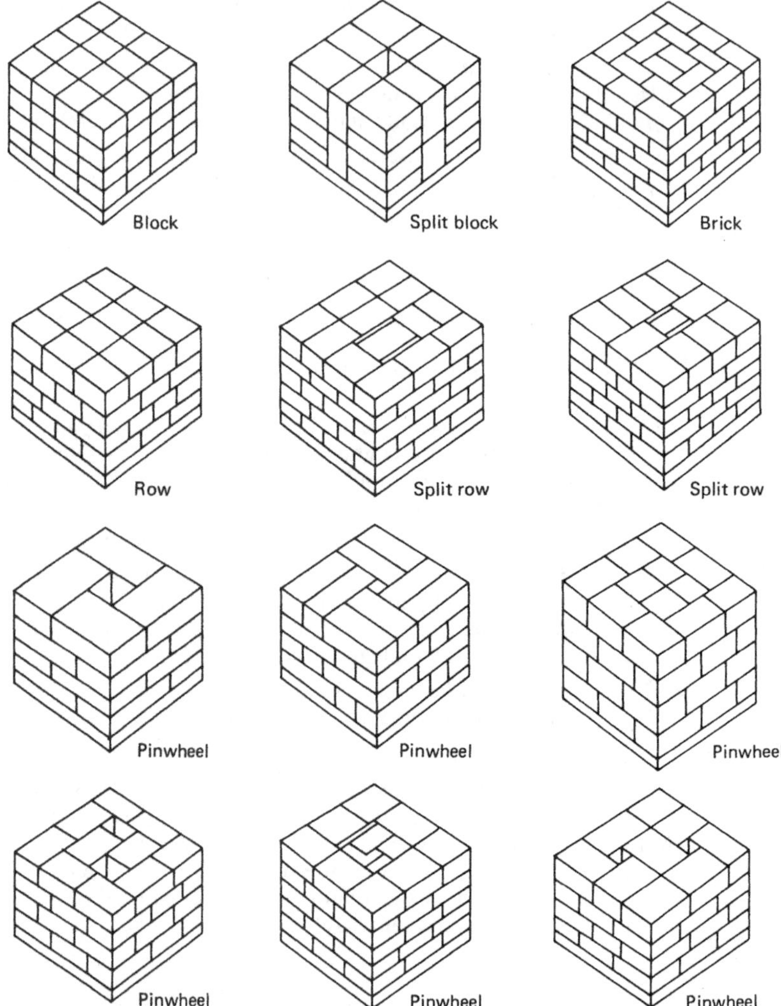

FIGURE 4.8 Typical pallet patterns. Paper or fiberboard binders are used between layers if necessary.

- Self-cleaning by shedding debris
- Material can sometimes be processed in the container in such processes as degreasing, cleaning, and air drying.

Wire-mesh container stacking is accomplished in two ways: (1) interlocking through the corner posts and (2) the arch (saddle) of the upper container resting on the rim of the container below.

Wrap-Around Collapsible Frame. This forms the most basic type of wire-mesh container by using a standard pallet as a base.

Collapsible Container. This type of container is constructed to permit the sides to fold down when not in use. The obvious advantage of this design, saving space, must be weighed against its shorter life and smaller payload capacity.

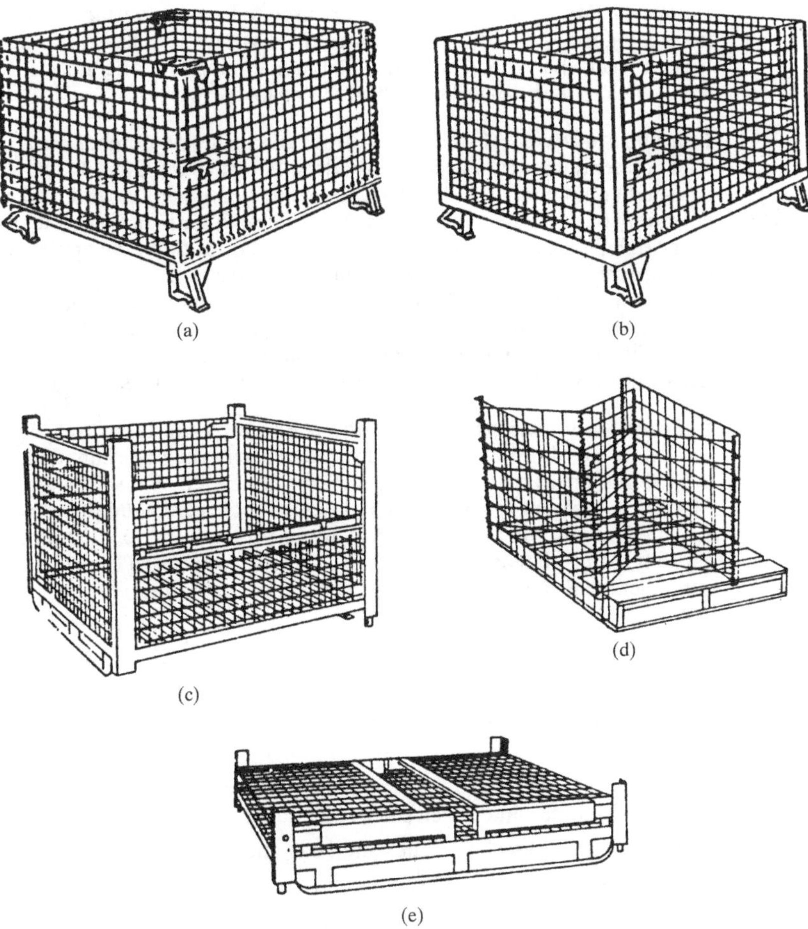

(a) (b) (c) (d) (e)

FIGURE 4.9 Kinds of welded-wire-mesh containers. (*a*) Collapsible wire container, (*b*) rigid wire container, (*c*) heavy-duty rigid, (*d*) wrap-around, (*e*) folded-down collapsible rigid.

Rigid Wire Container. This type is designed with additional vertical structural members that increase the strength of the containers, resulting in longer life and greater capacity than the collapsible types can provide. A heavy-duty rigid container has been developed to handle heavier loads in more strenuous environments. Additional horizontal bracing has been added to achieve the capability.

Collapsible Rigid Container. This is a heavy-duty version of the collapsible container. It is nearly equal in strength to the heavy-duty rigid type, but the higher original cost must be compared with savings on return freight costs.

Corrugated-Steel Containers. These are probably the strongest type of metal container available, since corrugation permits a longer surface of material to be used for a given size of a container than any other method of construction, and it is fabricated from hot-welded steels 0.105 to 0.179 in (2.7 to 4.6 mm) thick. This type of container has progressed from the one-type-only gondola bin to configurations that are almost virtually materials-handling systems in themselves. They can be considered as consisting of three basic *modules:* base, lifting and stacking aids, and container box, with options for specific applications.

Container Bases. Container bases are used to facilitate surface movement and can be essentially categorized as metal pallets since there are provisions for two- and four-way entry with forklift trucks and other materials-handling equipment. The general features of these bases were covered in the pallet section of this chapter.

Container Sections. Container sections can include a number of options (Fig. 4.10) to provide hopper dumping or accessibility, drop-bottom dumping, and end gates. Air- or hydraulic-actuated equipment and tilt stands are used in conjunction with, and as means to extend the versatility of, containers in assembly or machining stations. Container sections come in a wide range of heights; the deeper containers are used for dumping applications, and the shallower sizes are more suitable for manual handling of parts.

Several lifting and stacking aids can be designed or are available as part of the container system. Overhead movement can be accomplished by the addition of overhead crane lugs, chain-sling lugs, and bar-lift or sling-lift notches in the base. Stacking is accommodated by corner stacking angles, self-aligning hairpin brackets, and stacking lugs.

Industry Specifications. Specifications issued by the Industrial Metal Containers Product section of the Material Handling Institute specify the materials, design, construction, testing procedures, environmental considerations, utilization, and safety practices regarding the use of metal containers. The safety guidelines do not recommend the use of wire-mesh containers for applications where parts and material protrude or for dumping purposes. A load and capacity rating plate is used to indicate the load capacity of each container and the number of containers that can be safely stacked on top of each other.

Wood Containers

Wood containers are of three basic types.

Types

1. Bins with bases and closed sides and ends
2. Boxes with bases, closed sides and ends, and a top
3. Crates with open or slatted sides and ends

Wood containers with standard pallet bases are finding increasing use in applications where mechanized handling and storage are required for products ranging from solid materials of irregular shapes and sizes to granular materials. Pallet containers are now being used in the agricultural area, where fruits and vegetables are loaded into pallet containers when picked and then washed, transported, and placed in supermarkets in the same container.

The pallet container design may include collapsible sides, a feature which saves space when the container is not in use or when it is being returned empty.

Construction. The four major types of end-panel and side-panel construction in wood containers are:

1. *Solid.* Usually of plywood, this type of construction provides great strength as well as a smooth interior surface.
2. *Vertical-slat.* Used for deep containers and requires more bracing than the solid or horizontal-slat construction to prevent rocking.
3. *Horizontal-slat.* Used for shallower containers and requires less bracing than containers with vertical-slat construction.
4. *Wirebound.* Provides the economy of lightweight construction with the added strength of being wire-bound (Fig. 4.11). The main advantage of this construction is a high-strength container with low tare weight.

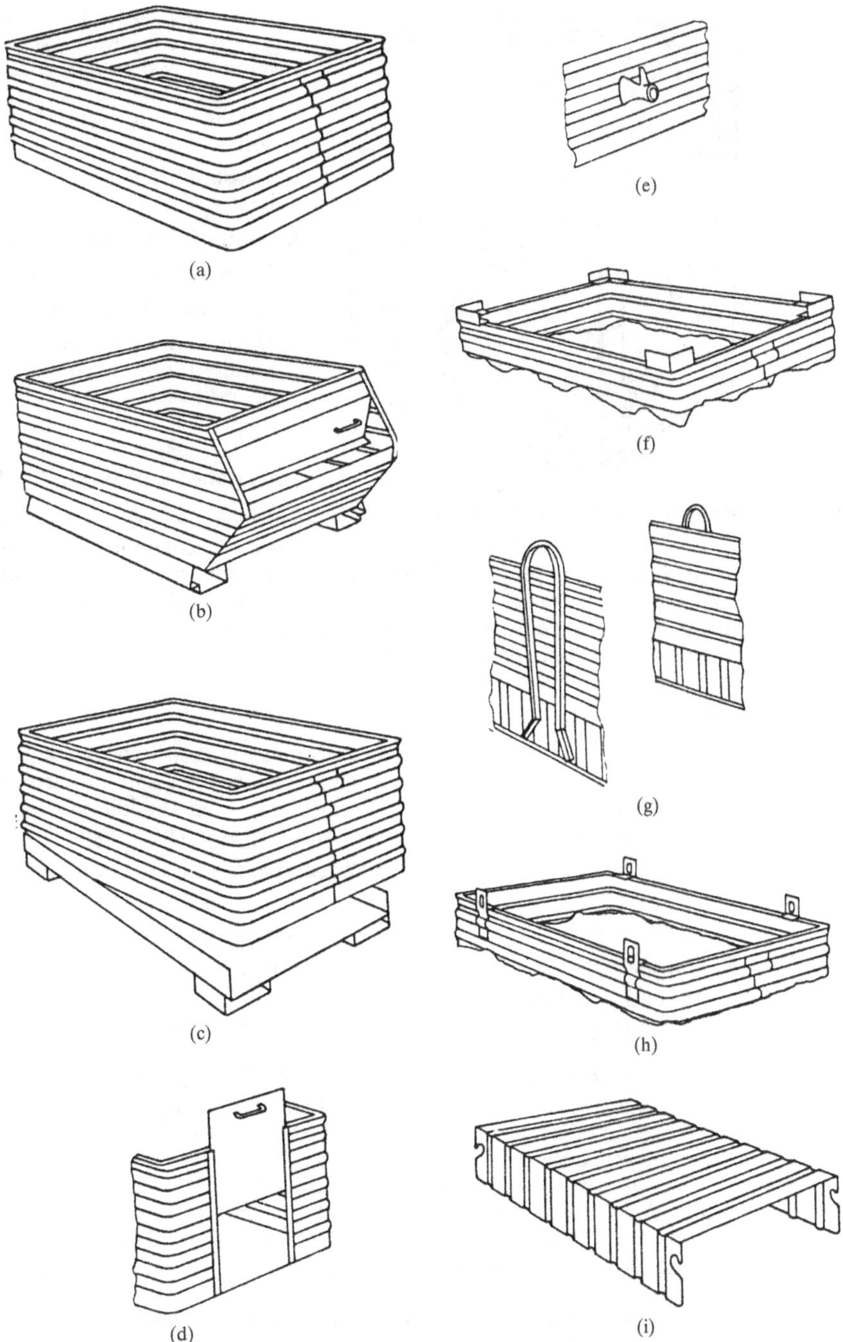

FIGURE 4.10 Corrugated-metal-container styles and lifting and stacking aids. (*a*) Basic corrugated box, (*b*) hopper-front pallet base container, (*c*) drop-bottom container, (*d*) end gate option, (*e*) overhead crane lugs, (*f*) corner-stacking lugs, (*g*) chain sling and stacking lugs, (*h*) bar lift, (*i*) sling lift notches.

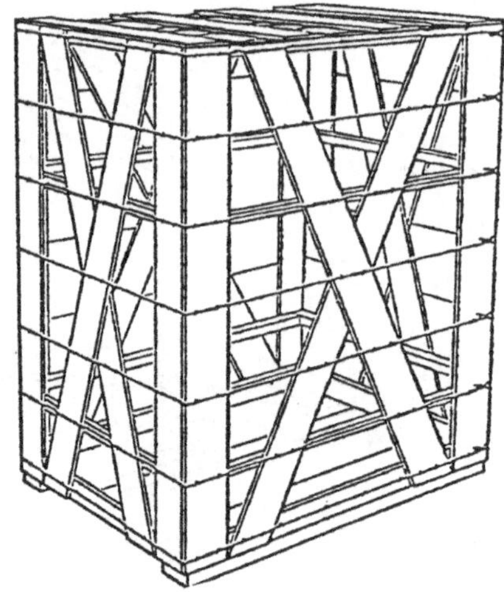

FIGURE 4.11 Typical wirebound container.

Selection of Wood. In the selection of the wood species to be used for wood containers, the following factors should be considered:

1. Intended usage and life requirements of the container
2. Pressure to be exerted on the bottom and sides by the goods
3. Permissible bulging of sides and edges
4. Degree of impact resistance required
5. Type of handling equipment used
6. Degree of weather and water resistance required for exposure to weather or cleaning operations

Corrugated-Cardboard Containers

Corrugated-cardboard containers offer a wide range of economical solutions for packaging or materials-handling problems. There are literally hundreds of unique designs of corrugated containers. Because of the relatively low cost, it is feasible to "tailor-make" a configuration that is an optimum design for a particular product and situation. This container may be one of, or a modification of, five common types.

Types

1. *Regular slotted container.* The most commonly used style. All flaps are of equal length, and other flaps meet when closed. Contents of the box are protected by one thickness of corrugation on the side and two thicknesses on top. If additional top and bottom protection is required, both outer top and bottom flaps can be designed to overlap.
2. *Telescope box.* A two-piece box designed so one part fits into the other.
3. *Five-panel folder.* A corrugated flat sheet, scored into five panels, which folds into a four-sided tube-type container that is closed by end flaps.
4. *One-piece folder or book fold.* Used for shipping books, catalogs, and wearing apparel.
5. *Gaylord.* A corrugated container that has a base, generally consisting of two or more wood runners that allow the container to be moved by forklift equipment.

Corrugated Material and Construction. The basic corrugated material is referred to as "double-face" or single-wall, consisting of an outer-facing, corrugated medium, and an inner facing, joined by adhesive. Corrugated material is also available in single-face, double-wall, and triple-wall designs.

Types of Board. These include Fourdrinier kraft linerboard, the highest-quality and -cost material made from virgin pulpwood, and cylinder linerboard which is generally made from a combination of reclaimed fibers and virgin pulp. The weight of a linerboard required for the contents of a corrugated container is specified in the Uniform Freight Classification and Motor Truck Classification.

Corrugating Medium. Straw, reclaimed fibers, or woods can be used as a corrugating medium. The combination of corrugated-medium thickness, weight, and flute configuration determines the strength and moisture-lockout properties of the container. Corrugating media are specified by thickness and weight per 1000 ft^2 (MSF).

FLUTE CONFIGURATION. This specifies the number of corrugations per lineal foot. The three most common flutes used are:

1. *A flute.* The highest flute with the least number of corrugations (36 per linear foot, 0.1875-in high) (12 per centimeter, 4.2 mm). When used in an upright position, A flute has the best stacking strength and greater capacity to absorb shock in the direction of the thickness.
2. *B flute.* Has the greatest number of corrugations per foot with the lowest flute height. It is stiffer and less shock-absorbent than A flute, but it has greater crush resistance to loads placed in the direction of thickness.
3. *C flute.* A compromise between A and B flutes.

Methods of Fastening Joints. The accepted methods are taping, stitching, or gluing. Common carriers publish detailed regulations governing the method to be used in regard to content type, weight, and other factors.

Corrugated boards may be specially treated to provide additional properties to the container, such as a coating and lamination, in order to:

- Retard slippage
- Inhibit mold
- Retain temperature
- Increase water or moisture resistance

Tote Boxes and Bins

Tote boxes are used for unit loads of smaller parts that can be moved manually through the operation or can be stacked in a larger container to become part of a unit load. Tote boxes are available mainly in metal and plastic and are also fabricated from other materials such as wood, cardboard, fiberboard, and Plexiglas.

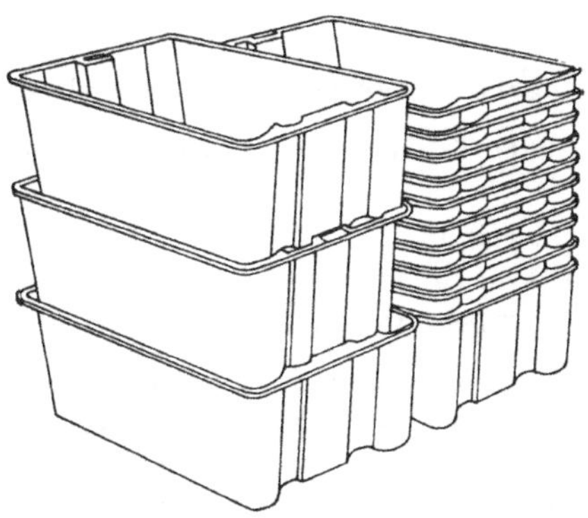

FIGURE 4.12 Typical tote boxes.

Plastic Tote Boxes. Plastic tote boxes have many applications where small and light parts are handled and in environments where protection from corrosive chemicals or a high degree of cleanliness is required. Plastic containers are easily cleaned without harm or deterioration to the material. Plastic totes are very cost-effective considering their durability and typical life cycle. Molding capabilities permit desirable features to be incorporated as an integral part of the container, permitting a number of provisions for nesting and stacking. A typical tote box is shown in Fig. 4.12. There are three major types:

Straight Nesting. This refers to tote boxes that can be nested when not used. This type requires the use of lids or covers if stacking is required but results in maximum product protection, since one box will not fall into another. The design of straight-nesting boxes is characterized by tapered sides which reduce cube utilization and should not be used for storing on shelves.

Straight Stacking. This method of stacking boxes is ideal, because of minimum tapered sides, for shelf storage and for use as an inner container of maximum cube utilization. Since the sides and ends support the loads, greater weights can be stacked than when using the straight-nesting variety.

Combination of Stack and Nest. This feature is available in some boxes and can be altered by the orientation of boxes in relation to each other.

The most common plastic materials and their major properties are indicated in Table 4.5.

TABLE 4.5 Plastic Materials Used in Tote Boxes

Material	General properties
ABS (acrylonitrile-butadiene-styrene)	High impact absorption, good compressive strength; more expensive than other thermoplastics
High-density polyethylene	Good to excellent stiffness, excellent temperature range, –40 to 150°F (–40 to 65.5°C); commonly used for food applications with USDA and FDA approval
High-impact polypropylene	More durable than polyethylene, not as stiff; tendency to crack at temperatures below 0°F (–18°C)
High-impact polystyrene	Extremely stiff, excellent compressive load strength, good temperature range; tendency to crack easily under high impact; readily attacked by solvents and oils
FRP (fiberglass-reinforced polyester)	Exceptional compressive load strength; can be heat-, fire-, and wear-resistant

DUNNAGE AND OUTER SECUREMENT OF CONTAINERS AND UNIT LOADS

Dunnage

Dunnage refers to inner package containment methods or material that is used to protect the contents of a container from damage. This is done in one of two ways, either by preventing the movement of the contents or by providing a cushioning medium to absorb shocks.

Plastic and other petroleum-based materials are used for dunnage because of their light weight and low bulk density.

Dunnage Materials. Polystyrene is used to cushion package contents in three general forms:

1. Loose foam strands are used to fill the airspace in the package and provide a cushioning barrier around the contents.
2. Polystyrene can be molded to the general form of the part in the container and actually becomes an inner case for the part.
3. Polystyrene is also used for corner forms to strengthen corrugated containers.

Bubblepack is two thin sheets of polyethylene with air entrapped within the "bubble" sections when laminated together. The contents of a container may be wrapped in bubblepack for protection during shipment or movement.

Corrugated board can be easily formed into many shapes to protect, support, and cushion products. The trend is to reduce the number of inserts by combining features into a single interior form, reducing the cost and inventory expense of packaging materials. Typical inner packings, portions, and sheets used for this purpose are shown in Fig. 4.13.

Outer Securements

Container closure can be achieved by sealing flaps with glue and tape, by stapling, and by strapping with plastic or steel bands.

The glue and tape sealing method lends itself to automation by use of case-sealing equipment, which automatically dispenses glue or tape close to the flaps. Stapling can also be automated by passing containers through, under, and between staple heads.

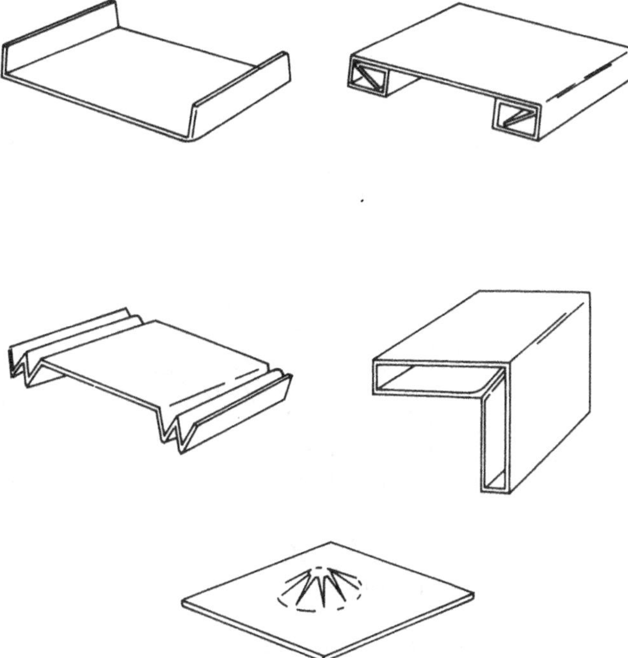

FIGURE 4.13 Typical inner packings, portions, and sheets.

Strapping. Strapping can be used to secure not only containers but stand-alone unitized loads and palletized unit loads as well. There are two major strapping classifications: steel and plastic. Steel strapping is dimensionally stable under all but extreme conditions and plastic strapping is more resilient.

Common cold-rolled-steel strapping, the least stretchable of all steel strapping, is good for strapping lightweight packages or pallet loads that are not subject to high impact or shock. Heavy-duty steel strapping, both hot- and cold-rolled steel, absorbs high impacts without breaking; however, it stretches under great stress and requires staples to keep it in place.

Plastic strapping continues to stretch under tensioned loads and therefore should not be used where continuous loads are present. However, its resilience keeps it tight on a package or load that shrinks or settles.

Steel strapping may be applied by manually operated tensioning tools which notch or crimp steels to retain the strapping. Plastic strapping can also be applied both manually and automatically in the same manner as steel strapping. Friction-welding and heat-sealing systems are also available.

Stretch Wrap. Stretch wrap is one method of confining unitized or pallet loads by wrapping a film of polyethylene material around the load. The stretch film is wound under light tension by rotating the load on a platform on horizontal- and spiral-type stretch film equipment. Horizontal-type equipment uses a full sheet wrap, while spiral-type equipment bands the load by dispensing several overlapping layers. Vertical wrapping equipment rotates the film around the load and is generally used when the load is not on a pallet.

Shrink Wrap. Shrink wrap uses a polystyrene film that can be a bag or flat sheet that is put over or around the unit load and then placed in an oven. Bag design should be selected to achieve the desired protection required or to optimize the holding power for the load configuration. While in the oven, the film reaches a temperature of 240°F (116°C). During the cool-

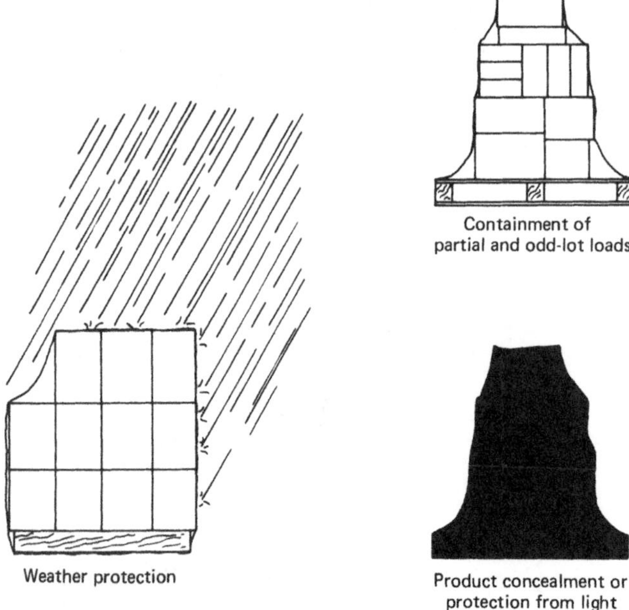

FIGURE 4.14 Other benefits from use of shrink wrap. (*Black Body Corp.*)

ing cycle, the film molecules try to return to their original compact orientation, shrinking the film tightly around the load.

Three types of ovens are used for this purpose, depending on the volume handled. Closet types can normally handle up to 20 loads per hour; bell-type ovens are for higher-volume operations, up to 20 loads per hour; and tunnel or feedthrough ovens are generally specified for automated lines and handle more than 75 loads per hour.

In addition to securing products in a unit load, there are extra benefits in using shrink wrap (Fig. 4.14). Irregular, hard-to-pack loads (like those in order-picking warehouses) can be secured with shrink wrap. Products can be protected from damage caused by weather, dirt, or moisture. Clear film allows for easy viewing for identification and inventorying of product. Opaque film can be used to conceal or protect product from light.

REFERENCES AND BIBLIOGRAPHY

1. "Storage and Materials Handling," Departments of the Army, Navy, Air Force, and U.S. Marine Corps, Washington, D.C., 1955.
2. "Unit Load Stretch Wrapping," *Modern Materials Handling,* October 1979.
3. Schumf, G., "Slip Sheets," *Material Handling Engineering,* July 1980.
4. "Pallets: Wood Isn't the Only Answer," *Material Handling Engineering,* November 1990.
5. James A. Cooke: "Survival Tactics," *Logistics Managment and Distribution Report,* May 2001.

CHAPTER 4.3
MATERIALS HANDLING: FIXED-PATH EQUIPMENT

K. W. Tunnell Co., Inc.
King of Prussia, Pennsylvania

INTRODUCTION

Conveyors, cranes, and hoists are generally considered fixed-path materials-handling equipment, since they often become a fixed part of the physical plant. Once in place, a considerable amount of time, disruption, and cost is needed to change the arrangement of the equipment. It is therefore very important to plan the installation of these pieces of equipment very carefully.

A complete materials-handling system can include a wide variety of fixed-path equipment for unit loads and bulk materials handling, as well as mobile handling equipment and storage racks. This further complicates the planning process, since the fixed-path equipment not only has to satisfy the requirements of the fixed-path handling but must also be compatible with the overall flow of the total handling system.

There are many considerations involved in planning fixed-path equipment installations; many of these considerations are unique to a specific type or class of equipment, but general areas that must be addressed in the planning and exploration stage are:

- *Flexibility of the system.* Must a wide range of unit-load sizes or bulk material be handled or conveyed?
- *State of materials to be handled.* Is it in a unit load or bulk form?
- *Weight, dimensions, and physical properties of the material being handled or moved.* Is it fragile, light, firm, or does it have other properties that require special attention?
- *Loading and unloading methods.* Is it handled manually or received from or delivered to other equipment such as lift trucks, palletizers, or packaging equipment?
- *Capacity of equipment.* Does the conveying speed match the speed or capacities of the equipment it is being interfaced with? Is there sufficient capacity or length to accumulate material when required?
- *Supporting-system requirements.* Is the material to be sorted, accumulated, weighed, or further processed while being handled or conveyed?
- *Environmental conditions.* Must provisions be made for dust, high or low temperature, high humidity, or other ambient conditions in the plant or outside?

- *Safety.* What special precautions must be taken to protect operating personnel or personnel working near the equipment? What provisions must be made to comply with regulatory requirements?
- *Maintenance.*
- *Facility restrictions.* Are overhead heights or floor loading capacities adequate for supporting and accommodating equipment? Is there sufficient plant area? Will the fixed-path system impede access to equipment and the flow of personnel and other materials within the plant?
- *Horizontal or vertical distances to be covered.* What hardware is required to negotiate inclines and declines throughout the system?
- *Power and energy requirements.*

There are many types and varieties of fixed-path equipment. Each of the major classifications of this type of equipment are discussed and described here, but no effort will be made to list all of the items that are contained in each classification. The classifications that are covered include:

- Conveyors and monorails
- Sorting, consolidating, and diverting devices
- Hoists and cranes
- Automatic guided vehicles
- Robots

CONVEYORS

Conveyors are gravity or power devices commonly used to move uniform loads continuously from point to point over fixed paths. The primary function of the conveyor is to move materials when the loads are uniform, and the routes do not vary. The movement rate and direction is usually fixed, although powered conveyors have the capability to alter the rate of speed. The major types of conveyor and related devices are chutes and wheel and roller conveyors.

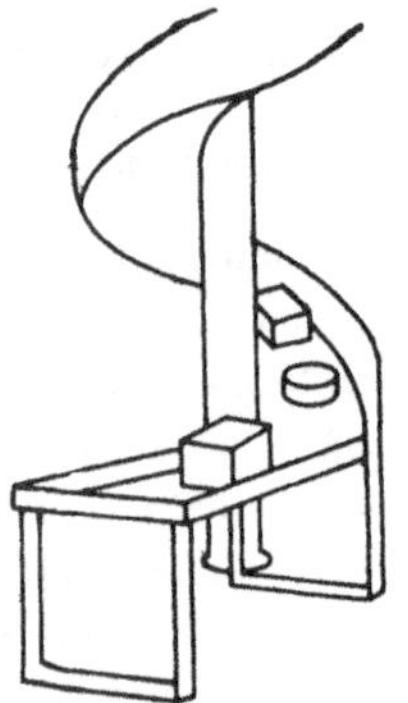

FIGURE 4.15 Spiral chute.

Chutes. Chutes are the simplest fixed-path devices that use gravity to convey bulk or unit loads down declines. Straight and spiral types are available. The spiral chute (Fig. 4.15) is a continuous trough over which bulk materials or discrete objects are guided in a helical path.

Wheel and Roller Conveyors. These depend on both gravity and power to move materials. Objects of various shapes can be handled by changing the cross section of the rolling surface or by aligning the objects in the conveyor framework. These conveyors are generally used to move materials horizontally.

Considerations for Chutes and Wheel and Roller Conveyors. The following sections discuss points that must be considered in specifying and designing both of these classes of conveyors.

Load Characteristics. These include maximum and minimum sizes of loads and the shapes and carrying surfaces of all units. The suitability of a load configuration to be handled on roller or wheel conveyor is important. Unsupported packages such as bags (Fig. 4.16) are not recommended for this type of equipment.

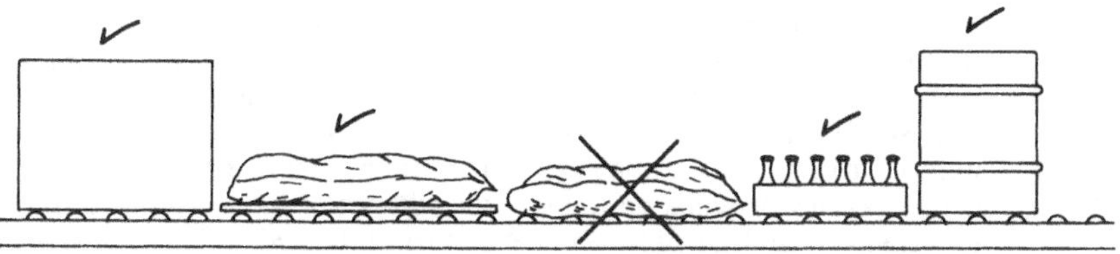

FIGURE 4.16 Can it be handled on a roller conveyor? (*Litton UHS.*)

Operating Conditions. These include the size and weight of the conveying surfaces, environmental conditions, and loading and unloading methods. These considerations determine the type and capacity of frame, roller, or wheel material and sizes, and the type of bearings that should be used.

Roller or Wheel Spacing and Pattern. This is determined by the size of the minimum package or unit load. (See Fig. 4.17) To determine roller centers, divide the minimum load length by three. Wheel pattern should be specified to provide a minimum of five wheels under the package. Other guidelines include:

1. A minimum of three rollers under a hard bottom surface
2. A minimum of four rollers under a flexible bottom surface

Roller and wheel capacity is determined by dividing the weight of the heaviest load to be handled by the minimum number of rollers or wheels under the carrying surface of the load. If special requirements must be considered, such as drop, shock, or side loads, then a roller with a higher load rating will have to be considered.

Conveyor Width; Wheel and Roller Setting. The width of conveyor is determined by the back-to-back frame dimension required to ensure sufficient clearance to carry the load

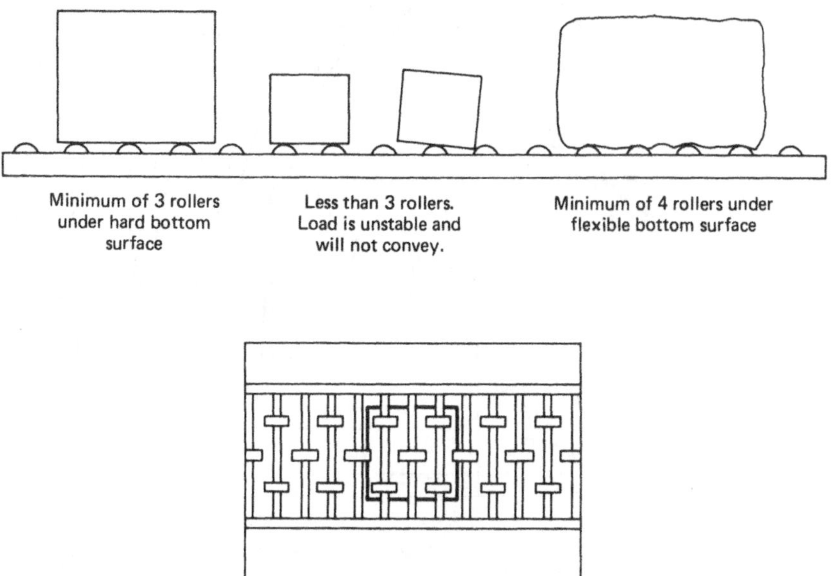

FIGURE 4.17 Roller and wheel spacing guidelines. (*Litton UHS.*)

around a 90° curve. The minimum clearance is dependent on the roller setting. If rollers are set high, the conveyor can handle loads up to 1.25 times the width of the conveyor. If the rollers are set low, a minimum of 1 in (2.5 cm) between frame and load must be allowed on each side. Package skew should also be considered in determining this dimension. Specification of curve sections can be calculated graphically from the chart in Fig. 4.18. The design of curve sections depends on the size and shapes of loads. Tracking of packages is important, especially when there are a number of turns involved and the skewing effect becomes cumulative. This effect can be minimized by using tapered rollers (Fig. 4.19) or a double-roller differential section.

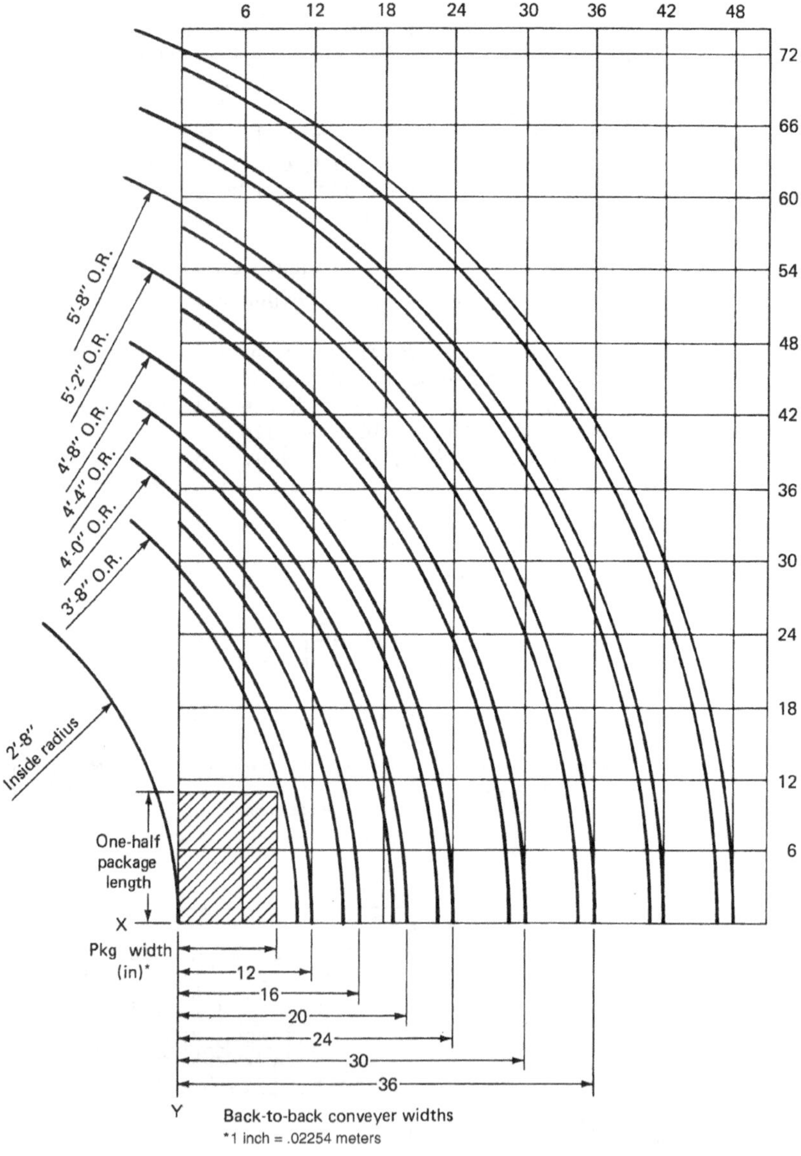

FIGURE 4.18 Curve-radius selection.

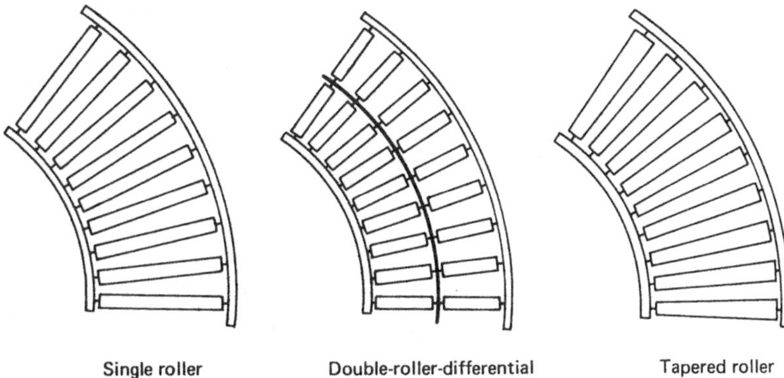

FIGURE 4.19 Roller-conveyor curve sections. (*Litton UHS.*)

Bearing Selection. This is dependent upon conveyor operating conditions. Plain ball bearings are used for indoor use where severe environmental conditions are not present. Dust-tight bearings, which are designed to run dry, are ideal in dusty atmospheres. Greased bearings require more force to turn and their use should be kept to a minimum on gravity-type conveyors.

Conveyor Pitch. The pitch that is required cannot be easily determined because it is dependent on a combination of many factors such as:

- Weight and stability of unit load
- Smoothness of bottom of load
- Firmness of load surface
- Length of rollers
- Number of rollers under load
- Length of runs
- Types of bearings

The suggested pitch for a number of types of unit loads is shown in Table 3-1 and should be used as a general guide for determining pitch.

Conveyor Support and Frame Capacity. Conveyor supports can be one of three types: permanent-floor, ceiling-hung, or portable. Supporting points (Fig. 4.20) should be located to handle the load equally. The design load is the weight of the conveyor section plus the maximum unit load for that section of the conveyor.

Special Considerations for Wheel Conveyors. Wheel conveyors are used for light-duty applications and have several advantages over roller conveyors for lightweight unit loads. Gravity wheel conveyors consist of a series of wheels which can be one of many different styles and materials, mounted on common axles and supported between two frames. They are generally less expensive and lighter in weight, making them ideal for portable applications. Light unit loads travel better on wheel conveyors because less pitch is needed and less force is required to start the wheels in motion (see Table 4.6). Another advantage inherent in the use of multicontact conveying surfaces is that the wheels provide a turning action which enables the package to maintain its original position.

Frames that support the wheel axles are either steel or aluminum. Low weight and corrosion resistance are properties that make aluminum attractive, but steel should be used where conditions require the conveyor to be more rugged.

TYPES OF WHEELS. There is a wide variety of both metal and plastic wheels now available, including:

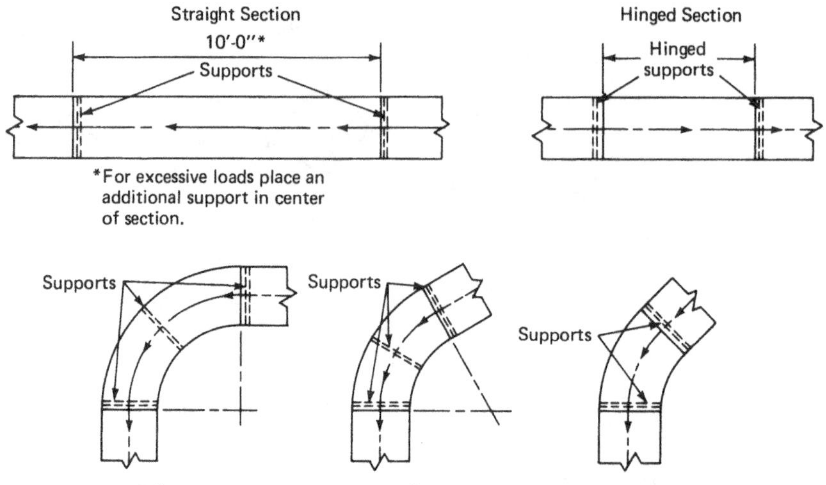

FIGURE 4.20 Suggested support locations.

- *Steel wheels with ball bearings.* The most rugged and commonly used wheels, these are used where long life is a requirement. Life potential of this wheel is 10 times that of aluminum. Steel wheels can be covered with *neoprene tires* and are used to reduce shock, prevent slippage, increase traction, prevent marring fragile surfaces, and reduce noise.
- *Aluminum wheels with steel or plastic ball bearings.* Used in applications where weight is a factor, particularly for portable conveyors. Plastic ball bearings should be used in corrosive atmospheres.
- *Nylon wheels.* Used where resistance to salt, water, and chemicals is required, as well as in applications where conveyors are cleaned frequently. Nylon wheels also do not mar or mark containers.
- *Polypropylene plastic wheels.* Have many properties which make them ideal for a wide range of applications. This material is highly impervious to a wide range of corrosive materials and is temperature-resistant from 230 to −30°F (110 to −34°C). The wheels do not absorb moisture and can be steam-cleaned repeatedly.
- *Hysteretic wheels.* Metal wheels surrounded by a tire formed by elastomeric material, these are used for line storage of heavy loads. The purpose is to absorb the energy of the initial load impact and to control the movement of the load to a safe speed.

LUBRICATION. Metal wheels can be greased, oiled, or dry. Nylon and plastic wheels are always furnished dry. Oiled or dry bearings should be used where high temperatures can cause thinning and leakage of grease. Dry bearings are recommended where temperatures are below 0°F (−18°C).

Special Types of Wheel Conveyors. These have been designed for handling specific products or for special industries. Samples of these are shown in Fig. 4.21.

Powered Conveyors. These are designed for continuous control of products on level surfaces, through inclines and declines, and around curves. Many powered conveyors are equipped with a computerized control system to provide tracking and diagnostics. Basic considerations for powered conveyors are the same as for gravity conveyors. Powered roller and belt conveyors are the major powered types used to move unit loads.

Powered Roller Conveyors. These are used mainly to accumulate loads because power-drive disengagement can be accomplished very simply when the forward motion of the unit

TABLE 4.6 Suggested Pitch for Gravity Conveyor

		Gravity roller conv., plain and dust-tight bearings			Gravity roller conv., pressure-lubricated bearings			Pi gravity wheel conv. or live rail conv.		
		Pitch per ft, pitch per 10′0″, and pitch per 90° curve								
Container being conveyed	Approx. cont. wt., lb	1″/ft	10′0″	90°	1″/ft	10′0″	90°	1″/ft	10′0″	90°
Cartons (fiber smooth bottom)	10	⅝	6¼	5				⅜	3¾	4
Cartons (fiber smooth bottom)	20	½	5	5				¼	2½	3½
Cartons (fiber smooth bottom)	45	⅜	3¾	3½	¾	7½	6	3/16	1⅞	3
Cartons (fiber smooth bottom)	100	¼	2½	3	⅝	6¼	5	¼–3/16		3
Fiber beverage carton empty	5							1½	15	10
Fiber beverage carton empty bottles	35	½	5	4–5	¾	7½	6	⅜	3¾	4
Fiber beverage carton filled bottles	45	⅜	3¾	3½	⅝	6¼	5	¼	2½	3½
Wood cases or boxes	5	⅝	6¼	5				⅜	3¾	5
Wood cases or boxes	10	⅜	3¾	4	¾	7½	6	¼	2½	5
Wood cases or boxes	25	¼	2½	4	⅝	6¼	5	¼	2½	4
Wood cases or boxes	50	3/16	1⅞	3	½	5	4	¼	2½	3½
Wood cases or boxes	100	3/16	1⅞	3	⅜	3¾	3½	⅛	1¼	2½
Steel drums	20	⅝	6¼	5½	¾	7½	6			
Steel drums*	55	7/16–½	4¼–5	5	⅝	6¼	5			
Steel drums	120	⅜	3¾	4	½	6	4½			
Steel drums	250	¼	2½	3	⅜	3¾	3½			
Steel drums*	550	¼–3/16	2½–1⅞	3	¼	2½	3			
Wood barrels	100	½	5	5	⅝	6¼	6			
Wood barrels	400	⅜	3¾	4	½	5	5			
Metal and fiberglass tote boxes	10	½	5	5				⅜	3¾	4
Metal and fiberglass tote boxes	25	⅜	3¾	4	¾	7½	6	¼	2½	3½
Metal and fiberglass tote boxes	50	¼	2½	3	⅝	6¼	5	3/16	1⅞	3
Milk cans empty	25	⅝	6¼	6	¾	7½	6½			
Milk cans full	105	⅜	3¾	4	½	5	5			
Milk crates empty	10	¾	7½	6	1	10	7½			
Milk crates empty bottles	45	½	5	5	⅝	6¼	5			
Milk crates full bottles	60	⅜	3¾	4	½	5	5			
Wood pallets smooth runner†	350	⅜	3¾		7/16	4¼				
Wood pallets smooth runner†	750	5/16	3⅛		⅜	3¾				
Wood pallets formica base	350	¼	2½		5/16	3½				
Wood pallets formica base†	750	⅛	1¼		3/16	2½				
Wood pallets pine plywood†	350	⅜	3¾		7/16	4¼				
Wood pallets pine plywood†	750	5/16	3⅛		⅜	3¾				
Wood beverage case empty	6	13/16	8⅜	6¾				¼	2½	3½
Wood beverage case empty bottles	30	½	5	5	¾	7½	6	⅛	1¼	2½
Wood beverage case full bottles	40	7/16	4¼	4	½	5	5	⅛	1¼	2½
Multiwall bags, firm	50							⅝	6¼	6
Multiwall bags, firm	100							½	5	5

* Varies with new or reconditioned drums.
† Depends on the way it is nailed and if banded or not. Bad nailing or banding will considerably increase the pitch and make a planned slope impractical.

load is stopped. Generally, power-drive disengagement is triggered when the unit load meets an obstruction, creating an opposite reaction which causes the carrying roller bushing to move up an angular slot, thus relieving the pressure and contact between the drive belt and rollers.

Powered roller conveyors are either chain- or belt-driven. Chain-driven units are used in heavy-duty applications and where oil or contaminants would have an adverse effect on the belting. The belt-drive-powered trains are designed for accumulation where the belt-to-roller pressure is very light or for transportation sections where belt-to-roller pressure is increased by center takeup rollers and the use of high friction drive belts.

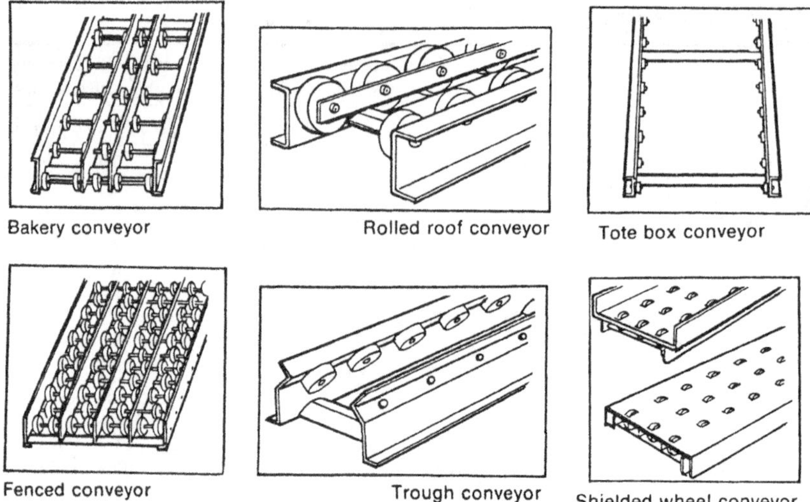

FIGURE 4.21 Special types of wheel conveyors.

Powered roller conveyors are not used for inclines greater than 5° since the contact force between the unit load and roller surface is not sufficient to overcome the gravity force due to a low coefficient of friction. This type of conveyor is not normally used over straight runs because of the higher cost as compared with belt conveyors.

Belt Conveyors. Belt conveyors consist of an endless moving belt which carries materials within a supporting frame. The belt can be made from a variety of materials and may or may not be equipped with cleats or other grabbing devices. The belt may be supported by a solid-slider-type bed of wood or metal or by rollers.

Conveyor manufacturers suggest friction surface belts on inclines up to 13°, and rough-top rubber belts should be used for inclines up to 25°. In applications where a steeper incline is required, heavily textured, nubbed or cleated surface belts can be used. Also, special requirements for belt surface material should be considered in applications where chemical or oil resistance is needed or for mandated cleanliness requirements.

Belt Conveyor Parameters. The parameters that must be defined before equipment is specified are belt speed, length, maximum load on belt at one time, tension loads, power requirements, and support and mounting hardware. Belt speed should be specified to be compatible with process equipment and other materials-handling hardware. Belt length should be adequate to accumulate the maximum expected product capacity. The maximum load on the belt at any one time can be calculated from the following formula:

$$P = \frac{K}{S \times 60 \text{ min}} \times C \text{ (ft)}$$

where P = maximum product weight
K = load per hour, lb
S = speed of belt, ft/min
C = center-to-center distance

For example, if

Load per hour in pounds = 10,000 lb

Speed of belt = 50 ft/min

Center-to-center distance = 36 ft

Then, the maximum product weight is

$$P = \frac{10{,}000}{50 \times 60} \times 36$$

$$= 1200 \text{ lb } (544 \text{ kg})$$

Considerations for Bulk-Materials Belt Conveyors. These considerations are similar to those for all conveyors; however, the properties of the materials to be moved affect the parameters of and the specification of the conveyor. The use of belt conveyors for bulk materials is limited by the characteristics of materials, some of which are:

- Stickiness, which may prevent materials from completely discharging from the conveyor or interfering with the power-train components.
- Temperatures that exceed 150°F (71°C) would cause deterioration or damage to most belt materials.
- Chemical reactions of conveyed materials with belt material. Some oils, chemicals, fats, and acids can damage belts.
- Large lump size becomes a factor and generally requires the system to be oversized for the amount of weight being moved.

Weight and friction are the common factors that determine the amount of incline that is possible for unit-load and bulk-materials-handling belt conveyors. Bulk-materials belt conveyors must also include materials characteristics such as size consistency, shape of lumps, moisture content, angle of repose, and flowability. The maximum angle of incline for various bulk materials is shown in Table 4.7. The ideal combination of belt width and speed (Table 4.8) is determined by the characteristics of the materials handled.

TABLE 4.7 Maximum Angle of Incline

Material carried	Maximum angle of incline, deg*	Material carried	Maximum angle of incline, deg*
Alumina, dry, free-flowing, ⅛″ lumps	10 to 12	Grain	8–16
Beans, whole	8	Ore (see stone)	15–20
Coal, anthracite	16	Packages	15–25
Coal, bituminous, sized,† lumps over 4 in	15	Pellets, depending on size, bed of material and concentricity (taconite, fertilizer, etc.)	5–15
Coal, bituminous, sized,† lumps 4 in and under	16	Rock (see stone)	15–20
Coal, bituminous, unsized†	18	Sand, very free-flowing¶	15
Coal, bituminous, fines, free-flowing‡	20	Sand, sluggish (moist)§	20
Coal, bituminous, fines, sluggish§	22	Sand, tempered foundry	24
Coke, sized	17	Stone, sized, lumps over 4 in	15
Coke, unsized	18	Stone, sized, lumps 4 in and under, over ⅜ in	16
Coke, fines and breeze	20	Stone, unsized, lumps over 4 in	16
Earth, free-flowing‡	20	Stone, unsized, lumps 4 in and under, over ⅜ in	18
Earth, sluggish§	22	Stone, fines ⅜ in and under	20
Gravel, sized, washed	12	Wood chips	27
Gravel, sized, unwashed	15		
Gravel, unsized	18		

* For ascending conveyors when uniformly loaded and with constant feed.
† See third footnote (‡) to Table 4.8 for definitions of "sized" and "unsized" as used in Material Carried columns of Table 4.7.
‡ Angle of repose 30 to 45°.
§ Angle of repose greater than 45°.
¶ Very wet or very dry, with angle of repose less than 30°.

TABLE 4.8 Recommended Belt Speed as Determined by Material Handled

Material		Recommended belt speed, ft/min*†												
		Belt width, in†												
Characteristics‡	Example	14	16	18	20	24	30	36	42	48	54	60	72	84
Maximum size lumps, sized or unsized														
Mildly abrasive	Coal, earth	300	300	400	400	450	500	550	600	600	650	650	650	650
Very abrasive, not sharp	Bank gravel	300	300	400	400	450	500	550	550	600	600	600	600	600
Very abrasive, sharp and jagged	Stone, ore	250	250	300	350	400	450	500	500	550	550	550	550	550
Half max. lumps, sized or unsized														
Mildly abrasive	Coal, earth	300	300	400	400	500	600	650	700	700	700	700	700	700
Very abrasive	Slag, coke, ore, stone, cullet	300	300	400	400	500	600	650	650	650	650	650	650	650
Flakes	Wood chips, bark, pulp	400	450	450	500	600	700	800	800	800	800	800	800	800
Granular ⅛″ to ½″ lumps	Grain, coal, cottonseed, sand	400	450	450	500	600	700	800	800	800	800	800	800	800
Fines														
Light, fluffy, dry, dusty	Soda ash, pulverized coal						220–250 ft/min							
Heavy	Cement, flue dust						250–300 ft/min							
Fragile, where degradation is harmful	Coke, coal											200–250 ft/min		
	Soap chips						150–200 ft/min							

* Normal for belts traveling horizontally on ball- or roller-bearing idlers. For picking belts, speed is usually 50 to 100 ft/min. Belts with discharge plows should not travel faster than 200 ft/min. For trippers, recommended speed is 300 to 400 ft/min. Trippers for higher-speed applications can be furnished. A speed of at least 300 ft/min should also be maintained for proper discharge when using 35 to 45° idlers, and also for materials tending to cling to belt.

† 1 in = .02254 m; 1 ft/min = .00508 m/s.

‡ *Unsized* means a uniform mixture of material in which not more than 10 percent are lumps ranging from maximum size to ½ maximum size, at least 15 percent are fines or lumps smaller than 1/10 maximum, and the remaining 75 percent are lumps of any size smaller than ½ maximum. *Sized* means a uniform mixture in which not more than 20 percent are lumps ranging from maximum size to ½ maximum size, and the remaining 80 percent are lumps no larger than ½ maximum size and no smaller than 1/10 maximum size.

Metal-Belt Conveyors. Similar in design to standard belt conveyors, these differ in that their surface is a belt of woven or solid metal. The materials include carbon steel, galvanized steel, chromium stainless steel, and other metals or special alloys that are required for a specific application and environment. Wire belts are also available for use where processing temperatures vary from 320 to 2500°F (160 to 1416°C). Wire-belt conveyors are used primarily to move product or unit loads through processes that include liquid or chemical treatment, heat treatment, or kiln firing operations. Wire belts can be cleaned or sterilized while in motion. Mesh openings in the belt permit circulation of water, gases, heat, and cooling air. Typical uses for metal-belt conveyors include operations such as spray-washing glass containers, moving baked goods to ovens, conveying cathode-ray tubes through various processes, and moving hot forgings from automatic die casting equipment.

Tracking of a wire-mesh belt is a problem since the belt is formed by a number of different sections joined together and a wide range of temperatures used in processing operations causes expansion and contraction of the belt material. The conveyor specification must frequently include one of the following features to compensate for these conditions and assure straight belt tracking.

- Multitooth sprocket belt drive
- Belt aligners, consisting of pulleys or rollers mounted on the supporting frame

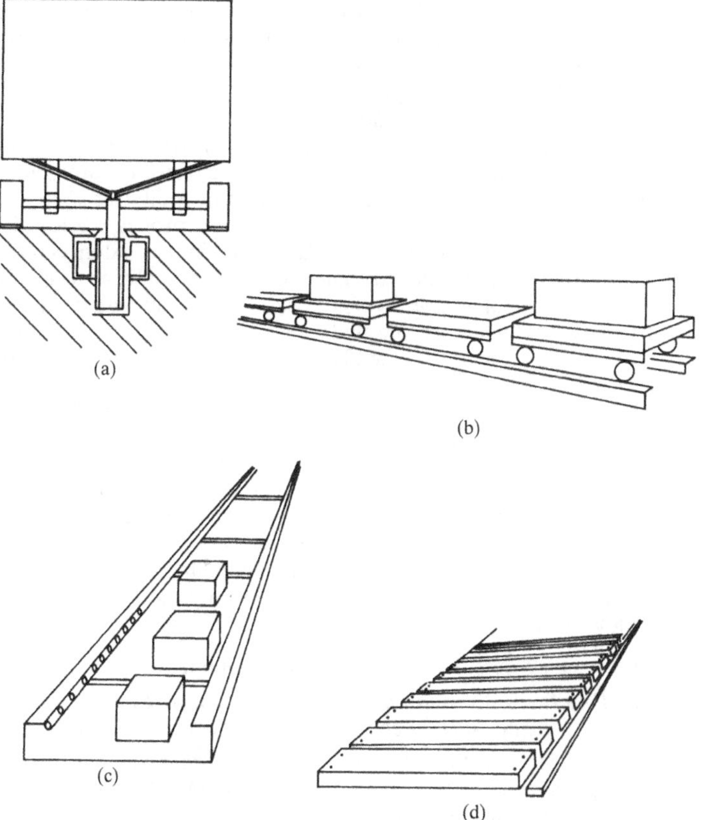

FIGURE 4.22 Types of conveyors. (*a*) In-floor towline, (*b*) trolley conveyor, (*c*) pusher bar conveyor, (*d*) slat conveyor.

- Self-tracking belt that has V-shaped wires on the underside that run through grooved driver drums

Surface Chain Conveyors. Surface chain conveyors (Fig. 4.22) include sliding chain, pusher bar, slat, and tow types and car-type trolley conveyors.

Sliding Chain Conveyors. These are the simplest type since they use the chain itself to convey packages down two sliding tracks. The conveyors are used to handle heavier loads than can belt conveyors, such as loaded pallets or unitized loads, but they have the same incline limitations as belt and powered-roller conveyors.

Pusher Bar Conveyors. These are able to convey loads up steeper inclines (to 45°) because the load is pushed by a car connected to the chain drives and this arrangement moves the load along a metal bed or trough. Pusher bar conveyors are generally used for floor-to-floor movement in multistory warehouses or facilities.

Slat Conveyors. These employ an endless chain to drive a conveyor surface of nonoverlapping, noninterlocking slats made of wood or metal. Slat conveyors can be used as moving work tables and to move heavy unit loads and are ideal for applications where the conveyor surface must be flush with a work station or with a floor surface. In the latter application, the installation will permit industrial trucks to cross over or be carried on the slat surface. Slat conveyors can be operated on inclines or declines, the angle of which is limited by the friction between slat surface and the load. Cleats may be added to support loads where steeper inclines are required.

Towline Conveyor. A towline conveyor uses an endless chain either supported from an overhead track or running in a track below the floor to tow trucks, dollies, or carts. The in-floor towline system is the type commonly found in warehouses. Existing building floors can be fit with towline by digging up the floor or mounting a rail on top, although it is preferable to install the system in the floor at the time the building is built. The recessed track allows the use of floor space for other equipment, but the track cannot easily be relocated once installed. It is truly versatile as it can be looped around storage areas, moved down aisles, and forked into spur sections for loading, unloading, and empty car storage. Switching onto spurs can be accomplished through the use of magnetic probes, radio-frequency signals, scanners, or mechanical switching. Carts and trucks used in this system can range from the ordinary pallet truck fitted with tow pins to engage with the chain drive to special carts or trucks.

Car-Type Trolley Conveyors. These employ an endless chain to pull a series of small cars or trolleys which carry the material to be moved. These often have fixtures to be used in assembly lines or contain molds for use in foundry processing operations.

Overhead Conveyors. Overhead conveyors include both trolley conveyors and power and free types of equipment. These conveyors are supported and function within a trolley track driven by a chain power drive to move parts or product. The path of the conveyor can be straight, inclined, declined, and around corners; it can make optimum use of building geometry and follow the general work flowpath within the limitations of building constraints and equipment design parameters. Conveyors can be supported independently or attached to existing beams and trusses, depending on the load factors involved.

In order to determine equipment design parameters, the following procedure should be followed.

1. From process flowcharts, determine all operations to be serviced by the conveyor.
2. Determine the path of the conveyor on a scaled plant layout (Fig. 4.23), showing all obstructions such as columns, walls, machinery, and work aisles.
3. Develop a vertical elevation view to determine incline and decline dimensions (Fig. 4.24). At this point a three-dimensional view could be prepared to give a multiplan view of the proposed installation.

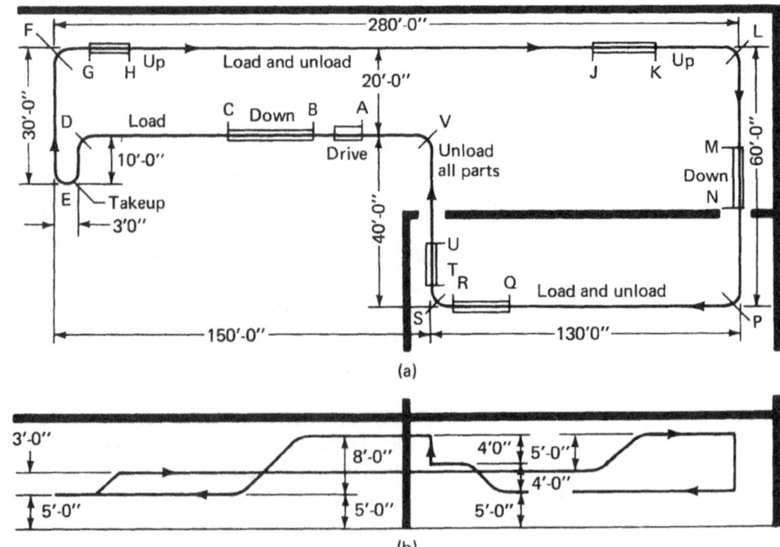

FIGURE 4.23 Plan and vertical elevation views of trolley conveyor system.

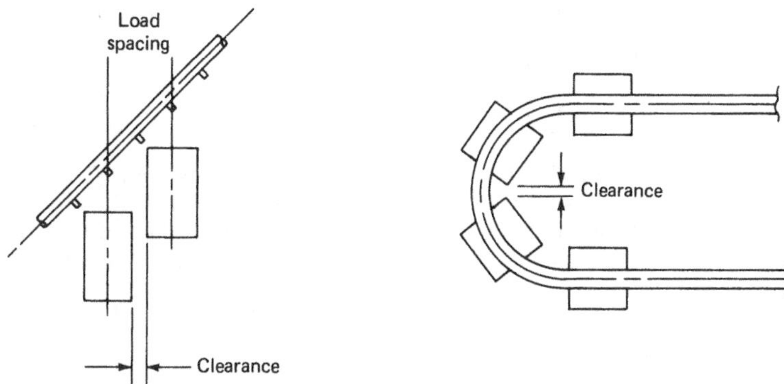

FIGURE 4.24 Load-spacing considerations for overhead conveyors.

4. Determine the movement rate, unit load size, spacing, and carrier design.
5. Modify turn radii to provide clearances that allow for desired clearance (Fig. 4.24) on turns.
6. Modify loading spacing to provide clearance on inclines and declines. As inclines and declines get steeper, load spacing has to be increased to provide a constant clearance or separation between loads. Table 4.9 shows the load spacing required for a given separation at various incline angles.
7. Redraw the conveyor path and vertical elevation views using new radii and incline information.
8. Compute the chain pull, which is the total weight of the chain, trolleys, and other components, as well as the weight of the carriers and load. For example, for a given system the tentative chain pull is calculated as follows:

TABLE 4.9 Load Clearance on Inclined Track for Overhead Conveyors

Load spacing, in	Incline angle, deg											
	5	10	15	20	25	30	35	40	45	50	55	60
	Horizontal centers, in*											
12	12	11⅞	11⅝	11¼	10⅞	10⅜	9⅞	9¼	8½	7¾	6⅞	6
16	15⅞	15¾	15½	15⅛	14½	13⅞	13⅛	12¼	11⅜	10⅜	9¼	8
18	18	17¾	17⅜	17	16⅜	15⅝	14¾	13⅞	12¾	11⅜	10⅜	9
24	24	23⅝	23¼	22⅝	21¾	20⅞	19¾	18⅜	17	15½	13¾	12
30	29⅞	29⅝	29	28¼	27¼	26	24⅝	23	21¼	19⅜	17¼	15
32	31⅞	31½	31	30⅛	29	27¾	26¼	24½	22⅝	20⅝	18⅜	16
36	35⅞	35½	34¾	33⅞	32⅝	31¼	29½	27⅝	25½	23⅛	20⅝	18
40	39⅞	39⅜	38⅜	37⅝	36¼	34⅝	32¾	30⅝	28¼	25¾	23	20
42	41⅞	41⅜	40⅜	39½	38⅛	36⅜	34⅜	32¼	29¾	27	24⅛	21
48	47⅞	47¼	46⅜	45⅛	43½	41⅝	39⅜	36¾	34	30⅞	27⅝	24
54	53⅞	53¼	52¼	50¾	49	46¾	44¼	41⅜	38¼	34¾	31	27
56	55⅞	55⅛	54⅛	52¾	50¾	48½	45⅞	42⅞	39⅝	36	32⅛	28
60	59¾	59⅛	58	56⅜	54⅜	52	49⅛	46	42½	38⅝	34½	30
64	63¾	63	61⅞	60⅛	58	55½	52½	49	45¼	41⅛	36¾	32
72	71¾	70⅞	69⅜	67¾	65¼	62⅜	59	55¼	51	46¼	41⅜	36
80	79¾	78⅞	77¼	75¼	72½	69⅜	65½	61⅜	56⅜	51½	45⅞	40

* 1 in = .02254 m.

$$\text{Total tentative chain pull} = 700 \times 60.0 \times 0.03 = 1260$$

where 700 = conveyor length, ft
0.03 = coefficient of friction, %
60.0 = 10.0 lb/ft (chain and trolleys) + 12.5 lb/ft (carriers) + 37.5 lb/ft (line load)

For this initial calculation, inclines and declines are assumed to be level sections if the number of declines balances out the number of inclines; however, for each additional incline, the weight has to be added to determine the total chain pull. If, in our example, a vertical incline that raises a load 8 ft is required, then the additional chain pull is

$$37.5 \text{ lb} \times 8\text{-ft lift} = 300 \text{ lb } (136 \text{ kg})$$

The total chain pull then becomes 1260 + 300 lb = 1560 lb (707 kg).

9. Select tentative conveyor size based on trolley load and chain pull.

10. Select vertical curve radii.

11. Determine power requirements and drive locations. This requires a point-to-point calculation of chain pull around the complete path of the conveyor, which is shown in Fig. 4.23. The following three formulas are used to compute point-to-point chain pull.

a. Pull for each straight horizontal run:

$$P_H = XWL$$

where $X = 0.02$ for standard ball-bearing trolleys
W = total moving weight, lb/ft (empty or loaded, as the case may be)
L = length of straight run, ft

b. Pull for each traction wheel or roller turn:

$$P_T = YP$$

where $Y = 0.02$ for traction wheel or roller turn and P = pull at turn, lb.

c. Pull for each vertical curve:

$$P_v = XWS + ZP + HW(1 + Z)$$

where $X = 0.02$ for standard ball-bearing trolleys
W = total moving weight, lb/ft
S = horizontal span of vertical curve, ft
H = total change of level of conveyor, ft (plus, when conveyor is traveling up the curve; minus, when conveyor is traveling down the curve).
$Z = 0.03$ for 30° incline; 0.045 for 45° incline; 0.06 for 60° incline; 0.09 for 90° incline
P = pull at start of curve, lb

Drive horsepower may be calculated from the following formula:

$$\text{Drive hp} = \frac{\text{drive capacity (lb)} \times \text{maximum speed}}{33{,}000 \times 0.6}$$

12. Design conveyor supports and superstructures.

13. Design guards which are required by federal, state, and other codes under high trolley runs, particularly over aisles and work areas. Guard panels are normally made from woven or welded wire mesh with structural angles and channels to suit the size and weight of the material being handled.

Power and Free Conveyors. These consist of two separate trolley systems: one moves and is powered by a chain drive; the other has a track under the powered track that accommodates a free-moving trolley containing a carrier from which a load is suspended (Fig. 4.25). In the

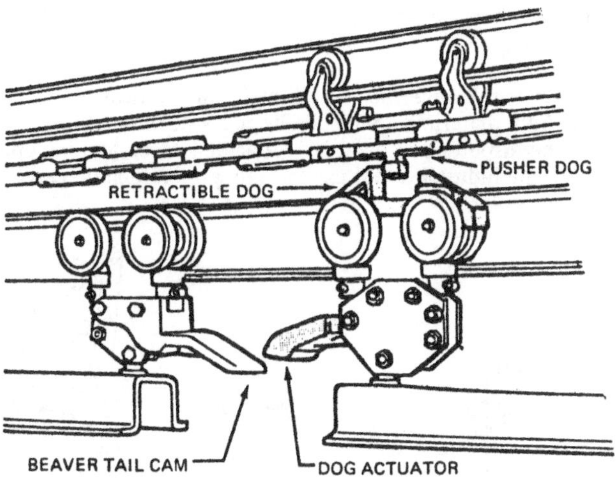

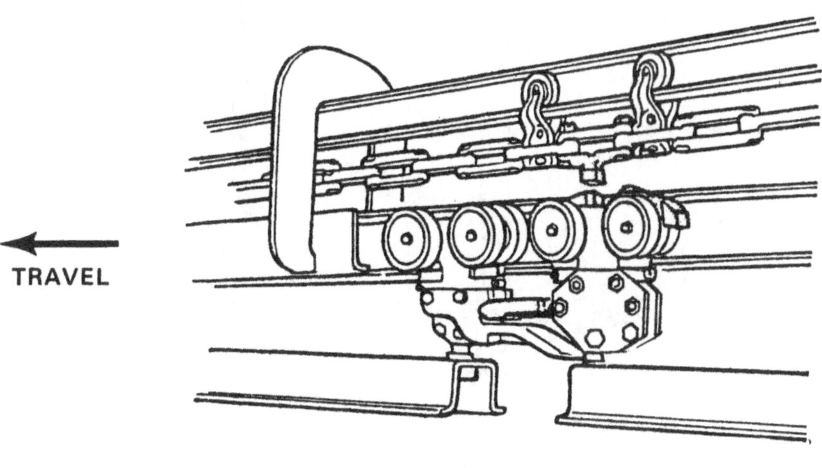

FIGURE 4.25 Power and free trolley.

powered mode, the powered trolley is engaged with the free trolley through contact of a pusher dog on the powered system to a retractable dog on the free system. Disengagement is accomplished by contact with another load or by actuating the dog actuator. The system is extremely flexible since each carrier can be stopped or started without interrupting the system. This conveyor can be utilized in a process where operation times vary, or where units need to be accumulated into a batch before the next operation begins.

Two variations of the power and free conveyor, the inverted and side-by-side, have been developed recently to overcome some limitations of the original. To decrease the amount of vertical space needed for the two trolleys, a side-by-side system was created, where the powered trolley is located next to the load trolley. The inverted power and free conveyor mounts the tracks on the floor, with the free track above the powered track. This configuration permits work to be moved at assembly levels.

The same design criteria apply to power and free conveyors as to other chain-driven trolley systems.

Bulk-Materials Vertical Conveyors. Bulk-materials vertical conveyors (Fig. 4.26) are generally used to lift bulk materials up to silos, hoppers, or other storage containers from which the material may be dispensed into a mixing, packing, truck-loading operation, or directly to a process. Some of the industries that use this equipment include glass, agricultural fertilizer, and powdered chemicals.

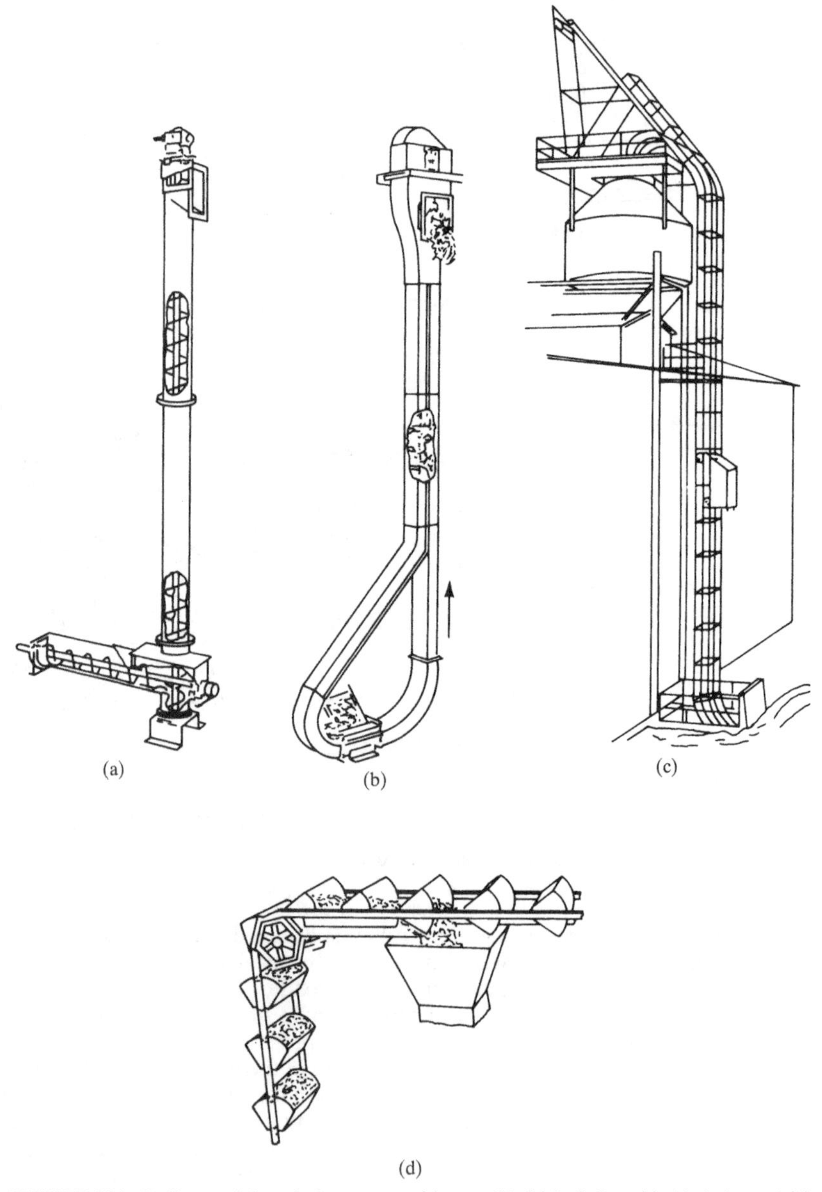

FIGURE 4.26 Bulk material vertical conveyors: (a) rotor lift, (b) bulk flow, (c) skip hoist, and (d) gravity-discharge conveyor elevator.

Skip Hoists. These are used to lift bulk materials handled in batches to very high points. A bucket which carries the material moves vertically in guides and is raised and lowered by a hoist-operated cable.

Gravity Discharge Conveyor-Elevators. These carry material in both horizontal and vertical paths. The buckets are rigidly mounted on two strands of chain running in tracks. Material is loaded into a bucket at the base of the equipment by feeding material into a lower trough, and discharge is effected when the bucket position changes in the horizontal run.

Bulk-Flos. These lift material by the use of flights attached to a chain drive which is contained in a dust-tight casing. Bulk-Flos are self-feeding and -discharging and lend themselves to continuous bulk-material processes.

Rotor Lifts. These are similar to screw conveyors but are mounted vertically to effect the lifting of bulk materials and are contained in a dust- and weatherproof casing. Screw feeders or conveyors are generally used to deliver material to rotor lifts.

Other Specialty Conveyors. There are innumerable variations on standard conveying systems, some of which are unique to individual industries. Six common examples are described below.

Screw Conveyor. This conveyor (Fig. 4.27) consists of a screw rotating in a stationary trough and the material moving along its length by rotation of the screw. This type of conveyor serves a dual purpose since it can also be used to perform processes such as blending and mixing of material while the material is being moved. The conveyor is generally enclosed

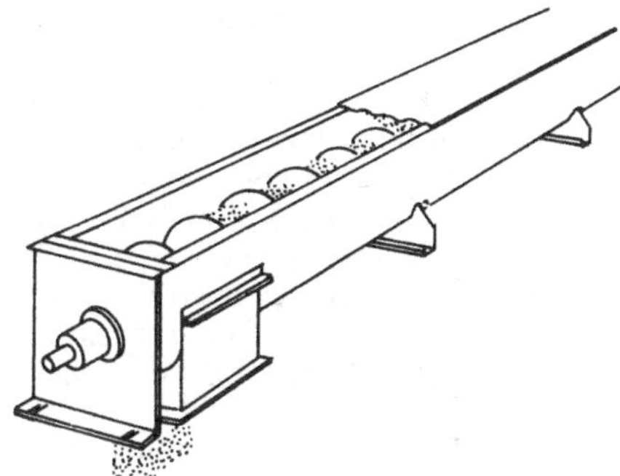

FIGURE 4.27 Screw conveyor.

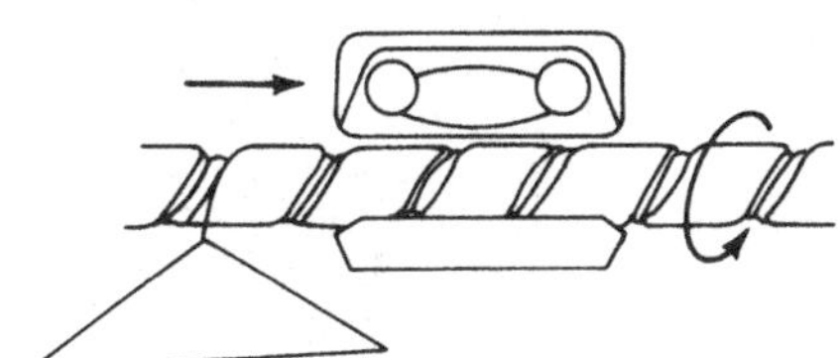

FIGURE 4.28 Spiral track conveyor.

to prevent dust or fumes from escaping and allow the conveyor to be cooled or heated. Loading or discharging can be located at any point along a conveyor.

Spiral Track Conveyors. These conveyors (Fig. 4.28) consist of a continuous spiral track with a power drive which turns the track, moving anything which is hung on it. It has wide application in the garment industry. It is generally used for items weighing less than 10 lb (5 kg). Interlocking nylon wafers can permit turns to be made in any direction on a radius of 18 in (46 cm).

Oscillating and Vibrating Conveyors. These use the natural frequency vibration of a trough to provide a conveying action to move material. Oscillating conveyors use a mechanically driven power train to move a trough carrying material against spring supports which provide a fast return and downward stroke, causing the trough to vibrate and convey the material. Vibrating conveyors utilize some form of magnetic pulsation to create this vibration motion. Wider variations of frequency are possible by simple control for vibrating conveyors, enabling speed changes compensating for material differences.

Application of both types of conveyors is growing in a number of different industries for uses such as: conveying light food products such as cereal in the food industry; moving, cooling, and breaking up lumps of casting sand in foundries; quenching and removal of glass cul-

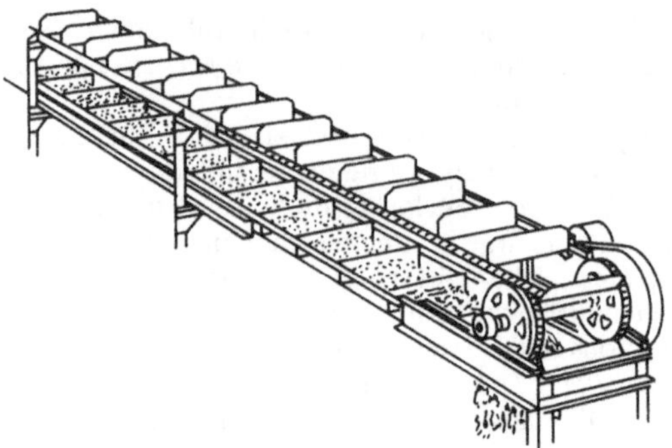

FIGURE 4.29 Flight conveyor.

let in water-filled troughs in the glass industry; removing ferrous from nonferrous materials in separation systems; and feeding small parts into automatic packaging or assembly equipment.

Flight Conveyors. These conveyors (Fig. 4.29) use scraper plates to push nonabrasive bulk material through a trough which can be horizontal or inclined.

Apron Conveyors. These conveyors use a series of interlocking apron pans supported in a stationary frame for conveying materials that are heavy, abrasive and lumpy, such as ore, stone, industrial refuse, and waste materials.

Pneumatic Tubes. These use a pressure or vacuum system to move materials or a container at relatively high speed. The major application is that of an internal mail carrier, although it can also be used to move certain types of high-volume fine particulate.

Monorails. A monorail is usually an overhead system on which carriers transport materials from one point to another on a track. Unlike trolley conveyors, each carrier is independent and the system can be powered or unpowered. A powered system, or automated electrified monorail (AEM), requires two rails like power and free, but each carrier is equipped with an electric motor that draws energy from an electrified rail. The monorail system can be designed with spurs, and carriers can travel forward or backward, eliminating the need for a closed loop. This equipment can also travel small vertical inclines without any assistance, and can be equipped with a drive chain to climb steeper slopes.

An AEM is suitable for use in most industries since monorail carriers can carry up to 10,000 lb (4545 kg) and travel up to 600 ft/min (3 m/s). For example, a transmission assembly plant uses monorails in all stages of assembly and testing, with assembly occurring directly on the carrier. In an AEM system, the carriers often have sensors or microprocessors on board to communicate to a central computer about its position, type of load, and even diagnostic data. Generally, monorails are given only the intelligence that is necessary so that factories beginning to automate can incorporate monorails into their system. As expansions are necessary, carrier and/or additional track can be added without disrupting the existing process.

SORTING, CONSOLIDATING, AND DIVERTING DEVICES

A materials-handling system must frequently have the ability at some point to identify, sort, and divert parts, products, or unit loads. Peripheral accessories and equipment do this, ranging from simple mechanical diverters to sophisticated optical recognition reading devices, which can actually read and identify alphanumeric characters and sort 20,000 items per hour and which are used mainly for check and mail handling. Whatever the complexity of the system, three basic elements must be considered: identification of the item to be sorted or consolidated, recognition of the item, and the command to activate the mechanisms to divert the item.

Simple Mechanical Sorting. Simple mechanical sorting utilizes inherent differences such as size, shape, weight, or other physical differences to identify or recognize items; it generally is

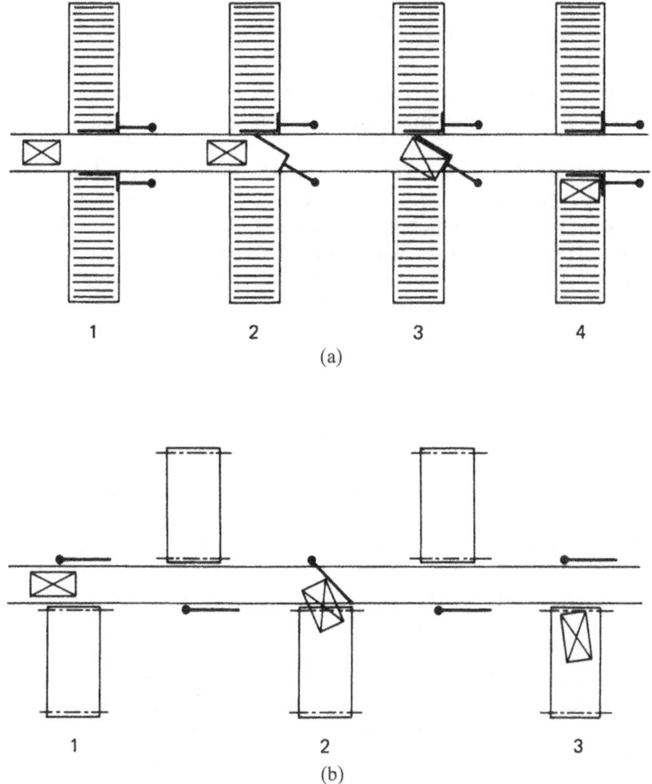

FIGURE 4.30 Diverting mechanisms.

contact sorting in which an item must make contact with a channel or feeler guides or discerns physical differences and contacts a cam or other simple mechanism to activate a diverter chute or other diverting device.

Diverting Mechanisms. Diverting mechanisms can be grouped into devices that deflect, push off, drive off, or tilt; many variations are included within each group (Fig. 4.30).

Electromechanical Sorting. Electromechanical sorting uses noncontacting identification devices that can sense both inherent differences and applied differences. These are identified on the load or package by a code that can be discriminated by a sensing or scanning device and that triggers a diverting mechanism.

Photosensors. These are the most commonly used sensing and scanning devices. A photoelectric control consists of a light source, photoreceiver, amplifier, and output. A beam of light from the light source activates the photosensitive elements of the photoreceiver which produces an electric signal, which drives a relay to activate the diverting mechanism. When used as a sensor, photoelectric controls can be used to:

Sense the presence or absence of containers or products on a conveyor

- Detect over- and undersized products
- Sort products by size
- Count items

Photoelectric controls can also be arranged in an array or ganged in a manner that a code on a container can be read. Each sensor that goes into the scanning device will detect a specific mark or blank space from the code which when scanned produces binary information into a logic function of a controller. This in turn supplies a signal for the diverting mechanism. Typical codes are ladderlike in format, and this allows a scanning device to read the code vertically.

Automatic Identification. Automatic identification refers to the ability to track materials in the factory through the use of semi- or fully automated technologies. Most common is bar code technology, which employs a scanner swept across the bar code and has been in use for 20 years. The bar code, which is a series of black bars of varying thickness and approximity, is used to identify the product. The human eye is not able to decipher the bar code, but a scanner can identify the product easily, since bar codes were designed for computers. Scanners can be handheld wands, laser guns, or a moving-beam laser scanner. A handheld wand must touch the bar code whereas several inches may separate the laser gun and code. Moving-beam laser scanners are ideal for reading bar codes on items being transported on a conveyor. The scanner is mounted in a location where it can emit a beam searching for the bar code.

Optical character recognition (OCR) is an alphanumeric character set which can be read by a special scanner, and can be read by humans. However, OCR technology is less reliable and more costly than bar coding. A compromise between the two is placing a label next to the bar code. An emerging technology is radio-frequency identification devices (RFIDs), which use a read-only tag fastened to a unit load to emit radio signals. When the signals are received by the computer, the contents of the load are identified. Some companies are experimenting with installing RFIDs on the carrier of a trolley and bar codes on the unit load to integrate tracking of materials and conveyor systems.

Palletizers. Palletizers receive individual packages, cases, or bags from a conveyor and automatically arrange them on a pallet in a predetermined pattern with the required number of tiers. Each tier need not have the same predetermined pattern. Generally, units to be stacked are received on a control belt at the entrance to the machine. At this point the unit will be counted and oriented depending on the patterns required. As each row is completed, a pusher moves the cases onto an apron. When the tier is completed, the apron is withdrawn, depositing the tier on the pallet or tier below. The operation is repeated until the pallet load is completed, when it is discharged and replaced with an empty pallet.

Large volumes of standard units are required, and it is estimated that palletizers become economical when approximately 900 units per hour require palletizing. The larger palletizers can handle in excess of 6000 cases per hour of certain products.

Depalletizers. Depalletizers are highly specialized pieces of equipment which automatically depalletize cartons and cases. Automatic squaring mechanisms permit the handling of loose pallet loads. Depalletizers generally operate in the range of 3500 cases per hour and are seen primarily in beverage distribution.

HOISTS AND CRANES

Hoists and cranes are materials-handling equipment used to move varying loads intermittently within a fixed area. The loads vary in size and weight and are not uniform. Most of the materials movement is devoted to raising and lowering loads, although some units are so constructed as to permit them to travel laterally over a specific area. The types of hoists, cranes, and attachments are listed below.

Hand and Powered Hoists. Hand and powered hoists (Fig. 4.31) are the most basic and economical lifting equipment which enables an operator to move a large load, up to 50 tons (45,360 kg), vertically by using some kind of mechanical advantage.

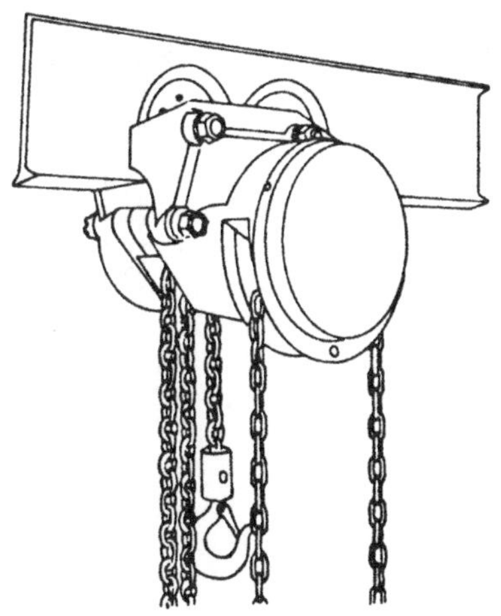

FIGURE 4.31 Hoist.

Jib Cranes. Jib cranes (Fig. 4.32) consist of a hoist that is mounted on a boom track. The hoist mechanism can be moved laterally in the track and the boom can be turned in an arc limited by the building restrictions or the mounting arrangement of the boom. Jib cranes are classified into basic groups of bracket jib, cantilever jib, and pillar jib. Load capacities range from small manually operated cranes to loading towers that exceed 300 tons (272,160 kg).

Bridge Cranes. Bridge cranes consist of a hoist mounted on a guider bridge which is supported by two trucks on each end and rides on runways supported by building members. Top-running bridges, where end trucks ride on top of runway tracks, are able to support a total bridge and load weight of hundreds of tons, but underhung or bottom-running bridges, where the trucks are suspended from the lower flanges of the runway track, normally are used for loads less than 20 tons (18,144 kg). Bridge cranes can be operated manually or powered or, in the cases of very large cranes, can be operated by remote control (Fig. 3-18).

Gantry Cranes. The gantry crane is very similar to a bridge crane except it is supported by self-contained vertical support members that travel in tracks on the floor surface and it is generally used where overhead runways are not feasible due to building restrictions. The gantry crane system also has the advantage of being usable in outdoor operations without the construction of an expensive supporting structure (Fig. 4.32).

Stacker Cranes. The stacker crane consists of a rigid mast suspended from an overhead bridge that travels laterally. A platform or a set of forks moves up and down on slider bars to lift and lower loads. The stacker crane is most commonly used to place or retrieve loads to and from racks from both sides of an aisle. In automatic storage and retrieval systems, the stacker crane is computer-controlled. The computer has the rack location of each item stored in the memory and is able to command the load-carrying platform to a specific location for storage and retrieval of a load.

Lifters. A lifter (Fig. 4.33) is an attachment suspended from the load hook of a hoist or crane that permits a load to be handled more easily or quickly than possible with a hook, and many load configurations cannot be handled with a hook. In many cases, lifters are designed for a specific application, but there are many standard types that are available for a wide range of applications. Lifters are categorized by the method in which the load is carried.

Supporting Lifters. These carry the load on the surface of the lifter, on bearing surfaces of cradles, or hooks and slings attached to lifters.

Clamping Lifters. These hold the load by surface friction or by squeezing load.

Surface-Attaching Lifters. These consist of both magnetic or vacuum types. Magnetic lifters can use either a permanent magnet that requires a strip-off device to release the load or an *on-off magnet* that can be activated by applying a voltage. Vacuum pads can be used to lift loads with nonporous and smooth surfaces and are commonly used to handle glass and aluminum.

Manipulating Lifters. These move the load through one or more axes for operations such as positioning or dumping.

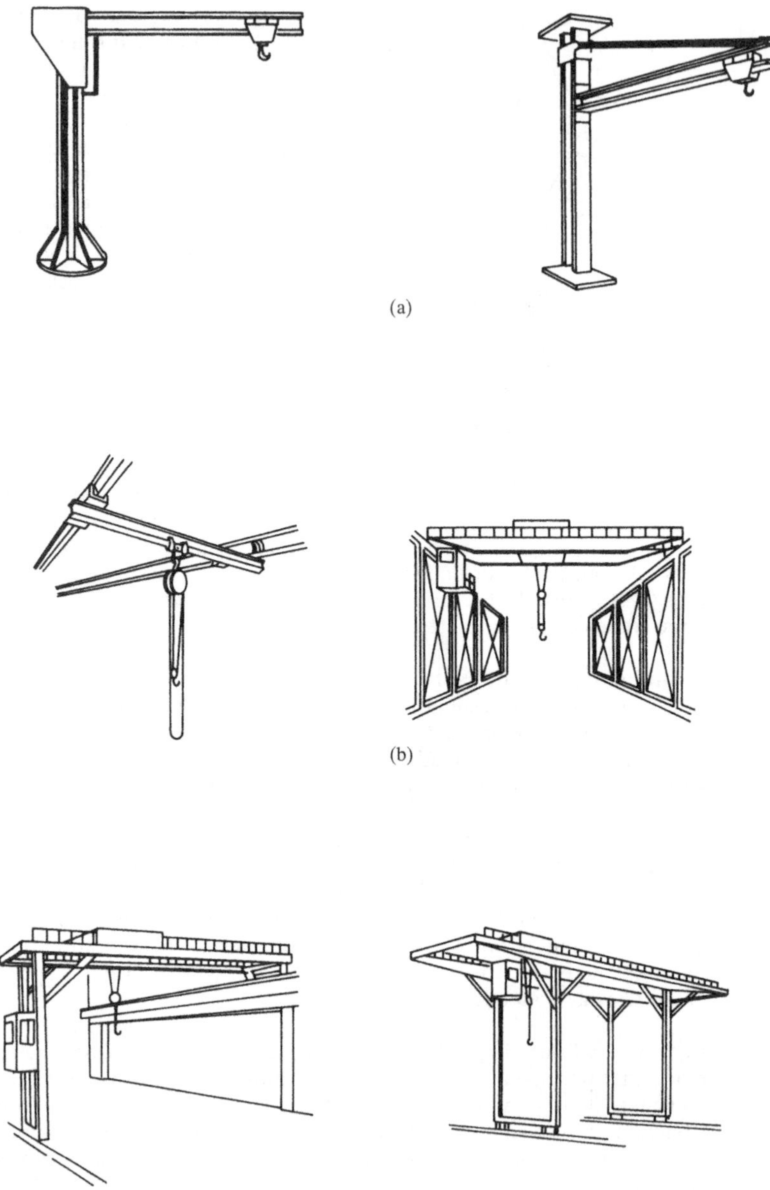

FIGURE 4.32 Types of cranes. (*a*) Jib crane, (*b*) bridge crane, (*c*) gantry crane.

AUTOMATIC GUIDED VEHICLES

Automatic guided vehicles move material over fixed paths but do not require the use of an operator or a mechanical drive train located below the floor surface or an overhead towline. They are useful when a variety of materials must be moved over long distances to and from a variety of fixed destinations. There are three identifiable types of vehicles: first, the driverless

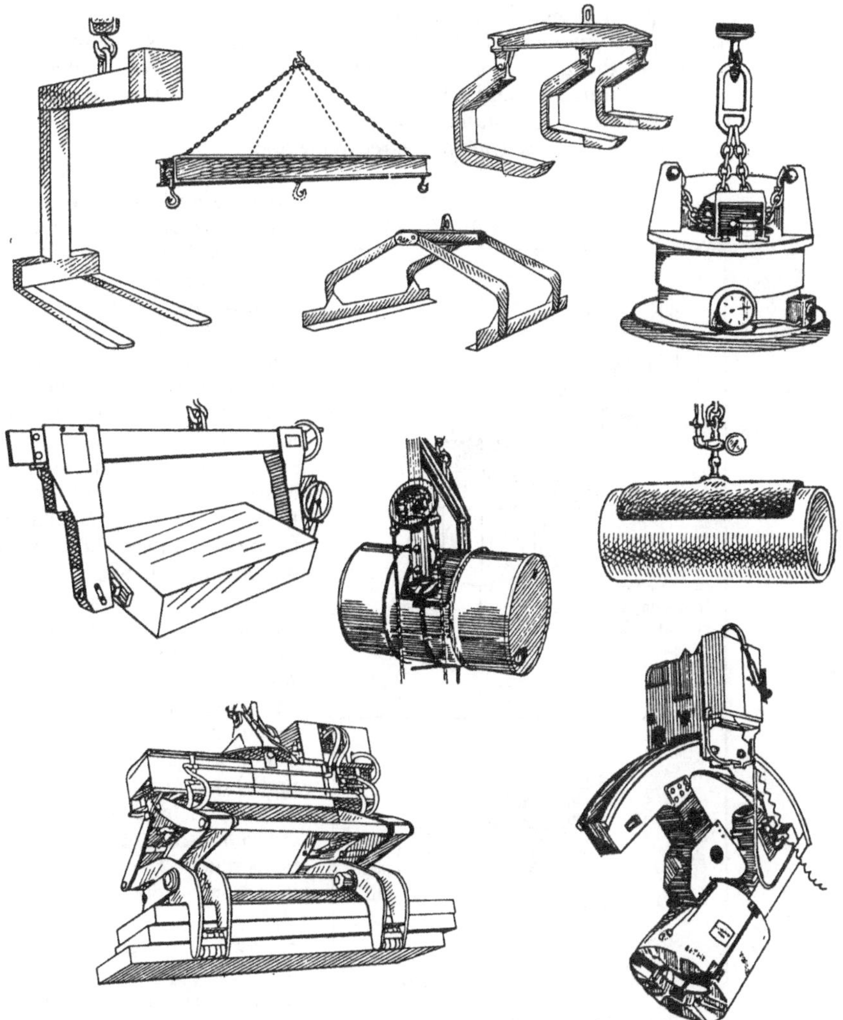

FIGURE 4.33 Types of lifters. (*Reproduced with permission from* Material Handling Engineering Handbook and Directory, 1977/1978, *published by Material Handling Engineering, Cleveland.*)

tractor (Fig. 4.34) which hauls trailers or cartloads of material; second, the individual unit-load or pallet mover (Fig. 4.35); and third, the multishelved self-contained vehicle. The last type is used primarily to move mail in office buildings or for food and supply deliveries in hospitals.

Guidance and Control Systems. Guidance and control systems are similar for all three systems. Two systems are used: optical, where the unit follows a line taped or painted on the floor surface; or magnetic, where a thin wire is set in a shallow channel sealed over in the floor. This latter system is less flexible and more costly to control but is not subject to obliteration or wear, which can be a problem in certain factory environments.

The driverless tractor, being unable to reverse on its own trailers, generally requires a closed-loop system. However, multiple-loop systems can be used. Unit-load movers are generally reversible and can operate on a spur.

The programming information which determines the paths and stops can be preset on the tractor programmer or can be controlled from a central dispatching point. These systems generally have the logic to allow the tractor to take the shortest route to the destination without traveling through the entire loop. Radio-control transmitters are often used to reposition the train within a loading station, eliminating unnecessary walking in operations such as order picking or loading the train at the receiving dock.

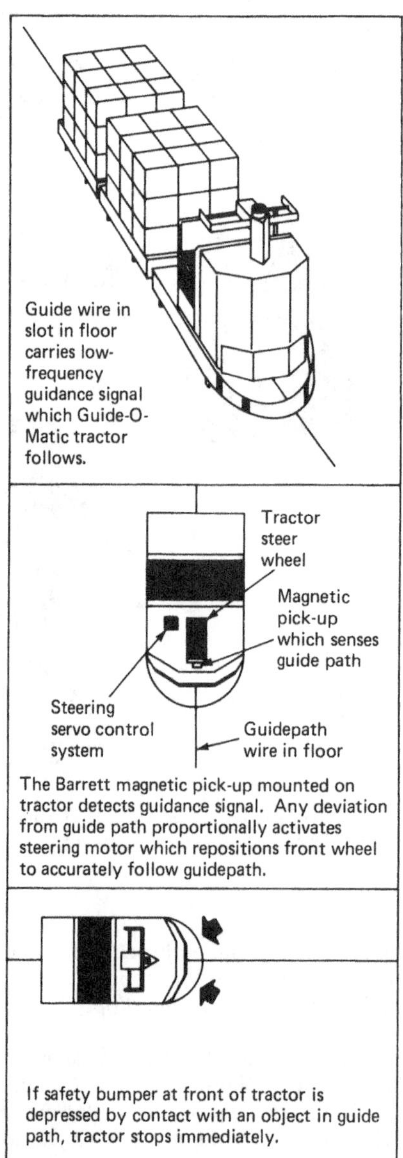

FIGURE 4.34 Typical features of a driverless tractor system.

Loading and Unloading. Although all vehicles can be loaded and unloaded with operator assistance, both tractors and unit-load movers can have automatic load and unload features. The tractor-trailer arrangement can have an automatic uncoupling option. More common are options whereby the trailers have rollers on the carrying surface and the loading-unloading stations where a pusher can be used to move the load. Similar systems can be used for unit-load movers, sometimes using powered roller systems. More common is the lifting device established in Fig. 4.35. This has particular potential in manufacturing operations where materials can be brought directly into the work station.

Routes are dependent on surface conditions. Cracked and broken slab can cause discontinuity in the tape or wire guides. Inclines and declines within a plant must be considered, in which case an acceleration or deceleration feature must be specified for the equipment. External routes linked with automatic door control, internal traffic lights, and automatic ramps to cover rail lines have been used. However, external use of this equipment is not widespread, and external surfaces must be prepared very carefully, especially in regions where snow and ice are involved.

Safety. Driverless tractors are available with many more safety options than any other automatic conveying system and include such features as encounter detection, sonic detectors, and optical detectors, which will all shut the tractor down if an object is detected in the path. Additional safety devices include a strobe light, siren, and panic buttons which can override all other controls. Using warning signs and placing mirrors at corners and blind spots are good preventive measures and so is keeping the tractor speed below 5 mi/h.

ROBOTS

Robots are programmable machines capable of automatically moving individual parts or objects over precise paths in space.[1] A robot can also be programmable so that it is able to move parts

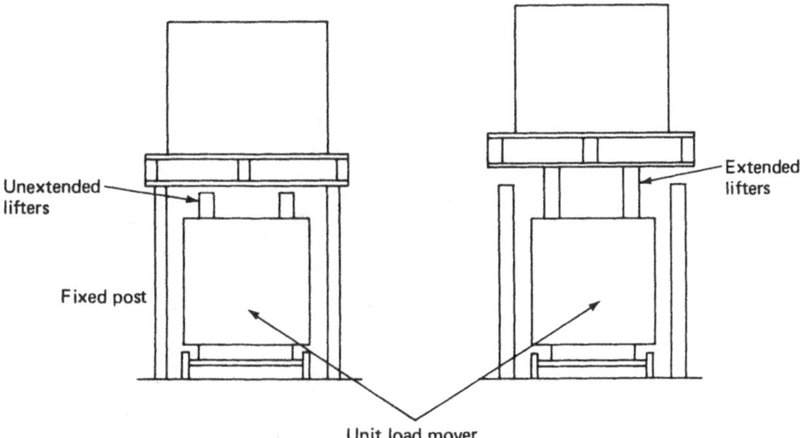

FIGURE 4.35 Individual unit load or pallet mover.

through different paths, capable of performing repetitive motions, able to duplicate the movements of the human arm by moving parts through four axes in space.

Applications. Present applications related to materials handling include machine loading and unloading, conveyor transfer, and pallet loading. The most practical applications for materials handling will be those areas that require repetitious manual operations, particularly those involving the interface between workers and machines. Robots are also ideal for these types of operations in poor working environments, such as those where heat, cold, fumes, or radiation exposure is present. Painting and welding are typical major potential application areas.

Design Components. Robots (Fig. 4.36) are available with a wide range of capabilities and in various design configurations. The major components include a manipulator which actually performs an operation and moves parts, a controller that stores data and directs the movements of the manipulator, and the energy source to power the robot.

A sophisticated robot with six axes of motion can perform many of the same movements as the shoulder, elbow, and wrist. Simpler, less expensive units with two degrees of freedom, called *put-and-place units,* are typically used for machine loading and should become widely used in the materials-handling field in the next decade.

Manipulator. The handling of objects by the manipulator is facilitated by the use of tools that give the robot "hand" capability. The general categories for this purpose are either grippers or surface-lift devices.

MECHANICAL GRIPPERS. These grippers (Fig. 4.37) are usually movable fingerlike levers paired to work in opposition to each other. They can be thought of as mechanical equivalents of the thumb and forefinger.

SURFACE-LIFT DEVICES. These can include simple forklift attachments, vacuum pickups (Fig. 4.38), hooks, or magnetic devices.

CONTROLLER. The controller initiates the motions of the manipulator through a sequence at the desired points and stops the motion when required. The controller can be programmed by adjustment of mechanical cams, stops, and limit switches on the simpler types of put-and-place robots. The more sophisticated robots can be "taught" a sequence of movements by an operator. In the teaching mode, the programmer manually moves the manipulator through the motions of the operation, and the coordinates of the path are stored in the controller memory.

Energy Sources. Nonservo, or *pick-and-place* robots operate through activation of a hydraulic or pneumatic system and are the simplest, lowest-cost units. They have limited flex-

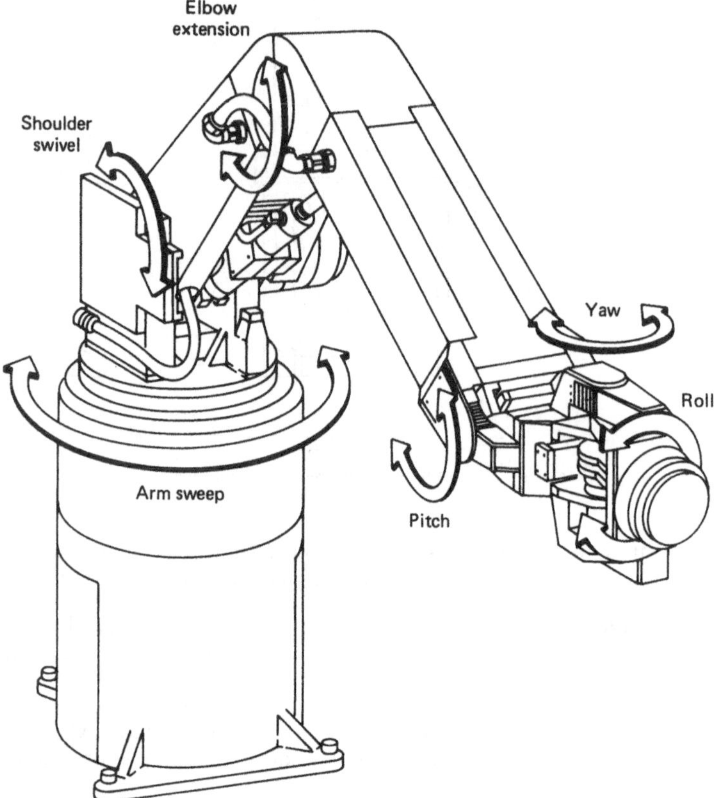

FIGURE 4.36 Robot with six axes of motion.

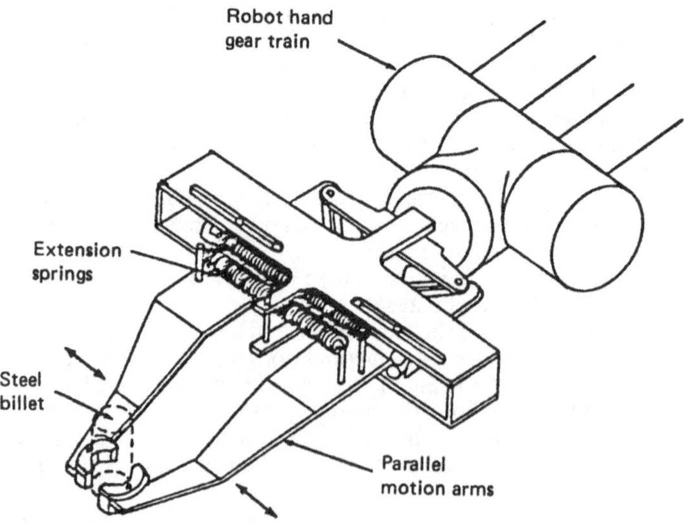

FIGURE 4.37 Robot grippers equipped with spring-loaded fingers.

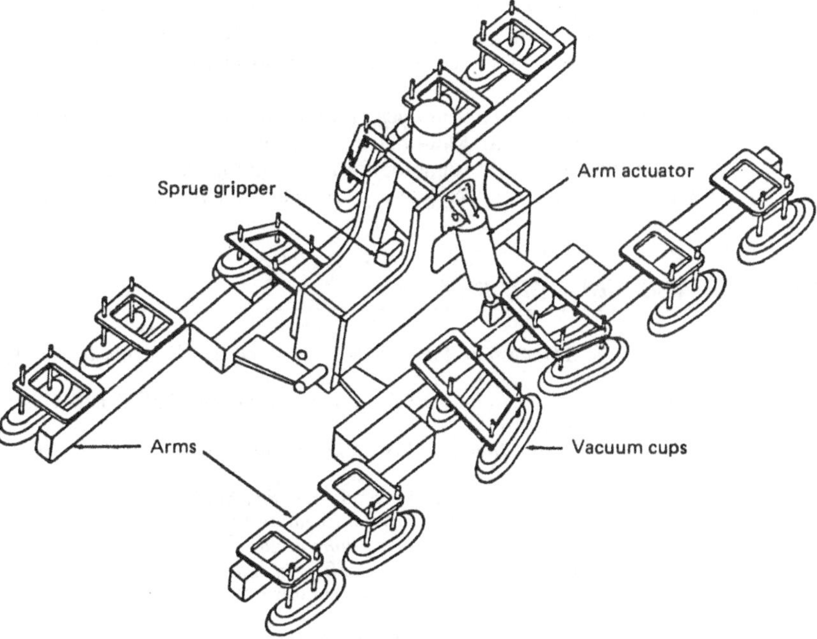

FIGURE 4.38 Vacuum pickup device for robot.

ibility in terms of program capability and positioning capability but are highly reliable. In the operation of this type of robot, as the sequence is indexed, the manipulator members move until the present limit of travel is reached. Since there are only two positions for each axis to assume, programming can be done by adjusting the end stops for each axis to establish the operation sequence.

Servo-type robots use servo motors or valves to move the manipulator members and can be further classified into either point-to-point or continuous-path types. *Point-to-point* servo robots are programmed or taught by feeding them manipulator-position data at discrete points and, in performing a task, they will internally select a path to that point. *Continuous-path* servo robots are programmed or taught to follow a precise path and are used for operations where movement is important, particularly in spray painting.

Future Developments. The technology of robots will be expanded in the future to include the capability to discriminate differences in objects by optical- or mechanical-sensing devices which would send a feedback signal to the controller which will make a decision to initiate a movement command to the manipulator. Further future developments include speech recognition for robot programming and three-dimensional optical-sensing devices. Also, while robots now in operation are generally large, floor-mounted units, future robots will also include table-mounted units able to assist in small subassembly and final assembly operations.

Planning Considerations for the Use of Robots. There are four points that must be considered when evaluating the feasibility of using a robot in materials handling. They are rate of handling, weight of the object, orientation of the object, and number of different items to be handled.

Rate of Handling. Robots are not high-speed handling equipment. If the handling rate is greater than 15 items per minute, another approach should be considered.

Weight of Object. The weight-handling capacity of robots is presently 500 to 2000 lb (227 to 907 kg), depending on the type of robot. The heavier the load, the lower the handling rate.

Orientation of the Object. Position of the object is important and should be consistent. A primary limitation of current robots is the precise orientation required of parts to be picked up by the robot and, hence, a feeding or positioning mechanism to the robot itself is often required.

Number of Items to be Handled. Setup time for product changes can be reduced by quick changeover grippers and automatic program selecting capability. In cases where dissimilar parts are handled in the same operation, a multipurpose gripper or "hand" should be used, along with a sensing device that can command the robot to switch to a preset program.

REFERENCES AND BIBLIOGRAPHY

1. Tanner, W. R.: *Industrial Robots,* Vols. 1 and 2, Society of Manufacturing Engineers, Dearborn, Mich., 1979.
2. *Automated Storage/Retrieval Systems Planning Guide,* Clark Handling Systems, 525 W. 26th St., Battle Creek, MI 49016
3. *Automated Storage/Retrieval Systems Justifications,* Clark Handling Systems 525 W. 26th St., Battle Creek, MI 49016.
4. "Industrial Robots," *Modern Material Handling,* April 1980.
5. Salvendy, G.: *Handbook of Industrial Engineering,* Wiley, New York, 1992.
6. "Towline Conveyors," *Material Handling Engineering,* February 1991.
7. "Power and Free Means Flexibility," *Material Handling Engineering,* April 1992.
8. "How to Evaluate and Plan for Warehouse Automation," *Material Handling Engineering,* March 1992.
9. "Monorails Deliver the Goods," *Plant Engineering,* March 7, 1991.
10. "Automated Electrified Monorails," *Material Handling Engineering,* April 1990.

CHAPTER 4.4
MATERIALS HANDLING: MOBILE EQUIPMENT

K. W. Tunnell Co., Inc.
King of Prussia, Pennsylvania

INTRODUCTION

The group of equipment that is described as mobile materials-handling equipment is made up of machines that essentially depend on a self-contained power source for movement and are independent in their movement route. The equipment, being self-contained material movers, provides a flexible, relatively inexpensive transportation link between plant activities. This broadly classified group of equipment includes devices and equipment from the simplest two-wheeled hand truck to highly sophisticated movers controlled by computer-based systems.

Within the mobile materials-handling equipment group there is a wide array of general-purpose and specialized material movers. Basically, there are two broad categories of mobile equipment. The powered equipment depends on a built-in power source for its operation. The unpowered device relies on a detachable prime mover, either a piece of powered equipment, or in many cases, the equipment operator. The least complex equipment provides transportation between two points without positioning or lifting capabilities. Other units lift or roughly position the load being transported as well as move the material. The multiple-axis movers transport the load; they also have a position capability along two or more axes to accomplish loading and unloading.

Generically, mobile materials-handling equipment falls into five groups, each of which will be discussed in this chapter:

1. Floor trucks and operator-powered movers
2. Powered lift trucks
3. Burden carriers
4. Tractors and tractor trains
5. Mobile industrial cranes

APPLICATION CONSIDERATIONS

Equipment Utilization and Selection

From available records, it appears that mobile equipment often has a low level of utilization. Powered equipment is often employed well beyond its economic life, generating penalty costs in spare-parts inventories, maintenance, and productivity.* Five to seven years has been calculated to be the average economic life of a powered vehicle, provided proper maintenance has been performed. Keeping a lift truck beyond the optimal time increases maintenance costs by 30 to 40 percent. Proper operator training, which is required by the U.S. Occupational Safety and Health Administration (OSHA), can improve efficiency and reduce maintenance costs as well. By tracking costs and utilization, it is possible to reduce the size of the lift truck fleet. Often, two old lift trucks can be replaced with one newer model.

Other general considerations in establishing equipment requirements include:

- Unit-load condition and size and center of load
- Terrain, environment, and aisle width in the movement area
- Length, type, and frequency of moves
- Positioning requirements of load(s)
- Operating economies and maintenance
- Standardization of equipment
- Critical nature of operation(s) serviced

Factors in Wheel Selection and Use

Solid Wheels. These are made in semisteel, forged steel, or molded plastic, hard rubber, and composite materials. They should be limited to small diameters and low-speed movement and should not be used to transmit power. They have low resistance to roll, but a short life span when overloaded or subjected to rough floor conditions. They will cause load vibration because of a lack of cushioning.

Rubber-Cushioned Tired Wheels. These consist of a metal wheel having a machined diameter onto which a rubber tire is pressed or molded. It has the lightest load-carrying capacity of those used on mobile equipment. Minimal power is required to move material, since rolling friction is minimized.

Oil-Resistant Tired Wheels. The tires are made of special oil-resistant rubber compounds which will resist the degrading effects of oil on rubber.

High-Traction Tired Wheels. The tires are made of rubber impregnated with abrasive or other materials to give additional traction on ice or in wet conditions.

Low-Power Tired Wheels. The tires are fabricated from rubber compounds that offer minimum roll resistance and have lower power requirements, causing less drain on battery-operated equipment.

Nonmarking Tired Wheels. The tires use a rubber compound filler other than carbon to avoid floor marking and contamination.

Conductive Tired Wheels. The tires avoid the chance of static sparking in hazardous or explosive environments by maintaining vehicle-to-floor conductivity.

Laminated Tired Wheels. The tires for these wheels are made up of sections of pneumatic tire carcasses threaded onto a steel band. Such tires are extremely tough, with a harsh ride. They are well suited to littered environments, such as scrap yards, and trash handling.

Polyurethane Tired Wheels. Though more expensive than rubber, these wheels have a significantly higher load-carrying capacity and are less susceptible to cuts than most rubber and rubber-compound wheels. Wheel hardness of polyurethane tires results in a harsher ride and increased plant floor damage.

---

* "Use the Optimum Economic Life to Help Cut Costs," *Modern Materials Handling,* February 1991.

Inflatable Tired Wheels. These wheels have vulcanized, reinforced rubber tires similar to automotive tires. The tires are both tube and tubeless. They generally carry a lower load rating for their size than solid-tire wheels. Their use will provide greater load cushioning, higher speed capability, easier maintenance, and less floor damage.

Factors in Internal-Combustion-Engine Selection and Use

Internal Combustion Engines. These are used in outdoor applications, in well-vented interiors, and in nonhazardous environments. They are generally powered by gas or liquid propane gas, although compressed natural gas (CNG) is a promising alternative. In anticipation of new government regulations, manufacturers are redesigning engines with reduced emissions and improved fuel efficiency.

Industrial Engine. Typically, this heavier engine is designed to operate in a lower rpm range than an automobile engine. It can be expected to give about 10,000 h of useful life before overhaul. At an equivalent operating speed of 20 mi/h (32 km/h) in an automobile, this would equate to 200,000 mi (321,800 km).

Automotive Engine. This is of lighter construction than the industrial engine and, because of the quantities in which it is produced, is of relatively lower cost. It generally operates most efficiently in a higher rpm range than the industrial engine and can be expected to give about 7000 h of useful life prior to overhaul. This life is equivalent to about 140,000 mi (225,260 km) of automobile travel. An advantage of this type of engine is the availability of replacement parts through automotive supply firms.

Air-Cooled Engine. This is restricted to lighter-duty applications where weight, size, and initial cost are the prime concerns. The absence of a separate cooling system is a distinct advantage, although this engine's life expectancy is a relatively short 1500 to 2000 h of operation.

Diesel Engine. Typically, this type is installed in large pieces of equipment where the additional size and cost is not significant. However, because of recent improvements in engine design, diesel engines are becoming more prominent in smaller trucks. This is largely due to the reduced need for periodic maintenance, greater fuel economy per hour of operation, and longer expected life—up to 20,000 h.

Compressed Natural Gas Engine. This engine design is ideal for indoor use due to its low noise and low emissions. CNG is a low-cost fuel and the truck can run a full shift before requiring fuel. Other benefits include fewer oil changes and lower maintenance costs. This truck is suited for most types of use and can accommodate loads up to 6000 lb (2700 kg).

Factors in Battery-Powered-Vehicle Selection and Use

Battery-Electric Equipment. This is mechanically simpler in design than engine-driven equipment. Typically, the high-torque dc electric-drive motor is coupled directly to the drive axle through a constant-mesh drive train. An electronic silicon-controlled rectifier (SCR) speed-control device regulates the motor's revolutions per minute through operator foot control. Direction is reversed electrically with a delay interlock to avoid reversing motor direction while in motion.

Storage Battery. These must be replenished frequently either by recharging or by exchanging them for fully charged batteries. Batteries used in a given piece of equipment should provide ample power to operate effectively for an 8-h day as determined by their ampere-hour (Ah) rates. The Ah rating, to some degree, limits the effective operating range of battery-operated equipment and requires that routine schedules for replenishment are followed. Also, because of the weight of a large storage battery, equipment application is sometimes adversely limited.

Advantages of Battery Vehicles. The advantages are low fume emission and heat contamination, quietness and cleanliness, and generally lower maintenance requirements.

Types of Batteries. The two primary types of batteries used are lead-acid and nickel-iron-alkaline. A lead-acid battery will provide 2.0 to 2.3 V per cell, while the nickel-iron-alkaline battery will provide 1.2 V per cell. Voltages used for modern battery-powered mobile equipment are 12, 24, 36, 48, and 72, with some higher voltages used in larger equipment.

Advantages. The advantages of the lead-acid battery are a lower initial cost, high ampere-hour capacity, and low resistance to self-discharge. The nickel-iron-alkaline battery is desirable because of its longer life expectancy, resistance to physical damage, noncorrosive electrolyte (KOH), and more rapid and less critical recharge rates.

Recharging Times. These are adjusted for different batteries by dividing the Ah rating of the battery by the 8-h Ah rating of the charger and multiplying by 8. For example, a battery having a 600-Ah rating and a 450-Ah charger will require

$$(600 \div 450) \times 8 = 10.64 \text{ h}$$

Battery Charging Area. Warehouses utilizing multishift operations require remote charging of the equipment batteries. These areas require hoists, conveyors, chargers, and required safety features (e.g., exhaust hoods and fans, shower and eyewash station, etc.).

FLOOR TRUCKS AND OPERATOR-POWERED MOVERS

This type of equipment is the most fundamental materials-handling aid available. The basic simplicity permits easy adaptation for single-purpose application. Standard catalogs indicate the wide variety available, often designed for specific industries. However, custom design may be specified with very little, if any, cost penalty.

Generally, floor trucks are described as follows.

Two-Wheeled Hand Trucks. Two-wheeled hand trucks (Fig. 4.39) are essentially levers on two wheels. The axle connecting the wheels serves as the fulcrum of the lever and carries up to 80 percent of the total load moved. The two-wheeled cart is normally used for short nonrepetitive moves of smaller loads over smooth floors. Carts are generally 48 to 64 in (1.2 to 1.6 m) high, and are designed to carry a variety of materials in bags, barrels, bales, boxes, and bins. Typical accessories include height extension, stair climbers, safety brake, spread clamps, and straps.

Dollies. Dollies are smaller-wheeled platforms upon which a load is placed for short distance and intermittent moves. Typically, dollies are fitted with caster-type wheels and are either pulled or pushed by an operator.

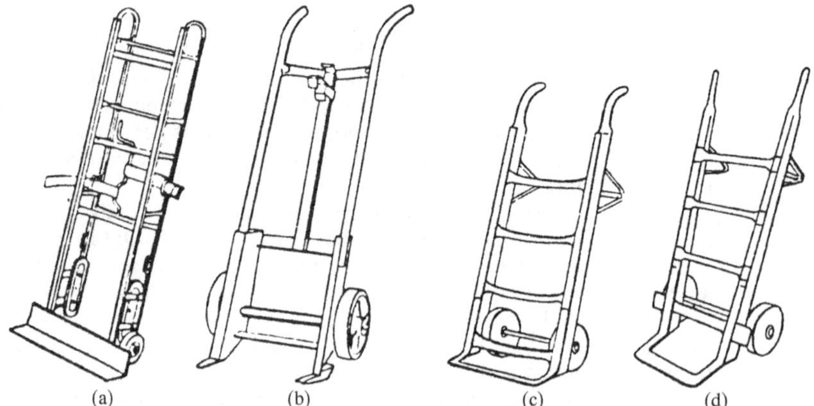

FIGURE 4.39 Two-wheeled hand trucks. (*a*) Appliance type, (*b*) drum and barrel mover, (*c*) general type with Western handle, (*d*) general type with Eastern handle. (*Reproduced with permission from* Material Handling Engineering Handbook and Directory, 1979–1980, *published by* Material Handling Engineering *magazine, Cleveland, Ohio.*)

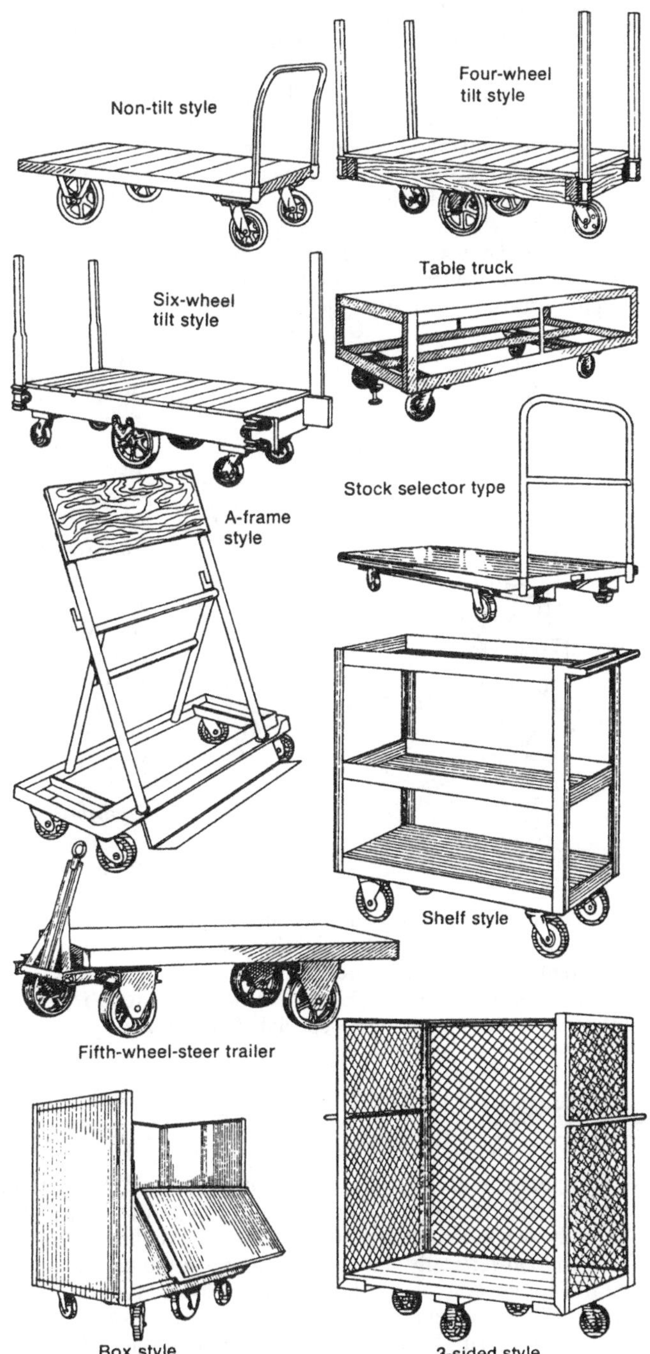

FIGURE 4.40 Factory trucks and wheel arrangement patterns. (*Reproduced with permission from* Material Handling Engineering Handbook and Directory, 1979–1980, *published by* Material Handling Engineering *magazine, Cleveland, Ohio.*)

Factory Trucks. Factory trucks (Fig. 4.40) are wheeled platforms or containers either moved by an operator or towed by detachable power units. There is a wide variety of devices in this group and an even wider variety of uses for materials movement and as mobile storage.

The hand factory truck is hand-powered, guided by the direction of the moving force, and closely related to the dolly. Several wheel-arrangement patterns are available with tradeoffs between maneuverability and stability.

The towed factory truck is connected to the prime mover by a tow bar which provides the steering direction. Both two-wheel and four-wheel steering are available on towed factory trucks. Two-wheeled steering is generally the least expensive and most commonplace. Because of the steering geometry involved, each truck will follow a turn of shorter radius than the preceding vehicle. As several of these units are connected in trains, the continual tightening of turns requires more space for maneuvering.

The four-wheel-steered truck, with properly adjusted steering, is capable of following the same path as the vehicle in front of it. Where long trains are economically justified and desirable, the four-wheel-steered devices may be used to minimize commitment of valuable manufacturing space to aisles.

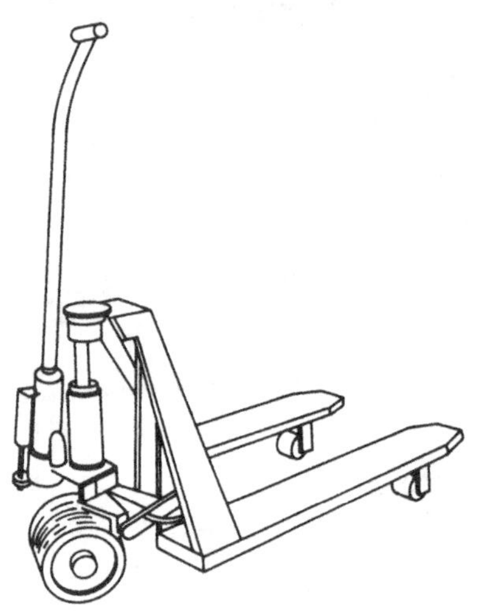

FIGURE 4.41 Hydraulic-lift truck.

The Semilive Skid. The semilive skid is a rectangular platform or box having two wheels on one end and two fixed supports on the other. The end having the fixed supports is also fitted with a heavy pickup pin to which a two-wheeled jack is attached. The jack and handle are used as the lifting device and tiller, allowing the skid to be maneuvered by the operator.

Hydraulic-Lift Trucks. Hydraulic-lift trucks (Fig. 4.41) are used for short distance moves at the workplace. They generally range in capacity from 2500 to 8000 lb (1130 to 3625 kg). These trucks require a minimum amount of maintenance, and can last for 20 years.[2] The trucks can be equipped with a jacklike manually operated hydraulic lift or pedal operated system to elevate a loaded pallet. Some units use an electrically driven hydraulic system to lift, often above the maximum 5 in of the manual system. These lift trucks generally use forks for lifting pallets or platforms for special containers and positioning heavy loads.

POWERED-LIFT TRUCKS

This equipment group represents what is probably the largest and most varied of equipment for materials handling. The powered-lift truck owes its popularity to its versatility, being able to easily pick up a unit load, transport it quickly in a variety of environments, and then position the load vertically at almost any point within the capability of the equipment. Depending on the volumes involved, they become less economical for moves over 300 ft (90 m) since rated speeds are generally between 5 and 10 mi/h (7 and 14 km/h). Powered lift trucks are usually fitted with lifting forks to carry a unit load, although a wide variety of special load-carrying attachments can be used in place of forks. Power for lift trucks is either by internal-combustion engine or battery electricity.

The various pieces of equipment in this group can be operated over a variety of terrains, depending upon the design and, specifically, the wheel and tire combination used. Load-carrying capacities from 1000 to over 40,000 lb (450 to 18,000 kg) are common. Large vehicles are available with capacities in excess of 100,000 lb (45,000 kg). The very large vehicles are generally used outside, particularly for the moving and stacking of shipping containers.

Establishing aisle widths and their relation to fork-truck selection are critical when significant storage areas are involved. Clearly, the narrower the aisles, the more rows of storage. Equipment manufacturers have been ingenious in designing specialty trucks to operate in narrow aisles. It should be noted that manufacturers specify equipment turning circles and, thus, aisles will require space to aid in fork-truck maneuverability. Specialty trucks designed to operate in narrow aisles permit better space utilization, but tend to trade off some aspect of performance, a factor to be considered in specifying specialty as opposed to general-purpose equipment.

Truck capacity is generally calculated as follows (see Fig. 4.42):

A = distance, in, from center of front axle to heel of fork

B = distance, in, from heel of fork to center of load

C = distance $(A + B)$ from center of front axle to center of load

D = length, in, of load on fork

W = weight of load, lb

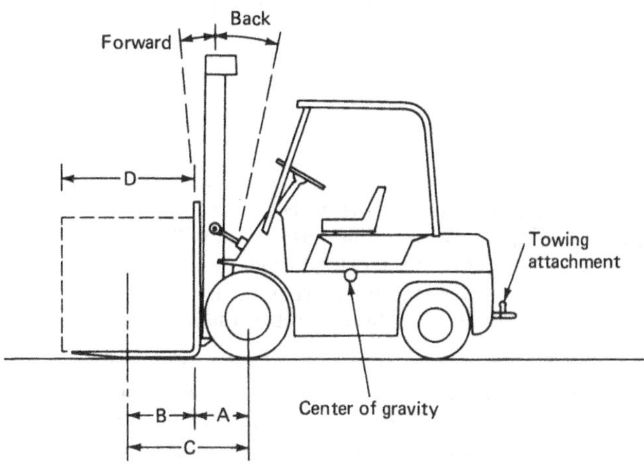

FIGURE 4.42 Rated truck capacity and counterbalanced truck.

1. *Inch · pound rating*

$$\text{Inch} \cdot \text{pound rating} = W \times C$$

2. *Maximum load length for given load*

$$C = \frac{\text{inch} \cdot \text{pound rating}}{W}$$

3. *Maximum load for given load length*

$$W = \frac{\text{inch} \cdot \text{pound rating}}{C}$$

A specific example is given to illustrate the actual calculations.

1. Truck is rated 3000 lb (W) at 20 in [3000-lb load which has a center 20 in (B) from heel of fork].
2. Distance from center of axle to heel of fork is 10 in (A).
3. Pallet load to be handled is 2000 lb:

$$C = A + B = 10 + 20 = 30 \text{ in}$$
$$\text{Inch-pound rating} = W \times C = 3000 \times 30 = 90{,}000 \text{ in} \cdot \text{lb}$$
$$C = \frac{\text{inch} \cdot \text{pound rating}}{W} = \frac{90{,}000}{2000} = 45 \text{ in}$$
$$B = C - A = 45 - 10 = 35 \text{ in}$$
$$D = 2 \times B = 2 \times 35 = 70 \text{ in allowable load length}$$

4. When selecting attachments, refer to the truck manufacturer to determine the amount of negative effect the attachment has on the truck's useful load-carrying capacity.

Aisle widths are generally established as follows:

A = aisle width
TR = turning radius of truck
L = load length
C = aisle clearance (total on both sides)
AX = distance from rear corner of load to centerline of axle:

$$A = TR + L + C + AX$$

The several varieties of powered-lift trucks are described below.

Counterbalanced Trucks. The counterbalanced trucks (Fig. 4.42) use their large, carefully positioned weight mass to offset (counterbalance) the moved load mass. These trucks are generally equipped with a tilting mast which will "tilt" the lifting mechanism rearward from the vertical lifting position and further counterbalance the load during movement. The load is positioned fully in front of the truck so that the truck structure does not interfere with adjacent stacks of material. This minimizes the aisle widths that are required.

Straddle Trucks. The straddle trucks (Fig. 4.43) differ from the counterbalanced type in that they do not depend on weight mass to counteract the weight of the load being handled. Instead, the straddle forklift positions the two main load-carrying wheels at or forward of the material load center. The truck is extremely stable as a result of this arrangement.

The straddle design is more compact and of lighter weight than the counterbalanced type. It is necessary, when negotiating loads into or out of racks, that either the straddle truck be equipped with an extending fork mechanism (pantograph) or the racks be positioned or constructed to allow the forward wheels of the truck to enter them.

Side-Loading Trucks. Side-loading trucks (Fig. 4.44) are a unique combination of a straddle-lift truck and a narrow-aisle truck. They are used where there are narrow aisles, where rapid transportation is called for, and where long narrow loads such as pipe and bar stock are handled. Side-loading trucks do not have to be turned to engage or place loads.

Nonrider Lift Trucks. Nonrider lift trucks (Fig. 4.45) are those where the operator walks along with the truck, directing the operation through a control unit attached to the truck. These units have basically the same features found in larger counterbalanced and straddle trucks. They are used for lifting and stacking light loads and moving these loads short distances.

Straddle Carriers. Straddle carriers (Fig. 4.46) are large-capacity, highly maneuverable, powered lift trucks. To load and unload, the vehicle is driven over the unit load(s). The actual loading and unloading is extremely fast, although precise positioning of loads requires other methods. Unit loads can be transported at rates approaching highway speeds.

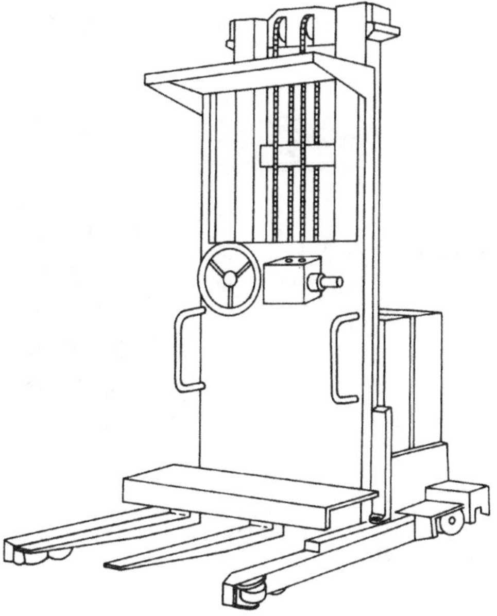

FIGURE 4.43 Straddle truck.

Order-Picker Trucks. Order-picker trucks have an elevated platform forward of the mast from which the truck and the platform can be operated. Typically, the trucks are used for picking partial loads in narrow aisles to heights of 36 ft, allowing for significant labor and space saving.

Materials-Handling Attachments

The most widely used attachments are the forks themselves. They can be set at various widths and generally range between 30 and 60 in (80 and 160 cm) in length. The forks should be at least two-thirds the length of the maximum load to be lifted.

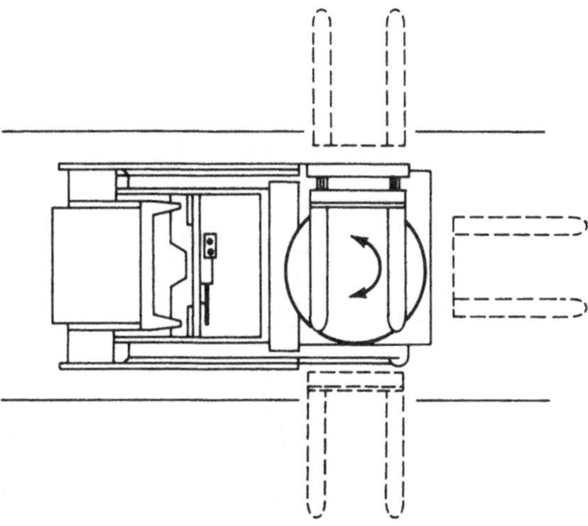

FIGURE 4.44 Side-loading truck.

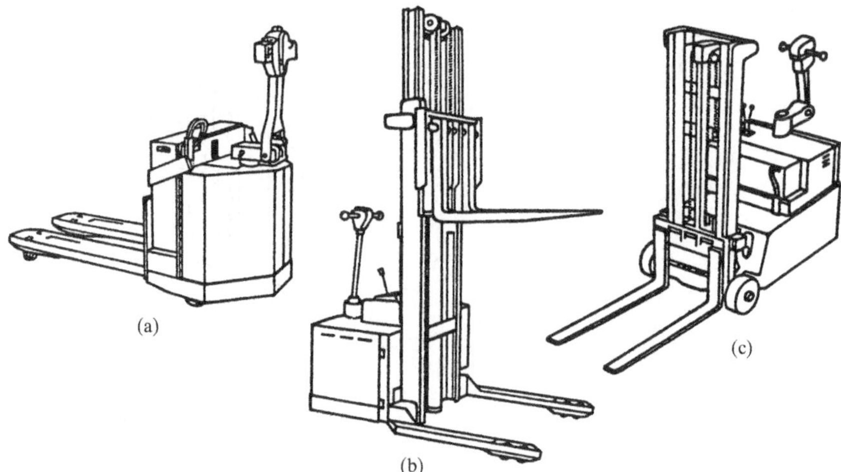

FIGURE 4.45 Nonrider lift trucks. (*Reproduced with permission from* Material Handling Engineering Handbook and Directory, 1979–1980, *published by* Material Handling Engineering *magazine, Cleveland, Ohio.*)

Standard two-stage uprights provide a lift height of approximately 18 ft (5.5 m), and three- and four-stage uprights provide heights to 20 ft (6 m). Certain specialty vehicles are designed to operate above 20 ft (6 m). The difference in fork height and total extended height is generally 4 ft (1.2 m), reflecting the height of the backrest. For low buildings, *free-lift* trucks should be specified. This feature permits the forks to be raised to lift loads to nearly half the total lift height without extending the uprights.

Frequently, a forklift truck will be fitted with an attachment or combination of attachments which allows the vehicle to perform special handling functions or simply allows it to operate more efficiently in a given situation. In some cases, these attachments replace the conventional forks for handling products which the forks cannot. In other instances, the attachments are used to augment the original fork function by giving the load-carrying forks additional motions.

When selecting attachments, it is always wise to consult with the truck manufacturer since attachments have a negative effect on a truck's useful load-carrying capability. When attachments are installed, the truck's information plate must be restamped, indicating the new effective truck capacity as required by OSHA 1910.178(4).

Attachments usually limit a forklift truck to a specialized function and, to some extent, limit its overall in-plant versatility. Some of the more simply designed attachments mount on the fork attachment rails and require only a few minutes to install or remove. The more complicated attachments, particularly those requiring hydraulic connections, should be considered as permanent conversions.

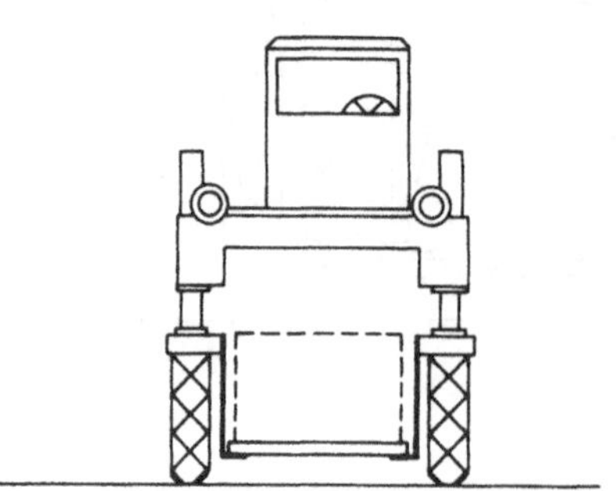

FIGURE 4.46 Straddle carrier.

The following is a list and a brief description of some of the more common attachments (Fig. 4.47):

1. *Ram.* A single projection, mounted in place of the forks for carrying coiled materials which can be easily entered horizontally. Rams have a variety of lengths and diameters to handle a variety of products, from steel coils to rolled carpet.

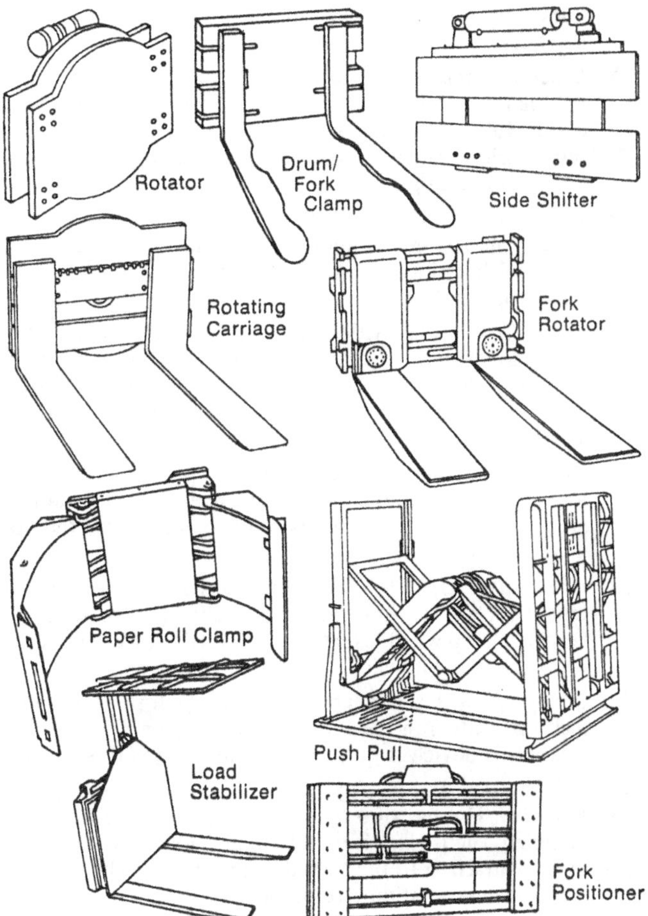

FIGURE 4.47 Common materials handling attachments. (*Reproduced with permission from* Material Handling Engineering 1979–1980, *published by* Materials Handling Engineering *magazine, Cleveland, Ohio.*)

2. *Barrel attachment.* Used to grasp the top seam of a steel drum and transport it in the vertical position.
3. *Concrete-block fork.* This and the similar brick fork are designed specifically for handling stacks of masonry products without pallets.
4. *Paper-roll clamp.* Specifically designed to carefully grasp and transport rolled materials in the vertical position. It is frequently combined with a rotator which allows the roll to be carried horizontally in the case of loosely rolled or easily damaged materials.
5. *Push-pull.* Uses a polished platen instead of forks to carry the load. Its purpose is to position loads in dense environments without the use of a pallet. In place of a pallet, a thin slip sheet is used under the load. This sheet is grasped by hydraulic clamps and pulled into the truck platen for loading and pushed off again into its next position.
6. *Bale clamp.* Used to grasp and carry baled materials and depends on hydraulic pressure to grasp the bales from the sides.
7. *Scoop.* Used to handle loose or granular bulk material and consists of a metal bucket mounted in place of the fork, with a dumping capability usually provided. Tilting the

bucket for loading and transport is accomplished by tilting the lift truck's mast forward and back.

8. *Squeeze clamp.* Used to grasp the sides of boxed products in a manner similar to the bale clamp, except that the grasping arms are smooth and deliver an even pressure to the carton to avoid damaging its contents. This device eliminates the need for pallets. It requires, however, additional side clearance on each side of the material moved to accommodate the clamps.
9. *Top-handling lift.* Used to handle folded cartons by hooking into the folded lip of the top carton. The most important advantage is that extremely high storage density can be accomplished since only minimum side clearances are required without the use of either pallets or slip sheets.
10. *Side shifter.* Used with almost any type of attachment as well as forks, the side shifter allows loads to be positioned accurately from right to left without relocating the truck. Its major function is to speed the positioning of loads and to minimize rack space between loads. The side shifter will also reduce wear on the truck itself by reducing repositioning.
11. *Adjustable forks.* Where a variety of pallet and load sizes are encountered, adjustable forks are used. While most fork arrangements are manually adjustable, the mechanically adjustable forks allow the operator to accomplish the operation while remaining in the driver's seat.
12. *Load stabilizers.* To assure that loosely arranged and unstable loads are firmly contained during transit, various load stabilizers are available. Such a device is essentially a vertical clamp which exerts a downward pressure on a load and thus holds it in position while it is being moved.
13. *Clamping forks.* Similar in design to the fork positioner, clamping forks can be used to pick up loads in a conventional manner or may be used to clamp loads between the forks. This device is quite commonly used with special notched forks for transporting drums.
14. *Rotator.* The use of a rotator allows a load to be rotated through 360°, generally for dumping. The rotator is used with unit-load devices that fully enclose the forks and thus remain attached to the fork during rotation. They are also used with various clamping devices when rotation is required.
15. *Extended-reach forks.* Commonly used with straddle forklifts to enable the truck to reach a load in the racks while the forward wheels remain outside of the rack space. The attachment also allows racked materials to be reached when two-deep storage is used. The reaching mechanism consists of a hydraulically operated pantograph system between the truck mast and the forks.

BURDEN CARRIERS

In the manufacturing process where sufficient volumes are involved, conveyor systems are often used to move materials from point to point. When smaller volumes or several moves of varying density are involved, however, a fixed-platform device is often used. These fixed-platform vehicles depend on an auxiliary loading and unloading method and are not tied to a specific unit-load module. Such devices are called *burden carriers.*

Burden carriers come in a wide variety of sizes and shapes. They are available in two basic types (Fig. 4.48). One is the walkie (nonriding) type and the other is the riding type. Both are available with battery-electric and internal-combustion power sources. They are usually limited in load-carrying capacity. High loads are generally handled by other types of handling equipment.

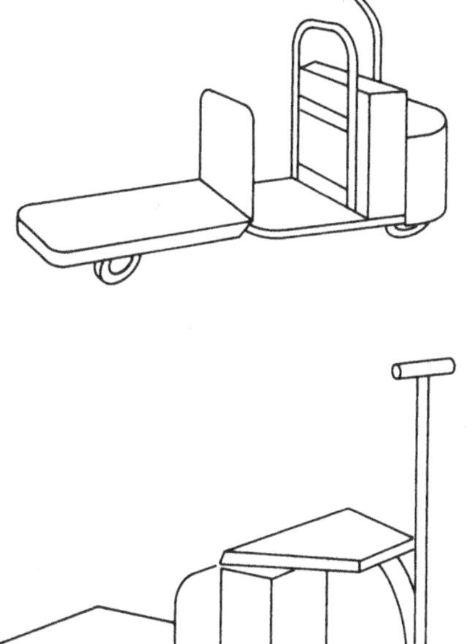

FIGURE 4.48 Types of burden carriers.

Walkie Burden Carrier. The walkie burden carrier is typically a three-point suspension hauler using battery-electric power, although some units are available that are powered by a small air-cooled engine. They are similar in design to the previously discussed walkie lift trucks, except that they have a fixed platform. Load ranges of 1000 to 3000 lb (450 to 1360 kg) are available and application is limited to noncontained loads. Loading is generally done by hand or, in the case of heavier loads, by hoists and cranes.

Rider Burden Carrier. The rider-type burden carrier is often tailored to a variety of special applications such as personnel carriers, fire trucks, and portable maintenance shops. In its simplest form, the truck provides a driver's seat and a flat load-carrying bed. In this configuration it serves most commonly as a miscellaneous hauler to deliver supplies and materials in-house for distances of more than 300 ft (90 m). The power source for the rider-type truck is fairly evenly divided between air-cooled gasoline engines and battery-electricity. Both three-point and four-point suspensions are common, with an operating suspension system being incorporated in many larger units, along with pneumatic tires. These vehicles are able to negotiate rougher terrains and may attain speeds of up to 20 mi/h (30 km/h).

TRACTORS AND TRACTOR TRAINS

The term *tractor* (Fig. 4.49) refers to a detachable power source supplying locomotion to one or a group of load-bearing vehicles not having on-board power. The tractor is a steerable mover which is directed by an operator. They are generally classified according to their drawbar pulling rating (DPR) into small, medium, and large sizes.

On all grades above 5 percent, the individual manufacturer should be consulted since a variety of other factors which must be considered vary with individual tractor designs. Minimum safety criteria for industrial tractors are covered in OSHA Standard Section 1910.178 (Powered Industrial Trucks) and should be referred to when equipment is being selected.

The main application for these vehicles is the movement of goods in volume over distances too long to be economically moved by fork trucks—approximately 300 ft (90 m). Since the tractor trains are not self-loading, a system of tractor loading stations and surplus tractors and trailers is required, involving constant hitching and unhitching of tractors and trailers. An alternative is to use a forklift as a tractor with the operator of the fork truck loading the trailers.

Apart from fork trucks, five types of tractors are used for most industrial applications.

Highway Tractors. Highway tractors are typically used in over-the-road applications and are relatively specialized to serve this purpose. They do, however, find application in large manufacturing complexes for the movement of materials between remote locations where

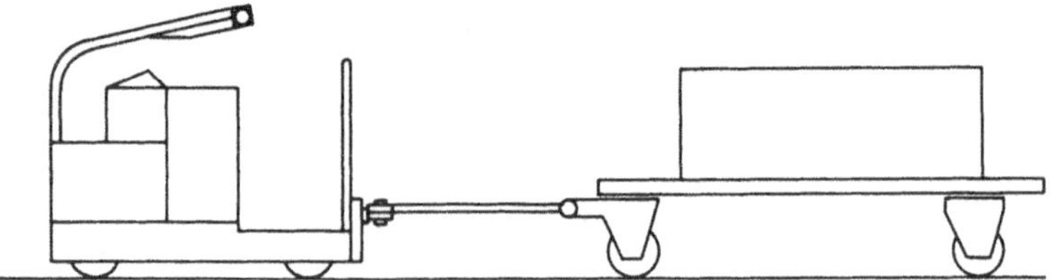

FIGURE 4.49 Tractor used in industrial applications.

warranted by the speed and density of materials flow. This type of tractor is also frequently used in factory shipping yards for the positioning of both loaded and unloaded semitrailers.

Walkie Tractors. Walkie tractors are the smaller variety of industrial tractors. These tractors are battery-electric with motive power, braking, and steering being provided by a single wheel or a close-coupled pair of wheels. The drive mechanism is tiller-controlled through hand controls, as in other walkie equipment, and deadman controls are provided. Two other wheels are provided for stability at the rear of the unit. A variety of coupling devices are available for attaching the tractor to trailers and semilive skids.

Walkie-Rider Tractors. The walkie-rider tractor is essentially a larger version of the walkie tractor. The major differences are that in a walkie-rider tractor a platform is provided for the operator to stand on during operation and two travel speeds are provided. A slow speed comparable to the operator's walking pace, on the order of 3 mi/h (5 km/h) and a higher speed of roughly 7 mi/h (11 km/h) is common in this type of equipment. Owing to higher operating speeds, these tractors have wider operating ranges. Because of the longer range of these tractors, larger-capacity batteries are used and, therefore, the units are heavier and larger than pure walkie tractors.

Rider Tractors. Rider tractors are available in both stand-up and sit-down configurations. The stand-up variety is more compact and generally applicable to more congested situations. The sit-down tractor is generally larger and is used where higher speed and longer distances, up to ½ mi (0.8 km) or more, are encountered. Battery-electric and internal-combustion engines are used as power sources in both versions; however, battery-electric power is more prevalent in the stand-up models, and the internal-combustion engine is the frequent choice in sit-down tractors.

Specialty Tractors. Specialty tractors are usually confined to very heavy load applications and are often built as an integral part of the load carrier itself. Two more common applications of these specialty tractors are large bulk-handling carriers for molten metals and granular materials and for spotting of railway cars.

MOBILE INDUSTRIAL CRANES

Mobile industrial cranes (Fig. 4.50) serve a variety of plant and production-related materials-handling functions. They are especially adaptable to loads of large or unusual size and where careful placement is required. In some applications, they are used only to position a given load, while in other applications they are used as both prime mover and positioner.

Mobile cranes differ from other plant hoisting equipment in that they operate independently of any supporting structure. The primary advantage of a crane is its ability to reach into places not normally accessible by other types of materials-handling equipment. With the exception of straddle cranes, the industrial crane depends on a boom for its reach and lift

FIGURE 4.50 Mobile industrial crane.

capability. It is the positioning of the boom that ultimately determines where a load will be placed and how large a load can be safely lifted.

The following text discusses the types of mobile cranes in use.

Portable Hand-Powered Crane. The portable hand-powered crane is similar in design to a small manual-lift truck, except that the load-carrying forks have been replaced by a boom and hook. This equipment is commonly used to move and position work pieces into and out of process equipment where volumes do not warrant a permanently installed hoisting system. It is also frequently found in maintenance and repair shops to assist in the disassembly and reassembly of in-plant equipment. Lifting is accomplished either through a hand winch and cable system or a manually operated hydraulic system. Typical lifting capacities are limited to 2000 lb (900 kg) or less.

Stevedore Crane. The stevedore crane is a nonswinging crane which requires that the hook be positioned by maneuvering the entire vehicle. This limits its use to relatively unobstructed areas. The boom may be extended outward by the operator to reach the load and returned back to a position closer to the vehicle for transport. The crane is a relatively fast vehicle which is used to pick up a load and transport it to a final destination.

The front, load-carrying wheels are also the powered wheels, with steering being achieved by the trailing wheels. Both three- and four-point suspensions are used, and the crane is often used to tow factory trucks while also loading and unloading them. Typical load capacities range from 2 to 4 tons (1800 to 3600 kg).

Swing-Boom Crane. The swing-boom crane is a larger-capacity crane than the stevedore crane and is used more for positioning loads than for transportation. The boom structure is constructed so that it can be rotated by the operator through 180°. Outriggers are provided for stability. They are usually powered by diesel or spark-ignition engines, with battery-electric power also being available.

Full Revolving Cranes. Full revolving cranes are capable of swinging a load through a full 360° and are generally the largest of mobile cranes. Their use is normally one of positioning loads as opposed to transportation. This type of crane will often be mounted on a truck-type chassis for rapid movement between jobsites. Power is provided by diesel or spark-ignition engines through direct hydraulic torque converters.

Load-lifting capacities at the boom's most upright position can be as high as 100 tons (90,000 kg) with reaches in excess of 100 ft (30 m) possible. Power is provided by diesel or spark-ignition engines through direct hydraulic torque converters. Industrial applications are almost totally limited to construction activities and maintenance of large structures.

Straddle Crane. The straddle crane has no boom but has a wheel-mounted framework on which are mounted two hoists. These hoists are capable of moving within the limits of the framework for precise load positioning. The straddle crane is related to the straddle carrier. It is a highly versatile crane, finding application as both positioner and mover of materials. In addition, its design is such that it is extremely stable and can move at relatively high speeds.

The framework consists of four vertical columns mounted above the vehicle's high flotation wheels, supporting two horizontal crane rails which carry the traveling hoists. Load-carrying capacity ranges from approximately 10 to 60 tons (9000 to 55,000 kg) per hoist for an aggregate capacity of approximately 20 to 120 tons (18,000 to 110,000 kg).

A variety of power systems are used, all of which are engine-driven. Hydraulic systems are those most commonly used for transport, hoisting, and positioning. However, one manufacturer employs an engine-driven electrical power plant to operate the crane's functions through electric motors.

The straddle crane is capable of operating in high-density areas and is highly maneuverable, since all four of its wheels can be turned and powered independently. Common applications are in steel storage yards, loading and handling of shipping containers, commercial concrete castings, truck and car loading, and boatyards. Special load-handling devices are easily adapted to this crane, increasing its versatility.

REFERENCES AND BIBLIOGRAPHY

1. "Better Maintenance Improves the Bottom Line," *Modern Materials Handling,* February 1991.
2. "Hand Pallet Trucks," *Material Handling Engineering,* July 1991.
3. "New Products and Services," *Logistics Management and Distribution Report,* May 2001.
4. "Not the Same Old Rack and Roll," *APICS—The Performance Advantage,* September 1999.

CHAPTER 4.5
MATERIALS HANDLING: WAREHOUSING AND STORAGE

K. W. Tunnell Co., Inc.
King of Prussia, Pennsylvania

INTRODUCTION

In the overall materials-handling system, warehousing provides the facilities, equipment, personnel, and techniques required to receive, store, and ship raw materials, goods in process, and finished goods. Storage facilities, equipment, and techniques vary widely depending on the nature of the material to be handled and the volume of orders to be shipped. Characteristics of materials, such as size, weight, durability, shelf life, and order lot size, are factors in designing a warehousing system and in solving warehousing problems.

Economics is also of great importance in the design of warehousing systems. Storage and retrieval costs are incurred, but add no value to the product. Thus, the investment in storage and handling equipment and in floor space must be based on minimizing unit storage and handling costs.

Other factors to be considered in designing warehousing systems include control of inventory size and location, provisions for quality inspection, provisions for order picking and packing, staging for receiving and shipping, appropriate numbers of shipping and receiving docks, and maintenance of records.

WAREHOUSING ACTIVITIES

Warehousing activities vary according to material amounts and characteristics. However, the activities associated with warehousing generally include the following procedures:

1. Unload inbound shipments.
2. Accumulate received material in a staging area.
3. Examine the quantity and quality and assign a storage location.

4. Transport the material to the storage area.
5. Place the material in the assigned storage location.
6. Retrieve the material from storage and place it in an order-picking line, if a picking line is used.
7. Fill orders, if applicable.
8. Sort and pack, if applicable.
9. Accumulate for shipping.
10. Load and check outbound shipments.
11. Label package and carrier selection.

WAREHOUSING INFORMATION AND DOCUMENT CONTROL

Associated with the physical handling and storage of materials is an information- and document-control system. The information- and document-control system provides for:

1. Acknowledging receipt of material for accounting purposes
2. Verifying the quality and quantity of received material
3. Updating the inventory records to reflect receipts
4. Locating all material in storage
5. Updating the inventory records to reflect shipments
6. Notification to the accounting function of shipments for billing purposes

Many information- and document-control systems are standard features of current warehouse management systems (WMSs). The cost-effectiveness of such systems over manual systems depends on such factors as:

1. The number of line items in storage
2. The number of customers served
3. The volume of material shipped
4. The speed and accuracy required to process orders

Generally, WMSs, technology, and automation are cost-effective for industries and distribution centers having many line items in storage, many customers, and a large volume of goods shipped. Distributors in many diverse industries often utilize WMSs, technology, and automated systems. Examples of technology include bar codes, radio-frequency (RF) tags, and RF mobile computer terminals. Examples of automated systems include automated storage and retrieval systems (AS/RSs), "pick-to-light" systems, high-speed sorting systems, and conveyor systems.

TYPES OF MATERIALS

Materials to be stored may be broadly classified as bulk materials or packaged goods. Bulk materials such as fuels, chemicals, minerals, and grain are stored in specialized storage facilities and transported in pipes, screw conveyors, power shovels, etc. In the many industries which handle and store bulk goods, each accomplishes these tasks with very specialized equipment and techniques. This discussion will be limited to warehousing packaged goods. The reader should consult specific publications that apply to bulk materials-handling industries for particulars in bulk handling.

Within the packaged-goods classification, materials are subdivided into categories according to their state of completion in the manufacturing process. Categories include raw materials, goods in process, and finished goods.

Raw materials. These vary widely in characteristics, depending on the industry. A few examples are raw foods and ingredients for food processors, thousands of small parts for electronics assemblers, engines and motors for manufacturers of vehicles, and wood and finishes for furniture manufacturers. Raw materials are the goods on which the manufacturing process will operate to produce salable products. Indeed, the finished goods of one manufacturer often become the raw materials of another.

Goods in process. This refers to goods which have completed some but not all of the manufacturing process. Typically, a manufacturing process involves several operations utilizing different equipment, skills, and materials. Goods in process are stored while awaiting the next manufacturing operation. They are often stored along the manufacturing process rather than in the warehouse proper.

Finished goods. These goods are those which have completed the manufacturing process and are stored in inventory to fill customer orders. Finished goods may be further subdivided into reserve and order-picking stock. Customer orders are filled from order-picking stock while the picking stock is replenished from reserve stock.

The amount of raw materials, goods in process, and finished goods to be handled and stored varies considerably from industry to industry. Industries having large inventories of raw materials usually are converters of bulk materials such as paper and steel. Manufacturers of highly complex equipment such as computers and automobiles require a significant amount of raw-materials storage for parts as well.

Industries having significant needs for goods in process handling and storage are those whose manufacturing process is not automated. Machine-shop and electronic-assembly operations are examples.

Finished-goods handling and storage capacity are a function of manufacturing volume and product bulk. Industries having high-volume and high-bulk output generally require a considerable handling capacity for finished goods. The paper conversion and bottling industries are examples.

CONSIDERATIONS IN WAREHOUSE PLANNING

The objective of warehouse planning is to provide space and equipment to hold and preserve goods until they are used or shipped in the most cost-effective manner. The efficient accomplishment of warehousing activities listed in Chap. 4.1 is dependent on thorough planning. The following sections discuss these considerations as a guide to the warehouse planner.

Type and Number of Materials

The type and number of materials to be stored and handled form the basis for warehouse planning. The physical characteristics of the material, to a great extent, determine materials-storage and -handling methods. Physical factors include dimensions, weight, shape, and durability. As a first step in warehouse planning, all materials to be stored must be identified and their physical characteristics listed.

The quantity of each material item to be stored must be established. The planner may require assistance from sales management for finished-goods inventory levels and manufacturing management for establishing levels of raw materials and goods in process. In establishing inventory levels, seasonality, changes in product mix, and expected turnover rate become factors.

With the inventory level of each item of stored material established, a storage unit is selected. A storage unit is the least number of an item which is stored as one unit. Examples include a single crated refrigerator, a pallet containing 20 cases of canned goods, and a bundle of pipe. The storage unit is usually selected according to the physical characteristics of the material, the available handling and storage equipment, the quantity, and the manner in which the material is received or shipped.

A storage unit may be larger than a shipping unit or a manufacturing unit. In this case, order-picking facilities are provided for items used or shipped in lots smaller than a storage unit. The service level of storage in an order-picking operation must be established as well.

Factors affecting order-picking stock levels include minimum order quantity, volume, and the physical characteristics discussed earlier. Sheet-metal screws, for example, might be in 3-months supply, while cased canned goods might be in only 8-h supply, and automotive components at an assembly plant might be in only 4-h supply.

Storage Equipment

Storage-equipment selection follows the establishment of the reserve and order-filling inventory storage units and levels.

In the case of selecting equipment for an existing building, the constraints of the building itself must be taken into consideration. Storage equipment must be compatible with floor loading capacity, clear height beneath sprinklers and structural steel, column spacing, and location of shipping and receiving docks, etc.

The characteristics of the storage unit, pallet, drum, bundle, etc., largely determine the type of storage equipment required. The inventory levels to be maintained determine the number of pieces of storage equipment. Materials characteristics and the volume of materials movement generally are deciding factors in selecting materials-handling equipment. Materials-handling equipment is discussed in Chaps. 4.3 and 4.4.

Storage equipment usually consists of general-purpose or specialty storage racks of varying height, depth, and load capacity. However, the warehouse floor may serve as all or part of the required equipment. Storage units such as pallets of cased canned goods, which have the rigidity and stability to support loads placed on top of them, are normally stored on the floor in stacks. Rolls of paper and coils of steel are frequently stacked on end. Storage units which have rigidity and are many in number lend themselves to floor-stacking techniques.

Heavy or bulky storage units which lack rigidity or which are few in number are generally better stored in storage racks. Storage units which are small, such as wristwatches or thumbtacks, are suitable for storage in shelving and bins. Containers used in conjunction with shelving or by themselves are discussed in Chap. 4.2.

Some types of available storage equipment are described below. Custom-designed special equipment is offered by many storage-equipment manufacturers.

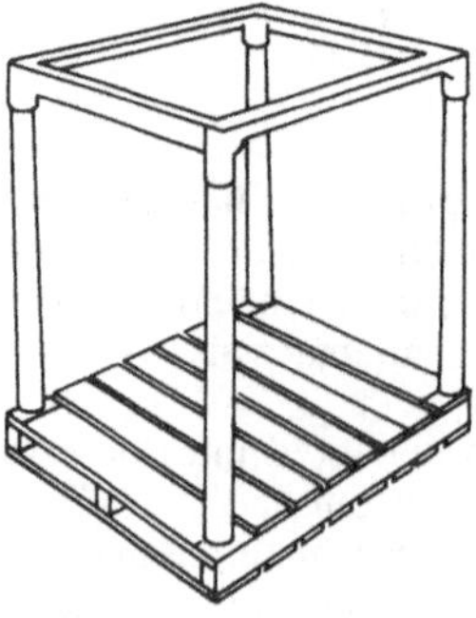

FIGURE 4.51 Pallet frame.

Pallet Frames. Pallet frames (Fig. 4.51) are useful where materials lack the rigidity or stability to be stacked on the floor and where there are a large number of storage units in inventory. The pallet frame attaches to the pallet and extends above the material. The frame acts as a structure on which another pallet is stacked. Pallets so stacked are often placed several stacks deep and thus conserve floor space as compared with pallet racks which require aisle access. The frames are removable for pallet loads not requiring support.

Pallet Racks. Pallet racks are the most commonly used storage aids and are available in many configurations adapted to particular materials characteristics and turnover rates. Pallet racks, for the purpose of this discussion, are classified into five groups.

One-Deep Pallet Racks. These are used when many items with small inventory quantities must be stored for ready accessibility. They may be configured to accommodate containers and other unit loads in addition to pallets. They are also used for order picking when it is most economical to pick directly from storage units.

One-deep pallet racks consist of vertical upright frames connected by horizontal crossbeams on which pallets and containers are placed one deep. Uprights are available in various heights and depths, and crossbeams are available in various lengths to accommodate most storage unit sizes. The load-carrying capacity for the upright and beam combinations is established by the manufacturers.

The normal storage height for this type of rack is from 20 to 24 ft (6 to 7 m) from the floor to the top of the top load. Lifting operations tend to be inefficient at greater heights because it becomes too difficult for the lift operator to accurately place the storage unit. Specialized equipment for heights greater than 24 ft (7 m) is available, however.

The horizontal crossbeams are adjustable so that the vertical height of the rack may be divided into as many storage levels as desired. The individual storage-opening height is tailored to suit the height of the storage unit. Clearance of 4 to 6 in (10 to 15 cm) from the top of the load to the bottom of the crossbeam above it is usually provided. In establishing the maximum height of the top load, the height of the fire-protection sprinklers must be considered. Most fire-protection codes and fire-insurance underwriters require a minimum clearance of 18 in (45 cm) between the top storage unit and the sprinklers. The warehouse planner should consult the local applicable fire code and the insurance underwriter.

The horizontal width of the storage opening is determined by two factors. These are the maximum weight and the maximum width of the loads to be stored in the opening. It should be noted that the load width may be larger than the pallet width because of load overhang. Normally, 4 in (10 cm) is provided horizontally between loads and between loads and uprights. Typically, two pallet loads are placed side by side in one opening. When the horizontal dimension of the opening has been determined, the rack manufacturer's catalog is consulted to select a compatible crossbeam length and weight capacity.

At this point in warehouse planning, the planner should calculate the floor load resulting from the fully loaded pallet rack. The floor loading will become a design parameter for new construction. In the case of an existing facility, the floor will be confirmed as adequate or the rack arrangement will be shown to be unfeasible.

The number of pallet racks required is determined by dividing the maximum number of storage units by the number of those units contained in one rack.

Two-Deep Pallet Racks. These racks (Fig. 4.52) are similar in design to the one-deep pallet rack except that two pallets, one behind the other, are stored in each position. Two-deep racks are used when there is insufficient floor space to accommodate the required number of one-deep racks. Two one-deep racks are normally placed side by side and require aisle access from each side of the two-rack combination. The two-deep rack requires access from only one side but stores the same amount of material as the two one-deep racks placed side by side. The ratio of aisle to storage is thus reduced by using two-deep racks.

Two-deep racks have some costs associated with them, however. Lift equipment must be fitted with extended reach capability in order to position loads in the rear storage position. The efficiency of storing and retrieving the load in the rear storage position is less than with single-position loading and unloading. Two-deep racks are often more expensive than their two, one-deep, side-by-side counterparts. Storage units are sometimes damaged when positioned in or retrieved from the rear position.

The manner of selecting the height, depth, width, and number of two-deep racks is similar to that for one-deep racks.

Drive-In or Drive-Through Pallet Racks. These racks (Fig. 4.53) are designed to provide storage several pallets deep. The racks consist of vertical uprights which are braced across the top of the rack. Angle-iron ledges are welded or bolted to the insides of the uprights to sup-

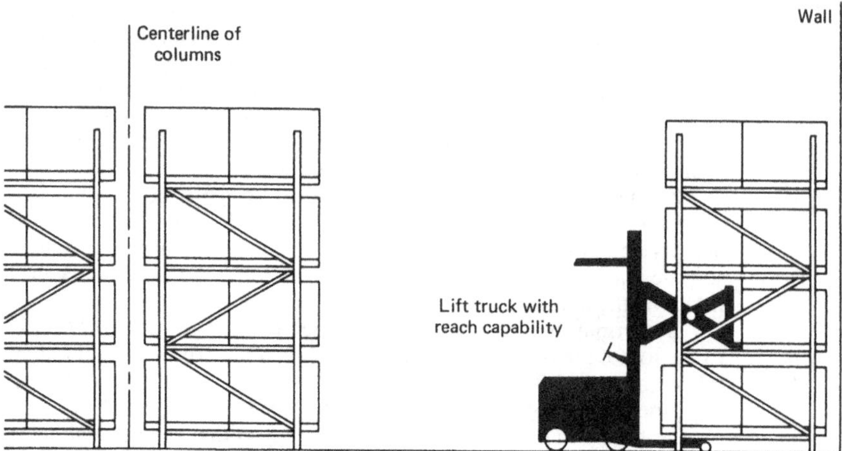

FIGURE 4.52 Two-deep pallet racks.

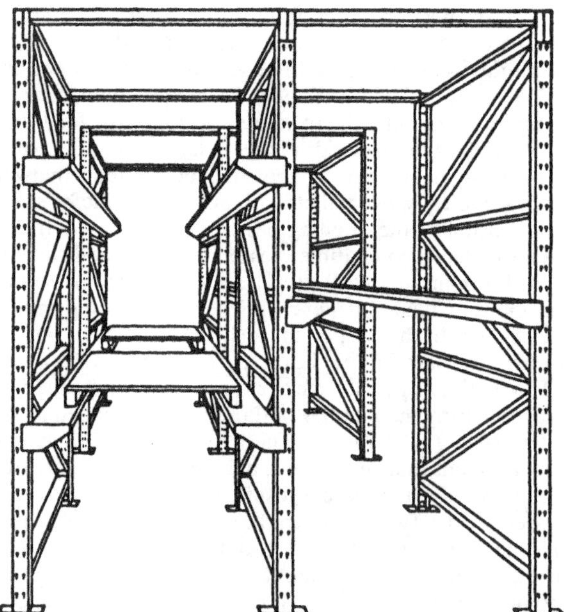

FIGURE 4.53 Drive-in or drive-through rack. (*Reproduced with permission from* Material Handling Handbook and Directory, 1977/1978, *published by* Material Handling Engineering *magazine, Cleveland, Ohio.*)

port the pallets. This arrangement allows the lift vehicle to enter the rack to place or retrieve a pallet.

Drive-in or -through pallet racks are used where floor space is limited and where there are many storage units of a particular item to be accommodated. Palletized items shipped or delivered by the entire truckload would be candidates for storage in this type of rack. The total number of positions of storage in one storage aisle in the rack could be designed to contain one truckload.

There are several limitations of drive-in or -through racks. Pallets stored in the rear or middle of a storage aisle cannot be retrieved until those in front are removed. This feature limits first-in–first-out inventory control except by loading or unloading entire storage-aisle lots one at a time. When entire storage-aisle lots are so treated, empty pallet positions are created if less than the entire aisle is filled or emptied. Storage efficiency is thereby reduced.

The lift operator must move and lift the storage unit in very confined spaces. Damage to the goods as a result of close tolerances is more frequent in this rack than in others. It is also clear that the lift operator's efficiency is reduced by the requirement of driving into the rack in confined spaces.

Since the ledges that support the pallets are fixed, storage units of uniform dimensions are required. Storage units having overhang on the side of the pallet generally do not lend themselves to this type of storage. Finally, the lift vehicle may be no wider than the distance between the ledges. In selecting drive-in or -through racks, the planner should keep in mind that the usual maximum height is again 20 to 24 ft (6 to 7 m) and that drive-in racks usually are no more than six pallets deep. Due to their depth and the relative lack of bracing of their own, higher drive-in and -through racks generally require bracing to the building structure.

Gravity-Flow Racks. These racks (Fig.4.54) are constructed to contain several pallets in depth and to support the pallets on inclined roller conveyors. Pallets are loaded on the high side of the roller conveyor and removed from the low side. As a pallet is removed from storage, the pallets behind it roll down toward the retrieval opening.

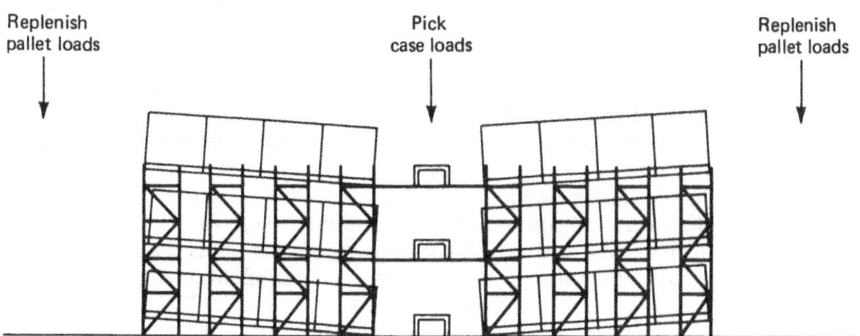

FIGURE 4.54 Gravity-flow racks.

Gravity-flow pallet racks and smaller versions for individual cartons and containers are commonly employed in order-picking operations. A continuous supply of an item is presented to the order picker without replenishment interference. Gravity-flow racks are also useful in maintaining first-in–first-out inventory control.

Gravity-flow pallet racks generally do not exceed six pallets in depth because of the high cost. However, the depth of the rack may be designed to contain a particular time period's supply. This configuration is appropriate when continuous replenishment is not employed.

The height of gravity-flow racks seldom exceeds 24 ft (7 m). The height of the rack is often limited to that conveniently reached by the order picker. In some cases two-level picking on the inside by personnel on foot is replenished by lift vehicles from the back side. See Fig. 4.54.

The height and width of storage or picking openings in a gravity-flow system are usually fixed with little or no adjustment conveniently possible. Instead, storage units are arranged to fit the gravity-flow configuration.

Gravity-flow racks may not be suited to very unstable storage units due to the shock of impact of movement and sudden stops on the inclined roller conveyor. Such difficulties are overcome by placing the unstable items in suitable containers.

Logic-Flow Racks. These are, in principle, designed to accomplish the same functions as gravity-flow racks. Instead of gravity providing the motive force to move full storage units to the picking opening, a powered conveyor does so. In most applications the order picker operates the powered conveyor with start-stop control. This arrangement eliminates the shock of impact experienced in gravity conveyors. Unstable and very delicate storage units are handled and stored in this manner.

In general, due to its high cost, the logic-flow rack is employed only for very specific, small storage situations. For the most part, these systems are manufactured from custom designs.

Bins and Shelving. Bins and shelving are widely used for the storage of goods in small lot sizes as raw materials, goods in process, and as finished goods, particularly in order-picking applications. They are available in many sizes, strengths, and degrees of closure. Indeed, pallet racks, previously discussed, may easily be converted to shelving.

In selecting shelf and bin storage, the planner determines for each storage item an appropriate shelf opening and depth or bin-drawer size. Shelves, bins, and drawers may be fitted with dividers to contain more than one item. The degree of protection from dust, light, theft, etc., determines the degree of closure required. Shelving and bin closures are available from totally open to totally enclosed and individually locked. Where many items require the same degree of protection, the shelving-bin system may be enclosed in a protective enclosure such as a clean room or refrigerated room.

Shelf, bin, and drawer arrangements can be obtained as separate units or in customized combinations. Customized combinations are more costly but may be justified in situations where stock may be advantageously stored in some picking order. Order-picking efficiency is maximized by reducing search and travel time on the part of the order picker.

Automatic Storage and Retrieval Systems. Automatic storage and retrieval systems (Fig. 4.55) are employed to achieve highly dense storage and very efficient placement and retrieval of materials. Many of the other activities listed under "Warehousing Activities" at the beginning of this chapter may also be mechanized and partially or fully automated as well.

The mechanization and automation of warehousing activities require a high capital investment and a very comprehensive feasibility study to justify the investment. The success of mechanized and automated warehousing also requires the *complete commitment by management* to support the planning, design, procurement, installation, and *especially* testing and implementation. In the past, planning to start up could take over 3 yr, but now manufacturers of these systems are providing preengineered proven components such as controls and racks, to reduce excessive start-up times.

Mechanized and automated warehousing systems may be considered by the planner if some or all of the following conditions exist:

- Many varieties of items in storage
- High-volume storage items
- High turnover in general
- Highly seasonal storage items
- High cost of land and floor space
- High labor costs
- Need for rapid customer service
- Random storage desirable
- Storage units uniform in size

Automatic storage and retrieval systems, whether automated or not, achieve their high density by storing goods at greater heights than in conventional racks. *High cube* warehousing from 20 to 100 ft (6 to 30 m) is in use. At heights above 20 ft (6 m) the system may become the structure of the building to which walls and the roof are attached.

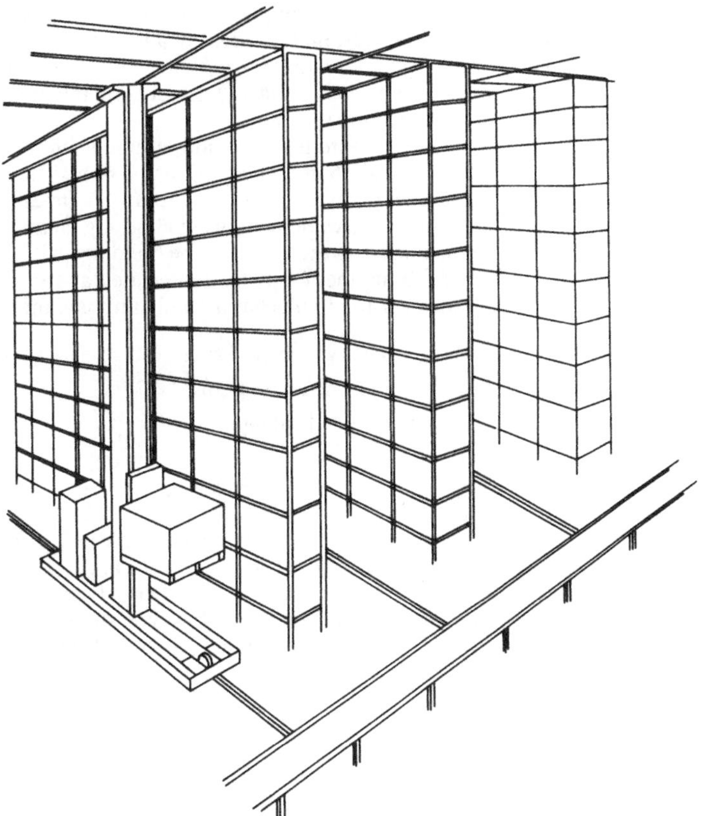

FIGURE 4.55 Automated storage and retrieval system.

The materials-handling equipment, referred to as storage and retrieval (S/R) machines, can be stacker cranes, turret trucks or automatic guided vehicles. The S/R machine travels on rails between the storage units and is guided by rails at the top of the storage units. It can operate in aisles as wide as itself. For example, turret trucks can accommodate up to 3300 pounds (1500 kg) in aisles only 66 in (1.7 m) wide. Each aisle has a dedicated machine which services both sides of the aisle, although systems can be designed to allow machine to transfer to other aisles. To determine the feasibility of aisle transfer machines, one should consider planned system utilization and required response time of storage and retrievals.

In a semiautomatic system, there is an operator on the S/R machine. In this type of system, the S/R machine can have the operator travel with it horizontally and vertically or just horizontally. The operator will select the bay and level on a keyboard, and the machine will position itself to perform the operation. In a fully automatic system, the operator may control the movements of the S/R machines from the computer console located at the pick and deposit station of the system. There are also systems in which a computer system will issue the commands to the S/R machines.

Goods to be placed in storage are often delivered to the S/R machine by conveyor; likewise, outgoing goods are sent by conveyor. Often, the conveyor is centrally located at the pick and deposit station and can be semi- or fully automated.

The degree of mechanization and automatic control of warehousing varies from user to user and from manufacturer to manufacturer. The planner should consider engaging consultants and equipment manufacturers in the planning process. Most manufacturers provide planning guides to assist in identifying requirements. See Refs. 2 and 3 in Chap. 4.3 for more information. Prior to or during the identification of requirements for a mechanized-automated system, the planner should also determine requirements for a comparable conventional warehouse system. The capital investment and the operating cost of each system are than analyzed for economic justification.

Typically the automated system will require a higher initial capital investment but incur lower annual operating costs than a conventional system. The automated system would be economically justified if its payback period and return on investment are satisfactory to management. There also may be tax implications in choosing an automated storage and retrieval system. Where the racking structure supports the building walls and roof, the structure may be considered equipment. Equipment may be depreciated at a faster rate than buildings. Other factors influencing the decision to mechanize and automate include:

- Competitive advantage in servicing customers
- Reliability and the need for backup systems
- Degree to which the market will change
- Time to become operational
- Availability of capital

Storage and retrieval for small lots can be accomplished with mechanized and automated arrangements. Horizontal (Fig. 4.56) and vertical carousels are frequently used, and can achieve picking rates of 100 picks/h. A vertical carousel uses less floor space and resembles a horizontal carousel rotated onto one end. However, the vertical carousel requires an automated retrieval system, since it can reach heights of 35 ft (10.6 m). In either carousel system,

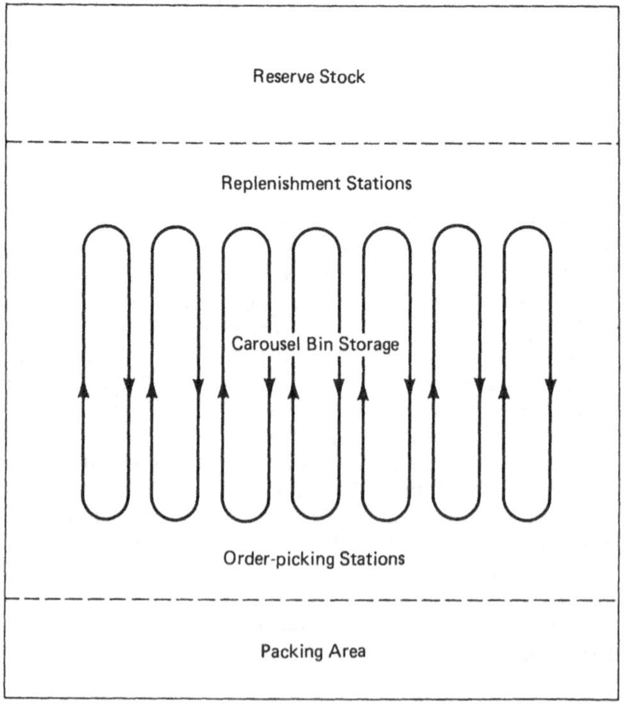

FIGURE 4.56 Schematic plan of automated bin storage.

bins of material are transported to a stationary order picker. The operator may select the picking face to be transported by an ON-OFF control panel in a semiautomatic configuration. Frequently, the picking list is prepared by computer in the order of storage to minimize moves. In automated systems, the computer automatically positions the next picking face as the order picker indicates completion of the last pick or that the item is out of stock. Robotic arms, guided by computer, may be used to perform the pick operations. Complete automation may also include provisions for automatic packing-list preparation, billing, and reordering.

Whether manual or mechanized, shelving and bin storage may be arranged in multilevels by the use of mezzanines. See Fig. 4.57. This configuration is appropriate when high bay space is available and high storage density is required. In general, high-volume and high-weight items are stored in lower levels, and lighter, slow-moving items in upper levels.

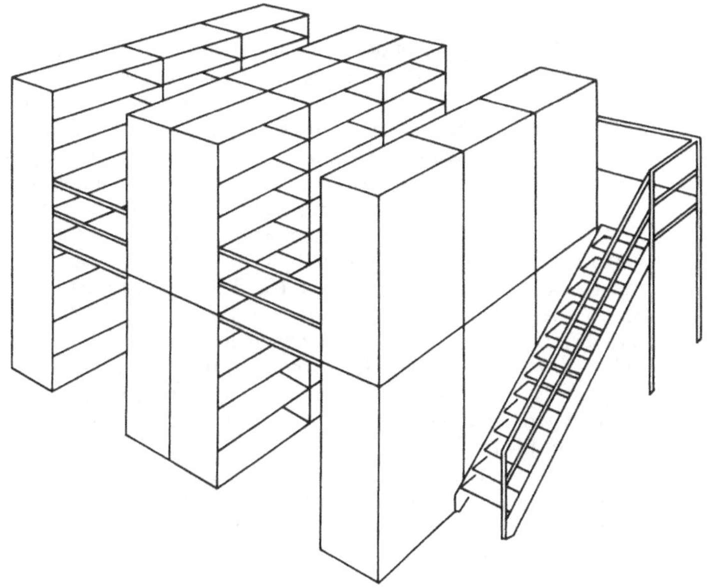

FIGURE 4.57 Multilevel shelving.

Shipping and Receiving Docks

Shipping and receiving docks are an important factor in warehouse planning because inefficient layouts may use too many lift trucks and personnel to achieve the necessary loading/unloading rates. Of prime importance is the number of docks and type of equipment needed. Recently, dock management software has been developed to track dock performance and interface with other software.

In warehousing situations involving very high inbound and outbound volume, the principles of queuing theory are applied to assess peak dock activity. The peak dock activity, using alternative numbers of docks, may then be simulated using computer models to determine the best combination of receiving and shipping docks. Suppliers of dock equipment will often assist in the planning.

In practice, warehouses having moderate to low inbound-outbound traffic are analyzed from historic or expected data. The peak rate of deliveries and shipments is determined and expressed in vehicles per hour. The rate at which vehicles are loaded and unloaded is also determined and expressed in vehicles per hour per dock. Dividing the vehicle rate of arrival by the rate at which the vehicles are loaded or unloaded results in the number of docks required.

EXAMPLE

Peak rate of truck deliveries = 10 trucks per hour

Rate at which trucks are unloaded = 2 trucks per hour per dock

$$\text{No. receiving docks required} = \frac{\text{rate of truck arrivals}}{\text{rate trucks are unloaded}} = \frac{10 \text{ trucks per hour}}{2 \text{ trucks per hour per dock}} = 5 \text{ docks}$$

The type of dock and the equipment provided at the dock are dependent upon the type of material to be handled, the need for specialized loading/unloading equipment, and the need for security or weather protection.

Types of Docks. Types of docks to be considered include the following.

Indoor Docks. These are designed to accommodate the delivery vehicle under the roof of the warehouse building. Accommodation for entire tractor-trailers, trailers only, small delivery trucks, and railcars are common. These arrangements are appropriate where there is need to contain heated or cooled air, where security is important, and where materials-handling systems such as bridge cranes load or unload the material.

Flush Docks. These are constructed flush with the warehouse floor and with the outer wall of the building. This type of dock is generally built at a height above the outside grade level to accommodate the height of the vehicles to be serviced. Where the terrain is flat, ramps down to the docks are usually provided. This type of dock is normally provided with dock seals for weather protection. Also typical are dock leveler installations to accommodate minor differences in the heights of vehicle load beds.

Open Docks. These are less expensive than those previously discussed. However, they offer the least protection from weather and pilferage. Open docks are appropriate in warmer climates when handling goods which are not weather-sensitive or pilferable. When installed, a canopy over the dock area is usually provided.

Sawtooth Docks. These are arranged at an angle to the face of the building wall. This configuration is applicable where maneuvering room for shipping and delivery vehicles is limited. The sawtooth arrangement may be fully enclosed, covered, or open, as described above.

Dock Equipment. This must accommodate the materials to be handled, the vehicle to be serviced, the loading and unloading equipment, and the need for weather and security protection. Some standard equipment includes the following.

Dock Doors. These should be of the overhead type, counterbalanced for ease of operation. The doors should ride vertically along the wall of the building. Doors with tracks curving inward, such as a household garage door, could be damaged by materials-handling equipment. Doors are sized to accommodate the loaded materials-handling equipment passing through them and/or the size and configuration of the delivery-shipping vehicle. Options available for dock doors include the material from which the door is made, insulation, windows, and mechanized door-opening or -closing operators.

Dock Levelers and Dock Boards. These provide a bridge between the delivery or shipping vehicle and the dock (Fig. 4.58). They also serve to accommodate differences in height between vehicles and the dock. Dock levelers are typically installed to be flush with the floor of the dock when retracted. Models which attach to the outside of the dock as a retrofit are also available. The levelers are activated or retracted by spring pressure or hydraulics. Dock boards are reinforced steel plates which are manually lifted into place to form the bridge.

Factors used in selecting the appropriate dock leveler or board include the weight of the heaviest load and materials-handling vehicle combination to cross it, the distance which the unit must span for bridging, and the combined width of the load and materials-handling vehicle. Manufacturers of this equipment offer a wide variety of weight capacities and sizes.

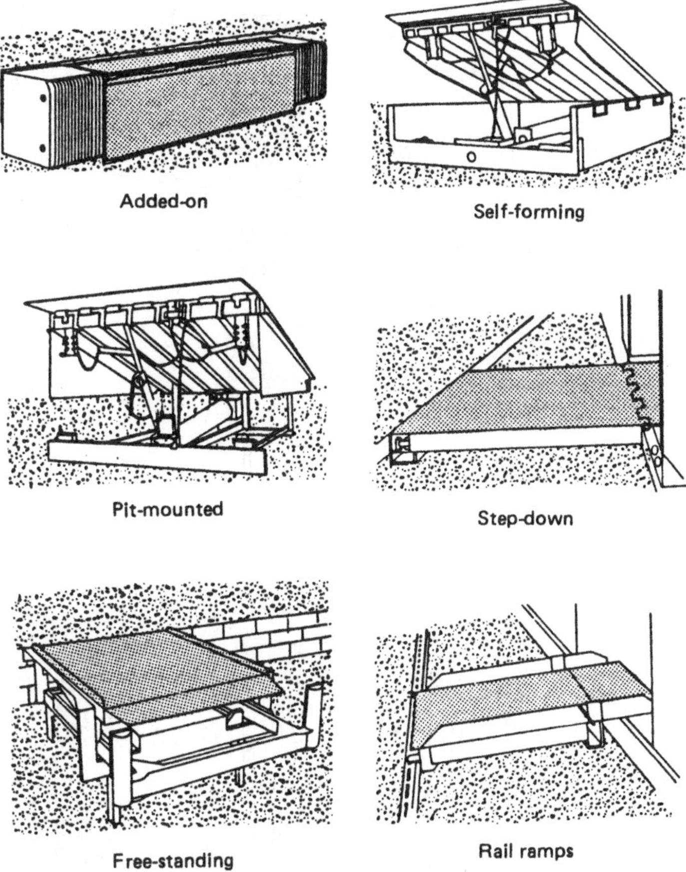

FIGURE 4.58 Types of dock levelers and dock boards. (*Reproduced with permission from* Material Handling Handbook and Directory, 1977/1978, *published by* Material Handling Engineering *magazine, Cleveland, Ohio.*)

Weather-Protection Equipment. This equipment (Fig. 4.59) consists of devices to seal or cover the loading-dock area. Truck-dock seals used in conjunction with flush docks are popular. They have flexible construction and are often inflated. The arriving truck backs into the seal surrounding the door to effect the seal. Also available are hood-type seals which are mechanically activated. Made of flexible material, the hood extends out from the dock and conforms to the shape of the truck or railcar.

Other devices in common use are inside docks (discussed previously) and canopies extending over the outside dock area for rain protection. Where high activity means that dock doors are seldom closed, weather curtains are indicated. The curtains consist of strips of clear flexible material covering the door opening. Materials-handling equipment can drive through the curtain. When there is no traffic through the curtain, the strips act, to some extent, to seal the door from weather.

Dock Lighting. Dock lighting is required for nighttime operations. Floodlights are typical for lighting outside driveways, rails, and maneuvering areas to facilitate spotting delivery and shipping vehicles. Lighting for open docks is required to facilitate loading- and unloading-

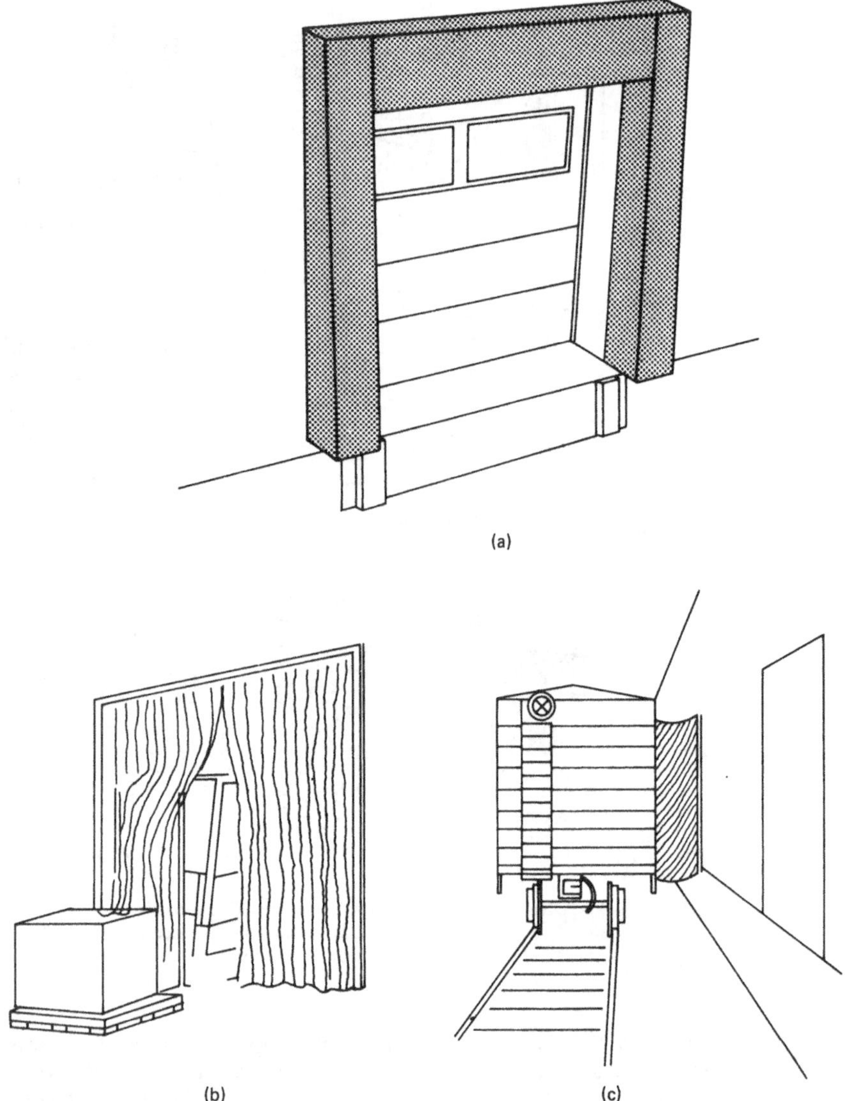

FIGURE 4.59 Types of dock weather protection.

vehicle movement. When dock seals or hoods are used, lighting may be required for the inside of the truck or railcar. Portable or adjustable fixed lighting for these purposes is available.

Warehouse Layout

The warehouse layout is the final and perhaps the most important step in the planning process. Prior to undertaking the layout, the planner establishes the activities to be completed and the type and number of materials to be stored and handled, storage and handling equipment, and docks. (See "Warehousing Activities" at the beginning of this chapter.) The warehouse layout should be planned to provide the space and arrangement that makes the best use of

- Storage cubes
- Efficiency of the flow of materials from activity to activity
- Effective communications between activities

Because thousands of combinations of types, sizes, weights, and volumes of materials have been observed, specific warehouse layout characteristics cannot be described in this book. However, general principles for warehouse design are discussed below.

Location in Storage. The location in storage (Fig. 4.60) of particular items is of importance. The following points should be considered.

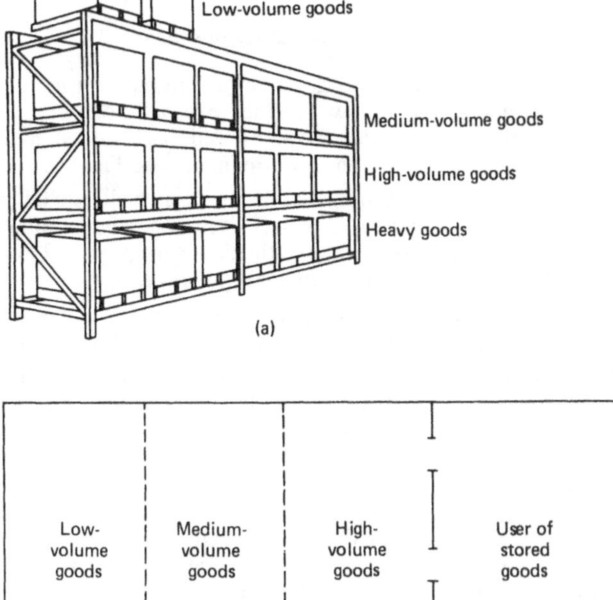

FIGURE 4.60 Location in storage according to volume.

- Items having a high turnover should be located near the user. The user may be a manufacturing operation, the shipping docks, or a quality-inspection area.
- Items having a high turnover should be stored and retrieved in the most convenient level vertically—slow movers high and fast movers low.
- Heavy and/or difficult-to-move items should be stored low.

- Where few items but large volumes of commodities are characteristic, individual loads of an item should be stored together in semidedicated areas.
- Where many items, but few of each, are encountered, random storage should be considered. A locator system is necessary.
- The nature of some storage items may require them to be stored in dedicated space. Some examples include hazardous materials, items of high value, and perishable goods.

Aisles

- Minimum aisle width is determined by the loaded maneuvering characteristics of materials-handling equipment. Determination of minimum aisle width is discussed in Chap. 4.4
- Aisle width may be reduced by imposing one-way traffic.
- Aisles should open from the supplying area and open to the user area for maximum efficiency.
- Aisles should not be located next to walls as only one storage face is presented.

Storage Equipment Location and Arrangement

- The arrangement may be effected by column spacing in existing buildings or may determine the column spacing in new facilities. Normally, one-deep, two-deep, and drive-in racks, as well as shelving, are placed end on end, back-side along column lines. This arrangement eliminates column interference with aisles. See Fig. 4.52.
- One-deep, two-deep, and drive-in racks, as well as shelving, are most effectively placed back-to-back in open floor areas. This arrangement minimizes access aisle requirements. In the case of racks, space between them must be provided for pallet overhang. Frequently, the width of a line of columns provides this space.
- Storage racks, except gravity-flow and logic-flow, in addition to shelving, are efficiently placed along walls with openings for doors and fire-protection equipment. See Fig. 4.52
- The height of the storage equipment is limited to that which provides no less than 18 in (45 cm) of clearance beneath fire-protection sprinklers. Local codes or fire underwriters may require a different clearance.

Docks

- Receiving and shipping docks are usually located to accommodate the flow of materials in the manufacturing process. The most common manufacturing flow patterns are straight-through and U-shaped. See Fig. 4.61.
- In the straight-through processes, raw materials are received and stored at the beginning of the manufacturing process. Finished goods appear at the end of the process and are stored and shipped from that location. Receiving and shipping areas which are so separated generally require more personnel and docks than an equivalent U-shaped arrangement.
- In U-shaped process flow, raw materials arrive at the same side of the building as that from which the finished goods are shipped. Shipping and receiving docks may be separated by no more than an imaginary line. This arrangement may offer economies in lower personnel requirements and in the number of docks since receiving and shipping personnel and equipment may be interchanged when necessary. The U-shaped process flow is also advantageous if high bay storage for both raw materials and finished goods is required.
- Spacing between docks is established to minimize interference between the docks during operations.
- Areas adjacent to docks for staging off-loaded material or material awaiting shipment are normally required.

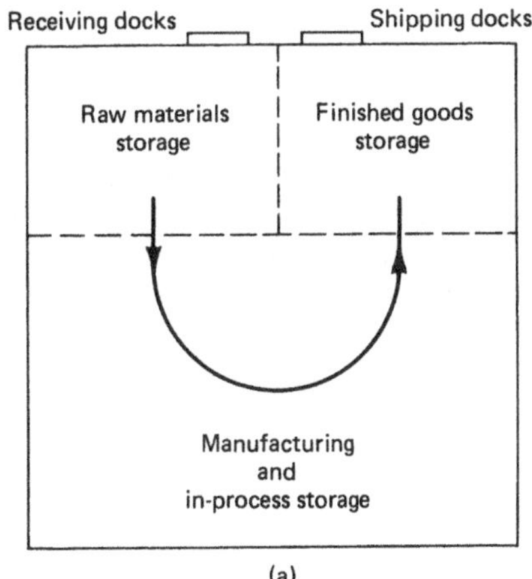

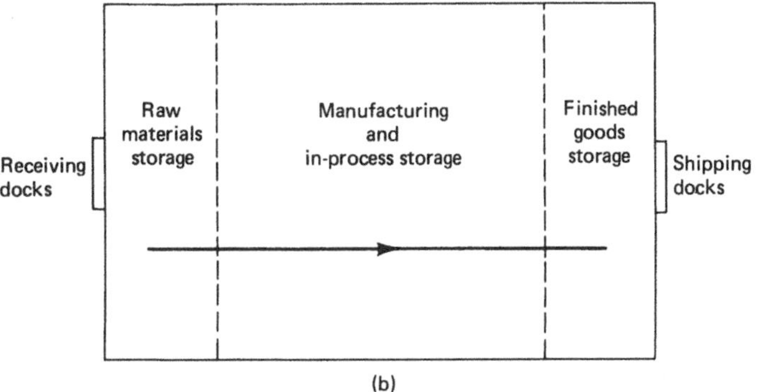

FIGURE 4.61 Materials flow patterns. (*a*) U-shaped materials flow, (*b*) straight-through materials flow.

- Provisions for enclosed space near docks may include administrative offices, personnel comfort facilities, and quality-inspection areas.

Building Characteristics

- Storage height may be limited not only by fire codes but also by local zoning restrictions.
- Building services such as piping and space heaters should be placed in aisles to avoid interfering with storage equipment and to be accessible for maintenance.
- Lighting is normally designed to aid materials-handling equipment operators and order-picking personnel in locating and identifying stored items. Architectural and engineering firms normally assist the planner in determining the number and location of lighting fixtures.
- The type and number of units of fire-protection equipment are governed by fire code and fire underwriter requirements. Generally, the flammability and the amount of material stored determines the requirements.
- Floors may be enhanced to increase durability and housekeeping qualities. Coatings are available to increase surface hardness and wearability and to reduce dust. Typically, 3- to 4-in (8- to 10-cm) lines are painted on the floor to mark traffic and storage aisles.

BIBLIOGRAPHY

"Home Grocery Delivery, A B2C Nightmare," *ID Systems,* March 2001.

"Investing in Inventory Accuracy," *APICS—The Performance Advantage,* July 2000.

"Not the Same Old Rack and Roll," *APICS—The Performance Advantage,* September 1999.

"Picking at the Speed of Light," *Warehousing Management,* April 2000.

"Survival Tactics," *Logistics Management and Distribution Report,* May 2001.

"What's Your Warehouse IQ?" *Warehousing Management,* April 2000.

CHAPTER 4.6
AIR POLLUTION CONTROL

Paul C. Siebert
Michael C. Carey
Roy F. Weston, Inc.
West Chester, Pennsylvania

REGULATORY REQUIREMENTS

Although considerable federal and state legislation concerning air quality was developed during the 1960s, the primary statutory framework now in place was established by the Clean Air Act (CAA) of 1970. This legislation was amended in 1974 to incorporate a program to ensure attainment of the ambient air-quality standards (AAQS) for certain pollutants, and emission or performance standards for certain industrial sources. In 1977 Congress enacted additional revisions, including the Prevention of Significant Deterioration (PSD) and nonattainment requirements applicable to new or modified sources located in pollutant attainment or nonattainment areas, respectively.

Reauthorization of the CAA was due in 1983; however, it was November 1990, after many years of intensive and acrimonious debate by Congress, before the statute was amended again. The CAA Amendments (CAAA) of 1990 were extensive and will double the annual costs to industry in order to achieve compliance over the next 10 yr or more. The major components of the preamendment clean air program are set forth here.

Pre-1990 Regulations and Standards

National Ambient Air-Quality Standards. The National Ambient Air-Quality Standards (NAAQS) define the quality of air that must be achieved to prevent adverse effects. The primary air-quality standards specify levels of pollution that cannot be exceeded without threatening adverse effects on human health. The secondary air-quality standards set limits that if exceeded could result in adverse effects on public welfare, including property and vegetation. For certain pollutants, both short- and long-term standards have been established. For each type of pollutant for which health- and welfare-based air-quality criteria have been established there is a set of these standards, each with its own regulatory program to limit atmospheric concentration levels at or below the levels set by its standards. The secondary standards

TABLE 4.10 National Primary and Secondary Ambient Air-Quality Standards

Pollutant	Type of standard	Averaging time	Compliance frequency parameter	Concentration µg/m³	ppm
Sulfur oxides (as sulfur dioxide)	Primary	24 h	Annual maximum*	365	0.14
	Primary	1 yr	Arithmetic mean	80	0.03
	Secondary	3 h	Annual maximum*	1,300	0.5
Particulate matter of 2.5 µm or less ($PM_{2.5}$)	Primary and secondary†	24 h	Three-year average of the 98th percentile daily value	65	—
	Primary and secondary†	1 yr	Three-year average	15	—
Particulate matter of 10 µm or less (PM_{10})	Primary and secondary	24 h	Annual maximum‡	150	—
		24 h	Annual arithmetic mean‡	50	—
Carbon monoxide	Primary and secondary	1 h	Annual maximum*	40,000	35
		8 h	Annual maximum*	10,000	9
Ozone	Primary and secondary	1 h	Annual maximum*	235	0.12
	Primary and secondary†	8 h	Three-year average of fourth-highest daily maximum	—	0.08
Nitrogen dioxide	Primary and secondary	1 yr	Arithmetic mean	100	0.05
Lead	Primary and secondary	3 mo	Arithmetic mean	1.5	—

* No more than one expected exceedance per year.

† A 1999 federal court ruling remanded these standards to EPA, which had proposed them in 1997, thus blocking their implementation. The U.S. Supreme Court has agreed to reconsider that decision.

‡ In 1997, EPA proposed to revise the compliance frequency parameters for PM_{10} to a 3-yr average of the 99th percentile 24-h average concentration; averaging of exceedances per year over a 3-yr period; and a 3-yr average of the annual arithmetic mean concentrations. A 1999 federal court ruling vacated these revisions, thus blocking their implementation. EPA has not contested this ruling.

are identical to these primary standards, except for sulfur dioxide (SO2), for which there is a 3-h, short-term secondary standard of 1300 mg/m3. Refer to Table 4.10, which presents the current NAAQS, including the particulate matter of 10 mg or less (PM10) standard that replaced the original total suspended particulate (TSP) standard in 1993. Whenever ambient air pollution levels exceed the standard (i.e., nonattainment areas), the thrust of the program is to mandate efforts to reduce air emissions to improve air quality.

State Implementation Plan. The State Implementation Plan (SIP) provides a scheme under which the NAAQS are expected to be achieved and maintained. These plans, which must be approved by the U.S. Environmental Protection Agency (EPA), provide comprehensive programs by each state to reduce pollution through a variety of specific abatement measures, some of which are delineated in facility-specific compliance programs. Once approved by EPA, these state-specific or facility-specific requirements become federally enforceable limitations. When deficiencies in the SIPs become apparent, modifications can be made, using specific legal procedures set forth by EPA. However, the process of issuing or modifying a SIP can be complicated and time consuming.

Air-Quality Control Regions. The Air-Quality Control Region (AQCR) is a basic delineation of each state into appropriate geographical airsheds. Within an AQCR, a state must define the attainment status for each pollutant. There are 247 AQCRs in the United States in which air quality is defined by ambient-air monitoring data, estimates made from mathematical dispersion analyses, or both. Each AQCR has its own program and schedule specified in the SIPs (some AQCRs are interstate), which have been developed to achieve and/or maintain acceptable air quality.

New Source Performance Standards. The New Source Performance Standards (NSPS) are an indirect way of attaining or maintaining air-quality standards by limiting emissions from certain new or modified sources. These standards now cover more than 70 basic industrial and process categories. A new plant is subject to the NSPS only if these standards are proposed by EPA for a certain source category before the plant is under construction. Thus, before proceeding with a new plant, a company should determine which categories are covered by NSPS. Industries affected by the NSPS are shown in Table 4.11. Industrial categories scheduled for NSPS development are listed in Table 4.12.

National Emission Standards for Hazardous Air Pollutants. The National Emission Standards for Hazardous Air Pollutants (NESHAPs) were designed as a means of limiting emissions from both new and existing sources of pollutants that impair human health. However, under the CAA, as amended in 1977, the development and promulgation of NESHAPs was required to follow a rather lengthy and arduous path that was subject to several general criteria that became the focus of regulatory and judicial challenges and interpretations, resulting in subsequent delays. Therefore, during the 20 yr prior to the CAAA of 1990, only 22 NESHAPs were promulgated for 8 hazardous air pollutants (HAPs) emitted from narrowly defined source categories (see Table 4.13). The main provisions of the act (in Section 112) that brought about this rather ponderous process were the following:

- The requirement that a pollutant had to go through a rather lengthy, and often contested, process of being listed as a HAP "which causes, or contributes to, air pollution which may reasonably be anticipated to result in an increase in mortality or an increase in serious, irreversible, or incapacitating reversible, illness." This listing process had to be completed before standards could be developed (which process in itself required numerous internal EPA, interagency, and public review steps).
- The requirement that NESHAPs "provide an ample margin of safety to protect the public health" without any other Congressional guidance so that the regulatory development process could not consider cost (as was permitted and required for the technology-based NSPSs).

PSD/New Source Review. A special feature of the clean air program, largely added by the 1977 amendments, applies to new or modified facilities that cause a significant increase in air emissions. The two primary features of this review process pertain to new or modified facilities located in areas where a significantly emitted pollutant is either in attainment or not in attainment of the applicable NAAQS. For those areas in attainment of the standard, the proposed industrial project must avoid a significant deterioration of the air quality by complying with the PSD requirements. In nonattainment areas, the project will be subject to New Source Review (NSR) nonattainment requirements designed to achieve a net improvement of the air quality. It is important to note that PSD and nonattainment requirements are applied separately to each regulated pollutant. Thus, a new plant to be located in an area classified as nonattainment for ozone and attainment for particulates and SO_2 could be required to comply with both the PSD and the NSR nonattainment requirements. The principal requirements of each program are presented here.

Prevention of Significant Deterioration. EPA's regulations for PSD of air quality in clean air areas represent one of the most complicated parts of the entire CAA. Their basic purpose is to prevent unlimited industrial growth from degrading air quality in those areas where the NAAQS are being met, to ensure preservation of cleaner air. The system designed to achieve that goal includes technology standards to ensure that any new major facility or modification incorporates the best available control technology (BACT), an incremental system to prevent

TABLE 4.11 Typical Categories for Federal New Source Performance Standards*

Subpart	Category 40 CFR 60	Proposed	Promulgated	Last revised
D	Fossil-Fuel-Fired Steam Generators—8/17/71–9/19/78	8/17/71	12/23/71	9/24/96
Da	Electric Utility Steam Generating Units—9/18/78	9/19/78	6/11/79	5/7/90
Db	Industrial-Commercial-Institutional Steam Generating Units (greater than 100 MBtu/h)	6/19/84	11/25/86	10/8/97
Dc	Small Industrial-Commercial-Institutional Steam Generating Units (10–100 MBtu/h)	6/9/89	9/12/90	5/8/96
E	Incinerators	8/17/71	12/23/71	2/14/90
Ea	Municipal Waste Combustors—12/20/89–9/20/94	12/20/89	2/11/91	12/19/94
Eb	Large Municipal Waste Combustors—9/20/94	9/20/94	12/19/95	8/25/97
Ec	Hospital/Medical/Infectious Waste Incinerators—6/20/96	6/20/96	9/15/97	—
F	Portland Cement Plants	8/17/71	12/23/71	2/14/89
G	Nitric Acid Plants	8/17/71	12/23/71	2/14/89
H	Sulfuric Acid Plants	8/17/71	12/23/71	2/14/89
I	Hot-Mix Asphalt Facilities	6/11/73	3/8/74	2/14/89
J	Petroleum Refineries	6/11/73	3/8/74	2/4/91
K	Storage Vessels for Petroleum Liquids—6/11/73–5/19/78	6/11/73	3/8/74	4/8/87
Ka	Storage Vessels for Petroleum Liquids—5/18/78–7/23/84	5/18/78	4/4/80	4/8/87
Kb	Volatile Organic Liquid Storage Vessels—7/23/84	7/23/84	4/8/87	10/8/97
L	Secondary Lead Smelters	6/11/73	3/8/74	2/14/89
M	Secondary Brass and Bronze Production Plants	6/11/73	3/8/74	2/14/89
N	Basic Oxygen Process (BOP) Furnaces—6/11/73	6/11/73	3/8/74	2/14/89
Na	Secondary Emissions from BOP Steelmaking Facilities—1/20/83	1/20/83	1/2/86	2/14/89
O	Sewage Treatment Plants	6/11/73	3/8/74	2/3/94
P	Primary Copper Smelters	10/16/74	1/15/76	2/14/89
Q	Primary Zinc Smelters	10/16/74	1/15/76	2/14/89
R	Primary Lead Smelters	10/16/74	1/15/76	2/14/89
S	Primary Aluminum Reduction Plants	10/23/74	7/25/77	10/7/97
T	Phosphate Fertilizer Industry (PFI): Wet-Process Phosphoric Acid Plants	10/22/74	8/6/75	2/14/89
U	PFI: Superphosphoric Acid Plants	10/22/74	8/6/75	2/14/89
V	PFI: Diammonium Phosphate Plants	10/22/74	8/6/75	2/14/89
W	PFI: Triple Superphosphate Plants	10/22/74	8/6/75	5/17/89
X	PFI: Granular Triple Superphosphate Storage Facilities	10/22/74	8/6/75	4/15/97
Y	Coal Preparation Plants	10/22/74	1/15/76	2/14/89
Z	Ferroalloy Production Facilities	10/21/74	5/4/76	2/14/90
AA	Steel Plants: Electric Arc Furnaces—10/21/74–8/17/83	10/21/74	9/23/75	5/17/89
AAa	Steel Plants: Electric Arc Furnaces—8/17/83	8/17/83	10/31/84	5/17/89
BB	Kraft Pulp Mills	9/24/76	2/23/78	2/14/90
CC	Glass Manufacturing Plants	6/15/79	10/7/80	5/17/89
DD	Grain Elevators	1/13/77	8/3/78	2/14/89
EE	Surface Coating of Metal Furniture	11/28/80	10/29/82	12/13/90
GG	Stationary Gas Turbines	10/3/77	9/10/79	6/27/89
HH	Lime Manufacturing Plants	5/3/77	4/26/84	2/14/89
KK	Lead-Acid Battery Manufacturing Plants	1/14/80	4/16/82	2/14/89
LL	Metallic Mineral Processing Plants	8/24/82	2/21/84	2/14/89
MM	Automobile and Light-Duty Truck Surface Coating Operations	10/5/79	12/24/80	10/11/94
NN	Phosphate Rock Plants	9/21/79	4/16/82	5/17/89
PP	Ammonium Sulfate Manufacture	2/4/80	11/12/80	2/14/89

TABLE 4.11 Typical Categories for Federal New Source Performance Standards* (*Continued*)

Subpart	Category 40 CFR 60	Proposed	Promulgated	Last revised
QQ	Graphic Arts Industry: Publication Rotogravure Printing	10/28/80	11/8/82	—
RR	Pressure-Sensitive Tape and Label Surface Coating Operations	12/30/80	10/18/83	12/13/90
SS	Industrial Surface Coating: Large Appliances	12/24/80	10/27/82	12/13/90
TT	Metal Coil Surface Coating	1/5/81	11/1/82	5/3/91
UU	Asphalt Processing and Asphalt Roofing Manufacture	11/18/80	8/6/82	6/27/89
VV	Equipment Leaks of VOCs in the Synthetic Organic Chemical Manufacturing Industry (SOCMI)	1/5/81	10/18/83	6/12/96
WW	Beverage Can Surface Coating Industry	11/26/80	8/25/83	12/13/90
XX	Bulk Gasoline Terminals	12/17/80	8/18/83	2/14/89
AAA	Residential Wood Heaters	7/1/88	2/26/88	6/29/95
BBB	Rubber Tire Manufacturing Industry	1/20/83	9/15/87	9/19/89
DDD	VOC Emissions from the Polymer Mfg. Industry	9/30/87	12/11/90	3/22/91
FFF	Coating of Flexible Vinyl and Urethane Coating and Printing	1/18/83	6/29/84	8/17/84
GGG	Equipment Leaks of VOCs in Petroleum Refineries	1/4/83	5/30/84	—
HHH	Synthetic Fiber Production Facilities	11/23/82	4/5/84	6/23/94
III	VOC Emissions from SOCMI Air Oxidation Unit Processes	10/21/83	6/29/90	9/7/90
JJJ	Petroleum Dry Cleaners	12/14/82	9/21/84	11/27/85
KKK	Equipment Leaks of VOCs from Onshore Natural Gas Processing Plants	1/20/84	6/24/85	1/21/86
LLL	Onshore Natural Gas Processing: SO_2 Emissions	1/20/84	10/01/85	2/14/89
NNN	VOC Emissions from SOCMI Distillation Operations	12/30/83	6/29/90	11/27/95
OOO	Nonmetallic Mineral Processing Plants	8/31/83	8/1/85	6/9/97
PPP	Wool Fiberglass Insulation Manufacturing Plants	2/7/84	2/25/85	2/14/89
QQQ	VOC Emissions from Petroleum Refinery Wastewater Systems	5/4/87	11/23/88	8/18/95
RRR	VOC Emissions from SOCMI Reactor Processes	6/29/90	8/31/93	11/27/95
SSS	Magnetic Tape Coating Facilities	1/22/86	10/3/88	11/29/88
TTT	Industrial Surface Coating: Surface Coating of Plastic Parts for Business Machines	1/1/86	1/29/88	6/15/89
UUU	Calciners and Dryers in Mineral Industries	4/23/86	9/28/92	7/29/93
VVV	Polymeric Coating of Supporting Substrates Facilities	4/30/87	9/11/89	—
WWW	Municipal Solid Waste Landfills	5/30/91	3/12/96	6/16/98

* For complete list, consult U.S. Environmental Protection Agency, Washington, D.C.

any single project from having undue impacts on air quality, and an intricate scheme of procedural and technical requirements to ensure compliance with the technological and incremental limitations. In brief, the system works as follows:

Best Available Control Technology (BACT). All new plants and modifications subject to PSD must install BACT. What constitutes BACT is determined on a case-by-case basis. During the late 1980s, EPA instituted a top-down system of BACT determination under which the applicant first identifies the most stringent state-of-the-art control technology that could possibly be used, and then places the burden on the applicant to justify any variations from such controls by demonstrating that they cannot be used with respect to the actual proposed facility based on technological, environmental, energy, or cost considerations. Top-down BACT, requiring consideration of the most stringent alternative first, has been set aside by the courts; however, it is still being implemented by many state agencies. EPA has established a BACT clearinghouse for each source category. Nevertheless, it is the responsibility of the applicant to define, support, and defend the selected BACT.

Increments of Air Quality. Even after satisfying the BACT controls, the applicant will not be permitted to proceed unless the remaining emissions can be accommodated within an

TABLE 4.12 Industrial Categories for Which New Source Performance Standards Are to Be Developed

Stationary fuel combustion	Basic chemical manufacture
Stationary internal combustion engines	Synthetic organic chemical manufacturing
	Borax and boric acid
Metallurgical processes	Hydrofluoric acid
By-product coke ovens	Phosphoric acid: thermal process
Foundries: gray iron	Potash
Foundries: steel	Sodium carbonate
Secondary aluminum	
Secondary copper	Chemical products manufacture
Secondary zinc	Ammonia
Uranium refining	Carbon black
	Charcoal
Mineral products	Detergent
Brick and related clay products	Explosives
Castable refractories	Fuel conversion
Ceramic clay	Printing ink
Gypsum	Synthetic fibers
Perlite	Synthetic rubber
Sintering: clay and fly ash	Varnish
Polymers and resins	Evaporative loss sources
ABS-SAN resins	Industrial surface coating: fabric
Acrylic resins	Industrial surface coating: paper
Phenolic resins	
Polyester resins	Petroleum industry
Polyethylene	Crude oil and natural gas production
Polypropylene	Gasoline additives
Polystyrene	Petroleum refinery
Urea-melamine resins	Transportation and marketing
Food and agricultural	Wood processing
Alfalfa dehydrating	Chemical wood pulping: acid sulfite
Ammonium nitrate fertilizer	Chemical wood pulping: neutral sulfite (NSSC)
Animal feed defluorination	Plywood manufacture
Starch	
Urea (for fertilizer and polymers)	Consumer products
Vegetable oil	Textile processing
Waste disposal	
Hazardous waste transfer, storage, and disposal facilities	

available *increment* of air quality. The incremental limitations rest first upon an *area classification system* under which all areas in the country meeting air-quality standards are classified as either Class I (to be kept in especially pristine condition) or Class II (where normal industrial growth is to be permitted). Whichever area is involved, emissions from the new facility or modification must not cause a projected degradation in preexisting air quality beyond the amount of the allowable increment. The idea is that in areas that today are much cleaner than required to satisfy the NAAQS, the air cannot be degraded so that it just barely satisfies the air-quality standards. In fact, in Class I areas, virtually no degradation will be allowed. It should be noted that these increments have been established for SO_2, nitrogen dioxide (NO_2), and particulates. Air-quality modeling analyses must be conducted to demonstrate compliance with applicable PSD increments and NAAQS. Depending on the geographical and top-

TABLE 4.13 Typical Categories for Federal Risk-Based National Emission Standards for Hazardous Air Pollutants*

Subpart	Category 40 CFR 61	Pollutant	Promulgated	Last revised
B	Radon from Underground Uranium Mines	Radon	12/15/89	—
C	Beryllium	Beryllium	4/6/73	11/7/85
D	Beryllium Rocket Motor Firing	Beryllium	4/6/73	11/7/85
E	Mercury	Mercury	4/6/73	9/23/88
F	Vinyl Chloride	Vinyl chloride	10/21/76	12/23/92
H	Radionuclides Other than Radon from DOE Facilities	Radionuclides (except radon)	12/15/89	—
I	Radionuclides from Federal Facilities Other than NRC Licensees and Not Covered by Subpart H	Radionuclides	12/15/89	12/30/96
J	Equipment Leaks (Fugitives) of Benzene	Benzene	6/6/84	—
K	Radionuclides from Elemental Phosphorus Plants	Radionuclides	12/15/89	12/19/91
L	Benzene from Coke By-product Recovery Plants	Benzene	9/14/89	12/19/91
M	Asbestos	Asbestos	4/5/84	6/19/95
N	Inorganic Arsenic from Glass Manufacturing Plants	Arsenic	8/4/86	5/31/90
O	Inorganic Arsenic from Primary Copper Smelters	Arsenic	8/4/86	5/31/90
P	Inorganic Arsenic from Arsenic Trioxide and Metallic Arsenic Production Facilities	Arsenic	8/4/86	10/3/86
Q	Radon from DOE Facilities	Radon	12/15/89	—
R	Radon Emissions from Phosphogypsum Stacks	Radon	6/3/92	—
T	Radon from the Disposal of Uranium Mill Tailings	Radon	12/15/89	7/15/94
V	Equipment Leaks (Fugitive Emission Sources)	Volatile HAP†	6/6/84	7/10/90
W	Radon from Operating Mill Tailings	Radon	12/15/89	—
Y	Benzene from Benzene Storage Vessels	Benzene	9/14/89	12/11/89
BB	Benzene from Benzene Transfer Operations	Benzene	3/7/90	10/31/90
FF	Benzene from Benzene Waste Operations	Benzene	3/7/90	1/7/93

* For complete list, consult U.S. Environmental Protection Agency, Washington, DC.
† Volatile HAPs currently include vinyl chloride and benzene.

ographical features, as well as the existing sources in the vicinity of the proposed facility, the modeling analysis can be extremely complicated and costly. A modeling protocol is typically required by regulatory authorities before any refined or detailed analysis is conducted.

Preconstruction Approval. Plants and modifications subject to PSD review cannot begin construction until a *permit* has been issued. To complete PSD reviews in areas where the air-quality baseline has not been established, which have become less common because PSD has been in effect for an extended period, an applicant generally must present *extensive monitoring data to establish the baseline* against which the increment will be calculated. Both meteorological and air-quality monitoring may be required for a period ranging from 4 to 12 months. These preconstruction requirements may delay any permit submission by a total of 7 to 18 months. A typical schedule of activities for a PSD-affected facility is shown in Fig. 4.62.

Plants subject to PSD review include plants within 28 specified industrial categories (see Table 4.14) if potential emissions of any regulated pollutant exceed 100 tons per year (tpy) and plants in other industrial categories if potential emissions exceed 250 tpy. These plants are defined as "major sources" under the PSD program. Modifications to major sources that

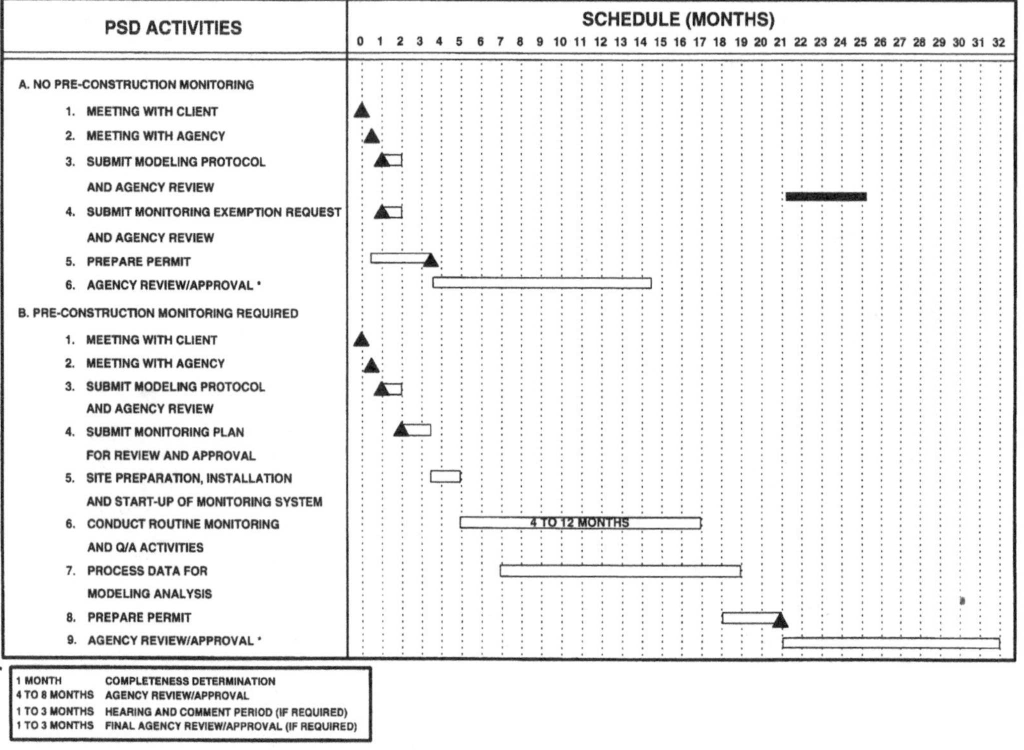

FIGURE 4.62 Typical schedule of activities for permitting PSD-affected facilities.

would result in a "significant net emissions increase" of any regulated pollutant also are subject to PSD review requirements (see Table 4.15). The significance levels vary for many of the regulated pollutants, but for SO_2, nitrogen oxides (NO_x), and volatile organic compounds (VOCs), an increase of 40 tpy is deemed significant.

New Source Review. In areas classified as nonattainment for any regulated pollutant, a different set of special rules restricts the construction of new sources. The basic thrust of these NSR rules is to require that such new sources install state-of-the-art control technology and also provide supplemental reductions in emissions from neighboring sources to more than offset whatever new emissions would result from the new source even after best controls are utilized. The principal requirements are as follows:

- *Lowest achievable emission rate (LAER).* This requirement deliberately imposes a technology-forcing standard of control. The 1977 amendments specified that it must reflect: (1) the most stringent emission limitation contained in any SIP (unless the applicant can demonstrate that such limitations are not achievable), or (2) the most stringent emission limitation achieved in practice within the industrial category, whichever is more stringent. In no event can LAER be less stringent than any applicable NSPS. Often it is comparable to the BACT standard, but in many cases it can be more stringent. Specific applications of this standard must be determined on a case-by-case basis.

TABLE 4.14 100-tpy PSD Source Categories*

1. Fossil-fuel-fired steam electric plants of more than 250 MBtu/h heat input
2. Coal cleaning plants (with thermal dryers)
3. Kraft pulp mills
4. Portland cement plants
5. Primary zinc smelters
6. Iron- and steel-mill plants
7. Primary aluminum ore reduction plants
8. Primary copper smelters
9. Municipal incinerators capable of charging more than 250 tons of refuse per day
10. Hydrofluoric acid plants
11. Sulfuric acid plants
12. Nitric acid plants
13. Petroleum refineries
14. Lime plants
15. Phosphate rock processing plants
16. Coke oven batteries
17. Sulfur recovery plants
18. Carbon black plants (furnace process)
19. Primary lead smelters
20. Fuel conversion plants
21. Sintering plants
22. Secondary metal production plants
23. Chemical process plants
24. Fossil-fuel boilers (or combinations thereof) totaling more than 250 MBtu/h heat input
25. Petroleum storage and transfer units with a total storage capacity exceeding 300,000 barrels
26. Taconite ore processing plants
27. Glass fiber processing plants
28. Charcoal production plants

* From EPA PSD regulations, 40 CFR 52.21(b)(1)(i)(a).

- *Emission offsets.* The key feature of offsets is that they must comprise legally enforceable reductions in emissions from other sources above and beyond those that would otherwise be required. These can be derived from installation of advanced controls producing extra reductions in emissions from existing sources or from the shutdown of such sources. Under the 1977 amendments, offsets can be traded on the market and, subject to certain limitations, can be "banked" for future use. Offsets must be measured against prior actual emissions and must be sufficient to cover potential emissions from the new source. They must also produce a "net air-quality benefit" in the vicinity of the new source.
- *Other requirements.* All other plants in the state owned by the same company must be in compliance with applicable regulations, and the state's SIP must also be carried out.
- *Sources covered.* Any new plant with potential emissions of 100 tpy or more of particulates, SO_2, NO_x, VOCs, or carbon monoxide (CO) was subject to the requirements of the 1977 amendments. Modifications to these major sources are subject to review if they result in significant increases, as with the PSD program. Under the CAAA, these levels have been lowered for the more seriously polluted ozone (O_3), CO, and PM_{10} nonattainment areas.

The most significant impact of these nonattainment requirements has been the restrictive effect of limitations on the availability of offsets. In earlier years, offsets could be more readily found, but as the regulatory requirements have become more stringent, especially in the more serious nonattainment areas, it has become increasingly difficult to find potential reductions in current emissions beyond those already demanded by regulatory controls.

TABLE 4.15 Significant Net Emissions Increase*

Pollutant	Increase, tpy
CO	100
NO_x (as NO_2)	40
SO_2	40
Particulate matter	25
TSP	25
PM_{10}	<15
Ozone	40 of volatile organic compounds
Lead	0.6
Asbestos	0.007†
Beryllium	0.0004†
Mercury	0.1†
Vinyl chloride	1
Fluorides	3
Sulfuric acid mist	7
Hydrogen sulfide	10
Total reduced sulfur	10
Reduced sulfur compounds	10
Municipal waste combustor (MWC)	
MWC acid gases	40
MWC metals	15
MWC organics	3.5×10^{-6}
Other CAA pollutants	>0‡

* From EPA PSD regulations, 40 CFR 52.21(b)(23)(i).

† Title III of the CAAA added a new section (112(b)(6)) that excludes these HAPs listed in Section 112(b)(1) of the revised act from federal PSD requirements. Current EPA policy (New Source Review Program Transitional Guidance, March 11, 1991) clarifies that states with approved PSD programs may continue to regulate these pollutants under state PSD regulations.

‡ EPA regulations, 40 CFR 52.21(b)(23)(ii), specify that any net emissions increase is considered significant for any CAA pollutants for which significant emission increase levels are not specified by 40 CFR 52.21(b)(23)(i) (e.g., arsenic, benzene, and radionuclides). However, current EPA policy (New Source Review Program Transitional Guidance, March 11, 1991) in response to the CAAA, specifies that HAPs under Section 112 of the CAA are no longer considered regulated pollutants under federal PSD. However, several other pollutants regulated under the CAA are still considered significant for any net emission increase because they are not regulated or listed as HAPs under Section 112.

Because of the procedural delays and difficulties of complying with PSD or nonattainment regulations, companies have sought to avoid these requirements by designing projects that cause little or no net increase in emissions. Federally enforceable permit conditions have been developed and approved by the permitting authority as a means of avoiding these requirements. A set of internal "netting" rules governing the assessment of net emissions increases has also allowed certain projects to avoid the NSR process entirely.

Post-1990 Regulations and Standards

The CAAAs have had a major impact on virtually every industrial and many commercial operations throughout the country. These are potentially onerous requirements that have affected the economic viability of some existing facilities and the design and scheduled operation of new or modified facilities. Fortunately, the requirements of many provisions of the

CAAA were implemented over a 10-yr or longer period. Unfortunately, the economic burden to industry over this period may represent an incremental cost increase on a national basis of \$20 to \$30 billion per year.

The primary provisions of the act are summarized here:

- *Nonattainment.* Title I of the act established a new classification of nonattainment areas for each pollutant and included deadlines and mandated degrees of emission reductions required to achieve attainment, based on the severity of present pollution levels.
- *Mobile sources.* As part of the O_3 and CO nonattainment programs, Title II directed further tightening of emission standards applicable to motor vehicles. The Title II provision also established new requirements for petroleum companies to produce alternative fuels for motor vehicles, and for the automotive industry to design and manufacture vehicles that are capable of using such alternative fuels, intended for sale to fleet operations in designated areas.
- *Air toxics.* Title III established requirements to limit routine emissions of HAPs and measures to prevent accidental releases of extremely hazardous substances (EHSs).
- *Acid rain.* Title IV provisions established a new set of requirements, primarily on coal-fired power plants, designed to cut emissions of SO_2 in half and to reduce emissions of NO_x. The amendments also provided market-driven mechanisms to facilitate achievement with the acid rain requirements.
- *Operating permit program.* Title V stipulated that each state develop and implement a comprehensive new operating permit program for all significant air emission sources. The program had to require rigorous monitoring, reporting, and compliance certification, as well as annual permit fees.
- *Stratospheric ozone.* Title VI established new requirements to restrict emissions of chlorofluorocarbons (CFCs) as a major step toward addressing problems of stratospheric ozone depletion and global warming.
- *Enforcement.* Title VII greatly strengthened enforcement authorities, particularly by adding new criminal sanctions. Unlike the other provisions of the act, the new enforcement requirements were effective the day the act was signed into law.

The three areas of primary concern to industry are the nonattainment, air toxics, and permit provisions of the CAAAs. The major requirements and planning considerations associated with each of these provisions are discussed in the remainder of this chapter. It is important to note that development and implementation of the regulations for some of these provisions have been delayed by EPA and various states. As a result, it is incumbent upon each company to track the status and schedule of new regulations, including penalties for noncompliance, and appropriately comment on proposed rules as they are developed.

Nonattainment Requirements. Currently, there are roughly 100 areas of nonattainment for O_3, over 40 for CO, and 70 for particulate matter. Also, there are a number of areas not in compliance with SO_2 standards. Thus, existing or proposed new or modified industrial facilities located in or near these areas will be subject to a variety of new requirements designed to limit existing and future emissions of these nonattainment pollutants. For certain existing sources, reasonably available control technology (RACT) requirements have or will be imposed. The construction of a new or modified source may require the installation of LAER technology and emission offsets.

RACT. The CAAAs have expanded the type and number of existing sources that will be subject to the RACT requirements. Previously, the requirement was generally applicable to major emission sources (i.e., with the potential to emit 100 tpy) of VOCs located in O_3 nonattainment areas. Up to this point, the O_3 nonattainment problem was addressed almost exclusively through the control of VOCs. The CAAAs extend applications of this basic RACT requirement to NO_x, with RACT now being applied to smaller sources in more seriously polluted areas. NO_x in the presence of VOCs and sunlight can contribute to the formation of ground-level O_3. Existing major sources subject to the new RACT requirement are shown in Table 4.16.

It is important to note that any existing source located in an 11-state area extending from the Washington metropolitan area to Maine, with the potential to emit 50 tpy of VOCs and 100 tpy of NO_x, will also be defined as a major source and be subject to RACT requirements. This area has been established by the CAAA and is known as the Northeast Ozone Transport Region.

A determination of the practices or procedures required to satisfy RACT regulations can or should be made on a case-by-case review of the circumstances at each individual facility. However, EPA has adopted the mechanism of issuing Control Technology Guidelines (CTGs) to provide a generic definition of RACT for specified industrial categories. Although this procedure has provoked criticism of EPA, it has, in effect, required states to apply the CTG requirements as though they have regulatory effect. EPA issued 29 CTGs prior to the CAAA and another 8 draft or final CTGs in the early 1990s. CTGs have been developed and issued only for VOC source categories. In addition, alternative control technique (ACT) documents were issued in the early 1990s for 3 additional VOC sources and 10 NO_x source or control device scenarios. Because RACT had to be defined, approved, and installed for existing applicable sources by May 31, 1995, EPA guidance was not available for all NO_x emission sources. As a result, categorical limits and/or presumptive requirements were established by various states for NO_x-related combustion sources.

For sources potentially subject to these regulations, it is important to quantify VOC and/or NO_x emissions over a range of operating conditions. It may be possible to accept permit conditions to limit emissions such that the source is not defined as a major source and thus not subject to the RACT requirements. In any event, it will be important to review the specific regulations and discuss the requirements with regulatory authorities prior to preparing any RACT determination. RACT determinations will vary greatly from state to state, especially for NO_x emission sources. Some states established presumptive RACT levels for certain sources, while other states implemented case-by-case RACT, often with the intent of granting greater flexibility to industry, but sometimes with the effect of greater uncertainly, cost of analysis, and inclusion of smaller sources.

NSR/LAER. One of the most significant impacts of the nonattainment requirements affecting American industry is on the construction of new facilities or modification of existing facilities. When any increase in emissions of a major source above the minimum amounts would result from an expansion or modernization project or certain other changes, the project will be subject to NSR requirements. These requirements under the CAAAs have expanded and complicated the permitting process. The principal requirements are: (1) a construction permit must be obtained, (2) technology standards reflecting LAER must be satisfied, and (3) offsets representing emission reductions from other sources must be provided.

The definition of *source applicability* or what constitutes a *major source* is the primary concern. The CAAAs reduced the previous definition (100 tpy) to lower levels in certain O_3, CO, and PM_{10} nonattainment areas. This will expand the reach of NSR to cover smaller sources. In

TABLE 4.16 Existing Source RACT Applicability Requirements for Each Ozone Nonattainment Area Category

Category of nonattainment area	Size of VOC or NO_x sources affected
Extreme	10
Severe	25
Serious	50
Moderate and marginal	100

Ozone attainment status and classification must be identified. Size is defined as the potential to emit VOCs or NO_x in tons per year.

addition, new or modified NO_x emission sources in O_3 nonattainment areas will be subject to the same threshold levels and NSR requirements as VOC emission sources. The CAAAs make no change in the LAER technology requirements. However, the CAAAs do tighten the offset requirements by increasing the ratio of offsets needed in areas with higher O_3 levels. The revised major source definition and offset ratios are presented in Table 4.17.

The enhanced NSR requirements, as well as the expanded regulations, will have a dramatic effect on new source construction and availability of emission offsets. Existing sources will apply RACT to control NO_x and/or VOC emissions. Unless an existing source "overcontrols" either VOC and/or NO_x emissions beyond RACT-related requirements or a source is permanently shut down, emission credits or offsets may not become available. Thus, any existing source would be well advised to "bank" or save any internally created emission credits. These internally derived emission-reduction credits may represent the only offsets available within certain geographic areas, especially for NO_x sources. However, the geographic area may expand considerably for existing sources or sources proposed in the Northeast Ozone Transport Region. With EPA's approval, it is the intent for states located in the region, as some already have, to allow intrastate emission trading and the establishment of a regional emission bank. This approach would significantly improve the likelihood for industrial growth because of an expanded geographic area and the opportunity to trade emission offsets on a regional basis.

One final aspect peculiar to new or modified major sources pertains to NO_x emissions. NO_x emission sources in certain ozone nonattainment areas can be subject to both PSD and NSR nonattainment requirements. For example, an existing major source (i.e., greater than 100 tpy of one of the 28 PSD source categories) located in a severe nonattainment area intends to install an NO_x emission source with the potential to emit 50 tpy. The source would be required to apply LAER (i.e., more stringent than BACT) and obtain offsets at a 1.3-to-1 ratio. However, a modeling analysis may be required to demonstrate that the PSD increment and/or the applicable NAAQS will not be exceeded. The applicant needs to be aware of these unusual and unique situations, which must be considered in developing an overall permitting strategy.

Air Toxics

Routine Release Conditions. The CAAAs include a substantially revised program to regulate HAPs. The revised program substantially streamlined NESHAP development by listing HAPs in the CAAAs and by setting standards based on available control technology rather than health effects. Residual health-effect risks after implementation of technology-based standards will later determine whether additional regulation is needed. HAPs were defined as 189 listed substances that may cause health and environmental effects at low concentrations and are not regulated as criteria pollutants (refer to Table 4.18).

The CAAAs define a major HAP source as any stationary source or group of stationary sources located within a contiguous area under common ownership that emits or has the

TABLE 4.17 Major Source Definitions and Offset Ratios in Ozone Nonattainment Areas

Category	Size of major source, tpy	Offset ratios
Marginal	100	1.1:1
Moderate	100*	1.15:1
Serious	50*	1.2:1
Severe	25*	1.3:1
Extreme	10	1.5:1
Ozone transport region not classified as serious, severe, or extreme	50 VOC, 100 NO_x	1.15:1

* States have the option of choosing a major source definition of 5 tpy (and accepting other conditions) in order to avoid complying with a requirement that emissions be reduced 15 percent over the first 6 yr. See Section 182(b)(1)(A)(ii).

TABLE 4.18 Section 112 Hazardous Air Pollutants[a]

Chemical Abstracts Service number	Pollutant	Chemical Abstracts Service number	Pollutant
75-07-0	Acetaldehyde	132-64-9	Dibenzofuran
60-35-5	Acetamide	96-12-8	1,2-Dibromo-3-chloropropane
75-05-8	Acetonitrile	84-74-2	Dibutyl phthalate
98-86-2	Acetophenone	106-46-7	1,4-Dichlorobenzene
53-96-3	2-Acetylaminofluorene	91-94-1	3,3′-Dichlorobenzidine
107-02-8	Acrolein	111-44-4	Dichloroethyl ether (bis[2-chloroethyl]ether)
79-06-1	Acrylamide		
79-10-7	Acrylic acid	542-75-6	1,3-Dichloropropene
107-13-1	Acrylonitrile	62-73-7	Dichlorvos
107-05-1	Allyl chloride	111-42-2	Diethanolamine
92-67-1	4-Aminobiphenyl	64-67-5	Diethyl sulfate
62-53-3	Aniline	119-90-4	3,3′-Dimethoxybenzidine
90-04-0	*o*-Anisidine	60-11-7	4-Dimethylaminoazobenzene
1332-21-4	Asbestos	121-69-7	*N,N*-Dimethylaniline
71-43-2	Benzene (including benzene from gasoline)	119-93-7	3,3′-Dimethylbenzidine
		79-44-7	Dimethylcarbamoyl chloride
92-87-5	Benzidine	68-12-2	*N,N*-Dimethylformamide
98-07-7	Benzotrichloride	57-14-7	1,1-Dimethylhydrazine
100-44-7	Benzyl chloride	131-11-3	Dimethyl phthalate
92-52-4	Biphenyl	77-78-1	Dimethyl sulfate
117-81-7	Bis(2-ethylhexyl)phthalate (DEHP)	NA	4,6-Dinitro-*o*-cresol (including salts)
542-88-1	Bis(chloromethyl) ether		
75-25-2	Bromoform	51-28-5	2,4-Dinitrophenol
106-99-0	1,3-Butadiene	121-14-2	2,4-Dinitrotoluene
156-62-7	Calcium cyanamide	123-91-1	1,4-Dioxane (1,4-diethylene-oxide)
105-60-2	Caprolactam (removed 6/18/96, 61 FR 30816)		
		122-66-7	1,2-Diphenylhydrazine
133-06-2	Captan	106-89-8	Epichlorohydrin (1-chloro-2,3-epoxypropane)
63-25-2	Carbaryl		
75-15-0	Carbon disulfide	106-88-7	1,2-Epoxybutane
56-23-5	Carbon tetrachloride	140-88-5	Ethyl acrylate
463-58-1	Carbonyl sulfide	100-41-4	Ethylbenzene
120-80-9	Catechol	51-79-6	Ethyl carbamate (urethane)
133-90-4	Chloramben	75-00-3	Ethyl chloride (chloroethane)
57-74-9	Chlordane	106-93-4	Ethylene dibromide (dibromo-ethane)
7782-50-5	Chlorine		
79-11-8	Chloroacetic acid	107-06-2	Ethylene dichloride (1,2-dichloroethane)
532-27-4	2-Chloroacetophenone		
108-90-7	Chlorobenzene	107-21-1	Ethylene glycol
510-15-6	Chlorobenzilate	151-56-4	Ethyleneimine (Aziridine)
67-66-3	Chloroform	75-21-8	Ethylene oxide
107-30-2	Chloromethyl methyl ether	96-45-7	Ethylene thiourea
126-99-8	Chloroprene	75-34-3	Ethylidene dichloride (1,1-dichloroethane)
1319-77-3	Cresol/cresylic acid (mixed isomers)		
95-48-7	*o*-Cresol	50-00-0	Formaldehyde
108-39-4	*m*-Cresol	76-44-8	Heptachlor
106-44-5	*p*-Cresol	118-74-1	Hexachlorobenzene
98-82-8	Cumene	87-68-3	Hexachlorobutadiene
NA	2,4-D (2,4-dichlorophenoxyacetic acid) (including salts and esters)	NA	1,2,3,4,5,6-Hexachlorocyclo-hexane (all stereoisomers, including lindane)
72-55-9	DDE (1,1-dichloro-2,2-bis (*p*-chlorophenyl) ethylene)		
		77-47-4	Hexachlorocyclopentadiene
334-88-3	Diazomethane	67-72-1	Hexachloroethane

TABLE 4.18 Section 112 Hazardous Air Pollutants[a] (*Continued*)

Chemical Abstracts Service number	Pollutant	Chemical Abstracts Service number	Pollutant
822-06-0	Hexamethylene diisocyanate	78-87-5	Propylene dichloride (1,2-dichloropropane)
680-31-9	Hexamethylphosphoramide	75-56-9	Propylene oxide
110-54-3	Hexane	75-55-8	1,2-Propylenimine (2-methylaziridine)
302-01-2	Hydrazine	91-22-5	Quinoline
7647-01-0	Hydrochloric acid (hydrogen chloride)	106-51-4	Quinone (*p*-benzoquinone)
7664-39-3	Hydrogen fluoride (hydrofluoric acid)	100-42-5	Styrene
123-31-9	Hydroquinone	96-09-3	Styrene oxide
78-59-1	Isophorone	1746-01-6	2,3,7,8-Tetrachlorodibenzo-*p*-dioxin
108-31-6	Maleic anhydride	79-34-5	1,1,2,2-Tetrachloroethane
67-56-1	Methanol	127-18-4	Tetrachloroethylene (perchloroethylene)
72-43-5	Methoxychlor	7550-45-0	Titanium tetrachloride
74-83-9	Methyl bromide (bromomethane)	108-88-3	Toluene
74-87-3	Methyl chloride (chloromethane)	95-80-7	Toluene-2,4-diamine
71-55-6	Methyl chloroform (1,1,1-trichloroethane)	584-84-9	2,4-Toluene diisocyanate
78-93-3	Methyl ethyl ketone (2-butanone)	95-53-4	*o*-Toluidine
60-34-4	Methylhydrazine	8001-35-2	Toxaphene (chlorinated camphene)
74-88-4	Methyl iodide (iodomethane)	120-82-1	1,2,4-Trichlorobenzene
108-10-1	Methyl isobutyl ketone (hexone)	79-00-5	1,1,2-Trichloroethane
624-83-9	Methyl isocyanate	79-01-6	Trichloroethylene
80-62-6	Methyl methacrylate	95-95-4	2,4,5-Trichlorophenol
1634-04-4	Methyl *tert*-butyl ether	88-06-2	2,4,6-Trichlorophenol
101-14-4	4,4′-Methylene bis(2-chloroaniline)	121-44-8	Triethylamine
75-09-2	Methylene chloride (dichloromethane)	1582-09-8	Trifluralin
101-68-8	4,4′-Methylenediphenyl diisocyanate (MDI)	540-84-1	2,2,4-Trimethylpentane
101-77-9	4,4′-Methylenedianiline	108-05-4	Vinyl acetate
91-20-3	Naphthalene	593-60-2	Vinyl bromide
98-95-3	Nitrobenzene	75-01-4	Vinyl chloride
92-93-3	4-Nitrobiphenyl	75-35-4	Vinylidene chloride (1,1-dichloroethylene)
100-02-7	4-Nitrophenol	1330-20-7	Xylenes (mixed isomers)
79-46-9	2-Nitropropane	95-47-6	*o*-Xylene
684-93-5	*N*-Nitroso-*N*-methylurea	108-38-3	*m*-Xylene
62-75-9	*N*-Nitrosodimethylamine	106-42-3	*p*-Xylene
59-89-2	*N*-Nitrosomorpholine		Antimony compounds
56-38-2	Parathion		Arsenic compounds (inorganic, including arsine)
82-68-8	Pentachloronitrobenzene (quintobenzene)		Beryllium compounds
87-86-5	Pentachlorophenol		Cadmium compounds
108-95-2	Phenol		Chromium compounds
106-50-3	*p*-Phenylenediamine		Cobalt compounds
75-44-5	Phosgene		Coke oven emissions
7803-51-2	Phosphine		Cyanide compounds[b]
7723-14-0	Phosphorus		Glycol ethers[c]
85-44-9	Phthalic anhydride		Lead compounds
1336-36-3	Polychlorinated biphenyls (Aroclors)		Manganese compounds
1120-71-4	1,3-Propane sultone		Mercury compounds
57-57-8	β-Propiolactone		Fine mineral fibers[d]
123-38-6	Propionaldehyde		Nickel compounds
114-26-1	Propoxur (Baygon)		

TABLE 4.18 Section 112 Hazardous Air Pollutants[a] (*Continued*)

Chemical Abstracts Service number	Pollutant
	Polycyclic organic matter[e]
	Radionuclides (including radon)[f]
	Selenium compounds

Note: For all listings above that contain the word *compounds* and for glycol ethers, the following applies. Unless otherwise specified, these listings are defined as including any unique chemical substance that contains the named chemical (i.e., antimony, arsenic, etc.) as part of that chemical's infrastructure. See August 14, 2000, memo from John Seitz, EPA, OAQPS on Guidance on the Major Source Determination for Certain Hazardous Air Pollutants for further information.

[a] This draft list from the EPA OAQPS Technology Transfer Network web site (as of November 2000) includes current EPA staff recommendations for technical corrections and clarifications of the hazardous air pollutants (HAPs) list in Section 112(b)(1) of the Clean Air Act. This draft has been distributed to apprise interested parties of potential future changes in the HAPs list and is informational only. The recommended revisions of the current HAPs list, which are included in this draft, do not themselves change the list as adopted by Congress and have no legal effect. EPA intends to propose specific revisions of the HAPs list, including any technical corrections or clarifications of the list, only through notice and comment rulemaking. Hydrogen sulfide, which had been inadvertently added to the original list of HAPs as published in the CAAAs, Section 112(b)(1) by a clerical error, was removed by a joint resolution of Congress that was approved by the president on December 4, 1991.

[b] X′CN where X = H′ or any other group where a formal dissociation may occur. For example, KCN or $Ca(CN)_2$.

[c] On January 12, 1999 (64 FR 1780), EPA proposed to modify the definition of glycol ethers to exclude surfactant alcohol ethoxylates and their derivatives (SAEDs). On August 2, 2000 (65 FR 47342), EPA published the final action. This action deletes each individual compound in a group called the surfactant alcohol ethoxylates and their derivatives (SAEDs) from the glycol ethers category in the list of HAPs established by Section 112(b)(1) of the CAA. EPA also made conforming changes in the definition of glycol ethers with respect to the designation of hazardous substances under the Comprehensive Environmental Response, Compensation, and Liability Act (CERCLA).

The following definition of the glycol ethers category of hazardous air pollutants applies instead of the definition set forth in 42 U.S.C. 7412(b)(1), footnote 2:

Glycol ethers include mono- and diethers of ethylene glycol, diethylene glycol, and triethylene glycol $R\text{-}(OCH_2CH_2)_n\text{-}OR'$

where $n = 1, 2$, or 3
R = alkyl C7 or less, or phenyl or alkyl substituted phenyl
R′ = H, or alkyl C7 or less, or carboxylic acid ester, sulfate, phosphate, nitrate, or sulfonate

The following definition is currently set forth in 42 U.S.C. 7412(b)(1):

Includes mono- and diethers of ethylene glycol, diethylene glycol, and triethylene glycol $R\text{-}(OCH_2CH_2)_n\text{-}OR'$

where $n = 1, 2$, or 3.
R = alky or aryl groups.
R′ = R, H, or groups that, when removed, yield glycol ethers with the structure $R\text{-}(OCH2CH)_n\text{-}OH$.

Polymers are excluded from the glycol category.

[d] Includes mineral fiber emissions from facilities manufacturing or processing glass rock or slag fibers (or other mineral-derived fibers) of average diameter of 1 μm or less. (This definition is currently under review.)

[e] Includes organic compounds with more than one benzene ring, and that have a boiling point greater than or equal to 100°C. (This definition is currently under review.)

[f] A type of atom that spontaneously undergoes radioactive decay.

potential to emit, considering controls, 10 tpy or more of any single HAP or 25 tpy or more of any combination of HAPs. Even seemingly minor facilities may be considered major sources of HAP emissions under these extremely stringent definitions. It should be noted that the EPA administrator has the discretionary power to establish lower-quantity threshold values for highly toxic or carcinogenic compounds. The triggering levels for these substances could be considerably less than 10 tpy.

The first step in determining the impact of the new air toxics programs is to simply compare the results of the facility emission inventory to the 10- and 25-ton thresholds. For certain operations, secondary by-products may be chemically created and exist in the effluent discharge stream. If they are triggered for any one or more of the 189 HAPs, then the following provisions need to be evaluated in detail.

The new HAP emissions control program requires the installation of state-of-the-art pollution control technology on the vast majority of emission sources. However, process modifications or product substitution may represent an acceptable alternative approach.

The HAP provisions regulate emissions in a two-step process:

1. Promulgation over 10 yr of emissions limits reflecting maximum achievable control technologies (MACTs) for 174 categories of industrial sources potentially emitting HAPs.
2. Residual risk determination to control categories of sources whose emissions still present health risks even after the application of MACTs.

In Step 1, EPA is mandated to develop a technology-based regulatory framework (the MACTs) to regulate source categories and limit HAP emissions by source categories over the 10 yr following passage of the CAAAs. HAP emissions will be reduced by setting limits on specific industrial emission sources. EPA identified for development of regulations 174 categories of industrial sources that emit substantial quantities of HAPs. Emissions levels reflecting MACTs are established for each regulated source category. These levels are used ultimately to set source-specific limits in facility operating permits. The entire process is similar to that for National Pollutant Discharge Elimination System (NPDES) permits in which effluent guidelines are used to set facility-specific limits. The identified source categories and the status of the MACT standards are given in Table 4.19.

The MACT determination process was outlined in the CAAA. MACTs were defined by Congress as the "maximum degree of reduction in emissions . . . taking into consideration the cost of achieving such emission reductions and any non-air quality, health and environmental impact and energy requirements . . . achievable for new and existing sources." The procedure for determining MACTs was also established by Congress. For new sources, the MACT for a source category is to be the control achieved in practice by the best-controlled similar source. In essence, for new stationary sources the HAP standards must be at least as effective as BACT requirements.

For existing sources, the MACT is required to be the average of the top 12 percent of the best-controlled sources in the source category, excluding those sources utilizing technology considered to meet the LAER, within 18 months of proposal or 30 months of promulgation. If there are fewer than 30 sources in the source category, the MACT for existing sources is to be the average emission level of the five best-performing sources. Thus, for existing major sources, HAP emission standards may be less stringent than the standards for new stationary sources in the same category, but may not permit less control than that achieved by the best-performing controls. The original MACT can also be achieved by process and material modifications in addition to the emission, design, equipment, work practice, or operational standards available for NESHAPs.

As previously described, step 1 in the HAP program is setting the technology-based control standards for HAP emissions sources. Step 2 is determining whether more stringent limits are required to protect public health and the environment with an ample margin of safety after setting the control standards. The issue of residual risk and its measurement and control has been studied by the government for several years. Within 8 yr after implementation of a particular standard, EPA must establish a second tier of emission standards for pollutants that present a cancer risk that is greater than 1 in 1 million, even when complying with the MACT emission limits established by this program. Thus, HAP emission sources may become subject to a second phase of increasingly stringent regulation. It is important to note that some states may retain their existing air-toxics regulations. These regulations may be in the form of ambient air concentration level guidelines. Thus, after the application of MACT, a source may be required to demonstrate compliance with a state's more stringent health-rated standard.

Accidental Release Conditions. The CAAAs include requirements for preventing and minimizing the consequences of accidental releases of hazardous substances. Part of the law required the Occupational Safety and Health Administration (OSHA) to promulgate a chemical process safety management standard to protect workers from hazards associated with accidental releases of extremely hazardous chemicals in the workplace. The final standard, called the Process Hazard Safety Standard, was published in the *Federal Register* on February 24, 1992.

The CAAAs also required EPA to promulgate the final form of the accidental release prevention regulations by November 15, 1993, for preventing and minimizing, if necessary, the consequences of accidental air releases from stationary facilities. The proposed risk manage-

TABLE 4.19 National Emission Standards for Hazardous Air Pollutants from Source Categories (MACT Standards)[a]

Industry group and source category	CAAA deadline	Proposed	Promulgated	40 CFR Part 63 subpart	Compliance date
Agricultural chemical production[b]					
Butadiene-furfural cotrimer (R-11) production	11/15/00	11/10/97	06/23/99	MMM[c]	06/30/02
Captafol production	11/15/97	11/10/97	06/23/99	MMM[c]	06/30/02
Captan production	11/15/97	11/10/97	06/23/99	MMM[c]	06/30/02
4-Chloro-2-methylphenoxyacetic acid production	11/15/97	11/10/97	06/23/99	MMM[c]	06/30/02
Chloroneb production					
Chlorothalonil production	11/15/97	11/10/97	06/23/99	MMM[c]	06/30/02
2,4-D salt and ester production	11/15/97	11/10/97	06/23/99	MMM[c]	06/30/02
Dacthal™ production	11/15/97	11/10/97	06/23/99	MMM[c]	06/30/02
4,6-Dinitro-o-cresol production	11/15/97	11/10/97	06/23/99	MMM[c]	06/30/02
Sodium pentachlorophenate production	11/15/97	11/10/97	06/23/99	MMM[c]	06/30/02
Tordon™ acid production	11/15/97	11/10/97	06/23/99	MMM[c]	06/30/02
Ferrous metal processing					
Coke by-product plants	11/15/00	(To be delisted)			
Coke ovens—charging, top side, and door leaks	12/31/92	12/04/92	10/27/93	L	Various
Coke ovens—pushing, quenching, and battery stacks	11/15/00	(09/00)	(09/01)	(CCCCC)	
Ferroalloy production	11/15/97	08/04/98	05/20/99	XXX	05/20/01
Integrated iron and steel manufacturing	11/15/00	(09/00)	(09/01)	(FFFFF)	
Iron foundries	11/15/00	(02/01)	(02/02)	(EEEEE)	
Steel foundries	11/15/00	(02/01)	(02/02)	(EEEEE)	
Steel pickling—HCl process	11/15/97	09/18/97	06/22/99	CCC	06/22/01
Fiber production processes					
Acrylic fiber/modacrylic fiber production	11/15/97	10/14/98	06/29/99	XX, SS, TT, UU, WW	06/29/02
Rayon production	11/15/00	08/28/00	(04/01)	UUUU	
Spandex production	11/15/00	(06/00)	(04/01)	(XXX)	
Food and agriculture processes					
Baker's yeast manufacturing	11/15/00	10/19/98	(12/00)	CCCC	
Cellulose food casing manufacturing	11/15/00	08/28/00	(04/01)	UUUU	
Vegetable oil production	11/15/00	05/26/00	(04/01)	GGGG	
Fuel combustion					
Engine test facilities	11/15/00	(02/01)	(05/02)	(DDDDD)	
Industrial boilers	11/15/00	(01/01)	(02/02)	(DDDDD)	
Institutional/commercial boilers	11/15/00	(01/01)	(02/02)	(DDDDD)	
Process heaters	11/15/00	(01/01)	(02/02)	(DDDDD)	
Stationary internal combustion engines	11/15/00	(09/00)	(11/01)		
Stationary turbines	11/15/00	(09/00)	(09/01)	YYYY	
Liquid distribution					
Gasoline distribution (Stage I)	11/15/94	02/08/94	12/14/94	R	12/15/97
Marine vessel loading operations[d]	11/15/97	05/13/94	09/19/95	Y	09/19/99
Organic liquid distribution (nongasoline)	11/15/00	(09/00)	(10/01)	(EEEE)	09/19/98 MACT; 09/19/98 RACT
Mineral product processing					
Alumina processing	11/15/00	(05/01)	(05/02)		
Asphalt/coal tar application—metal pipes	11/15/00	(05/01)	(05/02)		
Asphalt concrete manufacturing	11/15/00	(Expected to be delisted)			
Asphalt processing	11/15/00	(02/01)	(05/02)		

TABLE 4.19 National Emission Standards for Hazardous Air Pollutants from Source Categories (MACT Standards)[a] (*Continued*)

Industry group and source category	CAAA deadline	Proposed	Promulgated	40 CFR Part 63 subpart	Compliance date
Asphalt roofing manufacturing	11/15/00	(02/01)	(05/02)		
Chromium refractory production	11/15/00	(05/01)	(05/02)		
Clay product manufacturing	11/15/00	(05/01)	(05/02)		
Lime manufacturing	11/15/00	(05/01)	(05/02)	(AAAAA)	
Mineral wool production	11/15/00	05/08/97	06/01/99	DDD	06/01/02
Portland cement manufacturing	11/15/97	03/24/98	06/14/99	LLL	06/10/02
Taconite iron ore processing	11/15/00	(03/01)	(03/02)		
Wool fiberglass manufacturing	11/15/97	03/31/97	06/14/99	NNN	06/14/01
Wet-formed fiberglass mat production	11/15/00	05/26/00	(04/01)	HHHH	
Nonferrous metal processing					
Aluminum production, primary	11/15/97	09/26/96	10/07/97	LL	10/07/99
Aluminum production, secondary	11/15/97	02/11/99	03/23/00	RRR	03/23/03
Copper smelting, primary	11/15/97	04/20/98	(11/00)	QQQ	
Lead smelting, primary	11/15/97	04/17/98	06/04/99	TTT	06/04/02
Lead smelting, secondary	11/15/94	06/09/94	06/23/95	X	06/23/97
Magnesium refining, primary	11/15/00	(03/01)	(03/02)		
Petroleum and natural gas production and refining					
Oil and natural gas production	11/15/97	02/06/98	06/17/99	HH	06/17/02
Natural gas transmission and storage	11/15/97	02/06/98	06/17/99	HHH	06/17/02
Petroleum refineries—catalytic cracking (fluid and other) units, catalytic reforming units, and sulfur plant units	11/15/97	09/11/98	(09/00)	UUU	
Petroleum refineries—other sources not distinctly listed	11/15/94	07/15/94	08/18/95	CC	08/18/98
Pharmaceutical production processes					
Pharmaceutical production	11/15/97	04/02/97	09/21/98	GGG[c]	09/21/01
Polymer and resin production					
Acetal resin production	11/15/97	10/14/98	06/29/99	YY, SS, TT, UU, WW	06/29/02
Acrylonitrile-butadiene-styrene resin production	11/15/94	03/29/95	09/12/96	JJJ[c]	07/31/97
Alkyd resin production	11/15/00	(07/00)	(07/01)	(FFFF)	
Amino resin production	11/15/97	12/14/98	01/20/00	OOO	01/20/03
Boat manufacturing	11/15/00	07/14/00	(04/01)	VVVV	
Butyl rubber production	11/15/94	06/12/95	09/05/96	U[c]	07/31/97
Carboxymethylcellulose production	11/15/00	08/28/00	(04/01)	UUUU	
Cellophane production	11/15/00	08/28/00	(04/01)	UUUU	
Cellulose ether production	11/15/00	08/28/00	(04/01)	UUUU	
Epichlorohydrin elastomer production	11/15/94	06/12/95	09/05/96	U[c]	07/31/97
Epoxy resin production	11/15/94	05/16/94	03/08/95	W	03/03/98
Ethylene-propylene rubber production	11/15/94	06/12/95	09/05/96	U[c]	07/31/97
Flexible polyurethane foam production	11/15/97	12/27/96	10/07/98	III	10/08/01
Hypalon™ production	11/15/94	06/12/95	09/05/96	U[c]	07/31/97
Maleic anhydride copolymer production	11/15/00	(07/00)	(07/01)	(FFFF)	
Methylcellulose production	11/15/00	08/28/00	(04/01)	UUUU	
Methyl methacrylate–acrylonitrile–butadiene–styrene resin production	11/15/94	03/29/95	09/12/96	JJJ[c]	07/31/97
Methyl methacrylate–butadiene–styrene terpolymer production	11/15/94	03/29/95	09/12/96	JJJ[c]	07/31/97
Neoprene production	11/15/94	06/12/95	09/05/96	U[c]	07/31/97
Nitrile butadiene rubber production	11/15/94	06/12/95	09/05/96	U[c]	07/31/97
Nitrile resins	11/15/00	03/29/95	09/12/96	JJJ[c]	07/31/97
Nonnylon polyamide production	11/15/94	05/16/94	03/08/95	W	03/03/98

TABLE 4.19 National Emission Standards for Hazardous Air Pollutants from Source Categories (MACT Standards)[a] (*Continued*)

Industry group and source category	CAAA deadline	Proposed	Promulgated	40 CFR Part 63 subpart	Compliance date
Phenolic resin production	11/15/97	12/14/98	01/20/00	OOO	01/20/03
Polybutadiene rubber production	11/15/94	06/12/95	09/05/96	U[c]	07/31/97
Polycarbonate production	11/15/97	10/14/98	06/29/99	YY, SS, TT, UU, WW	06/29/02
Polyester resin production	11/15/00	(07/00)	(07/01)	(FFFF)	
Polyether polyol production	11/15/97	09/04/97	06/01/99	PPP	06/01/02
Polyethylene terephthalate resin production	11/15/94	03/29/95	09/12/96	JJJ[c]	07/31/97
Polymerized vinylidene chloride production	11/15/00	(07/00)	(07/01)	(FFFF)	
Polymethyl methacrylate resin production	11/15/00	(07/00)	(07/01)	(FFFF)	
Polystyrene resin production	11/15/94	03/29/95	09/12/96	JJJ[c]	07/31/97
Polysulfide rubber production	11/15/94	06/12/95	09/05/96	U[c]	07/31/97
Polyvinyl acetate emulsion production	11/15/00	(07/00)	(07/01)	(FFFF)	
Polyvinyl alcohol production	11/15/00	(07/00)	(07/01)	(FFFF)	
Polyvinyl butyral production	11/15/00	(07/00)	(07/01)	(FFFF)	
Polyvinyl chloride and copolymer production	11/15/00	(07/00)	(07/01)		
Reinforced plastic composites production	11/15/00	(08/00)	(08/01)	(WWWW)	
Styrene-acrylonitrile resin production	11/15/94	03/29/95	09/12/96	JJJ[c]	07/31/97
Styrene-butadiene rubber and latex production	11/15/94	06/12/95	09/05/96	U[c]	07/31/97
Production of inorganic chemicals					
Ammonium sulfate production—caprolactam by-product plants	11/15/00	(07/00)	(07/01)	(FFFF)	
Antimony oxide manufacturing	11/15/00	Delisted 11/18/99			
Carbon black production	11/15/00	(04/00)	(04/01)		
Chlorine production	11/15/00	(08/00)	(08/01)		
Cyanide chemical manufacturing	11/15/00	(06/00)	(04/01)		
Fume silica production	11/15/00	(02/01)	(02/02)		
Hydrochloric acid production	11/15/00	(02/01)	(02/02)		
Hydrogen fluoride production	11/15/00	10/14/98	06/29/99	YY, SS, TT, UU, WW	06/29/02
Phosphate fertilizer production	11/15/97	12/27/96	06/10/99	BB	06/10/02
Phosphoric acid manufacturing	11/15/97	12/27/96	06/10/99	AA	06/10/02
Uranium hexafluoride production	11/15/00	(05/01)	(05/02)		
Production of organic chemicals					
Ethylene processes	11/15/00	(06/00)	(04/01)	(XX)	
Quaternary ammonium compound production	11/15/00	(07/00)	(07/01)	(FFFF)	
Synthetic organic chemical manufacturing industry	11/15/92	12/31/92	04/22/94	F, G, H, I	05/14/01
Tetrahydrobenzaldehyde (butadiene dimers)	11/15/97	08/22/97	05/12/98	F	05/12/99, 5/12/98 (new sources)
Surface coating processes					
Aerospace industries	11/15/94	06/06/94	09/01/95	GG	09/01/98
Auto and light-duty truck (surface coating)	11/15/00	(02/01)	(02/02)	(IIII)	
Flat wood paneling (surface coating)	11/15/00	(12/00)	(12/01)	(QQQQ)	
Large appliance (surface coating)	11/15/00	(08/00)	(08/01)	(NNNN)	

TABLE 4.19 National Emission Standards for Hazardous Air Pollutants from Source Categories (MACT Standards)[a] (*Continued*)

Industry group and source category	CAAA deadline	Proposed	Promulgated	40 CFR Part 63 subpart	Compliance date
Magnetic tapes (surface coating)	11/15/94	03/11/94	12/15/94	EE	12/15/96 without new control devices; 12/15/97 with new control devices
Manufacture of paints, coatings, and adhesives	11/15/00	(07/00)	(07/01)	(FFFF)	
Metal can (surface coating)	11/15/00	(05/01)	(05/02)	(KKKK)	
Metal coil (surface coating)	11/15/00	7/18/00	(04/01)	SSSS	
Metal furniture (surface coating)	11/15/00	(08/00)	(08/01)	(RRRR)	
Miscellaneous metal parts and products (surface coating)	11/15/00	(02/01)	(02/02)	(MMMM)	
Paper and other webs (surface coating)	11/15/00	09/13/00	(06/01)	JJJJ	
Plastic parts and products (surface coating)	11/15/00	(12/00)	(12/01)	(PPPP)	
Printing, coating, and dyeing of fabrics	11/15/00	(04/01)	(04/02)	OOOO	
Printing/publishing (surface coating)	11/15/94	03/14/95	05/30/96	KK	05/30/99
Shipbuilding and ship repair (surface coating)	11/15/94	12/06/94	12/15/95	II	12/16/96
Wood furniture (surface coating)	11/15/94	12/06/94	12/07/95	JJ	11/21/97
Waste treatment and disposal					
Hazardous waste incineration	11/15/00	04/19/96	09/30/99	EEE	06/19/01
Municipal landfills	11/15/00	(08/00)	(09/01)	(AAAA)	
Off-site waste and recovery operations	11/15/94	10/13/94	07/01/96	DD, OO, RR, VV	07/01/00
Publicly owned treatment works emissions	11/15/95	12/01/98	10/26/99	VVV	10/26/02
Sewage sludge incineration	11/15/00	(Pending, possibly under Section 129)			
Site remediation	11/15/00	(03/01)	(03/02)	(GGGGG)	
Miscellaneous processes					
Aerosol-can-filling facilities	11/15/00	Delisted 11/18/99			
Benzyltrimethylammonium chloride production	11/15/00	(07/00)	(07/01)	(FFFF)	
Carbonyl sulfide production	11/15/00	(07/00)	(07/01)	(FFFF)	
Chelating agent production	11/15/00	(07/00)	(07/01)	(FFFF)	
Chlorinated paraffin production	11/15/00	(07/00)	(07/01)	(FFFF)	
Chromic acid anodizing	11/15/94	12/16/93	01/25/95	N	01/25/97
Commercial sterilization facilities	11/15/94	03/07/94	12/06/94	O	12/06/98
Decorative chromium electroplating	11/15/94	12/16/93	01/25/95	N	01/25/96
Dry cleaning (perchloroethylene)—dry-to-dry and transfer machines	11/15/92	12/09/91	09/22/93	M	09/23/96
Dry cleaners (petroleum solvent)	11/15/00	(May be delisted)			
Ethylidene norbomene production	11/15/00	(07/00)	(07/01)	(FFFF)	
Explosive production	11/15/00	(07/00)	(07/01)	(FFFF)	
Flexible polyurethane foam fabrication operations	11/15/00	(12/00)	(05/02)		
Friction product manufacturing	11/15/00	(12/00)	(12/01)		
Halogenated solvent cleaners	11/15/94	11/29/93	12/01/94	T	12/02/97
Hard chromium electroplating	11/15/94	12/16/93	01/25/95	N	01/25/97
Hydrazine production	11/15/00	(07/00)	(07/01)	(FFFF)	
Industrial process cooling towers	11/15/94	08/12/93	09/08/94	Q	03/08/95
Leather tanning and finishing operations	11/15/00	(06/00)	(04/01)	(TTTT)	

TABLE 4.19 National Emission Standards for Hazardous Air Pollutants from Source Categories (MACT Standards)[a] (*Continued*)

Industry group and source category	CAAA deadline	Proposed	Promulgated	40 CFR Part 63 subpart	Compliance date
OBPA (oxy-bisphenoxarsine)/1,3-diisocyanate production	11/15/00	(07/00)	(07/01)	(FFFF)	
Paint stripper users	11/15/00	(05/01)	(05/02)		
Photographic chemical production	11/15/00	(07/00)	(07/01)	(FFFF)	
Phthalate plasticizer production	11/15/00	(07/00)	(07/01)	(FFFF)	
Plywood/particle board manufacturing	11/15/00	(12/00)	(12/01)	(ZZZ)	
Pump and paper production					
Kraft, soda, sulfite, semichemical, mechanical pulping; and non-wood-fiber pulping	11/15/97	12/17/93	04/15/98	S	04/15/01 noncombustible; 04/16/01 nonchemical
Chemical recovery combustion sources	11/15/00	04/15/98	(12/15/00 court-ordered)	MM	4/16/01
Rocket engine test firing	11/15/00	(02/01)	(05/02)		
Rubber chemical manufacturing	11/15/00	(07/00)	(07/01)	(FFFF)	
Semiconductor manufacturing	11/15/00	(05/01)	(05/02)	(BBBB)	
Symmetrical tetrachloropyridine	11/15/00	(07/00)	(07/01)	(FFFF)	
production		12/31/92	04/22/94	H,I	05/12/99; 05/12/98 (new sources)
Tire production	11/15/00	(07/00)	(08/01)	(XXXX)	
Categories of area sources					
Chromic acid anodizing	11/15/94	12/16/94	01/25/95	N	01/25/97
Commercial sterilization facilities	11/15/94	03/07/94	12/06/94	O	12/06/98
Decorative chromium electroplating	11/15/94	12/16/93	01/25/95	N	01/25/96
Dry cleaning (perchloroethylene)—dry-to-dry and transfer machines	11/15/92	12/09/91	09/22/93	M	09/23/96
Halogenated solvent cleaners	11/15/94	11/29/93	12/02/94	T	12/02/97
Hard chromium electroplating	11/15/94	12/16/93	01/25/95	N	01/25/97
Lead smelting, secondary	11/15/97	06/09/94	06/23/95	X	06/23/97
Oil and natural gas production[e]	11/15/97	02/06/98	06/17/99	HH	06/17/02

[a] Dates and subpart identifiers in parentheses are those expected by EPA as of August 3, 2000, per the EPA OAQPS TTN web site. The following categories were also delisted on the date listed in parentheses: asbestos processing (10/30/95); electric arc furnace, stainless and nonstainless steel (06/04/96); chromium chemical manufacturing (06/04/96); nylon 6 production (02/12/98); wood treatment (06/04/96); cyanuric chloride production (02/12/98); and lead-acid battery manufacturing (5/17/96).

The following categories are identified on the upcoming standards list as of April 3, 2000, on the EPA OAQPS TTN web site:

Category	Expected proposal	Expected promulgation
Ceramics	05/01	05/02
Brick and structural clay products	02/01	02/02
Lightweight aggregate	05/01	05/02

[b] In the November 10, 1997 preamble to the proposed Subpart MMM MACT standards, EPA also noted its plan to combine the 11 individual agricultural chemical sources categories into a single category under this title, Pesticide Active Ingredient Production, as was done in the June 23, 1999, promulgation.

[c] Leak detection and repair (LDAR) provisions of the hazardous organic NESHAP (HON) rule of 40 CFR 63, Subparts H and I, apply to the emission of certain chemicals from facilities in this source category (see relevant subpart and 40 CFR 63.190 for details).

[d] Marine vessel loading source category was added November 12, 1993.

[e] EPA noted its plans to add this source category when it proposed the listed Subpart HH MACT standards.

Source: Compiled by Elsevier Science Inc.

ment planning regulations were issued on October 20, 1993, and the final regulations on January 31, 1994, as 40 Code of Federal Regulations (CFR) 68. Whereas the OSHA standard protects the worker from accidental releases, the EPA regulations protect the community and the environment, covering the off-site effects of accidental releases as well as on-site prevention and response. The accidental release program was developed in close coordination with the existing OSHA process safety management standard in order to minimize any unnecessary duplication of certain requirements.

EPA's accidental release prevention regulations required certain facilities handling more than a threshold amount of EHSs to develop detailed and integrated accidental release prevention and risk management plans, in most cases far beyond what has previously been expected of industry. EPA's initial proposed list of 162 EHSs was published in the January 19, 1993, *Federal Register.* The current list of toxic and flammable substances and their threshold quantities, as well as the reference and threshold quantities for explosives, are given in 40 CFR 68.130.

Operating Permit Program Requirements. The CAAAs established a massive new operating permit program to be administered by state agencies. EPA's formal requirements regarding permit programs were promulgated as 40 CFR Part 70 on July 21, 1992. The new operating permit program generally has been administered by state and local permitting authorities and has required nearly all sources of even minor amounts of air emissions to apply for and obtain an operating permit. As shown in Table 4.20, the federal operating permit program is substantially different from previous operating permit requirements.

Existing major sources as well as newly constructed sources will require permits. Even those facilities and operations previously grandfathered from permit requirements will now be required to obtain permits. After the permit program is in full force, a facility will not be able to operate without both obtaining a permit and meeting the requirements stated in the permit.

The operating permit program is generally modeled after the NPDES program under the Clean Water Act (CWA). Under the CAAAs, operating permits became the "centerpiece for compliance"—the new operating permits are intended to contain all of a facility's requirements for compliance with all air-quality regulations in one enforceable document. A typical operating permit for a facility is many pages in length and may require the collection and maintenance of reams of emissions data for each source or source grouping. Furthermore, industrial facilities are required to pay permit fees as a condition of the permit. The new operating permits are vastly different from previous operating permits.

Initially, all major sources of air pollution were required to obtain an operating permit. The definition of *major* varies, depending on the regulatory program and geographic location of the facility. Using the emission inventory data developed as part of the compliance review, Table 4.21 provides a summary of the types of sources required to apply for operating permits.

When an industrial operation is subject to the permit program as a major source for any one pollutant, all potential emissions of every regulated pollutant at the facility must be addressed in the permit application. The following section outlines the minimum operating permit program elements.

TABLE 4.20 Comparison of Pre- and Post-1990 Operating Permits

Pre-1990	Post-1990
Compliance required with nonspecific state regulations	Permit lists stack-by-stack allowable emissions
Emission monitoring rarely required	Continuous compliance monitoring
Annual emission reports (not compliance)	Annual emission reporting; annual compliance certification
	Five-year maximum term
	Significant permit fees

TABLE 4.21 Operating Permit Program Applicability

Source type	Major source threshold (potential to emit)
HAP	10 tpy* any single HAP or 25 tpy aggregate HAP emissions
VOC in an ozone nonattainment area	10–100 tpy VOC emissions
VOC in Northeast Transport Region	50 tpy VOC emissions
NO_x in an ozone nonattainment area	10–100 tpy NO_x emissions
CO in a serious nonattainment area	50 tpy
Particulate matter (PM_{10} nonattainment area)	50–70 tpy
All others	100 tpy any emissions
Sources subject to NSPS, National Emission Standards for Hazardous Air Pollutants (NESHAPs), PSD/NSR and acid rain regulations	No threshold quantity

* Tons per year.

At a minimum, state agencies must include certain specific elements in their operating permit programs for approval by EPA: requirements for permit applications, emission monitoring, compliance certification reporting, permit fee authority, personnel, funding, and authority to issue and process permits.

By regulation, facilities were required to submit permit applications by November 1995 at the latest. Although many states required permit application submittal before November 1994, due to delays in EPA regulations guidance and program approval, as well as extended schedules in some state programs, some initial applications had not been submitted by January 2001. To submit a timely and complete permit application, facilities must begin to prepare the information months or years before the submittal deadline. With dozens to hundreds of individual emissions units in a large industrial facility, accurate and well-documented information is essential. Unless a company negotiates its permits properly, after these emission limitations have been established, much of the operating flexibility of the past may be gone.

The permit application should include the following

- Potential emission of each regulated pollutant from a company's facilities
- Existing emission limits on a pollutant-by-pollutant basis for both hourly and annual emissions, whether by regulation or existing permit
- Any desired emissions of existing limits
- Potential emission rates in tons per year and in compliance terms (e.g., pounds per gallon, pounds per million Btu)

It is possible that the permit will cap both annual and short-term emissions for certain sources or groupings of sources. The permit application must cite and describe the regulatory requirements applicable to the facility. Requirements expected to become applicable during the term of a permit must also be identified in the application. EPA has issued several white papers that have clarified and narrowed the information required in the application, and are intended to facilitate operational flexibility.

Title V permits become the principal mechanism for detailing the specific requirements applicable to individual sources. It is extremely important to negotiate a permit that will allow substantial operating flexibility. Otherwise, the time required to modify air permits may seriously impede product development and facility planning processes.

If a facility is out of compliance at the time the permit application is submitted, a schedule for compliance, including enforceable milestones, is required. This compliance plan then becomes an enforceable part of the facility permit.

The new operating permit program establishes procedures for several classes of changes at industrial facilities (Table 4.22). In the past, many of these changes required no EPA involve-

TABLE 4.22 Minor Versus Significant Permit Applications

Minor modifications	Significant modifications
Comply with applicable requirements.	Violates existing applicable requirements.
Small emissions increases.	Not a minor modification or an administrative amendment; change is typically a Title I modification.
No change to certain types of permit conditions or case-by-case determinations.	Change to permit condition established to avoid regulatory applicability or change to previous case-by-case determination on emissions permits or ambient impacts.
Insignificant monitoring changes.	Significant monitoring changes.
Insignificant reporting or recordkeeping change.	Relaxation of reporting or recordkeeping permit conditions.
Facility can make change before modification is reviewed if state regulations accept EPA guidance.	Full review process before facility can make change.
No shield.	State may allow a shield.

ment. In the future, however, almost any change in method of operation will require EPA approval. Changes at sources are classified as: (1) administrative permit amendments, (2) minor permit modifications, or (3) significant permit modifications.

The new air permits summarize existing restrictions applicable to an operation. A permit change is not required to authorize a change in practice that is otherwise legal under the state's rules and regulations. Any change in practices or procedures that is within the operational flexibility provided by the permit and is otherwise legal may be made without a permit modification. Permit modifications, particularly significant ones subject to EPA involvement, could take from many months to more than a year for approval [40 CFR 70.7(e)(4)(ii) specifies that the review of the majority of significant permit modifications should be completed within 9 months]. Preparation of complete permit applications and their timely submittal are critical items in responding to the CAAA permitting requirements. The permit application and permit eventually crafted by the agencies based on the application will define a company's operations for many years to come. Applications require careful consideration of each item required in the permit application, as well as consideration of the potential for future process modifications or production increases, to ensure that the operating permit provides sufficient flexibility to allow the facility to run within a range of likely emissions and to avoid very lengthy permit modifications.

Compliance Assurance Monitoring. One substantial requirement related to the new federally mandated operating permit program is the Compliance Assurance Monitoring (CAM) rules that were promulgated as 40 CFR 64 on October 22, 1997. The CAM rules apply only to facilities subject to the new federal operating permit program. In particular, only sources that emit certain threshold levels of pollutants and utilize air pollution control devices must meet these monitoring, recordkeeping, and reporting requirements. If the new operating permit application was submitted prior to enactment of the final CAM rule, compliance need not be implemented until the permit is renewed. In order to address this delay, EPA issued *Periodic Monitoring Guidance for Title V Operating Permits* in September 1998. However, in April 2000 the federal appeals court ruled that the guidance must be set aside because it consisted of far more than interpretation of existing regulations.

Regulatory Information Sources. For information on the current status of air-quality regulations and related background documents, see the EPA Office of Air Quality Planning and Standards (OAQPS) Technology Transfer Network (TTN) web site at www.epa.gov/ttn/.

EMISSIONS SURVEY

The initial step in defining the emissions levels and determining the BACT or LAER technology (for a modification or a proposed new facility) or RACT or MACT requirements (to control atmospheric emissions from an existing plant) is an emissions survey. All applicable pollutant sources and quantities of emissions must be identified. The results of this survey will provide management with a representative and comprehensive overview and with sufficient detail with which to formulate abatement plans and design programs in order to file the permit application. The basic steps in a typical survey include the following:

- Cataloging emission sources
- Quantifying emissions
- Preparing a source identification file

Cataloging Emission Sources

General Plant Information. The first step in such cataloging activities is to develop general plant information. This information is required to provide a general background and overview of the proposed or existing plant and/or its modifications, and define applicable regulatory requirements. The following data should be provided:

- A schematic block flow diagram of the process(es) for the plant showing the flow of raw materials into and out of the process(es), and the design/operation of the air pollution control equipment. This diagram should identify sources of all potentially regulated pollutants, both process and fugitive emissions, which could be released into the atmosphere. Potential secondary or transformation by-product pollutants may need to be considered.
- A materials balance across the process(es) and across the air pollution control equipment.
- Airflows, either shown on a schematic block flow diagram for the process(es) and the control equipment, or provided separately.
- Dimensions of the plant buildings showing the length, width, and height of all structures, *present* and *proposed.*
- Details of existing and proposed stack(s): height, diameter, exit gas flow, and temperature.
- Details of the design characteristics of the air pollution control equipment.
- Details on the potential release of fugitive emissions from existing or proposed new operations.
- A facility plot plan, including all fenced-off areas, identifying all sources, buildings, and plant property lines.

Process Flowsheets. The second step is to prepare process flowsheets in sufficient detail to indicate the flow of all raw materials, additives, by-products, and waste streams. The flowsheet should identify all points of feed input, and all points at which atmospheric, liquid, and solid wastes are discharged. The engineer or supervisor responsible for each process should verify that the flowsheets identify all sources. Many plants have numerous sources, which must be identified correctly for tracking purposes.

Survey Data Sheets. Analysis of process flowsheets can be further verified by review of prior permit applications, process blueprints, photographs, and inspection manuals. With the aid of these and any other resources available, the emissions surveyor can then develop checklists in the form of survey data sheets that will be used during the plant tour to ensure complete and efficient gathering of pertinent data. These survey data sheets will pertain chiefly to process and feed data, and control equipment and emissions data. In addition, the survey form should identify the permit status (i.e., grandfathered, exempt, regulated, or not permitted) and the expiration dates of permits for each stack and/or vent. Figure 4.63 shows an example of a process survey data sheet, representing presurvey evaluation of a fuel-fired combustion source. Such an example can be modified to apply to most types of emission sources now in operation. The process survey data sheet should include the following:

- Detailed information on operating conditions for the process as designed
- Identification of normal and maximum throughput or processing rates during continuous, batch, or intermittent operation, with frequency of emission discharges for each
- Description of raw materials, products, and wastes
- Values for normal operating temperature, equipment performance ratings, flows, pressures, and similar data that are routinely monitored and/or recorded

POWER PLANT SURVEY FORM

Type of Heat Exchanger

	Primary	Standby
Coal-fired	☐	☐
Oil-fired	☐	☐
Gas-fired	☐	☐

If multiple-fired, check appropriate boxes

Rated Input Capacity__________Btu/hr
Maximum Operating Rate__________Btu/hr
Rated Steam Output__________lb/hr____(a)____Btu/hr steam
Maximum Steam Output__________lb/hr____(a)____Btu/hr steam
Furnace Volume width____ft x depth____ft x height____ft = ____cu ft
Operating Schedule__________ hr/day____ day/wk____wk/yr

Coal Firing

Type of Firing ☐ Grate ☐ Spreader stoker ☐ Pulverized coal ☐ Cyclone

Type __________

☐ Dry bottom ☐ Wet bottom

Fly Ash Reinjection ☐ Yes ☐ No

Soot Blowing

☐ Continuous
☐ Intermittent

Time Interval Between Blowing____ minutes

Duration____minutes

Outside Coal Storage ☐ Yes ☐ No

Maximum Amount Stored Outside____tons

FIGURE 4.63 Sample presurvey data sheet for fossil-fuel-fired steam generators.

Control Equipment Survey Information. The information given in a control equipment survey sheet is outlined in Tables 4.23 to 4.27. Table 4.23 shows information for a general survey evaluation form, while Tables 4.24 to 4.27 outline data requirements for control methods for particulates and acid gases.

Identical information and data must be developed and evaluated for any new or modified facility in order to select the control system that represents BACT or LAER technology.

Tour of Plant Facilities. The fourth step is a tour of the plant facilities. The tour should include discussions with the process engineer or supervisor in order to ensure identification of all sources, verify the process and control-equipment flowsheets, and account for any equip-

TABLE 4.23 General Control Equipment Data

A. At maximum continuous production rate (MCPR)
   1. Inlet and outlet absolute cubic feet per minute (acfm).*
   2. Inlet and outlet gas temperatures.
   3. a. Inlet and outlet percent of H_2O.
      b. Dew point.
   4. Inlet pollutant levels for:
      a. TSP,† lb/h, gr/scf,‡ etc.; also provide particulate size distribution.
      b. SO_2, lb/h, gr/scf, etc.
      c. NO_x, lb/h, gr/scf, etc.
      d. HC,§ lb/h, gr/scf, etc.
      e. CO, lb/h, gr/scf, etc.
   5. Expected and guaranteed efficiencies for each of the above pollutants.
   6. Expected and guaranteed pollutant levels at outlet for each of the above pollutants.
   7. If available, make, model, and type of control system(s).
   8. Explain basis for acfm and of selection for design. *Caution:* These may be different from data used in dispersion modeling.
   9. Material balance across control equipment.
   10. Block flow diagram of control equipment.
   11. Inlet particulate size distribution.
B. For system(s) that may have a lower efficiency at normal operating rates as compared to MCPR, provide all data outlined above at normal operating rate.

* Actual cubic-feet per minute.
† Total suspended particulates.
‡ Grains per standard cubic foot; 1 gr = 1/7000 lb.
§ Hydrocarbon.
Note: If available, provide specification sheets. Provide drawings of internals.

TABLE 4.24 Multiclone and Cyclone Data

1. Number of tubes for multiclones (multiple cyclones).
2. Length of tubes.
3. Fractional size efficiency vs. *P* curves.
4. Diameter of tubes for multiclone or diameter of cyclone.
5. Pressure drop in H_2O column.
6. Particulate size distribution into cyclone and/or multiclone.
7. Grain loading at inlet and/or outlet, gr/dscf*
8. ACFM at °F.
9. Design efficiency.
10. Disposal and handling of dry, collected dust.
11. Preventive maintenance program.

* Grains per dry standard cubic foot.

TABLE 4.25 Data for Scrubber for SO_2 and/or Other Pollutants

1. Design inlet and outlet volume flows, temperatures, percent moisture, and SO_2 loadings.
2. Type of scrubber: wet or dry? If wet, operational mode.
3. Is reheat necessary? Describe.
4. Percentage of exhaust gases? Describe.
5. Any prequench section? Describe.
6. Type of demister.
7. Type of internal construction.
8. Minimum number of isolatable modules.
9. Minimum liquid-to-gas ratio.
10. Minimum-maximum pH of scrubbing liquor.
11. Maximum gas face velocity.
12. Minimum ratio of reagent to SO_2 and chemical composition solution.
13. Pressure drop in H_2O column.
14. Is any bypass to be provided?
15. Outline any measures to be taken to produce more uniform SO_2 loading to the scrubber (e.g., precleaning, coal blending, etc.).
16. Design efficiency for PM and/or SO_2 and/or other pollutants.
17. Redundancy of control equipment.
18. Parameters for packed-bed scrubbers:
    a. Type of packing
    b. Packing size and/or shape
    c. Packing height
19. Parameters for spray scrubbers:
    a. Number of nozzles
    b. Nozzle droplet size
    c. Nozzle design and/or shape
    d. Nozzle pressure, pounds per square inch gauge (psig)
20. Materials of construction.
21. Disposal of sludge or wet dust.
22. Effluent to streams from sludge pond.
23. Percolation from sludge ponds to aquifers.
24. Preventive maintenance program.

TABLE 4.26 Fabric Filter or Baghouse Data

1. Type of filter media:
    a. Chemical composition
    b. Woven cloth or felt
    c. Porosity of new cloth
    d. Weight of cloth, oz/yd^2
    e. Napped or nonnapped fibers in woven bags
    f. Life of bags
2. Air-to-cloth ratio.
3. Cleaning mechanism.
4. Number of compartments. Can one compartment be shut down for maintenance while the rest of the baghouse continues operation? How is overall collection efficiency affected?
5. How is damage from gas dew point circumvented? What is dew point?
6. Overall collection efficiency.
7. Materials of construction.
8. Disposal of collected materials.
9. Preventive maintenance program.

TABLE 4.27 Electrostatic Precipitator (ESP) Data

1. Collection area, ft^2/1000 acfm.
2. Number of fields.
3. Number of compartments. Can one compartment be shut down for maintenance while rest of ESP continues operation? How is the overall collection efficiency affected during maintenance?
4. Gas velocity through ESP, ft/s.
5. Description of gas conditioning techniques.
6. Design of charging area: weighted wires, rigid frames, etc.
7. Temperature of operation.
8. Resistivity of fly ash or particulate matter at temperature of operation.
9. Inlet gas details: How is turbulence minimized? Any aerodynamic gas flow studies?
10. Rapping or cleaning details: provisions to minimize reentrainment when cleaning.
11. Design efficiency for preventive maintenance.
12. Design of collecting rappers to prevent bridging, plugging, or improper operation.
13. Materials of construction.
14. Disposal of collected materials.
15. Preventive maintenance program.

ment modifications already made. Sufficient data should be gathered to allow computation of material balances in order to have a basis for quantifying each emission source.

The plant tour starts at the files. There the surveyor will gather design specifications for each process and control device. Correspondence may also yield pertinent information relating to operations and maintenance (O&M) of process and control equipment, current status of compliance, comments of control agencies, public complaints, and the like. This kind of background information can enhance the understanding needed for a meaningful on-site inspection of each process and control device. If the particular (existing) emission source requires an air permit, most of the pertinent information on the operating and emission characteristics are contained in the permit. Under the new operating permit program, permits must be renewed every 5 yr.

Each air containment source generally has a duct that vents from the process to an outside chimney or stack. The exhaust gas is moved by a fan or, in some instances where heat is applied, by natural draft. For each operation, the ducting should be followed from the process to the point of entry to the atmosphere. In some instances, the exhaust gas stream is difficult to follow. The introduction of makeup air, splitting of gas streams, and ducting of several operations to a common stack complicate the overall exhaust system structure and require careful tracing to ensure that exhaust-gas paths are properly defined. Placement of fans and control devices must be noted along with the height and location of each stack. Air-conditioning, heating, and makeup vents must not be mistaken for process stacks. After defining all process sources, the surveyor should check the roof to identify any unaccounted-for emission points. Finally, potential sources of fugitive emissions should be confirmed.

Quantifying Emissions

All of the data obtained thus far from the process flowsheets, process survey forms, control equipment survey forms, stack survey forms, photographs, correspondence, discussions, and the plant tour can now be organized to develop an emission survey plan. This plan must indicate the quantity of emissions estimated from each source, with possible variations due to season, time of day, feed materials, and similar variables. The emissions characterization should identify all important parameters affecting control of the pollutants and possible sampling techniques. These data will be used to review the compliance status for each source. These programs will describe the plans that will be implemented by the company to achieve or maintain compliance, and should contain the following increments of progress or milestones:

- Date of submittal of the final compliance plan or permit to the appropriate air pollution control agency
- Date of issuance of an approved compliance plan or permit
- Date by which contracts for emission control systems or process modifications will be awarded, or date by which orders will be issued for purchase of component parts to accomplish emission control or process modification
- Date of initiation of on-site construction or installation of emission control equipment or process change
- Date by which on-site construction or installation of emission control equipment or process modification is to be completed
- Date by which final compliance or full operation is to be achieved

Figure 4.62 presents an example of the activities that must be completed before compliance can be achieved. Depending on the nature of the emission source and the complexity and size of the modifications required, the time requirements for compliance can range from a few months to several years.

Quantification techniques can then be applied in order to develop and utilize the emission survey information. Mass balances usually can be established around each process, particularly when the throughput rates and the composition of raw materials are known. The materials balance will indicate the extent of solid, liquid, and gaseous wastes.

A review of applicable air pollution control regulations will provide the basis for the mass balance. The control regulations state which pollutants are regulated and define each pollutant. The definition of each pollutant determines the conditions under which the pollutant is sampled and its chemical and physical makeup. For instance, because water vapor is not considered an air pollutant, it is not necessary to accurately account for it in the mass balance. Emissions of SO_2 or organic substances are usually regulated and must be estimated in the materials balance.

A materials balance for gaseous pollutants can be determined by analysis of raw materials, fuels, and products to give the gaseous pollutant potential of many of the compounds liberated during a combustion or chemical process. Knowledge of fuel composition is especially useful in estimating emissions because many compounds in the fuel become airborne gases after combustion (e.g., sulfur in fuel oil exhausts as SO_2). Other constituents, such as ash and volatile matter, directly affect the quantity of particulate emissions. Stack sampling tests may be the only acceptable way to quantify emissions of certain pollutants (e.g., VOCs and HAPs).

In reviewing a permit application, the regulatory authorities typically use an EPA publication titled "Compilation of Air Pollutant Emission Factors" (AP-42), in order to estimate the potential uncontrolled and controlled emissions of a particular pollutant. This document and all its supplements should be obtained by the plant engineer and used as reference material.

Preparing a Source Identification File

A source identification file provides a means of standardizing data for the emission survey. For each pollutant source, a standard identification form gives a description of the process, a summary of emission data, the current compliance status, and the proposed actions, if any are intended. A basic source identification form is shown in Fig. 4.64. Some sources involve several emission points with more than one pollutant at each point.

The source identification file should be indexed to provide easy access by any concerned party. The index should list all sources and identify the emission point for each source. Assignment of a number for each emission point will facilitate an alphanumeric search for emission points in the source identification file.

Emission Point No. ____________________

Emission Point Name ____________________

Date of Record ____________________

Source Name ____________________

Description of Source ____________________

Type of Permit ____________________

Date of Permit ____________________

Applicable Regulation(s) ____________________

Particulate Emissions ____________________ units ____________________

Allowable Emissions ____________________ units ____________________

Method of Determination ____________________

Gaseous Emissions ____________________

type ____________________ units ____________________

Allowable Emissions ____________________ units ____________________

Method of Determination ____________________

Compliance Status ____________________

Date Contract Awarded ____________________

Date Construction Began ____________________

Monitoring ____________________

ambient ____________________

stack ____________________

FIGURE 4.64 Source identification form.

EMISSION CONTROL METHODS

Emission control alternatives might be broadly grouped under the following categories: (1) process or raw material changes that reduce the quantity of pollutant, (2) changes that eliminate production of a particularly undesirable pollutant, and (3) collection and removal of the pollutant from the gas stream. Current designs employ some combination of these alternatives, with interaction between the alternatives. For example, using a scrubber to selectively remove a pollutant may lower an exhaust-gas temperature, which could significantly reduce the thermal buoyancy of the effective plume rise, thereby necessitating the requirement for plume reheat and/or increase in stack height. Disposal of an undesirable liquid or removal of solid waste from a gas stream may be a greater problem than abatement of gaseous pollutants. Clearly, there must be a compromise. The control system must be evaluated with regard to capital and operating costs, level of control, by-product generation, reliability, and other practical considerations.

The most important process parameters for selection of control equipment are the following:

Flue-gas characteristics

- Total flue-gas flow rate
- Flue-gas temperature
- Control efficiency required
- Composition of emissions
- Corrosiveness of flue gas over operating range
- Moisture content
- Stack pressure

Process or site characteristics (field survey)

- Reuse or recycling of collected emissions
- Availability of space
- Availability of additional electrical power
- Availability of water
- Availability of wastewater treatment facilities
- Frequency of start-up and shutdown

Controlling Particulate and SO_2 Emissions

Traditional and nontraditional air pollution control systems that primarily limit particulate and SO_2 emissions are discussed in this subsection. It is important to note that many HAPs are also considered to be particulate matter. Thus, controlling particulate emissions will limit HAP emissions. Some of the devices described here can be used in series or alone to reduce particulate and acid gas emissions to acceptable levels.

Mechanical Collectors. The most familiar and widely used mechanical collector is the cyclone separator (Fig. 4.65). Gas enters tangentially at the top of a cylindrical shell and is forced downward in a spiral of decreasing diameter in a conical section. Particles are centrifugally thrust outward and forced to spiral downward to the bottom, which is closed by an air lock. Because gas cannot escape at the bottom, it is forced to turn and travel, still whirling, back up the center of the vortex and exit at the top. Particles are discharged from the bottom through the air lock.

The tighter the spiral in which gas must flow, the greater the centrifugal force acting on a particle of given mass, and thus the more efficient the cyclone can be. Top diameters of cyclones range from more than 120 in (3 m) to as small as 24 in (60 cm), and capabilities reach 85 percent efficiency with particles as small as 10 μm at pressure drops from 0.5- to 3-in (13- to 77-mm) water gauge.

Cyclones alone are seldom adequate for pollution control, except where the load consists almost entirely of coarse particles, such as in woodworking shops, or where particle density is

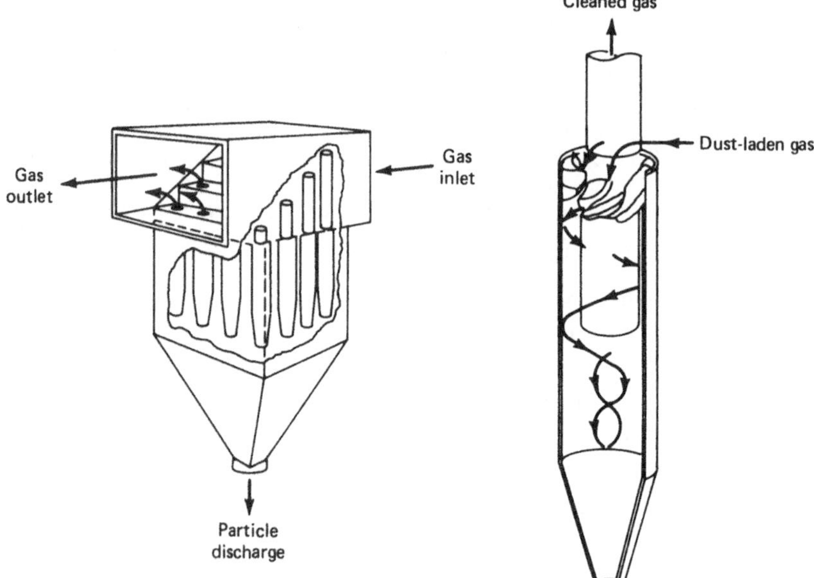

FIGURE 4.65 Mechanical collector cyclone.

unusually high. Cyclones are often used when there is a special reason to separate and collect reusable coarse particles from useless fine particles (for example, in fluid-bed catalyst regenerators through which the fine particles pass for subsequent separation by more efficient means). Cyclones are sometimes applied where gas cooling is necessary (for example, where the cyclone is followed by a fabric filter, the cyclone scalps the coarse fractions of the dust load and at the same time provides cooling).

Scrubbers. Scrubbers are collectors capable of separating solid or liquid particles from a gas stream. They are also used to separate a chemically reactive or soluble gas constituent from other gas constituents in a flue-gas stream. The most common use for gas separation involves its use in *flue-gas desulfurization* (FGD).

In the simplest application of scrubbing, liquid is sprayed in at the top of a column, and collision for particulate wetting or capture occurs as drops fall through a rising gas stream. Pressure drop is low, and application is limited to situations where 50 percent efficiency or less is acceptable and the percentage of particles smaller than 10 μm is low, or where coarse particles are to be scalped ahead of a precipitator or a more efficient scrubber. A spray column is often used primarily for quenching hot gas, with coarse particulate removal a useful but incidental effect.

Centrifugal Scrubbers. In a centrifugal scrubber, column design and directed sprays cause droplets and gas to mix in a rising vortex such that centrifugal force increases the momentum of collisions between particles and drops. Thus, smaller particles can be captured, and efficiencies as high as 90 percent can be achieved with particles as small as 5 μm, at pressure drops from 2- to 6-in (50- to 150-mm) water gauge.

Packed Scrubbers. A column may be fitted with impingement plates, wetted mesh, or fibrous packing, or packed with saddles, rings, or other solid shapes. In such scrubbers, typically at a 1- to 10-in (25- to 250-mm) water-gauge pressure drop, efficiencies can reach 95 percent with particles as small as 5 μm. Packed beds designed for gas absorption, however, are subject to fouling if the gas stream contains a significant fraction of solid particles. In designs that use sprays to wash the packing, or that are packed with small spheres agitated by the gas flow, the fouling problem is reduced. In some systems, a moist chemical-foam packing is used which drains slowly from the scrubber along with captured particulates and is then replaced.

Venturi Scrubbers. A venturi tube operating on the eductor or ejector principle, with scrubbing liquid as the motive fluid, can collect particles as small as submicrometer size with efficiencies as high as 95 percent, if grain loading is low. Gas-pressure drop is not needed for power input, and there can even be a gain in gas pressure across such a scrubber. The disadvantages are the requirement for substantial scrubbing-liquid flow at high pressure and the scrubber's inability to remove large particles because the induced gas-stream velocity is low.

Scrubbing a particulate- and acid-containing gas stream by bringing it into contact with a liquid is an effective means of removing both the dust and the acids. The collected particles, along with any acids (in either a "raw" or neutralized form), are disposed of in a wet form and present their own set of disposal problems. This is one of the major drawbacks to wet scrubbing and a reason that dry scrubbing systems are gaining wide acceptance.

When treating an acid-gas stream, an alkali compound (such as caustic soda or lime) is added to the scrubber water. The water is brought into contact with the gas stream through dispersed droplets (spray tower), contraction of the gas stream (venturi), or countercurrent flow (collision). This violent contact removes the particles and acids from the gas stream, but wet scrubbing is very dependent on size and concentration (for gases) at a potentially high energy cost. In general, the greater the degree of acid gas or the amount of particulate removal required (or the smaller the particle to be removed), the more energy must be expended to reach this goal. This energy usage is found in both the pumps to move the scrubbing liquor and the fan power to move the gas.

Semidry Scrubbers. Semidry scrubbing systems start with a wet scrubbing medium (usually a lime slurry or a soda ash solution) and produce a dry waste product. The central device of a semidry scrubbing system is the spray dryer, similar to equipment that has been used for years in industry to manufacture everything from powdered coffee and milk to paint pigments and detergents.

In a semidry scrubbing system, the solution or slurry is dispersed by nozzles or rotary atomization systems into a fine cloud of droplets. These droplets are brought into contact with a hot gas stream (and herein lies a disadvantage to semidry scrubbing systems) that proceeds to evaporate the water in the droplets. As the water evaporates, the acids in the gas stream react with the alkali material in the drying droplets and neutralize them, forming a fine powder. Most of this powder is removed from the bottom of the spray dryer, while the remainder is entrained in the gas stream and carried out to either a fabric filter or an electrostatic precipitator (ESP).

The necessity for a secondary emissions control device is another disadvantage of semidry scrubbing systems; this usually will result in a cost about 5 to 10 percent higher than for wet scrubbing systems. Two major advantages of the semidry scrubbing system are its relative simplicity to operate and the relative ease (compared to wet scrubbing systems) of the disposal of the dry waste product.

Another advantage of the semidry scrubbing systems is that, combined with a fabric filter, there is almost complete (95 percent for SO_2, 90 to 95 percent for HCl) removal of acid gases, particulates, metals, and, when used with activated-carbon injection, dioxins and furans.

Dry Injection Systems. Dry injection systems are effective in removing acids from gas streams. In operation, this method is very similar to the semidry scrubbing system in that a dry powder (usually lime or soda ash) is injected into a gas stream. In some applications this powder injection may be enhanced by a separate fine water spray to humidify the gas stream.

The acids will react with the injected powder and, again like the semidry systems, be removed from the gas stream (along with other particles) in either an ESP or a fabric filter. Two great advantages of the dry injection system are the simplicity of its operation and, depending on the layout and length of ductwork, its ease of retrofit into a manufacturing facility with an existing particulate control device.

One major disadvantage of the dry injection system is that without the aid of gas-liquid-solid mass transfer, as in the wet or semidry scrubbing systems, addition of alkali at a ratio much higher than stoichiometric is required for an equivalent removal.

Fabric Filters. Fabric filters collect solid particles by passing gas through cloth bags that most particles cannot penetrate. As the layer of collected material builds on the fabric, the pressure differential required for continued gas flow increases; consequently, the accumulated dust must be removed at frequent intervals. Fabric used to form the filter elements can range from nylon to wool to Teflon-coated fiberglass, shaped into cylindrical bags or envelopes with a roughly elliptical cross section.

Fabric filters are capable of 99+ percent collection efficiency with particles as small as submicrometer size. High efficiency is attained with moderate pressure drops, typically in the range from 2- to 4-in (50- to 100-mm) water gauge. Power input is thus comparable to that of multitube mechanical collectors, while the capability to collect fine particles is much greater. Operating costs, including maintenance, are somewhat higher, however, because some moving parts are involved and bags must be replaced periodically. Unlike wet scrubbers, the performance of fabric filters is relatively unaffected by variations in gas-flow rate. Fabric filters are sometimes preceded by mechanical dust collectors or settling chambers in the baghouse when excessive grain loading or abrasive, coarse particles, or both, are involved.

At certain times during operation, the pressure drop will increase to the point that necessitates removal of the particles from the fabric. Depending on the filter system design, the cleaning process is accomplished by shaking the bags, sending a pulse of compressed air into the filter bag, or sending a stream of clean gas in a reverse direction through the filter compartment.

As effective as fabric filters are, there are some drawbacks to their use. Excessive pressure drops may occur as the fabric becomes plugged with small particles or from water droplets in the gas stream; ripping or pinholing of the fabric will reduce its effectiveness; and retrofit of filters may be difficult, depending on the layout of the process and the facility.

Electrostatic Precipitators. Electrostatic precipitators (ESPs) are extremely efficient air pollution control devices that can remove more than 99 percent of the particles in a gas stream (Fig. 4.66). This high efficiency is possible because, unlike other pollution control devices, the precipitator applies the collecting force only to the particles to be collected, not to the entire gas stream. Thus, an extremely low, energy-saving power input [about 200 W per 1000 ft^3/min (0.5 m^3/s)] is required.

In operation, a voltage source creates a negatively charged area, usually by means of wires suspended in the gas-flow path. On either side of this charged area are grounded collecting plates. The high potential difference between these plates and the discharge wires creates a powerful electric field. As the polluted gas passes through this field, particles suspended in the gas become electrically charged and are drawn out of the gas flow by the collecting plates. They adhere to these plates until removed for disposal. Removal is accomplished mechanically by periodic vibration, rapping, or rinsing. The gas stream, now substantially free of particulate pollution, continues to flow for release to the atmosphere.

Because its collection mechanism depends on charging the particles in the gas stream, an ESP will be strongly affected by the characteristics of both the particles and the gas stream in which they are contained. Particles with too high a conductivity (low resistivity) will bleed their charge either before being collected or will be released back into the gas stream after impacting with the collecting electrodes.

Conversely, particles with very low conductivity (high resistivity) will not be easily charged and may require a very large precipitator to achieve an adequate collection efficiency. In specifying a precipitator, the user should have either a qualified testing company or a reputable vendor perform resistivity and particle size distribution measurements on the gas stream to be controlled to develop the needed design data. If this testing is not possible, then a qualified consultant or company experienced in precipitator design should be used.

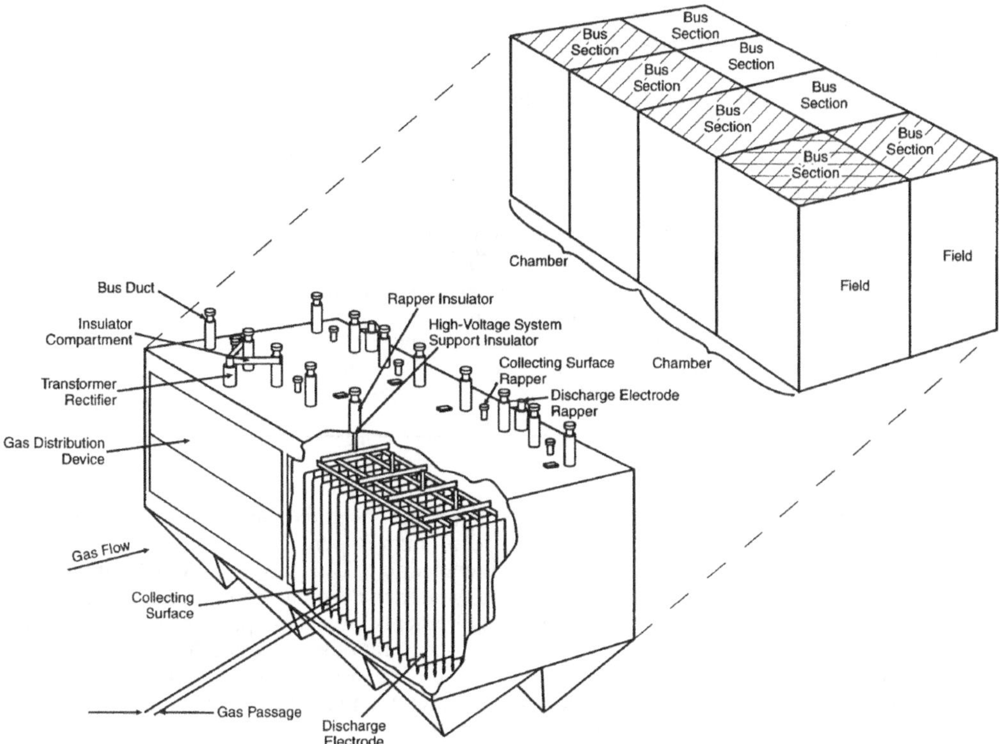

FIGURE 4.66 Electrostatic precipitator.

Controlling VOC Emissions

The air pollution control equipment marketplace offers many competing technologies for controlling emissions of volatile organic compounds (VOCs). It is important to note that over half of the HAPs are considered VOCs. Thus, controlling VOCs will limit HAP emissions. If any technology was economically and technically superior under all conditions, it would be the only one on the market. In fact, each technology used to control VOCs is superior under some set of conditions.

The reasons for choosing one control technology over another are situation-specific. Some general guidelines to VOC-control technologies and the situations where they may be appropriate are presented in this subsection. Table 4.28 summarizes the control technologies and their applications.

High-VOC Airstreams. Airstreams with a high VOC content are defined here as those containing several hundred parts per million (ppm) and greater.

Refrigerated Vapor Condensor. This removes VOCs from a gas stream by condensation at very low temperatures, typically around –112°F (–80°C). As an example, refrigerated condensation is used to remove gasoline from air displaced by fuel transfer operations. It is most applicable to gas streams with flow rates less than 1000 standard cubic feet per minute (scfm) and VOC concentrations of at least several hundred ppm.

Many hydrocarbons have significant vapor pressures even at –112°F (–80°C). This limits attainable recovery. For example, the vapor pressure of toluene at –112°F (–80°C) is 0.052 mmHg or 69 parts per million by volume (ppmv) at atmospheric pressure. If the initial toluene

TABLE 4.28 Control Technologies for VOCs

Airstreams with a high VOC content (>500 ppm)
• Refrigerated vapor condenser
• Solvent vapor adsorption
• Flare
Airstreams with a moderate VOC content (about 100 to 500 ppm)
• Thermal incinerator
• Catalytic incinerator
• Regenerative (ceramic) incinerator
• Regenerative carbon adsorber
Airstreams with low VOC content (<100 ppm)
• Once-through carbon adsorption
• Carbon adsorber-incinerator systems
• Catalytic oxidation
• Biological systems
• Reducing process airflow rates

concentration is 1000 ppmv, at least 90 percent removal is attainable. If the initial toluene concentration is 100 ppmv, recovery can be no better than 30 percent.

Everything passing through the system is cooled to –112°F (–80°C)—VOCs, air, inert gas, and water vapor. Water is converted to ice, which could plug the system. Therefore, high humidity must be avoided. Refrigeration requirements are a large part of the cost of operating the system, and refrigerating low VOC concentrations has less efficient recovery and higher operating costs. This restricts refrigerated condensation to relatively high VOC concentrations at relatively low flowrate conditions. Should the recovered VOC have value—for example, for reuse—the operating costs can be offset.

Alternatively, a chilled vapor condenser can be employed if the vapor pressure and concentration of the VOC will allow removal efficiencies meeting the targeted emission rate at operating temperatures higher than approximately 30°F (–1°C).

Solvent Vapor Adsorption. VOC-containing air passes through a nonvolatile solvent, which adsorbs most of the VOCs. VOCs are stripped from the loaded solvent with heat and a partial vacuum and are recovered by condensation. Recovered VOCs can be recycled, which can offset some or all of the costs of operating the system. The VOC concentration must be at least 100 ppmv. Skid-mounted package units might be used for fuel transfer and can handle up to 1000 scfm. Larger custom units can handle several thousand scfm and might be used for controlling chlorinated hydrocarbon emissions from degreasers.

Flares. Flares are considered an acceptable device for emission control. Instances where there is both a sustained high gas flow and a high VOC concentration are rare because the losses would be economically unacceptable. A flare would be effective for controlling intermittent short-term high-flowrate or high-concentration surges.

The flare has the advantages of relatively low capital cost, minimal operating cost, and the ability to handle VOC surges. VOC control of 95 to 98 percent can be assumed. In addition, flares are a combustion source and will result in emissions of NO_x and CO.

Moderate VOC Concentration. Airstreams with moderate VOC concentrations are defined here as ranging from 100 to 500 ppmv.

Thermal Incinerators. VOC-containing air is heated in a combustion chamber to a temperature of 1400° to 1800°F (760° to 982°C). A supplemental fuel is always required and natural gas is the usual choice. A temperature of 1436°F (780°C) is adequate for most VOCs and will ensure more than 99 percent destruction. The heat content of the VOC usually does not contribute significantly to the fuel requirements.

Heat exchangers can be built into the incinerator to use the stack gas to preheat the incoming VOC stream. Up to 65 percent heat recovery is attainable. The heat exchanger reduces the fuel at the expense of a higher capital cost. The incinerator may need to be made of more expensive metals if chlorides are present.

The thermal incinerator is a simple and effective system. The disadvantage is the cost of the supplemental fuel and the increase in emissions of combustion-related gases (i.e., NO_x and CO). Fuel cost is less of a concern for low airflow rates and where operation is not continuous. For example, a 2000-scfm incinerator would require about 3.5 MBtu/h. For a one-shift, 5-days-per-week operation, the incinerator would operate about 2200 h/yr (allowing for warm-up), creating an annual fuel requirement of about 7700 MBtu/h. Depending on natural gas prices, a $100,000 heat exchanger might not be justifiable. For the same airstream and a three-shift operation, a heat exchanger probably would be a good investment.

For a 20,000-scfm air stream, fuel requirements would be about 35 MBtu/h. Such a thermal incinerator should certainly have a heat exchanger; therefore, other VOC control methods should be considered.

Catalytic Incinerator. VOC combustion takes place on a catalyst surface at about 800°F (427°C). Fuel costs are lower than for a thermal incinerator because it takes less fuel to heat air to 800°F (427°C) than to 1436°F (780°C). Also, the heat value of the VOC may contribute significantly to the fuel requirements.

The cost of the catalyst is a significant part of the cost of a catalytic incinerator. This is offset because a catalytic incinerator can be made of less expensive materials due to the lower combustion temperature. The lower fuel cost means that catalytic incineration can be evaluated as an alternative to thermal incineration at airflow rates above a few thousand scfm. A catalytic incinerator will typically achieve greater than 98 percent destruction of VOCs. This is very acceptable for controlling VOCs from coating and printing operations.

Replacing the catalyst is expensive; therefore, catalytic incinerators should not be used when the catalyst could be contaminated. Burning chloride-containing VOCs produces HCl, which can attack the catalyst. Sulfur-containing VOCs can also attack some catalysts. Airstreams that contain suspended particulate matter or produce particulate matter upon combustion can cause the catalyst to be coated and rendered inert. If there is a VOC surge in the airstream, the VOCs' fuel value could cause an exothermic reaction, raising the combustion temperature above the design limit and rendering the catalyst inert ("burning" the catalyst). This could result from a drying oven being loaded with freshly coated surfaces, thus producing an initial rapid evaporation of the VOCs.

Regenerative (Ceramic) Incineration. The regenerative incinerator has a ceramic heat exchanger. Heat from burning the VOCs is transferred to the ceramic exchanger. This heat then is used to heat the incoming VOC stream. The system has dual exchangers, with one exchanger being heated while the other is giving up its heat. The incinerator achieves 98 to 99 percent destruction of VOCs at a combustion temperature of about 1400°F (760°C) with a thermal efficiency of up to 95 percent. The regenerative incinerator can be modified to include a catalyst in the regenerative ceramic beds, which allows the combined benefits of regenerative incineration with catalytic incineration. The result is lower combustion chamber temperature, further reducing fuel requirements while maintaining a high level of VOC destruction efficiency and heat recovery. The regenerative incinerator has the highest thermal efficiency of any incinerator, and therefore the lowest fuel cost. Capital costs, however, are higher than for other incineration systems. It is very effective in treating high-volume airstreams containing a VOC concentration up to a few hundred ppm.

The units typically range from 5000 to 100,000 scfm and weigh several tons. The size and weight restrict the location of the incinerator. The recuperative incinerator should be considered when airflow rates are several thousand scfm.

Regenerative Carbon Adsorption. VOC-containing air passes through a bed of activated carbon. The VOCs are adsorbed onto the carbon. When the bed is sufficiently loaded, VOCs are stripped from the bed using low-pressure steam followed by air drying. Steam and VOCs are recovered by condensation. For continuous operation, there are two or three carbon beds, with one bed being regenerated while the other beds are loading. Regenerative carbon adsorption requires on-site steam and compressed air.

Carbon adsorption works very well for water-insoluble organic compounds that are liquid at room temperature. Process economics benefit if the recovered VOCs can be recycled. A single-stage unit can achieve up to 99 percent control. Carbon adsorption is usable for VOC concentrations between 10 and several hundred ppm.

Carbon adsorption is less effective for the following conditions:

- Organic compounds that are gases or very low boiling-point liquids at ambient temperatures
- Airstreams above 100°F (38°C)
- Airstreams above 50 percent relative humidity

Under these conditions, VOCs are not tightly held on carbon, and control efficiencies can be unsatisfactorily low. Ketones can undergo exothermic polymerization on the carbon surface, coating the carbon with a solid polymer that is not removable by steam stripping, thus rendering the carbon nonadsorptive. The heat buildup can cause the carbon bed to ignite. If the VOCs are water-soluble, they are dissolved in the condensed steam and the condensate requires treatment prior to discharge. Using an alternative VOC-control method is usually preferable to treating the VOCs in the condensate.

Low VOC Concentration. Airstreams with a low VOC concentration are those below 100 ppm.

Once-Through Carbon Adsorption. This differs from regenerative carbon adsorption in that activated carbon is in canisters furnished by the carbon supplier. VOC-containing air passes through the canisters. Instead of regenerating the carbon, spent canisters are exchanged by the supplier for fresh canisters.

Advantages are: minimal capital investment is needed; steam and compressed air are not required, which makes canisters suitable for field sites; minimal labor and attention are required; and effective VOC loading is better in the low-ppm concentration range.

Activated carbon will load 0.3 lb of typical VOCs per lb of carbon at 100 ppm and 0.15 lb per lb of carbon at 5 ppm. Steam stripping will leave 0.2 and 0.13 lb of VOCs on the carbon, respectively. Therefore, for regenerative carbon the working margin is 0.1 lb of VOC per lb of carbon at 100 ppm (0.3 – 0.2) and 0.02 lb of VOC per lb of carbon at 5 ppm (0.15 – 0.13). Thus, at 5 ppm, once-through carbon can load 0.15 lb of VOC per lb of carbon, compared with 0.02 lb for regenerative systems.

The carbon canisters lack instrumentation and controls; therefore, the carbon could become overloaded unless it is carefully monitored. Fluctuating or less-than-design airflow rates could cause the carbon to form channels, reducing the effective capacity of the canister.

The VOC control cost is high for nonregenerative carbon. At a loading of 0.15 lb of VOC per lb of carbon and a carbon cost of $2.40 per lb, the VOC control cost for carbon alone is $32,000 per ton ($2.40/0.15 × 2000). Any VOC control strategy is expensive for 5-ppm streams.

Carbon Adsorber-Incinerator Systems. A combination of a carbon adsorber and an incinerator can be used to control a high-volume, low-VOC-content stream. For example, suppose it is necessary to control a 100,000-scfm stream having a 30-ppm VOC content and that incineration is the appropriate control method. Whichever mode of incineration was used, the fuel cost would be very high. However, the fuel cost could be greatly reduced by first passing the airstream through a carbon adsorber, followed by stripping the adsorber with warm air [approximately 250°F (121°C)], and then routing that airstream to the incinerator. The airstream out of the carbon adsorber could be 10,000 scfm with a VOC content of 300 ppm. In a catalytic incinerator, little or no supplemental fuel would be required. Capital costs for two control systems would be higher than for a single system, and two control systems would have more chance of being out of commission, but they would significantly reduce a very high fuel cost.

Catalytic Oxidation. A catalyst, typically a monolithic honeycomb type with a metal or ceramic substrate, is inserted into the process stream, typically a waste stream from a combustion source; VOCs are converted to H_2O and CO. The catalyst requires an operating temperature of 250 to 1200°F (121 to 680°C) to achieve a high destruction efficiency of up to 90 percent. Disadvantages of catalytic oxidation include the potential for catalyst poisoning by chemicals in the waste stream, the potential for plugging if particulate matter is present in the gas stream, and the potential for damage by a transient increase in VOC concentration, which can cause a temperature surge.

Biological Systems. Biological systems use microbes in an aerobic environment to biodegrade VOCs to CO_2 and H_2O. Biological systems can take the form of biofilters, trickling filters, or bioscrubbers. Bioscrubbers are not yet common in the United States and are not addressed in the following paragraphs.

VOC-removal efficiencies as high as 95 percent have been achieved with biofilters. Biofilters typically have low operating costs with minimal emissions of other pollutants, such as NO_x and CO. However, space requirements are higher and capital costs are moderate to higher compared to other VOC-abatement equipment. A number of factors must be considered in selection of a biofilter, including the following:

- Biodegradability of the VOCs, which differs by compound
- Compound solubility into the biomass
- VOC loading rate and concentration
- Temperature of the process stream being treated, because the optimum temperature range for microbial activity is 77 to 95°F (24 to 34°C)
- Acid by-product formation, typically from chlorinated VOCs, which affects biomass pH and effectiveness
- Acclimation period for the microbes to achieve optimum degradation, which can differ according to the VOCs present in the process stream to be cleaned
- Particulate matter or grease in the process stream, which can clog the biofilter media
- Operational characteristics of the process, because biosystems tend to operate best when the process stream to be treated is continuous-flow, with a constant composition
- Source of moisture to maintain biological activity

Reducing Process Airflow Rates. Regulatory agencies now look at making VOC controls more feasible by reducing process airflow rates. For example, suppose a coating operation has five stations, each with an airflow rate of 20,000 scfm and a VOC content of 80 ppm. The combined airflow rate is 100,000 scfm and the VOC content is still 80 ppm. VOC controls would be very expensive. However, if the airflow from the five stations was put in series (the air from Station 1 goes to Station 2, which goes to Station 3, etc.), then the airflow rate is 20,000 scfm with a VOC content of 400 ppm. If the lower flammability level of this VOC is 1.6 percent or 16,000 ppm, the VOC content is still well below 25 percent of the lower explosive limit (LEL) and may be economically controllable. When VOC controls are required, it may be technically and economically preferable to reduce the process airflow rate rather than buy a larger and more expensive control system.

Controlling NO_x Emissions

Nitrogen oxides (NO_x) are products of all conventional combustion processes. Nitric oxide (NO) is the predominant form of NO_x emitted by such sources, with lesser amounts of nitrogen dioxide (NO_2) and nitrous oxide (N_2O). The NO can further oxidize in the atmosphere to NO_2. The generation of NO_x from fuel combustion is a result of two formation mechanisms. Fuel NO_x is formed by the reaction of nitrogen chemically bound in the fuel and oxygen in the combustion air at high temperature in the combustion zone. Thermal NO_x is produced by the reaction of the molecular nitrogen and oxygen contained in the combustion air at high temperature in the combustion zone. The main factors influencing the NO_x reaction are combustion temperature, residence time within the combustion zone, amount of fuel-bound nitrogen, and oxygen levels present in the combustion zone.

For some combustion sources, RACT, BACT, and LAER have and will require add-on controls or combustion modifications to limit NO_x emissions. These add-on control technologies are relatively new and can represent high capital and maintenance costs, especially for industrial boilers and heaters. The low-cost combustion-modification techniques involve burner modifications with total capital cost under $50,000 per boiler.

Combustion Modification Techniques

Burners Out of Service (BOOS). On multiple-burner units, simply taking one or more burners out of service lowers the potential flame temperature (PFT) through several mechanisms. These include increasing gas flow to any particular remaining burner and reducing the flame temperature by reducing the flame-to-flame radiant exchange.

BOOS costs little to implement. This technique often is most effective in reducing NO_x for burners at middle to upper elevations in a boiler.

Boiler Derating. Boiler derating can be an acceptable procedure for industrial boilers under certain conditions. For example, steam demand may decrease when certain processes are phased out or replaced by a more energy-efficient process, or a plant's energy use can be reduced through efficiency planning. In such cases, the boilers may be derated to fire at lower loads.

Derating can be as simple as reducing the firing rate, or a facility may be permanently derated, for regulatory reasons, to take advantage of a higher NO_x limit for a lower-rated boiler. Even permanent derating can be accomplished inexpensively—for example, by installing a permanent restriction, such as an orifice plate, in the fuel line. If a boiler derating is impossible, often the burners themselves can be modified to reduce NO_x concentrations.

Burner System Modification (BSM). BSM is a relatively underused and low-cost strategy that can produce marked NO_x reductions. It can be used on ring and spud burners, the two main types. BSM may also consist of modifications to the air registers surrounding the burner. These modifications may require a level of expertise not usually found in the boiler house. However, many industrial facilities possess both a machine shop and fabrication capabilities. In such a case, the costs of BSM can be reduced by performing the work in-house under the guidance of a qualified combustion consultant.

Oxygen and Combustible Trim. Installation of oxygen or combustible trim controls increases boiler efficiency and reduces NO_x by limiting the amount of influent air admitted to the boiler. Systems costing less than $10,000 are available to monitor and automatically trim O_2 or combustibles to the desired levels. O_2 is usually reduced to about 3 percent. Combustible trim reduces combustible concentration to a target level, e.g., 200 parts per million by volume, dry (ppmvd) in the furnace. Combustibles are usually reduced to lower levels at the stack outlet due to their subsequent oxidation in the boiler. An alternative type of trim—the deliberate generation of combustibles—forces oxygen concentration to its minimum possible concentration, thereby reducing PFT. However, combustibles must not exceed certain levels because boiler corrosion can be accelerated by attendant flame impingement and a reducing atmosphere, or, in the case of unstable flames, by an alternating reducing and oxidizing atmosphere.

Steam Injection (SI) and Water Injection (WI). SI and WI are commonly applied to gas turbines to limit NO_x emissions. SI reduces the PFT by diluting oxygen near the burner front and directly removing heat from the burner flame. WI functions in a similar way, but removes even more heat from the burner flame due to its heat of vaporization. Both SI and WI lower boiler thermal efficiency—typically by no more than 1 or 2 percent. Most industrial boilers can handle these losses, although utility units cannot. SI or WI may be the lowest-cost option for reducing NO_x to less than 40 ppmvd, if simple combustion modifications are not successful.

Staged Combustion (SC). This technique deliberately partitions the air (air staging) or fuel (fuel staging) to create an initial fuel-rich zone, followed by an air-rich one to complete the combustion. SC uses a volume of air that is equal to or slightly greater than that for conventional combustion; however, the PFT is reduced because initial oxygen concentrations are reduced. SC is usually done in two separate sections of the boiler, and requires a large furnace volume. SC is sometimes adaptable to smaller boilers. Some low-NO_x burners stage combustion over short lengths to reduce NO_x.

The term *staging* means allowing fuel and air to react in multiple zones, or stages, rather than all at once. Compared with conventional combustion, SC tends to limit large excesses of temperature and oxygen, producing lower NO_x. This can be accomplished over small distances along the burner itself if the fuel, air, or both are injected at multiple locations along the burner.

Alternate Fuels. This term refers to fuels other than natural gas and fuel oil. If a boiler cannot meet mandated NO_x levels with its existing fuel, a switch to an alternate fuel may prove worthwhile. Capital costs for implementing alternate fuels range widely. For example, a facility containing equipment that can handle both fuel oil and natural gas can switch between fuels with minimal costs. However, if using an alternate fuel requires new handling and storage equipment, costs may exceed $50,000 per boiler.

Flue-Gas Recirculation (FGR). In this technique, a portion of the flue gas is withdrawn from the boiler stack and introduced with the combustion air. The oxygen availability is reduced, thus reducing the PFT and NO_x. Most applications recirculate roughly 20 percent of the flue gas to achieve less than 40 ppmvd NO_x. However, excessive FGR can produce unsafe, unstable flames and result in an increase in CO emissions.

Low-NO_x Burners (LNB). These reduce NO_x by replacing high-NO_x-producing burners. Most, but not all, LNBs are spud-type burners specifically designed to produce flames with lower PFT. This is usually done with bluff bodies and gas ports that induce recirculation zones and reduce local oxygen concentrations close to the burner (effectively producing staged combustion). Bluff bodies also increase flame stability, which is more critically needed when local oxygen concentrations are deliberately reduced. LNB is often combined with FGR, which, in tandem, can reduce NO_x emissions by 60 to 90 percent.

Postcombustion Techniques

Selective Catalytic Reduction (SCR). In the SCR process, NO_x is reduced to N_2 and H_2O by ammonia (NH_3) within a temperature range of approximately 540 to 840∞F (283 to 433∞C) in the presence of a catalyst, usually a base metal. The lower end of the operating temperature range is feasible when the acid gas impurity level is relatively low. NH_3 has been used as an acceptable reducing agent for NO_x in combustion gases because it selectively reacts with NO_x, while other reducing agents, such as hydrogen (H_2), carbon monoxide (CO), and methane (CH_4), also readily react with O_2 in the gases. In a typical configuration, flue gas from the combustion source is passed through a reactor, which contains the catalyst bed. Parallel-flow catalyst beds may be used in which the combustion exhaust gas flows through channels rather than ports to minimize blinding of the catalyst by particulate matter. NH_3 in vapor phase is injected into the flue gas downstream of the other control equipment that may be required for the particular combustion process (for removal of pollutants such as particulate matter and SO_2). The NH_3 is normally injected at a 1-to-1 molar ratio based on the NO_x concentration in the flue gas. Major capital equipment for SCR consists of the reactor and catalyst, NH_3 storage tanks, and an NH_3 injection system using either compressed air or steam as a carrier gas. Because of the toxic and explosive characteristics of NH_3, appropriate storage and handling safety features must be provided if anhydrous NH_3 is used. In addition, NH_3 slip is highly regulated by the agencies, and is generally limited in permits to 10 ppmv (dry, 15 percent O_2) or less. The two critical variables to ensure low NH_3 slip are controlling the NH_3 injection rate and ensuring proper mixing of the NH_3 in the gas stream prior to reaching the catalyst bed. NO_x-removal efficiencies approaching 9 percent have been reported when using SCR systems for boiler and gas-turbine applications.

Selective Noncatalytic Reduction (SNCR). This process involves NH_3, either anhydrous or in solution with water, or urea injection without a catalyst. A major commercially available SNCR system is the Exxon Thermal $DeNO_x$ NH_3 (anhydrous or aqueous) injection system.

Exxon Thermal $DeNO_x$. Exxon Thermal $DeNO_x$ ammonia injection, like SCR, uses the NO_x/NH_3 reaction to convert NO_x to molecular nitrogen. However, without catalyst use or supplemental hydrogen injection, NO_x reduction reaction temperatures must be tightly controlled between 1600 and 2200°F (871 and 1204°C) [between 1600 and 1800°F (871 and 981°C) for optimum efficiency]. Below 1600°F (871°C), and without hydrogen also being injected, NH_3 will not fully react, resulting in what is called *NH_3 breakthrough* or *slip.* If the temperature rises above 1800°F (981°C), a competing reaction begins to predominate:

$$NH_3 + 5/4\ O_2 \rightarrow NO + 3/2\ H_2O$$

As indicated, this reaction increases NO emissions. Therefore, the region within the boiler where NH_3 is injected must be carefully selected to ensure that the optimum reduction reaction temperature will be maintained.

For this noncatalytic NO_x reducing process, NH_3 must be injected at a 2-to-1 molar ratio (based on the flue-gas NO_x concentration). Therefore, there is some slip of NH_3 that does not react completely and can potentially cause odors. Also, fine particulate emissions that create a visible plume can be formed from the reaction of NH_3 and HCl (a solid-fuel combustion byproduct) downwind of the stack. Therefore, it is important to keep the NO_x injection rate to the minimum necessary.

$SCONO_x$™ System. The $SCONO_x$ process, employing catalytic combustion to reduce emissions of NO_x, is patented by Goal Line Environmental Technologies, LLC. A precious-metal catalyst is coated with an absorptive liquid. The catalyst oxidizes NO to NO_2. The liquid coating on the catalyst absorbs the NO_2. To prevent saturation of the absorptive liquid, the liquid is regenerated by passing a dilute hydrogen-reducing gas across the catalyst surface. The absorbed NO_2 is released as N_2 and H_2O. Unlike SCR or SNCR, no NH_3 is required, thus eliminating the concern about NH_3 slip emissions. The optimum operating temperature range for the system is 300 to 700°F (199 to 369°C). An additional benefit of the system is that catalyst also converts CO to CO_2. The N_2, H_2O, and CO_2 are exhausted to the atmosphere. The $SCONO_x$ system is equipped with a regeneration-gas-production module, which uses natural gas and air reactions to produce the regeneration gas. A disadvantage of this system is that the catalyst is subject to attack from sulfur oxides. Therefore, depending on the sulfur content of the fuel combusted, another system patented by Goal Line may be required to remove sulfur compounds from the combustion-source exhaust stream entering the $SCONO_x$ system. Most of the commercially available demonstrations of the $SCONO_x$ system have been conducted on combined-cycle cogeneration systems firing natural gas and comprised of a combustion turbine, duct burner, and heat-recovery steam generator. According to the manufacturer, NO_x and CO emissions can be reduced to 2 and 1 ppm, respectively. The manufacturer claims that the technology also can be applied to simple-cycle combustion turbines, boilers, and internal combustion engines if a heat exchanger is used to cool the turbine exhaust gases to 650°F (392°C).

Catalytic Combustion. Catalytic combustion is a process by which a lean fuel-air mixture is passed over a catalyst to flamelessly initiate combustion. The air and fuel are premixed upstream of the catalyst. A portion of the fuel is burned in the catalyst and the balance is burned downstream of the catalyst. This technique allows combustion temperatures to remain below 2730°F (1500°C), which is below the optimum temperature for production of thermal NO_x (the nitrogen and oxygen in the air react to form NO_x). The result is very low NO_x emissions, on the order of 3 ppm. The technology is being demonstrated on a gas turbine and has been deemed by EPA to have proven that the reductions have been achieved in practice.

BIBLIOGRAPHY*

1. EPA: "Compilation of BACT/LAER Determinations—Revised," EPA 450/2-80-70, May 1980.
2. EPA: "Compilation of Air Pollutant Emission Factors," AP-42, 5th ed., and Supplements A through F, Research Triangle Park, N.C., September 2000.
3. EPA, OAQPS: "Ambient Monitoring Guidelines for Prevention of Significant Deterioration (PDS)," EPA 450/4-80-012, November 1980.
4. EPA, OAQPS: *Guidelines for Air Quality Maintenance Planning and Analysis,* vol. 10, "Procedures for Evaluating Air Quality Impact of Near Stationary Sources," EPA-450/4-77-001, Research Triangle Park, N.C., October 1977.
5. EPA, OAQPS: "Guideline on Air Quality Models," EPA-450/2-78-027, Research Triangle Park, N.C., April 1978.
6. EPA, OAQPS, Region III: "Guidelines for Determining BACT," Philadelphia, December 1978.
7. EPA, Region III: "Permit Application Kit, Prevention of Significant Air Quality Deterioration," Air Programs Branch, Philadelphia, November 1978.
8. EPA, Technology Transfer: "Industrial Guide for Air Pollution Control," Contract 68-01-4147, by PEDCo Environmental, Inc., June 1978.
9. EPA: *Control Techniques for Particulate Emissions from Stationary Sources,* vol. 1, EPA 450/3-81/005a, Research Triangle Park, N.C., 1982.
10. EPA: *Handbook-Control Technologies for Hazardous Air Pollutants,* EPA 625/6-86/014, Center for Environmental Research Information, 1986.
11. *Federal Register,* 1977 Clean Air Act, PSD, SIP Requirements, Part 52, June 19, 1978.
12. *Federal Register,* 40 CFR Parts 51 and 52, September 5, 1979.
13. *Federal Register,* 40 CFR Parts 51 and 52, February 5, 1980.
14. Quarles, J. Jr.: "Federal Regulation of New Industrial Plants," P.O. Box 998, Ben Franklin Station, Washington, D.C. 20044, January 1979.
15. Rymarz, T. M., and D. H. Klipstein: "Removing Particulates from Gases," *Chemical Engineering Deskbook,* vol. 82, no. 21, October 6, 1975, pp. 113–120.
16. Corbitt, R. A.: *Standard Handbook of Environmental Engineering,* McGraw-Hill, New York, 1990.
17. Goal Line Environmental Technologies LLC web site, www.glet.com.
18. Air and Waste Management Association (AWMA): *Air Pollution Engineering Manual,* W. T. Davis, ed., John Wiley & Sons, New York, 2000.
19. Catalytica, Inc. web site, www.catalytic-inc.com.
20. Schimmoller, B. K.: "SCR Dominates NO_x Compliance Plans." *Power Engineering,* July 2000.
21. Giovando, C. A.: "Environmental Technologies Help Meet Strict Regulations," *Power,* May–June 2000.
22. Reynolds, F. E., Jr., and W. D. Grafton, III: "Biofiltration: An Old Technology Comes of Age," *Environmental Technology,* July–Aug. 1999.

* Due to the proliferation of new regulations and technology, the reader is urged to research the latest publications, government and private.

CHAPTER 4.7
LIQUID-WASTE DISPOSAL

Envirex Products
U.S. Filter
Waukesha, Wisconsin

GLOSSARY

Alkalinity Ability to neutralize acids—determined by the water's content of carbonates, bicarbonates, hydroxides, and borates, silicates, and phosphates, if present. Expressed in milligrams per liter of calcium carbonate ($CaCo_3$).

BOD_5 (biochemical oxygen demand) A measure of oxygen metabolized, in milligrams per liter, in 5 days by microorganisms that consume biodegradable organics in wastewater under aerobic (with air) conditions.

COD (chemical oxygen demand) The amount of oxygen, in milligrams per liter, needed to oxidize both organic and oxidizable inorganic compounds.

Effluent The liquid end product discharging from a process.

Floating matter Matter that passes through a 2000-μm sieve and separates by flotation in 1 h.

Settleable solids Solids larger than 0.01 mm in diameter settling in 2 h under quiescence.

Suspended solids Small filterable particles of solid pollutants in wastewater. The examination of suspended solids and the BOD_5 test constitute the two main determinations for water quality.

Total solids All dissolved, suspended, and settleable solids contained in a liquid.

Turbidity The amount of suspended matter in wastewater; quantity obtained by measuring its light-scattering ability.

DESCRIPTION OF THE PROBLEM

The passage of the Federal Water Pollution Control Act as amended in 1977 under Public Law 92-217 forced the plant engineer to become familiar with its many ramifications. Among the provisions of this act that are of direct interest to industry is the establishment of water quality standards typified by Table 4.29, column A, for maintaining aquatic life. These restrictions are enforced by the requirement that a permit be issued before discharges are permitted under the National Pollutant Discharge Elimination System (NPDES). Every holder of a NPDES permit is required to comply with monitoring sampling, recording, and reporting requirements.

In 1987, further amendments were incorporated into the Water Quality Act (WQA). These amendments were aimed principally to improve water quality in areas that lacked

TABLE 4.29 Maximum Discharge Limits

Constituent	A Direct discharge or recycle, mg/L	B* To POTW
Ammonia nitrogen (as N)	1.5	—
Arsenic (total)	1.0	—
Barium (total)	5.0	—
Boron (total)	1.0	1.0
Cadmium (total)	0.05	2.0
Chromium (total)	—	25.0
Chromium (total hexavalent)	0.05	10.0
Chromium (total trivalent)	1.0	—
Copper (total)	0.02	3.0
Cyanide (total)	0.025	@150°F & pH 4.5 = 2.0
Fluoride (total)	1.4	—
Iron (total)	1.0	50.0
Iron (dissolved)	0.5	—
Lead (total)	0.1	0.5
Manganese (total)	1.0	—
Mercury (total)	0.0005	0.0005
Nickel (total)	1.0	10.0
Fats, oils, and greases (FOG)	15.0	100.0 total
pH	5.0–10.0	4.5–10.0
Phenols	0.1	—
Phosphorus (as P)	1.0	—
Selenium (total)	1.0	—
Silver	0.005	—
Sulfate	500.0	—
Temperature	—	150°F (65°C)
Zinc (total)	1.0	—
Total dissolved solids	1000.0	No limit

* Units are milligrams per liter unless otherwise specified.

compliance with minimum national discharge standards. This is an ongoing, active regulatory area; the reader is urged to check the latest regulations before proceeding with plant modifications.

Many local ordinances have pretreatment requirements limiting high effluent concentrations of wastes and toxic materials which might adversely affect treatment processes of publicly owned treatment works (POTWs). A typical list of effluent limitations is shown in Table 4.29, column B. When discharging to POTWs, industry is expected to pay its proportionate share of capital cost of the POTWs collection and treatment equipment. These user fees are usually based on a multiplier of BOD_5, suspended solids, and liquid volume, and vary with each municipality.

For specific pollutants, effluent guidelines for specific industrial categories are published in the *Code of Federal Regulations* (40 CFR 401).

Different effluent levels are allowable depending on the following:

1. Industrial subcategory
2. Control technology required (e.g., best available technology economically achievable)
3. Existing or new source (new sources are more severely regulated than existing sources)
4. Where the effluents are discharged (effluent levels discharged into POTWs are different from direct discharges into navigable water)

For details pertaining to emissions by specific industry, the *Code of Federal Regulations* should be consulted. Because the promulgation of effluent guidelines is an ongoing process, the EPA should be contacted for the latest information.

All pollutants are classified as either conventional, toxic, or nonconventional. Conventional pollutants include BOD_5, TSS (total suspended solids), and pH. There are 129 priority pollutants that appear on the toxics list in the *Federal Register* 43(164)4108 (February 1978). Nonconventional pollutants are those that are neither toxic nor conventional, such as nitrogen, oil, and grease. Best conventional pollutant control technology was required for conventional pollutants by July 1, 1984. Best available technology economically achievable was also required for toxic and nonconventional pollutants by the same date.

PLANT SURVEY

The accurate measurement of flow volume and pollutants in the waste flow are essential in assessing any wastewater problem, and in designing a wastewater treatment system. Limitations on effluent (see Table 4.29) make it imperative to analyze flows and impurities quickly, accurately, and at reasonable cost.

The best approach is to make a comprehensive wastewater survey that will (1) determine the quantity of wastewater discharge, (2) locate the major sources of waste within the plant, (3) determine wastewater composition, (4) explore in-plant or process changes to minimize the waste problem, (5) establish the basis for wastewater treatment, and (6) evaluate effect of wastes on the receiving stream.

Composition of wastewater varies with the amount of impurities initially present in water and the chemical analysis of any pollutants that are added. While domestic sewage has a fairly uniform composition, industrial wastes have an almost infinite variety of characteristics, as shown in Table 4.30. Wastewaters should be analyzed for at least BOD_5, COD, color, total solids (suspended and dissolved), pH, and turbidity. Other impurities of interest will vary with the source and type of wastewater.

Concentration of pollutants must be correlated with average, minimum, and maximum flows encountered. The analysis program must also take effluent water quality standards into account (see Table 4.29) and the BOD_5 reduction required to meet them. Any toxic impurities in the wastewater that adversely affect water quality must be determined.

TABLE 4.30 Typical Process Discharge Volumes, BOD_5, and Suspended Solids for Industrial Wastewater Before Treatment

	Unit processed	Discharge per unit		BOD_5, mg/L	Suspended solids, mg/L
		Gallons	Liters		
Aluminum & copper	lb (kg)	12–13	45–50	N/A	300–500
Automotive	Car	10,800	40,900	190	215
Beverage, malt	bbl (L)	330	1250	390–1800	70–100
Canning					
Fruit	Case	20–40	75–150	300–1600	200–500
Vegetable	Case	50–100	190–380	700–2000	300–2000
Coal washing	ton (tonne)	125	138	N/A	2000–3000
Cooking	ton (tonne)	1500–2800	1650–3090	50–200	90
Dairy, milk	gal (L)	4–12	15–45	1800	560–4000
Electrical	kWh (kJ)	80	110	N/A	50–2000
Laundry					
Commercial wash	ton (tonne)	8600	36,000	600–1860	400–2200
Industrial	ton (tonne)	5000	20,000	650–1300	4900–8600
Manufacturing, gen. fabr.	ton (tonne)	700	3000	50–1500	200–15,000
Meat packing					
Cattle	Animal	400–2000	1515–7575	400–900	400–800
Chicken	Bird	8–9	30–34	150–2400	100–1500
Hogs	Animal	300–600	1136–2273	1000	650
Office building	Person	30–45	114–170	117	176
Paint, latex	gal (L)	3	11	2000–3000	15,000–60,000
Paper making	lb. (kg)	65	108	200–800	500–1200
Pharmaceutical	—	—	—	600–2500	500–1000
Phenolic resins	ton (tonne)	75,000	313,000	11,500	40
Railway maintenance	Locomotive	3000	11,360	500–800	200–600
Refining	bbl crude	770	2900	100–500	300–700
Rubber, synthetic	Car tire	500	1890	25–1600	60–2200
Steel					
Cold-rolled	lb (kg)	9	15	150	100–300
Hot-rolled	lb (kg)	18	30	80	500–2000
Tanning, hide	lb (kg)	8–12	30–45	900	6000
Textiles					
Synthetic	lb (kg)	12–25	45–95	1500–6000	500
Wool	lb (kg)	70	265	900	100
Vegetable oil	gal (L)	22	83	3050	900

Sampling and Flow Measurement

The starting point in any wastewater survey is an effective program of sampling and flow measurement. To be useful, a sample of wastewater must accurately represent the source from which it is taken and be large enough to run all the laboratory tests required. This means the method of sampling must be tailored to the type and kind of wastewater flow. A close check of each waste source will reveal whether flow is continuous or intermittent and any wide swings in flow rate.

It is also important to know if the concentration of pollutants changes drastically or is fairly constant. The presence of oil or excessive suspended solids may also cause problems.

An integral part of this program is the need to obtain flow information on various in-plant streams as well as the plant outfalls. Wastewater flows are measured for the following reasons:

1. To determine the quantity of water being discharged, as well as variations in the flow rate
2. To determine the number of pounds of constituents being discharged on the basis of the analytical data and the determined flow rate

3. To evaluate segregation possibilities
4. To determine the effect of the wastewater discharge on the receiving stream, if applicable

Measuring Rate of Flow. Rates of flow can be approximated by the methods discussed in the following paragraphs.

Water Meters on Influent Lines. Water consumption in the plant should be determined during a wastewater survey to check on wastewater-flow measurements and to compute a water balance for the plant. Meters can also be installed at particular water-using operations to obtain flow data.

Container and Stopwatch. The time required to obtain a given volume of water in a container is measured. Volume can be determined either by weight added or by a calibrated collection container. The weight of water added is divided by 8.34 lb/gal to determine the number of gallons collected. The flow is then determined by the formula

$$\frac{\text{Gal in container} \times 60}{\text{Time, s, to fill container}} = \frac{\text{gal/min}}{15.85} = \text{L s}^{-1}$$

If the container fills in less than 10 s, the accuracy of this method is questionable.

Weirs. A weir acts like a dam or obstruction, with the water flowing through the notch, which is usually rectangular or V-shaped.

To ensure accurate weir measurements:

1. The weir crest must be sharp or at least square-edged. Steel is the best construction material, but tempered wood is also used.
2. The weir must be ventilated. There must be air on the underside of the falling water.
3. Leaks around the weir plate must be sealed.
4. The weir must be exactly level.
5. Weirs should be kept clean.
6. The head on the weir should be measured at a distance of 2.5 times the head upstream from the weir.
7. The channel upstream from the weir should be straight, level, and free from disturbing influences. A stilling box may be used to quiet the water flow.
8. The weir should be sized after the flow is estimated by other methods. The head on any weir should be greater than 3 in (7.6 cm) but not more than 2 ft (61 cm).

The flow over V-notched (triangular) weirs and rectangular weirs can be taken from the nomographs shown in Fig. 4.67.

Parshall Flume. A Parshall flume (Fig. 4.68) can be used to measure flows in open channels at or near ground surface. This device is valuable when it is not possible to dam the water. It is also advisable for a permanent installation because it is self-cleaning.

Flow under submerged conditions can be calculated from readings taken at gauges. If the water surface downstream from the flume is high enough to retard the rate of discharge, submerged flow exists. When there is no backwater effect, water passing through the throat and diverging section assumes a level which corresponds to the floor of the channel. This pattern demonstrates free flow.

The flow of a free discharge from a Parshall flume is calculated by

$$Q = 4WH_a n$$

where Q = flow, ft^3/s
W = throat width, ft
H_a = head of water above level floor, ft
n = $1.522\ W^{0.026}$

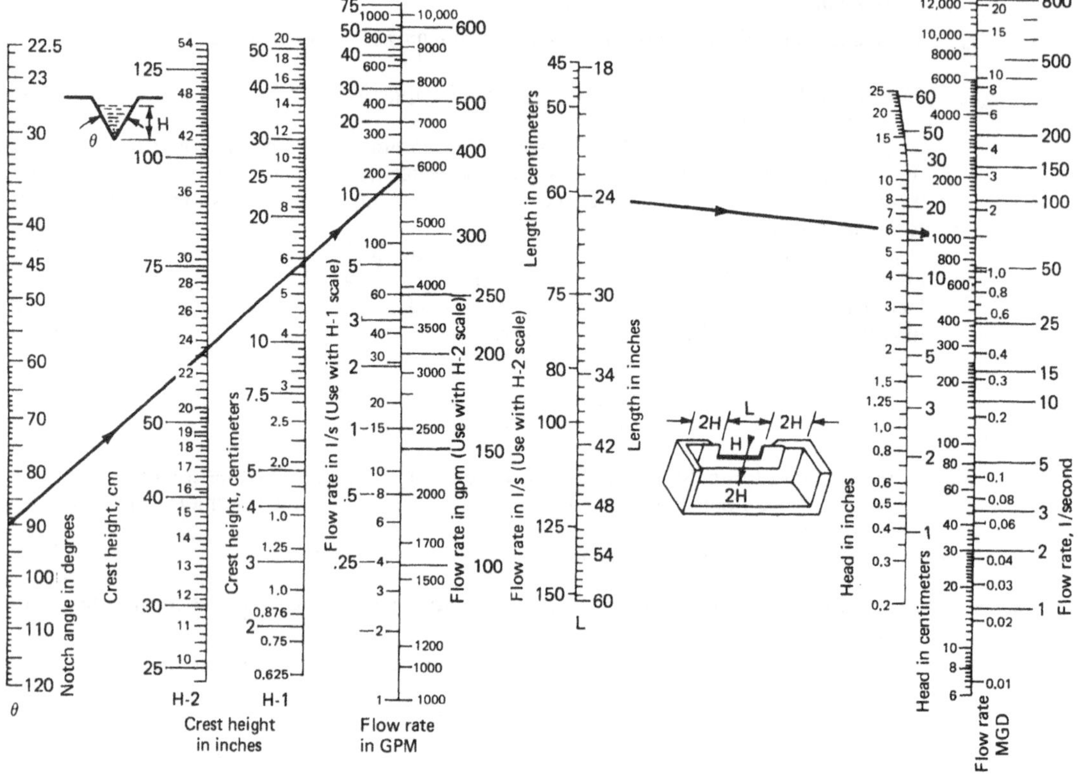

FIGURE 4.67 Nomographs for measuring flow over weirs.

or in metric units

$$Q = 8.52 \times 10^3(11.4 \log W)H_a(1.57 + 0.09 \log W)$$

where Q = flow, m^3/h
W = throat width, m
H_a = head of water above level floor, m

Flow under submerged conditions can be calculated from readings taken at gauges, one located at a point two-thirds the length of the converging section measured back from the crest of the flume H_a and one located near the downstream end of the throat section H_b. Degree of submergence is given by the ratio H_b/H_a.

Sample Collection and Analysis. Wastewaters are sampled and analyzed to identify those pollutants that require treatment and to select the proper treatment process.

Collecting Samples. Since a wastewater's characteristics can vary considerably, composite samples are collected to obtain a truer representation of the waste. Small samples are collected at frequent intervals during the sampling period. They are mixed together to form the composite sample.

Compositing Samples. Depending on plant operation, 8-, 16-, or 24-h composites can be collected. Daily sampling for 3 days generally constitutes a sampling program. Samples can be composited on the basis of the following criteria.

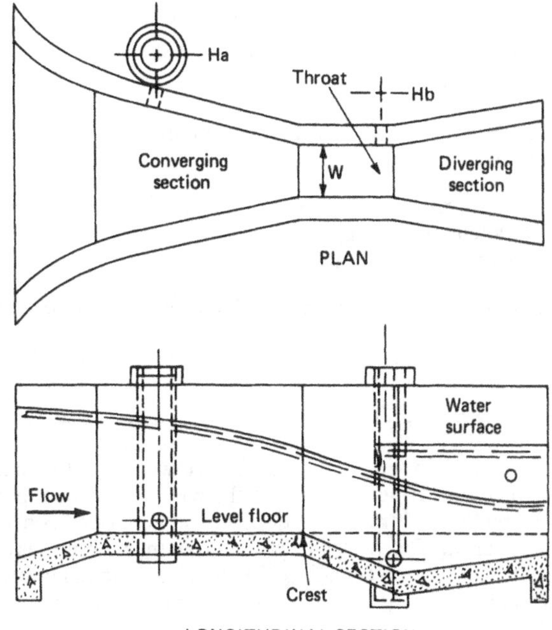

FIGURE 4.68 Parshall flume.

Flow. The amount of sample collected at any time during the sampling period is proportional to the flow of wastewater at that time.

Time. The same amount of sample is collected at every interval during the sampling period regardless of variations in wastewater flows.

Sample Size. The sample size collected at any one time should be at least 200 mL. Composite samples can be collected either manually or with automatic samplers. Automatic, battery-operated samplers are available for collecting composite samples on the basis of flow or time.

Amount of sample to be collected depends upon the laboratory tests to be run. The amount of sample required for each test to be performed should be determined before the sampling program is begun to ensure that sufficient sample is collected.

Analytical Determinations. Determinations that may be conducted on a wastewater sample are:

pH	Copper
Alkalinity or acidity	Nickel
Total hardness	Zinc
Chloride	Chromium, hexavalent
Sulfate	Chromium, total
Suspended solids	Iron
Volatile suspended solids	Manganese
Settleable solids	Solvent soluble (oil)
Total nitrogen	Phenol
Ammonia nitrogen	Biochemical oxygen demand (BOD_5)

Total phosphate	Chemical oxygen demand (COD)
Total solids	Total organic carbon (TOC)
Volatile total solids	Cyanide

Standard methods are available for conducting these determinations. Phases involved in a full pollution control program are:

I. Definition of the problem and development of an action plan (survey or feasibility study)
II. Detailed engineering
III. Construction and start-up

An outline of the steps involved in each of these phases is shown in Table 4.31.

TREATMENT

The necessity of effective industrial wastewater treatment must be considered an integral part of the manufacturing process, and the cost of treatment must be charged against the product.

The method of treatment depends on economic considerations and the degree of treatment required. The best alternative system for pollutant removal must be selected on the basis of a case-by-case study of efficiency and actual costs. It must be recognized that a complete system may involve several unit components and that pretreatment is required before tertiary treatment.

Figure 4.69 illustrates various treatment units combined to form a treatment system. Figure 4.70 shows the various treatment processes classified according to type, and illustrates how they can be combined to give the desired effluent quality.

Primary Treatment

The physical removal of combined chemical coagulation and physical removal of solids from wastewater is classified as primary treatment, especially if these processes are followed by biological treatment. pH adjustment is frequently used to convert soluble contaminants such as heavy metals to particulate matter, which can then be removed.

Oily waters are usually treated separately to remove the oil prior to mixing them with other waste streams. Chemicals can be used to enhance gravity separation of oil when emulsions are present.

Figure 4.71 presents a general guide for the design of a gravity separator based on parameters set forth by the American Petroleum Institute. It is desirable to have a minimum depth of 5 ft (1.5 m), a maximum horizontal flow-through velocity not exceeding 3 ft^3/min (0.015 L/s), and a minimum length-to-width ratio of 3:1 with a depth-to-width ratio of 0.3 to 0.5.

Alkaline or acidic waste streams must be neutralized before secondary treatment or discharge.

Secondary Treatment

Secondary treatment is used to remove soluble and residual suspended organic pollutants. Biological treatment is often a very cost-effective treatment method, depending on the nature of the organics and the flow and load. If the effluent quality required is higher than that which can be obtained by biological treatment, or if the organics are not biodegradable, tertiary processes are needed to remove the nondegradable fraction.

Commonly used biological treatment processes include anaerobic treatment for high-strength wastes (BOD > 2000 mg/L), and a wide variety of aerobic processes including acti-

TABLE 4.31 Pollution Control Program

- I. Survey or feasibility study
  - A. Fact finding:
    1. Develop a plant water balance for average and peak operating conditions.
    2. Inventory all industrial processes using water.
    3. Determine characteristics of the receiving waterway both upstream and downstream from plant's discharge.
    4. Determine chemical characteristics of waste streams.
    5. Study all operations using water and producing wastes.
    6. Determine local requirements with respect to pollution.
  - B. Analyze data to determine:
    1. Sources of offending contaminants.
    2. Feasibility of segregating contaminated wastes requiring treatment from dilute wastes which would be acceptable without treatment.
    3. Availability of "natural" dilution waters, that is, waters employed for useful purposes but not contaminated.
    4. Quality of effluent required for compliance with discharge standards.
    5. Whether treatment is necessary.
  - C. Exploit in-plant and/or process changes to minimize the problems by:
    1. Reducing wastes or waste volume at sources.
    2. Exploring the possibilities for reuse of process materials without treatment.
    3. Investigating recovery of valuable process materials.
    4. Reexamining the degrees of treatment required to meet standards.
    5. Reevaluating to decide whether treatment is necessary.
  - D. Detailed report on the engineering survey:
    1. Recommend a preliminary course of action.
    2. Advise management whether a waste-treatment plant is necessary.
    3. Describe the general type of plant required.
    4. Provide preliminary estimate of construction cost.
    5. Prepare preliminary estimate of operating costs.
- II. Detailed engineering
  - A. Process design and evaluation:
    1. Assign liaison and engineering personnel as required.
    2. Evaluate bench scale or pilot plant data.
    3. Translate the total evaluated data into process flow diagrams and functional specifications for the treatment plant.
    4. Prepare plot plan showing layout on plant site.
    5. Assemble an engineering report for review and approval.
    6. Obtain preliminary approval of regulatory agency.
  - B. Definitive engineering:
    1. Prepare detailed engineering flow diagrams which form the basis of final plant design.
    2. Obtain approval of overall plant design.
    3. Complete the definitive design.
    4. Obtain final approval and permit of regulatory agency.
- III. Construction and start-up
  - A. Procurement and scheduling:
    1. Prepare complete equipment specifications, bills of material, and preliminary timetable.
    2. Prepare item delivery and installation schedule.
    3. Use critical-path scheduling when warranted.
    4. Coordinate and inspect all phases of the work performed by fabricators.
  - B. Facilities erection and testing:
    1. Plan, supervise, and coordinate erection of the complete wastewater-treatment plant.
    2. Conduct unit tests, after assembly, to assure proper functioning of all related facilities.
    3. Inspect, adjust, and calibrate instruments and controls to conform to high accuracy standards, with engineers performing the work.

TABLE 4.31 Pollution Control Program (*Continued*)

C. Operator training:
   1. Prepare detailed operating manuals for all unit operations in the plant.
   2. Assemble vendors' manuals for use by plant personnel in maintaining, repairing, and replacing mechanical, instrument, and electric equipment parts.
   3. Assist with training of operating crews while construction work is in the final stages.

D. Start-up of treatment facilities:
   1. Initiate a control testing program.
      a. Operational
      b. Quality of effluent
   2. Initiate an efficiency testing program.
   3. Establish conditions for operations.
   4. Initiate a development program.
   5. Establish record-keeping procedures.

E. Supervise operation.

vated sludge, PACT, aerated lagoons, trickling filters, rotating biological contractors (RBCs), and biofilters.

Tertiary processes are selected based on the type of contaminants to be removed (suspended or dissolved, organic or inorganic) and the required quality of the treated final effluent.

Table 4.32 is indicative of removal efficiencies for unit treatment processes.

EQUIPMENT

Various unit items of equipment are combined to provide the degree of treatment required.

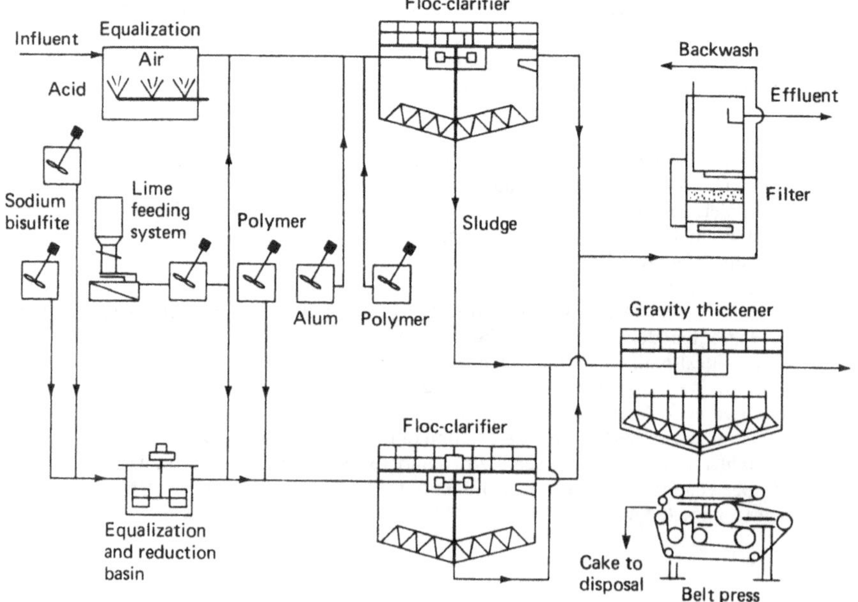

FIGURE 4.69 General manufacturing wastewater treatment.

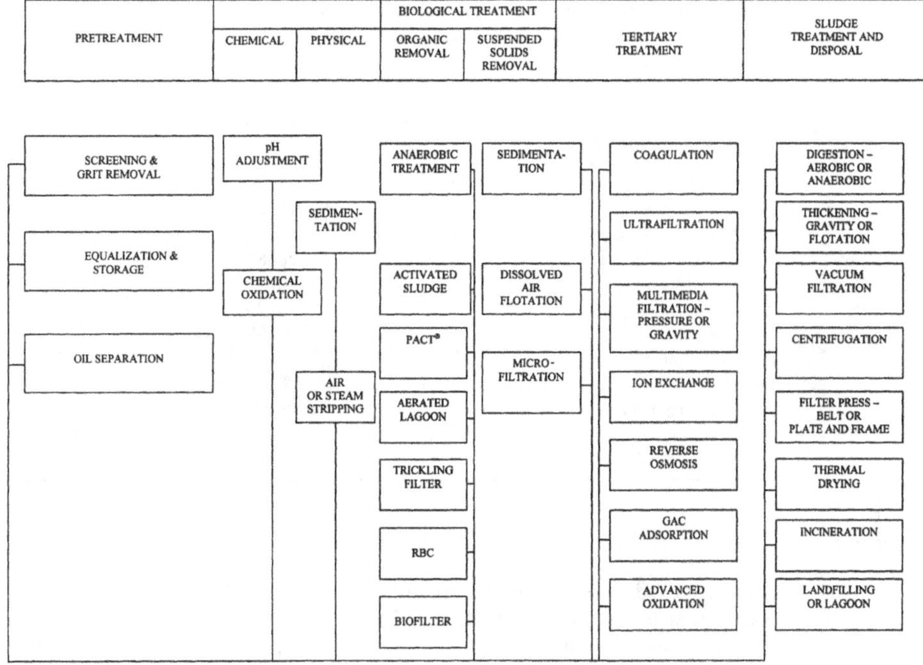

FIGURE 4.70 Alternatives for wastewater pollutant removal processes and how they may be combined in treatment programs. PACT = powder activated carbon treatment; RBC = rotating biological contactor; GAC = granular activated carbon.

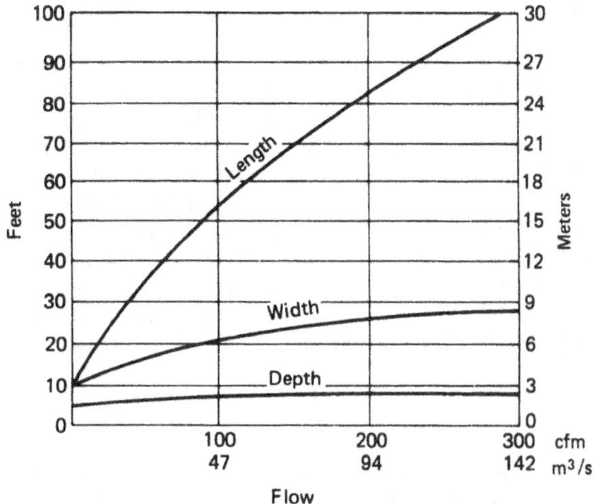

FIGURE 4.71 Design of gravity separator.

Bar Screens

A mechanically cleaned bar screen, like the one shown in Fig. 4.72, is the simplest tool for removing debris or suspended matter which could damage equipment or disrupt the treatment process. All solids with larger than ¾- to 2-in (2- to 5-cm) bar rack openings are trapped on the upstream side and removed by moving rakes.

TABLE 4.32 Treatment-Process Removal Efficiencies

Treatment method	Removal efficiency, %			
	BOD_5	Suspended solids	Total dissolved solids	Fats, oils, & greases
Screening	0–5	5–20	0	0
Sedimentation	5–15	15–6	0	5–15
Chemical precipitation	25–60	30–90	0–50	10–40
Dissolved-air flotation	10–30	70–85	10–20	80–95
Trickling filter	40–85	80–90	0–30	10–20
Nonaerated lagoon	30–70	30–70	30–80	0–40
Aerated lagoon	50–80	50–90	0–40	5–15
Activated-sludge	70–90	85–95	0–40	0–15
Filtration	80–85	30–70	10–60	0–10
Activated carbon	95–99	95–99	10	N/A
Ion-exchange	N/A	N/A	95–99	N/A
Reverse osmosis	N/A	N/A	99	N/A

Clarifiers

Clarifiers used for removal of settleable solids and readily floating oils and greases are either rectangular or circular basins (Figs. 4.73 and 4.74). Clarifiers are sized on the basis of settling rate (area) and detention time (volume). Inclined plates can be included to increase the effective settling area of a clarifier and reduce space requirements.

Typical overflow rates vary from 250 to 1400 gal/(day)(ft^2) [10 to 50 m^3/(day)(m^2)]. Detention time is in the range of 1 to 4 h.

Flocculation Systems

Flocculation is the agglomeration of finely divided suspended matter and floc caused by gently stirring or agitating the wastewater. The resulting increase in particle size increases the settling rate and improves suspended solids removal by providing more efficient contact between suspended solids, dissolved impurities, and chemical coagulants.

Mechanical flocculation uses paddles slowly rotating on a horizontal or vertical axis. Peripheral speed at the paddle tip is about 1 ft/s. Various other mechanical devices are used to achieve the same result. An air flocculation system has diffusers along one side of the basin near the bottom to produce a gentle rollover action perpendicular to flow.

The size of the required basin is determined by the detention time, which is normally in the range of 20 to 30 min at rated flow. In some cases involving industrial wastes, detention may be reduced to as little as 10 min.

Combined Equipment

Many clarifier designs, such as the solids contact unit, combine mixing, flocculation, and coagulation in one basin. This may have economic advantages and may produce better-quality effluent with shorter overall detention time than the approach using separate treatment units.

Flotation Systems

Flotation is sedimentation in reverse to remove floatable materials and solids with a specific gravity so close to that of water that they settle very slowly or not at all. The principle of air

FIGURE 4.72 Mechanically cleaned bar screen. (*Envirex.*)

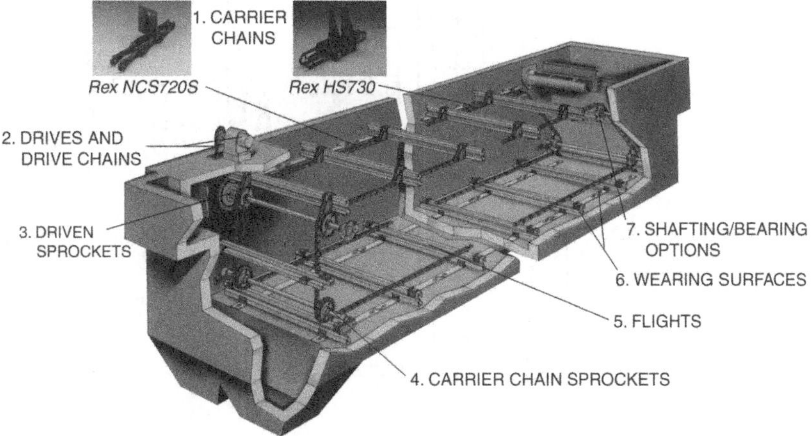

FIGURE 4.73 Clarifier for a rectangular basin. (*Envirex.*)

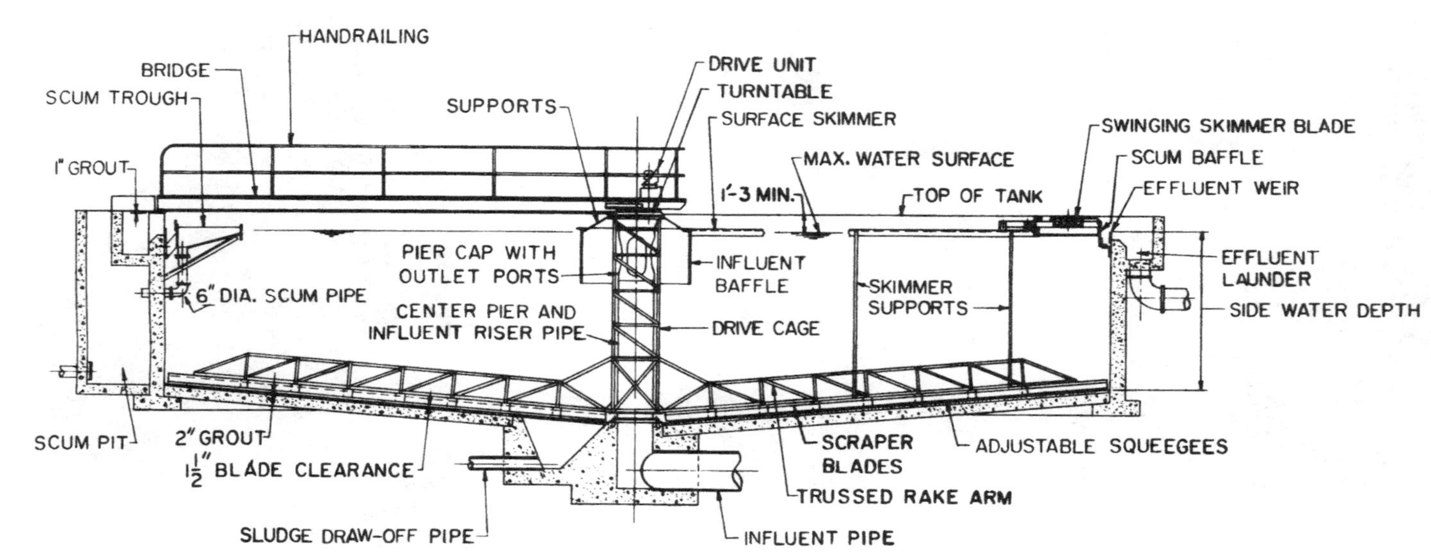

FIGURE 4.74 Clarifier for a circular basin.

flotation is based on the fact that when the pressure on a liquid is reduced, dissolved gases are released as extremely fine bubbles. These bubbles attach themselves to any suspended matter present and rapidly float them to the surface, where they concentrate and can be removed by skimming.

Pressure-flotation units dissolve air in the water under pressure and then release it to the atmosphere in the flotation tank.

Flotation equipment may be circular or rectangular.

Recycle Pressurization. The rectangular unit in Fig. 4.75 illustrates the use of recycle pressurization. Air is injected into the effluent recycle stream before it discharges into the inlet compartment of the flotation unit. There it is mixed with the incoming raw waste and releases the required air for flotation. The amount of effluent recycled varies from 25 to 50 percent of the forward flow. This approach has advantages when the raw waste is highly variable in composition or contains large amounts of solids.

Flotation units are normally designed for a feed-flow detention period of 15 min and an overflow rate of 1.5 to 4.0 gal/(min)(ft^2) [760 to 2025 L/(day)(m^2)]. Increased detention time is needed if floated sludge must be thickened since surface area is based on the loading rate of solids.

High-Rate Gravity Filters. High-rate gravity filters remove suspended solids and operate in the range of 5 to 10 gal/(min)(ft^2) [2532 to 5063 L/(day)(m^2)]. They may be vertical or horizontal and filled with a variety of media such as anthracite coal, sand, and gravel (Fig. 4.76).

PRODUCT RECOVERY

In many cases, industrial wastes contain valuable products such as high-value metals, acids, and other substances which can be used for manufacturing by-products, and these, when recovered, will yield high economic returns. Also obtainable are solvents, recovered with activated-carbon adsorption used for removal and recycling of solvents contained in the waste as vapors; these solvents include hydrocarbons, esters, alcohols, freons, ketones, and chlorinated or fluorinated organic compounds.

In the plating industry, rinse waters contain many of the following contaminants, usually in intolerable amounts: hexavalent chromium; sodium cyanide; complex cyanides of the heavy metals, such as cadmium, copper, zinc, and sometimes silver and gold; soluble nickel salts; strong mineral acids; and strong alkalis. Of these, the most toxic are the hexavalent chromium and cyanide ions.

Batch Treatment Method

There are several basic methods of treating wastewater or spent process solutions. One of the oldest is the batch treatment method whereby the waste is usually collected in a single vessel and chemically treated. The supernatant or clear treated portion of the batch is discharged and the settled sludge is sent to a sludge dewatering device, such as a filer press or vacuum filter. Heavy metals, such as copper, nickel, zinc, or lead, can be treated with sodium hydroxide or lime to form insoluble metal hydroxides. Hexavalent chromium is typically converted to trivalent chromium by the addition of an acid and a reducing agent, such as sodium metabisulfite or sodium hydrosulfite. The chromium is precipitated as a hydroxide by the addition of sodium hydroxide or lime. Toxic chemicals, such as cyanides, are oxidized to carbon dioxide and nitrogen by the addition of sodium hydroxide and bleach or chlorine gas. More difficult wastes containing complexing or chelating agents may need to be treated with a strong reducing agent, such as sodium borohydride, sodium sulfide, or an organosulfide such as dithiocarbomate (DTC).

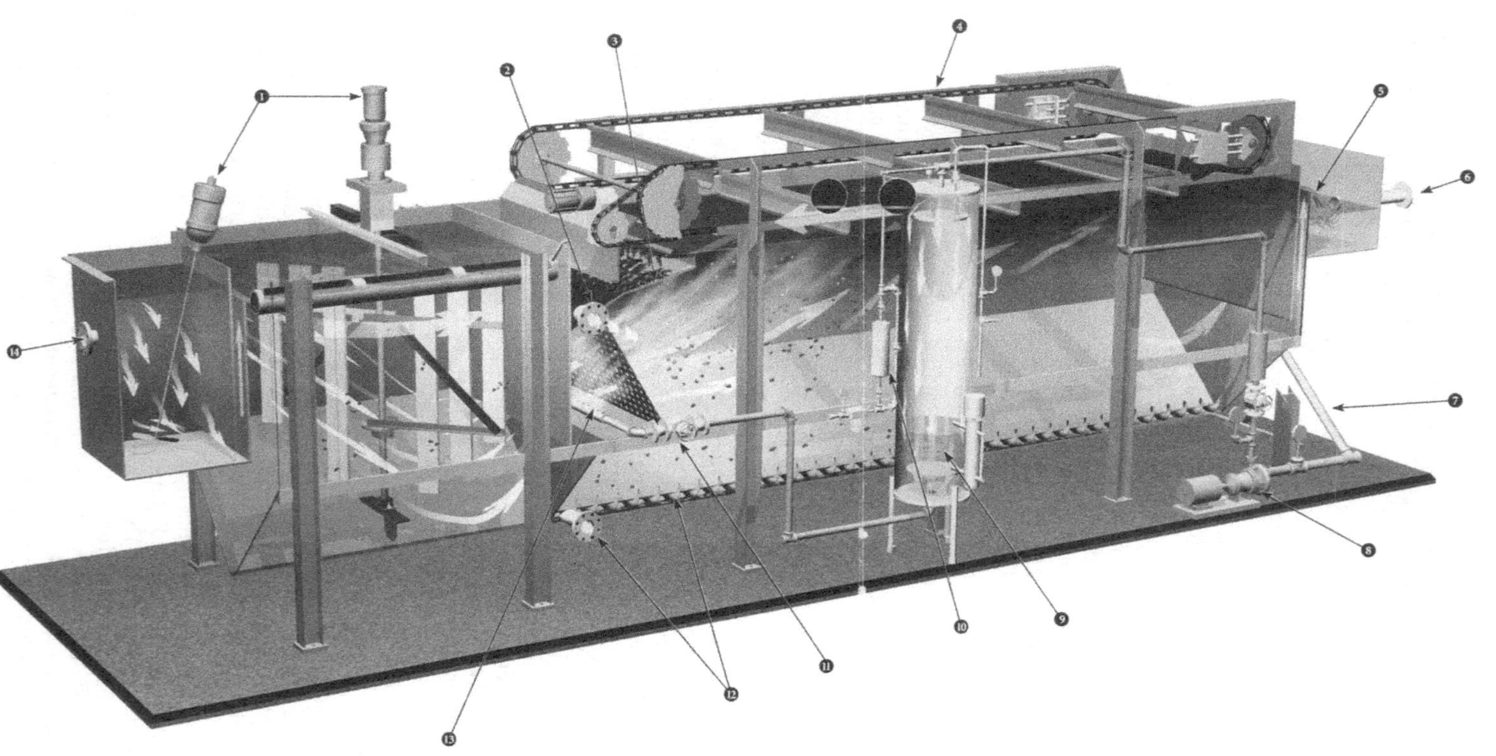

FIGURE 4.75 Rectangular flotation unit with recycle pressurization: (1) flash mixing and flocculation, (2) float discharge line, (3) float beach, (4) skimmer, (5) adjustable effluent weir, (6) effluent line, (7) recycled effluent, (8) recycle pump, (9) air-saturation tank, (10) air controls, (11) pressure-reduction valve, (12) settled solids removal and bottom solids discharge line, (13) influent and recycle blending and distribution system, and (14) influent. (*Envirex.*)

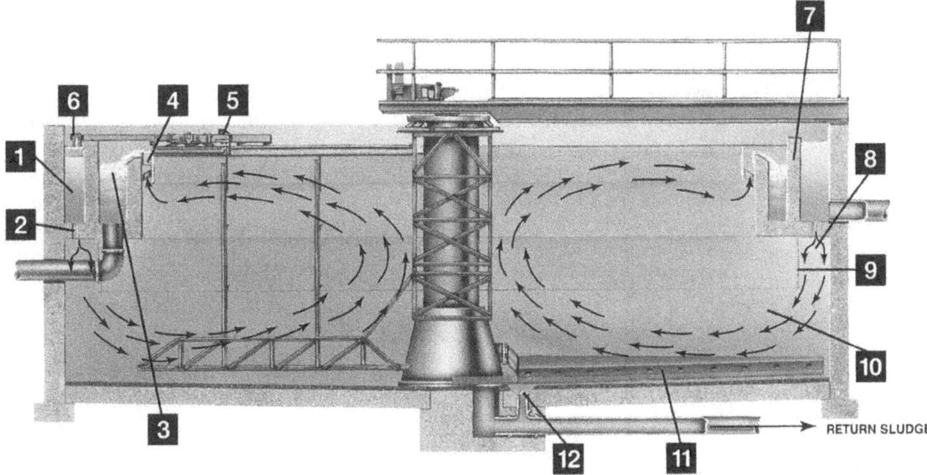

FIGURE 4.76 Solids contact clarifier: (1) influent channel, (2) inlet orifice, (3) effluent channel, (4) effluent weir and scum baffle, (5) full surface skimming, (6) scum removal, (7) common channel wall, (8) deflector baffle, (9) influent skirt baffle, (10) large inlet area, (11) sludge remover, and (12) tank drain. (*Envirex.*)

Continuous-Flow Method

Another approach to treatment of rinse water is the continuous-flow method. This approach incorporates a number of properly sized reactors and a solids separation device. To be effective, good instrumentation to monitor pH and reliable chemical reagent delivery systems are critical for proper operation. Chromium, cyanides, and heavy metals are effectively treated in a continuous-flow system. Depending on the discharge parameters, there are a number of options for solids separation. A conventional lamella clarifier can be employed if the allowable discharge of suspended solids is in the range of 10 to 20 mg/L. Additional solids polishing can be accomplished by filtering the effluent from the clarifier through a sand or multimedia filer. In place of the clarifier, more advanced filtration process, known as *microfiltration,* can be used. Instead of relying on gravity for solids settling in a clarifier, the microfilter uses polymeric membranes to separate the particles for the wastewater.

Usually associated with a continuous-flow treatment system is a sludge thicker or storage system and a filter press to dewater the sludge.

Ion Exchange

Ion exchange plays an important part in waste treatment, chemical reclamation, and water recycling. Ion exchange is a reversible, chemically driven process that sorbs hazardous ions onto specially treated porous organic materials (usually beads) and replaces them with nontoxic ions. The sorbed ions are then chemically stripped from the beads and recovered in the form of a more concentrated solution, called the regenerant.

Ion exchange is an ideal means for collecting low concentrations of ionic materials (such as metal salts) from dilute rinse waters. The metals in the regenerant can either be treated via conventional waste treatment to form metal hydroxides or recovered as a metallic by-product using electrowinning technology. If the total ionic loading is low enough (i.e., <500 mg/L), the effluent from the cation-exchange unit can be treated through another anion-exchange unit to completely purify the water. This deionized water can then be reused in the process. The purity of the water will usually be less than 1 mg/L of the total dissolved solids at a cost of roughly 5 cents per 1000 gal of treated wastewater.

If chemical reclamation is the objective, process segregation is necessary. This is particularly true when chrome is to be recovered from rinse waters. Cyanide rinses, copper and nickel plating rinses, and alkaline or acid rinses must be removed from the chrome rinse waters. The rinse waters containing the chrome will first pass through a cation-exchange unit where the chrome-bath impurities, such as copper, nickel, and trivalent chromium, are removed. The hexavalent chromium is removed in the anion-exchange unit. If so desired, since the water has been deionized, it can be recycled as high-purity rinse water.

Upon exhaustion, the cation exchanger is regenerated with acid and the anion exchanger is regenerated with caustic. The cationic regenerant solution will need to be treated for heavy metal reduction by either batch treatment or being bled into a continuous-flow treatment system. The anionic regenerant will contain sodium chromate and some caustic soda. This anionic regenerant can then be converted to chromic acid by passing through the cation exchanger. The resultant chromic acid can be used for bath replenishment.

Table 4.33 shows the recovery value of by-products from plating wastes using ion exchange and other recovery technologies.

Closed-Loop System

In the area of metal fabrication, parts are frequently washed or rinsed to clean away oils and other wastes. With a little care, improved wash operations cut downtime, reduce energy requirements, and cut soap costs. This is accomplished by a closed-loop washwater treatment system.

A typical closed-loop system (Fig. 4.77) is best described as a continuous-batch oil separator. It has dual compartments holding caustic wash solution, each equipped with an oil roll skimmer and separated by a waste tank. Piping leads from each compartment to a series of washers and back to a pump. Automated valves control flow from the pump to one of the two compartments.

One compartment continuously supplies caustic solution to a group of washers as the other stands for 24 h, allowing heavy materials to settle as oils float to the surface. Then, surface oils containing less than 0.1 percent water are skimmed off and drained into a waste tank; these may then be sold to an oil reclamation firm. While one wash solution in the first compartment undergoes treatment, the clean solution in the other compartment is circulated through the washers.

Another technology for treating aqueous alkaline cleaners uses ultrafiltration membranes. Polymeric membranes are typically used for low-temperature alkaline baths [<110°F (43°C)], whereas ceramic membranes are used for higher-temperature applications. The contaminated alkaline cleaner is pumped from the cleaning line to the membrane system. A feed valve, which is controlled by a level sensor, maintains a constant volume of cleaner in the process compartment. The contaminated cleaner is then pumped through a bag filter and into a settling compartment, where the free oils and sludge are removed. From the middle of the settling compartment, the fluid is pumped into the process chamber. The contaminants and emulsified oils are then concentrated.

TABLE 4.33 Economic Value of By-Product Recovery

Process	By-product	Concentration, mg/L	2000 value, $/1000 gal
Rinse water	Copper	100–500	1.7–8.5
	Nickel	150–900	2.3–19.8
	Zinc	70–350	1.1–6.8
	Cadmium	50–250	3.4–17.0
	Chromium	400–2000	9.0–42.6
	Tin	100–600	2.3–13.6
Aluminum, bright dip	Phosphoric acid	10%	1260

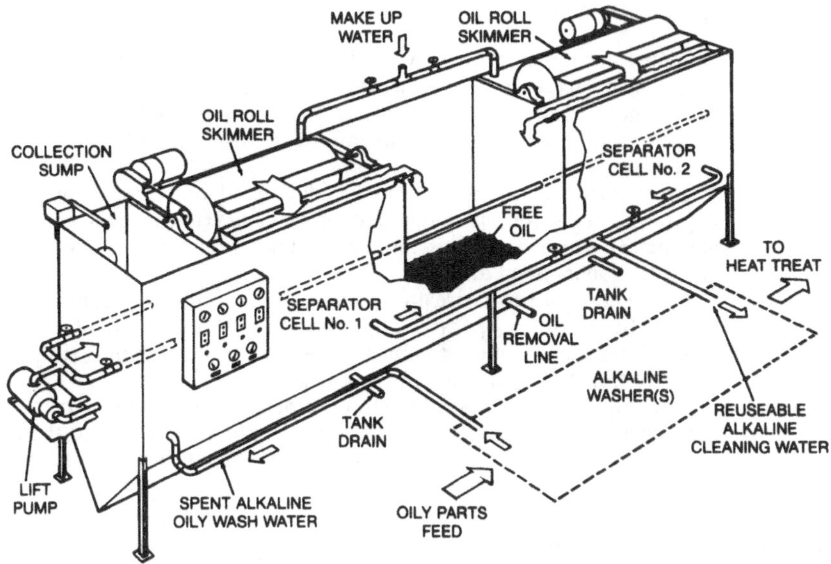

FIGURE 4.77 Alkaline wash system.

The solution is pumped through the membrane at a high flow rate. The emulsified oil and other contaminants are separated from the alkaline cleaner solution. The purified aqueous cleaner rapidly flows back to the cleaner line.

Treatment of Spent Coolants

Spent coolants can be treated for recovery of oil by acidification and centrifugation or by heat treatment at 223°F (106°C) for 22 h.

In the heat-treatment process, three distinct layers develop after there is water loss of approximately 12 percent through evaporation. The top layer consists of reusable oil, amounting to approximately 24 percent of the original scum volume; the middle layer, containing approximately 16 percent of the scum, is called the *rag layer* and consists of flocculated particulate matter; the bottom layer contains approximately 48 percent of the total scum and appears as relatively clear water. This bottom water layer is returned to the spent-coolant holding tanks for retreatment, and the rag, or intermediate, layer is combined with the swarfing dust and other heavy solids for removal by haulage.

A portion of the recovered oil can be used as the fuel for heating the floated scum, thereby making the system self-sustaining. The balance can be reprocessed for reuse as coolant. A simple product balance would show that for every 100 L of coolant treated, from 0.3 L to as much as 5 L of reusable oil can be recovered. A minimum flow of about 30,000 L of coolant per day might be selected as an economic break-even point.

At this minimum flow, taking, on an average, 3 L of recoverable oil for each 100 L of feed, 750 L per day of coolant would be recovered. Of this, approximately 50 L would be required to heat the scum to 223°F (106°C) for the 22-h break period, leaving a net total of 700 L of recovered oil.

Cost recovery can vary, depending on the type of coolant used. Caution should be exercised when this system is used where plated parts are machined, as there is the hazard of cyanide buildup in the oil after prolonged usage.

Water Recovery

Recycled water may ultimately be the major valuable product because of increasing water-supply costs, increasing water-treatment costs, and mounting charges for using municipal wastewater facilities. The recovery of product fines, usable water, and thermal energy are important methods of reducing overall waste-disposal costs and should be seriously considered in every case.

Frequently, waste streams can be eliminated or reduced by process modifications or improvements. A notable example of this is the use of save-rinse and spray-rinse tanks in plating lines. This measure brings about a substantial reduction in waste volume and frequently a net reduction in metal dragout.

Segregation of waste streams is a necessity at times, not from the product-recovery point of view, but from the operational point of view. An example of this is segregation of acidic metal rinses from cyanide streams to avoid the production of toxic hydrogen cyanide (HCN) and thus eliminate potential safety hazards.

The prime requirement of waste treatment, by-product recovery, and water reuse is that the principal product or products of the plant be satisfactory to the consumer, and the secondary requirement is that the operation of the plants be efficient and economical.

By-Product Recovery and Use

The urgent problems facing industries are how to recover by-products from the waste materials inherent in every industrial operation and what to do with the by-products. Confronted with the growing dangers of pollution, the anticipation of government regulation, and the loss of valuable materials through unprofitable waste-disposal methods, industries are forced to develop sophisticated refining methods for processing chemical and industrial by-products and even to develop markets for by-products.

Industry is becoming increasingly aware of the necessity for pollution abatement and product recovery, not only because of its effect upon the general welfare of the public, but also because of its own dependence upon rivers and streams for suitable water for manufacturing processes. Industries are also increasingly aware of the fact that the benefits accruing from pollution abatement through product recovery may be quite significant.

SOURCE CONTROL AND WATER REUSE AND RECYCLING

The plant layout and arrangement of process and manufacturing sequences must be considered with regard to wastewater pollution control. This means undertaking a complete engineering survey of water use to develop an accurate water balance for peak and average operating conditions. In effect, what is needed is a complete inventory of all plant operations that use water and produce wastes.

With these data in hand, a fresh look should be taken at plant and process operations to see if any changes can be made that will reduce the amount of water used or decrease the flow of wastewater produced. A simple process adjustment is often all that is required to lower the concentration of pollutants. Perhaps there are valuable chemicals that can be recovered; or there may be an alternative approach that might change the nature of the waste to make it easier to handle. Sometimes, segregating a contaminated process water from the rest of the waste discharge can reduce the size of the wastewater-treatment system.

The survey will uncover applications for water that can be recycled for repeated reuse. Some wastewater, now discharged to the sewer, may be well suited for cooling or boiler feed. The economics of using cooling towers to replace once-through cooling systems should be reconsidered. In some cases it may even pay to switch to air-cooled heat exchangers.

Machinery or operations having a common waste product can be connected to a centralized system of contaminant collection and treatment. With early involvement, the plant engineer can anticipate potential pollution problems and advise preventive measures. Where possible, the plant engineer should strive to prevent pollution at the source, thus minimizing the need to incorporate pollution control facilities. Typical methods of preventing pollution at the source include:

1. Substituting process materials
2. Modifying manufacturing procedures
3. Changing production equipment
4. Recycling process water

Examples of materials substitution are the use of chlorinated solvents of less toxicity for carbon tetrachloride and using a paint with reduced lead or zinc content.

Cascading or countercurrent water-use systems (where an operation that requires relatively low quality water uses wastewater from another operation) or recycling systems (where water is treated and returned to the same operation) can greatly reduce intake water requirements.

The quantities listed in column B, Table 4.29, should be considered the maximum pollutant values for reuse of process water.

BIBLIOGRAPHY

Corbitt, E. A.: *Standard Handbook of Environmental Engineering,* McGraw-Hill, New York, 1999.

Eckenfelder, W. Wesley, Jr.: *Industrial Water Pollution Control,* McGraw-Hill, New York, 1988.

Freeman, H. M.: *Standard Handbook of Hazardous Waste Treatment and Disposal,* McGraw-Hill, New York, 1998.

Lund, Herbert F.: *Industrial Pollution Control Handbook,* McGraw-Hill, New York, 1971.

Nemerow, Nelson L.: *Strategies of Industrial and Hazardous Waste Management,* 2d ed., John Wiley & Sons, New York, 1998.

CHAPTER 4.8
SOLID-WASTE DISPOSAL

C. Phillip Headley
Technical Specialist
Waste Equipment Technology Association (WASTEC)
Washington, D.C.

INTRODUCTION

One of the most significant factors in the successful operation of any industrial plant is the proper disposal of solid waste.

Obviously, such waste can be simply hauled away to an appropriate disposal site as it is generated, either by an outside contractor or by plant personnel. This chapter, however, discusses the alternative: employing an in-plant system to reduce the volume and weight of solid waste as generated so as to reduce haulage costs. However, hazardous material must be disposed of (usually with treatment) by specialists; this subject is discussed in the last portion of this chapter.

SYSTEM DESIGN

Due to large variations in the types of waste found in an industrial plant, each refuse-handling system should be custom designed to fit the needs of each plant. The key to a successful design is understanding current operations—that of the plant engineer as well as the operations of the outside solid waste managers (refuse collectors, equipment distributors, and manufacturers) servicing the plant. With this combination of knowledge, the most economical and efficient waste-handling system for a plant can be designed.

First, five essential factors must be analyzed:

1. The volume of waste produced
2. Its composition and characteristics
3. Special handling requirements
4. Location and other physical constraints
5. Requirements for safety and security

DISPOSAL METHODS

Having evaluated the five criteria, the plant engineer must determine the best method of disposal to be used. Two common methods employed in many industrial plants are (1) *compaction* and (2) *incineration.*

Compaction

Compaction is the method whereby a large volume of solid waste is reduced (squeezed) under high pressure, minimizing both container space and the number of times the container must be emptied. This reduces collection costs and permits a safe, clean method of handling in-plant refuse. The installation of an efficient compaction system, including chutes, conveyors, and large refuse-compactor containers, can reduce in-plant trash-handling costs by as much as 90 percent. Also, with an appropriate installation, rear-door pilfering can be practically eliminated.

Purchasers often want to specify *compaction ratio* or the density that can be produced by a compactor. Most manufacturers decline to guarantee specific compaction performance because it depends on both the specific machine employed and the material being compacted. Typically, however, compactors will reduce the volume of wastes by factors ranging from 2 to 5.

Stationary Compactors. If stationary compaction is chosen, several factors must be considered. The parameters of stationary compactors including charging-chamber volume, pressure of the packer ram, cycle time, penetration of the packing ram into the compaction container, and the compactor base size.

To assist purchasers of stationary compactors, the Waste Equipment Technology Association (WASTEC) of the Environmental Industry Associations (EIA) has developed a standardized method of rating compactors. Periodically (generally every 3 years), WASTEC publishes a book of standardized ratings for the equipment sold by its compactor manufacturer members.

Figure 4.78 shows a sketch of the most widely used style of stationary compactor, the horizontal detachable unit. Ratings are also provided for compactors with pivoting rams and self-contained compactors in which the compaction ram is integrally mounted within a refuse container.

Charging Chamber Volume. This must be large enough to accommodate, without any difficulty, the largest piece of refuse generated by the user.

Additional considerations must be taken into account when determining charging-chamber size. For example, the industrial plant may utilize in-plant trash carts of a specific size, so the charging chamber should be able to receive the entire contents of the full trash cart or a full container load. In many instances, this could mean that the charging area should be substantially greater than the largest single piece of refuse generated at the location.

Physical Dimensions of the Compactor. It may seem obvious that a compactor machine must fit properly into the space where it is intended to operate. However, errors made in this area are common, and utmost care is advised in planning. Conversely, a number of installation possibilities exist which may enable the utilization of a machine that, at first glance, would not appear to fit into the available space.

As a general rule, the placing of a stationary compactor should be calculated not only on the basis of the actual operating machine, but must include: (1) space for access to the charging chamber; (2) space for the container for the compacted wastes; and (3) space required for the pickup vehicle to maneuver, pull away the full container, replace an empty one, and then haul the full box away. Finally, sufficient room must be provided at the loading area to accommodate trash carts, containers, or even a conveyor chute leading to the hopper.

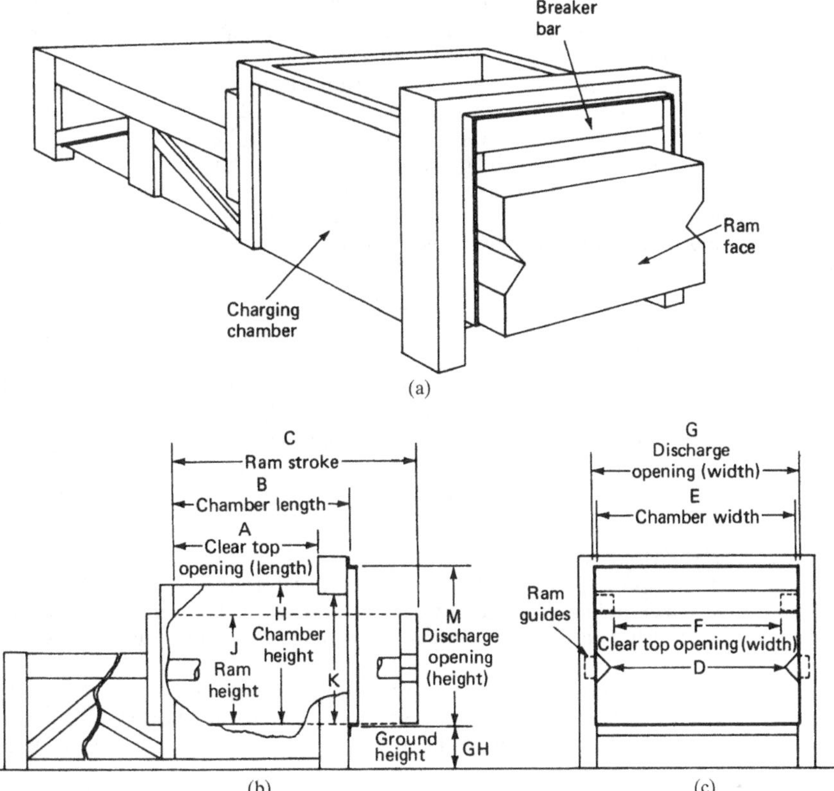

FIGURE 4.78 Horizontal commercial-industrial stationary compactor.

Dock space may be saved and plant security increased by installing the machine through the wall so that the charging area and the working mechanism remain inside the building, but the container is attached to the machine through an opening in the wall. Also, chutes from upper floors may run to the charging area, decreasing internal transportation costs and reducing the need for additional dock space.

Cycle Time. Cycle time is an important parameter of container specification. It refers to the time required for a fully retracted ram face to pack the refuse from the charging box into the container and return to its original fully retracted position ready for another load of refuse.

Cycle times run as short as 20 s to more than 1 min. Short cycle times are important if the application under consideration requires the capability of accepting refuse very quickly. A purchaser should carefully consider whether a compactor with a short cycle time is needed. Often, compactors built with a rapid cycle utilize a small-diameter hydraulic cylinder that consequently exerts a relatively low force on the ram face, thereby sacrificing compaction pressure and material density.

Pressure of the Ram Face. A third important parameter of a stationary-compactor specification is the pressure of the ram face. Pressure, in pounds per square inch, is more important than total force in determining compaction density.

The ram pressure is the total force exerted on the ram, divided by the area of the ram. For a hydraulically operated compactor, the total ram force is the product of the hydraulic pressure and the hydraulic cylinder area.

For example, a compactor with a ram face measuring 30 × 60 in (0.8 × 1.6 m) has an area of 1800 in^2 (1.3 m^2). If it is acted upon by a 5-in (12-cm) diameter hydraulic cylinder with a fluid pressure of 2000 lb/ft, the total ram force will be 39,200 lb (175,000 N), and the pressure exerted by the ram will be 21.7 lb/in^2 (140,000 N/m^2).

Most compactors have both a normal and a maximum pressure rating. The higher pressure is used when finally packing out a container to make it easy to detach the container and ensure that the waste remains within it.

Penetration of the Ram into the Container. The penetration of the ram face into the container is an important factor in the operation of a compactor. Essentially, it represents the position of the forward ram stroke available for final compaction of the refuse load into the container. As the container begins to fill up with compacted waste, there is a tendency for the portion of refuse closest to the ram to fall back into the charging area. This is of most concern when the container is detached because waste falling from the container will litter the environment. The further a ram penetrates into the compaction container, the easier it is to load and detach a container cleanly.

Base Size. Compactor specifications are summarized as a single parameter, the base size. This is defined as the volume of waste theoretically moved through the compactor in a single stroke. Purchasers commonly use the base size as the primary specification, adding the parameters listed earlier as necessary. Table 4.34 gives typical uses of compactors of various base sizes.

Rated stationary compactors manufactured by members of WASTEC carry the WASTEC Rating Seal. This seal assures performance conforming to established standards, certified by a registered professional engineer. Table 4.35 gives a hypothetical sample listing of rated compactors.

In addition to obtaining a compactor, the plant engineer will have to specify a container, which is usually provided by the refuse hauler. Before purchasing any equipment, compactor or container, the engineer should always consult the hauler to determine what type, size, and weight of containers can be handled. Some points to watch for are the following:

1. The container must be able to withstand the pressure of the waste without damage or distortion.
2. Roll-off and drop-off hoists that are used to remove containers have weight limits. The hoist will not be able to lift containers that exceed these limits.
3. State highway laws limit the amount of weight that can be carried on each chassis.
4. The length of the container affects the compaction ratio. The larger the container, generally the less the compaction.

TABLE 4.34 Typical Compactor Applications

Use	Base size	
	yd^3	m^3
High-rise apartment building	Up to 1	0.8
Commercial establishment	1 to 4	0.8 to 3
Industrial plants	1.5 to 7	1.2 to 5.5
Transfer stations	7 to 12	5.5 to 9

TABLE 4.35 WETA Commercial-Industrial Stationary Compactor Ratings

Manufacturer: Hypothetical Manufacturing, Inc. Date: August 1977

Model number	NSWMA base size, yd^3	Clear top opening length (L) width (W), in	Chamber length, in	Ram stroke, in	Ram penetration, in	Force rating normal (N) maximum (M), (ram lb/in^2 per force, lb)	System pressures normal (N) maximum (M)	Ram face, in	Volume displacement rate, yd^3/h	Rated motor size, hp	Cycle time, s	Discharge opening width (W) height (H), in	Ground height, in	Base unit weight, lb
A	0.38	$L = 24.0$ $W = 34.5$	24.5	30.0	5.5	$N = 25.3/18200$ $M = 27.8/20000$	$N = 1450$ $M = 1600$	$W = 36.0$ $H = 20.0$	39	1	35	$W = 37.0$ $H = 24.5$	15.5	1800
B	1.06	$L = 36.0$ $W = 51.5$	39.0	50.0	11.0	$N = 28.5/36200$ $M = 30.8/39100$	$N = 1850$ $M = 2000$	$W = 53.0$ $H = 24.0$	106	10	36	$W = 53.5$ $H = 32.5$	16.0	4000
C	1.60	$L = 40.0$ $W = 58.5$	41.0	55.0	14.0	$N = 21.6/38800$ $M = 24.3/43700$	$N = 2000$ $M = 2250$	$W = 60.0$ $H = 30.0$	144	10	40	$W = 61.0$ $H = 37.5$	15.5	5100
D	1.91	$L = 48.0$ $W = 58.5$	49.0	65.0	16.0	$N = 24.3/43700$ $M = 27.6/49600$	$N = 2250$ $M = 2550$	$W = 60.0$ $H = 30.0$	146	10	47	$W = 61.0$ $H = 37.5$	15.5	5900
E	3.43	$L = 84.5$ $W = 58.0$	89.0	110.0	21.0	$N = 55.8/100400$ $M = 61.4/110500$	$N = 2000$ $M = 2200$	$W = 60.0$ $H = 30.0$	225	20	55	$W = 61.5$ $H = 37.5$	16.5	13000

Note: The Waste Equipment Technology Association presents this technical information on commercial-industrial stationary compaction equipment to assist users in the selection of equipment. The Waste Equipment Technology Association does not present these data as a warranty for equipment performance. The ratings represent data supplied by manufacturers that have been computed according to the WETA criteria and certified for accuracy by a registered professional engineer, selected by the manufacturer.

There is no single formula to use in selecting a compactor for a plant. However, careful consideration of the tangible and intangible factors that have been outlined here will aid in avoiding costly mistakes.

In-Plant Incineration

An alternative method of disposal is *incineration.* Although incinerators are more expensive to build and install than stationary compactors, their use can provide additional savings in refuse-hauling costs. Energy recovery in the form of steam may also be considered as a source of additional savings.

When one is considering an incinerator, one should evaluate the following factors:

1. Any limitations imposed by air pollution emission limits in the area of the plant.
2. The physical constraints imposed by the dimensions of the plant.
3. The appropriateness in quantity and composition of the solid waste generated within the plant: a sufficiently high fired Btu value
4. Fuel requirements. Most incinerators require supplementary fuel, either gas or oil, to control air pollution. Fuel costs can be high, and fuel availability may become uncertain. However, energy recovery can offset some of the costs.
5. By-product disposal: adequate methods of disposal must be provided to handle the by-products of the incinerated waste.

Controlled-air incinerators (Fig. 4.79), the most frequently used incinerators today, were first commercially available in the United States in the early 1960s, but they were not really

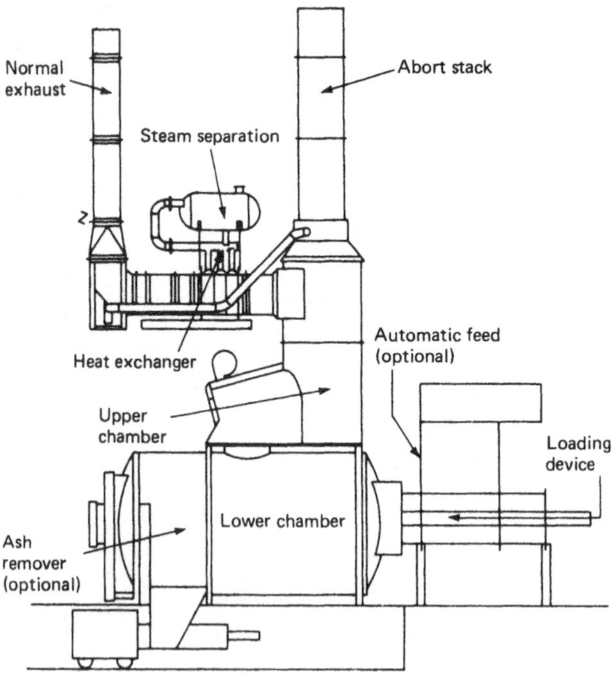

FIGURE 4.79 Controlled-air incinerator. (*Adapted with permission from the* American City & County Magazine.)

accepted until the late 1960s and early 1970s. This acceptance was mostly due to the increasing demands for high performance as measured by very low particulate emissions and a very high reduction ratio.

Most controlled-air incinerators employ two chambers. These chambers are designated lower, or primary, and upper, or secondary. Performance of the antipollution functions of the system depends on controlling the conditions within these two chambers. The lower chamber is required to operate at low interior gas velocities and under controlled temperature conditions. This is done by limiting the air introduced into the primary chamber to less than the amount required for complete combustion (hence, the process is called *starved-air incineration*). This gives the lower chamber the operating characteristics of a partial oxidation system.

The heat released in the lower chamber is controlled by limiting the introduction of combustion air to an amount which will give partial oxidation of the waste in the chamber. The heat is sufficient to sustain the partial oxidation reactions. The gases from the lower chamber pass into the upper chamber through a turbulent zone, where additional air is added and ignition takes place to complete the oxidation reactions. The noncombustible portion of the waste and carbonaceous residue from the reactions remain in the lower chamber. The noncombustibles are rendered sterile by the relatively high temperature, while the carbonaceous material is further oxidized by the incoming air. The result is a high-quality sterile ash.

The gas velocity in the lower chamber is influenced by several factors. The gas which evolves from the chamber is a result of the interaction of the air, the auxiliary fuel, and the oxidation and volatilization products from the waste. The quantity of gas from the waste can vary substantially depending on chamber conditions of the waste and could therefore alter the gas velocity in the lower chamber significantly. The airflow controls of the upper and lower chambers are integrated in order to minimize cycling and provide a uniform flow of gases. This is important for controlling pollution performance and especially so for an efficient energy-recovery system.

When volatilization proceeds at an excessive rate as a result of the high temperatures, two distinct adverse effects ensue. First, the velocity in the lower chamber will exceed the design velocity, and particles which are too large to be oxidized properly will be carried into the upper chamber. Second, the gases will flow to the upper chamber at a rate which exceeds the capacity of the chamber and can result in excessive particulate emissions or smoking.

The function of the upper chamber is to complete the oxidation reactions of the combustible products as they are received from the lower chamber. In order to accomplish this, conditions in the chamber must be controlled within a rather narrow band, from inputs which vary rather widely. The control system is designed to maintain the required conditions by modulating both air and fuel to the system. This in effect controls the air input, auxiliary fuel input, and gas flow from the lower chamber.

The gases pass from the lower chamber into the upper chamber through a turbulent mixing region in a controlled manner. Additional air is introduced into the system and the gases are ignited, again under controlled conditions. The gas temperature at this point is somewhat higher than the in the lower chamber, and the atmosphere is oxidizing (more than sufficient air for complete combustion). Temperatures in the upper chamber are limited to less than 2500°F (1400°C) in order to minimize production of nitrogen oxides and in the interest of equipment durability. On the lower-temperature side, it is recommended that at least 1800°F (1000°C) be maintained in order to stabilize an adequate reaction rate to complete the combustion process. The desired operating temperature point is adjustable but is factory preset for maximum performance. The primary means of controlling the temperature in the upper chamber is to control the quantity of combustion air. Air quantity is decreased when the temperature drops below the set-point and increased when the temperature rises above the set-point.

If heat is to be recovered, the gas temperature at the inlet to the heat exchanger is not allowed to exceed 1800°F (1000°C). This is done to protect the heat exchanger and is an addition to the normal safety controls. An overtemperature condition will automatically drive the hot-gas flow to the abort stack.

Compared with compaction, incineration is a more costly and more complex way of disposing of waste; however, the energy-recovery potential may in the long run prove to be a decisive factor for incineration in many plants.

REGULATIONS

Transportation, treatment, storage, and disposal of hazardous waste are regulated under the Resource Conservation Recovery Act (RCRA), enacted in 1976 and amended in 1984 (usually referred to as the Solid Waste Disposal Act).

The Hazardous Materials Transportation Act was enacted in 1974. In 1990 it was amended to incorporate tighter regulations on transportation of hazardous materials on highways, railways, and waterways. Authority to determine what is a hazardous material rests with the U.S. Department of Transportation.

To arrange for disposal of hazardous wastes, a plant engineer should contact a waste service company that is *fully permitted to transport and dispose of those wastes.* The company should provide a written statement certifying where the wastes are to be taken and how they are to be treated or disposed of. The disposal facilities should be inspected and the actual waste disposition verified.

The hazardous-waste generator will be requested to initiate a manifest for shipments of hazardous wastes going off-site for disposal and will also be required to designate where the wastes are taken.

This is an on-going, active regulatory area; the reader is urged to check the latest regulations before proceeding with plant modifications.

BIBLIOGRAPHY

Corbitt, Robert A.: *Standard Handbook of Environmental Engineering,* 2nd ed., McGraw-Hill, New York, 1998.

EPA-RCRA Orientation Manual, EPA/530-SW-86-001, Washington, D.C., 1986.

Freeman, H.M.: *Standard Handbook of Hazardous Waste Treatment and Disposal,* McGraw-Hill, New York, 1997.

Lindgren, G.F.: *Managing Industrial Hazardous Waste: A Practical Handbook,* CRC Press-Lewis Publishers, 1989.

Waste Equipment Technology Association, *Listing of Rated Stationary Compactors,* Washington, D.C., 2000.

CHAPTER 4.9

FIRE PROTECTION AND PREVENTION

Casey C. Grant, P.E.
Assistant Vice President, Codes and Standards Administration
National Fire Protection Association
Quincy, Massachusetts

GLOSSARY

Boiling point The temperature at which the liquid boils when under normal atmospheric pressure (14.7 psia). The boiling point increases as pressure increases and is dependent on the total pressure.

Combustible A material or structure that can burn is considered combustible. Combustible is a relative term; many materials that will burn under one set of conditions will not burn under others, e.g., structural steel is noncombustible, but fine steel wool is combustible. The term *combustible* does not usually indicate ease of ignition, burning intensity, or rate of burning, except when modified, as in *highly combustible interior finish.*

Fire prevention Measures directed toward avoiding the inception of fire.

Fire load The amount of combustibles present in a given situation, usually expressed in terms of weight of combustible material per unit area. This measure is employed frequently to calculate the degree of fire resistance required to withstand a fire or to judge the rate of application and quantity of extinguishing agent needed to control or extinguish a fire.

Fire point The lowest temperature of a liquid in an open container at which vapors are evolved fast enough to support continuous combustion.

Fire resistance A relative term, used with a numerical rating or modifying adjective to indicate the extent to which a material or structure resists the effect of fire, e.g., "fire resistance of 2 h."

Fire-resistive Pertains to properties or designs that resist the effects of any fire to which a material or structure may be expected to be subjected. *Fire-resistive materials* or assemblies of materials are noncombustible, but noncombustible materials are not necessarily fire-resistive; *fire-resistive* implies a higher degree of fire resistance than *noncombustible. Fire-resistive construction* is defined in terms of specified fire resistance as measured by the standard time-temperature curve.

Fire-retardant Usually denotes a substantially lower degree of fire resistance than fire-resistive and is often used to refer to materials or structures which are combustible in whole or in part, but have been subjected to treatments or have surface coverings to prevent or retard ignition or the spread of fire under the conditions for which they are designed.

Flame-resistant A term that may be used more or less interchangeably with flame-retardant.

Flame-retardant Materials, usually decorative, which due to chemical treatment or inherent properties, do not ignite readily or propagate flaming under small to moderate exposure.

Flammable A combustible material that ignites very easily, burns intensely, or has a rapid rate of flame spread. Flammable is used in a general sense without reference to specific limits of ignition temperature, rate of burning, or other property. *Flammable* and *inflammable* are identical in meaning. Flammable is used in preference to inflammable.

Flammable limits The extreme concentration limits of a combustible in an oxidant through which a flame will continue to propagate at the specified temperature and pressure. For example, hydrogen-air mixtures will propagate flame between 4.0 and 75 percent by volume of hydrogen at 21°C and atmospheric pressure. The smaller value is the lower (lean) limit, and the larger value is the upper (rich) limit of flammability. For liquid fuels in equilibrium with their vapors in air, a minimum temperature exists for each fuel above which sufficient vapor is released to form a flammable vapor-air mixture. There is also a maximum temperature above which the vapor concentration is too high to propagate flame. These minimum and maximum temperatures are referred to respectively as the lower and upper *flash points* in air. The flash-point temperatures for a combustible liquid vary directly with environmental pressure.

Flashover The phenomenon of a developing fire (or radiant heat source) producing radiant energy at wall and ceiling surfaces within a compartment. The radiant feedback from those surfaces heats the contents of the fire area so that all the combustibles in the space become heated to their ignition temperature.

Flash point The lowest temperature at which the vapor pressure of a liquid will produce a flammable mixture and resultant flame. The flame will not continue to burn at this temperature if the source of ignition is removed.

Glowing combustion and flame Combustion is the process of exothermic, self-catalyzed reaction involving either a condensed-phase or a gas-phased fuel, or both. The process is usually (but not necessarily) associated with oxidation of a fuel by atmospheric oxygen. Condensed-phase combustion is usually referred to as *glowing combustion,* while gas-phase combustion is referred to as a *flame.*

Ignition temperature (autoignition temperature, autogenous ignition temperature) The minimum temperature to which a substance in air must be heated in order to initiate, or cause, self-sustained combustion independently of the heating or heated element. The ignition temperature of a combustible solid is influenced by rate of airflow, rate of heating, and size and shape of the solid.

Latent heat Heat is absorbed by a substance when converted from a solid to a liquid and from a liquid to a gas. Conversely, heat is released during conversion of a gas to a liquid, or a liquid to a solid. Latent heat is the quantity of heat absorbed or given off by a substance in passing between liquid and gaseous phases (latent heat of vaporization) or between solid and liquid phases (latent heat of fusion). The high heat of vaporization of water is a reason for the effectiveness of water as an extinguishing agent.

Noncombustible Not combustible.

Nonflammable Not flammable.

Specific heat The heat, or thermal capacity, of a substance is the number of calories required to raise 1 g of a particular substance 1°C. The specific heats of various substances vary over a considerable range; for all common substances, except water, they are less than unity. Specific-heat figures are significant in fire protection as they indicate the relative quantity of heat needed to raise the temperature to a point of danger, or the quantity of heat that must be removed to cool a hot substance to a safe temperature.

Vapor density The weight of a volume of pure gas compared with the weight of an equal volume of dry air at the same temperature and pressure. A figure less than 1 indicates that a gas is lighter than air, and a figure greater than 1 indicates that a gas is heavier than air. If a flammable gas with a vapor density greater than 1 escapes from its container, it may travel at a low level to a source of ignition.

Vapor pressure Because molecules of a liquid are always in motion (with the amount of motion depending on the temperature of the liquid), the molecules are continuously escaping from the free surface of the liquid to the space above. Some molecules remain in space while others, due to random motion, collide with the liquid. If the liquid is in an open container, molecules (collectively called *vapor*) escape from the surface, and the liquid is said to evaporate. If, on the other hand, the liquid is in a closed container, the motion of the escaping molecules is confined to the vapor space above the surface of the liquid. As an increasing number strike and reenter the liquid, a point of equilibrium is eventually reached with the rate of escape of molecules from the liquid equals the rate of return to the liquid. The pressure exerted by the escaping vapor at the point of equilibrium is called *vapor pressure.* Vapor pressure is measured in millimeters of mercury (mmHg), or torr.

INTRODUCTION

Fire protection is an area in which most plant engineers can make a significant contribution to plant operations. At many facilities, the chief engineer may also serve as the *fire marshal* or *fire chief.* Even at larger plants, where there is a full-time safety or fire-protection engineer responsible for plant protection and loss prevention, the plant engineer should be familiar with the fire problem, fire-prevention methods, and fire-protection systems.

Fires in private industry have a great potential for significant economic and financial loss to both the industrial plant and the community. Recovery from an industrial fire includes not only replacement of equipment and facilities at higher costs, but also temporary and permanent lost business income, loss of skilled employees during the time the plant is closed, loss of profits on damaged finished goods, and extra expenses to restore operations. Many plants destroyed by fire do not reopen, contributing to local unemployment and disrupting the lives of employees.

This chapter is intended to give the plant engineer a basic understanding of fire protection and prevention, to provide some basic information about fire, and to identify other resources for the incorporation of fire safety in every aspect of plant operations.

In addition to the National Fire Protection Association (NFPA) in Quincy, Mass., and the Federal Emergency Management Agency (FEMA) in Washington, D.C., the following fire protection services are also available in most cities in the United States: the Society of Fire Protection Engineers (SFPE), fire protection consultants, equipment manufacturers, testing laboratories, companies specializing in special-hazard protection and control, fire investigation firms, and organizations that perform fire-protection equipment testing, installation, and servicing. There are also state and regional training facilities, and the resources of municipal fire departments should not be overlooked.

This chapter does not attempt to discuss management programs which may have significant impact on plant fire protection. Information regarding risk management, personnel training, plant fire brigades, plant emergency organization, and fire-prevention programs must be obtained from other sources.

THE NATURE OF FIRE

A simple method of visualizing what fire is and how burning takes place is to use a four-sided object, a tetrahedron. Each surface of the tetrahedron is used to represent one of the conditions necessary for fire to occur (see Fig. 4.80).

The four components of this simple fire model are fuel, heat, oxidant, and the chemical chain reaction. While this model does not provide a complete scientific description of fire, it is sufficient for explaining most fire-protection concepts. In chemical terminology, the *fuel* component may also be referred to as a *reducing agent*. During the fire reaction the reducing agent looses electrons. The *heat* component includes both the heat which causes the fire and the heat emitted by the fire which causes it to be self-sustaining. The *oxidant* required for fire is most often provided by the oxygen in the ambient air, approximlately 21 percent. Although

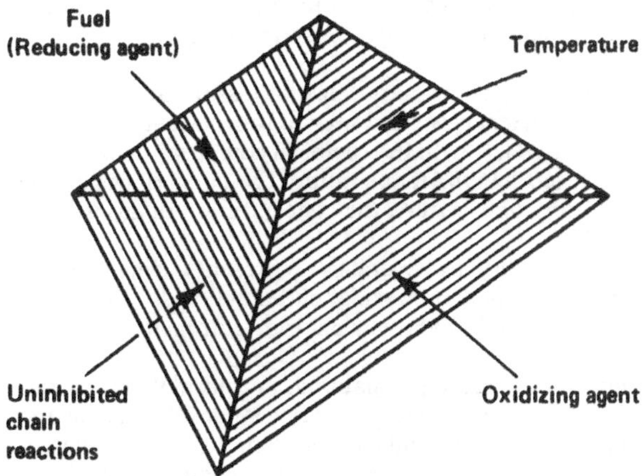

FIGURE 4.80 Tetrahedron fire model. (*From R. Tuve,* Principles of Fire Protection Chemistry, *NFPA, Quincy, Mass., 1998.*)

oxygen is the most common oxidizing agent and is usually necessary for fire to occur, there are some chemicals that release oxygen and some that can burn in an oxygen-free atmosphere.

Vapor State of Fuel

Before a fuel can be burned it must be in a vapor state. Therefore, flammable gases are most easily ignited. Even solids and liquids, such as wood and gasoline, must be vaporized before they will burn. The decomposition of matter due to heat which generates flammable vapors is called *pyrolysis.*

The initial vaporization of the fuel may be caused by heat from a source of ignition such as a chemical reaction, electric energy, or mechanical heat energy. Ambient heat may also be sufficient to vaporize fuels which are normally liquids.

Once sufficient vaporization has occurred, the combustible vapors can be ignited by an open flame of spark or, at a sufficiently high temperature, the vapor and oxygen mixture will ignite spontaneously.

After ignition of the vaporized fuel has occurred, the heat generated by the fire will cause further vaporization of the fuel and the intensity of the fire will increase. This process is known as *radiation feedback.*

Heat Transfer

The heat that causes the fuel to ignite and the heat generated by the resulting fire can be transmitted by one or all of the following three methods:

- *Conduction.* Heat transferred by direct contact from one body to another.
- *Radiation.* Energy travel through space or materials as waves.
- *Convection.* Heat transferred by a circulating medium, either a gas or a liquid.

Products of Combustion

There are four categories of products of combustion: (1) fire gases, (2) flame, (3) heat, and (4) smoke. All of these products are produced in varying degrees by each fire. The material or materials that are involved in the fire and the resulting chemical reactions produced by the fire determine the products of combustion.

Fire Gases. The primary cause of loss of life in fires is the inhalation of heated, toxic, and oxygen-deficient gases and smoke. The amount and kind of fire gases present during and after a fire vary widely with the chemical composition of the material burning, the amount of available oxygen, and the temperature. The effect of toxic gases and smoke on people will depend on the time of exposure, the concentration of gases in air, and the physical condition of the individual.

There are usually several gases present during a fire. Those that are commonly considered lethal are carbon monoxide, carbon dioxide, hydrogen sulfide, sulfur dioxide, ammonia, hydrogen cyanide, hydrogen chloride, nitrogen dioxide, acrolein, and phosgene.

Flame. The burning of materials in a normal oxygen-rich atmosphere is generally accompanied by flame. For this reason, flame is considered a distinct product of combustion. Burns can be caused by direct contact with flames or heat radiated from flames. Flame is rarely separated from the burning materials by an appreciable distance.

Heat. Heat is the combustion product most responsible for fire spread. Exposure to heat from a fire will affect persons in proportion to the length of exposure and the temperature of the heat. The dangers of exposure to heat from fire range from minor injury to death. Exposure to heated air increases the heart rate and causes dehydration, exhaustion, blockage of the respiratory tract, and burns. Fire fighters should not enter atmospheres exceeding 120°F (48°C) to 130°F (54°C) without special protective clothing and masks. The maximum survivable breathing level of heat from fire in a dry atmosphere for a short period has been estimated at 300°F (148°C). Any moisture present in the air greatly increases the danger and sharply reduces the time of survival.

Smoke. Smoke is matter consisting of very fine solid particles and condensed vapor. Fire gases from common combustibles (such as wood) contain water vapor, carbon dioxide, and carbon monoxide. Under the usual conditions of insufficient oxygen for complete combustion, methane, methanol, formaldehyde, and formic and acetic acids are also present. These gases are usually evolved from the combustible with sufficient velocity to carry droplets of flammable tars which appear as smoke. Particles of carbon develop from the decomposition of these tars; they are also present in the fire gases from the burning of petroleum products, particularly from the heavier oils and distillates.

Modes of Combustion

The combustion process may occur in two modes, *flaming* (including explosion) and *flameless* (including glow and deep-seated glowing embers). The flaming mode is characterized by relatively high burning rates. Intense and high levels of heat are usually associated with the flaming mode.

The flaming and flameless modes are not mutually exclusive; combustion may involve one or both modes. Often combustion may occur in the flaming mode and gradually make a transition to the flammable mode. At one point in this process both modes occur simultaneously.

Fire Control

Fire prevention and fire extinguishment can be described in terms of the fire model previously discussed. Fire prevention is generally a matter of keeping heat and fuel separated; or, in some processes, keeping heated fuel from combining with oxygen.

Fire extinguishment can be summarized by four methods:

1. Removal or dilution of air or oxygen to a point where combustion ceases
2. Removal of fuel to a point where there is nothing remaining to oxidize
3. Cooling of the fuel to point where combustible vapors are no longer evolved or where activation energy is lowered to the extent that no activated atoms or free radicals are produced
4. Interruption of the flame chemistry of the chain reaction of combustion by injection of compounds capable of quenching free-radical products

Removal of Oxygen. The amount of dilution of oxygen necessary to stop the combustion varies greatly with the kind of material that is burning. Ordinary hydrocarbon gases and vapors will not burn when the oxygen level is below 15 percent. Acetylene will continue to burn unless the oxygen concentration is lowered, but will continue to glow on the surface even if the oxygen level is as low as 4 to 5 percent.

A fire in a closed space can extinguish itself by consuming the oxygen. However, incomplete combustion, which takes place when the oxygen is consumed, usually results in considerable generation of flammable gases.

A commonly used method of putting a fire out by removing or diluting the oxygen is by flooding the entire fire area with carbon dioxide or with some other inert gas.

Fuel Removal. Fuel removal can be accomplished in a variety of ways. One of the most common examples is the practice of building a firebreak across the path of an advancing forest fire.

Fires in large coal or wood pulp piles can usually be controlled only by moving the pile of the fire zone. Fires in large oil storage tanks have been controlled by pumping the oil out of the burning tank into an empty tank. If a gas line is ruptured and the gas ignited, shutting off the supply of gas is the only way to stop the fire.

If it is not practical to remove the fuel, extinguishment can be accomplished by shutting off the fuel vapors or by covering the burning or glowing fuel. The use of fire-fighting foams and dry powder extinguishers are effective procedures for covering or coating a fire.

Cooling. For most common combustibles such as wood, paper, and cloth, the simplest and most effective means of removing the heat of a fire is through the application of water. Water application can be varied and will depend on the fire.

Applying water to the burning fuel cools the fuel until the rate of release of combustible vapors and gases is reduced and ultimately stopped. Heat developed by a fire is carried away by radiation, conduction, and convection. This helps reduce the amount of heat and makes the use of water more effective. Only a relatively small proportion of the heat evolved needs to be cooled by the water in order to extinguish the fire.

Effective use of water cannot be accomplished if the water cannot reach the burning fuel directly. For this reason, areas where firefighters cannot readily reach the fire with water streams, such as high-rise buildings and high-piled storage areas, must be provided with automatic sprinklers or other automatic fire-protection system.

Interruption of Chemical Reaction. Extinguishment by cooling, by oxygen dilution, and by fuel removal, is applicable to all classes of flaming and glowing fires. Extinguishment by chemical flame inhibition applies to the flaming mode only.

In the fourth method of extinguishment, the action occurs only during contact of the chemical agents with activated groups or with atoms being produced by the combustion process. In a sense this could be seen as a temporary extinguishment process, operating only when the agent particles are present in the flame. If activation energy continues to exist after withdrawal of the flame-inhibiting agent, the flame reaction will reestablish and will continue.

Summary

Combustion occurs under the following conditions:

1. An oxidizing agent, a combustible material, and an ignition source are essential for combustion
2. The combustible material must be heated to its ignition temperature before it will burn
3. Combustion will continue until:
   a. The combustible material is consumed or removed.
   b. The oxidizing agent concentration is lowered to below the concentration necessary to support combustion.
   c. The combustible material is cooled below its ignition temperature.
   d. Flames are chemically inhibited.

THE PLANT FIRE PROBLEM: CAUSES AND PREVENTION

In general, the cause of most plant fires is the exposure of a fuel to a source of heat. Where the fuel, such as accumulations of trash or debris, is not necessary plant operation, fires can be prevented by removal of the fuel. Where the exposed fuel, such as raw materials or finished products, is essential, the source of heat must be protected or controlled.

Some of the most common sources of heat and fuel that cause plant fires are heating and cooking equipment, smoking, electric equipment, burning, flammable liquids, open flames and sparks, incendiary (arson), spontaneous ignition, gas fires, and explosions. These sources of heat are summarized here.

Heating and Cooking Equipment

Defective or Overheated Equipment. This includes improperly maintained or operated furnaces, smoke pipes, vents, portable and stationary heaters, industrial commercial furnaces, and incinerators.

Chimneys and Flues. Fire can arise from ignition of accumulated soot or inadequate separation from combustible material.

Hot Ashes and Coals. These can cause problems when improper disposal or disposal in combustible containers or with combustible debris occurs.

Improper Location. This can mean installation too close to combustible or accumulation of combustibles near an appliance.

Electric Equipment

Wiring and Distribution Equipment. These include short-circuit faults, arcs, and sparks from damaged, defective, or improperly installed components.

Motors and Appliances. These include careless use, improper installation, and poor maintenance.

Flammable Liquids

Storage and Handling. These hazards include careless spills, leaking fuel, and overturned tanks.

Inadequate Safeguards. Fires can be started by improper storage containers or facilities, improper electrical equipment near open processes, or improper bonding and grounding of transfer processes.

Open Flames and Sparks

Trash and Rubbish. Burning trash and rubbish can furnish the fuel for accidental ignition; careless burning ignites other material.

Sparks and Embers. Problems include ignition of roof coverings by sparks from chimneys, incinerators, rubbish fires, locomotives, etc.

Welding and Cutting. Hazards include ignition of combustibles by the arc or flame itself, heat conduction through the metals being welded or cut, molten slag and metal from the cut, and sparks.

Friction, Sparks from Machinery. Friction heat or sparks resulting from impact between two hard surfaces are a hazard.

Thawing Pipes. Open-flame devices are a hazard when used in the dangerous practice of thawing pipes.

Other Open Flames. These include ignition sources such as candles, locomotive sparks, incinerator sparks, and chimney sparks.

Lightning. This includes building fires caused by the effects of lightning.

Exposure. Exposure fires are those originating in places other than buildings, but which ignite buildings.

Incendiary, Suspicious. These are fires that are known to be or thought to have been set, fires set to defraud insurance companies, fires set by mentally disturbed persons, and fires set by malicious persons.

Spontaneous Ignition. This means fires resulting from the uncontrolled spontaneous heating of materials.

Gas Fires and Explosions. These are fires and explosions that involve gas that has escaped from piping, storage tanks, equipment, or appliances and fires caused by misuse or faulty operation of gas appliances.

Smoking. The use of smoking materials in flammable or explosive atmospheres, or discarding smoking materials in combustible debris.

PLANT FIRE HAZARDS

Fire hazards in industrial and manufacturing plants are a mix of exotic hazards specific to particular industries and common hazards analogous to those found in any other type of property—electrical systems, heating, smoking, common appliances and tools, even cooking, and children playing with fire. Many texts and handbooks exist to address the methods for preventing, mitigating, and controlling fires involving each of the many special industrial hazards. These references are listed in the bibliography at the end of this chapter.

From a statistical perspective, structure fires in industrial and manufacturing properties are separable into two primary categories: (1) basic industry, utility, and defense properties;

and (2) manufacturing properties. This concept is illustrated in Table 4.36, which summarizes the structure fires that occurred in industrial and manufacturing facilities from 1993 to 1997.

Basic industry, utility, and defense properties include agriculture, forestry, mining, laboratories, energy production, communication facilities and defense sites. *Manufacturing properties* include those properties that make products of all kinds or are engaged in processing, assembling, mixing, packing, finishing, decorating, repairing, and similar operations. Only fires reported to public fire departments are included in these statistics. Tables 2 through 10 provide variations of other useful information on structure fires that have occurred in industrial and manufacturing facilities.

There are several noteworthy points of information in the statistical data of Tables 4.37 to 4.45. In summary, an average of 17,000 reported structure fires caused 16 civilian deaths, 631 civilian injuries, and $776.5 million in direct property damage per year. One-fifth of these structure fires occurred in facilities manufacturing metal or metal products.

TABLE 4.36 Industrial and Manufacturing Facility Structure Fires
Annual average of 1993–1997 structure fires.

Occupancy	Fires	Civilian deaths	Civilian injuries	Direct property damage, in millions
Industry, utility, and defense	4,200	2	75	$129.5
Energy production facility	200	1	4	22.5
Laboratory	300	0	23	5.1
Communication, defense, or document facility	200	0	5	7.6
Energy distribution property or utility facility	500	0*	12	12.9
Agricultural or farm production facility	2,300	1	13	55.7
Forest, fish hatchery, or hunting area	100	0	0	0.7
Mine or quarry	200	0	3	4.6
Nonmetallic mineral or mineral product manufacturing facility	300	0	12	12.7
Unclassified or unknown-type industry, utility, or defense	200	0	3	7.8
Manufacturing	12,800	15	556	647.0
Food product manufacturing	1,200	1	32	65.0
Beverage, tobacco, or related oil product manufacturing	200	0	9	2.9
Textile manufacturing	600	1	34	150.8
Wearing apparel or leather or rubber product manufacturing	400	0	21	37.7
Wood, furniture, paper, or printing product manufacturing	3,000	0	72	113.4
Chemical, plastic, or petroleum product manufacturing	1,300	6	116	58.0
Metal or metal product manufacturing	3,600	4	186	141.1
Vehicle assembly or manufacturing	600	1	31	24.0
Other manufacturing	1,000	0	30	30.6
Unclassified or unknown-type manufacturing	800	2	24	23.5
Total	17,000	16	631	$776.5

* Not zero, but rounds to zero.

Note: These are fires reported to U.S. municipal fire departments and so exclude fires reported only to federal or state agencies or industrial fire brigades. Fires are rounded to the nearest 100, deaths and injuries to the nearest 1, and direct property damage to the nearest $100,000. Sums may not equal totals due to rounding errors. Damage has not been adjusted for inflation.

Source: National estimates based on NFIRS and NFPA survey, NFPA, Quincy, Mass.

TABLE 4.37 Fire Protection Features Involved in Industrial and Manufacturing Structure Fires

Annual average of 1993–1997 structure fires.

Percentage of fires in buildings with smoke or other fire alarms present	32.2%
Percentage of fires in buildings having smoke or other fire alarms in which devices were operational	82.4%
Percentage of fires in buildings with operational smoke or other fire alarms (product of first two statistics)	26.6%
Percentage of fires in buildings with automatic suppression systems	41.2%
Deaths per 1000 fires with automatic suppression system present	1.1
Deaths per 1000 fires with no automatic suppression system present	1.2
Reduction in deaths per 1000 fires when automatic suppression system was present	10.8%
Average loss per fire when automatic suppression system was present	$19,871
Average loss per fire with no automatic suppression system	$53,646
Reduction in loss per fire when automatic suppression system was present	63.0%

Source: National estimates based on NFIRS and NFPA survey, NFPA, Quincy, Mass.

TABLE 4.38 Industrial and Manufacturing Facility Structure Fires, by Year

Year	Fires	Civilian deaths	Civilian injuries	Direct property damage, millions
1980	42,100	21	775	$ 608.9
1981	39,000	31	1322	659.3
1982	35,000	50	1118	449.6
1983	29,900	51	1091	727.1
1984	30,400	88	853	803.5
1985	30,900	37	779	569.0
1986	26,400	24	821	516.0
1987	25,800	44	991	542.3
1988	24,100	30	847	675.7
1989	21,700	60	783	1795.7
1990	18,700	79	830	688.7
1991	18,300	21	641	646.6
1992	17,500	11	545	538.6
1993	16,400	13	893	566.1
1994	18,000	16	648	567.4
1995	16,300	32	587	1253.8
1996	16,500	8	540	785.8
1997	17,800	16	485	709.4
1980–1997 annual average	24,700	35	808	$ 728.0
1993–1997 annual average	17,000	17	631	$ 776.5

Note: These are fires reported to U.S. municipal fire departments and so exclude fires reported only to federal or state agencies or industrial fire brigades. Fires are expressed to the nearest 100, deaths and injuries are rounded to the nearest 1, and property damage is rounded to the nearest $100,000. Property damage figures have not been adjusted for inflation.

Source: National estimates based on NFIRS and NFPA survey, NFPA, Quincy, Mass.

TABLE 4.39 Industrial and Manufacturing Facility Structure Fires, by Cause of Fire
Annual average of 1993–1997 structure fires.

Cause	Fires		Property damage, millions	
Other equipment	6,500	(38.4%)	$249.2	(32.1%)
Working or shaping machine	1,000	(5.6%)	22.0	(2.8%)
Furnace, oven, or kiln	900	(5.4%)	26.6	(3.4%)
Unclassified or unknown-type processing equipment	900	(5.2%)	47.6	(6.1%)
Unclassified or unknown-type special equipment	500	(2.8%)	30.4	(3.9%)
Heat-treating equipment	300	(1.8%)	10.7	(1.4%)
Separate motor or generator	300	(1.7%)	5.5	(0.7%)
Waste-recovery equipment	300	(1.5%)	4.5	(0.6%)
Casting, molding, or forging equipment	200	(1.4%)	12.1	(1.6%)
Conveyor	200	(1.0%)	6.9	(0.9%)
Open flame, ember, or torch	2,300	(13.8%)	107.0	(13.8%)
Torch	1,400	(8.1%)	82.5	(10.6%)
Open fire	300	(1.6%)	2.4	(0.3%)
Rekindle or reignition	200	(1.2%)	2.4	(0.3%)
Electrical distribution	1,700	(10.3%)	90.9	(11.7%)
Fixed wiring	500	(2.9%)	27.1	(3.5%)
Light fixture, ballast, or sign	200	(1.3%)	9.8	(1.3%)
Overcurrent protection device	200	(1.2%)	11.5	(1.5%)
Heating equipment	1,300	(7.4%)	34.1	(4.4%)
Fixed area heater	300	(1.8%)	12.0	(1.5%)
Central heating unit	200	(1.2%)	6.8	(0.9%)
Incendiary or suspicious	1,100	(6.6%)	174.5	(22.5%)
Natural causes	1,100	(6.6%)	41.9	(5.4%)
Spontaneous ignition or chemical reaction	700	(4.1%)	23.3	(3.0%)
Lightning	300	(1.5%)	10.4	(1.3%)
Appliance, tool, or air conditioning	800	(4.8%)	13.4	(1.7%)
Dryer	300	(1.8%)	4.0	(0.5%)
Other heat source	600	(3.7%)	12.5	(1.6%)
Cooking equipment	500	(2.9%)	22.8	(2.9%)
Exposure (to other hostile fire)	500	(2.7%)	13.8	(1.8%)
Smoking materials	400	(2.2%)	15.1	(1.9%)
Child playing	100	(0.5%)	1.3	(0.2%)
Total	17,000	(100.0%)	$776.5	(100.0%)

Note: These are fires reported to U.S. municipal fire departments and so exclude fires reported only to federal or state agencies or industrial fire brigades. Fires are expressed to the nearest 100, and property damage is rounded to the nearest $100,000. Property damage figures have not been adjusted for inflation. The 12 major cause categories are based on a hierarchy developed by the U.S. Fire Administration. Sums may not equal totals due to rounding errors. Unknown-cause fires have been allocated proportionally.

Source: National estimates based on NFIRS and NFPA survey, NFPA, Quincy, Mass.

TABLE 4.40 Industrial and Manufacturing Facility Structure Fires, Causes of Civilian Deaths and Injuries

Annual average of 1993–1997 structure fires.

Cause	Civilian deaths		Civilian injuries	
Other equipment	5	(30.9%)	310	(49.2%)
Unclassified or unknown-type processing equipment	2	(11.2%)	47	(7.4%)
Chemical processing equipment	1	(5.6%)	16	(2.5%)
Unclassified special equipment	1	(4.0%)	21	(3.3%)
Working or shaping machine	1	(3.1%)	38	(6.0%)
Natural causes	4	(25.0%)	58	(9.2%)
Spontaneous ignition or chemical reaction	3	(18.5%)	26	(4.2%)
Static discharge	1	(6.3%)	29	(4.5%)
Electrical distribution	3	(17.0%)	55	(8.7%)
Overcurrent protection device	1	(7.3%)	15	(2.3%)
Fixed wiring	1	(3.6%)	13	(2.1%)
Open flame, ember, or torch	2	(12.9%)	81	(12.9%)
Torch	2	(12.9%)	66	(10.5%)
Incendiary or suspicious	1	(7.1%)	13	(2.0%)
Heating equipment	1	(4.3%)	25	(4.0%)
Other heat source	0*	(1.4%)	24	(3.8%)
Cooking equipment	0*	(1.4%)	16	(2.5%)
Appliance, tool, or air conditioning	0	(0.0%)	32	(5.0%)
Smoking materials	0	(0.0%)	7	(1.2%)
Exposure (to other hostile fire)	0	(0.0%)	6	(0.9%)
Child playing	0	(0.0%)	3	(0.5%)
Total	17	(100.0%)	631	(100.0%)

* Not zero, but rounds to zero.

Note: These are fires reported to U.S. municipal fire departments and so exclude fires reported only to federal or state agencies or industrial fire brigades. Civilian deaths and civilian injuries are rounded to the nearest 1. The 12 major cause categories are based on a hierarchy developed by the U.S. Fire Administration. Sums may not equal totals due to rounding errors. Unknown-cause fires have been allocated proportionally.

Source: National estimates based on NFIRS and NFPA survey, NFPA, Quincy, Mass.

With respect to the overall fire problem, 2.9 percent of reported structure fires occurred in industrial or manufacturing properties. The 17,000 fires that occurred between 1993 and 1997 in industrial or manufacturing properties accounted for 2.9 percent of the 587,500 structure fires, 0.4 percent of the 3860 civilian structure-fire deaths, 2.8 percent of the 22,730 civilian structure-fire injuries, and 10.5 percent of the $7,382 billion in direct property damage per year during that time period.

Since 1980, fires in industrial and manufacturing properties fell 58 percent, from 42,100 in 1980 to 17,800 in 1997. From 1996 to 1997, however, structure fires in these occupancies increased 7 percent from 16,500 in 1996. In comparison, structure fires of all types declined 48 percent from 1980 to 1997. From 1996 to 1997, total structure fires fell 5 percent.

Industrial and manufacturing facility fires have a cause profile that reflects the wide variety of industrial processes used and activities conducted. Unlike most other nonresidential occupancies, incendiary and suspicious causes ranked fifth among the causes of fires in indus-

TABLE 4.41 Industrial and Manufacturing Facility Structure Fires, by Area of Origin
Annual average of 1993–1997 structure fires.

Area of origin	Fires		Civilian deaths		Civilian injuries		Property damage, millions	
Processing or manufacturing area	3,400	(19.9%)	5	(32.4%)	221	(35.0%)	$175.8	(22.6%)
Machinery room or area	1,700	(10.1%)	2	(9.5%)	79	(12.5%)	64.4	(8.3%)
Product storage area, tank, or bin	1,000	(5.9%)	3	(16.6%)	32	(5.0%)	87.0	(11.2%)
Maintenance shop or area	900	(5.1%)	1	(4.4%)	37	(5.9%)	42.8	(5.5%)
Duct	700	(4.1%)	0*	(1.0%)	26	(4.2%)	26.2	(3.4%)
Unclassified area of origin	700	(4.0%)	0*	(1.3%)	18	(2.8%)	21.6	(2.8%)
Attic or ceiling/roof assembly or concealed space	600	(3.7%)	0*	(1.0%)	11	(1.8%)	21.3	(2.7%)
Heating equipment room	600	(3.3%)	0	(0.0%)	20	(3.2%)	50.5	(6.5%)
Exterior wall surface	500	(2.9%)	0	(0.0%)	4	(0.6%)	25.3	(3.3%)
Exterior roof surface	400	(2.6%)	0	(0.0%)	5	(0.8%)	5.8	(0.7%)
Trash or rubbish area	400	(2.2%)	0	(0.0%)	7	(1.1%)	5.2	(0.7%)
Supply storage room or area	400	(2.2%)	0*	(1.3%)	14	(2.2%)	20.2	(2.6%)
Unclassified structural area	400	(2.1%)	1	(3.2%)	5	(0.8%)	13.3	(1.7%)
Unclassified service or equipment area	300	(2.0%)	1	(3.2%)	17	(2.7%)	15.8	(2.0%)
Unclassified storage area	300	(1.9%)	0*	(2.8%)	4	(0.6%)	11.1	(1.4%)
Wall assembly or concealed space	300	(1.7%)	0	(0.0%)	3	(0.5%)	9.7	(1.2%)
Kitchen	300	(1.5%)	0	(0.0%)	9	(1.4%)	4.4	(0.6%)
Switchgear area or transformer vault	200	(1.4%)	1	(7.8%)	13	(2.1%)	12.3	(1.6%)
Office	200	(1.4%)	0*	(1.3%)	3	(0.5%)	14.0	(1.8%)
Shipping, receiving, or loading area	200	(1.3%)	1	(5.0%)	8	(1.3%)	16.1	(2.1%)
Lawn, field, or open area	200	(1.3%)	0	(0.0%)	3	(0.4%)	2.3	(0.3%)
Electronic equipment room or area	200	(1.2%)	0	(0.0%)	3	(0.5%)	10.4	(1.3%)
Conveyor	200	(1.2%)	0	(0.0%)	3	(0.5%)	4.7	(0.6%)
Laboratory	200	(1.2%)	0	(0.0%)	23	(3.6%)	2.9	(0.4%)
Unclassified function area	200	(1.2%)	0	(0.0%)	12	(1.9%)	7.5	(1.0%)
Ceiling/floor assembly or concealed space	200	(1.2%)	0	(0.0%)	1	(0.2%)	22.4	(2.9%)
Laundry room or area	200	(1.0%)	0	(0.0%)	3	(0.4%)	2.0	(0.3%)
Other known service or equipment area	600	(3.5%)	0	(0.0%)	14	(2.2%)	10.3	(1.3%)
Other known function area	400	(2.3%)	0*	(2.8%)	9	(1.5%)	22.7	(2.9%)
Other known means of egress	200	(1.1%)	0	(0.0%)	2	(0.3%)	8.7	(1.1%)
Other known assembly area	200	(1.1%)	0*	(2.2%)	4	(0.6%)	8.4	(1.1%)
Other known storage area	200	(1.0%)	0	(0.0%)	4	(0.6%)	10.2	(1.3%)
Other known area	600	(3.3%)	1	(4.2%)	13	(2.1%)	21.5	(2.8%)
Total	17,000	(100.0%)	17	(100.0%)	631	(100.0%)	$776.5	(100.0%)

* Not zero, but rounds to zero.

Note: These are fires reported to U.S. municipal fire departments and so exclude fires reported only to federal or state agencies or industrial fire brigades. Fires are rounded to the nearest 100, deaths and injuries to the nearest 1, and direct property damage to the nearest $100,000. Sums may not equal totals due to rounding errors. Damage has not been adjusted for inflation. Unknown-cause fires have been allocated proportionally.

Source: National estimates based on NFIRS and NFPA survey, NFPA, Quincy, Mass.

trial and manufacturing properties. Almost one-fifth (38 percent) of the fires in these structures were caused by the wide variety of equipment and processes classed as "other equipment." Open flames, embers, or torches ranked second; and electrical distribution equipment ranked third.

Processing or manufacturing areas were the most frequent areas of origin. One of every five reported structure fires in industrial or manufacturing properties started in processing or

TABLE 4.42 Industrial and Manufacturing Facility Structure Fires, by Form of Material
Annual average of 1993–1997 structure fires.

Form of material	Fires		Civilian deaths		Civilian injuries		Property damage millions	
Unclassified form of material	2,100	(10.9%)	0	(2.5%)	108	(14.7%)	$62.4	(5.0%)
Dust, fiber, or lint	1,800	(9.0%)	2	(7.3%)	65	(8.9%)	34.6	(2.7%)
Electrical wire or cable insulation	1,300	(6.6%)	2	(3.3%)	38	(5.2%)	50.3	(4.0%)
Trash	1,200	(6.2%)	0	(0.0%)	27	(3.7%)	21.5	(1.7%)
Structural member or framing	1,100	(5.9%)	0	(2.2%)	13	(1.8%)	70.3	(5.6%)
Agricultural product	800	(4.1%)	0	(0.0%)	3	(0.5%)	20.4	(1.6%)
Gas or liquid in or from a pipe or container	700	(3.8%)	3	(12.3%)	102	(13.9%)	52.3	(4.2%)
Multiple forms	600	(2.9%)	5	(16.5%)	20	(2.8%)	43.0	(3.4%)
Exterior roof covering or finish	500	(2.7%)	0	(0.0%)	9	(1.2%)	25.8	(2.0%)
Fuel	500	(2.3%)	2	(4.3%)	32	(4.4%)	33.6	(2.7%)
Exterior sidewall cover or finish	400	(2.1%)	0	(0.0%)	3	(0.5%)	7.5	(0.6%)
Chips	400	(2.1%)	0	(0.0%)	15	(2.1%)	27.9	(2.2%)
Box, carton, or bag	400	(2.0%)	0	(0.0%)	14	(1.9%)	22.1	(1.8%)
Thermal or acoustical insulation	400	(1.9%)	0	(0.0%)	12	(1.5%)	11.7	(0.9%)
Interior wall cover or fixed items	400	(1.9%)	0	(0.0%)	9	(1.2%)	54.9	(4.4%)
Cooking materials	300	(1.6%)	0	(0.0%)	7	(0.9%)	4.0	(0.3%)
Atomized or vaporized liquid	200	(1.0%)	2	(7.8%)	26	(3.6%)	20.5	(1.6%)
Magazine or newspaper	200	(1.0%)	0	(0.0%)	3	(0.5%)	9.4	(0.7%)
Conveyor belt, drive belt, or V-belt	200	(1.0%)	0	(0.0%)	0	(0.0%)	16.8	(1.3%)
Other known form	3,500	(20.4%)	8	(47.1%)	194	(30.7%)	10.3	(24.1%)
Total	17,000	(100.0%)	17	(100.0%)	631	(100.0%)	$776.5	(100.0%)

* Not zero, but rounds to zero.

Note: These are fires reported to U.S. municipal fire departments and so exclude fires reported only to federal or state agencies or industrial fire brigades. Fires are rounded to the nearest 100, deaths and injuries to the nearest 1, and direct property damage to the nearest $100,000. Sums may not equal totals due to rounding errors. Damage has not been adjusted for inflation. Unknown-cause fires have been allocated proportionally.

Source: National estimates based on NFIRS and NFPA survey, NFPA, Quincy, Mass.

manufacturing areas; 10 percent started in machinery rooms or areas; 6 percent started in product storage rooms or areas; and 5 percent began in maintenance shops or areas.

The causes of structure fires resulting in civilian deaths and injuries indicate that "other equipment" caused 31 percent of the civilian deaths and 49 percent of the civilian injuries. As a point of historical reference, the deadliest U.S. fire or explosion on record in the industrial or manufacturing property class was the 1907 Monongha, West Virginia, coal-mine explosion that claimed 361 lives.

About one-fourth of reported fires occurred in properties with working smoke alarms. Only 27 percent of the reported fires in industrial and manufacturing properties occurred in facilities protected by working smoke or fire alarms during the 5-yr period from 1993 through 1997. Automatic suppression systems were present in 41 percent of the fires in these properties. Interestingly, the average estimated direct property damage was almost 3 times as high when no automatic suppression system was present.

Several specific fires or explosions related to the incidents in these statistics are in the all-time top 20 with respect to dollar loss. In the last several decades, four recent industrial or manufacturing fires and explosions rank among the costliest in U.S. history: the 1989 Pasadena, Texas, vapor-cloud explosion, which also killed 23 people (#5 in dollar loss after adjusting for inflation); the 1995 Methuen, Massachusetts, textile-mill fire (#9); the 1988 Norco, Louisiana, refinery fire (#12); and the 1987 Pampa, Texas, chemical-plant fire (#17).

TABLE 4.43 Industrial and Manufacturing Facility Structure Fires, by Type of Material
Annual average of 1993–1997 structure fires.

Type of material	Fires		Civilian deaths		Civilian injuries		Property damage millions	
Sawn wood	2,200	(11.0%)	0	(1.2%)	22	(3.0%)	$116.8	(8.9%)
Unclassified type of material	900	(5.6%)	1	(4.1%)	33	(4.7%)	28.6	(2.2%)
Wood shavings	800	(4.8%)	0	(1.0%)	16	(2.2%)	53.4	(4.1%)
Paper untreated or uncoated	800	(4.6%)	0	(0.0%)	27	(3.8%)	31.5	(2.4%)
Unclassified plastic	600	(3.6%)	0	(2.1%)	26	(3.6%)	30.9	(2.4%)
Grass, leaves, hay or straw	500	(3.1%)	0	(0.0%)	5	(0.6%)	15.6	(1.2%)
Rubber	500	(2.9%)	0	(2.7%)	16	(2.2%)	35.2	(2.7%)
Multiple types	500	(2.9%)	1	(4.0%)	16	(2.3%)	41.4	(3.2%)
Cotton, rayon or finished goods	500	(2.7%)	1	(3.6%)	8	(1.2%)	11.1	(0.8%)
Other plastic	400	(2.4%)	0	(1.2%)	25	(3.5%)	35.6	(2.7%)
Polyvinyl	400	(2.3%)	0	(0.0%)	23	(3.2%)	14.6	(1.1%)
Cardboard	400	(2.2%)	0	(0.0%)	16	(2.2%)	21.9	(1.7%)
Class IIIB combustible liquid	400	(2.2%)	0	(0.0%)	26	(3.7%)	41.1	(3.1%)
Manufactured fabric, fiber or finished good	300	(2.0%)	0	(0.0%)	17	(2.4%)	11.5	(0.9%)
Hardboard or plywood	300	(2.0%)	0	(0.0%)	3	(0.5%)	12.4	(0.9%)
Adhesive, resin or tar	300	(1.9%)	0	(0.0%)	21	(2.9%)	9.5	(0.7%)
Unclassified wood/paper	300	(1.8%)	0	(0.0%)	5	(0.7%)	14.9	(1.1%)
Other flammable/combustible liquid	300	(1.7%)	0	(0.0%)	16	(2.2%)	12.1	(0.9%)
Grain/natural fiber (pre-process)	300	(1.7%)	0	(1.5%)	7	(1.0%)	10.8	(0.8%)
Other wood/paper	300	(1.6%)	0	(0.0%)	13	(1.7%)	9.4	(0.7%)
Applied paint/varnish	200	(1.4%)	0	(0.0%)	11	(1.7%)	9.1	(0.7%)
Unclassified flammable or combustible liquid	200	(1.3%)	1	(3.9%)	19	(2.7%)	12.2	(0.9%)
Combustible metal	200	(1.2%)	0	(1.5%)	18	(2.5%)	3.9	(0.3%)
Fat or grease (food)	200	(1.2%)	0	(0.0%)	8	(1.1%)	5.2	(0.4%)
Fiberboard or wood pulp	200	(1.1%)	0	(0.0%)	3	(0.4%)	5.8	(0.4%)
Food or starch	200	(1.0%)	0	(2.7%)	6	(0.8%)	1.7	(0.1%)
Other known type	3,200	(18.8%)	13	(76.5%)	275	(43.6%)	4.8	(23.2%)
Totals	17,000	(100.0%)	17	(100.0%)	631	(100.0%)	$776.5	(100.0%)

* Not zero, but rounds to zero.

Note: These are fires reported to U.S. municipal fire departments and so exclude fires reported only to federal or state agencies or industrial fire brigades. Fires are rounded to the nearest 100, deaths and injuries to the nearest 1, and direct property damage to the nearest $100,000. Sums may not equal totals due to rounding errors. Damage has not been adjusted for inflation. Unknown-cause fires have been allocated proportionally.

Source: National estimates based on NFIRS and NFPA survey, NFPA, Quincy, Mass.

During the same 5-yr period from 1993 through 1997, an average of 52,300 outside and other fires per year in industrial or manufacturing properties caused an average of 7 civilian deaths, 173 civilian injuries, and $55.4 million in direct property damage per year. Also during this same period, an average of 5300 vehicle fires caused 18 civilian deaths, 51 civilian injuries, and $53.5 million in direct property damage per year.

FIRE HAZARDS OF MATERIALS

Virtually all matter can be changed by exposure to sufficient quantities of energy. Energy in the form of heat was previously discussed. The heat energy which causes or is produced by a

TABLE 4.44 Industrial and Manufacturing Facility Structure Fires, by Equipment Involved
Annual average of 1993–1997 structure fires.

Equipment involved	Fires		Civilian deaths		Civilian injuries		Property damage, millions	
No equipment involved	4,400	(25.9%)	2	(14.8%)	124	(17.8%)	$249.2	(18.6%)
Torch	1,100	(6.5%)	2	(9.3%)	62	(8.9%)	80.8	(6.0%)
Working or shaping machine	800	(4.6%)	0	(2.1%)	37	(5.2%)	32.5	(2.4%)
Furnace, oven, or kiln	800	(4.4%)	0	(0.0%)	39	(5.6%)	25.9	(1.9%)
Other processing equipment	500	(3.2%)	2	(9.9%)	40	(5.8%)	30.9	(2.3%)
Fixed wiring	400	(2.6%)	0	(2.7%)	12	(1.8%)	29.2	(2.2%)
Other special equipment	300	(1.9%)	0	(2.7%)	21	(2.9%)	27.5	(2.1%)
Dryer	300	(1.7%)	0	(0.0%)	4	(0.6%)	4.5	(0.3%)
Fixed local heating unit	300	(1.5%)	0	(1.5%)	4	(0.7%)	12.9	(1.0%)
Unclassified object or exposure to fire	300	(1.5%)	1	(3.1%)	16	(2.3%)	23.5	(1.8%)
Heat-treating equipment	300	(1.5%)	0	(0.0%)	13	(1.9%)	10.5	(0.8%)
Unclassified processing equipment	200	(1.5%)	0	(2.1%)	9	(1.3%)	25.4	(1.9%)
Separate motor or generator	200	(1.4%)	0	(0.0%)	4	(0.6%)	5.3	(0.4%)
Waste-recovery equipment	200	(1.3%)	0	(0.0%)	10	(1.4%)	4.7	(0.4%)
Casting, molding, or forging equipment	200	(1.1%)	0	(1.2%)	41	(5.8%)	13.1	(1.0%)
Power switch or overcurrent device	200	(1.1%)	1	(5.4%)	16	(2.1%)	12.9	(1.0%)
Light fixture, lampholder, or sign	200	(1.1%)	0	(2.1%)	4	(0.6%)	10.4	(0.8%)
Central heating unit	200	(1.1%)	0	(0.0%)	1	(0.2%)	7.4	(0.6%)
Other known equipment	4,300	(25.2%)	9	(52.9%)	218	(34.5%)	0.0	(21.9%)
Total	17,000	(100.0%)	17	(100.0%)	631	(100.0%)	$776.5	(100.0%)

* Not zero, but rounds to zero.

Note: These are fires reported to U.S. municipal fire departments and so exclude fires reported only to federal or state agencies or industrial fire brigades. Fires are rounded to the nearest 100, deaths and injuries to the nearest 1, and direct property damage to the nearest $100,000. Sums may not equal totals due to rounding errors. Damage has not been adjusted for inflation. Unknown cause fires have been allocated proportionally.

Source: National estimates based on NFIRS and NFPA survey, NFPA, Quincy, Mass.

fire usually causes undesired changes to the material involved. The relative fire risks and products of burning of different types of materials are presented in this section.

Wood

When in contact with sufficient heat, all wood or wood-based products will ignite, the time of ignition depending on the ignition source and the length of exposure.

Wood or wood-based products can be treated with fire-retardant chemicals. When so treated, the flammability of wood and wood-based products is reduced. The flammability of these products can also be reduced when they are used in combination with other materials, such as insulation.

Wood is made up primarily of carbon, hydrogen, and oxygen. Live wood cells retain considerable moisture. When the wood is dead, air replaces most of the water in the cellular structure of wood. Some of the concerns associated with wood are further addressed in NFPA 664, *Woodworking and Woodprocessing Facilities.*

The fire behavior of wood and other combustible solids of the same size and shape varies greatly with the moisture content. Wet wood is harder to ignite and will not burn as fast as dry wood. The burning rate is also influenced by the moisture content in materials. See Table 4.46

TABLE 4.45 Industrial and Manufacturing Facility Structure Fires, by Form of Heat
Annual average of 1993–1997 structure fires.

Form of heat	Fires		Civilian deaths		Civilian injuries		Property damage, millions	
Heat or spark from friction	1,200	(6.8%)	0	(1.0%)	41	(5.2%)	$38.3	(2.7%)
Heat from gas-fueled equipment	1,100	(6.3%)	0	(2.5%)	40	(5.1%)	$39.5	(2.7%)
Unspecified short circuit or arc	1,000	(5.8%)	0	(0.0%)	40	(5.1%)	$56.4	(3.9%)
Heat from properly operating electrical equipment	700	(4.2%)	0	(1.5%)	41	(5.2%)	$18.5	(1.3%)
Spontaneous ignition or chemical reaction	600	(3.5%)	3	(13.4%)	29	(3.7%)	$26.5	(1.8%)
Cutting torch operation	600	(3.4%)	2	(5.4%)	38	(4.8%)	$65.6	(4.6%)
Welding torch operation	500	(2.9%)	0	(1.2%)	26	(3.3%)	$31.5	(2.2%)
Molten or hot material	500	(2.8%)	0	(1.2%)	48	(5.9%)	$34.7	(2.4%)
Heat from overloaded equipment	400	(2.4%)	0	(0.0%)	13	(1.5%)	$14.3	(1.0%)
Unclassified heat from electrical equipment	400	(2.3%)	0	(0.0%)	23	(2.9%)	$37.5	(2.6%)
Heat from other hot object	300	(2.1%)	0	(0.0%)	19	(2.4%)	$16.0	(1.1%)
Spark or flame from gas-fueled equipment	300	(1.7%)	0	(0.0%)	18	(2.3%)	$23.1	(1.6%)
Open fire	300	(1.7%)	0	(0.0%)	4	(0.5%)	$3.9	(0.3%)
Short-circuit arc from defective or worn insulation	300	(1.7%)	0	(0.0%)	3	(0.4%)	$11.0	(0.8%)
Heat from improperly operating electrical equipment	300	(1.6%)	0	(0.0%)	15	(1.9%)	$8.1	(0.6%)
Cigarette	300	(1.6%)	0	(0.0%)	6	(0.8%)	$15.4	(1.1%)
Unclassified form of heat	300	(1.6%)	2	(3.1%)	10	(1.3%)	$28.6	(2.0%)
Match	200	(1.4%)	0	(0.0%)	6	(0.8%)	$16.7	(1.2%)
Arc or spark from operating equipment or switch	200	(1.4%)	0	(1.5%)	34	(4.2%)	$11.6	(0.8%)
Lightning discharge	200	(1.3%)	0	(0.0%)	3	(0.3%)	$11.8	(0.8%)
Hot ember or ash	200	(1.3%)	0	(0.0%)	6	(0.7%)	$3.6	(0.2%)
Unclassified heat from open-flame or spark	200	(1.3%)	0	(2.7%)	9	(1.2%)	$28.7	(2.0%)
Unclassified heat from hot object	200	(1.2%)	0	(0.0%)	13	(1.6%)	$11.4	(0.8%)
Torch operating (not cutting or welding)	200	(1.2%)	0	(2.7%)	15	(1.9%)	$3.0	(0.2%)
Heat from other open flame or spark	200	(1.1%)	2	(3.3%)	13	(1.5%)	$61.3	(4.3%)
Heat from liquid-fueled equipment	200	(1.1%)	0	(1.0%)	6	(0.8%)	$10.5	(0.7%)
Heat from equipment, fuel unknown	200	(1.1%)	0	(0.0%)	10	(1.2%)	$3.1	(0.2%)
Rekindle or reignition	200	(1.1%)	0	(0.0%)	0	(0.0%)	3.2	(0.2%)
Heat from direct flame or convection	200	(1.1%)	0	(0.0%)	3	(0.3%)	7.4	(0.5%)
Short circuit arc from mechanical damage	200	(1.1%)	2	(6.0%)	8	(1.0%)	16.3	(1.1%)
Heat from other electrical equipment	200	(1.0%)	2	(4.1%)	13	(1.6%)	10.1	(0.7%)
Electric lamp	200	(1.0%)	0	(2.1%)	5	(0.6%)	16.0	(1.1%)
Other known form of heat	2,000	(11.8%)	10	(58.8%)	189	(30.0%)	38.3	(12.0%)
Totals	17,000	(100.0%)	17	(100.0%)	631	(100.0%)	$776.5	(100.0%)

* Not zero, but rounds to zero.

Note: These are fires reported to U.S. municipal fire departments and so exclude fires reported only to federal or state agencies or industrial fire brigades. Fires are rounded to the nearest 100, deaths and injuries to the nearest 1, and direct property damage to the nearest $100,000. Sums may not equal totals due to rounding errors. Damage has not been adjusted for inflation. Unknown-cause fires have been allocated proportionally.

Source: National estimates based on NFIRS and NFPA survey, NFPA, Quincy, Mass.

for heat of combustion for various wood-based products composed with petroleum-based materials.

Even when exposed to a relatively high heat source for a prolonged period of time, ignition is generally difficult when the moisture content of wood (and similar fuels) is above 15 percent. Once ignition and resultant fire have begun, heat radiation and the rate of pyrolysis reduce the importance of the moisture factor.

TABLE 4.46 Heat of Combustion of Various Wood and Wood-Based Products and Comparative Substances

Substance	Heating value, Btu/lb
Wood sawdust (oak)	8,493*
Wood sawdust (pine)	9,676*
Wood shavings	8,248*
Wood bark (fir)	9,496*
Corrugated fiber carton	5,970*
Newspaper	7,883*
Wrapping paper	7,106*
Petroleum coke	15,800
Asphalt	17,158
Oil (cottonseed)	17,100
Oil (paraffin)	17,640

* Dry.

Source: Extracted from *Kent's Mechanical Engineers' Handbook,* 12th ed., H. B. Carmichael and J. K. Salisbury, eds., Wiley-Interscience, New York, 1950.

Plastics

There are thousands of plastic product formulations that are produced in a variety of shapes and sizes, such as solid shapes, films and sheets, foams, molded forms, synthetic fibers, pellets, and powders. They are classified into 30 major groups of plastics and polymers. In addition, most finished products contain additives such as colorants, reinforcing agents, fillers, stabilizers, and lubricants. These additives vary the chemical nature of the product still more.

Most plastics are combustible, but the degree of combustibility varies widely because of the range of chemical compositions and combinations. As a result, it is virtually impossible to assign a fire hazard or flammability limit to any general plastic group. The only method of determining the fire hazard of a particular plastic is to fire test the plastic under exact end-use conditions. See Tables 4.47 and 4.48.

The chemical composition, the physical form, and the manner and arrangement in which plastics are stored greatly affect the degree of fire hazard that is present. Large quantities of smoke are usually generated when stored plastics are involved in fire, a condition made more or less difficult by the amount of ventilation present or available in a given storage area.

Thermoplastics, such as polyethylene and plasticized polyvinylchloride, and thermosets, such as polyesters, present challenging fire hazards. Plastics in foamed material form present the most severe hazard of all. In a fire, thermoplastics will melt and break down and behave and burn like flammable liquids. Automatic sprinkler systems with high sprinkler-discharge densities are necessary for adequate fire protection.

Dusts

When some combustible solids are ground or rubbed into minute particles, the particles tend to mix with the air in much the same way that vapor or gas mixes with the air. The finer the dust particle, the more completely it will mix with the air and remain suspended in the air.

TABLE 4.47 Small-Scale Test for Combustibility of Plastics

Test method	Sample size, in	Position of sample	Ignition source, flame, in	Time and limit of exposure, s	Value reported, in/min	Usual material application
ASTM D 635	⅛ × ½ × 5	Horizontal	1	2–30	Burning rate	Rigid plastic
ASTM D 568	0.05 × 1 × 18	Horizontal	1	15	Burning rate	Films
ASTM D 229	½ × ½ × 5	Horizontal	1	30	Burning time for 4 samples	Electrical insulation
ASTM D 1692	½ × 2 × 6	Horizontal	1	60	Burning rate	Foam
UL 94*	⅛ × ½ × 5	Horizontal	¾	30	Burning rate	
	⅛ × ½ × 5	Vertical	¾	2–10	Extinguishment time	Rigid plastics Rigid plastics
NFPA 701	2¾ × 10	Vertical	1½	12	Length of char	Films
UL 214	2¾ × 10	Vertical	1½	12	Length of char	Films

* UL stands for Underwriters Laboratories Inc.
Source: NFPA, Quincy, Mass.

TABLE 4.48 Medium- and Large-Scale Tests for Combustibility of Plastics

Test for	Number
Surface burning of building materials	NFPA 255 UL 723 ASTM E 84
Fire tests of building construction and materials	NFPA 251 UL 263 ASTM E 119
Fire tests of roof coverings	NFPA 256
Radiant panel test for flame spread	ASTM E 162
Factory mutual calorimeter test	
UL and FM corner wall tests*	
Full-room burnouts—FM, UL	
Flame-retardant films	NFPA 701 UL 214

* UL stands for Underwriters Laboratories Inc. FM stands for Factory Mutual Systems.
Source: NFPA, Quincy, Mass.

Although dust particles from all combustible solids do not result in potentially explosive dust particles, a large number of combustible solids can yield explosive dust particles. See Table 4.49. A variety of NFPA standards deal directly with this subject. These include:

NFPA 61, *Fires and Dust Explosions in Agricultural and Food Products Facilities*

NFPA 650, *Pneumatic Conveying Systems for Handling Combustible Materials*

NFPA 654, *Fire and Dust Explosions from the Manufacturing, Processing, and Handling of Combustible Particulate Solids*

NFPA 655, *Prevention of Sulfur Fires and Explosions*

Metals

Nearly all metals will burn in air under certain conditions. See Table 4.50. Some oxidize rapidly in the presence of air or moisture, generating sufficient heat to reach their ignition tem-

TABLE 4.49 Common Combustible Solid Dusts Generating Severe Explosions

Type of dust	Maximum explosion pressure		Maximum rate of pressure rise	
	psig	bar	psig/s	bar/s
Corn (processing)	95	6.55	6,000	413.7
Cornstarch	115	7.93	9,000	620.5
Potato starch	97	6.89	8,000	551.6
Sugar (processing)	91	6.27	5,000	344.7
Wheat starch	105	7.24	8,500	586.0
Ethyl cellulose plastic molding compound	102	7.03	6,000	413.7
Wood flour filler	110	7.58	5,500	379.2
Natural resin	87	6.0	10,000	689.5
Aluminum	100	6.9	10,000	689.5
Magnesium (powder)	94	6.48	10,000	689.5
Silicon (powder)	106	7.31	10,000	689.5
Titanium (powder)	80	5.52	10,000	689.5
Aluminum magnesium alloy (powder)	90	6.20	10,000	689.5

Source: Extracted from U.S. Bureau of Mines Investigations and Reports, nos. 5753, RI 5971, RI 6561.

TABLE 4.50 Melting, Boiling, and Ignition Temperature of Pure Metals in Solid Form

Pure metal	Temperature: Melting point		Boiling point		Solid metal ignition	
	°F	°C	°F	°C	°F	°C
Aluminum	1220	660	4445	2452	1832*‡	555*‡
Barium	1337	725	2084	1140	347*	175*
Calcium	1548	842	2625	1440	1300	704
Hafnium	4032	2223	9750	5399	—	—
Iron	2795	1535	5432	3000	1706*	930*
Lithium	367	186	2437	1336	356	180
Magnesium	1202	650	2030	1110	1153	623
Plutonium	1184	640	6000	3315	1112	600
Potassium	144	62	1400	760	156*†	69†*
Sodium	208	98	1616	880	239‡	115‡
Strontium	1425	774	2102	1150	1328*	720*
Thorium	3353	1845	8132	4500	932*	500*
Titanium	3140	1727	5900	3260	2900	1593
Uranium	2070	1132	6900	3815	6900*§	3815*§
Zinc	786	419	1665	907	1652*	900*
Zirconium	3326	1830	6470	3577	2552*	1400*

* Ignition in oxygen.
† Spontaneous ignition in moist air.
‡ Above indicated temperature.
§ Below indicated temperature.
Source: From Arthur E. Cote, ed., Fire Protection Handbook, NFPA, Quincy, Mass., 1997.

peratures. Others oxidize so slowly that heat generated during oxidation is dissipated before they become hot enough to ignite. Certain metals, notably magnesium, titanium, sodium, potassium, calcium, lithium, hafnium, zirconium, zinc, thorium, uranium, and plutonium, are referred to as *combustible metals* because of the ease of ignition of thin sections, fine particles, or molten metal. The same metals in massive solid form are comparatively difficult to ignite.

Hot, burning metals may react violently with the extinguishants used on fires involving ordinary combustibles or flammable liquids. A few metals, such as uranium, thorium, and plutonium, emit ionizing radiations that can complicate fire fighting and introduce a contamination problem.

Temperatures in burning metals are generally much higher than the temperature in burning flammable liquids. Some hot metals can continue burning in nitrogen, carbon dioxide, or steam atmospheres in which ordinary combustibles or flammable liquids would be incapable of burning. NFPA standards that specifically address metals include:

NFPA 480, *Storage, Handling, and Processing of Magnesium Solids and Powders*

NFPA 481, *Processing, Handling, and Storage of Titanium Production*

NFPA 482, *Processing, Handling and Storage of Zirconium Production*

NFPA 651, *Machining and Finishing of Aluminum and the Production and Handling of Aluminum Powders*

Flammable and Combustible Liquids

The improper storage, handling, and use of flammable and combustible liquids has been the cause of many deaths, injuries, and disastrous fires. Various NFPA standards provide detailed requirements on this subject, including:

NFPA 30, *Flammable and Combustible Liquids Code*

NFPA 30A, *Automotive and Marine Service Station Code*

NFPA 30B, *Manufacture and Storage of Aerosol Products*

NFPA 31, *Installation of Oil Burning Equipment*

NFPA 395, *Farm Storage of Flammable and Combustible Liquids*

It is the vapor from the evaporation of a flammable or combustible liquid when exposed to air or under the influence of heat, rather than the liquid itself, which burns or explodes when mixed with air in certain proportion in the presence of some source of ignition. There is a flammable range below which the vapor mixture is too lean to burn or explode, or above which the vapor mixture is too rich to burn or explode. (see Table 4.51.)

For gasoline, one of the most common and widely used flammable liquid, the flammable range is 1.4 and 7.6 percent by volume. When the vapor-air mixture is near either the lower flammable limit (LFL) or upper flammable limit (UFL), the explosion is less intense than when the mixture is in the intermediate range. The violence of the explosion depends on the concentration of the vapor as well as the quantity of vapor-air mixture and the type of container. Thus, it is important in controlling the fire hazard to store a flammable liquid in the proper type of closed container and minimize the exposure to air. When exposed to heat from a fire, a tank or other container may rupture with dangerous results if properly designed vents are not provided or if the exposed tank or container is not cooled by hose streams. The principal fire and explosion prevention measures under such circumstances are: (1) exclusion of sources of ignition, (2) exclusion of air, (3) keeping the liquid in a closed container, (4) ventilation to prevent the accumulation of vapor in the flammable range, and (5) use of an atmosphere of inert gas instead of air.

TABLE 4.51 Flash Points and Flammable Limits of Some Common Liquids and Gases

	Flash point		
Liquid (or gas at ordinary temps.)	°F	°C	Flammable limits, percent by volume
Acetylene	(Gas)		2.5–81.0*
Benzene	12	−11	1.3–7.1
Ether (ethyl ether)	−49	−45	1.9–36.0
Fuel oil			
Domestic, no. 2	100 (min.)	38	None at ordinary temperatures
Heavy, no. 5	130 (min.)	54	None at ordinary temperatures
Gasoline (high test)	−36	−38	1.4–7.4
Hydrogen	(Gas)		4.07–75.0
Jet fuel (A & A-1)	110 to 150	43 to 65	None at ordinary temperatures
Kerosene (Fuel oil, No. 1)	100 (min.)	38	0.7–5.0
LPG (propane-butane)	(Gas)		1.9–9.5
Lacquer solvent (butyl acet.)	72	22	1.7–7.6
Methane (natural gas)	(Gas)		5.0–15.0
Methyl alcohol	52	11	6.7–36.0
Turpentine	95	35	0.8–(undetermined)
Varsol (standard solv.)	110	43	0.7–5.0
Vegetable oil (cooking, peanut)	540	282	Ignition temperature = 833°F

Source: NFPA, Quincy, Mass.

The following system of classifying liquids is generally recognized:

Flammable Liquids. Flammable liquids are any liquid having a flash point below 100°F (38°C) and a boiling point below 100°F (38°C) and having a vapor pressure not exceeding 2068.6 mmHg at 100°F (38°C). Class I liquids include those having flash points below 100°F (38°C) and may be subdivided as follows:

Class IA includes those having flash points below 73°F (23°C) and a boiling point below 100°F (38°C).

Class IB includes those having flash points below 73°F (23°C) and a boiling point at or above 100°F (38°C).

Class IC includes those having flash points at or above 73°F (23°C) and below 100°F (38°C).

Combustible Liquids. Liquids with a flash point at or above 100°F (38°C) are referred to as combustible liquids. They are:

Class II liquids include those having flash points at or above 100°F (38°C) and below 140°F (60°C).

Class IIIA liquids include those having flash points at or above 140°F (60°C) and below 200°F (93°C).

Class IIIB liquids include those having flash points at or above 200°F (93°C).

Some typical liquids would be classed as follows:

Denatured alcohol	Class IB
Fuel oil	Class II
Gasoline	Class IB
Kerosene	Class II

Peanut oil	Class IIIB
Turpentine	Class IC
Paraffin wax	Class IIIB

Proper storage and handling of flammable and combustible liquids are necessary to prevent fire or explosion. Ventilation to prevent accumulations of flammable vapors is of primary importance because there is the possibility of breaks or leaks in the storage and handling in a closed system. It is important to eliminate possible sources of ignition in an area where flammable liquids are stored, handled, or used.

Ventilation of an area where flammable liquids are manufactured or used can be accomplished by natural or mechanical means. Wherever possible, equipment such as compressors, stills, and pumps should be located in a spacious, open area. Most flammable liquids produce heavier-than-air vapors that flow along the ground or floor and settle in depressions. These can travel long distances and be ignited and flash back from a point remote from the origin of the vapors. NFPA 30, *Flammable and Combustible Liquids Code,* is the accepted national standard for fire protection of such liquids.

NFPA 30, *Flammable and Combustible Liquids Code,* specifies the construction, installation, spacing, venting, and diking of aboveground and underground storage tanks; container storage in buildings and warehouses; loading and unloading practices; and safeguards for dispensing of liquids. Transportation of liquids by truck, rail, vessel, and pipeline is under the jurisdiction of the U.S. Department of Transportation.

Table 4.52 provides storage limitations for flammable and combustible liquids in inside rooms, which are defined by NFPA 30 as storage rooms that have no exterior walls, hence no direct access from outside that can be used by emergency response personnel. Table 4.53 provides limitations for unprotected storage of flammable and combustible liquids in storage rooms that have access from outside, storage buildings, and warehouses. Section 4.8 of NFPA 30 includes storage requirements and limitations for flammable and combustible liquids that are protected by a properly designed and installed automatic fire sprinkler system or foam-water fire sprinkler system. The protection system design criteria, storage arrangements, class of liquid, and type of container vary considerably, and the information is too voluminous to be included here. Refer to Section 4.8 of NFPA 30 for this information.

TABLE 4.52 Storage Limitations for Inside Rooms

Total floor area, ft^2	Automatic fire protection provided?*	Total allowable quantity, gal/ft^2 of floor area
≤150	No	2
	Yes	5
>150 and ≤150	No	4
	Yes	10

For SI units, 1 ft^2 = 0.09 m^2; 1 gal = 3.8 L.

* The fire protection system shall be automatic sprinklers, water spray, carbon dioxide, dry chemical, or other approved system. (See Section 4.8 of NFPA 30.)

Source: NFPA 30-2000, *Flammable and Combustible Liquids Code,* Table 4.4.4.2, NFPA International, Quincy, Mass.

Gases

There are many kinds of materials that exist in the form of gas. In general, gases are thought of and described when the substance exists in a gaseous state at normal temperature and pres-

TABLE 4.53 Indoor Unprotected Storage of Liquids in Containers, Portable Tanks, and Intermediate Bulk Containers (IBCs)

	Container storage			Portable tank/ metallic IBC storage			Rigid plastic and composite IBCs		
Class	Maximum pile height, ft	Maximum quantity per pile, gal	Maximum total quantity gal	Maximum pile height, ft	Maximum quantity per pile, gal	Maximum total quantity, gal	Maximum pile height, ft	Maximum quantity per pile, gal	Maximum total quantity, gal
IA	5	660	660	—	NP*	—	—	—	—
IB	5	1,375	1,875	7	2,000	2,000	—	—	—
IC	5	2,750	2,750	7	4,000	4,000	—	—	—
II	10	4,125	8,250	7	5,500	11,000	7	4,125	8,250
IIIA	15	13,750	27,500	7	22,000	44,000	7	13,750	27,500
IIIB	15	13,750	55,000	7	22,000	88,000	7	13,750	55,000

For SI units, 1 ft = 0.3 m; 1 gal = 3.8 L.
* Not permitted.
Source: NFPA 30-2000, *Flammable and Combustible Liquids Code,* Table 4.4.4.1, NFPA International, Quincy, Mass.

sure 70°F (21°C) and 114.7 psia (101,430 N/m^2). NFPA standards that provide detailed information on this subject include:

NFPA 50, *Bulk Oxygen Systems at Consumer Sites*
NFPA 50A, *Gaseous Hydrogen Systems at Consumer Sites*
NFPA 50B, *Liquefied Hydrogen Systems at Consumer Sites*
NFPA 51, *Oxygen-Fuel Gas Systems, Welding and Cutting*
NFPA 51A, *Acetylene Cylinder Charging Plants*
NFPA 54, *National Fuel Gas Code*
NFPA 55, *Compressed and Liquefied Gases in Portable Cylinders*
NFPA 58, *Liquefied Petroleum Gas Code*

Classification by Chemical Properties. Gases can be broadly classified according to chemical properties, physical properties, or usage. Classification by chemical properties helps to define the hazards of gases to people and in fires.

Flammable Gases. Any gas that will burn in the normal concentrations of oxygen in the air is a flammable gas. Like flammable liquid vapors, the burning of this gas in air is in a range of gas-air mixture (the flammable range). See Table 4.54.

Nonflammable Gases. Nonflammable gases will not burn in air or in any concentration of oxygen. A number of nonflammable gases, however, will support combustion. Such gases are often referred to as *oxidizers* or *oxidizing gases.* A common oxidizer is oxygen or oxygen in a mixture with other gases.

Nonflammable gases that will not support combustion are usually called *inert gases.* Common inert gases are nitrogen, carbon dioxide, and sulfur dioxide.

Toxic Gases. Toxic gases endanger life when inhaled. Gases such as chlorine, hydrogen sulfide, sulfur dioxide, ammonia, and carbon monoxide are poisonous or irritating when inhaled.

Reactive Gases. Reactive gases react with other materials or within themselves by a reaction other than burning. When exposed to heat and shock, some reactive gases rearrange themselves chemically. Such gases can produce hazardous quantities of heat or reaction products. Fluorine is a highly reactive gas. At normal temperatures and pressures it will react with most organic and inorganic substances, often fast enough to result in flaming. Other examples of reactive gases are acetylene and vinyl chloride.

TABLE 4.54 Combustion Properties of Common Flammable Gases

Gas	Btu/ft³ (gross)	mJ/m³ (gross)	Limits of flammability, % by volume in air: Lower	Upper	Specific gravity (air = 1.0)	Air needed to burn 1 ft³ of gas, ft³	Air needed to burn 1 m³ of gas, m³	Ignition temperature: °F	°C
Natural gas									
High inert type*	958–1051	35.7–39.2	4.5	14.0	0.660–0.708	9.2	9.2		
High methane type†	1008–1071	37.6–39.9	4.7	15.0	.590–.614	10.2	10.2	900–1170	482–632
High Btu type‡	1071–1124	39.9–41.9	4.7	14.5	.620–.719	9.4	9.4	—	—
Blast furnace gas	81–111	3.0–4.1	33.2	71.3	1.04–1.00	0.8	0.8	—	—
Coke oven gas	575	21.4	4.4	34.0	.38	4.7	4.7	—	—
Propane (commercial)	2516	93.7	2.15	9.6	1.52	24.0	24.0	920–1120	493–604
Butane (commercial)	3300	122.9	1.9	8.5	2.0	31.0	31.0	900–1000	482–538
Sewage gas	670	24.9	6.0	17.0	0.79	6.5	6.5	—	—
Acetylene	1499	208.1	2.5	81.0	0.91	11.9	11.9	581	305
Hydrogen	325	12.1	4.0	75.0	0.07	2.4	2.4	932	500
Anhydrous ammonia	386	14.4	16.0	25.0	0.60	8.3	8.3	1204	651
Carbon monoxide	314	11.7	12.5	74.0	0.97	2.4	2.4	1128	609
Ethylene	1600	59.6	2.7	36.0	0.98	14.3	14.3	914	490
Methylacetylene, propadiene, stabilized§	2450	91.3	3.4	10.8	1.48	—	—	850	454

* Typical composition CH_4 71.9–83.2%; N_2 6.3–16.20%.
† Typical composition CH_4 87.6–95.7%; N_2 0.1–2.39%.
‡ Typical composition CH_4 85.0–90.1%; N_2 1.2–7.5%.
§ MAPPd gas.
Source: From Arthur E. Cote, ed., *Fire Protection Handbook,* NFPA, Quincy, Mass., 1997.

Classification by Physical Properties. Gases can also be classified by their physical properties. They can be compressed, liquefied, or cryogenic.

Compressed Gases. A compressed gas is at normal temperature inside a gas container and exists solely in the gaseous state under pressure; common compressed gases are hydrogen, oxygen, acetylene, and ethylene.

Liquefied Gases. Liquefied gases can be liquefied relatively easily and stored at ordinary temperatures at relatively high pressure. Liquefied gas exists in both liquid and gaseous states; at storage pressure, both the liquid and gas in the liquefied-gas container are in equilibrium and will remain so as long as any liquid remains in the container. Liquefied gas is more concentrated than compressed gas.

Cryogenic Gases. Cryogenic gases are stored in a completely liquid state. They must be maintained in their containers as low-temperature liquids at relatively low pressure. Cryogenic gases must be stored in special containers that allow the gas from the liquid to escape in order to prevent a pressure buildup caused by the production of the gaseous state within the container, which would result in container failure.

Classification by Usage. An understanding of gases as they are classified by usage is important to those involved in fire protection because the terms of these classifications are used in codes, standards, and general industrial and medical terminology (Table 4.54).

Fuel Gases. These gases are customarily used for burning with air to produce heat which in turn is used as a source of heat (comfort and process), power, or light. The principal and most widely used fuel gases are natural gas and the liquefied petroleum gases, butane and propane.

Industrial Gases. These are classified by chemical properties customarily used in industrial processes, for welding and cutting, heat treating, chemical processing, refrigeration, water treatment, etc.

Medicinal Gases. These are for medical purposes such as anesthesia and respiratory therapy. Cyclopropane, oxygen, and nitrous oxide gases are common medical gases.

Corrosive Chemicals

Corrosive chemicals are usually strong oxidizing agents that can increase fire hazards. Caustics, which are classified as water- and air-reactive chemicals, are also corrosive. Two types of corrosive chemicals that deserve consideration are inorganic acids and halogens.

Concentrated aqueous inorganic acids are not in themselves combustible; however, in addition to their corrosive and destructive effect on living tissue, their chief fire hazard results from the possibility of their mixing with combustible materials or other chemicals, resulting in fire or explosion. Almost all corrosive chemicals are strong oxidizing agents.

Halogens are salt-producing chemicals that are very active. They are noncombustible but will support combustion; presence of halogens causes turpentine, phosphorus, and finely divided metals to ignite spontaneously. The fumes are poisonous and corrosive.

Storage of corrosive chemicals should be provided with two considerations in mind: (1) protection against the damaging effect of corrosive chemicals on living tissue, and (2) guarding against any fire and explosion hazard that might be associated with the corrosive chemical.

Inorganic Acids. These should be stored in cool, well-ventilated areas that are not exposed to the sun or other chemical and waste materials; they should be protected from freezing temperatures. Water in spray form is the recommended procedure for fighting fires in inorganic acid storage areas; in fires involving perchloric acid, extra care should be taken, since this may mix with other organic materials and result in an explosion.

Halogens. Fluorine and chlorine should be stored in special containers; fluorine may be safely stored in nickel or Monel cylinders. Impurities or moisture in the cylinder may cause an explosive reaction; chlorine, a serious inhalation hazard, should be stored in areas where ventilation is a prime consideration. Where chlorine leakage is suspected, use a self-contained breathing apparatus; fluorine requires protection of the self-contained breathing apparatus and special protective clothing.

Very often, corrosive vapors are a by-product of an industrial or chemical process. Ducts must be used to carry these vapors safely from the area. The type of duct used is determined by the vapor. Heavier-gauge metal may be sufficient, although a protective coating or special lining may be required in the ducts. Stainless steel, asbestos cement, and plastic linings have been used with success, depending on the corrosive vapor.

Radioactive Materials

The main concern in fire protection of radioactive materials is to prevent the release (or control the release) of these materials during fire extinguishment. Although fire-protection operations are similar to those used when nonradioactive materials are involved, fire involving buildings or areas containing radioactive materials presents two additional considerations: (1) the presence of harmful radioactive materials might necessitate changing normal fire-fighting procedures and (2) because of the presence of radioactive matter, delay in salvage and resumption of normal operations may occur. NFPA standards that address radioactive materials include:

NFPA 801, *Facilities Handling Radioactive Materials*

NFPA 803, *Light Water Nuclear Power Plants*

DESIGN AND CONSTRUCTION FOR FIRE SAFETY

Fire-safety objectives must be determined before a facility is designed. The architectural design and the building methods and materials used for a structure will often determine how a fire will be confined or will spread. At times, code compliance is not sufficient to meet the level of risk acceptable to a particular organization.

Fire-safety design decisions are necessary in at least three objective areas: life safety, property protection, and continuity of operations. A fourth objective area that is gaining prominence in recent years is the protection of the environment, such as would be considered for a hazard like large outdoor used-rubber-tire storage yards. Building codes, fire codes, and, increasingly, systems-analysis techniques are used by fire-protection engineers to determine the level of protection to meet design objectives.

Building and Site Planning for Fire Safety

Two categories of decisions should be made in the design process to provide effective fire-safe design: interior building functions and exterior site planning. Building fire defenses, active and passive, should be designed so that the building assists in the suppression of fire while providing for safety of occupants.

Fire-Safety Planning for Buildings. Interior layout, circulation patterns, finish material, and building services are all important fire-safety considerations in building design. Building design also has a significant influence on the efficiency of fire-department operations.

Fire-fighting accessibility to building's interior includes access to the building itself as well as access to the interior of the building. Spaces in which fire-fighting access and operations are restricted because of architectural, engineering, or functional requirements should be provided with effective protection. A complete automatic sprinkler system is often the best solution, providing both life safety and property protection.

Ventilation is of vital importance in removing smoke, gases, and heat. Appropriate skylights, roof hatches, emergency escape exits, and similar devices should be provided when the building is constructed. NFPA has several design documents that address smoke control and ventilation.

Ventilation of building spaces performs the following important functions:

1. Protection of life by removing or diverting gases and smoke to allow for egress
2. Control of the spread and direction of fire by setting up air currents that limit fire movement.
3. Provision of a release for unburned, combustible gases before they acquire a flammable mixture, thus avoiding a backdraft explosion and reducing burning temperatures

NFPA 204, *Smoke and Heat Venting Guide,* recommends automatic smoke venting of large industrial buildings. NFPA 92A (*Smoke Control Systems*) and 92B (*Smoke Management Systems in Malls, Atria, and Large Areas*) present design methods for smoke-control systems in other types of buildings.

In large-area buildings, curtain boards are an important element, unless vented areas are subdivided by means of walls or partitions. The function of curtain boards is to delay and limit the horizontal spread of heat by providing the horizontal confinement needed to obtain the desired *stack* action. The depth of such curtain boards largely determines the height of the stack, which affects the capacity of the vent. If an area is protected by automatic sprinklers, curtain boards have added values: confinement of heat to speed up operation of sprinklers over the fire and obstructed lateral spread of heat to minimize the operation of an excessive number of sprinklers. See Fig. 4.81.

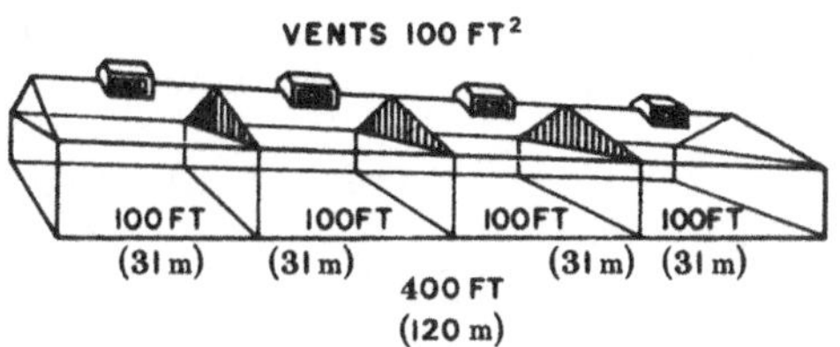

FIGURE 4.81 Curtain boards and roof vents. *(From* Fire Protection Handbook, *14th ed., G. P. McKinnon, ed., NFPA, Quincy, Mass., 1976.)*

Fire-Safety Planning for Sites. Proper building design for fire protection should include a number of factors outside the building itself. The site on which the building is located will influence the design. Among the more significant features are traffic and transportation conditions, fire-department accessibility, and fire water supply. Inadequately sized water mains and poor spacing of hydrants have contributed to the loss of many buildings.

Is the building truly accessible to fire apparatus? Ideal accessibility occurs where a building can be approached from all sides by fire-department apparatus; congested areas, topography, or buildings and structures located appreciable distances away from the street can cause difficulty and prevent effective use of fire apparatus. Inadequate attention to site details can place the building in an unnecessarily vulnerable position; if fire defenses are compromised by preventing adequate fire-department access, the building itself must make up the difference in more complete internal protection.

Are the water mains adequate, and are the hydrants properly located? The number, location, and spacing of hydrants and the size of the water mains are vital considerations; consult local standards and insurance requirements.

The Built Environment

Interior finish is defined as those materials that make up the exposed interior surface of wall, ceiling, and floor constructions. The common interior-finish materials are wood, plywood, plaster, wallboards, acoustical tile, insulating and decorative finishes, plastics, and various wall coverings.

Some building codes, and NFPA 101, *Life Safety Code,* include floor coverings under their definition of interior finishes that vary according to occupancy type.

Fire Tests

It is possible to estimate the damage that fire can cause to a building by studying: (1) the amount and kind of combustible materials in the building and (2) the way they are distributed throughout the building. These two factors help to indicate the rate of combustion, the duration of the fire, and the degree of difficulty in extinguishing the fire.

The effects of fire on the components of a building (such as the columns, floors, walls, partitions, and ceiling or roof assemblies) are tested against both time and temperature. Results of the tests, specific to the construction configuration, are recorded in hours or minutes and indicate the duration of fire resistance.

Ratings for flame spread of interior finish materials have been established with the use of the 25-ft (7.6-m) tunnel developed by A. J. Steiner at Underwriters Laboratories Inc. and defined in NFPA 255, *Surface Burning Characteristics of Building Materials.* These ratings are used in NFPA 101, *Life Safety Code,* and in other codes to indicate the area in which finishes of varying flame spread characteristics may be used (See Table 4.55). The classifications used in NFPA 101, *Life Safety Code,* are as follows:

Class	Flame-spread range
A	0–25
B	26–75
C	76–200

The higher the flame spread, the greater the hazard. Exit stairway enclosures typically require Class A interior finishes.

Materials are measured on a relative scale with inorganic-reinforced cement board rated 0 and red oak flooring rated at approximately 100. Some highly combustible wallboards received ratings as high as 1500.

One of the best sources of information showing the wide variety of building assemblies and giving the fire-resistance ratings of beams, columns, floors, walls, and partitions is the Underwriters Laboratories Inc. *Fire Resistance Directory,* published annually.

Confinement of Smoke and Fire

Design criteria for plant facilities are generally based on estimated fire severity (see Table 4.56). Specific industrial-hazard fire-severity data may be obtained from insurance organizations (see Table 4.57 and Fig. 4.82).

Fire doors are the most widely used and accepted means of protecting vertical and horizontal openings. Suitability of fire doors is determined by nationally recognized testing laboratories; doors that have not been tested cannot be relied on for effective protection. Doors are tested with the frame, hardware, wired-glass panels, and other accessories necessary to complete the installation to simulate expected installed exposure.

Most building codes reference NFPA 80, *Standard for Fire Doors and Fire Windows.* This standard establishes the minimum rating for the most commonly encountered types of openings in walls stated in hours. With regard to fire doors, the following is a description of various types:

Composite doors. Typically flush design and consisting of a manufactured core material with chemically impregnated wood-edge banding and untreated wood-face veneers, or laminated plastic faces, or surrounded by and encased in steel.

Hollow-metal doors. Typically flush and paneled design consisting of metal covered with steel of 24 gauge or lighter.

Sheet-metal doors. Typically formed no. 22 gauge or lighter steel and of the corrugated, flush, and paneled designs.

Rolling steel doors. These are of the interlocking steel slat design or plate-steel construction.

Tin-clad doors. Typically of two- or three-ply wood-core construction, covered with no. 30 gauge galvanized steel or terneplate (maximum size 14 by 20 in) or no. 24 gauge galvanized steel sheets not more than 48 in wide.

Curtain-type doors. Typically consisting of interlocking steel blades or a continuous formed-spring-steel curtain in a steel frame.

The suitability of a fire door is based on the fire-barrier resistance rating stated in time necessary life-safety or property-protection requirements.

NFPA 80, *Fire Doors and Fire Windows,* lists requirements for the installation of doors, windows, and shutters, and also specifies how the opening should be constructed and how the door or window should be mounted, equipped, and operated.

TABLE 4.55 Summary of Life Safety Code Requirements for Interior Finish

Occupancy	Exits	Access to exits	Other spaces
Assembly—new*			
Class A or B	A	A or B	A or B
Class C	A	A or B	A, B, or C
Assembly—existing*			
Class A or B	A	A or B	A or B
Class C	A	A or B	A, B, or C
Educational—new*	A	A or B	A or B C on low partitions**
Educational—existing*	A	A or B	A, B, or C
Day-care centers—new	A I or II	A I or II	A or B
Day-care centers—existing	A or B	A or B	A or B
Group day-care homes—new	A or B	A or B	A, B, or C
Group day-care homes—existing	A or B	A, B, or C	A, B, or C
Family day-care homes	A or B	A, B, or C	A, B, or C
Health care—new (A.S. mandatory)	A or B	A or B C lower portion of corridor wall**	A C in small individual rooms**
Health care—existing	A or B	A or B	A or B
Detention and correctional—new	A* I	A* I	A, B, or C
Detention and correctional—existing	A or B* I or II	A or B* I or II	A, B, or C
Residential, hotels, and dormitories—new	A I or II	A or B I or II	A, B, or C
Residential, hotels, and dormitories—existing	A or B I or II†	A or B I or II†	A, B, or C
Residential, apartment buildings—new	A I or II**	A or B I or II**	A, B, or C
Residential, apartment buildings—existing	A or B I or II†	A or B I or II†	A, B, or C
Residential, board and care‡	(See *Life Safety Code*, Chaps. 22 and 23)		
Residential, 1- and 2-family, lodging or rooming houses	A, B, or C	A, B, or C	A, B, or C
Mercantile—new*	A or B	A or B	A or B
Mercantile—existing Class A or B*	A or B	A or B	Ceilings—A or B Existing on walls —A, B, or C
Mercantile—existing Class C*	A, B, or C	A, B, or C	A, B, or C
Office—new	A or B I or II	A or B I or II	A, B, or C
Office—existing	A or B	A or B	A, B, or C
Industrial	A or B	A, B, or C	A, B, or C
Storage	A or B	A, B, or C	A, B, or C
Unusual structures	A or B	A, B, or C	A, B, or C

* Interior wall and ceiling finish in corridors, exits, and any space not separated from the corridors and exits by a partition capable or retarding the passage of smoke shall meet this interior finish classification.

** Refer to the *Life Safety Code* for details.

A.S. = automatic sprinklers

N.R. = not regulated

† Previous installed floor coverings may be continued in use, subject to the approval of the authority having jurisdiction.

‡ See Chap. 21 of NFPA 101.

Notes:

Class A interior wall and ceiling finish—flame spread 0–25, smoke developed 0–450.

Class B interior wall and ceiling finish—flame spread 26–75, smoke developed 0–450.

Class C interior wall and ceiling finish—flame spread 76–200, smoke developed 0–450.

Class I Interior Floor Finish—minimum 0.45 W/cm^2.

Class II Interior Floor Finish—minimum 0.22 W/cm^3.

Automatic Sprinklers—where a complete standard system of automatic sprinklers is installed, interior finish with flame spread rating not over Class C may be used in any location where Class B is normally specified and with rating of Class B in any location where Class A is normally specified; similarly, Class II interior floor finish may be used in any location where Class I is normally specified and no critical radiant flux rating is required where Class II is normally specified.

Exposed portions of structural members complying with the requirements for heavy timber construction may be permitted.

Source: From Arthur E. Cote, ed., *Fire Protection Handbook*, NFPA, Quincy, Mass., 1997.

TABLE 4.56 Estimated Fire Severity for Offices and Light Commercial Occupancies

Combustible content, total, including finish, floor and trim, lb/ft^2	Heat potential assumed Btu/ft^2*	Equivalent fire severity, approximately equivalent to that of test under standard curve for the following periods
5	40,000	30 min
10	80,000	1 h
15	120,000	1½ h
20	160,000	2 h
30	240,000	3 h
40	320,000	4½ h
50	380,000	7 h
60	432,000	8 h
70	500,000	9 h

* Heat of combustion of contents taken at 8000 Btu/lb up to 40 lb/ft^2; 7600 Btu/lb for 50 lb, and 7200 Btu for 60 lb and more to allow for relatively greater proportion of paper. The weights contemplated by the tables are those of ordinary combustible materials, such as wood, paper, or textiles.

Source: From Gordon P. McKinnon, ed., *Fire Protection Handbook,* NFPA, Quincy, Mass., 1976. Data applying to fire-resistive buildings with combustible furniture and shelving.

TABLE 4.57 Fire Severity Expected by Occupancy

(*See Fig. 4.82*)

Temperature curve A (slight)
Well-arranged office, metal furniture, noncombustible building
Welding areas containing slight combustibles
Noncombustible power house
Noncombustible buildings, slight amount of combustible occupancy
Temperature curve B (moderate)
Cotton and waste-paper storage (baled) and well-arranged, noncombustible building
Paper-making processes, noncombustible building
Noncombustible institutional buildings with combustible occupancy
Temperature curve C (moderately severe)
Well-arranged combustible storage, e.g., wooden patterns, noncombustible buildings
Machine shop having noncombustible floors
Temperature curve D (severe)
Manufacturing areas, combustible products, noncombustible building
Congested combustible storage areas, noncombustible building
Temperature curve E (standard fire exposure—severe)
Flammable liquids
Woodworking areas
Office, combustible furniture and buildings
Paper working, printing, etc.
Furniture manufacturing and finishing
Machine shop having combustible floors

Source: NFPA, Quincy, Mass.

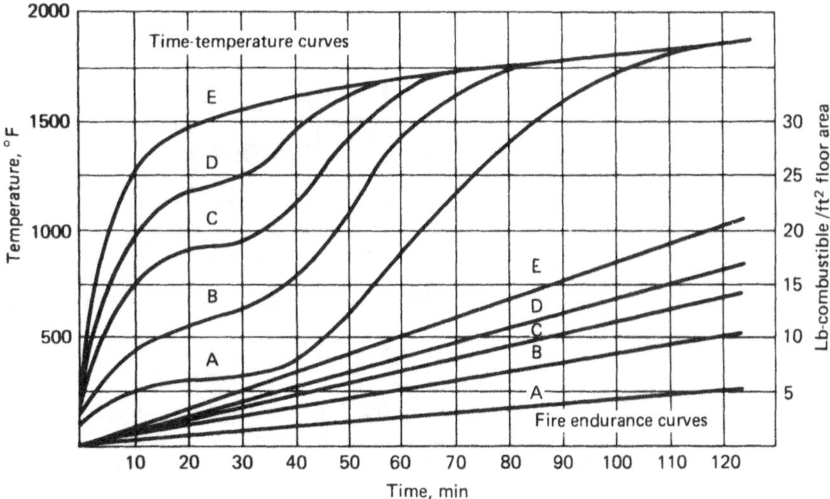

FIGURE 4.82 Possible classification of building contents for fire severity and duration. The straight lines indicate the length of fire endurance based on amounts of combustibles involved. The curved lines indicate the severity expected for the various occupancies (see Table 4.57). There is no direct relationship between the straight and curved lines, but, for example, 10 lb of combustibles per square foot will produce a 90-min fire in a C occupancy, and a fire severity following the time-temperature curve C might be expected. (*From R. Tuve,* Principles of Fire Protection Chemistry, *NFPA, Quincy, Mass., 1998.*)

FIRE-DETECTION AND -ALARM SYSTEMS

There are several general systems and many devices which can be effectively used to detect fire and transmit a warning. This section briefly describes this equipment. Details of this subject are addressed by NFPA 72, *National Fire Alarm Code.*

Heat Detectors

Heat-detection devices are categorized in two ways: (1) those that respond when the detection element reaches a predetermined temperature (fixed-temperature types), and (2) those that respond to an increase in heat at a rate greater than some predetermined value (rate-of-rise types). Some devices combine both principles. The same principles apply whether the devices are of the spot-pattern type, in which the thermally sensitive element is a unit, or the line-pattern type, in which the element is continuous along a line or a circuit.

Fixed-Temperature Detectors. Thermostats are one of the most widely used fixed-temperature heat detectors in signaling systems.

Bimetallic Thermostats. The common form of thermostat is the bimetallic type that utilizes the different coefficients of expansion of two metals under heat to cause a movement resulting in closing of electrical contacts (see Fig. 4.83).

Snap-Action Disk Thermostats. A metal disk goes from concave to convex when the temperature rating of the thermostat is reached. One special advantage of these thermostats is that when the temperature goes down, they are restored to their original condition.

Line Thermostats. The *thermostat cable* is a line type of thermostat. The cable is made up of two metals separated from each other by a heat-sensitive covering applied directly to the wires. When the rated temperature is reached, the covering melts and the two wires come in contact to initiate an alarm. The section of wire affected must be replaced after operation.

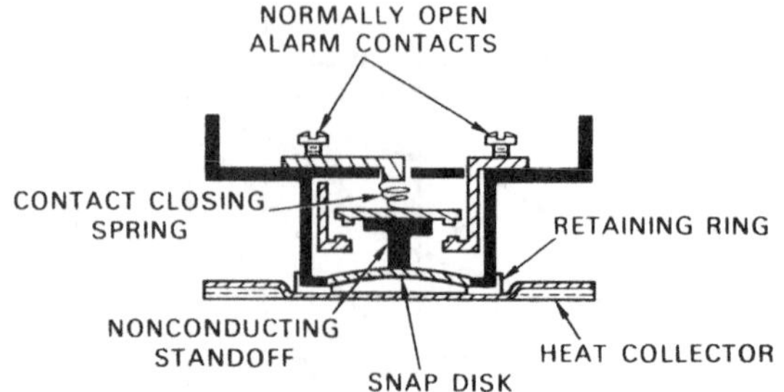

FIGURE 4.83 Spot-type, fixed-temperature snap-disk detector.

Other Types. Other forms of fixed-temperature heat detectors are the *fusible link,* occasionally employed to restrain operation of an electrical switch until the point of fusion is reached, and the *quartzoid bulb thermostat,* which depends on removal of the restriction by breaking the bulb. Both of these units require replacement after operation.

Rate-of-Rise Detectors. Fire detectors that operate on the rate-of-rise principle function when the rate of temperature increase exceeds a predetermined rate. Detectors of this type combine two functioning elements, one of which initiates an alarm on a rapid rise of temperature, while the other acts to delay or prevent an alarm on a slow temperature rise. Advantages of rate-of-rise devices are: (1) they can be set to operate more rapidly under most conditions than can fixed-point devices; (2) they are effective across a wide range of ambient temperatures; (3) they recycle rapidly and are usually readily available for continued service; (4) they tolerate slow increases in ambient temperature without giving an alarm. The disadvantages of rate-of-rise detectors for some applications are their susceptibility to false alarms where there is a rapidly increasing temperature and their possible failure to respond to a fire that propagates very slowly.

Pneumatic-tube detectors operate on the rate-of-rise principle. When the temperature increases at a certain rate, the air in the tube expands and causes a diaphragm to move and close a circuit, thus causing an alarm. The device will not cause an alarm if the temperature rise is slow.

Combined Rate-of-Rise and Fixed-Temperature Detectors. Thermostats have been developed to take advantage of the rate-of-rise feature to sense a fast-developing fire; the fixed-temperature part takes care of a fire whose growth is slow. The typical form of the rate-of-rise thermostat is a vented air chamber that heats up in a flexible diaphragm carrying electric contacts (see Fig. 4.84). Heat outside the chamber causes air within the chamber to expand. When such expansion exceeds the capacity of the vent to relieve pressure, the diaphragm is flexed, thus closing the electric contacts. Slow changes in ambient temperature near the chamber allow it to "breathe" through its vent, and the diaphragm is not moved sufficiently to cause an alarm.

Rate-Compensation (Anticipation and Differentiation) Devices. These provide an assured actuation at some predetermined maximum temperature and compensate for changes in rates of temperature rise (see Fig. 4.85). This essentially compensates for any expected thermal lag that would temporarily delay detector operation.

Smoke Detectors

There are four types of smoke detectors: (1) photoelectric, (2) beam-type, (3) ionization, and (4) sampling detectors.

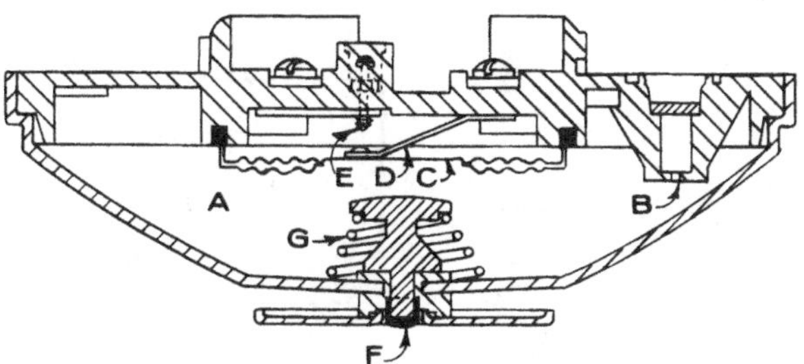

FIGURE 4.84 A spot-type combination rate-of-rise, fixed-temperature device. The air in chamber A expands more rapidly than it can escape from vent B. This causes pressure to close electrical contact D between diagram C and insulated screw E. Fixed-temperature operation occurs when fusible alloy F melts releasing spring G which depresses the diaphragm, closing the contact points.

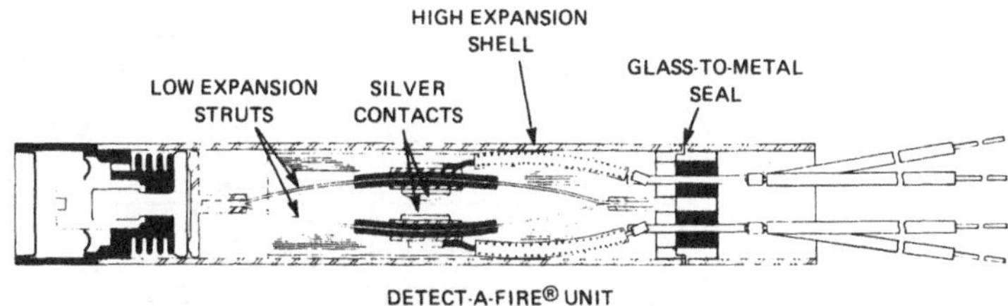

FIGURE 4.85 Rate-compensation heat detector. (*From* Fire Protection Handbook, *14th ed., G. P. McKinnon, ed., NFPA, Quincy, Mass., 1976.*)

Photoelectric Detectors. These detectors operate on a beam of light. The smoke either obscures a beam of light directly or enters a refraction chamber where the smoke reflects the light into the photocell. The change in electric current resulting from either partial obscuring of a photoelectric beam by smoke particles or the scattering of light onto a photosensitive device causes an alarm sound when the smoke reaches a sufficient density. (See Figs. 4.86*a* and 4.86*b,* respectively.)

Beam-Type Detectors. These employ a light beam that is carried between elements at extreme ends or sides of the protected area and crosses the area to be protected. The beam is projected into a photosensing cell. Smoke between the light source and the receiving photocell reduces the light that reaches the cell, activating the alarm. (See Fig.4.87.)

Ionization Detectors. These detectors consist of one or more ionization chambers and the necessary related amplification circuits. The ionization detector has as a sensing element, the ionization chamber, in which air is made electrically conductive (ionized) by a minute source of radioactive material. A voltage applied across the ionization chamber causes a very small electric current to flow as the ions travel to the electrode of opposite polarity. When smoke particles enter the chamber, they attach themselves to the ions and cause a reduction in mobility and thus a reduction in current flow. The reduced current flow increases the voltage on the electrodes which, when reaching a predetermined level, results in an alarm. (See Fig. 4.88.)

Sampling Detectors. These consist of tubing distributed from the detector unit into the area(s) to be protected. An air pump draws air from the protected area back to the detector. A high-intensity strobe, laser particle counter, or cloud-chamber smoke detector may be used

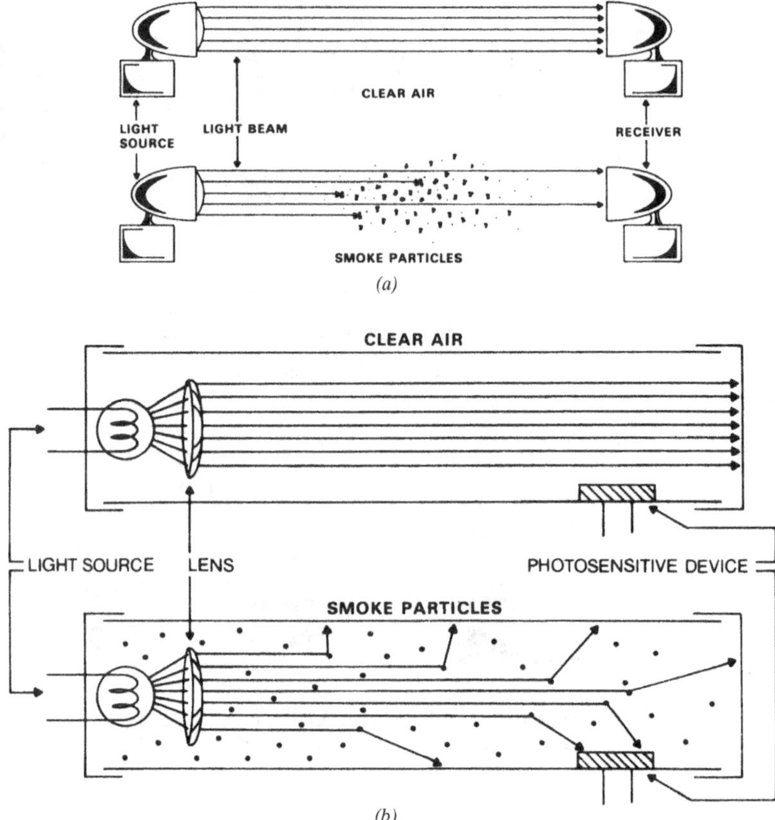

FIGURE 4.86 Principle of operation for (*a*) a photoelectric obscuration smoke detector and (*b*) a photoelectric scattering smoke detector.

as a sampling detector. The air pump draws a sample of air into a chamber for a highly sensitive analysis. These types of detectors are increasing in popularity.

Flame Detectors

There are four basic types of flame detectors: (1) infrared, (2) ultraviolet, (3) photoelectric, and (4) flame flicker.

Infrared and Ultraviolet. These detectors have sensing elements responsive to radiant energy outside the range of human vision.

Photoelectric. This type employs a photocell that either changes its electric conductivity or produces an electric potential when it is exposed to radiant energy.

Flame Flicker. This is a photoelectric type of detector that includes means to prevent response to visible light unless the observed light is modulated at a frequency characteristic of the flicker of a flame.

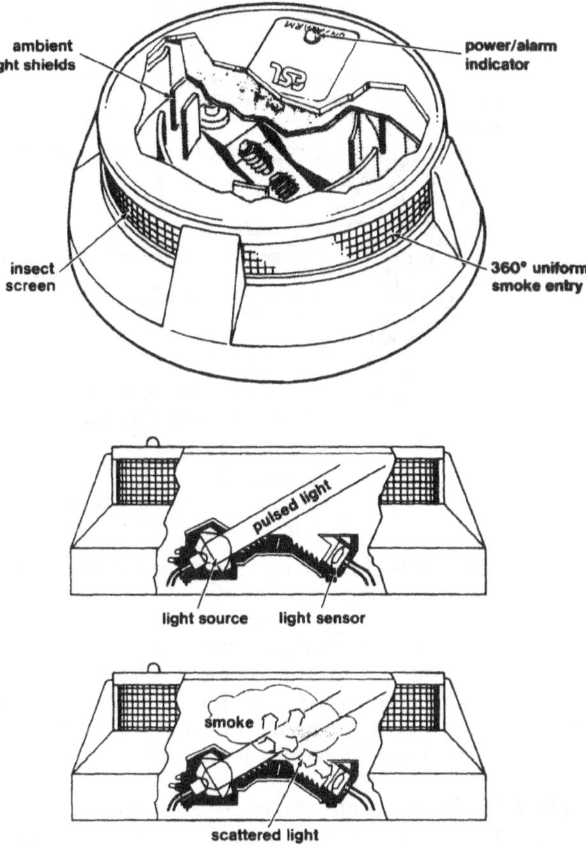

FIGURE 4.87 Cross-sectional view of a photoelectric light-scattering smoke detector. (*Electro Signal Lab., Inc.*)

Gas Detectors

Various detection devices will monitor the amount of flammable gases or vapors in an area. Portable gas detectors are used to detect the presence of combustible gas or vapor in basements, sewers, manholes, etc. Other devices will analyze the air samples brought into the device from various points. Gas and vapor testing equipment is valuable for preventing fires and explosions in petroleum and chemical plants and in industries where combustible vapors may be generated.

Fire-detection devices are usually installed in systems which combine manually activated fire-alarm stations and audible and visual warning devices. They may also be connected to fire-suppression systems in some hazardous areas.

Protective Signaling Systems

The detection and alarm systems in a plant building should be connected to a constantly supervised monitoring system. The most common systems are described in this section.

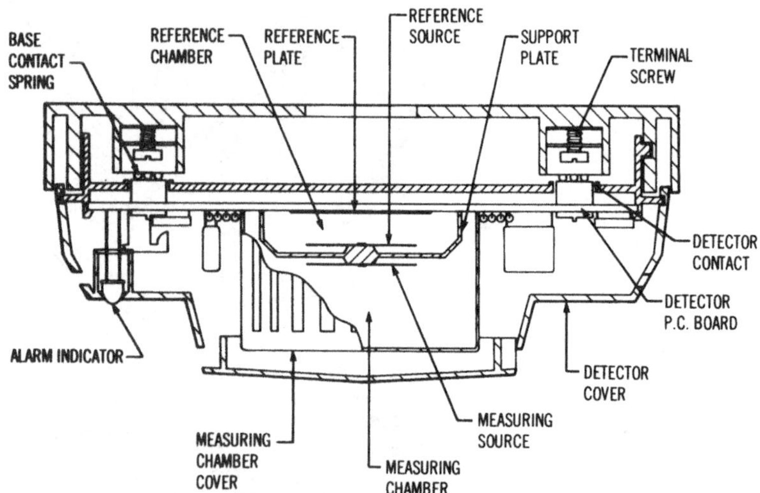

FIGURE 4.88 Cross-sectional view of an ionization smoke detector. (*Pyrotronics, Inc.*)

Local Systems. These systems produce a signal manually or automatically at the protective premises for an alarm of fire and for required supervisory services, including supervision of a security guard's rounds, supervision of sprinkler water-flow alarm service and of sprinkler systems, etc. Local systems are used for the protection of property and for the protection of life by indicating the necessity for evacuation of the building.

Proprietary Systems. Proprietary systems are used for individual properties where the system is under constant supervision by competent and experienced personnel in a central supervisory station at the property protected. Such systems are usually found in large industrial plants; signals are received at a central supervisory station where experienced operators are on duty at all times. The central supervisory station is under the control of the owner or occupant of the protected property and is usually on or near that property.

Remote-Station Systems. These are usually used to protect premises on which there is frequently no one present. The signal is received at fire-alarm headquarters or at the office of a communications agency, usually located at a distance from the protected property. Signals are transmitted and received on privately owned equipment; the agency receiving the signals may be a municipal fire department or a communications agency capable of receiving the signals and acting upon them.

Auxiliary Systems. This type connects devices in the protected plant with the municipal fire-alarm system. Alarms are received at fire-alarm headquarters on the same equipment and by the same alerting methods as alarms transmitted from municipal street boxes. Signals are recorded at a municipal fire department; connecting facilities between the protected property and the fire department are part of the municipal fire-alarm system. Devices in the protected plant are customarily owned and maintained by the property owner. Equipment that connects the devices to the city's circuits is owned and maintained by the municipality, or leased by it, as part of the municipal alarm system and limited to alarm service only.

Central Station Systems. Central station systems are operated by firms whose principal business is the furnishing and maintaining of supervised signaling service. The central station services properties subscribing to the service; alarm and signaling devices on the subscribers' property are connected to the central station where operators are on hand to receive the signal and take the appropriate action. Central station operators retransmit alarms to the fire department.

Standards for the installation and maintenance of fire detection and alarm equipment are listed in the last section of this chapter.

WATER-BASED FIRE-PROTECTION SYSTEMS

This section describes systems which suppress or control fires.

Water Supply

The most common type of fire-protection systems rely on water. Therefore it is essential that adequate supplies of water be provided and maintained. NFPA 24, *Installation of Private Fire Service Mains and their Appurtances,* provides requirements for on-site water supply systems.

The plant water-supply system or nearby public water supply will usually be the primary source used by the plant fire brigade or public fire department. Water must be provided with flows and pressure that are sufficient for supplying automatic sprinkler systems and fire hoses, in addition to normal plant requirements. When the public water supply is inadequate for plant protection, supplemental private supplies are necessary.

Pipe Networks. The minimum recommended pipe size for underground fire protection pipings is 6 in (15 cm). It is desirable to loop the pipes in a grid pattern (Fig.4.89) and minimize the friction loss wherever possible. Various pipe materials are available including steel, cast iron, and plastic for use underground. The selection of size and type of pipe should be based on the expected flows, soil conditions, and other similar factors.

Fire Hydrants. Fire hydrants are provided on public mains to allow the fire department to draw water with mobile pumpers to supply sprinkler and standpipe systems, as well as hose streams. Fire hydrants are provided on private mains to allow the fire brigade or fire department to supply hose streams, and to support sprinkler and standpipe systems with mobile pumpers.

Hydrants are available as wet-barrel (California, see Fig. 4.90) and dry-barrel (base valve, see Fig. 4.91) types. Dry-barrel hydrants are necessary where there is any chance of freezing.

Hydrants on plant pipe networks are normally located every 250 ft (76 m) and about 40 ft (12 m) from the buildings protected. They must be protected from damage by vehicles or machinery.

The available water flow for fire suppression is determined by flow testing the hydrant system. The water flow available at 20 psig (138 kPa) is determined, since this is the minimum pressure that usually must be maintained to satisfy water regulatory rules.

Valves in pipelines supplying fire-protection water are generally required to be indicating valves. These include underground gate valves with indicator post, underground butterfly valves with indicator post, and outside screw and yoke (OS&Y) gate valves.

Closed valves have been the primary cause of sprinkler systems' failing to control fires.

Fire Pumps. Fire pumps are essentially the same as typical water-supply pumps. Additional considerations for fire pumps are outlined in NFPA 20, *Installation of Centrifugal Fire Pumps.* Factors that should be considered with fire pumps include the following:

- Use of equipment listed for fire pumps
- Use of approved accessories

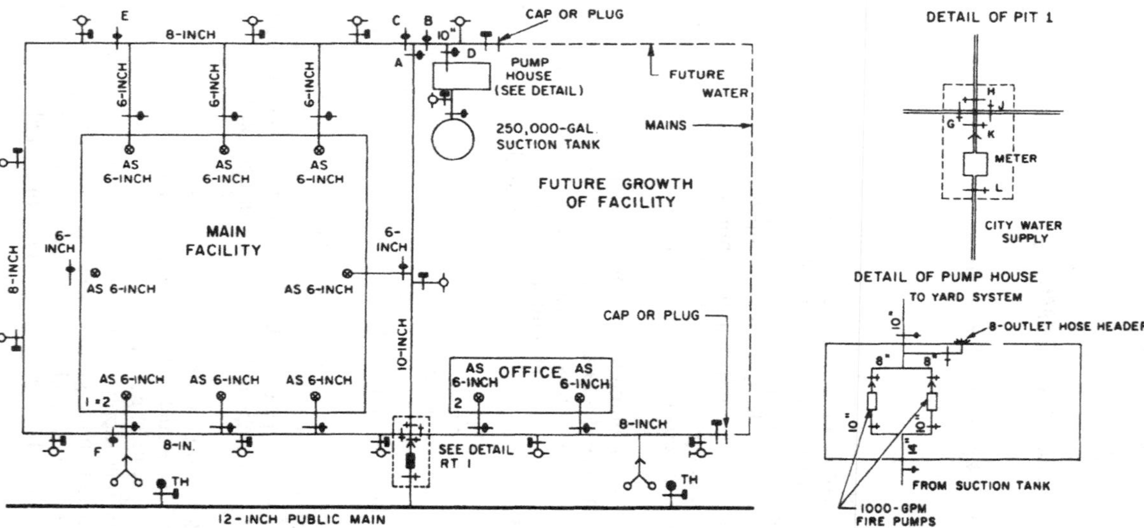

FIGURE 4.89 Water-pipe network. (*From Arthur E. Cote, ed.,* Fire Protection Handbook, *NFPA, Quincy, Mass., 1997.*)

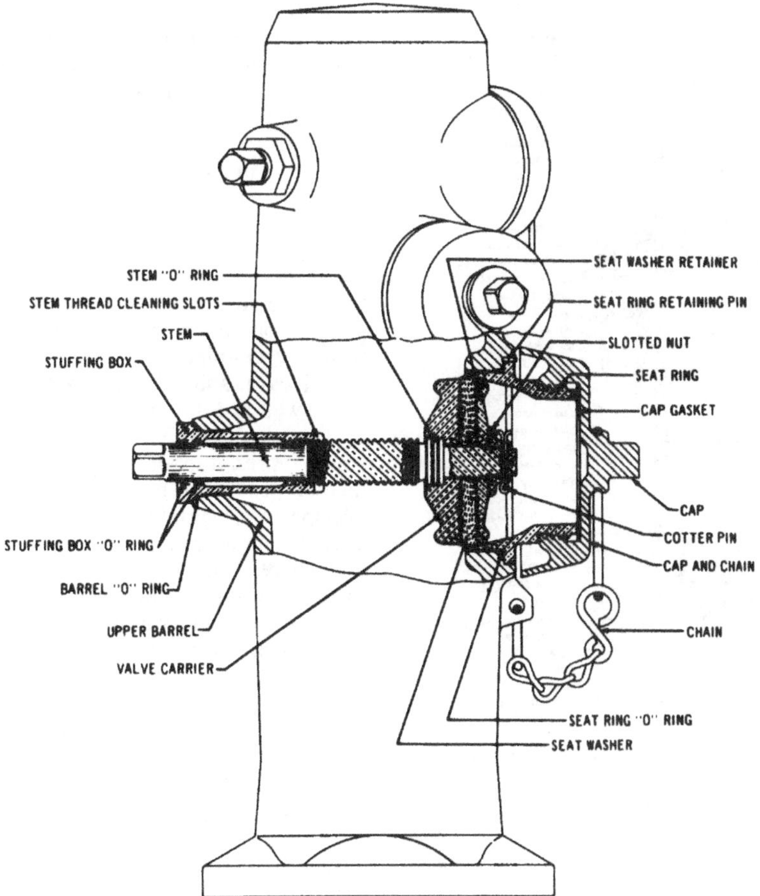

FIGURE 4.90 Wet-barrel hydrant. (*Mueller Co.; from Arthur E. Cote, ed.,* Fire Protection Handbook, *NFPA, Quincy, Mass., 1997.*)

- Adequate capacity to meet fire-flow demands
- Selection of fire pump driver based on reliability, adequacy, economy, and safety of power source
- Automatic operation
- Safe location for uninterrupted service
- Weekly and annual testing
- Maintenance

Sprinkler Systems

A sprinkler system, for fire-protection purposes, is an integrated system of underground and overhead piping designed in accordance with fire-protection engineering standards. The installation includes one or more water supplies. The portion of the sprinkler system aboveground is a network of specially sized or hydraulically designed piping installed in a building,

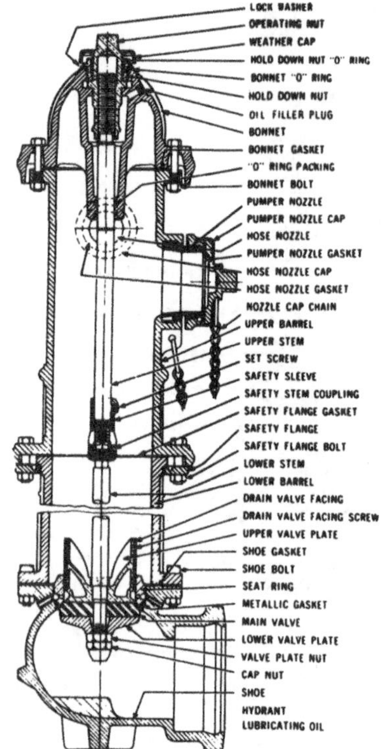

FIGURE 4.91 Dry-barrel hydrant. (*Mueller Co.; from Arthur E. Cote, ed.,* Fire Protection Handbook, *NFPA, Quincy, Mass., 1997.*)

structure, or area, generally overhead, and to which sprinklers are connected in a systematic pattern. The valve controlling each system riser is located in the system riser or its supply piping. Each sprinkler system riser includes a device for actuating an alarm when the system is in operation. The system is usually activated by heat from a fire and discharges water over the fire area. Further information on this subject is included in NFPA 13, *Installation of Sprinkler Systems.*

Wet-Pipe Systems. This type of system contains water under pressure at all times; water will be discharged immediately on operation of an automatic sprinkler. Water flowing from the wet-pipe sprinkler system actuates an alarm valve that gives off a signal. Figure 4.92 illustrates the total concept of the wet-pipe automatic sprinkler system.

The essential features of wet-pipe sprinkler systems, which represent about 80 percent of sprinkler installations, include provisions for water supplies, piping, and location and spacing of sprinklers. This system is generally used wherever there is no danger that the water in the pipes will freeze and wherever there are no special conditions requiring one of the other systems. Inspection of the wet-pipe sprinkler system at regular intervals is essential, and quarterly inspection of all water-control valves and water flow alarms is recommended.

Where subject to temperatures below 40°F (5°C), a wet-pipe system cannot be used. There are two recognized methods of maintaining automatic sprinkler protection in such locations: (1) through the use of systems where water enters the sprinkler piping only after operation of a control valve

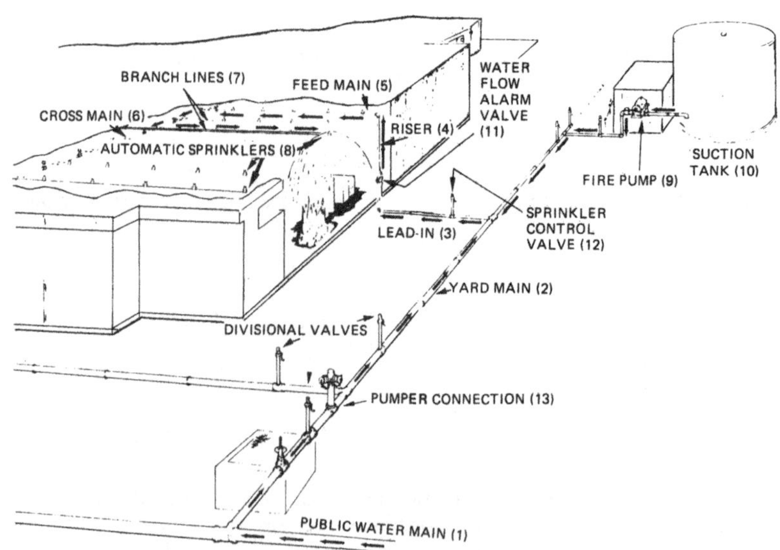

FIGURE 4.92 A typical sprinkler installation. (*From Arthur E. Cote, ed.,* Fire Protection Handbook, *NFPA, Quincy, Mass., 1997.*)

(dry-pipe, preaction, etc.) and (2) by the use of antifreeze solution in a portion of the wet-pipe system.

Dry-Pipe Systems. In locations where the building temperature cannot be maintained at 40°F (5°C), or higher a dry-pipe system is usually provided. In dry-pipe systems, the sprinkler piping contains air or nitrogen under pressure instead of water, and admission of the water into the system is controlled by a dry-pipe valve. When a sprinkler is opened by heat from a fire, the air pressure is reduced, the dry-pipe valve is opened by water pressure, and water travels to and flows out of any opened sprinklers.

Preaction Systems. Systems in which the air in the piping may or may not be under pressure are called *preaction systems.* These systems are designed primarily to protect properties where the danger of water damage from broken sprinklers or piping could be serious. (*Note:* Although inadvertent water discharging from a sprinkler system is extremely rare, this option is still viable.) The water supply valve is actuated independently of the opening of the sprinklers by an automatic fire-detection system; the valve is opened sooner than with the dry-pipe system, and the alarm is given when the valve is opened.

The preaction system has several advantages over a dry-pipe system. The valve is opened sooner because the detection system will usually respond to the fire prior to operation of the first sprinkler. Sprinkler piping is normally dry; thus, preaction systems are not subject to freezing and can be used when a dry-pipe system is required.

Deluge Systems. Deluge systems are reserved for use in some types of extra-hazard occupancies. All sprinklers are open at all times so that when the water comes on, the entire area is deluged with water; when heat from a fire actuates the fire-detecting device, water flows to and is discharged from all sprinklers on the piping system, thus "deluging" the protected areas. These systems are often used in airplane hangers and in areas where flammable liquids are handled or stored. Often, these systems use a foam solution, allowing faster control of the fire.

By using heat detection devices operating on the rate-of-rise or fixed-temperature principle, or other devices such as ultraviolet or infrared detectors designed for individual hazards, it is possible to apply water to a fire more quickly than with systems in which operation depends on opening of sprinklers only as the fire spreads.

Sprinklers

Sprinklers are designed with temperature ratings ranging from 135°F (57°C) to as high as 650°F (343°C). Ratings of 165°F (74°C) are usual for use in buildings that are maintained at normal, constant temperatures.

The location and spacing of sprinklers depends on the degree of hazard and type of construction. NFPA 13, *Installation of Sprinkler Systems,* provides detailed design and installation requirements. Table 4.58 provides a summary of spacing requirements for sprinkler installation.

Standpipes

The four generally recognized standpipe system concepts are described here. The design and installation of these systems is based on NFPA 14, *Installation of Standpipe and Hose Systems.* See Table 4.59.

1. A wet-standpipe system, having supply valve open and water pressure maintained at all times. This is the most desirable type of system.
2. A dry-standpipe system arranged to admit water to the system through manual operation of approved remote-control devices located at each hose station. The water-supply control mechanism introduces an inherent reliability factor that must be considered.

TABLE 4.58 Summary of Spacing Rules

	Light hazard	Ordinary hazard	Extra hazard[e]	High-piled storage[f]	Large-drop sprinklers[g]	Early suppression fast-response sprinklers[g,h]
Unobstructed construction[a]	225[b]	130	100	100	130	100
Noncombustible obstructed construction	225[b]	130	100	100	130	100
Combustible obstructed construction	168[c,d]	130	100	100	100	N/A

[a] Wood truss construction as defined in NFPA 13 is classified as obstructed construction for the purpose of determining sprinkler protection areas.
[b] For light hazard occupancies, the protection area per sprinkler for pipe schedule systems shall not exceed 200 ft² per sprinkler.
[c] For light combustible framing members spaced less than 3 ft on center, maximum spacing is 130 ft².
[d] For heavy combustible framing members spaced 3 ft or more on center, maximum spacing is 225 ft².
[e] For extra-hazard occupancies:

1. The protection area per sprinkler for pipe schedule systems shall not exceed 90 ft².
2. The protection area per sprinkler for hydraulically designed systems with densities below 0.25 gal/min·ft² may exceed 100 ft², but shall not exceed 130 sq ft².

[f] For high-piled storage occupancies:

1. The protection area per sprinkler may exceed 100 sq ft but shall not exceed 130 sq ft for systems hydraulically designed in accordance with NFPA 231 and 231C for densities below 0.25 gal/min·ft².
2. Where protection areas are specifically indicated in the design criteria of other portions of this standard or other NFPA standards, those protection areas shall be used.
3. For protection involving large-drop sprinklers use the large-drop sprinkler column in the table.

[g] For large-drop and ESFR sprinklers, the minimum spacing is 80 ft² per sprinkler.
[h] For special sprinkler protection areas, see 4-3.2 of NFPA 13.
N/A denotes data not available in current standard.
Source: From NFPA 13, *Standard for the Installation of Sprinkler Systems,* Quincy, Mass., 1991.

TABLE 4.59 Summary of Types of Standpipe

Type	Intended use	Size hose and distribution	Minimum size pipe	Minimum water supply
Class I	Heavy streams Fire department Trained personnel Advanced stages of fire	2½-in connections All portions of each story or section within 30 ft of nozzle with 100 ft of hose	4 in up to 100 ft 6 in above 100 ft (275 ft maximum unless pressure regulated)	500 gal/min first standpipe 250 gal/min each additional (2500 gal/min maximum) 30-min duration 65 psi at top outlet with 500 gal/min flow
Class II	Small streams Building occupants Incipient fire	1½-in connections (distribution same as class I)	2 in up to 50 ft 2½ in above 50 ft	100 gal/min per building 30-min duration 65 lb/in² at top outlet with 100 gal/min flowing
Class III	Both of above	Same as class I with added 1½-in outlets or 1½-in adapters and 1½-in hose.	Same as class I	Same as class I

Source: From NFPA 14, *Standard for the Installation of Standpipe, Private Hydrants, and Hose Systems,* Quincy, Mass., 2000.

3. A dry-standpipe system in an unheated building. The system should be arranged to admit water automatically by means of a dry-pipe valve or other approved device. The depletion of system air at the time of use introduces a delay in the application of water to the fire and increases the level of competency required to control the pressurized hose and nozzle assembly during the charging period.
4. A dry-standpipe system having no permanent water supply. This type is usually reserved for use in low-risk buildings. This type of system might also be used in buildings under construction, where allowed in lieu of the wet standpipe in unheated areas.

Water-Spray Fixed Systems

Water-spray fixed systems are generally used to protect flammable liquid and gas storage vessels; piping and equipment; electrical equipment such as transformers, oil switches, and rotating electrical machinery; and openings in firewalls and floors through which conveyors pass. The type of water spray required for any particular hazard depends, of course, on the nature of the hazard and the purpose for which the protection is provided.

NFPA 15, *Water Spray Fixed Systems for Fire Protection,* calls for nozzles, piping, valves, pressure gauges, and detection systems to be of an approved type. The spray nozzles generally used in these systems are open, and the pipes, especially outdoor ones that are subject to freezing temperatures, are usually dry.

Foam Extinguishing Systems

Foam extinguishing systems have been used extensively for many years, especially in the petrochemical industry, for the extinguishment of flammable liquid fires. The principal kinds of foam are chemical and mechanical (determined by how they are generated), though chemical foams are generally considered obsolete. These classes are further subdivided. NFPA standards that address this subject include the following:

NFPA 11, *Low Expansion Foam Extinguishing Systems*

NFPA 11A, *Medium and High Expansion Foam Extinguishing Systems*

NFPA 16, *Foam-Water Sprinkler Systems and Spray Systems*

Special compatible foam concentrates result in the generation of a foam that does not break down as readily as ordinary foam when mixed with dry chemical extinguishing agent. Other special foams are available for application on fires in alcohols, esters, ketones, and ethers (called *water-soluble* or *polar liquids*). This concentrate produces a foam that does not deteriorate like ordinary foam when in contact with water-miscible solvents.

Inspection, Testing, and Maintenance

NFPA 25, *Inspection, Testing and Maintenance of Water Based Fire Protection Systems,* provides requirements and procedures for ensuring that the systems discussed in the section will perform as intended. The standard provides requirements for proper inspection, test, and maintenance procedures for sprinkler, standpipe, and underground supply systems, as well as fire pumps, water storage tanks, spray systems, and foam-water sprinkler systems. Other chapters address record retention provisions, valve inspection procedures, and system impairment procedures.

NFPA 25 states the frequency of a procedure at the particular element (inspect, test, or maintain) and the method by which the procedure is carried out at that element. For purposes of NFPA 25, inspection is a visual examination, testing is a physical check of the component, and maintenance is work performed on a component to keep it operable.

SPECIAL AGENT SUPPRESSION SYSTEMS

Carbon Dioxide Systems

Carbon dioxide is a noncombustible gas that has been effectively used to extinguish certain types of fires. It acts to dilute the oxygen in the fire area to a point where it will no longer support combustion (Table 4.60). Because carbon dioxide is stored under pressure, it can readily be ejected from its storage container. Carbon dioxide is inert and will not conduct electricity. It can be used safely on energized electric equipment fires without causing damage to the equipment. NFPA 12, *Carbon Dioxide Extinguishing Systems,* provides details on these types of systems.

Because carbon dioxide does little or no damage to equipment or materials with which it comes in contact, it is very useful for protection of rooms with contents of high value and contents subject to water damage. Typical of such occupancies are rooms housing live electric equipment. Carbon dioxide is also widely used for extinguishing flammable liquid fires.

Halogenated Extinguishing Systems

A halon is a hydrocarbon (hydrogen and carbon) in which some of the hydrogen atoms have been replaced by such elements as bromine, chlorine, or fluorine, or by combinations of these (see Table 4.61). A number of halons are toxic, thus making them undesirable for general use; two of them, halon 1301 and halon 1211, have acceptable levels of toxicity and excellent flame extinguishment properties.

Halon 1211 and halon 1301 are the only two agents recognized by the NFPA Technical Committee on Halogenated Fire Extinguishing Agent Systems. The two standards handled by this Committee are NFPA 12A, *Halon 1301 Fire Extinguishing Systems,* and NFPA 12B, *Halon 1211 Fire Extinguishing Systems.* Both halon 1211 and 1301 are widely used for protection of electric equipment (both are nonconductors of electricity), airplane engines, and computer rooms. As both of these halons rapidly vaporize, they leave little corrosive or abrasive residue to clean up and do not interfere as much with visibility during fire fighting as foam or carbon dioxide. Halon 1211 is used in portable fire extinguishers and halon 1301 is used in total flooding systems.

Today, fire-protection halons are subject to international restrictions imposed by the Montreal Protocol on Substances that Deplete the Stratospheric Ozone Layer. Consequently, production of these fire protection agents has been phased out as of January 1, 1994. New halon replacements are now available. See NFPA 2001, *Clean Agent Fire Extinguishing Systems.*

Dry-Chemical Extinguishing Systems

Dry-chemical extinguishing agents consist of finely divided powders that effectively extinguish a fire when applied to the fire by portable extinguishers, hose lines, or fixed systems. The original dry powder was sodium bicarbonate (ordinary baking soda). Potassium bicarbonate and other chemical powders, with additives to make the powders free flowing and more moisture resistant, are now also in use. Dry chemical has been found to be an effective extinguishing agent for fires in flammable liquids and in certain types of ordinary combustibles and electric equipment, depending on the type of dry chemical used. Detailed installation requirements are included in NFPA 17, *Dry Chemical Extinguishing Systems,* and NFPA 17A, *Wet Chemical Extinguishing Systems.*

Dry-chemical extinguishing systems are used to protect flammable-liquid storage rooms, dip tanks, kitchen range hoods, deep-fat fryers, and similar hazardous areas and appliances. Because dry chemical is nonconductive, these systems are useful in the protection of oil-filled transformers and circuit breakers. Dry-chemical systems are not recommended for telephone-switchboard or computer protection. Dry chemicals are also widely used in portable fire extinguishers.

TABLE 4.60 Minimum Carbon Dioxide Concentrations for Extinguishment

Material	Theoretical min. CO_2 concentration, %	Minimum design CO_2 concentration, %
Acetylene	55	66
Acetone	27*	34
Aviation gas grades 115/145	30	36
Benzol, benzene	31	37
Butadiene	34	41
Butane	28	34
Butane—I	31	37
Carbon disulfide	60	72
Carbon monoxide	53	64
Coal gas or natural gas	31*	37
Cyclopropane	31	37
Diethyl ether	33	40
Dimethyl ether	33	40
Dowtherm	38*	46
Ethane	33	40
Ethyl alcohol	36	43
Ethyl ether	38*	46
Ethylene	41	49
Ethylene dichloride	21	34
Ethylene oxide	44	53
Gasoline	28	34
Hexane	29	35
Higher paraffin hydrocarbons $C_mH_{2m} + 2m - 5$	28	34
Hydrogen	62	75
Hydrogen sulfide	30	36
Isobutane	30*	36
Isobutylene	26	34
Isobutyl formate	26	34
JP-4	30	36
Kerosene	28	34
Methane	25	34
Methyl acetate	29	35
Methyl alcohol	33	40
Methyl butene—I	32	36
Methyl ethyl ketone	32	40
Methyl formate	32	39
Pentane	29	35
Propane	30	36
Propylene	30	36
Quench, lubricating oils	28	34

Note: The theoretical minimum extinguishing concentrations in air for the above materials were obtained from a compilation of Bureau of Mines limits of flammability of gases and vapors (Bulletins 503 and 627). Those marked with * were calculated from accepted residual oxygen values.

Source: From Arthur E. Cote, ed., *Fire Protection Handbook,* NFPA, Quincy, Mass., 1997.

Combustible-Metal Extinguishing Systems

A number of metals and metal powders found in industrial situations and in transport will burn. Some metals burn when heated to high temperatures by friction or exposure to external heat. Others burn from contact with moisture or in reaction with other materials. These metals and metal powders require special extinguishing agents and special fire-fighting techniques. Some result in explosions and very high temperatures, and some react violently with water. Still others give off toxic fumes when burning.

TABLE 4.61 Some Physical Properties of the Common Halogenated Fire-Extinguishing Agents

Agent	Chemical formula	Halon no.	Type of agent	Approx. boiling point, °F[†]	Approx. freezing point, °F[†]	Specific gravity of liquid at 68°F[†] (water = 1)	Approx. critical temp., °F[†]	Estimated pressure, psig[‡]		Latent heat of vaporization, cal/g water = 540 cal/g CO_2 = 138 cal/g
								At 130°F[†]	At critical temp.	
Carbon tetrachloride	CCl_4	104	Liquid	170	–8	1.595				46
Methyl bromide	CH_3Br	1001	Liquid	40	–135	1.73				62
Bromochloromethane	CH_2BrCl	1011	Liquid	151	–124	1.93				
Dibromodifluoromethane	CF_2Br_2	1202	Liquid	76	–223	2.28	389	23	585	29
Bromochlorodifluoromethane	CF_2BrCl	1211	Liquefied gas*	25	–257	1.83	309	75	580	32
Bromotrifluoromethane	CF_3Br	1301	Liquefied gas	–72	–270	1.57	153	435	560	28
Dibromotetrafluoroethane	$C_2F_4Br_2$	2402	Liquid	117	–167	2.17		3.8		25

* May be kept as a liquid at reduced temperatures.
[†] 5/9 (°F –32) = °C.
[‡] 1 psig = 6.895 kPa.
Source: From Arthur E. Cote, ed., *Fire Protection Handbook,* NFPA, Quincy, Mass., 1997.

Some combustible metal extinguishing agents' success in handling metal fires has led to the terms *approved extinguishing powder* and *dry powder.* Such terms have been accepted in describing extinguishing agents for metal fires, and should not be confused with the name *dry chemical,* which normally applies to an agent suitable for use on flammable-liquid and live electric equipment fires. Graphite powder, talc, and sand have all been used to smother metal fires.

PORTABLE FIRE EXTINGUISHERS

Portable fire extinguishers are required in most plants by local, state, and federal regulations and insurance companies. Where there are trained personnel available to use the proper extinguisher on a small incipient fire, extinguishers may prove useful in preventing a larger, more devastating fire.

The limitations of extinguishers, personal exposure to fire and smoke, capacity range, selectivity, and availability necessitate that training be provided if they are expected to be effective. Use of extinguishers should be simultaneous with notification of the fire brigade or department.

Types of Portable Fire Extinguishers

The kind and number of extinguishers needed for particular types of fires are specified in NFPA 10, *Portable Fire Extinguishers.* The most common types of extinguishers in use are the pressurized water, carbon dioxide, and multipurpose dry chemical. Other extinguishers commonly used are water pump tanks, halon 1211, and combustible-metal-type dry powder.

Application of Portable Fire Extinguishers

NFPA 10, *Portable Fire Extinguishers,* classifies fires in four ways:

Class A. Fires involving ordinary combustible materials (wood, cloth, paper, rubber and many plastics) (Table 4.62).

Class B. Fires involving flammable or combustible liquids, flammable gases, greases, and similar materials (Table 4.63).

Class C. Fires involving live electric equipment where safety to the operator requires the use of electrically nonconductive extinguishing agents. (*Note:* When electric equipment is de-energized, the use of Class A or B extinguishers may be used.)

TABLE 4.62 Fire Extinguisher Size and Placement for Class A Hazards

	Light-(low) hazard occupancy	Ordinary-(moderate) hazard occupancy	Extra-(high) hazard occupancy
Minimum rated single extinguisher	2-A	2-A	4-A*
Maximum floor area per unit of A	3000 ft^2	1500 ft^2	1000 ft^2
Maximum floor area for extinguisher	11,250 $ft^{2\dagger}$	11,250 $ft^{2\dagger}$	11,250 $ft^{2\dagger}$
Maximum travel distance to extinguisher	75 ft	75 ft	75 ft

* Two 2½-gal (9.46-L) water-type extinguishers can be used to fulfill the requirements of one 4-A rated extinguisher.
† See Appendix E-3.3 of NFPA 10, *Portable Fire Extinguishers.* For SI units: 1 ft = 0.305 m; 1 ft^2 = 0.929 m^2.
Source: From Arthur E. Cote, ed., *Fire Protection Handbook,* NFPA, Quincy, Mass., 1997.

TABLE 4.63 Fire Extinguisher Size and Placement for Class B Hazard Excluding Protection of Deep Layer Flammable Liquid Tanks

Type of hazard	Basic minimum extinguisher rating	Maximum travel distance to extinguishers, ft (m)
Low	5-B	30 (9)
	10-B	50 (15)
Moderate	10-B	30 (9)
	20-B	50 (15)
High	40-B	30 (9)
	80-B	50 (15)

Source: From Arthur E. Cote, ed., *Fire Protection Handbook,* NFPA, Quincy, Mass., 1997.

Class D. Fires involving certain combustible metals (such as magnesium, titanium, zirconium, sodium, and potassium) requiring a heat-absorbing extinguishing medium not reactive with the burning metals.

Figures 4.93 and 4.94 illustrate fire extinguishing agents, classifications, and symbols.

CODES AND STANDARDS

Fire Safety Standards-Making Organizations

ORDINARY

COMBUSTIBLES

1. Extinguishers suitable for "Class A" fires should be identified by a triangle containing the letter "A." If colored, the triangle shall be colored green.*

2. Extinguishers suitable for "Class B" fires should be identified by a square containing the letter "B." If colored, the square shall be colored red.*

3. Extinguishers suitable for "Class C" fires should be identifed by a circle containing the letter "C." If colored, the circle shall be colored blue.*

4. Extinguishers suitable for fires involving metals should be identified by a five-pointed star containing the letter "D." If colored, the star shall be colored yellow.*

FIGURE 4.93 Fire extinguisher identification.* Recommended colors per PMS (Pantone Matching System): green—*Basic Green,* red—192 *Red,* blue—*Process Blue,* yellow—*Basic Yellow.* (*From NFPA 10,* Standard for Portable Fire Extinguishers, *Quincy, Mass., 1998*).

American National Standards Institute (ANSI). ANSI sets public requirements for national standards and develops and publishes them on a wide range of subjects. In order to achieve uniformity in voluntary and mandatory state and federal standards, it coordinates voluntary standardization activities of concerned organizations.

ANSI standards cover a variety of products, materials, and equipment that is used both in highly specialized fields and in nearly all other areas of modern life. ANSI publishes standards on ceramic tiles, chemical process equipment, home appliances, electronics equipment, motion picture film and equipment, acids, refractory materials, oil burners, office machines and supplies, hospital supplies, and combustion engines.

American Society for Testing and Materials (ASTM). ASTM develops and publishes standards on finished products and on materials used in manufacturing and construction. Because some products and materials are used only within certain companies, industries, and government agencies, not all ASTM standards are developed by the full-consensus system. However, standards that deal with commodities used by the general public are developed by a full-consensus procedure, wherein all interested parties are fairly represented in the committee writing the standard. The standard committee is made up of anyone technically qualified or knowledgeable in the area of the committee's scope.

National Fire Protection Association (NFPA). Approximately 275 various NFPA codes and standards encompass the entire scope of fire prevention, fire protection, fire fight-

Typical Pictorial Extinguisher Marking Labels

*NOTE: Recommended colors, per PMS (Pantone Matching System):
(BLUE–*299*)
(RED–*Warm Red*)

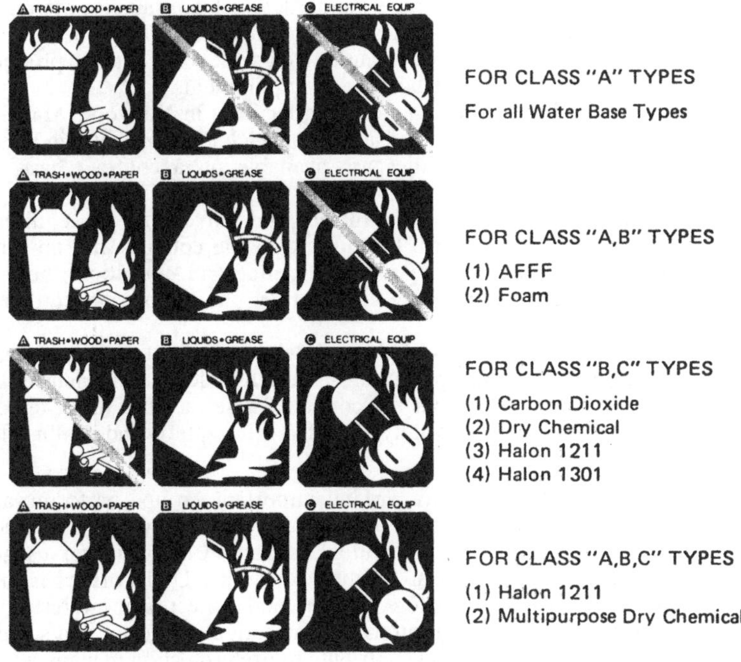

Color Separation Identification (picture symbol objects are white; background borders are white)

BLUE *	–	background for "YES" symbols
BLACK	–	background for symbols with slash mark ("NO")
RED *	–	slash mark for black background symbols

FIGURE 4.94 Fire extinguisher symbols. (*From NFPA 10,* Standard for Portable Fire Extinguishers, *Quincy, Mass., 1998.*)

ing, and fire hazards, ranging from the ***National Electrical Code,**** believed to be the most widely adopted set of safety requirements in the world, to codes or standards of specific limited areas.

Once a code or standard has been adopted by the NFPA, it becomes available for adoption by any organization or jurisdiction having enforcement authority. A number of NFPA standards are widely used and commonly referenced in fire legislation.

Fire Testing and Research Laboratories. There are many laboratories in the United States capable of performing, in varying degrees, fire tests of materials and/or equipment; many of these same laboratories, as well as other laboratories, have facilities for conducting fire-related research work. Generally, these laboratories can be classified into three categories: (1)

* National Electrical Code is a registered trademark of the National Fire Protection Association, Quincy, MA 02269.

private and industrial laboratories, (2) university laboratories, and (3) government laboratories. In the United States there are approximately 65 private and industrial laboratories that perform a wide range of fire tests. Space does not permit that each be described in detail. However, there are two, Underwriters Laboratories Inc. (UL) and Factory Mutual Laboratory Facilities (FM), whose work warrants particular emphasis.

Annually UL publishes lists of manufacturers whose products, when tested, have proved acceptable under appropriate standards which are subjected to one of the follow-up services provided by the laboratories as a countercheck. The work *listed* appears on UL labels attached to these products as authorized evidence that these products have been found to be in compliance with the laboratories' requirements.

Factory Mutual maintains testing facilities in Norwood, Mass., and also conducts large-scale applied research in its 1-acre, 60-ft-high FM test center in West Gloucester, R.I. Factory Mutual laboratory facilities are available on a contract basis through Factory Mutual Research.

More than 40 American colleges and universities are equipped with laboratories for fire testing and research. In addition to the colleges and universities that serve primarily as institutions for fire science training and education, there are others, both private and state-supported, whose engineering, physics, or science departments engage in such activities.

Several departments of the federal government—Agriculture, Air Force, Army, Commerce, Navy, and Transportation—as well as independent agencies also have research laboratories located throughout the country. These facilities are a direct result of an increasing national interest in fire safety as well as other safety- and health-related issues.

Insurance Organizations. Many important groups perform varied fire protection and inspection services on behalf of the insurance industry and its insureds. For example, the Association of Mill and Elevator Mutual Insurance Companies serves the mill and elevator industry's needs; the American Institute of Marine Underwriters is organized to serve marine underwriters and to promote, advance, and protect their interests.

There are five large insurance organizations, however, that serve a wide range of casualty and property insurers and contribute to fire protection in many ways. They are: (1) American Insurance Association, (2) American Mutual Insurance Alliance, (3) Factory Mutual System, (4) Industrial Risk Insurers, and (5) Insurance Services Office.

BIBLIOGRAPHY

Bryan, John L.: *Automatic Sprinkler and Standpipe Systems,* NFPA, Quincy, Mass., 1997.

Bryan, John L.: *Fire Supression and Detection Systems,* Glencoe Press, Beverly Hills, 1974.

Bugbee, Percy: *Principles of Fire Protection,* NFPA, Quincy, Mass., 1978.

Cote, Arthur E., ed.: *Fire Protection Handbook,* 18th ed., NFPA, Quincy, Mass., 1997.

Cote, Arthur E., ed.: *Industrial Fire Hazards Handbook,* 3d ed., NFPA, Quincy, Mass., 1990.

Factory Mutual: *The Handbook of Property Conservation,* Factory Mutual System, Norwood, Mass.

Factory Mutual: *Loss Prevention Data Books,* Factory Mutual System, Norwood, Mass.

Factory Mutual: *Property Conservation Workbook,* Factory Mutual System, Norwood, Mass., 1979.

Kimball, Warren Y.: *Fire Department Terminology,* NFPA, Quincy, Mass., 1979.

Magison, Ernest C.: *Electrical Instruments in Hazardous Locations,* 4th ed., Instrument Society of America, Triangle Park, N.C., 1998.

NFPA: *Guide to OSHA Fire Protection Regulations,* NFPA, Quincy, Mass.

NFPA: NFPA 600, *Industrial Fire Brigades,* NFPA, Quincy, Mass., 2000.

NFPA: *Introduction to Fire Protection,* NFPA, Quincy, Mass., 1982.

NFPA: *National Fire Codes,* NFPA, Quincy, Mass., annual.

Planer, Robert G.: *Fire Loss Control, A Management Guide,* Marcel Dekker, New York, 1979.

Roytman, M. Ya.: *Principles of Fire Safety Standards for Building Construction,* Amerind, New Delhi, India, 1975.

Tuck, Charles A., ed.: *NFPA Inspection Manual,* NFPA, Quincy, Mass., 1976.

Tuve, Richard C.: *Principles of Fire Protection Chemistry,* NFPA, Quincy, Mass., 1998.

U.S. Department of Labor: *General Industry Standards,* part 1910, title 29, Code of Federal Regulations, Occupational Safety and Health Administration.

Williams, C. A., Jr., and Heins, R. M.: *Risk Management and Insurance,* McGraw-Hill, New York, 1976.

Zajic, J. E., and Himmelmann, W. A., *Highly Hazardous Materials Spills and Emergency Planning,* Marcel Dekker, New York, 1978.

CHAPTER 4.10

TOXIC SUBSTANCES AND RADIATION HAZARDS

John McNary*
Senior Industrial Hygienist
Clayton Group Services, Inc.
Santa Ana, California

GLOSSARY

Aerosols Liquid droplets or solid particles dispersed in air that are of fine enough particle size (0.01 to 100 μm) to remain so dispersed for a period of time.

Alveoli Tiny air sacs of the lungs, formed by a dilation at the end of a bronchiole; through the thin walls of the alveoli, the blood takes in oxygen and gives up its carbon dioxide through respiration.

Anthrax A highly virulent bacterial infection picked up from infected animals and animal products.

Aplastic anemia A condition in which the bone marrow fails to produce an adequate number of red blood corpuscles.

* Revised and updated from second edition.

Asbestos A hydrated magnesium silicate in fibrous form.

Asbestosis A disease of the lungs caused by the inhalation of fine airborne asbestos fibers.

Asphyxia Suffocation from lack of oxygen. *Chemical asphyxia* is produced by a substance, such as carbon monoxide, that combines with hemoglobin to reduce the blood's capacity to transport oxygen. *Simple asphyxia* is the result of exposure to a substance, such as carbon dioxide, that displaces oxygen.

Bronchi The two main branches of the trachea that go into the right and left lung.

Bronchiole The smallest of the many tubes that carry air into and out of the lungs.

Byssinosis Disease occurring to those who experience prolonged exposure to heavy air concentrations of cotton dust.

Carcinoma Malignant tumors derived from epithelial tissues, that is, the outer skin, the membranes lining the body cavities, and certain glands.

Cesium 137 An isotope of the element cesium having an atomic mass number of 137; one of the important fission products.

Chloracne A disease caused by chlorinated polyphenyls and nephthalenes acting on sebaceous glands and the liver.

Chronic bronchitis An inflammation of the bronchial tubes lasting a long period of time or occurring frequently.

Cilia Tiny hairlike "whips" in the bronchi and other respiratory passages that aid in the removal of dust trapped on these moist surfaces.

Conjunctivitis Inflammation of the delicate mucous membrane (conjunctiva) that lines the eyelids and covers the front of the eyeball.

Contact dermatitis Dermatitis caused by a primary irritant.

Cristobalite A crystalline form of free silica. Quartz in refractory bricks and amorphous silica in diatomaceous earth are altered to cristobalite when exposed to high temperatures (calcined).

Curie A measure of radioactivity, or the rate at which a radioactive material throws off particles. The radioactivity of one gram of radium is a curie. One curie corresponds to 37 billion disintegrations per second or 37 becquerels. The becquerel (abbreviated Bq) is the SI unit that supersedes the curie (abbreviated Ci).

Cutie-pie A portable instrument equipped with a direct-reading meter used to determine the level of radiation in an area.

Cyclone As used in industrial-hygiene monitoring, a particle-size selector whose operation is based on imparting sufficient tangential velocities to relatively larger (heavier) particles sufficient to cause impaction on the walls of a conical chamber, while permitting smaller (respirable) particles to remain entrained in the air system.

Diatomaceous earth A soft, gritty, amorphous silica composed of small aquatic plants. Used in filtration and decolorization of liquids. Calcined and flux-calcined diatomaceous earth contains appreciable amounts of cristobalite.

Dose The total amount of a substance taken into the body by all routes of exposure during some recognized time period.

Dyspnea Shortness of breath, difficult or labored breathing.

Edema A swelling of body tissues as a result of being waterlogged with fluid.

Electromagnetic radiation The propagation of varying electric and magnetic fields through space at the speed of light, exhibiting the characteristics of wave motion.

Emphysema A lung disease, in which the walls of the alveoli have been stretched too thin and broken down; frequently accompanied by impairment of the heart action.

Epidemiology The science of correlating incidence and distribution of disease with causative factors or agents.

Etiology The study or knowledge of the causes of disease.

Fibrosis A growth of fibrous tissue in an organ in excess of that naturally present. A condition marked by increase of interstitial fibrous tissue.

Forced vital capacity (FVC) The maximum volume of air that can be expelled with maximum effort after a full inspiration.

Free crystalline silica Silicon dioxide with the SiO_2 molecule oriented in a fixed tetrahedral (crystalline) pattern, the most prevalent forms being quartz, cristobalite, and tridymite.

Hemoglobin The red coloring matter of the blood which carries the oxygen.

Ionizing radiation Refers to (1) electrically charged or neutral particles or (2) electromagnetic radiation which will interact with gases, liquids, or solids to produce ions.

Industrial hygiene The science of recognizing, evaluating, and controlling environmental stresses arising in or from the workplace.

Inertial impaction The forceful impingement on, or striking of, a particle on a surface with resulting adherence.

LD_{50} Abbreviation of lethal dose 50, the dose which is required to produce death in 50 percent of the exposed species. Death is usually reckoned as occurring within the first 30 days.

Leukemia A blood disease distinguished by overproduction of white blood cells. It may result from overexposure to radiation or it may generate spontaneously.

Lymph A clear, colorless fluid which circulates through the vessels of the lymphatic system.

Maser Microwave amplification by stimulated emission of radiation.

Metastasis Spread of malignancy from the site of primary cancer to secondary sites due to transfer through the lymphatic or blood system.

mg/m^3 Milligrams of substance per cubic meter of air, a common unit of exposure concentration.

Milliroentgen One one-thousandth of a roentgen. A roentgen is a unit of radioactive dose.

mppcf Millions of particles per cubic foot of air, a common unit of exposure concentration for mineral dusts.

Narcosis Stupor or unconsciousness produced by chemical substances.

Necrosis Destruction of body tissue.

Papilloma A small growth or tumor of the skin or mucous membrane.

Phagocyte A cell in the body that characteristically engulfs foreign material and consumes debris and foreign bodies, bacteria, and other cells.

Pharynx A part of the alimentary canal located between the mouth and the esophagus.

Pneumoconiosis A disease of the lungs caused by irritation of dusts and other particles.

Pulmonary function tests Measurement of ventilatory capacity of the lung by a series of tests, such as forced expiratory volume (FEV) and forced vital capacity (FVC).

Quartz The most prevalent, naturally occurring form of free silica, the basic raw material for the industrial sand industry.

Rad Standard unit of radioactive dose. It supersedes the roentgen.

Radioactivity Emission of energy in the form of alpha, beta, or gamma radiation from the nucleus of an atom.

Radiologist A specialist in the diagnostic and therapeutic use of x-rays and other forms of ionizing radiant energy.

Respirable Capable of penetrating into the lower respiratory tract, generally regarded as requiring a particle size of 10 μm or less.

Respirable mass That portion of total suspended particulate matter capable of penetrating into the lower respiratory tract.

Roentgen A unit of radioactive dose or exposure (abbreviated R). A roentgen is that amount of x- or gamma radiation that will produce one electrostatic unit of charge, of either sign, in one cubic centimeter of dry air at standard temperature and pressure. It is equivalent to 2.58×10^{-4} C/kg in air.

Roentgenogram The shadow picture formed on a sensitized film or plate by x-rays passing through a body.

Sensitizer A chemical that, at first exposure, may or may not cause irritation. After extended or repeated exposure, some individuals develop an allergic type of skin irritation called *sensitization dermatitis.*

Siderosis Lung disease resulting from inhalation of iron oxide.

Silicosis A lung disease resulting from fibrosis of the lungs due to inhalation of silica dust.

Spirometry The measurement of air movement in or out of the lungs with the use of a spirometer.

Synergism Combined action of substances whose total effect is greater than the sum of their separate effects.

Talc A hydrous magnesium silicate used in ceramics, cosmetics, paint, pharmaceuticals, and soap.

TLV Threshold-limit value. An exposure level under which most people can work consistently for 8 h/day, day after day, with no harmful effects. A table of these values and accompanying precautions is published annually by the American Conference of Governmental Industrial Hygienists (ACGIH); TLV is a registered trademark of ACGIH.

Time-weighted average concentration A calculated average obtained by dividing the sum of products of concentration times time for all activities by the total time of exposure.

Trachea The cartilaginous and membranous tube (windpipe) by which air passes to and from the lung.

Tridymite A vitreous, colorless form of free silica formed when quartz is heated to 870°C (1598°F). A form of crystalline silica rarely found in naturally occurring deposits.

X-ray diffraction Since all crystals act as three-dimensional gratings for x-rays, the pattern of diffracted rays is characteristic for each crystalline material. This method is of particular value in determining the presence or absence of crystalline silica in industrial dusts.

INTRODUCTION

Workplace safety and health are critical issues confronting employers. Exposure to an ever-increasing number of new chemicals in the workplace poses a continual challenge for the employer today. Moreover, the regulatory standards are continually being expanded and updated and, in most cases, result in lower and more stringent permissible exposure limits (PELs). These regulations have also heightened employee awareness of workplace hazards and have resulted in lawsuits against employers.

In January 1989, the U.S. Occupational Safety and Health Administration (OSHA) lowered 212 PELs and set new PELs for 164 substances not previously regulated by OSHA. The enormity of this step can be appreciated when one realizes that prior to January 1989, and during its 19 years of existence, OSHA issued only 24 such standards. The remaining 400-plus chemicals published in OSHA's original 1910.1000 regulation (the Z tables) were adopted from the 1968 American Conference of Governmental Industrial Hygienists (ACGIH) threshold limit values (TLVs), and had remained unchanged since 1970.

However, OSHA revoked the Final Rule Limits of January 1989 in response to the 11th Circuit Court of Appeals decision in AFL-CIO v. OSHA, effective June 30, 1993. The court ruled that OSHA had not given enough supporting data. On June 30, 1993, OSHA amended

29 CFR 1910.1000 to reflect the court's decision. The changes included a new Table Z-1 and the existing Tables Z-2 and Z-3 with minor corrections. Eleven states (Alaska, California, Connecticut, Maryland, Michigan, Minnesota, New Mexico, Tennessee, Vermont, Virginia, and Washington) and the Virgin Islands opted to continue enforcement of the 1989 PELs under their parallel state OSHA standards-setting authority.

OSHA continues to recommend that employees' exposure be limited to the more protective level of either the National Institute of Occupational Safety and Health (NIOSH) Recommended Exposure Levels (RELs) or the ACGIH TLVs. OSHA can issue citations for violation of the general duty clause [Section 654(a)(1)]. The general duty clause states that "each employer shall furnish to each of his employees employment and a place of employment which are free from recognized hazards that are causing or likely to cause death or serious physical harm to his employees."

Enactment of occupational health standards by OSHA has forced many plant engineers to become involved in the evaluation and control of industrial hygiene problems not only in existing facilities, but in the design of new facilities as well. For example, industrial hygiene considerations play a dominant role when considering ventilation systems (local and general), makeup air, enclosure and/or isolation of processes using toxic chemicals, handling and storage of toxic materials, and cleanup of exhaust emissions. Adequate disposal of toxic or hazardous wastes is also important to prevent environmental contamination and exposure of workers handling them. Plant engineers, therefore, need to be aware of toxic chemical and radiation hazards and how to effectively deal with them.

Industrial hygiene is defined by the American Industrial Hygiene Association (AIHA) as "that science and art devoted to the recognition, evaluation, and control of those environmental factors or stresses, arising in or from the workplace, that may cause sickness, impaired health, and well-being, or significant discomfort and inefficiency." An *industrial hygienist* is someone trained in "engineering, chemistry, physics, or medicine or related biological sciences (augmented by) special studies and training in all of the above cognate sciences" to practice industrial hygiene as defined above. A fully staffed industrial-health team also includes, at minimum, an *industrial physician,* an *industrial toxicologist,* an *analytical chemist,* and an *environmental* or *process engineer.*

TOXIC SUBSTANCES

Identifying Hazards: The Workplace Survey

Identifying potential health hazards requires knowledge of processes and the materials used. For example, it is safe to assume that petrochemical workers will probably be exposed to one or more of a variety of organic vapors, that sandblasters risk overexposure to respirable dusts (crystalline silica, etc.), and that radiation is a potential hazard affecting workers in nuclear power plants.

In many other situations, the presence of a hazard may only be suspected, as when employee complaints of disease symptoms fall into an identifiable pattern or when strange odors are detected in the workplace. These cases call for a more rigorous exploratory survey, both to identify the hazard and to measure its extent.

Whether the identity of the contaminant is known or not, a workplace survey should be conducted that includes an inventory of the chemicals and the processes in the work area or entire plant, the chemicals used in these processes, the raw materials, by-products, and so on. Generally, a comprehensive investigation requires the services of an experienced professional, such as a certified industrial hygienist. When the nature of the contaminant is indeterminate, the industrial hygienist should work with an industrial physician and a process engineer.

Many times in an investigation, visibility is a positive indicator of a hazard. The absence of visibility of dusts and fumes, however, does not mean that there are not dangerous levels of contaminants present, since many contaminants constitute a hazard at levels much too low to be visible.

In the same way, the investigator uses the sense of smell to pinpoint certain contaminants. For example, it is possible to distinguish between the haylike odor of phosgene and the fish-like odor of trimethylamine. Here again, however, certain precautions must be observed. For example, although hydrogen sulfide has a very distinct rotten-egg-like odor, it can also dull the sense of smell after prolonged exposure, and thus effectively mask heavy concentrations of other substances.

Tables 4.64 and 4.65 provide examples of typical occupational exposure to particulates and gaseous toxic agents.

Work and Process Inventory

The first step is to obtain descriptions of job functions within the facility. This information may be provided by the personnel department and/or manufacturing staff, and should include the following:

TABLE 4.64 Examples of Occupational Exposure to Particulate Toxic Agents

Contaminant	Physical state	Occupation
Asbestos	Dust, fiber	Fireproofers, insulation strippers, asbestos-cement workers, auto-garage mechanics, construction workers, shipbuilding and repair workers, gasket manufacturing workers, rubber compounders, vinyl-tile workers, maintenance workers, and asbestos abatement workers
Silica	Dust	Abrasive blast cleaners, pottery makers, glass makers, cement workers, coal miners, construction workers, enamellers, foundry workers, smelters
Lead	Dust, fume	Babitters, glassmakers, foundry workers, pottery makers, printers, paint manufacturers and sprayers, can and dye makers, and lead abatement workers
Chromic acid	Mist	Electroplaters, picklers, colored glass, ink, and refractory makers
Arsenic	Dust, fume, or gas (arsine)	Copper smelters, brass manufacturing workers, ceramic and glass makers, insecticide manufacturing workers, electroplaters
Beryllium	Dust, fume	Beryllium metal and alloy manufacturing workers, alloy machinists, glass and neon tube makers, rocket fuel manufacturing workers
Coal-tar pitch volatiles	Dust, fume	Metallurgical operations workers, metal casters, petroleum refinery workers, coking operations workers

TABLE 4.65 Examples of Occupational Exposure to Gaseous Toxic Agents

Contaminant	Physical state	Occupation
Carbon monoxide	Gas	Cokers, smelters, metal caster, forklift drivers, garage operators, heat treaters, coal conversion workers, pottery makers
Nitrogen oxides	Gas	Welders, electroplaters, forklift drivers, fertilizer manufacturing workers, explosive manufacturers, dye workers
Fluorocarbons	Gas or vapor	Food processors, storage workers
Sulfur dioxide	Gas	Brewers, copper smelters, ore roasters, petroleum refiners, glassmakers, powerplant operators, paper makers
Benzene	Vapor	Petroleum refiners, coke oven workers, organic chemical manufacturing workers, gasoline station attendants, ink makers, insecticide makers, lithographers, paint makers, rubber makers
Toluene	Vapor	Core makers (foundry operation), polyurethane foam makers and users, spray painters, ship welders

- Nature of the job
- Description of the process
- Work duration
- Nature of potential exposure (if known)

This information allows the investigator to group various workers according to similarity of exposure to chemical and physical stresses. This grouping is essential later in selecting representative workers who may need to be sampled for exposure levels.

Chemical Inventory

The investigator conducting the industrial hygiene survey should be knowledgeable about the chemicals that are used in the plant. In preparing a chemical inventory (preferably with the help of the purchasing department), all processes using chemicals are taken into consideration, including waste treatment, boiler operations, air conditioning, and any other pertinent sources including those involving raw materials, intermediate products, finished products, and cleaning compounds.

A major problem in recognizing chemical hazards arises from the variety of chemical formulations from different suppliers and manufacturers. Industry uses a large variety of materials sold under various trade names. Under OSHA's hazard communication standard (29 CFR 1910.1200) distributors are responsible for ensuring that their customers are provided with copies of the *material-safety data sheets* (MSDSs) for each hazardous chemical. The standard requires employers to have a list of all hazardous chemicals in the workplace and a readily available MSDS for each hazardous chemical. Employees must be provided with information and training prior to an initial assignment to work with a hazardous chemical. Updating of the MSDSs is recommended, using various relevant publications and fact sheets, including the hygiene guides prepared by the American Industrial Hygiene Association and the chemical-safety data sheets supplied by organizations such as the Chemical Manufacturers Association.

Consideration must also be given to toxic materials that may be present as impurities in relatively safe materials—diatomaceous earth, for example, which is supposedly 100 percent amorphous silica and thus generally considered a safe material. Diatomaceous earth has extensive applications, including use as a filtering aid and as a filler in cosmetics, detergents, and other household products. However, certain varieties of diatomaceous earth used as filtration aids have been found to contain significant quantities of crystalline silica, a widely recognized occupational health hazard.

Moreover, certain materials that apparently exist at subtoxic levels in the workplace can substantially appreciate in toxicity by undergoing chemical reactions within the body. For example, a metabolic reaction occurs in the case of benzidine dyes. The carcinogenic nature of benzidine has been recognized for many years, but benzidine-based dyes have been considered relatively safe on the assumption that they contain very little free or unreacted benzidine. Studies of analyses of urine samples collected from workers exposed to these dyes, however, indicate that metabolization of benzidine dye results in significantly elevated benzidine levels within the body.

Routes of Entry

Contaminants enter the body in three principal ways:

- Inhalation (through the respiratory tract)
- Skin absorption (through the skin)
- Ingestion (through the digestive tract)

Inhalation is by far the most common access for airborne contaminants to the body because of the continuous need to oxygenate the tissue cells and because of intimate contact with the body's circulatory system.

The effect of exposure to toxic agents is usually classified as *acute* or *chronic.*

Acute Exposure. Acute exposure is characterized by exposure to high concentrations of the toxic material over a short period. The exposure occurs quickly and can result in immediate damage to the body. For example, inhaling high concentrations of carbon monoxide gas or carbon tetrachloride vapors will produce acute poisoning.

Chronic Exposure. Chronic exposure occurs when there is continuous absorption of small amounts of contaminants over a long period. Each dose, taken independently, would have little toxic effect, but the quantity accumulated over a number of years can result in serious damage. Chronic poisoning can also be produced by exposure to small amounts of harmful material that produces irreversible damage to tissues and organs so that the injury rather than the poison accumulates. An example of such a chronic effect is silicosis, a disease produced by inhaling crystalline silica dust over a period of years.

Nature of Contaminants

Airborne contaminants can be present as liquids or solids, as gaseous material in the form of a true gas or vapor, or in combinations of both gaseous and particulate matter. Most often, airborne contaminants are classified according to physical state and physiological effect on the human body. Knowledge of these classifications is necessary for proper evaluation of the work environment. One must also consider the route of entry and action of the contaminant.

Physiological Classification of Toxic Effects

Irritants. Irritants cause inflammation of the moist mucous surfaces of the body. Irritants are corrosive, but inflammation of tissues may result from concentrations well below those needed to produce corrosion. Examples of irritant materials include aldehydes, alkaline and acid mists, and ammonia. Materials that affect both the upper respiratory tract and lung tissues are chlorine and ozone. Irritants that affect primarily the terminal respiratory passages are nitrogen dioxide and phosgene.

Asphyxiants. Asphyxiants deprive the tissues of oxygen. They are generally divided into two classes—*simple* and *chemical.*

Simple Asphyxiants. These are physiologically inert gases that deprive the tissue of oxygen by diluting the available atmospheric oxygen. Examples include nitrogen, carbon dioxide, hydrogen, helium, and aliphatic hydrocarbons such as methane.

Chemical Asphyxiants. These prevent either oxygen transport in blood or normal oxygenation of the tissues. Chemical asphyxiants are active far below the level required for damage from simple asphyxiants. Examples include carbon monoxide, hydrogen cyanide, and nitrobenzene.

Primary Anesthetics. Anesthetics depress the central nervous system, particularly the brain. Examples include ethylene and ethyl ether.

Systemic Poisons. Systemic poisons cause injury to particular organs or body systems. The halogenated hydrocarbons (such as carbon tetrachloride) can damage the liver and kidneys, whereas benzene, aniline, and phenol may cause damage to the blood-forming system. Examples of materials classified as neurotoxic agents include carbon disulfide, methyl alcohol, tetraethyl lead, and organic phosphate insecticides. Examples of metallic systemic poisons include cadmium, lead, manganese, and mercury.

Carcinogens in the Workplace

Concern about occupational carcinogens has increased with the promulgation of the OSHA hazard communication standard and the various state worker's-right-to-know laws. Management concerns have also heightened because of the potential for huge financial liabilities resulting from cancer-related litigation. Educating management and workers about actual or perceived carcinogenic risks is a delicate and difficult task for the industrial hygienist and the personnel or labor relations manager.

OSHA and ACGIH prepare the most commonly used workplace carcinogen lists. OSHA-regulated carcinogens are included in Table 4.66. The ACGIH publishes an annual list of threshold limit values (TLVs), including a list of human and suspect (i.e., animal) carcinogens. ACGIH TLVs and criteria are used extensively in the United States and many other countries.

Chemical carcinogens can cause tumors in mammalian species. Carcinogens may induce a tumor type not usually observed, or induce an increased incidence of a tumor type normally

TABLE 4.66 Individual Chemical Substance Standards

29 CFR 1910	Chemical name	Exposure limits		
		TWA	Ceiling	Action level
1000	Air contaminants			
1001	Asbestos	0.2 fibers/cm^3	1.0 fibers/cm^3	0.1 fibers/cm^3
1002	Coal tar pitch volatiles	0.2 mg/m^3		
1003	4-nitrobiphenyl	*		
1004	α-naphthylamine	*		
1005	Reserved			
1006	Methyl chloromethyl ether	*		
1007	3, 3′-dichlorobenzidine (and its salts)	*		
1008	bis-chloromethyl ether	*		
1009	β-naphthylamine	*		
1010	Benzidine	*		
1011	4-aminodiphenyl	*		
1012	Ethyleneimine	*		
1013	β-propiolactone	*		
1014	2-acetylaminofluorene	*		
1015	4-dimethylaminoazobenzene	*		
1016	N-nitrosodimethylamine	*		
1017	Vinyl chloride	1 ppm	5 ppm/15 min	0.5 ppm
1018	Inorganic arsenic	10 μg/m^3		5 μg/m^3
1025	Lead	50 μg/m^3		30 μg/m^3
1028	Benzene	1 ppm	5 ppm/15 min	0.5 ppm
1029	Coke oven emissions	150 μg/m^3		
1043	Cotton dust	200 μg/m^3 (yarn manufacturer)		100 μg/m^3
		500 μg/m^3 (textile mill wastehouse)		250 μg/m^3
		750 μg/m^3 (slashing and weaving)		375 μg/m^3
1044	1,2-Dibromo-3-chloropropane	1 ppb		
1045	Acrylonitrile	2 ppm	10 ppm/15 min	1 ppm
1047	Ethylene oxide	1 ppm	5 ppm/15 min	0.5 ppm
1048	Formaldehyde	1 ppm	2 ppm/15 min	0.5 ppm

* These carcinogen standards require that these chemicals be handled only in completely enclosed systems with exposures reduced to the lowest feasible level.

seen, or induce such tumors at an earlier time than would otherwise be expected. In some instances, the worker's initial stages of exposure to the carcinogen and the tumor appearance are separated by a latent period of 20 to 30 years. Examples of chemical carcinogens include benzo(α)pyrene, β-naphthylamine, vinyl chloride, and chromates.

Reproductive Hazards in the Workplace

Exposure to chemicals that affect the reproductive systems is an increasing concern among employees. California's Proposition 65 underscores these concerns shown by regulatory agencies regarding exposure of workers and the community to teratogens and other chemicals that display reproductive toxicity. The current professional and legal view is that reproductive health can no longer be separated from general occupational health and that workers do not have to choose between having a safe workplace and having a family.

Particulate Size

Airborne particulate matter varies in size from less than 0.01 to more than 25 μm (1 μm ≅ 1/25,000 in). These particles are invisible to the naked eye.

Nonrespirable particulates consist of particles that either are too large to escape the respiratory tract's defenses before they can reach the lungs or are too small to be retained in the lungs even if they get that far. Nonrespirable particulates present other types of health problems. Toxic fumes and nonrespirable dusts that are inhaled or ingested (or, less commonly, that come into contact with the skin) may eventually be absorbed into the bloodstream and cause systemic poisoning. Other nonrespirable dusts, inert "nuisance" substances like limestone and gypsum, can severely irritate the upper respiratory tract, endangering health as well as causing great discomfort.

Particulates that are small enough to find their way into the lungs and remain there, including dusts fine enough to be classified as *respirable dusts,* can produce serious chronic conditions such as the group of diseases known as *pneumoconiosis*—lung diseases like coal miners' black lung disease and silicosis.

In view of the preceding, the measurement of dust exposure in many cases needs to be limited to that fraction of inhaled particles that are small enough to be deposited in the alveolar spaces. Thus, a size-selective air-sampling method is required to separate the respirable fraction from the coarser material that is deposited in the upper respiratory tract. The size selector most commonly used in respirable dust sampling has the following characteristics:

Aerodynamic diameter, μm (unit density sphere)	% passing size selector
2.0	90
2.5	75
3.5	50
5.0	25
10.0	0

Measurement of Worker Exposure

A variety of techniques are used in measuring exposure of workers to toxic chemicals. Techniques used for measuring particulates differ somewhat from those used to measure gases and vapors.

Sampling to determine worker exposure to particulates is most often done by filtration, using a battery-operated pump that draws air through a filter, usually 37 mm in diameter and made of paper, fiberglass, or synthetic materials such as mixed cellulose-ester, polyvinyl chloride, or polycarbonate. A cyclone attached to a filter cassette is used to sample for respirable dust.

Time-Weighted Average Exposure

Threshold-limit values usually refer to time-weighted concentrations for a 7- or 8-h workday and 40-h workweek. Application of TLVs to workers with unusual work schedules requires careful evaluation in order to provide protection equal to that provided workers on standard work shifts. OSHA has proposed formulas for calculating equivalent PELs for unusual work schedules using either an hours-per-day or hours-per-week adjustment. These and other models for calculating equivalent PELs are discussed in *Patty's Industrial Hygiene and Toxicology,* 3d ed., vol. 3A (Wiley-Interscience, 1994).

Short-term exposures to concentrations above the threshold limit are permitted provided that they are compensated for by equivalent excursions below the limit during the workday. For example, if the permissible exposure limit is 10, a worker could be permitted to work in concentrations as high as 15 for 4 h, provided that the remainder of this 8-h shift did not result in exposure to concentrations above 5, thus yielding an average 8-h weighted concentration of 10.

Stated mathematically, the time-weighted average concentration, based on an 8-h shift, is defined as

$$\text{TWA} = \frac{1}{8}\sum_{i=1}^{N} \overline{C}_i t_i$$

where $\overline{C}$ is the average concentration of the contaminant at location i, N is the total number of work locations, Σ refers to the summation of the products of concentration $\overline{C}$ in percent, and t is the time in house spent at location i. Expressed otherwise,

$$\text{TWA} = \frac{(\text{exposure time})\ (\text{con. } A) + (\text{exposure time})(\text{con. } B) + \cdots}{\text{total work time per shift}}$$

Obviously, this approach requires reliance on personal-breathing-zone sampling or detailed job analyses for all classifications studied, and a comprehensive program of sampling sufficient to establish average airborne concentrations at all work sites.

Who to Sample

Sampling programs are based on statistical considerations, as it may be impractical and is often unnecessary to sample every worker's exposure. Thus, various statistical approaches have been devised to determine the minimum number of workers to be sampled to achieve adequate representation. A simple approach recommended by NIOSH is summarized as follows:

Number of employees exposed	Minimum number of employees whose individual exposures shall be determined
1–20	50% of the total number of exposed employees
21–100	10 plus 25% of the excess over 20 exposed employees
Over 100	30 plus 5% of the excess over 100 exposed employees

Indoor Air Quality

Over the past 40 years, significant changes have occurred in the construction and operation of buildings. Some of these changes have led to the generation of new classes of air contaminants. During the past 10 years, particularly in office buildings, the presence of indoor air pollutants has been associated with reports of new maladies known as *sick-building syndrome* and *building-related illness.*

Sick-building syndrome. Many occupants complain of headache; fatigue; and eye, nose, and throat irritation; as well as dry skin. Complaints are more intense in the afternoon. Relief is experienced shortly after occupants leave the building.

Building-related illness. One or more occupants develop a clinically diagnosed disease that may be related to the occupant's presence in the building. Examples include humidifier fever, asthma, hypersensitivity pneumonitis, Legionnaires' disease, and Pontiac fever.

Employee health may be adversely affected by indoor air contaminants including environmental tobacco smoke, volatile organic compounds, bioeffluents, microbial allergens, and *Legionella* (a bacterium). Some of these contaminants can cause discomfort; eye, nose, and throat irritation; humidifier fever; and hypersensitivity pneumonitis. *Legionella* can cause Legionnaires' disease and Pontiac fever. Some contaminants may be carcinogenic.

The action of building owners and office managers is important in preventing sick-building syndrome and building-related illness. Indoor air quality problems can be mitigated by such practices as purchasing of finishing materials with minimum volatile organic emissions and insisting on careful maintenance of the building's heating, ventilating, and air-conditioning (HVAC) system. See Chaps. 5.2 and 5.3, "Building Air Quality" and "HVAC System Control Equipment."

An essential aspect of solving most indoor air pollution problems is a thorough understanding of how the HVAC system moves air through the building. Whether a consultant is needed to solve an indoor air-quality problem depends on the expertise of the facilities engineering and environmental health staff. A considerable amount of literature is available for the guidance of in-house activities. Some aspects of an indoor air-quality evaluation, such as ventilation assessment and sampling for volatile organic compounds and microorganisms, often require the services of a consultant.

Occupational Safety and Health Standards

Measured results are compared to OSHA health standards or other widely accepted standards such as the American Conference of Governmental Industrial Hygienists threshold-limit values.

Federal Standards (29 CFR 1910 Subpart Z). The first compilation of health and safety standards promulgated by OSHA was based on the existing federal and national consensus standards.

Table 4.66 lists the standards in Sec. 1910.1001 through 1910.1052 of Title 29, Chap. XVII, Sec. 1910.1000. These standards are detailed sets of regulations for individual substances. In addition to setting exposure limits, they require exposure monitoring and medical surveillance of exposed employees.

Private Organizations. The first influential American organization in the field of occupational safety was the American Conference of Governmental Industrial Hygienists. The ACGIH publishes each year its revised and updated list of occupational exposure limits, commonly referred to as the TLVs. This listing is more extensive than the OSHA listing of air contaminants in CFR 29 1910.1000. It should be noted that the term *TLV* is a registered trademark of the AGGIH and should not be used to refer to OSHA's permissible exposure limits. Another organization involved in the promulgation of such standards, although its concerns extend into several other areas, is the American National Standards Institute (ANSI). Two prominent organizations in the occupational health field are the American Board of Industrial Hygiene (ABIH) and the American Industrial Hygiene Association (AIHA), both established in the 1950s; among the services they provide are the certification of industrial hygienists and accreditation of laboratories that analyze workplace samples.

Control Strategy—Reducing or Eliminating Hazards. A strategy for reducing or eliminating workplace hazards can include either or both of two approaches: *engineering* and/or *administrative controls.*

Engineering controls. Some examples are substituting relatively safe chemicals for more toxic ones, altering a process in such a way as to reduce worker exposure to contaminants, changing or upgrading ventilation systems, and isolating or enclosing contaminated areas.

Administrative controls. An example is adjusting work schedules so that workers receive only a fraction of their present exposure to contaminants.

As an example of an engineering control, if asbestos is used as insulation, it may be possible to replace it with fiberglass or calcium silicate, materials that do not present the same degree of hazard. Here again, however, it should never be assumed that such a substitution completely eliminates the hazard. The hazardous potential for fiberglass is not completely known, for example, while certain varieties of calcium silicate, generally regarded as a relatively safe substance, have been found to contain measurable amounts of asbestos fibers and/or free crystalline silica, another hazardous material.

A control strategy must take into account work *practices,* which can be a major factor in workplace exposure. Poor work practices result from lack of awareness of the potential hazards of a toxic material. For example, in one paint factory, workers were observed dusting off their clothes with a compressed-air line, starting from the shoes and proceeding upward. The raw material in the plant's primary process included, among other substances, lead chromate. Sampling measurements in the plant indicated that worker exposures to lead and hexavalent chromium (both highly toxic materials) were not excessive under normal working conditions. It was determined that *this dusting-off procedure with compressed air, however, exposed the workers to higher concentrations of lead and chromium during a 2- to 4-min period than they were exposed to during the remainder of the day* when handling these materials frequently. Vacuuming of dusty clothing (rather than blowing the dust off) could easily prevent this exposure. In such cases, employee-awareness programs can be highly effective in preventing work practices that result in unnecessary exposure.

The evaluation of existing and planned engineering controls is a major goal of an effective industrial-hygiene program. OSHA considers engineering controls to be the primary and most effective means of reducing or eliminating exposure to toxic substances in the workplace. The extent of engineering controls needed can be better determined after sampling measurements have been taken, other aspects such as work practices and housekeeping have been evaluated, and corrective actions have been instituted. Examples of poor engineering controls are all too plentiful.

Exposure Surveillance Programs

A comprehensive ongoing occupational health or industrial-hygiene program includes a monitoring program for surveillance of worker health in addition to determining compliance with the applicable regulatory standards. The *surveillance program* consists not only of ongoing sampling of worker exposure to airborne concentrations of chemicals but also monitoring for exposure to such physical agents as noise, radiation, and heat stress. The frequency of sampling will depend on the levels of various contaminants, the toxicity of the material, and the specific requirements under the applicable codes.

Biological monitoring of worker exposure to chemical hazards through analysis of samples of blood, urine, expired air, etc. is required in those cases where measurement of airborne concentration alone is not a reliable indicator of exposure hazard. In some instances, the metabolite (chemical produced within the body) of a toxic substance rather than the substance itself may be measured in the biological fluids.

Medical surveillance of exposed workers is also a necessary tool for protection against toxic exposures. Preemployment and periodic medical examinations can reveal the presence of toxic effects at an early stage, when a cure is often possible. Medical examinations should include specific organ functions to detect changes relative to the specific contaminants to which the worker is exposed.

RADIATION HAZARDS

Radiation is energy, which is emitted, transmitted, or absorbed in wave or energetic particle form. The electromagnetic (EM) waves consist of electric and magnetic forces. When these forces are disturbed, EM radiation results. Figure 4.95 is an arrangement of known EM radiations according to their frequency and/or wavelength. It includes microwave, infrared, visible, and ultraviolet radiation. The range of biological effects of exposure within the electromagnetic spectrum is, therefore, extremely broad and diverse.

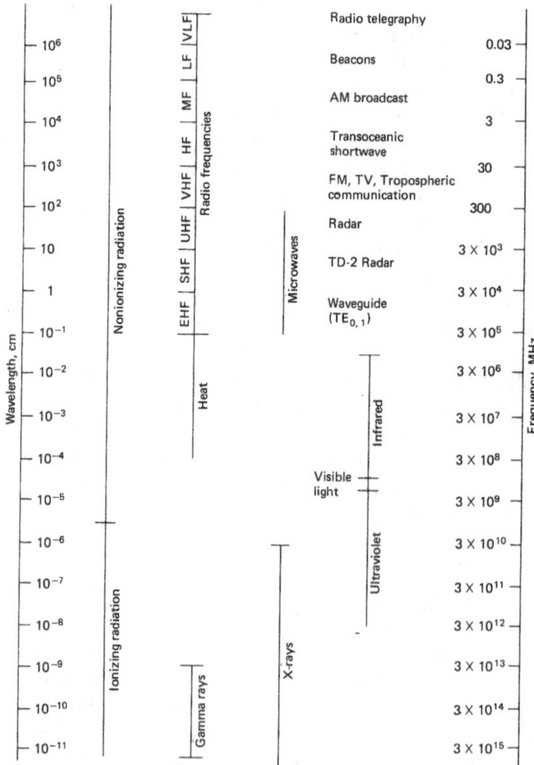

FIGURE 4.95 Electromagnetic spectrum.

Radiation is generally divided into two categories: *ionizing* and *nonionizing*. Ionizing radiation includes x-rays, alpha and beta particles, neutrons, and gamma rays. Nonionizing radia-

TABLE 4.67 Example of Occupational Exposure to Ionizing and Nonionizing Radiation

Type of radiation	Occupation
Ionizing (radioactive isotopes, x-rays)	Nuclear powerplant workers, food preservers, electron microscopists, biologists, food sterilizers, high-voltage repair workers, ceramic workers, drug makers
Ultraviolet	Meat curers, movie projectionists, pipeline workers, paint curers, nurses, bacteriologists, dentists, food preservers, lithographers, laboratory workers, welders, textile inspectors, plastic curers, printers
Infrared	Bakers, electricians, furnace workers, glassblowers, heat treaters, solderers, welders, firefighters, steel workers
Microwave radio-frequency	Radar workers, automotive workers, paper product workers, plastic heat-sealing workers, rubber-product workers, textile workers, tobacco workers, electronics workers, advertising sign workers
Laser	Medical technicians, surgeons, aerospace workers, semiconductor workers

tion includes ultraviolet, visible, infrared, microwave, and radio frequency. Laser radiations fall into most of these bands. Examples of occupational exposure to ionizing and nonionizing radiation are shown in Table 4.67.

Ionizing Radiation

Of the types of ionizing radiation, alpha particles are the least penetrating—paper and skin will ordinarily stop alpha particles.

Beta particles have considerably more penetrating power than alpha particles. X-rays and gamma rays both have very good penetrating power and require the use of heavy shielding material (lead). Neutrons are very penetrating and require shielding with materials of higher hydrogen-atom content rather than the use of mass alone.

Alpha emitters are an internal hazard because they do not have the ionizing ability to travel very long distances. One must take precautions against breathing or ingesting the alpha emitters. Beta emitters are also generally considered an internal hazard.

Regulations Regarding Ionizing Radiation. Jobs involving exposure to ionizing radiation fall under the provisions of Nuclear Regulatory Commission (NRC) standards (10 CER Part 20). OSHA standards (Sec. 1910.96) on ionizing radiation apply in cases where employees may not be protected under the NRC standards.

Table 4.68 summarizes the most frequent types of radiation encountered in health physics surveys and the type of detection most commonly used.

TABLE 4.68 Types of Detectors Used for Various Types of Radiation

Type of detector	Type of radiation
Proportional or scintillation counter	Alpha
Geiger-Mueller tube or proportional counter	Beta
Ionization chamber	X and gamma
Proportional counter	Neutron

A photographic film badge can be used for determining beta and gamma doses. Such film badges give a sufficiently quantitative indication of the integrated weekly doses of individuals. These should be recorded along with other records of radiation levels.

Nonionizing Radiation

Ultraviolet Radiation. The major source of ultraviolet radiation is the sun. Common constructed sources are mercury discharge lamps, welding and plasma torches, xenon discharge lamps, and lasers. The symptoms of overexposure to ultraviolet radiation are those characteristic of a severe sunburn.

Currently, there are no OSHA standards for exposure to ultraviolet radiation. OSHA Sec. 1910.97 concerning nonionizing radiation includes exposure levels and warning signs. However, this is an advisory and not a mandatory standard.

NIOSH has a recommended standard for occupational exposure to ultraviolet radiation. The American Conference of Governmental Industrial Hygienists has developed threshold-limit values for ultraviolet radiation in the spectral region between 180 and 400 nm and represent conditions under which it is believed that nearly all workers may be repeatedly exposed without adverse effect. The ACGIH urges caution in the use of the threshold values in that these values should be used as guides in the control of exposure to ultraviolet sources and should not be regarded as a fine line between safe and dangerous levels.

Infrared Radiation. Infrared radiation (ir) can be associated with the heat given off by all bodies that radiate heat and covers wavelengths from 0.75 μm to about 1 mm. Infrared rays are mostly absorbed by the skin and burn the skin in the same way the sun does. Like ultraviolet radiation, infrared radiation is invisible. Although infrared radiation may not be felt, over the years it can cause permanent eye damage. Water can be used as a barrier, as it can absorb most ir waves. Regular clothing protects the skin against ir, but goggles should be worn when there is a potential for overexposure.

Except for thermal burns (below wavelengths of 1.5 μm), infrared radiation is insignificant as a health hazard. However, when highly intense and compacted sources of radiant energy are being used, as with lasers, injury can occur in fractions of a second, before pain is evident.

Microwave Radiation. There has been a great deal of exposure to microwaves among people in the armed forces because microwave radiation is used for radar. With the advent of microwave ovens, however, this type of radiation is now encountered in the home. Microwaves have wavelengths of 1 m to 1 mm and their effect is related to power intensity and time of exposure as well as to the wavelength. Their ability to heat the body allows them to be used for medical treatment. This ability, however, can also be a hazard to the overexposed, unprotected worker. Microwaves penetrate deeply into the body and cause body temperature to rise; if the body temperature rises high enough, the person can go into a coma and die.

Before personnel are assigned to work in or about radar equipment, they should be given a complete physical including eye and blood examinations. Workers should be instructed never to look directly at a radar beam and should be given physicals periodically and whenever exposure is indicated. Most microwave-measuring devices are based on (1) bolometry, (2) colorimetry, (3) voltage and resistance changes in detectors, or (4) radiation pressure on a reflecting surface. The bolometry method is one of the most widely used in commercially available power meters.

One troublesome fact in the measurement of microwave radiation is that the near field (reactive field) of many sources may produce unpredictable radioactive patterns. Energy density rather than power density may be a more appropriate means of expressing hazard potential in the near field.

Radio-Frequency Radiation. Radio-frequency (rf) electromagnetic radiation obeys the general laws of electromagnetic radiation and is characterized by the following basic parameters: frequency f in hertz (1 Hz = 1 cycle per second), propagation time T, wavelength λ in meters, and velocity c, which in free space under normal conditions is equal to the velocity of light, i.e., 300,000 km/s. These quantities are interrelated according to the formula

$$\lambda = cT$$

Radio-frequency radiation covers wavelengths from 300 m to 1 mm. For measurement, either electric or magnetic field-strength meters or power meters are used.

In power meters, the rf energy causes a change in a temperature-sensitive element which is monitored by a bridge circuit and meter calibrated in units of $\mu W/m^2$ or mW/m^2.

All effects of rf radiation may be classified as either thermal or nonthermal, leading to heat stress or disturbance of the nervous system, respectively.

Nonionizing Radiation Instruments

Ultraviolet and Infrared Radiation Monitoring Devices. A variety of instruments are commonly used to measure ultraviolet and infrared radiation. They are classified according to the type of detector used, which is generally one of two types: thermal detectors or photoelectric detectors. Thermal detectors are those in which the absorbed radiation is degraded to heat and subsequently converted to an electric signal by changing the electric resistance of a filament. Photoelectric detectors are based on the principle that the absorbed photons eject electrons from a material. Most of these instruments are precalibrated by the manufacturer, but should be routinely checked prior to field use.

Microwave Monitoring Devices. Most microwave radiation detectors consist of the following components: a test antenna, an attenuator, a bolometer or thermistor, and a power meter. These detectors are usually in the form of a single integrated unit. The test antenna is specific to certain wavelengths and, therefore, interchangeable. These units must be calibrated at frequencies throughout the band to ensure accuracy.

Laser Monitoring Devices. A wide variety of laser radiation detectors are commercially available to fulfill the diversified needs resulting from different wavelengths, pulse durations, and power and energy densities of various laser instruments. Laser radiation detectors are generally based on either of two basic principles—photon or thermal detection.

Lasers

Laser technology has increased rapidly since its inception in 1960. Presently it involves virtually every scientific field in one way or another.

Laser is an acronym for light amplification by stimulated emission of radiation. A laser is a source of coherent energy that may appear in the near or far infrared, long or short ultraviolet, or visible regions. The properties of lasers are similar to those of other devices for various regions of the electromagnetic spectrum, except that the laser normally achieves great power densities. A laser beam does not rapidly diverge since a laser operates at specific frequencies in the electromagnetic spectrum. It travels in one direction in a straight line.

There are five types of lasers, classified by generating medium:

- Solid state (e.g., neodymium-yttrium-aluminum garnet)
- Gaseous state (e.g., helium-neon)
- Excimer (reactive gases such as chlorine and fluorine mixed with inert gases such as argon)
- Liquid-state, using organic dyes such as rhodamine 6G in liquid solution
- Semiconductor or junction

Some lasers operate continuously while others operate in pulses, sometimes as short as 10^{-11} s. Output levels range from milliwatts to kilowatts for continuous operation and up to gigawatts in pulse operations.

The two main areas of potential health effects from laser radiation are the eye and the skin. Associated electrical, chemical, and explosive hazards are also possible. Knowledge of and continued interest in laser safety are necessary in developing applications of lasers.

Laser Hazard Classes. Lasers are divided into four major hazard classifications based on a scheme of graded risk to the eye and skin.

- *Class I.* Cannot emit laser radiation at known hazard levels.
- *Class I.A.* Lasers not intended for viewing.
- *Class II.* Low-power visible lasers (aversion response provides protection).
- *Class IIIA.* Intermediate-power lasers (limit eye exposure).
- *Class IIIB.* Moderate-power lasers (no eye exposure).
- *Class IV.* High-power lasers (no eye or skin exposure; potential fire hazard).

Ocular Hazards. Protective eyewear should be worn whenever hazardous conditions may result from operation of a laser product. Laser radiation in the infrared region is highly absorbed at the surface of the cornea and could induce opacities in the cornea and destruction of the protective epithelial layer. In the ultraviolet region, exposure may result in extreme discomfort, but a moderate exposure is not thought to produce permanent damage. In the region of the spectrum from 320 to 1500 nm (the visible range is normally 380 to 760 nm), eye hazards are confined primarily to the retina and choroid. The most critical area for vision is the fovea. Thus, safety guidelines for laser radiation are designed to protect against foveal area damage. There is an often-undetected hazard from diffused laser radiation in the visible region due to the ability of the eye to focus such radiation to a very small spot on the retina. If the laser beam is incident upon a diffuse surface, the illuminated diffuse surface can serve as a secondary extended source. In this case, the retinal power density is unaltered regardless of viewing angel and how far away the surface is from the viewer.

Skin Hazards. The effects of laser radiation on the skin may vary from mild reddening (erythema), to blistering and charring, depending on the amount of energy absorbed, the wavelengths of the radiation, skin pigmentation, individual sensitivity, and duration of exposure.

Electrical and Explosion Hazards. Live parts of circuits and components with peak open-circuit potentials over 42.5 V are considered hazardous, unless limited to less than 0.5-mA circuit components of combustible materials. For example, transformers are potential fire hazards unless individual noncombustible enclosures are provided.

In the event of a tube or lamp failure, components such as electrolytic capacitors may explode if subjected to voltages higher than their ratings. A misdirected laser beam, in this case, could steam the coolant of a high-output laser system to trigger an explosion.

Chemical Hazards. OSHA currently has no occupational health standards for laser products. Threshold limits established by the American Conference of Governmental Industrial Hygienists are the recommended values to use. All laser products should meet specifications of the *National Electrical Code* (NFPA 70-1981), Arts. 300 and 400. Adherence to control procedures required under this standard should eliminate any significant probability of detrimental health effects from laser product operations. Therefore, an extensive, medical surveillance protocol for laser operators is not required under this standard.

BIBLIOGRAPHY

American Industrial Hygiene Association Journal, 2700 Prosperity Ave., Suite 250, Fairfax, VA 22031.

American Journal of Public Health, American Public Health Association, 1015 18th St. NW, Washington, DC 20036.

Archives of Environmental Health, Heldref Publications, 1319 18th St. NW, Washington, DC 20036-1802.

Cember, H.: *Introduction to Health Physics,* McGraw-Hill, New York, 1994.

Harris, R. L. (ed.): *Patty's Industrial Hygiene and Toxicology:* 5th ed., vols. 1–4, Wiley-Interscience, New York, 2000.

Harris, R. L., Cralley, L. J., and Cralley, L. V. (eds.): *Patty's Industrial Hygiene and Toxicology,* 3rd ed., vol. 3A, Wiley-Interscience, New York, 1994.

Cralley, L. V., and L. J. Cralley (eds.): *Industrial Hygiene Aspects of Plant Operations,* vol. 1, 1982, and vol. 2, 1983, Macmillan, New York.

OSHA Final Rule, *Air Contaminants, Permissible Exposure Limits:* Title 29 Code of Federal Regulations Part 1910.1000, www.osha-slc.gov/FedReg_osha_data/FED19930630.html.

Threshold Limit Value and Biological Exposure Indices 2000: American Conference of Governmental Industrial Hygienists, Cincinnati, Ohio, 2000.

CHAPTER 4.11
SANITATION CONTROL AND HOUSEKEEPING

Robert L. Bays
Scientist, Shared Technical Services
Ecolab Professional Products
St. Paul, Minnesota

GLOSSARY

Acid A compound that gives a pH below 7. Most soils including oils, greases, and waxes are acids.

Alkali A compound that gives a pH between 7 and 14. Hard-water films, carbonates, potash, and phosphates are alkaline.

Deodorize Destroying, masking, or modifying foul and unpleasant odors. May be accomplished by killing bacteria that make foul odors.

Detergent Any product that cleans. Usually a synthetic detergent but may be a soap or even an abrasive material.

Disinfectant Product used on hard inanimate surfaces which destroys microorganisms but not necessarily spores. Also called germicide.

Finish A protective coating used as a top coat.

Germicide See Disinfectant.

Hard floors Concrete, terrazzo, ceramic, quarry, slate, etc.

Polymer See Finish.

Resilient floors Vinyl, vinyl asbestos, asphalt, rubber, linoleum, etc.

Sanitary Relating to health or to the preservation of or restoration of health and hygiene.

Sanitation Use of sanitary measures to clean and maintain a building.

Sanitize Reduce bacterial counts to safe levels as determined by health requirements.

Sealer A product used to prevent excessive absorption of finish coats into porous surfaces. An undercoat.

INTRODUCTION

Proper housekeeping procedures play an important role in the total plant operation. Custodial employees work with a variety of chemicals, equipment, and procedures. They work in a building that has taken considerable money to construct, and improper chemicals and procedures can shorten the normal life of the building. In a relatively few years, the owners will have *spent more money on housekeeping and maintenance than the initial cost of the building and its furnishings.* Custodians should be well-trained to perform their tasks. Table 4.69 gives average cleaning times for various duties so that an estimate of efficiency can be determined. In addition to maintaining the appearance and prolonging the life of the building, good housekeeping provides immeasurable safety, hygienic, and sanitary benefits.

TABLE 4.69 Cleaning Operation Time Estimate*

Calculating cleaning times is a complex and frequently frustrating business. All cleaning times are estimates. There is no absolute time standard—nor can there be one. Two workers with the same tools, instructions, and area will clean at a different pace. A worker who must trek to the basement of a building to empty dirty cleaning solution will not clean as productively as a worker with convenient custodial closets and sinks on every floor.

When calculating cleaning times there are many other variables to consider. These include: age of the building, design of the building, climate, season, outside soil, placement of custodial closets, type of floors and walls, custodial training, day vs. night cleaning shifts, vandalism factor, etc.

To achieve a cleaning time standard there must be a set of uniform conditions. For example, each custodian must:

- Have sufficient materials, supplies and equipment to perform the work
- Be sufficiently trained to perform the task with the materials and supplies available
- Demonstrate the ability to perform the standard using the available supplies and materials
- Be aware of a cleaning time expectation
- Be regularly evaluated to provide for ongoing adherence to cleaning time standards.

General cleaning		
Dusting surfaces: dust with duster	150 ft^2	0.90 min
Dusting surfaces: dust with treated cloth	150 ft^2	1.80 min
Dusting surfaces: damp wipe with trigger sprayer and cloth	150 ft^2	2.88 min
Dusting surfaces: dust with hand-held duster vacuum	150 ft^2	1.35 min
Dusting surfaces: dust with tank/canister vacuum	150 ft^2	2.25 min
Dusting surfaces: dust with backpack vacuum	150 ft^2	1.62 min
Furniture, upholstered: vacuum with hand-held duster vacuum	25 ft^2	2.10 min
Furniture, upholstered: vacuum with tank/canister vacuum	25 ft^2	2.55 min
Furniture, upholstered: vacuum with backpack vacuum	25 ft^2	2.10 min
Glass door and hardware: clean using trigger sprayer and cloth (2 sides)	1 item	3.00 min
Glass panel/partition: clean using trigger sprayer and cloth	30 ft^2	3.42 min
Handrails/banister: damp-wipe with trigger sprayer and cloth	48 ft^2	0.86 min
Mats, walk-off: vacuum with upright vacuum	36 ft^2	1.08 min
Mats, walk-off: vacuum with tank/canister vacuum	36 ft^2	1.08 min
Mats, walk-off: vacuum with backpack vacuum	35 ft^2	0.95 min
Seating, upholstered: vacuum with hand-held duster vacuum	1000 ft^2	84.00 min
Seating, upholstered: vacuum with tank/canister vacuum	1000 ft^2	102.00 min
Seating, upholstered: vacuum with backpack vacuum	1000 ft^2	87.00 min
Telephone, desk: sanitize using trigger sprayer and cloth/cleaner–disinfectant	1 item	0.67 min
Telephone, wall: sanitize using trigger sprayer and cloth/cleaner–disinfectant	1 item	1.00 min

TABLE 4.69 Cleaning Operation Time Estimate* (*Continued*)

Trash removal		
Trash removal: empty trash/ash trays/pencil sharpener and wipe clean	2 items	1.00 min
Trash removal: empty trash/ash trays/pencil sharpener, wipe clean and reline basket	2 items	1.50 min
Trash pickup: pick up loose debris with lobby pan and porter broom/scrape up gum	1000 ft^2	18.00 min
Clean and polish		
Furniture, hard surface: clean/polish with trigger sprayer/chemical and cloth	100 ft^2	8.40 min
Stainless steel: clean/polish with trigger sprayer/chemical and cloth	10 ft^2	1.20 min
Wood paneling: clean/polish with trigger sprayer/chemical and cloth	100 ft^2	12.00 min
Restroom service		
Restroom pickup service: trash/replace supplies/touch-up, as needed	9 fixtures	9.02 min
Restroom service: trash/clean, disinfect/fixtures/wipe mirrors/replace supplies/sweep floor	9 fixtures	27.00 min
Restroom service: trash/clean, disinfect fixtures/wipe mirrors/replace supplies/dust mop floor	9 fixtures	21.60 min
Restroom service: trash/clean, disinfect fixtures/wipe mirrors/replace supplies/wet-mop floor	9 fixtures	44.98 min
Carpet care		
Protect from soiling using pump sprayer and soil protection chemical	1000 ft^2	10.20 min
Spot-remove by testing, applying spot remover and blotting	1 spot	4.00 min
Bonnet clean with immersion method using 17-in rotary floor machine	1000 ft^2	69.60 min
Bonnet clean with immersion method using 20-in rotary floor machine	1000 ft^2	60.00 min
Bonnet clean with immersion method using 24-in rotary floor machine	1000 ft^2	50.40 min
Bonnet-clean with spray-on method using 17-in rotary floor machine	1000 ft^2	54.00 min
Bonnet-clean with spray-on method using 20-in rotary floor machine	1000 ft^2	44.40 min
Bonnet-clean with spray-on method using 24-in rotary floor machine	1000 ft^2	34.80 min
Dry-clean, pretreat carpet with prespray chemical and pump tank sprayer	1000 ft^2	10.20 min
Dry-clean, spread dry cleaning compound	1000 ft^2	13.20 min
Dry-clean, agitate dry compound with 12-in revolving-brushes machine	1000 ft^2	34.80 min
Dry-clean, agitate dry compound with 24-in revolving-brushes machine	1000 ft^2	25.20 min
Dry-clean, vacuum up dry compound with 12-in upright vacuum	1000 ft^2	34.80 min
Dry-clean, vacuum up dry compound with 14-in twin-motor upright	1000 ft^2	31.00 min
Dry-clean, vacuum up dry compound with 16-in upright vacuum	1000 ft^2	30.00 min
Dry-clean, vacuum up dry compound with 18-in twin-motor upright	1000 ft^2	25.00 min
Dry-foam-clean using 12-in machine that requires separate foam pickup	1000 ft^2	34.80 min
Dry-foam-clean using 14-in machine that requires separate foam pickup	1000 ft^2	30.00 min
Dry-foam-clean using 18-in machine that requires separate foam pickup	1000 ft^2	25.20 min
Dry-foam-clean using 28-in machine that requires separate foam pickup	1000 ft^2	19.80 min
Dry-foam-clean using one-pass 13-in machine with simultaneous foam pickup	1000 ft^2	33.00 min
Dry-foam-clean using one-pass 24-in machine with simultaneous foam pickup	1000 ft^2	22.80 min
Extraction-clean using portable machine with hose and 12-in suction head	1000 ft^2	120.00 min
Extraction-clean using portable machine with hose and 16-in suction head	1000 ft^2	110.00 min
Extraction-clean using portable machine with hose and 12-in agitator power head	1000 ft^2	64.80 min
Extraction-clean using portable machine with hose and 16-in agitator power head	1000 ft^2	60.00 min
Extraction-clean using portable machine with hose and 17-in turbo rotating power head	1000 ft^2	15.00 min
Extraction-clean using 12-in self-contained, self-propelled machine	1000 ft^2	55.00 min
Extraction-clean using 16-in self-contained, self-propelled machine	1000 ft^2	29.00 min
Extraction-clean using 21-in self-contained, self-propelled machine	1000 ft^2	15.00 min
Rotary-shampoo with 175-r/min 17-in rotary floor machine	1000 ft^2	60.00 min
Rotary-shampoo with 175-r/min 20-in rotary floor machine	1000 ft^2	55.20 min

TABLE 4.69 Cleaning Operation Time Estimate* (*Continued*)

Carpet Care (*continued*)		
Rotary-shampoo with 350-r/min 17-in rotary floor machine	1000 ft^2	49.80 min
Rotary-shampoo with 350-r/min 20-in rotary floor machine	1000 ft^2	45.00 min
Rinse and extract shampoo using portable extractor with hose and 12-in suction head	1000 ft^2	60.00 min
Rinse and extract shampoo using portable extractor with hose and 16-in suction head	1000 ft^2	55.20 min
Scrub using one-pass machine with 12-in twin cylindrical brushes and wet pickup	1000 ft^2	27.00 min
Scrub using one-pass machine with 24-in twin cylindrical brushes and wet pickup	1000 ft^2	15.00 min
Vacuum with 12-in upright vacuum	1000 ft^2	22.80 min
Vacuum with 14-in upright vacuum	1000 ft^2	21.00 min
Vacuum with 14-in twin motor upright	1000 ft^2	17.00 min
Vacuum with 16-in upright vacuum	1000 ft^2	19.20 min
Vacuum with 18-in upright vacuum	1000 ft^2	17.40 min
Vacuum with 18-in twin motor upright	1000 ft^2	15.00 min
Vacuum with 20-in upright vacuum	1000 ft^2	15.60 min
Vacuum with 22-in upright vacuum	1000 ft^2	13.80 min
Vacuum with 24-in upright vacuum	1000 ft^2	12.00 min
Vacuum with 26-in large-area push-type vacuum	1000 ft^2	10.80 min
Vacuum with 28-in large-area push-type vacuum	1000 ft^2	7.50 min
Vacuum with 30-in large-area push-type vacuum	1000 ft^2	6.00 min
Vacuum with 32-in large-area push-type vacuum	1000 ft^2	4.00 min
Vacuum with 34-in battery-powered vacuum	1000 ft^2	6.50 min
Vacuum with backpack vacuum and 12-in orifice carpet tool	1000 ft^2	21.60 min
Vacuum with backpack vacuum and 14-in orifice carpet tool	1000 ft^2	19.80 min
Vacuum with backpack vacuum and 16-in orifice carpet tool	1000 ft^2	18.00 min
Vacuum with backpack vacuum and 18-in orifice carpet tool	1000 ft^2	16.20 min
Vacuum with backpack vacuum and 20-in orifice carpet tool	1000 ft^2	14.40 min
Vacuum with backpack vacuum and 22-in orifice carpet tool	1000 ft^2	12.60 min
Vacuum with backpack vacuum and 24-in orifice carpet tool	1000 ft^2	10.80 min
Vacuum with scrap-trap-type vacuum with 12-in carpet tool	1000 ft^2	12.00 min
Vacuum with scrap-trap-type vacuum with 16-in carpet tool	1000 ft^2	11.00 min
Vacuum with tank-type/canister vacuum and 12-in orifice carpet tool	1000 ft^2	24.00 min
Vacuum with tank-type/canister vacuum and 14-in orifice carpet tool	1000 ft^2	22.20 min
Vacuum with tank-type/canister vacuum and 16-in orifice carpet tool	1000 ft^2	20.40 min
Vacuum with tank-type/canister vacuum and 18-in orifice carpet tool	1000 ft^2	18.60 min
Vacuum with tank-type/canister vacuum and 20-in orifice carpet tool	1000 ft^2	16.80 min
Vacuum with tank-type/canister vacuum and 22-in orifice carpet tool	1000 ft^2	15.00 min
Vacuum with tank-type/canister vacuum and 24-in orifice carpet tool	1000 ft^2	13.20 min
Wet pickup with tank-type wet vacuum and 12-in orifice pickup tool	1000 ft^2	30.00 min
Wet pickup with tank-type wet vacuum and 14-in orifice pickup tool	1000 ft^2	28.20 min
Wet pickup with tank-type wet vacuum and 16-in orifice pickup tool	1000 ft^2	26.40 min
Wet pickup with tank-type wet vacuum and 18-in orifice pickup tool	1000 ft^2	24.60 min
Wet pickup with tank-type wet vacuum and 20-in orifice pickup tool	1000 ft^2	22.80 min
Wet pickup with tank-type wet vacuum and 22-in orifice pickup tool	1000 ft^2	21.00 min
Wet pickup with tank-type wet vacuum and 24-in orifice pickup tool	1000 ft^2	19.20 min
Hard-floor care		
Apply floor finish using mop	1000 ft^2	36.00 min
Apply floor finish using lambswool applicator	1000 ft^2	30.00 min
Apply floor finish using gravity-feed applicator	1000 ft^2	24.00 min
Apply floor seal using mop	1000 ft^2	36.00 min
Apply floor seal using lambswool applicator	1000 ft^2	30.00 min
Apply floor seal using gravity-feed applicator	1000 ft^2	24.00 min
Clean baseboards with manual swivel cleaning tool and handle	100 linear ft	6.60 min

TABLE 4.69 Cleaning Operation Time Estimate* (*Continued*)

Hard-floor care (*continued*)		
Clean baseboards with automatic rotary vertical brush machine	100 linear ft	3.00 min
Damp-mop with 12-oz mop head using single bucket and wringer	1000 ft^2	16.80 min
Damp-mop with 12-oz mop head using double bucket and wringer	1000 ft^2	15.60 min
Damp-mop with 16-oz mop head using single bucket and wringer	1000 ft^2	14.40 min
Damp-mop with 16-oz mop head using double bucket and wringer	1000 ft^2	13.20 min
Damp-mop with 24-oz mop head using single bucket and wringer	1000 ft^2	12.00 min
Damp-mop with 24-oz mop head using double bucket and wringer	1000 ft^2	10.80 min
Damp-mop with 32-oz mop head using single bucket and wringer	1000 ft^2	9.60 min
Damp-mop with 32-oz mop head using double bucket and wringer	1000 ft^2	8.40 min
Damp-mop with 18-in flat mop using single bucket and wringer	150 ft^2	2.52 min
Dry-buff/polish with 175-r/min 12-in rotary floor machine	1000 ft^2	40.20 min
Dry-buff/polish with 175-r/min 14-in rotary floor machine	1000 ft^2	34.80 min
Dry-buff/polish with 175-r/min 17-in rotary floor machine	1000 ft^2	30.00 min
Dry-buff/polish with 175-r/min 20-in rotary floor machine	1000 ft^2	25.20 min
Dry-buff/polish with 175-r/min 24-in rotary floor machine	1000 ft^2	19.80 min
Dry-buff/polish with 350-r/min 17-in rotary floor machine	1000 ft^2	19.80 min
Dry-buff/polish with 350-r/min 20-in rotary floor machine	1000 ft^2	15.00 min
Dry-buff/polish with 350-r/min 24-in rotary floor machine	1000 ft^2	10.20 min
Dry-buff/polish with 1000+ r/min 17-in rotary floor machine	1000 ft^2	7.20 min
Dry-buff/polish with 1000+ r/min 20-in rotary floor machine	1000 ft^2	6.60 min
Dry-buff/polish with 1000+ r/min 24-in rotary floor machine	1000 ft^2	5.40 min
Dry-buff/polish with 1000+ r/min 27-in rotary floor machine	1000 ft^2	4.80 min
Dry-burnish with 2000+ r/min 17-in rotary floor machine	1000 ft^2	6.60 min
Dry-burnish with 2000+ r/min 20-in rotary floor machine	1000 ft^2	6.00 min
Dry-burnish with 2000+ r/min 24-in rotary floor machine	1000 ft^2	4.80 min
Dry-burnish with 2000+ r/min 27-in rotary floor machine	1000 ft^2	4.20 min
Dust-mop with 12-in mop using dust treatment chemical	1000 ft^2	13.20 min
Dust-mop with 18-in mop using dust treatment chemical	1000 ft^2	9.00 min
Dust-mop with 24-in mop using dust treatment chemical	1000 ft^2	7.20 min
Dust-mop with 30-in mop using dust treatment chemical	1000 ft^2	6.00 min
Dust-mop with 36-in mop using dust treatment chemical	1000 ft^2	4.80 min
Dust-mop with 42-in mop using dust treatment chemical	1000 ft^2	3.60 min
Dust-mop with 48-in mop using dust treatment chemical	1000 ft^2	2.40 min
Dust-mop with 60-in mop using dust treatment chemical	1000 ft^2	1.80 min
Dust-mop with 72-in mop using dust treatment chemical	1000 ft^2	1.20 min
Scrub with 175-r/min 12-in floor machine that requires separate wet pickup	1000 ft^2	48.00 min
Scrub with 175-r/min 14-in floor machine that requires separate wet pickup	1000 ft^2	40.20 min
Scrub with 175-r/min 17-in floor machine that requires separate wet pickup	1000 ft^2	31.20 min
Scrub with 175-r/min 20-in floor machine that requires separate wet pickup	1000 ft^2	27.00 min
Scrub with 175-r/min 24-in floor machine that requires separate wet pickup	1000 ft^2	23.40 min
Scrub with 350-r/min 17-in floor machine that requires separate wet pickup	1000 ft^2	19.80 min
Scrub with 350-r/min 20-in floor machine that requires separate wet pickup	1000 ft^2	16.80 min
Scrub with 350-r/min 24-in floor machine that requires separate wet pickup	1000 ft^2	13.20 min
Scrub using one-pass machine with 12-in twin cylindrical brushes and wet pickup	1000 ft^2	12.00 min
Scrub using one-pass machine with 24-in twin cylindrical brushes and wet pickup	1000 ft^2	6.00 min
Scrub with 17-in automatic scrubber that includes wet pickup	1000 ft^2	9.00 min
Scrub with 21-in automatic scrubber that includes wet pickup	1000 ft^2	7.80 min
Scrub with 24-in automatic scrubber that includes wet pickup	1000 ft^2	6.00 min
Scrub with 27-in automatic scrubber that includes wet pickup	1000 ft^2	5.40 min
Scrub with 32-in automatic scrubber that includes wet pickup	1000 ft^2	4.20 min
Scrub with 36-in automatic scrubber that includes wet pickup	1000 ft^2	3.00 min
Spray-buff with 175-r/min 12-in rotary floor machine and finish restorer	1000 ft^2	45.00 min
Spray-buff with 175-r/min 14-in rotary floor machine and finish restorer	1000 ft^2	40.20 min
Spray-buff with 175-r/min 17-in rotary floor machine and finish restorer	1000 ft^2	34.80 min

TABLE 4.69 Cleaning Operation Time Estimate* (*Continued*)

Hard-floor care (*continued*)		
Spray-buff with 175-r/min 20-in rotary floor machine and finish restorer	1000 ft^2	30.00 min
Spray-buff with 175-r/min 24-in rotary floor machine and finish restorer	1000 ft^2	25.20 min
Spray-buff with 350-r/min 17-in rotary floor machine and finish restorer	1000 ft^2	25.20 min
Spray-buff with 350-r/min 20-in rotary floor machine and finish restorer	1000 ft^2	19.80 min
Spray-buff with 350-r/min 24-in rotary floor machine and finish restorer	1000 ft^2	15.00 min
Spray-buff with 1000+ r/min 17-in rotary floor machine and finish restorer	1000 ft^2	8.40 min
Spray-buff with 1000+ r/min 20-in rotary floor machine and finish restorer	1000 ft^2	7.80 min
Spray-buff with 1000+ r/min 24-in rotary floor machine and finish restorer	1000 ft^2	6.60 min
Spray-buff with 1000+ r/min 27-in rotary floor machine and finish restorer	1000 ft^2	6.00 min
Spray-buff with 2000+ r/min 17-in rotary floor machine and finish restorer	1000 ft^2	7.80 min
Spray-buff with 2000+ r/min 20-in rotary floor machine and finish restorer	1000 ft^2	7.20 min
Spray-buff with 2000+ r/min 24-in rotary floor machine and finish restorer	1000 ft^2	6.00 min
Spray-buff with 2000+ r/min 27-in rotary floor machine and finish restorer	1000 ft^2	5.40 min
Strip with 175-r/min 17-in rotary floor machine that requires separate wet pickup	1000 ft^2	79.80 min
Strip with 175-r/min 20-in rotary floor machine that requires separate wet pickup	1000 ft^2	75.00 min
Strip with 350-r/min 17-in rotary floor machine that requires separate wet pickup	1000 ft^2	52.80 min
Strip with 350-r/min 20-in rotary floor machine that requires separate wet pickup	1000 ft^2	45.00 min
Strip with mop-on chemical that requires separate wet pickup	1000 ft^2	18.00 min
Sweep with 8-in corn/synthetic broom	1000 ft^2	25.20 min
Sweep with 12-in push broom	1000 ft^2	15.00 min
Sweep with 16-in push broom	1000 ft^2	12.00 min
Sweep with 18-in push broom	1000 ft^2	10.80 min
Sweep with 24-in push broom	1000 ft^2	8.40 min
Sweep with 30-in push broom	1000 ft^2	6.00 min
Sweep with 36-in push broom	1000 ft^2	4.80 min
Sweep with 42-in push broom	1000 ft^2	3.60 min
Sweep with 48-in push broom	1000 ft^2	2.40 min
Sweep with 24-in push sweeper machine	1000 ft^2	6.00 min
Sweep with 30-in push sweeper machine	1000 ft^2	5.40 min
Sweep with 36-in rider power sweeper machine	1000 ft^2	3.00 min
Sweep with 42-in rider power sweeper machine	1000 ft^2	1.80 min
Sweep with 60-in rider power sweeper machine	1000 ft^2	0.60 min
Wet pickup with tank-type wet vacuum and 12-in orifice pickup tool	1000 ft^2	27.00 min
Wet pickup with tank-type wet vacuum and 14-in orifice pickup tool	1000 ft^2	25.20 min
Wet pickup with tank-type wet vacuum and 16-in orifice pickup tool	1000 ft^2	23.40 min
Wet pickup with tank-type wet vacuum and 18-in orifice pickup tool	1000 ft^2	21.60 min
Wet pickup with tank-type wet vacuum and 20-in orifice pickup tool	1000 ft^2	19.80 min
Wet pickup with tank-type wet vacuum and 22-in orifice pickup tool	1000 ft^2	18.00 min
Wet pickup with tank-type wet vacuum and 24-in orifice pickup tool	1000 ft^2	16.20 min
Wet-mop and rinse with 12-oz mop using single bucket and wringer	1000 ft^2	45.00 min
Wet-mop and rinse with 12-oz mop using double bucket and wringer	1000 ft^2	42.00 min
Wet-mop and rinse with 16-oz mop using single bucket and wringer	1000 ft^2	34.80 min
Wet-mop and rinse with 16-oz mop using double bucket and wringer	1000 ft^2	31.80 min
Wet-mop and rinse with 24-oz mop using single bucket and wringer	1000 ft^2	23.40 min
Wet-mop and rinse with 24-oz mop using double bucket and wringer	1000 ft^2	20.40 min
Wet-mop and rinse with 32-oz mop using single bucket and wringer	1000 ft^2	18.00 min
Wet-mop and rinse with 32-oz mop using double bucket and wringer	1000 ft^2	15.00 min
Stairways and landings		
Sweep with push broom	150 ft^2	4.50 min
Dust-mop using dust treatment chemical	150 ft^2	3.60 min
Damp-mop with mop bucket and wringer	150 ft^2	5.40 min
Vacuum with tank/canister vacuum	150 ft^2	4.50 min
Vacuum with backpack vacuum	150 ft^2	3.15 min
Vacuum with upright vacuum	150 ft^2	5.85 min

TABLE 4.69 Cleaning Operation Time Estimate* (*Continued*)

Overhead services		
Ceiling acoustical: clean with spray-on chemical and extension handle	1000 ft^2	84.00 min
Ceiling: wash manually with sponge using ladder and bucket	1000 ft^2	168.00 min
Light fixture diffusers: remove and clean in ultrasonic dip and return	4 items	12.00 min
Light fixtures: damp wipe with trigger sprayer and cloth using ladder	1 item	3.00 min
Overhead surfaces: damp-wipe with trigger sprayer and cloth using ladder	150 ft^2	9.90 min
Overhead surfaces: dust with backpack vacuum using ladder	150 ft^2	16.20 min
Overhead surfaces: dust with duster and extension handle	150 ft^2	0.90 min
Overhead surfaces: dust with tank/canister vacuum using ladder	150 ft^2	17.10 min
Overhead surfaces: dust with hand-held duster vacuum using ladder	150 ft^2	12.60 min
Vents: damp-wipe with trigger sprayer and cloth using ladder	1 item	0.50 min
Miscellaneous services		
Cubical curtain: remove and replace with clean curtain	1 item	3.00 min
Furniture, upholstered: shampoo with portable machine	25 ft^2	12.45 min
Garbage/trash cans: wash with pressure washer	1 item	2.00 min
Garbage/trash cans: wash with special can mounting sprayer system	1 item	1.00 min
Light bulbs/tubes: replace using ladder	1 item	3.00 min
Mats, fatigue: wash with pressure washer	36 ft^2	3.02 min
Mats, walk-off: wash with pressure washer	36 ft^2	3.02 min
Wall/partition, fabric: vacuum with backpack vacuum and 12-in-orifice tool	120 ft^2	3.24 min
Wall/partition, fabric: vacuum with tank/canister vacuum and 12-in-orifice tool	120 ft^2	3.96 min
Walls: wash manually with wall mop, extension handle, bucket and wringer	120 ft^2	23.98 min
Walls: wash manually with sponge, bucket and wringer using ladder	120 ft^2	36.00 min
Walls: wash with wall washing machine using ladder	120 ft^2	12.02 min
Windows		
Exterior: wash with brush, squeegee, and bucket	100 ft^2	10.02 min
Exterior: wash with high-rise extension tools	100 ft^2	13.20 min
Exterior: wash with trigger sprayer and cloth	100 ft^2	11.40 min
Interior and exterior: wash with trigger sprayer and cloth	200 ft^2	22.80 min
Interior: wash with brush, squeegee, and bucket	100 ft^2	10.02 min
Interior: wash with trigger sprayer and cloth	100 ft^2	11.40 min
Multiple pane: wash with trigger sprayer and cloth	1 pane	0.07 min
Walkways and steps		
Sweep with 8-in corn/synthetic broom	300 ft^2	7.56 min
Sweep with 12-in push broom	300 ft^2	4.50 min
Sweep with 16-in push broom	300 ft^2	3.60 min
Sweep with 18-in push broom	300 ft^2	3.24 min
Sweep with 24-in push broom	300 ft^2	2.52 min
Sweep with 30-in push broom	300 ft^2	1.80 min

* Reprinted with permission from International Sanitary Supply Association. These cleaning time estimates represent average cleaning times. Layout, obstacles, maintenance level desired, environmental conditions, etc., will affect cleaning time.

Safety is a key reason for good custodial procedures. The Occupational Safety and Health Administration of the U.S. Department of Labor states in its Standards and Interpretations, 1910.22(a) Housekeeping:

> (1) All places of employment, passageways, storerooms, and service rooms shall be kept clean and orderly and in a sanitary condition.
>
> (2) The floor of every workroom shall be maintained in a clean, and so far as possible, a dry condition. Where wet processes are used, drainage shall be maintained, and false floors, platforms, mats, or other dry standing places should be provided where practicable.

(3) To facilitate cleaning, every floor, working place, and passageway shall be kept free from protruding nails, splinters, holes, or loose boards.

Other benefits from the results of a well-trained custodial work force are:

1. A clean, sanitary environment contributes to the health of all the personnel.
2. Training improves the morale and mental attitude of each worker.
3. Each worker can have a feeling of pride in a job well done.
4. A well-maintained building is easier to keep in shape, thus saving time and money.
5. A well-maintained building and a crew of employees that enjoy their work contribute to goodwill and public relations.
6. The life of the various parts of the building and the building itself can be greatly extended by proper housekeeping procedures.

TRAINING TIPS

Some key tips to use in training are:

1. The longer a soil, stain, or coating is allowed to remain on the surface, the harder it will be to remove.
2. Quality products made by reputable manufacturers are usually best in the long run.
3. Chemicals should be measured and applied properly when required.
4. Two thin coats of a finish are better than one thick coat.
5. Always follow directions on labels or given in training, for safety as well as effectiveness.
6. Do not depend on perfumed products to cover up odors. Clean thoroughly and kill germs with disinfectants or sanitizers and there will not be putrefaction odors.
7. Surfaces vary considerably and aging often changes them.
8. Use products and procedures that will do the best job and that are made for that particular surface.
9. Clean and maintain equipment in proper working order.
10. Learn how to use equipment to the best advantage.
11. Use products with the best blend of properties to do the job intended.
12. Read supplier directions, literature, trade books, and magazines for generating ideas and becoming aware of new products.
13. Talk with fellow custodians about common problems and situations.
14. Attend local, regional, or state training programs where possible.
15. Practice cooperating with other departments, and they will cooperate with you.

PLANT DESIGN

The actual design of the plant can play a key role in contributing to simpler maintenance. Too often, those responsible for maintaining the building are never consulted during the early planning of the building design. If given a chance, many experienced custodians can provide some good ideas in planning or rearranging plant areas. Following are some basic suggestions:

1. Custodial service closets should be provided on every floor and be centrally located to save walking time.

2. The closet should have sufficient shelving, floor space, a floor-level sink, hard surface floors and wall racks for mops and brooms, and other facilities that fit the needs.
3. Keys for doors, cabinets, machines, etc., should be planned and organized to make locking and unlocking by the custodians quick and easy as well as secure.
4. Access through one room, closet, or office to another should not be permitted.
5. Restrooms should be planned to make use and cleaning easy. (Wall-hung fixtures, placement of fixtures for sequential use, glazed walls, ceramic floors, and floor drains are good choices.)
6. Kick plates and push plates can help the appearance and cleanability of the custodial closet, restroom, and other doors.
7. Wall and furniture surfaces should be easily washable.
8. Entrance mats should be either recessed or integrated so that tracked-in soils are reduced.
9. Light-colored carpeting should not be laid where soiling will show quickly.
10. OSHA states in Standards and Interpretations 1910.22(b) aisles and passageways:

> (1) Where mechanical handling equipment is used, sufficient safe clearances shall be allowed for aisles, at loading docks, through doorways and wherever turns or passage must be made. Aisles and passageways shall be kept clear and in good repair, with no obstruction across or in aisles that could create a hazard.
>
> (2) Permanent aisles and passageways shall be appropriately marked.

EQUIPMENT

Cleaning equipment may vary somewhat according to type of surfaces, size of area, and other factors. Though it is tempting to order large versions of all equipment to apparently save time in covering large surface areas, there are many occasions when the larger equipment version will be too clumsy, heavy, and hard to adapt to the inevitable smaller areas. When it is impractical to own a variety of sizes to fit every need, consider a compromise size. Many units have attachments to broaden their usage in various areas; they may combine two or more functions and may be of the rider type. Details and demonstrations should always be obtained from equipment suppliers prior to commitment. All equipment should be cleaned after each use and kept in working order. Standardization of purchases can make training easier, requires stocking of fewer replacement parts, and makes possible larger quantity discounts. Following are some of the basic items:

1. Rotary *floor machines* for use in scrubbing, stripping, scouring, and polishing resilient and hard-surface floors as well as shampooing carpets. A 20-in model is most efficient.
2. Wet or dry *vacuums* are available in a number of different varieties. They may use a wand or squeegee, be lightweight backpack models, be extra quiet, contain filters, and be of various sizes or suction powers. Sweeper versions may be of the rider type.
3. *Automatic scrubber vacuums* to scrub soils and pick up the scrub solution.
4. Pressure sprayers for fast and effective cleaning.
5. Wall-washing equipment with many variations.
6. *Carpet extractors* for cleaning and rinsing on location.
7. *Compactors* for trash disposal.
8. Mops, buckets, wringers, carts, pads, sprayers, brushes, dispensers, and many other miscellaneous items.

TABLE 4.70 Solving Floor Finish Problems*

Problems															
Black marking	Conspicuous formation of traffic lanes	Detergent resistance (poor)	Dirt pickup	Discoloration	Dullness in side aisles	Excessive foaming of polish or seal	Floor will not come clean during stripping	Gloss (poor initial)	Mop marks in finish	Gloss does not last	Pad marks in finish	Polish not leveling well	Polish not adhering to floor	Powdering or dusting	Scratching
■		■	■	■		■		■	■			■	■	■	
								■				■			
■	■				■			■							
						■									
				■								■	■	■	
		■			■			■	■				■	■	
			■	■	■			■		■		■	■	■	
					■				■		■				■
								■		■		■			
		■	■	■	■	■		■	■	■	■	■	■	■	
	■	■		■											
			■	■											■
			■	■											■
			■	■											■
								■		■		■			
		■						■		■		■	■	■	
■								■					■	■	
		■	■	■				■							
							■								
	■				■				■						
			■	■											
		■	■	■	■				■						
										■				■	
	■														
		■		■									■		
		■			■				■						
			■	■							■				
■	■		■	■	■				■						
		■			■	■		■		■		■	■	■	
				■			■								

* Most floor finishes are quality formulations, yet many outside factors can cause the finest products to do a poor job. This chart explains some of the problems and their solutions.

Source: Huntington Laboratories Inc., Huntington, Indiana.

Problems

Scuffing excessively	Slipperiness	Streaked appearance	Tackiness	Uneven or dull film	Unpleasant odor	Water resistance (poor)	Whitening of polish film	The solutions
■		■	■	■	■	■		Do not pour used polish or seal back into container
				■				Buff clean dry floor before applying polish or seal
				■				Use seal
								Do not use mop wringer
■				■				Remove all previous coatings or factory finish
■		■				■		Use floor neutralizer after stripping
■		■	■	■	■			Check for product freezing damage
								Use fine or medium scrubbing pads rather than coarse
		■	■					Allow adequate drying time between coats
■		■		■	■	■		Use clean equipment and applicator
■								Remove excess wax buildup on surface
								Remove dust and loose soil more frequently
							■	Make sure adequate walkoff mats are available
								Use automatic scrubber daily
		■	■					Check for proper ventilation
■	■	■	■			■	■	Check for and remove soap or oil film from floor
■			■					Unstable plasticizer in tile—allow curing time
		■			■			Clean equipment properly after use
								Use stronger stripper and/or allow more dwell time
■								Areas not being buffed often enough—check schedule
	■	■						Areas not being cleaned properly—check schedule
						■	■	Mix cleaner according to directions
								Use a less aggressive buffing pad
				■				Feather new coats into old at edge of traffic lane
						■		Check for use of metal drum pump in polish—use plastic pump
	■		■		■			Check for foreign materials being used
■			■			■	■	Check for the use of too strong a cleaner
■			■					Check for over use of polish
	■	■		■				Make sure you are recoating and using restorer as recommended
■		■				■		Rinse floor thoroughly after stripping, deep scrubbing and cleaning
				■				Floor may not be properly stripped—restrip

CHEMICALS

Understanding some of the basic properties of the products used for cleaning, sanitation, and/or disinfecting should become one of the most important goals of maintenance personnel. Each product will perform certain tasks if used properly but have the potential to harm surfaces or personnel if improperly applied. Use only where and in the manner recommended. Observe the cautions on labels to preclude dangerous situations. Use the mildest abrasive, coupled with the mildest chemical possible, to complete the task and protect the surface.

A cleaning product may be either all-purpose for a variety of tasks or highly specialized. Mild *alkaline cleaners* will remove oily, greasy soils from any surface not harmed by water alone. Special alkaline cleaners are recommended for specific tasks such as stripping of floor coatings, window washing, carpet cleaning, wall washing, etc. Strong alkaline cleaners and degreasers are for heavy industrial soil buildups. *Acid cleaners* remove lime and hard-water deposits or rust stains from hard surfaces such as toilet bowls, metals, ceramics, and concrete. *Solvents* are generally used solely for spot removers in housekeeping.

Abrasive or mechanical action of some sort is usually necessary for the best cleaning procedures. Scouring powders, abrasive hand pads, abrasive floor pads, brushes, pressure sprayers, and simple "elbow grease" rubbing are supplements to the chemicals.

Germicides combined in certain cleaners can help control harmful microorganisms and unpleasant odors. This results in a healthful as well as a pleasant environment.

Floor coatings are of two types, sealer and finish. Sealers are generally designed for porous surfaces to smooth them out and make them easier to maintain. Concrete and terrazzo sealers stop dusting and may be used to provide a good base for a finish. Wood sealers bring out the natural beauty of wood, prevent penetration of stains, and provide gloss, nonslip, and other desirable properties. Resilient-tile sealers are mainly for porous tile and provide a good base for a finish. Note that some products may perform both as a sealer and finish.

Finishes generally are water-based and easily removed with a detergent stripping solution. The first finishes were made of natural waxes that required much dry buffing to retain a smooth, glossy appearance. The relatively recent chemical revolution has brought about many new coatings. Table 4.70 outlines specific solutions to floor finish problems.

UNDERSTANDING OSHA'S RIGHT-TO-KNOW LAW

The original right-to-know law was passed by Congress in 1970. The law said that chemical manufacturers and importers must determine the physical and health hazards of each product they make or import. Then they must pass the information about these hazards on to users through material safety data sheets and container labels. As of May 23, 1988 employers in the nonmanufacturing sector were included and must be in compliance with provisions of the OSHA standards.

Most chemicals in use have been diluted. They are not nearly so dangerous as the concentrate form when they are put to use by the user employee. There are notable exceptions, and there is the possibility of dangerous situations occurring because of lack of knowledge or carelessness. A serious emphasis on communication and training will help people to know and then reduce the risk by sensible handling of the chemicals.

What Does the OSHA Standard Require?

Three particular groups are mentioned by the OSHA standard: chemical manufacturers, distributors, and importers. They are required to evaluate each of their chemicals for possible physical and health hazards and then communicate the information from the research on each chemical by material safety data sheets and product labels. Employers who use chemicals in their business are required to: develop a written hazard communication program, inform employees about hazardous chemicals at the time of initial assignment, and, whenever a new hazard is introduced to the area, maintain a list of hazardous chemicals used and an available file of material safety data sheets (MSDSs) for reference when needed. They are also required to conduct employee training that includes at least the following:

1. Methods and observations that may be used to detect the presence of a hazardous chemical in the work area.
2. Recognition of the physical and health hazards of chemicals in the work area.
3. Measures employees can take to protect themselves from these hazards.
4. Details of the hazard communication program developed by the employer—including an explanation of the labeling system and material safety data sheets.
5. Documentation of the training program and those in attendance.

It stands to reason that if hazardous chemical precautions are going to function, the employee must read labels and follow the directions and information presented and know where the MSDSs are filed and refer to them if an unusual situation requires it.

What Is a Hazardous Chemical?

Hazardous chemical means any chemical which is a physical hazard or a health hazard. The warning required relative to a hazardous chemical could be words, pictures, symbols, or combination of these appearing on a label or other appropriate form of warning which convey the hazards of the chemical in the container.

What Is on a Label?

Labels are the direct link in communication from the chemical source to the employee who will use the product. Seldom will it be necessary for the employee to refer to the MSDS. Although regulation requires ready access to the MSDS, it would be only an emergency situation where such reference would be helpful. All the important, practical information is on the label.

Labeling problems sometimes occur when large containers of product are shipped and transfer is made into smaller unlabeled containers. If it is important to transfer products, labels with complete information are available for easy application to the new container. Even dispensers with silk-screened information are available for product when appropriate.

Labels provide the information needed for safe use of the chemical. They must include:

1. Identification of product
2. Active ingredients
3. Use information
4. Physical hazards
5. Health hazards
6. First-aid information
7. Storing and handling instructions
8. Basic protective clothing, equipment, and procedures that are recommended when working with this chemical
9. Name and address of the manufacturer

The labels must also:

1. Not be removed or defaced
2. Identify the material that may be hazardous
3. Warn of hazards in a consistent language:
   a. *Caution* is the least hazardous and indicates slight irritation from eye and skin contact with the concentrated chemical product.
   b. *Warning* is the next most hazardous and indicates moderate eye and skin irritation from the concentrated chemical.
   c. *Danger* is the most hazardous and indicates severe eye and skin irritation potential from the concentrated product.

Material Safety Data Sheets

The material safety data sheets (Fig. 4.96) must be available and are required from chemical manufacturers and importers for each hazardous chemical they produce or import. Employers in the manufacturing sector are required to have a material safety data sheet for each hazardous chemical which they use. Employers in the nonmanufacturing sector were required to be in compliance with the provisions for communication on hazardous materials by May 23, 1988.

The material on the MSDS will not be as helpful to the employee using the chemical as the product label. Nevertheless, the MSDS must be on file and accessible. In time of need for information not included on the label, the MSDS should be available for referral. The MSDS will:

1. Identify the source of the chemical
2. List the hazardous components
3. Give physical and chemical characteristics
4. Warn about possible physical hazards
5. Note the stability or possible reactivity of the product
6. List possible health hazards
7. Give precautions for safe handling
8. Suggest control measures and protective equipment and what to do in an emergency

Note that MSDSs are required, must be accessible to the worker, and must give all the information you need to work safely with the chemical.

It is essential to follow through in maintaining safety in the work place. A key element to safety and well-being in the work place is the employee. Reading the label does not ensure safety. Doing what it says will. The law means that employees have a right to know about chemical hazards where they work. But hazard communication can protect only if they:

1. Read labels and, when appropriate, material safety data sheets
2. Know where to find information about their chemicals
3. Follow warnings and instructions
4. Use the correct protective clothing and equipment when handling hazardous substances
5. Learn emergency procedures
6. Practice sensible, safe work habits

Our concern for hazards in the work place can, to a great extent, be minimized if we conform to the following simple rules:

1. Use common sense.
2. Read the label.

Huntington.

CUSTOMER SEE REVERSE SIDE

MATERIAL SAFETY DATA SHEET

DATE 12/25/91 | MSDS DATE 10/01/91 | PRODUCT NO. 102915
CUSTOMER NO. ______ INVOICE NO. ______

SECTION I - IDENTITY

MANUFACTURER'S NAME	EMERGENCY TELEPHONE NO.	ADDRESS (Number, Street, City, State and Zip Code)
Huntington Laboratories, Inc.	219/356-8100	970 East Tipton Street, Huntington, Indiana 46750

CHEMICAL NAME AND SYNONYMS: N/A
TRADE NAME AND SYNONYMS: BLUE BLAZES
CHEMICAL FAMILY: DETERGENT
FORMULA: N/A

SECTION II - HAZARDOUS INGREDIENTS

CAS NO.	PRINCIPAL HAZARDOUS COMPONENTS	%	ACGIH TLV	OSHA PEL
141-43-5	MONOETHANOLAMINE	1-10	3 PPM	3 PPM
	OTHER INGREDIENTS > OR = 3%:			
7732-18-5	WATER			

SECTION III PHYSICAL DATA

APPEARANCE AND ODOR: BLUE LIQUID, CLEAN AND FRESH ODOR. | pH VALUE: 10.2-11.2
SPECIFIC GRAVITY (H_2O=1): 0.99 | PERCENT, VOLATILE BY VOLUME (%): >99 | REACTIVITY IN WATER: NONE | EVAPORATION RATE (WATER = 1): 1
BOILING POINT (°F.): 212 | SOLUBILITY IN WATER: 100% | VAPOR DENSITY (AIR=1): <1 | VAPOR PRESSURE (mm Hg.): 17.5

SECTION IV - FIRE AND EXPLOSION DATA

FLASH POINT METHOD USED: NONE, TCC | FLAMMABLE LIMITS: UPPER N/A, LOWER N/A | AUTO-IGNITION TEMP.: UNK | SPECIAL FIRE FIGHTING PROCEDURES: NONE
CONT: | EXTINGUISHING MEDIA: AS FOR SURROUNDING FIRE.
UNUSUAL FIRE AND EXPLOSION HAZARDS: NONE

SECTION V - PHYSICAL HAZARDS

STABILITY: UNSTABLE ___ STABLE X | CONDITIONS TO AVOID: NONE
INCOMPATABILITY (MATERIALS TO AVOID): STRONG ACIDS, OXIDIZERS
HAZARDOUS DECOMPOSITION PRODUCTS: THERMAL DECOMPOSITION MAY YIELD OXIDES OF CARBON AND NITROGEN.
HAZARDOUS POLYMERIZATION: MAY OCCUR ___ WILL NOT OCCUR X | CONDITIONS TO AVOID: NONE

SECTION VI - HEALTH HAZARD DATA

THRESHOLD LIMIT VALUE: SEE SEC II
EFFECTS OF OVEREXPOSURE: 1. INHALATION: NO ADVERSE REACTION EXPECTED.
2. EYES: CAUSES IRRITATION. 3. SKIN: MAY CAUSE IRRITATION.
4. INGESTION: MAY BE HARMFUL.

CHEMICAL LISTED AS CARCINOGEN OR POTENTIAL CARCINOGEN: NATIONAL TOXICOLOGY PROGRAM: YES ___ NO X | I.A.R.C. MONOGRAPHS: YES ___ NO X | OSHA: YES ___ NO X

EMERGENCY AND FIRST AID PROCEDURES:
1. INHALATION: MOVE TO FRESH AIR.
2. EYES: IMMEDIATELY FLUSH EYES WITH PLENTY OF WATER FOR AT LEAST 15 MINUTES. CALL A PHYSICIAN.
3. SKIN: FLUSH SKIN WITH PLENTY OF WATER. REMOVE CONTAMINATED CLOTHING AND WASH BEFORE REUSE. CALL A PHYSICIAN IF IRRITATION PERSISTS.
4. INGESTION: GIVE VICTIM A GLASS OF WATER. CALL A PHYSICIAN. NEVER GIVE ANYTHING BY MOUTH TO AN UNCONSCIOUS PERSON.

SECTION VII - SPILL OR LEAK PROCEDURES

STEPS TO BE TAKEN IN CASE MATERIAL IS RELEASED OR SPILLED: CONTAIN SPILL. DO NOT CONTAMINATE FOOD, FEED, OR WATER.
WASTE DISPOSAL METHOD: DISPOSE OF IN ACCORDANCE WITH ALL LOCAL, STATE, AND FEDERAL REGULATIONS.

SECTION VIII - SPECIAL PROTECTION INFORMATION

RESPIRATORY PROTECTION (SPECIFY TYPE): NONE REQUIRED.
VENTILATION: LOCAL EXHAUST ___ MECH X | SPECIAL ___ | OTHER ___
PROTECTIVE GLOVES: RUBBER | EYE PROTECTION: SAFETY GLASSES | OTHER: NONE

SECTION IX - SPECIAL PRECAUTIONS

PRECAUTION TO BE TAKEN IN HANDLING AND STORING: AVOID PRODUCT CONTACT WITH EYES, SKIN, AND CLOTHING. WASH THOROUGHLY AFTER HANDLING.

OTHER PRECAUTIONS: KEEP OUT OF REACH OF CHILDREN.

FIGURE 4.96 Sample material safety data sheet.

3. Follow the directions.
4. Be considerate of others.

Your Right to Know—A Summary

Chemical manufacturers and importers are required to:

- Evaluate products and determine whether there are any health hazards associated with using them.
- Communicate their findings by providing labels and material safety data sheets for each product they manufacture or import.

Employers are required to:

- Establish a written hazard communication program explaining how workers will be informed about hazards and how to handle them. The written program should be available for workers to read or review at all times.
- Label products appropriately.
- Obtain and keep available material safety data sheets for all products with physical or health hazards. These documents should be kept in a place where both workers and leadership can easily refer to them.
- Train employees to identify and deal with hazardous materials and make them aware of any new hazards introduced into the work area.

Employees should:

- Read labels, follow directions, and be knowledgeable about MSDS.
- Identify any hazardous materials and obtain the proper equipment to work with them safely.
- Use proper techniques to perform tasks and be familiar with appropriate emergency procedures.
- If they have questions, ask their supervisor.

The Hazard Communication Standard says you have a right to know about chemical hazards where you work, but remember, hazard communication can protect you only if you:

- Read labels and follow the directions.
- Know where to find information about your chemicals.
- Follow instructions and respect warnings.
- Use proper protective clothing and equipment when handling hazardous substances.
- Know what to do in an emergency.
- Use common sense and work safely.

OSHA COMPLIANCE

Understanding and Complying with OSHA's *Occupational Exposure to Bloodborne Pathogens: Final Rule*

Industries where workers are in contact with or handle blood and other potentially infectious materials *will* be affected by OSHA's *Occupational Exposure to Bloodborne Pathogens: Final Rule* (29 CFR 1910.1030). Over 500,000 establishments and 5,000,000 workers are estimated to be governed by this rule. OSHA has calculated that *occupational exposures* are responsible for 5800 to 6600 cases of hepatitis B (HBV) every year and believes that the majority of these cases can be avoided by following this standard. Compliance with this standard is estimated to prevent approximately 8800 occupational and nonoccupational cases of HBV infections per year. These cases could result in an estimated 190 deaths.

What Is the Final Rule? On December 6, 1991 OSHA published its final rule on *occupational exposure* to bloodborne pathogens. This standard marks the end of a four-year rulemaking process and produced the first comprehensive and specific OSHA standard to deal with HBV, AIDS, (HIV), and other bloodborne pathogens.

The standard specifies universal precautions, engineering and work practice controls, personal protective equipment, and housekeeping, combined with HBV vaccinations/postexposure follow-up, hazard communication labels/signs, record keeping and training, to reduce *occupational exposure* for all employees exposed to blood and potentially infectious materials. Meeting these requirements is *not optional* to employers or employees: *it is now required by law.*

Who Is Affected? Since there is no population free from the risk of HIV or HBV infection, *any employee* who has *occupational exposure* to blood or other potentially infectious material will be included in this standard. *Twenty-four industries* were identified for the regulatory impact and flexibility analysis, but the scope of this standard is in *no way limited* to employees in those job classifications.

States with state plans are required to adopt standards (within 6 months after the publication of the final standard) at least as effective as federal OSHA standards. The 23 states and 2 territories with state plans are: Alaska, Arizona, California, Connecticut, Hawaii, Indiana, Iowa, Kentucky, Maryland, Michigan, Minnesota, Nevada, New Mexico, New York, North Carolina, Oregon, Puerto Rico, South Carolina, Tennessee, Utah, Vermont, Virginia, the Virgin Islands, Washington, and Wyoming. In Connecticut and New York, the plans cover only state and local employees. Until the state plan is promulgated, interim protection is provided by the federal OSHA standard.

In states with *no* state plan, the public employees are not covered by the federal OSHA standard. Although state, county, and municipal employees are *not* covered by the federal OSHA standard, OSHA has urged these states to extend the protection of their requirements to public employees who have exposure to blood and other infectious materials. OSHA will refer complaints or inquiries to the state public health agency having jurisdiction over these health care facilities.

Exposure Control Plan. The exposure control plan is a pivotal provision of the bloodborne standard. It requires the employer to identify the employees who will receive the training, protective equipment, vaccinations, and other provisions of the standard. The employer must identify the positions and task procedures where *occupational exposure* to blood and other potentially infectious materials can occur.

Methods of Compliance. Methods of compliance shows the employer how to protect employees from the hazards of bloodborne pathogens and comply with the standard through universal precautions, engineering and work practice controls, personal protection equipment, housekeeping, and the handling of regulatory waste.

Vaccinations/Postexposure and Follow-Up

Hepatitis B vaccinations, postexposure evaluations, and follow-up are required by the employer for all employees who have occupational exposure or an exposure incident. This is to ensure that *all* employees are protected from infection. It also makes sure the employee receives appropriate medical follow-up after *each* exposure incident.

Hazard Communication. Hazard communication requires that employees receive warning through labels, signs, and training in order to eliminate or minimize their exposure to bloodborne pathogens. Warning labels are to be fluorescent orange or orange-red with lettering or symbols in contrasting color and include the *biohazard* legend.

Record Keeping

The records must be kept for *each* employee covered by the Occupational Exposure of Bloodborne Pathogens Standard. Training records, medical evaluations and treatment are important to the employer in determining vaccination status and follow-up involving exposure.

Availability of Records. The employer shall ensure that all records required to be maintained shall be made available on request to the assistant secretary of labor for Occupational Safety and Health Administration, the director of the National Institute for Occupational Safety and Health, U.S. Department of Health and Human Services, or designated representatives for copying and examination.

Compliance

It is *not* the responsibility of either the federal OSHA department of the state OSHA department to individually *notify public or private employers/employees of this standard.* Notification of the standard is done via press releases and publication in the *Federal Register.* Ignorance of the law is no excuse for not being in compliance.

BIBLIOGRAPHY

Bays, Robert L.: "Changes and Choices in Cleaning and Disinfecting," *Infection Control Today*, April 1998, pp. 12–18.

"Bloodborne Standard Training Requirements" International Executive Housekeeping Association, Lincolnwood, IL, August 1992.

Federal Register, vol. 52, no. 163, Monday, Aug. 24, 1987, Rules and Regulations.

Federal Register, vol. 56, no. 235, Friday, Dec. 6, 1991, Rules and Regulations.

Feldman, Edwin B.: *Building Design for Maintainability,* McGraw-Hill, New York, 1975.

Feldman, Edwin B.: *Housekeeping Handbook for Institutions, Business and Industry,* Frederick Fell Publishers, New York, 1978.

General Industry Standards and Interpretations, U.S. Department of Labor, Occupational Safety and Health Administration, 1977.

"Is Your Facility Affected by the OSHA Bloodborne Standard?" *Executive Housekeeping Today*, September 1992.

Meyers, Earl M.: "Standardized Housekeeping Program Improves Utilization of Manpower and Materials," *Maintenance Engineering,* October 1972.

"Pesticide Assessment Guidelines," PB83-153294, Environmental Protection Agency, Washington, D.C., pp. 51–53.

Ruhlin, Robert R.: "Work Control: A Sure Way to Improve Building Maintenance," *Buildings,* February 1974.

"292 Cleaning Times," International Sanitary Supply Association Inc. (ISSA), Lincolnwood, Ill.

CHAPTER 4.12
NOISE CONTROL

Douglas H. Sturz
Senior Acoustical Consultant
Acentech Incorporated
Cambridge, Massachusetts

Jeffrey L. Fullerton
Acoustical Consultant
Acentech Incorporated
Cambridge, Massachusetts

INTRODUCTION

Noise produced inside industrial plants is most often controlled to provide a healthy, safe, and comfortable work environment. Controlling excessive workplace noise can reduce the occurrence of occupationally induced hearing loss among employees, reduce future payments associated with hearing loss compensation claims by employees, improve speech communication and the audibility of warning signals, improve productivity by reducing employee fatigue and discomfort, and help to enhance job satisfaction. Noise produced outside industrial plants is most often controlled to promote good neighborhood relations, to avoid degrading the local environment, and to meet local or state regulations. To be most effective, steps taken to control excessive noise must avoid or minimize impact on production, inspections, maintenance, and safety.

This chapter includes basic information about the measurement, evaluation, and control of noise at industrial plants. It does not provide "cookbook" solutions for controlling noise. Rather, engineering noise control information is based on the authors' experience gained from acoustical consulting practice.

Readers interested in specific or advanced noise control information are referred to the end of this chapter. Listed there are many texts, manuals, standards, and publications specializing in noise control that the authors have found to be useful.

OCCUPATIONAL NOISE EXPOSURE LIMITS

OSHA Safety and Health Standards (29 CFR 1910.95) provide maximum daily occupational noise exposure limits that give both the level and the duration of the noise to which employees may be exposed while in the workplace. This Standard allows 8-h exposures to a daily average noise level of up to 90 dBA. For shorter durations, the maximum limits are greater. A 4-h exposure is allowed to a daily average noise level of 95 dBA *or* a 2-h exposure is allowed to a daily average noise level of 100 dBA. Exposure to continuous noise levels as great as 115 dBA is permitted for 15 min or less per day. It is the employee's time-weighted average (TWA) exposure or noise dose received while at work that is subject to the basic OSHA noise limits. These noise exposure limits and the method to calculate TWA exposures are listed in Table 4.71. In addition, the Standard also limits exposure to impulsive or impact noise—for example, from a punch press—to no greater than a 140-dB peak sound pressure level.

TABLE 4.71 Permissible Noise Exposures*

Daily duration, h	Sound level, dBA
8	90
6	92
4	95
3	97
2	100
1.5	102
1	105
0.5	110
0.25 or less	115

* When the daily noise exposure is composed of two or more periods of noise exposure of different levels, their combined effect should be considered, rather than the individual effect of each. If the sum of the fractions: $C_1/T_1, + C_2/T_2 + C_n/T_n$ exceeds unity, then the mixed exposure is considered to exceed the limit value. C_n indicates the total time of exposure at a specified noise level, and T_n indicates the total time of exposure permitted at that level.

OCCUPATIONAL HEARING CONSERVATION REQUIREMENTS

In addition to the Table 4.71 noise exposure limits, OSHA Safety and Health Standards provide the requirement that employers administer an effective hearing conservation program to protect employees who may receive 8-h TWA noise exposures exceeding 85 dBA (a noise dose exceeding 50 percent). The 8-h TWA of 85 dBA is considered the "action level" at which an effective hearing conservation program becomes mandatory. No requirements are imposed when 8-h TWA noise exposures are less than 85 dBA (a noise dose less than 50 percent). The Standard describes in detail the many elements that must be included in an effective hearing conservation program. Some of these elements are summarized in the following text. Readers responsible for establishing or administering such programs are referred to the Standard itself and the many professionals who specialize in this field.

Feasible administrative, scheduling, or engineering controls should be utilized to reduce TWA noise exposures to within the limits of Table 4.71. Hearing protection must be provided and used to achieve the TWA noise exposure limits of Table 4.71 whenever administrative, scheduling, or engineering controls fail to do so.

In plant areas where 8-h TWA noise exposures might exceed 85 dBA, monitoring must be performed to document noise levels or noise exposures. The monitoring must be repeated as necessary to ensure that results are kept up to date. Records of the monitoring results must be maintained and monitoring instruments must be calibrated.

Notification is required for all employees likely to receive TWA noise exposures greater than 85 dBA.

In addition, observation of the measurements must be allowed by employees likely to receive TWA noise exposures greater than 85 dBA and audiometric testing must be made available to these employees.

Hearing protectors must be made available at no cost to employees likely to receive TWA noise exposures greater than 85 dBA. A suitable variety of hearing protectors that are comfortable, well maintained, and effective must be provided, along with training in the fitting, care, and use of the hearing protectors. Effective hearing protectors must be worn by employees with a TWA noise exposure exceeding 90 dBA, and by some employees with a TWA noise exposure greater than 85 dBA. Although not required by the Standard, it is suggested that employees be required to wear hearing protectors every time they enter noisy work areas and encouraged to purchase hearing protectors for use at home when shooting or operating noisy equipment. For hearing protectors to be effective they must fit well, be comfortable and well maintained, and, of course, be worn.

Earplugs and earmuffs are the two most common types of hearing protectors used in industrial plants. Earplugs are generally of the formable or premolded type. Formable plugs made of an expandable closed-cell foam material are commonly used at many plants and can be effective at reducing noise at the user's ears when inserted properly. Premolded plugs made of soft silicone, rubber, or plastic can also be effective. Failure to achieve a correct and snug fit will limit the noise attenuation of any plug. Workers should be instructed in how to straighten the ear canal and insert the plug correctly. While such instruction may appear unnecessary, many workers may need several tries before achieving a fit snug enough to get the best possible noise attenuation.

Earmuffs are also commonly used in many industrial plants and can be effective at reducing exposure when worn properly. Earmuffs consist of two plastic cups with interior sound-absorptive foam and soft seals that cover the entire external ear. They are held in place with a headband or are sometimes attached to a hard hat. Again, workers should be provided instruction in the proper use of earmuffs. Eyeglasses or long hair can cause a gap at the muffs' seals and reduce the noise attenuation.

Several types of effective earmuffs and/or earplugs should be made available to employees. They should be selected based on the type of noise to be attenuated and the degree of noise attenuation needed. They should also be comfortable to wear and easy to use. It is important that plant management support and encourage their use whenever workers enter noisy areas. Hearing protectors only work when worn properly!

Hearing protector attenuation must be evaluated for effectiveness in the specific noise environment in which the protector will be used. Hearing protectors provided to employees must attenuate noise exposures to an 8-h TWA of 90 dBA or less. For employees with significant hearing loss, hearing protectors must be provided that attenuate noise exposures to an 8-h TWA of 85 dBA or less. Manufacturers of hearing protectors provide a noise reduction rating (NRR) for their protectors based on laboratory testing. The published NRR is often much greater than the actual noise attenuation that is achieved by typical employees in real-world industrial working environments. A conservative estimate of the actual real-world attenuation provided by hearing protectors in typical industrial work environments is approximately equal to one-half of the NRR reduced by 7 (0.5 [NRR-7]). For example, a hearing protector with an NRR of 39 could be expected to reliably reduce typical industrial noise exposures by about 16 dBA. In a plant work area where the noise level is 100 dBA, a hearing protector rated at 39 NRR would reduce the noise to about 84 dBA at the worker's ear. For some work environments the actual noise attenuation might be less. This illustrates the importance of purchasing high-quality and high-performance hearing protectors.

Training programs must be provided at least annually to employees with 8-h TWA noise exposures greater than 85 dBA. These programs must be up to date and must inform employees of at least the following: the effects of noise on hearing; the purpose of hearing protectors; the performance of the various types of hearing protectors; and instructions on selection, fitting, use, and care of the hearing protectors. Employees must also be informed of the purpose and procedures of the audiometric testing.

The OSHA Noise Standard must be made available to all affected employees and a copy of it must be posted in the workplace.

Accurate records of all employee exposure measurements must be maintained for at least two years. It is suggested that the records be well organized and maintained for a longer period in case of future hearing-loss compensation claims. Records of all audiometric test results must be maintained for the duration of an employee's employment. Again, it is suggested that they be well organized and maintained for a longer period in case of hearing-loss compensation claims. Copies of these records must be available to employees upon their request.

The Standard also includes many appendices providing detailed information on the required audiometric testing of employees, including computations, measurements, calibrations, record keeping, and qualifications. People responsible for administering hearing conservation programs must be familiar with and understand the requirements contained in the appendices.

For a hearing conservation program to be effective, it must be well supported by plant management. Without management-level support and encouragement, the program is not likely to succeed at protecting employees' hearing. The use of hearing aids is not a good answer to hearing loss, just as the use of wooden legs, mechanical hands, and false teeth is not a good answer to industrial accidents. The best answer is *prevention*—occupational hearing loss is preventable!

CONCEPTS AND VOCABULARY OF NOISE CONTROL

Sound Noise is simply unwanted sound, which is a series of vibrations in the air. Human ears are sensitive to these vibrations; they sense them and pass them on to the brain to decipher. Acoustics—the branch of physics that deals with the production, transmission, and control of sound—has its own concepts and vocabulary. Several key terms and concepts needed by engineers working in noise control are defined in the following text.

Frequency Frequency f is the number of oscillatory cycles a sound completes in one s. Units of frequency may be expressed in cycles per second (cps) or Hertz (Hz).

The audible frequency range extends from about 20 to 20,000 Hz for a young person with ideal hearing, but for adults this range is frequently narrower due to aging and noise exposure effects. For engineering analysis purposes, the audible frequency range is often divided into a series of octave bands. Just like an octave on a piano keyboard, an octave in sound analysis represents the frequency interval between a given frequency and twice that frequency. For ease of use and to develop a consistent basis for communication, standard octave bandwidths have been defined and are built into measurement instrumentation.

Each of the standard bands is identified by the frequency at the geometric mean of the frequencies at the extremes of the band. The center frequencies and approximate cutoff frequencies are listed in Table 4.72.

TABLE 4.72 Center and Approximate Frequency Limits for the Standard Set of Contiguous Octave Bands Covering the Audio Frequency Range

Lower band limit	Octave band center frequency	Upper band limit
11	16	22
22	31.5	44
44	63	88
88	125	177
177	250	355
355	500	710
710	1,000	1,420
1,420	2,000	2,840
2,840	4,000	5,680
5,680	8,000	11,360
11,360	16,000	22,720

Velocity of sound Sound waves in air at normal temperatures and pressures (about 68°F [20°C] and 1 atmosphere pressure) travel at a velocity c that is approximately 1127 ft/s (344 m/s). The velocity of sound waves in air changes somewhat with temperature and pressure, but for most practical applications in plant engineering it can be considered a constant.

Wavelength The distance that a sound wave travels in completing one cycle is the wavelength λ and can be calculated at any frequency by:

$$\lambda = \frac{c}{f}$$

Decibel The decibel (dB) is a dimensionless unit for expressing the ratio of two numerical values on a logarithmic scale. It is convenient to use decibels in dealing with sound power, intensity, or pressure because of the tremendous range of values of these quantities that can be perceived by the ear. Audible intensities range from 10^{12} to 1 W/m^2.

Sound power and sound power level Sound power describes the total acoustical energy emission of a sound source in watts (W). The sound power level (L_w or PWL) is the designation in decibels of the ratio of two sound powers expressed as follows:

$$L_w = 10 \log \frac{W}{W_r}, \text{ Customarily } L_w = 10 \log \frac{W}{10^{-12}} \text{ dB}$$

These terms can be applied to the entire frequency spectrum or to a narrow bandwidth. These quantities can be thought of as being analogous to the total light energy emitted by a light bulb. They are independent of the environment and the distance from the source. The customary reference sound power W_r in use today and in this chapter is 10^{-12} W. Originally, 10^{-13} W was used as the standard reference sound power. When the reference sound power is 10^{-13}, the sound power level can be converted to a sound power level referenced to 10^{-12} by subtracting 10 dB. To avoid ambiguity, the reference sound power should always be stated with sound power level data.

Sound pressure and sound pressure level Sound waves produce small changes in the density of air as they travel through it. These changes in the density cause pressure fluctuations around the ambient static pressure. The magnitude of the pressure fluctuations above and below the ambient pressure is the sound pressure p.

The unit approved by the International Standards Organization (ISO) for measuring sound pressure is the pascal (Pa), though the terms microbars (μbar), dynes per square centimeter (dyn/cm^2), and newtons per square meter (N/m^2) have all been used.

Conversion factors are:

$$1 \text{ bar} = 10^5 \text{ Pa}$$
$$1 \text{ μbar} = 1 \text{ dyn/cm}^2$$
$$1 \text{ dyn/cm}^2 = 10^{-1} \text{ Pa}$$
$$1 \text{ N/m}^2 = 1 \text{ Pa}$$

The sound pressure level (L_p or SPL) is the designation in decibels of the ratio of the square of two sound pressures expressed as:

$$L_p = 10 \log \left(\frac{p}{p_r}\right)^2 \quad \text{or} \quad L_p = 10 \log \left(\frac{p}{2 \times 10^{-5}}\right)^2 \quad \text{dB}$$

The customary reference sound pressure p_r is 2×10^{-5} Pa and should be indicated with all data to avoid ambiguity. This reference sound pressure is set to be approximately equal to the threshold of hearing at 1000 Hz. So a sound pressure level of 0 dB at 1000 Hz corresponds to the approximate threshold of hearing.

The sound pressure level can be thought of as analogous to the light level in foot-candles from a light bulb in a room. Just as the light level is a function of the distance from the bulb and the color of the walls of the room, so the sound pressure level in a room is a function of the distance from the source and the acoustical characteristics of the room. The distance from the source, and the acoustical characteristics of the space in which the data were measured, or are to apply, should be stated with sound pressure level data.

Sound intensity and sound intensity level Sound intensity I is the sound power radiated in a specified direction through a unit area normal to the direction of propagation in units of W/m^2. The sound intensity level (IL) is the designation in decibels of the ratio of two intensities and is expressed by:

$$IL = 10 \log \left(\frac{I}{I_r}\right)$$

When indicating IL, it is customary to use a reference intensity of 10^{-12} W/m^2, which should be indicated to avoid ambiguity.

$$IL = 10 \log \frac{I}{10^{-12}} \quad \text{dB}$$

IL is a function of the distance from the source but, as typically used, does not take the environment into account. It is analogous to the amount of light energy from a light bulb that falls on your hand in a room at a particular distance from the bulb due to the direct emission from the bulb only and discounting any light energy that may fall on your hand due to reflections from the room surfaces. Sound intensities are more difficult to measure than sound pressures, and prior to recent developments in instrumentation this was not practically possible in the field. Even the latest instrumentation devices still measure pressures only (using two microphones instead of one), then internally calculate the associated intensities based on the phase relationship of the data from the two microphones. Measuring sound intensity can be useful in investigating a sound source in a reverberant space because associated with the magnitude is a direction of wave propagation. Sound pressure level measurements do not include information about the direction of sound propagation.

A-weighted sound level There are many single-number schemes for evaluating sounds according to people's response to how loud a noise is perceived to be. One of the simplest and most useful of these single-number schemes is the sound level in dBA. The **A** means that the frequency spectrum has been weighted by a specific electrical network in the sound-measuring equipment (or manually) before the single-number level is derived. It is simple to measure the A-weighted sound level, and, fortunately, this single number often correlates with people's perceptions as well as or better than most other single-number noise rating schemes. The frequency response of the A-weighting network in standard octave bands is shown in Table 4.73. To determine the A-weighted sound level from unweighted octave band levels, simply apply these factors to the levels in the various bands and add the weighted octave band levels as decibels.

TABLE 4.73 A-Scale Weighting Factors

Octave band center frequency, Hz	Approximate A-weighting relative response, dB
31.5	−39
63	−26
125	−16
250	−9
500	−3
1000	0
2000	+1
4000	+1
8000	−1

Table 4.74 shows some typical A-weighted sound levels in industrial environments.

TABLE 4.74 Typical A-Weighted Sound Levels in Industrial Plants, Near Various Equipment

Equipment	dBA
Synthetic spinning machine	95
Rock crusher	101
Letterpress	96
Hammer mill	96
Hand-held sand blaster	95
Plastic extruder	97
Candy wrapper	90
Fly shuttle loom	102
Can filling/seamer	98
Ring twister	94
Chipper	105
Wood chipper	109
Billet heater	102
Buffing machine	101
Punch press	102
Wire-drawing machine	96
Molding machine	98
Concrete-block machine	106

Addition of decibels Since decibels are logarithmic units, they are not added arithmetically. Decibels are added by converting them to power, intensity, or pressure; adding these quantities arithmetically; and then converting them back to decibels. The mathematics involved may be somewhat complex for people not accustomed to working with logarithms, so charts have been developed for convenience. See Fig. 4.97 for one example of such a chart. To add the sound levels of 80 dB and 74 dB that are produced individually by

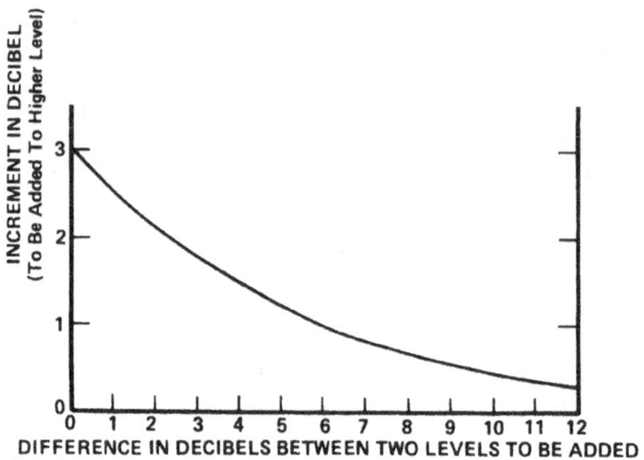

FIGURE 4.97 Chart for combining sound levels.

two separate sources, note that the difference between the two levels is 6 dB. Enter the chart in Fig. 4.97 at the 6 on the horizontal axis and draw a line straight up to the curve. From the intersection of the vertical line and the curve, draw a horizontal line over to the vertical axis and read the decibel increment (1 dB) to be added to the higher of the two levels being added (80 dB). The total level is thus 81 dB. Similarly, a 90-dB sound level added to another 90-dB sound level results in 93 dB.

Sound propagation outdoors—the inverse square law Under *free field* conditions—away from any nearby surfaces to reflect or disturb the sound field—sound radiates spherically away from a source. The acoustical energy is radiated to progressively larger spherical shells as it moves outward from the source. The acoustical energy per unit area (the intensity) is thus inversely proportional to the surface area of the spherical shell and this area is a function of the square of the radius of the shell. Hence, sound levels observed at any distance from a source are said to follow an inverse square relationship.

Two forms of the inverse square relationship that are useful for analyzing everyday acoustical problems are shown below in equation form. The first estimates the sound pressure level at distances away from a source of known sound power level under free field conditions:

$$L_p = L_w + 10 \log \left(\frac{Q}{4\pi r^2} \right) + C \text{ dB}$$

where Q = the directivity of the source, which is equal to the inverse of the portion of a sphere to which the sound is radiated, i.e., for spherical radiation, $Q = 1$; for hemispherical radiation, $Q = 2$; for radiation of a quarter of a sphere, $Q = 4$, etc.

C = 0 if r is expressed in m, and 10.5 if r is expressed in ft.

This relationship applied in reverse is also useful for estimating the sound power level of a source from sound pressure levels that are known at some distance from the source:

The second equation allows calculation of the sound pressure level at one distance from a source, given the sound pressure level at some other known distance.

$$Lp_1 = Lp_0 - 10 \log(r_1^2/r_0^2) = Lp_0 - 20 \log(r_1/r_0)$$

From both these equations, it can be seen that the sound pressure level in the free field falls off at the rate of 6 dB per doubling of distance from the source.

Absorption coefficient The ability of a material to absorb or dissipate sound energy incident on its surface is of interest in assessing and treating acoustical problems. The sound absorption coefficient, α, indicates the percentage, in decimal form, of the incident sound that is absorbed (removed from the sound wave) through interaction with the surface. Sound absorption coefficients range from 0, for no absorption, to 1, for complete absorption. Some laboratory test data give absorption coefficients greater than 1, indicating that the surface absorbed more sound energy than was incident upon it. Clearly this cannot be; the data are the result of acoustical complexities and the standard laboratory test methodology. Absorption coefficients of materials are functions of frequency, the mounting condition of the material, and sometimes the angle of incidence of the sound upon the surface. Tables 4.75 and 4.76 show typical published data for random incidence sound. Note also that the percentage of sound that is reflected from a surface is $1 - \alpha$.

Room absorption The total amount of sound absorption in a room can be determined by summing the products of the area of each type of surface material in a room times its absorption coefficient. Often this must be done separately for all frequency bands that are of interest. The term that is typically applied to this quantity for noise control purposes is the room

TABLE 4.75 Representative Sound Absorption Coefficients of General Building Materials and Furnishings*

Materials	Octave band coefficients					
	125	250	500	1000	2000	4000
Brick, unglazed	0.03	0.03	0.03	0.04	0.05	0.07
Brick, unglazed, painted	0.01	0.01	0.02	0.02	0.02	0.03
Carpet, heavy, on concrete	0.02	0.06	0.14	0.37	0.60	0.65
Same, on 40-oz hair felt or foam rubber	0.08	0.24	0.57	0.69	0.71	0.73
Same, with impermeable latex backing on 40-oz hair felt or foam rubber	0.08	0.27	0.39	0.34	0.48	0.63
Concrete block, coarse	0.36	0.44	0.31	0.29	0.39	0.25
Concrete block, painted	0.10	0.05	0.06	0.07	0.09	0.08
Fabrics						
Light velour, 10 oz/yd², hung straight, in contact with wall	0.03	0.04	0.11	0.17	0.24	0.35
Medium velour, 14 oz/yd², draped to half area	0.07	0.31	0.49	0.75	0.70	0.60
Heavy velour, 18 oz/yd², draped to half area	0.14	0.35	0.55	0.72	0.70	0.65
Floors						
Concrete or terrazzo	0.01	0.01	0.01	0.02	0.02	0.02
Linoleum, asphalt, rubber, or cork tile on concrete	0.02	0.03	0.03	0.03	0.03	0.02
Wood	0.15	0.11	0.10	0.07	0.06	0.07
Wood parquet in asphalt on concrete	0.04	0.04	0.07	0.06	0.06	0.07
Glass						
Large panes of heavy plate glass	0.18	0.06	0.04	0.03	0.02	0.02
Ordinary window glass	0.35	0.25	0.18	0.12	0.07	0.04
Gypsum board, ½ in nailed to 2 × 4's 16-in o.c.	0.29	0.10	0.05	0.04	0.07	0.09
Marble or glazed tile	0.01	0.01	0.01	0.01	0.02	0.02
Openings						
Stage, depending on furnishings			0.25–0.75			
Deep balcony, upholstered seats			0.50–1.00			
Grills, ventilating			0.15–0.50			
Plaster, gypsum or lime, smooth finish on tile or brick	0.013	0.015	0.02	0.03	0.04	0.05
Plaster, gypsum or lime, rough finish on tile or brick	0.14	0.10	0.06	0.05	0.04	0.03
Same, with smooth finish	0.14	0.10	0.06	0.04	0.04	0.03
Plywood paneling, ⅜-in thick	0.28	0.22	0.17	0.09	0.10	0.11
Water surface, as in a swimming pool				0.02	0.02	0.02
Air, sabins per 1000 ft³ at 50% RH	0.01	0.01	0.01	0.9	2.3	7.2

* Complete tables of coefficients of the various materials that normally constitute the interior finish of rooms may be found in the various books on architectural acoustics. The following short list will be useful in making simple calculations of the reverberation in rooms.

TABLE 4.76 Representative Sound Absorption Coefficients of Common Acoustic Materials

Materials*	Octave band coefficients					
	125	250	500	1000	2000	4000
Fibrous glass (typically 4 lb/ft³) on solid backing						
1 in thick	0.07	0.23	0.48	0.83	0.88	0.80
2 in thick	0.20	0.55	0.89	0.97	0.83	0.79
4 in thick	0.39	0.91	0.99	0.97	0.94	0.89
Polyurethane foam (open cell) on solid backing						
¼ in thick	0.05	0.07	0.10	0.20	0.45	0.81
½ in thick	0.05	0.12	0.25	0.57	0.89	0.98
1 in thick	0.14	0.30	0.63	0.91	0.98	0.91
2 in thick	0.35	0.51	0.82	0.98	0.97	0.95

* For specific grades, see manufacturer's data.

constant R.* The units of R are designated sabins and are in ft² or m². Take care to distinguish between English unit sabins or metric unit sabins when applying various formulas.

$$R = \sum S_1\alpha_1 + S_2\alpha_2 \ldots\ldots S_n\alpha_n \quad \text{m}^2\ (\text{ft}^2)$$

where S_1, S_2 and S_n are the areas of the individual materials.

Sound propagation indoors The sound pressure level in a room due to a particular sound source can be estimated using the following relationship:

$$L_p = L_w + 10 \log\left(\frac{Q}{4\pi r^2} + \frac{4}{R}\right) + C \text{ dB}$$

where R is the room constant in units consistent with the units of r, and C is the appropriate conversion term for the units of r and R (0 dB if metric, 10.5 dB if English).

The first quantity within the parentheses is the direct sound field and is identical to the equation given previously for spherical spreading of sound in the free field (without reflecting surfaces nearby). The second quantity within the parentheses is the reverberant sound field, which relates to how much sound absorption there is in the room and hence how much the sound is reflected within the room. The resulting total sound level is the decibel sum of the direct and reverberant sound levels. Note that the reverberant portion of the sound field is not a function of the distance from the source; it is essentially constant throughout the room. In typical rooms, the sound level due to a source diminishes at the rate of 6 dB per doubling of distance from the source until the direct field sound level equals the reverberant field sound level. Beyond this distance, the sound level remains essentially constant. Actually, in the transition region between the direct and reverberant fields the two levels add, resulting in a smooth transition between the regions. The relationship for the sound pressure level indoors is shown graphically in Figure 4.98.

Noise reduction by sound absorption Sound that originates at a source in an enclosed factory space will spread until it reaches the surfaces, where it will either be absorbed or reflected. If the room surfaces are hard (reflective), sound will reverberate, intermittent sounds will be mixed together, and steady sounds will add up. The result may be a relatively noisy space. If room surfaces are soft (absorptive), the space will be less noisy, because the

*The room constant R is technically the sum of $S_n\alpha_n/1 - \alpha_n$, but for many practical applications and especially for rooms with relatively little sound absorption, R is closely approximated by the sum of the products of the surface area times its respective absorption coefficient.

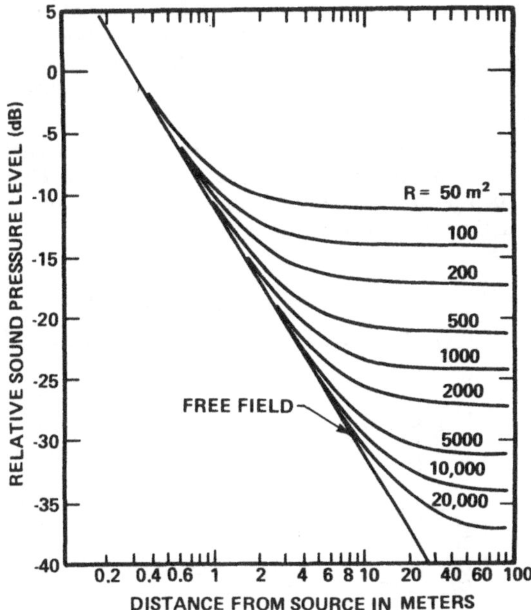

FIGURE 4.98 Drop-off of sound pressure level as a function of distance from sound source and room constant.

aforementioned effects will be minimal. The noise level in the reverberant portion of the room, well away from any particular sound source, is a function of how much sound absorption is in the room. The relative noise reduction in the reverberant field that can be obtained through the introduction of sound absorption to the space can be estimated as follows:

$$NR = 10 \log \left(\frac{R_1}{R_0} \right) \qquad \text{dB}$$

where R_0 is the original amount of sound absorption in the room and R_1 is the new amount of sound absorption.

For each doubling of the sound absorption in a room, the reverberant field noise level is reduced by 3 dB.

Transmission coefficient The ability of a material to block the transmission of sound is of interest in analyzing and treating acoustical problems and is given by the sound transmission coefficient τ, which is the fraction of incident sound energy that is transmitted through a barrier material. Thus:

$$\tau = \frac{W_2}{W_1}$$

where W_1 is the incident sound energy in watts and W_2 is the sound energy transmitted in watts.

Transmission loss The sound transmission loss TL of a barrier material is defined as the ratio of the transmitted sound energy to the incident sound energy.

In decibel form, this ratio is expressed as:

$$TL = 10 \log \frac{W_1}{W_2} = 10 \log \frac{1}{\tau} \qquad \text{dB}$$

To calculate in detail the sound transmission loss for even simple constructions is difficult. As a result, engineers usually rely on laboratory test data. The transmission loss of a particular material has to do with the material only and is only part of the equation to determine what noise level will occur on the opposite side of a barrier. See the paragraph on noise reduction on the following page.

Mass law For each doubling of the weight of a single-material construction, the average sound transmission loss increases by about 6 dB. Figure 4.99 shows a graphic presentation of the transmission loss of a single wall. The empirically determined curve (solid) is lower than the theoretical curve (dashed).

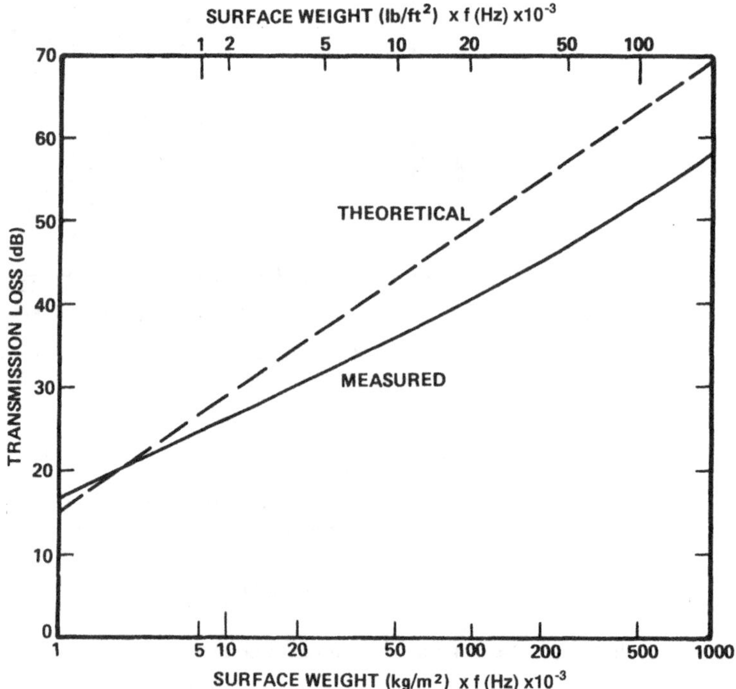

FIGURE 4.99 Average transmission loss of single homogeneous walls in the 100- to 3150-Hz frequency range.

Composite sound transmission loss The composite sound transmission loss of a multi-element barrier is estimated from the transmission coefficients for each part. The sound power transmitted by the elements with a common incident sound power is:

$$W_{\text{trans}} = W_{\text{inc}}\,(\tau_1 S_1 + \tau_2 S_2) \qquad \text{W}$$

where τ_1 and τ_2 are the transmission coefficients for the individual parts, and S_1 and S_2 are the areas of the individual parts in ft^2 (m^2).

The composite transmission loss becomes:

$$TL = 10 \log \frac{S_1 + S_2}{\tau_1 S_1 + \tau_2 S_2} \qquad \text{dB}$$

Noise reduction The difference in sound pressure levels between two rooms is called the noise reduction (*NR*). *NR* accounts for the total amount of energy being radiated through the common construction by way of the area term and the attenuating effect of sound absorption in the receiving room.

This noise reduction can be expressed as:

$$NR = TL - 10 \log \frac{S}{A_2} \qquad \text{dB}$$

where S = area of common wall, ft^2 (m^2)
A_2 = total sound absorption in the receiving room, ft^2 – sabin (m^2 – sabin)

HUMAN RESPONSE TO NOISE

Human response to noise is as varied as people and the range of potential noise sources. The noise of a mosquito or dripping sink may prevent one from sleeping, but the louder noise of surf at the beach or rain on the roof may induce sleep. The noise produced by operating machinery may be acceptable at a manufacturing plant, but intrusive and unacceptable at a nearby residential area. One person might hardly notice the noise produced by an ultrasonic welder and another person might find it unacceptable. Exposure to mid-frequency and high-frequency noise causes far more damage to hearing than exposure to low-frequency noise of the same level. Investigators have spent decades studying human response to noise, and many books have been written on the subject. This section provides a brief overview.

The pressure oscillations of the sound in the air cause the eardrum and small bones of the middle ear to vibrate. These vibrations are transmitted to the fluid-filled cochlea, the inner ear's sensory organ. Sensory hair cells that line the cochlea translate these vibrations into nerve impulses that are transmitted to the brain, where they are perceived and interpreted. Exposure to loud sound levels of industrial noise may cause temporary hearing loss by overstressing the sensitive hair cells in the cochlea. With time away from the noise, normal hearing typically returns. Long-term exposure to high levels of industrial noise destroys some of the sensory hair cells and causes permanent hearing loss because there are fewer functioning sensory cells. No treatment has yet been found to repair the cells and restore noise-induced hearing loss.

A faint sound level is in the range of 10 to 20 dB. A moderate sound level is about 50 to 60 dB, and a loud sound level is about 90 to 100 dB. Sound levels greater than about 130 dB often cause pain in the ear.

When the magnitude of a noise is increased (or decreased) by 10 dB, this is often perceived as a doubling (or halving) of the loudness.* Changing the magnitude of a noise by 5 to 6 dB is generally considered significant, and is clearly noticeable. Changing the magnitude of a noise by less than 2 to 3 dB is usually just noticeable.

The frequency of a noise is analogous to its tonal quality or pitch. The fundamental frequency of middle C on a piano keyboard, for example, is 262 Hz. A tuning fork produces sound at a single frequency, often called a *discrete tone.* Transformers produce sound at several discrete frequencies that are even multiples of line frequency. In the United States, transformer noise is concentrated at 120, 240, 360, 480, and 600 Hz. However, most sounds include a composite of many frequencies and are characterized as random or broadband. Rotating equipment such as fans and motors usually produces both broadband and discrete tonal noise.

The normal frequency range of human hearing extends from a low-frequency rumble at about 20 to 250 Hz up to a high-frequency hiss at about 4,000 to 12,000 Hz. People have different hearing sensitivity to different frequencies and generally hear best in the mid-frequency range that is common to human speech, from about 500 to 4000 Hz.

The noise environment in most plants and communities varies from place to place and varies with time at any given location due to the composite of many noise sources. The noise

*Provided the character of the sound is not appreciably changed.

environment may include intermittent high-noise single events and slowly varying background noises. The frequency and amplitude statistics of most noise environments are rather complex.

At any one location, a complete physical description of the noise environment might include its noise level at various frequencies as a function of time. It is common practice to simplify this multidimensional description by eliminating the frequency variable and measuring the A-weighted noise level, as observed on a standard sound level meter. The A-weighting filter emphasizes the mid-frequency components of the noise and demphasizes the low-frequency components in order to approximate the sensitivity of the human ear.

For certain applications, it is also common practice to further simplify this multidimensional description by eliminating the temporal variable and measuring the equivalent sound level (Leq) or OSHA sound level (Losha) time average sound level, as observed on standard dosimeters and modern sound level meters. Leq is often used in community noise analysis. It is also used during the analysis of employee noise exposures in many countries other than the United States. Losha is often used in employee noise exposure analysis in the United States.

These simple measures are convenient and sufficient for many applications. However, a more sophisticated frequency analysis is often necessary to evaluate fully a community's response to intruding noise from a plant. Detailed frequency analysis is almost always necessary for engineering design or specification of noise control treatments.

Many acoustic rating scales and rating procedures have been developed by investigators interested in assessing people's reactions to intruding noise and evaluating adverse health effects from noise exposures. Several of the scales and procedures that are currently popular for assessing or evaluating plant noise follow.

Overall sound pressure level (Lp in dB)
A-weighted sound level or noise level (LA in dBA)
C-weighted sound level or noise level (Lc in dBC)
Energy equivalent sound level or noise level (Leq in dB)
OSHA equivalent sound level or noise level (Losha in dB)
Day-night sound level or noise level (Ldn in dB)
Speech interference level (SIL)
Noise criterion curves (NC)
Room criterion curves (RC)
Balanced noise criterion curves (NCB)

Definitions and uses of these acoustic rating scales and rating procedures can be found in acoustical references and textbooks.

MATERIALS SELECTION

The most commonly used materials for control of noise in industry are sound-absorbing and barrier materials for airborne sound, and vibration isolators and dampers for structureborne sound. Selection of materials is often affected by factors other than acoustics. Factors include:

- Vibration
- Corrosion
- Temperature
- Erosion by fluid flow
- Cost

- Appearance
- Structure
- Weight
- Clogging
- Restrictions on materials near food processing lines
- Requirements for materials not to be damaged by disinfecting
- Firebreak requirements on ducts, pipe runs, shafts
- Flame spread rate limits
- Fire endurance limits
- Restrictions on shedding of fibers in air by acoustically absorbing materials
- Elimination of uninspectable spaces in which vermin may hide
- Requirements for secure anchoring of heavy equipment
- Restrictions on hole sizes in machine guards (holes can reduce radiated noise of vibrating sheets)

CONTROL OF PLANT NOISE

Noise is controlled inside plants to provide a safe, healthful, and comfortable work environment. Noise is controlled outside plants to avoid degrading the neighborhood environment. Noise can be controlled directly at the *source,* such as a fan or transformer, where it is produced. Noise can also be controlled along the *paths* that it travels from the source to the receiver, for example, by the use of a barrier or a muffler. Noise can also be controlled at the *receiver,* for example, by earmuffs or earplugs, both of which were discussed earlier. This section addresses control of plant noise at the source and along propagation paths.

Some common methods used to control plant noise are barriers, enclosures, absorption, mufflers, lagging or wrapping, damping, isolation, equipment selection, and improved maintenance. An additional, newer method that is available to reduce plant noise is active noise control. Each of these methods is discussed subsequently. Further information about noise control techniques and costs is available from noise control material manufacturers and publications. Some manufacturers have simple computerized selection programs available.

Mufflers

Mufflers, also called silencers or attenuators by their manufacturers, are located in pipes, ducts, and passageways used to convey gas, liquid, or solid materials to or from a machine, process, or area for the purpose of reducing the propagation of noise from the source to the receiver. Mufflers are commonly used at the intake and/or exhaust of engines, fans, and compressors and at the outlets of high-pressure gas vents.

The muffler must be designed, specified, and selected such that its dynamic acoustic insertion loss is adequate to meet the required noise attenuation design goal established for the specific application. In addition, noise that radiates from the outer shell of the muffler must be adequately controlled, which is often accomplished by constructing it of sufficiently heavy material or by wrapping it with an acoustic lagging. Achieving high noise attenuation through a muffler usually requires that flexible connections be installed between the muffler and the noise source to reduce structureborne flanking and minimize shell-radiated noise. Also, noise caused by high-velocity gas flow at the muffler discharge (self-noise) must be adequately controlled for the specific application. Gas flow speed at the discharge should usually be low enough to achieve self-noise levels at least 10 dB less than the required design goal. The temperature of the gas within the muffler must be considered because the wavelength of sound increases as temperature increases and muffler performance varies with wavelength.

In harsh environments, mufflers must be designed and constructed to withstand erosion, corrosion, thermal stresses, and sometimes relatively high vibration caused by the noise source or turbulent gas flows. Mufflers installed in dirty air flows must be designed to avoid becoming clogged. In these applications, it is often necessary to specify and custom design open-cavity-type, nonclogging, tuned-dissipative mufflers. Mufflers installed in a fluid stream often must also be designed, constructed, and installed so as to avoid causing excessive pressure losses and the consequential impact on flow.

Dissipative mufflers, often in the form of parallel baffles used at the inlet and outlet of a noisy fan, provide noise attenuation through the use of a fibrous sound-absorptive material, such as glass-fiber insulation. Sound waves passing through the muffler enter the fibrous material and are absorbed or dissipated as movements of air molecules are converted to heat through friction effects. Dissipative mufflers are most often in the form of (1) a rectangular duct section containing absorptive parallel baffles or (2) a circular duct section containing absorptive material on the outer wall and/or an absorptive element ("bullet") on the duct center line. Figure 4.100 illustrates the approximate insertion loss achievable with 90-cm-long mufflers designed with different spacing between the baffles and baffle thickness. Note that the insertion loss of these dissipative mufflers is greatest at mid-frequencies and least at low and high frequencies.

The glass fiber material within dissipative mufflers must be protected with woven glass cloth and/or steel screen when exposed to high-velocity gas flows. When used in wet environments or environments that need to be especially clean, the glass fiber material is often encapsulated in thin plastic bags to provide protection.

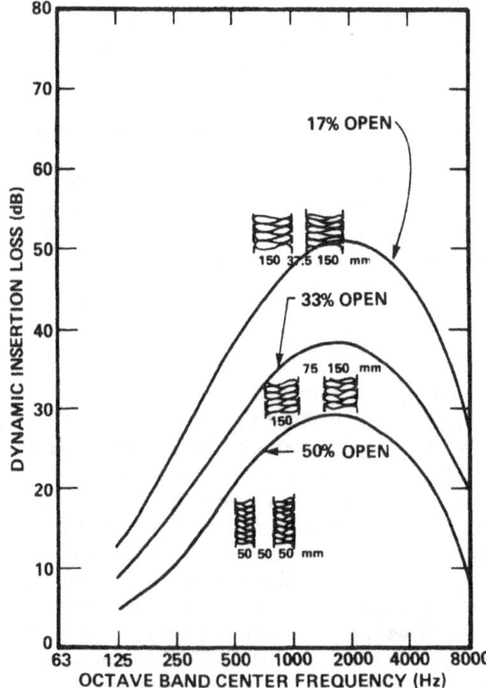

FIGURE 4.100 Acoustical performance of 90-cm (3-ft)-long mufflers with different spacing and thickness of baffles.

Reactive mufflers, such as those used at the outlets of reciprocating engines, provide noise attenuation through the use of one or more internal expansion chambers interconnected with perforated pipes. The expansion chambers serve to reflect acoustic energy in the gas stream back toward the source rather than along the pipe to the exit.

Reactive/dissipative mufflers include both reactive expansion chambers and dissipative insulation elements. They are commonly used when a large degree of noise attenuation is required at the exhaust of reciprocating or turbine engines and compressors. One advantage of reactive/dissipative mufflers used to control noise from large fans is that they can be designed to provide both broadband noise attenuation and additional attenuation of the tonal noise produced by the fan at its blade-passing frequency.

While not actually mufflers, diffusers formed with porous or sintered metal are sometimes installed at the outlets of small gas vents to reduce jet noise.

A wide variety of dissipative and reactive mufflers are available as predesigned catalog items from many manufacturers. Simple dissipative and reactive mufflers will typically attenuate low-frequency noise by approximately 3 to 6 dB and mid-frequency noise by approximately 10 to 30 dB. Such mufflers are often well suited for common or simple noise control applications without special requirements. Many muffler manufacturers have simple computer programs available to help with the selection of such silencers. When significantly greater noise attenuation is needed, for complex installations or critical applications where low noise levels are required, and when low pressure loss is necessary, it is suggested that an experienced professional be consulted to advise in the design, specification, and selection of appropriate mufflers for the specific application.

Equipment Specification, Selection, Layout, Maintenance, and Operation

The specification, selection, layout, maintenance, and operation of plant equipment can influence plant noise levels and employee noise exposures. Each of these items is worth considering when evaluating noise control requirements and available options. For those companies that are moving office workers out onto the factory floor, these considerations are essential.

A wide variety of equipment is now available directly from manufacturers with noise levels reduced by as much as 5 to 15 dBA compared to some standardly produced equipment. Examples include compressors, motors, valves, transformers, cooling towers, and reciprocating and turbine engines. Entire manufacturing and assembly lines can be designed and installed to have reduced noise levels. To achieve this, a realistic understanding of the plant's noise-level/noise-exposure design goals and a well-written technical specification that defines the noise requirements are necessary. When purchasing replacement equipment, adding additional equipment, or designing a new plant, the specification of low-noise equipment is a worthwhile investment, yielding an improved work environment and less hearing loss among employees. In some cases, low-noise equipment also operates more efficiently than standard high-noise equipment. Quieting the source is often the most cost-effective way to reduce plant noise.

It is obvious that equipment layout and spacing within a plant will affect both work-area noise levels and employee noise exposures. Sometimes logical adjustments to equipment layout during the design of a plant addition or new plant will yield an improved work environment without loss in production efficiencies. In some cases all that is needed is a fresh look at plant requirements rather than simply repeating previous layouts. Relocating equipment such as fans, pumps, engines, and metal shears away from conference rooms and office areas should be considered to avoid excessive intrusive noise. Questions should be asked when considering the layout of a new plant or plant addition: Is it necessary to locate several motor pump sets so close together that maintenance workers inspecting or repairing one unit are exposed to excessive noise from the adjacent operating units? Can instrument panels and operator workstations be relocated somewhat farther away from particularly noisy equipment? Can a booth or shelter be incorporated into an employee workstation? Is it possible to locate workstations away from noisy equipment and noise-reflecting walls? Can very noisy operations, such as riveting, chipping, or metal grinding, be located away from work areas that would otherwise be less noisy?

Well-maintained equipment tends to operate with less noise than does poorly maintained equipment. One obvious example is a small leak in the packing of a valve passing high-pressure steam and causing high noise levels nearby. Old engine exhaust mufflers and equipment enclosures are other examples. Loose guards and worn bearings are sometimes the source of unnecessary noise as are slipping belt drives and fans with distorted inflow conditions.

Equipment operation also affects noise. Some fans and pumps produce higher noise levels when operated at low load rather than at full-rated load. For motor-driven units, variable-speed electronic drives are available that reverse this, resulting in less noise during low-load operation. Modifying the cutoff of high-pressure centrifugal fans may help to reduce the generation of tonal noise. Ventilation systems produce less noise when well balanced and operating at the lowest acceptable pressure. It is well known that excessive noise is one result of metal cutting at incorrect speeds. Pump cavitation also causes excessive noise. Plant operation with open doors for additional ventilation during hot summer weather sometimes results in excessive community noise and poor public relations. Exercising emergency engine generator sets during third shift, rather than during the daytime, can also result in poor community relations. The nighttime use (and occasional abuse) of outdoor paging systems, rather than radios or beepers, is sometimes the source of community noise complaints. The volume of indoor paging systems is sometimes set far higher than is necessary for adequate communication and contributes to employee noise exposures. All-night idling of trucks, particularly refrigerated trucks, at warehouses has been the source of many community noise problems. The noise caused by releasing high-pressure steam to the atmosphere can be reduced significantly by the use of an operator-controlled valve with low noise trim.

Active Noise Control

Considerable research efforts have been undertaken to develop reliable and cost-effective means to actively (or electronically) control noise produced by equipment. Though it is not a new concept, modern theory and hardware for adaptive digital signal processing have recently made the commercial use of active noise control worth considering for certain applications with special requirements.

One basic application of active noise control is the reduction of low-frequency tonal noise generated by fans in long ducts of relatively small cross section. An input microphone installed in the duct is used to generate a signal that is proportional to the unwanted noise. The signal is processed electronically and fed to a power amplifier that drives a loudspeaker mounted in the duct. The electrically generated sound from the loudspeaker is used to destructively interfere with and cancel a portion of the unwanted noise. In more complex applications, several microphones and loudspeakers must be employed. In most real-life applications, various physical conditions, such as temperature, flow velocity, and frequency, with temporal and spatially varying system characteristics, must be considered. Adaptive active control systems that include one or more error-measuring microphones are being employed to improve noise attenuation performance in applications with varying system characteristics.

The use of active noise control should be considered when passive noise control methods are expensive or difficult to install. The control of very-low-frequency noise in a duct is an example where passive noise control may be difficult. In low-frequency applications where space is limited and no additional pressure is available, active noise control becomes even more attractive. However, it is important to recognize that the microphones, electronic circuits, and loudspeakers will require at least some ongoing maintenance, particularly when installed in hot or corrosive environments.

Active noise control has also been applied inside earmuffs and communication headsets to reduce noise exposure and improve speech intelligibility. Earmuffs and headsets with built-in active noise control elements are available for use in high-noise environments where conventional hearing protectors would not provide adequate noise attenuation.

For special applications requiring significant attenuation of low-, mid-, and high-frequency noise, it might be most practical to consider the use of both active and passive noise control methods. A passive noise control treatment, such as a small muffler or duct lining, can be installed to attenuate the mid- and high-frequency noise and active noise control might be employed to attenuate the low-frequency noise.

Additional information on active noise control and its applications can be found in sources listed at the end of this chapter.

Surface Damping

Machinery housings often consist of large areas of flexible metal plate; if they are set into vibration, such plate areas may become significant radiators of airborne noise. Vibration is caused by a forcing mechanism, such as internal oscillating, rotating, or reciprocating components, that may excite the machine's surfaces at their resonant (or natural ring) frequency. Application of damping (vibration-energy-absorbing) material to the surfaces will reduce the amplitude of vibration at the resonant frequency. Adding more material can stiffen and change the natural frequency so that it is no longer so easily excited.

Damping (viscoelastic) materials can be applied as either a free layer or a constrained layer. (See Figs. 4.101*a* and *b*.) The material can be troweled on, glued and baked on, or (for intermittent use) attached by a magnetic layer. Practical concerns regarding the specific material include ease of cleaning, toxicity, durability, and temperature-independent efficiency in the required frequency ranges. As a rule of thumb, for a damping material to be effective it must be at least as thick as the material to be damped.

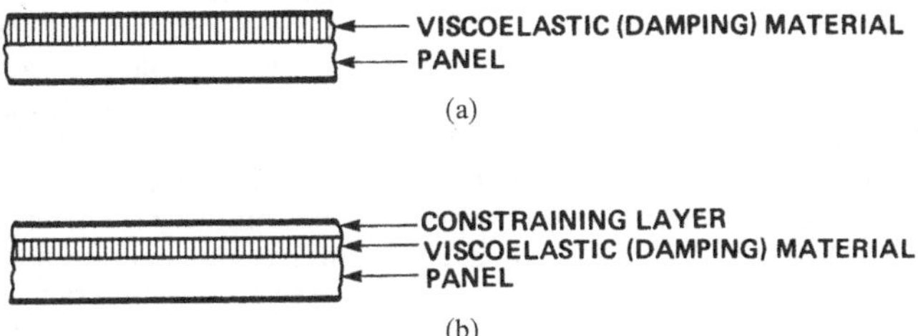

FIGURE 4.101 (*a*) Panel with free layer of viscoelastic (damping) material. (*b*) Panel with constrained layer of viscoelastic (damping) material.

Addition of Sound-Absorptive Materials to a Room

One commonly used method for noise reduction in a room is to clad significant portions of the interior surfaces (usually the ceiling and walls) with highly sound-absorptive materials. Adding such materials to a room provides only minor noise reduction near the sources, but at greater distance, in the reverberant field, significant noise reduction can be achieved. If sound-absorptive material is added to a space that initially has little absorption in it, the reverberant field noise reduction may be as high as 5 to 8 dB; however, in spaces that already have a modest level of absorption, the effect will be less. Reverberant-field noise reduction is approximately 3 dB per doubling of the amount of absorption in the room. For many spaces, a practical limit for noise reduction by adding absorptive material is about 5 dB. The first addition of absorption in a room provides the greatest reduction of noise and subsequent additions of absorption provide progressively less noise reduction (on a percentage basis). In some spaces, the cost of absorption to achieve the needed noise reduction may be higher than that of other methods to reduce the noise.

Enclosures

Enclosures around noise sources or receiving personnel can provide effective noise reduction. They consist of materials with substantial sound transmission loss and are arranged to surround the noise source or receiver.

Equipment enclosures are structures that surround the noise source and thus contain the sound it generates. However, enclosures can cause a buildup of high-level acoustic energy within themselves. Enclosures therefore usually consist of a wall with surface weight chosen to provide the required attenuation and an inner lining of porous material to dissipate the buildup of acoustic energy. In some cases where complete enclosures are built, the machines inside may require placement on vibration-isolator devices that prevent the transmission of structureborne noise to the outer surfaces of the enclosure.

If there are gaps in the enclosure, sound can escape. The greater the percentage of open area in an enclosure, the smaller the reduction in radiated sound. Table 4.77 gives an indication of the effects of openings in otherwise well-designed enclosures.

TABLE 4.77 Noise Level Effects of Openings

Percentage of open area in enclosure	Maximum average noise reduction, dB
50	3
25	6
10	10
1	20

In a practical sense, enclosure designs often require some accommodation. For ease of maintenance, enclosures can be constructed to rise upward and away from the machinery by overhead cranes. Access openings can be provided by tunnels lined with acoustically absorbent material (in effect, mufflers). Access for controls can be designed with hinged covers that lift easily, or the controls can be relocated. Some enclosure panels can be lifted automatically at the correct point in the machine cycle to provide access. Ducts can be provided with small ventilating fans to produce a controlled cooling airstream and, if combined with filters, can allow a controlled environment for some operations.

Employees exposed to high-level noise from a number of sources can be protected by acoustic booths. These can range from small open-fronted telephone-booth-sized cabinets (into which the operator steps while observing the operation of a semiautomatic machine) to completely enclosed control consoles. In many cases, acoustic booths for personnel can provide an island of protection; an employee exposed to high-level noise during part of the workday can be protected well enough to reduce total exposure during normal working hours to within the permissible exposure limit.

The design of worker enclosures requires suitable walls to produce the necessary sound reduction and also some internal sound absorption to prevent reverberant buildup of transmitted sound. The design and location of such booths require careful study and measurement in the field.

Wrapping and Lagging

Noise-control wrappings (lagging) can be thought of as tight-fitting enclosures installed and supported directly on the surface of equipment such as ducts, pipes, valves, and machines to reduce noise radiation. For high-temperature noisy equipment, thermal insulation can often be adapted also to serve as effective noise control wrapping. To serve both functions, the inner insulation layer must be porous and resilient (e.g., a glass fiber blanket), so that it is a good sound absorber and is resilient enough to avoid transmitting vibration from the equipment surface to the outer noise barrier layer. The outer protective layer should be impervious, preferably limp, and isolated from the noise-radiating equipment surface. Most often the outer layer (jacket) is aluminum or steel sheet metal, or mass-loaded silicone or vinyl material.

Noise control wrappings can be thought of as a spring-mass isolation system, where the porous insulation forms the spring and the impervious jacket forms the mass. At frequencies equal to and below the system resonance, the noise attenuation (insertion loss) is zero or negative, sometimes providing an increase in low-frequency noise radiation. At higher frequencies—above about 200 Hz for most practical wraps—noise attenuation increases with increasing frequency.

Noise attenuation of acoustical wrappings increases as (1) the surface weight of the impervious outer jacket is increased, (2) the stiffness per unit area of the porous insulation is reduced, (3) the thickness of the porous layer is increased, and (4) frequency is increased. Well-designed noise control wrappings routinely provide mid- to high-frequency noise attenuations of 10 to 30 dB. To achieve substantial noise attenuation, it is essential that the outer jacket not contact the vibrating machine surface and that the insulation layer be resilient. Rigid or impervious thermal insulations such as calcium silicate, glass foam, and closed cell foams are generally not suited for noise control applications.

Flexible blanket insulations have proven to provide substantial noise attenuation, be easy to remove and reinstall during equipment maintenance, and be cost effective in many applications requiring ready access to the equipment. Information about thermal-acoustic flexible blanket insulation is available from many manufacturers of industrial blanket insulation and in the sources listed at the end of this chapter.

Vibration Isolation

Large vibrating machinery housings or sections of framing may radiate large amounts of sound, even when the vibration is almost imperceptible. For noise control, the surfaces can be acoustically "isolated" from the vibrating drive mechanism by vibration isolation mounts, breaks, or pads installed between the vibrating source and the radiating surface. (See Chap. 4.13, "Vibration Control.")

Barriers

A barrier is a solid wall used to shield a receiver from the direct radiated sound of a machine. To be effective, it must be sufficiently massive to provide the required reduction in the sound transmitted through the barrier, and it must be sufficiently wide and high to prevent the sound defracted around the edges from becoming significant. Barriers should be placed close to either the source or the receiver for maximum effectiveness. A barrier positioned halfway between the source and the receiver will require the greatest dimensions to produce a given noise reduction.

Barriers can also be used to shield a group of workers not working on noisy machines from an area containing noisy machines. Of course, the effectiveness of such barriers can be significantly reduced by reflected sound, for example, from a ceiling or nearby sidewalls. Therefore, barriers may need to be combined with ceiling and possibly wall treatments to reduce the sound transmitted by alternative paths. Figure 4.102 shows the average mid-frequency noise reduction provided by a barrier. At low frequencies the reduction will be less and at high frequencies the reduction will be greater than shown in Fig. 4.102.

Barriers can be provided with acoustically absorptive material on the side facing the source to avoid a reflected buildup of sound near the machine. Without absorptive material, the barrier will tend to increase the noise level on the machine side of the barrier.

Simple small barriers can be located on some machinery to provide individual operator protection. In this case, a transparent material may allow visual monitoring. On small

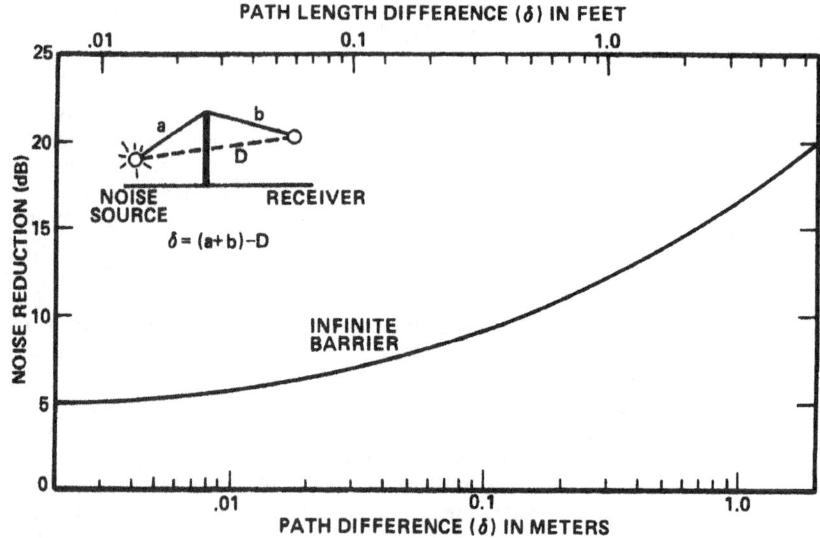

FIGURE 4.102 Average noise reduction of acoustical barrier of infinite length.

machines the controls may still be reached under or around the barrier. Such arrangements can also act as safety shields.

Although a barrier is designed to stop the sound from reaching a given receiver, the barrier does not necessarily have to impede the passage of materials and products. The use of overlapping entrances to produce a visual and acoustic blockage can allow easy access for forklift trucks and the locations of conveyors.

Barriers are often useful for blocking noise emissions from rooftop equipment to the adjacent community. The barriers must block the line of sight from source to receiver by a suitable amount to produce the desirable attenuation. Such barriers must be made of solid materials; they cannot be porous screens. To avoid unwanted sound transmission under barriers, they often must extend down close to the roof.

COMPUTER MODELING

In the past decade, several acoustical modeling programs have been introduced. These programs have the capability of modeling interior or exterior acoustical environments using common acoustical algorithms. They perform a vast number of calculations over large areas, the results of which can be presented in a graphical form. For a plant engineer or manager, the interior acoustical models can be helpful for planning layouts of loud or noise-sensitive activities, locating sound-isolating constructions, and anticipating areas for hearing conservation programs. Exterior acoustical models can be employed to make predictions in the community for compliance with noise codes, aid in the selection of suitably quiet equipment, and assess potential noise control options.

The benefits of such programs include their ability to perform repeat calculations of various conditions relatively quickly. In addition, they facilitate the assessment of more complex conditions, where hand calculations are too numerous to be feasible, Ultimately, the colorful plots can be a useful tool for helping people who are not familiar with acoustics to visualize the impact of noise sources and the resulting environment.

The drawbacks of these programs include the substantial effort required to properly configure and verify the model, the complexity of some output data, and the potential for unrecognized errors (garbage in, garbage out). There is also the issue that the precise results that are generated are perceived as an unarguable, accurate prediction; this is not always true. The precision of the model (an extreme of which is taken to several decimal places) is often no better than good hand calculations and judgment. It is highly advisable for the user of these programs to be experienced with basic acoustical calculations, and to have a thorough understanding of the details of the modeling program. Anything that can be done to verify the results of the computer modeling is worthwhile.

MEASUREMENT AND INSTRUMENTATION

Before a noise measurement program is started, its objective should be clearly understood. The objective could be to determine the approximate level of the noise during a short period of time or to find out whether the noise environment is hazardous to the health and welfare of the worker. The objective also might be to learn if the noise exceeds some locally adopted sound level limits. Depending on the purpose and the required accuracy, a wide range of instrumentation is available and the instruments may be used in a variety of ways. A basic program to assess noise throughout an entire plant may require many measurements. A noise control investigation of a particular machine may require extensive, detailed information. Sometimes monitoring systems are set up to detect excessive levels.

The simplest instrument available to measure sound levels is a sound level meter. There are four different classifications of portable sound level meters:

1. Precision, used for laboratory work or other measurements requiring extreme accuracy
2. General purpose, used for most industrial applications requiring normal accuracy
3. Survey, used for rapid surveys to determine approximate levels; typically measures A-weighted levels only
4. Special purpose, used for special purposes such as impulse noise measurements

The precisions and tolerances of indicating meters and weighting networks vary significantly for the various types of sound level meters.

A sound level meter must be calibrated on a regular schedule if it is to provide meaningful data. Most equipment is battery operated; the batteries must be fresh and capable of supplying the instrument with sufficient power. A battery check is followed by a field calibrator check. The calibrator produces a pure tone at a known sound pressure level, which will allow appropriate adjustments to the meter.

Sound level meters measure noise only at a given point at the time of observation. If the noise being measured is constant in both space and time, meters will give an accurate representation of the situation. However, if the sound level changes with time and location (for instance, as an operator moves around), it will be necessary to record either the sound level manually using short time intervals (5 to 10 s) or the noise data for later analysis of the time history of the noise. The second approach is preferable when a worker's noise exposure is related to duty cycles or product flow. In this case, extrapolations can be made on the basis of total day production to determine the noise exposure of an employee over a full day.

In industry there are situations in which time pressure is great, more than one person must be monitored, and duty cycles are not easily definable. In such cases, audiodosimeters can be used to measure employees' noise exposure. Audiodosimeters, devices about the size of a deck of cards, are worn by employees to record the noise exposure of the wearers wherever they go. The microphone can be fixed to the unit or detached from the unit and placed in the hearing zone of the wearer. Audiodosimeters are available with an internal circuit that integrates the sound level and time in accordance with various noise exposure regulations.

BIBLIOGRAPHY

Textbooks, Manuals, and Guides

Listed following are textbooks, manuals, and guides containing technical information that are useful in the application of noise control treatments at industrial and commercial facilities.

American Society of Heating, Refrigeration, and Air-Conditioning Engineers, Inc: *1999 ASHRAE Handbook: Heating, Ventilating, and Air-conditioning Applications,* Inch Pound ed., Chap. 46, Atlanta, Ga., 1991.

Beranek, Leo L., and István Vér: *Noise and Vibration Control Engineering Principles and Applications,* John Wiley & Sons, New York, 1992.

Bies, David A., and Colin H. Hansen: *Engineering Noise Control Theory and Practice,* Unwin Hyman Ltd., London, 1988.

Edison Electric Institute: *Electric Power Plant Environmental Noise Guide:* Vols. I and II, 2nd ed., Washington, D.C., 1984.

Harris, Cyril M.: *Handbook of Acoustical Measurements and Noise Control,* 3rd ed., McGraw-Hill, New York, 1991.

Jensen, Paul, Charles R. Jokel, and Laymon N. Miller: *Industrial Noise Control Manual,* U.S. Department of Health, Education, and Welfare, NIOSH, Cincinnati, Ohio, 1978.

Kinsler, Lawrence E., Austin R. Frey, Alan B. Coppens, and James V. Sanders: *Fundamentals of Acoustics,* 3rd ed., John Wiley & Sons, New York, 1982.

Peterson, Arnold P. G.: *Handbook of Noise Measurements,* 9th ed., Gen Rad, Concord, 1980.

Suter, Alice H., and John R. Franks: *A Practical Guide to Effective Hearing Conservation Programs in the Workplace,* U.S. Department of Health and Human Services, U.S. Government Printing Office, Washington, D.C., 1990.

U.S. Department of Health, Education, and Welfare, NIOSH: *Compendium of Materials for Noise Control,* Cincinnati, Ohio, 1975

Vér, István, and Eric J.W. Wood: *Induced Draft Fan Noise Control Technical Report and Design Guide,* Empire State Electric Energy Research Corp., ESEERCO Report No. EO82-15, New York, 1984.

Journals

Numerous journals dealing with acoustics and noise control engineering are available throughout the world. Three published in the United States have been selected and are listed here.

Sound & Vibration
Acoustical Publications, Inc.
27101 East Oviatt Road
PO Box 40416
Bay Village, OH 44141
216-835-0101

Noise Control Engineering Journal
Institute of Noise Control Engineering
PO Box 3206 Arlington Branch
Poughkeepsie, NY 12603
914-462-4006

Journal of the Acoustical Society of America
Acoustical Society of America
500 Sunnyside Boulevard
Woodbury, NY 11797
516-349-7800

Noise Control Equipment Manufacturers

Sound & Vibration magazine (listed in the previous section) publishes an annual list of companies that produce and sell hardware for the control of noise and vibration.

Acoustic Instrumentation

Sound & Vibration magazine (listed previously) publishes an annual list of companies that produce and sell instrumentation for the measurement of noise and vibration.

Noise Control Consultants

National Council of Acoustical Consultants, 66 Morris Avenue, Springfield, NJ 07081, (voice) 201-379-1100, (fax) 201-379-6507.

Institute of Noise Control Engineering
PO Box 3206 Arlington Branch
Poughkeepsie, NY 12603
914-462-4006

Professional Organizations

Professional organizations engaged in acoustics and noise control engineering are located in more than 30 countries. Two organizations in the United States are listed here.

Acoustical Society of America
500 Sunnyside Boulevard
Woodbury, NY 11797
516-349-7800

Institute of Noise Control Engineering
PO Box 3206 Arlington Branch
Poughkeepsie, NY 12603
914-462-4006

Standards

Standards dealing with acoustics and noise control engineering applicable to industrial plants have been prepared by and are available from the following organizations, most of which are located in the United States. Many excellent standards are available from similar organizations located in other countries.

Acoustical Society of America (ASA)

Air Conditioning and Refrigeration Institute (ARI)

Air Movement and Control Association (AMCA)

American Boiler Manufacturers Association (ABMA)

American Gear Manufacturers Association (AGMA)

American National Standards Institute (ANSI)

American Society of Heating, Refrigeration and Air-Conditioning Engineers (ASHRAE)

American Society for Testing and Materials (ASTM)
Compressed Air and Gas Institute (CAGI)
Cooling Tower Institute (CTI)
Diesel Engine Manufacturers Association (DEMA)
Industrial Silencer Manufacturers Association (ISMA)
Institute of Electrical and Electronic Engineers (IEEE)
International Electrotechnical Commission (IEC)
International Organization for Standardization (ISO)
National Electrical Manufacturers Association (NEMA)
National Fluid Power Association (NFPA)
Society of Automotive Engineers (SAE)

CHAPTER 4.13
VIBRATION CONTROL

Eric E. Ungar
Chief Engineering Scientist
Acentech, Inc.
Cambridge, Massachusetts

Jeffrey A. Zapfe
Senior Consultant
Acentech, Inc.
Cambridge, Massachusetts

CHARACTERIZATION OF VIBRATIONS

Vibration refers to oscillatory (back-and-forth) motions of structures, mechanical systems, or components of these. A vibration generally is characterized by the displacement, velocity, or acceleration measured at one or more points on the item of interest in specific directions of interest (e.g., perpendicular to a floor, wall, or shaft).

The time variation of a vibration sometimes appears approximately like the idealized curve shown in Fig. 4.103*a*, obtained from a sensor and displayed on an oscilloscope or chart recorder. Such a regular curve, which corresponds mathematically to a sine or cosine, is called *sinusoidal* or *simple harmonic*. Note that it deviates from zero (the middle position) equally in both directions; the maximum excursion from zero in *one* direction is called the *amplitude* A, the total excursion in *both* directions is called the *double amplitude* $2A$ (or sometimes the peak-to-peak value). Amplitudes may be given in units of displacement, velocity, or acceleration, depending on how the vibration is measured.

The time interval T between successive peaks is called the *period* and usually is measured in seconds. The number of vibration cycles (i.e., the number of periods) that occur per second is called the *frequency* f and is generally measured in hertz (Hz). Hertz is the internationally standardized name for cycles per second (cps).

In practice one rarely obtains a simple signal like that of Fig. 4.103*a*. One is more likely to obtain a signal that looks like Fig. 4.103*b*, which consists of a basic sinusoid like that of Fig. 4.103*a*, to which are added one or more sinusoids with higher frequencies (shorter periods) and generally smaller amplitudes. Fig. 4.103*b* is said to represent a *multifrequency* or *complex* vibration. The component with the lowest frequency (greatest period) is called the *fundamental* component. Components that occur at frequencies that are integer multiples of the fundamental one are called *harmonics*.

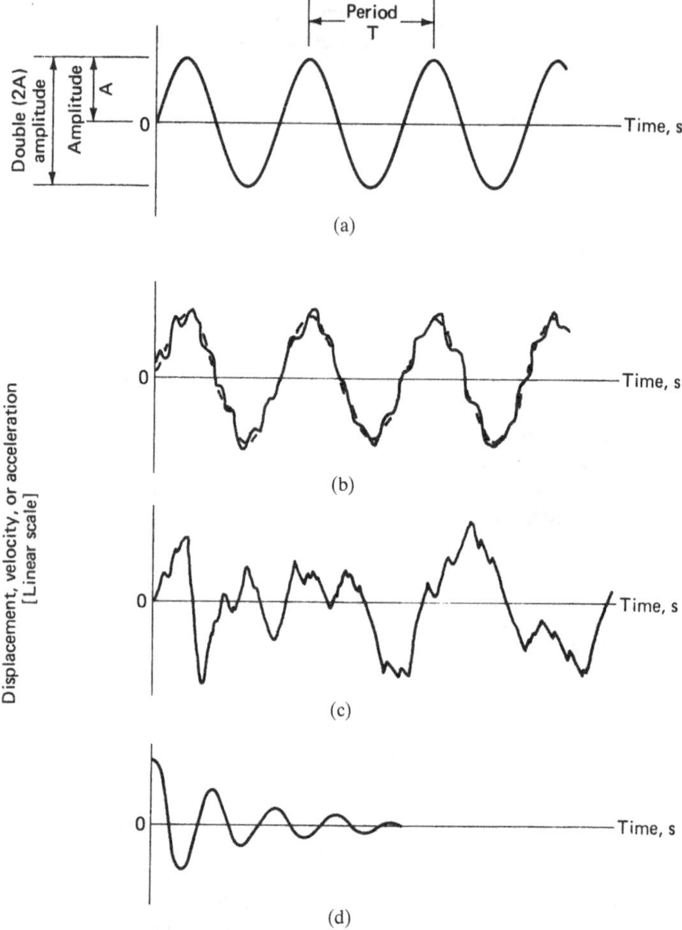

FIGURE 4.103 Typical vibration records: (*a*) steady sinusoidal or simple harmonic vibration; (*b*) steady multifrequency vibration; (*c*) irregular (nonperiodic) vibration; (*d*) decaying transient single-frequency vibration.

Vibrations like those illustrated in Fig. 4.103*c,* which have no well-defined period or amplitude—i.e., where the signal essentially never repeats itself—are called *nonperiodic* or *irregular.*

A vibration that has essentially the same amplitude over an extended time period is called *steady,* whereas a vibration with time-varying amplitude is called *transient.* Figures 4.103*a, b,* and *c* illustrate steady vibrations, whereas Fig. 4.103*d* illustrates a typical decaying transient containing a single-frequency component; similar transients with multiple-frequency components or with irregular behavior are usually encountered in practice.

It is often useful to describe a vibration in terms of its frequency components. A plot of amplitude vs. frequency is called a *spectrum.* There are available a variety of instruments called *spectrum analyzers* that automatically calculate the frequency components of a vibration signal and provide amplitude-frequency displays.

Knowing the amplitude at a given frequency in terms of any one of the three motion quantities (displacement d, velocity v, acceleration a), one can calculate the amplitude in terms of the other two from:

$$a = 2\pi f v = (2\pi)^2 f^2 d$$

$$v = a\backslash 2\pi f = 2\pi f d \tag{1}$$

$$d = a\backslash(2\pi)^2 f^2 = v\backslash 2\pi f$$

where *a, v,* and *d* always contain the same length units and seconds and *f* is in hertz. For example, to *d* given in mils there corresponds *v* in mils/s and *a* in mils/s^2. For $v = 10$ ft/s and 20 Hz, $a = 2\pi(20)(10) = 1256$ ft/s^2 and $d = 10/2\pi(20) = 0.080$ ft. Acceleration is often expressed in units of gravitational acceleration *g,* where $1g = 32.2$ ft/s^2 = 386 in/s^2 = 9.80 m/s^2.

Figure 4.104 is a representation of Eq. (1) and a convenient chart for the approximate conversion between motion quantities.

CAUSES OF VIBRATIONS

Vibrations are always caused by time-varying forces or torques, that is, by forces or torques that may be oscillatory in magnitude or direction or by forces that are suddenly applied or

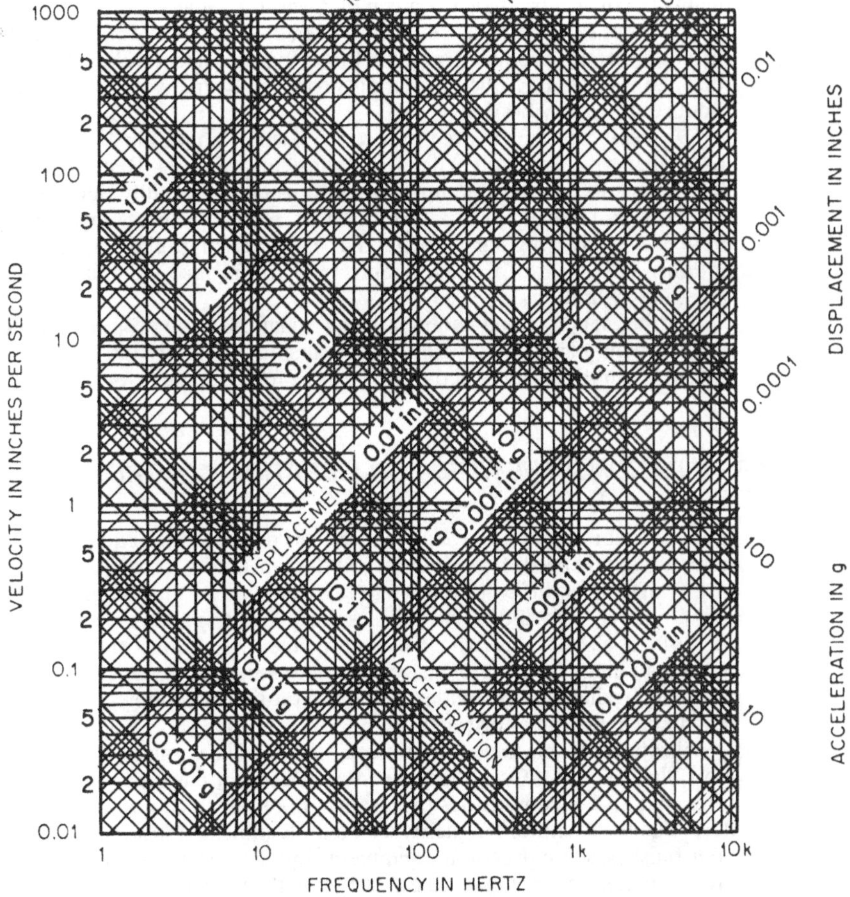

FIGURE 4.104 Chart for conversion between displacement, velocity, and acceleration. Reproduced with permission from *Handbook of Acoustical Measurements and Noise Control,* C. M. Harris (ed.), 3d ed., McGraw-Hill, New York, 1991.[1]

released. These forces need not be due to mechanical causes; electromagnetic, aerodynamic, or fluid-related forces also are often encountered in practice.

Unbalances in rotating machines produce net centrifugal forces that change direction in space as the machine rotates. For a machine with a horizontal shaft, such a force acts upward at one instant and downward a half-rotation later, thus producing a force that acts on the floor vertically at a frequency that corresponds to the *shaft rotation frequency* f_r (hertz) = $N/60$, where N denotes the shaft rotation speed in revolutions per minute (rpm). As such a machine comes up to speed, it produces transient vibrations that increase in frequency until steady operating conditions are reached; when such a machine is turned off and coasts toward a stop, it produces decaying transient vibrations with decreasing frequency.

Reciprocating machines also produce unbalanced inertia forces that are transmitted to the machine housing and supports. The primary components of these forces occur at the crankshaft's rotational frequency and at the first few integer multiples of that frequency. These forces generate vibrations along the direction of piston travel, as well as perpendicular to that direction, in the plane of the crank, and also produce vibratory moments or couples; the relative magnitudes depend on the cylinder arrangement and the degree of dynamic balance.

Fans, blowers, and pumps tend to generate steady vibrations, due to both unbalances and repetitive fluid pulses. The latter occur primarily at the *blade passage frequency,* which is the frequency with which blades pass a fixed point. For a rotor with n blades, rotating at N rpm, the blade-passage frequency is given by f_b (hertz) = $nN/60$.

Turbulent flows of water or air in a duct, or flows from a blower or air jet impinging on a surface, typically produce irregular forces on the structural surfaces. Similarly, irregularly repeated impacts, e.g., due to footfalls produced by many people walking on a floor, tend to produce irregular vibrations.

Single impacts, such as are produced by a single operation of a punch press, generate force pulses that typically result in decaying transient vibrations. Repetitive impacts result in repetitive transients, but if these impacts are repeated so rapidly that the vibrations due to one impact do not decay much before the next impact occurs, then the vibrations tend to have more of a steady, irregular character. For example, continuing impacts of materials against the bottoms and sides of a gravity chute tend to produce essentially continuous nonperiodic (irregular) vibrations.

Rattling, slippage, and nonlinearities (deviations from proportionality between forces and displacements) associated with large excursions generally introduce higher-frequency components than those associated with the basic forces and motions.

EFFECTS OF VIBRATIONS: THE NEED FOR VIBRATION CONTROL

Excessive vibrations can have adverse effects on personnel, equipment, and structures. Vibrations can annoy people, can interfere with their ability to perform or concentrate on mental tasks, can make it difficult for them to carry out precise movements or to make accurate readings of instruments, and in extreme cases can lead to physical disabilities and injuries. Vibrating surfaces also act somewhat like loudspeaker membranes in that they radiate sound, which again may range from annoying to painful to injurious, depending on its intensity. Vibrations may also produce such secondary effects as rattling of windows or motion of lights, which also tend to be annoying or distracting.

Vibration of a machine may reduce the life of its components, particularly those that are most highly loaded. Oscillatory stresses induced in machine parts, supports, building structures, and also in connections (hold-down bolts, pipes, cables) may produce failures of these items due to structural fatigue. Machine tools subjected to excessive vibrations may produce poor finishes; some precision equipment (optical systems, microscopes, gauges, microassembly equipment) cannot be used effectively in the presence of significant vibrations.

Vibration control is needed wherever adverse effects due to vibration exist or are expected. How much vibration reduction is required depends on the existing conditions and on the magnitude of vibration that is acceptable. Zero vibration is as impossible as an immovable object or an irresistible force.

In many instances vibration limits can be approximated in terms of a frequency-independent velocity value. This value typically is adequate for preliminary evaluation, but for more reliable assessment one needs to take account of the appropriate frequency-dependent vibration criteria. A person who has any part of his or her body in direct contact with a vibrating surface typically can feel vibrations with velocities in excess of 0.008 in/s (0.0001 m/s). It is advisable that the vibrations of floors in operating rooms and critical working areas not exceed half this value, which corresponds to the threshold of perception of the most sensitive persons. In offices, twice the aforementioned value is acceptable; in workshops, four times this value.[2,3,4] Vibration limits pertaining to task performance proficiency, comfort, health, and safety, are considerably higher.[5] A vibration velocity of 2 in/s (0.05 m/s) commonly has been considered as a safe limit for buildings, although minor damage occasionally has occurred at half this value.[6,7]

The suppliers of sensitive equipment generally specify limits on the vibrations to which the equipment may be exposed. In absence of specific information, 0.002 in/s (50 μm/s) may be taken as an appropriate limit for microscopes up to 400× magnification, precision balances, and metrology laboratories. One may take half this value as a limit for microscopes at higher magnification and for optical equipment that is mounted on isolation tables. A velocity limit of 0.0005 in/s (12 μm/s) may be used for electron microscopes at magnifications up to 30,000× and for magnetic resonance imaging systems, as well as for some mass spectrometers. More stringent limits may apply for electron microscopes at greater magnifications, for certain equipment used in the manufacture of microelectronics, and for specialized research equipment.[8]

By measuring the vibrations at the bearing housings of rotating equipment, one frequently can judge whether this equipment is vibrating excessively. Vibration velocities in excess of 0.2 in/s (0.005 m/s) typically indicate rough operation; velocities in excess of 0.5 in/s (0.012 m/s) tend to be indicative of severe problems.[9] Velocity values for evaluating the vibration severity of small and medium-sized electrical machines are given in Ref. 10 and reproduced in Ref. 11. The causes of excessive vibrations in rotating machinery may often be determined from the predominant frequencies of these vibrations. For example, vibrations at the shaft rotation frequency f_r are indicative of unbalance, misalignment, or a bent shaft; vibrations at twice the rotation frequency tend to indicate mechanical looseness; other frequencies can be related to bearing faults, problems with belt drives, electrical problems, and aerodynamic effects from fans and blowers.[12]

DIAGNOSING A VIBRATION PROBLEM

It is usually convenient to consider any vibration problem in terms of (1) the *source* of the undesirable vibrations; (2) the *receiver,* i.e., whatever is adversely affected by vibrations and requires protection; and (3) the *path* along which vibrations from the source reach the receiver. In practical situations, many sources often contribute to the vibrations experienced by a single receiver, and vibrations from a given source often reach a single receiver via several paths.

The most convincing approach for identifying the vibration source responsible for a given problem consists of turning off all possible sources, then turning them on one at a time while observing the resulting effects at the receiver of interest. Similarly, the predominant paths can best be identified by interrupting one at a time or by disabling all and then reestablishing one at a time. These procedures can only rarely be carried out in practice to their full extent, but they can often be carried out sufficiently to provide valuable partial, if not full, insight into the problem.

Useful information can often be gained by comparing the vibration spectra measured at various sources with those measured at the receiver of concern. Sources that produce significant vibrations only at frequencies at which the receiver experiences no adverse effects generally cannot be responsible for these effects. Similarly, paths that do not transmit vibrations at frequencies at which the receiver is subject to problems generally are not the dominant paths along which vibrations that cause these problems are transmitted. For example, a source that produces vibrations only at 30 Hz will not cause adverse effects at a receiver that is sensitive only to vibrations at 50 and 80 Hz, and paths that do not transmit 50- or 80-Hz vibrations are not likely to play a role in producing these adverse effects.

Sources and paths responsible for vibration of a receiver frequently can be identified by correlation measurements. These involve simultaneous measurement of the vibrations at the receiver and a source (or path) and calculation of the corresponding statistical correlation coefficient or coherence. For sources or paths that contribute significantly to the vibration at the receiver, high correlation (or a coherence value near unity) is obtained, whereas for sources or paths that make insignificant contributions, low correlation is obtained. Most modern spectrum analyzers are equipped to carry out correlation or coherence calculations automatically.

In dealing with any vibration problem, one must keep in mind the phenomenon called *resonance.* Any mechanical system or structure has a number of frequencies at which it can be set into vibration very easily; these are called *natural frequencies.* The lowest of these, called the *fundamental natural frequency,* is often most easily excited and of greatest importance. Resonance occurs if a system is subjected to a vibratory force or motion at one of its natural frequencies; large vibrations can then result *even with small inputs.* The natural frequencies of a system can be determined readily; if a system is deflected and released, or if it is struck, it will vibrate at one or more of its natural frequencies. For observation of the natural frequencies, however, it is usually necessary to have all relevant vibration sources turned off, so that the natural frequencies will not be masked by the excitation frequencies.

A *stroboscope* is often useful for investigating vibrations with large displacements at relatively low frequencies. This consists of a light that is made to flash at precisely timed intervals. The light is aimed at a vibrating part and the flashing frequency is adjusted manually until the vibrating part appears to stand still; the frequency of the vibration can then be read from the instrument. Then, by changing the flashing frequency slightly, one can observe the vibration in apparent slow motion to see where the largest excursions occur.

VIBRATION CONTROL STRATEGY

It is generally best to control a vibration at its source, because this approach avoids problems at all potential receivers. However, in cases where only a limited number of receivers are of concern and where control at the source(s) is not feasible, control at the receiver(s) may be preferable. Reduction or elimination of vibrations at the source typically involves improving the dynamic balance of rotating or reciprocating equipment, substituting items with lesser vibrations for those with more (e.g., centrifugal pumps for reciprocating pumps), or changing operating speeds to eliminate resonance conditions. Reduction of the adverse effects of vibrations at the receiver generally involves substitution of less-vibration-sensitive items or processes, or adding stiffening or mass judiciously in order to eliminate resonances, if any are present. Balancing and other machine adjustments generally should be left to specialists.

Vibration *isolation* most often turns out to be the most practical and cost-effective means for vibration control in industrial plants. Isolation involves insertion of soft, flexible elements in the propagation path so as to reduce the transmitted forces and motions. Because of the multitude of paths that can begin at any source or terminate at any receiver, isolation is best accomplished near a source or receiver.

Some other vibration control methods that are useful only under certain specific circumstances are described at the end of this chapter.

Vibration Isolation at the Source

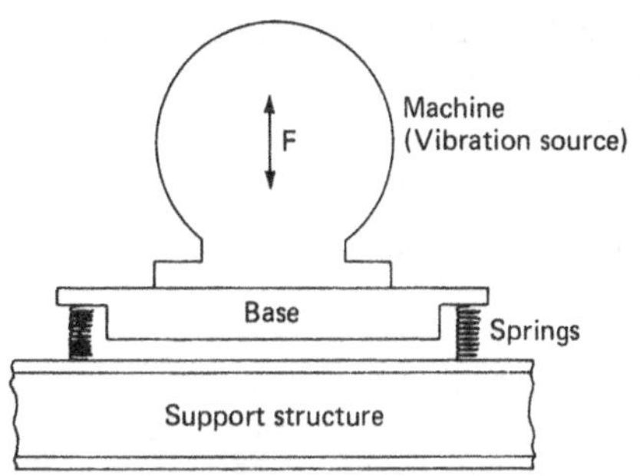

FIGURE 4.105 Conceptual sketch of isolation of vibration source.

Basic Principles. The basic concepts of vibration isolation can be understood with the aid of the schematic sketch of Fig. 4.105, which shows a machine that generates a vertical oscillatory force of amplitude F (e.g., due to an imbalance) rigidly attached to a base. This base, which may represent a machine housing or frame, a foundation, or an inertia base, is mounted atop a supporting structure via a series of springs or other resilient elements generally called *isolators.*

The oscillatory force produces motion by accelerating the combined mass m of the machine and the base. This motion of the base produces oscillatory compression (and extension) of the springs (superposed on the static compression due to the weight they support), which in turn gives rise to oscillatory forces on the support.

If the force varies slowly, i.e., at low frequencies, then the inertia of the mass offers little opposition to the motion, and the force essentially acts directly to compress the springs. The machine-base mass here moves just enough for the total spring force to match the externally applied force, as it would if the force were applied statically, and thus the entire applied force is transmitted to the support structure. On the other hand, if the force F varies rapidly, i.e., at high frequencies, then the inertia of the mass of the machine and base opposes the motion to a much greater extent than do the springs. The springs then compress and extend very little, and only the spring forces resulting from these small spring deflections are transmitted to the support structure.

The ratio of the amplitude F_s of the total force that the springs exert on the support (assumed rigid) to the amplitude F of the exciting force is called the *transmissibility* T. The transmissibility also is equal to the factor by which the force that acts on the support is reduced if the machine is supported on the given isolators instead of being fastened rigidly to the support structure. Very often this factor is at least approximately equal to the factor by which the vibratory motion of the support structure is reduced when the isolators are used in place of rigid connections. The *isolation efficiency* E, which is defined by $E = 1 - T$, indicates what fraction of the exciting force is prevented from acting on the support and approximately by what factor the motion of the support is reduced due to use of the springs or springlike elements. For example, if a given isolation system results in a transmissibility of 0.05, then its isolation efficiency is 0.95, indicating that the use of the isolators reduces the vibratory force on the support structure by 95 percent.

To produce significant vibration reduction, isolators must be soft enough to produce transmissibility T of 0.1 or less. The total stiffness k (that is, the sum of the stiffness of all of the isolators that support the machine and base) of the isolators that isolate a mass m (of the machine and base) at a transmissibility T may be estimated from

$$k \approx mf^2T/C_1 \tag{2}$$

where f represents the disturbing frequency in hertz and $C_1 = 10$ for k in units of lb/in and m in units of lb. ($C_1 = 25$ for k in units of N/mm and m in kg.) Better isolation (smaller transmissibility) is associated with smaller stiffness (softer isolators).

Many isolators, notably helical coil springs of steel, have straight-line force-deflection curves in their rated load range. For such isolators the slope of the force-deflection curve is equal to the stiffness and the stiffness is simply related to the *static deflection s* that the isolator experiences due to the load it supports. The static deflection needed to obtain a given transmissibility may be estimated from

$$s \approx C_2 \backslash f^2 T \tag{3}$$

where $C_2 = 10$ for s in units of inches and $C_2 = 250$ for s in mm. For a given disturbing frequency f, total isolator stiffness k, and total supported mass m, the transmissibility T and isolation efficiency E may be determined from

$$T = 1 - E \approx C_1 k \backslash m f^2 \approx C_2 \backslash s f^2 \tag{4}$$

Note that better isolation (smaller transmissibility and greater isolation efficiency) is obtained for smaller isolator stiffness and greater mass.

Practical Considerations. As is evident from the foregoing expression, the lowest disturbing frequency f corresponds to the greatest vibration transmission. Therefore, an isolation system must be designed for the lowest disturbing frequency of concern.

In general, vibration transmission can be reduced in two ways: (1) by using softer resilient elements (reducing the total stiffness k and thus increasing the static deflections) or (2) by increasing the supported mass m. It should be noted that the use of softer springs (beyond those needed to achieve a transmissibility of 0.1) leads to greater static deflection and reduced vibration transmission, but has little effect on the vibratory motion of the machine and base. If the supported mass is increased (e.g., by the addition of a heavy "inertia" base) and the total spring stiffness is not changed, then there results greater static deflection, reduced vibration transmission, and a reduction in the vibratory motion of the machine. In practice, where springs or other isolators often are selected to carry as much load as they can support safely, the addition of mass requires that more and/or stiffer isolators be used, so that the total spring stiffness is increased by approximately the same factor as the mass; in this case the static deflection and vibration transmission essentially remain unchanged, but the vibratory motion of the mass is reduced.

It is important to note that useful isolation can be achieved only if the resilient elements (springs) are *considerably softer* than the supporting structure, i.e., only if the resilient elements deflect considerably more under a given load than does the support structure. Otherwise, the support structure provides the predominant resilience and the springs merely serve to transmit the forces to the support essentially without attenuating them.

Figure 4.106 is a convenient chart for estimating the major isolation system parameters needed to achieve a desired vibration reduction or transmissibility. The dashed line in the figure illustrates the case where there exists a disturbing frequency of 3000 cpm (corresponding to 50 Hz), at which it is desired to reduce the vibration by 99.9 percent, i.e., to 0.01 times its original value. The chart shows that here a value for k/m of about 2.5 (lb/in)/lb is needed; thus, for a 2000-lb mass, a total stiffness of $2.5 \times 2000 = 5000$ lb/in is required. (A lesser stiffness would provide better isolation than that prescribed, whereas a greater stiffness would result in poorer isolation.) One may also read from the chart that to $k/m = 2.5$ (lb/in)/lb there corresponds a static deflection of about 0.4 in and a resonance frequency of about 300 cpm or 5 Hz.

Selection of an isolation system must account for the fact that an excitation frequency f that is produced by a machine usually varies with the machine's rotational speed. As a machine is brought up to speed or slowed to a stop, a speed at which the exciting frequency matches the natural frequency of the machine on its resilient supports may be encountered. This speed in revolutions per minute (rpm) may be estimated from

$$N = C_3\, s = C_4\, k/m$$

where $C_3 = C_4 = 188$ for s in in, k in lb/in, and m in lb ($C_3 = 945$ and $C_4 = 302$ for s in mm, k in N/mm, and m in kg). At this speed, intense vibrations may occur. Their magnitude depends on how fast the machine passes through this resonance speed and on the damping characteristics of the isolation system. If the machine accelerates or decelerates rapidly, vibrations do not have time to build up. For machines that accelerate or decelerate slowly, the magnitude of the vibration produced at resonance is inversely proportional to the damping in the system.

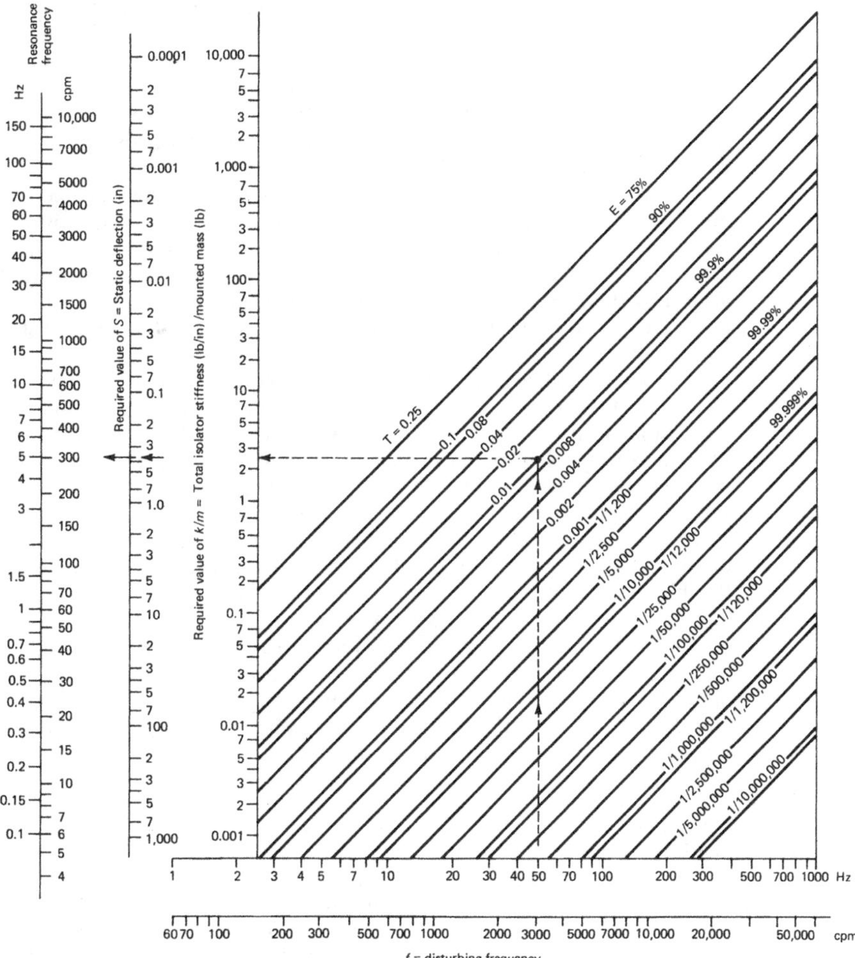

FIGURE 4.106 Chart for estimation of isolation system requirements. (1 in = 25.4 mm; 1 [lb/in]/lb = 0.386 [N/mm]/kg.)

Damping refers to a spring's or a structure's capability of dissipating oscillatory energy. A bell or a steel spring rings (i.e., vibrates) for a long time after it is struck; that is, it takes a relatively long time to dissipate the vibratory energy imparted to it. On the other hand, a rubber or cork rod vibrates only briefly after an impact; it dissipates vibratory energy rapidly and thus is highly damped. For machines that accelerate or coast down slowly, isolation elements of highly damped materials (e.g., rubber or cork) should be used, or energy-dissipation devices (e.g., friction pads or hydraulic dampers like those in automobile shock absorbers) should be added in parallel with the resilient members. In some cases, the incorporation of snubbers in the isolation systems may suffice to limit the excursions of the base as the machine passes through resonance. Such snubbers may, for example, be in the form of rubber cones that are mounted so that the base bumps against them when its excursion exceeds a given amount. Snubbers are incorporated in many commercial isolator assemblies.

If slow acceleration or coast-down is not a problem, then the type of isolator material and the isolator configuration essentially make no difference; the only parameter that counts is the total stiffness k under the expected load conditions and operating frequencies. For many resilient elements, particularly metal springs, the stiffness is practically independent of frequency and also of load within the design load range. Whenever vibrations in more than one direction (e.g., in the horizontal as well as the vertical direction) are of concern, however, the proper resilience for isolation in all directions must be provided. In this case, appropriately selected and aligned springs, specially configured rubber pads, or suitably chosen commercial isolators should be employed.

Because of the beneficial effect of increased mass of the machine base, it usually is desirable to mount several machines on the same base. In this way, the vibration transmitted from each machine is attenuated by the mass of the base and also by the masses of the other machines. If such a common base is to be effective, it must not have any resonances of its own at or near the operating speeds of all machines mounted on it; ideally, its fundamental resonance should occur above all operating frequencies. This implies that such bases should be as stiff as possible.

Attention must also be given to avoidance of excessive rocking vibrations of isolated bases. For this purpose it is usually useful to employ wide bases, so that the springs act with large moment arms. It is also beneficial, particularly for bases supporting several items, to have the springs distributed so that they "pick up" the loads locally, e.g., by using stiffer or more closely spaced springs near the heavier items and less stiff or more widely spaced springs near the lighter items.

In practice it is usually necessary to make provisions to avoid the transmission of vibrations via paths that bypass the isolated base. For this purpose it usually is desirable to provide flexible sections of piping, electrical conduit, cable, and ducts between the isolated machines and the unisolated surroundings. (Appropriate devices especially designed for vibration isolation are commercially available; devices such as loops and bellows that are primarily designed to accommodate thermal expansion rarely are adequate for vibration isolation.) Similarly, hold-down bolts through isolators, and similar arrangements that in effect serve as vibration "short-circuit" paths around the isolators, must be removed or themselves isolated, e.g., with soft rubber sleeves and washers.

Isolation of Vibration-Sensitive Items

Basic Principles. The fundamental concepts pertaining to isolation of items to be protected from vibrations may be visualized with the aid of the schematic sketch of Fig. 4.107. This shows a vibration-sensitive item attached to a base, which is fastened to a vibrating structure (e.g., a floor of a building housing vibration sources) via resilient elements, represented in the sketch by springs. In reality, the base may consist of a machine frame or foundation, an inertia block, or a floated (isolated) slab or floor.

If the sensitive item is attached to the vibrating structure directly, without springs, then it vibrates with the same amplitude as the vibrating structure. The same is true if the attachment springs are stiff. However, if the springs are soft, then the oscillatory deflection of the vibrating structure leads to only relatively small forces acting on the base; the inertia of the mass of the base resists acceleration and thus keeps the resulting motion small. This inertia effect is more pronounced for larger accelerations, i.e., higher frequencies, and therefore better isolation is obtained at higher disturbance frequencies, all other things being equal.

The ratio of the amplitude of the motion V_I (displacement, velocity, or acceleration) of the sensitive item to that of the vibrating structure V (measured in terms of the same quantity as is V_I) is called the *motion transmissibility* T_v, or just *transmissibility.** This transmissibility also is equal to the factor by which the amplitude of the vibratory motion of the sensitive item is

* It is somewhat unfortunate that the same word, *transmissibility,* is commonly used to refer to the force transmissibility T associated with source isolation and to the motion transmissibility T_v associated with receiver (sensitive item) isolation. However, because T and T_v obey the same relation, this usage tends to cause little difficulty.

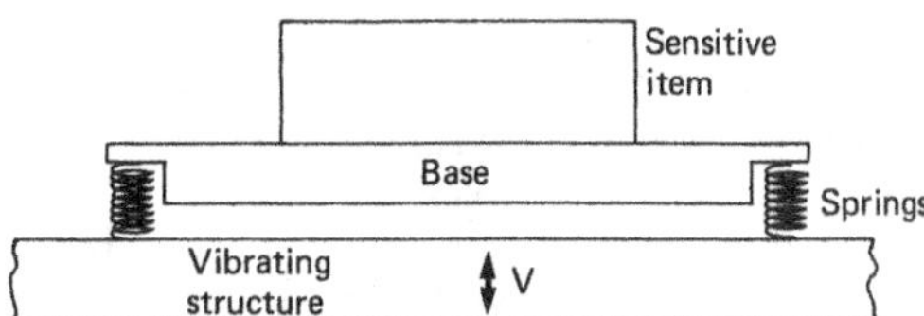

FIGURE 4.107 Conceptual sketch of isolation of a sensitive item.

reduced if it is supported on the given isolators instead of being fastened rigidly to the vibrating structure.*

The *isolation efficiency*† E_v is defined by $E_v = 1 - T_v$ and thus indicates what fraction of the motion of the vibrating structure is kept from the sensitive item. For example, if $T_v = 0.01$, then the isolation efficiency is 99 percent; the sensitive item vibrates only 1 percent as much as it would if it were rigidly attached to the vibrating structure.

Equations (2) to (5) and the discussions accompanying them also apply if T and E are replaced by T_v and E_v and if m is taken to represent the mass of the isolated item and its base.

Practical Considerations. Because the lowest disturbing frequency f corresponds to the greatest transmissibility T_v, an isolation system must be designed with respect to the lowest disturbing frequency of interest.

Figure 4.106 is a convenient chart for estimating the major isolation system parameters needed in order to achieve a desired vibration reduction or transmissibility. The dashed line in the figure illustrates the approach for a case where there exists a disturbing frequency of 3000 cpm (corresponding to 50 Hz), at which frequency it is desired to reduce the vibration by 99.9 percent, i.e., to 0.01 of its original value. The chart shows that here a value of k/m of about 2.5 (lb/in)/lb is needed; thus, for a 2000-lb mass, a total stiffness of $2.5 \times 2000 = 5000$ lb/in is required. (A lesser stiffness would provide better isolation than that prescribed, whereas a greater stiffness would result in poorer isolation.) One may also read from the chart that to $k/m = 2.5$ (lb/in)/lb there corresponds a static deflection of about 0.4 in and a resonance frequency of about 300 cpm or 5 Hz. Improvement in the isolation (reduction in the transmissibility) can be obtained by (1) using softer springs and/or (2) increasing the mass of the base. Reducing the spring stiffness by a given factor has the same effect on reducing the motion of the isolated item as increasing the combined mass of the item and its base by the same factor. Use of softer springs is usually less difficult and less costly than obtaining significantly increased base mass. However, softer springs are subject to greater static deflection and also tend to permit the isolated item to vibrate or to rock more due to motions of its own parts (e.g., of pistons, slides, spindles). In order to minimize this rocking and secondary vibration, it is often useful to balance the isolated items and to distribute the springs so as to oppose the dynamic loads directly (by placing more or stiffer springs where the greater loads act and having the spring force axes aligned with the direction of the loads).

In practice, isolators often are loaded to their full capacity. In such a case, an increase in the mass supported by the isolators requires that more and/or higher capacity (and therefore stiffer) isolators be used, which results in the static deflection and transmissibility remaining essentially unchanged. A mass increase then has practically no effect on the transmitted vibrations, but it does reduce the vibration of the isolated item due to the aforementioned motions of its own parts.

In all cases, in order to be able to provide useful isolation, the resilient elements (springs) must be more flexible than the vibrating structure to which they are attached and the base

* This equality holds only if the motion of the vibrating structure is unaffected by the magnitude of the reaction forces that act on that structure.

† Again, *isolation efficiency* is common usage, although *motion isolation efficiency* is more precise.

supporting the sensitive item. That is, for a given force applied at the center of gravity of the isolated item, the resilient elements must deflect more than either one of the structures to which they are attached. The base of the sensitive item, in addition to being as heavy as possible, should be as stiff as possible.

It is often advantageous to have several sensitive items mounted on the same base; then the total mass of all items and the base acts to assist in the isolation of each item. In this case it is important that the base be as rigid as possible, so that all masses are well interconnected dynamically. In selecting items to be mounted on a common base, it must be kept in mind that items rigidly fastened to the same base can transmit vibrations and rocking motions to each other; thus it is generally inadvisable to mount extremely sensitive items together with items that may cause dynamic disturbances.

Any connections between the isolated item or its base and the vibrating structure must be less stiff than the resilient (spring) elements if these connections are not to transmit more vibration than the springs. Thus, care must be taken to use flexible bellows, tubing, hoses, conduits, cables, etc., with coils and loops where appropriate, to reduce the vibration transmission via these connections.

Some Specialized Control Techniques

Air Spring Isolators.[13,14] Air springs obtain their resilience primarily from the compressibility of confined volumes of air. Such springs, which may take the form of piston-in-cylinder arrangements or of air-filled rubber bladders, can be made to have very low stiffness. They typically are employed where very good isolation is needed and can provide system natural frequencies of 3 Hz or less. Because of their low stiffness, air springs deflect considerably if there is change in the loads they support. However, such deflections can be counteracted by adjustment of the air pressure. Air spring isolation arrangements often are provided with control systems that automatically adjust the air pressure so as to maintain the position and orientation of the isolated item.

Pendulum Isolators. Pendulum arrangements can be made to have low natural frequencies for motions in the horizontal plane, and thus can provide good isolation of horizontal vibrations. The natural frequency of a pendulum arrangement varies inversely with the square root of the pendulum length. A system with a pendulum that is 4 in (10 cm) long has a natural frequency of about 1.6 Hz; one with a 16-in (40-cm)-long pendulum has a natural frequency of about 0.8 Hz. Because pendulum systems only provide isolation in the horizontal plane, they must be combined with vertically acting resilient elements (such as air springs, soft helical steel springs, or resilient rubber pads) if good isolation of vertical vibrations is also needed.[13]

Two-Stage Isolation.[14,15,16] In two-stage isolation, each isolator shown in Fig. 4.105 or 4.107 is replaced by two isolators in series, with an auxiliary mass between them. The auxiliary mass may be included in each isolator arrangement or it may consist of an extended secondary machine base that has isolators below and above it. A two-stage isolation system can provide better isolation at high frequencies that can readily be obtained by use of conventional simple isolation systems, but in a range of low frequencies a two-stage system performs more poorly than a simple system. Generally, the greater the auxiliary mass, the greater the vibration reduction at high frequencies; the lower the frequency range in which the two-stage system performs poorly, the lower the frequency at which the benefit from the auxiliary mass comes into play. Careful design of two-stage systems is required to ensure that the aforementioned poor low-frequency isolation performance does not occur in the frequency range of concern in a particular situation.

Detuning and Structural Damping. The large vibrations that occur if any system or structure vibrates at a resonance may best be reduced by *detuning* (changing the excitation or the system so as to avoid operation at resonance). If such changes cannot be made—either for operating reasons, or because a system is subject to multifrequency excitation, or because it may have a multitude of resonances at closely spaced frequencies—then increasing the damping constitutes essentially the only means of obtaining useful vibration reductions.

For thin metal panels, increased damping can usually be obtained most conveniently by means of layers of viscoelastic (tacky, rubbery) damping material that may be sprayed, troweled, or glued onto the metal panels. A variety of such materials is commercially available, and suppliers of these materials have experience and data pertaining to desirable material thicknesses.

For thick plates of metal or concrete, or for beams, it is often useful to employ a viscoelastic damping material as the middle layer of a sandwich configuration, with the outer layers consisting of the structural material. Careful design of such damped sandwich configurations, based on well-known but somewhat intricate specialized procedures, is required if these are to work properly.[15,17]

In cases where large structures vibrate severely, it is sometimes useful to fasten boxes filled with sand or some other granular material at the locations where the greatest motions occur. Agitation of grains—and their "rattling" against each other and against the sides of the box—extracts energy from the vibration, thus reducing it.

Vibration Absorbers.[15] Vibration absorbers, often also called *dynamic absorbers* or *tuned dampers*, are useful primarily where fixed-frequency vibration is of concern. Such an absorber consists of a mass that is attached to a vibrating system via a spring, with the mass and spring values chosen so that the resonance frequency of this absorber system coincides with the frequency of concern. When this mass-spring system is attached to the base or housing of the vibrating machine or to the vibrating structure, it acts to reduce the vibratory motion of the item to which it is attached. The motion of the added mass is opposite that of the structure to which it is attached; thus, the mass makes the spring push down on the vibrating structure when the latter moves upward, and vice versa.

For a vibrating component or structure with an attached absorber, the product of the absorber mass and its excursion amplitude is equal to the product of the mass of the component and its excursion. Thus, absorbers with small masses need to operate at large excursions; trading off between absorber masses and excursions is part of the design process.

Active Vibration Control.[13,18] Fully active vibration control systems employ sensors to measure the vibration of concern, actuators to provide forces that act to reduce this vibration, and signal processors to provide appropriate control signals to the actuators. The actuators, which may be electrodynamic or piezoelectric, may react against a support (often advantageously in parallel with conventional isolators) or against an inertial mass. In semiactive vibration control systems, the characteristics of some elements of the vibrating system—such as the stiffness of isolators, the resonance frequency of a dynamic absorber, or the positions of masses—are adjusted automatically on the basis of sensed vibrations.

The advent of fast digital signal processing has in recent years given impetus to research and development of active vibration control systems. They have been shown to be advantageous at low frequencies, where conventional "passive" systems are impractical or involve great weight penalties, but they require power, involve added complexity, and need more maintenance. Unlike passive systems, active control systems are not available off the shelf and generally must be developed specially for any given application.

REFERENCES

1. Harris, C. M. (ed.): *Handbook of Acoustical Measurements and Noise Control,* 3d ed. McGraw-Hill, New York, 1991.
2. Beranek, L. L., and I. L. Ver: *Noise and Vibration Control Engineering,* John Wiley & Sons, New York, 1992.

3. von Gierke, H. E., and W. D. Ward: "Criteria for Noise and Vibration Exposure," Chap. 16 of Ref. 2.
4. American National Standard ANSI S3.29-1983: "Guide to the Evaluation of Human Exposure to Vibration in Buildings," Acoustical Society of America, New York, 1983.
5. American National Standard ANSI S3.18-1979: "Guide for the Evaluation of Human Exposure to Whole Body Vibration," Acoustical Society of America, New York, 1979.
6. von Gierke, H. E., and W. D. Ward: "Criteria for Noise and Vibration Exposure," Chap. 26 of Ref. 2.
7. Hall, W. J.: "Vibrations of Structures Induced by Ground Motion," Chap. 24, part I, of C. M. Harris (ed.), *Shock and Vibration Handbook,* 3d ed. McGraw-Hill, New York, 1987.
8. Ungar, E. E., D. H. Sturz, and C. H. Amick: "Vibration Control Design of High Technology Facilities," *Sound and Vibration 24,* 20–27, July 1990.
9. Maten, S.: "Velocity Criteria for Machine Vibration," Chap. 9 of M. P. Blake and W. S. Mitchell (eds.), *Vibration and Acoustic Measurement Handbook,* Spartan Books, New York, 1972.
10. International Standard ISO 2373: "Mechanical Vibration of Certain Rotating Electrical Machinery with Shaft Heights between 80 and 400 mm—Measurement and Evaluation of Vibration Severity," American National Standards Institute, New York.
11. Tímár-Peregrin, L.: "Noise and Vibration of Electrical Machinery," Chap. 20 of Ref. 2.
12. Courrech, J.: "Condition Monitoring of Machinery," Chap. 39 of Ref. 1.
13. Horning, R. W., and D. W. Schubert: "Air Suspension and Active Vibration-Isolation Systems," Chap. 33 of C. M. Harris (ed.), *Shock and Vibration Handbook,* 3d ed. McGraw-Hill, New York, 1987.
14. Ungar, E. E., and D. W. Sturz: "Vibration Control Techniques," Chap. 28 of Ref. 1.
15. Mead, D. J.: *Passive Vibration Control,* John Wiley & Sons, Chichester, UK, 1988.
16. Ungar, E. E.: "Vibration Isolation," Chap. 11 of Ref. 2.
17. Ungar, E. E.: "Structural Damping," Chap. 12 of Ref. 2.
18. Fuller, C. R., S. J. Elliott, and P. A. Nelson: *Active Control of Vibration,* Academic Press, London, 1996.

CHAPTER 4.14
ENERGY CONSERVATION

Peter M. Fairbanks, P.E.
Bluestone Energy Services, Inc.
Braintree, Massachusetts

INTRODUCTION

There are opportunities in almost every facility to reduce energy consumption and energy cost. Although facilities vary greatly in their use of energy and in the potential for reducing energy consumption, one or more of the following conditions generally apply at any facility that creates opportunities to conserve energy:

- Systems are designed for worst-case design conditions and operate most of the time at less than design conditions.
- System designers add margin on top of design conditions to ensure systems will perform. Many times, energy efficiency is not the first priority in system design.
- Space usage or system requirements have changed over time.
- Equipment is purchased on a lowest-first-cost basis rather than a life-cycle basis.
- Higher-efficiency equipment has evolved since existing equipment was installed.
- Equipment is worn and is operating at low efficiency.
- Controls have failed and equipment is operating inefficiently.
- Personnel are unaware of the energy use incurred from unnecessary operation of equipment and systems.

Varying degrees of energy conservation can be achieved in facilities, depending on the existing potential and the level of facility commitment to this goal. Some measures, such as shutting off lights when not needed, require little analysis or management support to implement. Other measures, such as chiller plant or refrigeration system upgrades, may require metering and significant analysis to demonstrate life-cycle cost-effectiveness in order to obtain commitment of project funding.

There is insufficient space here to provide a comprehensive energy conservation treatment that can sufficiently address all of the variables regarding facilities and energy measures. Various facilities will require various levels of documentation of savings and costs. Implementation of one measure may affect the savings and viability of another measure. Various caveats apply regarding implementation of more sophisticated measures, and there are site-specific issues that will affect feasibility of measures. Many measures will require careful integration into existing systems to avoid adverse effects on operation. In order for a comprehensive energy conservation program to be implemented effectively at most facilities, it should ideally be performed on a site-specific basis under the direction of or with the assistance of engineers skilled in such projects. The benefits from such a program can be significant. For facilities where energy conservation has not been addressed, a comprehensive program will typically save from 10 to 30 percent of facility energy costs with projects that have simple paybacks of 5 yr or less. If there are utility energy conservation programs in the area that provide incentives for energy conservation, qualifying projects will typically have paybacks of 3 yr or less. A 3-yr payback project has an internal rate of return of 20 percent. There are few opportunities where a facility can receive this level of return for such a relatively low-risk investment. In many instances energy conservation measures provide additional benefits such as reduced maintenance costs, improved working conditions, or increased production. Improvements to the environment also accrue, as less energy use means less pollution.

The focus of this chapter is to provide information that will be helpful in assessing the potential for energy conservation in your facility and organizing a comprehensive energy conservation program, as well as to provide information on evaluation techniques for specific energy conservation opportunities (ECOs) and identification of some of the potential problems to be avoided.

Energy Conservation Program Preparation

In order to develop a successful comprehensive energy conservation program for your facility, the following preparation is recommended.

- Identify if there are applicable electric or gas utility energy conservation programs that would supply technical assistance in identifying ECOs and provide incentives to install approved measures (contact your utility representatives). *Do this first.* If there are programs with significant incentive payments, you will need to follow the utility's protocol to qualify for their assistance. Utility programs can be a great source of technical assistance and project funding; however, it should also be understood that the assistance provided is usually oriented toward that utility's energy and may not result in optimum energy conservation for your facility. For example, an electric utility may provide engineering analysis and incentives for optimizing an electric chiller upgrade, whereas a gas-fired two-stage absorption chiller could be the most cost-effective option for the facility.
- Identify those in-house resources that can be committed to the program, or whether outside technical assistance is needed.
- Identify the threshold of return and sophistication of analysis necessary for your organization to obtain commitment of funds for project implementation. For many facilities, go/no-go decisions for less sophisticated measures are made on a simple payback basis (project invested money recouped in savings over two yr, three yr, etc.). For higher-cost projects, projects with alternative installation options, or projects involving benefits occurring on different timelines, a more sophisticated life-cycle analysis may be appropriate. With this analysis, the value of avoided energy costs (and other avoided costs such as maintenance, labor, etc.) over the expected useful life of the installation is discounted to present worth by using an appropriate interest rate that represents rate of return for alternative investment of the project money or avoided borrowing cost of the project money. On an "apples to apples" present-worth basis, the benefits of the project can be compared to the installed cost or to alternative projects.
- Identify the level of documentation and savings verification necessary. At a minimum, this typically includes the following:
  - Description of existing conditions and proposed change
  - Savings analysis
  - Cost estimate for project installation

For higher-cost projects, more sophisticated projects, or facilities that require a higher comfort level, preinstallation metering of operating parameters may be required. This could include spot measurement of amps, volts, kilowatts, kilowatt hours (kWh), flow (gallons per minute [gpm]), temperatures, British thermal units (btu), units of oil or gas consumption as applicable with assumptions made to address parameter variation as a function of time of day, outside conditions, or production level. The highest level of metering and demonstration of savings potential would involve recorded measurement, over several weeks or a month's time, of applicable parameters as a function of an appropriate variable such as time of day or outside temperature. The operating performance can then be reasonably extrapolated over a year using historical weather data for the variable parameter(s).

- Identify level of savings verification required. Utility programs or in-house management may require postinstallation savings verification in the form of either one-time measured parameters or recorded metered data collected over a period of time to verify savings calculations.

After completion of the preparation groundwork, an energy conservation program can be implemented that meets the requirements of your facility. A program structure that is recommended and that has been successful at many facilities consists of the following phases:

- Determining the energy conservation potential of the facility
- Refining the program
- Implementation of approved measures
- Maintaining savings measures

This format allows the program to start slowly (minimizing time and cost) and allows management "buy-in" on an informed basis. For example, Phase I (see the following text) requires relatively minimal investment. At its completion, the overall potential of the program and potential ECOs will be defined. Phase II, a more in-depth analysis of measures, will require a larger investment of time and resources; decisions regarding the scope and investment commitment for this phase can be made on an informed basis with the information developed in Phase I. At the end of Phase II, the feasibility of measures will be identified with associated annual energy savings and any associated positive (or negative) operating and maintenance impacts, potential utility incentive payments, and associated installed costs will be accurately defined. Decisions can now be made regarding measure implementation and priority of measure implementation in Phase III. Phase IV involves protection of the investment through savings maintenance.

PHASE I—DETERMINING ENERGY CONSERVATION POTENTIAL

A useful first step in a comprehensive energy conservation program is determining the facility's potential for becoming more energy efficient. This is a preliminary analysis that takes a macro look at the facility energy usage profile supplemented with walk-through identification of likely ECOs. In order to understand how systems are operating, it may be necessary to perform some preliminary diagnostics using spot measurement of amps, kilowatts, flows, and temperatures, and observing equipment operation in response to control manipulation. (More discussion of diagnostic technique is included later.) Depending on the size and complexity of a facility, a list of potential ECOs and the facility's potential for cost-effective energy conservation (5 to 10 percent of total energy cost, 20 to 30 percent of total energy cost, etc.) can be determined by a knowledgeable energy engineer in several days to several weeks. Ideally this analysis is performed with a holistic approach that, with input from operating and maintenance personnel, identifies operating and maintenance problems, indoor air quality issues, equipment at the end of useful life, etc. The preliminary analysis consists of the following process.

Gathering Facility Energy Consumption Data

Information is obtained on energy consumption and cost for the major types of energy (electricity, gas, oil) used at the facility. Many facilities track these data on a monthly basis for budgeting and performance monitoring purposes. If the data are not being tracked to show demand, energy consumption, and cost, billing data should be obtained to demonstrate these. Ideally, several years of utility bills will be available. The desired result is to obtain sufficient data to provide an annual profile of energy use and cost for each type of energy. The billing structure for each type of energy should be identified (on-peak energy cost, off-peak energy cost, applicability of "demand ratchet" where the highest monthly demand for electricity is charged for the entire year, etc.). For purposes of the preliminary analysis of electricity-conserving ECOs, unless there are significant pricing mechanisms such as demand ratchet that should be accounted for, using average cost of energy on a dollar-per-kilowatthour basis is usually sufficiently accurate. The seasonal average cost for fossil fuels is also sufficiently accurate for preliminary analysis of fossil fuel–conserving ECOs.

Energy Use Reconciliation

As a filter and a guide, it is sometimes useful to roughly reconcile end use of energy with annual facility energy billings. For some facilities—especially for large, complex facilities—the value of the reconciliation may not be worth the effort to accurately achieve it and the time may be better spent analyzing discrete systems for preliminary ECO potential. If it is decided that a reconciliation is desirable, and metered data, submetered data, operator log data, etc. are available, these can be used to provide a higher degree of accuracy. However, for purposes of this preliminary analysis, simple calculations based on operating system assumptions will generally suffice. Reconciliation of end energy use with total annual energy use based on billing should be achieved to an accuracy of ± 10 percent. If this level of accuracy is not attained, operating data should be refined and assumptions should be changed regarding loads, efficiencies, etc. until reconciliation is achieved. This exercise can result in a basic understanding of how energy is used in the facility and can help to prioritize where the best opportunities for conservation exist.

End user energy consumption should be broken down into major categories by utilizing the following assumptions and calculations.

Electricity. Convert all usage to billing units (kWh).

- *Lighting*—Multiply the wattage of light fixtures by rough counts of lighting and hours of operation.
- *Electric chiller plants*—Multiply the average chiller load (tons) by efficiency at that load (0.5 to 0.7 kW/ton typical) multiplied by annual hours of operation. Associated chiller pump and cooling tower fan energy is determined by motor nameplate horsepower multiplied by assumed load factor (85 percent) multiplied by 0.746 kW/hp multiplied by annual hours of operation.
- *Air handling units*—Multiply the motor nameplate horsepower by the assumed load factor (85 percent) multiplied by 0.746 kW/hp multiplied by annual hours of operation. Associated direct expansion cooling energy is determined by average cooling load multiplied by efficiency (1 to 1.2 kW/ton typical) multiplied by annual hours of cooling load operation.
- *Production line motorized equipment*—Multiply the motor nameplate horsepower by the assumed load factor (85 percent) multiplied by 0.746 kW/hp multiplied by annual hours of operation.
- *Production line electric heating equipment*—Multiply the kilowatt rating by the average load multiplied by annual hours of operation.

- *Production line electric cooling equipment*—Multiply the average cooling load tonnage by the efficiency (1 to 1.2 kW/ton typical) multiplied by annual hours of operation.

Fossil Fuel. Convert all usage to billing data units (therms gas, gallons oil, etc.) or convert all energy into common units such as megabtu (Mbtu).

- *Building skin loss heating load*—Based on building construction and insulation characteristics, determine the average heating load on an average heating season day and multiply by heating season hours. Publications from the American Society of Heating, Refrigeration and Air Conditioning Engineers (ASHRAE) (www.ashrae.org) as well as many other heating, ventilating, and air conditioning (HVAC) manuals provide information on typical building construction heat transfer factors, average winter temperatures in various parts of the country, and formulas for heat load calculation. Divide heat load by boiler system, or, if individual heaters are used, by burner efficiency (typically 60 to 80 percent) to determine the quantity of purchased heat energy.
- *Outside air heating load*—Identify the amount of outside air brought into the building through analysis of drawing schedule data for air-handling units, observation of air handling unit operation, assumptions based on ft^2 (m^2) of air-handling unit intakes, and typical design airflow velocities (150 to 800 ft/min [250 m/min] at intake, 500 ft/min [150 m/min] at coil, etc.). Using a differential temperature based on the average heating season outdoor temperature and average air-handling unit discharge air temperature, and annual hours of heating season operation, calculate the amount of heating btu required using the following formula:

$$1.08 \times \text{CFM} \times \text{differential temperature} \times \text{annual hours}$$

  If the building has more exhaust than make-up air and pressure is negative, determine the quantity of infiltration air as a percentage of the shortfall and calculate its heat load as well. Adjust for boiler/burner efficiency.
- *Steam absorption chiller*—Multiply the average chiller load (tons) by the efficiency (20 to 30 lb/ton for a single-stage absorber, 11 to 14 lbs/ton for a two-stage absorber) multiplied by annual hours of operation. Adjust for boiler/burner efficiency.
- *Production line heating equipment*—Multiply the thermal rating by the average load multiplied by annual hours of operation. Adjust for boiler/burner efficiency.

Identification of Potential ECOs

In the preliminary analysis, the focus is on identifying all significant ECOs. There may be additional smaller opportunities that can be explored in Phase II or III, but the Phase I emphasis is typically to "get the most bang for the buck" and identify major opportunities. For each ECO, a cursory estimate of savings and installed cost should be developed with assumptions made as necessary regarding operating conditions, load, etc. Candidate ECOs that have typically been found to be cost-effective based on "normal" operating conditions and New England energy pricing are listed in the following sections. The basis for normal operating conditions is office building systems operating 3000 plus h/yr, commercial facilities operating 4000 h/yr, small manufacturing facilities with single-shift operation, and large manufacturing facilities operating two or three shifts. The energy cost basis is electricity at an average cost of $0.07 per kWh, and fossil fuel costs range from $3.00 per Mbtu for larger users with No. #6 oil and interruptible gas contracts to $7.00 per Mbtu for facilities burning firm gas and No. 2 oil.

Lighting (See Also Chap. 5.9). Retrofitting lighting is usually one of the first and best ways to conserve energy in a facility. It is not uncommon to be able to reduce lighting energy use by 40 percent or more while maintaining acceptable light levels. With reduction in lighting energy, there is associated reduction in air-conditioning load and a lesser negative impact on building heating load. Because newer technology lighting fixtures such as T-8 lamps with electronic ballasts have higher light output and better color rendition than older T-12 lamps with magnetic ballasts, it is possible to significantly reduce lighting energy while improving light quality. Many areas, especially those with office/cubicle task lighting and computer use, are overlit. It is useful to spot-meter with a foot-candle meter to determine existing light levels and potential for delamping. Lens treatment as part of the retrofit, although adding expense, may be desirable to avoid computer screen glare or to eliminate aged prismatic lenses. It may be feasible to improve the efficiency of a fluorescent fixture by installing a reflector and reducing its energy level by delamping. Decisions regarding retrofitting existing fixtures versus installing replacement fixtures (higher cost) will depend on issues such as lens treatment and fixture efficiency. Higher levels of operating hours will make more expensive treatment more feasible. It is recommended that, as part of Phase II, sample lighting retrofits be installed and light levels measured before installation and after installation with sufficient time allowed for new lamp burn-in. Lighting energy conservation is very site specific; however, the following are typical measures that have been found feasible and cost effective:

- Standard 2- × 4-ft four-lamp fixtures with T-12 (1.5-in-diameter) lamps and magnetic ballasts should be changed to T-8 (1-in-diameter) lamps and electronic ballasts. Delamp as appropriate to three lamps with or without reflector, two lamps with or without reflector, or new three- or two-lamp fixtures.
- Standard 2- × 2-ft T-12 U-tube fixtures should be changed to T-8 U-tube/electronic ballast or three 2-ft T-8 lamps and electronic ballast, with or without reflector.
- High-output fluorescent fixtures should be replaced with new fixtures containing T-8 lamps/electronic ballasts and reflectors.
- Two-lamp strips containing 4-ft or 8-ft T-12 lamps should be retrofitted with reflector and a single strip of 4-ft T-8 lamps or replaced with new fixtures containing single 4-ft T-8 lamp strips with electronic ballasts and reflectors.
- Incandescent fixtures, recessed and surface, 150 W or less, should be retrofitted with hard-wired compact fluorescent 13-W or 26-W retrofit kits (some allow dimming if required) or new compact fluorescent fixtures.
- Incandescent fixtures, 200 W or more, should be replaced with T-5/electronic ballast fluorescent fixtures.
- Mercury vapor, high-pressure sodium, or metal halide fixtures should be replaced with T-5/electronic-ballast fluorescent fixtures.
- For office lighting redesign, replace three- to four-lamp troffers with pendant-mounted direct/indirect two-lamp fixtures.

Lighting Controls. Lighting control strategies that have been found effective include:

- Motion sensor control for spaces with sufficient lighting wattage controlled (200 W minimum) manual on, automatic off, ultrasonic sensor.
- Photocell control for outside lighting or lighting near window walls or under skylights.
- Energy management system (EMS) control of building lighting on an area or floor basis allowing scheduling of lighting. Local override and emergency lighting circuits would be required.

HVAC (See Also Chaps. 3.5, and 5.2, and 5.3). Heating, ventilating, and air-conditioning energy conservation strategies include the following.

Controls

- Install an EMS that provides the following control strategies for HVAC equipment: scheduling control, setback control, optimum start/stop, air handling unit discharge temperature reset, air side economizer control.
- For buildings with electric heat, install CO_2 sensor control of outside air dampers through the EMS.

Air Handling Systems

- On variable-air-volume (VAV) systems, install variable-frequency drives (VFDs) to control supply fan and return fan speed as a function of system pressure set-point in place of inlet-vane control. This is applicable for airfoil or backward-curve fans, not forward-curve fans, which have relatively efficient unloading with inlet-vane control.
- Convert dual-duct systems to cooling-only VAV systems if separate perimeter heating is installed.
- Convert primary electric heat in large air-handling units to hot-water coils served by new gas-fired condensing boiler(s).
- Eliminate any simultaneous heating and cooling (lock out reheat in cooling season, isolate steam heat and reheat in cooling season, etc.).

Hydronic Systems

- Install VFDs on heating, reheating, or cooling pump sets (10 hp [7.5 kW] and larger) that serve end users with two-way valves. The VFD controls pump speed as a function of system pressure in place of pressure bypass control.
- Install outside temperature–based reset control, especially on systems without VFD control.

Chilled Water Plant

- Replace inefficient chillers with efficient chillers. As a rule of thumb, cooling from a single-stage steam absorption chiller costs 2 to 3 times as much per ton as cooling from a new efficient electric centrifugal chiller or a two-stage steam absorption chiller. An electric centrifugal chiller that was installed before 1990 would have an efficiency on the order of 0.65 to 0.8 kW/ton, compared to 0.5 kW/ton or better (depending on size) for a new electric centrifugal chiller. If the chiller is inefficient, the boiler is inefficient, and the facility is heated with hot water, a direct-fired two-stage steam-absorption chiller could be cost-effective.
- Install a VFD on a chiller with inefficient unloading characteristics.
- Consolidate chilled-water systems and schedule chillers to use the most efficient chiller first.
- Install a plate and frame "free-cooling" heat exchanger for facilities with cooling towers and cold-weather chilled-water loads (data centers, process, absence of air side economizer capability).
- Reduce chilled-water pumping energy by conversion to primary/secondary pumping, or, if compatible with chillers and chiller controls, variable pumping through chillers. This measure requires two-way valve control at the end user.
- Install a VFD(s) on the condenser water pump(s) responding to chiller head pressure or differential condenser water temperature.
- Install a VFD(s) on the cooling-tower fan(s) in place of single-speed cycling fan control.

- Reset condenser water temperature lower as a function of wet-bulb temperature down to the chiller manufacturer's recommended minimum (typically 65° F [25°C]).

Heating System

- Upgrade boiler/burner controls. On larger boilers (100 hp [75 kW] plus), replace jackshaft "fixed" fuel-air relationship controls with distributed digital/servomotor controls and exhaust-gas analysis systems (O_2, CO, CO_2). With this type of control system, excess air can be precisely controlled.
- If boiler stack temperature exceeds 400°F (200°C) and boiler heat-transfer surfaces are not fouled, install heat recovery to heat feedwater or combustion air.
- If tubes are fouled, remove scale and improve water treatment to avoid scale deposits. If present, remove soot and deposits on the fire side of tubes.
- Insulate exposed steam system piping and valves.
- Install a VFD on the boiler feed pump.
- Install VFDs on forced-draft fans and induced-draft fans to control fan speed in lieu of damper or vane control.
- Replace steam traps with steam venturis where applicable. Venturi devices are engineered and must be accurately sized for the application. The benefit is that there are no moving parts to wear out, reducing maintenance and steam losses.
- Reduce boiler steam pressure if compatible with system requirements.
- Install gas-fired radiant heating for large open space heating in place of heating and distributing air.

Process

Compressed-Air Systems. Recent electric utility research conducted in New England indicates that 6 to 17 percent of industrial electric energy is used in compressed-air systems and that 30 percent of this energy can typically be conserved.

- Fix system leaks.
- Lower the system pressure to the lowest required by end users.
- Install pressure regulators at end users.
- Install solenoid valves that isolate compressed air to machines when machines are off-line.
- Modify the distribution system piping to reduce pressure drop if it is found to be excessive.
- Install engineered vortex nozzles and air knives for blow-off applications.
- Use mechanical vacuum pumps in place of compressed-air venturi vacuum devices.
- Install blowers for low-pressure applications, such as tank agitation, in place of using compressed air.
- Convert air-driven mixers to motor-driven mixers where feasible.
- Install motor-driven hoists in place of air hoists.
- Install adequate storage with intermediate pressure control for applications where air demand is variable and single-stage rotary screw compressors with load/no-load control are not fully blowing down during periods of low air demand.
- For variable air-demand systems, replace single-stage rotary screw compressors equipped with inlet-valve modulation control (or other control types [load/no-load, turn valve, etc.] not efficiently following load variations) with variable-frequency drive or reluctance-motor-drive air compressors.

Cooling

- Use cooling-tower/plate-and-frame free cooling in place of chilled water for winter operation or for processes that can tolerate 60 to 70°F (16 to 21°C) cooling water.

Heating/Drying

- Install gas or electric infrared heat in place of electric resistance heat or gas-fired convection heating for drying or curing applications.

Vacuum Systems

- Install a cooling tower and reduce the seal water temperature on liquid-ring vacuum pumps, allowing resheaving and reduction in horsepower to achieve the same vacuum level.
- Replace liquid-ring vacuum pumps with efficient blowers (equipped with water separators if water is being removed from the process).

Pumping Systems

- Install VFDs on pumping applications where flow is throttled and recirculated back to the suction side of the pump or pressure is simply allowed to build up.

Make-up Air/Exhaust Systems

- Install controls to coordinate makeup air and exhaust with process requirements. Reduce airflows by using contaminant sensor/VFD control.

Refrigeration Systems

- Install floating head pressure condenser controls.
- Install evaporator fan controls.
- Install hot gas defrost controls based on demand.
- Install VFD condenser fan controls.
- Optimize ammonia refrigeration systems with thermosyphon compressor cooling, liquid overfeed evaporators, pumped refrigerant circulation, and automatic air purgers.

General

- Optimize processes requiring serial heating/cooling. Eliminate simultaneous heating and cooling.

PHASE II—REFINING THE PROGRAM

Based on the information gathered in Phase I, decisions can be made regarding the scope of budgeting for ECOs, the relative attractiveness of identified opportunities, and the subsequent extent of Phase II analysis. This decision-making process will result in a list of candidate ECOs for further analysis in Phase II. The intent of Phase II is to prove or disprove assumptions made in Phase I regarding measures and to perform sufficient engineering analysis, metering, and information gathering so that measure costs, savings, and benefits will be accurately defined to support implementation decisions in Phase III. Issues regarding refrigerant selection, appropriate disposal of hazardous waste, safety issues, potential VFD harmonic distortion, VFD/motor compatibility, etc. should be addressed. Sufficient engineering design should be performed to obtain contractor pricing or highly accurate cost estimates. Analyses should be performed based on metered data as necessary to verify savings. Historical weather data are often used for such analysis of system operation that is a function of outside conditions. These are engineering weather data that have been collected for as long as 30 years at representative locations across the country. One source for these weather data is the Department of Defense Engineering Weather Data, USAF Handbook 32-1163, recently updated and available in draft form at the Air Force Civil Engineer Support Agency Web site (www.afcesa.af.mil) and soon to be available for purchase in CD-ROM format from the National Climate Data Center (www.ncdc.noaa.gov). Data are provided for 511 U.S. locations and 292 international locations and are in a bin-temperature format that summarizes on a monthly basis the average number of hours each location experiences a specific dry bulb temperature range (e.g., 80 to 84°F [27 to 29°C]) with mean coincident wet-bulb temperature for the range at specific time periods (e.g., 5 to 8 A.M.). Another source of engineering weather data is ASHRAE's updated Weather Year for Energy (WYEC-2) software and data, which allows typical monthly weather profiles to be developed for 77 U.S. cities. This is available as a CD-ROM or as a download from ASHRAE (www.ashrae.org).

Following are some diagnostic procedures that have been found useful in verifying existing operation of equipment and systems.

Diagnostics

- An amp probe, or preferably a kilowatt-measuring device, can be used to measure amps and volts or kilowatts of existing motors, verifying existing energy levels.
- An ultrasonic flow meter allows flow measurement of water and other liquids from the outside of piping.
- A multichannel recorder capable of measuring and recording variable parameters over a period of time provides information on equipment operation as a function of a variable. Typical parameters measured include kilowatts, kWh, temperatures, and, with the use of the ultrasonic flow meter, flow. With this equipment, for example, it is possible to record chiller load and energy use as a function of outdoor air temperature by recording chilled-water supply and return temperature, chilled-water flow, chiller kilowatts, and outdoor air temperature.
- A multichannel recorder is also useful for metering the kilowatt load of air compressors over a period of time. Using compressor manufacturer performance data and observed maximum and minimum load, unloading point, etc., the facility air-load profile can be established. If it is determined to be necessary, a more accurate profile of energy use and air use can be established by incorporating an airflow meter, but this requires the time and expense of installation in the piping.
- A portable CO_2 meter is useful for determining the potential for reducing outside air in a facility.
- A multimeter that measures dry-bulb temperature and wet-bulb temperature is useful in diagnosing air handling system performance. With this device one can establish percentage of outside air, performance of heating and cooling coils, presence of simultaneous heating and cooling, etc.
- An infrared temperature recorder is used to obtain spot measurements of hot or cold surface temperatures to verify operation.

PHASE III—IMPLEMENTATION

With the completed analyses of Phase II, it should now be clear which ECOs meet the facility criteria for project implementation and what order of implementation is desirable. Many of the ECOs will require final engineering to address typical design issues. Depending on the scope of the ECO and the facility preference, decisions can be made regarding method of implementation—full engineered plans and specifications, design/build, third-party performance contract, etc. Issues of phasing the work in and around facility operation to minimize disruption should be addressed. Any savings verification requirements should be clearly identified and incorporated into the project before the work is started. Operating requirements to maintain savings levels should be clearly defined and documented.

PHASE IV—SAVINGS MAINTENANCE

After the facility has incorporated energy conservation measures that meet its criteria of cost and value, the measures must be maintained. Periodic reviews of performance should be conducted to verify that set-points have not changed, schedules are still appropriate, and equipment is maintained and not operating in "temporary" off-design condition. Over time, if the energy conservation measure operating requirements are not vigilantly maintained, savings will typically erode. It is also usually possible, after gaining experience with the new systems and equipment, to actually tweak performance and increase savings.

Going forward, energy efficiency should be a component of the decision-making process for replacing equipment that is at the end of its useful life or purchasing equipment for facility or process expansion. Motors that wear out should typically be replaced with premium efficiency motors. For more sophisticated equipment, such as chillers or boilers, it may be desirable to perform a life-cycle analysis and identify the present value of the energy savings for each increment of efficiency improvement over the life of equipment. As an example, consider a proposed chiller installation with an expected useful life of 20 yrs; an average load of 300 tons over a 2500-h cooling season; electricity cost of $0.07 per kWh, escalating at 3 percent; and value of money at 8 percent per year. The life cycle energy savings for each 0.01-kW/ton chiller efficiency improvement under this scenario would be $6431. This value can be listed as an evaluation criterion in the purchase specification, and a fair evaluation can be made of a lower-first-cost chiller with a lower efficiency compared to a higher-first-cost chiller with a higher efficiency. With this type of purchasing analysis, a performance guarantee with penalties equal to the present worth value of life-cycle efficiency should be specified with efficiency demonstrated by a witnessed performance test.

SECTION 5

BASIC MAINTENANCE TECHNOLOGY

CHAPTER 5.1*

ELECTRIC MEASURING INSTRUMENTS

Simpson Electric Company
Elgin, Illinois

INTRODUCTION

Electrical measuring instruments are important to all aspects of plant maintenance and control. Some tasks require only simple yes or no checks, e.g., continuity, presence or absence of

* Revised for this third edition.

line power voltage, and current flow. Instruments for these tests may be rugged and inexpensive. Yet they must be safe, reliable, and accurate. Whether you are working on the production line, in the laboratory, or servicing equipment, measurements can yield cost savings proportional to instrument accuracy. Electrical measuring equipment is available in three basic configurations: handheld, bench-mounted, and panel-mounted. The following paragraphs in this chapter provide: (1) overviews of uses for electric and electronic measuring instruments, (2) guides to the safe operation of these instruments, and (3) criteria for selecting particular instrument features.

THE MULTIMETER

There is a need for electrical measurement everywhere in the plant:

- Incoming component inspection
- Production line assembly
- Quality control
- Plant maintenance and repair
- Power monitoring and conservation

Popular with manufacturing and service professionals alike, the most versatile test instrument is the multimeter, combining measurement of ac and dc voltage, resistance (ohms), and current (amperes). Usually compact, battery-powered, and portable, two types of multimeter are widely available: the analog volt-ohm-milliammeter (VOM) and the digital multimeter (DMM).

VOLT-OHM-MILLIAMMETER (VOM)

The analog volt-ohm-milliammeter (Fig. 5.1) is basically a test instrument, with several ranges for measuring volts, ohms, and amperes. The VOM consists of:

- An electromechanical meter movement and precision meter face
- Selection switches with resistors and rectifiers that allow the meter to measure various voltage, current, and resistance ranges

The analog display used by VOMs is well-suited to displaying quick trend information, nulling and peaking, ranges, and go/no-go readings. The dB scale can be used in conjunction with the ac voltage measurement to give an indication of the power gain of an amplifier. A zero (dB) power level must be chosen, typically 1 mW in 600 Ω.

The power gain is defined as

$$\text{dB} = 10 \log_{10} \frac{P_2}{P_1}$$

Today's high-quality analog models feature taut-band construction which can operate for more than 20 million cycles with a repeatability error of less than ±0.02 percent. This corresponds to about 70 yr of normal use. Taut-band construction also withstands moderate shock and vibration and operates on limited power of less than 5 μW.

The VOM is the best all-around electronic instrument for industrial plant maintenance. Various electrical tests can be made on rotating electrical equipment, such as motors and generators, to check all operating conditions. For example, to check continuity of shunt field and armature coils.

FIGURE 5.1 Analog volt-ohm-milliammeter (VOM).

Measurement of rotating equipment depends, in part, on its operating characteristics. DC motors, for example, can be series, shunt, combination series-shunt, or compound. In the series type, where the field winding is in series with the armature, the field strength changes with armature current, and starting torque is high. An increased load reduces the speed and the torque increases. In the shunt type, the field winding is in parallel with the armature and the field strength does not vary with armature current. Starting torque is lower and speed varies little with load changes. In the compound type of motor, there are two sets of field windings, one set in parallel with the armature and the other set in series. Characteristics of the compound motor for both speed and load can be varied by adjusting relations of the two sets of windings. See Chaps. 5.7 and 5.8, "Electric Motors" and "Electric Motor Controls."

A wide range of accessories such as high-voltage probes, temperature and high-frequency probes, and ac or ac/dc ampere clamp-ons have greatly increased the measurement capabilities of the analog multimeter, well beyond its VOM designation.

DIGITAL MULTIMETER (DMM)

Digital multimeters (Fig. 5.2) have been evolving rapidly. The electronics of a present-day digital multimeter are extremely complex. In the simplest terms, it consists of an integrated analog-to-digital converter which measures voltage by comparing it to an internal reference voltage. In measurement of current, a resistance is placed in series with the load, and the DMM measures the voltage drop across the standardized resistance, converting it to a current measurement. Like its analog counterpart, the DMM measures voltage, current, and resist-

FIGURE 5.2 Digital multimeter (DMM).

ance, but displays measurements directly on a digital display. Digital multimeters today offer the kind of features that you would expect from integrated circuits (ICs) with recent advances in digital technology. Expanded capabilities include the following:

- Memory
- Frequency measurement sensing
- Peak hold and data hold
- Logic probes
- Continuity beeper and diode test
- Capacitance measurement
- Analog bar graphs

Automatic range selection, and the features mentioned above, make the DMM more versatile and easy to use than ever before. Accessory adapters have also been developed to measure insulation resistance, watts, power factors, and ac line frequency. Some very recent designs in DMMs incorporate RS-232C interface capability. Here the multimeter is connected to a PC, ready to download stored readings for hard copy to document production test records or to build a database for future analysis or presentation via PC software.

METER SELECTION CRITERIA/FEATURES

When purchasing a panel, bench, or handheld meter, the most important characteristics to select are:

- Root-mean-square (rms) sensing
- Accuracy and resolution
- Range

RMS Sensing

Root-mean-square sensing measures the power of an ac signal (i.e., of alternating currents and voltages, the effective current or voltage applied). It is that value of alternating current or voltage that produces the same heating effect as would be produced by an equal value of direct current or voltage. For a sine wave, it is equal to 0.707 times the peak value. For measuring perfect sine waves (Fig.5.3), less expensive meters that average a rectified signal are sufficient. If your applications have varying (nonsinusoidal) wave amplitude and frequency (Fig. 5.4), select "true rms" capability for voltmeters and clamp-on ammeters.

Accuracy and Resolution

Accuracy and resolution indicate how fine and how precise are the readings on the instrument. Accuracy is the maximum deviation to be expected between the meter reading and the actual value being measured. Accuracy is usually expressed as a percent of full scale for analog instruments and as a percent of reading for digital instruments. A high-accuracy analog voltmeter may have a specification of ±2 percent at full scale. This indicates that, when set to the 100-mA scale, for example, expect the reading to be within 2 percent of 100, or ±2 mA.

Resolution is the degree to which nearly equal values of a quantity can be discriminated. In analog meters, resolution is the difference between adjacent scale divisions. In digital meters, resolution is the value represented by a one-digit change in the least-significant digit.

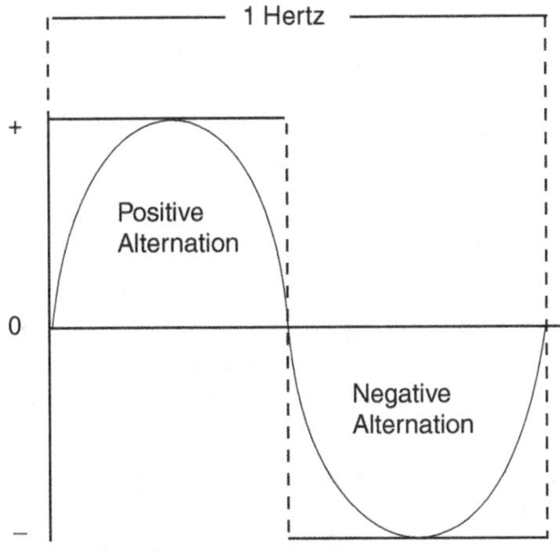

FIGURE 5.3 Sine wave.

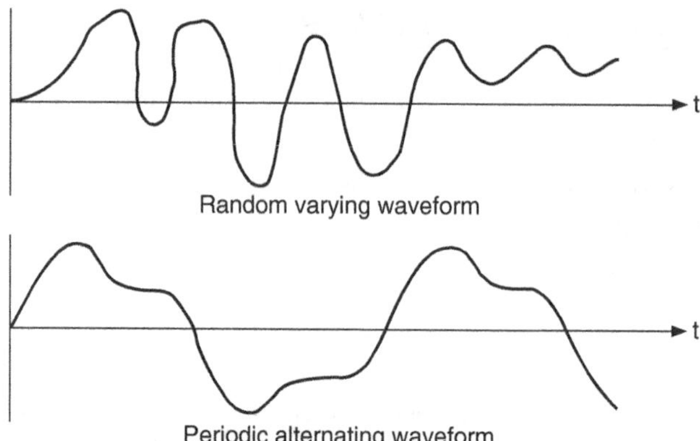

FIGURE 5.4 Varying wave intensity and frequency.

Published accuracy and resolution specifications are useful in selecting a meter that meets your requirements economically.

Range

Range indicates the highest reading that can be safely measured with the meter. Several resistors internal to VOMs and DMMs allow readings over a very broad range.

Ammeters measure the current in a circuit. Attempting to measure current above the range of the meter can damage the movement or burn out internal components. Ammeters must be selected for the current range to be tested. Accessories such as shunts and amp-clamps extend the range of an ammeter.

Many designers overlook the value of the *suppressed-zero meter* (Fig. 5.5), in which the significant portion of the range is expanded to occupy the entire scale, while the lower portion is suppressed. Accuracy may be increased in proportion to the scale expansion. For example, a

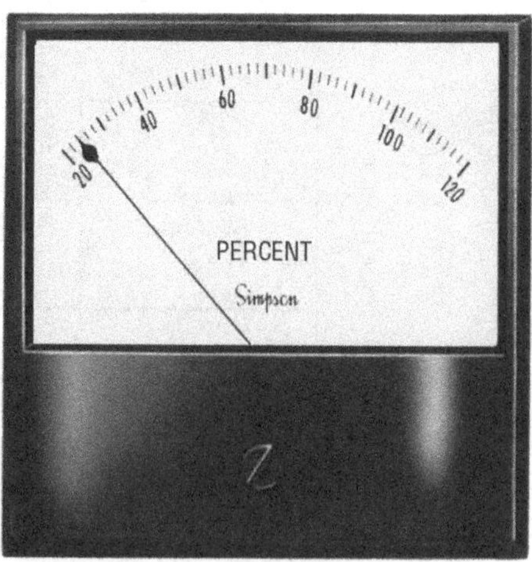

FIGURE 5.5 Suppressed-zero meter.

FIGURE 5.6 250°-swing meter.

meter with an acuracy of 2 percent of the full scale can have a maximum error of 2 V if the range is 0 to 100 V, but will have a maximum error of only 1 V if the range is 50 to 100 V.

Another useful style of meter is the *250°-swing meter* (Fig. 5.6), which provides a wider range.

VOLTMETERS

A voltmeter measures the electrical potential (voltage) between two points of a circuit. Depending on the application, voltmeters may be used to read fractions of a volt in a solid-state circuit one day and a 440-V power line the next. Applications such as cathode-ray tube troubleshooting require measurements in the kilovolt range.

Voltmeter Selection Criteria/Features

Aside from having the required characteristics of rms sensing, accuracy, resolution, and range, a voltmeter must also have the correct internal resistance for the application.

Many solid-state and logic circuits operate on very small currents. Putting a low-resistance meter in parallel with a solid-state circuit may increase current enough to damage sensitive components. In addition, the current change may be sufficient to bias the PN junction in solid-state devices, altering the circuit current paths. Generally, a meter is considered high resistance when the input resistance is 10 MΩ or higher.

Sensitivity and Meter Loading

Sensitivity is a ratio of the response of a measuring device (meter) to the magnitude of the measured quantity (volts, ohms, amperes, etc.). Voltage-measuring devices are rated in ohms per volt (Ω/V). On any particular range, it is obtained by dividing the resistance of the instrument (in ohms) by the full-scale voltage value of that range. A meter with a 100-kΩ resistance on the 0- to 100-V range has a 1000-Ω/V rating. A meter with a 2-MΩ resistance on the 0- to 100-V range has a 20-kΩ/V rating.

The loading effect of a voltmeter can be demonstrated by using these two meters (Fig. 5.7). First the 1000-Ω/V meter is used to measure the voltage across the 100-kΩ resistor in the circuit of Fig. 7.5*a*. As soon as the meter is placed across R2, a parallel combination of resistance (R2 and the 100-kΩ internal resistance of the meter) is formed. This forms a total resistance of 50 kΩ. By Ohm's law ($V = IR$) the voltage drop across R2 should be 40 V, the circuit current (0.4 mA) times the 100-kΩ resistance. However, multiplying the 50-kΩ parallel bank by the circuit current of 0.4 mA gives 20 V, an error of 50 percent.

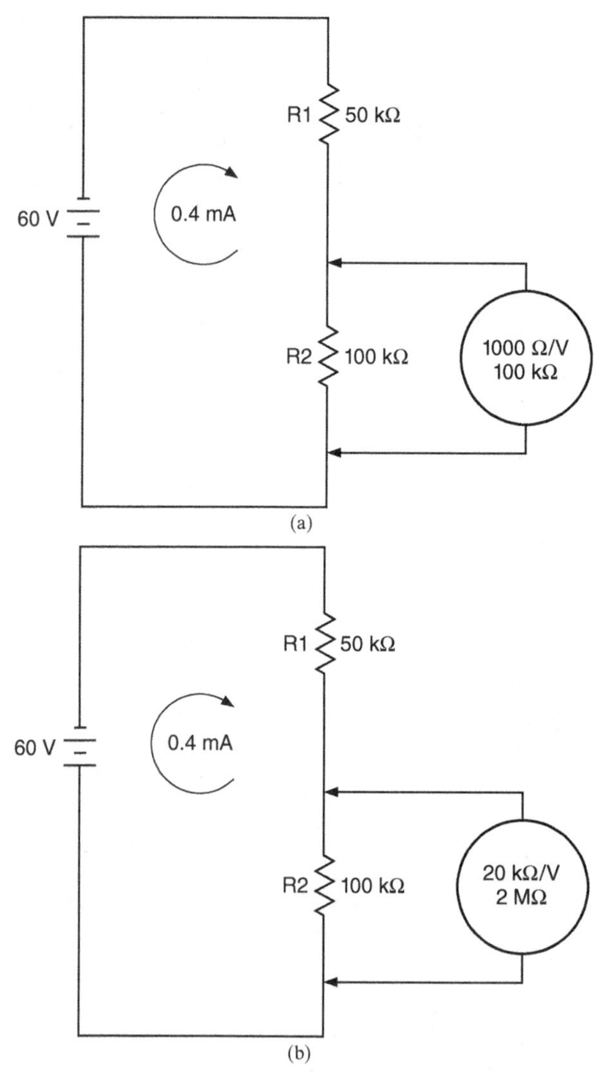

FIGURE 5.7 (*a*) Loading effect of a voltmeter. (*b*) Correct reading with a 2-MΩ meter.

FIGURE 5.8 Connect the voltmeter in parallel with the load.

In Fig. 5.7*b*, the 2-MΩ resistance meter measures the same voltage drop, only this time the parallel resistance of R2 and the meter (100 kΩ/2 MΩ) equals 95,239 Ω. Now the circuit current of 0.4 mA times the parallel bank of 95,239 Ω equals approximately 38 V. It is obvious that meter 1 is loading down the circuit. For circuits with very high impedance, another form of meter is necessary, such as the DMM.

Voltmeter Operation

When operating a voltmeter, take precautions to prevent damage to the meter and the circuit under test:

- Always connect the meter in parallel with the load (Fig. 5.8).
- Select a voltage range that provides a midscale reading. When testing a 5-V circuit, select the 10-V scale. If circuit voltage is unknown, select the highest scale and work downward.
- Observe the correct polarity, particularly with analog meters. A reversed-polarity signal can bend the meter needle, damage the movement, or burn calibration inside the instrument.
- Select a meter with the proper resistance for the circuit being tested.

Voltmeter Accessories

Accessories can expand the range and versatility of the voltmeter. Some common accessories are:

- High-voltage probe allows a voltmeter to read voltages above 1 kV, which are common to cathode-ray tubes.
- Temperature probe uses a voltmeter scale to display temperature. Use for heating, ventilating, and air conditioning (HVAC), refrigeration, or machine operating temperatures. Infrared-sensitive noncontact temperature probes are available for fragile or moving objects.

AMMETERS

An ammeter measures the current flowing past a single point in a circuit. Ammeters may be required to read microamperes (μA) through more than 1000 A in welding and manufacturing equipment. Ammeters measure the current flowing in a circuit. Therefore, an ammeter is useful in determining varying loads on motors and other machinery. There are two major types of ammeter. The *direct-wire* type connects directly to the circuit. *Clamp-on* types have jaws that close over the conductor, taking a reading without interrupting the circuit. Clamp-on ammeters generally have a higher current range than direct-wire types.

Ammeter Selection Criteria/Features

The most important characteristics to select from are:

- Range
- Direct wire or clamp-on

Range

Range is the highest amperage the meter can safely read without being damaged. Typical handheld instruments may have a maximum capacity of 2 A. Some bench models range up to 10 A. Accessories such as shunts (for dc) and amp-clamps (for ac) allow meters to handle much higher current ranges.

Direct Wire or Clamp-on

Multimeters are connected in *series* with a circuit when measuring current. This requires that the circuit be turned off and disconnected, and the meter probes connected to complete the circuit again. This is practical for small circuits and bench operations, but may not be acceptable for high-current or continuous-usage machinery.

AC above 1 A may be measured by a clamp-on meter. The complete ammeter measurement and display function is contained in the handle of the meter (Fig. 5.9). Clamp the jaws of the meter over the current-carrying conductor, and internal circuitry selects the proper range and displays the reading. Clamp-on meters are most useful in:

- High-current applications
- Circuits where current cannot be interrupted

Ammeter Operation

When operating an ammeter, take precautions to prevent damage to the meter and the circuit.

Direct-Wire Ammeter

- Always connect the ammeter in series with the load under test, Fig. 5.10. Observe polarity.
- Before reapplying power to the circuit, select a current range that provides a midscale reading. When testing a 2-A circuit, select the 5-A scale. If circuit voltage is unknown, select the highest scale of the meter and work downward.
- If the current is above the range of the meter, select a current shunt or amp-clamp.

Clamp-on Ammeter. The ac clamp-on ammeter is actually a special application of a current transformer. Clamp-on ammeters can be used to determine the current loading and to check that current densities are not being exceeded for the cables and wiring involved. Even though circuits are initially in balance, they tend to get out of balance as a result of machines being added to the system.

Depending on the size of the instrument, ranges covered may be as high as 2000 A. More common ranges are 500, 250, 100, 50, and 10 A ac. Some clamp-on meters also provide ac voltage measurement capability through use of separate plug-in leads. Others include analog outputs so that measurements can be sent to chart recorders and displayed on oscilloscopes. DC clamp-on meters are also available, but they utilize a different operating principle.

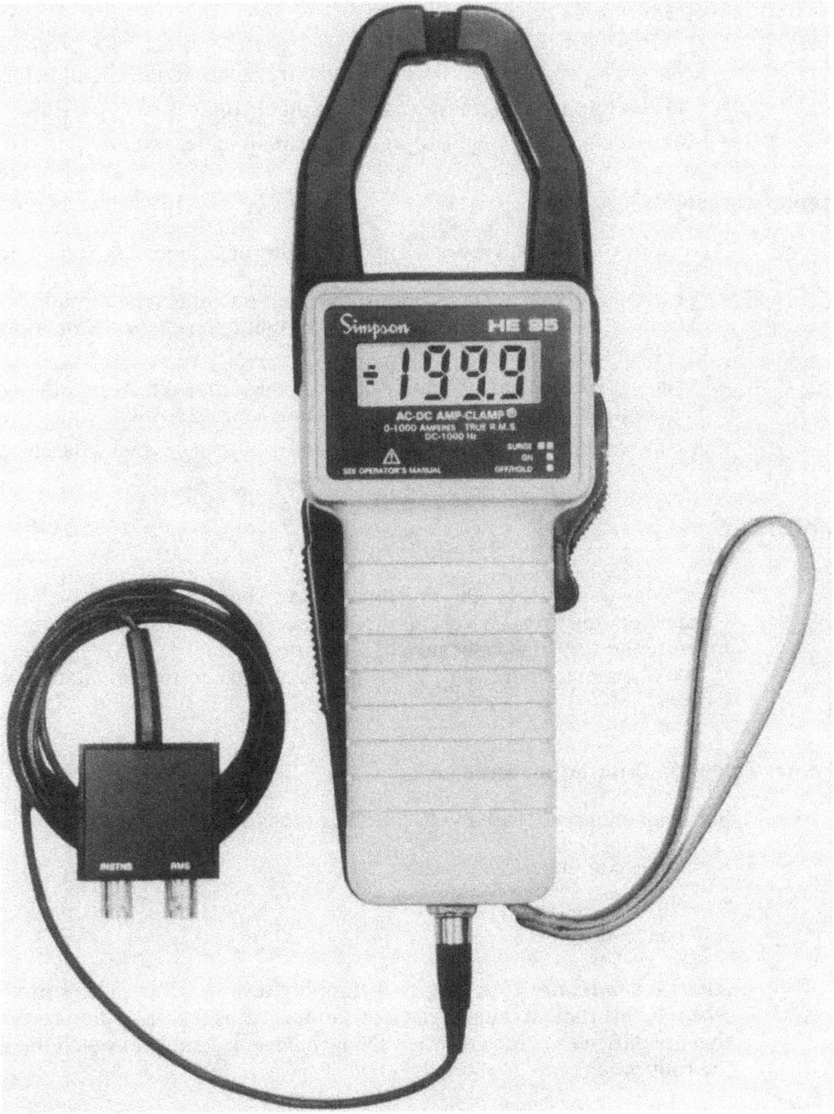

FIGURE 5.9 The ammeter display is contained in the handle of the meter.

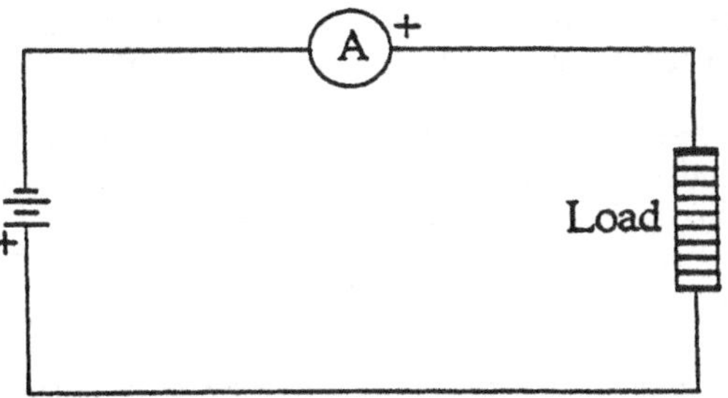

FIGURE 5.10 Series ammeter connection.

- Select an instrument appropriate for the frequency in the circuit being tested.
- Use as far away from transformers and other conductors as possible.
- Make sure only the conductor from the circuit under test is within the jaws of the clamp.

Ammeter Accessories

Accessories can expand the range and versatility of an ammeter. Some common accessories are:

- *A current multiplier* is used only with clamp-on ammeters to multiply the amps reading of extremely low current ac circuits. Current multipliers typically multiply the current reading by a factor of 10.
- *An amp-clamp* extends the range of direct-wire ammeters by permitting clamp-on operation. The amp-clamp replaces the direct-wire ammeter leads, then clamps over the conductor.
- *A shunt* extends the range of the ammeter by diverting some of the current around the meter.

OHMMETERS

An ohmmeter measures the resistance of any component or circuit between the test probes. Ohmmeters may read a very low resistance in conductors, or, using a special instrument, measure the very high resistance of insulators.

An ohmmeter operates by applying low voltage to the circuit, then measuring the circuit current.

Ohmmeter Selection Criteria/Features

Select an ohmmeter with the best features for your application, such as:

- Audible continuity checking
- Testing voltage
- Diode testing

Audible Continuity Checking. To simplify testing, some meters produce an audible tone when the instrument detects circuit continuity. This feature is most useful for testing circuits that are difficult to reach or when the technician is unable to watch the display while making the connection.

Testing Voltage. The voltage applied by the ohmmeter to the circuit affects how the circuit operates. Most general-use ohmmeters apply 1.5 V to the circuit. Meters designed for use with solid-state circuits may apply just a few microvolts. Voltage at this level can detect even the small resistance caused by corroded terminals that can disrupt circuit operation.

High-voltage testers, using either a battery or a hand-cranked generator, are used in testing insulation. Testing voltages may exceed 5000 V.

Diode Testing. A diode testing function measures the forward bias voltage of a diode or semiconductor. Instead of displaying ohms, the meter displays the voltage required to bias (turn on) the PN junction of a solid-state device.

Ohmmeter Operation

- Connect the ohmmeter only to an *unenergized* circuit.
- Always zero the meter before each use to compensate for battery condition.

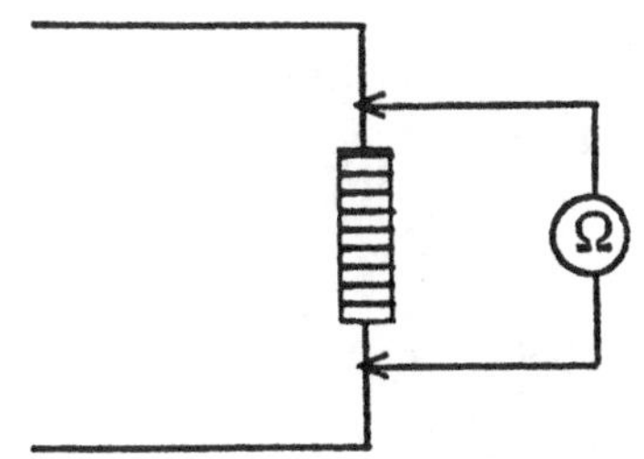

FIGURE 5.11 Parallel ohmmeter connection.

- Connect the leads in parallel with the item under test (Fig. 5.11).
- Select a resistance range that will provide a midscale reading. If you are using an 11-kΩ circuit, select the 20-kΩ scale.

PANEL METERS

Analog

Analog panel meters come in many standard models and ranges as well as in custom versions. The two classic meter styles are round and rectangular. The rectangular type provides maximum space for the scale arc. For minimum use of panel space, edgewise reading meters are available, but they often introduce parallax problems and are limited to short angular pointer swings (see Fig.5.12).

With proper calibration, panel meters can be mounted horizontally or vertically. This subjects them to environmental factors which must be considered, such as ambient temperature, magnetic panels, stray fields, shock and vibration, and moisture and dirt. Ruggedized construction, case sealants, and self-shielding movements are available in certain models to withstand extreme environmental conditions.

Analog panel meters are available that measure dc voltage and current, ac voltage and current, temperature, motor load, elapsed time, and frequency. Analog meters are preferred where readings are constantly changing or fluctuating.

FIGURE 5.12 Panel meter styles.

Digital

A digital panel meter (DPM) can be as simple as a dc measuring unit or as complex as a microprocessor-based controller (Fig. 5.13). DPMs differ from analog in that they have their own power supply and are generally more accurate. Controller/indicators are available for monitoring a variety of process variables: ac/dc voltage and current, TRMS ac voltage and current, 4 to 20 mA dc, 1 to 5 V dc, resistance, three-wire potentiometer, frequency, tachometer, resistance-temperature detector (RTD), and thermocouples.

Built-in displays are available as liquid-crystal displays (LCDs), light-emitting diodes (LEDs), or planar gas-discharge units. Accuracies are generally from 0.1 to 1.0 percent. Analog-to-digital circuitry is kept to a minimum to reduce cost.

Digital panel meters overcome the problems of parallax, polarity, and reading error often encountered with analog meters.

Analog meters, however, provide a better means for observing fluctuations, trends, and approximations. Misreading a DPM by even one digit can result in significant error.

Both analog and digital meters have their place in the field of test and measurement.

FIGURE 5.13 (*a*) Digital dc measuring meter; (*b*) microprocessor-based controller/indicator.

POWER METERS

As demand for energy increases, the cost of energy continues to rise. Plant engineers can save their companies thousands of dollars in enery costs, as well as prevent power outages and equipment damage, by applying power meters. The power meter measures voltage, current, power, and other parameters of power quality, such as power factor and phase. Many power meters have ports that can be connected to a computer (Fig. 5.14). Some have data-logging capability.

The power meter can be used for several energy management applications:

- Demand control
- Load profiling
- Power quality
- Alarm notification/remote metering

Demand Control

Individual plant areas can be billed internally, identifying costly processes that require additional analysis. Power requirements for each area can be compared over time to assess the

FIGURE 5.14 Power meter.

effectiveness of energy-savings measures on a per-building basis. Historical data can be analyzed to track each area's energy usage from season to season and year to yar. Staff personnel can use this information to predict future energy and financial requirements when planning budgets, expansions, or additional energy-savings strategies.

Load Profiling

The plant engineer can conduct an energy audit using a power-monitoring system that can quickly and easily analyze the plant's electrical system to accurately identify areas of excessive loading. Data collected for each load can be combined and displayed to create a detailed model of the plant's electrical usage for a typical working day. The plant engineer can then use this information to identify areas where power is being wasted, and to reduce that waste to improve energy efficiency. Load profiling can prevent overloads and outages by ensuring that each component of the electrical system is running well within its maximum rated levels.

Power Quality

An energy management system can be put in place to monitor and record a full range of harmonic values and phase unbalances at the building level to help staff personnel to minimize power-related problems and identify areas in need of conditioning. Parameters such as time of use and peak demand can be configured to accommodate virtually any utility pricing structure. This information can be used to help engineers efficiently distribute loading to avoid exceeding predicted peak demand. The energy management system can be optimized to avoid costly peak demand penalties. Comprehensive power-quality monitoring identifies disturbances associated with loss of production. Harmonic levels can be displayed and recorded for anlysis at the building level or for the entire system.

Alarm Notification/Remote Metering

When used in conjunction with an energy management system, a power meter can offer alarm notification of excessive voltage, phase, or harmonic disturbances. When used with a computer, an option to initiate a predetermined response can be activated.

OTHER METERS

Other meters and measuring devices can be useful in the plant. Although not all of these meters are strictly electrical measuring devices, they are often used by the same assembly or maintenance personnel who use multimeters. These devices are often manufactured by the same companies that make multimeters. They include:

- Microwave leakage detectors
- Frequency counters
- Sound/noise measuring instruments

Microwave Leakage Detectors

More and more processes use microwaves for heating and drying. Maintenance personnel need a handheld microwave leakage detector to test seams, joints, and door seals to locate microwave leakage. The unit of measurement of microwave radiation is milliwatts per square centimeter (mW/cm^2).

Frequency Counters

A frequency counter measures the frequency of electrical signals. The frequency counter is often used during troubleshooting and assembly procedures for computers and other electronic devices as well as any device using oscillators. A frequency counter with a lowpass filter eliminates incorrect signals due to high-frequency noise.

Sound/Noise Measurements

Occupational Safety and Health Administration (OSHA), Walsh-Healy, and other applicable regulations require documentation of dosage levels of workplace noise. Select a sound/noise meter that is compatible with a sound dosimeter or strip chart if you require a permanent record.

Sound/noise meters also require a specialized tool for calibration. Calibrators may provide a single output or several ranges for calibrating instruments. (See also Chap. 4.12, "Noise Control.")

CONCLUSION

Electrical test equipment comes in a wide variety of configurations to meet the needs of plant maintenance, assembly, and other operations. Selecting meters to economically meet your exact needs can be done by studying instrument specification sheets. Proper maintenance and handling ensures long life of the instrument. Keep in contact with leading manufacturers to stay up to date in this rapidly changing field.

CHAPTER 5.2
BUILDING AIR QUALITY*

Travis West
President, Building Air Quality
The Woodlands, Texas

THE ISSUE OF INDOOR AIR QUALITY

Concerns over indoor air quality (IAQ) have propelled it into the forefront of the workplace environment, and it appears it will become a dominant issue confronting facility management and maintenance personnel well into this century. As the public has become more aware of the health and comfort implications of IAQ, pressure has increased on commercial facilities to maintain an acceptable level of air quality.

Like many other building issues, IAQ problems can result in tensions between the occupants affected and those with the responsibility for the building's management. We are now learning that successfully resolving IAQ problems is based as much on a person's understanding that something is being done to resolve the problem as it is on the actual expenditure of resources leading to the solution. Consequently, a rapid and well-thought-out response to calls for assistance begins to build the foundation for a positive relationship.

Before discussing the issue of maintenance of mechanical systems and its effects on IAQ, it is important that the reader understand the variety and range of factors which can affect IAQ. This chapter discusses the types of sources which can create a concern for IAQ and how

* Adapted from Chap. 20.1 in Robert C. Rosaler (ed.), *HVAC Maintenance and Operations Handbook,* McGraw-Hill, New York, 1998; used with permission.

pollutants are distributed and ultimately can be controlled. It also discusses the impact that building occupants can have on IAQ. In addition, information regarding current legislation and proposed guidelines, as well as a table offering suggested preventive maintenance scheduling, is included.

THE IMPACT OF INDOOR AIR QUALITY

Much information has been learned about the relationship of IAQ and the maintenance of mechanical systems in commercial buildings. Although now widely recognized by IAQ consultants, many property owners and managers are only recently beginning to accept the following facts:

- The expense and effort required to address IAQ in a proactive manner is almost always less than the expense and effort required to address a problem after it has occurred.
- Many IAQ problems can be prevented by making sure that the facility's management staff and operations personnel understand the many factors which can create these problems. In addition, they must be well trained in the operations of their equipment and must have budgets and support staff to allow them to remain diligent in preventive maintenance activities.

IAQ problems that occur in high-rise buildings are very similar, in nature and cause, to those problems that occur in one-story strip centers and other similar buildings. Consequently, the design and use of the building does assume the same level of importance as the building's maintenance and operations activities.

FACTORS AFFECTING INDOOR AIR QUALITY

The air quality found inside a building is the result of many factors which can affect it continually during a day's activities. There are factors that can change rather quickly, such as occupant loading; a change in space conditioning; and even odor or contaminant releases, which can occur in an uncontrolled manner. In addition, there are other factors which can develop into air-quality problems over extended periods of time. Problems of this nature include the development of biological growths, the deterioration of building components, and even gradual modifications to the tenant makeup, affecting how and when the building and its mechanical systems are being used.

Experience has shown, however, that for an IAQ problem to occur there need to be four specific factors present:

- A contaminant source
- A pollutant pathway
- HVAC systems
- Building occupants

While any of these factors can play an important part in the overall IAQ complaint, all four must be present for a complaint to occur. Consider, for example, that a source such as an odor is present in a building, but away from any occupied spaces. The building's occupants are located at the other end of the building. In this instance, even with the odor occurring and the building occupied, if there is no transport mechanism present (HVAC system) an IAQ complaint is not likely to occur. Consider a second example in a building that clearly has a contaminant source and a fully functional HVAC system, but no building occupants. In this instance the chance of an IAQ complaint is also limited. However, in any instance where an

HVAC system, contaminant source, and building occupants are present concurrently, the likelihood of an IAQ complaint increases.

CONTAMINANT SOURCES

The source is considered the area or product which is creating the IAQ contaminant. This can occur as an odor or fume brought into the building through the outside air supply. This can also occur within a building from deteriorating components or off-gassing of fixtures or furnishings, or it can be generated by a building occupant's activity (such as gluing, soldering, etc.).

As previously discussed, IAQ contaminant sources can occur *outside* of the building or can be generated *inside* the building. Whenever these sources are uncontrolled, the potential for an IAQ problem is increased dramatically. Recent IAQ training programs developed by the EPA have identified a variety of areas in which sources can occur.

Sources Outside of the Building

Sources found outside of the building can include outdoor air which has been contaminated by a variety of things, both naturally occurring and resulting from human activity. The naturally occurring ones are usually anticipated in the forms of pollen, dust, and spores, which are generated by trees, grasses, and other living things. Pollutants from human activity can include gases, dusts, and fumes from industrial activities.

In addition, outside-air fans can bring in other outside sources, such as exhaust from vehicles parked at building loading docks, from idling traffic in parking lots, and from nearby congested freeways.

Mechanical Equipment Sources

Sources of IAQ contaminants related to the HVAC system can include dirt and dust which is found in ductwork or on other components, but is most often seen as microbiological growth found in drip pans, humidifiers, ductwork coils, and other moist portions of the system units. In addition, IAQ investigators commonly look for improper venting of combustion products such as carbon monoxide from water heaters, steam boilers, and gas appliances.

Building Occupant Activities

Activities which can create contaminant sources in buildings come from a variety of areas. Building management firms must be apprised of the fact that housekeeping activities as well as maintenance activities can create potential IAQ contaminant sources. In addition, occupant activities have been shown to be a source for a variety of IAQ contaminants. These contaminants include the obvious, such as tobacco smoking, and may also include cooking, cosmetic odors, and even the use of certain office supplies, which may affect more allergic individuals. Building housekeeping and maintenance crews must also be made aware of the fact that the materials used for cleaning, pest-control activities, maintenance, and lubrication can also generate odors, vapors, and fumes which can be considered potential IAQ contaminant sources.

Building Components and Furnishings

Building components and furnishings can create IAQ problems. Such products can act as collectors for dust or fallen airborne fibers and other particles. These furnishings include actual hard-surface fixtures such as cabinets, desks, and bookshelves. In addition, carpeting has been found to be one of the largest collectors of dust and other indoor air contaminants. This problem increases as carpeting ages because the buildup of dirt and other contaminants can never fully be removed through cleaning processes. Other types of unsanitary conditions can be created through water damage which may occur in the occupied space. Water leaks or spills which come in contact with items such as cellulose ceiling tiles, plasterboard walls, and even the aforementioned carpeting present a tremendous potential for the growth of micro- or macrobiological organisms of concern.

When products, fixtures, and furnishings are new, the potential for IAQ concerns also exists. Research has shown that the dyes and glues used in the manufacture and installation of items such as wallpaper, carpeting, and even drapery systems can offgas harmful chemicals into a building's environment. Because of those concerns, many manufacturers have had air-quality laboratories test their fixtures and furnishings for the level and type of offgassing that they can create. Prior to purchasing new fixtures or furnishings for a tenant space, the buyer should ask for the results of these laboratory tests to ensure that no unusual levels of offgassing can occur.

Other Unusual Sources

There are many other sources of IAQ contaminants which are considered unusual, yet can be found in most any building. The most common of these is the small kitchenette area found in many office suites. Odors are the primary cause of IAQ complaints where these are found. In addition, odors from a building or tenant cafeteria can create IAQ complaints.

Other areas which can generate unique though not unusual problems include laboratory areas, print shops, gymnasium or health spa areas, beauty salons, smoking lounges, and even office areas with high-volume copy machines.

POLLUTANT PATHWAYS

A fourth and sometimes critical factor affecting indoor air quality involves pollutant pathways and naturally occurring driving forces. Pollutant pathways can most commonly be referred to as *pressure differentials* which occur throughout the inside of a building.

These differentials can be caused from an oversupply of conditioned air being provided to one room while limited quantities of return air are being allowed. When this occurs, a positive pressure differential happens. If this room as described were a laboratory, kitchen, or other source of potential contaminants, those contaminants could then be delivered through a pollutant pathway to other occupied areas in the building.

Driving forces describe the activity, which generally occurs through natural forces exerted on the outside of a building. These forces can include high winds on the north side creating a low-pressure area on the south side of the building or can be in the form of the phenomenon known as *stack effect,* which can occur most dramatically during winter months. During stack effect, warm air rises inside a building, causing an accumulation of warm air into the upper floors while at the same time generating an obvious lack of warm air on the lower floors. As heated air rises, it is replaced by cold air that is sucked into the building at lower floor levels.

Even when a building is designed to maintain a positive pressure, there will always be an area or two which will periodically be under negative pressure. As this occurs, there is the potential for a source to affect the low-pressure areas. The interaction between pollutant pathways and intermittent or variable driving forces can lead to the problem of a single source causing IAQ complaints in a variety of sometimes distant areas within a building.

HVAC SYSTEM

Although the HVAC system is not able to control the generation of air contaminants at the source, it can be used to affect pressure relationships and even dilute (through the addition of outside air) IAQ contaminants. For the sake of this discussion it's important to note that while the IAQ contaminant source can in fact be the HVAC system, the system is most commonly merely the carrier of the contaminants from the source to the building occupants.

BUILDING OCCUPANTS

Without a person or persons who are impacted by the presence of a contaminant source, there would not be an IAQ complaint.

STANDARDS GOVERNING INDOOR AIR QUALITY

Although at present there is no federal legislation that sets standards for acceptable IAQ, a number of states have enacted legislation requiring them. The scope and nature of IAQ regulation varies on a state-by-state basis. However, a good rule of thumb is to include what legal experts refer to as "providing standard practice of care" when determining the level of IAQ.

However, during the design, construction, and even retrofit of commercial buildings there are recommended trade association standards which should be followed. The standards such as those offered by the American Society of Heating, Refrigerating and Air-Conditioning Engineers (ASHRAE) were created to address many different aspects of commercial building management which can affect indoor air quality.

ASHRAE Standard 62-1989

The standard recommending guidelines for acceptable IAQ is referred to as Standard 62-1999: "Ventilation for Acceptable Air Quality."

ASHRAE 62-1999 is intended to assist professionals in the proper design of ventilation systems for buildings. The purpose of the standard is to specify minimum ventilation rates and indoor air quality that will be acceptable to human occupants and minimize the potential for adverse health effects. The standard applies to all indoor or enclosed spaces that people may occupy, and even discusses the release of moisture in residential kitchens, bathrooms, locker rooms, and swimming pools.

Readers are urged to obtain a copy from ASHRAE, 1791 Tullie Circle, N.E., Atlanta, GA 30329-2305, or the ASHRAE web site, www.ashrae.org.

ASHRAE Standard 55-1992

The second of the two ASHRAE standards which can have an effect on indoor air quality is referred to as standard 55-1992: "Thermal Environmental Conditions for Human Occupancy."

ASHRAE 55-1992 covers several environmental parameters, including the temperature, radiation, humidity, and air movement designed within a building's structure.

Since the perception of being too warm or cold, too humid or dry, can sometimes be misinterpreted as an IAQ issue, thermal comfort within the occupied space becomes important.

Some of the important features of ASHRAE 55-1992 include the following:

- A definition of acceptable thermal comfort
- Recommendations for summer comfort zones and winter comfort zones, clearly defined in graphs
- Guidelines for making adjustments in air delivery depending on occupant activity levels

PROGRAM MANAGEMENT

Once the construction of a building has been completed, the responsibility for preserving good IAQ rests with the building operations and maintenance functions. A comprehensive facilities program should exist within each occupying firm or building manager. This program can be used to manage and monitor acceptable IAQ—including a scheduled testing and analysis phase, repairs, and minor alterations to facilities—as well as to guide the reaction to unscheduled breakdown repairs and maintenance.

To implement an IAQ program, three elements must be provided:

- Management support and commitment from all levels of the organization
- Staffing by trained and competent personnel
- Budgetary allocations to provide the fiscal resources necessary to perform the comprehensive services required to maintain good indoor air quality

The following pages cover areas that are shown to have an effect on the indoor air quality of a building.

Facility Inventory

An audit of the condition of all buildings should be conducted as early as possible to establish a baseline for the physical plant. In addition to anticipating potential IAQ problems, an audit has value to all other aspects of operations and maintenance. Results of an effective (and periodic) audit can also be used as due diligence groundwork in instances of lawsuits related to building IAQ. Such audits can be performed by properly prepared and knowledgeable mechanical and environmental engineers.

The major building components that should be considered for evaluation include the following:

- Roof
- HVAC (including local exhaust systems in specialty areas)
- Plumbing
- Electrical
- Subfloor and crawl-space areas
- Other areas—loading docks, food services, duplicating and copy rooms, laboratories, storage facilities, industrial and manufacturing areas, and so forth

Systems Description

Providing a description of each system is a particularly important but often overlooked program element. It is important that both existing and new staff personnel in the building understand the capacities and capabilities of the systems, to ensure that the building is performing at its optimum level.

Review your existing equipment lists and mechanical plans. Compare the as-built drawings to the equipment installed. This helps to ensure that components are receiving regular attention. Also, equipment that has been installed in inaccessible locations is often overlooked during routine maintenance. By verifying the status of all equipment noted on the mechanical drawings, you are forced to confirm its presence and condition.

Operations Plan

An operations manual should describe how to run the building systems. The plan should provide a detailed step-by-step procedure covering operation from start-up to shutdown. This document, an essential reference for day-to-day activities, will be a useful training document for new employees. A vital part of this plan should be a compilation of all available manufacturers literature and manuals.

Comprehensive Preventive Monitoring Plan

Preventive monitoring is incorporated into all building IAQ programs. The objective of this monitoring is to prevent small deficiencies from blossoming into major costly outages or repairs. For example, regularly oiling a bearing on a fan system will extend the useful lifetime of the unit and prevent the potential loss of makeup or exhaust air that is necessary for proper IAQ. Table 5.1 lists the preventive monitoring tasks that are necessary in any program related to IAQ.

Scheduling periodic audits by consultants can be an important means of determining IAQ status. Both frequency and timing are important. For example, biological growths of fungal and bacterial organisms have often been found in condensation pans. In some environmental studies, these organisms have been shown to cause disease. Since HVAC systems run throughout the year, they will accumulate biological growths as the year progresses and may disseminate biologically active aerosols and odors to the building's occupants all of the time. *If building management waits until a chosen annual date to clean and disinfect these units, it may pose a greater risk to building occupants than if the units were inspected on a regular schedule and cleaned and disinfected as problems begin to develop.*

Scheduled maintenance and any preventive monitoring must also be tied to a building's utilization schedule. Painting, coil cleaning, pest control, or other projects involving the use of chemicals should be scheduled when there are few, if any, occupants in the immediate vicinity. The work area should be well ventilated, using fans or supplying large quantities of outside air wherever possible. In contrast, the monitoring phase must be scheduled when the building is occupied to allow for retrieval of "occupied" and/or potential exposure analysis figures.

Mechanical Systems. Pulleys, belts, bearings, dampers, heating and cooling coils, and other mechanical systems must be checked periodically. These are included in Table 5.1; consequently, there is no reason that these should be the source of an IAQ problem if the staff is following the table schedules.

Pulleys and belts should be tightened as needed and changed prior to failure. Bearings should be lubricated or repacked to prevent major failure of vital system components. Air dampers and baffles should be cleaned and cleared of debris periodically. Failure to perform these activities will result in an increase in resistance, which causes a decrease in air supply.

The air-conveyance system (ductwork) should have access ports available to allow for periodic sampling of IAQ at various points in the system. These data, when retrieved, can offer accurate information on the presence of mold, mildew, bacteria, and dust, allowing building management to address a problem before any concerns are raised.

TABLE 5.1 Preventive Monitoring Tasks

Task	Recommended frequency
Air-handling system	
Heat/cool/vent	
Clean or replace filters	Quarterly or as needed
Clean coils and spray with disinfectant	Annually
Clean and flush condensate pans with disinfectant as needed	Quarterly or as needed
Check filters; clean if needed	Monthly
Check condensate pan drains	Monthly
Check and clean air intake, intake louvers, etc.	Monthly
Check air-distribution boxes and overall duct system for cleanliness, leaks, collapse	Quarterly
Clean heat and cool coils	Annually
Check and clean fan and blower blades of dirt and trash buildup	Quarterly or as needed
Return-air fan	
Check and clean air intake screening, louvers, etc.	Monthly
Check overall duct system for cleanliness, leaks, collapse	Monthly
Clean fan and blower blades	Annually
Unit heaters	
Check and clean coil and fan blades	Quarterly or as needed
Fan cooling units	
Clean and flush condensate pans with disinfectant as needed	Quarterly
Check and clean filters as needed	Quarterly
Check and clean coil with spray and disinfectant	Annually
Ceiling fans	
Check and clean fan blades	Monthly
Rooftop units	
Heat/cool/vent	
Check operation of roll filter	Monthly
Clean or replace filters	Quarterly or as needed
Clean coils and spray with disinfectant	Quarterly or as needed
Clean and flush condensate pans with disinfectant as needed	Quarterly or as needed
Clean all coils	Annually
Drainage systems	
Roof drains	
Check and clean roof or floor drains in area	Weekly
IAQ audit	
Independent testing and analysis of mechanical systems and tenant areas for acceptable air-quality levels	Quarterly

Building ventilation networks are systems that serve multiple locations. The common practice of arbitrarily adjusting the dampers or baffles to accommodate complaints from one area must be avoided. By changing the airflow in one area, the system balance is shifted and distribution throughout the entire network is affected. If there are distribution complaints, a test of the building's air-handling system should be performed to confirm that the HVAC system and distribution network are in balance and are adequate.

Condensate Pans. Components that are exposed to water, such as condensation drain pans, require scrupulous maintenance to prevent biological growth and the entry of undesired biological contaminants to the indoor airstream. This is an item in Table 5.1. Therefore, there is no excuse if a pan is found to be the source of an IAQ problem. The need to use an algicide is an indication of inadequate maintenance.

Filters. Mechanical equipment used for heating, ventilating, and air conditioning usually contains a filter medium or screen that is designed to minimize the system's exposure to particulate material. Air filters were originally added to HVAC systems to control contamination and reduce the amount of daily maintenance required. Over the years, however, as the issue of IAQ in buildings has developed, filters have become a first line of defense against distributing pollutants to the occupied spaces.

Filtration for large-building systems is available in a variety of media, ranging from spun fiberglass on the low end to carbon-impregnated filter media on the higher-efficiency (and more expensive) end of the scale. Choosing the right filter medium for your type of building, regional climate, and tenant activities is important in controlling the IAQ.

Regardless of the filter medium you choose, it is important that periodic checks of filter loading be done to ensure that problems do not develop. The recommended period for checking and changing air filters is covered in Table 5.1.

Vacuuming. Normal industrial vacuums emit particles and fibers in their exhaust. An improvement in performance can be obtained if the vacuum can be fitted with a high-efficiency particulate air (HEPA) filter. HEPA vacuums should be used in areas that might have spores or microorganisms. HEPA vacuum filtration ensures that potentially toxic or harmful aerosols are not dispersed while responding to a problem. Dry sweeping in problem areas should be avoided.

Roofing. Poorly designed or improperly drained roofs may be a potential source of poor IAQ. Flat roofs will invariably collect water and cause pooling. Stagnant, standing water on roofs can support the growth of algae, bacteria, and possibly viruses that can be drawn into building air systems. Leaks in roofs result in water damage or accumulation in ceiling tile, carpeting, or internal wall spaces. Fungi and bacteria that develop in this moisture have been found to be responsible for allergies and respiratory disease. Consequently, when roofs are sloped inadequately or roof repairs are postponed, IAQ can easily be compromised. Water-damaged materials must be removed and replaced in a timely fashion before they serve as a breeding ground for biological growth.

Pipe Leaks. Pipe leaks can occur through corrosion, mechanical failure, or because of the age of the facility. In any case of leakage, repair or replacement of the damaged pipe section must be performed immediately, and the contaminated water must be quickly removed and disposed of by pouring it into an appropriate drain. It is prudent to have wet vacuums, submersible pumps, squeeze brooms, and mops available to handle water emergencies. Water-damaged ceiling tiles, rugs, insulation, or walls must also be moved and replaced in a timely fashion to prevent mold from growing. Following storms, it is good practice to inspect the building for discolored ceiling tiles or leaks as signs of water problems. Pipes that may be subjected to situations causing condensation should be insulated to prevent the problem.

Drains. The antisiphon P-traps in sinks must contain water to prevent noxious odors from the sanitary sewer lines from migrating back into the indoor air spaces. Sinks and drains that are used infrequently can dry out, allowing a path for sewer gases to enter. Cup sinks in laboratory fume hoods and on benches frequently dry out and have often been found to be the sources of odors. This problem can be resolved and prevented by periodically running water in these drains, plugging unused drains with a rubber stopper, or using a nontoxic liquid with a low vapor pressure to fill the P-trap. Drains located in mechanical or air-handling rooms are designed for condensate pan runoff. If no moisture is evident, these lines can dry out (this occurs most frequently in dry climates and during the winter months in humid climates). Dry P-traps allow a path for noxious odors from sewer lines to enter. These odors can then be unknowingly distributed throughout the building by the air-distribution system.

Drains in laboratories must be kept clear and in working order. Sediment in drain traps can provide conditions that support the growth and accumulation of biological organisms.

Materials Management Program

It is important to understand that most chemicals used in housekeeping, maintenance, operations, pest control, and cafeteria services can affect air quality. Their use in the building should be properly managed. Material safety data sheets (MSDSs) are obtained from the manufacturers or distributors and should be on file in the building where any potentially hazardous material is to be stored or used. MSDSs for all chemical products used should be kept in an easily accessible area for employee access. A master file should also be kept in the offices of the personnel responsible for the chemical's use. A building's population should not be subjected to unknown and potentially harmful effects of any material because of the absence of an MSDS.

The chemical composition of hazardous ingredients must be evaluated along with the hazardous reaction potential. Wherever possible, materials, chemicals, and reagents that present the lowest toxic potential should be used. There are many authoritative sources available to help in this determination, including vendor salespeople, vendor chemists, and even the local or state health department.

Less toxic materials should be substituted for more toxic materials. In general, water-soluble materials should be given preference over organic-solvent systems. Materials that are higher in flash point (ignition) and/or have a lower vapor pressure are also preferred.

Building Use Changes. Special care must be exercised when building space utilization is changed. Renovation, redesign, or changes in building use can create situations that may lead to compromises in IAQ. For example, if a copy machine is installed in a small closet or other unventilated space, chemical emissions such as ozone, carbon-black dust, or ammonia may become problems. Further, the addition of blueprint processes, paper-bursting activities, or even microwave cooking can add unexpected odors, dusts, or gases to an unprepared or improperly engineered area. The effect on heat and noise levels must also be anticipated if new equipment is added to an area.

A common renovation problem arises when additional personnel need to be accommodated in a space. Office or instructional areas are often partitioned and additional furniture and equipment is installed. Anticipate the need to modify the air distribution in these situations. Conversely, when partitions are removed, creating new spaces, the ventilation distribution and balance must be revised. Care must be taken to ensure that, in the final layout, air-supply grilles are located far enough away from the returns so that complete air balance and mixing does occur. Otherwise, stagnant areas will develop.

Evaluation of Building Materials Prior to Construction or Renovation. New materials used during construction or renovation present the potential for occupant exposure to emissions (offgassing) from those materials. Many IAQ problems can be avoided by selecting building materials that are less likely to pollute. In cases where this is not possible, the responsibility for addressing the issue of product offgassing falls to building management. The most likely avenue to pursue is dilution of the concentration of emissions. Chemical content, chemical emission potential, and the potential for toxicity and irritation are all issues that should be considered in construction or renovation.

Most manufacturers are now supplying MSDSs with the delivery of their products. If the product is a carpet, wallboard, paneling, or material, some vendors can now supply product emission statistics on these items. In cases where the product emission rates are not available, most vendors will employ staff or consultants who can relate information about the products' effects on IAQ.

The simplest approach to material evaluation is to identify materials that can raise IAQ concerns, and select the lowest-emitting and least-toxic products.

Ventilation and Isolation During Construction or Renovation. During renovation projects, special care must be taken to isolate the area being remodeled from all occupied areas. The contaminants can spread via the HVAC system, by airflow through corridors, up and down elevator shafts, and through ceiling plenum areas. Contamination of occupied areas during remodeling can lead to tenant complaints ranging from dry eyes, itchy skin, and burning throats to nausea and even vomiting. Individual effects almost always include lower productivity, which is used to support claims of tenant move-out.

There should be an effort to supply sufficient exhaust-only ventilation to areas that are involved in this construction or renovation process. Further, adjoining (occupied) areas should be kept under a slight positive pressure to avoid cross-contamination. This positive pressure in occupied areas, along with exhaust ventilation at the source of the contaminants, will help to isolate and remove most contaminants.

After the construction or renovation project has been completed, other strategies fall into play that can have a major impact on the IAQ of the facility. Much has recently been written about "bake-out" or "flush-out" procedures.

Bake-out is a process of overheating a building or space to artificially age the materials that are sources of contaminants. Research on the effectiveness of this process is limited, and any attempts at performing this process should be guided only by IAQ consultants.

Pesticides. Building management should be committed to providing the building with a pest-managed environment through the implementation of preventive hygienic methods and chemical strategies when necessary. Integrated pest management (IPM) emphasizes the use of nonchemical techniques for the management of pests, relying on the use of pesticides only when nonchemical strategies are not effective.

Each building should receive a scheduled inspection for the purpose of identifying existing or potential problems that may contribute to harboring, feeding, or population growth of pests. Inspectors should recommend nonchemical pest-control measures whenever possible. Finally, the inspector will list options for chemical treatments should the nonchemical measures prove to be unsuccessful.

Chemical control treatment should be applied after building hours with the exception of emergency situations. All contract personnel will be required to possess a valid pest-control applicators license, and such license must be on file in the building management office.

Building management has a responsibility to notify all tenants about the pending treatment for pests. Management should also assign personnel to accompany the application technician, and to monitor the use and location of all pesticides.

Record Keeping

While there is a clear record-keeping requirement for asbestos, other potential contaminants have no set procedure. Therefore, it is recommended that records on issues related to IAQ be kept on a scheduled, periodic basis. These records should contain information related to the scheduled preventive monitoring program, as well as engineering reports showing the IAQ health of the building. This IAQ survey will involve testing for specific contaminants related to IAQ and should be performed every 3 months. If no IAQ problems are encountered during the first year, consideration can be given to extending the schedule to a 6-month basis.

The minimum testing required should include review of the operations of the HVAC system and testing and evaluations for the presence of carbon monoxide, carbon dioxide, temperature and humidity (as a comfort indicator), pressure differential of various rooms and air-conditioning zones, and in some instances samples for the presence of excessive levels of mold, mildew, fungi, and bacteria. Other tests which can be performed include sampling for the presence of ammonia, nitrogen dioxide, formaldehyde, ozone, and respirable suspended particles (RSPs), such as glass fibers and inhalable dusts.

This record keeping should include records developed during an initial baseline audit, laboratory field tests (if any), reinspections (every 3 months), and other additional testing. Action plans and any maintenance or operation activities that have been used to mitigate problems, and results of work performed by any outside consultants or contractors, should be kept on file as well.

PREVENTIVE MONITORING TASKS GUIDELINES

The preventive monitoring tasks guidelines (Table 5.1) are designed to offer guidance on daily, weekly, or monthly procedures that can have a dramatic effect on IAQ. These items are offered for guidance only, and should not be treated as suggested requirements. Specific information for each building depends on a number of factors, including occupancy levels, location, number of heating days versus cooling days, ambient air conditions, type of HVAC system, recommended manufacturer maintenance guidelines, and other unique factors not addressed in this chapter. The purpose of these guidelines is to begin the thought process regarding items that can exhibit the potential for the development of an IAQ problem.

Additional information on the issue of IAQ can be obtained by visiting the U.S. Environmental Protection Agency's web site, www.epa.gov/iaq. Additional information on IAQ and building maintenance and operations issues relative to IAQ can be found at www.iaqpro.com.

CHAPTER 5.3

HVAC SYSTEM CONTROL EQUIPMENT*

Terry Hoffmann
Johnson Controls, Inc.
Milwaukee, Wisconsin

INTRODUCTION

In order to discuss the maintenance and operations of system control equipment, it is important to understand their purpose. Fundamentally, a control system does four things: (1) establishes a final condition, (2) provides safe operation of the equipment, (3) eliminates the need for ongoing human attention, and (4) assures economical operation.

Maintaining a set final condition is often considered the only purpose of the system because it has a direct impact on occupants of the facility. The comfort level of each individual is the end product of all control efforts, yet ASHRAE design standards are set to satisfy only 90 percent of all occupants. Well-designed systems that are operated correctly and maintained to their original delivered condition are capable of doing much better than this.

Safety pertains to both individuals and property. The most obvious safety-related HVAC controls are pressure and temperature cutout switches that shut off equipment to prevent damage to both the device and the operator. It is interesting to note that levels of safety consciousness vary throughout the world. While the general pattern would seem to indicate more concern for safety in economically developed countries, there are variances from country to country based on culture and operational practices. For example, it is quite normal for a motor

* Revised from Chap. 21 in Robert C. Rosaler (ed.), *HVAC Maintenance and Operations Handbook,* McGraw-Hill, New York, 1998; used with permission.

control system installed in Germany to monitor each position of a hand-off-auto switch, as well as all electrical overload devices. As we develop into a completely global society based on a single set of standards, which are then tailored to the needs of a particular geographic area, it will be important to understand these differences and adapt our operational practices to meet them.

Elimination of the need for human attention is the cornerstone of the modern control system. In fact, we often call it an automatic temperature control system or an automated building control system. The impact here is that the whole concept of operation is less tangible because the device or system is designed to operate itself without the aid of a person. The first thermostat, dating back to the 1880s, was not designed for the *comfort* of occupants, but for their *convenience* so they did not have to notify the boiler room staff whenever they wanted more or less heat. A good automatic control system reduces human attention to making sure that commissioning has been correctly accomplished, and occasionally making set-point adjustments to better meet the current situation if it is outside of normal design conditions. These variances might include extremes of heat and cold, as well as building upgrades and retrofits that affect control variables either temporarily or over the long term.

Assurance of economical operation has become another important factor due to two important global trends. First, the need to conserve our global natural resources places a responsibility on each individual and organization to utilize them in the most efficient manner. If these resources are depleted without the development of suitable alternatives, it would have a tremendous impact on the world economy. Second, the advent of global competition has made productivity a central piece of the strategic operation of both manufacturing and service industries. Automatic control systems provide the vehicle for this conservation and productivity. It is notable that many specifications refer to the controls as an energy management and control system.

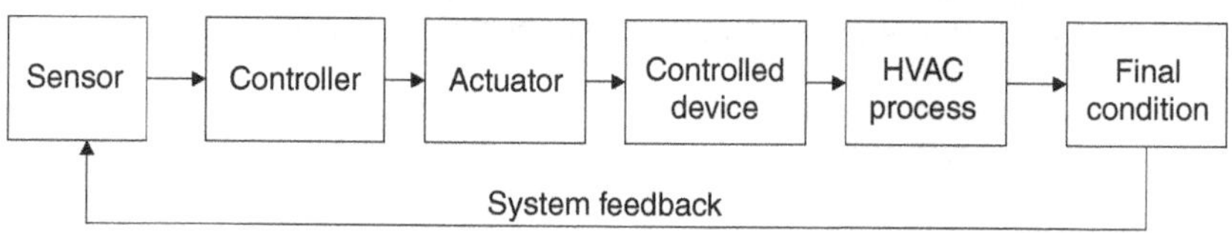

FIGURE 5.15 System feedback.

FUNDAMENTAL CONTROL DEVICES

By looking at a functional block diagram of the fundamental control loop, we can easily identify the individual pieces of equipment common to HVAC control systems (see Fig. 5.15).

Sensors at the beginning of the diagram sense the controlled variable (i.e., the final condition we are attempting to impact). They send a signal to the controller's input. For HVAC controls, these devices usually sense temperature, humidity, flow (velocity), speed (percent or rpm), and position or state (on/off, high/low).

The *controller* analyzes the sensor's input and makes a comparison against the desired final condition—the set-point. The controller then outputs a signal to the controlled device assembly to maintain the set-point value. In networked systems, it is also responsible for reporting information to a higher level controller on the network, which then aggregates the information and interfaces to a higher-level information system. In turn, this information can be accessed on a personal computer interface or a dedicated display device.

The *actuator* is often part of a controlled device assembly, and is responsible for taking the output from the controller and transforming it—with the aid of some energy source—into a change of position in the controlled device. Actuators are usually powered electrically or pneumatically.

The *controlled device* regulates the flow of steam, water, or air according to the commands of the controller as applied by the actuator. The regulation of these fluids helps to balance the load, gains, or losses in the HVAC system. These devices include motors, valves, and dampers.

The *HVAC process* refers to the actual mechanical equipment, ducts, coils, fans, pumps, and so on, that impact the *final condition* in accordance with mechanical design.

In summary, we are therefore concerned with the operation and maintenance of sensors, controllers, actuators, valves, dampers, motors, the infrastructure connecting them to the mechanical equipment, and finally, when installed, all related network components that feed into the human operator of a computerized management and control system.

CONTROL SYSTEM TYPES

HVAC control systems are usually categorized by energy source:

Self-contained control systems combine the sensor, controller, and controlled device into one unit. In this system, a change in the controlled variable is used to actuate the controlled device. A self-contained pressure relief valve is an example of this type of control device. If the pressure in the line exceeds the maximum set-point—which may be fixed or adjustable—it is vented to relieve pressure and protect the end devices, along with ensuring the safety of the operator and occupants.

Pneumatic control systems use compressed air to modulate the controlled device. In this system, the air is applied to the controller at a constant pressure, and the controller regulates the output pressure to the controlled device according to the desired rate of change. Typical air pressure for an HVAC control system is 20 psi (136 kPa).

Electromechanical control systems use electricity to power a mechanical device. These systems may have two-position action (on/off) in which the controller switches an electric motor, solenoid coil, or heating element. Or, the system may be *proportional* so that the controlled device is modulated by a motor.

Electronic control systems use solid-state components in electronic circuits to create control signals in response to sensor information. Greater reliability of electronic components and reduced power requirements have allowed electronic systems to supplant electric systems in all but the simplest applications—and where digital control is not required.

Digital control systems [direct digital control (DDC)] utilize computer technology to detect, amplify, and evaluate sensor information. This evaluation can include sophisticated logical operations such as rule-based and fuzzy logic operations. The resulting digital output signal is then converted to an electrical or pneumatic signal, which is capable of operating the controlled device. Often this transducer is included in the actuator. Table 5.2 compares the different types of systems.

TYPICAL APPLICATIONS

Since this book is about HVAC control, we are interested primarily in the following applications, or subsets, of HVAC control:

- Temperature controls
- Flow controls
- Pressure controls

While the mechanical equipment may vary greatly in these applications, the control elements are largely the same. In the case of electric or electronic controls, the only difference

TABLE 5.2 Control System Types

Classification	Source of power	Output signal
Self-contained	Vapor Liquid filled	Expansion as a result of pressure
Pneumatic	15 psi (104 kPa) 20 psi (136 kPa)	0–15 psi (0–104 kPa) 0–20 psi (0–136 kPa)
Electric	24 V ac 120 V ac 220 V ac	0–24 V ac 0–120 V ac 0–220 V ac
Electronic	5, 12, or 15 V dc 12 V ac 24 V ac	0–5 V dc (typical) 0–12 V ac 0–24 V ac
Digital	24 V ac 120 V ac 220 V ac	0–20 milliamps 0–10 V dc direct digital

between each application might be the sensing element. With regard to digital control systems, there is flexibility to handle any of these applications—and in many cases a single controller assumes responsibility for temperature, flow, and pressure in order to control a single piece of mechanical equipment. For a detailed description of each application and system design fundamentals, refer to the *1995 ASHRAE Applications Handbook* or the *Handbook of HVAC Design*) (see Bibliography).

IMPACT OF SYSTEM FAILURE ON SYSTEM PERFORMANCE

Failure of the control system has an obvious impact on the performance of the HVAC system. The most noticeable impact is on occupant comfort due to the loss of temperature, humidity, or ventilation control. Older pneumatic and electronic systems are designed for a single "best alternative" failure mode. In the case of temperature control, failure usually defaults to a *full heat* setting to prevent freezeups in the northern climates, and to a *no flow* setting where freezeups are not a problem.

Either alternative is concerned strictly with safety, not comfort. Modern digital control systems are designed for fail-safe or fail-soft operation to lessen the impact of such component failures. For example, if a sensor is determined to be out of range or nonoperational, the digital controller can continue to operate using the last good reading that it had obtained prior to the failure—or it can be programmed to fail to a particular output. It might also be programmed to look at another input, such as outside air temperature, and make the best decision possible given that information. Any of the three alternatives is probably superior to the older "on/off" scenario.

Modern digital controllers, with their built-in failure modes, also have eliminated much of the domino effect that plagued less sophisticated systems. The chances of melting all of the chocolate bars in the cafeteria because temperature control failure defaulted to full heat over the weekend are greatly diminished. By the same token, it is less likely that any failure in a system will result in the failure or destruction of components in another, complementary system. That's because operators can preprogram and select a sophisticated array of checks and balances.

Another significant step toward the elimination of catastrophic failures is the growing trend of system manufacturers to include not only an adjustable alarm limit for critical values, but also a pre-alarm limit that notifies the operator of an impending failure. This form of predictive feedback fits well with the concept of both predictive maintenance and reliability-centered maintenance. For older systems with less built-in intelligence, it is the operator's responsibility to spot these anomalies.

HISTORICAL EXPERIENCE CONCERNING FAILURE RATES

When comparing the mean time between failures (MTBF) of a digital HVAC control system to that of a pneumatic or electric system, it is appropriate to examine the system as having four levels of technology, each with its own characteristics (see Table 5.3).

Since the control and monitoring system requires all of the layers to be operating correctly, system reliability is equal to that of the least reliable component—which often is the *personal computer.* Therefore, it is important to make sure that a networked system is configured with multiple PCs having duplicate functionality. A recommended plan is to have access to a spare monitor, and to maintain a daily backup of the system on a second hard drive or a high-density removable backup device (tape or disk).

Actuators also present concerns because they are often the next least-reliable devices. Most buildings have older style actuators that are not very reliable, although many high-quality actuators being built today have one-year reliability rates of over 99 percent. It is reasonable to ask the system manufacturer for this information as part of the original purchase decision. The information also is valuable when determining which spares—and how many—to have on hand. There is a high probability that any manufacturer maintaining ISO 9000 certification of its design and manufacturing facilities will be able to provide these data.

The *wiring network* is not noted as having a high probability of failure. However, this is only true if the network does not rely on a PC-based server and if the network is not shared with other information technology organizations within the facility. Network maintenance is best handled by a computer network management specialist, so it lies beyond the scope of HVAC controls. The only way to make sure that network failures do not adversely impact system operations is to design and maintain a system with no single point of failure. In HVAC applications, the impact of a single failure can be minimized by using distributed control techniques and increasing the level of intelligence as far down the control ladder as possible. When adding devices to an existing system, always evaluate the impact of a controller failure on the operation of the system given its new tasks. In many cases, it is better to add a new controller than to overburden an existing one.

TABLE 5.3 Mean Time Between Failures (MTBF)

Layer	MTBF characteristics
Sensors/actuators	Very long MTBF for sensors (passive components) Medium MTBF for actuators (mechanical components)
Digital controllers (microprocessor based)	Long MTBF (failures usually relate to power supply)
Wiring network	Very long MTBF (passive components)
Personal computers	Short MTBF (mechanical components in monitors and disk drives)

OPERATOR TRAINING REQUIREMENTS

Depending on the size of the facility and the complexity of the HVAC control system, it is possible to have vast differences in the educational requirements for those who operate and maintain the equipment. Anyone with responsibility for the system should have a working knowledge of HVAC basics, as well as familiarity with the system itself and the technology behind it. In addition, the operator should be able to accomplish the following tasks:

- Program/change set-points and schedules as required to meet the operational needs of the organization.
- Optimize energy usage and efficiency.
- Ensure occupant comfort.
- Understand the impact of system changes and maintenance on the core business functions of the organization.
- Assist with building systems safety, health, and environmental standards, requirements, and codes.
- Assist with planning for system upgrades and improvements designed to increase safety, improve reliability, and improve occupant comfort levels, thereby increasing productivity.

In order to accomplish these tasks, the operator will require training above and beyond that which is available on the job. The following recommendations are to be viewed as consecutive and incremental. Each level builds on the next, leading to the ultimate goal of an informed, productive control system manager.

HVAC Basic Training

This course of study should include the following information:

- HVAC system types and piping systems
- Fan systems and fan characteristics
- Dampers and damper actuators
- Valves and valve actuators
- Boilers and related equipment
- Fundamentals of refrigeration
- Chiller types and applications
- Cooling towers
- Heat exchangers and miscellaneous equipment

Sources for this training include local technical colleges and universities, ASHRAE professional development seminars, mechanical systems and control systems manufacturers, and self/group teaching via books, videos, and study guides available from a variety of sources.

Control Systems Fundamentals

A typical course outline should include:

- Basic control theory
- Proportional control fundamentals

- Proportional plus integral control
- Proportional plus integral plus derivative control
- Economizer systems
- Mixed air systems control
- Variable-air-volume control systems control
- Building pressurization control strategies
- Fan capacity control methods
- Control fundamentals for water systems

Sources for this training are similar to those for HVAC systems, with a heavier emphasis on materials obtained from control systems manufacturers.

System-Specific Operation and Maintenance (Pneumatic)

If pneumatic systems or devices are prevalent in the facility, it is important to understand:

- Pneumatic controls and air supply systems
- Air supply maintenance
- Thermostats: single and dual set-point
- Controllers
- Valve and damper actuators
- Valve and damper actuator service
- Auxiliary pneumatic devices
- Temperature, pressure, and humidity transmitters
- Pneumatic receiver/controllers
- Calibration and adjustment or receiver/controllers
- Safety devices and procedures

Hands-on courses are best suited for this training, and are available from local technical colleges and control system manufacturers.

System-Specific Operation and Maintenance (Digital)

The differences between the pneumatic and digital courses are technology- and terminology-based:

- Control systems evolution
- Computers, programming, and direct digital control (DDC)
- Systems terminology
- Basic hardware, software, and user input devices
- Analog and binary sensor types and applications
- Output types, actuators, and transducers
- Software data types
- Control loop types
- Programming direct digital controllers
- Communication to automation systems

Training of this nature should be provided directly from the manufacturer of the digital controls or from an alternate source familiar with the equipment.

Energy Management Concepts

This course of study ensures the understanding of basic energy management concepts and features relating to heating, cooling, air distribution, electrical, and lighting systems:

- Energy management fundamentals
- The building envelope
- Modeling the building environment
- Opportunities for conservation
  Cooling systems
  Heating systems
  Air distribution systems
  Water distribution systems
  Electrical systems
  Lighting systems
- Performing an energy audit

The Association of Energy Engineers (AEE) has courses available that lead to examination and granting of a Certified Energy Manager (CEM) certificate. This line of study is suitable for both operators and aspiring managers. ASHRAE also offers professional development seminars on this topic, as do local colleges and universities. Manufacturers of mechanical and control systems are likely to offer courses in applying their products to manage energy usage.

Building Automation and Control Systems Operation and Maintenance

With a firm foundation of mechanical, systems, and energy knowledge in place, it is possible to comprehend and apply the features and functions of the building automation system to the needs of the enterprise:

- Basic system hardware configurations
- Communications network architecture
- Operator workstations and I/O devices
- System software
  On-line help, point types, menus
  Passwords, commands, summaries
  Reports and alarm management
  Scheduling points and summaries
  Point history, trending, operator transactions
  Energy management features
  Defining/creating systems and points
  Programming control loops and functions
- Hardware maintenance

It is strongly suggested that this course be taken on site using the installed equipment, or at a well-equipped lab operated by the control system manufacturer.

Given the breadth and depth of training required to properly maintain a complex HVAC control system, a decision must be made to provide training in-house or to augment the operation and maintenance team with an external control systems specialist. There are three ways to obtain this assistance:

- Include maintenance of the control system in a service contract with the mechanical contractor or controls company. This provides a limited amount of time each month to perform important tasks and observe system trends that may impact operation.
- Outsource a mechanical systems and controls specialist from an organization that has experienced staff available. The advantage here is that the individual is full-time and blends into the organization, while not burdening the organization with the requirements for training and development.
- The operation and maintenance of the mechanical systems and controls can be handled by an outside supplier as a task under a total facility management contract. This allows management to concentrate on planning and developing strategies for improvements in operation and productivity, while ensuring competent operation and maintenance at a price that is constant for a fixed period of time.

PRODUCT PREVENTIVE/PREDICTIVE MAINTENANCE PROCEDURES

The following tasks are recommended for a comprehensive inspection and calibration of the HVAC control system. This may be accomplished on a rotating basis or several times throughout the year, depending on seasonal conditions, winter shutdown, spring changeover, and so on. Many of the tasks are system-specific and may not be required in a particular facility.

It must also be noted that the type of maintenance program adopted in the facility may have a significant impact on the frequency of many tasks. For example, a program that includes predictive maintenance information may reduce the frequency of tasks requiring hardware checks for wear or accuracy. Likewise, a good reliability-centered maintenance (RCM) program (see Chap. 2.2) will reduce the frequency of most tasks while increasing the frequency of a lesser number of tasks performed on mission-critical equipment.

Finally, the word *clean* is used sparingly in the following procedures to describe situations where dirt has a direct adverse effect on the operation of the device. It is good general policy to maintain all control equipment to a high standard of cleanliness. The interval between general cleaning can be evaluated on a yearly basis and extended if it is determined to be acceptable.

Control Air Compressor

- Drain the tank and check all traps
- Change compressor oil and check the oil pressure
- Check the belts, bearings, and sheaves, visually or by vibration analysis
  Replace worn or damaged equipment
  Ensure proper sheaf alignment to avoid premature failures
- Change the suction filter on a regular basis
- Inspect the unloader and check valve
- Inspect the high-pressure safety valve
- Lubricate the motor(s) and analyze operating conditions
- Check the PE switch, starter, and alternator as required
- Measure and record the compressor runtime under loaded conditions during building occupancy
- Where applicable, record the oil carryover rate

Refrigerated Air Dryer

- Check and record the refrigerant temperature
- Check and record the refrigerant pressure
- Clean all grilles and coils as required
- Check and operate the drain trap and bypass valves

Filter and Pressure-Reducing Station

- Inspect the coalescent filter and change as required
- Inspect the charcoal filter and change as required
- Check and adjust the pressure-reducing valve, record all settings

Boilers, Chillers, Pumps, and Zone Control

- Check all controllers for correct calibration and adjust as necessary
- Calibrate all transmitters and set gauges as required
- Check all PE switches
- Check all control valves for operation and flow
- Check all pilot positions for accuracy
- Check auxiliary control devices

Fan Systems and HVAC Unit Controls

- Review the sequence of operation for conformance to original specification and resolve differences
- Check operation and calibration of all dampers, lubricate as required
- Check all pilot positioners for accuracy
- Check all control valves and actuators
- Calibrate all controllers as required
- Calibrate all transmitters
- Set receiver gauges as required
- Check all solenoid valves, PE switches, and air valves where installed and applicable
- Check all auxiliary controls

Room-Terminal Unit Controls

- Check all thermostats and calibrate as required
- Check all control valves for operation and leakage
- Check operation of unit coil steam traps
- Check operation of all dampers, lubricate
- Check all solenoid valves, PE switches, and air valves where installed and applicable
- Check all auxiliary controls

Terminal Units

- Boxes—mixing and variable air volume
  Inspect connection to ductwork
  Lubricate and adjust dampers and linkage
  Verify operation of control
- Electric duct heaters
  Inspect coil for damage, clean if accessible
  Inspect isolators for damage or cracks
  Torque the heating terminals
  Verify operation and staging of control
- Induction units
  Visually inspect coil and clean
  Check and clean drain pans
  Clean discharge grille

Check and clean strainers
Check steam traps and hand valves

- Reheat coils
  Inspect coil for leaks/damage
  Dust coil, remove debris
  Check and clean strainers
  Verify operation of steam trap if applicable
  Verify operation of control
- Radiation
  Visually inspect fins/cast iron
  Check and clean strainers
  Check valve for leakage and operation
  Clean as required

Note: Always refer to the manufacturers' written documentation for calibration and maintenance procedures where available.

Stand-Alone Digital Controller, or Application-Specific Controllers of a Building Automation System

On a Scheduled Basis

- Verify that the device is being controlled at the appropriate set-points and schedules
- Change one set-point value; verify smooth transition and stable control at the new set-point
- Return set-point to original value
- Repeat for each additional control loop, if any
- Verify that controlled valves and dampers will stroke fully in both directions, sealing tightly where appropriate
- Verify the proper operation of critical control processes and points associated with this unit, and make any necessary adjustments

As Required

- Verify/calibrate other points associated with these units where the need for possible corrective maintenance is indicated

Network-Level Controllers and Interface Devices of a Building Automation System

On a Scheduled Basis

- Check meter and LED indications to verify proper ac and dc power levels and appropriate transmit and receive activity on the trunks, and to check for possible error code and diagnostic indications
- Inspect wiring for signs of corrosion, fraying, or rapid discoloration
- Check voltage level of battery
- Cycle power to initiate self-test diagnostic, if appropriate for device
- Remove excessive dust from heat sink surfaces
- Clean faceplate and input pad, if present
- Clean transparent window in door, if appropriate
- Clean enclosure exterior surfaces
- Verify the proper operation of critical control processes and points associated with this unit, and make any necessary adjustments

As Required

- Verify/calibrate other points and control processes, where the need for possible corrective maintenance is indicated

Computerized Operator Workstations of a Building Automation System

On a Daily Basis

- Review workstation for *critical, follow-up,* and *off-line* status indications
- Review workstation for *override, disabled,* and *lockout* status indications
- Review system event log
- Review building automation system operational concerns
- Perform on schedule corrective maintenance procedures as appropriate to resolve situations noted in the preceding reviews

On a Scheduled Basis

- Check monitor for clarity, focus, and color
- Clean read/write heads of removable disk drive(s)
- Cycle power, listen for unusual motor/bearing noise
- Verify proper system restart (check system date, time, and hardware status)
- Clean exterior surfaces
- Save/copy workstation database, including graphics and resident controller databases
- Install appropriate workstation software refinement and problem correction revisions ("minor revs") as they become available

Note: Major revisions to the building automation system workstation software that add new features and capabilities—or significantly enhance existing features—are not included in maintenance procedures outlined here.

Network Services for a Building Automation System

On a Scheduled Basis

- Reset the system diagnostic counters
- Allow data to tabulate in the diagnostic registers, if appropriate
- For each operator workstation and network connected unit:
  List all diagnostic statistics
  Analyze the error rate for each network node
  Analyze the transmission rate for each network node
  Determine the network performance ratio
- For each application specific device level trunk:
  List all diagnostic statistics
  Analyze the error rate for each network node
  Analyze the transmission rate for each network node
  Determine the network performance ratios
- Provide a report summarizing network analysis results

As Required

- Perform network analysis tasks as appropriate to verify or discount suspected communications or network throughput problems
- Perform network analysis tasks as appropriate to evaluate the impact on network performance of various configuration options—as part of a proposed system expansion or modification

TROUBLESHOOTING AND REPAIR TIPS

The following information is not designed to be a complete summary of troubleshooting techniques, but may be helpful in sorting out basic problems and identifying the cause of persistent failures.

The *control air compressor* is often the source of problems in pneumatic systems. Most compressors are designed for best performance when operating between 33 percent and 50 percent on-time. More than 50 percent on-time can result in objectionable levels of oil carry-over if proper filtration procedures are not followed. Oil in the control air lines can destroy certain pneumatic devices or make them inoperable. Excessive on-time also may result in a significant energy loss.

Common causes of excessive on-time include:

- Air leaks in the system
- Water in the compressor tank
- Extra control devices added to the system that require capacity above the original design specifications
- Inefficient compressor operation

Checking the efficiency of the compressor is easy if you have the documentation that comes with the unit. Simply close off the compressor from the control lines, bleed off the air in the tank, and record the pump-up time to pressurize the tank to 90 psi (620 kPa). Compare this to the manufacturer's specifications. If the number is too high, the compressor may require attention. If it is satisfactory, you must check the system for leaks or extra equipment.

Room controllers are basically maintenance-free devices as long as they receive accurate input signals, have clean uninterrupted power or supply air, and are not abused. Once installed, they usually require little calibration unless they have been tampered with or application requirements change. It is still a good idea to check the operation of a controller by reading the output pressure or voltage for a known condition in the space.

In actuality, the first evidence of a problem with an HVAC system is usually a complaint from an individual who is uncomfortable or bothered by drafts. Experience has shown that, in most cases, the problem lies in a misunderstanding of the capabilities and application of the control system as opposed to a malfunction of the equipment. There are several common causes of occupant discomfort that do not involve the system:

Sun load. Direct sunlight on a thermostat will cause overcooling or underheating in a zone. Also, direct sunlight on occupants will cause them to feel warmer than the actual space temperature.

Zone control. A person outside a controlled zone may feel uncomfortable because the air distribution system is not delivering the full controlled medium to the space. Also, different zones may be controlling at different temperatures due to location or installed equipment requirements, so it may take a few minutes to adjust to the differences when moving from zone to zone. People working in an office atmosphere might not be comfortable in a high-tech manufacturing environment where constant air movement is required for cleanliness.

Covering of grilles. Occupants frequently will cover part or all of a discharge diffuser, causing improper heating or cooling. The imbalance in the air distribution also can have an impact on other occupants of the same zone.

Occupant location. If occupants are located immediately adjacent to an outside wall or window, they may be subject to cold air leakage or to radiant cooling or heating from the walls.

People and equipment. Overheating may occur if more people or equipment occupy an area than it was originally designed for. This can occur when a meeting is held in an area not designed for the purpose, when an office area is converted to a makeshift laboratory, or when an area's computer usage increases significantly. In addition to the problems associated with overheating, it is also possible that the system is no longer delivering sufficient outside air to each person to ensure a healthy environment.

Drafts. To some people, even slight air movement is uncomfortable. Since most systems use air as a means of heating and cooling, there must be movement for the system to operate. This results in the need to relocate workstations or install new diffusers to solve minor problems. In extreme cases, the air delivery system may need to be rebalanced.

Even if none of these conditions exists, it is still possible for the controller to be operating properly. If the controller appears to be operating correctly, check this list before reprogramming or recalibrating the unit:

- Are the actuators operating correctly and coupled to the controlled device (damper or valve)?
- Is the control agent (water or air) sufficiently hot or cold to maintain the desired setpoint?
- Is the pressure sufficient to deliver enough of the control agent to maintain the desired setpoint?
- Are there unusual loads in the space that may overpower the efforts of the controller?
- Is there leakage in water lines or ductwork?

If the answer to all of the above is no, it is time to refer to the manufacturer's instructions and recalibrate or reprogram the controller.

Valves are another occasional source of problems in HVAC control systems. The most common valve failures are sticking, leakage, and failure to close completely. Leakage around the valve stem is easily identified and corrected by replacing the packing. If there is persistent leaking of multiple valves, it is possible that the water supply is dirty and causing premature failures.

Valves that are not controlling correctly may be sticking. To check this, stroke the valve from one extreme position to the other, and check for travel that is smooth and even. If the stem sticks, replace the packing. Failure of a valve to shut off tightly also can cause comfort problems. To check for leakage through the valve, stroke it fully closed and check the temperature of the outlet piping. If the outlet remains hot on a heating valve or cold on a cooling valve, there is some leakage through the valve. Check with the control valve manufacturer to see if a repair kit is available for the disk, or replace the entire valve.

Networked systems present an entirely different set of challenges to the operator. Far and away the most likely scenario for a network failure that can be treated by a layperson is a faulty connection. Check all connections by carefully disconnecting, inspecting, and reconnecting the network to the device that is not responding, and then to the next device in line. Pay special attention to shields and grounds, as a faulty shield may cause a segment of wiring to act like an antenna. Once the connections have been checked and reconnected, power down and repower the device that is not responding—and check again for a response.

Beyond these simple troubleshooting measures, it is time to bring in a network specialist with the correct hardware and software tools to thoroughly test the network. *Always perform a complete backup of all information on computer hard drives before attempting to fix any software problems on the system—or allowing anyone else to touch the hardware.* Rebuilding a network from scratch can take weeks—and all because someone neglected to back up the system or erased the existing backup medium to use it for another purpose. It is not too severe to recommend keeping the backup under lock and key.

Additional suggestions for troubleshooting and repair of HVAC control systems are available directly from manufacturers and are specific to their products. Ask for a copy of the operating manual for the type of system or equipment you are operating, as well as any publications dealing with maintenance and troubleshooting.

CRITERIA FOR REPAIR VERSUS REPLACEMENT

Replacement parts for most HVAC controls are available directly from the manufacturers or their authorized distributors. The decision regarding repair versus replacement of the device is easier for controls than it is for mechanical equipment in an HVAC system because controls usually are much less expensive—and the troubleshooting process often recommends replacement. In general, there are three scenarios for inoperable or damaged equipment:

- If the manufacturer, or a third party, offers a repair kit, it is usually safe to assume that repair is cost-effective and easy to accomplish.
- If the manufacturer offers a repair and return service, or exchange credit, there is benefit to taking advantage of the reduced pricing. This is usually limited to large controllers or complex electronic devices where a repair station has been established to handle the exchange process.
- If a repair parts kit is not available and there is no repair and exchange policy, the part must be replaced. Controllers and auxiliary equipment should not be repaired locally, as testing and adjustment of most items cannot be adequately accomplished without sophisticated laboratory testing.

If the troubleshooting and replacement process is unsuccessful and there is no service contract in place with the controls vendor, then a suitable third party must be called in to determine the next step in the repair process.

HOW MUCH SHOULD MAINTENANCE ON MY CONTROLS COST?

There are a lot of answers to this question based on the vintage of the controls, the technology applied, the company of manufacture, and the level of investment in staff training.

In general, though, you should expect to pay the same price for a good control technician as you would for any other professional in your geographic area. For mechanical and electrical control systems, an hourly rate would be comparable to that of a journeyman steamfitter or electrician. For computerized systems that are networked within the building or across a campus, expect to pay the same hourly rate as your department does for a certified network specialist. This assumes that the individual you are paying for is either factory trained or has taken part in a company program that includes formal training packages in conjunction with on-the-job mentoring. It is also recommended that you ask your service provider for training plans, certificates, and/or complete resumes of all persons who will touch your system(s).

It sounds like service can be expensive if you purchase it on a failure-mode basis. But, there are ways to reduce the fiscal cost of repairs and maintenance, including: (1) the negotiation of terms with your provider on a yearly basis, (2) multiyear contracts, which include improvements that reduce overall costs in subsequent periods, and (3) the inclusion of HVAC controls maintenance in a comprehensive operations and maintenance package, which includes all low-voltage electrical applications and the mechanical equipment itself. The more options that your service vendor can provide, the better opportunity you have to come up with a cost-effective solution for your facility.

The cost of controls maintenance also depends on the maturity of your system technology. For example, pneumatic controls are still reliable, safe, and efficient, but are increasingly expensive to maintain. Spare parts are still available from distributors or directly from manufacturers, but the number of people qualified to maintain these systems is in short supply. To make matters worse, the older the device, the more tuning it requires to maintain set-point or to ensure accuracy of control loop operation. Training is available for pneumatic systems and should be part of an ongoing maintenance plan if they are going to remain part of the control strategy.

Likewise, electronic and digital controls will demand more attention, depending on their vintage. Self-tuning algorithms were available in most systems in the late 1980s, but there have been improvements in the past few years, making them more reliable and less prone to performance problems at the extremes of their control band. Better controls mean less maintenance and repair costs.

Maintenance contract costs for building automation systems, which include interfaces to other building functions, are even more difficult to predict because there are so many differences in terms of computing hardware and network design. If your company is already paying for network maintenance and your system is operational over standard Ethernet/IP, you should talk to your internal networking group to determine the best solution for each part of the system. The downside to a split responsibility situation for a single system is the continuing saga of "It's not my problem, it's theirs." Single source responsibility is usually best and cheaper in the long run if the organizational costs of downtime to the enterprise are factored in. A long-standing rule of thumb for computerized automation systems maintenance is to start at a yearly cost equal to 10 percent of the purchase price for a comparable system in today's dollars. Adjust the cost up or down based on complexity, geographic factors, level of satisfaction, and the cost of downtime if the system fails. Maintenance of automated systems is difficult because they are expected to last for decades, while we replace our computing equipment every 3 to 5 years. Think of the half-life of today's computing equipment and the pace at which network technology is changing.

Hopefully, these guidelines and suggestions will help to target and tune your maintenance budget and aid in decision making regarding replacement of existing equipment.

SUMMARY

The importance of well-maintained HVAC system control equipment to the delivery of comfort for building occupants cannot be overstated. Insufficient attention to controls can undermine the efforts—and the cost advantages—of keeping the building's mechanical equipment in peak condition. Control equipment failure can cause extensive damage to HVAC systems, which may in turn create unsafe conditions.

Effective maintenance and operations of HVAC system control equipment depends on a number of factors, including the type of equipment installed in the facility, the maintenance system that is being employed, and—most important—well-trained people who understand the complexities of building controls.

INTERNET RESOURCES

The following links were active as of January 1, 2001, and may be of assistance to HVAC maintenance and operations professionals with an emphasis on controls and automation.

XREF Publishing Company
HVAC/R information source
Unbiased cross-reference database for the HVAC/R industry
www.xrefpub.com

HVAC Controls
Meeting place for the HVAC industry
Forums, links, downloads
www.hvac-controls.com

Naval Facilities Engineering Service Center
Navy's center for specialized facility engineering
Publications and links to many other government sites
www.nfesc.navy.mil

ASHRAE
American Society of Heating, Refrigerating and Air-Conditioning Engineers
Learning institute, publications, standards
www.ashrae.org

AEE
Association of Energy Engineers
Books, seminars, links, news
www.aeecenter.org

Automated Buildings
Online magazine for automated buildings
Products, systems, articles
www.automatedbuildings.com

The HVAC Mall
HVAC/R directory, index, resource
Products, contracts
www.hvacmall.com

Johnson Controls
Controls manufacturer's home page
Products, training courses, articles
www.johnsoncontrols.com/cg

Facilities Net
Home page for *Building Operating Management* and *Maintenance Solutions* magazines
Products, articles, forums
www.facilitiesnet.com

As links change on a regular basis due to technology upgrades and corporate acquisitions/mergers, the best way to update these references is to use a search engine and an appropriate string such as "HVAC maintenance."

BIBLIOGRAPHY

1995 ASHRAE Applications Handbook, Chapters 8 and 42, American Society of Heating, Refrigerating and Air-Conditioning Engineers, Inc., Atlanta, Ga., 1995.

Alliance Planned Service, Johnson Controls, Inc., Milwaukee, Wis., 1992.

ANSI/ASHRAE 55-1992 ASHRAE Standard, American Society of Heating, Refrigerating and Air-Conditioning Engineers, Inc., Atlanta, Ga., 1992.

Building Environments—HVAC Systems, Chapters 16 and 17, Global Learning Services, Johnson Controls, Inc., Milwaukee, Wis., 1997.

Grimm, Nils R., and Robert C. Rosaler: *Handbook of HVAC Design,* McGraw-Hill, New York, 1990.

Installation and Service Manual for PureFlow™ Air Compressors, 1st ed., Johnson Controls, Inc., Milwaukee, Wis., 1993.

Johnson Controls Institute Training Programs, Johnson Controls, Inc., Milwaukee, Wis., 1996.

CHAPTER 5.4
MECHANICAL EQUIPMENT: GEARING AND ENCLOSED GEAR DRIVES

William A. Bradley III
American Gear Manufacturers Association (AGMA)
Alexandria, Virginia

GEARS AND GEAR DRIVES

The major uses of gears and gear drives are to transmit power, with reduction or increase of speed and multiplication of torque. In addition, gears are used to transmit rotary motion from one shaft to another in a uniform manner, and for positioning. Gear drives are usually classified as reducers, increasers, high-speed drives, or drives for the accurate positioning of loads.

Gear Drive Types

Speed Reducer. Economically, it is normally better to use a small, high-speed prime mover and gear-reducer combination than a larger, low-speed power source.

Speed Increaser. In some instances, it is impractical to operate a prime mover at a speed high enough to suit requirements of the driven equipment. For such applications, gears may be used as a speed increaser.

Positioning. It may be necessary to index or position objects with a shaft that does not normally make a complete revolution. The relative load to drive size may make it practical to use gearing or gear drives.

Shaft Orientation. Gear drives are furnished with various shaft orientations and speed ratios. Some common arrangements that are available include: in-line, parallel shaft, and right angle. Miter gears (1:1-ratio bevel gears), for example, serve the specific purpose of providing a 90° shaft orientation. Other angles can be supplied by specially designed gears of several types.

Gear Types

The most common types of gears are illustrated in Figs. 5.16 to 5.25. Other available types are generally modifications of the basic gears shown. Gear nomenclature and definitions can be found in ANSI/AGMA 1012-F90, *Gear Nomenclature, Definitions of Terms with Symbols.*[1]

Spur gears A spur gear has a cylindrical pitch surface and teeth that are parallel to the axis. Spur gears operate on parallel axes (Fig. 5.16).

Spur rack A spur rack has a plane pitch surface and straight teeth that are at right angles to the direction of motion (Fig. 5.16).

Helical gears A helical gear has a cylindrical pitch surface and teeth that are helical. Parallel helical gears operate on parallel axes. Mating external helical gears on parallel axes have helices of opposite hands. If one of the mating members is an internal gear, the helices are of the same hand (Fig. 5.17).

Single-helical gears Gears have teeth of only one hand on each gear (Fig. 5.18).

Double-helical gears Gears have both right-hand and left-hand teeth on each gear. The teeth are separated by a gap between the helices (Fig. 5.19). Where there is no gap, they are known as *herringbone gears.*

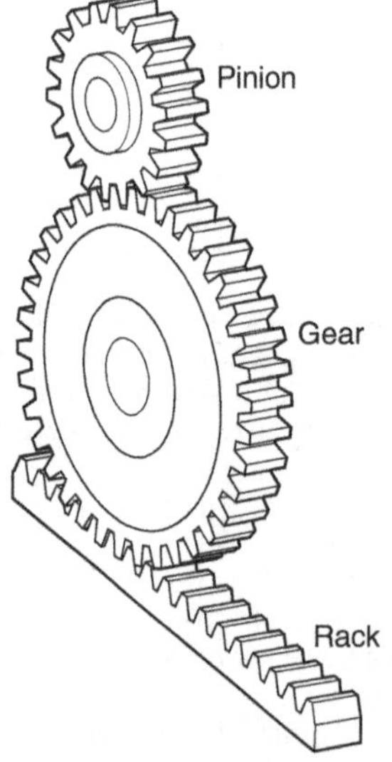

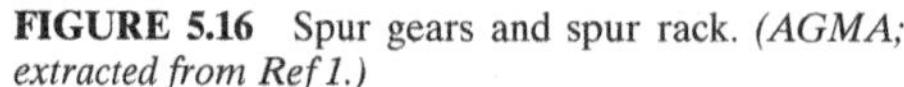

FIGURE 5.16 Spur gears and spur rack. *(AGMA; extracted from Ref 1.)*

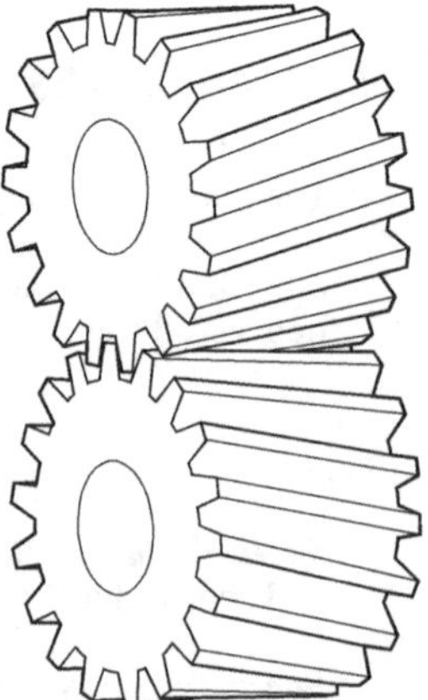

FIGURE 5.17 Parallel helical gears. *(AGMA; extracted from Ref 1.)*

FIGURE 5.18 Single helical gears. *(AGMA; extracted from Ref 1.)*

FIGURE 5.19 Double helical gears. *(AGMA; extracted from Ref 1.)*

Wormgearing Includes worms and their mating gears. The axes are usually at right angles (Fig. 5.20).

Wormgear (wormwheel) The gear that is the mate to a worm. A wormgear that is completely conjugate to its worm has a line contact and is said to be *enveloping.* An involute spur gear or helical gear used with a cylindrical worm has point contact only and is said to be *nonenveloping* (Fig. 5.20).

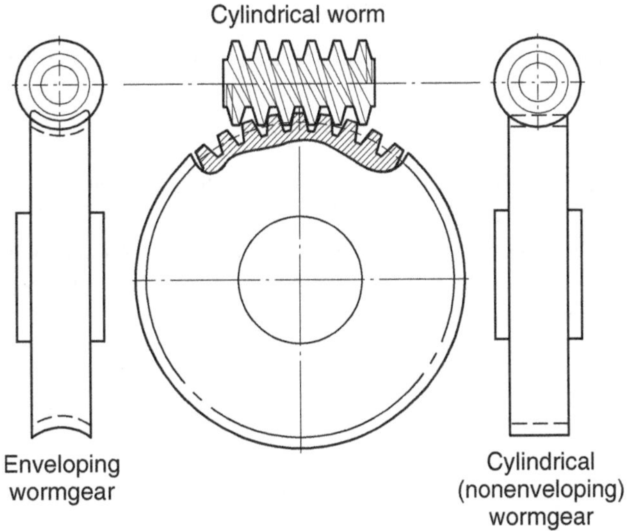

FIGURE 5.20 Cylindrical, single-enveloping wormgearing. *(AGMA; extracted from Ref 1.)*

Cylindrical worm A worm that has one or more teeth in the form of screw threads on a cylinder.

Enveloping worm (hourglass) A worm that has one or more teeth and increases in diameter from its middle portion toward both ends, conforming to the curvature of the gear (Fig. 5.20).

Double-enveloping wormgearing This is comprised of enveloping (hourglass) worms mated with fully enveloping wormgears (Fig. 5.21).

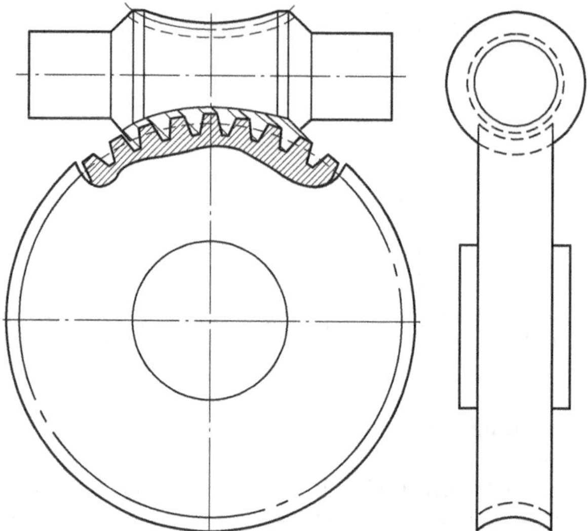

FIGURE 5.21 Double-enveloping wormgearing. *(AGMA; extracted from Ref 1.)*

Bevel gears These are gears that have conical pitch surfaces and operate on intersecting axes that are usually at right angles (Fig. 5.22).

Miter gears These are mating bevel gears with equal numbers of teeth and with axes at right angles (Fig. 5.23).

Straight bevel gears These have straight tooth elements which, if extended, would pass through the point of intersection of their axes (Fig. 5.24).

Spiral bevel gears These have teeth that are curved and oblique (Fig. 5.24).

Hypoid gears Similar in general form to bevel gears, hypoid gears operate on nonintersecting axes (Fig. 5.25).

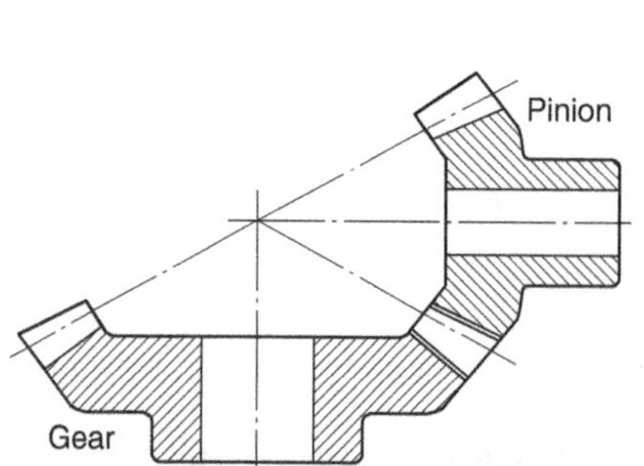

FIGURE 5.22 Bevel gears. *(AGMA; extracted from Ref 1.)*

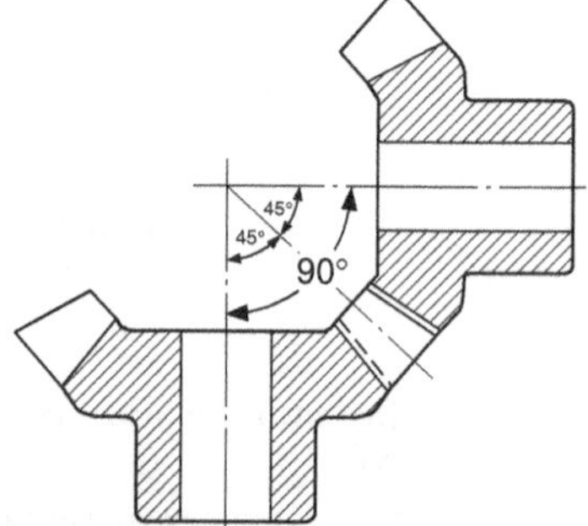

FIGURE 5.23 Miter gears. *(AGMA; extracted from Ref 1.)*

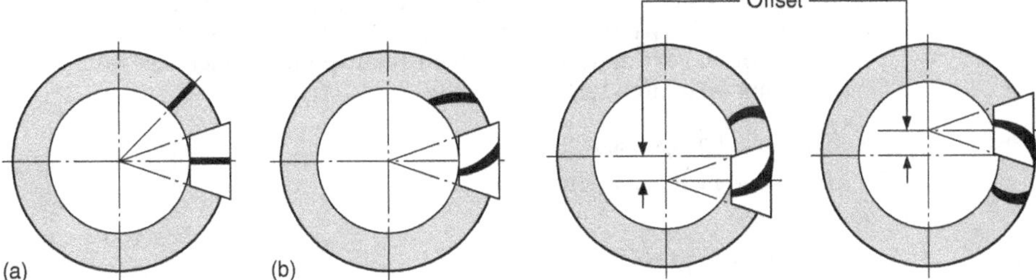

FIGURE 5.24 Bevel gears: (*a*) straight, and (*b*) spiral. *(AGMA; extracted from Ref 1.)*

FIGURE 5.25 Hypoid gears. *(AGMA; extracted from Ref 1.)*

Gear Geometry

Gear-Tooth Action. Gears of all types have the common objective of transmitting theoretically smooth motion through the engagement of successive teeth. Certain design parameters need to be maintained to ensure that this occurs, such that one or more pairs of teeth are engaged before a preceding pair disengages.

Standard Tooth Forms. The involute tooth form is almost universally used for spur and helical gears. In principle it is applied to bevel gears and to some wormgears. The involute form can provide accurate transmission of motion, even when there is some slight change in center distance between shafts. It also offers a number of manufacturing advantages. In some worm and other noninvolute gearing, conjugate forms and accurate manufacturing processes must be employed to ensure smooth transmission.

Modified and Special Tooth Forms. Common modifications to involute gear geometry include tip and root relief, crowning, and profile shift. These are applied to optimize gears for a given application. See AGMA 913-A98, *Method for Specifying the Geometry of Spur and Helical Gears*[2] for additional design information. Some noninvolute tooth forms are used to achieve higher capacity. Normally, very accurate manufacturing and mounting must be maintained to realize this advantage.

Gear Materials and Heat Treatment

Common Materials. Gears may be made of a wide range of materials that may be ferrous, nonferrous, and nonmetallic. For industrial applications of spur, helical, and bevel gears, steel is most commonly used, with iron and other metals occasionally used. Wormgear pairs generally are made up of bronze or iron for the wormwheel and steel for the worm. Use of plastics is generally limited to light applications, particularly those in which minimal lubrication is available.

Through-Hardened Versus Surface-Hardened Material. Since a gear set's load capacity is determined in part by the surface hardness of the teeth, heat treating is often used to transmit increased load in a smaller space. Pinions, whose teeth must endure more load cycles, if carburized and hardened, are made the same hardness as or slightly harder than their mating gears. Surface hardening adds to the manufacturing cost, particularly where an additional finishing operation is required to correct distortion. For many applications, through-hardened gears may prove more economical. This can happen when the equipment arrangement does not permit use of smaller surface-hardened gears.

Material Selection. Information regarding material selection and properties is contained in ANSI/AGMA 2004-B89, *Gear Materials and Heat Treatment Manual.*[3] Many factors influence the selection of gear materials, and their relative importance will vary. These factors include, but may not be limited to, the following:

- Mechanical properties
- Grade and heat treatments
- Cleanliness
- Availability and cost
- Machinability and other manufacturing characteristics

For detailed lists of metallurgical characteristics related to load capacity, see the material sections of ANSI/AGMA 2001-C95, *Fundamental Rating Factors and Calculation Methods for Involute Spur and Helical Gear Teeth*[4] or ANSI/AGMA 2003-B97, *Rating the Pitting Resistance and Bending Strength of Generated Straight Bevel, Zerol Bevel, and Spiral Bevel Gear Teeth.*[5]

Specific material and heat-treatment combinations should be based on experience and an analysis of the overall application requirements and conditions.

The following factors should be considered when making a material or heat-treatment selection:

- Safety factor, loading, duty cycle, mounting, gearing enclosure, lubrication, and operating ambient atmospheric conditions.
- For replacement gearing, the life obtained from the previous gearing should be evaluated. If satisfactory, replace with similar material. If longer life is required, selection of a heat-treatment specification yielding a higher hardness and, if necessary, a better material may provide the desired improvement.
- Annealed carbon steels, bar stock, or forgings are usually satisfactory for pinions and gears for uniform or moderate shock loads when the size of the gearing is not an important factor. Annealed carbon-steel pinions with cast-iron gears are sometimes used.
- Alloy steel pinions are used for higher loads or when greater life is desired. They may be used with cast-iron or annealed (forged or cast) steel gears, usually with larger gears or with higher reduction ratios.
- Heat-treated alloy steel pinions and gears should be used with higher hardness ranges when space limitation is a factor, i.e., where smaller center distances and face widths may be necessary.
- Steel pinions and gears which are machined and have teeth cut after heat treatment should have a pinion minimum hardness higher than that of the gear minimum hardness.
- Steel pinions and gears hardened to 400 Bhn or surface hardened higher after cutting are generally specified with the same hardness.
- Where impact loads exist, the use of high alloys and the lowering of hardness are recommended for surface-hardened gears and pinions.

When accelerated wear is encountered in service, heat-treated gearing providing greater hardness will, in most cases, help to alleviate this condition. Consult with an AGMA company member for appropriate recommendations.

Manufacturing Processes

Generating. The most common method of producing gear teeth is by a generating process that uses a rack-shaped tool and relative motions between the tool and the workpiece similar to the gear's intended operation. Examples are hobbing, shaping, rolling, grinding, and shaving.

Forming. The forming process produces gear teeth by a direct-replication method. Examples are casting, powder metal molding, and broaching.

Finishing. Finishing operations remove a relatively small amount of material from gear teeth. They may be used to improve accuracy and surface finish. Shaving is used to improve profile and finish on relatively soft gears. Grinding or honing is employed for harder gear surfaces.

Size Limitations. General limitations on gear diameters are listed in Table 5.4. A specific company may have the capability of producing larger gears of a particular type. Spur gears of very large diameters for low-speed application have been cut in arc sections and also fabricated from rack sections, then bent to the proper radius of curvature.

TABLE 5.4 General Limitations on Gear Diameters

	Approximate maximum diameter	
Gear type	in	m
Spur	400	15
Helical	400	15
Straight bevel	100	4
Spiral bevel	90	3.5
Wormgear	150	6

Accuracy Requirements

Control of gear accuracy may be necessary for many reasons. Proper operation, particularly at higher speeds, and low dynamic loads are probably the most important. Depending on the application, specifications on sound level, transmission smoothness, and accuracy of motion are often controlling.

AGMA 2000-A88, *Gear Classification and Inspection Handbook—Tolerances and Measuring Methods for Unassembled Spur and Helical Gears;*[6] ANSI/AGMA 2009-A98, *Bevel Gear Classification, Tolerances, and Measuring Methods;*[7] and ANSI/AGMA 2011-A98, *Cylindrical Wormgearing Tolerance and Inspection Methods,*[8] contain classification systems covering tolerances for common types of gears. In these systems, tolerances are tabulated for a series of *accuracy grades* that range from 3 to 15. When selecting an accuracy grade number, care must be taken as the newer international convention standards use lower numbers to represent more precise tolerances. In older AGMA standards, the higher numbers represent more precise tolerances. The accuracy grade standards cover the gears themselves and do not generally apply to gears in enclosed drives.

By far the majority of applications can be satisfied by gears in the midrange, around grade 8. Rarely does an application require a gear in the most precise grade classification. The selection of the proper accuracy grade for a specific application requires good judgment and experience. Care should be used in making a final selection to meet the specific conditions, as more precise gears are more expensive to manufacture.

Precise Motion Transmission. Many applications use gears as a means for precise transmission of motion. Usually, more precise accuracy grades are required for these gears. Examples of such applications are navigational tracking, telescopes, printing presses, and index devices for machine tools.

High-Speed Considerations. Gear inaccuracies become more critical as operating speed increases. The most obvious problem may be high noise level. In addition, the dynamic loads on the teeth caused by inaccuracies may be a substantial part of the total transmitted load.

Accuracy Versus Cost. The cost of manufacturing and inspection escalates rapidly with increasing accuracy. Therefore, the user should use care to avoid specifying higher grades than required for the application. Table 5.5 compares relative first cost to gear type.

TABLE 5.5 Relative First Cost Versus Gear Type

Gear type	First cost*
Spur	Low to moderate
Wormgear	Moderate
Helical and bevel	Moderate to high

* Cost per unit hp (kW) or torque.

Mounting of Gears. Close attention must be given to the mounting of gears and the maintenance of alignment under operating conditions. Obviously, the accuracy and stiffness of mounting must be commensurate with the precision of the gears themselves to obtain optimum results. Some types of gears require close axial positioning of either or both members of a pair to obtain proper operation. Examples are bevel gears (both), cylindrical wormgears (gear only), and double-enveloping wormgears (both). The gearshaft positioning must be supported by bearings and bearing mounts of suitable capacity to accommodate all radial and thrust loads involved.

Additional Considerations. When specifying the quality of a given gear, there are additional or special considerations that should be reviewed.

- Backlash allowances in tooth thickness
- Matching pinion and gear mating modifications
- Replacement gearing

Enclosed Gear Drives

Advantages. Enclosed gear drives marketed by gear manufacturers offer several advantages over open power-transmission devices.

- Safety—protection from moving parts
- Retention of lubricant
- Protection from environment
- Economics of quantity manufacture
- Availability

Types and Features. Enclosed gear drives generally are classified by the principal type of gearing used. They may have a single set of gears or additional gear sets of either the same or different types in the same enclosure. Figures 5.26 to 5.33 provide illustrations of common enclosed gear drives. Table 5.6 covers typical application ranges of each. ANSI/AGMA 6010-F97, *Standard for Spur, Helical, Herringbone, and Bevel Enclosed Drives,*[9] is a typical standard that provides criteria for assembled gear drives, such as unit rating, lubrication, components, thermal rating, storage, and installation.

Mounting. Gear drives may be designed for base, flange, or shaft mounting. The last type makes use of a hollow output shaft for direct mounting on the driven shaft (refer to Figs. 5.32 and 5.33). A reaction arm or similar device is required to secure the unit against rotation.

FIGURE 5.26 High-speed single-reduction helical gear drive. *(Philadelphia Gear Corporation.)*

FIGURE 5.27 Double-reduction helical gear drive. *(Flender Corporation.)*

Gearmotors. A gearmotor is an integral drive unit incorporating an electric motor and gear reducer, with the frame of one supporting the other. Some designs use motors with special shaft ends and/or mountings, while others adapt standard motors (refer to Fig. 5.33). ANSI/AGMA 6009-A00, *Standard for Gearmotor, Shaft Mounted and Screw Conveyor Drives,*[10] contains information for gearmotors with spur, helical, bevel, and worm gears.

High-Speed Drives. AGMA standards for enclosed gear drives used in industrial service define a high-speed drive as having a shaft speed greater than 4000 r/min for helical and bevel units and 3600 r/min for cylindrical wormgears. An additional qualification is when the gear set pitch line velocity exceeds 7000 ft/min (35 m/s) for helical and bevel units and 6000 ft/min (30 m/s) sliding velocity for wormgears. Above these limits, special consideration should be given to such items as gear accuracy grade, lubrication, cooling, bearings, etc. See ANSI/AGMA 6011-H98, *Specification for High Speed Helical Gear Units.*[11]

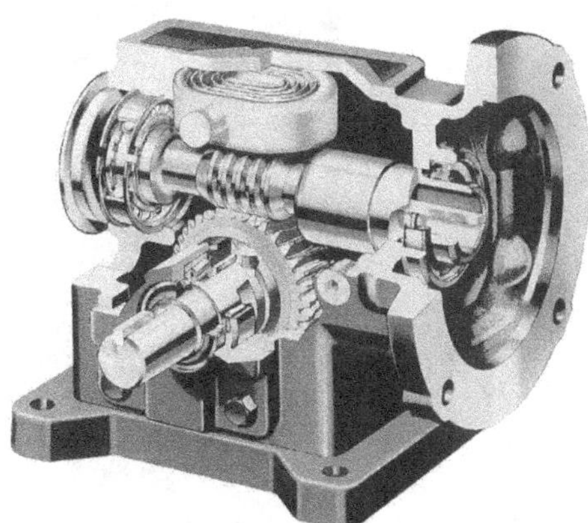

FIGURE 5.28 Single-reduction wormgear drive. *(Rockwell Automation.)*

FIGURE 5.29 Double-reduction wormgear drive. *(Boston Gear Division of IMO Industries, Inc.)*

FIGURE 5.30 Single-reduction spiral bevel gear drive. *(Falk Corporation.)*

INSTALLATION AND MAINTENANCE

The variety of types and sizes of gears and enclosed gear drives makes it impractical to cover installation and maintenance in specific detail. The user should refer to the manufacturer's literature, nameplate data, and warning tags. Such information should take precedence over the generalized comments that follow.

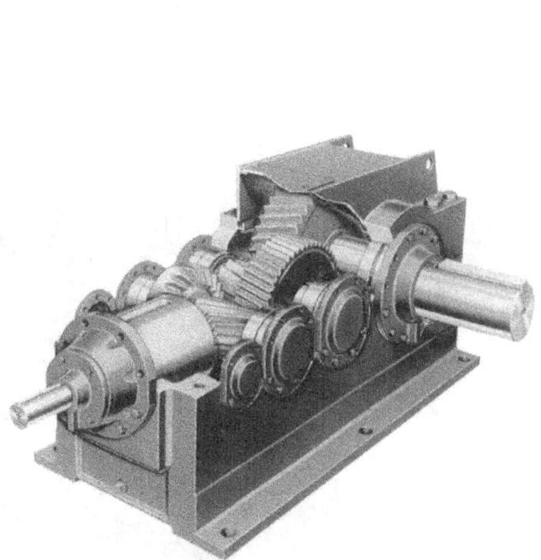

FIGURE 5.31 Triple-reduction spiral bevel–helical gear drive. *(Falk Corporation.)*

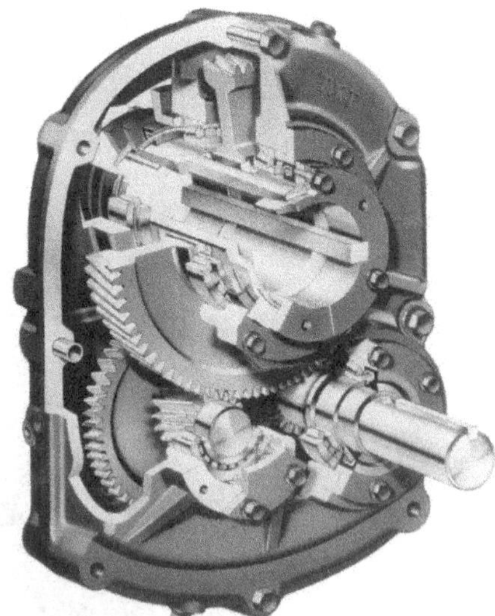

FIGURE 5.32 Shaft-mounted double-reduction gear drive. *(Rockwell Automation.)*

FIGURE 5.33 Triple-reduction helical-bevel-helical shaft-mounted gearmotor. *(Rockwell Automation.)*

Mounting of Gears and Enclosed Drives

To ensure long service and dependable performance, gears and enclosed gear drives must be rigidly supported and shafts accurately aligned. Reference 9 describes the minimum precautions required to accomplish this end. Many different types of units for widely varying applications are covered by this standard. Table 5.7, which is arranged as a checklist, is intended to act as a guide. Specific items should be applied as appropriate for a particular gear drive and the specific application.

Provision must be made for adequate lubrication of open, semienclosed, or enclosed gears. See Chap. 5.12, "Lubricants," for information on gear lubrication. ANSI/AGMA 9005-D94, *Industrial Gear Lubrication,*[12] also provides information on lubricant types and methods of lubrication.

TABLE 5.6 Enclosed Gear-Drive Units—Features (Refer to Figs. 5.26 to 5.33)

Type of unit	Ratio range*	Horsepower*	kW	Efficiency, %†	Fig. no.
Helical & double helical					
Single reduction	to 7:1	to 20,000	14,920	96–98	5.26
Double reduction	5:1–40:1	10,000	7,460	95–97	5.27
Triple reduction	20:1–300:1	3,000	2,238	93–95	5.31
Quadruple reduction	150:1–1000:1	1,000	746	91–93	—
Bevel, helical	5:1–40:1	3,000	2,238	95–97	—
Bevel, helical, helical	20:1–300:1	2,000	1,492	93–95	5.31
Bevel					
Single bevel	to 9:1	2,000	1,492	96–98	5.30
Wormgear					
Single reduction	5:1–70:1	300	224	50–96	5.28
Double reduction	25:1–4900:1	100	74	20–92	5.29
Helical, worm	20:1–300:1	150	112	50–94	—

* The information on ratio and horsepower ranges is approximate and is for the product usually offered.
† Efficiency is based on transmission of full rated power.

TABLE 5.7 Installation and Start-Up Checklist

Step	Instruction
Storage and handling	Observe manufacturers' instructions for unpacking and handling of gear drive. Store the unit properly until installation. Preserve the unit properly until placed into service. Handle the unit properly. Safety of personnel comes first. Lift only at adequate lifting points. Protect the mounting surface from damage.
Foundation	Provide adequate foundation commensurate with size and type of unit. Provide level mounting surfaces unless drive is designed otherwise.
Installation	Install the unit properly on an adequately supported foundation. Bolt the unit securely into place. Level the unit properly so as not to distort the gear case. Install any auxiliary components and connections, such as cooling water. Properly install couplings, sprockets, etc. suitable for the application. Ensure accurate alignment with other equipment. Install adequate machinery guards to protect operating personnel. Ensure that driving equipment rotates in the correct direction before coupling to a gear drive designed to operate in a specific direction.
Accessibility	Provide adequate access space to permit future maintenance.
Lubrication	Observe manufacturers' instructions, using the specified lubricant and amount. (*Note:* Most gear drives are shipped without lubricant.)
Start-up	Ensure that switches, alarms, heaters, coolers, and other safety and protection devices are installed and operational for their intended purposes. On a unit equipped with a separately driven lubrication pump, run the pump and check the lubrication system prior to starting the unit. Fill the unit or sump to proper level with correct lubricant before starting. Refill as necessary immediately after starting the unit. Ensure that all grease points have received the proper amount of grease. Check for proper rotation and freedom from obstructions before the start of full operation.
Operation and maintenance	Operate the equipment as it was intended to be operated. Do not overload. Run at correct speed. Inspect for oil leaks and unusual noise or vibration immediately after start-up. Check oil temperature after several hours [temperatures of 180–200°F (82–95°C) are usual for most gear drives]. Maintain lubricant in good condition, at the proper level. Dispose of used lubricant in accordance with applicable laws. Apply proper amount of grease to specified locations at prescribed intervals. Recheck alignment and tightness of all fasteners, fittings, and plugs after 1week of operation. Perform periodic maintenance of the gear drive as recommended by the manufacturer.

Installation, Start-Up, and Maintenance of Enclosed Gear Drives

The handling, installation, and maintenance of new gears or an enclosed gear drive deserve close attention to avoid damage, ensure proper operation, and provide adequate protection of personnel safety. A checklist of important items is provided in Table 5.7.

Lubrication of Enclosed Gear Drives. Improper lubrication is one of the major causes of failure of gear drives. The gear manufacturer's instructions must be followed to ensure proper operation.

Reference 12 provides detailed information on recommended lubricants and their maintenance procedures. Gear type, size, and speed, along with ambient temperature range, are major influencing factors on the proper lubricant selection.

The gear unit should be drained and cleaned with a flushing oil after 4 weeks of initial operation. When refilling, either the filtered original lubricant which has been tested for correct properties or new lubricant should be used. For normal operation, oil changes should be made after every 2500 h of service.

Periodic checks must be made on oil levels, oil cups, and grease fittings, and the lubricant must be replaced when required. Care should be taken not to overpack grease fittings. When a pressurized lubrication system is used, proper functioning of pump, filter, and cooler should be frequently audited.

Seals and Breathers. It is recognized that gear drives applied in certain industries and atmospheric conditions should be equipped with special oil seals and breathers. Extremely dusty or corrosive environments as well as severe moisture and vapor-laden atmospheres have special requirements. Applications subject to direct or indirect washdown, such as paper, food, and drug industries, may preclude use of breathers. In these cases, expansion chambers may be used.

Troubleshooting

Alertness to changes in operating characteristics, such as increased temperature rise over ambient, increased noise and vibration, and discovery of oil leakage, can prevent costly shutdowns. Table 5.8 provides a checklist to diagnose various problems in operation.

Failure Analysis

The proper analysis of a gear failure or incipient failure can be very valuable in preventing future expensive downtime. Gear wear is frequently subjective. Whether the gears should be considered usable or not depends on how much wear is tolerable (Fig. 5.34). Gears might not be usable if the wear causes excessive noise or vibration. Failures are more obvious, such as when several gear teeth break and transmission ceases. To find the cause(s) of wear or failure, the differences between primary and secondary problems and effects may have to be determined. Destructive wear varies in significance. For example, contact fatigue is often less serious than bending fatigue. This is because contact fatigue can progress slowly, starting with a few pits that may increase in size and number. If the teeth continue to wear, an increase in noise or vibration can warn of the impending failure (Fig. 5.35). Bending fatigue, however, can break a tooth without warning. ANSI/AGMA 1010-E95, *Appearance of Gear Teeth—Terminology of Wear and Failure*,[13] defines gear-tooth failure modes with 75 helpful illustrations, such as shown in Figs. 5.34 to 5.36.

APPLICATION OF GEARING AND ENCLOSED GEAR DRIVES

Gear Load Capacity

AGMA has many standards which can be used for gear capacity calculations of most gear types and enclosed drives. The capacities determined from these standards are intended for

TABLE 5.8 Trouble Chart

Trouble	What to inspect
Heating	Is unit and fan assembly covered with dirt? Is unit overloaded? Has recommended oil level been exceeded or is level too low? Are couplings in alignment? Have bearings been properly adjusted? Are oil seals or stuffing boxes the cause? Is oil clean or is sludge content high? Have oil filters been cleaned? Is oil pump functioning?
Shaft failure	Check alignment; many shafts fail owing to misalignment. Some troubles are caused by use of rigid couplings. Is an overhung load beyond the capacity of unit? Is unit subject to high dynamic loads or extreme repetitive shocks not previously considered?
Bearing failure	Rust formation caused by high humidity or the entrance of water. Unsuitable lubricant or the lack of a sufficient amount of lubrication. Abnormal loading causing excessive deflection which could result in flaking, cracks, and fractures. Improper adjustment causing abnormal loading if bearings are excessively preloaded, or abnormal bearing and gear wear if bearings have excessive freedom. Depending on bearing type, proper mounting and setting could be critical.
Oil leakage	Check oil seals and replace if worn. Check for housing pressure buildup due to a blocked breather. Check stuffing boxes and adjust or replace packing. Check tightness of drain, level, and other plugs or fittings.
Gear wear	Excessive loads. Incorrect lubrication. Insufficient lubrication. Misalignment due to worn bearing. Backlash may be insufficient. Lubricant carrying foreign matter, e.g., abrasive dirt or particles of worn metal teeth. Excessive temperature. Excessive speeds.
Noise or vibration	Bad alignment. Loose or worn bearings. Insufficient lubrication. Excessive lubrication. Gear wear.

specific applications where all the operating and environmental conditions are known. The capacities (*equivalent power* ratings) that are normally tabulated in manufacturers' catalogs are for a given life at the stated speeds and loads of a very uniform nature, operated for less than 10 h/day, and provided the lubricant bulk temperature does not exceed a temperature of 200°F (95°C).

Service Factors

Before selecting an enclosed gear drive from a catalog for an application, the power to be transmitted is multiplied by a service factor to determine an *equivalent power* as listed in a

FIGURE 5.34 Moderate gear-tooth wear. *(AGMA; extracted from Ref. 13.)*

FIGURE 5.35 Teeth with progressive pitting. *(AGMA; extracted from Ref. 13.)*

catalog. Service factors have been developed from the experience of manufacturers and users to allow for the nature of the transmitted load and duration of service. The selection of a service factor less than 1.0 is not recommended. Most AGMA standards for enclosed gear drives provide tables for specific applications with recommendations for their service factor. Reference 9, annex A provides such a tabulation for enclosed speed reducers or increasers using spur, helical, and spiral bevel gears. Table 5.9 illustrates a range of subjective service factors.

Product Selection

After the equivalent power has been determined, selection of an enclosed gear drive can be made by comparing this value with the catalog rating. It is necessary that the product selected

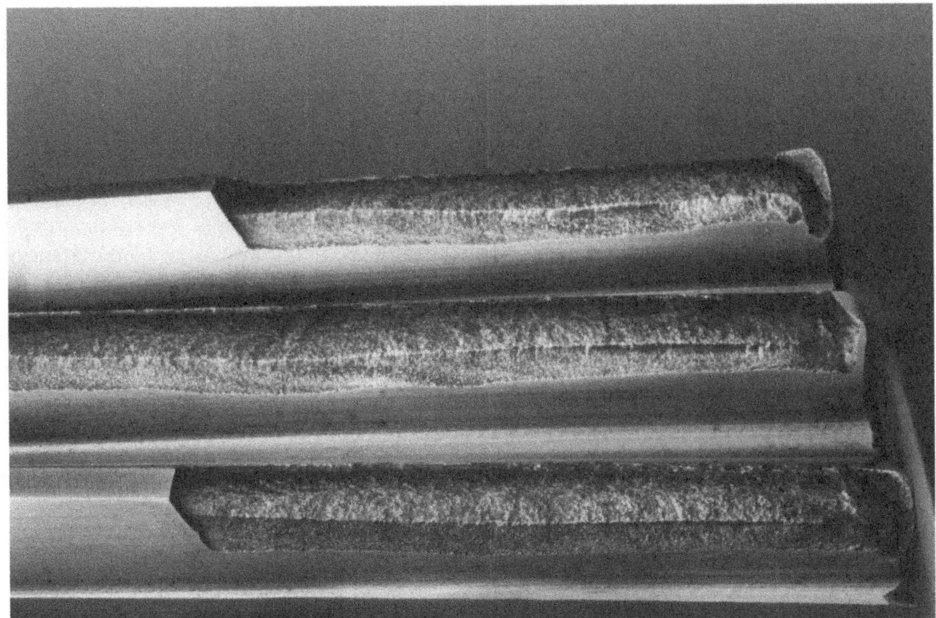

FIGURE 5.36 Case-core separation breakage. *(AGMA; extracted from Ref. 13.)*

TABLE 5.9 Typical Subjective Service Factors for Gear Drives

Prime mover	Duration of service, h/day	Driven machine load classification		
		Very uniform	Light shock	Heavy shock
Induction motor or steam turbine	≤10	1.00	1.25	1.75
	>10	1.25	1.50	2.00
Multicylinder engine	≤10	1.25	1.50	2.00
	>10	1.50	1.75	2.25
Single-cylinder engine	≤10	1.50	1.75	2.50
	>10	1.75	2.00	2.75

have a rated load capacity equal to or in excess of the equivalent power. An enclosed gear drive must also usually be checked for thermal rating. *Thermal rating* is the power that can be transmitted continuously without causing the gear drive temperature to increase more than 100°F (38°C) above ambient. Should this limitation prevail, several alternatives are available, such as auxiliary cooling systems, oil pans to reduce churning, or selection of a larger unit.

Systems Considerations

An essential phase in the design of a system of rotating machinery is the analysis of the dynamic response of the system to excitation forces. System torsional and lateral dynamic response can impose vibratory loads that are superimposed on the mean running load. Depending on the dynamic behavior of the system, this additional loading could lead to failure of system components. In a gear drive, these failures could occur as pitting of the gear teeth, tooth breakage, shaft breakage, bearing failures, etc. A dynamic system analysis becomes more important as driving and driven inertia becomes large, or rotating speeds increase, or both. A dynamic analysis must consider the complete system, including prime mover, gear drive, driven equipment, couplings, and foundations. The dynamic loads imposed on a gear unit are the result of the dynamic behavior of the total system and not of the gear unit alone.

For further information, see Ref. 11, annexes C and D.

Gear Sound and Vibration Control

The two greatest concerns regarding gear sound and vibration are their contribution to environmental pollution of personnel health and that they may be symptomatic of abnormal wear and impending failure. Refer to Chaps. 4.12 and 4.13, "Noise Control" and "Vibration Control," for additional information; or see AGMA 299.01, *Gear Sound Manual:* Sec. 1, *Fundamentals of Sound as Related to Gears;* Sec. 2, *Sources, Specifications, and Levels of Gear Sound;* Sec. 3, *Gear-Noise Control.*[14]

REFERENCES AND BIBLIOGRAPHY

AGMA standards and information sheets are continually being updated. For an up-to-date listing of all document abstracts see the AGMA web site, www.agma.org.

1. ANSI/AGMA 1012-F90, *Gear Nomenclature, Definitions of Terms with Symbols.*
2. AGMA 913-A98, *Method for Specifying the Geometry of Spur and Helical Gears.*
3. ANSI/AGMA 2004-B89, *Gear Materials and Heat Treatment Manual.*

4. ANSI/AGMA 2001-C95, *Fundamental Rating Factors and Calculation Methods for Involute Spur and Helical Gear Teeth.*
5. ANSI/AGMA 2003-B97, *Rating the Pitting Resistance and Bending Strength of Generated Straight Bevel, Zerol Bevel, and Spiral Bevel Gear Teeth.*
6. AGMA 2000-A88, *Gear Classification and Inspection Handbook—Tolerances and Measuring Methods for Unassembled Spur and Helical Gears.*
7. ANSI/AGMA 2009-A98, *Bevel Gear Classification, Tolerances, and Measuring Methods.*
8. ANSI/AGMA 2011-A98, *Cylindrical Wormgearing Tolerance and Inspection Methods.*
9. ANSI/AGMA 6010-F97, *Standard for Spur, Helical, Herringbone, and Bevel Enclosed Drives.*
10. ANSI/AGMA 6009-A00, *Standard for Gearmotor, Shaft Mounted, and Screw Conveyor Drives.*
11. ANSI/AGMA 6011-H98, *Specification for High Speed Helical Gear Units.*
12. ANSI/AGMA 9005-D94, *Industrial Gear Lubrication.*
13. ANSI/AGMA 1010-E95, *Appearance of Gear Teeth—Terminology of Wear and Failure.*
14. AGMA 299.01, *Gear Sound Manual:* Sec. 1, *Fundamentals of Sound as Related to Gears;* Sec. 2, *Sources, Specifications, and Levels of Gear Sound;* Sec. 3, *Gear-Noise Control.*
15. ANSI/AGMA 2008-B90, *Assembling Bevel Gears.*
16. ANSI/AGMA 6000-B96, *Specification for Measurement of Linear Vibration on Gear Units.*
17. ANSI/AGMA 6001-D97, *Design and Selection of Components for Enclosed Gear Drives.*
18. ANSI/AGMA 6025-D98, *Sound for Enclosed Helical, Herringbone, and Spiral Bevel Gear Drives.*
19. ANSI/AGMA 6034-B92, *Practice for Enclosed Cylindrical Wormgear Speed Reducers and Gearmotors.*
20. ANSI/AGMA 9001-B97, *Flexible Couplings—Lubrication.*

CHAPTER 5.5

MECHANICAL EQUIPMENT: ROLLING ELEMENT BEARINGS

Industrial Division Engineering Staff
The Torrington Company
Torrington, Connecticut

GLOSSARY*

Aligning bearing A bearing which, by virtue of its shape, is capable of considerable misalignment.

Antifriction bearing A term given to ball and roller bearings to distinguish them from non-rolling-element bearings, i.e., sliding bearings.

Average life See median life.

Axial internal clearance The measured maximum possible movement parallel to the bearing axis of the inner ring in relation to the outer ring.

Axial load See thrust load.

Ball bearing An antifriction bearing with balls as rolling elements.

Basic dynamic radial (axial) load rating (BDLR/BDALR) The constant stationary radial load (constant centric axial load) that a rolling bearing can theoretically endure for a basic rating life of 1 million (10^6) revolutions.

*Based upon ANSI/ABMA Standard 1—1990.

Basic static-load rating (BSLR) The static load that corresponds to a total permanent deformation of ball and race, or roller and race, at the most heavily stressed contact (0.0001 in [0.00254 mm] of the ball or roller diameter).

Bearing runout Radial or axial displacement of the surface of a bearing relative to a fixed point when one raceway is rotated with respect to the other raceway.

Bearing series A graduated dimensional listing of a specific type of bearing.

Bore The area of the bearing making contact with the bearing seat on the shaft.

Boundary dimensions Dimensions for bore, outside diameter, width, and corners.

Cage A bearing component that partly surrounds all or several of the rolling elements and moves with them. Its purpose is to space the rolling elements and generally also to guide and/or retain them in the bearing.

Cage pocket An aperture or gap in a rolling bearing cage to accommodate one or more rolling elements.

Combined load A combination of all radial and axial forces on a bearing.

Cone An inner ring of a tapered roller bearing.

Conrad ball bearing A deep-groove, radial non-filling-slot type ball bearing in which each ring has uninterrupted raceway grooves.

Cup An outer ring of a tapered roller bearing.

Diametral clearance See radial internal clearance.

Double-row bearing A rolling bearing with two rows of rolling elements.

End play See axial internal clearance.

Equivalent load The calculated, constant, stationary load which, if applied to a bearing, would give the same life as attained under actual conditions of load and rotation.

Filling-slot bearing A deep-groove, radial ball bearing having a loading slot in one shoulder of each ring to permit the insertion of a larger number of balls.

Fixed bearing A bearing that positions a shaft against axial movement in both directions.

Flanged housing A bearing housing with a radial flange with bolt holes for mounting on a support perpendicular to the bearing shaft.

Floating bearing A bearing so designed or mounted as to permit axial displacement between shaft and housing.

Full-complement bearing A rolling bearing without a cage in which the sum of the clearances between the rolling elements in each row is less than the diameter of the rolling elements and small enough to give satisfactory function of the bearing.

Housing bearing seat The part of the housing bore which contacts the outside diameter of the bearing.

Housing fit The amount of interference or clearance between the bearing outside surface and the housing bearing seat.

Inch series bearing A rolling bearing that conforms to an inch series of a standardized dimension plan.

Internal clearance See radial and axial internal clearance.

Lateral travel See end play.

Life For an individual rolling bearing, the number of revolutions (or hours at some given constant speed) that the bearing runs before the first evidence of fatigue develops in the material of either the rings (or washer) or any of the rolling elements.

Load rating See basic dynamic radial (axial) load rating.

Lubrication groove A continuous recess in a bearing or housing to permit relubrication.

Lubrication hole A hole in the bearing rings to provide lubricant to the rolling elements.

Median life The life attained or exceeded by 50 percent of a group of apparently identical rolling bearings operating under the same conditions (L_{50}).

Metric series bearing A rolling bearing that conforms to a metric series of a standardized dimension plan.

Minimum life See rating life.

Multirow bearing A rolling bearing with more than two rows of rolling elements supporting load in the same direction. It is preferable to specify the number of rows and type of bearing, for example "four-row cylindrical roller radial bearing."

Needle roller A cylindrical roller of small diameter with a large ratio of length to diameter. It is generally accepted that the length is between 3 and 10 times the diameter, which does not usually exceed 5 mm. The ends of a needle roller may be any one of several shapes.

Needle roller bearing An antifriction bearing with needle rollers as rolling elements.

Nonseparable bearing A rolling bearing in which neither bearing ring can be freely separated.

Outside diameter The outer ring outside diameter (OD) of a radial bearing or the housing washer outside diameter of a thrust bearing.

Pillow block An assembly comprising a radial-type bearing and a bearing housing which has a base with bolt holes for mounting on a support parallel with the bearing shaft.

Pitch diameter, rolling elements The diameter of the pitch circle generated by the center of a rolling element as it traverses the bearing's axis of rotation.

Pocket See cage pocket.

Preload Force applied on a bearing, for example, by axial adjustment against another bearing, before the "useful" load is applied (external preload), or force induced in a bearing by raceway and rolling element dimensions resulting in "negative clearance" (internal preload).

Raceway The surface of a load-supporting part of a rolling bearing, suitably prepared as a rolling track for the rolling elements.

Radial bearing A rolling bearing designed to support primarily radial load, having a nominal contact angle between 0 and 45° inclusive. Principal parts are inner ring, outer ring, and rolling elements with or without a cage.

Radial internal clearance (RIC) For a single-row, radial contact bearing, this is the average outer-ring raceway diameter, minus the average inner-ring raceway diameter, minus twice the rolling element diameter.

Radial load A load acting in a direction perpendicular to the bearing axis.

Radial play See radial internal clearance.

Rating life The rating life of a group of apparently identical bearings is the life in millions of revolutions that 90 percent of the group will complete or exceed (L_{10}).

Retainer See cage.

Ring An annular part of a radial rolling bearing incorporating one or more raceways.

Roller bearing An antifriction bearing with rollers as rolling elements.

Sealed bearing A rolling bearing that is fitted with a seal on one or both sides of the bearing.

Self-aligning bearing A bearing with built-in compensation for shaft or housing deflection or misalignment.

Self-contained bearing See nonseparable bearing.

Separable bearing A bearing assembly that may be separated completely or partially into its component parts.

Separator See cage.

Shaft bearing seat The portion of the shaft on which the bearing is mounted.

Shaft fit The amount of interference or clearance between the bearing inside diameter and the shaft bearing seat diameter.

Shield A circular part fixed to one bearing ring to cover the interspace, but not in contact with the other ring.

Single-row bearing A bearing having only one row of rolling elements.

Special bearing A bearing not meeting the requirements of standard or established line bearings.

Spherical (radial) roller bearing An internally self-aligning, radial rolling bearing with convex or concave rolling elements to accommodate misalignment.

Spherical roller thrust bearing An antifriction thrust bearing using spherical rollers as rolling elements.

Standard bearing An antifriction bearing conforming to ABMA (American Bearing Manufacturers Association), "General Boundary Plans of Metric and Inch Dimensions."

Static equivalent radial (axial) load The calculated static radial or static centric thrust load that, if applied to a bearing, would cause the same total permanent deformation at the most heavily stressed rolling element and raceway contact as that which occurs under an actual condition of loading.

Static load A load acting on a nonrotating bearing.

Tapered roller A roller with one end smaller than the other. The general shape is that of a truncated cone.

Thrust bearing A rolling bearing designed to support primarily axial load, having a nominal contact angle of greater than 45° up to and including 90°. Principal parts are shaft washer, housing washer, and rolling elements with or without cage.

Thrust load A load acting in a direction parallel with the bearing axis.

Washer (thrust bearing) An annular part of a thrust rolling bearing incorporating one or more raceways.

INTRODUCTION

Antifriction (rolling-element) bearings are used wherever the reduction of friction is required at the interface of dynamic and static components in machinery. A principal advantage of these bearings is their ability to operate at friction levels considerably lower than those of plain or oil-film bearings, while maintaining coefficients of friction at start-up that are close to those in normal operation. Favorable operating characteristics resulting from these lower friction levels are higher operational speeds, lower operational temperatures, and reduced operational torque (thus reduced power consumption).

The complement of rolling elements in antifriction bearings comprises balls or rollers (or even a combination of both in some special designs). In concept, the balls or rollers are arranged within a bearing primarily to support either pure radial or pure thrust loading, but they sometimes have the capability to accomplish both.

Bearings usually consist of four essential components: the inner ring and outer ring with their raceways; a complement of rolling elements (balls or rollers); and a cage, retainer, or separator. In operation, the hardened and ground surfaces of the raceways form the track for the rolling elements (load-supporting members) to follow while transmitting load from the dynamic members of an assembly to the stationary members. Many bearings are also designed with integral shields or seals to allow the bearings to be prelubricated and to protect the rolling elements from contamination. Bearings are often supplied in housed units for simplified mounting.

APPLICATION INFORMATION

Bearing Life

The life of a bearing is expressed as the number of revolutions or the number of hours at a given speed for which the bearing will operate before any evidence of fatigue develops on the rolling elements or the raceways. Life may vary from one bearing to another, but stabilizes into a predictable pattern when considering a large group of the same size and type of bearings. The *rating life* of a group of such bearings is the number of hours or revolutions (at a given constant speed and load) that 90 percent of the tested bearings will exceed before the first evidence of fatigue develops. This is called L_{10} life or *minimum life.*

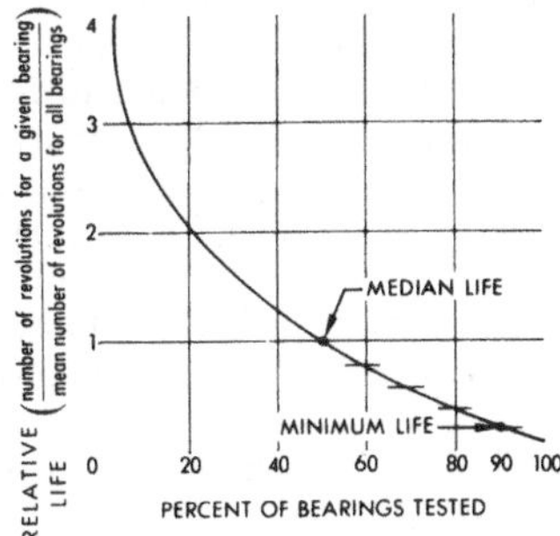

FIGURE 5.37 Median life curve.

Median Life. The results of testing a large group of ball or roller bearings may be graphically illustrated. The distribution curve shown is obtained by plotting relative life vs. percent of bearings tested (see Fig. 5.37). From the curve in Fig. 5.37 it is apparent that the median L_{50} life is approximately five times the minimum life. About 50 percent of the bearings will exceed the median life. Since it is not possible to predict the exact life of a single bearing, a safety factor must be allowed to minimize the chances of early failure.

The cost of replacing a bearing plus the expense of machine downtime may greatly exceed the relatively low cost of the bearing. Therefore, most designers prefer to use the L_{10} minimum life as a basis for design. In some applications where safety or maintenance economy is not critical and low initial bearing cost is desirable, the median-life value may be used.

Life-and-Load Relationship. Empirical calculations and experimental data point to a predictable relationship between bearing load and life. This relationship may be expressed by formulas. In these empirical formulas the bearing life is found to vary inversely as the applied load to an exponential power. The assigned value of the exponent depends upon the basic type of rolling element. For all types of roller bearings the formula is

$$\text{Life} = \left(\frac{\text{basic dynamic capacity}}{\text{load}}\right)^{3.33} 10^6 \text{ revolutions}$$

For all types of ball bearings the formula is

$$\text{Life} = \left(\frac{\text{basic dynamic capacity}}{\text{load}}\right)^{3} 10^6 \text{ revolutions}$$

Ring Rotation Factors (RF). The basic dynamic load rating, often referred to as *dynamic capacity,* of a radial bearing is based on the inner ring rotating with respect to the applied load. The bearing dynamic capacity does not have to be downgraded when the outer ring rotates. If the inner ring or outer ring rotates with respect to load, the rotation factor is 1.0.

Effect of Load. It is also evident from the exponential character of the basic life-load relationships that, for any given speed, a change in load may have a substantial effect on the life. For a roller bearing, if the load is doubled, the life is reduced to one-tenth its former calculated duration. Similarly, if the load is halved, the life is increased tenfold. For a ball bearing, if the load is doubled, the life is reduced to one-eighth its former value. Likewise, if the load is reduced one-half, the life is increased eightfold.

Effect of Speed. The preceding expressions are independent of the speed of the bearing. If bearing life is measured in hours, an increase in speed results in a decrease in hours of life. The number of revolutions per unit time determines the hours of life available before the fatigue limit of the bearing is reached. If the speed is doubled, the hours of life are halved. Conversely, if the speed is reduced by 50 percent, the hours of life will be doubled.

The limiting speed of a bearing is dependent on factors such as bearing type, application loads, mounting arrangement, and lubrication conditions. This will be discussed in a later section.

Selection of Bearing Type

Selection of bearing type is made after the general design concept of the machine has been established and the magnitude of the loads and speeds estimated. Special conditions can directly affect bearing operation and must be considered. These include ambient or localized temperatures, shock or vibration, dirt or abrasive contamination, difficulty in obtaining accurate alignment, space limitations, need for shaft rigidity, and reliability of lubrication.

Selection of the proper type of bearing is not an exact science. The fields of application for many types of bearings overlap, and the value of experience in bearing applications cannot be overemphasized. Each type of bearing, however, has inherent features, which determine its relative suitability for a specific application. Careful analysis of these features and familiarization with the fundamental characteristics of each type of bearing will help in selecting the proper bearing.

As an aid to experienced designers and inexperienced bearing users alike, the similarities and differences of ball and roller bearings are outlined. Table 5.10 shows the relative operating characteristics of each cataloged bearing type.

Ball Bearings versus Roller Bearings. Using balls as the rolling elements in bearings offers certain performance advantages. Most of the advantages of ball bearings are derived from the small areas of contact between the ball and raceways.

Ball bearings may be operated at higher speeds, with less internal friction and less heat generation. They have a greater inherent ability to accommodate slight misalignment. Under certain conditions of combined loading, ball bearings occupy less space than required for roller bearings of the same bore size.

Rollers are not limited to a single geometric shape. There are several types, such as tapered rollers, spherical rollers, and cylindrical rollers. For a given load and diameter of rolling element, rollers transmit load through a larger contact area than do balls. This allows roller bearings to support greater loads and accommodate far more shock than ball bearings of equivalent size. For a given applied load, contact area stresses for roller bearings are lower than for ball bearings, and, therefore, they produce lower elastic deformation. Since the larger contact areas create more friction, permissible operating speeds for roller bearings are lower than those for ball bearings.

Selection of Bearing Size

Once the designer selects a suitable type of bearing for a set of specific conditions, the size of the bearing is determined. In many cases more than one bearing type will satisfy the operating conditions. In these instances the designer should determine the most suitable size of each type considered and make the final selection on the basis of mounting simplicity, space considerations, and overall economy.

The basic parameters affecting the choice of bearing size are radial load, thrust load, speed, required life, ring rotation, and shock or vibration conditions. Other factors such as misalignment, abnormal temperature, contamination, and poor lubrication have the potential to seriously reduce service life. It is difficult to predict their individual or combined negative effects on performance, so these factors should be eliminated by proper mounting and system design rather than by attempting to estimate their effect on bearing life.

TABLE 5.10 Relative Operating Characteristics

Bearing type/style		Radial capacity	Thrust capacity	Limiting speed	Resistance to deflection	
					Radial	Axial
Ball radial						
Conrad		M	M2	H	M	L
Maximum-capacity		M+	M1, L1	M	M+	L+
Angular-contact		M	M1+	H	M	M
Roller radial						
Cylindrical	2 flanges	H	N	M+	H	N
	3 flanges	H	L2	M+	H	NR
	4 flanges	H	L2	M+	H	NR
Journal	No flange	H+	N	L	H+	N
Spherical		H	M2	M	H–	M
Tapered roller						
Single-row		H–	M1+	M	M–	M
Single-row, steep angle		M+	H1	M–	M	H
Double-row		H	M2+	M	H	M
Double-row, steep angle		M+	H2	M–	M	H
Four-row		H+	M2+	M–	H+	M
Thrust						
Angular-contact ball		L+	H1–	M	L	H–
Ball		N	H1	M–	N	H
Cylindrical roller		N	H1+	L	N	H+
Spherical roller		N	H1+	L	N	H+
Tapered roller		N	H1+	L	N	H+

NR Not Recommended
N None
L Low
M Moderate
H High
1 One Direction
2 Both Directions

Limiting Speeds. The ability of a bearing to operate at high speeds is dependent upon the rate at which generated heat is dissipated. Maximum speed is governed by bearing type, size, and load; ambient temperature; and type of lubricant.

The geometric design of a bearing and the method of positioning the rolling elements basically determine the coefficient of friction. Since frictional losses are proportional to the peripheral speed, it follows that the smaller the bearing the greater the speed at which it may operate. A common presentation of limiting speeds used by many bearing manufacturers is either the P_dN value, where P_d is the bearing design pitch diameter in mm and N is the operating speed in revolutions per minute, or a dN value, where d is the bearing bore in mm and N is the speed in rpm. Typical values are shown in Table 5.11.

Elastic deformation of the raceways and rolling elements is increased by heavy loads, and this creates additional heat, thereby limiting the allowable speed.

Ambient temperature may affect the rate of heat dissipation. Applications having a high ambient temperature require careful selection of the method of lubrication.

The type of lubricant is a basic criterion for establishing limiting speeds. Lubricants with high viscosity offer more frictional resistance; therefore, oil is preferable to grease for higher speeds. Modern grease formulations, however, are allowing operational speeds in the 1 to 2 million P_dN range in selected applications, notably high precision spindles. The use of circulating oil or oil mist can allow higher speed limits than oil-bath or grease lubrication. The bearing manufacturer should be consulted for limiting speeds of bearings in specific applications.

TABLE 5.11 Limiting Speeds

Radial ball and roller bearings P.D. × N values (PD in millimeters × R.P.M.)*

Bearing type series	Cage type	ABEC-1 RBEC-1		ABMA Class 2,4		ABEC-3			ABEC 5 and 7		
		Grease	Oil†	Grease	Oil	Grease	Oil†	Oil mist	Grease	Circulating oil	Oil mist
		Ball bearings									
Single row											
Nonfilling slot	Ball piloted molded nylon (PRB)	250,000	300,000	—	—	250,000	300,000	—	300,000	300,000	300,000
9300K, 9100K, 200K	Pressed steel, bronze	300,000	350,000	—	—	300,000	350,000	—	350,000	400,000	450,000
300K	Ring piloted molded reinforced nylon (PRC)	350,000	400,000	—	—	350,000	450,000	—	400,000	550,000	650,000
XLS and variations	Composition (CR)										
Filling slot	Ball piloted molded nylon (PRB)	250,000	250,000	—	—	—	—	—	—	—	—
200W and variations, 300W and variations	Pressed steel	250,000	300,000	—	—	—	—	—	—	—	—
Angular contact	Ball piloted pressed steel, molded nylon (PRB)	250,000	300,000	—	—	300,000	350,000	—	—	—	—
7200WN	Ring piloted bronze (MBR), ball piloted bronze (MBR)	300,000	400,000	—	—	—	—	—	—	—	—
7300WN	Ring piloted molded reinforced nylon (PRC)	350,000	400,000	—	—	350,000	400,000	—	—	—	—
Angular contact—extra precision											
2M9300WI, WO, 2M200WI, 2M300WI, 2M9100WI	Ring piloted composition (CR or PRC)	350,000	400,000	—	—	750,000	1,000,000	1,200,000			
2MM9300WI, WO, 2MM9100WI 2MM200WI, 2MM300WI									1,000,000	1,400,000	1,700,000
Double row											
5200	Ball piloted molded nylon (PRB), pressed steel	250,000	300,000	—	—	—	—	—	—	—	—
5300	Ball piloted bronze (BR)										

		Cylindrical roller bearings									
RU, RIU RN, RIN RJ, RIJ RF, RIF RT, RIT RP, RIP	Bronze	150,000	300,000	—	—	—	—	—	—	—	—
5200WS	Roller piloted pressed steel	150,000	300,000	—	—	—	—	—	—	—	—
		Spherical roller bearings									
239, 230, 240 231, 241, 222 232, 213, 223 233	Pressed steel Bronze Molded reinforced nylon	150,000	300,000	—	—	—	—	—	—	—	—
		Tapered roller bearings									
TS, TSS TDI, TDO, TDOC, TDOD	Pressed steel Bronze	—	—	150,000	300,000	—	—	—	—	—	—

$* \ \dfrac{\text{Bore} + \text{O.D.}}{2}$

† For oil bath lubrication, oil level should be maintained between ⅓ and ½ from the bottom of the lowest ball.
Note: Single or double normal contact (P or PP) sealed bearings should not exceed 350,000 PN.

Fits of Shaft and Housing. To ensure the full utilization of bearing capacity under operating conditions, it is important to have the proper fit between inner ring and shaft, and outer ring and housing. The tolerances to which the bearing is made are standardized, so desired fits may be obtained by controlling the dimensions and tolerances for the shaft and housing.

Normally, the problem in fit determination is to make the rotating ring of the bearing and its associated shaft or housing rotate as a single unit by using an interference fit. The fit of the nonrotating ring should be loose, with minimum clearance, for ease of assembly and axial movement in the housing.

The amount of interference fit employed should not create in the bearing rings excessive stress that might result in early fatigue failure. Under conditions of light load, the interference fit can be small. As the loads increase or shock loading is introduced, the interference must be increased so that no clearance exists and none can be induced by the load. This is the only effective means of preventing "creep." As a rule, axial clamping cannot be relied on to prevent creep since the clamping force must be excessively high. Thus, the heavier the load, the tighter the fit.

The degree of fit is designated in the American Bearing Manufacturers Association (ABMA) tolerance system. This tolerance system is also in accordance with that adopted by the American National Standards Institute (ANSI). These various fits apply to all ball thrust bearings and radial bearings except tapered roller bearings. (For tapered roller bearings, an adaptation of the recommended ABMA fitting practice has been made for the convenience of designers.)

Special Materials for Bearings. The standard materials for ball and roller bearings are usually ANSI 52100 or equivalent "ball bearing steel" or case carburized steels. Industrial demands for special bearings to meet abnormal service requirements spur the continual search for new and improved bearing materials. High temperatures, corrosive atmospheres, massive size, marginal lubrication, complex design, and space and weight limitations are typical abnormal requirements.

Conventional bearing steels are often inadequate when these problems are present. Sustained high operating temperatures reduce hardness, wear resistance, yield strength, and, therefore, bearing life. Conventional bearing steels also lack resistance to the oxidation that takes place at elevated temperatures.

Materials such as 440-C stainless or corrosion-resistant coated steels may be required for severely adverse environments. For extremely high speeds and high temperatures, special alloy steels and materials such as metallic carbides and ceramics are used.

The combination of bearing size, complexity of design, and space and weight limitations can be a governing factor in the selection of bearing material. For example, a large-diameter, thin-section bearing with integral gear teeth and bolt holes would require a material that could be selectively hardened.

Monel and beryllium copper are not as hardenable as bearing steels and are nonmagnetic and resist saltwater corrosion. These qualities make them excellent materials for marine applications.

Although low-carbon steels are the most popular, a variety of other materials is also used for bearing cages. Molded nylon, synthetic resin-impregnated fabrics, and bronze/brass are popular in the normal temperature ranges. High-performance polymers, carbon steel, certain stainless steels, and iron-silicon bronze are used for higher temperatures.

Mounting Design

Mounting design varies widely, depending upon the type of bearing used and the requirements of the application. Selection of bearing type and mounting design are closely related since many cases exist where selection of bearing type is influenced by mounting design considerations.

Most applications require the use of more than one bearing on a shaft. Two identical bearings or a combination of different types and sizes may be used on a common shaft. The advantages of each combination should be evaluated by the designer in selecting bearings.

A *fixed*-bearing mounting locates the shaft and carries any thrust loads that exist in the application. A *float*-bearing mounting accommodates relative axial movement between the shaft and housing. Various combinations of these two basic mountings are used:

1. Fixed-float mounting
2. Fixed-fixed (or opposed) mounting
3. Float-float mounting

The fixed and float arrangement may be necessary: (1) when thermal expansion or contraction of the shaft with respect to the housing occurs, (2) when separate housings are required for two or more bearings on a common shaft, or (3) to optimize bearing loading.

The fixed and float combination offers many desirable features for heavy industrial equipment. The upper half of Fig. 5.38 shows a typical arrangement. The axial location is accomplished through the fixed bearing since it is clamped rigidly to the shaft and the housing. A two-row tapered roller bearing is shown in the fixed position. However, other radial bearings capable of taking thrust loads in both directions and applied radial loads may be used. For the float position the cylindrical roller bearing shown supports relatively heavy loads and permits free axial displacement. A type TDO or TNA tapered roller bearing or a spherical roller bearing may be used if the sliding pressures between the outer ring and the housing bore are not excessive.

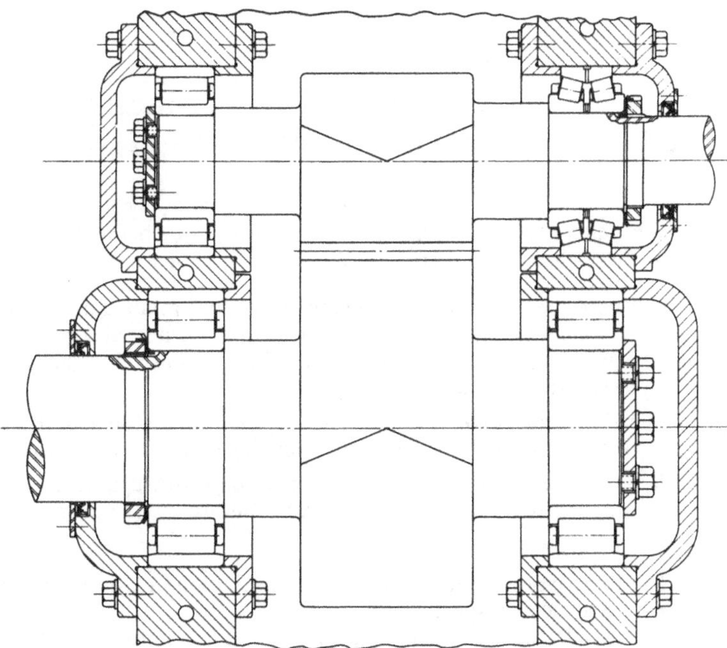

FIGURE 5.38 Typical fixed- and float-bearing arrangements.

The lower half of Fig. 5.38 demonstrates the float-float mounting. A pair of cylindrical roller bearings permits the entire shaft to float. A herringbone gear is used and the floating shaft allows the gears to mesh properly. The arrangement shown permits manufacturing economies since gear axial alignment is assured without very accurate machining or shimming of the bottom housing cover and end plate. A pair of type TDO or TNA tapered roller bearings or a pair of spherical roller bearings, not fixed axially in their housings, may be used if the sliding pressures are not excessive.

Figure 5.39 illustrates a typical mounting arrangement for a rotating housing where the bearings will be subjected to a radial load with a requirement for axial rigidity. A gap is left between the clamping ring and the face of the inner ring to provide for tolerance accumulation. Use of a spacer between the faces of the outer rings allows through-boring of the housing. Although maximum-capacity (types BH or BIH) ball bearings are shown, Conrad (types BC or BIC) ball radial bearings could also be used when there are moderate loads. Applications such as flywheels or spur gear hubs require a tight fit of the outer ring in the housing with a loose fit of the inner ring on the shaft.

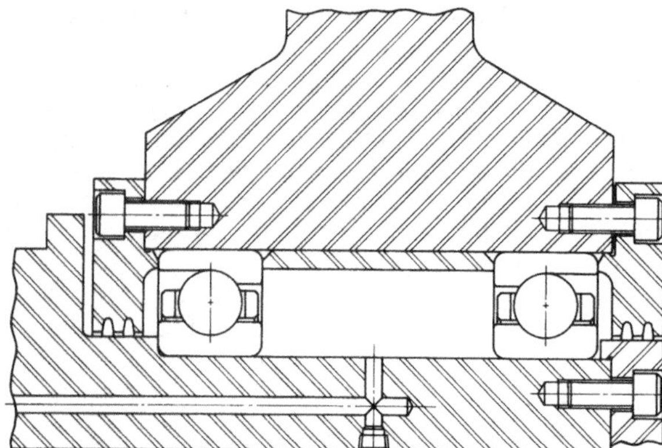

FIGURE 5.39 Typical fixed-fixed bearing arrangement.

Seals and Closures. Seals are used to protect the bearing from contamination as well as to retain the lubricant. There are three basic types of seals:

1. *Contact seals.* The most common types for prelubricated bearings are integral rubber lip or felt contact seals. For open-type bearings, external commercial cartridge seals or packings are mounted outboard of the bearings. These are usually standard components made by several manufacturers.
2. *Shields* and *labyrinth seals.* These are noncontact seals with slight clearance between stationary and rotating members depending on the lubricant to effect a frictionless closure.
3. *Slinger seals.* External types depend on centrifugal force to fling foreign matter away from the shaft. Internal slinger-type seals are used to distribute lubricant within the housing.

The basic types of integral bearing seals and end closure arrangements are shown in Figs. 5.40*a* and 5.40*b*, respectively.

Types DD and LL Integral shields and labyrinth seals. These constructions are usually pressed-steel members, which are designed with close clearances between the fixed and rotating parts of the bearing. Both designs provide lower bearing torques than contact seals and are usable for high-speed and high-temperature applications.

Type PP Integral contact seals. These constructions usually consist of a synthetic rubber lip seal or a felt seal held by a pressed-steel member that is fixed in the bearing outer ring. The rubber lip seal is a popular design for prelubricated standard-width ball bearings.

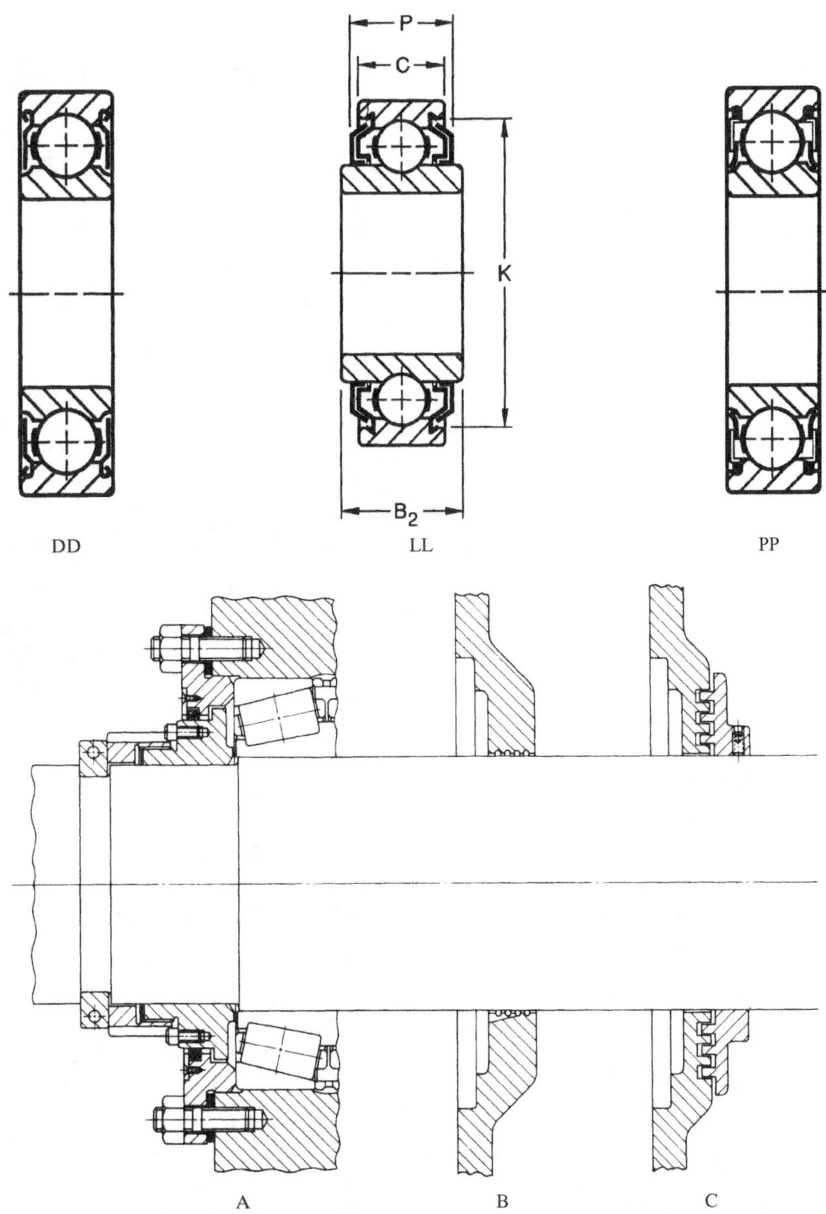

FIGURE 5.40*a* Integral seals and shields.

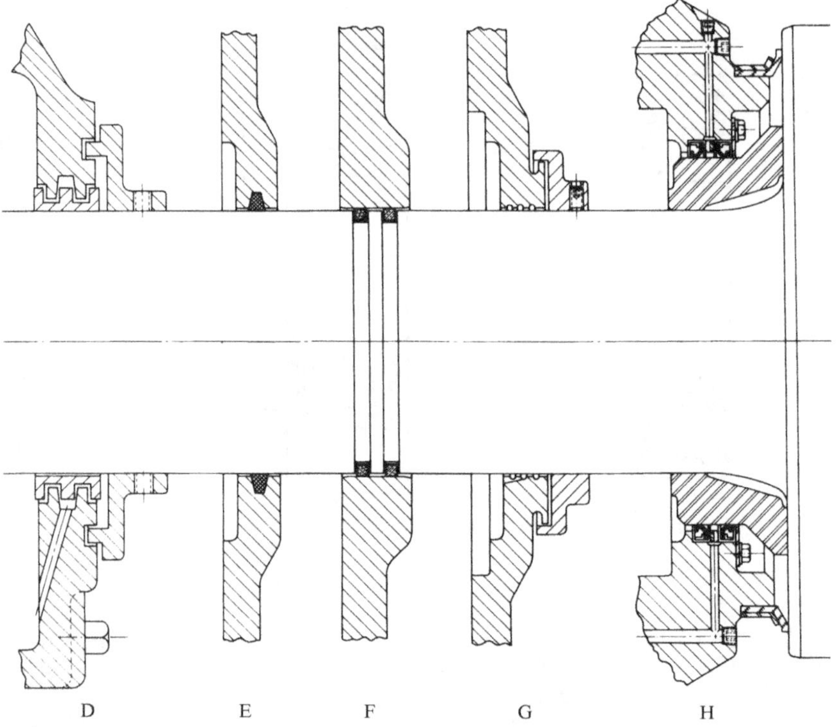

FIGURE 5.40*b* Basic sealing arrangements.

Type A Commercial seals. The contact lip may be synthetic rubber or leather, spring-backed for more positive sealing. Consult manufacturers' literature for shaft finish and limiting speeds. May be used for either grease or oil.

Type B Annular grooves. Shown here with drain slot at bottom, these may be in either the shaft or the housing, or in both. The effectiveness is increased by keeping the running clearance small and by using multiple grooves. Used for oil or grease lubrication.

Type C Axial labyrinth seals. Do not require a split housing. Clearance must be allowed for axial movement. Effective for abrasive environment. Use for oil or grease.

Type D Radial labyrinth seals. Used with a split housing or end cap. Bore of grooved sealing ring is slightly larger than shaft, allowing it to float axially. Angular surface of housing groove reduces pumping action. Suitable for oil or grease.

Type E Felt seals. Provide medium effectiveness at low speeds but lose their effectiveness at high speeds and high temperatures. In most cases, it functions as a contact seal, but after "wearing in" often functions as a simple close-clearance annulus seal. Not suitable for an abrasive environment. Use for grease lubrication.

Type F Piston ring seals. Modification of the labyrinth seal. The piston rings are stationary and are mounted under radial compression in the housing. The split rings have rabbeted joints. Clearances between the rings and grooves are slight. This seal accommodates axial displacement. It is easy to install or remove, and can be used with a spacer or in shaft grooves as shown. Accepting deflection and misalignment, it offers a relatively positive closure augmented by a grease annulus. A particularly effective seal for an abrasive environment, it is suitable for grease and oil.

Type G Combination annular groove-axial labyrinth seals. Annular grooves retain lubricant in the housing. The external flinger serves as a shield and flings contaminants away from the seal. It is suitable for grease and oil.

Type H Triple combination of lip contact seals. Two commercial seals are mounted and opposed with a spacer in between to allow relubrication of the contact surfaces. A face contact seal is also used to prevent the entrance of contaminants. It is used primarily for grease.

PHYSICAL DESCRIPTION

Ball Bearings

Radial Ball Bearings. There are three major types of radial ball bearings with a metric bore range of 10 to 320 mm and an inch bore range of 5⁄32 in to 40 in normally listed in bearing catalogs.

Radial
- Conrad (see Fig. 5.41)
- Maximum capacity (loading groove) (see Fig. 5.42)

Angular contact
- Single row (see Fig. 5.43)

Wide inner
- Locking collar (see Fig. 5.44*a* and 5.44*b*)
- Setscrew (see Fig. 5.44*c*)

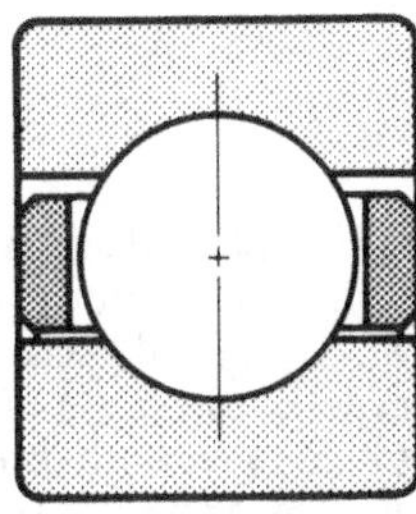

FIGURE 5.41 Conrad-type ball bearing.

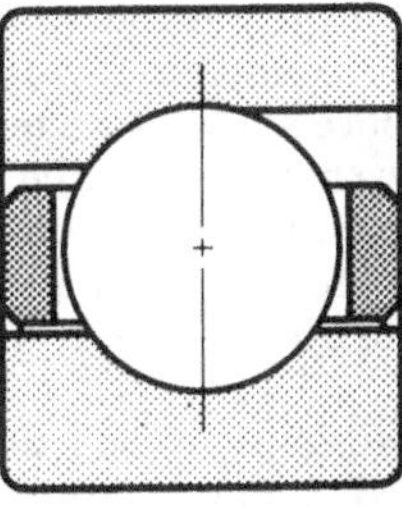

FIGURE 5.42 Maximum-capacity-type ball bearing.

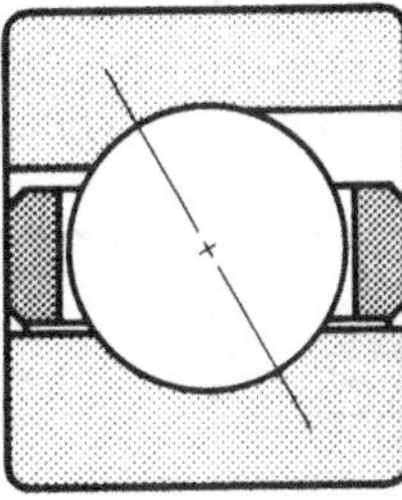

FIGURE 5.43 Angular-contact-type ball bearing.

Selection of the proper type of ball bearing is determined by consideration of the direction and magnitude of the bearing load. Ball bearings are particularly suited to high-speed operation because they have a lower coefficient of friction than roller bearings.

Figure 5.41 shows the Conrad, or deep-groove radial, ball-bearing type. Although it is primarily a radial bearing, it is capable of handling moderately high thrust loads from either direction and operating at relatively high speeds. The bearing rings have symmetrical, deep-grooved raceways without ball-loading grooves or a counterbore. The raceways are precision-ground to conform closely to ball curvature, consistent with minimum friction, maximum capacity, and practical manufacturing techniques. Balls are selected for uniformity to ensure optimum internal load distribution. The bearing utilizes the maximum number of balls which can be inserted between the raceways by eccentrically displacing the inner and outer rings. Balls are spaced by a cage (a ball-piloted cold rolled steel design is shown here). For higher speeds, and when other

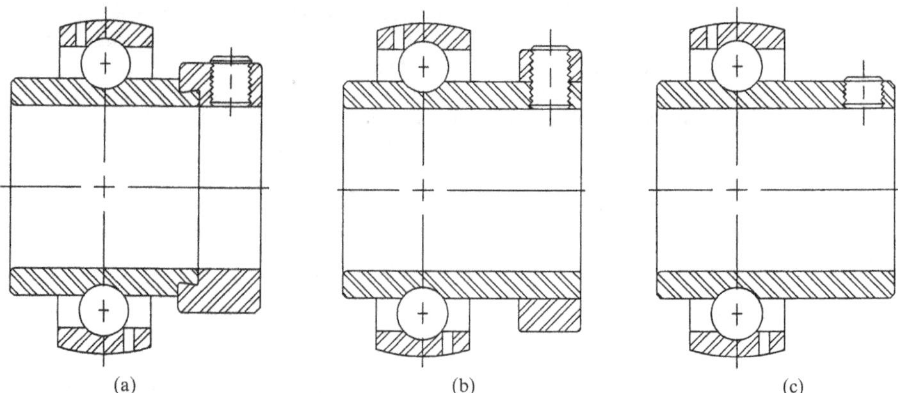

FIGURE 5.44 Bearing with (*a*) eccentric locking collar, (*b*) concentric locking collar, (*c*) set-screws in the inner ring.

operating conditions warrant, the cage may be ring-piloted and made of other materials such as bronze or polymers. The nonseparable bearing construction facilitates handling.

The maximum-capacity ball bearing shown in Fig. 5.42 has the greatest radial capacity obtainable in a single-row ball bearing with cage, but because of the ball-loading grooves in the inner and outer rings, it has limited thrust loading in both directions. This type of bearing may be identical to a deep-groove radial (Conrad) bearing, but the ball-loading grooves allow an increased number of balls to be inserted into the bearing. This increased number of balls results in a higher theoretical radial load rating compared to a comparably sized Conrad bearing, thus leading to the term "maximum capacity."

The design of angular-contact ball bearings, as shown in Fig. 5.43, is similar to the maximum-capacity type, but the outer ring is counterbored; this substantially reduces the raceway shoulder on one side. The counterbore allows a maximum complement of balls in a one-piece, machined-bronze cage to be assembled in the bearing. The outer ring is thermally expanded and slipped over the cage, ball, and inner-ring assembly. After cooling, the bearing is nonseparable. This design allows the bearing to resist heavier thrust loads and minimizes axial deflection under load in one direction only against the full shoulder side of the bearing.

Angular-contact bearings can be furnished in matched pairs for duplex mounting (Fig. 5.45). When the bearings supplied for opposed mounting are placed together, there will be a small gap between the inner rings (type DB) or outer rings (type DF) before clamping. After clamping together, an internal preload is introduced in the set which increases the axial and radial rigidity of the duplexed pair. When supplied for tandem mounting (type DT), very heavy unidirectional thrust loads may be almost equally distributed among the bearings of the set. Details of application for DB and DF pairs should be submitted to the engineering department of the bearing manufacturer for preload recommendations.

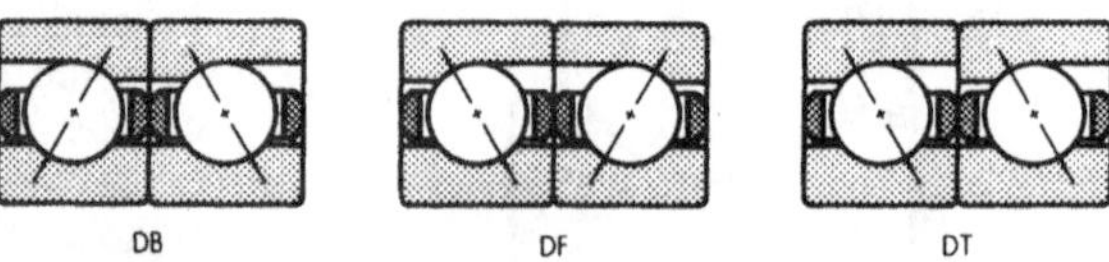

FIGURE 5.45 Duplex mounting of matched-pair angular-contact ball bearings.

Radial Ball Bearings with Locking Devices. The internal constructions of these bearings are basically the same as deep-race, single-row radial types with ability to carry radial, thrust, and combined loads. However, the bearings have a wide inner-ring design that allows them to be slip-fitted and locked on straight shafts without shoulders, lockouts, or adapters. The most widely used locking devices are the cam (eccentric) locking collar, the concentric setscrew locking collar, and setscrews in an extended end of the bearing inner ring (see Fig. 5.44).

These bearings are made with either straight or spherical OD outer rings, are usually prelubricated with grease, and have double integral contact or labyrinth seals or shields.

Radial Ball Bearing Housing Units. Many types of housing units with wide inner-ring bearings are available. The four most common basic housing constructions are pillow blocks, flanged cartridges, cylindrical cartridges, and take-up units (see Fig. 5.46). These units are usu-

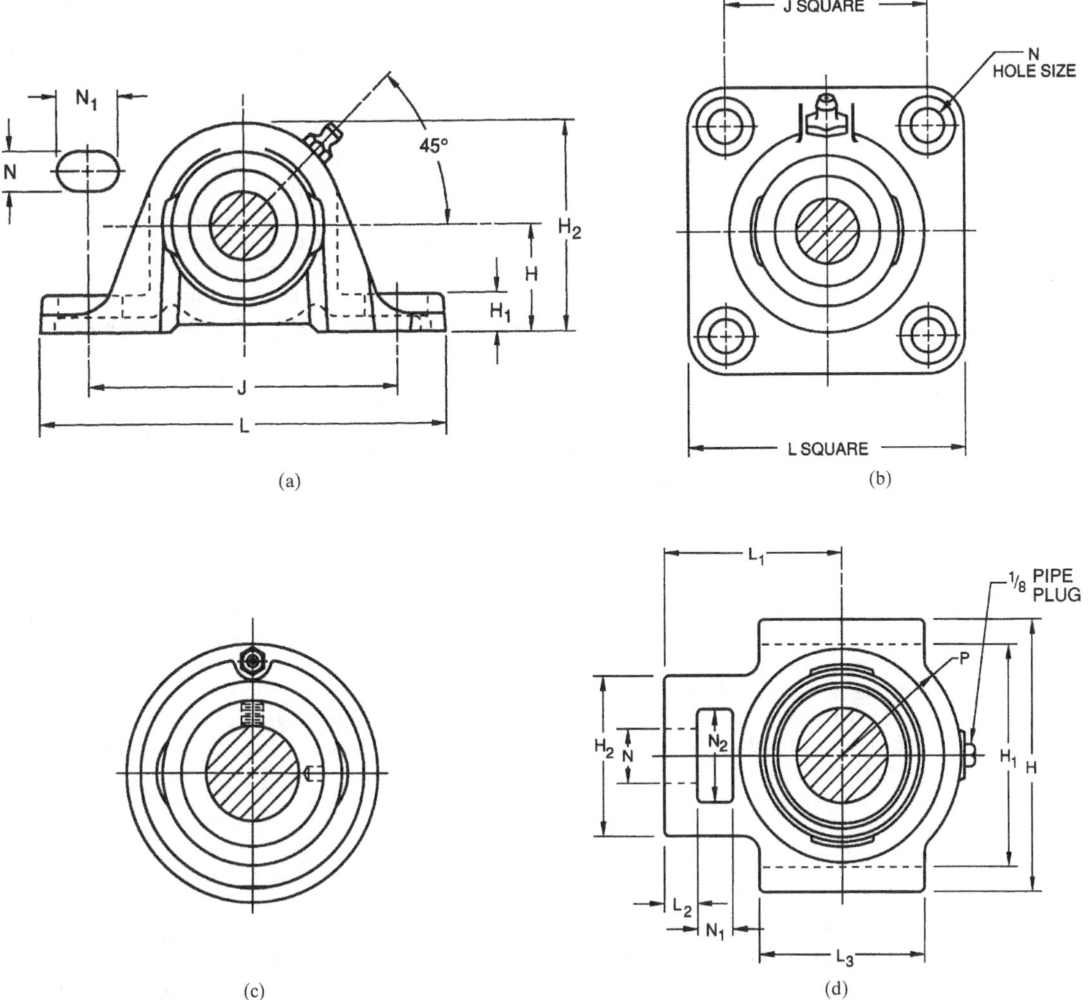

FIGURE 5.46 Ball bearing housing units. (*a*) Pillow block; (*b*) flange cartridge; (*c*) cylindrical cartridge; (*d*) take-up unit.

ally assembled with prelubricated, double-sealed, wide inner bearings that incorporate a self-aligning feature to allow the bearing to be aligned at mounting. The housings are generally made of cast gray or malleable iron, pressed steel, rubber, or other materials and are usually designed to permit bearing relubrication.

Ball Thrust Bearings. Ball thrust bearings are used for lighter loads and higher speeds than roller thrust bearings.

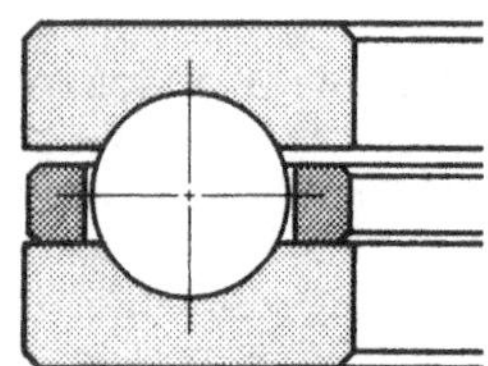

FIGURE 5.47 Ball thrust bearing.

Type TVB ball thrust bearings (Fig. 5.47) are separable, and consist of two hardened and ground steel washers with grooved raceways and a cage that separates and retains precision-ground and -lapped balls. The standard cage material is bronze, but this may vary according to the requirements of the application. Type TVB bearings provide axial rigidity in one direction, so using them to support radial loads is not recommended. They are very easily mounted. Usually the rotating washer is shaft mounted. The stationary washer should be housed with sufficient outside diameter clearance to allow the bearing to assume its proper operating position. In most sizes both washers have the same bore and outside diameter. The housing must be designed to clear the outside diameter of the rotating washer, and it is necessary to step the shaft diameter to clear the bore of the stationary washer.

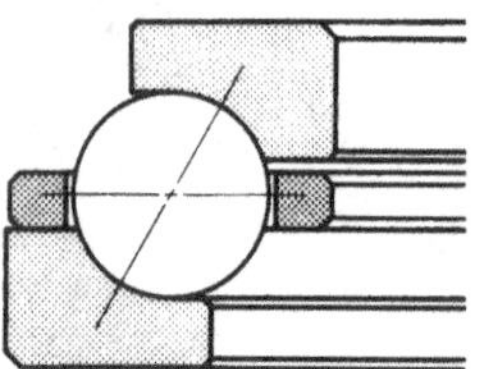

FIGURE 5.48 Angular contact ball thrust bearing.

Type TVL (Fig. 5.48) bearings are separable angular-contact ball bearings designed primarily for unidirectional thrust loads. The angular-contact design, however, will accommodate combined radial and thrust loads since the loads are transmitted angularly through the balls. The bearing has two hardened and ground steel rings with ball grooves and a one-piece bronze cage that spaces the ball complement. Although not strictly an annular ball bearing, the larger ring is called the outer ring, and the smaller the inner ring. Usually, the inner ring is the rotating member and is shaft mounted. The outer ring is normally stationary and should be mounted with outside diameter clearance to allow the bearing to assume its proper operating position. If combined loads exist, the outer ring must be radially located in the housing.

A type TVL bearing should always be operated under thrust load. Normally, this presents no problem as the bearing is usually applied on vertical shafts in oilfield rotary tables and machine-tool indexing tables. If a constant thrust load is not present, it should be imposed by springs or other built-in devices.

Low friction, cool running, and quiet operation are advantages of the type TVL bearing, which may therefore be operated at relatively high speeds. The bearing is also less sensitive to misalignment than other types of rigid thrust bearings.

Roller Bearings

Tapered Roller Bearings. Tapered roller bearings are generally considered to offer the best support for combinations of heavy radial and thrust loads at moderate speeds.

A single-row tapered roller bearing consists of an inner ring called a *cone,* an outer ring called a *cup,* a bronze or steel cage, and a complement of controlled-contour rollers. In multiple-row tapered roller bearings one or more cones, cups, and cage assemblies may be used.

Tapered rollers and raceways are designed on the geometric principle of a cone (Fig. 5.49). Extensions of the lines of contact between the rollers and the raceways all meet at a common point on the axis of the bearing. The design assures true geometric rolling. The large ends of the tapered rollers are spherically ground to match the spherically ground face on the guiding cone rib. Under load, the nominal pressure exerted between these two ground surfaces accurately positions the rollers within the load zone.

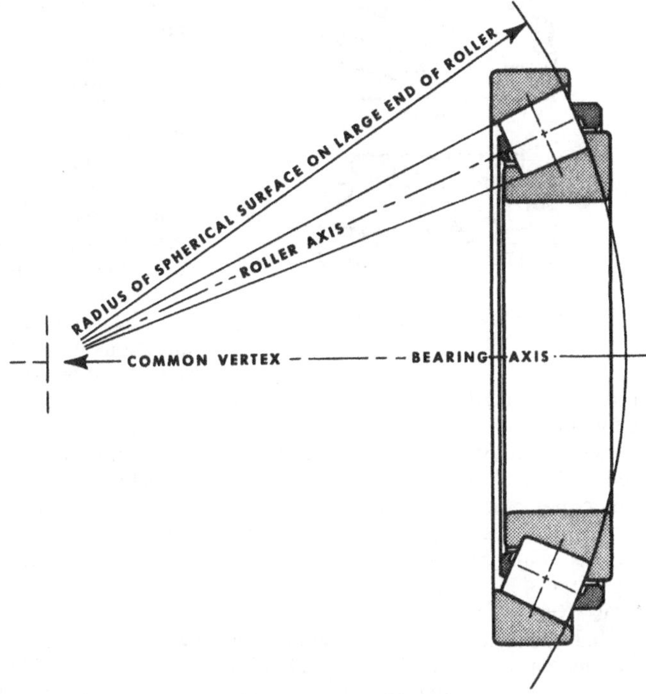

FIGURE 5.49 Geometric principle of tapered roller bearing.

Single-Row Tapered Roller Bearings. There are three types of single-row tapered roller bearings: TS, TSF, and TSS (Fig. 5.50). Each has a cup and a cone with a cage and roller assembly. Type TS serves as the basic design for the others.

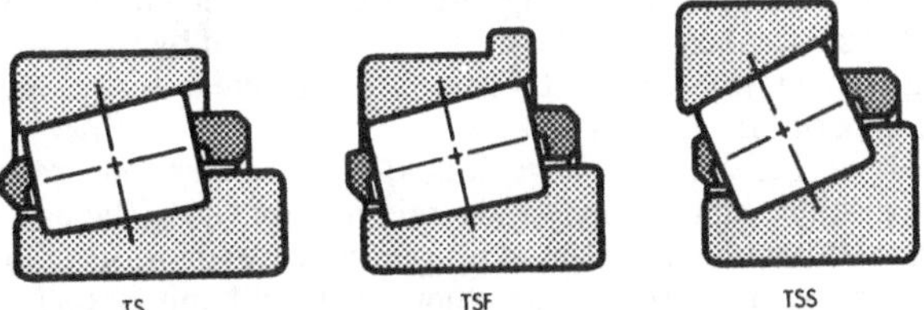

FIGURE 5.50 Single-row tapered bearings.

Since a single-row tapered roller bearing supports thrust loads from one direction only, the preferred mounting is in opposed pairs. The proper internal clearance for the two bearings may be obtained by axial adjustment at the time of assembly.

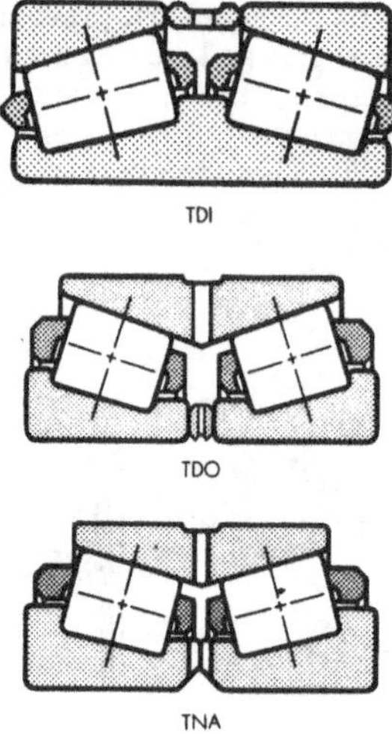

FIGURE 5.51 Two-row tapered roller bearings.

Type TSF bearings are identical with type TS except that the cup of the TSF type incorporates an external flange which, in some mountings, facilitates location and permits economies in design. When through-boring of the housing is advantageous, the use of the type TSF bearing is suggested.

Type TSS bearings are similar to type TS, but have a steeper angle of contact. These bearings are recommended for applications where the thrust load is predominant.

Two-Row Tapered Roller Bearings. Three basic types of two-row tapered roller bearings are available: TDI, TDO, and TNA (Fig. 5.51). Two-row tapered roller bearings have twice the radial capacity of single-row bearings of the same series and are used in positions where radial loading is too severe for single-row bearings. They have the further advantage that a two-row bearing can take thrust loads in both directions, thus allowing all applied thrust and shaft location to be taken at one position. This often simplifies design and reduces the danger of bearing clearance changes due to axial shaft expansion.

Similar to single-row tapers, steep-angle versions offer greater thrust capacity with reduced radial capacity. These are widely used as backup thrust bearings in rolling mills and other applications where heavy thrust loads are encountered.

The TDI design is such that the contact angles converge as they approach the axis of rotation (see Fig. 5.51). Consequently, use of these bearings will not appreciably increase the rigidity of the shaft mounting, and they should not be used singly on a shaft since they will not resist overturning moments.

In the TDO and TNA, the contact angle lines diverge as they approach the axis of rotation, thus increasing the rigidity of the shaft mounting (see Fig. 5.51). Therefore, these bearings are suited for resisting overturning moments. Due to the increased bearing rigidity, housing bore alignment is somewhat more critical than with TDI styles.

Cylindrical Roller Bearings. Six standard types of cylindrical roller bearings are shown in Fig. 5.52. All six types have the same roller complements and, consequently, the same capacity for a given size. All types can be mounted with interference fits on either the inner or outer ring, or both. In the latter case, a bearing with increased internal clearance must be specified to provide proper running clearance.

For convenience, bearings are listed according to bore, with both metric and inch bearings in the same figure. Inch bearings are identified by the letter "I" in the type code of the bearing number; thus where RN denotes a particular type of metric bearing, RIN denotes an inch bearing of the same type.

Types RU and RIU have double-ribbed outer and straight inner rings. Types RN and RIN have double-ribbed inner and straight outer rings. The use of either type at one position on a shaft is ideal for accommodating nominal expansion or contraction. The relative axial displacement of one ring to the other occurs with minimum friction while the bearing is rotating. These bearings may be used in two positions for shaft support if other means of axial location are provided.

Types RJ and RIJ have double-ribbed outer and single-ribbed inner rings. Types RF and RIF have double-ribbed inner and single-ribbed outer rings. Both types can support heavy radial loads, as well as light unidirectional thrust loads up to 10 percent of the radial load. The thrust load is transmitted between the diagonally opposed rib faces in a sliding action rather

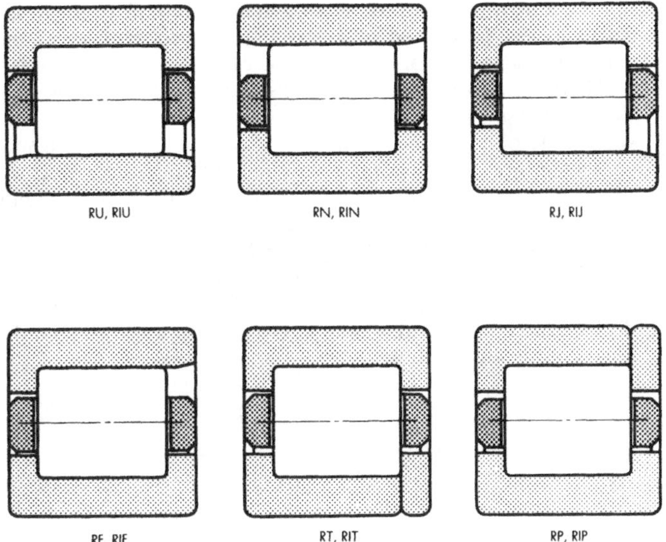

FIGURE 5.52 Standard types of cylindrical roller bearings.

than a rolling action. Thus, when limiting thrust conditions are approached, lubrication can become critical. When thrust loads are very light, these bearings may be used in an opposed mounting to locate the shaft. In such cases, shaft end play should be adjusted at time of assembly.

Types RT and RIT have a double-ribbed outer ring and a single-ribbed inner ring with a loose rib, which allows the bearing to provide axial location in both directions. Types RP and RIP have a double-ribbed inner ring and a single-ribbed outer ring with a loose rib.

Types RT and RP (RIT and RIP) provide heavy radial capacity and light thrust capacity in both directions. Factors governing the thrust capacity are the same as for RF and RJ (RIF and RIJ) bearings.

A type RT or RP (RIT or RIP) bearing may be used in conjunction with a type RN or RU (RIN or RIU) bearing for applications where axial shaft expansion is anticipated. In such cases the fixed bearing is usually placed nearest the drive end of the shaft to minimize alignment variations in the drive. Shaft end play (or float) is determined by the axial clearance in the bearing.

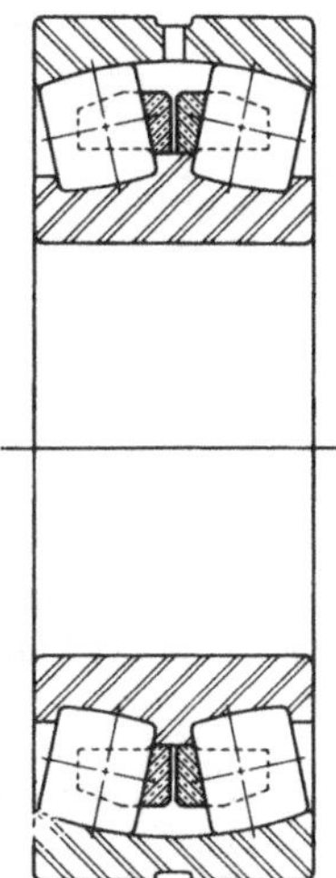

FIGURE 5.53 Self-aligning spherical roller bearing.

Spherical Roller Bearings. The self-aligning spherical roller bearing (Fig. 5.53) is a combination radial and thrust bearing designed for taking misalignment under load. When loads are heavy, alignment of housings difficult, and shaft deflections excessive, the use of spherical roller bearings assures best service life results.

With the spherical rollers operating on the spherically shaped outer race, the assembly of the inner ring, retainers, and rollers may take up to ±1½° of misalignment and continue to function with full capacity. High radial load capacity is secured by a large area of roller-to-race contact. Double-direction thrust capacity results from angular location of rollers relative to bearing axis.

Shaft deflections and housing distortions caused by shock loads are compensated for by the internal self-alignment of the free-rolling bearing elements. The binding stresses that limit service life of non-self-aligning bearings cannot develop in spherical roller bearings. Various cage styles are used, depending on the operating parameters and bearing manufacturer. Unit design and construction make the spherical roller bearing simple to handle at assembly point or during removal for maintenance.

A disassembled view of a spherical roller bearing pillow block is shown in Fig. 5.54. This consists of the pillow block, bearing, adapter, locknut, lockwasher, stabilizing ring, and triple labyrinth seals. The split design allows for ease of inspection and installation. Housings are designed to accommodate either grease or oil lubrication and are easily adapted to circulating oil lubrication systems.

FIGURE 5.54 Spherical roller bearing pillow block.

Because the bearings are self-aligning, shaft misalignment is not considered critical. Standard split pillow blocks are designed to provide static and dynamic alignment. A variety of sealing arrangements are available in addition to the labyrinth seals for more demanding requirements.

Roller Thrust Bearings. Cylindrical roller thrust bearings withstand heavy loads at relatively moderate speeds. Special design features can be incorporated into the bearing and mounting to attain higher operating speeds. Because loads are usually high, extreme pressure lubricants should be used with roller thrust bearings. Preferably, the lubricant should be introduced at the bearing bore and distributed by centrifugal force.

The type TP cylindrical roller thrust bearing (Fig. 5.55) has two hardened and ground steel washers, with a cage retaining one or more rollers in each pocket. When two or more rollers are used in a pocket they must be of different lengths and be placed in staggered positions in adjacent cage pockets to create overlapping roller paths. This prevents wearing grooves in the raceways and prolongs bearing life.

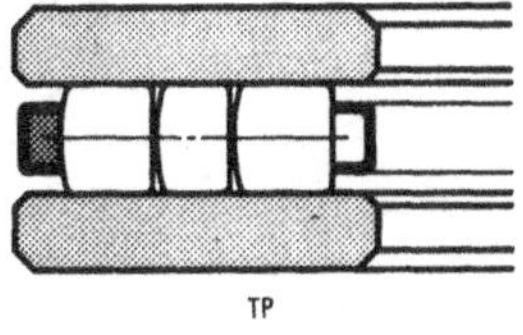

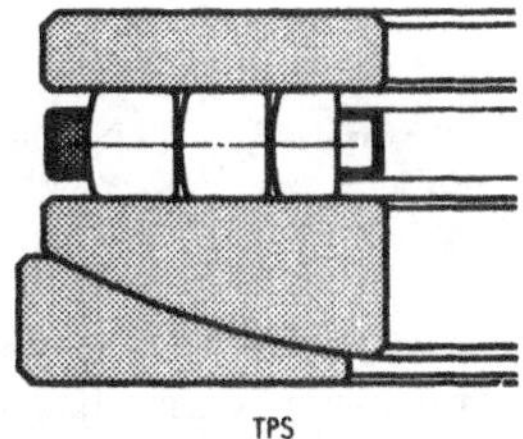

FIGURE 5.55 Cylindrical roller thrust bearings.

Because of the simplicity of their design, TP bearings are economical. Since minor radial displacement of the raceways does not affect the operation of the bearing, its application is relatively simple and often results in manufacturing economies for the user. Shaft and housing seats, however, must be square to the axis of rotation to prevent initial misalignment problems.

The type TPS bearing (Fig. 5.55) is the same as type TP, except one washer is spherically ground to seat against an aligning washer, thus making the bearing adaptable to initial misalignment. Its use is not recommended for operating conditions where alignment is continuously changing (dynamic misalignment).

The type TTHD tapered roller thrust bearing (Fig. 5.56) has an identical pair of hardened and ground steel washers with conical raceways, and a complement of tapered rollers equally spaced by a cage.

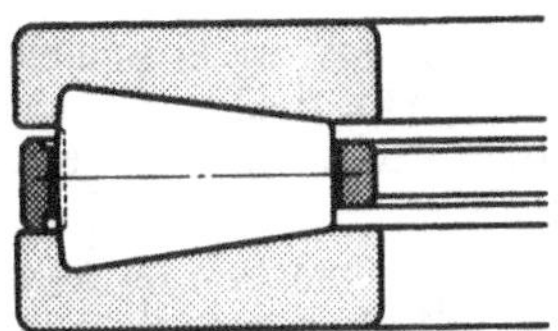

FIGURE 5.56 Tapered roller thrust bearing.

TTHD bearings are well suited for applications such as crane hooks where extremely high thrust loads and heavy shock must be resisted and some measure of radial location obtained. For very low speed, extremely heavily loaded applications, these bearings are supplied with a full complement of rollers for maximum capacity.

The TSR spherical roller thrust bearing design (Fig. 5.57) achieves a high thrust capacity with low friction and continuous roller alignment. The bearings can accommodate pure thrust loads as well as combined radial and thrust loads. This design allows higher speeds than do any of the other roller thrust bearing designs. Typical applications are air regenerators, centrifugal pumps, and deep-well pumps.

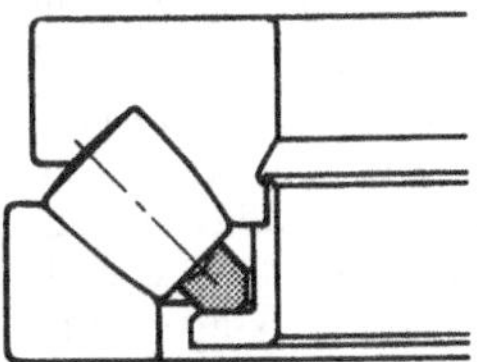

FIGURE 5.57 Spherical roller thrust bearing.

Because the spherical roller thrust bearing must always carry some thrust load, it cannot be used as a float bearing. Most applications require the use of two or more bearings on a shaft: a fixed bearing and a float bearing.

Needle Roller Bearings. Size for size, needle bearings (Fig. 5.58) have more rollers and more lines of contact, and thus generally have higher capacities than other roller bearings. This is particularly true under static, slow-rotating, or oscillating conditions. Both types have a far higher capacity in less radial space than other antifriction bearings.

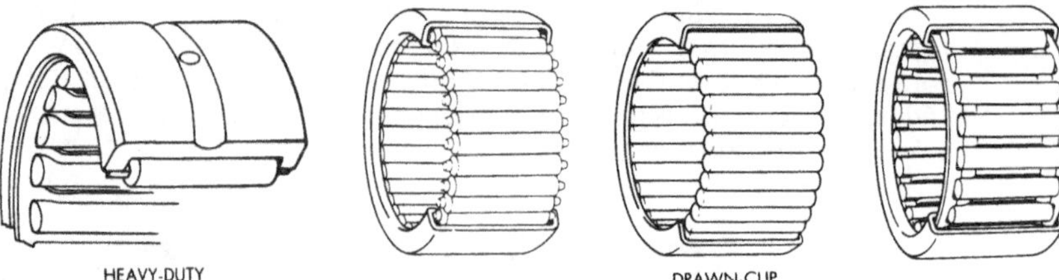

FIGURE 5.58 Needle roller bearings.

The heavy-duty bearing has an outer race that is deeply hardened, while the race of the drawn-cup bearing has a thin, hardened case over a relatively ductile core. Thus, the heavy-duty bearing can withstand heavier shock or more continuous loads than the drawn-cup bearing, which should never be dynamically loaded, even momentarily, beyond the load limit given in the manufacturer's tabular data. The smaller, more compact cross section is provided by the drawn-cup bearing with its smaller roller diameter and thin-steel outer race, as opposed to the larger roller diameter and thick outer race of the heavy-duty bearing.

For moderate speeds, both types are satisfactory. However, the cage in the roller bearing allows higher speeds for a given shaft size.

MOUNTING AND MAINTENANCE

General Installation Rules

Depending on the size of bearing and the application, there are different methods for mounting bearings. In all methods, however, certain basic rules must be observed.

Maintain Cleanliness. Choose a clean environment. Work in an atmosphere free from dust or moisture. If this is not possible (and sometimes in the field it isn't), the installer should make every effort to ensure cleanliness by the use of protective screens, clean cloths, and the like.

Plan the Work. Know in advance what you are going to do and have all necessary tools at hand. This reduces the amount of time for the job and lessens the chance for dirt to get into the bearing.

Preparation and Inspection. All component parts of the machine should be on hand and thoroughly cleaned before proceeding. Housings should be cleaned, even to blowing out the oil holes. *Do not use an air hose on bearings.* If blind holes are used, insert a magnetic rod to remove metal chips that might have become lodged during fabrication.

Shaft shoulders and spacer rings contacting the bearing should be square with the shaft axis. The shaft fillet must be small enough to clear the radius of the bearing.

On original installations, all component parts should be checked against the detailed specification prints for dimensional accuracy. Shaft and housing should be carefully checked for size and roundness.

Mounting Straight-Bore Bearings

Heat-Expansion Method. Most applications require a tight interference fit on the shaft. Mounting is simplified by heating the bearing to expand it sufficiently to slide easily onto the shaft. Two methods of heating are in common use:

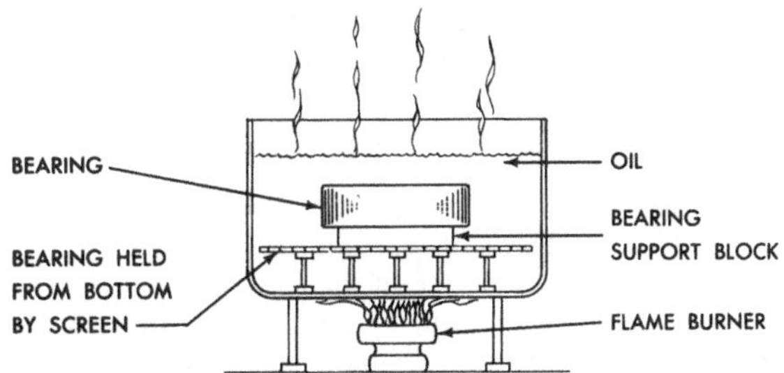

FIGURE 5.59 Typical arrangement for heat expansion method.

1. Immersion in a tank of heated oil (Fig. 5.59)
2. Induction heating

The first is accomplished by heating the bearing in a tank of oil having a high flash point. The oil temperature should not exceed 250°F (120°C). A temperature of 200°F (93°C) is sufficient for most applications. The bearing should be heated at this temperature, generally for 20 or 30 min, until it is expanded sufficiently to slide onto the shaft very easily.

The induction-heating method is particularly suited for mounting small bearings in production-line assembly. Induction heating is rapid, and care must be taken to prevent bearing temperature from exceeding 200°F (93°C). Trial runs with the unit are usually necessary to obtain the proper timing. Thermal crayons which melt at predetermined temperatures can be used to check the bearing temperature.

While the bearing is still hot, it should be positioned squarely against the shoulder. Lockwashers and locknuts or clamping plates are then installed to hold the bearing against the shoulder of the shaft. As the bearing cools, the locknut or clamping plate should be tightened.

The oil bath is shown in Fig. 5.59. The bearing should not be in direct contact with the heat source. The usual arrangement is to have a screen several inches off the bottom of the tank. Small support blocks separate the bearing from the screen. It is important to keep the bearing away from any localized high-heat source that may raise its temperature excessively, resulting in race hardness reduction.

Flame-type burners are commonly used, but may have to be replaced with some other heat source to comply with local safety regulations. An automatic device for temperature control is desirable. If safety regulations prevent the use of an open heated-oil bath, a mixture of 15 percent soluble oil in water may be used. This mixture may be heated to a maximum temperature of about 200°F (93°C) without being flammable. The bath should be checked from time to time to ensure its proper composition as the water evaporates. The bath leaves a thin film of oil on the bearing sufficient for temporary rust prevention, but normal lubrication should be supplied to the bearing as soon as possible after installation. Be sure all of the oil-in-water solution has been drained away from the bearing.

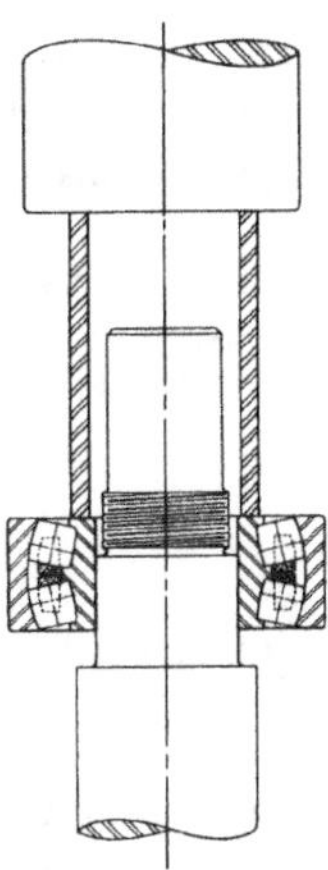

FIGURE 5.60 Arrangement for typical arbor press mounting.

Arbor Press Method. The alternative method of mounting, generally used only on smaller sizes (Fig. 5.60), is to press the bearing onto the shaft or into the housing. This can be done by using an arbor press and a mounting tube. The tube can be of soft steel.

To press-fit the bearing on its shaft seat, the inside diameter of the tube should be slightly larger than the shaft. The outside diameter of the tube should not exceed the maximum shoulder height given in the tables of dimensions. The tube should be faced square at both ends, thoroughly clean inside and out, and long enough to clear the end of the shaft after the bearing is mounted.

If the outer ring is being pressed into the housing, the outside diameter of the mounting tube should be slightly smaller than the housing bore, and the inside diameter should not be less than the recommended housing-shoulder diameter in the tables of dimensions.

Coat the shaft or housing bore with light machine oil to reduce the force needed for the press fit. Carefully place the bearing on the shaft, with a smooth blended OD corner, making sure it is square with the shaft axis. Apply steady pressure from the arbor ram to drive the bearing firmly against the shoulder. Never attempt to make a press fit on a shaft by applying pressure to the outer ring, or to make a press fit in a housing by applying pressure to the inner ring, as internal bearing damage (race brinelling) may result.

Straight-Bore-Bearing Removal

Bearing pullers of various types and designs are available from several manufacturers. These pullers are useful in removing bearings up to about 10-in (25-cm) bore. The preferred method utilizes a split ring which is placed behind the inner ring. The pulling device is assembled so as to cause the split ring to push the bearing off the shaft. If machine elements interfere with this method, a prong-type puller may be used with a roller bearing that has side flanges on the inner ring. Insert the prongs of the puller behind the flange and pull the bearing off the shaft.

Hydraulic Method. Remove lockplate or locknut. Provide a means for pulling the bearing off the shaft. A bearing puller applied to the outer ring may be used without fear of brinelling the races since only a nominal force is required for removal by this system. Or, the shaft may be placed in a vertical position, simply allowing gravity and slight manual force to remove the bearing. Pull the bearing all the way off the shaft quickly to prevent freezing after the loss of oil pressure as the bearing clears the oil groove.

Mounting Bearings in Housings

To facilitate installing the bearing in the housing and to minimize fretting corrosion during operation, coat the housing with a light-grade machine oil. Make sure the outer ring is square with the housing bore before inserting it. If the outer ring becomes misaligned and sticks, do not attempt to force it farther into the housing. Use a soft brass or steel bar and tap the outer ring until it becomes free and is realigned. An assembly plate for holding the outer ring in alignment with the inner ring is often useful for applications in which conditions tend to result in misalignment.

In case of outer-ring rotation, where the outer ring is a tight fit in the housing, the housing member can be expanded by heating. Split housings present no alignment problems during installation, but the housings should be checked carefully for size and roundness.

Internal Clearance

Internal clearance is the amount of radial play within a bearing. This clearance accommodates the effects of tight shaft and housing fits, radial thermal expansion, speed, and other operating conditions.

The internal clearance in radial bearings other than tapered roller bearings is defined as *diametral clearance.* The diametral clearance is the amount of radial play built into the bearings (see Figs. 5.61 and 5.62).

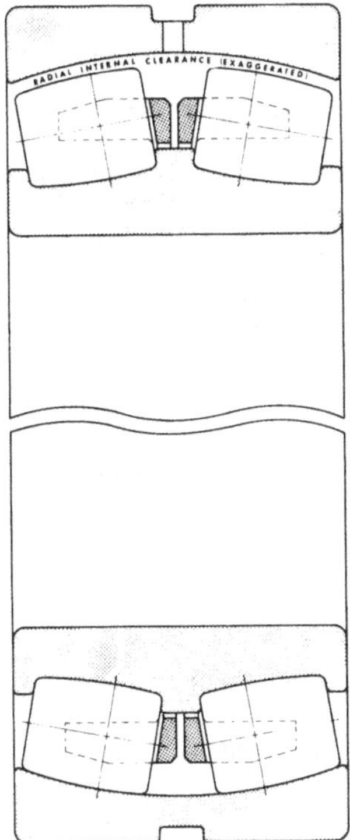

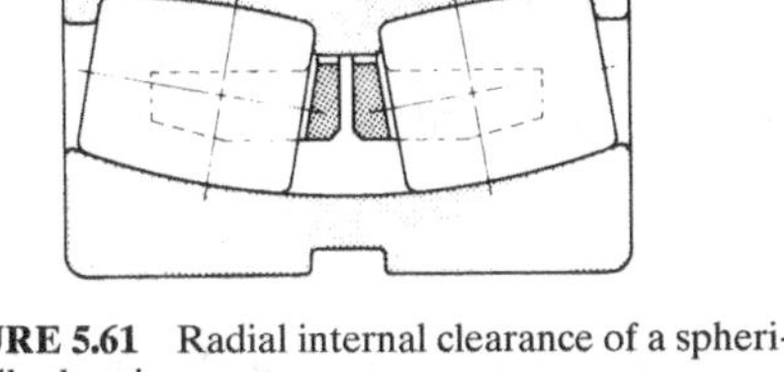

FIGURE 5.61 Radial internal clearance of a spherical roller bearing.

FIGURE 5.62 Radial internal clearance of a cylindrical roller bearing.

Straight-bore or tapered-bore bearings installed with interference fits will result in a reduction of internal or diametral clearance. Several factors influence the reduction of internal clearance. When the bearing inner ring is pressed onto a solid steel shaft, reduction of internal clearance is approximately 80 percent of the shaft interference fit for a standard series radial bearing. For an outer ring pressed into a heavy-section steel or cast-iron housing, the reduction of internal clearance is about 60 percent of the housing-interference fit. If the shaft is hollow, the housing walls are thin, or materials other than steel are used, clearance reduction may vary considerably from the above percentages.

Tapered-bore spherical roller bearings require a slightly greater interference fit with the shaft than straight-bore bearings of corresponding size. The resultant fit cannot conveniently be determined by direct shaft measurement. It can be checked by the distance the bearing is pressed onto a carefully gauged tapered shaft, or more easily by the effects of radial expansion of the inner ring as it is pushed onto the taper.

Lubrication (See Also Chap. 5.12)

Since the lubricant affects bearing life and operation, selecting the proper lubricant is an important design function. The purpose of lubrication in bearing applications is to:

1. Minimize friction at points of contact within the bearings
2. Protect the highly finished bearing surfaces from corrosion

3. Dissipate heat generated within the bearings
4. Remove foreign matter or prevent its entry into the bearings

Two basic types of lubricants used with antifriction bearings are oils and greases. Each has its advantages and limitations.

Since oil is a liquid, it lubricates all surfaces and is able to dissipate heat from these surfaces more readily. Because oil retains its physical characteristics over a wider range of temperatures, it may be used for high-speed and high-temperature applications. The quantity of oil supplied to the bearing may be accurately controlled. Oil lubricants can be circulated, cleaned, and cooled for more effective lubrication.

Grease, which is easier to retain in the bearing housing, aids as a sealant against foreign matter and corrosive fumes.

Oil Lubrication

1. Oil is a better lubricant for high speeds or high temperatures. It can be cooled to help reduce bearing temperature.
2. Oil is easier to handle, and with oil it is easier to control the amount of lubricant reaching the bearing. It is harder to retain in the bearing. Lubricant losses may be higher than with grease.
3. As a liquid, oil can be introduced into the bearing in many ways, such as drip feed, wick feed, pressurized circulating systems, oil bath, or air-oil mist. Each is suited to certain types of applications.
4. Oil is easier to keep clean for recirculating systems.

Grease Lubrication

1. This type of lubrication is restricted to lower-speed applications within the operating-temperature limits of the grease.
2. Grease is easily confined in the housing. This is important in the food, textile, and chemical industries.
3. Bearing enclosure and seal design is simplified.
4. Grease improves the efficiency of mechanical seals to give better protection to the bearing.

For all new applications, a competent bearing or lubrication engineer should be consulted to recommend the specific lubricant and method of lubrication for the specific bearing's operating and ambient conditions.

BIBLIOGRAPHY

For additional or more detailed information on a specific bearing type or application, refer to the bearing manufacturers' catalogs and/or one of the following:

ABMA Standards, The American Bearing Manufacturers Association, Inc.*

SKF Industries, Inc., Front Street and Erie Avenue, P.O. Box 6731, Philadelphia, PA 19132.

The Timken Company, 1835 Dueber Avenue S.W., Canton, OH 44706.

The Torrington Company, 59 Field Street, Torrington, CT 06790.

*Recently changed from the Anti-Friction Bearing Manufacturers Association (AFBMA).

CHAPTER 5.6

MECHANICAL EQUIPMENT: SHAFT DRIVES AND COUPLINGS

PART 1

FLEXIBLE SHAFTS

B. W. Elliott Manufacturing Co., LLC
Binghamton, New York

INTRODUCTION

Flexible shafting is a direct, mechanical method for transmitting rotary power or motion between two points through a simple or complex path. In its many applications, flexible shafting has simplified and/or replaced conventional methods of power or motion transmittal such as gears, belts, pulleys, and universal joints.

The benefits of using (Fig. 5.63) flexible shafting are becoming increasingly apparent to design engineers familiar with its capabilities. Flexible shafting provides a high degree of freedom in locating transmission system components and dramatically reduces or eliminates the need for precise alignment between driving and driven shafts. It is relatively unaffected by vibration, highly efficient (80 to 95 percent) and, being intrinsically safe, does not require expensive safety guards. It solves space- and weight-related problems and operates where temperatures, chemical and biological contaminants, or nuclear radiation is a primary design consideration.

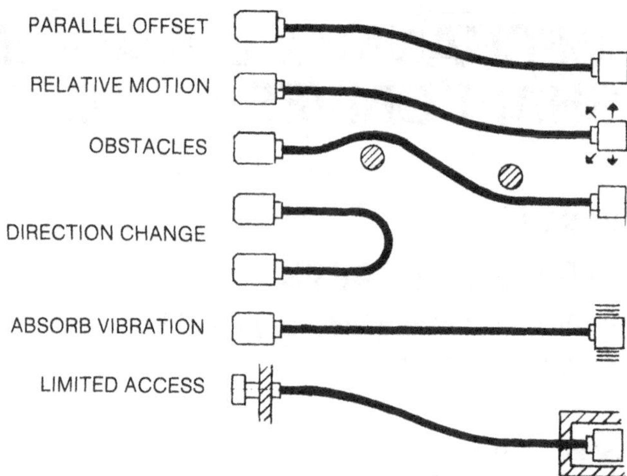

FIGURE 5.63 Typical benefits of flexible shafts.

SHAFT ALIGNMENT

When mounting a motor to drive a pump, a gearbox, or any device with self-contained bearings, a conventional coupling can absorb only small amounts of misalignment [typically less than 2° in angularity and 0.005 in (0.13 mm) in parallel misalignment]. The large tolerance for misalignment inherent in the flexible shaft allows the production engineer to do away with time-consuming and expensive realignment requiring dial indicators every time a motor is removed and replaced.

RELATIVE MOTION BETWEEN SHAFTS

Use in printing machinery and paper-coating equipment, or whenever power must be transmitted from a stationary motor to a moving carriage, are natural applications for flexible shafting. The shafting will absorb the relative motion between the driving and the driven member without resorting to constant-velocity universal joints.

MORE THAN ONE ANGLE IN POWER-TRANSMISSION PATH

When the power must be transmitted through several angles or a complex path to the driven member, a flexible shaft is much easier to apply than a series of universal gearboxes. When using gears, an angle other than 90° will require two gearboxes with their attendant alignment and expense problems.

POWER TRANSMISSION AT ANGLES OTHER THAN 90°

Flexible shafting can be used for small angles and angles other than the standard 90° available with gearboxes. As a side benefit, the flexible shafting does not have to be mounted as a gearbox needs to be.

SHOCK AND VIBRATION ABSORPTION

The flexible shaft will act as a heavily damped spring in the system to absorb the shock of sudden machinery stops and starts. It will also tend to smooth out vibration in the drive system. This heavy internal damping can cause a temperature rise in the shaft if it is used with a continuously pulsating driver, such as a single-cylinder engine. To reduce the heating, the shaft should be oversized in applications like this.

The engineer can apply flexible shafting in many applications where solid shafting and/or gearboxes would provide an expensive system that is difficult to align and maintain. *The shafting manufacturer should be consulted as early as possible in the design process to allow use of predesigned shafting whenever possible.* Consultation in the early design stage will eliminate use of nonstandard sizes, which, in turn, will result in lower final cost and shorter delivery cycles.

WHAT IS A ROTARY-MOTION FLEXIBLE SHAFT?

The flexible shaft is made of a flexible core in casing. The core is made up of layers of wire wound helically over one center wire (which acts as a mandrel) with each layer being wound in the opposite direction. The core is then covered by a flexible casing, which acts as a bearing surface for the core and at the same time acts as a support for the core. The casing also acts as a cover to hold in lubricant, keep out dirt, and provide safety for anyone in the immediate area.

Flexible shafting can be supplied in a large variety of design configurations and materials in response to customer needs. There are two primary types of flexible shafting: *power drive* and *remote control.*

POWER DRIVE (UNIDIRECTIONAL)

Flexible shafting for power drive applications is designed for continuous and intermittent operation where torque must be carried in one direction of rotation only. Power drive flexible shafts are specified as either right hand for clockwise rotation, or left hand for counterclockwise rotation. (See core selection Table 5.12 and Fig. 5.64.)

REMOTE CONTROL (BIDIRECTIONAL)

Flexible shafting for remote-control applications is designed for continuous and intermittent operation where torque must be carried in both clockwise and counterclockwise directions. (See core selection Table 5.13 and Fig. 5.64.).

FLEXIBLE COUPLINGS

These are short lengths of core that are supplied with couplings that interconnect the driving and driven elements and are available for both power drive and remote control operation. Flexible couplings under 16 in (40 cm) long do not require casings. (See Fig. 5-64.)

Since flexible shafting is relatively stiff in torsion and compliant in bending, flexible couplings inherently possess a unique combination of characteristics that permit them to accommodate both lateral and angular misalignment while being subjected to torsional loading.

TABLE 5.12 Power Drive Core Selection (Unidirectional)

Nominal diameter, in	Actual diameter, in	Part no.*	Weight/100 ft	Max. static torque capacity, lb·in†	Max. continuous rpm	Min. radius of operation, in	Horsepower rating (Max. dynamic torque, lb·in) Radius of operation, in									
							3	4	6	8	10	12	15	20	25	50
1/8	0.125–0.130	A-211 10170	3.3	12	20,000	3	0.04 (0.5)	0.08 (1.0)	0.12 (1.6)	0.14 (1.8)	0.15 (1.9)	0.15 (1.9)	0.16 (2.0)	0.16 (2.1)	0.17 (2.3)	0.18 (2.5)
5/32	0.145–0.150	A-211 10168	4.5	24	20,000	4		0.10 (1.5)	0.24 (3.5)	0.28 (5.0)	0.32 (5.5)	0.34 (5.8)	0.37 (6.0)	0.38 (6.2)	0.40 (6.5)	0.44 (7.0)
3/16	0.181–0.185	A-211 10630	6.8	48	15,000	4		0.19 (4.0)	0.34 (5.5)	0.40 (7.0)	0.46 (7.5)	0.49 (8.0)	0.53 (8.5)	0.57 (9.0)	0.60 (9.5)	0.66 (12.0)
1/4	0.241–0.245	A-211 8145	12.5	96	15,000	5			0.44 (12.0)	0.50 (14.0)	0.60 (15.0)	0.65 (16.0)	0.70 (18.0)	0.75 (20.0)	0.80 (22.0)	0.87 (24.0)
1/4	0.248–0.252	A-211 8552	12.4	88	20,000	5			0.46 (13.0)	0.52 (15.0)	0.58 (15.5)	0.62 (16.0)	0.66 (17.0)	0.70 (18.0)	0.75 (20.0)	0.80 (22.0)
5/16	0.309–0.312	A-211 9723	19.2	190	10,000	6			0.51 (18.0)	0.62 (20.0)	0.74 (24.0)	0.82 (26.0)	0.91 (30.0)	1.00 (32.0)	1.05 (35.0)	1.10 (40.0)
3/8	0.370–0.374	A-211 8149	28.2	250	10,000	6			0.62 (30.0)	0.88 (42.0)	1.10 (48.0)	1.20 (52.0)	1.40 (55.0)	1.50 (65.0)	1.60 (70.0)	1.70 (75.0)
7/16	0.437–0.441	1A-211 8134	38.5	380	6,000	6			0.74 (40.0)	1.00 (50.0)	1.20 (60.0)	1.40 (65.0)	1.60 (75.0)	1.80 (80.0)	2.00 (85.0)	2.10 (90.1)
1/2	0.500–0.496	1A-211 8585	49.6	500	6,000	6			0.80 (50.0)	1.10 (60.0)	1.30 (74.0)	1.50 (88.0)	1.80 (95.0)	2.10 (110)	2.30 (130)	2.60 (150.0)
5/8	0.621–0.626	A-211 8586	77.6	700	4,000	8				1.40 (100.0)	2.00 (132)	2.40 (158)	2.80 (180)	3.10 (190)	3.30 (205)	3.60 (235.0)
11/16	0.677–0.681	A-211 9263	91.2	860	3,750	10					2.20 (150)	2.60 (172)	3.00 (198)	3.40 (215)	3.60 (232)	3.90 (260)

* Standard remote-control cores are wound in the right-hand direction. Add a –1 to part number for right-hand or –2 for left-hand winding.

† Each core will either break or helix under this load. For short-term overloads do not exceed 75 percent of this value.

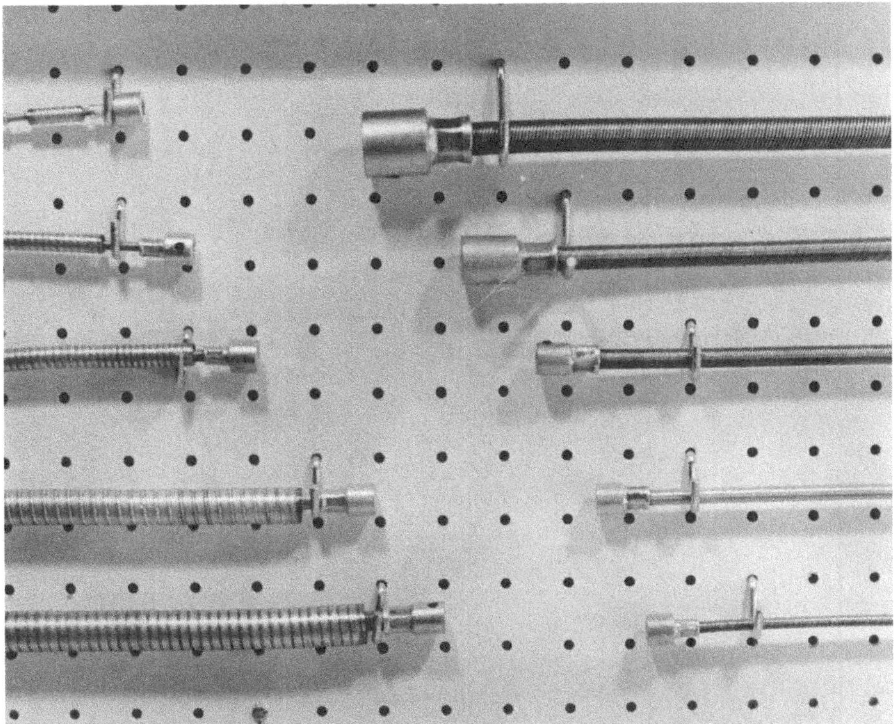

FIGURE 5.64 Typical flexible-shaft flexible couplings with setscrews used to hold them in place.

Available flexible shaft diameters range from 0.050 to 1.625 in (1.3 to 41 mm) and power drive shafts are capable of transmitting up to 12 hp (9 kW) (Fig. 5.65). For manual remote control, torques as high as 3500 in · lb (4040 cm · kg) can be transmitted.

CRITICAL PROPERTIES

Typically, the plant engineer is not expected to design a flexible shaft; that can be left to the engineering department of the flexible shaft manufacturer. *However, the engineer must know what is expected of the flexible shaft and must provide the manufacturer with certain basic data so that a suitable selection or design can be provided.* Some of the critical factors are:

- Type of shaft (i.e., remote control or power)
- Manual or dynamic operation
- Maximum torque to be transmitted
- Horsepower of driving unit
- Peak allowable rpm
- Direction(s) of rotation
- Total permissible angular deflection
- Cycling time or duty-cycle rest periods available
- Minimum bend radius

TABLE 5.13 Remote-Control Core Selection (Bidirectional)

Nominal diameter, in	Actual diameter, in	Part no.*	Weight/100 ft	Max. torque capacity, lb·in†	Max. rpm, intermittent operation	Max. torsional deflection			Torque rating for both directions of operation, lb·in‡									
						Torque, lb·in	Deflection, °/ft		Radius of operation, in									
							Wind	Unwind	3	4	6	8	10	12	15	20	25	50
⅛	0.124–0.128	A-212 10171	3.4	10	30,000	1	14	17	3.0	3.6	5.5	6.5	7.0	7.5	7.5	7.5	7.5	7.5
5/32	0.145–0.150	A-212 10169	4.8	20	20,000	1	7	9	6.0	7.0	8.0	9.0	10.0	11.0	12.0	12.0	12.0	12.0
3/16	0.181–0.185	A-212 8144	6.9	45	20,000	1	3.5	4		14.0	16.0	18.0	22.0	24.0	26.0	26.0	26.0	26.0
¼	0.245–0.249	A-212 8146	12.8	95	20,000	5	5	6		28.0	32.0	36.0	44.0	48.0	55.0	55.0	55.0	55.0
5/16	0.307–0.311	A-212 8148	19.7	150	20,000	10	6	7.5		56.0	64.0	72.0	88.0	96.0	110.0	110.0	110.0	110.0
⅜	0.374–0.378	A-212 8150	28.8	220	20,000	10	3.5	5			102.0	116.0	124.0	132.0	140.0	140.0	140.0	140.0
½	0.496–0.499	A-212 8296	54	340	10,000	100	10	13				200.0	220.0	240.0	260.0	280.0	280.0	280.0
⅝	0.614–0.618	A-212 8172	75	550	7,000	100	2.5	3.5					275	300	330	370	380	400
¾	0.740–0.747	A-212 8587	113	670	5,000	100	1.5	1.6							440.0	480.0	500.0	500.0
1	0.990–0.997	A-212 8588	200	1300	5,000	960	7	7								920.0	950.0	975.0
1¼	1.240–1.247	A-212 8559	310	1900	2,500	1900	5	5								1000	1200	1500
1.300	1.292–1.299	A-212 8660	340	1920	2,500	1920	5	5								1100	1300	1600
1⅝	1.618–1.611	A-212 8663	600	3000	1,750	3000	4	4									2500	2700

* Standard remote-control cores are wound in the right-hand direction. Add a –1 to part number for right-hand or –2 for left-hand winding.

† Each core will either break or helix under this load. For short-term overloads do not exceed 75 percent of this value.

‡ Each core can transmit this torque in both directions of operation for remote-control applications (less than 100 rpm) and intermittent power drive applications (no longer than it takes to raise core temperature 70°F (39°C) above ambient with rest duration allowing core to cool within 30°F of ambient). For continuous power drive applications in both directions, use only 30 percent of these torques.

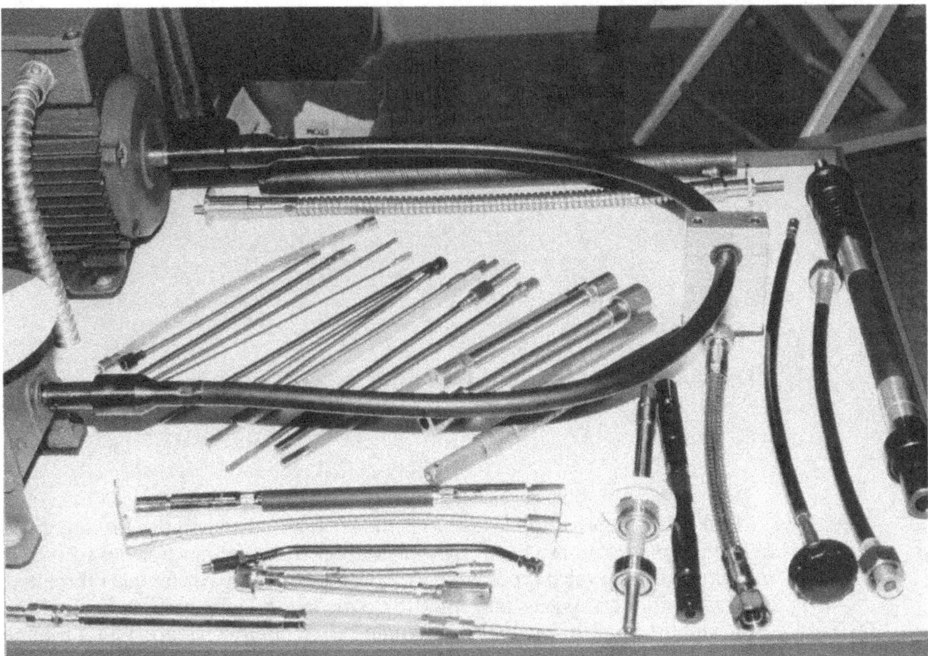

FIGURE 5.65 Wide range of sizes and types of flexible shafts with a variety of end fittings, some of which are shown here.

- Length of shaft
- Ambient temperature
- Unusual service conditions

Note that not all these data will apply in all instances, but the more background the manufacturer can be given, the more suitable and economical the flexible shaft design will be.

In many instances, a predesigned shaft or a flexible shaft/flexible coupling will solve the plant engineer's power transmission problem. However, if the requirements are unusual, a special shaft may have to be designed. Obviously, whenever possible, predesigned shafts should be selected because they are the less costly solution.

STANDARD POWER DRIVE FLEXIBLE SHAFTS

Standard power drive shafts are available with built-in bearings and couplings on the end that can be bored out and held in place with setscrews to suit the particular application (see Fig. 5.66).

CASINGS AND FITTINGS

In many applications, a bare flexible shaft equipped with suitable fittings, as previously described, is perfectly adequate for the task at hand. In other cases, however, a flexible shaft should be used with special fittings and a casing.

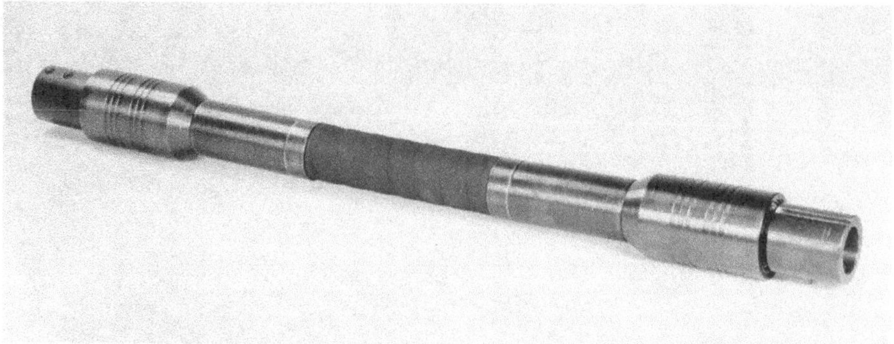

FIGURE 5.66 Standard power drive flexible shaft with built-in bearings and couplings; this offers a very simple hookup.

End fittings are installed on both the casing and the shaft core itself. The casing fittings keep the shaft assembly fixed in place; neither these fittings nor the casing rotates. Casing fittings may be of several types, for example, a loose male (or female) threaded coupling nut or a quick disconnect. With shaft-end fittings, even more choices are available: integrally formed drive square, tang fitting, male or female spline, hollow square, panel mounting, or setscrew. The choice of fittings depends to a large extent on the application, how accessible the shaft is, and how often it is expected to be serviced. If a shaft is to rotate at high speeds for prolonged periods and has to be lubricated regularly to compensate for this severe service, it is recommended that the assembly be designed for easy withdrawal of the shaft from the casing. In that event, one end of the shaft is usually swaged square so that the drive does not permanently lock the shaft into the casing at both ends.

A casing provides the following benefits:

- It keeps the flexible shaft properly lubricated for longer periods of time.
- It offers intermediate points of attachment when shaft length exceeds 18 in (46 cm).
- It keeps plant personnel from coming into contact with a high-speed rotating member.
- It protects the flexible shaft against hostile environments.

Casings can be classified into two types: *metallic* and *covered*. Metallic casings are neat, strong, durable, and bendable. Although they will retain grease, they are not oil- or watertight.

Covered casings are more expensive because their construction is more elaborate. Following an inner liner that may be either metallic or plastic, the casing is reinforced with layers of steel braid and finished off with a final outer sheath of rubber or plastic. This type of casing permits the flexible shaft to operate under water, hydraulic fluid, or other liquid, so long as the fluid does not attack the elastomeric outer covering.

SHAFTS THRIVE ON SPEED

Power shafts function most efficiently when operated at high speed. The reason stems from basic engineering principles: transmitted horsepower is directly proportional to speed.

$$\text{hp} = KTN$$

where T is torque, N is speed, and K is a constant to reconcile the units of torque and peripheral speed being used in the equation.

Typically, if torque is expressed in foot pounds and speed in revolutions per minute, hp = 0.00019 × ft · lb × rpm. For torque expressed in inch pounds, hp = 0.000016 × in · lb × rpm. Therefore, for a given horsepower requirement, the higher the shaft speed, the less the torque on the shaft and therefore, the smaller the shaft needed to accomplish the task. This factor should be kept in mind during the early stages of design, because quite often the insertion of a suitable speed-reduction device at the correct location in the drive train may serve to reduce the overall cost of the assembly.

A power shaft is meant to rotate at the highest permissible speed; it is, therefore, important to consider this factor in the design. If a speed reducer is necessary, insert it at the correct end of the shaft. Thus, if motor speed is 5000 rpm and final operating speed is to be 500 rpm, the speed reducer should be inserted at the output, not at the motor end. Conversely, if a speedup is desired, the speed increaser should be inserted at the motor end.

SHAFT RADIUS OF CURVATURE

Although rotary-motion flexible shafts are almost snakelike in their ability to reach into inaccessible areas, there is a limit to how much they may be bent without damaging them. This is the minimum operation radius (MOR). Each shaft construction (diameter, number of wires per layer, and number of layers) leads to a different MOR, which is always available from the manufacturer. The MOR is established by calculations based on the diameter and material of the mandrel wire.

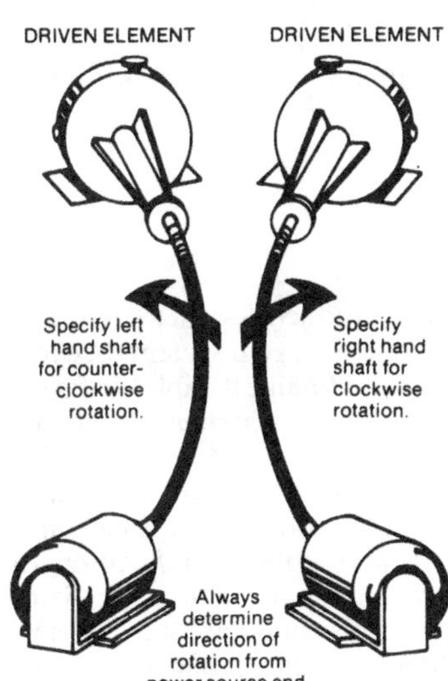

FIGURE 5.67 Direction of rotation.

It should come as no surprise that the amount of torque load a flexible shaft can comfortably manage is in inverse proportion to the MOR. The more a flexible shaft is curved, the higher the internal friction, the greater the heat generated, and the smaller the permissible continuous load. Here again, these figures are known for each construction and should be rigidly adhered to if overload or early failure of the shaft is to be avoided. Table 5.12 shows typical values for a range of high-tensile-steel power shafts, including such basic parameters as diameter, radius of curvature versus torque capacity, torsional breaking load, and weight per foot.

In using Table 5.12, remember that the torque-carrying capacities are shown for shafts rotating in the direction that tightens up the outer layer. If the same shaft is rotated so as to unwind the outer layer—clockwise for a left-hand flexible shaft or counterclockwise for a right-hand shaft—its torque capability is reduced by up to 50 percent. See Fig. 5.67 for an example of right- and left-hand shaft application.

SHAFT LIFE

As a rule of thumb, a power shaft running under no load at the MOR can be expected to have a useful life in excess of 10^8 revolutions. Such factors as service temperature, number of shaft bends, torque output, speed, shock loading, and operating radius can reduce shaft life.

SHAFT SELECTION

To select the proper shaft, determine the following items:

- Torque requirement
- Shaft radius of curvature
- Length of path between driving and driven systems
- Operating speed
- Acceptable torsional deflection or backlash

Determining the torque requirement on a shaft may not be easy. Remote-control shaft requirements can be determined by attaching a lever of known length to the device to be turned and pulling the end of the lever with a spring scale. Use the highest torque as the required torque. Measuring torque on a power shaft is more difficult because the requirement is dynamic. The best way is to instrument-load with a torque cell and measure the torque under operating conditions. Component manufacturers usually have these data available. A shaft efficiency of about 90 percent should be attainable for most applications; therefore, increasing the output torque by about 10 percent leads to the input torque.

The shaft path should then be examined. A drawing should be made that shows the path in a true view. The MOR should be determined from this drawing. A full-scale prototype can be used. The length of the shaft can also be determined at this time. For power shafts, the operating speed is generally given. For remote-control shafts, in which the speed is essentially zero, the amount of torsional deflection acceptable between the turning device and the turned device is the important consideration.

DOS AND DON'TS

Used intelligently, flexible shafts are the product designer's allies. They are more flexible than universal joints; they are more versatile than gear systems because they are totally unaffected by the exact angle or offset necessary; finally, flexible shafts offer an inherent shock-absorption capability, ease of installation, and maintenance unmatched by other forms of rotary-motion transmission.

However, flexible shafts do have to be treated carefully if they are to provide the service life built into them. They cannot be bent completely out of shape and must be used at radii equal to or greater than the MOR specified by the manufacturer. They should be secured approximately every 18 in (46 cm) to prevent the possibility of helixing. They must be lightly lubricated as preventive maintenance at regular intervals. Replacement must match the original design because a remote-control shaft is not interchangeable with a power shaft. The type of service expected must be clearly specified. Two differently built shafts, even of the same diameter and length, are not interchangeable. Above all, flexible shafts must be designed for the task they are to perform. For this reason, early consultation between the user's engineering department and the shaft manufacturer is highly recommended.

Follow these simple, basic suggestions, and the design flexibility of a rotary-motion flexible shaft will be applied to utmost advantage.

TYPICAL APPLICATIONS

Flexible-shaft flexible couplings in lengths of 6 in (15 cm) or less are ideal for connecting a driven member to its power source when a small amount of angular, parallel, or combined misalignment must be accommodated. Not only does a motor-type coupling readily compensate for the misalignment, but it will also handle a certain amount of shock loading and attenuate vibration.

Longer segments of flexible shaft, from 1 to 100 ft (0.3 to 30.5 m), are ideal for transmitting rotary motion to very distant locations, either for remote control or for power transmission.

Flexible shaft applications in industry are wide ranging, so it would be quite a task to list them all. However, a few typical cases may help to demonstrate how rotary-motion problems have been avoided or solved with a flexible shaft:

An operating device on a swinging oven door has to rotate freely, no matter how much the angle changes as the door opens and closes.

Two rotating devices have to be synchronized through a central gearbox, even though one of the devices moves laterally while the other stays fixed in place.

Several door-actuating devices have to be moved simultaneously, all by one driving motor and each one at a different angle to it.

A small, buried control has to be readily turnable from "outside"; a valve has to be closed from a remote location (Fig. 5.68); a potentiometer has to be adjusted in an "unreachable"

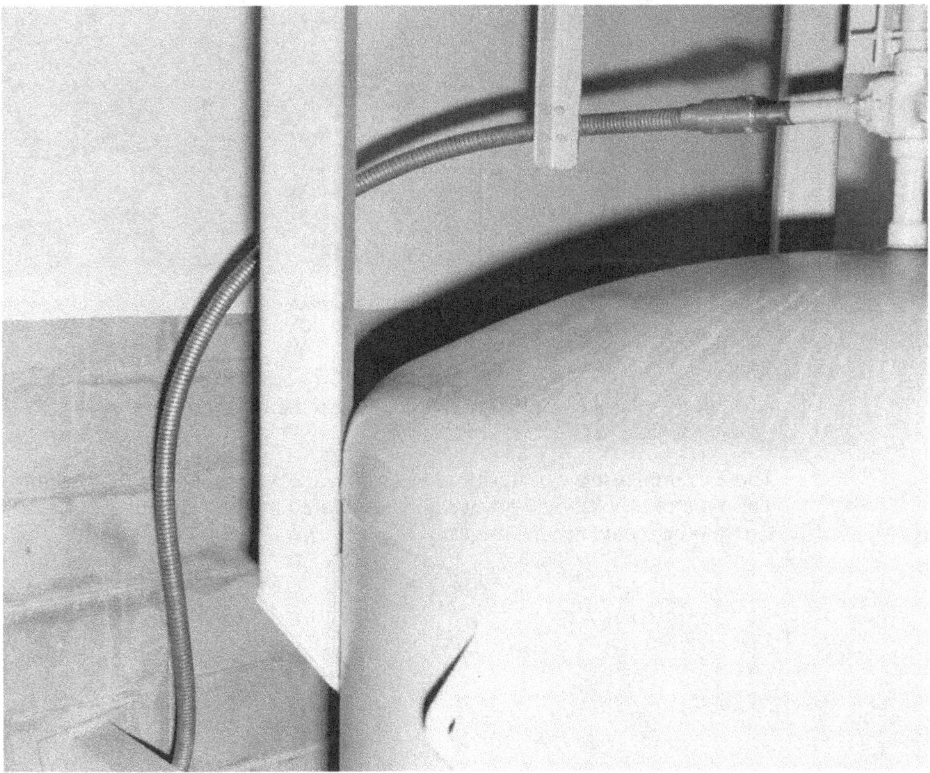

FIGURE 5.68 Flexible shaft operating valve in remote location underneath a tank.

FIGURE 5.69 ⅜ in (9.5 mm) flexible shaft drives eccentric inside of steel shell to vibrate concrete; operation speed up to 13,000 rpm.

spot; a switch has to be operated from far away; and a rotating vibrator head has to be driven while in concrete (Fig. 5.69).

These examples serve to illustrate the wide variety of problems that flexible shafts can solve. This type of drive should always be investigated by the plant engineer as a possible alternative to other conventional drives.

CHAPTER 5.6
SHAFT DRIVES AND COUPLINGS

PART 2
BELT DRIVES

Brent Oman
Application Engineer
Power Transmission Product Applications Department
Gates Rubber Company
Denver, Colorado

BELT DRIVES

Belt drives are the most widely used method of flexible power transmission. Improvements in materials and methods of manufacture have allowed the introduction of new belts with much broader application capabilities.

Belt drives basically fit into four types: flat, V, V-ribbed, and synchronous. Although each of these basic belt types are more suitable in specific application areas, most applications can be successfully designed with more than one type of belt.

Flat belts, one of the earliest forms of flexible power transmission, are generally more suited to high-speed, low-horsepower applications. At low speeds and high loads, flat-belt drives usually become too large to be cost-effective.

V belts are the most commonly used today, and they are the only belts that can be used on variable pitch and variable speed drives.

V-ribbed belts are described by some as guided flat belts. Their thinner cross section makes them suitable for operation on smaller diameters at higher speeds.

Synchronous (timing) belts are specifically designed as alternatives to roller chains and gears on drives, which require exact speed ratios and synchronization between the driver and driven machines. They are also widely used today in low-maintenance, energy-efficient applications.

FLAT BELTS

Flat belts are still widely used for power transmission. Their thin, flexible cross section allows them to operate over small diameters and, in some cases, at very high speeds. Many different sizes and constructions are available for a wide variety of uses. Flat belts are made either of fully molded or woven construction and may or may not have a tensile member.

One significant disadvantage of flat belts is that they depend entirely on friction in order to transmit power. Thus they require higher belt tension to do the same work, which results in higher shaft and bearing loads. The need for higher tension may cause more belt stretch, causing the belt to slip more easily than V belts.

V BELTS

The problems of high tension with flat belts led to the development of V belts. Unlike flat belts, which depend only on friction, V belts have deep V-shaped cross sections that wedge into the sheave groove to provide added horsepower capacity. Because of the wedging action, V belts are highly stable and can operate at tensions considerably lower than those needed by flat belts. Thus, V-belt drives can be more compact and allow for smaller shafts and bearings.

The load on a banded or covered V belt is carried by a tensile section located near the top of the belt. This section contains one or more layers of cord, depending on the method of manufacture (Fig. 5.70). Generally, most V belts today use polyester tensile cords, which provide significantly improved capacity compared to older tensile types for V belts.

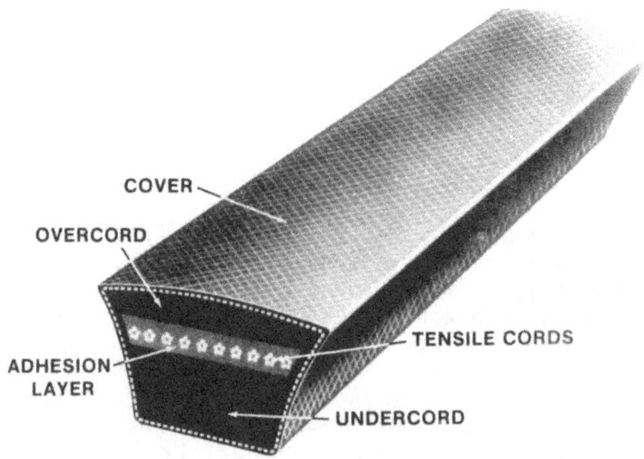

FIGURE 5.70 Typical V-belt construction.

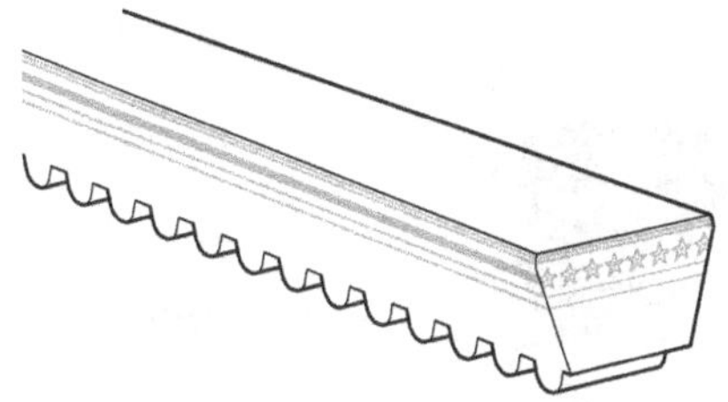

FIGURE 5.71 Molded-notch V belt.

A widely used variation of the banded V belt is the bandless molded notch or molded cog belt (Fig. 5.71).

The molded notches in these belts provide greater flexibility allowing the belt to operate on smaller diameter sheaves. This provides for even more compact drives where space is limited.

V belts operate over a wide range of speed and provide a broad choice of center distance. While V belts can operate at very high speeds, standard sheaves are normally limited to a 6500 ft/min (33 m/s) rim speed maximum for safety purposes. Beyond this speed, special materials and balancing may be required.

There are four basic types of V belts: heavy-duty industrial, light-duty industrial, agricultural, and automotive. The first two are described in the following paragraphs; the last two are outside the scope of this book.

HEAVY-DUTY V BELTS

Classical

This line of belts includes five standard cross sections—A, B, C, D, and E. These range in width from ½ in (13 mm) for an A section to 1½ in (32 mm) for an E section (Fig. 5.72). Also available are the molded-notch cross sections—AX, BX, CX . Classical belts can be used in single or multiple belt drives, can transmit hundreds of horsepower continuously, and absorb reasonable intermittent shock loads. Operating temperature limits range from –30 to +140°F (–34 to +60°C). These belts also provide excellent oil resistance.

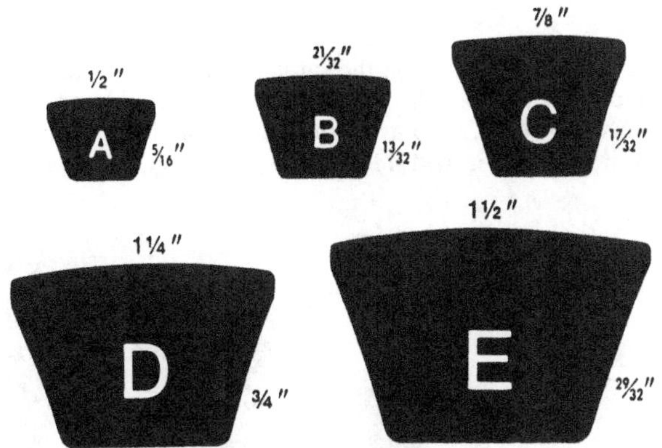

FIGURE 5.72 Classical V-belt cross sections.

Classical banded and bandless molded-notch belts are also available in a joined configuration, where two or more belt strands are connected across the top by a tie band (Fig. 5.73). The joined belt improves lateral stability, making it almost impossible for individual belts to turn over in the groove or jump out of the sheave. The tie band rides above the sheave outside diameter and does not interfere with the normal wedging action of the individual belts. Joined

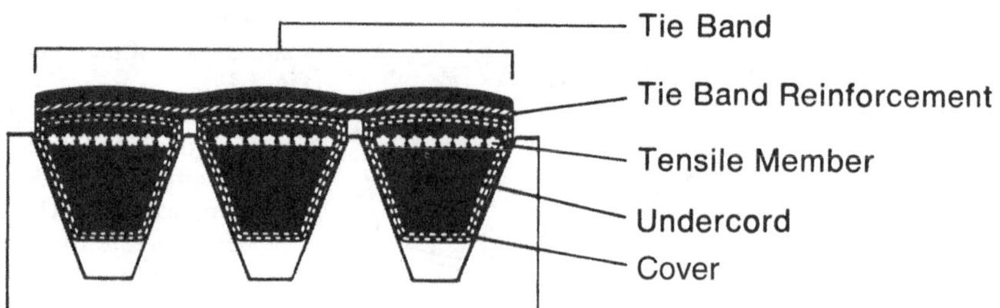

FIGURE 5.73 Joined V-belt tie band configuration.

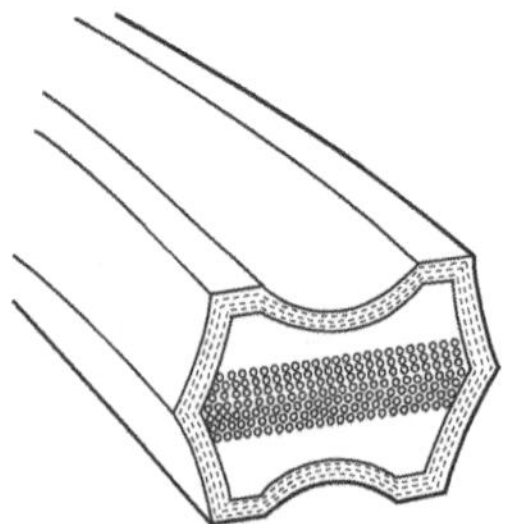

FIGURE 5.74 Double-V belt.

belts do not carry any higher horsepower rating, but should be designed using published single belt horsepower ratings.

Another variation of the classical belt is the double-V or hexagonal-V belt (Fig. 5.74). Double-V belts are used when power input or power takeoff is required on both sides of the belt. This is necessary when the driver and driven sheaves must rotate in opposite directions. (Drives with several counter-rotating shafts are called *serpentine* drives.) Double-V belts are available in AA, BB, CC, and DD cross sections.

Narrow

These belts can be used on the same applications as the classical belts, but allow for a lighter, more compact drive. Three cross sections—3V, 5V, and 8V—replace the five classical cross sections (Fig. 5.75). Also available are the bandless molded notch cross sections—3VX, 5VX. These belts range in width from ⅜ in for the 3V to 1 in for the 8V.

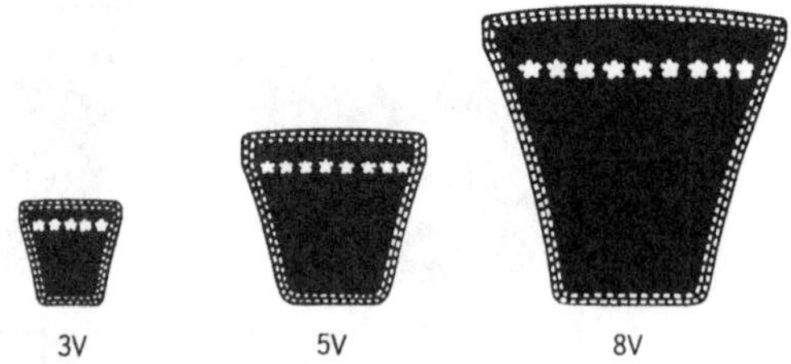

FIGURE 5.75 Narrow V-belt cross section.

Narrow belts, because of their relatively higher horsepower capacity, usually provide substantial space and weight savings compared to classical belts. For instance, narrow belts can transmit the same horsepower loads in ½ to ⅔ the space of a classical belt drive. Narrow belts often allow the use of greater speed ratio between driver and driven machine.

In addition to exhibiting excellent oil and heat resistance, most classical and narrow industrial V belts meet industry standards for static conductivity. This allows for safe operation in potentially hazardous environments.

Variable Speed

Within certain limits, V belts are well suited to drives that must run at varying input or output speeds. These are common in the air-moving industry and are referred to as *variable-pitch* drives. These drives must incorporate special sheaves. Speed ratio on these drives is controlled by moving one sheave sidewall relative to the fixed sidewall so that the belt can ride at different pitch diameters. Variable-pitch drives using a single variable-pitch sheave and classical cross-section belts will yield only about 1.4:1 overall speed variation.

Wide range *variable-speed* drives can easily allow for overall speed variations of up to 10:1 and transmit up to 100 horsepower (75 kW). Generally the higher the horsepower load, the lower the speed ratio. Variable-speed belts have a wide, thin cross section in comparison to other V belts (Fig. 5.76).

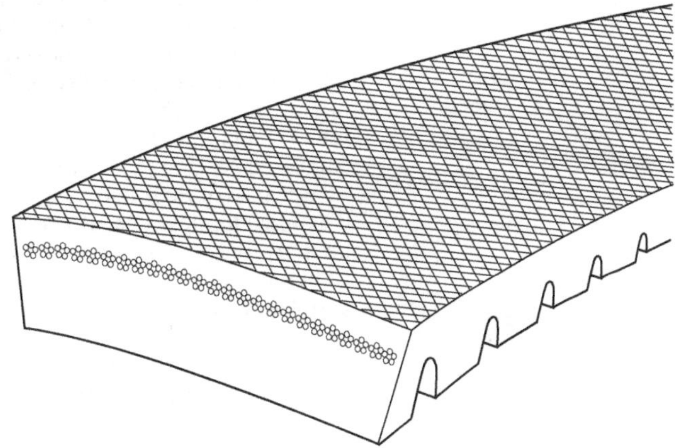

FIGURE 5.76 Variable-speed belts.

Unlike a gear transmission where speed variation is limited to finite discrete steps, variable-speed belt drives offer an infinite selection of speed ratios within the design range. In simple applications, speed is varied manually by moving the sliding motor base in and out or by adjusting an idler. On other drives, the adjustable or movable sheave sidewall is controlled through a mechanical linkage while the center distance is fixed. The driven sheave is generally spring loaded to provide adequate belt tension. In applications such as recreational vehicles (e.g., snowmobiles), the speed ratio is controlled by sensing speed and/or torque requirements of the drive, which in turn control the shifting mechanism.

The variable-speed operation can generate high frictional heat buildup and sidewall stresses. The wide, thin section is more easily subject to collapsing under high loads. In the past few years, manufacturers have developed improved materials and constructions to withstand the heat and high belt sidewall stresses. As a result, variable-speed belts are moving into more demanding applications.

These belts are used in many industrial, agricultural, automotive, and recreational vehicle applications where speed variation is a requirement.

LIGHT-DUTY V BELTS

These belts are similar to classical belts, except they have a slightly thinner cross section. This makes them ideally suited for the smaller sheaves typically found on many fractional-horsepower drives. Since they are generally intended to be used on single-belt drives requiring less than one horsepower, they are often referred to as fractional-horsepower belts. Standard cross sections are 2L, 3L, 4L, and 5L and range in width from ¼ to ⅝ in (6 to 16 mm).

Usually, light-duty belt drives operate on an intermittent basis. Service requirements may vary from 1–3 h/week for a power lawn mower to 40 or more h/week for office machines or commercial power tools.

FIGURE 5.77 V-ribbed belt.

V-RIBBED BELTS

Ribs on the bottom side of V-ribbed belts (Fig. 5.77) mate with corresponding grooves in the sheave. This tracking guides the belt and makes it more stable than a flat belt. Like V belts, V-ribbed belts operate in grooved sheaves, but they do not have the full wedging capability of V belts. As a result, they must operate at tensions slightly higher than V belts, but significantly lower than flat belts. A variation used more often today uses a truncated rib (Fig. 5.78), which provides more wedging action.

Because of their lightweight and thin construction, V-ribbed belts develop less centrifugal and bending tension as they are bent around small-diameter sheaves. Compared to a V belt, V-ribbed belts have a continuous tensile section across the entire belt width, providing additional load capacity.

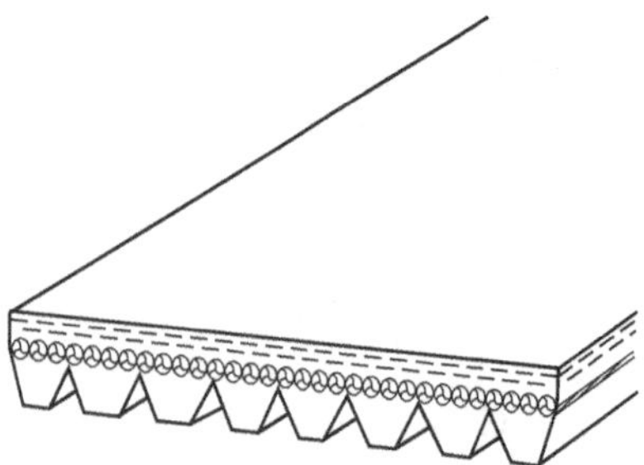

FIGURE 5.78 Truncated-rib V belt.

V-ribbed belts are used in applications requiring a high-speed ratio such as clothes dryers or on drives such as small power tools that require very small drives. V-ribbed belts are also more common on automotive front-end engine applications such as serpentine drives. V-ribbed belts require closer alignment than V belts to minimize stress on the ribs.

ROUND BELTS

The use of round belts is diminishing in power transmission today in favor of today's improved V belts. Because of their round design, they are still used often in drives where the shafts are in different planes. Consequently, they are still used in roller conveyors that must run around curved corners. Although different sizes and constructions may be available, the most common are ½-in (12.5-mm) and 9⁄16-in (14-mm) diameters.

SYNCHRONOUS BELTS

Synchronous belts, sometimes referred to as *timing* or *positive drive* belts, are especially suited to applications requiring a synchronized input and output speed (Fig. 5.79). Synchronous belts are designed to overcome the "creep" or slip of V-belt and flat belt drives. These belts eliminate slip by transmitting power through the positive engagement of the belt teeth against the pulley or sprocket teeth. Thus, precise speed ratios and synchronization are possible in such applications as automotive camshafts, machine tool indexing heads, computer printers, and robotics.

Synchronous drives have an advantage over gears and chains in that they can transmit high loads over a wide range of speeds with low noise and no lubrication requirements. Synchronous belts, like V-ribbed belts, require closer alignment tolerances in order to avoid premature wear or failure.

FIGURE 5.79 Synchronous belt.

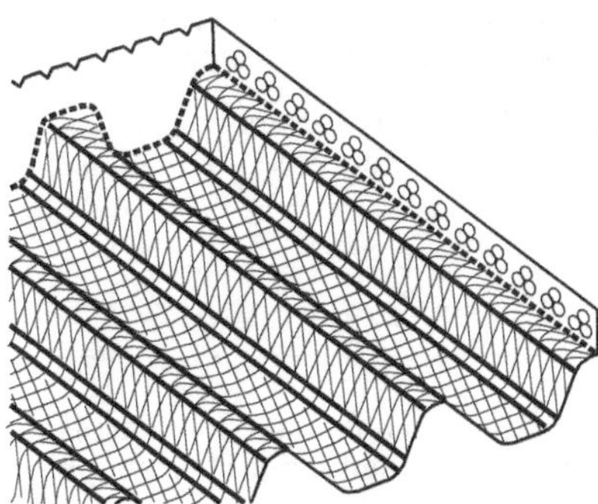

FIGURE 5.80 High-torque drive belt.

There are several situations in which synchronous belts have an advantage over V belts. These belts use an extremely high modulus, low stretch fiberglass or aramid tensile cord to maintain uniform spacing between the belt teeth. This low growth under load results in minimal need for center distance adjustment. Therefore, in addition to being low-maintenance items, synchronous belts are often used on drives with limited space for center distance adjustment. Since they do not slip, they exhibit a consistently high level of drive efficiency—generally about 98 percent in power transmission.

There are three basic tooth configurations associated with synchronous belts. The standard trapezoidal shape was introduced in the 1940s. In the 1970s the *high torque drive* (HTD) belt was introduced with a round-shaped tooth. In more recent years a number of modified curvilinear tooth belts have been introduced with a tooth shape similar to gear teeth (Fig. 5.80). It is very important that each belt type be run in compatible pulleys or sprockets.

BELT SELECTION

To aid in the selection of the proper belt for each application, manufacturers provide technical and performance data about their belts. In addition, the Rubber Manufacturers' Association (RMA) and the Mechanical Power Transmission Association (MPTA) have worked together to publish engineering standards and bulletins for most types of belts and drive hardware (see Bibliography). These publications contain information that supplements design catalogs.

There are four basic questions that need to be answered in the drive design:

1. What horsepower is required of the drive?
2. What is the speed (rpm) of the driver shaft?
3. What is the speed (rpm) of the driven shaft?
4. What is the approximate desired center distance?

In addition to the basic elements, there are a number of special drive characteristics that may require consideration. These might include:

- Special environmental conditions such as abrasives, chemicals, and so on.
- Overhung load (OHL) considerations for gear motors and reducers
- Driven pulley inertia (WR^2) requirements for equipment such as piston compressors, crushers, and so on.
- Special drive characteristics such as shock loads, inherent misalignment, clutching requirements, and so on.

Also, while selecting and evaluating the drive, consider the following points:

- Selecting larger diameter pulleys will keep drive face width to a minimum.
- Selecting larger diameter pulleys will keep drive tensions and shaft load at a minimum.
- Larger diameter pulleys will often give a more economical drive, but should not be so large that multiple V-belt capability is sacrificed.
- If space is limited, consider using the smallest diameter drive. However, pulleys on electric motors must be at least as large as the National Electric Manufacturers Association (NEMA) minimum recommended standards.
- When the drive is between two belt cross-section sizes, the larger section will usually be more economical. However, it is always recommended to check drives in both cross sections.

Selecting an optimum belt drive involves many factors, but drive selection can be readily accomplished using manufacturers' design literature. *Many manufacturers also offer computer programs for drive design selection.*

MAINTENANCE AND SAFETY

Once a belt drive has been selected it requires a minimum of maintenance, but certain procedures can help reduce equipment downtime, extend service life, and increase safety.

Installing Belts

When installing belts on any multiple-belt drive, *always replace all the belts* because older belts naturally become worn and stretched from use. If old belts are mixed with new belts, the new belts will be tighter, carry more than their share of the load, and probably fail before their time.

Be sure to use a set of belts from a single manufacturer. If brands are mixed, the belts may have different characteristics and could work against each other. This could result in unusual strain and reduce the life of the belts.

There are two ways to assure a well-matched drive system, depending on the method of matching used by the belt manufacturer.

Most manufacturers today use a no-match or single match system. Because of improved manufacturing systems and inspection procedures, manufacturers like Gates Rubber Co. can build belts to overall length tolerances that are within the RMA matching tolerances specified in the appropriate standard.

Some belt manufacturers, and some specialty belts, may still use the older, conventional matching system. Each belt is measured under industry standard tension in V sheaves and marked with a "punch" or match number designating the small increment of length within the overall belt length tolerance. With 50 as ideal, these match numbers generally run from 47 to 53. Belts using match numbers are grouped in sets within the limits given in Table 5.14.

TABLE 5.14 Belt Matching Limits

Belt length		
Inches	Millimeters	
Up to 63	Up to 1,600	Use only one number
63–150	1,600–3,810	Within two consecutive match numbers only
150–250	3,810–6,350	Within three consecutive match numbers only
250–375	6,350–9,525	Within four consecutive match numbers only
375–500	9,525–12,700	Within five consecutive match numbers only
Over 500	Over 12,700	Within six consecutive match numbers only

Belts intended to be run as single belts, such as light-duty, V-ribbed, and synchronous belts do not require a matching system.

After the belts have been properly selected, it is time to put them on the drive. The most important rule is *never to pry or roll the belts onto the pulley.* Prying the belt may cause invisible damage to the tensile cords, shortening the belt's life. The proper way to install the belt is to use the drive adjustment, generally the motor base, to move the pulleys closer together, allowing the belt to be easily placed onto the pulleys. A sturdy pry bar will help move the motor. Keep the rails free of dirt, grit, and rust; lubricate them frequently. This will make the belt change easier and safer.

Be sure to properly align the motor and driven pulleys. V-belt drives should be aligned to within ½° or ¹⁄₁₆ in/ft of center distance. V-ribbed and synchronous belt drives should be installed and aligned to within ⅓° or ¹⁄₁₀ in/ft of center distance. Misaligned drives can result in V-belt turnover and premature and unusual belt wear leading to shorter belt life.

The final installation procedure involves properly tensioning the belt or belts for long, trouble-free service. Here are three helpful tensioning tips:

- The best tension for the belt is the lowest tension at which the V belt will not slip, or synchronous belts will not ratchet (jump grooves). Too much tension shortens belt life and may also overload shafts and bearings.
- Tension the drive and allow it to run-in for at least 15 min. The more lightly loaded the drive, the longer the run-in period. Check and retension the drive to the recommended tension.
- Check the V-belt tension periodically and retension after the initial run-in period. Synchronous belts generally should not be retensioned. Most belt manufacturers publish tensioning procedures in their design catalogs. Some manufacturers, such as Gates, also make available computer-aided design and tensioning programs.

Belt Safety

Safety is a critical factor in the efficient operation of belt drives. The maintenance personnel can take several positive steps to help ensure drives run smoothly and safely:

- Keep belt drives properly guarded. Regulatory agencies [particularly the Occupational Safety and Health Administration (OSHA)], insurance companies, and other safety authorities require that drives be completely guarded. The guard should allow proper ventilation but should not have gaps that allow workers to reach inside and become caught in the drive. The guard also prevents debris from entering into and damaging the belt drive.
- Always turn the equipment off before working on the drive. Lock and tag the control box to indicate maintenance personnel are working on the drive.

- Check position of all components. Make sure machine components such as flywheels are in a safe position and that clutches are in neutral position so as to avoid accidental movement or start-up.
- Wear proper clothing. Loose or bulky clothing, such as neckties, loose sleeves, or lab coats may become entangled in the drive. Wear gloves when inspecting pulleys to avoid being cut by nicks or burrs in the metal.
- Keep the area clean. Debris and loose tools or other obstructions near the drive may result in tripping or falling. Keep floors clean.
- Use proper tools and procedures. Never attempt to roll or pry belts onto the pulleys. In addition to causing unseen damage to the belt tensile cords (leading to early failure), you may painfully injure fingers or hands. Follow recommended installation procedures as discussed previously.

Belt Inspection

If proper belt inspection procedures are followed, 90 percent of all maintenance problems can be eliminated. A properly installed belt drive is a remarkably trouble-free piece of equipment. But, to assure continued trouble-free service, quick, periodic inspections of the belt drive should be made a part of the routine preventative maintenance schedule.

The inspection period may be influenced by a number of factors. These include, but are not limited to, drive operating conditions, critical nature of the equipment, temperature extremes in the drive vicinity, other environmental factors, and accessibility of the equipment.

Look, listen, feel, and smell. Watch the drive operate. Remove the guard *temporarily* if necessary. Look for excessive or unusual belt vibration. Listen for unusual noise. Feel the driver and driven machine for excessive vibration. Unusual odors may suggest badly slipping belts or worn or damaged bearings. Stop the drive, tag the control box for safety, and carefully inspect the drive components.

Belts Immediately, but carefully, grasp or touch the belt. Your hand can safely tolerate up to about 140°F (60°C) surface temperature. If the V belt is uncomfortably hot, it may be slipping excessively due to inadequate tension. Check belt tension and retension as necessary. Never use a belt dressing to quiet a slipping V belt. It causes the belt to collect dirt and grit and damages the belt. Inspect the belt for any damage or unusual wear. Cracking, fraying, excessive wear, or loss of teeth (on a synchronous belt) requires belt replacement. Replace all the V belts in the set with a matched set of new belts from one manufacturer.

Pulleys Inspect the pulleys for unusual damage or excessive wear on V-belt sheaves. Small nicks or burrs may be smoothed over with a file or emery cloth. Manufacturers' groove gauges may be used to check for excessive wear on V-belt sheaves. If more than $\frac{1}{32}$ in (0.8 mm) of wear can be seen, the sheaves should be replaced. Visible wear on synchronous sprockets always suggests replacement.

Guards and Other Components Check guards for damage and repair or replace as necessary. Check shafts, brackets, and bearings for damage and correct as necessary.

Reassemble the drive, turn it on, and again look and listen for unusual signs to ensure the equipment is operating properly.

TROUBLESHOOTING

Occasionally belts may fail prematurely or in an unusual manner. Troubleshooting a belt-drive problem requires that you identify the causes. Proper procedures can make the investigative process easier for you and your belt supplier.

1. Describe the problem.
   - What is wrong?
   - When did it happen?
   - How often does it happen?
   - What is the drive application?
   - Have the machine operations or output changed?
   - What kind of belt(s) are you using?
   - What are your expectations for belt performance in this application?

2. Identify symptoms and record any unusual observations. These might include any of the following:
   - Premature belt failure
   - Severe or abnormal belt wear
   - Banded (joined) belt problems
   - V belt turns over or jumps off sheave
   - Belt stretches beyond take-up
   - Belt noise
   - Unusual vibration
   - Problem with pulleys
   - Problems with drive components
   - Hot bearings
   - Performance problems

Belt Failure

The most common symptoms, causes, and corrections of V-belt and synchronous-belt failures are shown in Tables 5.15 and 5.16.

TABLE 5.15 Why V Belts Fail

Trouble area and observation	Probable cause	Corrective action
Worn side patterns	Constant slip	Retension drive until belt stops slipping
	Misaligned sheaves	Align drive
	Worn sheave	Replace sheave
	Incorrect belt	Replace belts
Bottom of belt cracking	Belt slipping, causing heat buildup and gradual hardening of undercord	Install new belt, tension to prevent slip
	Idler installed on wrong side of belt	Refer to a V-belt installation manual
	Improper storage	Follow proper storage procedures
Bottom and sides burned	Belt slipping under starting or stalling load	Replace belt and tighten drive until slippage stops
	Worn sheave	Replace sheave
Belt turnover	Foreign material in grooves	Remove material, shield drive
	Misaligned sheaves	Align drive
	Worn sheave	Replace sheave
	Tensile member broken through improper installation	Replace with new belt(s) and follow proper installation procedure
	Incorrectly aligned idler pulley	Carefully align idler, checking alignment with drive loaded and unloaded
Belt pulled apart	Extreme shock load	Remove cause of shock load or redesign drive for increased capacity
	Belt came off drive	Check drive alignment, foreign material in drive; ensure proper tension and drive alignment

TABLE 5.16 Why Synchronous Belts Fail

Trouble area and observation	Probable cause	Corrective action
Excessive belt edge wear or cracking	Flange damage	Repair flange or replace sprocket
	Belt too wide	Use proper width belt and sprocket
	Misaligned drive	Align drive
	Rough flange surface	Repair flange
	Improper tracking	Correct alignment
	Belt hitting drive guard or bracketry	Remove obstruction or use inside idler
Tensile break	Excessive shock loads	Redesign drive for increased capacity
	Subminimal diameter	Redesign drive using larger diameters
	Improper handling or storage prior to installation	Follow proper handling and storage procedures
	Extreme sprocket run-out	Replace sprocket
	Misaligned drive	Align drive
Belt cracking	Subminimal diameter	Redesign drive using larger diameters
	Backside idler	Use inside idler or larger diameter backside idler
	Extreme low temperature at start-up	Pre-heat drive environment
	Extended chemical exposure	Protect drive
Premature tooth wear	Too low or high belt tension	Correct belt tension
	Belt running off unflanged sprocket	Align drive
	Incorrect belt profile for sprocket (i.e., HTD, GT, etc.)	Use proper belt/sprocket combination
	Worn sprocket	Replace sprocket
Tooth shear	Excessive shock loads	Redesign drive for increased capacity
	Less than 6 teeth-in-mesh	Redesign drive
	Worn sprocket	Replace sprocket
	Incorrect belt profile for sprocket	Use proper belt/sprocket combination
	Misaligned drive	Align drive
	Belt undertensioned	Correct belt tension

BIBLIOGRAPHY

The following are publications of the Rubber Manufacturers Association, Washington, D.C.:

IP-3 *Power Transmission Belt Technical Bulletin* The complete set of IP-3-1 through 3-13 is listed below (also available separately).

IP-3-1 *Heat Resistance of Power Transmission Belts (1989).*

IP-3-2 *Oil & Chemical Resistance of Power Transmission Belts (1999).*

IP-3-3 *Static Conductive V Belts (1995).*

IP-3-4 *Storage of Power Transmission Belts (1999).*

IP-3-6 *Use of Idlers with Power Transmission Belt Drives (1999).*

IP-3-7 *V-Flat Drives (1999).*

IP-3-8 *High Modulus Belts (1999).*

IP-3-9 *Joined V Belts (1999).*

IP-3-10 *V Belt Drives with Twist & Nonalignment (1999).*

IP-3-13 *Mechanical Efficiency of Power Transmission Belt Drives (1999).*

IP-3-14 *Drive Design Procedure for Variable Pitch Drives (1999).*

IP-20 *Specification: Joint MPTA/RMA/RAC Classical V Belts (1988).* A, B, C, and D Cross Sections.

IP-21 *Specifications: Joint RMA/MPTA Double V Belts (1991).* AA, BB, CC, DD Cross Sections.

IP-22 *Specification: Joint MPTA/RMA/RAC Narrow Multiple V Belts (1991).* 3V, 5V, and 8V Cross Sections.

IP-23 *Specification: Joint RMA/MPTA Single V Belts (1997)*. 2L, 3L, 4L, and 5L Cross Sections.

IP-24 *Specification: Joint MPTA/RMA/RAC Synchronous Belts (1997)*. MXL, XL, L, H, XH, and XXH Belt Sections.

IP-25 *Specification: Joint MPTA/RMA/RAC Variable Speed Belts (1991)*. Twelve Cross Sections.

IP-26 *Specification: Joint MPTA/RMA/RAC V-Ribbed Belts (1977)*. H, J, K, L, and M Cross Sections.

IP-27 *Specification: Curvilinear Toothed Synchronous Belts,* 8M & 14M Patches *(1997)*.

CHAPTER 5.6

SHAFT DRIVES AND COUPLINGS

PART 3

CHAIN DRIVES

R. B. Curry
S. E. Winegardner
Senior Engineers
Rexnord Corporation
Indianapolis, Indiana

INTRODUCTION

Chain drives are one of the most efficient methods used to transmit mechanical power between two or more parallel rotating shafts that cannot be directly coupled. A typical chain drive consists of a series of assembled links (commonly referred to as the chain) and two or more sprockets (Fig. 5.81). Power is transmitted through the chain by positive engagement

FIGURE 5.81 A roller chain drive operates multiple rolls in a heat-treating furnace.

with a driven sprocket to one or more drive sprockets that are keyed to rotating shafts. Roller chain, engineered chain, and silent (inverted tooth) chain are the three principal types of chain used in industrial drive applications.

Some benefits of a chain drive, compared to a belt drive, are:

- Shaft center distances are relatively unrestricted.
- Chains are easily installed and maintained.
- Chain drives provide a highly efficient, positive drive and a fixed ratio means of providing power.
- Chain drives operate in adverse environmental conditions such as high temperatures and corrosive, dusty, and generally dirty surroundings.
- Chain drives do not require pre-tension between shafts for power transmission, thereby reducing bearing loads.

Some benefits of a chain drive, compared to a gear drive, are:

- The load in a chain drive is distributed over a number of sprocket teeth simultaneously.
- Chain drives are generally more compact.
- All components rotate in the same direction, eliminating the need for additional components to maintain that feature.

This chapter encompasses basic design, application, use, and maintenance of roller chain. Additional information on this subject and other chain applications, along with sprocket selection procedures, can be obtained from manufacturers' catalogs. *The reader is cautioned that all calculations are only examples to aid in understanding concepts.* All design work should be done by those versed in this subject and/or with the cooperation of the manufacturer.

ROLLER CHAIN DESIGN

Roller chain drives are used in a wide range of power transmission applications for all basic industries such as food processing, materials handling, oil field equipment, construction, agricultural equipment, and machine tools.

Fourteen standard sizes of single- and multiple-width roller chain are listed in ANSI Standard B29.1 (American National Standards Institute, New York; see Figs. 5.82 and 5.83. Table 5.17 shows the chain number, corresponding pitch size, and the ANSI minimum ultimate tensile strength for single-width chain. This standard includes those roller chains that are intended for use in power transmission and conveying applications. These chains *are not intended* for use in overhead hoists or lifting applications. For specific information on hoist and leaf chains, see ANSI B29.24 and B29.8, respectively.

FIGURE 5.82 Single-width roller chain.

FIGURE 5.83 Multiple-width roller chain.

The chain number can be broken down to determine the pitch and general type of roller chain. The rightmost digits indicate the type, "1" indicates a lightweight chain, "5" indicates a bushing only (with no rotating roller), and "0" designates a standard roller chain. The remaining digit(s) are multiplied by ⅛ in (3.2 mm) to obtain the pitch in in (mm). Thus, a #35 roller chain is a 0.375-in (9.6-mm) pitch roller chain with an integral roller/bushing design.

The minimum ultimate tensile strength refers to the minimum allowable tensile strength that the American Chain Association and ANSI have determined that manufacturers of roller chain must meet. Reputable manufacturers can easily meet or exceed this value. The minimum ultimate tensile strength for multiple-width roller chain is equal to the minimum ulti-

TABLE 5.17 Standard Single-Width Roller Chain Size and Strength

	Chain pitch		Minimum ultimate strength	
Chain number	in	mm	lb	N
25	0.250	6.35	780	3,470
35	0.375	9.52	1,760	7,825
41	0.500	12.70	1,500	6,672
40	0.500	12.70	3,125	13,900
50	0.625	15.88	4,880	21,710
60	0.750	19.05	7,030	31,720
80	1.000	25.40	12,500	55,600
100	1.250	31.75	19,530	86,870
120	1.500	38.10	28,125	125,100
140	1.750	44.45	38,280	170,270
160	2.000	50.80	50,000	222,400
180	2.250	57.15	63,280	281,470
200	2.500	63.50	78,125	347,500
240	3.000	76.20	112,500	520,400

mate tensile strength of single-width roller chain times the number of strands in the multiple-width roller chain.

ANSI standard roller chain generally can be identified by three critical dimensions (Fig. 5.84). The first is pitch, which is the distance between the centers of adjacent pins. The second

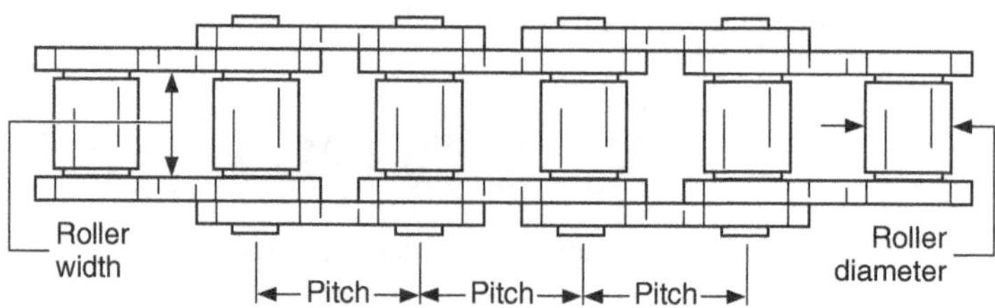

FIGURE 5.84 The three critical dimensions.

is the roller diameter, which is the outside diameter of the roller. The third is the chain width, which is the minimum distance between roller link plates. Note that the chain width is not the overall width. Both the chain width and roller diameter are approximately ⅝ of the pitch.

Basic component part nomenclature, as recognized by the American Chain Association, is shown in Fig. 5.85.

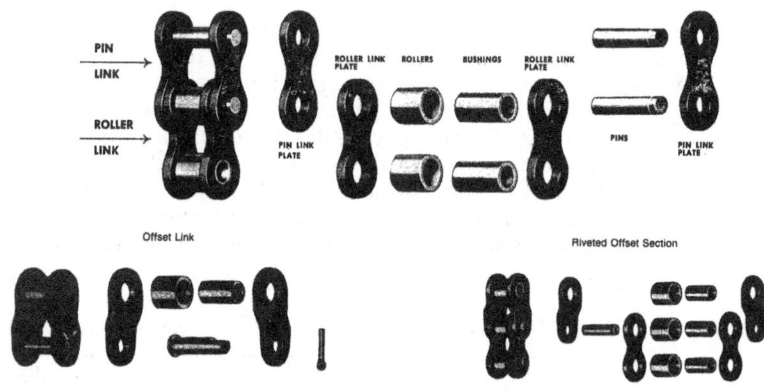

FIGURE 5.85 Basic part nomenclature.

Interchangeability between manufacturers is achieved by conformance to ANSI B29.1M. While standard roller chains from different manufacturers will intercouple and fit over the same sprockets, it is not a recommended practice. Small differences in material, heat treatment, and manufacturing specifications can contribute to premature failure of the drive.

Standard roller chain strength and life is very dependent on the quality of materials and heat treatment of component parts and press fits. Standard roller chain utilizes press fits at the bushing-to-roller link plate joints and the pin-to-pin link plate joints. These press fits are what hold the chain together during operation. Proper press fits [generally 0.003 to 0.006 in (0.08–0.16 mm)] contribute significantly to enhancement of fatigue life.

Some manufacturers go beyond a simple press fit by cold working the roller link plate and pin link plate pitch holes before assembly. This imparts residual compressive stresses around the hole and improves the geometry of the hole. Cold working can be done before or after heat treatment. Post-heat treatment yields the greatest return because the residual stresses are retained. This results in improved pin plate and bushing plate fatigue life. This is a particularly important feature to look for when the chain drive is subjected to a combination of low speed and high loading.

Speed and horsepower ratings are the prime considerations in engineering a chain drive. The ratings are normally listed for the smaller sprocket, regardless of whether it is the drive or driven member. Chain manufacturers should be consulted when special conditions such as composite duty cycles, idlers, or more than two sprockets are involved in the drive cycle.

The speed and horsepower ratings are based on approximately 15,000 h of service life at full-load operation and a service factor of 1.0. Operating conditions that establish the service factor are shown in Table 5.18. Strand factors for multiple-width chains are given in Table 5.19. Note that the strand factors are not equal to the number of widths in a multiple-width chain.

TABLE 5.18 Service Factors

	IC engine w/hydraulic drive	Electric motor or turbine	IC engine w/mechanical drive
Uniform load	1.0	1.0	1.2
Moderate shock load	1.2	1.3	1.4
Heavy shock load	1.4	1.5	1.7

TABLE 5.19 Multiple-Strand Factors for Multiple-Width Chains

Number of strands	Multiple-strand factor
2	1.7
3	2.5
4	3.3
5	4.1
6	5.0
7 or more	Consult manufacturer

Specialty roller chains have been developed for particular applications. For example, double pitch chain is an economical choice for slower speed drives on relatively long centers. Heavy series roller chain is used when conditions demand additional capacity to withstand the shortened sideplate fatigue life imposed by low-speed/high-load applications. Flexible joint type chain has been designed for smaller horsepower drives where the chain follows a guided serpentine path. Lubricated joint chains have oil-impregnated bushings to provide cleaner and longer operating life where either lubrication is restricted or external lubrication is absent.

Standard roller chain made of stainless steel is recommended for applications where high resistance to corrosive attack is required. When stainless is considered, specific attention must be paid to the corrosive environment and compatibility with manufacturers' offerings. It should be noted that most stainless chains have ⅓ to ¼ the horsepower capacity of conventional carbon steel roller chains.

SILENT CHAIN DRIVES

Silent (inverted tooth) chain is a drive chain constructed of sidebars (links), joint components, pins, and bushings that are unique to each manufacturer (Fig. 5.86). These chains do not have rollers, and the joint design of each manufacturer will prevent sections of chains of different manufacturers from being coupled together. The sidebars are designed to mesh with sprocket

FIGURE 5.86 Silent chain drive.

teeth in a gear-type engagement. Generally a silent chain drive is selected for high-speed/high-load applications and for smooth, quiet operations in industrial services such as electric generating plants, automotive test stands, machine tools, and ventilating systems.

There is a wide range of silent chain-link configurations available from various manufacturers. Most silent chain manufacturers produce chains of their own design that will transmit more horsepower at higher speeds than ANSI-standard chains. For this reason, silent chains cannot be interchanged on different manufacturers' sprockets. However, the SC series of silent chains shown in ANSI Standard B29.2 is interchangeable on sprockets among manufacturers.

ENGINEERING STEEL CHAIN DRIVES

Engineering steel chain drives are especially suited for heavy-duty applications. The normally offset sidebar chain (Fig. 5.87) can handle speeds up to 1000 ft/min (305 m/min) and power requirements as high as 500 hp (370 kW). These chains are commonly used in elevator drives, large conveyor drives, drum drives, and applications with poor operating conditions.

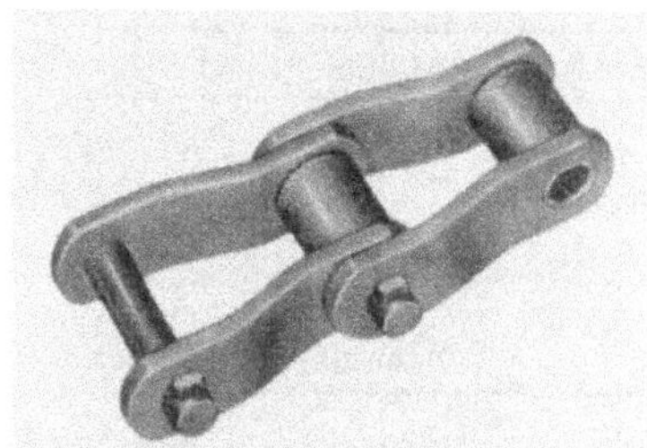

FIGURE 5.87 Engineering steel chain drive.

The eight sizes of engineering steel chain available are listed in ANSI Standard B29.10. Table 5.20 shows the chain number, the pitch size, and the minimum ultimate strength for each of the eight chains. As is the case with roller chain, speed and horsepower ratings are the prime considerations in selecting a chain drive. Normally, the ratings are listed for the smaller sprocket, regardless of whether it is the driver or driven member. Chain manufacturers should be consulted for proper drive selections when special conditions are encountered.

TABLE 5.20 Engineering Steel-Chain Size and Strength

	Chain pitch		Minimum ultimate strength	
Chain number	in	mm	lb	kn
2010	2.500	63.5	57,000	254
2512	3.067	77.9	77,000	342
2814	3.500	88.9	106,000	471
3315	4.073	103.4	124,000	552
3618	4.500	114.3	171,000	701
4020	5.000	127.0	222,000	987
4824	6.000	152.4	315,000	1401
5628	7.000	177.8	425,000	1890

DESIGN SUGGESTIONS FOR ENGINEERING A DRIVE

This section is intended to provide general suggestions or guidelines for evaluating and determining some of the mechanical design details in the drive design process.

Center Distance

Adjustable centers should be included in the initial design to allow compensation for wear. Adjustment range should be at least 1½ pitches. A center distance of 30 to 50 pitches is preferred. Minimum center distance should equal the diameter of the large sprocket plus half the diameter of the small sprocket. These proportions assure the suggested minimum chain wrap of 120° on the small sprocket. Absolute miminum center distance must be enough to provide clearance between sprocket teeth. Maximum center distance should be limited to 80 pitches. If maximum center distance exceeds 80 pitches, the catenary should be supported by one or more idlers or guides.

When sprocket centers cannot be adjusted, make a conservative drive selection by using a larger service factor than indicated. Providing good lubrication will have a positive impact on drive life, particularly drives with fixed centers. Fixed centers should have an automatically or manually adjustable idler as a method for taking up chain slack.

Chain Tightener

The purpose of a chain tightener is to remove control slack without creating excess tension. Chain tightening devices are often used for the following purposes:

- To control chain slack when drives are on fixed centers
- To control chain slack and pulsations on drives with vertical centers
- To eliminate the whipping action sometimes found in drives with long centers

The most common type of chain tightener is an idler sprocket mounted on an adjustable bracket, either manually adjusted or spring loaded. Rollers, shoes, and vibration dampeners are also used for regulating chain excess tension. The idler design and installation should be capable of providing enough adjustment to remove two full pitches of chain. This avoids the need for a less desirable offset link.

An idler sprocket should have a minimum of 17 teeth and should be located adjacent to the driver sprocket on the non-load-carrying span of chain. At least three teeth of the idler should be in full engagement. The tightener sprocket can contact the chain on either the inside or the outside. Some typical arrangements of chain tighteners are shown in Fig. 5.88.

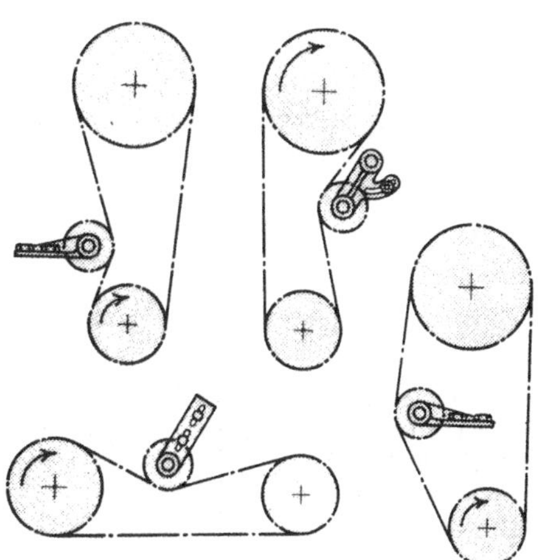

FIGURE 5.88 Chain tighteners used to control tension.

Low-Speed Drives

Ratings are not shown in the horsepower tables for extremely low speeds. For operation at these speeds, select chain on a chain strength basis. The ratio of ultimate tensile strength to working load should be at least 6:1.

Drive Ratio

The drive ratio is determined by the speeds of the driving and driven shafts. Normally, the speed ratio should be limited to approximately 7:1. Properly engineered, drives with ratios up to 10:1 will perform satisfactorily. However, double-reduction drives with smaller ratios have better operating characteristics and are often more economical than large ratio, single-reduction drives. Minimum chain wrap on the small sprocket always should be at least 120°.

Relatively large diameter sprockets should be selected for 1:1 and 2:1 ratio drives, especially if the drives required to operate on long fixed horizontal centers with the slack strand on top. This assures adequate distance between the two spans of chain and prevents them from striking as wear accumulates.

Chain Pitch

Use the smallest pitch chain that will handle the horsepower and load requirements. Single-strand chains satisfy most requirements and are usually more economical. Use small pitch, multiple-strand chains for high speed drives or when noise reduction is required. This allows a larger number of teeth in the driver sprocket (otherwise, typically small) and results in smoother operation. When the lowest possible noise generation is required, choose a smaller pitch, a wider chain, and a driver sprocket with at least 25 teeth.

Chordal Action

The rise and fall of each pitch of chain as it engages a sprocket is termed *chordal action* and causes repeated chain speed variations. As illustrated by Fig. 5.89, chordal action and speed variation decrease as the number of teeth in the small sprocket are increased.

Number of Teeth for Small Sprockets

The minimum number of teeth recommended for the small sprocket used with roller chain varies with operating conditions. The following are general guidelines:

- 50 to 100 ft/min (15–30 m/min), 15 teeth
- 100 to 300 ft/min (30–90 m/min), 17 teeth
- 300 to 500 ft/min (90–140 m/min), 19 teeth
- Over 500 ft/min (140 m/min), 23 teeth

Whenever possible, an odd number of teeth should be selected for the small sprocket. This causes the chain to continually seat on different sprocket teeth on each revolution, thereby creating a more even wear pattern on the sprocket and roller chain.

Number of Teeth for Large Sprockets

The number of teeth in the large sprocket has an appreciable effect on the amount of joint wear (or pitch elongation) that can be accommodated by the chain before it tends to jump or ride over the teeth. This is shown in Fig. 5.90. Generally, a roller chain has reached the end of its useful wear life when the elongation per pitch is in the range of 2 to 3 percent of pitch.

Drive Arrangements

Illustrated in Fig. 5.91 are the drive arrangements recommended for optimum drive life. The preferred direction of rotation is indicated, although arrangements *A, B,* and *C* will operate

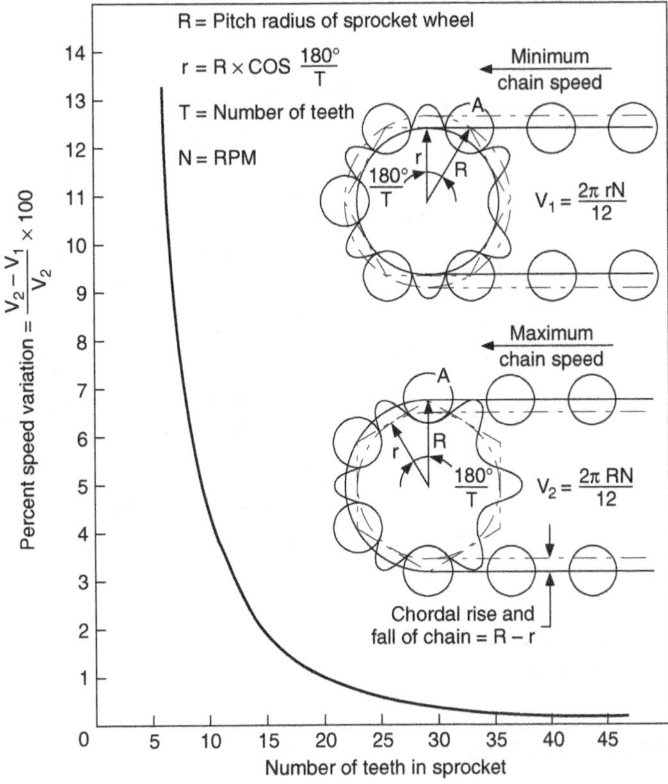

FIGURE 5.89 Variations in chain speed due to chordal action.

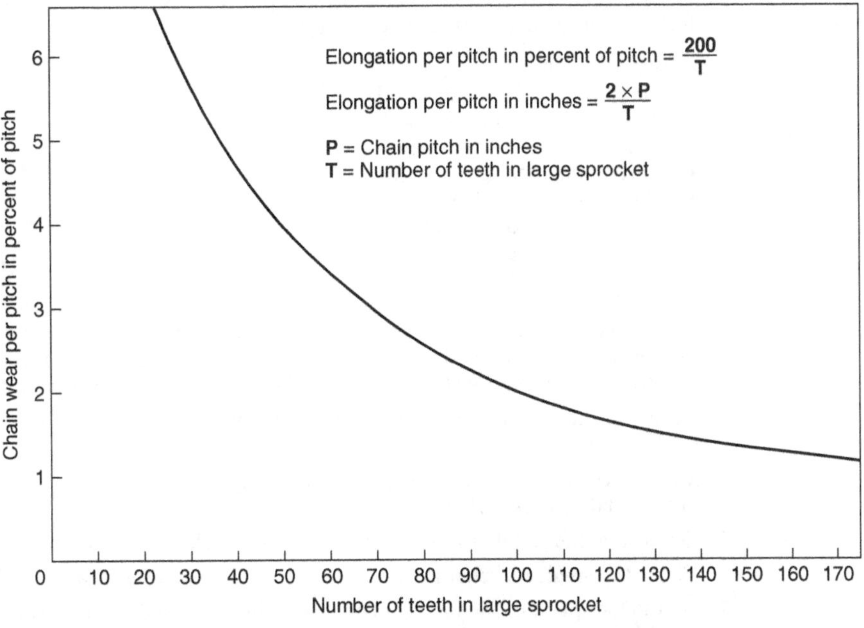

FIGURE 5.90 Variations in useful chain life based on pitch elongation and number of teeth in large sprocket.

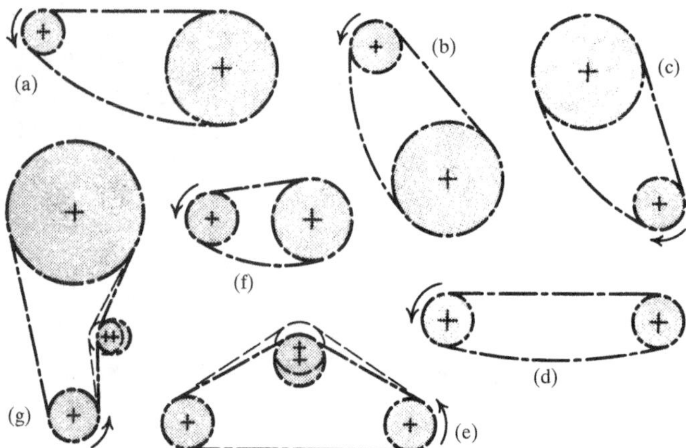

FIGURE 5.91 Preferred drive arrangements.

satisfactorily in either direction. Consult the chain manufacturer for approval on other drive arrangements.

DRIVE SELECTION PROCEDURE

The proper steps for selecting a chain drive are listed below:

1. Although horsepower and speed are the prime considerations for selecting a drive, the following information is also necessary:
   - Source of power
   - Horsepower to be transmitted
   - Size and speed of driving shaft
   - Size and speed of driven shaft
   - Approximate center distance
   - Relative positions of shafts
   - Type of driven equipment
   - Operating conditions
   - Space limitations
2. Establish the service factor from the actual operating conditions. The service factor is a factor by which the transmitted horsepower is multiplied to compensate for drive conditions. The composite or final service factor is the product of the separate service factors (see Table 5.18).
3. Calculate the factored horsepower value by multiplying the horsepower to be transmitted by the final service factor.
4. Make a trial chain selection based on the design horsepower and revolutions per minute of the small sprocket (Fig. 5.92).
5. Determine the number of teeth in the small sprocket from the speed-horsepower charts.
6. Check the small sprocket for bore capacity, number of teeth, and availability.
7. Divide the speed of the faster-turning shaft by the speed of the slower shaft to determine the drive ratio.
8. Calculate the chain length and exact sprocket centers.

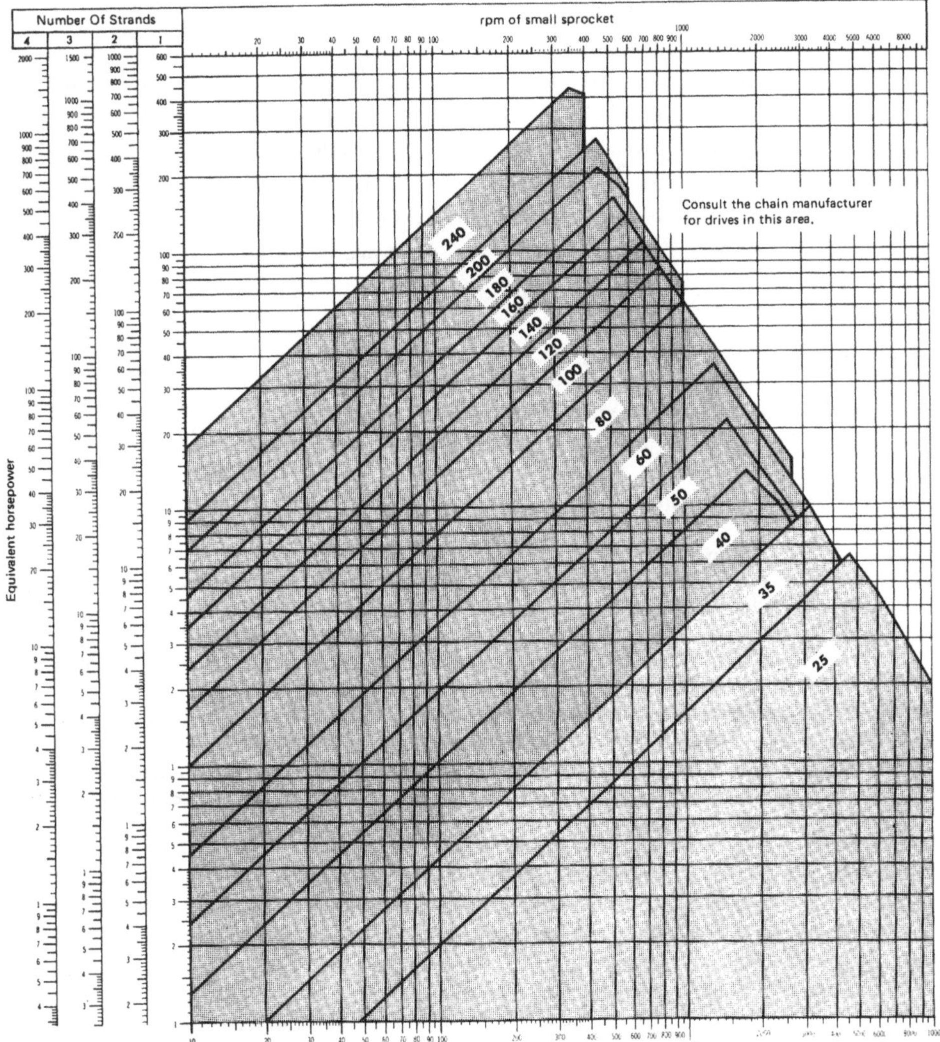

FIGURE 5.92 Trial selection chart for ANSI standard roller chains.

9. Determine the chain sag. The chain sag should be approximately 2 percent of the distance between shaft centers at the initial installation.

10. Determine the method of lubrication.

DRIVE SELECTION EXAMPLE

Application

Select a roller chain drive for the following conditions:

Source of power	Gearmotor
Horsepower to be transmitted	10 hp (7.5 kW)

Size of driving shaft	2.438 in (62 mm) diameter
Speed of driving shaft	100 r/min
Driven equipment	Uniformly fed coal elevator for power plant
Size of driven shaft	2.938 in (75 mm) diameter
Speed of driven shaft	42 r/min
Approximate center distance	24 in (610 mm)
Relative position of shafts	On same horizontal plane
Space limitations	None

Solution

Service Factor. The service factor listed in Table 5.18 for a uniformly fed elevator driven by a gear motor is 1.0.

Equivalent Horsepower. The equivalent horsepower is $10 \times 1.0 = 10$ hp.

Trial Chain. In Fig. 5.92 note that the intersection of the 100 r/min vertical line and the 10 hp (7.5 kW) single-strand horizontal line falls in the area for No. 100 chain. Thus, the trial chain is No. 100 single strand.

Small Sprocket. In Table 5.21 for No. 100 roller chain, the 100 r/min column lists 10.3 hp (7.68 kW), which corresponds closely to the equivalent horsepower of 10 required for this application. This rating is for single-strand chain when used with a 17-tooth sprocket.

Check the Small Sprocket. As shown in the rating table, the maximum bore of a 17-tooth No. 100 sprocket is larger than the 2.438-in (62-mm) bore required; therefore, the selection is satisfactory.

Drive Ratio. The drive ratio equals:

$$\frac{100 \text{ r/min}}{42 \text{ r/min}} = 2.38:1$$

Number of Teeth in Large Sprocket. The number of teeth in the large sprocket equals $2.38 \times 17 = 40.5$ teeth. Use a 40-tooth sprocket.

Center Distance and Chain-Length Computations. The chain length is a function of the number of teeth in each sprocket and the center distance between shaft centerlines. In addition, the chain must consist of an integral number of pitches, with an even number preferred in order to avoid the use of an offset assembly.

The most convenient method to obtain this information is by using center distance tables. If these tables are not available, then the values must be calculated.

A pitch diameter PD for any sprocket can be calculated when the chain pitch P and the number of teeth n or N are known.

$$\text{PD} = \frac{P}{\sin\left(\dfrac{180}{n \text{ or } N}\right)}$$

$$P = 1.250 \qquad n = 17 \qquad N = 40$$

TABLE 5.21 Speed-Horsepower Ratings for No. 100 Roller Chain

Number of teeth in small sprocket	Maximum bore, inches	Horsepower for single strand chain *																			
		rpm of small sprocket																			
		25	50	100	200	300	400	500	600	700	800	900	1000	1200	1400	1600	1800	2000	2200	2400	2600
11	2.000	1.85	3.45	6.44	12.0	17.3	22.4	27.4	32.3	37.1	32.8	27.5	23.4	17.8	14.2	11.6	9.71	8.29	7.19	6.31	1.29
12	2.250	2.03	3.79	7.08	13.2	19.0	24.6	30.1	35.5	40.8	37.3	31.3	26.7	20.3	16.1	13.2	11.1	9.45	8.19	7.19	0
13	2.500	2.22	4.13	7.72	14.4	20.7	26.9	32.8	38.7	44.5	42.1	35.3	30.1	22.9	18.2	14.9	12.5	10.6	9.23	8.10	0
14	2.813	2.40	4.48	8.36	15.6	22.5	29.1	35.6	41.9	48.2	47.0	39.4	33.7	25.6	20.3	16.6	13.9	11.9	10.3	9.05	0
15	3.250	2.59	4.83	9.01	16.8	24.2	31.4	38.3	45.2	51.9	52.2	43.7	37.3	28.4	22.5	18.4	15.5	13.2	11.4	10.0	0
16	3.500	2.77	5.17	9.66	18.0	26.0	33.6	41.1	48.4	55.6	57.5	48.2	41.1	31.3	24.8	20.3	17.0	14.5	12.6	11.1	0
17	3.813	2.96	5.52	10.3	19.2	27.7	35.9	43.9	51.7	59.4	63.0	52.8	45.0	34.3	27.2	22.3	18.7	15.9	13.8	0.79	0
18	4.188	3.15	5.88	11.0	20.5	29.5	38.2	46.7	55.0	62.3	68.6	57.5	49.1	37.3	29.6	24.2	20.3	17.4	15.0	0	
19	4.563	3.34	6.23	11.6	21.7	31.2	40.5	49.5	58.3	67.0	74.4	62.3	53.2	40.5	32.1	26.3	22.0	18.8	16.3	0	
20	4.875	3.53	6.58	12.3	22.9	33.0	42.8	52.3	61.6	70.8	79.8	67.3	57.5	43.7	34.7	28.4	23.8	20.3	17.6	0	
21	5.250	3.72	6.94	13.0	24.2	34.8	45.1	55.1	65.0	74.6	84.2	72.4	61.8	47.0	37.3	30.6	25.6	21.9	19.0	0	
22	5.625	3.91	7.30	13.6	25.4	36.6	47.4	58.0	68.3	78.5	88.5	77.7	66.3	50.4	40.0	32.8	27.5	23.4	20.3	0	
23	5.813	4.10	7.66	14.3	26.7	38.4	49.8	60.8	71.7	82.3	92.8	83.0	70.9	53.9	42.8	35.0	29.4	25.1	7.74	0	
24	6.000	4.30	8.02	15.0	27.9	40.2	52.1	63.7	75.0	86.2	97.2	88.5	75.6	57.5	45.6	37.3	31.3	26.7	0		
25	6.125	4.49	8.38	15.6	29.2	42.0	54.4	66.6	78.4	90.1	102	94.1	80.3	61.1	48.5	39.7	33.3	28.4	0		
28	7.000	5.07	9.47	17.7	33.0	47.5	61.5	75.2	88.6	102	115	112	95.2	72.4	57.5	47.0	39.4	33.7	0		
30	7.625	5.47	10.2	19.0	35.5	51.2	66.3	81.0	95.5	110	124	124	106	80.3	63.7	52.2	43.7	10.0	0		
32	8.250	5.86	10.9	20.4	38.1	54.9	71.1	86.9	102	118	133	136	116	88.5	70.2	57.5	45.2	0			
35	9.125	6.46	12.0	22.5	42.0	60.4	78.3	95.7	113	130	146	156	133	101	80.3	65.8	55.1	0			
40		7.46	13.9	26.0	48.5	69.8	90.4	111	130	150	169	188	163	124	98.1	80.3	0				
Lubrication type†		A	B					C													

* Ratings are based on a service factor of 1.
The ratings tabled above apply directly to lubricated, single strand, standard roller chains.
† Type A: Manual or drip (maximum chain speed 500 fpm)
Type B: Bath or disk (maximum chain speed 3500 fpm)
Type C: Forced (pump)

$$PD_{17} = \frac{1.250}{\sin\left(\frac{180}{17}\right)} \qquad PD_{40} = \frac{1.250}{\sin\left(\frac{180}{40}\right)}$$

$$PD_{17} = \frac{1.250}{0.18375} \qquad PD_{40} = \frac{1.250}{0.07846}$$

$$PD_{17} = 6.803 \qquad PD_{40} = 15.932$$

The center distance (CD = 24.00 in) has to be converted to chain pitches for the remaining calculations.

$$C = \frac{\mathrm{CD}}{P}$$

$$C = \frac{24.00}{1.250}$$

$$C = 19.2$$

Referring to Fig. 5.93, another calculation needed to determine the chain length required for this application is angle *a,* in degrees.

$$C = 19.2 \qquad n = 17 \qquad N = 40$$

$$a = \sin^{-1}\left(\frac{\dfrac{1}{\sin\dfrac{180}{N}} - \dfrac{1}{\sin\dfrac{180}{n}}}{2 \times C}\right)$$

$$a = \sin^{-1}\left(\frac{\dfrac{1}{\sin\dfrac{180}{40}} - \dfrac{1}{\sin\dfrac{180}{17}}}{2 \times 19.2}\right)$$

$$a = 10.964°$$

The entire length of chain in pitches L becomes

$$L = 2\,(BE + ME + KB)$$

and can be calculated by

$$L = 2\left[C\cos a + \frac{N+n}{4} + \frac{a}{360}\,(N-n)\right]$$

$$L = 2\,[(19.2 \times .982) + 14.25 + 0.700]$$

$$L = 67.61 \text{ pitches (round up to 68)}$$

As stated in the opening paragraph, chain length must be a whole number with an even number preferred, so this calculation was rounded to 68 pitches.

An approximation of chain length can be calculated by the formula:

$$L = 2C + \frac{N+n}{2} + \frac{(N-n)^2}{4\pi^2 C}$$

$$L = 38.4 + 28.5 + 0.698$$

$$L = 67.598$$

When the calculated chain length is rounded to the next even number of pitches, the center distance must be revised to reflect the longer or shorter chain length. By solving the following formula, a corrected center distance can be calculated.

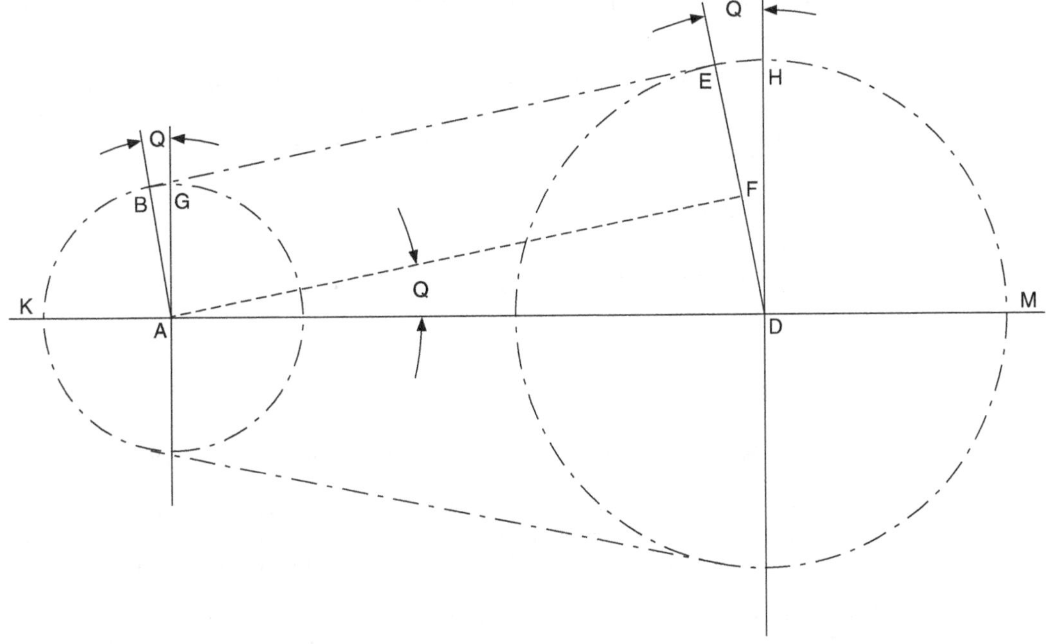

FIGURE 5.93 Calculation for chain length.

$$C = \frac{L - n\dfrac{90 - a}{180} - N\dfrac{90 + a}{180}}{2\cos a}$$

$$C = \frac{68 - 7.465 - 22.436}{1.963}$$

$$C = 19.404 \text{ pitches of chain}$$

Remember this value has to be converted from pitches of chain to a center distance in inches.

$$AD = C \times 1.250 \text{ (RC 100 pitch)}$$

$$AD = 19.404 \times 1.250$$

$$AD = 24.255 \text{ in}$$

In designing a chain drive, it is important that the theoretical center distance (as obtained from center distance tables or calculations) should not be exceeded. This means that center distances based on table or calculated dimensions require negative tolerances. When a plus tolerance is permitted, there is risk of the new chain being too tight around the sprockets, which could result in chain damage either in installation or during the break-in period.

If chain drive centers are not adjustable and chain length has to be rounded to the next larger even number, the amount of sag in the chain can be calculated by the following formula:

$$Z = \sqrt{0.375UE}$$

where Z = sag, in
U = unsupported horizontal length of catenary, in
E = excess chain, in

When extreme accuracy of chain length or center distance is required, consult a chain manufacturer.

Lubrication. The No. 100 rating table specifies type A manual or drip lubrication for a drive operating under these conditions. However, it is suggested that a casing with bath or disk lubrication be considered to extend the life of the chain and sprockets.

The drive selected for this application consists of:

17-tooth RC 100 driving sprocket

40-tooth RC 100 driven sprocket

68 pitches of RC 100 chain

LUBRICATION

Lubrication is the most important factor in maintaining high chain efficiency and providing a long service life.

The primary purpose of chain lubrication is to provide a clean film of oil at all load-carrying points where relative motion occurs. The lubrication method recommended in the speed-horsepower selection charts should be used to ensure proper lubrication at all times for a drive chain selection.

Several methods of lubrication have been developed to fit a particular range of horsepower, chain speed, and relative position of shafts. Manual or drip lubrication is used for open running drives which operate in a nonabrasive atmosphere. These methods should be confined to low-horsepower drives with a chain speed under 600 ft/min (183 m/min). Bath lubrication is the simplest automatic method of lubricating encased chain drives and is highly satisfactory for low or moderate speeds. Disk lubrication is used for moderately high-speed drive arrangements unsuitable for oil-bath lubrication. Forced lubrication or oil-pump lubrication is recommended for large-horsepower drives, heavily loaded drives, high-speed drives, or where oil-bath or oil-disk lubrication cannot be used (Fig. 5.94).

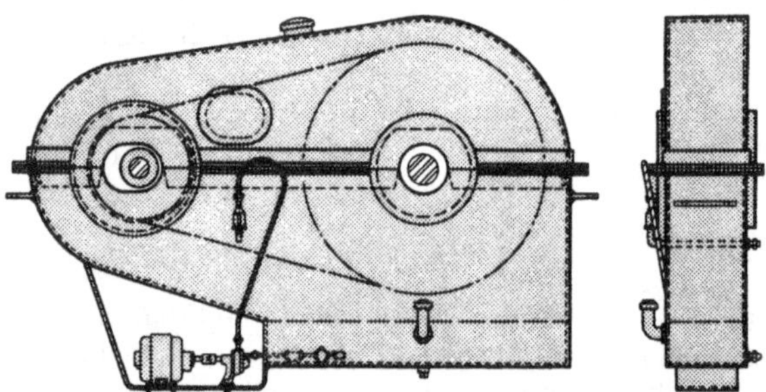

FIGURE 5.94 Forced lubrication. An oil pump is used to provide a continuous spray of oil to the inside of the lower span of chain.

Chains can be lubricated with any neutral grade of straight mineral oil in the 20 to 50 viscosity range, depending on the temperature. For difficult operating conditions, such as high temperature or an abrasive atmosphere, consult a lubricant manufacturer for proper lubricant selection.

CASINGS

Casings are important considerations in providing adequate lubrication to prolong the efficiency and service life of a chain drive. Casings contain the lubricant, exclude foreign materials from the lubricant, and function as a safety guard. There are four basic types of casings used with chain drives.

The oil-bath or oil-disk casing is used with the bath and disk methods of lubrication, with variations from the standard casing available depending on the chain speed and shaft center distances (Fig. 5.95). The second type of casing is used in outdoor applications when water,

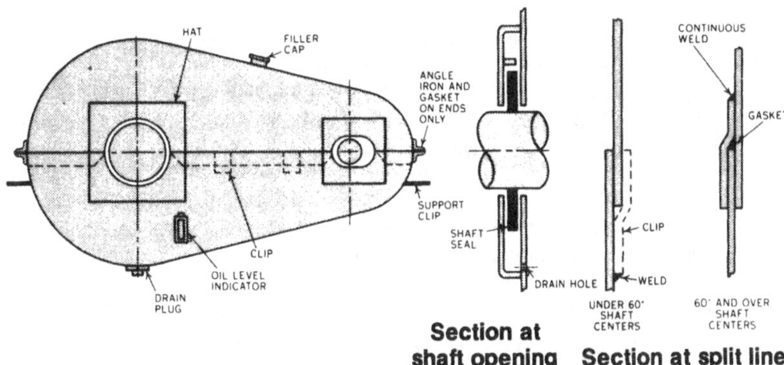

FIGURE 5.95 Oil-bath or oil-disk casing.

dust, and other contaminants are present, when extra protection against oil leakage is needed, or when forced lubrication is used (Fig. 5.96). The weather-resistant casing is the third type of construction, and it is used when additional precaution against contamination is desired. Guard-type casings are used primarily as a safety device rather than a lubrication accessory and are employed when other types of casings are not compulsory.

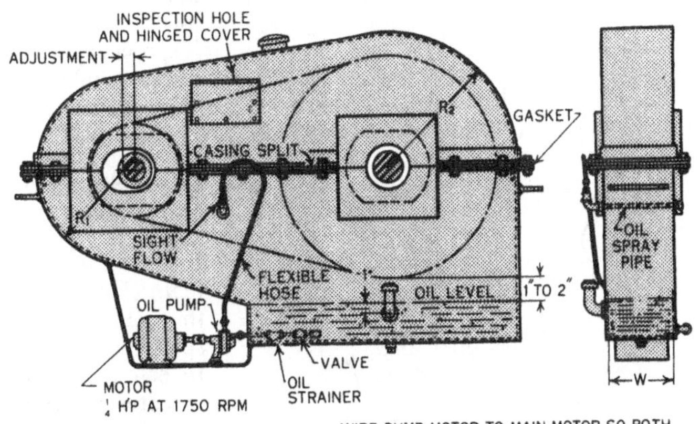

FIGURE 5.96 Casing for forced lubrication.

MAINTENANCE

Proper care of chain drives includes establishing regular periods of inspection. All drives should have an initial inspection after the first 100 h of operation, a second inspection after the next 500 h, and periodically thereafter. The periodic inspections of vertical center drives and drives subject to shock loading, reversal of rotation, or dynamic braking should be made at more frequent intervals. During the inspection, the chain tension should be carefully tightened.

Inspection of chain drives should include the following review.

Check the Drive Components

The chain and sprockets should be checked for accumulation of dirt or foreign materials. Foreign materials packed into the chain or sprockets may cause premature drive failure. A thorough visual inspection for cracked, broken, deformed, or corroded components should be completed. If any are found, the entire chain and sprocket assemblies should be replaced since other sections of the chain may be near failure as well. If an old section of chain is connected to a new section of chain, the drive will run rough and noisy and may ultimately damage the drive.

Inadequate lubrication is the most common cause of drive failure. Inadequate lubrication can be identified by one or more of the following:

- Chain joints with brown or rusty color indicate fretting or corrosive attack.
- Stiff or frozen chain joints indicate either broken components and/or significant lack of lubrication.
- Grooving or galling of pins.
- Blue or purple tint to pins indicates frictional heating. Properly lubricated chain will not show brownish color at the joints, and pins will be highly polished and smooth.

Check Sprocket Alignment

If wear is apparent on the inner surface of the roller link sidebars and on the sides of the sprocket teeth, then sprockets and/or shafts are misaligned. This condition should be corrected immediately to prevent undue wear of chain and sprocket teeth. ***Caution***: Tighten setscrew in sprocket hub after adjustment has been made. If sprockets are well aligned, check setscrews in sprocket hubs and retighten if necessary.

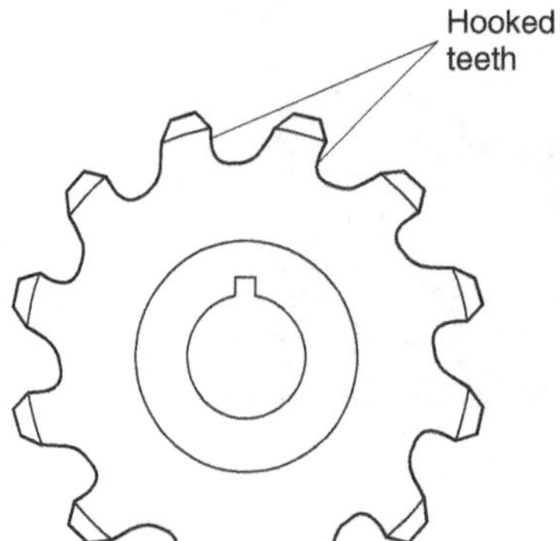

FIGURE 5.97 Worn sprocket.

Check Sprocket Tooth Wear

Excessive tooth wear will result in hook-shaped teeth (Fig. 5.97). When sprockets are found in this condition, they should be replaced. If new sprockets are not immediately available, then reverse the sprocket if physically possible. This will present the unworn tooth faces to the chain. If the worn sprocket is in a reversing drive, the sprockets must be replaced.

New chain should not be installed on worn sprockets as the unworn chain pitch will not match the worn pitch pattern of the sprockets. If new chain is installed on worn sprockets, the results will be excessive break-in wear, followed by reduced wear life and/or tensile failure of the chain due to uneven load distribution on the sprockets.

Check Chain Tension

Adjust sprocket centers to take up excess slack and provide proper chain tension. This is particularly important at the initial 100-h check since some slack will be apparent as a result of the initial seating of chain joint components. A sudden increase in slack indicates permanent yielding of components and would be the result of excessive overloading or shocks. See the "Drive Installation" section later in this chapter.

Chain elongation up to 3 percent over original chain length indicates that the chain has reached the end of its useful wear life and is riding near its limit of allowable height on the sprocket teeth. Another factor to be considered when determining allowable chain wear is the number of teeth in the large sprocket. If there are more than 67 teeth, the percentage of allowable wear then becomes less than 3 percent, or 200 divided by the number of teeth.

If the chain centers are vertical or fixed and nonadjustable, it is especially important to control the excess chain to prevent the chain from sagging away from the lower sprocket and jumping sprocket teeth. This can happen when the chain elongates approximately ½ chain pitch. If these values are exceeded, the chain *and* sprockets should be replaced.

Check Oil Level

If manual lubrication is being used, make sure that the maintenance schedule is being followed and that the oil is being properly applied. Proper application includes using the thickest oil that will still penetrate through the side plates and reach the pin and bushing bearing surface. Application of oil to the roller only is not sufficient. Grease or heavy gear oils are specifically not recommended, as they will not penetrate to the pin and bushing bearing surface.

The oil level should be checked when the drive is not operating, allowing enough time for the lubricant to accumulate in the casing sump. Also check to be sure there is no sludge accumulation. Any oil addition should be made when the drive is not operating.

Check Oil Flow

If the oil system includes a pump, the function of the pump should be checked by viewing oil flow through a site tube or flow meter. If this is not possible, the lubrication system should be operated with the drive off and checked by looking through an inspection door to make sure that adequate oil is discharging from the nozzle(s) and the discharge is striking the chain.

It is the preferred practice to start the oil system before the drive is operational to assure oil delivery to contact surfaces before loading occurs. This will further improve the wear life of drive components.

Change Oil

After the first 500 h of operation, drain the lubricant and refill the casing with fresh oil. If the system includes a filter, it should be removed and cleaned or changed if a disposable filter is used. The lubricant should be changed after each 2500 h of operation if operating conditions do not cause contamination. If at any time the lubricant is found to be contaminated, the casing and pump system should be flushed and the chain carefully recleaned. If the chain is found thoroughly dry, remove the chain and immerse in oil and reinstall.

DRIVE INSTALLATION

Accurate alignment, proper chain tension, good lubrication, and periodic inspection are required to obtain maximum chain and sprocket life. To assist you in safely installing the chain and sprockets, careful attention should be given to the following instructions.

Caution: Shut off and lock out all power to the equipment so that it cannot be started accidentally during these installation steps. Failure to do so can result in serious personal injury.

- *Shaft alignment.* Mount the sprockets on their respective shafts. As illustrated in Fig. 5.98, align the shafts horizontally with a machinist's level and adjust the shafts for parallel alignment with a vernier, caliper, or feeler bar. The distance between shafts on both sides of the sprockets must be equal. When shafts have been accurately aligned, the motor, bearings, and the like should be bolted securely in place so that alignment will be maintained during operation.
- *Sprocket alignment.* Sprockets must be in axial alignment for correct chain and sprocket tooth engagement. Apply a straight edge or heavy cord to the machined sprocket surfaces as shown in Fig. 5.99. When a shaft is subject to end float, the sprocket should be aligned for the normal running position. Tighten setscrews in sprocket hubs to guard against lateral movement and to hold key in position.
- *Chain installation.* Inspect chain to make sure it is free from dirt or grit before it is installed. Fit the chain around both sprockets, bringing the free ends together on one sprocket as shown in Fig. 5.100. Insert connecting link and secure in place.

Caution: Due to their flexibility, chains can be somewhat difficult to handle. When installing the chain, pick it up by the end links to avoid the possibility of pinching fingers or hands. Failure to do so may result in personal injury.

- *Chain tension.* Adjust drive centers for proper chain tension, as outlined in the following. Normally, horizontal and inclined drives should be installed with an initial sag equal to approximately 2 percent of sprocket centers. Vertical center drives, and those subject to shock

FIGURE 5.98 Aligning shafts.

FIGURE 5.99 Aligning sprockets.

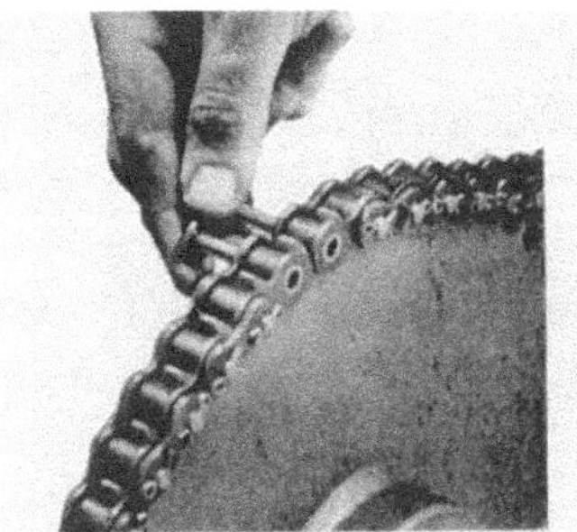

FIGURE 5.100 Inserting connecting link.

FIGURE 5.101 Determining chain sag.

loading, reversal of rotation, or dynamic braking, should be operated with both spans of chain almost taut. Periodic inspection of such drives should be made to avoid operation with excessive slack and to maintain proper chain tension.

To determine the amount of sag, pull one side of the chain taut, allowing all the excess chain to accumulate in the opposite span. As illustrated in Fig. 5.101, place a straight edge over the slack span and, pulling the chain down at the center, measure the amount of sag. If necessary, adjust drive centers for proper sag that will result in correct chain tension. (See Table 5.22.)

TABLE 5.22 Sag Based on 2% of Sprocket Centers

Shaft centers	in	20	30	40	50	60	70	80	90	100	125	150
	mm	508	762	1016	1270	1524	1778	2032	2286	2540	3175	3810
Sag	in	.50	.63	.88	1.00	1.25	1.50	1.63	1.88	2.00	2.50	3.00
	mm	12.7	16.0	22.4	25.4	31.8	38.1	41.4	47.8	50.8	63.5	76.2

Warning: The drive should be enclosed in an oil retaining casing or safety guard. Failure to enclose a drive could result in serious personal injury.

CHAPTER 5.7
ELECTRIC MOTORS

John C. Andreas
Consultant
MagneTek
St. Louis, Missouri

GLOSSARY OF COMMON MOTOR TERMS

Ambient temperature The temperature of the surrounding cooling environment.

Amperes The unit of intensity of electric current flowing in a conductor produced by the applied voltage.

*In updating this chapter for this third edition, the editor consulted with the Fasco Motors Group, Eldon, Missouri, www.fasco.com.

Full-load amperes (FLA): The current drawn by the motor when the motor is delivering its rated horsepower at its rated speed (full-load speed) and with rated voltage and frequency applied to the motor.

Locked-rotor amperes (LRA) (starting current per ampere): The current drawn by the motor with the rotor locked (zero speed) and with rated voltage and frequency applied to the motor.

Service-factor amperes (SFA): The current drawn by the motor when the motor is delivering its service-factor horsepower with rated voltage and frequency applied to the motor.

Efficiency How well a motor turns electric energy into mechanical energy, expressed as the ratio of mechanical power output (watts) to electric power input (watts).

$$\text{Efficiency \%} = \frac{\text{hp} \times 746}{\text{input watts}} \times 100$$

Alternatively, efficiency may be defined as

$$\text{Efficiency} = \frac{\text{mechanical energy out}}{\text{electrical energy in}}$$

But

$$\text{Mechanical energy out} = \text{electrical energy} - \text{motor losses}$$

Frequency The number of complete cycles of current per second made by alternating current. While sometimes called cycles per second, the preferred terminology is hertz. The standard frequency in the United States is 60 Hz; however 50 and 25 Hz can be found in the United States as well as in other industrial nations.

NEMA The National Electrical Manufacturers Association (NEMA), an organization which establishes voluntary standards and represents general practice in industry. NEMA has defined motor standards that include nomenclature, construction, dimensions, tolerances, operating characteristics, performance, quality, rating, and testing.

Phase The number of circuits over which electric power is supplied. In a single-phase motor, power is supplied in a single circuit or winding. In a three-phase system, power is provided over three circuits, each circuit reaching corresponding cyclic values at 120° intervals. AC motors are typically qualified as single-phase or polyphase.

Poles The number of magnetic poles in a motor, determined by the location and connection of the windings:

$$\text{Synchronous speed} = \frac{\text{frequency} \times 120}{\text{number of poles}}$$

Power *Mechanical power* is the rate of doing work, usually expressed in terms of horsepower.

$$\text{Power (ft} \cdot \text{lb/min)} = \frac{\text{work (ft} \cdot \text{lb)}}{\text{time (min)}}$$

$$\text{Horsepower (hp)} = \frac{\text{ft} \cdot \text{lb/min}}{33{,}000} = \frac{\text{watts}}{745.7}$$

$$\text{Horsepower (metric)} = \frac{\text{m} \cdot \text{kg/min}}{4500} = \frac{\text{watts}}{735.5}$$

Electric power is measured and expressed in watts, kilowatts (1 kW = 1000 W), or megawatts (1 MW = 10^6 W).

In *dc circuits,*

$$P = VI$$

where P = power in watts, V = line voltage in volts, and I = line current in amperes.

In *single-phase ac circuits,*

$$P = VI \times \text{PF}$$

where PF is the power factor.

For *three-phase ac circuits,*

$$P = 3\ VI \times \text{PF}$$

If the voltage and current are not in phase (as in most magnetic circuits), the volt-amp product is not actual power but apparent power (voltage times total line current) measured in voltamperes (VA) or kilovoltamperes (kVA).

Power factor The cosine of the phase angle between line voltage and current in ac circuits. The phase angle is determined by the electrical characteristics of the load. See Chap. 3.2 for a detailed discussion of power factor.

Rotor The rotating part of an electric motor.

Rated temperature rise The NEMA allowable rise in temperature for a given insulation system above ambient, when operating under maximum allowable load (i.e., service-factor load).

Stator The stationary part of an electric motor.

Service factor A multiplier which indicates what percent higher than the nameplate horsepower can be accommodated continuously at rated voltage and frequency without injurious overheating (i.e., exceeding NEMA allowable temperature rise for given insulation systems).

Slip The percentage reduction in speed from synchronous speed to full-load speed (known as *percent slip*). All ac induction squirrel-cage motors have slip.

$$\text{Percent slip} = \frac{(N_s - N_l) \times 100}{N_s}$$

where N_s = synchronous speed, r/min, and N_l = load speed, r/min.

Speed The rotational velocity of the motor shaft, measured in terms of revolutions per minute (r/min).

Full-load speed: Motor speed at which rated horsepower is developed.

No-load speed: Motor speed when allowed to run freely with no load coupled.

Synchronous speed: The synchronous speed of an ac motor is that speed at which the motor would operate if the rotor turned at the exact speed of the rotating magnetic field. However, in ac induction motors, the rotor actually turns slightly slower. This difference is the slip and is expressed in percent of synchronous speed. Most induction motors normally have a slip of 1 to 3 percent.

Squirrel cage A term used to describe the construction of one type of induction motor. The rotor is made of an iron core mounted on a concentric shaft. Copper, brass, or aluminum bars run the entire length of the core in slots on the core. These bars act as conductors and are fastened on each end of the rotor to end rings in order to form a complete short circuit within the rotor.

Torque The turning effort of a motor, normally expressed in ounce-feet (for fractional horsepower motors) or pound-feet (for integral horsepower motors). A motor developing 15 lb · ft of torque develops a force of 15 lb at the end of a 1-ft-radius lever arm.

Accelerating torque: The difference between the torque developed by the motor and torque required by the load at any given speed. This excess torque accelerates the motor and load.

Breakdown torque: The maximum torque developed by the motor at rated voltage and frequency, without an abrupt drop in speed.

Locked-rotor torque (also called *starting torque, static torque, breakaway torque*): The minimum torque developed at rest for all angular positions of the rotor with rated voltage applied at rated frequency.

Pull-in torque: The torque of a *synchronous motor* that brings the driven load into synchronous speed. (There is no corresponding term for induction motors.)

Pull-out torque: The maximum torque of a *synchronous motor* developed at synchronous speed with rated frequency and excitation.

Pull-up torque: The minimum torque developed by the motor during the period of acceleration from zero speed (rest) to the speed at which breakdown occurs.

Voltage A unit of electromotive force. One volt applied to a conductor offering 1 Ω of resistance will produce a current in that conductor of 1 A.

Wound-rotor induction motor An induction motor in which the secondary circuit consists of a polyphase winding or coils whose terminals are either short-circuited or connected to an external circuit.

INTRODUCTION

Electric-motor application for the plant engineer is the common-sense matching of load requirements with motor characteristics. Motor types, styles, sizes, mountings, and enclosures vary greatly. So the first step in using electric motors correctly is to understand them and the terminology the motor industry uses to describe them.

To help achieve such an understanding, the first part of this chapter presents a glossary of common motor terms essential to motor use and application. After that, motors and the many variables by which they are classified are discussed in terms of National Electrical Manufacturers Association (NEMA) standards, the unifying doctrine within the motor industry.

There are many ways to classify motors. But whichever one might choose, a familiarity with NEMA classifications and standards will at some point be necessary. Thus NEMA standards are as good a basis as any on which to organize a discussion of motors for the plant engineer.

NEMA is a nonprofit trade organization whose voluntary standards have been widely adopted by motor manufacturers and users alike. The NEMA standard "Motors and Generators" (MG1-1993) is designed to eliminate misunderstandings between manufacturer and purchaser and to assist the purchaser in selecting and obtaining the proper product for his or her particular needs.

Motors are classified by size, application, electrical type, NEMA design letter, and environmental protection and cooling methods. They are rated for special standard environmental and operating service conditions by performance and mechanical configuration: voltage and frequency, locked-rotor kVA, service factor, horsepower, speed, torque, locked-rotor current, performance, temperature rise, duty cycle, and frame size.

Because of the large number of motor types and configurations, this chapter is limited to motors of most interest to plant engineers. Subfractional horsepower and special-use motors (such as small-instrument, small-fan, stepping, and timing motors, for example) are only rarely specified for plant-engineering applications.

Motor types and classifications are followed by a brief guide to motor selection, the essentials of picking the right motor for either a new or replacement application. Following this application guide are discussions of two of the more practical concerns that face the electric motor user: troubleshooting and energy efficiency.

The section ends with a list of references and sources of additional information for each of the areas discussed.

MOTOR CLASSIFICATION

NEMA standards for electric motors cover frame sizes and dimensions, horsepower ratings, service factors, temperature rises, and performance characteristics. Such standards provide greater availability, more convenience in use, a basis for accurate comparison, faster repair service, shorter delivery times, and maximum mechanical and electrical interchangeability from motor to motor.

Motors are classified by size, application, electrical type, design letter, environmental protection, and cooling methods.

Classifying by Size

Virtually all electric motors used by the plant engineer can be classified as either fractional or integral horsepower. Despite the obvious distinction of horsepower, frame size (discussed later in this chapter) actually determines to which category a motor belongs.

A fractional-horsepower (FHP) motor is a motor built in a frame which is designated either by a two-digit frame number or by a three-digit frame number up to 140. An integral-horsepower (IHP) motor is one built in a frame which has a three-digit frame number from 140 to 680.

Classifying by Application

NEMA also classifies electric motors by application as general-purpose, definite-purpose, or special-purpose. A *general-purpose motor* is an induction motor which has a continuous rating, service factor, and temperature rise in accordance with NEMA standards. General-purpose motors are built in quantity in standard ratings with standard operating characteristics and mechanical construction for a wide variety of common applications.

A *definite-purpose motor,* on the other hand, is designed for specific service conditions and applications. It differs from the general-purpose motor with respect to rating, service factor, and temperature rise, one or all of which have limits much narrower than those of general-purpose motors. Definite-purpose motors conform to established NEMA standards, are produced in high volume, and are often low in cost compared with general-purpose motors of the same ratings. However, use of a definite-purpose motor for a duty other than that for which it was intended must be carefully considered.

Special-purpose motors incorporate specialized operating characteristics and/or mechanical construction to serve one-of-a-kind applications not satisfied by general- and definite-purpose motors.

Because of both the limited scope of this discussion and the fact that the great majority of plant applications require general-purpose motors, only general-purpose motors are discussed here.

Classifying by Electrical Type

Figure 5.102 illustrates the family of ac and dc general-purpose motors which together can serve virtually all needs of the plant engineer. This family is organized according to the characteristics of the electric power driving the motor and the variations in motor winding and rotor configuration.

AC Motors

Single-Phase Induction Motors. Alternating current motors fall into three major categories: single-phase, polyphase, and universal (ac-dc). Single-phase induction motors are inherently unable to start themselves. Thus these motors are classified by their means of starting as well as by their basic design, either induction or synchronous. Induction motors fall into two further categories: squirrel-cage and wound-rotor.

The squirrel-cage motor consists of a wound stator and laminated, cylindrical iron-core rotor. Cast-aluminum conductors imbedded within the rotor and short-circuiting end rings form a "squirrel-cage" configuration.

Split-Phase Motors. Split-phase motors use two windings—a main winding and a start winding (Fig. 5.103). The high resistance of the start winding creates a phase shift and induces a torque that causes initial motor rotation and acceleration. When a predetermined speed (the cutout point) is attained, a centrifugal mechanism opens the start-winding circuit. The motor then accelerates with only the main winding energized and runs as an induction motor. In some designs a current-sensing relay is used instead of a centrifugal switch.

Capacitor-Start Motors. Capacitor-start motors are similar to split-phase motors except that a capacitor is placed in series with the start winding to produce greater starting and accelerating torque (Fig. 5.104). After the start winding is removed from the circuit by a centrifugal or electronic switch, performance is identical to that of split-phase motors.

Permanent Split-Capacitor Motors. Permanent split-capacitor motors also have a start winding with a capacitor (Fig. 5.105). Because the capacitor and start windings are continuously energized, these motors operate at a higher power factor than other designs, although at the expense of a lower locked-rotor torque. Since no centrifugal switch is needed, the motor is usually shorter and often more reliable than other single-phase designs. These motors are usually used to drive fans or pumps with low starting torque requirements.

Two-Value-Capacitor Motors. Two-value-capacitor motors have both a switched-start capacitor and a run capacitor to improve full-load current, starting torque, and power factor (Fig. 5.106). Both are connected in parallel to the start winding, with the start capacitor disconnecting as the motor accelerates. These motors provide good overall torque characteristics and are quiet-running.

Shaded-Pole Motors. Instead of a start winding, shaded-pole motors have a continuous solid-copper loop around a small portion of each salient pole (Fig. 5.107). This shading coil causes the reaction necessary to start the motor, but produces rather low starting and accelerating torque. Because of their low starting torque, shaded-pole motors are best suited to light-duty applications such as direct-drive fans and blowers. Efficiency and power factor are also lower than that of other single-phase motors.

Wound-Rotor Motors. Wound-rotor motors have a stator winding connected to the power source and a rotor winding connected to a commutator.

Unlike squirrel-cage induction motors, wound-rotor motors have controllable speed and torque. Their application is considerably different from squirrel-cage motors because of the accessibility of the rotor circuit. Various performance characteristics can be obtained by inserting different resistances in the rotor circuit.

Wound-rotor motors may be used as constant-speed or as adjustable-speed motors. They are frequently used where high locked-rotor and accelerating torque with low starting current are required.

Repulsion Motors. Repulsion motors have an armature winding, commutator, and brushes (Fig. 5.108). Brushes are short-circuited and shifted to give the effect of two stator windings: a field winding at right angles to the brush axis and an induction winding along the

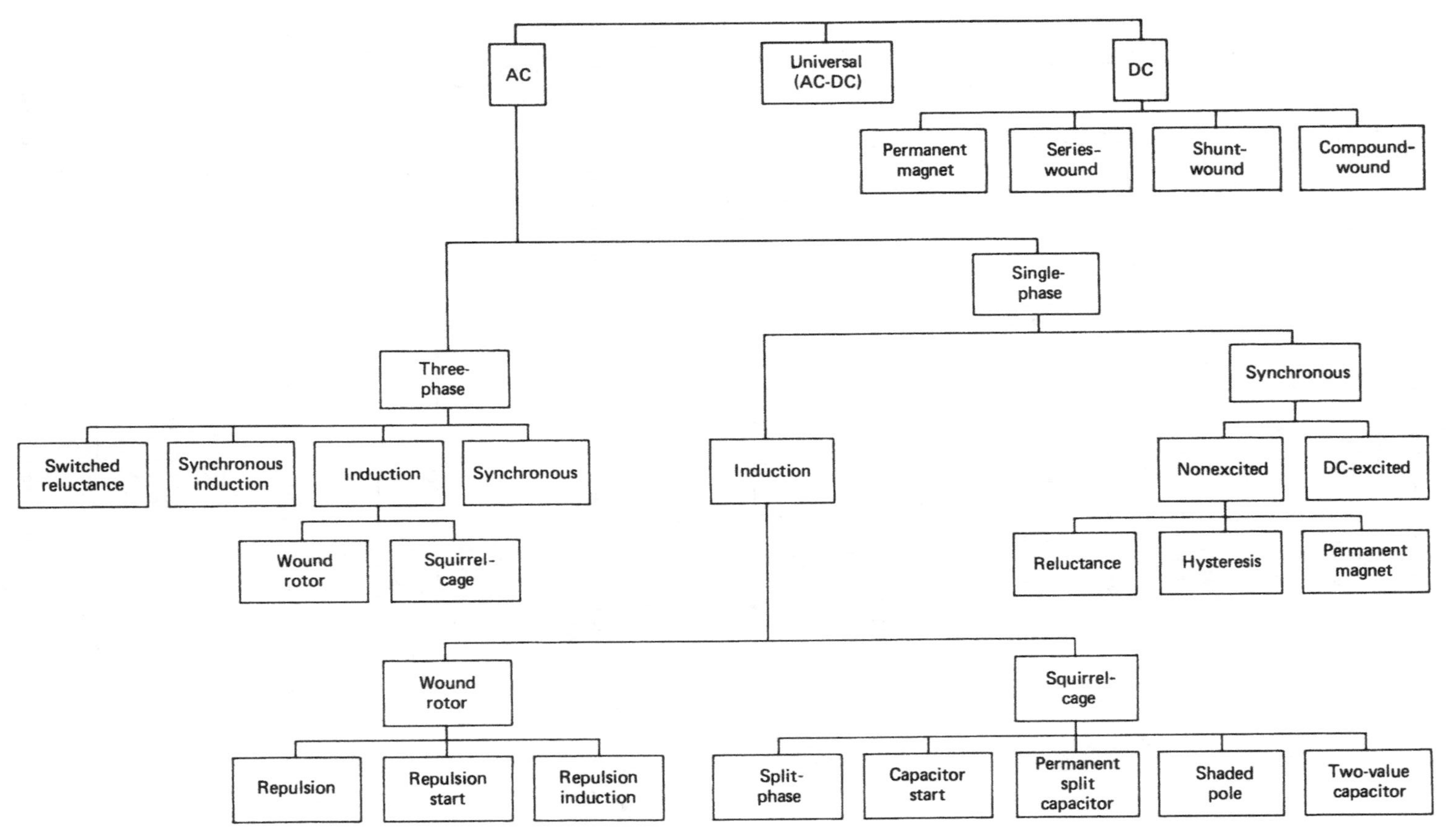

FIGURE 5.102 Electric motor family for plant engineers.

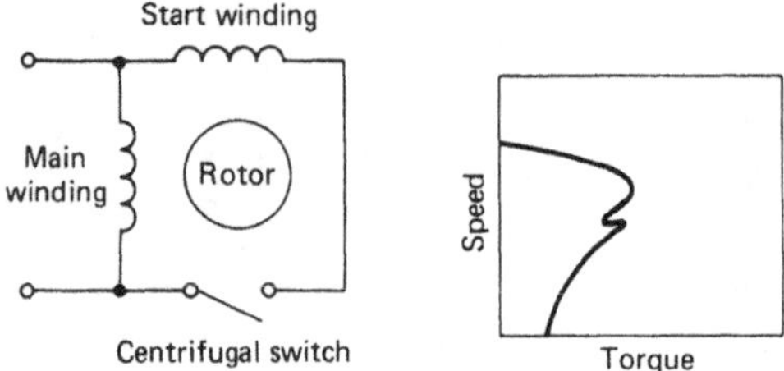

FIGURE 5.103 Schematic diagram and speed vs. torque diagram—ac split-phase motor.

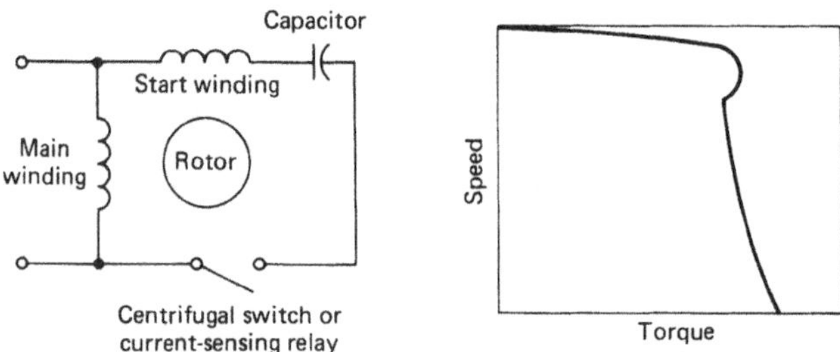

FIGURE 5.104 Schematic diagram and speed vs. torque diagram—ac capacitor-start motor.

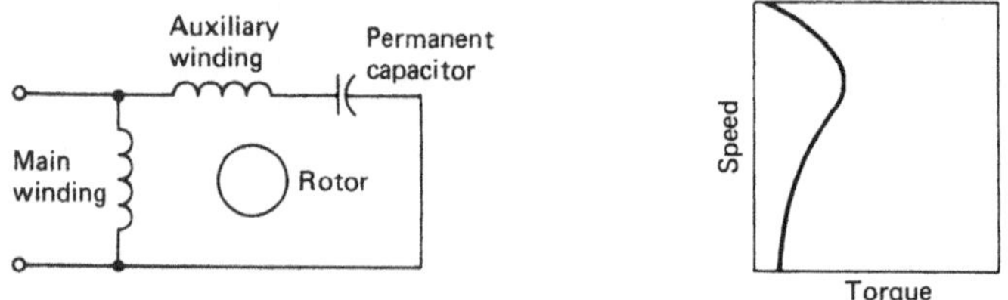

FIGURE 5.105 Schematic and speed vs. torque diagram—ac permanent split-capacitor motor.

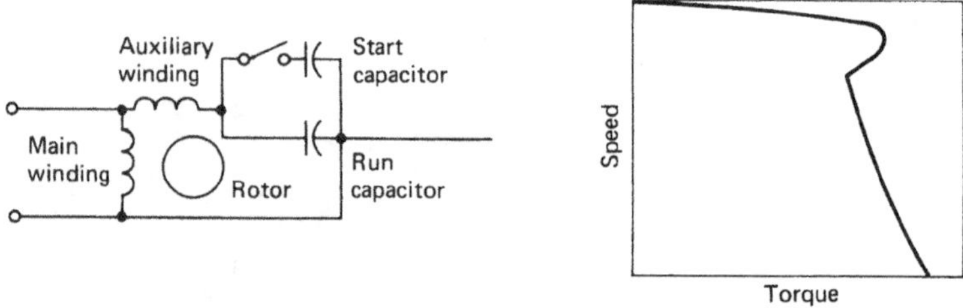

FIGURE 5.106 Schematic and speed vs. torque diagram—ac two-value-capacitor motor.

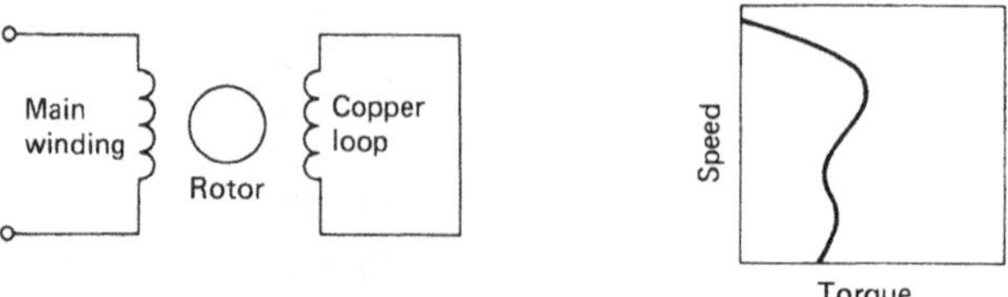

FIGURE 5.107 Schematic and speed vs. torque diagram—ac shaded-pole motor.

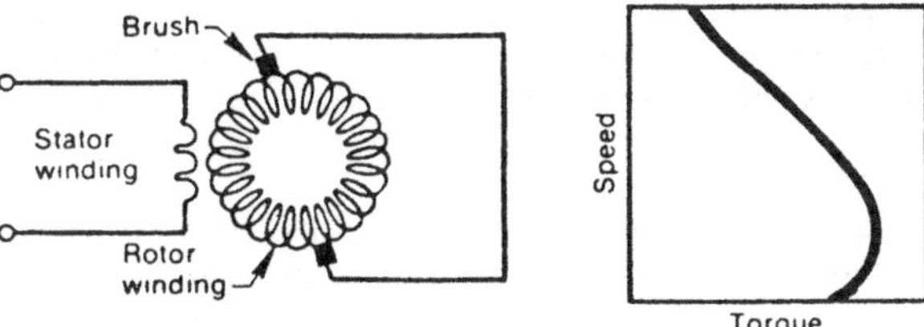

FIGURE 5.108 Schematic and speed vs. torque diagram—ac repulsion motor.

brush axis. The induction winding induces current in the armature winding that reacts with the magnetic field set up by the field winding to produce starting torque. Repulsion motors feature good starting characteristics and are often used for heavy, hard-to-start loads.

Repulsion-Start Motors. Repulsion-start motors are repulsion motors with a centrifugal switch (Fig. 5.109). At about 75 percent of synchronous speed, the switch short-circuits the commutator bars and the motor performs like a squirrel-cage motor. Repulsion-start motors are expensive and no longer widely used in industry.

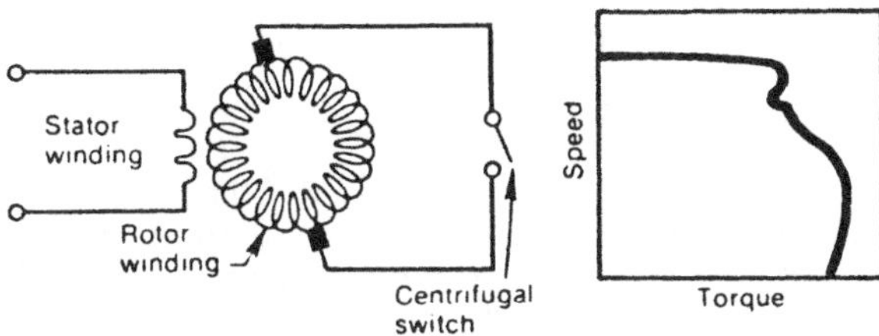

FIGURE 5.109 Schematic and speed vs. torque diagram—ac repulsion-start motor.

Repulsion-Start Induction-Run Motors. Repulsion-start induction-run motors are simply repulsion-start motors with the addition of a squirrel-cage rotor winding to improve speed regulation (Fig. 5.110). At a predetermined speed, a centrifugal switch shorts the commutator and the motor operates as a squirrel-cage induction motor. These motors are ideal for applications requiring high starting torque and low starting current.

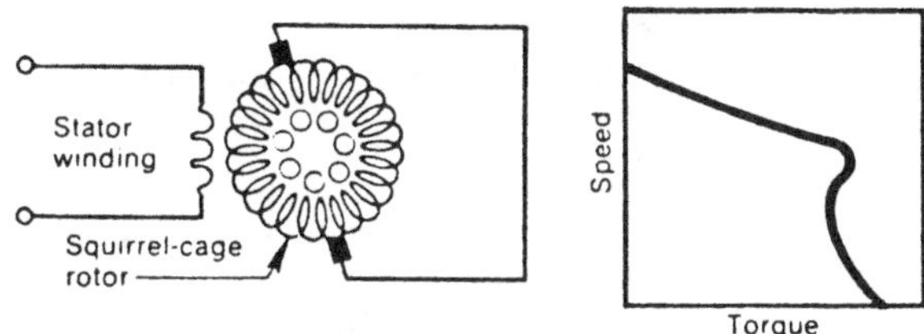

FIGURE 5.110 Schematic and speed vs. torque diagram—ac repulsion-induction motor.

Single-Phase Synchronous Motors. Single-phase synchronous motors are constant-speed motors that operate in synchronism with line frequency. As with squirrel-cage induction motors, speed is determined by the number of pairs of poles and is always a ratio of the line frequency.

Synchronous motors range from subfractional self-excited units to large horsepower, dc-excited motors for industrial drives. In the single-phase fractional-horsepower range, synchronous motors are primarily used where precise constant speed is required.

Like single-phase induction motors, synchronous motors cannot start themselves. They employ self-starting circuits.

Nonexcited Reluctance Motors. Reluctance synchronous motors have squirrel-cage construction with salient poles (Fig. 5.111). The rotor has one cutout for each pole, which together cause magnetic reluctance to be greater between poles than along the axis. The motor locks into synchronism in less than one cycle of applied voltage.

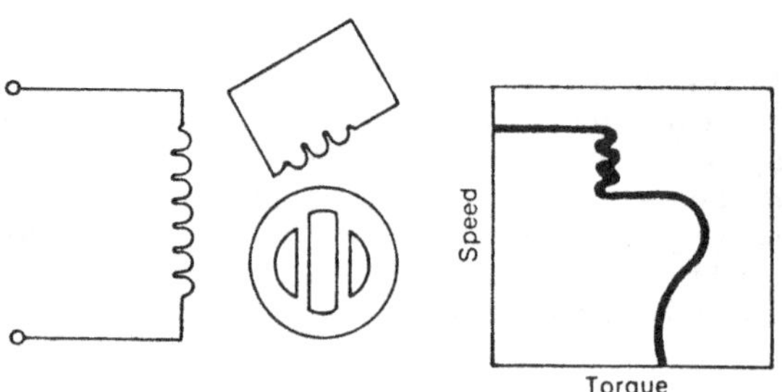

FIGURE 5.111 Schematic and speed vs. torque diagram—ac reluctance motor.

Efficiency and power factor are lower than for dc-excited synchronous or squirrel-cage induction motors. However, the motor is inexpensive, simple, and suitable for light loads.

Nonexcited Hysteresis. Hysteresis motors have no physical pole arrangement on their rotors but develop fixed magnetic poles in some random angular position as they reach synchronous speed (Fig. 5.112).

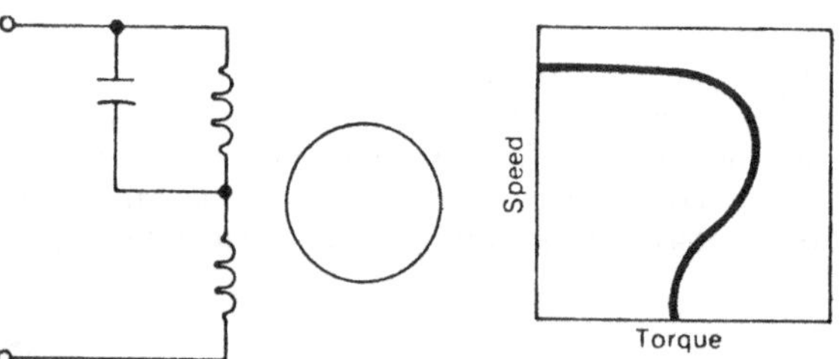

FIGURE 5.112 Schematic and speed vs. torque diagram—ac hysteresis motor.

Used as timing motors or for applications requiring precise constant speed, they are of secondary interest to plant engineers.

Nonexcited Permanent-Magnet Motors. Permanent-magnet motors have permanent magnets embedded in a squirrel-cage-type rotor (Fig. 5.113). They produce fixed poles that lock into step with the armature field. Because of its relatively high efficiency and power factor, this motor is popular in the fractional- and lower-integral-horsepower range.

Three-Phase Motors. The rotating magnetic field provided by three-phase ac power permits a simple and low-cost means of constructing an electric motor. In general-purpose use, three-phase motors require no start windings, switches, or starting or running capacitors, thereby eliminating some major sources of failures in single-phase motors.

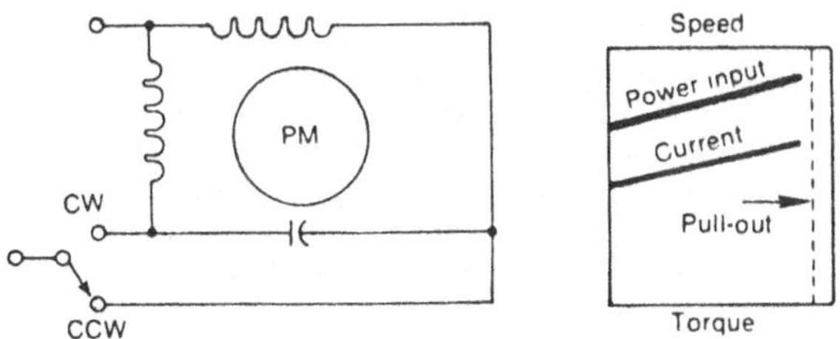

FIGURE 5.113 Schematic and speed vs. torque diagram—ac permanent-magnet motor.

The horsepower of three-phase motors ranges from ½ to 2500 or more. Starting current required is low to medium, about 5 to 7 times full-load current.

Three-phase motors can be easily reversed electrically, making them useful for applications involving control of direction of rotation or remote positioning. Different combinations of speed-torque characteristics are also available so that motor performance can be specifically matched to an application.

Three-Phase Induction Squirrel-Cage Motors. Three-phase squirrel-cage motors are basically constant-speed machines, although operating characteristics can be varied to some degree by modifying the rotor design. These variations produce predictable changes in torque, current, and full-load speed.

Evolution and standardization within the motor industry have resulted in five fundamental types of three-phase induction motors known by NEMA design letters (discussed later in this section).

Three-Phase Synchronous Induction Motors. The synchronous induction motor has a conventional three-phase stator winding. The rotor is modified to provide cutouts for a specific number of poles on the rotor face to match the polarity of the stator winding and flux guiding barriers are included to improve the pull-in and pull-out torques. Figure 5.114 shows a typical rotor configuration for a two-pole synchronous induction motor. Similar configura-

FIGURE 5.114 Synchronous induction motor rotor configuration.

tions are used for four-pole and six-pole synchronous induction motors. These motors start as an induction motor and at operating speed pull into synchronism and run at synchronous speed. The locked rotor current is considerably higher than the locked rotor current for the same horsepower induction motor. The power factor and efficiency are lower. However, they have the advantage of operating at synchronous speed. A typical speed-torque curve is shown in Fig. 5.115. The pull-in torque is a function of the total mechanical inertia of the driven system. The motors have been built in ratings up to 100 hp. They are usually applied to multimotor drives, where all of the motors in the system must stay in synchronism.

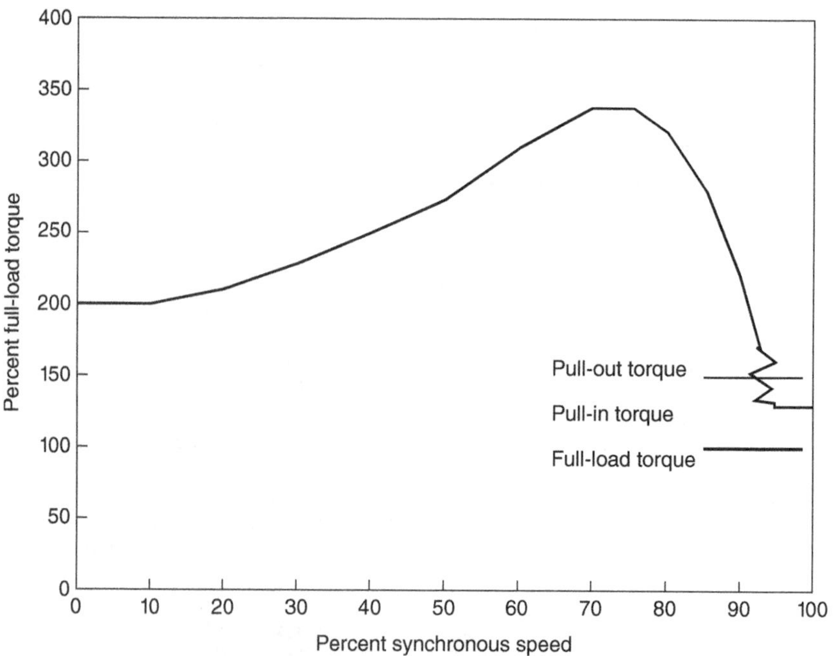

FIGURE 5.115 Synchronous induction motor speed-torque curve.

Three-Phase Induction Wound-Rotor Motors. Wound-rotor motors offer more speed and torque control than squirrel-cage induction motors, as well as the major advantage of accessibility of the rotor circuit. Performance characteristics can be varied merely by inserting different values of resistance in the rotor circuit.

Wound-rotor motors may be used as either constant-speed or adjustable-speed motors. With full load, the speed may be reduced by as much as 50 percent of synchronous speed for certain fixed loads such as fans or compressors. These motors are frequently used when high locked-rotor and accelerating torque with low starting current are required. They are also used where heavy or delicate loads must be accelerated gradually and smoothly, as in hoists and elevators. A variety of solid-state control systems are available for use in the rotor circuit of wound-rotor motors.

AC Three-Phase Synchronous Motors. Like single-phase motors, three-phase synchronous motors cannot start by themselves. One of two starting methods is used in most motors: dc excitation or reluctance.

Synchronous motors employing dc excitation, although offered in sizes as small as 20 hp, are primarily used in applications requiring 50 to several thousand horsepower. High-speed synchronous motors (from 514 to 1800 r/min) are normally used for the same applications as NEMA design A, B, or F squirrel-cage motors. Low-speed synchronous motors (below 450 r/min) are usually used as direct-connected drives for compressors and pumps, where they are more economical than induction motors with gear, chain, or belt drives.

These synchronous motors usually have pole-face squirrel-cage-type windings for both starting and damping. The motor is started as an induction motor with the dc field shorted, and at a predetermined speed the field is excited and the motor pulls into synchronism.

Switched Reluctance Motors. The switched reluctance motor (SRM) is a motor with double saliency with an unequal number of rotor and stator poles. The torque is produced by the tendency of the rotor poles to align with the poles of the excited stator phase and is independent of the direction of the phase current. The motor is singly excited from stator windings that are concentric coils wound in series on diagonally opposite stator poles. Figure 5.116 is a simplified diagram of an SRM drive. Shown in this figure is the phase winding and the switching circuit for one of the phases. Continuous rotation of the rotor is obtained by exciting the stator phases sequentially, the rotor stepping around in a direction opposite to that of the stator phase excitation. Output torque and speed depend on the control switching system including frequency, conduction angle, and ignition angle. Figure 5.117 illustrates the power output for different ignition angles and conduction angles. Because of their simple construction and relatively simple control system, these motors are finding applications in adjustable-speed drive systems.

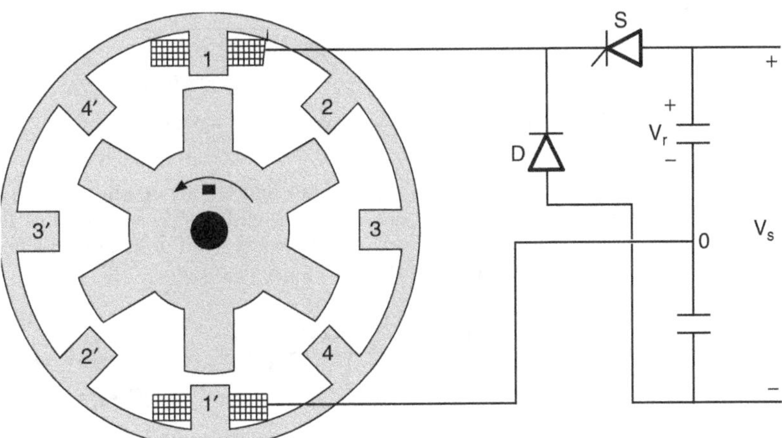

FIGURE 5.116 Simplified SRM motor drive system.

DC Motors. DC motors see a wide variety of industrial applications because their speed-torque relationships can be varied to almost any useful form—for both motor and regeneration applications and in either direction of rotation. Many dc motors can be operated continuously over a speed range of 8:1. Speed control down to zero for short durations or for driving reduced loads is also common.

AC motors lose speed rapidly and sometimes stall at loads above twice their rated torque. DC motors, by contrast, are often applied where they momentarily deliver 3 or more times their rated torque. And in emergency situations, dc motors can supply over 5 times rated torque for a limited time without stalling if the required power is available.

DC motor speed can be regulated smoothly down to zero, immediately followed by acceleration in the opposite direction without power circuit switching. DC motors also respond quickly to changes in control signals due to their high torque-to-inertia ratio.

Wound-field dc motors are classified by the type of motor field: shunt-wound, series-wound, and compound-wound. Permanent-magnet types are also popular, normally as fractional-horsepower motors.

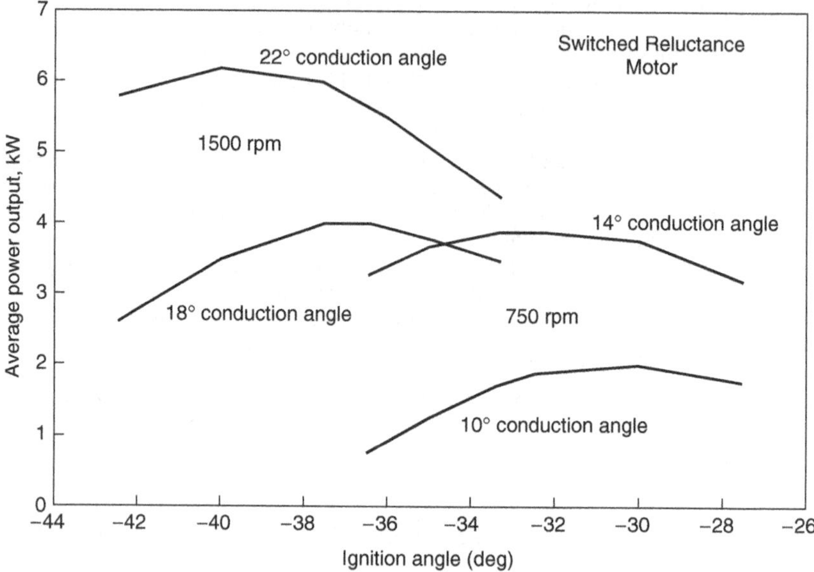

FIGURE 5.117 Switched reluctance motor output power.

Shunt-Wound Motors. Shunt-wound and stabilized shunt-wound dc motors can supply both constant speed at any control setting and a wide speed range that is field-controllable (Fig. 5.118). Most shunt motors are operated from adjustable voltage power supplies and, therefore, do not need auxiliary starting provisions.

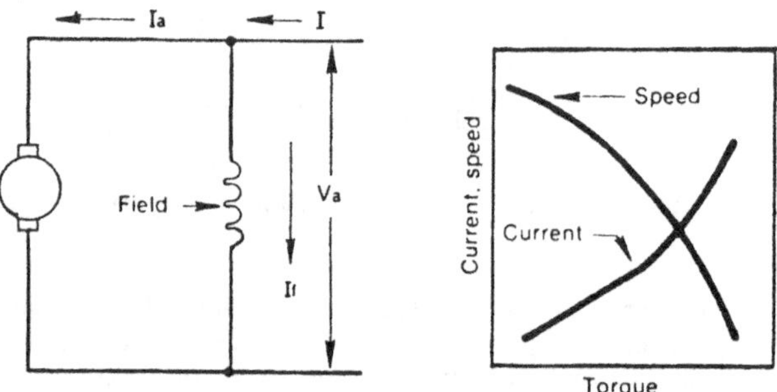

FIGURE 5.118 Schematic and speed vs. torque diagram—dc shunt-wound motor.

A stabilizing winding helps prevent speed increases as the load increases at weak field settings. This winding has disadvantages in reversing applications, however, because it must be reversed with respect to the shunt winding when the armature voltage is reversed. Reversing contactors are normally used.

The shunt winding can either be connected to the same power supply as the armature (self-excited) or be separately excited. Care must be taken never to open the field of a shunt-wound motor that is running unloaded. The loss of field flux causes motor speed to increase to dangerously high levels.

Series-Wound Motors. In series-wound motors, the field flux is created by coils that are electrically in series with the armature (Fig. 5.119). When the motor starts, the current and, consequently, the magnetic field are at maximum values, producing a large starting torque. As the motor speeds up and the current is reduced, the field flux also becomes smaller. With no external load on the shaft, the field flux drops nearly to zero and motor speed becomes dangerously high. For this reason, series-wound motors should be used only where the load is directly connected or geared to the shaft.

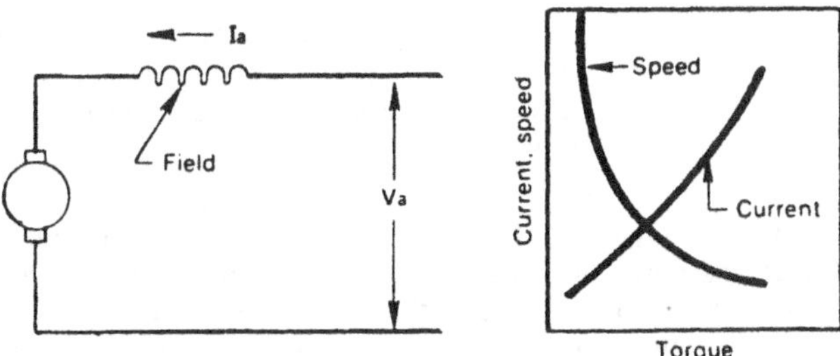

FIGURE 5.119 Schematic and speed vs. torque diagram—dc series-wound motor.

Compound-Wound Motors. Compound-wound motors combine both series and shunt fields (Fig. 5.120). The disadvantage of series-motor overspeeding at light loads is avoided since there is so little current in the series field at no load that speed is determined by the shunt field alone. At higher loads, speed depends on the sum of the two fields, making speed reduction similar to that of a series motor.

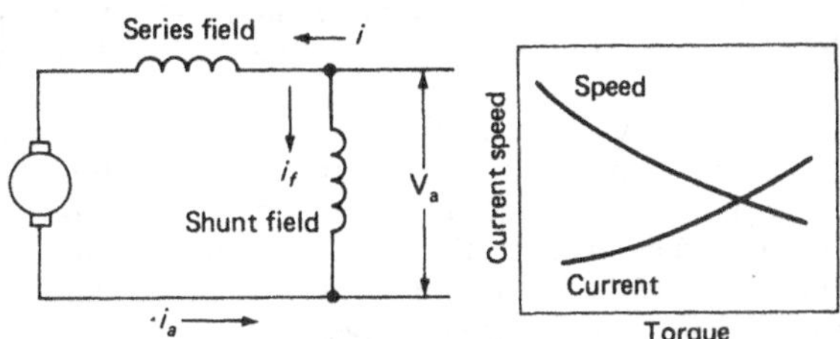

FIGURE 5.120 Schematic and speed vs. torque diagram—dc compound-wound motor.

Compound motors have high starting torques and fairly flat speed-torque characteristics at rated load. Because of the elaborate circuits needed to control compound motors, however, only large bidirectional types are built.

Permanent-Magnet Motors. Permanent-magnet motors have fields supplied by permanent magnets (Fig. 5.121). Those fields create two or more poles in the armature by passing magnetic flux through it. The magnetic flux causes the current-carrying armature conductors to move, creating a torque. This flux remains basically constant at all motor speeds; speed-torque and current-torque curves are linear.

Permanent-magnet motors, available in fractional- and low-integral-horsepower sizes, have several advantages over field-wound types. Excitation power supplies and associated wiring are not required, and reliability is improved. Efficiency and cooling are also improved by elimination of the power loss associated with an excited field.

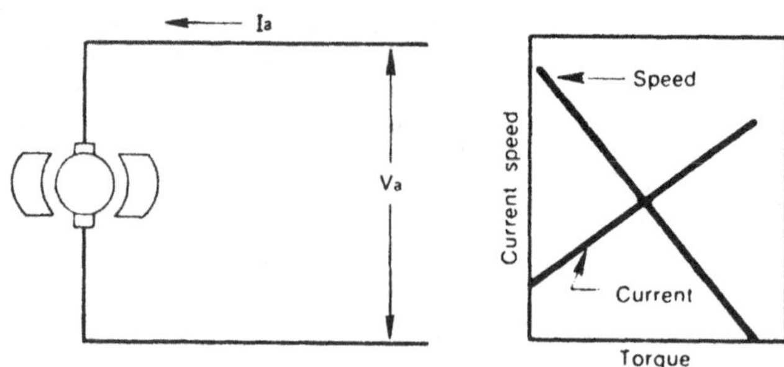

FIGURE 5.121 Schematic and speed vs. torque diagram—dc permanent-magnet motor.

Disadvantages are the absence of field control and special speed-torque characteristics. Overloads may also cause partial demagnetization that changes motor speed and torque characteristics until magnetization is fully restored.

Permanent-magnet motor performance is actually a compromise between compound-wound and series-wound motors. It has better starting torque but approximately half the no-load speed of a series-wound motor. In applications where compound motors are traditionally used, the permanent-magnet motor can offer slightly higher efficiency and greater overload capacity. In series-motor applications, permanent-magnet motors provide a cost advantage.

Brushless DC Motors. The brushless dc motor consists of a rotor with permanent magnets mounted on it—usually permanently bonded to the rotor structure—and a stator with windings similar to an ac induction motor. In addition, an electronic controller is provided to control the switching of the stator winding. You might consider this as an electronic commutator. The advantage of this type of motor is the simplicity of the ac motor type of construction with the excellent performance of a dc drive system. This eliminates the mechanical commutator and its maintenance. The disadvantage, compared to a dc motor drive system, is the cost. The advantage is low maintenance and superior performance. This type of system, with proper ventilation, can provide precise full load torque down to zero speed.

Universal Motors. Universal motors are essentially series motors that operate with nearly equivalent performance on direct current or alternating current up to 60 Hz. They differ from dc series motors in that they have different winding ratios and thinner iron laminations. A dc series motor runs on alternating current, but inefficiently. A universal motor, however, runs on direct current with essentially equivalent ac performance, but with poorer commutation and brush life than an equivalent dc series motor.

A universal motor has the highest horsepower per pound ratio of any ac motor because of its ability to operate at much greater speeds than any other 60-Hz motor. It is ideally suited for operation at a rated output, where an occasional overload or intermittent heavy load occurs. Stall torque may be as much as 10 times the continuous rated torque. The motor may even be operated in a stalled condition for short periods of time.

High starting torque, adjustable-speed characteristics, small size, and economy are all advantages of the universal motor. Universal motors are not more widely used, however, because their operating life is shorter, their size range is limited [to about 2 hp (1.5 kW)], and their very high speeds limit their applications.

Universal motors can be built to deliver speeds ranging from 4000 to 24,000 r/min and rated power from 0.1 to 1 hp (75 to 750 W). Efficiency varies from 30 percent for small sizes to 75 percent for large sizes.

Metric Motors. There is an increasing trend to express electric motor output in kilowatts rather than horsepower. This is consistent with the International Electrotechnical Commission (IEC) standards for rotating electrical machines including induction motors.

Table 5.23 shows a comparison of the units of measure in the International System (metric) and the American customary system, as related to induction motor rating. This table also shows the induction motor kilowatt output equation for both systems of measurement and the conversion from one system to the other.

TABLE 5.23 Comparison of International System (SI) Units (Metric) and American Customary System Units

Quantity	International System units (metric)	American customary units
Linear measure	Millimeter	Inch
Angular measure	Degree	Degree
Mass	Kilogram	Pound
Force	Newton	Pound-force
Torque	Newton meter	Pound-force foot
Power	Kilowatt	Horsepower

Output equation:

$$\text{kW output} = \frac{T \times n}{K \times 1000}$$

where T = torque
n = speed, r/min
K = 9.549 if torque is in newton meters
= 7.043 if torque is in pound-force feet

Conversion of torque:

1. Pound-force feet to newton meters

$$\text{Newton meters} = 1.356 \times \text{pound-force feet}$$

2. Newton meters to pound-force feet

$$\text{Pound-force feet} = 0.7376 \times \text{newton meters}$$

Conversion to horsepower:
In American customary units

$$\text{Horsepower} = \frac{\text{kW}}{0.7457}$$

In metric units

$$\text{Horsepower} = \frac{\text{kW}}{0.7355}$$

Table 5.24 shows the IEC preferred kilowatt output ratings converted to equivalent horsepower and the NEMA standard horsepower ratings.

In order to compare the physical dimensions of the IEC and NEMA induction motors Table 5.25 shows the comparison of the IEC standard induction motor shaft heights H and the nearest NEMA standard shaft height D, for the T frame line of NEMA induction motors. The shaft height dimension H in millimeters for the IEC standard and the D dimension in inches for the NEMA standard motors are defined as the distance from the centerline of the shaft to the bottom of the mounting feet. For other mounting dimensions, the specific metric motor should be determined.

Classifying by Design Letter

Three-phase, squirrel-cage, IHP motors are the most widely used ac induction motors in industry. Classification of performance requirements has results in NEMA standardized designs that

TABLE 5.24 Comparison of IEC (Metric) Preferred Output Ratings and NEMA Standard Horsepower Ratings

IEC preferred rating, kW	Equivalent horsepower, rounded	NEMA standard horsepower
0.06	0.080	1/12
0.09	0.121	1/8
0.12	0.161	1/6
0.18	0.241	1/4
0.25	0.335	1/3
0.37	0.496	1/2
0.55	0.737	3/4
0.75	1.005	1
1.1	1.475	1.5
		2
2.2	2.95	3
3	4.02	
3.7	4.96	5
4	5.36	
5.5	7.37	7.5
7.5	10	10
11	15	15
15	20	20
18.5	25	25
22	29	30
30	40	40
37	50	50
45	60	60
55	74	75
75	101	100
90	121	125
110	147	150
132	177	175
150	201	200
160	214	
185	248	250
200	268	
220	295	300

satisfy torque, horsepower, speed, and current requirements for a large number of applications. The classifications are distinguished by a NEMA design letter (A, B, C, and D).

Single-phase motors also have NEMA design letters, but they do not specify the performance characteristics to the same extent that the three-phase design letter does. The letters N and O are used for FHP motors and L and M for IHP motors.

Design A. NEMA design A motors are general-purpose motors with high starting (locked-rotor) currents and normal locked-rotor torques (Fig. 5.122). Slip is less than 5 percent, except for motors with 10 or more poles, which have a slightly greater slip. These motors are suitable in applications where load inertia is small and starts are infrequent. They are normally used where high breakdown torque (compared to NEMA design B motors) is required.

Design B. NEMA design B motors are general-purpose motors with normal starting torques and currents and relatively high breakdown torques (Fig. 5.123). Pull-up torque normally available allows rapid acceleration to full load speed. Slip is the same as design A.

TABLE 5.25 Comparison of IEC (Metric) Motor Shaft Heights and NEMA Shaft Heights for Induction Motors

IEC standard dimensions			Nearest equivalent NEMA dimensions	
Frame number	*H*, shaft height, mm	Equivalent inches	NEMA frame	*D*, shaft height, in
56	56	2.20		
63	63	2.48		
			42	2.62
71	71	2.80		
			48	3.00
80	80	3.15		
60	60	2.36		
			56	3.50
100	100	3.94		
112	112	4.41		
			180T	4.50
132	132	5.20		
			210T	5.25
160	160	6.30	250T	6.25
180	180	7.09	280T	7.00
200	200	7.87		
			320T	8.00
225	225	8.86		
			360T	9.00
250	250	9.84		
			400T	10.00
280	280	11.02	440T	11.00
315	315	12.40		
			500	12.50
355	355	13.98		
			580	14.50
400	400	15.75		

Design B motors are the most popular in industry. As the classification *general purpose* implies, the NEMA design B motors can be used to drive most loads such as pumps, fans, and conveyors.

Design C. NEMA design C motors have high starting torques with normal starting currents (Fig. 5.124). Breakdown torques are normal, though slightly less than design B motors. Although slip is less than 5 percent, it is higher than the slip of a design B motor. These motors are typically used in applications where breakaway loads are high and where starting torques higher than those available from design B motors are required. The motors have good running characteristics, although efficiency is somewhat poorer than that of design B motors.

Design D. NEMA design D motors have very high starting torques, with moderate starting currents and high breakdown torques (Fig. 5.125). Slip is high, 5 percent or more at rated load. For design D motors with full-load slip in the 8 to 13 percent range, the maximum torque is usually at locked rotor and so does not exhibit a breakdown torque. Thus speed can fluctuate significantly with changing loads. For practicality these designs have been subdivided into several groups in terms of slip (5 to 8 percent, 8 to 13 percent, etc.). These motors are typically found in applications where heavy loads are suddenly applied or removed at frequent intervals.

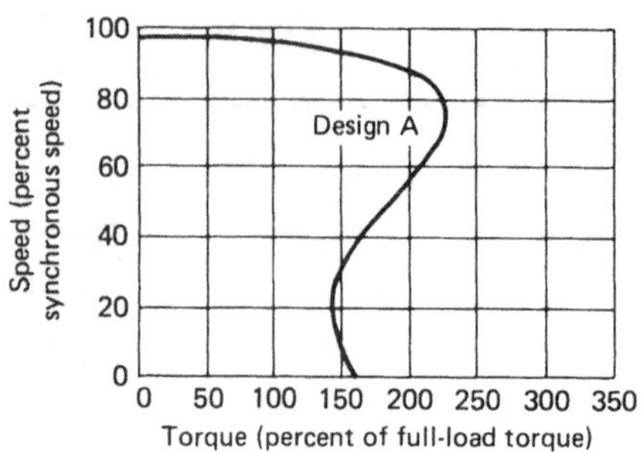

FIGURE 5.122 NEMA Design A.

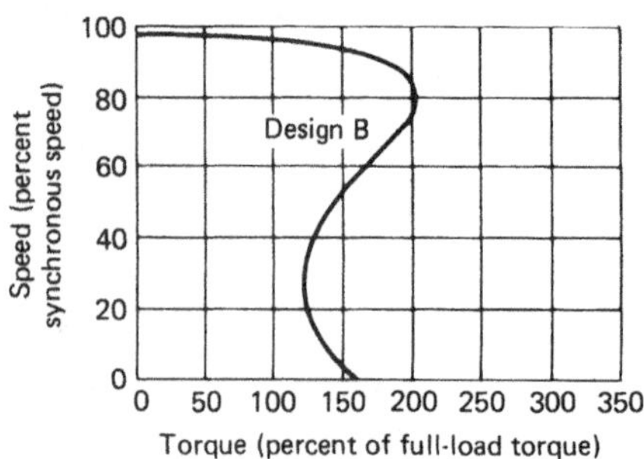

FIGURE 5.123 NEMA Design B.

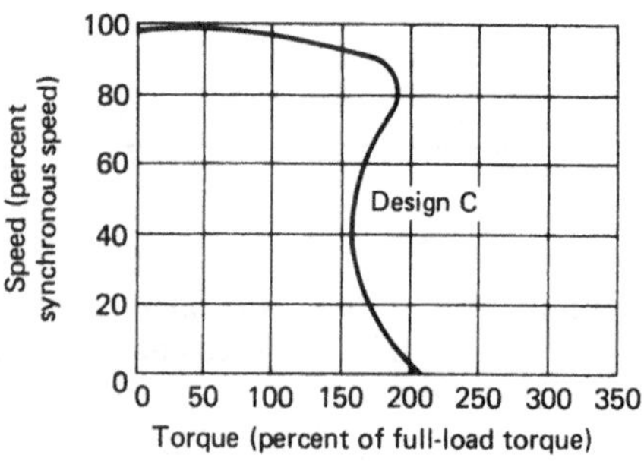

FIGURE 5.124 NEMA Design C.

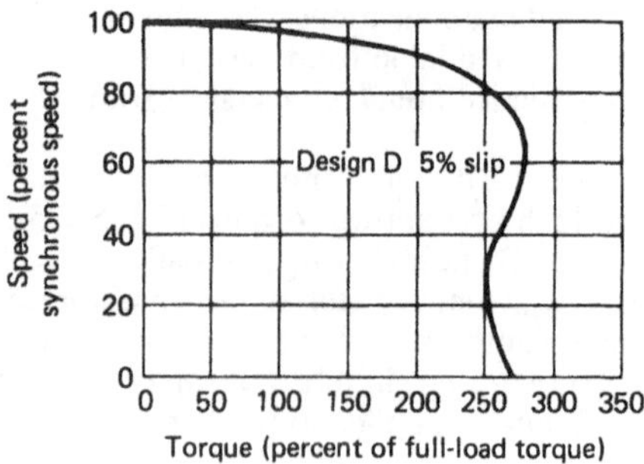

FIGURE 5.125 NEMA Design D.

Classifying by Environmental Protection and Cooling Methods

The two general NEMA classifications for motor enclosures are open (O) and totally enclosed (TE). An open machine is one having ventilating openings which permit passage of external cooling air over and around the windings of the motor. A totally enclosed machine is constructed so as to prevent the free exchange of air inside and outside the motor, but not sufficiently enclosed to be termed airtight. These enclosures are further designated by the degree of protection they provide (Table 5.26).

MOTOR RATINGS

The NEMA rating of a motor consists of the output of that machine along with any other characteristics assigned to it by the manufacturer. These characteristics include but are not limited to speed, voltage, current, and service factor.

TABLE 5.26 Motor Enclosures

Types	Characteristics
	Open
Drip-proof (ODP)	Will not allow dripping liquids or solids to enter the motor when falling on the motor at an angle not greater than 15° from vertical.
Splash-proof	A splash-proof motor is similar to a drip-proof motor, except it is constructed to exclude liquids and solids falling on it at any angle not more than 100° from the vertical.
Guarded	Motor openings are limited in size. Openings giving access to live or rotating parts do not permit the passage of a rod ¾ in (2 cm) in diameter and are at least 4 in (10 cm) away from those parts.
Semiguarded	Top half of the motor with limited-size openings as defined under guarded type.
Externally ventilated	Motor cooled by air circulated by a separate motor-driven blower. Normally used on large low-speed motors where rotor fans cannot move sufficient air to properly cool the motor.
Pipe ventilated	Motor end shields are constructed to accept pipe or ducts to supply ventilating air. Air can be supplied from a source remote from the motor, if necessary, to supply clean air for cooling.
Weather protected, type 1	Ventilating passages minimize entrance of rain, snow, and airborne particles. Passages are less than ¾ in (2 cm) in diameter.
Weather protected, type 2	Motors have, in addition to type 1, passages to discharge high-velocity particles blown into the motor.
	Totally enclosed
Nonventilated (TENV)	Enclosed motor not equipped for cooling by external means.
Fan-cooled (TEFC)	Cooled by external integral fan mounted on the motor shaft.
Explosion-proof	Enclosed motor which withstands internal gas explosion and prevents ignition of external gas.
Dust-ignition-proof	Excludes ignitable amounts of dust and amounts of dust that would degrade performance.
Waterproof	Excludes liquids and airborne solids except around shaft.
Pipe-ventilated	Openings accept air inlet and/or exit ducts or pipe for air cooling.
Water-cooled	Cooled by circulating water.
Water–air-cooled	Cooled by water-cooled air.
Air-to-air cooled	Cooled by air-cooled air.
TEFC guarded	Fan cooled and guarded by limited-size openings.
There are two other classes of protection for both open and totally enclosed motors:	
Encapsulated windings	Machine having random windings filled with resin.
Sealed windings	Machine with form wound coils with windings and connections sealed against contaminants.

Voltage and Frequency

The standard voltages for FHP dc motors, universal motors, and 60-Hz, single-phase ac motors are 115 and 230 V. Three-phase ac motors at 60 Hz have standard voltages of 115 [15 hp (11 kW) and smaller], 230, 460, 575, 2300, 4000, 4600, and 6600 V.

Motors must operate successfully under running conditions within the limits specified for voltage and frequency variation by the NEMA operating-service conditions listed below. Successful operation within the variations specified, however, does not necessarily mean that the motor will be able to start and accelerate the load to which it is applied.

Service Conditions

Usual environmental service conditions are defined by NEMA as follows:

1. Ambient temperatures in the range of 32 to 105°F (0 to 40°C) or, when water cooling is used, in the range of 50 to 105°F (10 to 40°C)
2. Barometric pressure corresponding to an altitude not exceeding 3300 ft (1000 m)
3. Installation on a rigid mounting surface
4. Installation in areas or supplementary enclosures which do not seriously interfere with ventilation

Usual operating service conditions are as follows:

1. Voltage variation up to 6 percent of rated voltage for universal motors and 10 percent of rated voltage for ac and dc motors
2. Frequency variation not more than 5 percent above or below rated frequency
3. Combined voltage and frequency variation not more than 10 percent above or below the rated voltage and frequency
4. V-belt drive, flat-belt, chain, and gear drives, in accordance with NEMA standards

Locked-Rotor kVA

Every ac motor (except three-phase wound-rotor types) rated 1/20 hp and larger has a letter designation for locked-rotor kVA per horsepower (Table 5.27). It is calculated as follows:

TABLE 5.27 Locked-Rotor kVA/hp Code Designations

Code letter	Locked-rotor kVA/hp	Code letter	Locked-rotor kVA/hp
A	0–3.15	L	9.0–10.0
B	3.15–3.55	M	10.0–11.2
C	3.55–4.0	N	11.2–12.5
D	4.0 –4.5	P	12.5–14.0
E	4.5 –5.0	R	14.0–16.0
F	5.0 –5.6	S	16.0–18.0
G	5.6 –6.3	T	18.0–20.0
H	6.3 –7.1	U	20.0–22.4
J	7.1 –8.0	V	22.4 and up
K	8.0 –9.0		

$$\text{kVA/hp} = \frac{IE}{\text{hp} \times 1000} \qquad \text{for single-phase motors}$$

$$\text{kVA/hp} = \frac{IE\sqrt{3}}{\text{hp} \times 1000} \qquad \text{for three-phase motors}$$

where I is the locked-rotor amperage at rated voltage E.

For motors with dual ratings, the following rules determine the code letter to be used:

Motor type	Code letter corresponds to kVA/hp for
Multispeed	
Variable torque	Highest speed
Constant torque	Highest speed
Constant horsepower	Highest kVA/hp
Wye delta, starting on wye	Wye connection
Dual voltage	Highest kVA/hp
Dual frequency, 60/50 Hz	60-Hz kVA/hp
Part-winding start	Full-winding kVA/hp

Service Factor

NEMA service factor is a multiplier which indicates what percent higher than the nameplate horsepower can be accommodated continuously at rated voltage and frequency without injurious overheating (i.e., exceeding NEMA allowable temperature rise for given insulation systems). Service factors for general-purpose ac motors are shown in Table 5.28.

Level of Performance

NEMA also rates motors by level of performance, specifically horsepower, speed, torque, and locked-rotor current for each category of motor type and size. Tables specifying acceptable limits for each are too detailed to include here, but may be found in NEMA standard MG1-1993.

TABLE 5.28 Service Factors

	Synchronous speed, r/min							
hp	3600	1800	1200	900	720	600	514	
1/20	1.4	1.4	1.4	1.4				Fractional-horsepower motors
1/12	1.4	1.4	1.4	1.4				Fractional-horsepower motors
1/8	1.4	1.4	1.4	1.4				Fractional-horsepower motors
1/6	1.35	1.35	1.35	1.35				Fractional-horsepower motors
1/4	1.35	1.35	1.35	1.35				Fractional-horsepower motors
1/2	1.35	1.35	1.35	1.35				Fractional-horsepower motors
1/2	1.25	1.25	1.25	1.15*				Fractional-horsepower motors
3/4	1.25	1.25	1.15*	1.15*				Fractional-horsepower motors
1	1.25	1.15*	1.15*	1.15*				Integral-horsepower motors
1/2–125	1.15*	1.15*	1.15*	1.15*	1.15*	1.15*	1.15*	Integral-horsepower motors
150	1.15*	1.15*	1.15*	1.15*	1.15*	1.15*		Integral-horsepower motors
200	1.15*	1.15*	1.15*	1.15*	1.15*			Integral-horsepower motors

* In the case of three-phase squirrel-cage integral-horsepower motors, these service factors apply only to design A, B, and C motors.

Temperature Rise

All standard motors unless otherwise stated are designed for use in a maximum ambient temperature of no greater than 104°F (40°C). The allowable temperature rise for the various enclosures and classes of insulation are shown in Table 5.29. For motors with a service factor of 1.15 or higher the temperature rise shall not exceed the values given in Table 5.29.

Duty Rating

The duty or time rating of an electric motor is determined by the electrical design, the enclosure, the method of cooling, and the class of insulation system used. This rating is the length of time the motor may operate without causing overheating or reducing the normal life of the motor.

An intermittent-duty motor is intended to operate for short periods of time totaling no more than 1 or 2 h/day. Motors with short-time ratings are designed to operate for no longer than the period shown on the nameplate. They must be shut off and allowed to cool to room temperature before being reactivated. Typical short-time duty ratings are 5, 15, 30, and/or 60 min. A motor with a continuous duty rating can be run indefinitely at rated load and voltage without injurious overheating or reduction in motor life.

Frame Designations and Assignments

The NEMA system for designating frame dimensions of motors consists of a series of numbers (frame number) in combination with letters. For convenience, manufacturers are allowed to use letters of the alphabet preceding the frame number for their own identification. However, such letters have no reference to the standards in Table 5.30 and vary in meaning from manufacturer to manufacturer.

The NEMA frame numbering system provides a useful relationship between motor horsepower and speed ratings and motor size. Although no such relationship currently exists for FHP motors, IHP motor horsepower and speed ratings and their assigned motor frames are supplied in MG13-1984 (Reaffirmed 1990).

APPLICATION AND SELECTION

Motor application begins by matching load requirements with motor characteristics. A correctly applied motor must be able to start the load, bring it up to operating speed, and run as long as necessary through all expected variations in the load.

TABLE 5.29 Allowable Temperature Rise

	Temperature, °C			
Class of insulation	A	B	F	H
1. Windings—fractional-horsepower motors				
a. Open motors other than those given in parts 1*b* and 1*d*	60	80	105	125
b. Open motors with 1.15 or higher service factor	70	90	115	
c. Totally enclosed nonventilated and fan-cooled motors, including variations	65	85	110	135
d. Any motor in a frame smaller than the 42 frame	65	85	110	135
2. Windings—integral-horsepower motors				
a. Motors other than those given in parts 2*b*, 2*c*, 2*d*, and 2*e*	60	80	105	125
b. All motors with 1.15 or higher service factor	70	90	115	
c. Totally enclosed fan-cooled motors, including variations	60	80	105	125
d. Totally enclosed nonventilated motors, including variations	65	85	110	135
e. Encapsulated motors with 1.0 service factor, all enclosures	65	85	110	

TABLE 5.30 Frame Designations

FHP motors	
The following letters immediately follow the frame number and denote specific variations.	
O	Face mounting
G	Gasoline-pump motors
H	A frame having an F dimension larger than that of the same frame without the suffix letter H
J	Jet-pump motors
K	Sump-pump motors
M	Oil-burner motors
N	Oil-burner motors
Y	Special mounting dimensions (consult manufacturer)
A	All mounting dimensions are standard except the shaft extension

IHP motors	
The following letters immediately follow the frame number and denote specific variations.	
C	Face mounting on drive end (When the face mounting is at the end opposite the drive, the prefix F is used, making the suffix letters FC.)
CH	Face mounting dimensions different from those for the frame designation having the suffix letter C (The letters CH are considered to be one suffix and should not be separated.)
D	Flange mounting on drive end (When the flange mounting is at the end opposite the drive, the prefix F is used, making the suffix letters FD.)
E	Shaft extension dimensions for elevator motors in frames larger than the 326U frames
HP and HPH	Vertical solid-shaft motors having dimensions in accordance with MG1-18.625 (The letters HP and HPH are to be considered as one suffix and should not be separated.)
JM	Face-mounted close-coupled pump motor having antifriction bearings and dimensions in accordance with NEMA standards (The letters JM are to be considered as one suffix and should not be separated.)
JP	Face-mounted close-coupled pump motor having antifriction bearings and dimensions in accordance with NEMA standards (The letters JP are to be considered as one suffix and should not be separated.)
LP and LPH	Vertical solid-shaft motors having dimensions in accordance with NEMA standards (The letters LP and LPH are to be considered as one suffix and should not be separated.)
P and PH	Vertical solid-shaft motors having dimensions in accordance with NEMA standards
R	Drive-end tapered-shaft extension having dimensions in accordance with NEMA standards
S	Standard short shaft for direct connection
T	Included as part of a frame designation for which standard dimensions have been established
U	Previously used as part of a frame designation for which standard dimensions had been established
V	Vertical mounting only
VP	Vertical solid-shaft motors having dimensions in accordance with NEMA standards (The letters VP are to be considered as one suffix and should not be separated.)
X	Wound-rotor crane motors with double shaft extension.
Y	Special mounting dimensions (Consult manufacturer.)
Z	All mounting dimensions standard except the shaft extension(s) (Also used to designate machine with double shaft extension.)

Suffix letters are added to the frame number in the following sequence.

Suffix letters	Sequence
A	1
T, U, HP, HPH, JM, JP, LP, LPH, and VP	2
R and S	3
C, D, P, and PH	4
FC and FD	5
V	6
E, X, Y, and Z	7

The first step is determining the load characteristics to which the motor will be matched: power, torque, speed, and duty cycle. Starting torque, as well as running torque, must be considered. It may vary from a small percentage to a value several times the full-load torque. The greater the excess torque than that required to start the load, the more rapid the acceleration.

Power is the product of torque times speed. The term *load* usually refers to the horsepower required to drive a machine:

$$\text{Horsepower (hp)} = \frac{\text{torque} \times \text{r/min}}{5252}$$

Duty cycle (how much is asked from a motor how often and for how long) is the final parameter. The higher the duty cycle (i.e., horsepower load above rated and time of operation) at that load, the more care should be taken in the selecting the motor size.

Having established what load the motor must drive, the correct motor can be selected by considering the following factors. The speed-versus-torque curve tells much about the performance characteristics of a motor (Fig. 5.126).

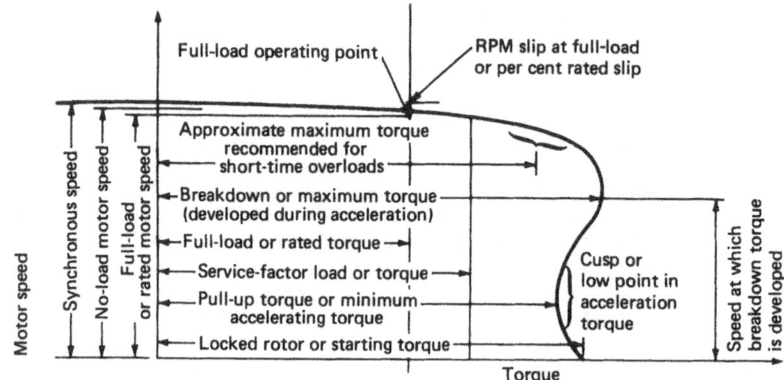

FIGURE 5.126 Speed-torque characteristics.

A standard general-purpose polyphase induction motor can be used when:

1. The momentary overload does not exceed 75 percent of the motor breakdown torque.
2. The root-mean-square (rms) value of the motor losses over an extended period of time do not exceed the losses at the service factor rating of the motor.
3. The duration of any overload does not raise the momentary peak temperature above a value safe for the motor's insulation system.

Power Supply

Power supply will be either single- or three-phase. A single-phase motor of the proper voltage can be used on a three-phase system if properly connected. But a three-phase motor cannot be used on single-phase supply. Three-phase motors generally cost less, perform better, and last longer than single-phase motors of the same size.

Voltage

Motor rating must match the nominal voltage and frequency of electricity supplied. Most motors are available in several standard voltages, but generally, the highest available voltage gives the lowest installation cost.

Horsepower

Mechanical power available at the motor shaft is the nominal horsepower rating at rated revolutions per minute. When installing a motor on new equipment, match the motor to the required horsepower computed when determining the load characteristics. When replacing a worn-out motor, review the application. Is the application overmotored? Could a lower horsepower rated motor be used? Many applications are overmotored as much as 50 percent. An energy-efficient motor should be considered as the replacement if the horsepower load and operating hours justify it.

Type

Type of motor best suited to a particular application is a major question because of the wide variety of motor types available. Tables 5.31 and 5.32 provide some guidelines, the first one by motor characteristics, the second by application.

Coupling

If motor speed matches the input shaft speed, a simple mechanical coupling can be used. But if it turns at a speed different from that recommended or calculated for the equipment, a speed-conversion drive is needed. It includes pulley and belt, gear, or chain and sprocket.

Enclosures

The drip-proof, general-purpose motor is suitable for dry, clean, and ventilated locations. Wet, dirty, or explosive conditions require other enclosures, such as totally enclosed fan-cooled or totally enclosed nonventilated motors in standard, severe duty, or explosion-proof construction. Table 5.33 matches enclosure to environment.

TABLE 5.31 Single-Phase Motors—Selection by Characteristics

Type	Horsepower ranges	Load-starting requirement	Starting current	Characteristics	Electrically reversible
Split-phase	1/20 to 1/2	Easy	High	Small, inexpensive, simple construction; nearly constant speed	Yes
Capacitor-start	1/8 to 10	Hard	Medium	Simple construction, long service; nearly constant speed	Yes
Two-value capacitor	1/4 to 20	Hard	Medium	Simple construction, long service; nearly constant speed	Yes
Permanent-split capacitor	1/20 to 1	Easy	Low	Inexpensive, simple construction; speed reduced by lowering voltage	Yes
Shaded-pole	1/400 to 1/2	Easy	Medium	Inexpensive for light duty	No
Wound-rotor (repulsion) types	1/6 to 10	Very hard	Low	Larger than other equivalent single-phase motors	No
Universal or series	1/150 to 2	Hard	High	High speed, small size; speed changes with load variations	Yes, some types
Synchronous	Very small fractional			Constant speed	

TABLE 5.32 Motor Selection by Application*

Application	Single-phase		Three-phase	
	Small, hp	Large, hp	Small, hp	Large, hp
Compressors				
Air	CS	CS,CP	B	B
Refrigeration	CS	CS,CP	B	C
Centrifugal	SP,CS	CS,CP	B	B
Reciprocating				
Loaded	CS	CS,CP	B	C
Unloaded	SP,CS	CS,CP	B	B
Conveyors and elevators				
Unloaded	CS,WR	CS,WR	B	B
Loaded	CS,WR	CS,WR	B	C
Cooling towers	CS	CS	B	B
Dryers	CS	CS,CP	B	B
Fans and blowers				
Centrifugal	SP,C	CS,CP	B	B
Propeller	SP,C,P	CS,CP	B	B
Unit heaters	SP,C,P	CS,CP	B	B
Machine tools				
Lathes	CS	CS	B	B
Milling machines	CS	CS	B	B
Drill presses	CS	CS	B	B
Grinders	CS	CS	B	B
Oil burners	SP	CS	B	B
Pumps				
Reciprocating	SP,CS	CS,CP,WR	B	B
Centrifugal	CP,CS	CS,CP,WR	B	B
Heavy oil	WR	CP,WR	B	D
Saws				
Metal, band saw	CS,U	CS	B	B
Wood, circular	CS,U	CS	B	B

* SP, split phase; CS, capacitor start; CP, two-value capacitor; C, permanent split capacitor; P, shaded pole; WR, wound rotor; U, universal; B, NEMA design B; C, NEMA design C; D, NEMA design D.

ADJUSTABLE SPEED SYSTEMS

The polyphase induction motor is basically a fixed-speed device, or at best a multiple-speed device with fixed speeds based on the stator winding configuration. The no-load or synchronous speed of the induction motor is

$$\text{Speed (synchronous)} = \frac{120 \times f}{p} \quad \text{r/min}$$

where f = power input frequency, Hz, and p = number of magnetic poles in the motor stator winding.

The use of an adjustable-frequency power supply (ac inverters) in conjunction with the induction motor results in an adjustable-speed drive system. The speed of the motor is adjusted by controlling the output frequency of the ac inverter. Thus the induction motor can be used on many adjustable-speed applications. The speed-torque motor performance at different input motor frequencies is illustrated by Fig. 5.127*a*. This permits the application of induction motor adjustable-speed systems to loads such as fans and pumps for flow control with considerable power savings over fixed-speed systems with dampers or valves for flow control. Figure 5.127*b* shows the performance of a 10-hp motor adjustable-frequency system driving a fan load.

TABLE 5.33 Motor Enclosures

		Standard		
Condition or application	Drip-proof	Totally enclosed	Severe-duty	Explosion-proof
Atmospheric				
Dry	x	—	—	—
Humid	x*	—	—	—
Outdoor, mild	x*	—	—	—
Outdoor, severe	x*	x*	x	—
Chips				
Metal or plastic	—	x	x	—
Wood	x	x	x	—
Dust				
Abrasive, nonexplosive	—	x	x	—
Abrasive, explosive	—	—	—	x
Carbon, coal, or coke	—	—	—	x
Flour	—	—	—	x
Metal, nonexplosive	—	x	x	—
Metal, explosive	—	—	—	x
Sand	x	x	x	—
Sawdust	x	x	x	—
Textile fibers	—	x	—	—
Fumes				
Explosive	—	—	—	x
Nonexplosive	—	x	x	—
Corrosive	—	x*	x	—
Liquids				
Acid or alkali	—	x*	x	—
Dripping water	x*	—	—	—
Explosive	—	—	—	x
Nonexplosive	—	x	x	—
Paint	—	—	—	x
Petroleum, oil	—	—	—	x
Splashing water	x*	x*	x	—
Solvents				
Corrosive, nonexplosive	—	x*	x	—
Noncorrosive, nonexplosive	—	x	x	—
Noncorrosive, explosive	—	—	—	x

* Depending on concentration and/or severity, additional protection may be necessary.

TROUBLESHOOTING

Table 5.34, a chart of motor problems and possible causes, can serve as a guide in identifying and correcting motor-system malfunctions.

ENERGY EFFICIENCY

Energy efficiency is a growing issue in the motor industry, and for good reasons. In addition to scarce energy resources, a great potential for energy conservation and reduced operating costs exists with the electric motor.

Today, energy-efficient three-phase induction motors, both open and TEFC, are available in ratings from 1 to 500 hp (0.7 kW to 373 kW). In many locations the cost of electric power is high and energy-efficient motors can be justified on loads operating a high percentage of the time.

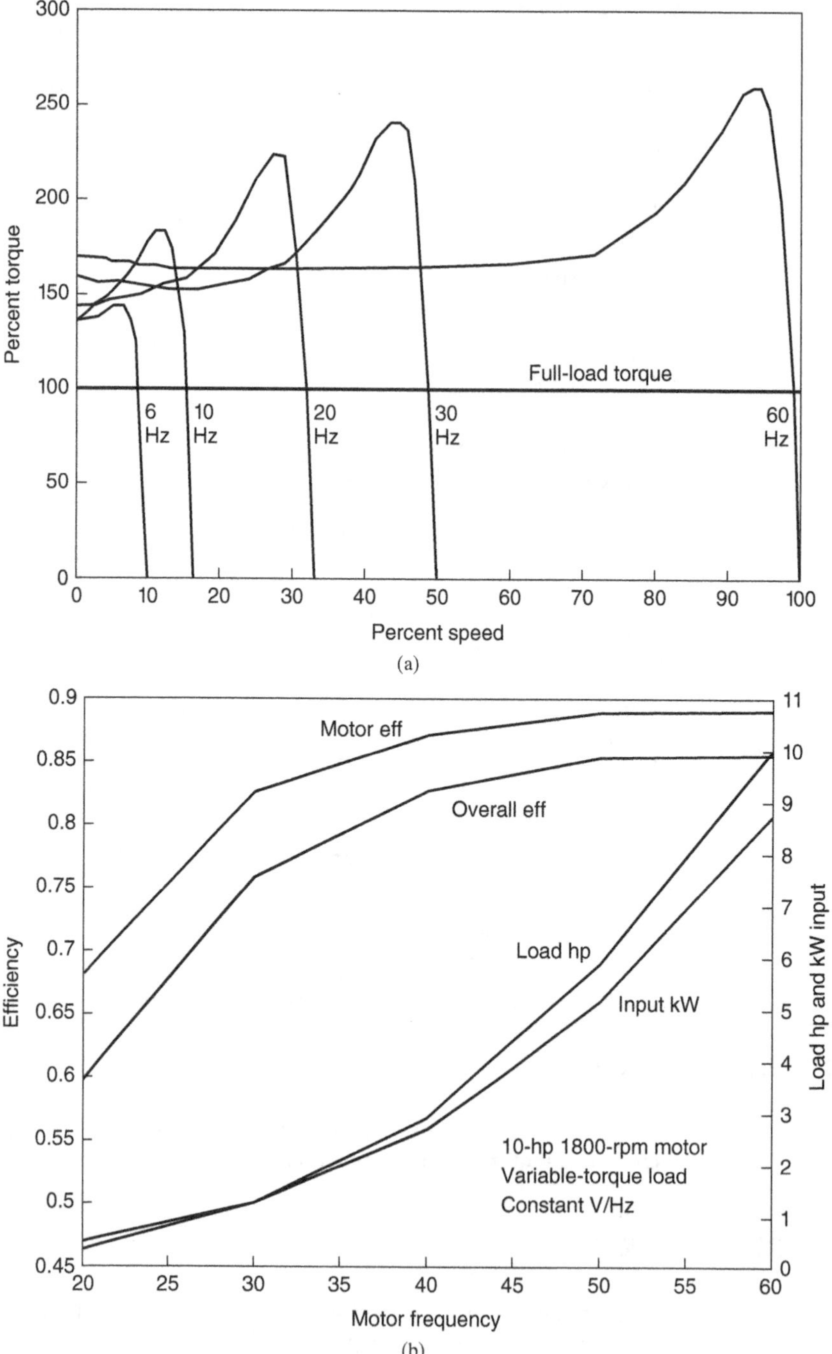

FIGURE 5.127 Induction motor adjustable-frequency system. (*a*) Speed-torque curves. (*b*) Driving a fan load.

TABLE 5.34 Troubleshooting Guide

Trouble	Cause	What to do
Motor fails to start	Blown fuses	Replace with time-delay fuses matched to nameplate amperes. If motor has service factor greater than 1.0, use fuse size up to 12.5% of motor amperes. Check for grounded winding.
	Improper power supply	Check to see that power supplied (voltage-frequency phases) agrees with motor nameplate.
	Less than 208 V on a 208-V system	Use a 200-V motor.
	Low voltage	Inadequate wiring. Long or inadequate extension cords.
	Improper line connections	Check connections against diagram supplied with motor.
	Overload (thermal protector) tripped	Check and reset overload relay in starter. Check heater rating against motor nameplate current rating. Check motor load. If motor has a manual reset thermal protector, check whether tripped.
	Open circuit in winding or starting switch	Indicated by humming sound when switch is closed. Check for loose wiring connections, whether starting switch inside motor is closed, or if capacitor is defective.
	Mechanical failure	Check to see if motor and drive turn freely. Check belts, bearings, and lubrication.
	Short-circuited stator	Indicated by blown fuses and/or high no-load line current. Motor must be rewound or replaced.
	Poor stator coil connection	Repair or replace.
	Rotor defective	Look for broken bars or end rings.
	Motor may be overloaded	Reduce load; increase motor size.
	If three-phase, one phase may be open	Indicated by humming sound. Check lines for open phase. Check voltage with motor disconnected from line; one fuse may be blown.
	Defective capacitor	Check for short-circuited grounded, open, or low-value capacitor. Replace if necessary.
Motor stalls	Wrong application	Change type or size of motor. Consult motor service firm.
	Overloaded motor	Reduce load or increase motor size. Belt may be too tight.
	Low motor voltage	See that nameplate voltage is maintained.
Motor runs and then dies down	Power failure	Check for loose connections to line, to fuses, and to control. Check if thermal protector tripped, fuses blown, check overload relay, starter, and push-buttons.
Motor does not come up to speed	Not applied properly	Consult motor service firm for proper type and size of motor. Use larger motor.
	Voltage too low at motor terminals (because of line voltage drop or voltage drop in wiring to motor)	Use higher-voltage tap on transformer terminals; increase wire size.
	Starting load too high	Check load capability of motor.
	Shorted or weak capacitor	Replace capacitor.
	Low temperature (below 0°F)	Replace capacitor with one of higher value. Check ball bearing lubricant—use low-temperature grease.
	Broken rotor bars	Look for cracks near the end rings or broken rotor bars. A new rotor may be required as repairs are usually temporary.

TABLE 5.34 Troubleshooting Guide (*Continued*)

Trouble	Cause	What to do
Motor takes too long to accelerate	Excess loading; tight belts; high inertia load	Reduce load; increase motor size. Loosen belts.
	Inadequate wiring	Check for high resistance; increase wire size.
	Defective rotor	Replace with new rotor.
	Applied voltage too low	Check incoming voltage drop in wiring to motor. Power may have to supply higher voltage.
	Weak or shorted capacitor	Replace capacitor.
	Low starting torque	Replace with larger motor.
Motor overheats while running under load	Overload	Reduce load; increase motor size; belts may be too tight.
	Insufficient airflow over shell of "air-over" motor	Modify installation or change to a self-cooled motor.
	May be clogged with dirt to prevent proper ventilation of motor.	Good ventilation is apparent when a continuous stream of air leaves the motor. If it does not after cleaning, check with manufacturer.
	Motor may have one phase open (three-phase motors)	Check to make sure that all leads are well-connected and a fuse is not blown in one line.
	Grounded or shorted coil	Repair or replace the motor.
	Unbalanced terminal voltage (three-phase motors)	Check for faulty leads, connections, and transformers. Excessive single-phase loads on one circuit.
	Faulty connection	Clean, tighten, or replace.
	High or low voltage	Check voltage at motor with voltmeter; should not be more than 10% above or below rated.
	Rotor rubs stator bore	If not poor machining, replace worn bearings.
Motor vibrates after corrections have been made	Motor misaligned	Realign.
	Coupling out of balance	Balance coupling.
	Driven equipment unbalanced	Rebalance driven equipment.
	Defective ball bearing	Replace bearing.
	Balancing weights shifted	Rebalance rotor.
	Polyphase motor running single phase	Check for open circuit or blown fuses.
	Excessive end play	Adjust bearing or add shims to eliminate excess end play.
	High voltage	Correct.
Unbalanced line current on polyphase motors during normal operation	Unequal terminal volts	Check leads and connections. Check transformers.
	Single-phase operation	Check for open contacts, blown fuses.
	Unbalanced supply	Check line-to-line voltage.
Scraping noise	Fan rubbing air shield	Repair.
	Fan striking insulation	Repair.
	Bent shaft	Straighten shaft or replace rotor.
Noisy operation	Air gap not uniform	Check and correct end shield fits or bearing.
	Rotor unbalance	Rebalance.
	High voltage	Reduce line voltage by changing power taps.
Hot bearings, general	Loose on mounting surface	Tighten holding bolts.
	Bent shaft	Straighten or replace shaft or rotor.
	Excessive belt pull	Decrease belt tension.
	Pulleys too far away from bearing nose	Move pulley closer to motor bearing.
	Pulley diameter too small; slipping	Use larger pulleys (both motor and load). Check belt tension.
	Misalignment	Correct by realignment of drive.
Hot bearings, sleeve	Oil grooving in bearing obstructed by dirt	Remove end shield, clean bearing housing, and oil grooves; renew oil.

TABLE 5.34 Troubleshooting Guide (*Continued*)

Trouble	Cause	What to do
	Oil too heavy	Use recommended oil.
	Oil too light	Use recommended oil.
	Too much end thrust	Reduce thrust induced by drive, or supply external means to carry thrust.
	Dry bearing	Add oil.
	Badly worn bearing	Replace bearing.
	Feeder wick not touching shaft	Repair or replace wicking.
Hot bearings, ball	Insufficient grease	Maintain proper quality of grease in bearing. Replace sealed bearing.
	Deterioration of grease or lubricant	Remove old grease, wash bearings thoroughly in clean kerosene, and replace with new grease.
	Contaminated water in bearing	Replace bearing if sealed type. Eliminate source of moisture.
	Excess lubricant	Reduce quantity of grease. Bearing should not be more than ½ filled.
	Overloaded bearing	Check alignment, side, and end thrust.
	Broken ball or rough races	Replace bearing. First clean end shield housing thoroughly.

Factors to consider in the selection of energy efficient induction motors are as follows:

1. Electric power cost savings and life-cycle cost comparison to standard motors
2. Improved ability to perform under adverse conditions such as abnormal voltage variations
3. Lower operating temperatures
4. Lower noise levels
5. Ability to accelerate higher inertia loads
6. Higher operating efficiencies at all load points

Efficiency and Power Factor

Motor efficiency is simply a measure of mechanical work output over electric power input. A motor's power factor is a measure of how well the motor uses the current it draws.

Total line current is made up of two components: real and reactive. Power factor is the ratio of *real power* to *apparent power,* or the cosine of the angle between the two. Technically, it is input kilowatts divided by the input kilovoltamperes.

Figure 5.128 shows efficiency and power factor as functions of loading for a typical small integral-horsepower polyphase motor. Efficiency is relatively stable over a wide range of loading conditions. But power factor drops off fast as the motor is unloaded—about a 15-percent spread between half- and full load. Proper application and sizing are thus important to maintain good power factor. A high power factor means that the motor requires less total current, and the resulting lower line current means that less energy is wasted in all feeder circuits serving the motor.

Any inefficiency in the process of converting electric into mechanical energy occurs as heat, often referred to as *watt losses,* or simply losses.

To increase efficiency, it is necessary to minimize these losses. The rotors of energy-efficient motors are built with additional aluminum to reduce losses resulting from current flowing in the aluminum rotor bars. Additional copper is used in the stators to reduce losses in the motor. Most motors have 100 percent copper stator windings: few are aluminum-wire-wound. More steel, together with special processing, is also used in the motors to reduce the

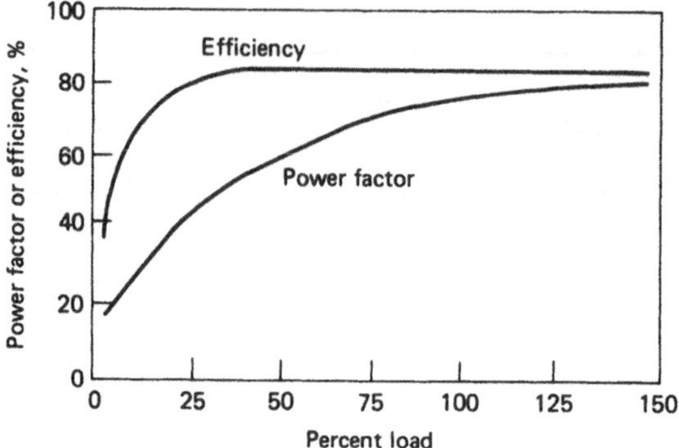

FIGURE 5.128 Efficiency and power factor vs. load.

stator and rotor losses. This specially processed steel includes thinner laminations as well as silicon electric steels.

Windings in the most energy-efficient motors are designed for optimum winding distribution. The air gap is also optimized for the best power factor and efficiency performance. Stator and rotor slots are usually designed for optimum performance and are not compromised to serve other uses.

Table 5.35 compares the nominal full-load efficiencies of standard motors and energy-efficient motors for both open and TEFC 1800 r/min motors. Energy-efficient induction motors carry a premium price over standard motors. However, based on the cost of electric

TABLE 5.35 Nominal Efficiency of Energy-Efficient Motors vs. Industry Average Standard Motors, 1800 r/min*

	Open motors		TEFC motors	
Motor hp	Standard motor	Energy-efficient motor	Standard motor	Energy-efficient motor
2	79.1	84.2	81.0	85.1
3	80.9	86.8	81.3	87.7
5	82.5	87.7	83.6	88.3
7.5	84.5	89.8	85.2	90.1
10	85.4	90.6	86.6	90.7
15	86.8	91.7	87.0	92.2
20	87.5	92.2	88.0	92.6
25	88.9	92.6	88.9	93.2
30	89.6	93.0	89.8	93.4
40	90.3	93.9	90.7	93.8
50	90.9	94.1	91.4	94.0
60	91.3	94.3	91.4	94.5
75	91.6	94.8	92.0	94.9
100	92.0	95.1	92.0	94.8
125	92.6	95.0	92.1	95.1
150	92.2	95.3	92.9	95.3
200	92.8	95.5	93.8	95.8
250	93.0	95.8	93.9	95.8
300	93.0	95.8		

* At full load.

TABLE 5.36 Annual Energy Savings in Dollars for Open, 1800-r/min Drip-Proof Motors*

hp	Standard motor efficiency, %	Energy-efficient motor efficiency, %	Annual kWh saving	Annual dollar saving
2	79.09	84.23	461	28
3	80.90	86.83	756	45
5	82.47	87.70	1,079	65
7.5	84.50	89.79	1,560	94
10	85.37	90.56	2,003	120
15	86.80	91.72	2,766	166
20	87.51	92.16	3,438	206
25	88.86	92.64	3,428	206
30	89.60	92.98	3,632	218
40	90.26	93.88	5,103	306
50	90.89	94.08	5,574	334
60	91.31	94.34	6,288	377
75	91.64	94.80	8,133	488
100	91.97	95.12	10,740	644
125	92.57	95.00	10,301	618
150	92.18	95.25	15,633	938
200	92.75	95.50	18,529	1,112
250	93.00	95.80	23,445	1,407
300	93.00	95.80	28,134	1,688

* Based on 6 cents/kWh power cost and 4000 h/yr operation.

power and annual operating hours, this premium price is often paid back in a matter of months. The simplest payback formula for energy-efficient motors is

$$\text{Payback (months)} = \frac{\text{price premium (dollars)}}{\text{kW saved} \times \text{cost per kWh} \times \text{annual running time (h)}} \times 12$$

There are other, more detailed, payback formulas available based on life-cycle costing; however, the above formula is adequate when the payback is 2 yr or less.

Table 5.36 has been developed to illustrate the energy cost savings for open 1800-r/min energy-efficient motors. This table can be used as a guide by adjusting the electric power costs and operating hours to suit your installation.

BIBLIOGRAPHY

Andreas, J. C.: *Energy Efficient Electric Motors,* 2d ed., Marcel Dekker, New York, 1992.

Beaty, H. W. and J. C. Kirtley Jr.: *Electric Motors Handbook,* McGraw-Hill, New York, 1998.

Cochran, P. L.: *Polyphase Induction Motors,* Marcel Dekker, New York, 1989.

Fink, D. G., and H. W. Beaty: *Standard Handbook for Electrical Engineers,* 13th ed., McGraw-Hill, New York, 1993.

Yeadon, W. H., and A. W. Yeadon: *Handbook of Small Electric Motors,* McGraw-Hill, New York, 2000.

ACKNOWLEDGMENTS

The following publications are acknowledged as partial sources for this section.

Form 5S1248, "Answers to Common Electric Motor Problems," Dayton Electric Manufacturing Company, Chicago.

Machine Design, Electrical and Electronics Reference Issue, sec. 7, "Motors," May 17, 1979, pp. 7–42.

NEMA Standards Publication MG1-1993, *Motors and Generators,* National Electrical Manufacturers Association. Washington, D.C., 1993.

NEMA Standards Publication MG13-1984 (Reaffirmed 1990), *Frame Assignments for Alternating-Current Medium Induction Motors,* National Electrical Manufacturers Association, Washington, D.C., 1984.

CHAPTER 5.8
MOTOR CONTROLS

James R. Wright, P.E.
Manager of Codes and Standards
Siemens Energy & Automation, Inc.
Batavia, Illinois

GLOSSARY

Definitions essential to an understanding of ac and dc motor control are provided. Most definitions are in accordance with National Electrical Manufacturers Association (NEMA) and American National Standards Institute (ANSI).

Accelerating relay A relay used to aid in motor starting or accelerating from one speed to another. It may function by: current-limit (armature-current) acceleration; counter-emf (armature-voltage) acceleration; or definite-time acceleration.

Across-line starting A method that connects the motor directly to the supply line on starting.

Actuator The cam, arm, or similar mechanical piece used to actuate a device.

Ambient conditions Conditions of the atmosphere adjacent to the electric apparatus. The specific reference may apply to temperature, contamination, humidity, etc.

Antiplugging protection A control function that prevents application of counter-torque until the motor speed has been reduced to an acceptable value.

Brake An electromechanical friction device employed to stop and hold a load. When the brake is set, a spring pulls the braking surface into contact with a braking wheel which is directly coupled to the motor shaft.

Branch circuit The portion of a wiring system extending beyond the final overcurrent device protecting the circuit.

Blowout coil Electromagnetic coil used in contactors and starters to deflect an arc when a circuit is interrupted.

Circuit breaker A device designed to open and close a circuit by nonautomatic means and to open the circuit automatically at a predetermined overload of current without injury to itself when properly applied within its rating (ampere-, volt-, and horsepower-rated).

Closed-circuit transition (as applied to reduced-voltage controllers) A method of motor starting in which power to the motor is not interrupted during the normal starting sequence.

Combination starter A magnetic starter having a manually operated disconnecting means built into the same enclosure. The disconnect may be a motor-circuit switch (with or without fuses) or a circuit breaker.

Contactor A device for repeatedly establishing and interrupting an electric power circuit. Definite-purpose contractors are those designed for specific applications such as air-conditioning, heating, and refrigeration equipment control.

Controller service The specific application of the controller. General purpose: standard or usual service. Definite purpose: specific application other than usual.

Control, three-wire Control function incorporating a momentary contact pilot device and a holding circuit contact to provide undervoltage protection.

Control, two-wire Control function utilizing a maintained contact pilot device to provide undervoltage release.

Controller A device, or group of devices, that governs in a predetermined manner the electric power delivered to the apparatus to which it is connected.

Controller function To regulate, accelerate, decelerate, start, stop, reverse, or protect devices connected to an electric controller.

Controller, drum A manual switching device having stationary contacts connected to a circuit by the rotation of a group of movable contacts.

Current-responsive protector Devices that include time-lag fuses, magnetic relays, and thermal relays (normally located in the motor or between the motor and controller) that provide a degree of protection to the motor, motor control apparatus, and branch-circuit conductors from overloads or failures to start.

Dash pot Device employed to create a time delay. It consists of a piston moving inside a cylinder filled with a liquid or gas which is allowed to escape through a small orifice in the piston. Moving contacts actuated by the piston close the electric circuit.

Drop-out (voltage or current) The voltage or current at which a device will return to its de-energized position.

Duty Specific controller functions. Continuous: constant load, indefinitely long time period. Short time: constant load, short or specified time period. Intermittent: varying load, alternate intervals, specified time periods. Periodic: intermittent duty with recurring load conditions. Varying: varying loads, varying time intervals, wide variations.

Dynamic braking The process of disconnecting the armature from the power source and either short-circuiting it or adding a current-limiting resistor across the armature terminals while the field coils remain energized.

Electromechanical Term applied to any device which uses electric energy to magnetically cause mechanical movement.

Electronic control Term applied to define electronic, static, precision, and associated electronic control equipment.

Electronic protector A monitoring system in which current in the motor leads is sensed and the thermal characteristic of the motor reproduced during both heating and cooling cycles. Other circuits compensate for copper and iron losses and detect a lost phase.

Electropneumatic controller An electric controller having its basic functions performed by air pressure.

Faceplate controller Controller having multiple switching contacts mounted near a selector arm on the front of an insulated plate. Additional resistors are mounted on the rear to form a complete unit.

Feeder The circuit conductors between the service equipment and the branch-circuit overcurrent device.

Float switch A switch responsive to the level of liquid.

Foot switch A switch designed for operation by an operator's foot.

Frequency The number of complete variations made by an alternating current per second, expressed in hertz.

Fuse An overcurrent protective device with a circuit-opening fusible member that is severed by the heat developed from the passage of overcurrent through it.

Field weakening A method of increasing the speed of a wound field motor by reducing field current to reduce field strength.

Instantaneous A qualified term applied to the closing of a circuit in which no delay is purposely introduced.

Interlock An electric or mechanical device, actuated by an external source, and used to govern the operation of another device.

Interrupting capacity The highest current at rated voltage that a device can interrupt.

Inverse time A qualifying term indicating that a delayed action has been introduced. This delay decreases as the operating force increases.

Jogging (inching) The intermittent operation of a motor at low speeds. Speed may be limited by armature series resistance or reduced armature voltage.

Latching relay Relay that can be mechanically latched in a given position when operated by one element and released manually or by the operation of a second element.

Limit switch A device that translates a mechanical motion or a physical position into an electric control signal.

Locked-rotor current Steady-state current taken from the line with the rotor locked and with rated voltage (rated frequency in the case of ac motors) applied to the motor.

Locked-rotor torque The minimum torque a motor will develop at rest for all angular positions of the rotor with rated voltage applied at the rated frequency.

Low-voltage protection (magnetic control only) The opening of the motor circuit by the controller upon a reduction or loss of voltage. Manual restarting is required when the voltage is restored.

Low-voltage release (magnetic control only) The effect of a device, operative on the reduction or failure of voltage, to cause the interruption of power supply to the equipment, but not preventing the reestablishment of the power supply on return of voltage.

Microprocessor controller Motor controller that utilizes feedback control and a computer complete with processor, memory, I/O, keyboard, and display.

Motor circuit switch A switch, rated in horsepower, capable of interrupting maximum operating overload current of a motor of the same horsepower rating at rated voltage.

Motor control circuit The circuit that carries the electric signals directing controller performance but does not carry the main power current. Control circuits tapped from the load side of motor branch circuits' short-circuit protective devices are not considered to be branch circuits and are permitted to be protected by either supplementary or branch-circuit overcurrent protective devices.

Nonreversing A control function that provides for motor operation in one direction only.

Normally open or closed Terms used to signify the position of a device's contacts when the operating magnet is deenergized (applies only to nonlatching-type devices).

Off-delay timer A device whose output is discontinued following a preset time delay after the input is deenergized.

Open-circuit transition A method of reduced-voltage starting in which power to the motor is interrupted during normal starting sequence.

Operating overload The overcurrent to which electric apparatus is subjected in the course of normal operating conditions that it may encounter. *Note:* Maximum operating overload is considered to be 6 times normal full-load current for ac industrial motors and control apparatus and 4 to 10 times normal full-load current, respectively, for dc industrial motors and control apparatus used for reduced- or full-voltage starting. It should be understood that these overloads are currents that may persist for a very short time only, usually a matter of seconds.

Operator's control (push–button) station A unitized assembly of one or more externally operable push–button switches, sometimes including other pilot devices.

Overcurrent Any current in excess of equipment or conductor rating. It may result from overload, short circuits, or ground faults.

Overload Operation of equipment in excess of normal full-load rating. A fault such as a short circuit or ground fault is not an overload.

Phase-failure protection Protection provided when power fails in one wire of a polyphase circuit to cause and maintain the interruption of power in all wires of the circuit.

Phase-lock servo A digital control system in which the output of an optical tachometer is compared to a reference square wave to generate a system error signal proportional to both shaft velocity and position.

Phase-reversal protection The prevention of motor energizing under conditions of phase sequence reversal in a polyphase circuit.

Pilot device A low-current status indicating or initiating device such as a pilot light, push button, and limit or float switch.

Plugging Motor braking by reversal of the line voltage or phase sequence in order to develop a countertorque which exerts a retarding force.

Programmed control A control system in which operations are directed by a predetermined input program consisting of cards, tape, plug boards, cams, etc.

Proximity switch A device that reacts to the presence of an actuating means without physical contact or connection.

Pull-up torque (ac motors) The minimum torque developed by the motor during the period of acceleration from rest to the speed at which breakdown occurs.

Push button A master switch having a manually operable plunger or button for actuating the switch.

Rating, continuous The substantially constant load that can be carried for an indefinite time.

Rating of a controller Designation of operating limits based on power governed and the duty and service required.

Rating, eight-hour The rating of a magnetic contactor based on its current-carrying capacity for 8 h without exceeding established limitations. Rating considerations include new, clean contact surfaces, free ventilation, and full-rated voltage on the operating coil.

Rating, make or break The value of current for which a contact assembly is rated for closing or opening a circuit repeatedly under specified operating conditions.

Reactor, saturable An inductor having the means to change the degree of magnetic saturation of its core(s) to control the magnitude of alternating current supplied to a load.

Regenerative braking In ac motors, it results from the motor's inherent tendency (through a negative slip) to resist being driven above synchronous speed by an overhauling load. In shunt-wound dc motors, it occurs when driven by an overhauling load, when shunt field strength is increased, or when armature voltage is decreased (in adjustable-voltage drives).

Relay A device operated by a variation in the conditions in one electric circuit to effect the operation of other devices in the same or other circuits. Examples include current, latching, magnetic control, magnetic overload, open-phase, low or undervoltage, and overload.

Reset A manual or automatic operation that restores a mechanism or device to its prescribed state.

Resistance starting A form of reduced-voltage starting employing resistances that are short-circuited in one or more steps to complete the starting cycle.

Reversing Changing the operation of a drive from one direction to the other.

Rod-and-tube (rate-of-rise) sensor A thermostat consisting of an external metal tube and an internal metal rod that operates as a differential-expansion element. This element actuates a self-contained snap switch.

Service of a controller The specific application in which the controller is to be used—either general-purpose or definite-purpose (crane and hoist, elevator, machine tool, etc.).

Starter An electric controller for accelerating a motor from rest to normal speed. *Note:* A device designed for starting a motor in either direction includes the additional function of reversing and should be designated a reversing controller.

Starting, slow-speed A control function that provides for starting an electric drive only at the minimum speed setting.

Static control A system that may contain electronic components that do not depend on electronic conduction in a vacuum or gas. The electrical function is performed by semiconductors or the use of otherwise completely static components such as resistors, capacitors, etc.

Static controller A controller in which the major portion of all of the basic functions are performed through the control of electric or magnetic phenomena in solids such as transistors, etc.

Synchronous motor controller A controller consisting of a three-pole starter for the ac stator circuit, a contactor for the dc field circuit, an automatic synchronizing device to control the dc field contactor, and a cage-winding protective relay to open the ac circuit without synchronizing, in order to start a synchronous motor, accelerate it to synchronous speed, and synchronize it to supply frequency.

Switch A device for making, breaking, or changing the connections in an electric circuit. In controller practice, a switch is considered to be a device operated by other than magnetic means.

Switch selector A manually operated multiposition switch for selecting alternative control circuits.

Temperature-responsive protector A protective device for assembly as an integral part of a motor that provides a degree of protection to the motor against dangerous overheating due to overload and failure to start.

Test, dielectric The application of a voltage higher than the rated voltage for a specified time to determine the adequacy of insulating materials and spacings against breakdown under normal conditions.

Tests, application Tests performed by a manufacturer to determine those operating characteristics not necessarily established by standards but which have application interest.

Thermal cutout An overcurrent protective device that contains a heater element and renewable fusible member which open the motor circuit. It is not designed to interrupt short-circuit currents.

Thermistors Devices that sense temperature through changes in resistance. Signals from a thermistor may be amplified to interrupt the contactor holding coil to provide a degree of protection against motor locked-rotor conditions and running overloads.

Threading Signifies low-speed operation similar to jogging, but for longer periods with interlocked control.

Time, accelerating The time to change from one specified speed to a higher or lower speed while operating under specified conditions.

Time delay A time interval purposely introduced into the performance of a function.

Time response An output, expressed as a function of time, resulting from the application of a specified input under specified operating conditions.

Torque A turning or twisting force that tends to produce rotation.

CONTROL OF AC MOTORS

The control of an ac motor includes motor starting and stopping; governing the motor speed, torque, horsepower, and other characteristics; and protecting personnel and equipment.

Types of AC Motor Starters

Motor starters can be divided into three basic types (manual, magnetic, and electronic operation) and three categories (full voltage/across the line, reduced voltage, and multispeed). Starters consist of a contactor to switch the electrical load and an overload relay which provides protection for the motor.

Control Selection Considerations

The selection of a specific motor control system also requires the consideration of a number of factors. Depending on the particular type, size, and application of the motor to be controlled and the particular characteristics of the driven load, motor control selection may be simple or complex.

Special Considerations. Selection of the proper motor control system also involves several other key factors. These include operator-versus-automatic machine starting, expected starting requirements, continuous or intermittent machine operation, and special functions, if any, required during operation.

Separate from these special functions are requirements that specify the need to reverse direction or stop the motor, and the types and number of protective devices necessary to assure proper and continued operation.

Manual Motor Starters

Manual starters are typically used on small motors in applications requiring infrequent starting. In general, comments applying to magnetic controls also pertain to manual motor controls.

Full-Voltage Starting. Full-voltage manual starters and contactors provide direct control for applications not requiring remote control and which permit automatic restarting.

Fractional-Horsepower Starting. The simplest type of manual starting switch is the one- or two-pole fractional-horsepower toggle switch consisting of an on/off snap-action mechanism. This method is generally applied to single-phase motors with ratings up to a maximum of 1 hp (0.75 kW) at 120 or 240 V, where only infrequent starting and stopping are required.

Magnetic Motor Starters

Three-Phase Magnetic Starting. Three-phase magnetic starters are designed for full-voltage starting of squirrel-cage induction motors when full starting torque and current surge are permitted. They are also used for primary circuit control for wound-rotor (slip-ring) motors that have provision for manual starting and speed control in their secondary circuits.

Wound-Rotor Controls. To control starting, accelerating, and regulation, a variable resistance is added to the rotor circuit. Full rotor resistance is used during motor starting. As the motor begins to accelerate, the resistance is reduced in steps. When the motor is connected to full line voltage (all resistance shorted out), it acts as a squirrel-cage motor.

The basic control circuit consists of a full-voltage starter and a balanced, adjustable three-phase resistor, wye-connected in the rotor circuit. Speed can be established for a given load by adjusting rotor resistance; once set, speed will vary with load conditions.

A static controller may be used to control the operating speeds of wound-rotor motors. In one method, a controlled saturable reactor is placed in the rotor circuit with the accelerating resistance. For fixed operating speeds, reactor saturation (which controls the motor speed) can be varied using a control resistor. Static controllers can also be used to reverse the direction of motor rotation by placing saturable reactors in the motor primary circuit rather than reversing contactors. Controlled reactor saturation directs the reversal of the motor rotation.

Synchronous Motor Starting. This method is used for power-factor correction of heavy concentrations of induction motors. It is also used for constant-speed, slow-speed industrial drive applications and for maximum efficiency on continuous heavy loads in excess of 75 hp (55 kW). Three-phase ac power is connected to the stator and dc to the rotor (which has both a field and a squirrel-cage winding).

A full-voltage magnetic contactor connects the ac motor winding to the line, and the rotor winding is closed through a starting and discharge resistor. The motor starts and comes up to speed like a squirrel-cage motor. At the correct rotor speed, a polarized-field frequency relay and reactor automatically apply dc excitation to the field to synchronize the motor with maximum synchronizing torque, while drawing minimum line current.

Speed-Torque Characteristics. Three-phase ac motors are designed to operate at speeds directly proportional to the frequency of the voltage applied to the stator field. However, while the motor's synchronous speed is directly proportional to the applied frequency, it is inversely proportional to the number of motor poles.

Since induction motors rely on rotor bars or windings to cut the flux of the rotating field to turn the rotor, they will operate at a speed slightly less than synchronous speed.

In constant-horsepower designs, output torque can vary inversely with motor speed. For applications requiring constant output torque, however, the air-gap flux must be held constant over the entire speed range of the motor.

Types of Motor Starters

Starters are rated or commonly described as being one of three types. These are traditional NEMA, definite purpose, and International Electrotechnical Commission (IEC) starters. Definite-purpose contactors and starters are an original equipment manufacturer (OEM) component found primarily in heating, ventilating, and air conditioning (HVAC), refrigeration, computer and power supply applications (Fig. 5.129). The plant engineer will ordinarily only come into contact with these devices on a repair or replacement basis. These contactors and starters are rated for motor loads in full-load and locked-rotor amperes and in amperes for resistive loads such as electric heating.

Some Basic Differences. The primary difference between IEC and traditional NEMA starters is one of concept. A 10-hp (7.5-kW) NEMA starter can be used in virtually any 10-hp (7.5-kW) motor application. This is not the case for IEC control, which must be derated for difficult applications.

(a)

(b)

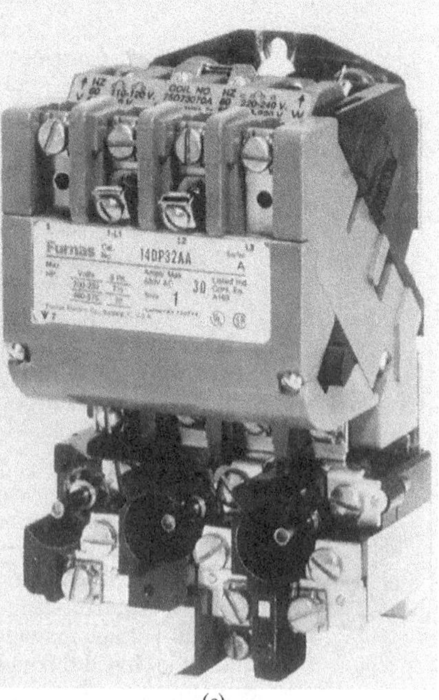

(c)

FIGURE 5.129 (*a*) Typical definite-purpose starter. (*b*) Typical NEMA starter. (*c*) Typical IEC starter. (*Furnas Electric Co.*)

Not all applications require a heavy-duty industrial starter (NEMA type). In applications where space is limited and minimum maintenance is required, IEC devices represent a very cost-effective solution.

IEC Starter Ratings. Utilization categories are used with IEC devices to define the typical duty cycle for a starter or contactor. The IEC utilization categories are:

AC1	Noninductive or slightly inductive loads
AC2	Starting of slip-ring motors
AC3	Starting of squirrel-cage motors and switching off only after the motor is up to speed [make at load running amperes (LRA), break at full-load amperes (FLA)]
AC4	Starting of squirrel-cage motors with inching and plugging duty; rapid start/stop (make and break at LRA)
AC11	Auxiliary (control) circuits

NEMA Starter Ratings. NEMA-type starters are rated for motor loads as shown on the NEMA ratings chart, Table 5.37. Some manufacturers also supply custom or motor-matched sizes which do not have standardized ratings, but may offer an economical alternative to the user.

Maximum-Horsepower Plugging and Jogging. The starters must be derated for applications requiring repeated interruptions of stalled motor current or repeated closing of high transient currents encountered in rapid motor reversal, involving more than 5 openings or closings per minute and more than 10 in a 10-min period, such as plug stop, plug reverse, or jogging duty.

Ballast-Type, Tungsten, and Other Discharge-Type Lighting Loads. Contactors may also be used for controlling tungsten and other discharge-type lighting loads. Contactors are specifically designed for such loads and are applied at their full rating.

Resistance Heating Loads. Contactors may be employed to switch a load at the utilization voltage of a resistance heat-producing element with a duty which requires continuous operation of not more than five openings per minute.

TABLE 5.37 Motor Starter Sizes

Single-phase horsepower		Three-phase horsepower			
115 V	230 V	200 V	230 V	460–575 V	Starter size*
⅓	1	1½	1½	2	00
1	2	3	3	5	0
2	3	7½	7½	10	1
3	5	10	10	15	1P, 1¾
3	7½	10	15	25	2
		15	20	30	2½
		25	30	50	3
		30	40	75	3½
		40	50	100	4
		50	75	150	4½
		75	100	200	5
		150	200	400	6

* NEMA Standard and custom sizes included.

Reduced-Voltage Starters (Electromechanical)

Unless prohibited by local utility restrictions, any size ac motor operating at any voltage can be started at full voltage. In actual application, however, when full voltage is applied to the motor terminals, its locked-rotor current may range anywhere from 6 to 10 times the value of normal running current. By design, this current may not harm the motor; however, it may damage the driven load because of the motor's high starting torque. Starters include overload relays which help provide protection to motor windings from harmful currents and resultant temperature rise that may be caused by overloading the motor, sustained low line voltage, or stalled rotor conditions.

Under these conditions, the application of a reduced-voltage starter could eliminate such potential problems.

Definition. A reduced-voltage starter reduces inrush line current and/or starting torque to a polyphase squirrel-cage motor in one of three ways:

1. It reduces voltage applied to the motor during starting.
2. It uses only part of the motor windings during starting.
3. It changes motor winding connections.

Reduced-voltage starters are used when limitations exist for the amount of current that can be drawn from the electrical service.

The starting torque developed by the associated driven equipment is also reduced when a reduced-voltage starter is used. Each type has different characteristics that determine where it may best be used.

Primary Resistor Starting. Using this method, series resistance is added in each conductor to the motor. Control is used to gradually short out the resistance as the motor comes up to speed, until the motor is connected to the full line voltage.

Primary Reactor Starting. This motor-starting method is similar to the primary resistor method except that reactors are substituted for the resistors.

Autotransformer Starting. At starting, wye-connected autotransformer coils reduce motor terminal voltage in each conductor to 50, 65, or 80 percent of line voltage. After a timed interval, a manual switch or a contactor connects the motor across the line and bypasses the autotransformer coils, employing either an open or closed transition.

Part-Winding (Increment) Starting. Though technically not reduced-voltage starting, part-winding controllers apply voltage through one starter to one motor winding followed by the second starter which connects voltage to the second winding. Three-step starters incorporate a series resistance with one motor winding to increase the motor terminal voltage gradually as the voltage drops across the resistor.

Wye-Delta Starting. Although technically not reduced-voltage starting, a wye-delta starter energizes the motor windings through electric contacts that form a wye connection giving about 33 percent of full line voltage across each winding. After a set time delay, the motor windings are connected in a delta configuration. Wye-delta starting can be used in applications requiring a low starting torque when supplying full starting current would cause significant voltage drops.

Open versus Closed Transition. Two terms are used when discussing reduced-voltage starting: *open-transition* and *closed-transition* starters. These terms are used to describe circuit continuity during the starting sequence.

Any open-transition starter momentarily disconnects the motor from the line at the transition from one step to another. A closed-transition starter never disconnects the motor from the line during the starting sequence. Closed-transition starters provide a smoother start and a lower peak current flow than open-transition starters. An open-transition starter may have fewer contactors and is less expensive than its closed-transition equivalent.

SOLID-STATE MOTOR CONTROL FOR AC INDUCTION MOTORS

Solid-state reduced-voltage controllers and starters fall into a class of motor controllers known as *variable-voltage controllers.* These units do not change the frequency of the power applied to the motor, only the magnitude of the applied rms voltage. Solid-state starters provide an alternative method of reduced-voltage starting for various induction motors. Solid-state starters are able to give both a physical (mechanical) and an electrical "soft" start to the load.

Electromechanical versus Solid-State Soft Starting

Electromechanical reduced-voltage starting is softer than across-the-line starting in both electrical and mechanical terms but it does not qualify as true soft starting. In Fig. 5.130, notice the abrupt changes in motor current and output torque that occur when a reduced-voltage autotransformer starter makes the transition from reduced to full voltage. These discontinuous points cause shocks to the electrical and mechanical systems. Other drawbacks of this method of starting include a lack of adjustability, special motor requirements in some cases, complicated control schemes, large size, and uncertain starting characteristics under varying load conditions. In the case of a reduced-voltage autotransformer starter, the user is confined to the transformer tap settings to determine the amount of voltage first applied to the motor. Wye-delta starters will only work with wye-delta motors and part-winding starters require part-wind motors to give an effective reduced-voltage start.

Solid-state reduced-voltage starters (soft starts) apply an adjustable reduced voltage to the motor and then gradually increase the applied voltage until the motor is receiving full voltage (see Fig. 5.131). In this case there are no sudden changes in the voltage applied to the motor. As shown in Fig. 5.132, this prevents the sudden changes in current and torque that occur with reduced-voltage starting. Reducing the motor torque also reduces the accelerating torque and the mechanical stress on the system, and the load comes up to speed more slowly.

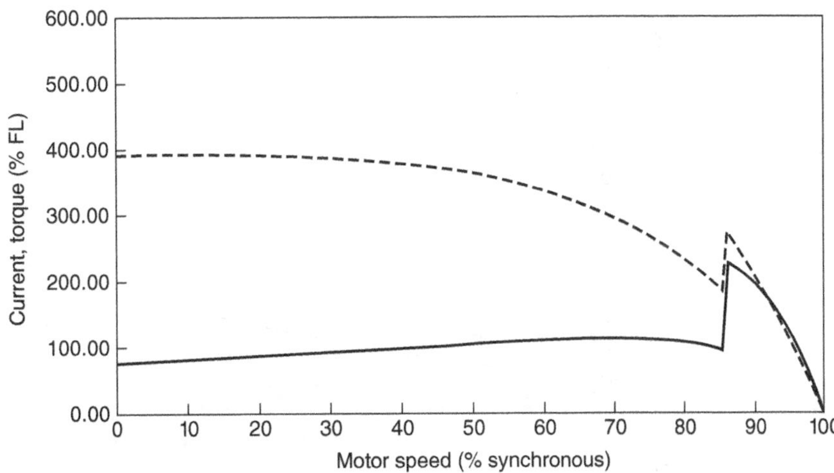

FIGURE 5.130 Starting current and torque, autotransformer starter.

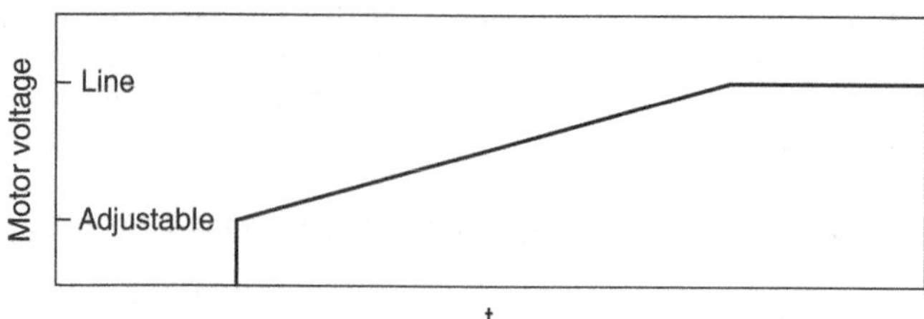

FIGURE 5.131 Solid-state soft start.

Selection of Solid-State Motor Controls

Types of Solid-State Reduced-Voltage Motor Controls

Solid-State Motor Controller. A solid-state motor controller is a device consisting of an electronic control assembly and one or more solid-state power assemblies. It is used with an electromechanical motor starter. A controller can be used for single-speed, multispeed, and reversing applications when the proper electromechanical device is also used. A controller typically has no means of motor overload protection or ability to completely remove power from the motor.

For a single-speed nonreversing application, a motor controller is wired in series between a motor starter and motor. The motor starter provides the power isolation and overload protection while the controller switches the motor current on and off.

Solid-State Motor Starter. A solid-state motor starter is a device consisting of an electronic control assembly, one or more solid-state power assemblies, and motor overload protection. The main difference between a controller and starter is that the starter can be used without any other motor control device.

Solid-state starters usually incorporate additional features and functions providing additional motor control. These may include such things as overvoltage, overcurrent, phase-loss, phase-reversal, and overtemperature protection.

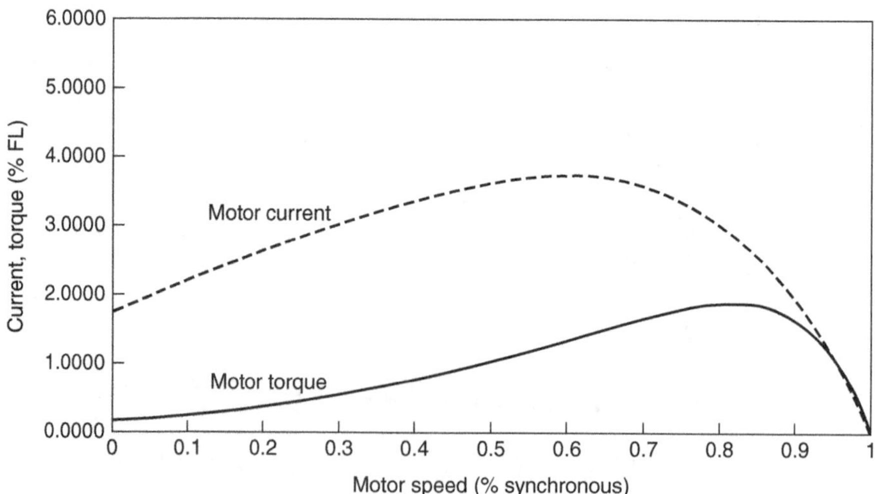

FIGURE 5.132 Starting current and torque for solid-state soft start.

Starting Methods for Solid-State Motor Controls

Voltage Ramp Soft Start. Voltage ramp soft-start units apply a reduced voltage to the motor on starting and then ramp the voltage up to line voltage in a more or less linear fashion (see Fig. 5.133). The *initial torque* or *initial voltage* adjustment determines the amount of voltage applied to the motor when the start signal is given. This voltage is usually adjusted so that the motor will develop just enough torque to overcome the mechanical friction in the system. On applications involving little friction (pumps, large inertial loads, fans) the initial voltage adjustment can be set relatively low. Applications involving much friction or demanding high locked-rotor torque (conveyors, positive displacement pumps) will require a higher torque setting. Ramping the voltage more quickly will cause the motor to accelerate more quickly and result in higher peak inrush currents.

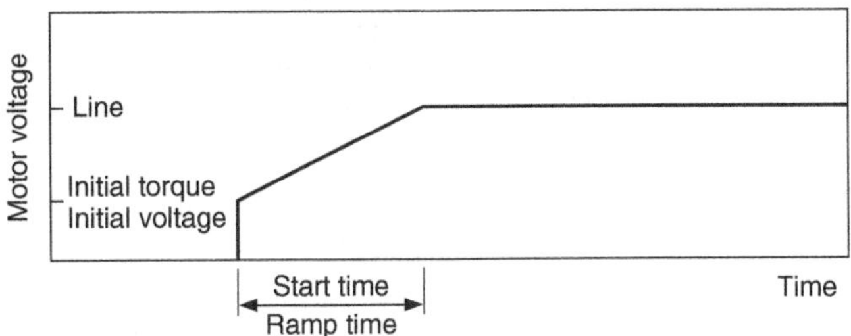

FIGURE 5.133 Voltage-ramp soft-start adjustments.

The key features of the voltage-ramp soft-start unit are that it can softly start varying loads, it does not limit motor torque, and it is easily adjusted. In general, it is desirable to bring the load up to speed as quickly as possible given the mechanical and electrical power distribution system constraints.

Applications for voltage-ramp-type soft-start controls generally fall into two categories: equipment protection and process protection. In terms of process protection, voltage ramp units are used to prevent spillage on conveying and packaging lines that results from the sudden acceleration of across-the-line starting. Electrical soft start is also a form of process protection, since power loss due to breaker tripping or generator shutdown would bring the process to a halt. Voltage-ramp units can be applied on variable-torque, constant-torque, high-inertia, and varying loads, and they give both physical and electrical soft start. As with any type of reduced voltage or soft starting means, it is necessary to evaluate some high-starting-torque applications to determine feasibility. Some loads demand such high torque at locked rotor that any kind of reduced voltage will not allow the motor to start.

Reduced-Voltage Current-Limit Soft Start. Current-limit soft-start units limit the current drawn by the motor during starting. They allow the motor to draw as much current as it needs up to a user-defined limit and hold the motor to that amount of current until the motor comes up to speed or the unit either shuts down or removes the limit. The current is applied to the motor in one of two different fashions as shown in Fig. 5.134. The *current-limit amps* adjustment determines the maximum amount of current the motor will be allowed to draw during a normal start. The *ramp-time* adjustment determines how quickly the motor current ramps up to the limit amount. For units with a ramp-time adjustment, a shorter ramp time results in a harder physical start, higher starting currents (up to the limit amount) for a shorter time, and a shorter acceleration time.

The key features of the current-ramp soft-start unit are that it limits the motor inrush current to a specific amount, it can provide physical soft start, and it limits motor torque. These units are used mainly for electrical soft start (reducing inrush current). The shutdown of production is a

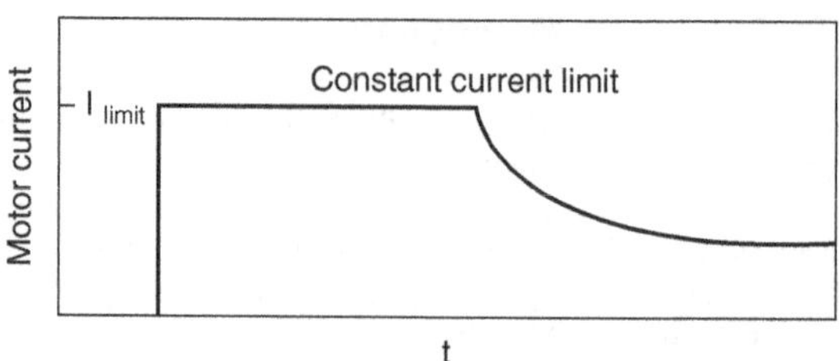

FIGURE 5.134 Current-limit units can apply the amount of current immediately or ramp up to the limit.

possibility with a current-ramp unit, since it limits the motor torque. Set the current-limit adjustment low enough to start the motor without light flicker, excessive voltage drop, or generator shutdown. If the unit has a ramp-time adjustment, it should be set to accelerate as quickly as possible given the mechanical system constraints.

Current-ramp units can be applied on variable-torque, some constant-torque, and high-inertia loads. They are not recommended for loads that vary considerably since they may not be able to start the load sometimes. As with any type of reduced-current or soft-starting means, it is necessary to evaluate high-starting-torque applications to determine feasibility. Some loads demand such high torque at locked rotor that any kind of reduced current will not allow the motor to start.

Environmental Conditions for Solid-State Motor Controls

Ambient Temperature. When selecting a mounting location, observe the specified ambient temperature range. The ambient temperature is the temperature of the air surrounding the control. Operate the control at an ambient temperature as close to the recommended minimum temperature as possible, since some semiconductor devices may cease proper functioning temporarily if overheated, while others may be damaged permanently.

Avoid mounting locations near heat sources, such as equipment that raises the ambient air temperature after a period of operation (e.g., power supplies, transformers, and starters); heat conducted into the control through the mounting surface; or radiated heat where a control is exposed by line of sight to the sun, extremely hot materials, or open flames.

When a mounting location with high ambient temperatures cannot be avoided, consider employing special cooling methods. Observe the following guidelines:

1. Position heat sinks such that air may move vertically between cooling fins across the heat sink surface. This allows ease of airflow for convection cooling.
2. In large systems, distribute components carrying large currents evenly throughout an enclosure to avoid localized hot spots.
3. Ventilated enclosures are effective, but take precautions to avoid air contamination by dust and conductive particles. Use air filters with fan cooling, since dust accumulation on heat sinks reduces their thermal efficiency.
4. Extreme situations may require heat exchangers or air conditioning of enclosures. Also, consider locating the entire control system in a separate room adjacent to the controlled machinery.

Humidity. During operation, heat is generated, and moisture seldom condenses on the equipment. A moisture problem is most likely to develop during inoperative periods or during storage. If necessary, apply humidity control measures during these periods.

Also, avoid dry conditions to prevent static electricity problems. Static electricity applied directly to module or circuit terminals can cause erratic control operation and, in some instances, can produce voltage breakdown failures in semiconductor devices.

Location. A vented Type 1 enclosure is acceptable in indoor locations where the location is dry and relatively dust-free. Type 1 enclosures allow air to enter the enclosure and cool the components inside. Any moisture present or an excessive amount of dust, dirt, or other airborne contaminants allowed to enter a Type 1 enclosure could cause damage to the motor control.

In situations where these conditions exist, a motor control must be mounted in a fully enclosed environmental enclosure. When installed in a sealed enclosure it is often necessary to add additional equipment to minimize or eliminate the heat generated by the motor control while it is operating. Consult the control manufacturer for recommendations.

Shock and Vibration. Shock and vibration can have a potentially damaging effect on solid-state motor controls. Special mounting provisions may be required to prevent such damage.

When determining the mounting location for a reduced-voltage motor control, attempt at all costs to avoid mounting the device to a piece of equipment that constantly vibrates or has the tendency to shake during its normal operation. If this situation exists it may be necessary to mount the motor control off the machine, on a building support or other solid object.

Wiring Considerations for Solid-State Motor Controls

Low-Level-Signal Wiring. Coupled electrical noise can be viewed as an unwanted signal originating in other wiring. Solid-state control lines should be separated from noise sources by routing them through their own separate conduit.

When possible, ac lines or lines carrying high currents or inductive loads should not run close to and parallel to sensitive low-level-signal wiring. Besides being separated, they should intersect only at right angles.

Shielded Cable. Using twisted-pair shielded cable for low-level-signal wiring provides a very effective shield against electrostatic and magnetic coupling. The cable should have approximately 12 twists per foot.

The shield must be connected to control ground at only one point and shield continuity must be maintained over the entire length of the cable. Be certain the shield is insulated over its entire length to assure the one-point ground connection. When possible, route the cable around, rather than through, high-noise areas.

Note: For device grounding, attach the shield directly to a large grounded metal object such as a machine tool. Some devices achieve better noise immunity by grounding the shield at the input circuit rather than at the device. Also, in some cases, the shielding may be more effective by connecting the shield to signal common instead of ground.

Grounding. In solid-state control systems, the grounding practices employed have a significant effect on noise immunity. Each ground must be connected to its respective reference point by no more than one wire (one-point grounding). Two or more systems must never share a common single ground wire, either equipment ground or control common. For proper grounding practices, observe the following:

1. Minimum wire sizes, color coding, safety practices, etc., must comply with the *National Electrical Code* and any local codes and regulations.*
2. Observe the recommendations specified on the data sheets for control system products.
3. For further information on grounding practices, refer to IEEE Standard 142-1972, "Recommended Practices for Grounding of Industrial and Commercial Power Systems."

Impedance coupling produced by ground sharing is a source of conducted noise that must be avoided. Guidelines for minimizing impedance coupling include the following:

**National Electrical Code* is a registered trademark of the National Fire Protection Association, Inc., Quincy, MA 02269.

1. Use individual signal return wires.
2. Where signal paths are shared, use conductors of large diameter (e.g., the conductor could be a heavy busbar).

Use a separate dc power supply or separate ac branch circuit to power each control system. When a power source is shared, the line impedance (ac supply) or current capability (dc supply) must be sufficient to maintain rated voltage under all loading conditions.

Using Power-Factor-Correcting Capacitors. The capacitive load of a power-factor-correcting capacitor has a very damaging effect on solid-state motor controls if used incorrectly. As shown in Fig. 5.135, when a capacitor is used it must be wired to some point on the line side of the motor control.

If the capacitor is wired to the output terminals of the solid-state motor control, damage will result to its silicon-controlled rectifier (SCR) power devices.

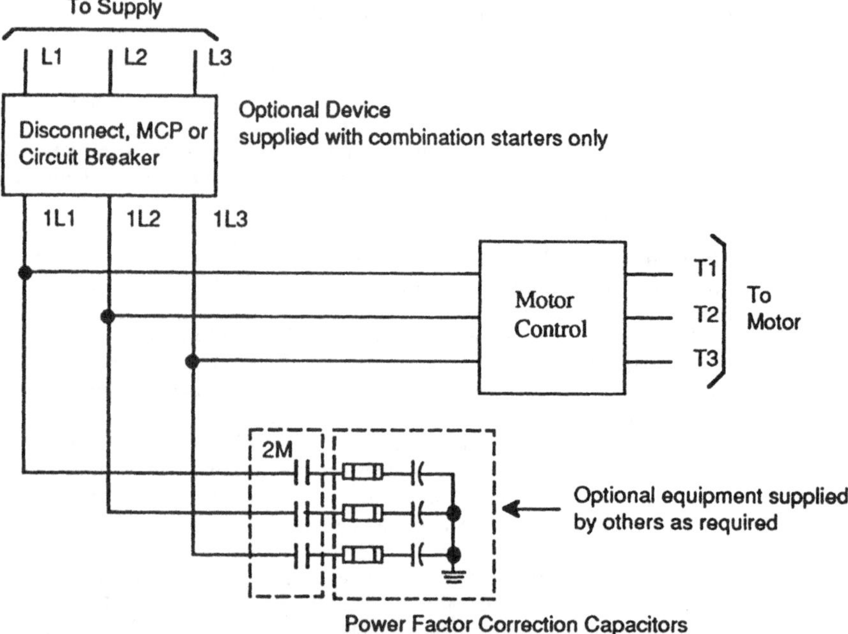

FIGURE 5.135 Wiring power-factor-correcting capacitors.

Cost Considerations

The most significant cost to consider when purchasing a reduced-voltage soft-start motor control is the purchase price. But there are several other costs which must also be considered. In both new and existing installations, it is necessary to figure in the cost of installing the device. Depending on the location selected, the cost to install and wire the device could exceed the cost of the device itself.

Other factors to consider include product life and the elimination of potential downtime. Solid-state devices, when used properly, offer a much longer life than their electromechanical counterparts.

SPEED CONTROL OF ELECTRIC MOTORS

The term *speed control,* as applied to electric motors, covers a wide range of control functions including motor starting, control of motor speed during normal operation, and reversing and stopping of motors.

Operating requirements for specific motor applications must also be considered and generally include constant, variable, adjustable, and multispeed operation.

Constant Speed. Motors of this type may be designed with speed ratings from 80 r/min and horsepower ratings ranging up to 5000 hp (3700 kW). A typical application is water pumps.

Variable Speed. Variable-speed motors slow as the applied load is increased and speed up as the load is decreased. Applications include cranes and hoists.

Adjustable Speed. Adjustable-speed motors can be varied over a wide speed range while the motor is running. Motor speed remains almost constant once set, even with load applied. A typical application is machine tools.

Multispeed. Multispeed motors are designed to operate at two or more definite speeds. However, once adjusted to a particular speed, the motor speed remains nearly constant regardless of applied load changes. A typical application is turret lathes.

Multispeed Magnetic Starters

Applications. Multispeed magnetic starters automatically reconnect multispeed motor windings for the desired speed in response to a signal received from push-button stations or other pilot devices.

Consequent-pole multispeed motors having two speeds on a single winding (consequent pole) require a starter which reconnects the motor leads to half the number of effective motor poles at the high speed point. In this type of motor, the low speed is one-half the high speed.

Separate winding motors, having separate windings for each speed, provide more varied speed combinations in that the low speed need not be one-half the high speed. Starters for separate winding motors consist of a starter unit for each speed.

Multispeed motor starters are available for constant-torque, variable-torque, and constant-horsepower motors.

Constant Torque. These motors maintain constant torque at all speeds. Horsepower varies directly with speed. This type of motor is applicable to conveyors, mills, and similar applications.

Variable Torque. These motors produce a torque characteristic which varies as the square of the speed. This type of motor is applicable to fans, blowers, and centrifugal pumps.

Constant Horsepower. This type of motor maintains constant horsepower at all speeds and therefore torque varies inversely with speed. This type of motor is applicable where the same horsepower is required at all speeds. The higher current required at low speed requires derating on starters for constant-horsepower applications. This type of motor is applicable to metalworking machines such as drills, lathes, mills, bending machines, punch presses, and power wrenches.

Operation. Magnetic starters for multispeed applications select the desired speed in accordance with the pilot control.

The shock to machinery on the reduction of speed is greater than when the speed is increased. Therefore, the pilot control should be wired so that the stop button must be depressed before dropping to a lower speed. Time delays should be used for applications requiring full automatic operation.

These controls may be modified for compelling or acceleration pilot control.

Compelling control requires that the motor always be started at the lower speed and that the push buttons be operated in speed sequence to the next higher speed. To change to a lower speed, the stop button must be depressed and then the push buttons are pressed in speed sequence until the desired speed is reached.

Acceleration control provides that the motor be accelerated automatically with timers by progressively energizing the controls at the push button station from the lowest to the highest speed. To change to a lower speed, the stop button is depressed and then it is necessary to proceed as if starting from rest.

Deceleration control provides that the motor be decelerated automatically with a timer when going from high speed to low speed. The timer allows the motor to decelerate from high speed to a lower speed before automatically restarting the motor in low speed.

Motor Braking

Changing the speed of, or stopping, an electric motor can be accomplished using (1) electrically operated mechanical brakes or (2) dynamic or regenerative braking or plugging (either with an electromechanical starter or a solid-state starter), or by a combination of the two. Brakes are used for two purposes: (1) to provide the means of stopping a driven load quickly and accurately and (2) to hold the load in place after stopping is accomplished.

Dynamic Braking. When very quick or accurate stopping is not a necessity, the motor may be used to stop itself. AC motors can be dynamically braked by removing the ac power source and reconnecting a dc power source (supplied either from batteries or rectified ac power). The motor then acts like a dc generator with a short-circuited armature. Energy is dissipated in the form of rotor heat.

Regenerative Braking. In ac motors, regenerative braking is developed by the motor's natural tendency to resist being driven above synchronous speed by an overhauling load. As this situation occurs, a negative slip is developed. The energy absorbed in slowing down is put back into the power supply. Regenerative braking generally is not used with rectified power supplies because this process requires the reversal of armature current.

SOLID-STATE VARIABLE-SPEED AC MOTOR CONTROLS

One method of controlling the speed of ac motors is with a pulse-width-modulated (PWM) variable-frequency drive (VFD). In the simplest terms, the VFD converts ac line voltage into variable voltage and frequency, which is then used to control the speed of ac squirrel-cage motors. As a controller, start/stop, hand-off-auto, and various other control schemes may be applied.

This type of equipment is almost limitless in its use in very diverse applications such as conveyors, HVAC (air handlers), irrigation systems (pumps), machine tools, food processing, overhead cranes, textile mills, chemical processing, and automotive manufacturing.

The benefits derived by the use of a VFD are closer and more responsive control of motor speed; less mechanical wear on systems because of inherent soft start; substantial energy savings achieved on variable-torque loads (e.g., fans and pumps); and elimination of noisy mechanical speed-control systems and their associated maintenance problems. Some of the main considerations in applying a drive are the following:

Environment. This will determine the enclosure type. The open-chassis type is also available for installation in other enclosures.

Cooling. If the ambient temperature is not within the manufacturer's rating, typically 104°F (40°C), steps must be taken so that the operational temperature is not exceeded.

Location. The location should be convenient.

When sizing the VFD to match the connected mechanical load, the foremost consideration is *inertia,* which is a measure of an object's resistance to changes in speed, whether that object is at rest or moving at a constant speed. This speed can either be linear (feet per minute, ft/min) or rotational (revolutions per minute, r/min).

The term for inertia is WK^2, usually given as a numerical value, in units of lb · ft². (This is normally calculated by the mechanical engineer or a machine design company). This then directly affects the torque required.

Torque is the action of a force tending to produce or producing rotation. Torque may even exist when no movement or rotation occurs. Work, unlike torque, exists only when movement occurs.

Because most power transmission is based on rotating elements, torque is extremely important as a measurement of effort to produce work.

The amount of amperage required to produce a given starting torque may be higher than the output rating of the VFD; therefore, the drive must be sized to the calculated starting amperage for the given acceleration time.

Estimating Accelerating Torque (Rotational)

English

$$T = \frac{WK^2 \times dN}{308 \times t}$$

Metric

$$T = \frac{J \times 2dN}{60 \times t} = \frac{J \times dN}{9.55 \times t}$$

where T = acceleration torque, lb · ft (N · m)
WK^2 = total system inertia, lb · ft² (i.e., the equivalent WK plus rotor inertia)
J = total system inertia, kg · m
dN = change in speed, r/min
t = time to accelerate, s

Note: In some cases inertia may be expressed by a metric value analogous to the English WK^2 whose dimension is $N \cdot m^2 = J \times g$, where $g = 9.8\ m/s^2$.

After torque has been calculated, the horsepower (or watts) can then be estimated:

English

$$hp = \frac{T \times N}{5250}$$

Metric

$$W = \frac{T \times 2N}{60} = \frac{T \times N}{9.55}$$

Estimate motor current if required:

English

$$I_m = \frac{hp \times 746}{1.73 \times E \times Eff_m \times PF_m}$$

Metric

$$I_m = \frac{W}{1.73 \times E \times Eff_m \times PF_m}$$

Efficiency (Eff) and power factor (PF) of the motor are expressed as a decimal, typically 0.85 and 0.81, respectively.

These formulas are examples only. Gear and belt reductions, which can make great differences in the sizing of the drives, must also be taken into consideration. In the case of pumps and fans, full-load ampere nameplate ratings of the motors are usually all that is required.

In comparison with average across-the-line starters, initial cost can seem quite high. As compared to soft starters, the cost differential is reduced, but still higher. The previously mentioned advantages can offset the difference in price.

CONTROL OF DC MOTORS

The control of dc motors, as with ac motors, includes motor starting and stopping; governing motor speed, torque, horsepower, and other characteristics; and ensuring the safety of personnel and equipment. The methods of controlling dc motors, however, vary considerably.

DC Motor Starting and Braking

Starting Considerations. Starting of dc motors rated 2 hp (1.5 kW) or less can generally be accomplished by manually operated, full-voltage starters. For dc motors rated above 2 hp (1.5 kW), reduced-voltage starting is usually recommended to avoid commutator damage. A common method of limiting starting current is by the addition of resistance (which is decreased in steps). The final step (larger motors or those with smooth starting requirements use several steps) is to connect the motor to full line voltage.

Braking. Like ac motors, dc motors may be brought to a stop quickly by electromechanical brakes. Dynamic or regenerative braking schemes can also be used with dc motors.

Speed Control

DC motor speed is directly proportional to the applied armature voltage and inversely proportional to the field flux. Shunt field control and armature voltage control are, then, the two basic methods used to accomplish speed adjustment.

Shunt Field Control. With the addition of a rheostat to the shunt field circuit, control is obtained by adjusting the rheostat to weaken the shunt field current. This, in turn, increases the motor speed and reduces the output torque.

The maximum standard speed range using field control is 3:1; specially designed motors can achieve speed ranges of 4:1 or more. For a specific motor, the nameplate rating should be checked or the manufacturer consulted if the allowable range of motor speed is not known.

Armature Voltage Control. This method requires the application of a variable-voltage power supply with a capacity equal to that of the motor, while holding the shunt field current constant. The resulting speed is proportional to the motor counter emf. The torque remains constant at rated current, regardless of motor speed. In applications where a speed range of greater than 3:1 is required, armature voltage control should be used.

DC Drive Characteristics

Adjustable-Voltage Drive. DC motor armature voltage can be varied to obtain wider ranges of speed control than are available with standard dc motors. An ac-to-dc motor-generator set controls the dc motor speed by increasing the generator voltage while holding the motor field at full strength. When generator voltage reaches full value, the motor runs at base speed. If the speed control is increased further, the motor field is weakened while generator voltage is held constant, causing the motor speed to increase above its base.

Chopper Drives. The high-speed switch (chopper) characteristics of phase-controlled rectifiers (thyristors) can be used to create adjustable-speed dc drives. A constant-voltage rectifier or a dc bus is used as the incoming supply. When the first thyristor is turned on by a reference-level signal, the dc motor is connected to the constant-voltage bus. When motor speed rises above the reference level, a second thyristor is turned on, shutting off the first, and the motor is driven from power stored in an inductor. When the motor speed drops below the reference level, the first thyristor is turned on, turning off the second which again supplies power from the bus. Current transfer between thyristors is regulated by capacitor discharge through a transformer. Dynamic braking can be added by turning on a third thyristor that dissipates energy into a dynamic braking resistor. This type of drive is often used in applications such as adjustable-voltage hoists because power supplies can be subjected to wide voltage deviations.

PROTECTION OF MOTOR CONTROLLERS

Electrical Enclosure Types and Specifications

Electrical enclosures should be selected and specified according to NEMA Standard Publication No. 250. A brief description follows:

Type 1. Indoor use primarily to provide a degree of protection against contact with the enclosed equipment in locations where unusual service conditions do not exist.

Type 3. Outdoor use primarily to provide a degree of protection against windblown dust, rain, and sleet and to resist damage by the formation of ice on the enclosure.

Type 3R. Outdoor use primarily to provide a degree of protection against falling rain and to resist damage by the formation of ice on the enclosure.

Type 4. Indoor or outdoor use primarily to provide a degree of protection against windblown dust and rain, splashing water, and hose-directed water and to resist damage by the formation of ice on the enclosure.

Type 4X. Indoor or outdoor use primarily to provide a degree of protection against corrosion, windblown dust and rain, splashing water, and hose-directed water and to resist damage by the formation of ice on the enclosure.

Type 6. Indoor or outdoor use primarily to provide a degree of protection against the entry of water during occasional temporary submersion at a limited depth.

Type 6P. Indoor or outdoor use primarily to provide a degree of protection against the entry of water during prolonged submersion at a limited depth.

Type 7. Indoor use in locations classified as Class I, Groups A, B, C or D, as defined in the *National Electrical Code.**

Type 9. Indoor use in locations classified as Class II, Groups E or G, as defined in the *National Electrical Code.*

Type 12. Indoor use primarily to provide a degree of protection against dust, falling dirt, and dripping noncorrosive liquids.

Type 13. Indoor use primarily to provide a degree of protection against dust, spraying of water, oil, and noncorrosive coolant.

Motor Overload (Running) Protection

Overload relays are intended to protect motors by "tripping" if heat generated in the motor windings is approaching damaging levels. This interrupts the control circuit and causes the contactor to open, disconnecting current to the motor.

Traditional NEMA designs use either melting alloy or bimetal strips, and have interchangeable heaters. NEMA has assigned classes in reference to the designed protection level provided by the individual overload relay. Class 10 overload relays are designed to protect artificially

**National Electrical Code* is a registered trademark of the National Fire Protection Association, Inc., Quincy, MA 02269.

cooled motors such as submersible pump motors or other motors of low thermal capacity. Class 20 overload relays are designed to protect standard American T-frame industrial motors.

IEC designs use a bimetal-type triggering device with integral (nonreplaceable) heaters. IEC-type overload relays are set for trip speed equivalent to NEMA Class 10.

Melting-Alloy-Type Overload Relays. The triggering mechanism in the melting-alloy type is a spring-loaded lever which is held in place by a solid plug of eutectic solder. When enough heat is generated, the solder melts, releasing the spring-loaded mechanism and opening the contact. Once the melting alloy cools enough to return to solid state, the spring-loaded triggering mechanism must be manually reset by turning it against a ratchet.

These overload relays are highly resistant to vibration. They can be adjusted only by changing heaters, cannot be modified for automatic reset, and are inherently sensitive to ambient temperatures.

Bimetal Overload Relays. Traditional NEMA-style bimetal overload relays share the heater interchangeability of the melting-alloy type. They may also have the ability to adjust up or down the equivalent of one heater size, or about ±15 percent. IEC overloads are bimetal type, but have nonreplaceable integral heaters. To compensate for this, these units can typically be adjusted ±20 to 25 percent. Bimetal devices use the heat generated to cause deflection of one or more bimetal strips. Once this deflection is great enough, the trip mechanism is operated.

Bimetal overloads begin movement toward the trip point the moment current begins to pass through the heaters. Thus, as they near the trip point, they may be more sensitive to shock and vibration.

Both bimetal types may be provided with means of compensating for ambient temperature, which may be desirable if the motor and control are in dissimilar environments. For example, a motor in a deep well pump is in a constant ambient temperature, while its surface control is in a variable ambient. Where motor and heater are located in the same ambient, compensation may actually be harmful. The overload relay is supposed to track the thermal conditions in the motor windings, and a compensated overload would ignore the ambient portion of the total heat present.

Difficulties can arise when an IEC overload is used with a typical American industrial T-frame motor. There, the quick trip action may result in nuisance tripping under normal starting loads. To avoid this nuisance tripping, users often must set the built-in adjustment to increase the trip current level. While this does prevent nuisance tripping, it can be a misadjustment and may negate the overload relay's protection at running speed. In this situation, a heavily loaded motor could see excessive currents and a corresponding temperature rise (see Fig. 5.136). One engineering rule of thumb states that each 10°C increase in temperature reduces motor insulation life by 50 percent. Increasing the trip current setting to avoid nuisance trips may well lead to early motor failure.

Solid-State Overload Relays. These overload relays are of two types, self-powered and externally powered. The self-powered relay (Fig. 5.137) uses currents induced by the current flowing to the motor for power, while the externally powered units contain an electronic power supply and require a separate source of control power to operate. The advantage of the self-powered units is that they are simpler, require no extra wiring, and are self-protected from short circuits.

High-Accuracy Trip Curve Adjustments. Certain aspects of an overload relay's trip curves are critical for real motor protection. These include time-to-trip on locked rotor, motor overload conditions, and phase loss.

Thermal overloads are limited from approaching the ideal motor protection curves for these critical conditions. Because they do not directly measure current loading, but instead approximate it through use of heaters within the overload, they are subject to thermal overshoot (continued heat absorption after locked-rotor current is removed), residual heat remaining after reset, mechanical variables, and other difficulties in placing the various curves in optimum relationship with each other.

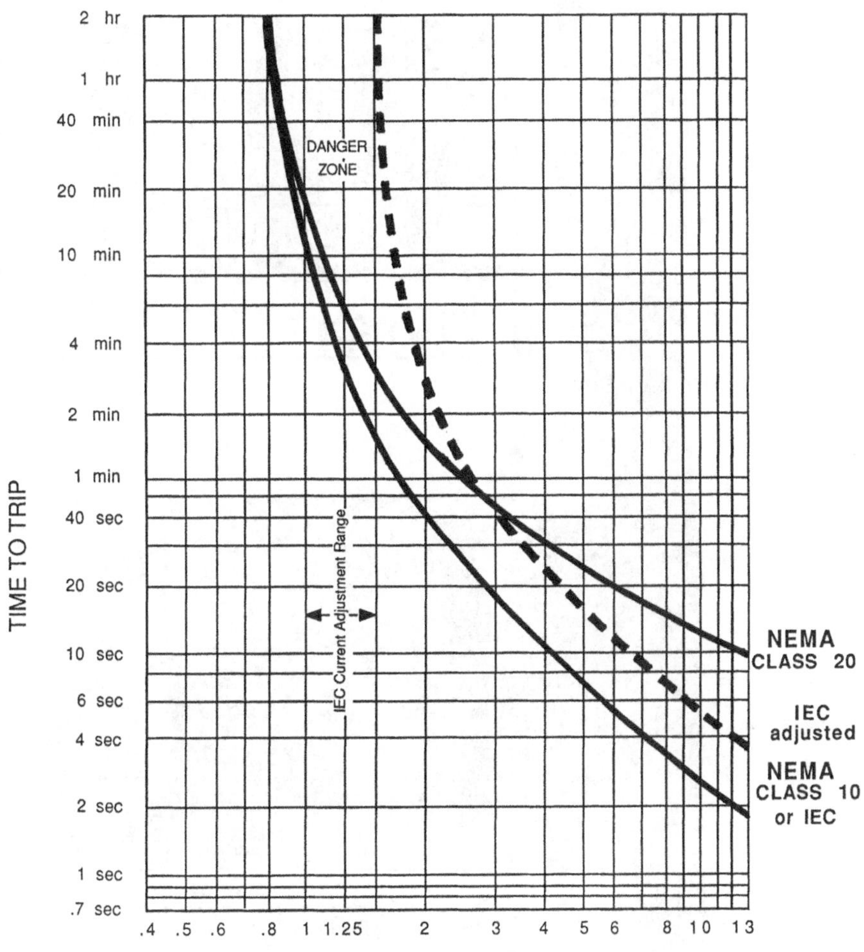

FIGURE 5.136 Graph illustrating the dangers of adjusting an overload relay above the full-load current of the motor to prevent nuisance tripping on start-up.

The solid-state overload has no such limitations. Optimum curves have been designed in (see Fig. 5.138).

Short-Circuit Self-Protection. In the event of a short circuit, the thermal overload heaters often burn out and can consequently cause damage to the overload relay.

The self-powered solid-state overload is self-protecting in short-circuit conditions. Current sensors automatically saturate, and sensed current is well below the safe maximum limit of the electronics. This same protection isolates the electronics of the solid-state overload from damage due to spikes or transients in line current.

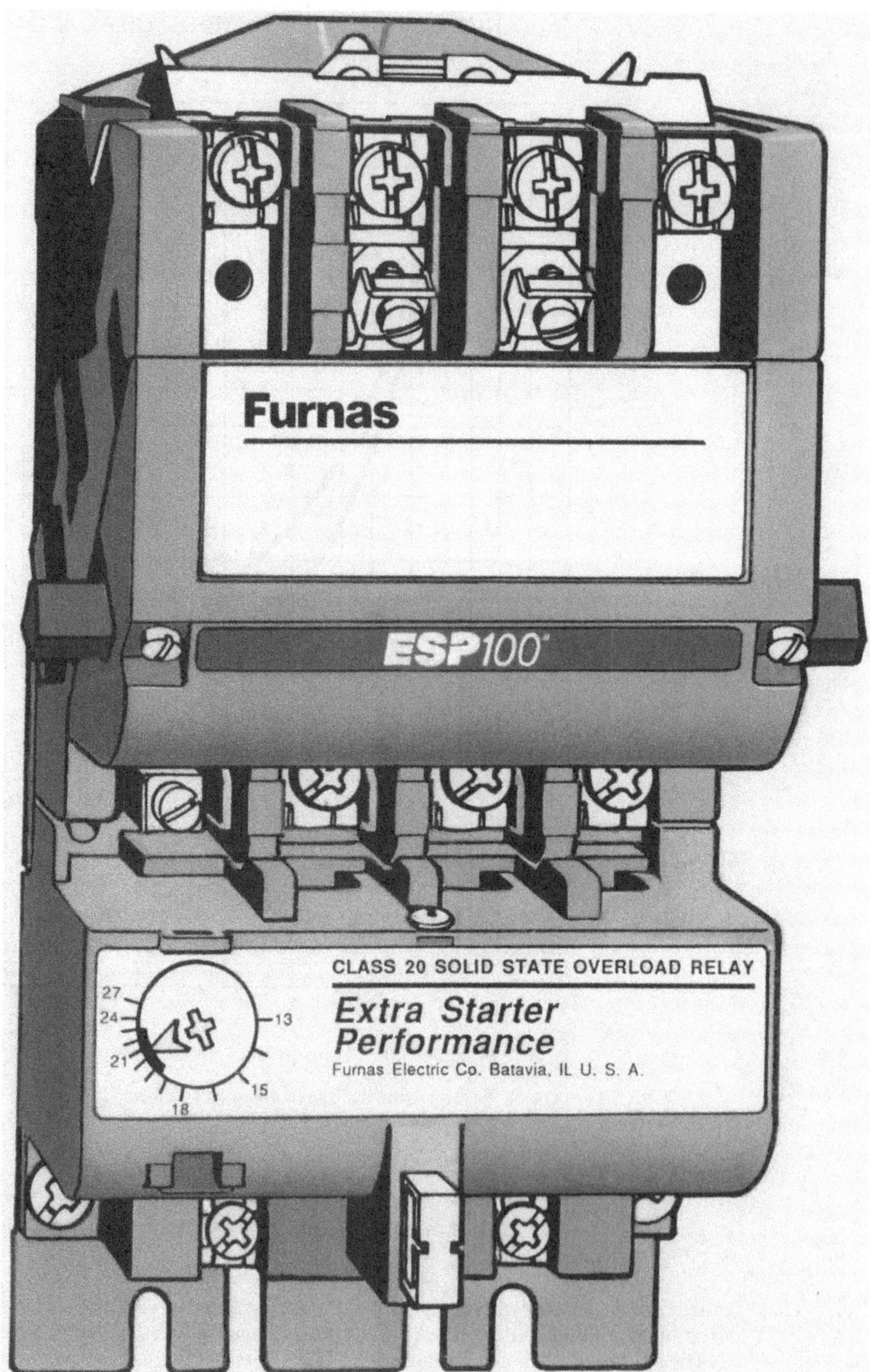

FIGURE 5.137 Starter with self-powered solid-state overload relay. (*Furnas Electric Co.*)

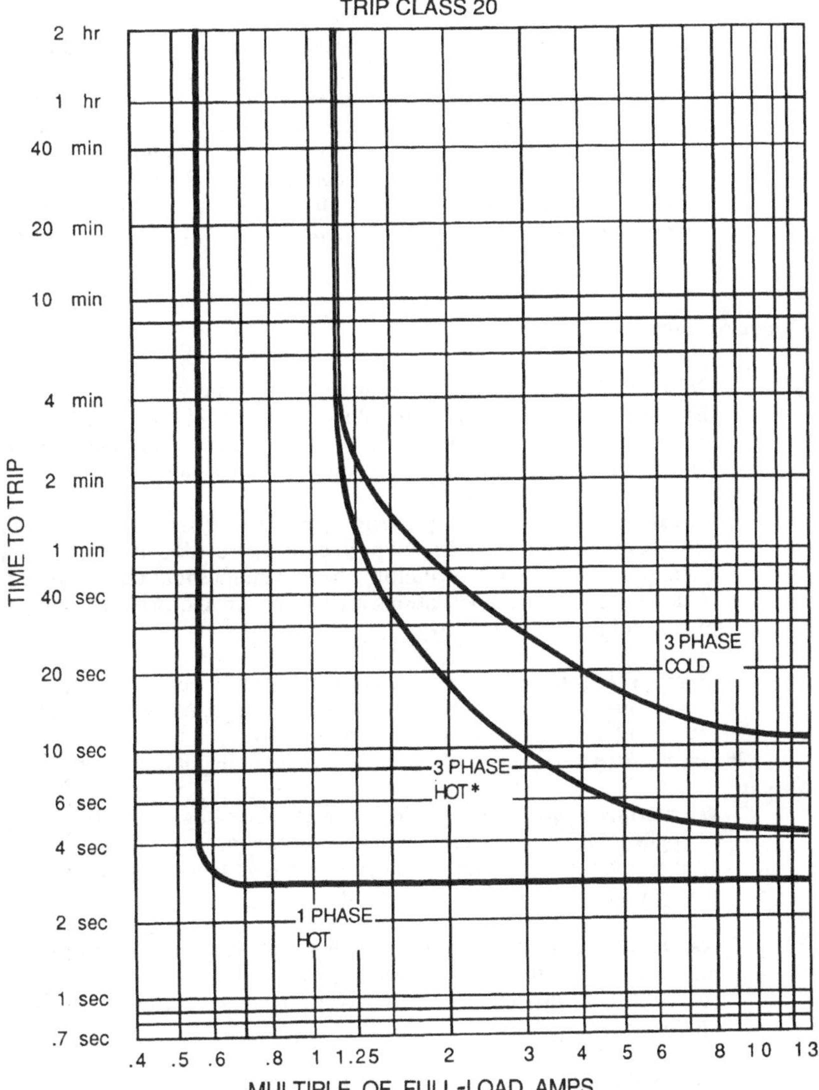

FIGURE 5.138 Solid-state overload relay trip curve.

Open-Phase Protection. Phase failure may be caused by a blown fuse, an open connection, or a broken line. If phase failure occurs while the motor is stopped, stator current will rise to a very high value while the motor remains stationary, and excessive heating may damage the windings.

Thermal overload relays, to varying degrees, are or can be made sensitive to single phasing. Melting-alloy-type overload relays are somewhat sensitive to phase loss, in that sensing of current in all three phases is independent. Loss of one phase will increase current in the other two. If the motor is fully loaded, this increase may be enough to cause the overload to trip. On the other hand, phase loss with a lightly loaded motor may never cause tripping with this type of overload.

Many such overload relays are constructed with a differential trip mechanism, and are phase-loss sensitive. A phase-loss-sensitive overload relay, compensated for temperature variations, and with two poles energized at 115 percent of full-load amperage, shall trip in "less than 2 h."

Neither this nor the sensitivity of the melting-alloy type to phase loss should be seen as true single-phasing protection. If single-phasing protection is a concern, complete protection may be provided by an electronic voltage monitor or solid-state overload relay.

The solid-state overload relay, by contrast, will trip within seconds under the same conditions. This provides enough of a delay to avoid nuisance tripping in momentary outages, but eliminates the potential for extended operation in phase loss condition under load.

Other Types of Motor Protection

Open-Field Protection. DC shunt and compound-wound motors can be protected against the loss of field excitation by installing field-loss relays in the shunt field circuit. Larger dc motors may race dangerously with the loss of field excitation, while other motors may not race because of friction and the fact that they are small.

Overspeed Protection. To prevent damage to a driven machine, materials in the industrial process, or the motor, overspeed protection is provided by controlling the power-supply frequency in ac motors and by limiting the maximum shunt field resistance or armature voltage in dc motors. Typical applications include paper and printing plants, steel mills, processing plants, and the textile industry.

Overtravel Protection. In applications where precise positioning is required, control devices (such as limit switches, cams, photoelectrics, etc.) are used to govern the starting, stopping, and reversal of electric motors. These devices can be used to control regular operation or as emergency switches to prevent the improper functioning of machinery.

Overvoltage Protection. Devices such as overvoltage relays provide a reduction in the level of applied voltage or a maintained interruption of power to the motor circuit when excessive voltage levels are experienced. High induced voltages may occur as a result of the interruption of inductive motor control circuits.

Reversed-Current Protection. Rectifiers are used to protect dc controllers in dc or three-phase ac systems that can be subject to damage when experiencing phase failures and phase reversals. It is also important to provide reversed-current protection for battery-charging equipment.

Reversed-Phase Protection. Phase-failure and phase-reversal relays are used to prevent the operation of elevators and industrial machinery to protect motors, machines, and personnel from potential hazards when phase reversal or loss occurs. Interchanging two phases of the supply of a three-phase induction motor will reverse its direction of rotation.

Undervoltage Protection. Undervoltage relays and three-wire control are employed to initiate the opening of a motor circuit and maintain it open on a reduction or failure of voltage. Undervoltage protection generally is applied where uncontrolled motor starting could result in potential hazards.

Undervoltage Release. Undervoltage relays and two-wire control are employed to initate the temporary opening of a motor circuit on the reduction or failure of voltage. On restoration of full voltage, the motor is automatically restarted.

Short-Circuit Protection

For motors with greater than fractional horsepower ratings, devices such as fuses, circuit breakers, and self-protected combination controllers must be installed ahead of the motor control apparatus to protect branch-circuit conductors, motor control apparatus, personnel, and the motor itself against fault conditions that may be the result of short circuits or grounds.

Definition. Short-circuit current is the current that flows when a fault in the electrical system occurs—for example, when phase conductors are accidentally shorted out. Faults are commonly referred to as either *arcing* or *bolted* faults. The bolted fault does not involve an arc. Such a fault can result from incorrect wiring, conductors that make contact and are welded together during an arcing fault, or foreign objects such as a wrench or screwdriver inadvertently left across conductors during maintenance work. In an arcing fault, a complete connection is not made across the phases, and part of the voltage is consumed across the arc. The total fault current is, therefore, somewhat smaller than for a bolted fault.

When a fault occurs, the overcurrent device must safely open and interrupt the fault. Overcurrent devices such as fuses, circuit breakers, and self-protected combination controllers, which interrupt fault currents, must have an *interrupting* rating equal to or greater than the available short-circuit current at their line-side terminals. Control devices, such as motor starters and overload relays, must have a *short-circuit current* rating equal to or greater than the available short-circuit current. That is to say, they must be able to withstand the fault current for the time it takes the overcurrent device to interrupt the fault. In practice, the actual current which flows during a fault is less than the available short-circuit current, and it is this current which the control device must withstand.

The available short-circuit current at any point in an electrical system is the worst-case or maximum short-circuit current that can flow under fault conditions. To safely apply overcurrent devices within their interrupting ratings, it is necessary to calculate the available short circuit currents at various points of application in an electrical distribution system. The procedure to calculate the magnitude of the fault current is described in various publications available from circuit protective device manufacturers.

Overcurrent devices and motor controls must be provided with marking which indicates their interrupting or short-circuit current ratings, as appropriate. For available currents above 10,000 A, discrete components, such as motor starters, must also be provided with marking which indicates the proper protective device which must be provided ahead of the component. This enables the user to properly coordinate the protective device and motor controls with the available short-circuit current.

Circuit Protective Devices. There are various types of short-circuit protective devices available for electrical circuits and equipment rated 600 V or less.

A disconnect switch or a nonautomatic circuit interrupter can be used to provide an electrical circuit disconnect function. Manual motor controllers marked "Suitable as Motor Disconnect" may also be used. Short-circuit protection is provided for these devices when fuses are added to the disconnect switch or thermal and/or magnetic trip units are added to the nonautomatic circuit interrupter. Self-protected combination controllers may be used to provide both the disconnect function and the short-circuit protection for motor circuits.

Fuses. Several types of fuses are in use: standard one-time, time-delay (or dual-element), and current-limiting. The standard one-time fuse has no time-delay feature. The time-delay dual-element fuses provide time delay where a heavy overload current might exist for a short time (normal motor starting). This type of fuse will not open the circuit during this brief time; however, if the overload continues, the fuse will cause the circuit to open. In a short-circuit condition, this type of fuse will open the circuit in an extremely short time.

Current-limiting-type fuses are used where extremely high short-circuit currents are available. They react to open the circuit in a fraction of a cycle, thereby limiting the actual current that can flow.

Circuit Breakers. Types of circuit breakers now in use are thermal magnetic, magnetic only, and solid state.

A circuit breaker is opened thermally by a bimetal element in each breaker pole. The function and operation of the bimetal is the same as the overload relay on magnetic starters.

In a short-circuit condition, magnetic action, accomplished by an electromagnet in each breaker pole, causes the breaker to trip instantaneously. In a magnetic-only breaker, the thermal sensing bimetal is left out. Solid-state breakers utilize current transformers and solid-state circuitry in place of the thermal and magnetic trip units.

Sizing of the protective device should be in accordance with the *National Electrical Code* (NEC)*, local code requirements, and equipment manufacturer specifications.

Self-Protected Combination Controllers. A self-protected combination controller operates on the same principal as the thermal-magnetic circuit breaker. In addition to providing the short-circuit protection and disconnecting function, the self-protected combination controller also provides the motor control function.

Protection of Solid-State Motor Controls

Solid-state reduced-voltage starters must be protected against faults and short circuits in the same manner as electromechanical reduced-voltage starters, in accordance with the NEC. Within this range there are two levels of protection—Type 1 and Type 2 protection. The NEC requirements normally afford Type 1 protection to solid-state reduced-voltage starters.

Type 1 protection. When Type 1 protected, the equipment is damaged (i.e., SCRs are shorted) and needs to be repaired, but the wiring and enclosure are intact. Type 1 protection is usually given by circuit breakers and class H, K, R, and RK-5 fuses.

Type 2 protection. When Type 2 protected, the equipment is not damaged during a fault/short circuit, and the equipment is protected by semiconductor, rectifier, and in some instances RK-1 fuses. Note that the semiconductor and rectifier fuses do not qualify as Underwriters Laboratories (UL) or NEC branch-circuit protection devices. However, occasionally a control manufacturer will use SCRs that can be protected by RK-1 fuses. Consult the manufacturer to be certain.

Solid-state reduced-voltage starters must also be protected against voltage transients. This is usually accomplished by installing metal oxide varistors (MOVs) line-to-line on line and load, or by connecting them from line-to-load across the SCRs. If MOVs are not provided, the SCRs will likely short-circuit due to voltage transients.

COMBINATION STARTERS

A combination starter (Fig. 5.139) combines a disconnect switch with or without fuses or a circuit breaker together with the starter and its accessories into one compact package. It may be provided as an enclosed unit or as an open unit which must be mounted in the appropriate enclosure.

The components are closely coordinated to work together, and the unit is rated for a particular value of short-circuit interrupting current. The unit may contain a thermal-magnetic or magnetic-only circuit breaker and a starter, or a fusible or nonfusible switch and a contactor or starter. The suitability of the enclosure (either provided with the combination starter or into which the combination starter is mounted) to its environment is an important consideration.

* *National Electrical Code* is a registered trademark of the National Fire Protection Association, Inc., Quincy, MA 02269.

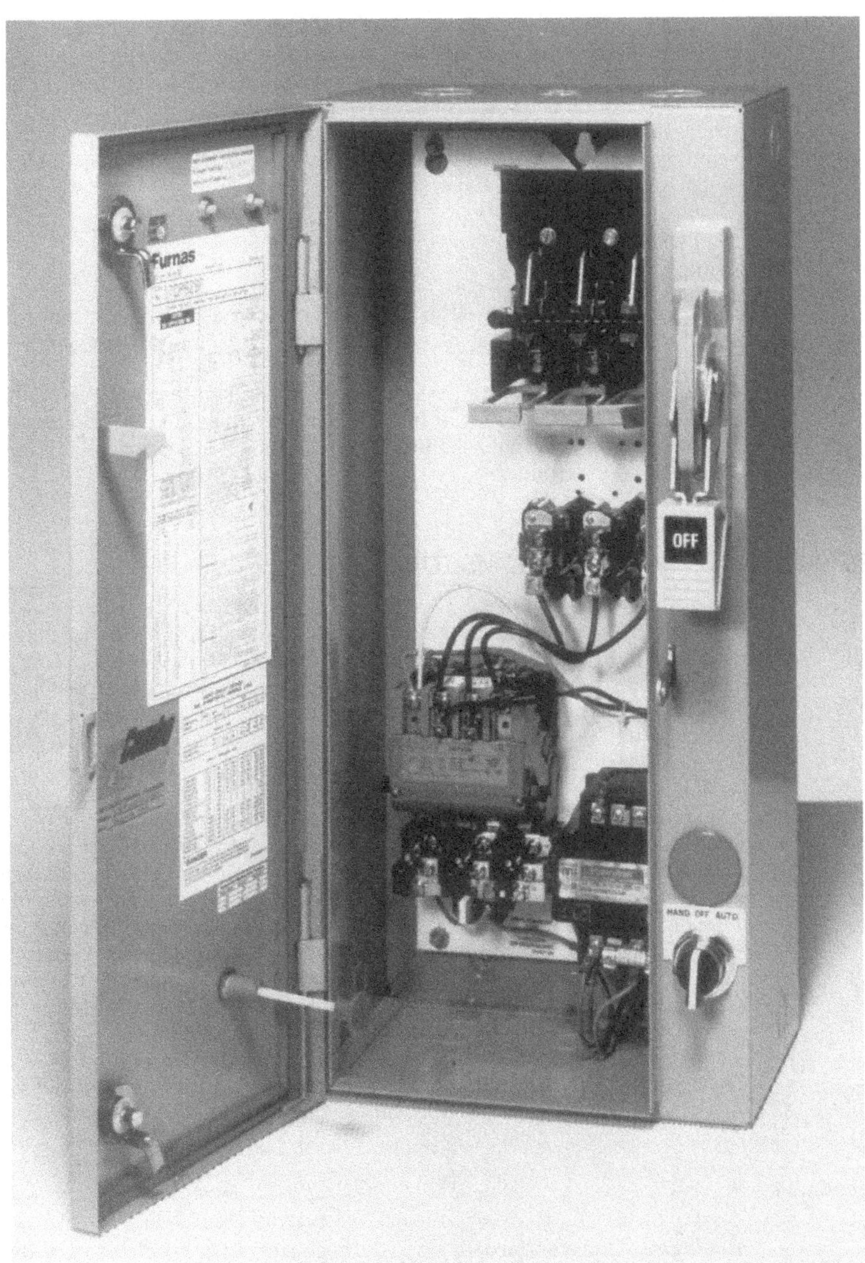

FIGURE 5.139 Combination starter. (*Furnas Electric Co.*)

A self-protected combination starter differs from the combination starter in that it combines the disconnecting means, short-circuit protection, and starter into one unit. The self-protected combination may include a magnetic controller function, or a manual controller function, or both. The self-protected combination controller that contains a manual-only controller may also be used with a separate magnetic or solid-state controller. In such a case, the separate controller is marked to indicate the proper self-protected combination controller which must be provided ahead of the controller.

Interrupting Rating

A combination starter or self-protected combination controller must be capable of interrupting the total symmetrical available short-circuit current in amperes, at the rated voltage, at the point of installation.

In a combination starter, the magnetic or solid-state contactor or starter is coordinated with a circuit breaker or set of fuses so that the equipment will successfully complete a fault interruption up to its published rating. The same coordination is built into the self-protected combination controller.

TROUBLESHOOTING ELECTRIC MOTOR CONTROLS

Varying factors such as temperature, humidity, and atmospheric contamination may adversely affect the performance of motor controls. Misapplication of a control may also lead to serious trouble and is often regarded as the major cause of motor control problems. Visual inspection every 6 months or so, and less-frequent electrical checks with the proper instruments, will help to ensure that production will not be interrupted because of a starter failure that could have been prevented.

It is important to make a complete mechanical check of motor control equipment before and after installation. Damaged or broken parts can usually be found easily and quickly, and replaced if necessary. Visual checks should be made with the aid of a flashlight, air hose, and a small brush. Debris and dirt can be brushed from contacts and other areas of the switch; light rust and dirt on pole faces can be removed with compressed air and brush. Never use a file or abrasive of any kind on pole faces since this can upset the precise fit between core components. A simple tightening of terminal screws should be sufficient to correct many motor controller problems.

It is recommended that the following general procedures be observed by qualified personnel in the inspection and repair of motor controller involved in a fault. Manufacturer's service instructions should be consulted for additional details.

CAUTION: *It must be understood that all inspections and tests are to be made on controllers and equipment which are deenergized, disconnected, and isolated so that accidental contact cannot be made with live parts and so that all plant-safety procedures will be observed.*

Procedures

Enclosure. Substantial damage to the enclosure, such as deformation, displacement of parts, or burning, requires replacement of the entire controller.

Circuit Breakers. Examine the enclosure interior and the circuit breaker for evidence of possible damage. If evidence of damage is not present, the breaker may be reset and turned on. If it is suspected that the circuit breaker has opened several short-circuit faults or if signs of possible deterioration appear within the enclosure, the test described in paragraph AB 1-2.38 of the NEMA standards publication, "Molded Case Circuit Breakers," Publication AB 1, should be performed before restoring the breaker to service.

Disconnect Switch. The external operating handle must be capable of opening the switch. If it fails or if visual inspection after opening indicates deterioration beyond normal, such as overheating, contact blade or jaw pitting, insulation breakage, or charring, the switch must be replaced.

Fuse Holders. Deterioration of fuse holders or their insulating mounts requires their replacement.

Terminals and Internal Conductors. Indications of arcing damage and/or overheating, such as discoloration and melting of insulation, require the replacement of damaged parts.

Contactor. Contacts showing heat damage, displacement of metal, or loss of adequate wear allowance require replacement of the contacts and, where applicable, the contact springs. If deterioration extends beyond the contacts, such as binding in the guides or evidence of insulation damage, the damaged parts of the entire contactor must be replaced.

Overload Relays. If burnout of the current element of an overload relay has occured, the complete overload relay must be replaced. Any indication that an arc has struck and/or any indication of burning of the insulation of the overload relay also requires replacement of the overload relay. If there is no visual indication of damage, the relay must be electrically or mechanically tripped to verify proper functioning of the overload-relay contacts.

Final Check

Before returning the controller to service, checks must be made for the tightness of electric connections and for the absence of short circuits, grounds, and leakage. All equipment enclosures must be closed and secured before the branch circuit is energized.

For these and other complex problems, the manufacturer's wiring and schematic diagrams should be reviewed before attempting repairs. A listing of standard wiring diagram symbols is shown in Fig. 5.140. Also, to aid in troubleshooting, a listing of many of the possible motor control problems and their probable causes and solutions are presented in Table 5.38.

TROUBLESHOOTING OF SOLID-STATE MOTOR CONTROLS

Repairs and Service

The repair of solid-state devices is as varied as are the designs. Before attempting to service any solid-state device, caution should be exercised when removing any printed-circuit boards or solid-state power devices. While designed for use in an industrial environment, they often contain small components or circuits which could be damaged if improperly serviced. Consult the appropriate documentation supplied by the control manufacturer for specific information pertaining to the servicing of a solid-state motor control.

Tools Required for Troubleshooting

Typically, no special tools are required to troubleshoot solid-state reduced-voltage motor controls. A digital multimeter or volt-ohmmeter and clamp-on ammeter are usually sufficient. The multimeter can be used to verify incoming and outgoing voltages or measure resistance or voltage drop across the SCRs, while the ammeter is used to measure motor current.

ELECTRICAL SYMBOLS

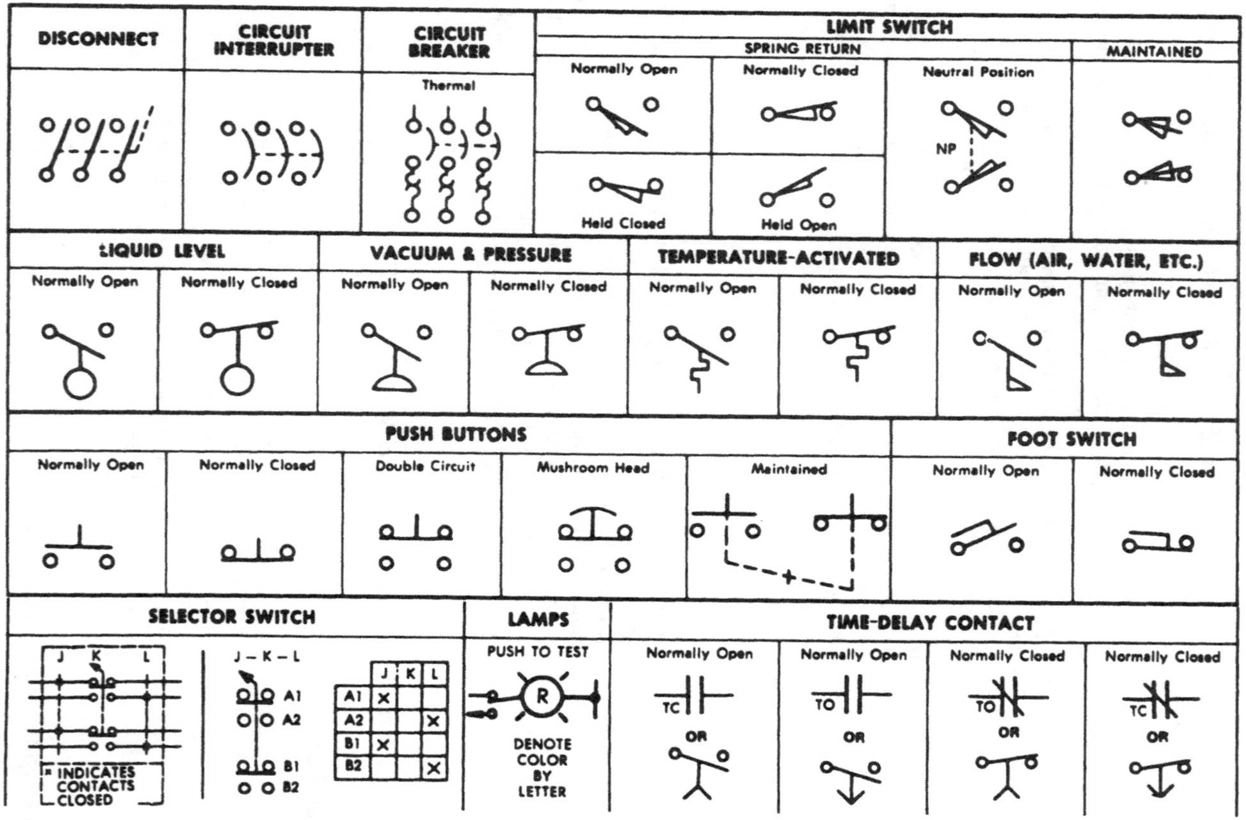

DISCONNECT

CIRCUIT INTERRUPTER

CIRCUIT BREAKER
Thermal

LIMIT SWITCH
SPRING RETURN
Normally Open
Normally Closed
Neutral Position
NP
Held Closed
Held Open
MAINTAINED

LIQUID LEVEL
Normally Open
Normally Closed

VACUUM & PRESSURE
Normally Open
Normally Closed

TEMPERATURE-ACTIVATED
Normally Open
Normally Closed

FLOW (AIR, WATER, ETC.)
Normally Open
Normally Closed

PUSH BUTTONS
Normally Open
Normally Closed
Double Circuit
Mushroom Head
Maintained

FOOT SWITCH
Normally Open
Normally Closed

SELECTOR SWITCH
J K L
× INDICATES CONTACTS CLOSED
J – K – L
A1
A2
B1
B2

	J	K	L
A1	X		
A2			X
B1	X		
B2			X

LAMPS
PUSH TO TEST
R
DENOTE COLOR BY LETTER

TIME-DELAY CONTACT
Normally Open
TC
OR
Normally Open
TO
OR
Normally Closed
TO
OR
Normally Closed
TC
OR

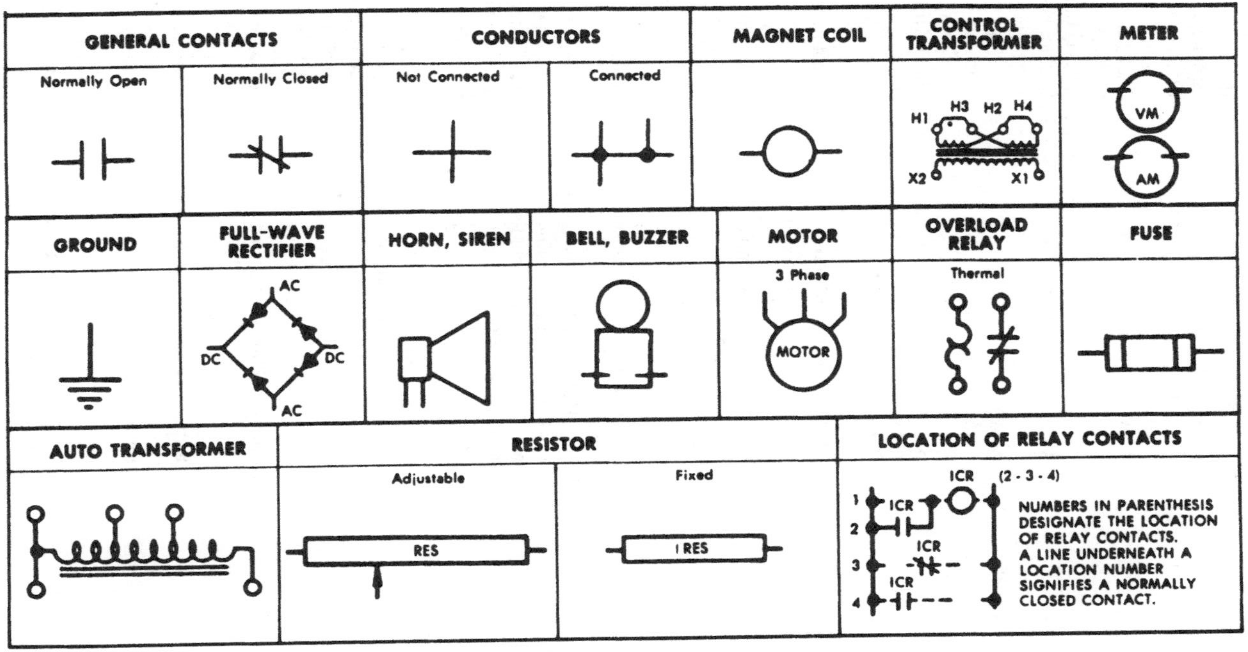

FIGURE 5.140 Wiring diagram symbols.

TABLE 5.38 Troubleshooting Motor Controls

Problem	Possible cause	Solution
I. Magnetic and mechanical parts		
Noisy magnet (humming)	1. Misalignment or mismating of magnet pole faces	1. Realign or replace magnet assembly.
	2. Foreign matter on pole face (dirt, lint, rust, etc.)	2. Clean (do not file) pole faces; realign if necessary.
	3. Low voltage applied to coil	3. Check system and coil voltage. Observe voltage variations during start-up time.
Noisy magnet (loud buzz)	4. Broken shading coil	4. Replace shading coil and/or magnet assembly.
Failure to pick up and seal in	1. Low voltage	1. Check system and coil voltage; watch for voltage variations during starting.
	2. Wrong magnet coil or wrong connection	2. Check wiring, coil nomenclature, etc.
	3. Coil open or shorted	3. Check with an ohmmeter, and when in doubt replace.
	4. Mechanical obstruction	4. Disconnect power and check for free movement of magnet and contact assembly.
Failure to drop out or slow dropout	1. Gummy substance on pole faces or magnet slides	1. Clean with nonvolatile solvent or degreasing fluid.
	2. Voltage to coil not removed	2. Shorted seal-in contact (exact cause found by checking coil circuit).
	3. Worn or rusted parts causing binding	3. Clean or replace worn parts.
	4. Residual magnetism due to lack of air gap in magnet path	4. Replace any worn magnet parts or accessories.
	5. Mechanical interlock binding (reversing starters)	5. Check interlocks for free pivoting. New bushing or light lubrication may be required.
II. Contacts		
Contact chatter (source probably from magnet assembly)	1. Broken shading coil	1. Replace assembly.
	2. Poor contact continuity in control circuit	2. Improve contact continuity or use three-wire control.
	3. Low voltage	3. Correct voltage condition. Check momentary voltage dip on starting.
Welding	1. Abnormal inrush of current	1. Use larger contactor; check for grounds, shorts, or excessive load current.
	2. Rapid jogging	2. Install larger jogging-rated device or caution operator.
	3. Insufficient tip pressure	3. Replace contact springs; check contact carrier for deformation or damage.
	4. Low voltage preventing magnet from sealing	4. Correct voltage condition. Check momentary voltage dip on starting.
	5. Foreign matter preventing contacts from closing	5. Clean contacts with nonvolatile solvent. Low-current or -voltage contactors, starters, and control accessories should be cleaned with solvent and acetone to remove solvent residue.

TABLE 5.38 Troubleshooting Motor Controls (*Continued*)

Problem	Possible cause	Solution
	6. Short circuit	6. Remove short fault and check for correct fuse or breaker size.
Short contact life or overheating	1. Filing or dressing	1. Do not file silver contacts. Rough spots or discoloration will not harm or impair their efficiency.
	2. Interrupting excessively high currents	2. Install larger device or check for grounds, shorts or excessive motor currents.
	3. Excessive jogging	3. Install larger device rated for jogging or caution operator.
	4. Weak contact pressure	4. Replace contact springs; check contact carrier for deformation or damage.
	5. Dirt or foreign matter on contact surface	5. Clean contacts with nonvolatile solvent.
	6. Short circuits	6. Remove short fault and check for correct fuse or breaker size.
	7. Loose connection	7. Clean and tighten.
	8. Sustained overload	8. Install larger device; check for excessive load current.
	9. Excessive wear	9. Higher than normal voltage may result in mechanical wear and bounce.
Contacts, supports, discoloring	1. Loose connections	1. Tighten hardware or replace.*
III. Coils		
Open circuit	1. Mechanical damage	1. Handle and store coils carefully.
Cooked coil (overheated)	1. Overvoltage or high ambient temperature	1. Check application and circuit. Coils will operate satisfactorily over a range of 85 to 110% of rated voltage.
	2. Incorrect coil	2. Check rating; replace with proper coil if incorrect.
	3. Shorted turns caused by mechanical damage or corrosion	3. Replace coil.
	4. Undervoltage, failure of magnet to seal in	4. Correct system voltage.
	5. Dirt or rust on pole faces increasing air gap	5. Clean pole faces.
	6. Sustained low voltage	6. Remedy according to local code requirements, low-voltage system protection, etc.
IV. Overload relays		
Nuisance tripping	1. Sustained overload	1. Check for equipment grounds and shorts and excessive motor currents due to overload. Check motor winding resistance to ground.
	2. Loose connections	2. Clean connections and tighten. This includes load wires and heater-element mounting screws.
	3. Incorrect heater	3. Check heater sizing and also check ambient temperature.
Failure to trip out (causing motor burnout)	1. Mechanical binding, dirt, corrosion, etc.	1. Clean or replace.
	2. Incorrect heater or jumper wires used or heaters omitted	2. Recheck ratings and heater size. Correct if necessary.

TABLE 5.38 Troubleshooting Motor Controls (*Continued*)

Problem	Possible cause	Solution
	3. Wrong calibration adjustment	3. Consult factory. Calibration adjustment is normally not recommended unless under factory supervision. It is customary to return units to factory for check and calibration.
V. Manual starters		
Failure to operate (mechanically)	1. Mechanical parts, including springs, worn or broken	1. Replace parts as needed.
	2. Welded contacts due to application or other abnormal cause	2. Replace contacts and recheck operation.
Trips out prematurely	1. Motor overload, incorrect heaters, or misapplication	1. Check conditions; replace or adjust as needed.
VI. Timers		
A. Pneumatic		
Erratic timing	1. Foreign matter in valve	1. Clean if at all possible, or replace timing head completely and exchange unit with factory.
Contacts do not operate	1. Adjustment incorrect on time-actuating screw	1. Follow service bulletin instructions for desired adjustment.
	2. Worn or broken parts in switch assembly	2. Replace defective parts.
B. Electronic relay		
Erratic timing	1. Loose connections	1. Check over unit visually.
	2. Timing relay worn out	2. Plug in new tested relay.
	3. Defective components	3. Check and replace if necessary.
Timer stops operating	1. Mechanical relay	1. Substitute good relay.
	2. Defective circuit components	2. Check over unit visually and with multimeters. Replacement of circuit board probably better than repair if relay normal.
C. Electronic (solid-state)		
Erratic timing	1. Loose connections	1. Visually inspect all connections.
	2. Check external connections	2. Check functions with multimeters in accordance with prescribed service instructions.
Timer will not time out	1. External connections	1. Systematically check system.
	2. Check power supply to timer	2. Fuses, etc.
	3. Initiation circuit open	3. See step 1.
	4. Contacts dirty	4. See step 1; clean if necessary.
VII. Limit switches		
Broken parts	1. Excessive overtravel of actuator	1. Use resilient actuator or operate within device tolerance.
Inoperative	1. Switch actuator out of position or broken	1. Inspect, repair, or replace.
	2. Lack of contact continuity	2. Clean contacts; replace contact block if necessary.
VIII. Drum controls		
Poor contact	1. Dirty rotor contacts and fingers	1. Inspect contacts; if copper, burnish with 4-0 sandpaper until clean; if silver, use a suitable solvent. Check for approximately $\frac{3}{64}$-in finger movement.
	2. Dirt or other foreign matter on horizontally mounted units	2. Clean systematically with contact cleaner and air.

TABLE 5.38 Troubleshooting Motor Controls (*Continued*)

Problem	Possible cause	Solution
Sluggish or hard operating switch	1. Dry bearings 2. Worn parts	1. Lubricate bearings sparingly. 2. Inspect carefully; replace worn parts.
IX. Pressure switches		
Pressure switch inoperative	1. Foreign matter in pressure-sensing area 2. Contacts burned	1. Remove switch and clean opening. 2. Clean contacts; replace if necessary.
Erratic operation	1. Worn parts 2. Diaphragm faulty	1. Inspect, adjust, or replace. 2. Replace diaphragm.
Very frequent operation	1. Likely due to a waterlogged system	1. Drain part of water from pressure tank and, if possible, pump in about 4 lb air.
X. Push buttons		
Button inoperative (mechanical) (electrical)	1. Shaft has dirt or residue binding 2. Contact board spring broken 3. Contaminated contacts and corrosion	1. Check, clean, and clear. 2. Replace contact board. 3. Clean.
XI. Push button, pilot lamp		
No light	1. Bulb out or burned out 2. Broken parts, wire, or transformer 3. Short life of bulb due to excessive high voltage	1. Replace with proper unit. 2. Inspect, repair, or replace. 3. Replace with next higher voltage pilot lamp (brillance may be reduced slightly).

* Any contact replacement should include a complete set of replacement including the support springs, screws, etc.

Preventive Maintenance

Solid-state reduced-voltage motor controls require very little preventive maintenance. About the only maintenance required is ensuring tight electrical connections, keeping filters and cooling fans clean and free to turn, and verifying that voltage transient protection is intact. The filter cleaning operation should be performed as often as the environment dictates.

Troubleshooting

Solid-state reduced-voltage starters and controllers typically comprise only two main sections, power and logic. Faults to the motor control will occur in one of these two places. When a problem occurs in a motor control system, it is important to consider all aspects including the motor control, control wiring, sensing switches, programmable control equipment, etc.

Power Circuit, SCR. Failures in the power section usually result in a shorted SCR. This failure may cause the motor to growl and rumble during starting or it may cause circuit breakers to trip during starting. Shorted SCRs can be determined by measuring the resistance across the motor control with an ohmmeter. (***Caution:*** Disconnect the power prior to measuring the resistance.) A shorted SCR will read 0 Ω. Most controls have built-in shorted SCR detectors that will light an indicating light-emitting diode (LED) or open an alarm contact.

Power Circuit, Protective Devices. A failed voltage-transient protective device must be checked visually unless an indicator is provided. The voltage-transient protection should be inspected any time a shorted SCR occurs.

Logic Circuit. Failures in the logic section usually prevent the motor control from doing anything or result in low voltage being applied to the motor (make certain this low voltage is not the result of an incomplete start sequence or some other adjustment). Logic failures usually require factory repair or replacement.

Control Circuits. Should a controller fail to start, even though the logic and power sections are in working order, it may be that the proper control commands have not been received by the unit. Consult the unit's instruction manual to determine the required control inputs and check the control circuit for tripped overload relays, open circuits, open disconnects, etc.

Troubleshooting Solid-State Variable-Speed (AC) Drives

Repairs, if required, can be lengthy because with electronic equipment, training of employees to make repairs may be necessary. Repair costs may also be greater than for nonelectronic equipment. They will always be greater if factory personnel are required for on-site repairs. When there are a large number of units in place at one facility, training of the maintenance staff is most cost-effective.

Most controls have built-in diagnostics to expedite troubleshooting. For any other testing, a good analog multimeter is normally all that is required. The decision to repair a piece of equipment or replace it depends on the individual production schedules. For high-volume production lines, it may be best to keep spare drives on hand to swap out with a problem unit. Repairs can then be made under less stressful conditions.

In most cases, VFDs require very little maintenance. The only moving parts are relays and input/output contactors which may wear out mechanically. It is rare that contacts need to be replaced because arcing is normally nonexistent.

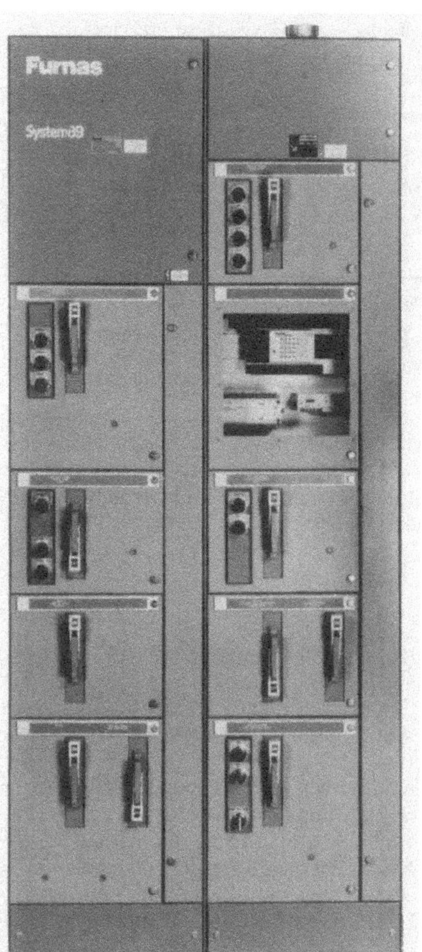

FIGURE 5.141 Motor control center. (*Furnas Electric Co.*)

MOTOR CONTROL CENTERS

Motor control centers are floor-mounted assemblies of one or more enclosed vertical sections having a horizontal common power bus and principally containing combination motor control units. The units are mounted one above the other in the vertical sections. These sections may incorporate vertical buses connected to the common power bus, thus extending the common power supply to the individual units. Units may also connect directly to the common power bus by suitable wiring (definition as stated by NEMA).

The common enclosure design (Fig. 5.141) and the use of combination starters offer both economy and ease of installation in multiple motor control installations. In addition, motor control centers (MCCs) provide proper coordination between short-circuit protective devices and the controller. Since MCCs are engineered systems, the components are closely coordinated to work together, and the unit is rated for a particular value of short-circuit interrupting duty at the point of its installation. MCCs may contain a molded-case circuit breaker and starter, or a fused switch and a starter.

MCCs centralize all the electric control apparatus for a given installation in housings which are convenient for easy field installation and maintenance. Use of MCCs can minimize the total amount of floor and wall space required by isolated motor control apparatus. The individual units generally are interchangeable and easily removable as well. With proper planning, MCCs can be designed with provisions to allow for future system expansion.

Vertical sections generally are 20 in (50.8 cm) wide by 90 in (228.6 cm) high, and most designs can accommodate up to six motor starter units per section.

Motor Control Center Specifications

Usually the following information is required to process an MCC order properly.

1. Service voltage
2. Configuration
3. Size of main horizontal bus
4. Bus bracing
5. Incoming service
6. Main protective device
7. Wiring NEMA class
8. Wiring NEMA type
9. Metering
10. Branch protective devices

1. Service Voltage. Low-voltage ratings for industrial control apparatus are based on utilization voltage per NEMA ICSI-112.22 voltage rating and are as follows for 60-Hz alternating current, multiphase: 115, 200, 230, 460, and 575 V. Corresponding system voltages are 120, 208, 240, 480, and 600 V.

Enter the proper values for hertz, phase, and number of wires used in system.

2. Configuration. The MCC should be of dead-front, indoor design and fabricated from code-gauge steel, with all sections joined to form a single assembly. All side, front, and rear cover plates should be field-removable.

The MCC enclosure should be (unless otherwise noted):

- NEMA 1 indoor construction
- Non-walk-in—front accessible

Main and branch units should be front-connected. The MCC should have space or provisions for future expansion, as noted on the plans.

The MCCs are to be constructed in accordance with the latest NEMA-ICS and UL 845 standards.

Individual sections are to be front-accessible, not less than 15 (38.1 cm) deep, and the rear of all sections should align. All bolts used to join current-carrying parts should be installed so as to permit servicing from the front only so that no rear access is required. An unobstructed conduit entry area is to be provided at the top and bottom of each standard control section.

3. Size of Main Horizontal Bus. Mark as required. Busbar is to be UL rated for ____600 A ____800 A ____1200 A ____1600 A and ____2000 A.

4. Bus Bracing

Main Bus. The MCC should be bused with rectangular busbars made of ____tin ____silver plated copper, and braced for ____42,000 A ____65,000 A ____85,000 A ____100,000 A.

The through bus on the end section should be extended and predrilled to allow the addition of future sections with standard busbar splice plates.

Ground Bus. Specify amperage or size. Specify location if specific location is desired. Recommend always mounting in bottom front unless full length neutral bus is also required. Specify lug size required or ground cable size.

Neutral Bus. If required, shall be ____300 A ____600 A ____800 A, located in the ____incoming line section only ____full length, mounted in the bottom. Specify lug size required for neutral cable or actual neutral cable size.

5. Incoming Cable Entry. ____Lugs ____crimp-type connectors to terminate ____copper ____aluminum cable, should be furnished as detailed on plans for ____top ____bottom entry.

Bus Duct Entry. The MCC is to be fed by ____copper ____aluminum, ____A, ____bus duct, as detailed on plans, ____ and other sections of the specification. The MCC manufacturer is to be responsible for coordination, proper phasing, and internal busing to the incoming bus duct. Detailed plans of the incoming bus duct is to be supplied to the manufacturer of the MCC.

Unit Terminals. ____A (None) ____Bd (control terminals only) ____Bt (control and power terminals through Size 3)

Each protective device should have an individual door over the front, or common door for dual-mounted units, equipped with one or more voidable interlocks that prevent the door from being opened when the switches are in the on position, unless the interlock is purposely defeated by activation of the defeater mechanisms. All breakers or fusible switches should be removable from the front of the MCC without disturbing adjacent units. The MCC should have space or provision for future units.

6. Main Protective Device. The main protective device, to be installed in the service section, is as indicated below.

Molded-case circuit breaker of the quick-make, quick-break, trip-free, ____ heavy duty, ____ extra heavy duty, ____ energy limiting, ____ solid-state type. It should be an ____ frame ____ two-pole ____ three-pole, 600-V breaker with a trip current rating of:

____400 A ____1000 A ____2000 A
____600 A ____1200 A ____2500 A
____800 A ____1600 A ____3000 A

and an interrupting capacity of not less than ____A rms symmetrical at the system voltage.

The following accessory features are to be included: ____ shunt trip, ____ electrical operator, ____ ground-fault relaying, ________________________________ (other).

Fusible switch of the quick-make, quick-break type. It should be a ____ two-pole ____ three-pole, ____ 240-V ____ 600-V unit with a continuous current rating of ____ 400 ____ 600 ____ 800 ____ 1200 A, and with ____A class ____ fuses, suitable for application on a system with ____A symmetrical available fault current.

Bolted pressure switch of the quick-make, quick-break type. It should be a ____ two-pole ____ three-pole ____ 240-V ____ 480-V unit with a continuous current rating of:

____800 A ____1600 A ____2500 A
____1200 A ____2000 A ____3000 A

and with ____A class L fuses, suitable for application on a system with ____A symmetrical available fault current.

The following accessory features are to be included: ____ shunt trip, ____ electrical operator, ____ ground-fault relaying, ________________________________ (other).

Power circuit breaker with a ____ stationary ____ drawout frame, and a current rating of:

____600 A ____2000 A
____800 A ____3000 A
____1600 A

It is to be ____ manually ____ electrically operated with a(n) ____ electromechanical ____ solid-state trip device, and an interrupting capacity of ____ A rms symmetrical at the system voltage.

The following accessory features are included: ____ short time delay, ____ ground-fault relay (trip), ____ shunt trip (M.O.C/B only), ____ control power transformer, ________________________________ (other).

7. Wiring NEMA Class. Indicate class I or class II as required. If class II, elementary or schematic drawings should be supplied with the order.

8. Wiring NEMA Type. Mark as required. If C terminals are mounted in bottom wiring space section, additional unit space will be required for more than one row of terminals.

Unit Terminals (B or C Wiring). If B or C wiring, NEMA class is required indicate type of terminals desired.

Pull-Apart Terminals. Indicate where required.

9. Metering. The following customer metering equipment should be furnished as shown on the plans.

Main Bus

____ Voltmeter, with ____-phase transfer switch
____ Ammeter with ____-phase transfer switch
____ Watthour meter(s) (two) (three)-element (with) (without) demand attachment
____ ________________________________ Current transformer(s),
________________________________ /5 or suitable rating
____ ________________________________ Potential transformer(s), suitable rating

Branch Circuits

____ Ammeter(s), with ____-phase transfer switch
____ ________________________________ Current transformer(s),
________________________________ /5 or suitable rating

10. Branch Protective Devices. All molded-case circuit breakers, fusible switches, and/or motor starter units used as a protective device in a branch circuit should meet the requirements of the appropriate paragraph below.

Each protective device should have an individual door over the front, equipped with a voidable interlock that prevents the door from being opened when the switch is in the on position, unless the interlock is purposely defeated by activation of the defeater mechanism.

Molded-case circuit breakers should be of the quick-make, quick-break, trip-free, ____ motor circuit protector (MCP), ____ thermal-magnetic type ____ solid-state, with frame, trip, and voltage ratings, either two-pole or three-pole, as indicated. All breakers should have an interrupting capacity of not less than ____ A rms symmetrical at the system voltage. All breakers should be removable from the front of the MCC without disturbing adjacent units. The MCC should have space or provision for future units.

Fusible switches should be quick-make, quick-break units and conform to the ratings shown on the plans.

All switches should have externally operated handles. Switches should be equipped with ____ fuse holders and class ____ fuses of ampere rating and type as indicated. Suitable for application on a system with ____ A symmetrical available fault current.

NEMA or motor matched magnetic starters are to be furnished of the type and horsepower ratings as indicated. Thermal overload relays on starters should be ____noncompensated melting-alloy type ____bimetal ____bimetal ambient compensated ____self-powered solid-state type. Three overload elements should be furnished on each starter. The overload heater elements or solid-state overload should be sized from the actual motor nameplate data.

TABLE 5.39 Checklists for Installing and Maintaining Motor Control Centers

Installation quick checklist

Receiving

- ❑ Inspect package for damage.
- ❑ After unpacking, inspect equipment for damage in transit.
- ❑ If damaged or incomplete, substantiate claims against shipper with identification of parts, description of damage, and photographs.

Handling

- ❑ Simplify handling by leaving equipment on shipping skid.
- ❑ Use the lifting means provided for moving the equipment.
- ❑ Take care to use the proper method of moving a motor control center.

Storage

- ❑ Store in a clean, dry space at moderate temperature.
- ❑ Cover with a canvas or heavy-duty plastic cover.
- ❑ If storage area is cool or damp, cover equipment completely and heat to prevent condensation.

Location Selection

- ❑ Flat and level floor
- ❑ Overhead clearance
- ❑ Accessibility front and rear.
- ❑ Protection from splash and drip, dust, and heat.
- ❑ Space for future expansion.
- ❑ If bottom conduit entry is used, conduit should be in place and stubbed up before equipment is installed.

Installation Method

- ❑ Grout into the foundation.
- ❑ Weld channel sills to steel leveling plates.
- ❑ Embedded anchor bolts in floor.

Field Assembly

- ❑ Remove hardware and horizontal bus connecting links from shipping splits.
- ❑ Install first shipping split.
- ❑ Remove end cover plates of structures to be joined (if required).
- ❑ Carefully align second split with first. Bolt structure together per manufacturer's instructions.
- ❑ Remove horizontal wireway barrier or units to expose horizontal bus.
- ❑ Connect horizontal buses with bus links. Torque bolts to manufacturer's recommendations.
- ❑ Grommet top and bottom horizontal wireways.
- ❑ Install heater coils (check selection against motor nameplate data).
- ❑ Install fuses.

Conduit Entry at Top

- ❑ Remove top plates from structure.
- ❑ Cut conduit entry holes in top plates.
- ❑ Reinstall top plates
- ❑ Install conduits

Incoming Line Connections

- ❑ Choose the shortest, most direct route from remote mains.
- ❑ If cables cannot be directly routed to terminals, provide adequate space for clamping the cables.
- ❑ Torque incoming lines to main lugs only per manufacturing recommendations.
- ❑ Torque all incoming connections to main circuit breakers and fusible disconnects as per the breaker or disconnect manufacturer's recommendations. The torque requirements will be found on a label located on the disconnecting device.

Outgoing Power and Control Wiring

- ❑ Disengage plug-in unit stabs from vertical bus. Connect control and power wiring to units.
- ❑ Use stranded wire.
- ❑ Leave enough slack to permit partial withdrawal of unit for test position maintenance.
- ❑ Pull wiring between units through vertical and horizontal wireway securing wires in the vertical wireway with wire ties provided.
- ❑ Route wiring between sections through the top or bottom horizontal wireways.
- ❑ Reinsert plug-in units to engage stabs.

Preoperation Checks

- ❑ Test insulation resistance of all circuits with the control center as ground.
- ❑ Remove restraining devices from contacts and shunts from current transformers. Make sure that all parts of magnetic devices operate freely.
- ❑ Check electrical interlocks for proper contact operation. Make sure that each motor is connected to its proper starter.
- ❑ Check all heater elements for proper installation.
- ❑ Check all timers for proper time interval setting and contact operation.
- ❑ Check fusible disconnect starters for proper fuse size. Clean the control center. Rid it of all extraneous material. Use a vacuum cleaner, not compressed air.
- ❑ Check all connections for mechanical and electrical tightness.
- ❑ Close all access plates and doors.

Energizing Motor Control Centers

- ❑ Make sure all unit disconnect handles and control center mains are turned to OFF.
- ❑ Turn on remote mains. Turn on motor control center main circuit breakers or fusible disconnects.
- ❑ Turn on unit disconnect handles one by one.
- ❑ Jog motors to check for proper rotation
- ❑ Adjust MCPs

Insulation Test (Megger)

- ❑ Measure phase to phase and phase to ground. Resistance measurements should be taken before a motor control center is placed into service and after installation or maintenance. When performing resistance measurements on motor control centers use an insulation tester (Megger) with a potential of 500–1000 V.

TABLE 5.39 Checklists for Installing and Maintaining Motor Control Centers (*Continued*)

WARNING: Before performing installation or maintenance, turn off electrical power to the control unit to avoid electrical shock.

WARNING: The main disconnect must be in the OFF position during all megger testing of the motor control center.

CAUTION: Devices such as solid-state components, capacitor units or any other devices which are not designed to withstand Megger voltage should be disconnected before testing the rest of the motor control center.

Take readings between each phase and from each phase to ground. This should be done with the branch disconnects OFF and again with the branch disconnects ON.

Branch Disconnects OFF:
Typically readings taken with all disconnects in the OFF position should be between 5–20 MΩ.

New equipment which was stored in a damp area may register lower on initial start-up. If readings are above 1 MΩ during start-up the following procedure may be observed to help dry the motor control center.

Energize several individual control units. If additional readings are above 1 MΩ, energize additional units.

After the equipment has been in operation for 48 h, the readings should be in the 5–20 MΩ range.

If at any time Megger readings are below 5 MΩ (1 MΩ during start-up) consult your local motor control center representative.

Branch Disconnects ON:
Before taking reading with the branch disconnects ON, disconnect all devices completing circuits between phases or between phases and neutral such as control transformers. Readings observed may be slightly lower than the OFF readings, but the start-up 1-MΩ lower limit still applies.

Record and keep a record of the Megger readings. Abrupt changes in resistive values may be an indication of potential failure. Even sudden changes within the 5–20-MΩ range may be an advance signal of insulation failure. The early detection of faulty insulation components can save costly repairs and downtime.

Maintenance quick checklist

Scheduling
- ❑ Schedule maintenance appropriate to the severity of service.
- ❑ Consider environment (dampness, heat, and dust), severity of operations, and the importance of the machinery being controlled.
- ❑ Control unit maintenance should coincide with inspection of the motor being controlled.
- ❑ Buswork inspection entails shutting down the entire control center.

Cleaning
- ❑ Use a vacuum cleaner, not compressed air.
- ❑ Excess deposits of foreign materials signify faulty gasketing.
- ❑ Pay particular attention to conductive deposits.

Loose Connections
- ❑ Periodic checking of tightness of connections promotes reliability and reduces heating.
- ❑ Overheating and discolorations signify loose connections.
- ❑ Torque horizontal bus bolts to manufacturer's recommendations.
- ❑ Torque incoming line connections to main lugs only to manufacturer's recommendations.
- ❑ Torque all incoming connections to main circuit breakers and fusible disconnects as per the breaker or disconnect manufacturer's recommendations.
- ❑ The torque requirements will be found on a label located on the disconnecting device.

Disconnect Operating Handle Adjustments
- ❑ Remove unit from center or place in test position.
- ❑ Adjust handles per manufacturer's recommendations.

Contacts
- ❑ Make sure that all contacts are free from extraneous materials, excess pitting or burning.
- ❑ Check for spring pressure.

Locking in Engaged Position
- ❑ To lock in ON, drill out the indentations on the disconnect operating handle and insert a padlock.
- ❑ To lock in OFF, as many as three padlocks may be inserted in the disconnect operating handle.

Field Additions of Sections
- ❑ For field additions of sections, follow the same procedure as for the field assembly of shipping splits.

Addition and Replacement of Control Units
- ❑ Deenergize motor control center. Follow manufacturer's recommendations.

Adding to a Blank Unit Space
- ❑ Follow manufacturer's recommendations.

Installation Test (Megger)
- ❑ Same as previous

The following accessory features should be furnished on each starter:

____Individual control power transformers

____Pilot light(s)

____Auxiliary contacts

____NO ____NC

Push buttons, selector switches, and other pilot devices shall be furnished as indicated.

The checklist shown in Table 5.39 is provided to assist the plant engineer in installing and maintaining MCCs.

APPLICABLE CODES AND STANDARDS

American National Standards Institute (ANSI). Standards on a wide variety of electrical equipment, published by the American National Standards Institute.

Canadian Standards Association. Canadian standards on a wide variety of electrical equipment, published by the Canadian Standards Association, 178 Rexdale Boulevard, Rexdale, Ontario M9W1R3.

Machine Tool Electrical Standards. Standards of the National Machine Tool Builders Association, 7901 Westpark Drive, McLean, VA 22101.

National Electrical Code. Standard of the National Fire Protection Association, Quincy, MA 02269.*

National Electrical Manufacturers' Association (NEMA). Standards on motors and control published by the National Electrical Manufacturers' Association, 2101 L Street N.W., Washington, D.C. 20037.

Underwriters Laboratories Inc. (UL). An independent organization that tests and makes recommendations based on safety and fire hazard conditions relating to the tested equipment. Standards are published by Underwriters Laboratories Inc., 333 Pfingsten Rd., Northbrook, IL 60062.

BIBLIOGRAPHY

Alerich, W. N.: *Electric Motor Control,* Van Nostrand Reinhold, New York, 1975.

Chestnut, H.: *Systems Engineering Tools,* John Wiley & Sons, New York, 1965.

DC Motors-Speed Controls-Servo Systems, Electro-Craft Corp., Hopkins, Minn., 1972.

Heumann, G. W.: *Magnetic Control of Industrial Motors,* John Wiley & Sons, New York, 1961.

Kintner, P. M.: *Electronic Control Systems in Industry,* McGraw-Hill, New York, 1968.

McPartland, J. F.: *Motor and Control Circuits,* McGraw-Hill, New York, 1975.

Millermaster, R. A.: *Harwood's Control of Electric Motors,* 4th ed., John Wiley & Sons, New York, 1970.

National Electrical Manufacturing Association: *ICS-1,* Washington, D.C., 1994.

National Electrical Manufacturing Association: *ICS-2,* Washington, D.C., 1994.

National Fire Protection Association: *National Electrical Code,* Quincy, Mass., 1999.

Siskind, C. S.: *Electrical Control Systems in Industry,* McGraw-Hill, New York, 1963.

CHAPTER 5.9
LIGHTING*

Gary Smith
Director, Product Management Venture Lighting Technologies, Inc.
Solon, Ohio

DETERMINE LIGHTING CRITERIA
DETERMINE LIGHT SOURCE REQUIREMENTS
UTILIZE MULTIPLE DESIGNS

Proper plant lighting can be one of the most significant factors in improving your company's bottom line. Study after study has shown that proper lighting increases worker productivity and decreases absenteeism, which will improve profitability.

In addition, companies can improve lighting while decreasing energy cost. Pennsylvania Power and Light (PP&L) upgraded its lighting and produced an energy savings of 69 percent. PP&L subsequently found that they received a 13 percent increase in productivity and a 25 percent decrease in absenteeism. This gave PP&L a 540 percent return on investment (ROI) on the $8362 spent on the project.

To properly choose, plan, and implement a successful lighting system, you must assess your needs, determine the light source, and consider creating multiple designs. Boeing decreased lighting power usage by 90 percent with a lighting retrofit, creating an ROI of 53 percent. This does not take into account the 20 percent improvement in defect rates that were encountered after the lighting upgrade.

DETERMINE LIGHTING CRITERIA

How much and what type of light do you need? That depends on the type of work that is being performed. A quality-control inspection area typically needs far higher light levels than a raw material storage area. Don't fall into the trap of lighting your entire plant at a single level. This is the easiest thing to do, but it will often end up costing you more for the overilluminated areas, and be detrimental to your production in the underilluminated ones. As mentioned, a well-lit area can decrease scrap and absenteeism and at the same time increase production.

Different industries or tasks require different levels of lighting. For a complete list refer to the Illuminating Engineering Society of North America (IESNA) *Lighting Handbook.* If you don't have access to this publication, your local lighting representative should be able to supply you with a copy of the pertinent data.

Once you have determined the lighting guidelines, take out a blueprint of your entire plant and identify what tasks are being performed where. It is usually a good idea to highlight in different colors the work areas requiring different light levels (foot-candles).

How effectively should your light source render colors? The higher the color rendering index (CRI) of a lamp, the better individual colors will be rendered. The CRI compares how closely a light source will render colors to a standard. The standard is usually an incandescent lamp, which is, by definition, a black body and has a CRI of 100, the highest rating (Table 5.40). Metal halide lamps usually have a CRI between 65 and 85. By comparison, high-pressure sodium lamps have a CRI of 22.

* Reprinted with permission from *Facilities Engineering Journal,* March/April 2000.

TABLE 5.40 Light Sources for Industrial Plants

Source	LPW*	CRI†	Color	Life, h	Overall rating
Incandescent	5–15	100	2800K	50–3000	Bad
Halogen	15–25	100	2900K	2000–5000	Poor
Fluorescent	40–80	62–90	2700–6000K	6000–20,000	Fair
Mercury	30–60	15	5700K	24,000+	Poor
High-pressure sodium	80–125	22	2000K	Up to 24,000	Good
Metal halide	80–100	65–70	2700–4000K	Up to 20,000	Better
Pulse-start metal halide	100–110	65–75	3000–4000K	Up to 30,000	Best

* Lumens per watt.
† Color rendering index.

What about light color (in degrees Kelvin) and CRI? Does any job area require hand-eye coordination, reading, or differentiating between parts? If so, you probably need a white (4000K) light source and a high CRI (65 or higher). If an area is simply for storage, you may be able to get by with a low color temperature and lower CRI. Be careful though. Lower-CRI lamps, such as high-pressure sodium, can contribute to mistakes made in part selection or shipping errors.

Once you have determined your facility's needs, you can choose a light source that will achieve the results you require.

DETERMINE LIGHT SOURCE REQUIREMENTS

When determining light source requirements, four factors should always be considered: system lumens per watt (LPW), lamp color, CRI, and the rated life. In industrial settings, you want high LPW, a crisp white color (4000K), high CRI, and a long life rating.

Incandescent and Halogen. Both of these lamps have excellent CRI and a generally pleasing color. However, they make poor sources for industrial lighting because they are so inefficient. These lamps consume most of their power by generating heat, not light. As a general rule of thumb, incandescent lamps take 5 times as much energy to generate the same amount of light as do more efficient sources (i.e., metal halide). The other disadvantage of incandescent and halogen lamps is their relatively short lamp life. Users of these products have to relamp frequently.

Fluorescent. These lamps are a better choice for industrial lighting than incandescents, but they still have several disadvantages. At a first glance, fluorescent sources seem to have all the desirable qualities previously mentioned (high LPW, high CRI, etc.) In fact, that is why in the past so many factories used these light sources. Their main disadvantage, however, is that they are inherently low-wattage sources. Think of this; it would take 35 to 45 ft of fluorescent lamps to generate the same light output as a single 1000-W high-output metal halide or HPS lamp. Fluorescent technology typically requires too many bulky fixtures to generate adequate illumination levels in industrial plants.

Mercury. While mercury lamps used to be the standard in high-intensity discharge lighting, their time has passed. This technology has too many drawbacks: low LPW, low CRI, poor lumen maintenance, and a blue light that has questionable appeal. In the next 5 to 10 yr, expect sales of mercury lamps to virtually evaporate.

High-Pressure Sodium. High-pressure sodium (HPS) lamps are very efficient in the production of raw lumens (high LPW), and they also have long life ratings. Their downfall, however, is poor CRI and an unattractive yellow-orange light. In many applications, such as storage facilities, this color may be acceptable, but in the workplace it is far less desirable. The low CRI makes it very difficult to differentiate colors and perform functions requiring eye-hand coordination. It is even claimed that yellow HPS light may have adverse psychological effects on certain people. At the least, visibility is extremely limited by this lamp of type.

Standard Metal Halide. Metal halides are generally viewed as the best source of light in industrial applications. The reasons are high efficiency (up to 100 LPW), respectable CRI, and long life. One of the most desirable attributes of metal halides is their ability to be manufactured to burn at different color temperatures. Most industrial plants, however, will probably prefer the crisp white 4000K color temperature.

Pulse-Start Metal Halide. These lamps combine the efficiencies of new metal halide arc tube lamp designs with the widely accepted ignitor-based ballast systems introduced with HPS. The resulting energy efficiencies (up to 110 LPW), high CRI (65 to 70), long life (up to 30,000 h), and superior lumen maintenance (80 percent) make these lamps the new standard for industrial, warehouse, and commercial applications.

UTILIZE MULTIPLE DESIGNS

Unless you or your staff have considerable experience in specifying and/or designing lighting layouts, now is the time to get help. This can be accomplished by hiring the services of a professional lighting designer or by contacting several lighting manufacturers' representatives. These are more than salespeople; they typically have years of application experience and constitute an excellent design resource.

Lighting representatives today are qualified individuals who have received hundreds of hours of training. This, combined with cutting-edge computer programs and years of real-life experience, makes them extremely qualified to recommend a variety of lighting designs.

If you do not know who your local reps are, it is easy to find out by calling your electrical distributors and asking for their recommendations. Another alternative is to use a lighting representative locator service offered by some of the larger national master distributors.

When you contact the reps (usually three is a good number), indicate to them the lighting levels you require and your light source preference. The lighting representative will want to visit your location to get as many details as possible to create proper design criteria and prepare a proposal. This is where you will appreciate having trained professionals involved in your project. They will take multiple factors into consideration to design a suitable system: reflectance of walls, floors, and ceilings, dirt depreciation factors, ballast parameters, mounting heights, and fixture efficiencies, just to name a few.

When the rep has collected the necessary information, you should request that several proposals be prepared using different lamp and system combinations. There are definite differences in performance and economics depending on which system is selected. You should also ask your rep to submit an ROI statement for each of the systems that are proposed.

There are many variations in similar products, each of them offering different performance. As you consider them, make sure that you do not sacrifice light quality for energy savings. With the products available today, you can obtain energy-efficient, high-quality lighting.

CHAPTER 5.10
COMPUTERS

Thomas E. Wiles
The Golf System, Inc.
Plano, Texas

OVERVIEW

There are three basic maintenance requirements for computers: *hardware maintenance, operating system maintenance,* and *software maintenance.* Each of these areas will be addressed individually. The modern microcomputer has an effective life span of about 10 yr. Capacitors drift over time. It takes about 10 yr for critical capacitors to lose enough of their capacitance to adversely affect the reliability of the machine. Computers older than 8 yr are living on borrowed time. Other critical components also fail with time (microprocessor, chipset, etc). While the actual life of these components cannot be accurately predicted, the incidence of failure increases very rapidly after 10 yr.

HARDWARE MAINTENANCE

Production facilities are generally very marginal environments for computers. The two problems are environmental contaminants and power. Power is addressed in Chap. 3.6. Environmental contaminants consist of dirt, dust, insects, heat, and chemicals.

Dirt, Dust, and Insects

Keeping the computer clean is the cornerstone of reliable operation. The best way to clean a computer is to blow the contaminants out with compressed air. This should be done on a regular basis, depending on the contaminant buildup in the computer. In a high-contaminant environment, the computer should be cleaned once a month. In a relatively clean environment, once a quarter is acceptable. When the machine is cleaned all cards fitting in edge connectors, plus the computer's memory, should be reset. This prevents film buildup on these connectors, which can cause the computer to run erratically. Some consideration should be given to a conductivity enhancer. A product such as Stabalant-22 can materially increase both the reliability and longevity of a computer.

Hard Drive

Ten years ago, computer hard drives rotated at 3000 r/min. Today, the slower drives rotate at 5600 r/min, and the faster drives rotate at 7400 r/min or 10,000 r/min. These parts rotate considerably faster than the engine in a car, and are expected to run at these speeds for 200,000 h without any form of maintenance or lubrication. As you might expect, a powered-up drive gets quite hot, and then cools when the computer is turned off. To allow for expansion inside the drive's case, there is a small breathing hole. This allows air to enter the case when the drive cools and leave the case when the drive heats. The condition of the filter on this hole is critical. This filter can be neither replaced nor cleaned. If it becomes clogged, particles can enter the drive case, causing friction, heat, and the eventual failure of the drive. Studies made during the 1980s indicated that *hard drive* life could be extended from a few months to about 6 yr in certain contaminated environments (by allowing the machine to run continuously). If the computer is not going to be used for 48 h, turn it off; otherwise, leave it running. Always disable the *hard drive* power-saving features in the operating system and the BIOS, as these power-saving features can shorten the life of a computer in a contaminated environment to a few months.

Floppy-Disk Drive

A good floppy-disk drive costs about $15 retail. These devices are robust and should last the life of the computer. Unfortunately, they collect dirt and other contaminants. The only way to keep these devices running reliably is to use them frequently. Floppy-disk drives tend to destroy diskettes when they are contaminated. Never put a diskette that you cannot afford to lose into a disk drive. If you need to use the drive with an important diskette, put a new diskette into the drive and format it. If the diskette formats cleanly, the drive is safe to use with valuable data.

Important data should be loaded on the file server first, then downloaded to the target computer.

A malfunctioning floppy-disk drive can prevent a computer from turning on. If you have a sudden computer failure, try disconnecting the floppy-disk drive and see if the machine will boot into the BIOS. If the machine passes the power-on self-test (POST), then replacing the drive will solve the problem. Also, removing the drive and disabling the drive in the BIOS will immediately return the machine to functionality.

CD-ROM Drive

The CD-ROM drive is the most delicate component in the computer. When a drive becomes contaminated it will not only destroy itself, it will totally destroy the CD placed in it. Do not put software distribution CDs in the target computer. All updates should be done through the file server (which should be in a clean room and get regular use). We recommend that CD-ROM drives be removed from, or never installed on machines located in production facilities. Map the file server's CD-ROM for all CD access.

Upgrading the Computer

In general, computers should not be upgraded. Upgrading an older computer is expensive, and the result is almost always unsatisfactory. By the time most computers have been upgraded sufficiently to perform their new tasks properly, the cost of the upgrade exceeds the cost of a new machine. Upgrading older hardware almost always results in unreliable computer performance.

OPERATING SYSTEM MAINTENANCE

Do not maintain the computer's operating system. The operating system that comes on the computer should last for the lifetime of the computer. Upgrading a Microsoft operating system is almost always a disaster. Most enterprise-level software packages replace some key elements of the Microsoft operating system. Upgrading the operating system can replace these files and render your software inoperable (most commonly on Windows 9X machines). If the computer is running your software reliably, leave the operating system alone. If a software update requires that the operating system be upgraded, perform the following steps:

1. Back up all data.
2. Fully erase the hard drive.
3. Do a complete installation of the new operating system using an original equipment manufacturer's version of the software. Install all drivers in the recommended order.
4. Reinstall the software packages.

The ROM BIOS on the motherboard of the computer can also be upgraded. On modern machines the BIOS can be *flashed* (written directly from a floppy disk). Because hardware manufacturers tend to ship motherboards with out-of-date BIOSs, flashing the motherboard BIOS is a fairly common practice. Our recommendation is not to flash the BIOS unless there is a known reason to do so. Most BIOS updates extend the capabilities of the motherboard (such as supporting faster microprocessors). While BIOS upgrades are generally safe, there have been documented cases in which the computer ran erratically after the BIOS flash. If you are going to flash the BIOS, try to put the computer on an uninterruptible power supply first. If power is lost during the BIOS flash, the motherboard will have to be replaced. (A replacement motherboard will probably not run with the operating system installed on the computer's hard drive.)

SOFTWARE MAINTENANCE

Software updates are normally essential and mandatory. Software vendors are very careful about the robustness of their updates—they do not stay in business if they are not. When possible, update a few machines and test the reliability of the update. Always back up all machines prior to the update as insurance in the event of a disaster. The largest danger related to software updates is the possibility of a corrupted file on the update media. This can result in the entire system coming down if it is not detected early. The basic rule is to *test before you run live.*

CHAPTER 5.11

POWER DISTRIBUTION SYSTEMS*

National Electrical Contractors Association, Inc.
Washington, D.C.

* Updated for the second edition by Challenger Electrical Equipment Corp., Roseville, Calif., and by the Editor-in-Chief.

GLOSSARY

Frequently used terms are defined here. More complete definitions are available in the latest edition of the *National Electrical Code,* published by the National Fire Protection Association.*

Ampacity Current-carrying capacity of electric conductors, expressed in amperes.

Ampere The unit of measure of electric current. Electric current is measured by the number of electrons that flow past a given point in a circuit in 1 s.

Branch circuit The conductors between the final overcurrent device protecting the circuit and the outlets or utilization equipment.

Bus(es) Metal conductors, usually copper or aluminum, of large size utilized to transmit large blocks of power.

Capacitor A device capable of storing electric energy. It is basically constructed of two conductor materials separated by an insulator.

Circuit breaker A device designed to open and close a circuit manually, and to open the circuit automatically (trip) on a predetermined overcurrent without injury to itself when properly applied within its rating.

Connected load The sum of the continuous loads of the connected power-consuming apparatus.

Controller A device, or group of devices, that serves to govern, in some predetermined manner, the electric power delivered to the apparatus to which it is connected.

Current The movement of electrons through a conductor material.

Demand The peak rate at which energy is consumed, specified usually in kilowatts.

Electromagnet A magnet in which the magnetic field is produced by an electric current. A common form of electromagnet is a coil of wire wound on a laminated iron core, such as the potential element of a watthour meter.

Electromotive force (emf) The force which tends to produce an electric current in a circuit. The common unit of electromotive force is the volt.

Electron A negatively charged particle which revolves about the nucleus of an atom.

Equipment A general term including material, fittings, devices, appliances, fixtures, apparatus, and the like used as a part of, or in connection with, an electric installation.

Fault Any system problem, but usually a short between phase conductors or a short to ground.

Frequency The number of cycles of an alternating current completed in a certain period of time, usually 1 s.

Fuse A protective device made up of a conductor which melts and opens when the current through it is more than the ampere rating of the fuse.

Ground A conducting connection, either intentional or accidental, permitting current to flow between an electric circuit or equipment and the earth.

Ground-fault circuit interrupter A device whose function is to interrupt the electric circuit to the load when a fault current to ground exceeds some predetermined value that is less than that required to operate the overcurrent protective device of the supply circuit.

Impedance The total vector sum of resistance and reactance opposing current flow in an ac system.

Inductance The property of a coil or any part of a circuit which causes it to oppose any change in the value of the current flowing through it. The unit of measure of inductance is the henry (H).

---

* *National Electrical Code®* is a registered trademark of the National Fire Protection Association, Inc., Quincy, MA 02269.

Kilo A prefix meaning thousand, or 10^3.

Load The equipment or appliance which is operated by electric current. Also, the current drawn by such a device.

Load factor The ratio of average current to the maximum demanded.

Mega A prefix meaning million, or 10^6.

Ohm The unit of electric resistance. A circuit has a resistance of 1 Ω when 1 V applied to it produces a current of 1 A in the circuit (Ohm's law).

Outlet A point on the wiring system at which current is taken to supply utilization equipment.

Overcurrent Any current in excess of the ampacity of equipment or conductor. It may result from overload, short circuit, or ground fault.

Overload Operation of equipment in excess of normal, full-load rating or of a conductor in excess of rated ampacity.

Panelboard A single integral enclosed unit including cabinet buses and automatic overcurrent protective devices, with or without manual or automatic control devices, for the control of electric circuits; designed to be accessible only from the front.

Peak load The maximum demand on an electric system during any particular period. Units may be kilowatts or megawatts.

Power factor The relationship between the active power (watts) and the voltamperes in any particular ac circuit. It is defined as the ratio of the total active power to the total voltamperes. It is also numerically equal to the cosine of the angle of phase difference between the total circuit voltage and current.

Relay A switch operated by means of electromagnetism.

Resistance The tendency of a device or a circuit to oppose the movement of current through it. The unit of resistance is the ohm.

Switchboard An integrated, factory-coordinated combination of circuit protective devices, control devices, meters, relays, busbars, and wireways enclosed in a single, preplanned unit designed to be a self-contained center. Protective devices may be individually compartment-mounted (switchgear) or group-mounted in barriered compartments. The entire structure is designed and built to operate as a coordinated unit.

Switching device A device designed to close and/or open electric circuits either manually or automatically.

Transformer A device which transfers electric energy from one coil to another by means of electromagnetic induction.

Utilization equipment Equipment which converts electric energy to mechanical work, chemical energy, heat, or light, or performs similar conversions.

VAR The term commonly used for voltamperes reactive.

Volt The practical unit of electromotive force, or potential difference. One volt will cause 1 A to flow when impressed across a 1-Ω resistance.

Voltage (of a circuit) The measured potential difference between any two circuit conductors or any conductor and ground.

Voltampere Voltamperes are the product of volts and the total current which flows because of the voltage. In dc circuits and ac circuits with unity power factor, the voltamperes and the watts are equal. In ac circuits at other than unity power factor, the voltamperes equal the square root of (watts squared plus reactive voltamperes squared).

Watt The practical unit of active power which is defined as the rate at which energy is delivered to a circuit. It is the power expended when a current of 1 A flows through a resistance of 1 Ω. $P_w = I^2R$

Watthour The unit volume of electric energy which is expended in 1 h when the power is 1 W.

INTRODUCTION

Approximately 36 percent of all energy consumed in the United States each year is used by industry. About one-third of the energy is used in the form of electricity.

Some 80 percent of the electricity used by industry is applied for electric drives which are elements of electromechanical systems. These systems are used to *form* (extrude, roll, cast, press, and spin), *shape* (mill, ream, drill, hone, and tap), and *transport* (conveyors, elevators, brakes, fans, pumps, and compressors). The remaining 20 percent of industry's electrical use is applied for electrolytic processes, process heating, lighting, and comfort conditioning.

The larger a plant is, the more important is its electric distribution system. It must be capable of meeting the needs of all electric equipment, from the point of entrance of the power company's service (or the plant generating powerhouse) to the terminals of the utilization equipment. If electric power is not available when and where needed, the owner's investment in both plant and inventory becomes idled.

The electric distribution system which is best for a given plant depends on the value assigned to dependability and flexibility. For example, if electricity is needed to manufacture a product whose design is frequently changed, a flexible, easily changed system is best. When continuity of service is essential, as in some chemical processes when a batch could be ruined by a power failure, an extremely reliable system is best.

Flexibility and reliability are only two concerns affecting electric distribution systems. There are several others, not the least of which is cost, including initial cost and life-cycle cost.

The way in which an industrial power distribution system is designed, installed, and maintained has considerable influence on virtually all aspects of system performance. For optimum performance, those involved with operation of the electric distribution system require at least a basic understanding of the factors involved in the generation, transformation, distribution, and utilization of electricity.

Basics of Electricity

Direct Current. *Direct current* is current which flows through the circuit in the same direction at all times. A dc flow in which the level is always constant is often called a continuous current.

Direct current is used in some industrial electrolytic processes, in almost all vehicles, and for certain motors. However, most of the electric energy used in America is generated as alternating current.

Alternating Current. An *alternating current* is one which passes through a regular succession of changing positive and negative values by periodically reversing its direction of flow. Total positive and negative values of current are equal.

If a typical alternating voltage is plotted against time, it will resemble the curve shown in Fig. 1-1.

The curve is called a sine wave because it has the same shape as the curve described by the equation

$$e = E_{max} \sin \theta$$

where θ is an angle and e is the instantaneous voltage.

Figure 1-1 shows the variation in ac voltage through two cycles. Voltage is zero at the beginning of the cycle, rises to a maximum value, and then falls back to zero halfway through the cycle. In the second half of the cycle, the voltage achieves a maximum negative value and then returns to zero at the end of the cycle. The number of cycles which the voltage goes through in 1 s is called *frequency.* Frequency is expressed in *cycles per second,* also known as *hertz.* Most common ac power supplies in the United States have a frequency of 60 cycles per second (cps), or 60 hertz (Hz).

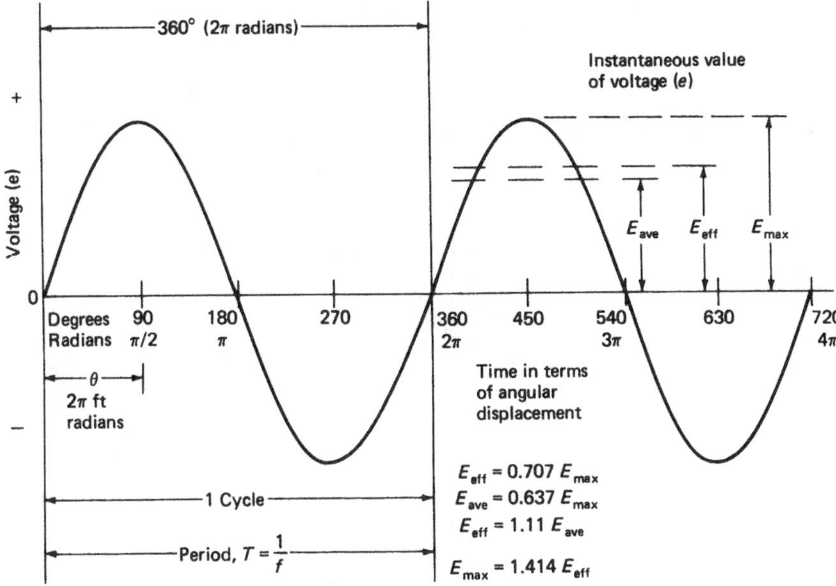

FIGURE 5.142 Sine-wave relationships.

Figure 5.142 shows that a cycle covers a definite period of time and is completed in 360° of the rotation of the armature of the generator. Accordingly, time may be expressed in terms of an angle of rotation:

$$\text{Period } T \text{ in seconds} = 360° \text{ or } 2\pi \text{ rad}$$

To express time, angle θ becomes

$$\theta = 2\pi ft \text{ rad}$$

where t is the time in seconds and f is frequency in hertz.

As such, the equation for the sine wave shown in Fig. 5.142 may be expressed as

$$e = E_{max} \sin 2\pi ft$$

or

$$e = E_{max} \sin \omega t$$

where $\omega = 2\pi f$ or

$$e = E_{max} \sin \theta$$

A voltage sine wave can be expressed in three values by:

1. Maximum or peak value E_{max}.
2. Average value E_{av}, which is equal to average value of e for the positive half or negative half of the cycle.
3. The root mean square (rms) or effective value (E_{eff}), a value of current which gives the same heating effect in a given resistor as the same value of direct current. Unless another description is specified, mention of alternating currents or voltages refers to the effective (rms) value.

Resistance is the only factor which opposes flow of current in a dc circuit.

Resistance also opposes flow in an ac circuit, but so do two other qualities: reactance from circuit *inductance* and *capacitance.*

As alternating voltage rises and falls, ac circuit amperage also rises and falls. If the circuit contains resistance only, voltage and current cycles are "in phase"; both voltage and current rise and fall at the same time, as shown in Fig. 5.143.

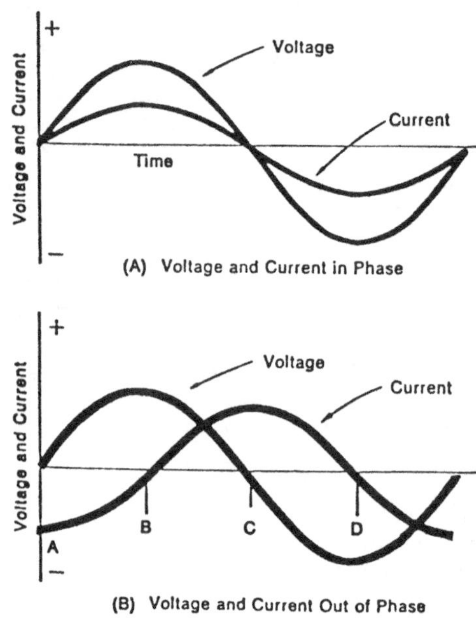

FIGURE 5.143 Voltage and current graphs.

Inductance is produced typically when a coil of wire is connected into an ac circuit. It opposes changes in current flow as voltage changes during a cycle, causing voltage and current to be out of phase, as shown in Fig. 5.143.

Capacitance typically is produced when a capacitor or condenser is inserted into an ac circuit. A capacitor or condenser also opposes voltage changes produced by a generator. As a result, voltage and current go out of phase, but in a direction which is opposite to that caused by inductance. In a purely inductive circuit, voltage leads current by 90°. In a purely capacitive circuit, voltage lags current by 90°.

Since all circuits contain resistance, total opposition to flow of ac current is dependent on the vector sum of the resistance, inductive reactance, and capacitive reactance in the circuit. This vector sum is called *impedance,* and is measured in ohms.

Ohm's law can be applied to ac circuits by substituting impedance for resistance:

$$I = \frac{E}{Z}$$

where I is the current, E is in volts, and Z is in impedance in ohms. In dc circuits, the power in watts is simply voltage times current, $P = EI$.

When ac current is used to operate a magnetic device, current lags applied voltage creating an out-of-phase relationship.

In Fig. 5.144 for example, the current wave is said to be θ degrees out of phase with the voltage wave, because it lags the voltage wave by the angle θ. It reaches its peak value θ degrees of angular displacement after the voltage wave reaches its peak. Thus, $P = EI \cos \theta$.

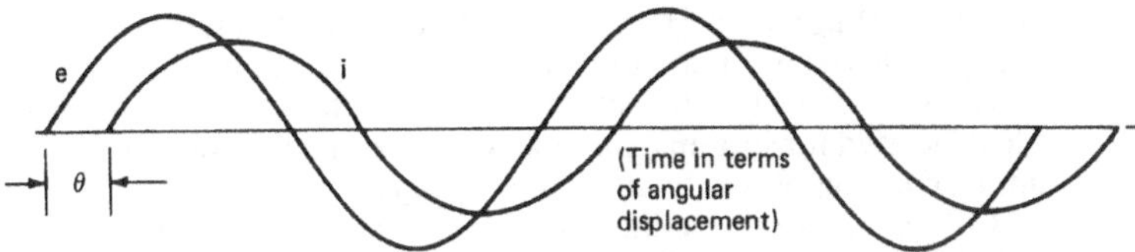

FIGURE 5.144 Current wave lagging voltage wave.

The cosine of the angle between the current and the voltage is known as the *power factor.* It is a measure of reactive power which produces additional heat in electric system components, *but performs no real work.* This relationship is illustrated mathematically in Fig. 5.145.

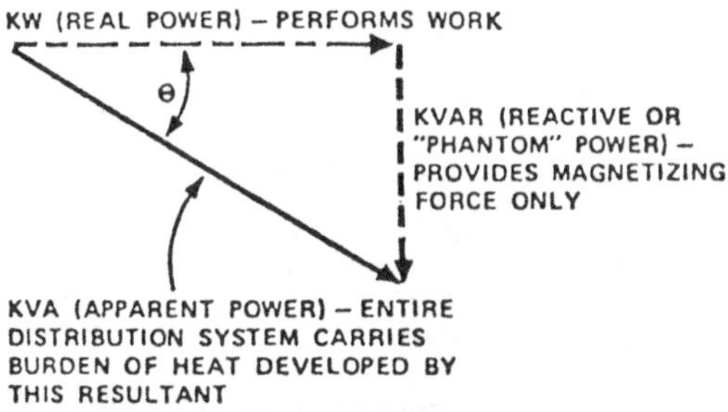

FIGURE 5.145 Power factor relationships.

Buying and Conserving Electric Energy

Purchased energy represents an increasing share of manufacturing costs. Most utilities use somewhat complicated methods to charge their customers for energy use. The basic charge is for actual kilowatthours used in a billing period. This charge is computed on a scale that recognizes blocks of energy use at different rates. The more energy used, the less is paid for each unit. This reflects the fact that overall utility company costs go down in proportion to the quantity of constant energy supplied by the system.

Most utilities also impose a charge for demand, expressed in kilowatts. This is a charge for maximum power requirements. It is determined by dividing energy consumed in a short (15- or 30-min) time period by the length of the time period. Thus, if 100 kWh is consumed in a 15-min demand interval, demand is 400 kW (100 kWh ÷ 0.25 h). Modern computer technology makes it possible to select the one demand interval of the billing period during which maximum wattage demand occurs.

A third charge sometimes imposed is a penalty for poor power factor. This charge is made to cover the cost of generating the unusable portion of electricity consumed in reactive power losses.

A fourth component of most utility bills, over which users have no control, is a *fuel adjustment charge.* This is added to cover the cost of fuel not included in the basic rate structure.

Careful examination of plant electric systems, utilizing a one-line diagram, can identify many steps to reduce energy consumption and electrical costs. Manual or automatic shutoff devices can be placed on short-term or rarely used equipment. Load-limiting and demand-limiting control devices can be installed. Power factor can be improved by the installation of capacitors or synchronous motors. Management can also investigate the feasibility of installing multifunction programmable controllers or centralized automated, computerized control to limit power and energy within the facility.

One-Line Diagram. There is no industrywide standard for the graphic description of electric systems, but most designers use a one-line diagram.

A one-line diagram of a plant's basic electric system should be developed and maintained. It can be used for system maintenance and as a management tool which can provide "instant" information on system loading and capacity.

Commonly used one-line diagram symbols are shown in Fig. 5.146; they are defined in IEEE Standard 315-1975, "Graphic Symbols for Electrical and Electronics Diagrams," ANSI Y32.2-1975.

The devices in switching equipment are referred to by numbers, according to the functions they perform. These numbers are based on a system which has been adopted as standard for automatic switchgear by IEEE. Numbers and their meanings are shown in Table 5.41.

A typical power-distribution diagram for an industrial facility is shown in Fig. 5.147. Note that the diagram is schematic. Locations shown refer to system relationships and not geography within the plant. One-line diagrams should be kept up to date as changes occur.

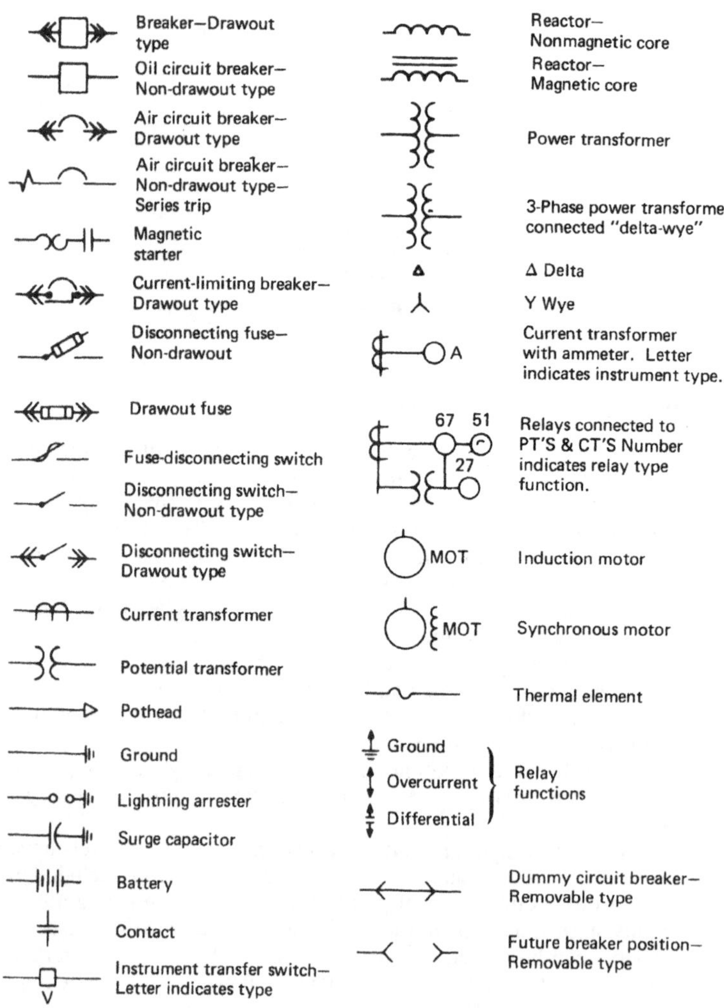

FIGURE 5.146 Commonly used symbols for one-line electrical diagrams.

TABLE 5.41 IEEE Device Numbers and Functions

Device no.	Function	Device no.	Function
1	Master element	51	AC time overcurrent relay
2	Time-delay starting or closing relay	52	AC circuit breaker
3	Checking or interlocking relay	53	Exciter or dc generator relay
4	Master contactor	54	High-speed dc circuit breaker
5	Stopping device	55	Power-factor relay
6	Starting circuit breaker	56	Field application relay
7	Anode circuit breaker	57	Short-circuiting or grounding device
8	Control power disconnecting device	58	Power rectifier misfire relay
9	Reversing device	59	Overvoltage relay
10	Unit sequence switch	60	Voltage balance relay
11	(Reserved for future application)	61	Current balance relay
12	Overspeed device	62	Time delay stopping or opening relay
13	Synchronous-speed device	63	Liquid or gas pressure level or flow relay
14	Underspeed device	64	Ground protective relay
15	Speed or frequency matching device	65	Governor
16	(Reserved for future application)	66	Notching or jogging device
17	Shunting or discharge switch	67	AC directional overcurrent relay
18	Accelerating to running transition contactor	68	Blocking relay
19	Starting to running transition contactor	69	Permissive control device
20	Electrically operated valve	70	Electrically operated rheostat
21	Distance relay	71	(Reserved for future application)
22	Equalizer circuit breaker	72	DC circuit breaker
23	Temperature control device	73	Load resistor contactor
24	(Reserved for future application)	74	Alarm relay
25	Synchronizing or synchronism check	75	Position-changing mechanism
26	Apparatus thermal device	76	DC overcurrent relay
27	Undervoltage relay	77	Pulse transmitter
28	(Reserved for future application)	78	Phase-angle measuring or out-of-step protective relay
29	Isolating contactor	79	AC reclosing relay
30	Annunciator relay	80	(Reserved for future application)
31	Separate excitation device	81	Frequency relay
32	Directional power relay	82	DC reclosing relay
33	Position switch	83	Automatic selective control or transfer relay
34	Motor-operated sequence switch	84	Operating mechanism
35	Brush-operating or slip-ring short-circuiting device	85	Carrier or pilot-wire receiver relay
36	Polarity device	86	Locking-out relay
37	Undercurrent or underpower relay	87	Differential protective relay
38	Bearing protective device	88	Auxiliary motor or motor generator
39	(Reserved for future application)	89	Line switch
40	Field relay	90	Regulating device
41	Field circuit breaker	91	Voltage directional relay
42	Running circuit breaker	92	Voltage and power directional relay
43	Manual transfer or selector device	93	Field-changing contactor
44	Unit sequence starting relay	94	Tripping or trip-free relay
45	(Reserved for future application)	95	(Reserved for special applications)
46	Reverse-phase-balance current relay	96	(Reserved for special applications)
47	Phase-sequence voltage relay	97	(Reserved for special applications)
48	Incomplete sequence relay	98	(Reserved for special applications)
49	Machine or transformer thermal relay	99	(Reserved for special applications)
50	Instantaneous overcurrent or rate-of-rise relay		

Source: ANSI C37.2-1970.

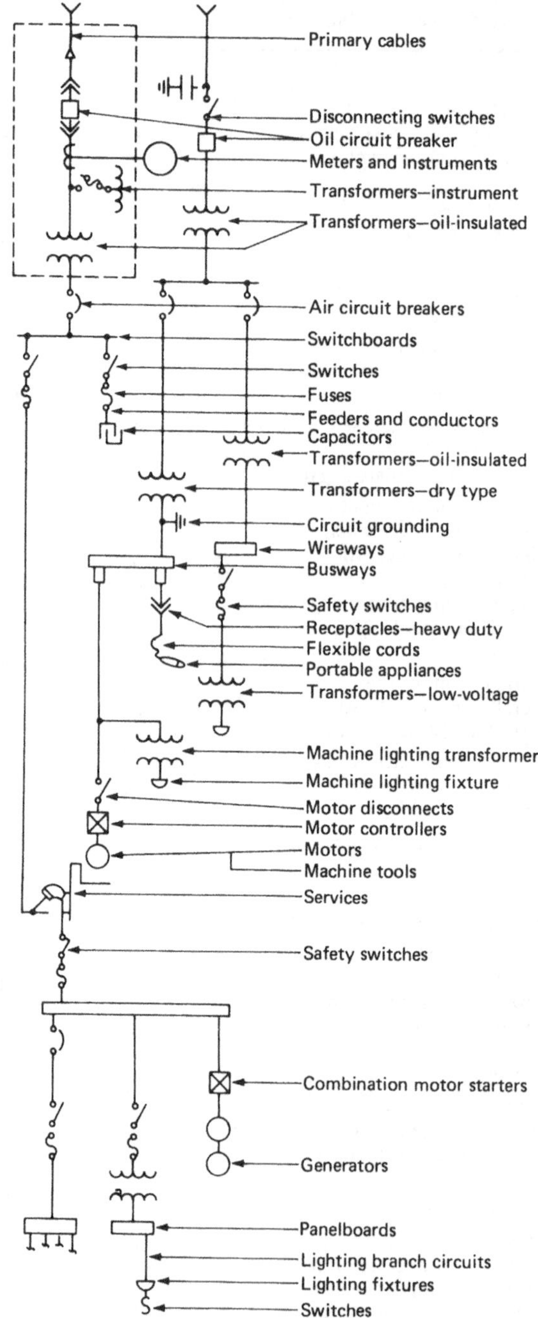

FIGURE 5.147 Typical electrical distribution one-line diagram.

Standards

Standards relate to requirements, including agreed-upon definitions of terms, measurement and test procedures, and equipment dimensions and ratings. Some of the more commonly used standards relating to electric distribution systems follow.

National Electrical Manufacturers Association Standards. National Electrical Manufacturers Association (NEMA) standards establish dimensions, ratings, and performance requirements for electric equipment, regardless of manufacturer. NEMA standards are used widely in specifications.

National Fire Protection Association Standards Documents. National Fire Protection Association (NFPA) standards specify requirements for fire protection and safety.

NFPA's National Electrical Code*. The NFPA's *National Electrical Code* (NEC) is recognized as the minimum standard for the "practical safeguarding of persons and property from hazards arising from use of electricity." The NEC may be supplemented by local requirements and ordinances pertaining to electric systems requiring more stringent safety measures. In general, the NEC is nationally recognized. This Code is under constant review and is updated every 3 years. All plant engineers should have working knowledge of this Code.

Underwriters Laboratories, Inc., Standards. Underwriters Laboratories (UL), Inc., develops safety testing standards for electric equipment. Only equipment which complies with those requirements may be listed or labeled.

American National Standards Institute. The American National Standards Institute (ANSI) promotes, coordinates, and approves as American National Standards documents which have been prepared in accord with ANSI regulations. American National Standards carry the identification numbers both of ANSI and originating organizations. The sponsoring organization is responsible for keeping a standard up-to-date.

Occupational Safety and Health Act. The Occupational Safety and Health Act (OSHA) requires that employers provide a safe and healthful workplace for all employees.

Local Codes

Some large cities develop their own codes, but these generally are based on the *National Electrical Code*.

SYSTEM COMPONENTS

From point of service to point of use, electric power is directed, protected, and modified by segments of the system whose function, performance, and efficiency are vital to the utilizing equipment. To operate and maintain an industrial electric distribution system properly and to understand system planning fully, it is necessary to comprehend the function of each system component and its place in the overall system. Continued reference to Fig. 5.147, the one-line diagram, will aid in putting the various elements of the system into proper perspective.

* *National Electrical Code®* is a registered trademark of the National Fire Protection Association, Inc., Quincy, MA 02269.

Note that some system components have no normal operation and maintenance requirements. For such a device, it is necessary to have only a basic understanding of its appearance, construction, and function.

One such device is the *pothead,* used for connecting medium- and high-voltage cable systems to the internal distribution system. It is a terminator which permits termination of a complex cable at one end and attachment of cable lugs at the other. Its body usually is cast and can be filled with an insulating compound like petroleum jelly.

Outdoor high- and medium-voltage substations usually employ *oil-immersed circuit breakers* as a system protective device. These devices are equipped with direct-acting internal current transformers and trip coils enclosed in a jacket filled with a mineral oil. The oil jacket provides rapid cooling for contacts which become heated due to arcing action created by each switching operation. The oil-filled jacket is a better heat dissipator than air, so the device can be smaller than one which is air-cooled.

An oil-immersed circuit breaker must be located with care. The oil is flammable; a serious explosion or fire could result from a spark coming into contact with the cooling medium. In addition, in case of a leak a means for disposing of the oil, usually in the form of a gravel-filled drain, must be provided.

The oil must be examined periodically. Inspection ports are provided for this purpose. If the oil becomes contaminated, it must be replaced or drained and filtered.

Transformers

Transformers make possible the use of the high distribution and utilization voltages found in industrial electric systems. They are used to transform one primary voltage to a second primary level, to step from primary down to secondary voltage (not more than 600 V), and to step a secondary distribution voltage to a secondary utilization level.

The names used to categorize transformers relate to their applications:

General-purpose transformers. Dry-type units rated 600 V or less. They are used for local step-down from a secondary distribution voltage to a utilization level, serving lighting and appliance loads. They also are called *general power* and *light* transformers or *lighting transformers.* Ceiling-suspended and floor-standing units are available.

Load-center transformers. Either dry-type or liquid-filled units, primary rated from 2400 to 15,000 V. They are used for both indoor and outdoor applications to step down to a voltage of 600 V or less. Load-center transformer units may be used separately, in combination with separate protective and switching devices and secondary distribution switchboards, or in combination with primary and secondary switching and protection in a packaged unit called a substation or load-center substation. They are base-mounted, free-standing units.

Distribution transformers. Single-phase and three-phase oil-immersed, pole-mounted, or platform-mounted units. They are primary rated from 480 to 15,000 V, with step-down to a secondary level (or to a lower high-voltage level for units over 10-kV primary). Distribution transformers in capacities up to 167 kVA are used for pole-line distribution. Other outdoor wiring systems use platform units rated up to 500 kVA.

Substation transformers. Oil-immersed units which are primary rated from 2400 to 67,000 V, with secondary ratings ranging from less than 600 to 15,000 V. Substation transformers are used in utility distribution and industrial substations. Power transformers are available for over 67 kV.

A transformer's nameplate indicates its kVA rating.

Transformer Ratings

Insulation classifications include liquid and dry types. Liquid insulation can be subclassified by type of liquid. Dry-type transformers can be subclassified as ventilated or sealed gas-filled.

kVA rating includes both self-cooling and forced-draft (fan-cooled) ratings for a specified temperature rise. The self-cooled rating should not exceed the peak demand by more than 25 percent to achieve the most efficient operation. A significant increase in transformer capacity can be obtained by the application of proper fan cooling. This permits transformers to be applied to present load conditions with an optimum load factor while still being able to serve expanded loads.

Transformers are designed to withstand short-duration overloads without any damage except a shortening of the useful life. The permissible overload and its duration vary with ambient temperature, preloading condition, and duration.

Transformer voltage ratings include primary and secondary continuous duty levels at specified frequencies, along with each winding's basic impulse level (BIL). The primary winding's continuous rating is the nominal line voltage of the system. The secondary voltage rating is the value under no-load conditions. Secondary voltage change experienced under load is termed *regulation.* It is a function of the system including the transformer impedance and the load power factor. Good regulation can often be achieved by tap adjustments on the primary side.

The *basic impulse level* rating for a transformer winding identifies the transient overvoltage withstand capability of its insulation.

Voltage Taps. Voltage taps are used either to compensate for small changes in primary supply to the transformer or to vary secondary voltage level with changed load requirements. A manually adjustable no-load tap changer is a common standard arrangement since tap changing under load is a highly complex problem.

Connections. Delta primary and wye secondary connections such as those shown in Fig. 5.148 are the most common transformer connections utilized today. In older plants, such variations as the grounded wye secondary and delta-delta connections to provide power to three-phase equipment are frequently encountered.

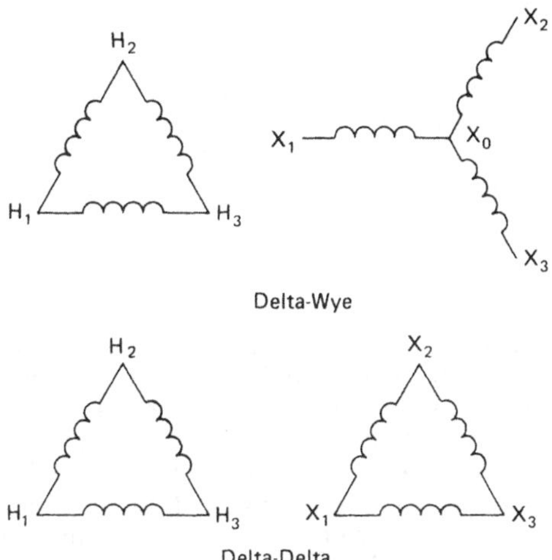

FIGURE 5.148 Three-phase transformer configurations.

Impedance. A transformer's impedance is its opposition to current flowing through it with the secondary short-circuited. Impedance is expressed as the percentage of normal-rated primary voltage which must be applied to cause full-rated load current to flow in the short-circuited secondary. For example, when a 480- to 120-V transformer is impedance-rated at 5 percent, it means that 5 percent of 480 V, or 24 V, must be applied to the primary to cause rated load current to flow in the shorted secondary. When 5 percent of primary voltage causes such current, 100 percent of primary voltage will cause 20 times full-load-rated secondary current to flow through a solid short-circuit on the secondary terminals.

The lower a transformer's impedance, the higher the short-circuit current it can deliver. The impedance values of general-purpose transformers generally range between 3 and 6 percent.

Circuit Breakers (600 V and Below)

Two types of circuit breakers are commonly used for applications at 600 V and less: *power circuit breakers* and *molded-case circuit breakers.*

Power circuit breakers are open-construction assemblies on metal frames for service in switchgear compartments and similar enclosures. Molded-case circuit breakers also are used in switchgear. While their protective value is high, they offer few control options.

Circuit breakers can be subcategorized in a variety of ways, based on different characteristics, as follows:

Adjustable. Indicates that the overcurrent device of the circuit breaker can be set to trip at different values of current and/or time within a predetermined range.

Instantaneous trip. Indicates that no delay is purposely introduced in the tripping action of the circuit breaker. Automatic circuit breakers may have both time overcurrent and instantaneous tripping devices.

Inverse time. Indicates that a delay in the tripping action of the circuit breaker is purposely introduced. The delay decreases as the magnitude of the current increases.

Nonadjustable. Indicates that the circuit breaker does not have any adjustment to alter the value of current at which it will trip or the time required for its operation.

Generally speaking, circuit breakers are rated by a frame size which determines the maximum overcurrent trip setting and an interrupting capacity indicating the maximum short-circuit current which can be interrupted safely without damage to the circuit breaker.

Circuit breakers are sometimes used with integral current-limiting fuses to provide for interrupting current requirements up to 200,000 A, symmetrical rms.

New Developments in Circuit Breakers

Circuit Breakers with Internal Electronics. Maintenance and industrial electricians will find that many new circuit breakers now contain electronic circuits which allow for adjustability of trip characteristics to fit a broad range of protection requirements. This adjustability allows for the customizing of the circuit breaker trip curve to coordinate with other distribution equipment or to offer additional protection for individual circuits. Adjustment features are placed on the face of the breaker. Qualified design engineers or technicians can customize these trip settings to provide enhanced system protection. Also, ground fault capabilities can be installed in the circuit breaker and a level of adjustability to the ground fault protection can be included. Multiple circuit breakers with ground fault protection can be used sequentially or cascaded so that the smallest branch breaker in the sequence would clear the fault first, without tripping the main circuit breakers or other feeder breakers in the circuit. Manufacturer's instructions should always be followed when altering characteristics of the circuit breaker.

High-Fault-Current Circuit Breakers. In modern distribution facilities, very high levels of fault current are available, up to or exceeding 200,000 A. In the case of a fault (a short circuit), there may be in excess of 200,000 A passing through the distribution equipment and the circuit breaker to the fault. Circuit breakers which are not designed to interrupt these high levels of fault current may be damaged or destroyed as they try to open or clear the fault condition.

In the past, a fuse or a circuit breaker–fuse combination was required in order to offer fault current protection as high as 200,000 A. Many of today's circuit breakers are capable of opening at these very high fault levels without the assistance of fuses. Utility companies can advise users what the available fault current is. Facility engineers and service personnel should be aware of the fault current ratings of equipment within their facility and ensure that existing equipment is capable of safely clearing faults. When replacing existing equipment or adding new distribution equipment, care should be taken that new circuit breakers or distribution equipment meets or exceeds the ratings required to safely clear an electrical fault.

Current-Limiting Circuit Breakers. There are also special types of circuit breakers which limit the amount of current that can flow through them during a short-circuit condition. These are called *current-limiting circuit breakers.* In the case of very high levels of fault current being available from the power company, they restrict the amount of current that would pass through in the case of a fault and reduce the possibility of damage to downstream equipment.

SERIES (CASCADED) RATINGS. Underwriters Laboratories recognized the short-circuit rating of a combination of devices, that is, two or more devices used in series. A combination of circuit breakers, where circuit breakers of a lower-fault-current capability could be used in conjunction with higher-fault-current circuit breakers on high-fault-current systems, or combinations of fuses and circuit breakers, are also recognized.

In the case of a system that has 200,000 A fault current available, a 200,000-A main circuit breaker might be used, and downstream from the 200,000-A main circuit breaker, there may be a 100,000-A feeder breaker, then a 65,000-A breaker, down to a 42,000-A breaker, protecting branch circuits, to the minimum circuit breaker available today, which is a 10,000-A interrupting capacity (AIC) breaker. All circuit breakers are labeled on the face with their AIC rating—the amount of fault current that can safely be interrupted by the breaker.

With the UL recognition of integrated or coordinated ratings, there is a requirement that distribution equipment be labeled with the coordination information of the circuit breakers installed in the distribution equipment. Service personnel must be mindful of the service rating of the equipment they are servicing and ensure that replacement or add-on equipment is rated equal to or higher than the markings on the distribution equipment call for. Circuit breakers should be rated higher than the available fault current. Or *use the series rating of equipment,* which is provided by the manufacturer and states what circuit breakers can be used in series. This will limit and prevent damage to equipment.

CRITICAL CIRCUITS. In situations where series-rated equipment is used, caution should be taken that critical power circuits are not interrupted. In the case of a series-rated circuit breaker system and a high-fault-current situation, two circuit breakers would trip simultaneously and limit the amount of fault current passing to the lower-rated breaker in the system. Thus two circuit breakers trip simultaneously; this terminates the power in the circuit in which the fault occurred as well as the feeder circuit feeding the branch.

Care should be taken in the design of systems such that critical circuits are not used with series-rated systems. In the case of our 200,000-AIC main, if there were a high current fault at a downstream sight, the 100,000-AIC circuit breaker would interrupt and it would also cause the 200,000-AIC main circuit breaker to interrupt simultaneously, which would terminate power to the entire distribution system. If this were used in, say, a hospital, you would not want a short circuit in a panel board or a downstream distribution switchboard to also cause the main circuit breaker to trip, taking all power out of the hospital. In those types of critical situations, the minimum circuit breaker AIC rating should be no less than the available system fault current delivered by the utility.

SPECIAL CAUTIONARY NOTE: All circuit breaker manufacturers recommend that their circuit breakers be "exercised" annually via the "trip test" button. Exercising the circuit breaker actuates the trip mechanism. The trip mechanism of the circuit breaker is separate from the on-off handle mechanism, and annual operation of the trip mechanism ensures that the breaker will trip when called on to operate. This is also a test indication: if the breaker fails to trip, it has malfunctioned and should be replaced.

Switchboards and Switchgear

Switchboards and switchgear include buses, conductors, control devices, protective devices, circuit switching devices, interrupting devices, interconnecting wiring, accessories, supporting structures, and enclosures.

Switchgears can serve as main secondary service equipment, as main primary service equipment, and as load center equipment when located near load concentrations. The assembly and its devices provide for the control and distribution of electricity to utilization or subdistribution equipment. Most switchgear manufactured since 1955 has been metal-enclosed, dead-front, free-standing type, with its circuit protective devices enclosed each in its own compartment. For certain types of load applications, the protective devices are group-mounted in separate cubicles instead of individual compartments.

Older switchboards may be live-front type with the devices wired and mounted to slate panels with the wiring exposed on the rear of these panels.

A preventive maintenance plan should be conducted for all plant switchgear. Normal maintenance includes checking and tightening cable connections, testing air circuit-breaker operations, and checking relay and meter calibrations.

Types and Functions of Switchgear. The three most common types of switchgear used in industrial plants are:

1. Medium- and high-voltage metal-clad circuit breaker
2. Low-voltage power circuit breaker, either drawout or stationary (molded-case breakers are being used more frequently for switchboard applications)
3. Load interrupter switch and fused switchgear

Metal-Clad Switchgear. This is metal-enclosed power switchgear characterized by removable circuit switching and interrupting devices. Major parts of the primary circuit are enclosed by grounded metal barriers. Instruments, secondary control devices, meters, relays, and their wiring also utilize grounded metal barriers to isolate them from primary circuit elements. All live parts are enclosed in grounded metal compartments, bus conductors are insulated, and mechanical interlocks are provided.

Low-Voltage Power Circuit Breaker Switchgear. This is metal-enclosed and generally consists of low-voltage power circuit breakers; bare or insulated bus and connections; control instruments, meters, and relays; and control wiring and accessories, together with cable and busway termination facilities. Circuit breakers may be drawout type, stationary air circuit breakers, or molded-case circuit breakers, usually of the solid-state, adjustable trip variety.

Metal-Enclosed Load Interrupter Switch and Fuse Switchgear. This consists of fused load interrupter switches, bare or insulated bus and connections, power fuses, instruments for control, and control wiring and accessories. Both stationary and removable interrupter switches and power fuses are used.

Protective Relays. Protective relays consist of an operating element and a set of movable contacts. In today's solid-state relays it is difficult to distinguish each of these elements visually. The operating element receives power from a control power source within the switchgear. It obtains input from a sensing element in the circuit. This input can be in the form of either a voltage or current variation. The relay measures the variation against its standard settings and, when a preset limit has been exceeded, it initiates an action by causing its contacts to assume a new configuration. This action is used to perform some other control function such as sounding an alarm or tripping a circuit breaker.

It is essential for all protective relays to be maintained in proper working order well within prescribed tolerances. Relay manufacturers provide complete maintenance and setting instructions including test procedures for all their relays. Such instructions should be kept in an overall operating manual in the plant engineer's office.

Directional Relays. Directional relays are used to initiate an action when current flows in a direction other than normal. Such relays will initiate actions preventing the reverse flow of current through equipment.

Differential Relays. Differential relays are so named due to their current-balance characteristic. If the current entering the protected section is not balanced by the current leaving it, the relay initiates an action. These relays are used most frequently for protection of transformers and substation buses where they permit or prevent casing of the breakers.

Voltage Relays. Voltage relays are fundamentally similar to overcurrent relays, except voltage is used to activate the operating element. Some voltage relays are activated by overvoltage, others by overcurrent, and still others by a combination of the two conditions.

Ground Relays. Ground relays are instantaneous relays that respond to unwanted ground current. As such, they are unaffected by load currents and so may be set to operate for single-phase-to-ground currents that are smaller than full-load currents.

Control power for relay use is supplied generally by the use of current and potential transformers. Low-voltage equipment operating at less than 1000 V to ground generally requires no potential transformers because relays and instruments are designed to function at normal low voltages.

Current Transformers. Current transformers are transformers in which the switchgear bus usually forms the single turn of the primary winding while the secondary is a combination of many turns. As such, the ratio of the current flowing in the main bus to that which flows in the secondary of the transformer is directly proportional to the bus rating. All relays and instruments are calibrated to read directly based on this ratio. The full secondary current is usually 5 A. Maintenance personnel must be careful not to open-circuit the secondary side of a current transformer. The resulting voltage across the open terminals is the transformer turn ratio times the primary voltage and is very dangerous.

Safety Switches and Disconnecting Means

Each piece of utilization equipment located remote from its source of power or overcurrent protective device requires a disconnecting means. In most cases, a safety switch is provided for plant machinery which is not built with its own internal on-off switch designed to completely disconnect the device from the electrical system. A safety switch may be either fused or unfused. The disconnecting means is provided to permit equipment maintenance without the possibility of some remote-control device functioning to turn the machine on while the maintenance person is in contact with some dangerous part. Some of the commonly used switch types include safety switches and toggle, or snap, switches. A *safety switch* is a spring-loaded, quick-make, quick-break, device enclosed in a metal housing. Also known as service switches or disconnect switches, they are rated in amperes and/or horsepower. *Toggle,* or *snap, switches* are also used for disconnect purposes. They are designed for outlet box mounting and are rated by voltage and current. Another such device is the thermal motor switch or manual motor starter, intended as the overcurrent protective device for small motors. It is a combination of motor circuit protector circuit breaker and thermal overload device, built as an integrated unit designed for outlet box mounting.

Fuses

Many types of fuses are used in switches and other protective devices. The National Electrical Manufacturers Association lists standardized fuse types.

A fuse protects a circuit by means of a "link" or internal element which melts because of the heat caused by excessive current, thereby opening the circuit.

Fuses are selected on the basis of voltage, current-carrying capacity, and interrupting rating, in accordance with NEMA standards.

Spare part planning should include provisions to have the plant stock at least one complete set of each fuse in use in the plant.

Power Fuses (Over 600 V). The NEMA E rating for power fuses requires that all fuses must carry their current rating continuously. Two types of power fuses are in common use: *current-limiting* and *expulsion type.*

Current-Limiting Type. These power fuses are designed in such a way that the melting introduces high arc resistance into the circuit during the first half-cycle's peak current, thus restricting short-circuit current.

Their current-forcing action produces transient overvoltages. Surge-protective apparatus may be needed to compensate.

Expulsion Type. These fuses are generally used in distribution system cutouts or disconnect switches. An arc-confining tube with a deionizing fiber liner and fusible element provide fault current interruption. Production of pressurized gases inside the fuse tube extinguishes the arc by expelling it from the open end of the fuse.

Low-Voltage Fuses (600 V and below). There are two styles of low-voltage fuse design, *plug* and *cartridge.*

Plug. These fuses are of three basic types, all rated 125 V or less to ground and up to 30 A maximum. Although they have no interrupting rating, they are subjected to short-circuit test with an available current of 10,000 A. The three types are: Edison base without time delay and all ratings interchangeable, Edison base with time delay and interchangeable ratings, and type S base available in three noninterchangeable current ranges: 0 to 15, 16 to 20, and 21 to 30 A. These last two types normally have a time-delay characteristic.

Cartridge. These fuses are either renewable or nonrenewable. Nonrenewable fuses are factory-assembled and must be replaced after operating. Renewable fuses can be disassembled and the fusible element replaced. Consult the NEMA standards and the code for fuse class definitions.

Wires and Cable

Wire and cable used for the distribution of electric power consists of a conducting medium usually enclosed within an insulating sheath and sometimes further protected by an outer jacket. The conducting medium is generally either copper or aluminum. The insulating medium can be made of any one of several materials depending on the ambient characteristics in which the wire is to be applied. Modern building wire is usually insulated with some form of plastic insulation. The various insulating materials are described in and defined by the *National Electrical Code,* Tables 310 to 313.*

Outer cable jacket materials range from cross-linked polyethylene to copper armor. Each jacket type has its own unique set of rules for application.

The cable protection method, conduit, cable tray, or bare exposed, depends on many design considerations. Cables require no basic maintenance except to assure that they are not applied beyond their ampacity. Cable and wire ampacities are published in the *National Electrical Code* and by the manufacturers who make cables. All cables used for normal electrical usage are sized in the United States in accordance with the American Wire Gauge (AWG) or circular mils.

Terminations of aluminum cables require periodic examination to ensure that "cold flow" has not loosened the connection. Lugs connecting aluminum cable must be tightened in exact

---

* *National Electrical Code*® is a registered trademark of the National Fire Protection Association, Inc., Quincy, MA 02269.

accordance with the recommendations of the equipment manufacturer. Maintenance personnel should possess a set of torque wrenches for aluminum cable maintenance. A program for terminal examination should be included in maintenance manuals.

Busways and Busbars

Busways and busbars perform the same functions as wires and cables, but their construction features and applications are far different.

Busbars can be cut, welded, bent, punched, drilled, plated, and insulated to meet specifications. They are most practical when configurations must be precise, terminal points inflexible, and installation locations confined.

Standard busway types, ampacities, components, and accessories are so numerous that almost any routing plan can be created. They can be enclosed or ventilated, installed indoors or out, and used either as point-to-point power feeders, as multipoint power sources, or as continuous power takeoff routes.

Motor Controllers and Motor Control Centers

All motor-operated devices require some form of motor controller to provide start-stop, reversing, and other functions. Sometimes these control elements are built into the machine and the controls are machine-mounted. More often than not the control of a building utility or process system motor is from a remote motor controller.

Motor controllers—commonly called *starters*—are usually magnetically operated devices with thermal overload protection built in through the application of melting alloy links. A preventive maintenance program is essential for motor starters if they are to continue to function properly. Remote controls with operating lights indicate only that the controls functioned. Therefore, it is necessary to make periodic inspections of these devices.

It is often more economical to group-mount motor controllers rather than individually mount them. A group of starters collected together into one integrated unit, is called a *motor control center.* NEMA has well-defined standards for type and class of wiring and the degree of interconnecting and interwiring present in any given center. These data can be obtained from the manufacturer and should be filed in the plant maintenance manuals together with complete detailed instructions for a periodic maintenance program. For further discussion, see Chap. 5.8.

Panelboards

Panelboards are assemblies of switching and overcurrent devices enclosed in a cabinet, usually wall-mounted, and protected by a cover or trim which can have a hinged door.

Panelboards provide for local protection and control of apparatus and lighting. They are usually composed of integrated groups of circuit breakers and fused switches providing protection for circuits terminating in their immediate area.

All of the maintenance and operating considerations previously discussed for circuit breakers and switches apply to panelboards. Unless the circuit breakers in a panelboard bear a *switching duty rating* and the panel nameplate or breaker is so marked, they are not designed for such service on a constant basis. Constant use for switching will shorten the life of a standard circuit breaker. Occasional use will not have a negative effect.

Panelboards serving lighting and appliance circuits which must be switched on at fixed times and off at other times can well be equipped with magnetic contactors. These devices operate in a manner similar to motor controllers except they contain no overload devices. They permit remote switching of panelboards without using the breakers as a switching device.

PRIMARY SERVICE METHODS

There are a number of different distribution methods used in modern industrial systems. In all of these, electric power is received from a public utility or in-house generating plant at some convenient primary voltage and passed through a control and distribution system to the point of utilization. There is no one standard system of industrial plant distribution. Each system is tailored to meet conditions specified at time of design. Prime factors in the type of system chosen are the required utilization voltage and the distances involved in the distribution. Because the number of plants in many areas has increased and the plants have moved further from the utility generators, utilities have been using higher voltages to transmit electricity to the plants. This technique also minimizes line losses and provides good voltage regulation. As a result, many utilities are distributing at up to 230 kV and some at 350 kV. In many cases the rates paid for power at higher voltages are low enough to generate savings which repay higher installation costs for a main substation within several years.

Systems utilizing primary power as their source require some point at which this power is received and transformed to a useful form. The point of connection is normally the plant substation. If the primary voltage is in the high voltage classification (34,500 V and above), chances are that the service is aerial. In these cases, service enters the substation from an overhead structure containing a series of disconnect switches and lightning arresters. The substation requires oil-filled circuit breakers on each feeder. Liquid-filled transformers change the incoming voltage to some more usable level and provide a method (either cable or bus) for distribution to the main secondary switchboard.

Main Substations

If the purchased power voltage can be used for the plant primary system without transformation, a plant main bus can serve the same purpose as a main substation.

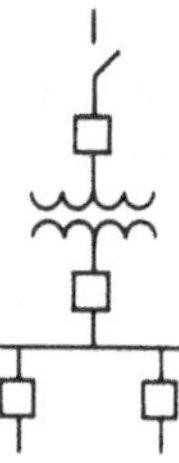

FIGURE 5.149 A typical main substation arrangement used by an industrial plant.

The principal functions of a main substation are indicated in Fig. 5.149. This is a simple arrangement which meets the requirements of many smaller plants. More complex arrangements are needed when there is more than one incoming line, or more than one power transformer, or one of a number of other bus arrangements. For large plants with heavy loads in widely separated areas, substations may require transmission voltage feeders connected to the incoming-line bus.

SECONDARY SERVICE METHODS

Many plants are switching to higher in-plant voltages to feed power over long distances throughout the plant without excessive line losses or loss of regulation. The most commonly used utilization voltages in new plants are 480 and 240 V, with 480 V becoming more popular. The best overall secondary utilization voltage is 480 V. It costs less and provides fewer line losses and less voltage drop.

Most new plants use load-center distribution, with unit substations being close to the loads and primary distribution being made at 2400, 4160, or 13,800 V. Primary-distribution voltage between master substation and load center should be selected based on size of the load and the distance to be covered. In general, 4160 V is used for loads under 10,000 VA and 13.8 kV for loads over 20,000 VA. For loads between 10,000 and 20,000 VA, 4160 V is used when plant layout is compact and 13.8 kV for long, rambling layouts.

The standard method of receiving and distributing secondary power is through use of radial systems.

Radial Systems

Conventional Simple Radial System. A conventional simple radial system (Fig. 5.150) receives power at the utility supply voltage. Voltage is stepped down to the utilization level by a transformer.

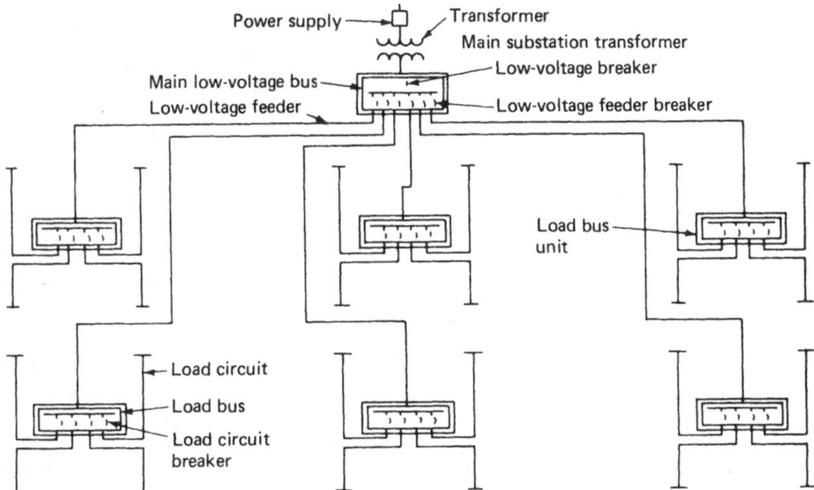

FIGURE 5.150 Conventional simple radial system.

Because the full building load is served from a single incoming substation, diversity among loads can be used to full advantage; installed transformer capacity can be minimized. The system's drawbacks include poor voltage regulation and poor service reliability.

Modern Load-Center Simple Radial System. The modern load-center simple radial system (Fig. 5.151) distributes incoming power at the primary voltage to power-center transformers located in building load areas. These transformers step the voltage down to utilization levels.

Each transformer must have enough capacity to handle the peak load of its specific load area. Combined transformer capacity requirements, therefore, may exceed those of a conventional simple radial system. This approach results in reduced losses, improved voltage regulation, reduced cost of feeder circuits, and no need for large low-voltage feeder circuit breakers.

Unit Substations

A unit substation contains one or more sections of each of three main components.

A *primary section* provides for connection of incoming medium- or high-voltage circuits, usually with disconnecting and circuit protective devices such as switches and circuit breakers. A *transformer section* includes one or more transformers. A *secondary switchboard section* provides for connection of secondary distribution feeders, each with a circuit protective switching and interrupting device.

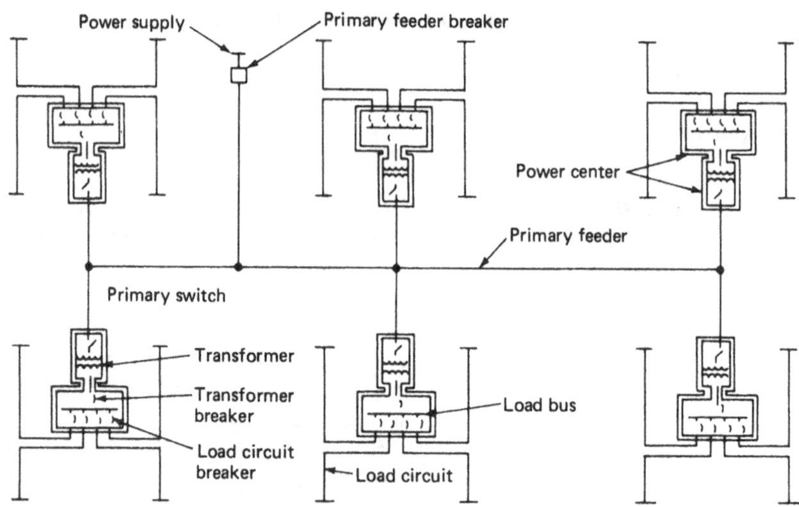

FIGURE 5.151 Modern simple radial system.

Unit substation sections usually are subassemblies designed for field connection into an integral single unit. Numerous types are available for both indoor and outdoor applications and are described fully in manufacturers' catalogs.

Unit substations can be either single-ended, fed by a single primary feed, or double-ended, fed from either end by a separate primary feed. Double-ended substations are usually designed so that either transformer can assume two thirds of the load in the event of a failure of one primary feeder. This is accomplished by the inclusion of a normally open tie circuit breaker in the switchgear lineup. The application of double-ended substations to industrial plants increases the reliability of the system and allows partial operation in the event of partial power failures. Plant operators must have a plan for immediate dumping of nonessential loads before closing the tie circuit breaker. This plan will keep essential processes functioning while power is down.

System Voltage. The voltage class of both primary and secondary distributions is called *nominal system voltage*. This term identifies the basic voltage normally utilized, such as 120/208 or 277/480 V. The actual voltage of each nominal system may be a slight variation such as 125/216 or 265/460 V. Each utility company uses its own selected secondary system. When a plant buys or generates primary power, the secondary voltage can be set very close to the nominal system rating.

Ranges for standard nominal system voltages are covered in ANSI C84.1-1977. Nominal system voltages most often found in the United States are indicated below.

120 V. A single-phase, two-wire system. Used for convenience outlets and incandescent lighting.

120/240 V. A single-phase, three-wire system. Nominal voltage between the two-phase conductors is 240 V. Nominal voltage from each phase conductor to ground is 120 V; used for power equipment, power outlets, electric heating processes, and in some cases high-intensity discharge lighting.

240 V. A three-phase, three-wire system, delta-connected, with 240 V between phase conductors and no ground or neutral conductor. Used for motor and three-phase power loads. This system is gradually being replaced by the more modern 120/208-V system with grounded neutral.

120/208 V. A three-phase, four-wire system, wye-connected with 208 V between phase wires and 120 V between phase and ground. This system permits single-phase, three-wire circuits to be taken from the system as well as 120-V single-phase circuits. The system is in general use for all types of loads. Recently, larger buildings have been designed to use a higher utilization voltage, but it often is converted to 120/480 V for convenience outlets and incandescent lighting.

277/480 V. A system similar in use and characteristics to 120/208 V for direct operation of motors, process equipment, and all forms of discharge lighting including fluorescent. Transformers are required to convert 277/480 to 120/208 V for uses described under that system.

4160 or 2400/4160 V. Sometimes used in large plants for internal distribution or for direct power operation on motors in excess of 250 hp.

Medium- and high-voltage systems are rarely used for distribution within plant buildings. In multibuilding sites higher distribution voltages may be found running between plant areas and between substations. A detailed discussion of medium- and high-voltage distribution or transmission systems is beyond the scope of this chapter.

Conductor Sizing and Load Growth

Wire and cables are usually sized exactly for the loads they serve. The sizes of feeders running to panelboards or switchboards usually allow for all spare circuits built into these devices, computed as if they were half-loaded.

Feeders serving grouped loads can usually have some additional load added to them because they are protected from overload by the natural diversity of equipment operation. Some study must be made of the actual use pattern of the equipment to determine the extent of the excess capacity available through this diversity.

All feeders serving either lighting, receptacle, or motor loads are sized to include a 25 percent factor to account for heating due to continuous loading. This factor must be maintained even when maximum loading is desired.

Most modern designers utilize only half a circuit's normal capacity during initial design; thus a 20-A receptacle circuit is loaded to only about 10 A initially. This approach permits addition of another 5 A of continuous load.

GROUNDING

The subject of electric system grounding is broad and complex, and is discussed here in brief overview.

Contemporary approaches hold that all power systems should have a grounded neutral included in the system. It is imperative for life safety that all metal elements of electric systems remain at ground potential at all times. Grounding should also take into account that buildings and equipment can build up a hazardous static charge of far greater magnitude than that encountered in winter when walking across a carpet and touching a doorknob. For tall structures or buildings located in isolated surroundings without other construction around, lightning can pose a potential hazard, which must be considered.

Ungrounded Systems

For many years industrial plants relied on an *ungrounded system,* essentially a delta-connected system without a grounded neutral. In this system, a single line-to-ground fault does not cause automatic circuit tripping. However, a second undetected ground on such a system can cause continuous nuisance tripping and even equipment burnout on devices not connected to the affected circuits. This is especially true on higher voltage systems. Ungrounded systems also pose the problem of transient overvoltages caused by grounded circuits. For all of these reasons there are now relatively few of these systems being installed, and existing ungrounded systems are being converted.

System Grounding

There are several types of grounded systems generally used in today's industrial plants. These are described briefly in the following paragraphs.

Resistance-Grounded Systems. Resistance-grounded systems are characterized by a resistance connection between system neutral and ground. This system introduces impedance in the ground path which tends to limit the current flow to ground. This technique also limits overvoltages caused by an intermittent-contact line-to-ground short circuit. Resistance is used to provide a ground-fault current which can be used for protective relaying operation.

Solidly Grounded Systems. Solidly grounded systems permit better control of overvoltages than any other scheme, but ground-fault currents can be higher.

These systems are used at operating voltages up to 600 V. The low line-to-neutral voltage reduces the risk of dangerous voltage gradients. A large magnitude ground-fault current helps attain optimum performance of phase-overcurrent protective devices.

Equipment Grounding

Equipment grounding is provided by a system of conductors utilized to maintain the metallic housings of electric system devices at ground potential. By grounding equipment in this manner, the system provides life safety and severely limits the fire hazard of short-circuit currents by providing a simple path to ground. This system should be periodically inspected and tested to maintain it in proper working order.

In general, it is an essential of a safe electric system that everything which might come into contact with a live system be maintained at ground potential. Further safety is gained by providing the system with a grounded neutral leg in the system.

System grounds can be derived from a ground attachment to a cold-water pipe ahead of the water meter or by a system of ground electrodes or some combination of these. A single ground point for the system should be established and periodically checked for continuity. Provision should be made to disconnect the system neutral during tests.

FAULT PROTECTION AND SYSTEM COORDINATION

Fault protection also is a complex subject. The principal types of faults which plague electric systems are three-phase short circuits, line-to-ground faults, and intermittent ground faults.

A three-phase bolted short circuit can be caused by any accident. The instant voltage fluctuations and the large overcurrent flowing in the system, coupled with the rapid decay of the system voltage, cause the circuit breaker to trip or the fuses to blow at the nearest point. The elapsed time for a circuit breaker to clear such a fault is from three to eight cycles depending on the size of the breaker. It is here that the design of the system is tested, because the circuit breaker is forced to open a circuit of far greater current than its trip rating. If breaker interrupting ratings are selected properly, clearing of three-phase bolted faults is simple. A fuse operates in the first one-half cycle. Fig. 5.152 illustrates a typical operating characteristic of current-limiting during a high-fault current interruption.

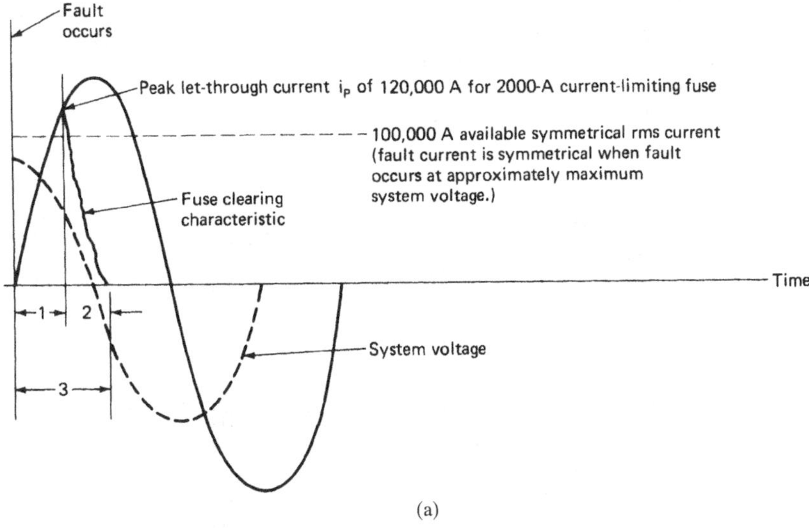

(a)

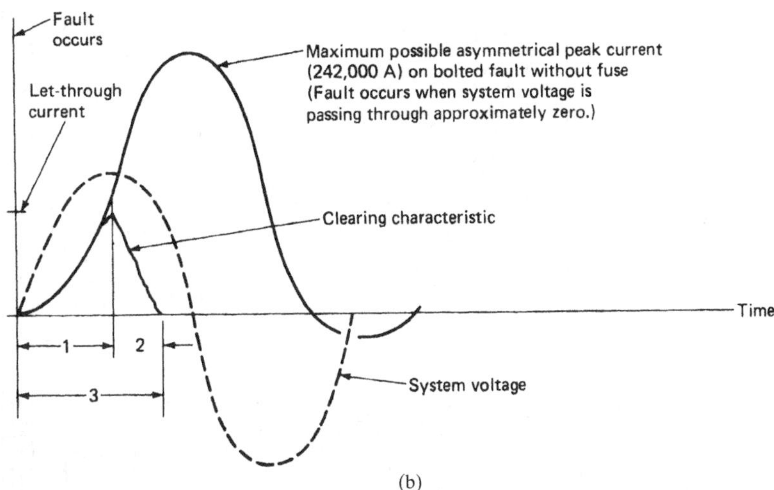

(b)

FIGURE 5.152 Typical current-limitation characteristics showing peak let-through and maximum prospective fault current as a function of the time of fault occurrence (100-kA available symmetrical rms current): (*a*) Fault occurring at peak voltage. (*b*) Fault occurring at zero voltage: 1 = melting time, 2 = arcing time, and 3 = total clearing time. *(Source: IEEE Standard 141-1976.)*

A solid line-to-ground fault causes system protective devices to react as they do for a three-phase short circuit.

The intermittent ground fault is the most difficult and therefore the most dangerous type of system fault. At no time does an overcurrent flow for a period long enough for the protective device to detect the problem and react. Intermittent ground faults are detected best by a *zero sequence* protection system. This system measures any current flow in the ground path and uses this current to operate a relay to cause the system protective device to operate and clear the fault.

Protective devices are usually coordinated so that the unit closest to the fault opens first. If the first unit in the system fails to clear the fault, the next one acts, and so on until the main opens, shutting the whole system down.

All circuit breakers, fuses, and most relays come with, or have available from the manufacturer, time-current operating curves. See Fig. 5.153 for comparative time-current curves of a

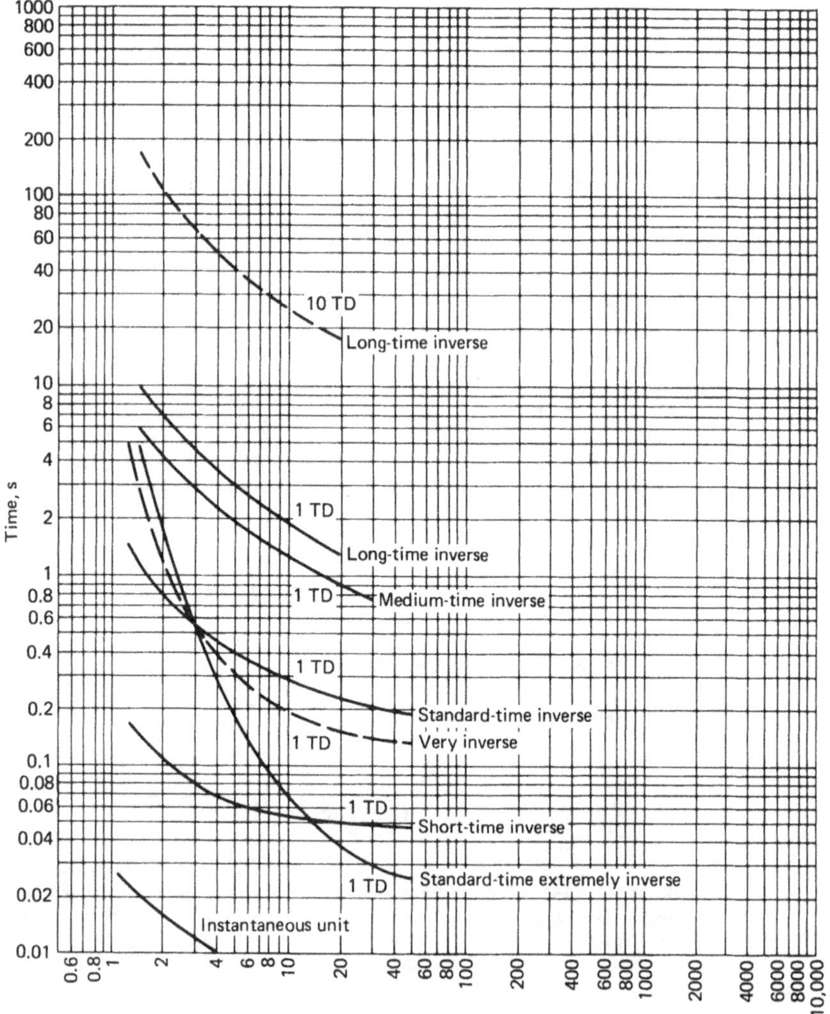

FIGURE 5.153 Comparative time-current curves of a typical induction overcurrent relay plus instantaneous trip.

typical induction overcurrent relay, plus instantaneous trip. By overlaying these curves and selecting services with coordinated operating times, it is possible to achieve the sequential operation needed.

Primary-system and even secondary-system protective relays are set to achieve this same system of sequential tripping.

It is incumbent on the operating personnel to never simply reclose a protective device that has opened on fault without first ascertaining the cause of the fault and either correcting it or removing the offending piece of equipment from the circuit.

SWITCHGEAR LOCATION AND WIRING

The best location for secondary switchboards is near the center of load. Frequently these are located on mezzanines above the plant floor, to isolate them from unauthorized personnel as well as to have them located near the load without taking valuable floor space.

In most industrial applications, a special enclosure is provided for a switchboard to isolate it from unauthorized personnel.

Wiring to equipment should be kept as short as possible, making load-center location of switchgear advantageous. Power switchboards are passive devices and require almost no operation. They are reset only if they trip under fault conditions.

Metering of utility company services can be done best at the single point of entry of service to the plant site. Metering of individual circuits for the purpose of monitoring energy use or some other function can be done anywhere in the plant. All such meters can be gathered in a single location with appropriate transmitters and telemetering equipment.

WIRING METHODS

Primary and Secondary Feeders

Two methods of running primary and secondary feeders in industrial plants are used: *overhead* and *underground.*

Overhead Distribution. Overhead distribution is used to interconnect buildings of the same facility which are separate from one another, when plants are operating on a temporary basis, or when a plant is more or less in a state of continual change.

Underground Distribution. Underground distribution is used for short distances, congested plant areas, or when the importance of high reliability justifies the extra expense.

A variety of raceways is available for distribution wiring. Rules governing the application and installation of raceways are discussed in various sections of the *National Electrical Code.** Adhering to these rules will help assure a safe, reliable installation.

Main feeders and *distribution feeders* for localized switchboards and panelboards usually are run in rigid metal conduit. Large power blocks are distributed in busways. Where plant floor use or equipment is changed frequently, it often is advantageous to distribute all main feeders in the form of interlocked armored cable run in cable trays.

The method of power distribution which is best for a given plant depends on characteristics of the plant itself. In some, workstations are fixed. In others, the production system is fluid and electrical demands change from product to product.

In the case of a fixed production system, electricity can be brought to the point of utilization in rigid conduit or by any of the other means already discussed. Plants which require more flexible power distribution often rely on a grid of busways which permits circuit-breaker-protected power takeoffs at any point.

* *National Electrical Code*® is a registered trademark of the National Fire Protection Association, Inc., Quincy, MA 02209.

Busways

Busway systems are widely used for feeders, plug-in subfeeders, and plug-in and trolley-type branch circuits.

They permit frequent changes in load layout, because they can be easily adapted to production-line rearrangement without seriously disrupting production.

Busway systems also have the benefit of controlled voltage drop characteristics, permitting them to meet different voltage stability requirements.

The types of busway commonly employed for industrial applications follow.

Feeder Busway. Low-impedance-type busway is frequently used for all types of high-capacity feeders and risers. Typical applications include transformer-vault-to-switchboard service runs, switchboard-to-load-center panel feeders, and welder feeders.

Plug-in Busway. Plug-in distribution busway is extremely flexible because it has easily accessible plug-in openings along its length, permitting tap-ins. It can be run from a switchboard or tapped from a run of feeder busway to carry power to closely spaced machines and other loads. Branch circuits to motors also can be tapped in.

UTILIZATION EQUIPMENT CONTROLS

Small equipment and hand tools are run from plug-in-type receptacles. These receptacles should be grounded and, for safety, equipped with ground-fault interrupter devices. Locking-type receptacles should be used for tools or equipment which are subject to extensive movement during use.

All equipment and machinery should be supplied with a disconnect device within sight of the machine, preferably within easy reach of the operator. This is a code requirement because it is mandatory for maintenance of a safe work space.

Utilization equipment, such as fans, pumps, oil burners, and lighting fixtures on branch circuits, can be controlled by contactors such as push-button switches, toggle or tumbler switches, and rotary snap-action switches.

Magnetic switches are widely used for motor control and where it is necessary to operate the switch contacts from remote push-button stations.

Small pieces of equipment and hand tools usually have controls built into them.

A variety of controls can be added to an existing plant to increase safety and/or reduce energy consumption and expense.

Microprocessor-based programmable controllers are available in different sizes, offering much diversity. Among the most popular is the so-called demand-based multifunction controller. Typically, this device provides demand control, remote start-stop, duty cycling, and optimal start-stop. The number of loads which can be controlled depends on the nature of the controller involved. In many cases they are modular, enabling addition of loads as needed. Devices such as these can be used for machinery, lighting, comfort cooling and heating, and other purposes. Each of the functions provided by a multifunction device generally can be performed by a smaller device which performs one function only, such as remote start-stop.

Timers and time-clock controls have been used for many years. Their function is generally to start and stop a certain operation at a predetermined time, for example, outdoor lighting.

Photocell controls are applied to lighting. As ambient lighting conditions fall to a certain predetermined point, the photocell activates lighting. When ambient lighting rises above that point, lighting is deactivated.

Dimming controls are available for fluorescent as well as mercury vapor lamps. They reduce light output to what is required for the tasks involved. The percentage of light reduction is close to the percentage of energy consumption reduction. In no case should lighting levels be reduced to less than what is required. Worker productivity will suffer; safety and security hazards may be created.

HAZARDOUS LOCATIONS

The dust and fumes put into the air by many industrial processes create an environment which is classified as electrically hazardous. Although the airborne substances may not be toxic, they can create an atmosphere which is subject to explosion and fire when heated by an electrical switching operation.

Electrical hazards are classified into three groups by the *National Electrical Code,* in the chapter titled "Hazardous Locations."* Each group is further subdivided into divisions by degree of hazard and type of substance present.

The basic classifications are as follows:

Class I. Air is contaminated by hazardous fumes or vapors including gases and airborne chemicals. This is subdivided into two divisions.

Division 1 Locations where hazardous vapors and gases can or do exist under normal operations.

Division 2 Areas where flammable vapors or gases are handled in proper containers but where hazardous concentrations are normally prevented by forced ventilation.

Class II. Similar to class I except it includes atmospheres containing or likely to contain combustible dust. The divisions of this class resemble those of class I.

Class III. This class of hazard deals with the presence of combustible materials in the manufacturing process and the presence of airborne particles created by the process. The divisions deal with the manufacturing site and the storage sites.

In addition to class and division of hazard, electrical hazards are further defined in each segment by group and designated by a letter such as "A," "B," etc.

The *National Electrical Code* accurately defines each class, division, and type of hazard. The methods of wiring required are also defined in the same Code.

The plant operator must be able to identify those areas of the plant presenting identifiable hazards and must maintain wiring systems which conform to the probable hazard present. Great skill is needed to keep to a minimum the electric devices actually located in the highest hazard areas. Careful arrangement keeps control and switching devices in the area with the lowest classification. Explosion-proof wiring is expensive. It is often less expensive to remove or reduce the hazard than it is to provide the appropriate electric system.

It is essential for the plant engineer to become familiar with the rules of the Code and also with other publications of the National Fire Protection Association that further define the steps which can reduce the hazards.

BIBLIOGRAPHY

1. Beeman, D. L. (ed.): *Industrial Power Systems Handbook,* McGraw-Hill, New York, 1955.
2. "Constructing Electrical Systems" (Special Report), *Electrical Construction and Maintenance,* May 1979, pp. 59–90, 164–170.
3. G.E. *Specifier's Guide,* General Electric Co., Schenectady, New York.
4. Gonen, T.: *Electric Power Distribution System Engineering,* McGraw-Hill, New York, 1986.
5. *IEEE Recommended Practice for Electric Power Distribution for Industrial Plants,* The Institute of Electrical and Electronics Engineers, Inc., New York, 1976.
6. *Industrial and Commercial Power Distribution Course,* The Electrification Council, New York, 1975.
7. Kurtz, E. B., and Shoemaker, T. M.: *The Lineman's and Cableman's Handbook,* 7th ed., McGraw-Hill, 1986.
8. McPartland, Joseph F., and William J. Novak: *Electrical Equipment Manual,* 3d ed., McGraw-Hill, New York, 1965.

9. *National Electrical Code*, National Fire Protection Association, Quincy, Massachusetts, 1981.

10. NECA Electrical Design Library: *Fault Detecting and Disconnecting Devices,* National Electrical Contractors Association, Washington, D.C., June 1977.

11. NECA Electrical Design Library: *Power Distribution Systems,* National Electrical Contractors Association, Washington, D.C., March 1977.

12. NECA Electrical Design Library: *Specifying Electrical Conductors,* National Electrical Contractors Association, Washington, D.C., June 1978.

13. *NECA Wiring Symbols Standard,* National Electrical Contractors Association, Washington, D.C., May 1976.

14. *Recommended Practice for Electrical Equipment Maintenance,* NFPA 70-B, National Fire Protection Association, Quincy, Massachusetts, 1977.

15. Smeaton, R. W.: *Switchgear Control Handbook,* 2d ed., McGraw-Hill, New York, 1987.

16. *System Neutral Grounding and Ground Fault Protection,* Westinghouse Relay Instrument Division, Westinghouse Electric Corporation, Coral Springs, Florida, January 1978.

17. Thumann, Albert: *Electrical Design, Safety, and Energy Conservation,* Fairmont Press, Atlanta, Georgia, 1976.

18. *Westinghouse Construction Specification,* 5th ed., Westinghouse Electric Corporation, Pittsburgh, Pennsylvania, 1978.

---

* *National Electrical Code*® is a registered trademark of the National Fire Protection Association, Inc., Quincy, MA 02269.

CHAPTER 5.12
STANDBY AND EMERGENCY POWER: ROTATING EQUIPMENT SYSTEMS

Jim Iverson
Manager, Technical Marketing
Industrial Business Group
Onan Corporation
Minneapolis, Minnesota

GLOSSARY

Automatic sequential paralleling system (ASPS) An emergency power supply system that operates multiple engine-generator sets in parallel. It allows the first set to reach acceptable speed after receiving the start command to be immediately connected to critical loads. It automatically synchronizes and connects remaining sets. It is often applied in health-care facilities with a 10-s transfer requirement.

Automatic transfer switch (ATS) An electric switching device that alternately connects the load to the normal or emergency power source as required without operator involvement. It may also include controls that start and stop the engine-generator set, monitor voltage and frequency of both power sources, and perform other timing and logic functions. See also nonautomatic transfer switch.

Nonautomatic transfer switch An electric switching device that alternately connects the load to the normal or emergency source as determined and initiated by an operator.

Prime mover The engine that drives the generator through permanent coupling. It may be either spark-ignited, diesel cycle, or gas turbine.

Paralleling switchgear Switchboard dedicated to control of engine-generators. Contains start sensor, synchronizer, reverse-power relay, and other controls. It closes generator output to the emergency bus through a fast-acting circuit breaker.

Paralleling load transfer equipment Switchboard to transfer load from the utility to a single generator set and back again in a nonload break operation using fast-acting circuit breakers. Contains start sensor, synchronizer, protective relaying, load sharing, and other controls. Paralleling load transfer equipment is often applied in facilities to control a standby generator set used for load curtailment or in conjunction with interruptible rate utility programs.

INTRODUCTION

The emergency power supply system (EPSS) encompasses a wide variety of equipment. Equipment under the EPSS heading ranges from a simple, self-contained battery light to a complex, highly engineered, multiple engine-generator set system with a capacity of several megawatts. These two extreme examples suggest a division of these systems based on the source of emergency power. The first example uses a stored-energy device, a battery, as the power source while the second example uses rotating equipment, engine-generator sets, as the power source. Stored-energy systems and rotating-equipment systems are the two categories of emergency power supply systems. The subject of this part is rotating-equipment systems.

SYSTEM DESCRIPTION

When the normal power source fails, the EPSS functions to supply electric power to specific, select loads. The equipment that composes the system is largely determined by the characteristics and requirements of the loads being served. Two equipment groups, the emergency power source and the electric switching equipment, subdivide the system based on function. While the two equipment groups have independent functions, the groups are interrelated and both serve the common purpose of the complete system.

Emergency Power Source

The function of this equipment is to generate electric power. The engine-generator set is the principal part of the group. The generator is permanently coupled to and driven by a prime mover which may be a diesel, gasoline, or gaseous engine or a gas turbine. Figure 5.154 illus-

FIGURE 5.154 Engine-generator set. (*Onan Corporation.*)

trates a typical diesel engine-generator set. Included in this group are an independent fuel supply with storage and the engine-generator set(s) with supporting equipment such as the governor, voltage regulator, exciter, cooling system, ventilation equipment, exhaust system, and engine control with meters and alarms.

Electric Switching Equipment

The function of the electric switching equipment is to interconnect the generator power with the utilization equipment. Included in this group are transfer switches, illustrated in Fig. 5.155, either automatic or nonautomatic. The transfer switch is interlocked to prevent simultaneous closing to both the normal and emergency power source. Automatic transfer switches also monitor both sources and initiate starting of the system. Other electric switching equipment includes bypass-isolation switches if needed, paralleling load transfer equipment, and in the case of parallel operation of multiple engine-generators, paralleling and totalizing switchboards.

SYSTEM CLASSIFICATION

Recognizing the diversity of applications for emergency power supply systems, the National Fire Protection Association (NFPA) Committee on Emergency Power Supplies (NFPA Standard 110) has adopted the following system definitions based on type, class, category, and level.

Type

Response time is the criterion for determining the system type. Types range from uninterruptible power supply systems that "float" on-line to power systems with no response time

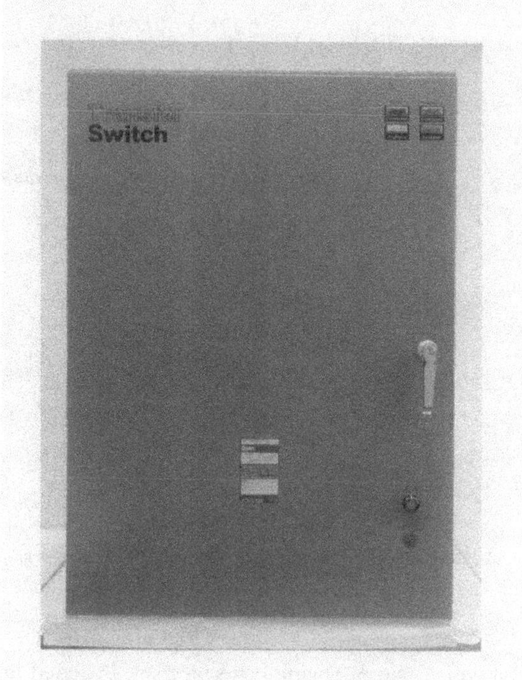

FIGURE 5.155 Transfer switch. (*Onan Corporation.*)

requirement. Systems with response times of 60 s or shorter generally are automatic systems, while systems with no response time limit are often manual-starting or portable.

Class

Systems are placed in a class based on the length of full-load operation possible without refueling or recharging. The extremes of class are those of very short duration, 5 min, to those systems with indefinite duration based on the user's needs. Properly designed rotating-equipment systems have the capability of unlimited operation.

Category

The NFPA Committee on Emergency Power Supplies defined two categories. One category is stored-energy systems that receive energy from the normal power source. The requirements for stored-energy systems are contained in NFPA Standard 110A. The second category is rotating-equipment systems that use engine-generator sets as the power source, and is covered in NFPA Standard 110.

Level

The critical nature of the load being served by the EPSS determines the system level. For example, a system that supplies loads which support human life and safety is the most critical level. The lowest level defines a system that supplies loads which would result in economic loss when without power. The level of a legally required system greatly influences the requirements the equipment must meet.

SYSTEM JUSTIFICATION

Loss prevention, safety, and legal requirements are three common justifications for an EPSS. A fourth justification, energy economics, is becoming more common as the cost of energy rises. Many electric utilities offer interruptible or load curtailment rates, which can justify the cost of an engine-generator set and either a transfer switch or paralleling load transfer equipment. The significant savings on insurance also justifies the cost.

Loss Prevention

An EPSS in an industrial plant can prevent several kinds of measurable loss. Some examples are the loss of wages during production downtime, the loss of in-process product, the loss of data processing, and the loss of refrigerated storage. These examples, of course, are not exhaustive.

Safety

Loss of electric power can directly threaten personnel safety. Industrial processes that present a hazard when without power are one example. Emergency power is also needed to operate elevators, fire pumps, fire alarms, communication networks, and other safety-related equipment.

Legal Requirement

Most states and some major cities have adopted codes that require EPSSs in specific buildings. The classification of the required system is determined by the building occupancy. Because legal requirements change often and are different from state to state, check the current local regulations. A good source of additional information is the local inspector.

Energy Economics

Peak shaving, interruptible or load curtailment rate programs, cogeneration, and heat reclamation are methods for saving energy costs that are becoming feasible as the cost of energy rises. Peak shaving uses the emergency system to reduce utility demand charges by assuming loads during periods of peak demand. Cogeneration uses heat from the engines to make steam or hot water for industrial processes. Heat reclamation captures waste heat from the engine through heat exchangers for hot water or space heating or for cooling with absorption chillers. Many utilities offer interruptible rate programs which pay customers to either curtail load or switch to on-site generation to relieve demand on the utility.

SYSTEM CONSIDERATIONS

Most system considerations can be grouped under three headings: reliability, capacity, and quality.

Reliability

System reliability is the capability of the EPSS to fulfill its intended purpose predictably without problems. The degree of reliability necessary for an emergency system is closely related to the critical nature of the load being served. Many factors contribute to system reliability; those given here are preinstallation considerations. After installation, effective maintenance, exercise, and experienced operators contribute to reliability.

1. *Compatibility.* Since all the equipment in the system is interrelated and must function toward a common purpose, it is important that each piece of equipment be coordinated with the rest of the system. A single design source for the complete system helps to ensure compatibility.
2. *Testing.* Prototype testing during the development process of the engine-generator set ensures the fitness of the equipment for its intended purpose. Prototype testing helps to ensure performance under a variety of conditions such as overloads, surges, short circuits, and others that may occur in actual service. Testing of the complete system for compatibility and performance as specified before installation also increases reliability.

Capacity

One of the most important considerations for an EPSS is capacity, or the amount of power the system can generate. Determination of capacity requires careful study of the load equipment. List the voltage, current, and power requirements of each piece of load equipment. Include pertinent information about the load such as inertia, power factor, starting method, and other data which would affect capacity. Table 5.42 shows a load tabulation example.

Choose between full protection or selected protection. Full protection means a larger system, and selected protection means less capacity and special wiring costs. If needs increase in the future, consideration given to expansion can pay off.

Quality

The EPSS should generate power that is essentially equal in quality to the power of the normal source. Requirements such as frequency regulation, voltage regulation, waveform deviation, harmonic content, and noise interference define the quality of the emergency power.

REPRESENTATIVE SYSTEM

Illustrated in the simplified block diagram, Fig. 5.156, is a typical automatic sequential paralleling emergency system. This system, designed to meet NFPA-99 and -110 requirements, consists of engine-generator sets, paralleling switchboards, totalizing board, and automatic transfer switches. In this system, three 400-kW engine generators furnish 1200 kW of power. The loads, 1200 kW total, are divided into three priorities of 400 kW each, matching the capacity of one engine-generator. If utility power is interrupted, an automatic transfer switch initiates starting of all three engine-generator sets simultaneously. When the first of the three engine-generators reaches operating voltage and frequency, the corresponding paralleling switchboard closes its circuit breaker, connecting the generator power to the emergency bus. With one generator on-line, the priority control enables the first priority transfer switches and the most critical loads to receive power. The shaded line in Fig. 5.156 represents this power distribution. The second and third engine-generators are automati-

TABLE 5.42 Emergency Electric Power Supplies—Load Tabulations for Example Location or Building (Onan-Fridley); Nominal Voltage 120/208, Frequency 60 Hz

Load (block) functional description	Load				Phase	Circuit no.	Phases used	Static loads			Running loads				Motor starting data				Starting sequence	Notes
	Type	Class	Category	Level				kW	PF*	A	hp output	kW input	PF	A	NEMA motor code	Locked rotor kVA	Locked rotor PF	Locked rotor kW		
Incandescent lights, exit—evacuation	10	2	A	1	1	5	A-N	2	1.0	–17									1	
Fluorescent lights, office—general	60	8	B	2	1	7	A-B-C	5.5	0.95	16									3	
Fire pump, 1	10	2	B	1	3	2	A-B-C				40	35	0.85	114	F	212	0.4	84.8	1	Across-line starting
Pump chiller, 1	60	8	B	3	3	7	A-B-C				40	35	0.85	114	F	212	0.4	84.8	3	Reduced-voltage starting, series reactor, 80%, closed transition
Elevators, 3 each	10	8	B	1	3	4	A-B-C				20	17.5	0.85	57	F	106	0.4	42.4	1	SCR controlled, one elevator at a time
Air compressor, 1	60	8	B	2	3	6	A-B-C				20	17.5	0.85	57	F	106	0.4	42.4	2	Compression relief during starting, low-inertia load

* PF = power factor

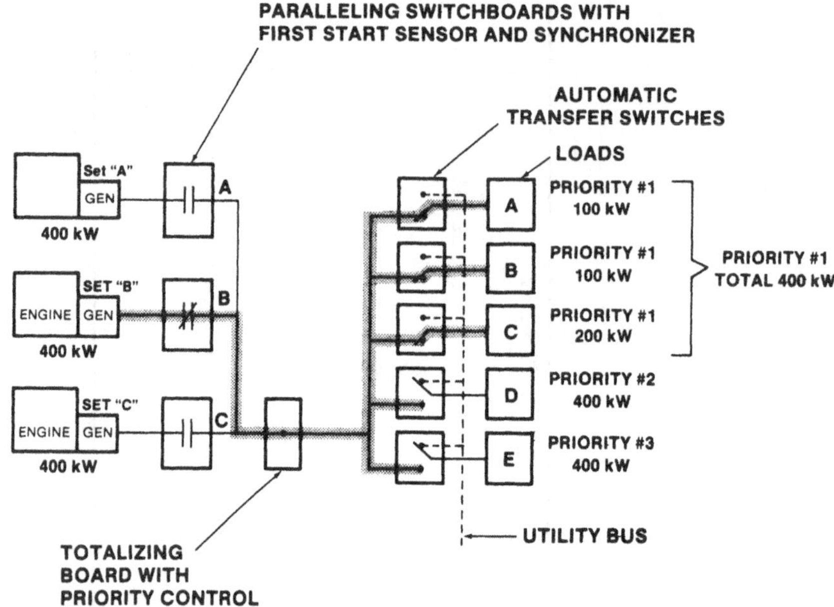

FIGURE 5.156 Automatic sequential paralleling system. (*Onan Corporation.*)

cally synchronized with and closed to the emergency bus. As each successive generator is closed to the bus, the priority control enables the second- and third-priority transfer switches to operate.

COMPARISON OF PRIME MOVERS

Table 5.43 reviews the advantages and disadvantages of various types of systems.

Standard Units

Major components of an EPSS include a generator set, transfer switch control, and if applicable, paralleling switchgear. Units discussed are representative only of the sizes and features generally available from EPSS manufacturers.

Generator sets suitable for use in EPSSs include diesel units ranging from 12 to 1500 kW, and spark-ignited units from 5 to 100 kW; rated at 1800 r/min. Three-phase generators are available throughout the range, and single-phase generators are available up to 125 kW typically. Standard cooling systems are set-mounted radiators, with air-cooled models available at the low end of the range. For this type of service, a generator set should have at least a battery-charging alternator, battery-charge-rate ammeter, oil-pressure gauge, coolant-temperature gauge, and a run-stop switch on the engine control (Fig. 5.157). Desirable features, some of which may be required by codes, include an engine-cranking limiter, low-oil-pressure shutdown, high-coolant-temperature shutdown (water cooling), and running-time meter. Water-cooled engines will generally have available as options heat exchanger, city water, or remote radiator systems, and water-cooled exhaust manifolds.

Common sizes of transfer switch controls range from 40 to 3000 A. An automatic transfer switch offers completely automatic, unattended operation and can include time delays for

TABLE 5.43 Advantages and Disadvantages of Various Types of Prime Movers

Advantages	Disadvantages
Gasoline reciprocating	
Lower initial cost than diesel Quick starting, especially in low ambients Lightweight	Fuel storage Gasoline deteriorates over time Lower thermal efficiency than diesel
Gaseous reciprocating	
Lower initial cost than diesel Fuel does not deteriorate over time More efficient combustion than gasoline Lower maintenance needs Easy starting	Requires high Btu content (1100 Btu/ft^3) or derating necessary May not be permitted in seismic risk areas without backup fuel storage
Diesel reciprocating	
Low maintenance requirements Easy fuel storage Low operation costs Good thermal efficiency Availability in wide range of kW ratings	Higher initial cost than gasoline or gaseous
Gas turbine	
Lightweight Smaller than equivalent diesel Low vibration, low noise No cooling water required Adaptable for cogeneration Multiple fuel capability Low maintenance	Higher initial cost Longer start time Poor partial load efficiency

engine starting, load transfer to the emergency power source, retransfer of load to the normal power source, and stopping of the engine (Fig. 5.158). It often has voltage sensors for sensing either undervoltage or overvoltage conditions of the normal source voltage and usually for sensing undervoltage conditions of only the emergency source voltage. Operation indicator lamps and meters are also available, if not standard features. An exerciser which automatically test-starts the generator set on a regularly scheduled basis is a popular option.

Paralleling switchgear includes, for each generator set of the paralleling system, an ac ammeter, ac voltmeter, frequency meter, wattmeter, synchronizing lights, a circuit breaker for connecting generator output to the bus, and voltage and frequency adjustment controls. An ac ammeter, an ac voltmeter, and a wattmeter are also connected to the bus for readings of the total paralleling system output. While manual paralleling or automatic paralleling systems are available, the use of automatic systems is more prevalent. (Automatic paralleling switchgear has provisions for manual paralleling if necessary.)

SYSTEM PREPAREDNESS

Once an EPSS is correctly installed and correctly interconnected, system failure results usually from lack of maintenance support. Whether manual or automatic system start-up, preparedness depends on a maintenance program, system exercise, and competent maintenance personnel.

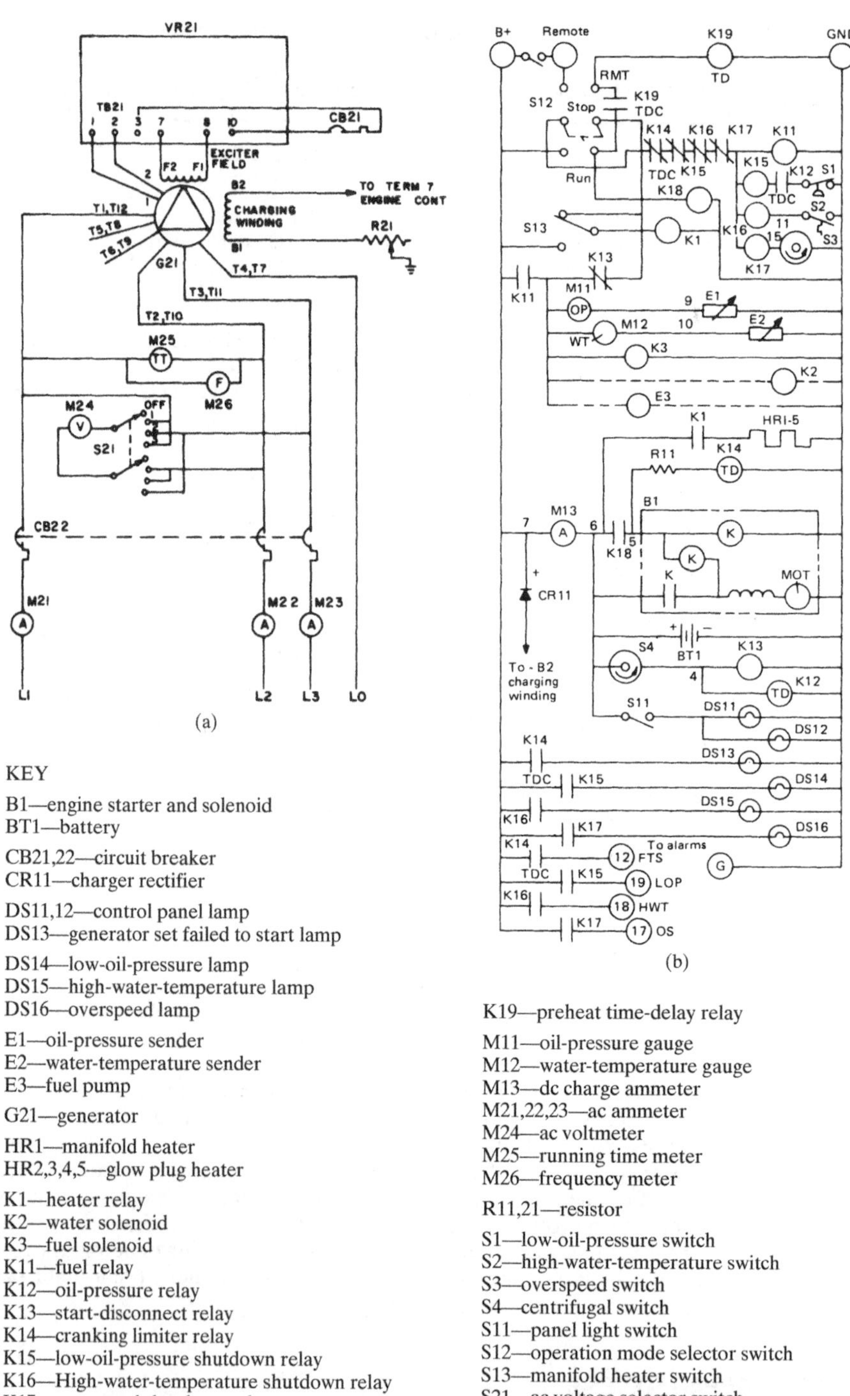

KEY

B1—engine starter and solenoid
BT1—battery

CB21,22—circuit breaker
CR11—charger rectifier

DS11,12—control panel lamp
DS13—generator set failed to start lamp

DS14—low-oil-pressure lamp
DS15—high-water-temperature lamp
DS16—overspeed lamp

E1—oil-pressure sender
E2—water-temperature sender
E3—fuel pump

G21—generator

HR1—manifold heater
HR2,3,4,5—glow plug heater

K1—heater relay
K2—water solenoid
K3—fuel solenoid
K11—fuel relay
K12—oil-pressure relay
K13—start-disconnect relay
K14—cranking limiter relay
K15—low-oil-pressure shutdown relay
K16—High-water-temperature shutdown relay
K17—overspeed shutdown relay
K18—starter pilot relay

K19—preheat time-delay relay

M11—oil-pressure gauge
M12—water-temperature gauge
M13—dc charge ammeter
M21,22,23—ac ammeter
M24—ac voltmeter
M25—running time meter
M26—frequency meter

R11,21—resistor

S1—low-oil-pressure switch
S2—high-water-temperature switch
S3—overspeed switch
S4—centrifugal switch
S11—panel light switch
S12—operation mode selector switch
S13—manifold heater switch
S21—ac voltage selector switch

VR21—voltage regulator

FIGURE 5.157 Schematic diagram of generator set. (*a*) AC generator set. (*b*) DC generator set control. All components are shown in the deenergized position. FTS—failed to start; LOP—low oil pressure; HWT—high water temperature; OS—overspeed. (*Onan Corporation.*)

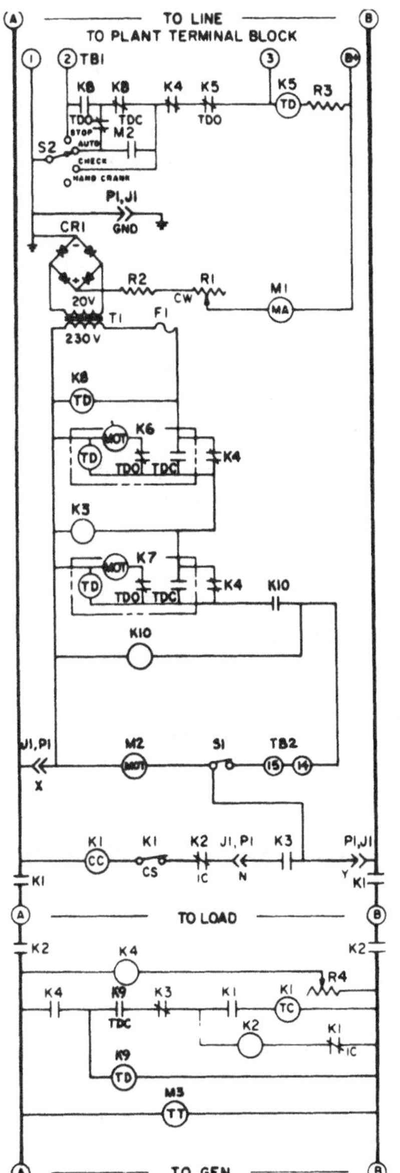

FIGURE 5.158 Schematic diagram of transfer switch. (*Onan Corporation.*)

Maintenance Program

Personnel establishing the maintenance program must prepare a maintenance performance schedule for the complete EPSS. Next, they must establish a means to document the system history of maintenance and service performed.

Although maintenance personnel prepare the maintenance schedule, they must include manufacturer's recommendations of maintenance items and schedules. Due to the infrequent nature of operation for these systems, some maintenance items require service intervals of operational hours while other items require time intervals of days, weeks, months, or years. Some items require both. (The EPSS should have a running time meter to indicate hours of operation.)

The following gives a representative listing of maintenance items and maintenance intervals. Use only as a guideline for recognizing particular system maintenance needs and for establishing a schedule. A health-care facility, for example, might require additional items.

Every 8 Operational Hours

1. Check coolant level (water cooling).*
2. Check crankcase oil level.* Wait 15 min after shutdown for accuracy.
3. Check oil sump level (turbine).
4. Visually inspect generator set. Look for fuel, oil, or coolant leaks. Check exhaust if possible with generator set running. Note security of hardware and fittings.
5. Check fuel level.

Every 50 Operational Hours

1. Check air cleaner.* Perform more often in extremely dusty conditions. Replace if necessary.
2. Inspect governor and carburetor-injector-pump linkage. Clean if necessary. Perform more often in extremely dusty conditions.
3. Drain fuel filter sediment.

Every 100 Operational Hours

1. Clean and inspect crankcase breather.*
2. Change engine crankcase oil.* Change oil at least every 3 months, more often in extremely dusty conditions.
3. Replace engine oil filter element.* Coincide with engine oil changes.
4. Clean engine-cooling fins (air cooling).*

* Reciprocating engine only.

Every 250 Operational Hours

1. Replace fuel filter element.* For diesel fuel systems with two filters, the second filter from main fuel tank usually needs replacement after several thousand hours.
2. Inspect battery charging alternator.*
3. Replace the ignition points and spark plugs; time ignition (spark ignition).*
4. Check water filter (if equipped).*

Every 500 Operational Hours (Turbine Engine Only)

1. Replace fuel filter.
2. Clean and inspect fuel drain valve.
3. Replace oil filter element.
4. Clean and inspect combustion chamber liner assembly (liquid fuel).
5. Clean and inspect fuel nozzle assembly (liquid fuel).
6. Change oil. Turbine engine manufacturer may allow longer oil change periods if oil sample tests are performed and oil meets engine manufacturer's specifications.

Every 1000 Operational Hours

1. Check generator brushes (if applicable). Brushes must not stick in brush holders.
2. Clean generator. Blow out with low-pressure, filtered, compressed air.

Every 2500 Operational Hours

1. Clean and inspect combustion chamber liner assembly (gaseous fuel).
2. Clean and inspect fuel nozzle assembly (gaseous fuel).
3. Clean and inspect igniter plug.

In contrast to operational hours, the following items usually require inspection or maintenance on a regular basis. Time intervals are week, month, half-year, or year.

Every Week

1. Main fuel tank level—keep full as much as possible.
2. Day tank fuel level.
3. Coolant level (water cooling)—coolant should have rust inhibitor and antifreeze, if applicable.
4. Fan and alternator belts.
5. Hoses and connections.
6. Coolant heater operation (if applicable).
7. Oil heater operation (if applicable).
8. Batteries—check cleanliness, electrolyte level, and cable connections.
9. Battery charger—note charge rate.
10. Exhaust condensation trap—drain out water.
11. Emergency power supply area—note general cleanliness. Wipe down entire system. For an exceptionally clean area, longer intervals could be used.

---

* Reciprocating engine only.

12. Running tests. Start the generator set and note following (load is preferable):
    a. Fuel system—check operation of fuel solenoid auxiliary fuel pump (if applicable) and general fuel system operation.
    b. Lubrication system—note engine oil pressure and record.
    c. Exhaust system—inspect for tight connections and leaks. Note condition of muffler, exhaust line, and exhaust support.
    d. Cooling system—note operating temperature and record (engine must run long enough to warm up).
    e. Battery charging—note charge rate of generator set.
    f. Meters—note general operation.
13. System documentation—check that operation manual, wiring diagram, maintenance schedule, and log are accessible to maintenance personnel.

Every Month

1. Cooling system (water cooling)—inspect for adequate water flow. Remove any material which interferes with radiator airflow, etc.
2. Ventilation—air inlets and outlets should have unrestricted airflow. Check security of duct work. Check operation of any motor-operated louvres.
3. Fuel system—drain water from main fuel tank and day tanks if applicable. Check fuel tank vents.
4. Battery—check specific gravity of electrolyte. Clean battery terminals.
5. System operation indicator lamps—test lamps with test switch, if equipped.
6. Transfer switch and paralleling switchgear (if applicable); inside cabinets should be clean and free of foreign objects. Check appearance of wiring insulation and color of terminals.

Every 6 Months

1. Cooling system (water cooling)—check for rust and scale. If necessary, flush out system and replace coolant.
2. Engine alarm shutdown devices.
3. Transfer switch control—inspect components and check settings of time delays, voltage sensors, and exerciser, if applicable. Clean cabinet with low-pressure, filtered, compressed air.
4. Generator set control—clean interior with low-pressure, filtered, compressed air.
5. Paralleling switchgear (if applicable)—inspect components, bus bars, and feeder connections. Clean cabinet with low-pressure, filtered, compressed air.

Every Year

1. Generator—measure insulation resistances of windings with a Megger®. Record readings.
2. Paralleling switchgear (if applicable)—perform insulation tests and record.

Maintenance Records

Records of maintenance and service performed on the emergency power supply system have two main benefits. They help to ensure that maintenance procedures were performed, and they provide an excellent system history. A maintenance record form should have entry provisions for the date, maintenance work performed, personnel involved, and general comments (form could also include provisions for labor and parts costs). It will show if schedules were met and if authorized personnel performed the maintenance.

Maintenance history can point out repeating problems or symptoms of a problem in the system. It can become a communication reference with a manufacturer for troubleshooting, for repair, or for warranty purposes.

System Exercise

Most engines left idle for long periods of time will have difficulty starting. For this same reason, the EPSS needs a regularly scheduled exercise program to promote operation readiness. The program might utilize either an automatic exercise feature of a transfer switch control or manually initiated testing.

Frequent system use especially benefits the generator set. It evaporates water from the lubrication system and generator windings and coats internal moving parts with a film of oil.

Exercise, with load if possible, for at least 30 min should evaporate water in the lubrication system and minimize engine carbon buildup and exhaust-system fouling. System exercise should occur at least once a week.

An automatic exerciser can have settings for the length and number of unattended exercise periods. Manually initiated testing gives the same system benefits as automatic exercising, except that it does require the presence of maintenance personnel. However, personnel can use this time to increase their own system familiarity, to train other personnel, or to perform system inspection.

Competent Maintenance Personnel

One popular method to obtain competent maintenance personnel is through maintenance contracts. The manufacturer or manufacturer's representative offering the contract service usually ensures that personnel performing the maintenance are trained for the equipment and that maintenance procedures meet agreed schedules. If plant personnel perform maintenance procedures, they should receive prior equipment training. (Some manufacturers of emergency power supply systems offer such service schools or training sessions.)

REFERENCE STANDARDS

"Emergency and Standby Power Systems," NFPA 110.

"Electrical Power Distribution for Industrial Plants," IEEE Standard 141-1976.

"Recommended Practice for Emergency and Standby Power Systems," IEEE Standard 446-1987.

"Motors and Generators," ANSI/NEMA Standards Publication MG1-1978.

"Stationary Combustion Engines and Gas Turbines," NFPA 37.

"*National Electrical Code*," NFPA 70.*

"Essential Electrical Systems for Health Care Facilities," NFPA 99.

"*Life Safety Code*," NFPA 101.*

BIBLIOGRAPHY

Stromme, Georg: "Coordination Procedures for On-Site Power Generators," *Specifying Engineer,* May 1977, pp. 116–119.

Stromme, Georg: "Emergency and Standby Power Systems," *Building Operation Management,* July 1977.

* *National Electrical Code*® and *Life Safety Code*® are registered trademarks of the National Fire Protection Association, Inc., Quincy, MA 02269.

CHAPTER 5.13

STANDBY AND EMERGENCY POWER: BATTERIES*

General Battery Corporation
Reading, Pennsylvania

GLOSSARY

Alkaline cell Primary cell with excellent leakage protection capable of higher energy output than carbon-zinc cells.

Ampere-hour A unit of electricity (symbol A·h) equal to the current flowing past any point in a circuit for 1 h at a constant amperage.

Ampere-hour capacity The number of ampere-hours which a cell delivers under specified conditions of discharge rate, temperature, initial specific gravity, and final voltage.

Carbon-zinc cell Low-cost primary cell with moderate leakage protection and low energy output.

Cycle A discharge and its subsequent recharge.

Cycle service A type of battery operation in which a battery is repeatedly discharged and recharged during the life of the battery.

Equalizing charge An extended charge given to a cell or battery to ensure the complete restoration of the active materials within a cell to the fully charged condition.

---

* Updated for the second edition by the Editor-in-Chief.

Final voltage A prescribed voltage at which a discharge is to be terminated, usually chosen to realize optimum useful capacity without overly discharging the battery.

Finish rate The maximum value of current at which a charged or nearly charged cell or battery may be charged without causing excessive gassing or heating equal to 5 A per 100 A·h of 6- or 8-h rating.

Float service A method of battery operation in which a battery is continuously connected to a bus whose voltage is set slightly higher than the open-circuit voltage of the battery. Under these conditions the battery will either charge or discharge into the load, according to the fluctuations in the bus voltage occasioned by varying load conditions. The bus voltage is set to maintain the battery during normal operation in a fully charged condition with a minimum of overcharging.

Gassing The evolution of gases from either the positive or negative plate of a cell.

Level indicators A float or visible reference mark used to indicate the electrolyte level within a cell.

Nickel-cadmium cell A secondary cell commonly used in portable power applications.

Nominal voltage The voltage rating of a cell or battery arbitrarily assigned to a cell type for the purpose of establishing the operating voltage range. For example, the nominal voltage of a lead-acid cell is 2 V and the nominal voltage of a nickel-cadmium cell is 1.2 V.

Specific gravity The ratio of the density of the electrolyte to the density of water at standard conditions. The specific gravity (spgr) of battery electrolyte is usually measured with a hydrometer. Temperature, water loss, and state of charge will all affect the specific gravity of a cell.

INTRODUCTION

A *battery* is a device consisting of one or more cells that store chemical energy which can be converted into electric energy on demand. The unit of measure of this electric energy is the kilowatthour (kWh), but battery output is often rated in ampere-hour capacity (A·h) because it is an easily measured quantity used to indicate the work capability. A *cell* is the smallest unit a battery can consist of. The minimum components of a cell are two dissimilar electrodes, an electrolyte, a means of conducting the electric power from the cell, and a container. Other components such as separators, covers, and vents are added to improve performance, life, or usage of the cell. The capacity of a battery depends on the internal construction of the cells. The voltage of the battery is the sum of the voltage of each cell connected in series. Cell voltage is a function of the electrode and electrolyte material.

A *primary cell* is a cell that cannot be easily recharged because the electrochemical reaction is nonreversible. The major types of primary cells are carbon-zinc (CZn) and alkaline. These batteries are best used in applications where long shelf life, low current draw, infrequent use, and low initial cost are important. The disadvantages of these cells include low output current, high voltage drops at high current, and the inability to be recharged. The selection of the proper type will depend upon cost, energy output, current draw, frequency of use, and amount of leakage protection required. Primary cells can be found in flashlights, instrumentation, alarm systems, cameras, and many portable low-power devices.

Secondary cells are fully reversible. The chemical energy can readily be restored by supplying electric energy to the cell in a process called *recharging.* Cells can store only a limited amount of energy. No amount of overcharging will store additional energy. The lead-acid battery is the most common form of secondary cell; it provides most of the traction, stationary, and engine-starting requirements of industry. Advances in technology have resulted in sealed lead-acid batteries being used more frequently in portable power applications. The other type of secondary cell used in industrial plants is the nickel-cadmium (NiCd) battery. It is occasionally used in traction, stationary, and engine-starting applications when cost is not a major factor. The nickel-cadmium battery is frequently used in portable power applications because of its ability to provide lightweight, high-current output in repeated-cycle service for a reasonable cost.

BATTERY CLASSIFICATION BY USAGE

An industrial plant usually has many different types of batteries in a great variety of sizes and shapes. Batteries can range from a single cell weighing a few ounces to a large battery that fills a room and weighs many tons. The electric output of batteries can range from a few milliwatts to hundreds of kilowatts. Industrial plants frequently use batteries for portable power, motive power, engine cranking, and/or stationary applications.

Portable Power

These cells and batteries are designed to be easily carried and supply energy requirements for portable lighting, power tools, instrumentations, communications, and alarm signals. They are sealed to prevent leakage and can be either nonrechargeable (primary cell) or rechargeable (secondary cell) depending on load and usage. These batteries usually provide intermittent or steady low-current power for long periods of time. The life can range from a few hours to many years depending on load and cycle conditions. Table 5.44 lists typical ratings used in the selection of portable power cells.

TABLE 5.44 Typical Range of Ratings of Portable Power Cells

	Primary	Secondary
Weight, lb (kg)	0.1–5.0 (0.045–2.3)	0.1–5.0 (0.045–2.3)
Volume, in^3 (cm^3)	0.5–200 (6–2500)	1–200 (12–2500)
Voltage per cell	1.3–1.5	1.2–2.0
5-h capacity, A · h	0.1–6.0	0.1–10.0
No. of cycles	None	100–1000
Cost, dollars per cell*	1–20	2–40
General types	Carbon	Nickel-cadmium
	Mercury	Lead-acid (GEL)
	Alkaline	Lead-acid (sealed)

* 1993 dollars.

Motive Power

These batteries are designed for repeated-cycle service supplying the energy to propel and operate electrically powered industrial trucks, sweepers, scrubbers, personnel carriers, mine equipment, and over-the-road electric vehicles. Lead-acid industrial, golf cart, and automotive batteries are the most frequently used types for this service. These batteries usually provide intermittent moderate and high-current power for 3 to 10 h to and 80 percent depth of discharge between recharges. They are discharged 3 to 10 times a week with a life ranging from 1 to 7 yrs. Table 5.45 lists data on and ratings of typical motive power batteries. Figure 5.159 shows the internal construction of a typical motive power cell. This diagram points out some of the important design criteria used to achieve good performance and long life in motive power service.

TABLE 5.45 Typical Range of Ratings of Standard Motive Power Batteries

	Minimum	Maximum
Weight, lb (kg)	30 (14)	7,000 (3,200)
Length, in (cm)	10 (25)	60 (150)
Width, in (cm)	7 (0.8)	45 (120)
Height, in (cm)	9 (23)	36 (90)
Voltage per battery, V	6	96
No. cells	3	48
6-h capacity, A · h	50	2,000
Cycles	1000	2,000
Life, years	1	7
Cost, dollars per battery*	100	15,000
General types:		
Industrial truck, mine, golf cart, automotive, electric vehicle		

* 1993 dollars.

Engine Cranking

These batteries are designed to furnish the electric energy requirements of internal combustion engines used in vehicular and stationary applications. These requirements include starting, lighting, and ignition (SLI). Some nickel-cadmium batteries are used in this type of service but most are either automotive or industrial lead-acid storage batteries. These batteries usually provide high-current power for a very short period of time during the engine-starting process. This results in a very shallow depth of discharge and very many cycles. Table 5.46 lists information and ratings of typical engine-starting batteries.

TABLE 5.46 Typical Range of Ratings of Standard Engine-Cranking Batteries

	Minimum	Maximum
Weight, lb (kg)	5 (2.3)	1,600 (700)
Length, in (cm)	4 (10)	45 (115)
Width, in (cm)	3 (7.5)	30 (76)
Height, in (cm)	3 (7.5)	20 (50)
Voltage per battery, V	6	64
No. cells	3	32
Cranking rate, A	50	3,500
Life, years	2	8
Cost, dollars per battery*	30	12,000
General types:		
Motorcycles, automotive, truck, marine, aircraft, diesel locomotive, and stationary diesel engine		

* 1993 dollars.

Stationary

These batteries are designed to be permanently installed on supporting racks and operated in float service as the backup or emergency energy source for communication systems, switchgear and control equipment, emergency dc power for lighting and essential equipment,

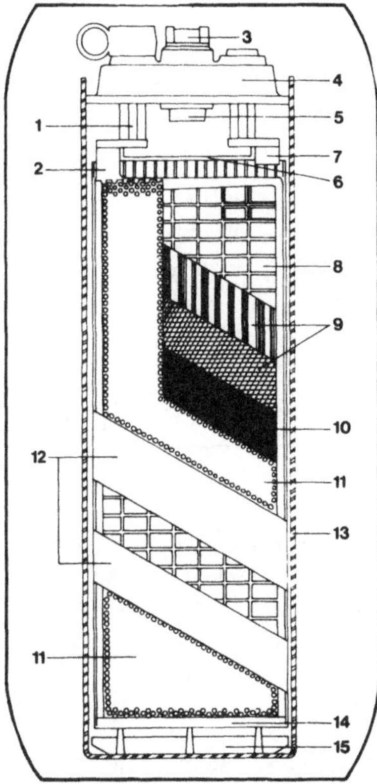

FIGURE 5.159 Internal construction of typical motive power cell. Key: 1, terminal post; 2, positive grid; 3, vent cap; 4, cover; 5, acid level indicator; 6, separator protector; 7, negative grid; 8, active material; 9, positive active material retention material; 10, positive active material retention material; 11, positive active material retention material; 12, separator; 13, container or jar; 14, positive active material retention material; 15, plate rest and sediment area.

and uninterrupted power supplies (UPS). The most common types of stationary batteries are antimony and calcium lead-acid. A few nickel-cadmium and Edison batteries are also found in this service. The power requirements of stationary batteries vary depending on the service. Carefully selecting a battery to match the load and cycle requirements of a particular operation is very important in achieving the desired life. Stationary batteries are often custom-designed to match the usage requirements. Table 5.47 lists information about and ratings of typical stationary cells.

BATTERY CHARGING

There are many ways to charge a battery. Modern chargers are becoming more and more automated, reducing the maintenance needs and increasing the life of the battery. The charging methods used will depend on the type of service and battery. Battery recharge should always be conducted to return the correct amount of charge. Both *overcharging* and *undercharging* are detrimental to battery life.

Constant Current

This is a charge conducted at a constant rate. The rate is usually at or below the finish rate. A *trickle charge* is a low-rate constant-current charge given to maintain the battery at a fully charged condition. This should be used only when the charging rate is matched to the battery. Some portable power and a few small-size stationary batteries use the trickle-charge method.

Two-Step Charging

This is a motive power charge technique commonly used on motor generators. The type of charge consists of a high-rate charge followed by a lower finishing rate charge. The finish rate charge is often initiated by a temperature voltage relay (TVR) which detects the gassing voltage. A timer to limit the length of the finish rate charge is often started by the TVR in order to limit overcharging.

Modified Constant Potential

This is a charging method frequently used in motive power service to automatically regulate the charging rate throughout a recharge. Ferromagnetic circuits are commonly used to control the initial rate, taper, and finishing rate of the recharge over a broad range of conditions.

Constant Voltage (Potential)

This charging method is frequently used with engine cranking and stationary batteries employing a charger with a voltage which is maintained at a constant value. The charging cur-

TABLE 5.47 Typical Range of Ratings of Standard Stationary Cells

	Minimum	Maximum
Weight per cell, lb (kg)	10 (4.5)	1800 (800)
Length per cell, in (cm)	3 (7.5)	19 (48)
Width per cell, in (cm)	3 (7.5)	18 (46)
Height per cell, in (cm)	6 (15)	60 (150)
No. cells	3	120
Voltage per battery, V	6	250
8-h capacity, A · h	10	8000
Life, years	5	20
Cost, dollars per cell*	100	5000
General types:		
Communication, utility, emergency lighting, uninterrupted power supplies		

* 1993 dollars.

rent is dependent on battery needs once the battery reaches the fixed potential. Stationary batteries frequently use a *float charge* which is a constant-voltage charge in which the voltage is set at a value slightly greater than the open-circuit potential to maintain all internal losses without overcharging the battery.

Charger Selection

Battery-charger selection is an important part of achieving good battery life and recharge efficiency. The following items should be considered when selecting a charger: automatic charging rate control; provisions for equalizing charge; fail-safe design (component failure will not harm battery); low electrolyte temperature rise during recharge; automatic over- and undercharge protection; automatic charger termination; charger protected from shorted and open-circuit output; charger protected from reverse polarity; polarity- and voltage-keyed connectors; high electrical efficiency and power factor; and low ac line draw.

BATTERY MAINTENANCE

It is a good practice to keep batteries clean and dry. Cleaning reduces losses due to contact resistance and prevents shorting or grounding through conductive dirt films. The heat dissipation is improved on clean batteries, which helps reduce operating temperature. Batteries are commonly operated between 40 and 120°F (6 and 71°C). Freezing should be avoided because it can permanently damage a battery. High-temperature storage or operation will effectively reduce life. For optimum performance and life, the manufacturer's recharge and maintenance instructions should be closely followed.

Portable power batteries require little maintenance beyond being kept clean, dry, and cool. Primary cell maintenance is limited to replacement when leaking or discharged. Secondary portable power cells require charging when discharged and replacement when leaking. Since most secondary portable power cells are nickel-cadmium, recharging should be conducted after a full discharge. Repeated charging after partial discharge will reduce the capacity.

Motive power, engine cranking, and stationary batteries require cleaning, watering, and charging. When cleaning, electrolyte on the cell covers or connectors should be neutralized and rinsed to prevent corrosion and shorting. Water that is approved for batteries or distilled water should be added as required to keep the electrolyte between the high and low level. The electrolyte level should always be above the plates. Gas bubbles created during charging displace volume causing the electrolyte level to increase. Therefore, water should be added when the battery is on charge and gassing. If water is added after the gas has dissipated from the electrolyte, room should be left for expansion.

Battery connections and charging equipment should be checked at least once a month. A loose or dirty connection can reduce performance or cause an explosion. Malfunctions in charging equipment can result in over- or undercharging which will reduce the life of the battery.

Maintenance records of batteries are useful in scheduling periodic maintenance functions such as checking charging equipment, keeping the battery clean, and maintaining levels. Records can also be used to locate problem areas.

BATTERY SAFETY

Caution should be used when storing, operating, or repairing a battery because of the chemical, explosive, and electrical safety hazards associated with all batteries.

Chemical

Batteries contain corrosive liquids which can be harmful on contact. Always wear protective clothing when exposed to corrosive liquids.

Explosive

Some batteries present an explosive safety hazard because of hydrogen gas released during charging. This hazard is controlled by ventilation and preventing ignition by sparks or open flame in charging areas.

Electrical

High-voltage batteries should be treated like any other high-voltage source for protection from shock hazards. Precaution should be taken to keep metal objects away from battery connectors and terminals to prevent shorting. A battery stores a large amount of energy which can be released rapidly when shorted.

Batteries are safe when proper safety practices are followed. All personnel working with batteries should be trained in the operation and safety practices provided by the battery manufacturer to prevent injury and damage to equipment.

SOURCES OF INFORMATION

Federal Specifications

Superintendent of Documents, U.S. Government Printing Office, Washington, DC 20402.

NEMA Standards

National Electrical Manufacturers Association, Suite 300, 2101 L Street, N.W., Washington, DC 20037.

BCI Standards

Battery Council International, 111 East Wacker Drive, Chicago, IL 60601.

IEEE Standards

Institute of Electrical and Electronic Engineers, 345 East 47th Street, New York, NY 10017.

BIBLIOGRAPHY

Fink, D.: *Standard Handbook for Electrical Engineers,* 13th ed., McGraw-Hill, New York, 1990.

Meurer, M.: *Sealed Battery Selection for Designers and Users,* McGraw-Hill, New York, 1990.

CHAPTER 5.14
PLASTICS PIPING

Stanley A. Mruk
Executive Director, Plastics Pipe Institute
A Division of the Society of the Plastics Industry
Washington, D.C.

INTRODUCTION

Outstanding chemical resistance is a principal reason for the growing use of plastics piping in practically every phase of U.S. industry including the pharmaceutical, chemical, food, paper and pulp, electronics, oil and gas production, water and waste treatment, mining, power generation, steel production, and metal-refining industries. With the general recognition of its other features (e.g., it is easy to work with, durable, and economically advantageous), plastics piping has also become widely accepted for a broad range of applications for other reasons than just its chemical inertness. Current major uses include water mains and services, gas distribution, storm and sanitary sewers, plumbing (drain, waste, and vent piping and hot- and cold-water piping), electrical conduit, power and communications ducts, chilled-water piping, and well casing.

The diversity of plastic pipe—reflecting the many available materials, wall constructions, diameters, and techniques for joining—is so great and the attendant technology is evolving at such a rate that it is difficult to present in a concise reference all the pertinent information available. This chapter, therefore, offers only basic and general information. In applying plastic pipe, the reader is advised to ensure that all appropriate code requirements and government safety regulations are complied with. Sources and references for additional information are provided for this purpose at the end of the chapter.

DESCRIPTION, CLASSIFICATION, AND TYPICAL USES

"Plastic" pipe is as indefinite a term as "metal" pipe. As with metal products, plastic pipes are made from a variety of materials. Plastics used for pipe exhibit a wide range of properties and characteristics. The variabilities in properties of plastics are derived not only from the chemical composition of the basic synthetic resin, or polymer, but are also largely determined by the kind and amount of additives, the nature of reinforcement, and the process of manufacture. For example, it is possible to formulate mixtures of polyvinyl chloride (PVC) resins plus appropriate additives that range from a clear, soft, and pliable product (such as that used for laboratory tubing and upholstery) to a rigid and strong product (such as for pressure pipe). Another example is the construction of reinforced thermosetting resin pipe (RTRP) by using glass-fiber winding techniques that can adjust to the desired value the ratio of the circumferential (or hoop) strength to the axial strength.

Plastics are divided into two basic groups, thermosetting and thermoplastic, both of which are used in the manufacture of plastic pipe. Thermoplastics, as the name implies, soften upon the application of heat and reharden upon cooling: they can be formed and reformed repeatedly. This characteristic permits them to be easily extruded or molded into a wide variety of useful shapes, including pipe and fittings. Because thermoplastics are shaped, by a die or mold, while in a "molten" state their properties are essentially isotropic (i.e., independent of direction). In some processes and under some conditions, some anisotropy (i.e., direction dependence) may result.

Thermosetting plastics, on the other hand, form permanent shapes when cured by the application of heat or a "curing" chemical. Once shaped and cured during the manufacturing process, they cannot be reformed by heating. The excellent adhesion properties of thermosetting resins permit their utilization in composite structures by which strength and stiffness can be greatly enhanced through the use of reinforcements and fillers. The greater strength and higher temperature limits of these composite structures permit RTRP to handle fluids at temperatures and pressures beyond the limits for thermoplastic pipe. Piping systems now available are capable of operating at temperatures in excess of 300°F (148°C). All important commercial constructions of thermosetting pipe utilize some form of reinforcement and/or filler. By orienting the reinforcement, RTR pipes can be given directionally dependent properties.

Thermoplastic Pipe

Most thermoplastic pipes and fittings are made from materials containing no reinforcements, although fillers are occasionally used. Pipe is manufactured by the extrusion process, whereby molten material is continuously forced through a die that shapes the product. After being formed by the die, the soft pipe is simultaneously sized and hardened by cooling it with water. Fittings and valves are usually produced by the injection-molding process, in which molten plastic is forced under pressure into a closed metal mold. After cooling, the mold is opened and the finished part is removed. Some items, especially larger-sized fittings for which there is insufficient demand to justify construction of injection-molding tooling, are fabricated from pipe sections, or sheets, by utilizing thermal or solvent cementing fusion techniques. To compensate for the lower strength, the fitting may either be made from a heavier wall stock or reinforced with a fiberglass-resin overwrap. The engineer designing a pressure-rated system should make sure that the pressure ratings of the selected fittings are adequate.

There is some thermoplastic pipe made of a cellular-core construction (for example, ASTM* F 628) in which the pipe wall consists of thin inner and outer solid skins sandwiching

* American Society for Testing and Materials.

a high-density foam (Fig. 5.160). The primary benefit of such construction is improved ring and longitudinal (beam) stiffness in relation to the material used. Because the foam-wall structure results in some loss of strength, applications for cellular-core pipe are in nonpressure uses, such as for above- and below-ground drainage piping, which can take advantage of the more material-efficient ring and beam stiffness.

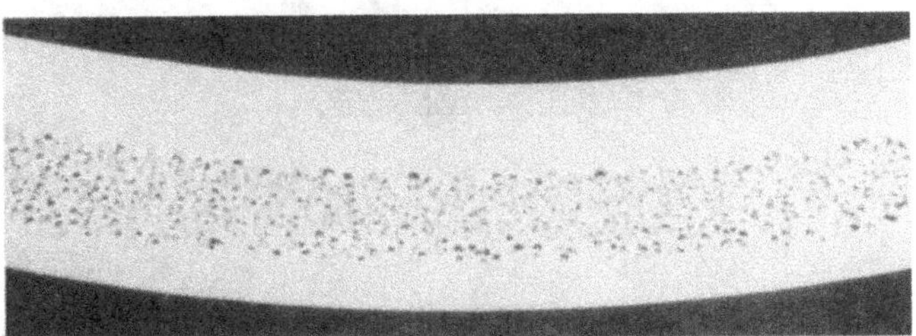

FIGURE 5.160 ABS cellular-core construction. (*Borg-Warner Chemicals.*)

For buried nonpressure applications, a composite pipe (ASTM D 2680) is produced that consists of two concentric tubes that are integrally braced with a truss webbing. The resultant openings between the concentric tubes are filled with a lightweight concrete. This construction increases both the ring and the beam stiffness. Composite pipe is used only for nonpressure buried applications such as sewerage and drainage.

Several other processes for improving the radial (i.e., ring) stiffness of thermoplastic pipe for buried applications have in common the formation of some type of rib reinforcement. A well-established technique is forming corrugations in the pipe wall. Corrugated polyethylene pipe (ASTM F 405) in sizes from 2 to 12 in (5 to 30 cm) is widely used for building foundations, land, highway, and agricultural drainage, and communications ducts. Ribbed pipe also is commercially made by the continuous spiral winding of the plastic over a mandrel of a specially shaped profile. Adjacent layers of this profile are fused to each other to form a cylinder that is smooth on the inside and has ribbed reinforcements on the outside (Fig. 5.161). The smooth inside diameter is preferable for many applications, such as sewerage, because it creates no flow disturbances. Pipes with ribbed construction are available in PVC and polyethylene (PE). PE pipes, which are made with hollow ribs to minimize material usage, are available in sizes from 18 to 120 in (45 cm to 3 m) in diameter.

The distinctive characteristics of the principal thermoplastic piping materials are discussed in the following pages.

Polyvinyl Chloride. Polyvinyl chloride (PVC) piping is made only from compounds containing no plasticizers and minimal quantities of other ingredients. To differentiate these materials from flexible, or plasticized PVCs (from which are made such items as upholstery, luggage, and laboratory tubing) they have been labeled rigid PVCs in the United States and unplasticized PVC (uPVC) in Europe. Rigid PVCs used in piping range from Type I to Type III, as identified by an older classification system that is still much in use. In this system, the type designations are supplemented by grade designations (e.g., Grade 1 or 2) which further define the material's properties. Type I materials, from which most pressure and nonpressure pipe is made, have been formulated to provide optimum strength as well as chemical and temperature resistance. Type II materials are those formulated with modifiers that improve impact strength but that also somewhat reduce, depending on modifier type and quantity, the aforementioned properties of Type I materials. There is little call for Type II pipe, as the impact strength of the stronger Type I pipe is more than adequate for most uses. Type III materials contain some inert fillers which tend to increase stiffness concomitant with some lowering of both tensile and impact strength and chemical resistance. Some nonpressure PVC piping, such as that used for conduit, sewerage, and drainage, is made from Type III PVCs.

The currently used classification system for rigid PVC materials for piping and other applications is described in ASTM D 1784, "Standard Specification for Rigid Polyvinyl Chloride and

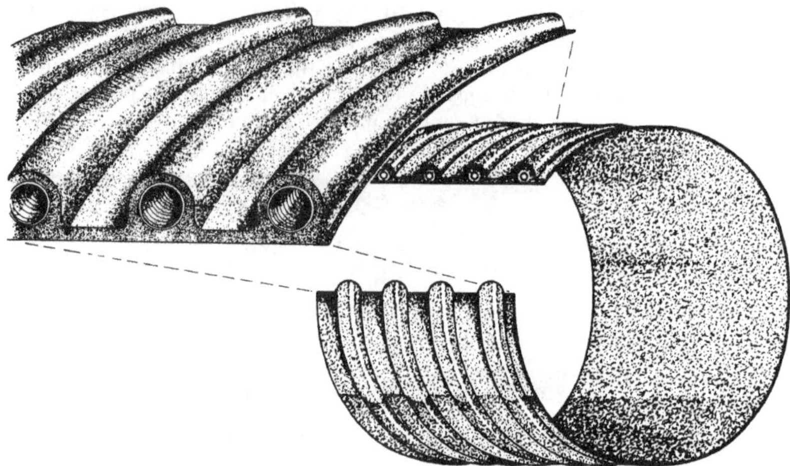

FIGURE 5.161 Typical profile of hollow-ribbed polyethylene pipe. (*Chevron Chemical Co.*)

Chlorinated Polyvinyl Chloride Compounds." This specification categorizes rigid PVC materials by numbered cells that designate value ranges for the following properties: impact resistance (toughness), tensile strength, modulus of elasticity (rigidity), deflection temperature (temperature resistance), and chemical resistance. The following table cross-references the designations of the principal PVC materials from the older to the newer classification system.

By cell classification system of ASTM D1784 minimum cell class	By older system	
	Type and grade	Designation
12454-B	Type I, Grade 1	PVC 11
12454-C	Type I, Grade 2	PVC 12
12433-D	Type II, Grade 1	PVC 21
13233	Type III, Grade 1	PVC 31

Because (as expanded in the discussion on properties) short-term properties of plastic materials are not a reliable predictor of long-term capabilities, those PVC materials that have been formulated for long-term pressure applications are also designated by their categorized maximum recommended hydrostatic design stress (RHDS) for water at 73.4°F (23°C) as determined from long-term pressure testing. The most commonly used designation system for PVC pressure-piping materials is based on the above older designation system with two added digits that identify, in hundreds of pounds per square inch, the maximum recommended design stress.* For example: PVC 1120 is a Type I, Grade 1 PVC (minimum cell class 12454-B) with a maximum recommended HDS of 2000 lb/in^2 (13.8 MPa) for water at 73.4°F (23°C); PVC 2110 is a Type 2, Grade 1 PVC (minimum cell class 14333-D) with an RHDS of 1000 lb/in^2 (6.9 MPa). Other PVC material designations available for pressure pipe are listed in Table 5.51. Most pressure-rated PVC pipe is made from PVC 1120 materials.

The combination of good long-term strength with higher stiffness explains why PVC has become the principal plastic pipe material for both pressure and nonpressure applications. Major uses include: water mains; water services; irrigation; drain, waste, and vent (DWV) pipes; sewerage and drainage; well casing; electric conduit; and power and communications

*Since the maximum recommended hydrostatic design stress is for continuous water pressure at 73°F (23°C), it is up to the designer to determine the extent, if any, by which this stress should be reduced to account for any departure from these conditions and the need for a suitable margin of safety against other considerations. See the discussion on design.

ducts. A much broader range of fittings, valves, and appurtenances of all types is available in PVC than in any other plastic.

Chlorinated Polyvinyl Chloride. As implied by its name, chlorinated polyvinyl chloride (CPVC) is a chemical modification of PVC. CPVC has properties very similar to PVC but the extra chlorine in its structure extends its temperature limitation by about 50°F (28°C), to nearly 200°F (93°C) for pressure uses and about 210°F (99°C) for nonpressure applications. ASTM D 1784, the rigid PVC materials specification, also covers CPVC which it classifies as Class 23477-B. By the older designation system, it is known as Type IV, Grade 1 PVC. CPVCs for pressure pipe are designated CPVC 4120 [i.e., Type IV, Grade 1 CPVC with a maximum recommended hydrostatic design stress of 2000 lb/in^2 (13.8 MPa) for water at 73.4°F (23°C)]. At 180°F (82°C) the maximum recommended hydrostatic design stress* for CPVC is 500 lb/in^2 (3.4 MPa).

Principal applications for CPVC are for hot or cold water piping and for many industrial uses which take advantage of its higher temperature capabilities and superior chemical resistance.

Polyethylene. Polyethylene (PE) is the best-known member of the polyolefin group—plastics that are formed by the polymerization of straight-chain hydrocarbons—known as olefins—that include ethylene, propylene, and butylene. Polyethylene plastics are tough and flexible even at subfreezing temperatures. They are generally formulated with only an antioxidant (for protection during processing) and some pigment (usually carbon black) or other agent designed to screen out ultraviolet radiation in sunlight which over long-term exposure could be damaging to the natural-color polymer.

ASTM D 1248, the PE molding and extrusion materials specification, classifies these materials into three types depending on the density of the natural resin. Type I consists of lower-density materials which are relatively soft and flexible and have low heat resistance. Type II PEs are of medium density, slightly harder, more rigid, and more resistant to elevated temperatures; they also have better tensile strength. Type III materials show maximum hardness, rigidity, tensile strength, and resistance to the effects of increasing temperature. Pipe is made almost exclusively from Type II and Type III PEs. ASTM D 1248 also provides for grade designations to further classify PEs according to other physical characteristics. PE piping materials for pressure piping are classified by a designation system that combines the type and grade coding with that for the maximum RHDS* for water at 73.4°F (23°C). PEs utilized for pressure piping, and so designated, are listed in Table 5.51.

The more recently issued standard, ASTM D 3350, classifies PE piping materials according to broader physical property criteria, including long-term strength, and it is expected to become the primary PE piping material standard.

Outstanding toughness and relatively low flexural modulus, which permits coiling of smaller-diameter pipe, are large factors in PE's prominence in gas-distribution and water-service piping. Other features, such as heat fusibility, good abrasion resistance, and availability in large diameters [up to 120 in (3 m)], account for the use of PE piping for chemical transfer lines, slurry transport, sewage force mains, intake and outfall lines, power ducts, and renewal (by insertion into the old pipe) of deteriorated sewer, gas, water, and other pipes.

Cross-Linked Polyethylene (PEX). PEX pipe and tubing is made from essentially the same materials and by the same extrusion process used to manufacture PE pipe. The difference is that during or directly after extrusion the polymer molecules are cross-linked to form a thermosetting material. The cross-linking may be effected by the addition of cross-linking compounds to the PE, or by subsequent electron or high-frequency radiation.

The primary benefit of cross-linking is the raising of PE's upper operating temperature limit. PEX can be used for pressure applications up to 200°F (93°C).

Principal applications for PEX and tubing are for underfloor heating systems, melting of ice and snow, and hot-cold water piping.

*Since the maximum recommended hydrostatic design stress is for continuous water pressure at 73°F (23°C), it is up to the designer to determine the extent, if any, by which this stress should be reduced to account for any departure from these conditions and the need for a suitable margin of safety against other considerations. See the discussion on design.

Polybutylene. Polybutylene (PB) is a unique polyolefin. Its stiffness resembles that of low-density PE, yet its strength is higher than that of high-density PE. However, its most significant feature is that it retains its strength better with increasing temperature. Its upper temperature limit is higher than that for any PE: nearly 200°F (93°C) for pressure uses and somewhat higher for nonpressure applications.

PB piping materials are covered by ASTM D 2581. Materials for pressure applications are designated PB 2110; the last two digits signify a maximum recommended hydrostatic design stress* of 1000 lb/in^2 (6.9 MPa) for water at 73.4°F (23°C). At 180°F (83°C) this design stress is 500 lb/in^2 (3.4 MPa).

Major applications of PB pipe tend to take advantage of its improved temperature resistance and its toughness. They include hot or cold water piping and industrial uses such as hot effluent lines. Because of its excellent abrasion resistance, PB pipe is also used for slurry lines.

Polypropylene. Polypropylene (PP) is a polyolefin similar in properties to Type III PE but is slightly lighter in weight, more rigid, and more temperature-resistant. PPs are classified by ASTM D 2146 into two types: Type I covers homopolymers that generally have the greatest rigidity and strength but offer only moderate impact resistance; Type II covers copolymers of propylene and ethylene, or other olefins, which are less rigid and strong but have much improved toughness, particularly at lower temperatures.

Although on the basis of its short-term properties PP shows better resistance to temperature, this material is inherently somewhat more sensitive to thermal aging than PE. To overcome this sensitivity, specially formulated grades containing appropriate heat-stabilizer systems have been developed. For pressure uses, the adequacy of long-term heat stability is evaluated by means of long-term pressure tests of the PP formulation, conducted at various higher temperatures. A PP material containing flame-retardant additives is available for drainage-piping applications.

Polypropylene piping is used for chemical waste and drainage and various other industrial uses that take advantage of excellent chemical resistance, good rigidity and strength, and higher-temperature operating limits. One feature that accounts for its use in laboratory and industrial drainage is its superior resistance to many organic solvents.

Acrylonitrile-Butadiene-Styrene. Acrylonitrile-butadiene-styrene (ABS) is a family of materials formed from three different monomers (chemical building blocks), acrylonitrile, butadiene, and styrene. The proportions of the components and the way in which they are combined can be varied to produce a wide range of properties. Acrylonitrile contributes rigidity, strength, hardness, chemical resistance, and heat resistance; butadiene contributes toughness; and styrene contributes gloss, rigidity, and ease of processing.

ASTM D 1788 classifies ABS plastics into numbered cells that designate value ranges for each of three properties: impact strength (toughness), tensile stress at yield (strength), and deflection temperature under load. ABS pipe materials are categorized into types and grades in accordance with established minimum cell requirements for each type and grade. Like the other major thermoplastics, ABS materials for pressure pipe are designated by a coding that identifies both short-term properties and long-term strength. For example, ABS 1316 is a Type I, Grade 3 (minimum cell classification 3-5-5 per D 1788) material with a maximum RHDS* of 1600 lb/in^2 (11.2 MPa) for water at 73.4°F (23°C). Other ABS pressure-pipe materials are listed in Table 5.51.

An advantageous combination of toughness with good strength and stiffness largely accounts for the most common uses for ABS piping, i.e., for drain, waste, and vent (DWV) applications as well as for sewers, well casings, and communications ducts. One especially tough formulation of ABS is utilized to manufacture piping for compressed-air service. Most other thermoplastic materials are not recommended for above-ground compressed-air service because, should they fail, the pipe failure mechanism would sometimes produce flying fragments which could be injurious.

*Since the maximum recommended hydrostatic design stress is for continuous water pressure at 73°F (23°C), it is up to the designer to determine the extent, if any, by which this stress should be reduced to account for any departure from these conditions and the need for a suitable margin of safety against other considerations. See the discussion on design.

Polyvinylidine Fluoride. Polyvinylidine fluoride (PVDF) is a fluoroplastic, i.e., its chemical composition includes the element fluorine. Fluoroplastics are distinguished by their exceptional chemical and solvent resistance, excellent durability, and broad working-temperature range. To these features PVDF adds good strength and toughness and radiation resistance (which explains its use for conveying radioactive materials). The ASTM Specification covering PVDF molding and extrusion materials is D 3222. A standard for PVDF pressure piping is under development at ASTM. Standard recommended hydrostatic design stresses for PVDF have not yet been established, and an effort to develop these values is underway. Manufacturers of PVDF piping systems should be consulted for recommended design-stress values.

PVDF piping is utilized primarily for chemical processing and other industrial applications that are beyond the reach of the more common thermoplastics. These include the handling of active materials such as chlorine and bromine and piping materials that subject it to higher temperatures. Because of its immunity to radiation, PVDF piping is also used in reprocessing nuclear wastes.

Other Thermoplastics. Some thermoplastics of lesser current commercial importance are either used for special reasons or were more popular some time ago. Among these is *chlorinated polyether,* a tough rigid material of outstanding chemical resistance and useful to about 220°F (104°C). PVDF has largely displaced it as a specialty material. *Cellulose acetate butyrate* (CAB) was extensively used in petroleum production for conveying saltwater and crude oil as well as for natural gas distribution. However, because of its relatively low strength and moderate resistance to temperature and chemicals, it has been largely displaced by other materials. Nylon, a tough, strong, and heat-resistant material that is also very resistant to aromatic and chlorinated solvents, has some special uses. One of these is for coiled small-diameter tubing for compressed-air service. Nylon insert (with barbs) fittings are also made for use with polyethylene pipe.

Polyacetal, a strong, hard plastic with relatively good temperature resistance, is being increasingly used for various hot- or cold-water plumbing components such as fittings, valves, and faucets.

Reinforced Thermosetting Resin Pipe

Reinforced thermosetting resin pipe (RTRP) is a composite largely consisting of a reinforcement imbedded in, or surrounded by, cured thermosetting resin. Included in its composition may be granular or platelet fillers, thixotropic agents, pigments, or dyes. The most frequently used reinforcement is fiberglass, in any one or a combination of the following forms: continuous filament, chopped fibers, and mats (Fig. 5.162). While reinforcements such as asbestos or other mineral fibers are sometimes used, fiberglass-reinforced pipe (FRP) is by far the most popular. One form of FRP, called reinforced plastic mortar pipe (RPMP), consists of a composite of layers of thermosetting resin–sand aggregate mixtures that are sandwiched by layers

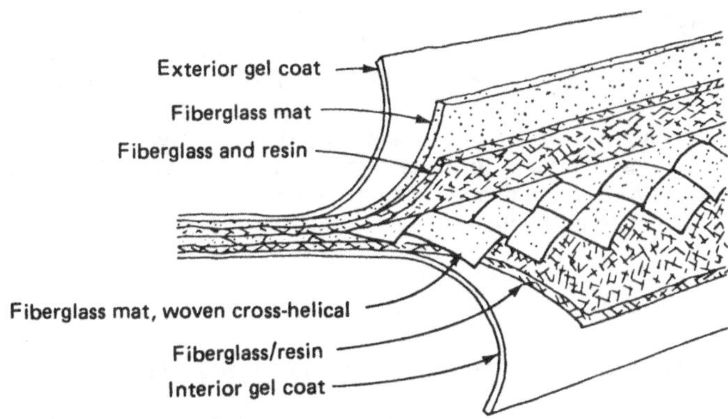

FIGURE 5.162 Fiberglass-reinforced pipe.

of resin-fiberglass reinforcements. In another construction, the sand is replaced by glass microspheres. The high content of reinforcements in RTRP, which may run from 25 to 75 percent of the total pipe weight, and the specific design of the composite wall construction are the major determinants of the ultimate mechanical properties of the pipe. The resin, although also influencing these properties somewhat, is the binder that holds the composite structure together, and it supplies the basic source of temperature and chemical resistance. Glass fibers, as well as many other reinforcements, do not have high resistance to chemical attack. For enhanced chemical and/or abrasion resistance, RTRP construction may include a liner consisting of plastic (thermosetting or thermoplastic), ceramic, or other material. The outer surface of the pipe—especially that of the larger diameter sizes—may also be made "resin rich" to better resist weathering, handling, and spills. Reinforced thermosetting resin pipe is available in a variety of resins, wall constructions, and liners with diameters ranging from 1 in (2.5 cm) to more than 16 ft (5 m). Stock and specially fabricated fittings are readily available.

Filament Winding. Pipe is produced by machine-winding, under controlled tension and in predetermined patterns, of glass reinforcement—which may consist of continuous-filament strands, woven roving, or roving tape—onto the outside of a mandrel. The reinforcement may be saturated with a liquid resin or preimpregnated with partially cured resin. After the pipe has been formed and cured, the mandrel is removed. The dimension of the inside diameter is set by the mandrel; that of the outside diameter by the thickness of the wall.

A high glass content and a precise machine-controlled fiber orientation which permits control of the radio of circumferential to axial strength make this pipe more economically suited for certain uses including higher-pressure applications. Since for a given pressure rating the axial strength of filament-wound pipe tends to be lower than that for pressure-molded pipe, it may have to be offset by more frequent support spacing. The use of automatic machines results in pipes of closer dimensional tolerance and in mechanical properties that are very consistent. To provide a stronger barrier against corrosion of the fiberglass filaments, a resin-rich liner, 0.02 to 0.1 in (0.5 to 2.5 mm) thick, is usually deposited on the inside of the pipe. In one process a thick, highly chemically resistant liner is produced by overwrapping a PVC, CPVC, or PVDF pipe (the only materials used at this time) with resins that can bond tightly to the thermoplastic.

The matched bell-and-spigot socket and the butt-and-strap methods are those primarily used to join filament-wound pipe.

Centrifugal Casting. In this process, resin and reinforcements (such as chopped fibers) are applied to the inside of a rotating mold. After the pipe has cured by action of heat or a catalyst, it is removed from the mold. The outside diameter of the pipe is set by the mold dimensions: the inside diameter is determined by the amount of material introduced into the mold. This type of pipe is almost fully machine-made and it provides very consistent mechanical properties and the closest tolerances. Because its glass fiber content is lower than that of filament-wound pipe it offers generally better chemical resistance. For additional resistance it may be made with a corrosion-resistant liner. The strength properties of centrifugally cast pipe are less direction-dependent than those of filament-wound pipe. Fittings are normally of the socket type for bell and spigot joining. The joint may be overwrapped if added strength is desired.

The smaller-diameter fittings for RTRP are usually produced by a molding (compression or injection) process that is similar to that used with thermoplastics. Fittings may also be made by the filament-winding or contact-molding process whereby the part is first fabricated on a steel or fiberglass form, then cured and removed. Centrifugal casting is also sometimes employed. By all these procedures (except filament winding) random orientation of the reinforcement is obtained.

Resins Used for Piping

Brief descriptions of the more important resins used for piping are given in the following text.

Epoxies. Epoxy resins are strong and have good resistance to solvents, salts, caustics, and dilute acids. Epoxies are cross-linked by curing agents which become an integral part of the polymer and affect the thermal, chemical, and physical properties of the polymer. For instance, the maximum service temperature of epoxy pressure pipe cured with anhydrides is 180°F (83°C), and it has little resistance to caustics; that cured with aromatic amines can be used at temperatures above 225°F (107°C), and it has good caustic resistance.

The major use of epoxy pipe is in oil fields, where its resistance to corrosion and paraffin buildup makes it preferable to steel pipe for crude-collection and saltwater-injection lines. Other uses are in the chemical process industry, in heating and air conditioning, in food processing, for gasoline and solvents, and in mining applications (including abrasive slurry transport, communications ducts, and power conduits).

Polyesters. Although typically not as strong as the epoxies, polyesters offer good resistance to mineral acids, bleaching solutions, and salts. The most commonly used polyester resins for pipe are isophthalic polyesters and bisphenol A fumarate polyesters. Isophthalics have poorer resistance to caustics and oxidizers. Bisphenol A fumarates have improved resistance to these materials and are widely used in paper mills for bleach lines.

Isophthalic resin pipe is used in waste-treatment and power plants in services where corrosive conditions are not severe. Maximum operating-temperature limits for pressure pipe vary, depending upon the specific material, but are generally below 200°F (93°C).

Vinyl Esters. These resins include chemical features of both epoxies and polyesters. Vinyl ester resins offer better chemical resistance, somewhat higher temperature limits, and better solvent resistance than ordinary polyesters but generally do not compare to epoxies in these properties. Vinyl ester resins are preferred over polyesters because they are more chemical-resistant than the isophthalics and less brittle than the bisphenol A fumarates. Typical services are in fertilizer plants (acid lines), chlorine plants (chlorine-saturated brine lines), and paper mills (caustic and black-liquor lines).

Furans. Furan resins offer very good chemical, solvent, and temperature resistance—up to about 300°F (150°C). Because they extend the limitations of the other resins, they are often selected for use in the processing industries in place of exotic metal piping.

Desirable Qualities of FRP Resins

The performance criteria that usually determine the choice of one or more FRP resins over another include chemical resistance, mechanical strength, heat resistance, and (for the manufacturer) processability. A qualitative summary of FRP resin performance is presented in Table 5.48. For the final choice of the appropriate FRP resin(s) and liner combination for a given service (chemical environment, weathering exposure, abrasion resistance, etc.) more detailed information, including case history data, should be obtained.

PROPERTIES

Compared with other materials, pipes made of plastic have less strength and rigidity and are more temperature-sensitive. However, they offer these essential properties sufficiently to satisfy the performance requirements of most industrial piping applications. Moreover, they have good to excellent strength-to-weight ratios; are durable and easy to install and maintain; and have outstanding chemical resistance. The thermoplastics are lower than thermosets in strength, rigidity, and maximum operating temperature. However, their chemical resistance tends to be superior. Thermosets, in contrast, are capable of handling corrosive fluids at pressures and temperatures well beyond the service limits of most thermoplastic pipe.

Physical and Mechanical

Typical physical, mechanical, and thermal properties of the major thermoplastic piping materials are presented in Table 5.49. Those for thermosets are given in Table 5.50. The actual val-

TABLE 5.48 Qualitative Summary of FRP Resin Performance*

Resin	Chemical resistance: Acids	Bases	Solvents	Other properties: Processability	Strength	Heat resistance
Polyester resins	Fair–good	Poor	Fair	Good	Fair–good	Fair–good
Isophthalic acid based	—	O	+	+	+	—
Het acid based	—	O	+	+	+	+
BPA/fumarates	+	—	—	+	+	—
Vinyl ester terminated polyesters	—	—	—	+	+	+
Vinyl ester resins	Good	Fair–good	Fair–good	Good	V. good	Good–v. good
BPA/ECH epoxy derived						
$n = 0$	+	—	+	+	++	+
$n = 2$	+	+	—	+	++	—
Phenolic-Novolac epoxy derived	+	—	+	—	+	+
Epoxy resins	Fair	Good	V. good	Fair	V. good	Good–v. good
Aliphatic amine cured	—	—	+	—	+	+
Aromatic amine cured	—	+	++	—	++	++
Anhydride cured	—	O	+	+	+	+
Lewis acid cured	+	+	+	—	+	++
Furan resins	Fair	Good	V. good	Poor	Good	V. good
Furfuryl alcohol derived	—	+	++	O	+	++

* ++ = Very Good, + = Good, — = Fair, and O = Poor.

Source: Based on table from M. B. Launikitis, "Chemically Resistant FRP Resins," *Proceedings of 1977 Plastics Seminar,* National Association of Corrosion Engineers, Houston, Texas.

TABLE 5.49 Typical Physical Properties of Major Thermoplastic Piping Materials*†

	ASTM Test no.	ABS I	ABS II	PVC I	PVC II	CPVC	PE II	PE III	PB	PP	PVDF
Specific gravity	D 792	1.04	1.08	1.40	1.36	1.54	0.94	0.95	0.92	0.92	1.76
Tensile strength, lb/in² (× 10³)	D 638	4.5	7.0	8.0	7.0	8.0	2.4	3.2	4.2	5.0	7.0
Tensile modulus, lb/in² (× 10⁶)	D 638	0.3	0.3	0.41	0.36	0.42	0.12	0.13	0.06	0.2	0.22
Impact strength, Izod, ft · lb/in notch	D 256	6	4	1	6	1.5	>10	>10	>10	2	3.8
Coeff. of linear expansion, in/(in)(°F)(× 10⁻⁵)	D 696	5.5	6.0	3.0	5.0	3.5	9.0	9.0	7.2	4.3	7.0
Thermal conductivity, (Btu)(in)/(h)(ft²(°F)	C 177	1.35	1.35	1.1	1.3	1.0	2.9	3.2	1.5	1.2	1.5
Specific heat, Btu/(lb)(°F)	—	0.32	0.34	0.25	0.23	0.20	0.54	0.55	0.45	0.45	0.29
Approx operating limit‡											
°F, nonpressure	—	180	180	150	130	210	130	160	210	200	300
°F, pressure	—	160	160	140	120	200	120	140	200	150	280

* The properties of a piping material may vary from one commercial material to another. The pipe manufacturer should be consulted for specific properties.

† Consult Table 5.51 for values of long-term strength at 73°F.

‡ Exact operating limit may vary from each particular commercial plastic material (consult manufacturer). Effects of environment should also be considered.

TABLE 5.50 Typical Physical Properties of Glass-Fiber-Reinforced Thermosetting Resin Pipe*†

Property at 75°F	Test no.	Polyester	Filament wound epoxy	Filament wound vinyl ester	Filament wound reinforced plastics mortar
Specific gravity	D 792	1.6	1.9	1.9	1.7 to 2.2
Tensile strength, lb/in² (× 10³)	D 2105				
Hoop direction		35	20 to 50	20 to 50	15 to 60
Axial direction		25	2 to 10	2 to 10	3 to 20
Tensile modulus, lb/in² (× 10⁶)	D 2105				
Hoop direction		1.4	1.5 to 4	1.5 to 4	1.4 to 3.6
Axial direction		1.4	0.8 to 2.0	0.8 to 2.0	0.8 to 1.8
Coeff. of linear expansion, in/(in)(°F)(× 10^{-5})	D 696				
Hoop direction		1.3	0.5 to 0.7	0.5 to 0.8	1.0
Axial direction		1.3	1.0 to 1.8	1.0 to 1.8	1.5
Thermal conductivity, (Btu)(in)/(h)(ft²)(°F)	C 177	0.9	1.5 to 4.2	1.5 to 4.2	1.3
Approximate temperature limit‡					
°F, nonpressure		240	300	250	200
°F, pressure		200	240	220	140

* The properties of a piping material may vary from one commercial material to another. The pipe manufacturer should be consulted for specific properties.

† Consult Table 5.512-4 for values of long-term strength at 73°F.

‡ Exact operating limit may vary from each particular commercial plastic material (consult manufacturer). Effects of environment should also be considered.

ues for any pipe will vary according to the specific material(s) used and, in the case of composite products, such as the thermosets, will also depend on the specific wall construction.

Because the properties of plastics are influenced by duration of loading, temperature, and environment, data-sheet values for mechanical properties such as those presented in Tables 5.49 and 5.50 are not satisfactory for design purposes. The stress-strain, and strain-stress, responses of plastics reflect their viscoelastic nature. The viscous, or fluidlike, component tends to damp or slow down the response between strain and stress. For example, if a load is continuously applied on a plastic material, it creates an instantaneous initial deformation that then increases at a decreasing rate. This further deformation response is known as *creep.* If the load is removed at any time, there is a partial immediate recovery of the deformation followed by a gradual creep recovery. If on the other hand the plastic is deformed (i.e., strained) to a given value that is then maintained, the initial load (stress) created by the deformation slowly decreases at a decreasing rate. This is known as the *stress-relaxation response.* The ratio of the actual values of stress to strain for a specific time under continuous stressing, or straining, is commonly referred to as the *effective creep modulus,* or the *effective stress-relaxation modulus.* In the case of thermoplastics this effective modulus is significantly influenced by time. For continuous loading of 20-yr duration, it can be from one-quarter to one-third the value of the short-term modulus. In the case of reinforced thermosets the viscous response is of lower order and the long-term effective modulus tends to be at least three-quarters of the short-term values. Most pipe manufacturers are prepared to provide values of effective moduli for specific materials and loading conditions.

The effective strength of plastics is influenced by time, temperature, and environment. For example, the breaking point for a thermoplastic material under short-term tensile testing is reached only after considerable material deformation has taken place, at least 10 percent, and in some cases over 100 percent (the ultimate elongation for reinforced thermosets is lower than for thermoplastics). Under long-term continuous loading, material failure (or an unacceptable level of material damage) will occur at much lower deformations than in tensile testing. With thermoplastics the strain levels at failure can be as low as 3 percent, and with some reinforced thermosets they may be below 0.5 percent. Material damage, and not creep or excessive deformation, represents the durability limit for plastics subject to long-term loading. These durability limits, are time-, temperature-, and environment-dependent.

Durability Under Continuous Loading. To establish its longer-term hydrostatic strength, plastics pipe is tested in accordance with ASTM D 1598, "Time to Failure of Plastic Pipe Under Constant Internal Pressure." After obtaining a sufficient number of stress vs. time-to-fail points that must span a testing time from 10 to 10,000 h, the data are extrapolated to determine the estimated average 100,000-h strength. The extrapolating procedures are those of ASTM method D 2837, "Obtaining Hydrostatic Design Basis for Thermoplastic Materials," or procedure B of ASTM D 2992, "Obtaining Hydrostatic Design Basis for Reinforced Thermosetting Resin Pipe and Fittings." The extrapolated long-term hydrostatic strength (LTHS) so determined is then rounded off into the appropriate hydrostatic design basis (HDB). The HDBs are design-stress categories represented by a series of preferred numbers (i.e., 2000, 2500, 3150, 4000, 5000, etc.) that ascend in steps of 25 percent. The following relationship, known as the ISO (International Standards Organization) plastics pipe hoop stress equation, is used to relate the test pressure and pipe dimensions to the resultant circumferential stress in the pipe wall:

$$S = \frac{p}{2}\,\frac{D_m}{t} \tag{1a}$$

where S = hoop stress, lb/in^2 (N/m^2)
p = internal hydraulic pressure, lb/in^2 (N/m^2)
D_m = mean pipe diameter, in (cm)
t = minimum pipe wall thickness, in (cm)

Once the HDB is established and then reduced to a hydrostatic design stress (HDS) by the application of an appropriate service (or design) factor, the same formula, rewritten as follows, may be used to compute the pipe pressure rating:

$$p = 2\text{HDS}\,\frac{t}{D_m} \tag{1b}$$

where HDS = hydrostatic design stress, lb/in^2 (N/m^2)
= HDB × SF
SF = service factor

The selected service factor considers two general groups of conditions: The first encompasses the normal variations in material and pipe manufacture and the second those of the pipe's application and use (environment, hazards, life expectancy). The general practice has been to use service factors of not more than 0.5 for static water pressure at 73°F (23°C). Smaller factors are used for more demanding conditions.

The HDBs for static water pressure at 73°F (23°C) for the major thermoplastic pipe materials are presented in Table 5.51. As indicated by this table, the ASTM material designations for pressure thermoplastics code the material according to its short-term properties and its maximum RHDS. The RHDS is determined by applying a factor of 0.5 to the material's HDB for water at 73°F (23°C). Values of HDB for other environments and temperatures will be different, even for materials of the same ASTM designation. For example, not all PE 3408s will show the same long-term strength at 120°F (49°C). For other than the standard HDB rating conditions the pipe manufacturer should be consulted for data or recommendations on the specific pipe material.

In the case of thermosets, because of their composite construction, their HDB is determined not only by the properties of the materials used but also by the specific wall construction and manufacturing process. Typical HDBs for some machine-made RTRPs for water service at 73°F (23°C) are given in Table 5.52. The pipe manufacturer should be consulted for HDBs for the specific pipe construction under consideration.

TABLE 5.51 Hydrostatic Design Basis* (Strength Categories) for Thermoplastic Pipe Compounds Determined with Water at 73.4°F (23°C)

Material designation†	HDB lb/in² (MPa), 73.4°F (23°C)‡
PE 1404	800 (5.5)
PE 2406	1250 (8.6)
PE 3406	1250 (8.6)
PE 3408	1600 (11.2)
PEX 0006	1250 (8.6)
PVC 1120	4000 (27.6)
PVC 1220	4000 (27.6)
PVC 2116	3150 (21.7)
PVC 2120	4000 (27.6)
CPVC 4120	4000 (27.6)
ABS 1208	1600 (11.2)
ABS 1210	2000 (13.8)
ABS 1316	3150 (21.7)
ABS 2112	2500 (17.2)
PB 2110	2000 (13.8)
PFA 1008	1600 (11.2)
PVDF 2020	4000 (27.6)

* Per ASTM D 2837.

† The last two digits code the maximum recommended hydrostatic design stress (RHDS), expressed in hundreds of pounds per square inch. RHDS = HDB × 0.5, where 0.5 is the generally accepted maximum value for the design factor. Lower values than 0.5 may be justified by certain operating and safety considerations. The first two digits code the material according to short-term properties.

‡ Since thermoplastics, even though of the same ASTM designation, may be affected differently by increasing temperature, HDBs at higher temperatures must be established for each specific commercial product. A number of such products have HDBs for temperatures as high as 180°F (82°C). Consult the most current Plastics Pipe Institute (1275 K Street NW, Suite 400, Washington, DC 20005) Technical Report TR-4 for latest listing of HDBs of commercial pipe compounds.

Durability Under Cyclic Loading. The higher shorter-term strength of plastics permits them to easily tolerate relatively infrequent short-lived excursions in pressure beyond those established for static pressure conditions. However, under repeated cyclic stressing, as in pressure surging, plastics tend to fatigue and their long-term strength is reduced. The fatigue sensitivity varies from plastic to plastic, and for each material it is dependent on various factors including the amplitude and the frequency of the cyclic stressing. The reduction in limiting strength of thermoplastics may range from slight to as much as one-half the value under static conditions. For services with only moderate and infrequent surging, as is the case in many water piping systems, selecting the pipe pressure rating solely on the basis of the maximum static pressure has been demonstrated to be effective. The American Water Works Association Standard for PVC Water Piping (AWWA C-900) establishes pipe pressure classes with a built-in consideration of cyclic pressure.

Reinforced thermosetting plastic pipes are somewhat more sensitive to fatigue than thermoplastic pipes. Accordingly, they are often pressure-rated for surge service on the basis of their HDB as determined by Procedure A of the aforementioned ASTM D 2992 from test

TABLE 5.52 Hydrostatic Design Basis for Machine-Made Reinforced Thermosetting Resin Pipe

ASTM pipe specification*	Type	Classification per ASTM D 2310†			HDB, lb/in²‡	
		Grade	Class	Designation	Static	Cyclic
D 2517 (gas pressure pipe)	Filament-wound	Epoxy, glass-fiber-reinforced	No liner	RTRP-11AD	—	5000
				RTRP-11AW	16,000	—
D 2996	Filament-wound	Epoxy, glass-fiber-reinforced	No liner	RTRP-11AD	—	5000
				RTRP-11AW	16,000	—
	Filament-wound	Epoxy, glass-fiber-reinforced	Reinforced epoxy liner	RTRP-11FE	—	6300
				RTRP-11FD	—	5000
	Filament-wound	Polyester, glass-fiber-reinforced	Reinforced polyester liner	RTRP-12EC	—	4000
				RTRP-12ED	—	5000
				RTRP-12EU	12,500	—
	Filament-wound	Polyester, glass-fiber-reinforced	No liner	RTRP-12AD	—	5000
				RTRP-12AU	12,500	—
D 2997	Centrifugally cast	Polyester, glass-fiber-reinforced	Polyester liner	RTRP-22BT	10,000	—
				RTRP-22BU	12,500	—
	Centrifugally cast	Epoxy, glass-fiber-reinforced	Epoxy liner	RTRP-21CT	10,000	—
				RTRP-21CU	12,500	—

* See Table 5.55 for the abbreviated title of the pipe standard.
† "Standard Classification for Machine-Made Reinforced Thermosetting Resin Pipe," ASTM D 2310. (The first three symbols following RTRP designate type, grade, and class; the last symbol codes the HDB.)
‡ Multiply by 6894 to convert to pascals.

data obtained in accordance with ASTM D 2143, "Test for Cyclic Pressure Strength of Reinforced Thermosetting Plastic Pipe." Typical HDB values for cyclic pressure service for certain RTRPs are also shown in Table 5.52.

Durability Under Continuous Straining. Stress relaxation gradually reduces the initial stress level that is first generated when a plastic material is strained to a given level and then maintained there. As a consequence of this reduction in stress, plastics can tolerate somewhat larger strains under constant-strain (i.e., stress-relaxation) conditions than the ultimate strain that ensues in a constant-load condition (under which there is no stress relaxation). At present there is no unanimously preferred method for establishing limiting strain values for conditions of stress relaxation. Appendix X2 to ASTM D 3262, "Standard Specification for Reinforced Plastic Mortar Sewer Pipe," includes a method, not officially part of the standard, by which one may evaluate the strain limits of the subject pipe in an acid environment. The strain limits for RTRPs tend to be significantly lower than those for thermoplastics piping. In fact, under conditions of continuous straining, limiting strain is not a design constraint for most thermoplastics. When dealing with situations such as in pipe bending or in deflection of pipe under earth loading (in which limiting strain is a possible consideration), recommended design values for maximum permissible strain should be obtained for the specific material and end-use conditions.

Durability Under Cyclic Straining. Both ultimate strength and allowable strain may be reduced by fatigue. Guidance should be sought from the pipe manufacturer.

Other Mechanical Properties. In addition to the test for long-term strength, a number of special tests have also been established for evaluating other relevant properties of plastic pipe:

ASTM D 2444 "Impact Resistance of Thermoplastic Pipe and Fittings by Means of a Tup (Falling Weight)"

ASTM D 2105 "Longitudinal Tensile Properties of Reinforced Thermosetting Pipe and Tube"

ASTM D 2925	"Measuring Beam Deflection of Reinforced Thermosetting Pipe Under Full Bore Flow"
ASTM D 2412	"External Loading Properties of Plastic Pipe by Parallel-Plate Loading"
ASTM D 2924	"External Pressure Resistance of Reinforced Thermosetting Resin Pipe"

Flow Properties

Plastic pipes offer minimal resistance to fluid flow. Tests indicate that they may be characterized as hydraulically smooth conduits according to the well-established rational flow formulas. The Hazen-Williams equation, a frequently used engineering approximation formula, yields good correlations for water flow when a C_H factor of 150 to 155 is used. For calculating water flows in open channels, or nonfull flow in conduits, Manning's equation is often used. The Manning n factor for plastics for clean water is approximately 0.0085 to 0.0095. With sewerage, values of $n = 0.010$ to 0.012 are used since they provide margin for flow-disturbing influences such as sedimentation and slime growth.

Since plastics do not corrode, their original flows do not deteriorate with time. Plastics are also easier to clean than other materials.

Chemical Resistance

Plastics do not corrode in the sense that metals do. Being nonconductors they are immune to galvanic or electrochemical effects. If they are affected by an environment, it is generally through direct chemical attack, solvation, or strain corrosion. In addition to differences in chemical resistance due to the nature of the plastic and pipe wall construction, the extent of resistance may also depend on time and temperature of contact and, in some cases, on the presence of an externally applied stress.

In *direct chemical attack,* the molecules of either the polymer or the reinforcement are altered, and the alteration leads to a gradual deterioration of properties. Such attack can be brought about by strong oxidizing and reducing agents and by ultraviolet and other radiation. Thermoplastics as a group tend to be inherently more resistant than thermosets to chemical attack. Good protection against ultraviolet radiation (which might be required with the more ultraviolet-sensitive pipes that are to be used in continuous exposure to the weather) is afforded by incorporation into the formulation of a finely divided carbon black, titanium dioxide, or some other opaque pigment.

Solvation is the absorption of an organic solvent by the plastic. Its effect may range from a slight swelling and softening, with minor effects on properties, to a complete solution. The solvent cementing of ABS and PVC pipe is based on solvation. By the use of selective solvents that evaporate after their task is completed, solvent cementing makes it possible to create a monolithic joint that retains the properties of the base material. Thermosets, because of their cross-linked chemical structure, tend to have superior resistance to solvation.

In *strain corrosion,* damage will occur only under the combined action of strain (i.e., stress) and environment. In the case of thermoplastics this form of attack is called *environmental stress cracking.* The mechanism, although not fully understood, is essentially the development and ensuing slow growth and propagation of cracks by the combined action of stress and a sensitizing agent. Stress-cracking agents tend to be materials such as detergents and alcohols that have a surface-wetting tendency. Stress cracking may be controlled by selecting stress-crack-resistant grades of material or by creating designs that ensure that the stress, or strain, is below the threshold value necessary to set this mechanism in action.

In the case of thermosetting materials, the formation of crazes or microcracks exposes the glass fiber, or other reinforcement, to possible chemical attack. To protect against this, a tough resilient liner is used, and the stress, or strain, level is limited to established safe values which will preclude the formation of crazing under the specific exposure conditions.

Table 5.53 presents a broad guide to the chemical resistance of the major thermoplastic piping materials. Table 5.48 includes a general guide to FRP resin performance. Final selection of material and pipe wall construction for chemical or corrosive service should follow consultation with more detailed chemical-resistance information. Because ultimate resistance is affected by stress, time, and temperature, it may not be reliably predicted by shorter-term "soak" tests. Successful previews in similar service are the best guide. Lacking these, new applications would best be evaluated by actual service testing. An advantage of service testing is that it is sure to include some minor (but often overlooked) contaminant which could influence the final result.

In the case of RTRP piping, the liner is considered the first line of defense and therefore a critical factor in corrosive service selection. Special resin liners, which may be deposited in extra thicknesses, are available for extra protection. Pipes with thermoplastic liners are also available.

Inertness to Potable Water and Other Fluids

Nearly all of the base materials from which plastic pipes are made are inert to potable water. It is possible, however, that through the addition of ingredients such as stabilizers, catalysts, modifiers, and pigments, the final formulation may render a pipe inadequate for potable water

TABLE 5.53 Thermoplastic Piping Materials: Chemical-Resistance Guide for Ambient Temperatures*

Attacking chemicals	ABS	PVC		CPVC	PE	PB	PP	PVDF
		I	II					
Inorganic compounds								
Acids, dilute	G	G	L	G	G	G	G	G
Acids, concentrated 80%	L	L	L	G	L	L	L	G
Acids, oxidizing	L	P	P	L	P	P	P	G
Alkalies, dilute	G	G	G	G	G	G	G	G
Alkalies, concentrated 80%	L	G	L	G	G	G	G	G
Gases, acid (HCl and HF), dry	L	L	L	L	G	G	G	G
Gases, acid (HCl and HF), wet	L	G	L	G	G	G	G	G
Gases, ammonia, dry	L	G	L	G	G	G	G	G
Gases, halogens, dry	L	L	L	L	L	L	P	G
Gases, sulfur gases, dry	P	G	L	G	G	L	P	G
Salts, acidic	G	G	G	G	G	G	G	G
Salts, basic	G	G	G	G	G	G	G	G
Salts, neutral	G	G	G	G	G	G	G	G
Salts, oxidizing	L	L	L	L	G	G	G	G
Organic compounds								
Acids	G	G	G	L	G	G	G	G
Acid anhydrides	L	L	L	P	L	L	L	L
Alcohols, glycols	L	G	L	G	L†	G	G	G
Esters, ethers, ketones	P	P	P	P	L	L	L	L
Hydrocarbons, aliphatic	L	L	L	G	L	L	L	G
Hydrocarbons, aromatic	P	P	P	L	P	P	P	G
Hydrocarbons, halogenated	L	L	L	L	P	P	P	L
Natural gas (fuel)	G	G	G	G	G	G	G	G
Mineral oil	G†	G	G	G	L†	G	G	G
Oils, animal and vegetable	G†	G	G	G	L†	G	G	G
Synthetic gas (fuel)	L	L	L	L	L	L	L	G

* G, good; P, poor; L, limited knowledge: determination requires precise knowledge of individual conditions.
† Stress-crack-resistant grade should be used.

in terms of its effect on the toxicological safety and the taste and odor quality of the water. To safeguard against this, most potable water piping standards require that the pipe be evaluated for this purpose by a laboratory recognized by the public health profession. A very commonly used laboratory is NSF International, Ann Arbor, Michigan. NSF evaluates, and lists as acceptable, those plastic pipes that satisfy the requirements of their standards for potable water service.

Because they do not contaminate fluids with metallic ions, plastic pipes are often utilized in the transport of pure materials, including deionized water. For food service there are pipes available that have been made from materials approved by the Food and Drug Administration.

COMMON METHODS FOR JOINING PLASTIC PIPES

Plastic pipe may be joined by a variety of methods (see Table 5.54), the choice of which is influenced in some cases by the properties of the basic material. For example: the polyolefins (PE, PB, and PP) may not be solvent-cemented; the vinyls (PVC and CPVC) and ABS heat-fuse with relative difficulty; and the thermosets may not be either heat-fused or solvent-cemented. Of the available choices, the one that is selected will depend upon pipe performance, installation, and maintenance requirements as well as availability of fitting and joining equipment. Heat fusion (of PE, PP, PB, and PVDF), solvent-cementing (of ABS, PVC, CPVC), and adhesive joining (of reinforced thermosetting resin pipe), all methods which produce monolithic joints, are preferred for applications requiring maximum strength and optimum chemical resistance (Fig. 5.163). Bell (sometimes referred to as socket) and spigot connections are utilized to join pipe by these three techniques. RTRP may also be joined by the butt-and-strap method whereby two pieces of pipe are butted together and overlays of a laminate are then applied over the butted section and allowed to cure (Fig. 5.164). Larger-diameter polyolefin pipe is joined by the heat fusion of the pipe butt ends.

Flanged connections are often used in industrial applications, particularly when making transitions to other materials, such as when connecting to a metal valve or to a tank outlet, and when it is advantageous to provide for easy removal of a pipe section or other component

TABLE 5.54 Techniques for Joining Plastic Pipe

	Thermoplastic pipe							
Method of joining	ABS	PVC	CPVC	PE	PP	PB	PVDF	RTR pipe
Adhesive	—	—	—	—	—	—	—	o
Solvent cements	o	o	o	—	—	—	—	—
Heat fusion	—	—	—	o	o	o	o	—
Threading*	o	o	o	o	o	—	o	o
Flanged connectors†	o	o	o	o	o	o	o	o
Grooved joints‡	o	o	o	o	o	—	o	o
Mechanical compression§¶	o	o	o	o	o	o	o	o
Elastomeric seal¶	o	o	o	o	o	o	o	o
Flaring	—	—	—	o	—	o	—	—
Insert	—	—	—	—	—	o	—	—

* Molded thread adapters are available for attachment on the pipe by another technique. Threads may not be cut in thermosetting pipe. Some thermosetting threaded connections may be adhesive-bonded for extra strength. For threading, the wall thickness of thermoplastic pipe should be not less than Schedule 80.

† Flanged adapters are applied on pipe by heat fusion, solvent-cementing, or threading.

‡ Minimum wall thicknesses are prescribed depending on the pipe material.

§ With thinner-walled pipe, stiffening inserts must be used.

¶ Many designs of elastomeric seal and compression fittings provide no thrust restraint and therefore may be used only in situations, such as buried pipe, in which the pipe is restrained from pullout. Elastomeric seal and compression fittings are available in special designs that incorporate end restraint.

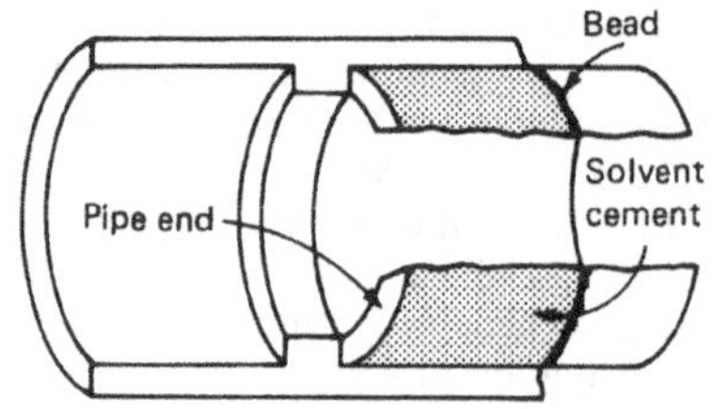

FIGURE 5.163 Threadless joint in PVC coupling (inside view).

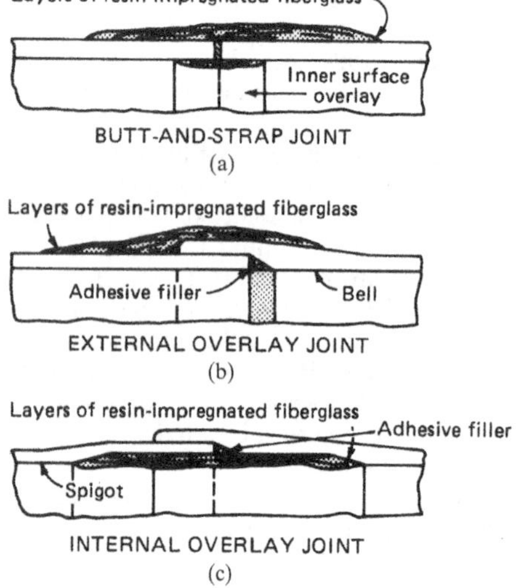

FIGURE 5.164 Butt and strap joints.

from the system for cleaning, maintenance, or other purpose.

Threading is also used with plastic pipe. However, molded threads are preferred for thermoplastics and are required for most thermosetting pipe. Molded threaded adapters are available and may be applied by solvent-cementing or with adhesive, whichever is applicable. Threads may not be cut on most thermosetting pipe for they may damage the structural integrity of the pipe wall. Thermoplastic pipe may be threaded, provided its wall thickness is not less than a prescribed minimum, normally at least that of Schedule 80 pipe.

For installations not excluding the use of elastomeric sealants such as neoprene or red rubber, there are mechanical-compression as well as bell and spigot connectors which incorporate such sealants into their design. Much thermoplastic and thermosetting piping specifically made for buried water and sewer lines is available with integral elastomeric-seal bell and spigot connectors. Such connectors greatly facilitate pipe construction, partly because of the ease of making the connection (a stab fit) and partly because the connection may be made under almost any weather or field condition.

Sometimes connectors utilizing grooved pipe are used with standard grooved-end systems such as Victaulic or Gustin-Bacon. With thermoplastic pipe of sufficient wall thickness, the grooves may be cut or rolled in some cases. Cutting is not permitted with thermosetting pipe. Grooved adapters are available for both thermoplastic and thermosetting pipes.

COMMERCIALLY AVAILABLE PRODUCTS

Plastics piping is manufactured in an imposing array of materials, constructions, diameters, wall thicknesses, lengths, and fitting types. The more important products are listed in Table 5.55 which also reports for each product the available size range and its important end uses. Whenever the product is covered by a major standard, the applicable document is identified. Because of the dynamic rate at which new plastic piping standards are currently being written (and older ones revised), the reader is advised to check with the major standards issuing organizations for the most current listing.*

Fittings for larger-diameter pipes are not listed in Table 5.55 because they are often custom-fabricated rather than available from stock. Also not listed in Table 5.55 are piping components such as valves and flanges. In diameters ranging from ⅜ to 4 in, valves made from PVC, CPVC, PP, and PVDF are available in a great variety of different styles including check, ball, diaphragm, globe, gate, and needle. They are available with socket (for solvent-cementing or heat-fusion

*The Plastics Pipe Institute, a Division of the Society of the Plastics Industry, 1275 K Street NW, Suite 400, Washington, DC 20005, regularly updates its TR-5, "Standards for Plastics Piping," which lists standards for thermoplastics piping issued by all major U.S., Canadian, and international standards organizations.

joining), butt, flanged, or threaded ends. Valves are also available in "Tru" union style and with multiports. A number of models may be obtained with pneumatic or electric actuators for automatic valve positioning. In the ⅜- to 4-in (0.95- to 10-cm) sizes a large array of other piping components, such as strainers, expansion joints, roof and floor drains, and line tapping fittings, is also available. Many of these items may be obtained up through about 8-in (20-cm) diam. Larger-sized components are sometimes specially fabricated, or metallic products are used which are connected to the plastic pipe by means of flanges or some other suitable connector.

Standard and special design manholes and access holes are available in both thermoplastic and thermosetting materials for sewerage and drainage applications. ASTM D 3753, "Specification for Glass Fiber Reinforced Manholes," is, to date, the only adopted standard for such products.

PRODUCT STANDARDS, CODES, AND APPROVALS

Standards

The primary source of standards on plastics piping is the American Society for Testing and Materials (ASTM), 1916 Race Street, Philadelphia, PA 19103. (See Table 5.55.) There are a number of other organizations that also develop such standards, generally on products related to their particular interest or activities. These include the following: American Water Works Association (AWWA), 6666 West Quincy Avenue, Denver, CO 80235, which issues standards for water distribution; the American Petroleum Institute (API), 300 Corrigan Tower Building, Dallas, TX 75201, on piping for oil and gas production; the National Electrical Manufacturers Association (NEMA), 2101 L Street, N.W., Washington, DC 20037, on conduit and ducting; and the Underwriters Laboratories (UL), 333 Pfingston Road, Northbrook, IL 60062, on conduit and ducting. In addition, the federal government and some state agencies have also issued standards on plastic pipe, generally for projects in which they exercise regulatory or financial functions, or for purchases on their own behalf. Typical of these are the standards issued by the U.S. Department of Agriculture (USDA), the U.S. Department of Defense (DOD), the Federal Housing Administration (FHA), and the General Services Administration (GSA). Most of these documents parallel the basic requirements of ASTM and other listed standards.

The most frequently used dimensioning scheme for setting the outside diameter (OD) of plastic pipe is the traditional iron-pipe size (IPS) system of commercial wrought-steel pipe (ANSI B36.10). Most thermoplastic pipes are available with standard IPS outside diameters. Some PE and PB pipes which are designed to be joined with insert-type fittings are sized to the same inside diameters as Schedule 40 wrought-steel pipe. In the case of thermoset pipes their outside diameters may be exactly, or approximately, equal to the reference dimension depending on whether the pipe is sized from the outside in (as in centrifugal casting) or from the inside out (as in filament-winding over a mandrel). Other diameter systems that are utilized include:

- *Copper tubing size (CTS).* Based on standard outside diameters of copper tubes, CTS pipe is used for water and gas services and for hot- or cold-water plumbing.
- *Cast-iron (CI) pipe size.* PVC and RTR water-main piping is made to this outside-diameter basis as well as to the IPS system.
- *International Standards Organization (ISO) sizes.* Some of the larger polyolefin pipes are made with these internationally set outside diameters.

Much pipe, especially in the larger sizes (for which compatibility with traditionally sized plastic piping components such as valves and fittings is not necessary), is made to fit into special diameter-sizing systems determined by the specific product standard or by the pipe manufacturer. Most of these systems have established diameter dimensions of such proportions that the resultant size of the inside bore is close to the pipe nominal diameter.

The wall thicknesses of most thermoplastic pipe of solid and homogeneous wall construction are defined in accordance with the standard dimension ratio (SDR) concept, whereby the

TABLE 5.55 Principal Commercially Available Plastic Piping Products

Pipe material	Product standard*	Title (abbreviated) of standard or brief product description	Diameter range, in	End use: Water supply and distribution							Sewer and drain					Industrial			Duct	
				Mains	Services	Drop pipe, wells	Well casing	Various: industrial, coml.	Distributing: cold only	Distributing: hot & cold	Collecting system	Building connections	Drainage	Drain, waste, & vent	Natural gas distribution	Corrosives and abrasives	Compressed gases†	Liquid fuels	Conduit (above ground)	Duct (below ground)
PVC, PE & PB	ASTM D 2513	Thermoplastic Gas Pressure Pipe & Fittings	¼–24												X		X			
ABS, PVC & SR	ASTM F 480	Thermoplastic Water Well Casing	2–16				X													
ABS, PVC & PP	ASTM D 3311	DWV Plastic Fittings Patterns	1¼–6											X						
ABS & PVC	ASTM D 2680	ABS & PVC Composite Sewer Pipe	6–15								X	X	X							
	ASTM F 409	ABS & PVC Accessible & Replaceable Tube and Fittings	3–12											X						
ABS	ASTM D 1527	ABS Plastic Pipe, Sch. 40 & 80	⅛–12	X	X	X		X	X			X				X	X			
	D 2282	ABS Plastic Pipe, SDR-PR	⅛–12	X	X	X		X	X			X				X	X			
	D 2468	ABS Plastic Pipe Fittings, socket, Sch. 40	¼–8	X	X	X		X	X			X				X	X			
	D 2661	ABS DWV Pipe and Fittings	1¼–6										X	X						
	D 2750	ABS Utility Conduit & Fittings	1–6																	X
	D 2751	ABS Sewer Pipe and Fittings	1–6									X	X							
	F 628	ABS Foam Core DWV	1¼–6										X	X						
PVC	ASTM D1785	PVC Plastic Pipe, Sch. 40–80 & 120	⅛–24	X	X	X		X	X			X				X	X			
	D 2241	PVC Plastic Pipe, SDR-PR	⅛–36	X	X	X		X	X			X				X	X			
	D 2464	PVC Plastic Pipe Fittings, threaded sch. 80	⅛–6	X	X	X		X	X			X				X	X			
	D 2466	PVC Plastic Pipe Fittings, socket, sch. 40	⅛–8	X	X	X		X	X			X				X	X			
	D 2467	PVC Plastic Pipe Fittings, socket, Sch. 80	⅛–8	X	X	X		X	X			X				X	X			
	D 2665	PVC DWV Pipe & Fittings	1¼–6											X						

	D 2672	PVC Plastic Pipe, Bell End	⅛–8	X	X	X		X	X			X				X	X			
	D 2729	PVC Drain Pipe & Fittings	2–6										X							
	D 2740	PVC Plastic Tubing	½–1¼		X	X		X												
	D 2949	3-in PVC Thin Wall DWV Piping	3											X						
	D 3034	PVC Sewer Pipe & Fittings, type PSM	4–15								X	X	X							
	F 512	PVC Conduit for Buried Installation	2–6																	
	F 679	PVC Sewer Pipe & Fittings	18–36								X		X							
	F 758	PVC Underdrain Piping	4–8										X							
	F 789	PVC Sewer Pipe, Type 46	4–15								X	X	X							
	F 794	PVC Sewer Pipe, Ribbed Wall	4–48								X		X							
	F 891	PVC Pipe with a Cellular Core	1½–18								X	X	X	X						
	F 949	PVC Corrugated Pipe, Smooth I.D.	4–18								X	X	X							
	F 1336	PVC Sewer Gasketed Fittings	4–27								X		X							
	AWWA C900	PVC Pressure Pipe for Water	4–12	X				X												
	C 905	PVC Water Transmission Pipe	14–36	X				X												
	C 907	PVC Pressure Fittings	4–8	X				X												
	UL 514	Electrical Outlet Boxes & Fittings	½–6																X	
	651	Rigid Nonmetallic Conduit	½–6																X	
NEMA	TC-2	Electrical Plastic Tubing & Conduit	½–6																X	
	TC-3	PVC Fittings for Conduit & Tubing	½–6																X	
CPVC	ASTM D 2846	CPVC Hot Water Distribution Systems	⅜–2							X										
	F 437	CPVC Plastic Pipe Fitting, threaded, Sch. 80	¼–6					X		X						X	X	X		
	F 438	CPVC Plastic Pipe Fittings, socket, Sch. 40	¼–6					X		X						X	X	X		
	F 439	CPVC Plastic Pipe Fittings, socket, Sch. 80	¼–6					X		X						X	X	X		
	F 441	CPVC Plastic Pipe, Sch. 40 & 80	¼–12					X		X						X	X	X		
	F 442	CPVC Plastic Pipe, SDR-PR	¼–12					X		X						X	X	X		
	F 443	CPVC Bell End Pipe	⅛–8					X		X						X	X	X		

* Issuing agency identified in discussion on standards.

† Thermoplastic piping is not normally recommended for above-ground service because of safety considerations should the pipe fail. A special grade is available (see ABS piping).

TABLE 5.55 Principal Commercially Available Plastic Piping Products (*Continued*)

Pipe material	Product standard*	Title (abbreviated) of standard or brief product description	Diameter range, in	End use																
				Water supply and distribution							Sewer and drain					Industrial			Duct	
				Mains	Services	Drop pipe, wells	Well casing	Various: industrial, coml.	Distributing: cold only	Distributing: hot & cold	Collecting system	Building connections	Drainage	Drain, waste, & vent	Natural gas distribution	Corrosives and abrasives	Compressed gases[†]	Liquid fuels	Conduit (above ground)	Duct (below ground)
PE	ASTM																			
	D 2104	PE Plastic Pipe, Sch. 40	½–6	X	X	X		X	X		X	X				X	X	X		
	D 2239	PE Plastic Pipe, SDR-PR	½–6	X	X	X		X	X		X	X				X	X	X		
	D 2447	PE Plastic Pipe, OD-based, Sch. 40 & 80	½–12	X	X	X		X	X		X	X				X	X	X		
	D 2609	Plastic Insert Fittings for PE Pipe	½–4		X	X		X									X			
	D 2683	PE Fittings, socket-fusion type for OD-based pipe	½–4	X	X	X		X	X		X	X				X	X	X		
	D 2737	PE Plastic Tubing	½–2		X	X		X	X								X	X		
	D 3035	PE Pipe, OD-Based, SDR-PR	½–24	X	X	X		X	X		X	X	X			X	X	X		
	D 3261	PE Fittings, Butt-fusion Type	½–10	X	X	X		X	X		X	X				X	X	X		
	F 405	PE Corrugated Tubing & Fittings	3–8										X							
	F 667	PE Corrugated Tubing & Fittings, Larger Diam.	8–24										X							
	F 714	PE Plastic Pipe, SDR-PR larger diam.	3–63	X				X			X	X	X			X	X	X		
	F 771	PE Irrigation Piping	½–6					X												
	F 810	PE Pipe for Drainage & Absorption Fields	3–6										X							
	F892	Corrugated PE Pipe with Smooth Interior	4										X							
	F 894	PE Profile Wall Pipe	18–120								X		X							
	F 1055	PE Fittings, Electrofusion Type	½–8												X					
	AWWA C901	PE Pipe, tubing & fittings for water	½–3		X															
	C 906	PE Water Transmission Pipe	4–63	X				X								X				
	API 5LE	PE Line Pipe	½–12					X	X							X	X	X		

PB	ASTM-																			
	D 2662	PB Plastic Pipe, SDR-PR	½–6	X	X	X		X		X						X	X	X		
	D 2666	PB Plastic Tubing	½–2		X	X		X		X						X	X	X		
	D 3000	PB Plastic Pipe, SDR-PR, OD-controlled	½–6	X	X	X		X		X						X	X			
	D 3309	PB Plastic Hot-Water Distributing Systems	¼–2		X					X										
	F 809	PB Plastic Pipe, SDR-PR, larger diam	3–42	X	X						X	X	X			X	X	X		
	F 845	Plastic Insert Fittings for PB Tubing	⅜–¾																	
	AWWA C902	PB Pipe, tubing & fitting for water	½–3		X															
PP	ASTM F412	PP Chemical Drainage Pipe & Fittings	1½–4										X	X						
RTRP	ASTM																			
	D 2517	RTR Pipe & Fittings for Gas	2–12												X		X			
	D 2996	Filament Wound RTR Pipe	1–16	X	X	X		X		X						X	X	X		
	D 2997	Centrifugally Cast RTR Pipe	1–14	X	X	X		X	X							X	X	X		
	D 3262	Filament Wound RTR Pipe & Fittings	8–144								X	X	X							
	D 3517	Filament Wound RTR Pressure Pipe	8–144					X			X					X	X	X		
	D 3754	Filament Wound RTR Sewer & Industrial Pipe	8–144					X			X	X	X			X	X	X		
	D 3840	RTR Fittings for Nonpressure	8–144								X	X	X							
	D 4024	RTR Flanges	½–24	X	X			X			X	X	X	X		X	X	X		
	F 1173	Epoxy Fiberglass Piping for Marine Applications	½–24					X								X				
	AWWA C 950	Fiberglass Pipe for Water	1–144	X	X				X							X	X	X		
	API 15 LR	Lower Pressure Fiberglass Pipe	2–12			X	X									X	X	X		
	15 HR	Higher Pressure Fiberglass Pipe	1–8			X	X									X	X	X		
	15 AR	Fiberglass Tubing	1–4½			X	X									X	X	X		
	MIL-P-28584	Steam Condensate Lines, RTRP Filament Wound	2–6					X												
	MIL-P-22245	Pipe and Pipe Fittings, glass-fiber-reinforced plastic	2–16	X	X	X	X	X		X						X	X	X		

* Issuing agency identified in discussion on standards.

† Thermoplastic piping is not normally recommended for above-ground service because of safety considerations should the pipe fail. A special grade is available (see ABS piping).

ratio of *average* outside diameter to *minimum* wall thickness is a constant value for each SDR pipe series over the entire range of pipe diameters. The standard diameter ratios that have been adopted by ASTM and other standard-writing organizations represent a series of preferred numbers that increase in steps of 25 percent, as follows: 11, 13.5, 17, 21, 26, 32.5, 41, etc. The advantage of establishing wall-thickness categories for thermoplastic pipe according to a constant ratio of diameter to wall thickness is evident from inspection of Eq. (1*b*): within each SDR category the pressure rating is the same for all pipe sizes. There is some pipe made to other than the established SDRs; for such pipe the actual diameter to wall-thickness ratio is simply referred to as the diameter ratio (DR).

Thermoplastic pipes specifically intended for industrial uses are often made to the IPS Schedule 40, 80, and 120 system which sets not only the outside diameter but also the wall thickness for each nominal size in each schedule. Since in a given pipe schedule the ratio of diameter to wall thickness tends to decrease with increasing diameter, so too does the pipe pressure rating. Schedule 80 wall thickness, or greater, is required whenever pipe is threaded. The pressure rating of thermoplastic threaded pipe is reduced to half that for unthreaded pipe.

The wall thicknesses of thermoset pipe are not defined in the same way as for thermoplastic pipe because key properties such as strength and stiffness depend not only on material but also on exact wall construction, which can vary not only among manufacturers but even with pipe diameter. The resultant pipe wall thickness will generally be set by the performance requirements of the pipe. Some standards set minimum values for wall thickness. Thermosetting pipe may not be field-threaded. If the pipe is to be joined by threading, it is available with factory-applied molded threads. Threaded adapters are also available.

Codes

The use of piping for plumbing, fire protection, and for the transport of hazardous materials may be subject to the provisions of a code and/or to those of local, state, federal, or other regulations. All the major model plumbing codes which have become adopted, or referenced, by state and local jurisdictions permit and prescribe to a varying but fairly extensive degree the use of plastics piping for hot-cold water lines; water services; drain, waste, and vents (DWV); sewerage; and drainage. Plastics piping is also covered by other codes, such as the following which are of interest to industrial users:

American National Standards Institute Codes

- ANSI B31.3 Chemical Plant and Petroleum Refinery Piping
- ANSI B31.8 Gas Transmission and Distribution Piping Systems
- ANSI Z223.1 National Fuel Gas Code

Department of Transportation, Hazardous Materials Board, Office of Pipeline Safety Operations

- Code of Federal Regulations (CFR), Title 49, Part 192, Transportation of Natural Gas and Other Gas by Pipeline: Minimum Federal Safety Standards
- Code of Federal Regulations (CFR), Title 49, Part 195, Transportation of Liquids by Pipeline, Minimum Federal Safety Standards

The National Fire Protection Association (Quincy, Mass.) Model Codes

- NFPA 30 Flammable and Combustible Liquids Code
- NFPA 54 National Fuel Gas Code
- NFPA 70 *National Electrical Code**
- NFPA 70A Electrical Code for One and Two Family Dwellings
- NFPA 34 Outdoor Piping

---

* *National Electrical Code*® is a registered trademark of The National Fire Protection Association, Quincy, MA 02269.

Approvals

Some standards and various jurisdictions and authorities require that before a pipe may be used for certain applications it first must be approved for that use by a recognized, or specifically designated, organization. Organizations with listing and approval programs for plastic pipe include the following:

- *For Potable Water* NSF International, NSF Bldg., P.O. Box 1468, Ann Arbor, MI 48105; Canadian Standards Association, 178 Rexdale Boulevard, Rexdale, Ontario, Canada M9W 1R3
- *For Drain, Waste, and Vent* NSF International and Canadian Standards Association (see above)
- International Association of Plumbing and Mechanical Officials, 5033 Alhambra Ave., Los Angeles, CA 90032
- *For Meat- and Food-Processing Plants* U.S. Department of Agriculture, 14th and Independence S.W., Room 0717 South, Washington, DC 20250
- *For Underground Fire Protection Systems* Underwriters Laboratories, Inc., 333 Pfingston Road, Northbrook, IL 60062; Factory Mutual Research Corporation, 1151 Boston-Providence Turnpike, P.O. Box 688, Norwood, MA 02062
- *For Underground Gasoline and Petroleum Lines* Underwriters Laboratories Inc. (see above)

DESIGN AND INSTALLATION

Standard piping products offered for specific uses such as cold water; hot or cold water; drain, waste, and vent; sewerage; and drainage are largely predesigned. For example, CPVC and PB hot- or cold-water tubing systems made in accordance with their respective standards ASTM D 2846 and D 3309 are pressure-rated at 100 lb/in^2 (690 kPa) for water at 180°F (83°C). Design and installation recommendations are included in an appendix to these documents. Since most standards for products dedicated to a specific application contain design and installation recommendations, such documents should be consulted by designers and installers. The installation of plastic plumbing piping products is often regulated by the applicable plumbing code. More detailed information on design and installation may be obtained from most manufacturers and plastic pipe trade associations (see "Additional Information").

When design is conducted to meet special requirements of a given application, particular attention must be given to the effects of solvents and corrosives on piping properties. Temperature and unusual loadings (such as cyclic pressure and vibration) must also be considered. Most manufacturers of industrial piping products can provide performance data and design and installation recommendations for special as well as ordinary conditions. Various references that provide design and installation information are listed under "Additional Information."

Fundamentally, the principles of the design and installation of plastic piping are the same as those applying to steel piping. However, because of differences in their properties, certain aspects of the design and installation of plastics may require different emphasis and solutions. The following paragraphs briefly discuss these more important aspects. Detailed design and installation recommendations for each specific product should be followed.

Pressure Rating

Continuous Pressure. The pipe pressure rating should be based on the design stress [see Eq. (1*b*)] that is established by taking into account the anticipated effect of time, temperature, and environment on the pipe strength properties. It should be recognized that because of material and fabrication differences, plastic fittings and joints may have a pressure rating lower than that for the pipes. The design stress should include adequate margin for safety considerations. With this in mind it should be recognized that most thermoplastics, excepting those specially formulated for this purpose, are not suitable for conveying compressed gases in above-ground service. In the event of accidental pipe failure, the large potential energy stored in compressed gases could precipitate a catastrophic-type failure mechanism that sometimes produces dangerous flying debris. Thermosets and certain specially formulated thermoplastics resist such failure and are suitable for this application.

Vacuum or External Pressure. The service capabilities of thinner-walled pipes made of less rigid materials may be determined in some cases not by internal pressure but by vacuum conditions created by transients (surges) or by external hydrostatic pressure loading. The buckling resistance of plastic pipes may be estimated using Timoshenkos's classic elastic buckling equation including the following adaptation which gives consideration to the effect of pipe ovality:

$$P_c = \frac{2E}{1-\mu^2}\left(\frac{t}{D_m}\right)^3 C \tag{2}$$

where P_c = collapsing pressure of unconstrained pipe, lb/in^2
E = effective modulus of elasticity of pipe material, lb/in^2
μ = Poisson ratio (approximately from 0.35 to 0.45 for thermoplastics for short-term loading)
t = pipe-wall thickness, in
D_m = pipe mean diameter, in
C = factor correcting for pipe ovality = $(r_o/r_i)^3$, where r_i is the major radius of curvature of the ovalized pipe, and r_o is the radius assuming no ovalization

For short-term loading conditions, the values of E and μ as obtained from short-term tensile tests yield reasonable correlations. For long-term loading, appropriate values as determined from long-term loading tests should be employed.

Cyclic Pressure. The shock load or high-pressure surge created by sudden closure of valves could exceed the pressure capabilities of a pipe and, if the pipe is not properly anchored, could result in fitting failure by overstraining of joints. Excessive surging should be eliminated by control of the rate of valve closure or the installation of accumulators. All plastic pipe can tolerate some surging in excess of working pressure, and because of its greater flexibility and viscoelastic properties the pressures generated by surging are of lower order than those for metal piping and are more quickly damped. However, under frequent and continuous surging the long-term strength of plastic pipe tends to be reduced by fatigue. Under these conditions an appropriately lowered value of hydrostatic design stress should be used.

Considerations for Above-Ground Uses

Supports, Anchors, and Guides. Most manufacturers supply information on support spacings. Typical recommendations are presented in Tables 5.56 and 5.57. Values are usually given for either single or continuous spans and for a given liquid specific gravity. A most important consideration is to ensure that the span distance between supports is based on the maximum system temperature. Depending on the piping material, pipe dimensions, and application, the support span may be dictated by the permissible deflection [generally about ½ in (1.3 cm) at midspan] or maximum allowable stress. The allowable stress should provide allowance for stresses generated by fluid pressure and by thermal and other loadings. Vertical pipe runs can be supported either in compression or tension. Long runs should be checked to ensure that

TABLE 5.56 Typical Recommended Maximum Support Spacing, in Feet, for Thermoplastic Pipe for Continuous Spans and for Uninsulated Lines Conveying Fluids of Specific Gravity up to 1.35

Pipe dimension		PVC			CPVC				PVDF				PP			
Nominal diam, in	Wall schedule	60°F	100°F	140°F	60°F	100°F	140°F	180°F	80°F	100°F	140°F	160°F	60°F	100°F	140°F	180°F
½	Schedule 40	4½	4	2½	5	4½	4	2½	3¾	3½	2	Continuous support recommended	1¾	1¾	1½	1¼
¾		5	4	2½	5½	5	4	2½	4	3¾	2½		2	2	1¾	1¾
1		5½	4½	2½	6	5½	4½	2½	4¼	4	2½		2	2	2	1¾
1¼		5½	5	3	6	5½	5	3	—	—	—		2½	2¼	2	2
1½		6	5	3	6½	6	5	3	4½	4¼	2½		2½	2½	2¼	2
2		6	5	3	6½	6	5	3	4½	4½	2¾		3	2¾	2½	2¼
3		7	6	3½	8	7	6	3½					3½	2¾	3	2¾
4		7½	6½	4	8½	7½	6½	4					4	3¾	3½	3
6		8½	7½	4½	9½	8½	7½	4½								
8		9	8	4½												
½	Schedule 80	5	4½	2½	5½	5	4½	2½	4½	4½	2½	Continuous support recommended	2	2	2	1½
¾		5½	4½	2½	6	5½	4½	2½	4½	4½	3		2½	2½	2¼	2
1		6	5	3	6½	6	5	3	5	4¾	3		2½	2½	2¼	2
1¼		—	—	—	—	—	—	—	—	—	—		3	2¾	2½	2½
1½		6½	5½	3½	7	6½	5½	3½	5½	5	3		3	3	2¾	2½
2		7	6	3½	7½	7	6	3½	5½	5¼	3		3½	3¼	3	2¾
3		8	7	4	9	8	7	4					4	4	3½	3½
4		9	7½	4½	10	9	7½	4½					4½	4½	4	3½
6		10	9	5	11	10	9	5								
8		11	9½	5½												

TABLE 5.57 Typical Recommended Maximum Support Spacing, in Feet, for Fiberglass-Reinforced Pipe for Temperatures up to 150°F (65°C) for Uninsulated Lines Conveying Fluids of up to 1.25 Specific Gravity

Nominal pipe size, in	Continuous span	Single span
1	7.5	6.3
1½	8.5	7.2
2	9.9	8.4
3	11.2	9.4
4	11.9	10.0
6	14.2	11.9
8	15.7	13.2
10	17.4	14.6
12	18.9	15.9

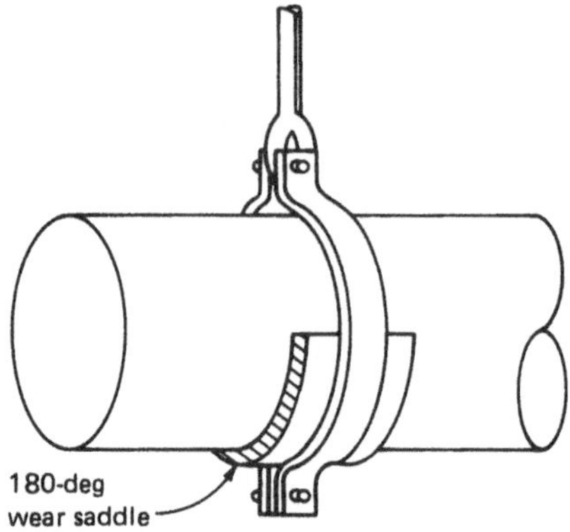

FIGURE 5.165 Wear-saddle prevents damage from standard hanger.

the tensile or compressive load does not exceed the permissible design value. Standard strap, sling, clamp, clevis, and saddle supports providing at least 120° of contact are generally recommended (Fig. 5.165). Supports offering narrow or point contact should be avoided. Valves and other heavy piping components should be individually supported.

Anchors, which divide a pipe system into sections, must positively restrain the movement of pipe against all applied and developed forces, particularly dynamic loading. Anchors should generally be employed near changes of direction, when transitioning to another piping material or when there is a change in line size. In long, straight runs, anchor spacing is generally recommended at about 200 to 300 ft (60 to 90 m). Anchor spacing and location will also be determined by the selection of anchoring for the control of expansion and contraction.

When the pipe is restrained against expansion and contraction it should be guided to prevent buckling. The guides should encircle the pipe but be loose enough to allow it to move freely in its axial direction.

Control of Expansion and Contraction. Because of the inherent flexibility of plastic piping, it is generally possible to design a pipe system so that no expansion joints are necessary. Their use should be avoided where possible, since they are expensive, and they remove the ability of the pipe to carry longitudinal loads. Offsetting this load with anchors can be an added problem with larger pipes. Preferred techniques for dealing with expansion and contraction are: (1) anchoring and guide spacing, (2) changing direction (offset legs), and (3) using expansion loops.

Although plastics expand more than steel, their relatively low modulus of elasticity results in significantly lower end load for the same temperature change. The smaller thermal forces can generally be readily relieved by changes in direction. However, when using directional change to absorb thermal forces, neither the pipe nor any of its components should be subjected to a bending stress (or strain) in excess of that recommended by the manufacturer. The stress may be controlled by placement of anchors not closer than a calculated distance from the point of change of direction. The length of the legs of expansion loops should similarly be determined.

Oscillations and Vibrations. Because plastic pipe is so much more flexible than steel pipe, oscillations due to changes in velocity of fluid set up more easily and tend to be of greater amplitude. In long runs this is generally no problem but they could damage connected piping by subjecting it to excessive stresses or strains. The solution is to restrain the pipe by the use of anchors. High-amplitude vibrations from connected equipment, such as pumps, should be isolated from the piping by the use of flexible connectors.

Considerations for Below-Ground Uses

In terms of their underground performance all plastic pipes are classified as flexible, which signifies that when they are properly installed they are capable of developing sufficient diametrical deformation, without incurring material failure, to fully activate soil support forces. Soil-assisted flexible pipes can easily support earth loads that would crush stronger rigid pipes for which total load bearing ability is almost entirely dependent on their own strength. To activate soil support, flexible pipes must be embedded in soils that are stable and that have been properly placed and densified around the pipe. ASTM documents D 2321, "Underground Installation of Flexible Thermoplastic Sewer Pipe," and D 3839, "Underground Installation of Flexible Reinforced Thermosetting Pipe and Reinforced Plastic Mortar Pipe," present detailed recommendations on the proper installation of flexible plastic pipe designed for nonpressure uses. Nonpressure pipe, which is generally of thinner wall construction and not rounded by internal pressure, requires somewhat more care in installation than heavier-wall pressure pipe. Recommendations for installation of thermoplastic pressure pipe are given in ASTM D 2774, "Underground Installation of Thermoplastics Pressure Piping."

The basic principle of the installation of buried plastic piping is to embed it in a soil of such quality that the resultant ultimate pipe deflection is controlled to an acceptable value that is limited by either the pipe performance requirements or the pipe material capabilities. The former generally permits deflection up to about 10 percent (some engineers may set conservatively lower values) while the latter is determined by the maximum allowable stress, or strain, in the pipe wall for the given pipe material and construction. The pipe supplier can provide the limiting deflection values for a given pipe material and construction. These values may depend on the fluids being handled. With thermoplastic pipe of solid wall construction deflection will seldom be limited by material performance constraints.

The extent to which a flexible pipe will deflect when embedded in a given quality of soil may be estimated by a variety of methods. One of the better-known relationships, sometimes called the Iowa equation, was developed for flexible metal conduits at Iowa State University. A modification of this equation is

$$\frac{\Delta x}{D_i} = \frac{L_D K P}{EI/r^3 + 0.061\,E'}$$

where Δx = horizontal deflection of the pipe, in (For relatively small deflections, the change Δy in vertical diameter of a circular section deforming elliptically is equal to 1.10 Δx. As an approximation, it is often assumed $\Delta y = \Delta x$.)

D_i = pipe inside diameter prior to loading, in

L_D = deflection lag factor compensating for the time dependence of soil deformation, dimensionless

K = bedding constant which varies with the angle of bedding (i.e., bedding support), dimensionless (The bedding constant ranges from 0.110 for a point support on the bottom of a pipe to 0.083 for full support. For plastic pipe, the typical value is taken as 0.10.)

P = total vertical pressure acting on the pipe, lb/in^2

r = pipe radius, in

E = modulus of elasticity of pipe material, lb/in^2

I = moment of inertia of pipe wall per unit of length, in^4/in (For round pipe $I = t_a^3/12$, in which t_a is the average wall thickness.)

E' = modulus of passive soil resistance, lb/in^2

TABLE 5.58 Bureau of Reclamation Values of E' for Iowa Formula for Initial Average Deflection of Flexible Pipe

Soil type for pipe embedment material per ASTM D 2321	Soil type description (United Classification System, ASTM D 2487)	E', lb/in² for degree of compaction of embedment (proctor density, %)*			
		Dumped	Slight (>85%)	Moderate (85–95%)	High (>95%)
I	Manufactured angular, granular materials (crushed stone or rock, broken coral, cinders, etc.)	1000 (+4%)	3000 (+4%)	3000 (+3%)	3000 (+2%)
II	Coarse-grained soils with little or no fines	N.R.†	1000 (+4%)	2000 (+3%)	3000 (+2%)
III	Coarse-grained soils with fines	N.R.	N.R.	100 (+3%)	2000 (+2%)
IV	Fine-grained soils	N.R.	N.R.	N.R.	N.R.
V	Organic soils (peats, mulches, clays, etc.)	N.R.	N.R.	N.R.	N.R.

* Values in parentheses give the approximate limit of deflection beyond the average deflection that is computed by using the given E' values. These limits are for pipe of relatively low stiffness. As pipe stiffness increases, the limit is narrowed.

† N.R. indicates use not recommended by ASTM D 2321.

As a result of extensive field investigations of the load versus deflection characteristics of various flexible pipes, the U.S. Bureau of Reclamation* has developed a series of soil reaction E' values for use in the Iowa equation under the assumption that $K = 0.1$ and $D_L = 1.0$. These E' values may be used to estimate a pipe's initial average deflection. To assist in estimating the initial maximum (i.e., acceptance) deflection as a consequence of both soil loading and installation factors, the Bureau of Reclamation has also reported the observed upper limits of deflection values. Both these limits, which primarily apply to pipes of lower ring stiffness, and the values of E' are shown in Table 5.58 as a function of the embedment materials recommended by D 2321.

In actual practice, it is seldom necessary to go through the Iowa equation calculation, for if the recommended installation practices in D 2321 are followed, initial installed deflections can quite readily be held to approximately 5 percent and less. Burial of the thinner-walled, more flexible pipes may also require consideration of the adequacy of the pipe's wall compressive strength as well as its buckling stability (Fig. 5.166).

Recommendations for the design and installation of buried plastic pipe may be obtained from pipe manufacturers, trade associations, or from the listed information given under "Additional Information."

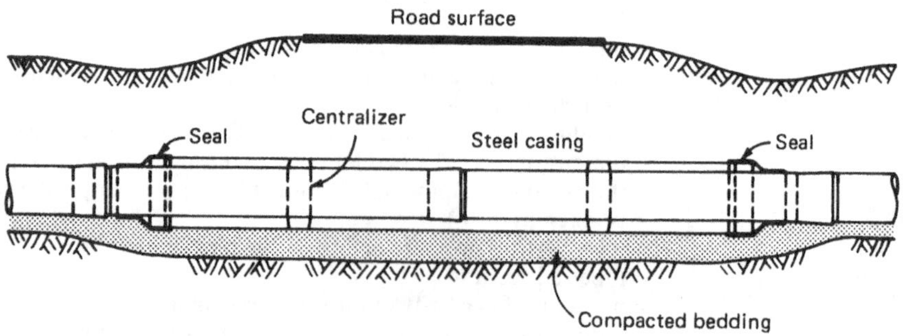

FIGURE 5.166 Steel casing protects pipe from concentrated loadings.

ADDITIONAL INFORMATION

The various plastic pipe trade associations issue reports, manuals, and lists of references on design and installation of their members' products. A list of current reports may be obtained by contacting each organization as follows:

- *Reinforced Thermosetting Piping* Fiberglass Pipe Institute/Composites Institute, a Division of the Society of the Plastics Industry, 355 Lexington Avenue, New York, NY 10017; The Materials Technology Institute of the Chemical Process Industries, Inc., 12747 Olive St. Rd., Suite 203, St. Louis, MO 63141-6269
- *Thermoplastics Pipe (Industrial, Gas Distribution, Sewerage, Water, and General Uses)* The Plastics Pipe Institute, a Division of the Society of the Plastics Industry, Inc., 1275 K Street, NW, Suite 400, Washington, DC 20005
- *Thermoplastics Pipe (Plumbing Applications)* Plastics Pipe & Fittings Association, 999 North Main Street, Glen Ellyn, IL 60137
- *PVC Piping (Water Distribution, Sewerage, and Irrigation)* Uni-Bell Plastics Pipe Association, 2655 Villa Creek Drive, Suite 164, Dallas, TX 75234

BIBLIOGRAPHY

The following references include useful information on plastics piping:

Britt, William F., Jr.: "Design Considerations for FRP Piping Systems," *Proceedings of the 1979 Conference on Managing Corrosion with Plastics,* National Association of Corrosion Engineers, Houston.

Cheremisinoff, Nicholas P., and N. Paul: *The Fiberglass-Reinforced Plastics Deskbook,* Ann Arbor Science, Ann Arbor, Michigan, 1978.

Cooney, J. L.: "Guidelines for the Inspection and Maintenance of FRP Equipment and Piping," *Proceedings of the 1979 Conference on Managing Corrosion with Plastics,* National Association of Corrosion Engineers, Houston.

Escher, G. A.: "Transition to FRP, Basic Guidelines for Piping Designers and Users," *Proceedings of the 1975 Conference on Managing Corrosion with Plastics,* National Association of Corrosion Engineers, Houston.

Escher, G. A., and W. B. MacDonald: "Chemical and Mechanical Properties of Butt and Strap Joints," *Proceedings of the 1975 Conference on Managing Corrosion with Plastics,* National Association of Corrosion Engineers, Houston.

Greenwood: "Buried FRP Pipe—Performance Through Proper Installation," *Proceedings of the 1975 Conference on Managing Corrosion with Plastics,* National Association of Corrosion Engineers, Houston.

Kutschke, C. T.: "Use of Plastic Pipe for Industrial Applications," *Proceedings of the 1975 Conference on Managing Corrosion with Plastics,* National Association of Corrosion Engineers, Houston.

Launikitis, M. B.: "Chemically Resistant FRP Resins," *Proceedings of the 1977 Conference on Managing Corrosion with Plastics,* National Association of Corrosion Engineers, Houston.

Mallison, John H.: *Chemical Plant Design with Reinforced Plastics,* McGraw-Hill, New York, 1969.

Mruk, Stanley: "Thermoplastics Piping: A Review," *Proceedings of the 1979 Conference on Managing Corrosion with Plastics,* National Association of Corrosion Engineers, Houston.

Petroff, Larry J., and Luckenbill, Michael: "Flexibility of the Design of Fiberglass Pipe," *Proceedings of the 1981 International Conference on Plastic Pipe,* American Society of Civil Engineers, New York.

Plastics Piping Manual, The Plastics Pipe Institute, 1275 K Street, NW, Suite 400, Washington, DC 20005, 1976.

Proceedings of the International Conference on Underground Plastic Pipe, March 30–April 1, 1981, New Orleans, La., American Society of Civil Engineers, New York.

"PVC Pipe, Design and Installation," AWWA Manual No. M23, American Water Works Association, Denver, 1980.

Rolston, Albert: "Fiberglass Composite and Fabrication," *Chemical Engineering,* January 28, 1980, pp. 96–110.

Rubens, A. C.: "Designing RTRP Systems Utilizing Published Engineering Data," *Proceedings of the 1979 Conference on Managing Corrosion with Plastics,* National Association of Corrosion Engineers, Houston.

Schrock, B. J.: "Thermosetting Resin Pipe," Preprint 3088, American Society of Civil Engineers, New York, 1977.

"Standard for Reinforced Thermosetting Resin Pipe," AWWA C-950, American Water Works Association, Denver. (Appendix to standards includes much design and installation information.)

CHAPTER 5.15
PIPE INSULATION

Michael R. Harrison
Vice President and General Manager
Manville Mechanical Specialty Insulations
Schuller International, Inc.
Denver, Colorado

Ricardo R. Gamboa
Manager, Engineering and Technical Services
Manville Mechanical Specialty Insulations
Schuller International, Inc.
Denver, Colorado

HEAT-TRANSFER FUNDAMENTALS

Heat energy is transferred from one location to another by three different mechanisms: *conduction, convection,* and *radiation.* In insulation design theory, the objective is to minimize the contribution of each mode in the most efficient and economical manner. As temperatures vary, the relative importance of each transfer mechanism also varies, making different insulation designs appropriate for various applications.

Conduction

Energy transfer by conduction is a result of atomic or molecular motion. As molecules become heated, their vibration increases, and energy is transferred to surrounding molecules. Conduction occurs in all three forms of matter: gas, liquid, and solid. Within most insulation, solid conduction is minimized by using an open-pore structure and a minimum amount of solid material. Conduction within a gas is more difficult to control; to achieve much greater insulation efficiency, this mode must be limited. One method employs a vacuum, thus eliminating the gas from the system. This is very effective but costly, since the vacuum seal must be maintained in order to assure adequate performance. The second method of controlling gas conduction is to replace the air in the insulation by a heavier gas such as Freon®. Again, the seal must be maintained to avoid eventual air and moisture migration back into the cell structure.

Convection

Convective currents are established when a hot fluid (gas or liquid) rises from a heat source and is replaced by a cooler fluid, which in turn is heated and rises, carrying the energy with it. In an insulation structure with many small cells, convection is minimized since the gas cannot freely pass through the structure. Most insulations are of sufficient density and formation to almost eliminate this mode within them, but convection plays a very important part in transferring energy from the insulated surface to the surrounding environment.

Radiation

As temperature increases, electromagnetic radiation gains in significance with regard to the total amount of energy transferred. Radiation occurs in a vacuum as well as in a gaseous environment, and its magnitude is dependent on the emittance of the radiating and receiving surfaces and the temperature difference between them. To control radiant flow, low-emittance surfaces are used in conjunction with absorbers and reflectors within the insulation itself. The mass density of the insulation is very important with a higher density reducing the level of radiation transfer.

There are obvious tradeoffs that must be made in controlling the various heat-transfer mechanisms. Figure 5.167 illustrates the contribution to total conductivity of each mechanism at three different temperatures. The most efficient insulation design, both thermally and economically, will vary depending on the application conditions. References 1 and 2 are basic texts on heat transfer for further study.

Heat Flow

The level of heat flow to or from a system is directly proportional to the difference between the system and ambient temperatures, and inversely proportional to the thermal resistance placed in the heat flow path:

$$\text{Heat flow} = \frac{\text{temperature difference}}{\text{resistance to heat flow}}$$

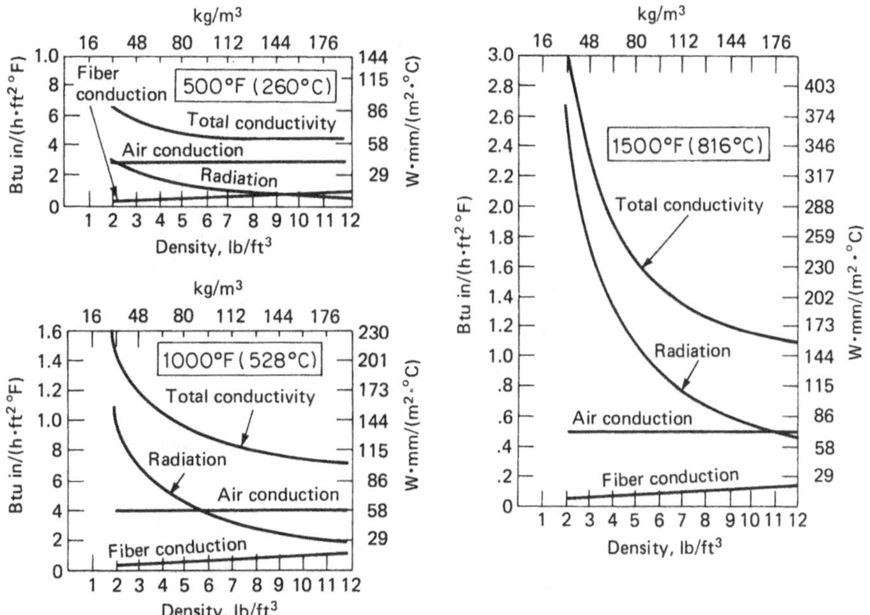

FIGURE 5.167 Contribution of each mode of heat transfer.[3]

In this light, the temperature difference is the forcing function; as long as there is a differential, energy will flow. No amount of thermal resistance can completely stop the heat transfer; it can only slow the rate at which it occurs.

Total thermal resistance is generally composed of two distinct types of resistances: insulation and surface. Insulation resistance for a homogeneous material is determined by dividing the insulation thickness *tk* by its thermal conductivity.

$$R_I = tk/k$$

The thermal conductivity, or k value, is an experimentally measured property of homogeneous material indicating the time rate of steady state heat flow induced by an unit temperature gradient in a direction perpendicular to that unit area. k are expressed in in-lb units (Btu)(in)/(h)(ft_2)(°F) and in SI units (W/m · °C). In the case of nonhomogeneous thermal insulation, materials that exhibit thermal transmission by several modes of heat transfer (resulting in property variation with specimen thickness or surface emittance), an apparent thermal conductivity is assigned.

The accurate measurement of thermal conductivity is very important since materials are often compared on this basis. The American Society for Testing and Materials (ASTM) has developed standardized test methods for measuring thermal conductivity. For pipe insulation, the standard test is C 335.[4]

Since thermal conductivity increases with temperature, it is important to use the k value at the insulation mean temperature rather than the value at either the operating temperature (too high) or the ambient temperature (too low).

The other type of thermal resistance is surface resistance:

$$R_s = 1/f$$

where f represents the surface film coefficient. Surface emittance, air velocity across the surface, and temperature difference all influence the value of R. Table 5.59 lists various R_s values for three common surface types, temperature differentials, and wind velocities. These values will be used in subsequent heat-transfer calculations.

TABLE 5.59 Values for Surface Resistance, R_s, (h)(ft²)(°F) and Btu (m² · °C/W)

A. Values for still air

$t_s - t_a$		Plain, fabric dull metal	Aluminum	Stainless steel
°F	°C	ε = 0.95	ε = 0.2	ε = 0.4
10	5	0.53 (0.093)	0.90 (0.158)	0.81 (0.142)
25	14	0.52 (0.091)	0.88 (0.155)	0.79 (0.139)
50	28	0.50 (0.088)	0.86 (0.151)	0.76 (0.133)
75	42	0.48 (0.084)	0.84 (0.147)	0.75 (0.132)
100	55	0.46 (0.081)	0.80 (0.140)	0.72 (0.126)

B. R_s Values with wind velocities

Wind velocity		Plain, fabric dull metal		Aluminum		Stainless steel	
mi/h	km/h	mi/h	km/h	mi/h	km/h	mi/h	km/h
5	8	0.35	0.06	0.41	0.07	0.40	0.07
10	16	0.30	0.05	0.35	0.06	0.34	0.06
20	32	0.24	0.04	0.28	0.05	0.27	0.05

* For heat-loss calculations, the effect of R_s is small compared with R_I, so the accuracy of R_s is not critical. For surface-temperature calculations, R_s is the controlling factor and is, therefore, quite critical. The values presented in Table 5.60 are commonly used values for piping and flat surfaces. More precise values based on surface emittance and wind velocity can be found in the referenced texts.

Source: Schuller International, Inc., Ref. 5.

Since thermal resistances are additive, they are very convenient to work with in calculating heat transfer. From the basic definition,

$$Q = \frac{\Delta t}{R_I + R_s} = \frac{\Delta t}{tk/k + R_s} \tag{1}$$

becomes the basic equation for heat transfer through insulation. Use of this equation is illustrated later.

Insulation Effectiveness

Before leaving the fundamentals, the importance of insulation should be illustrated. Table 5.60 shows the amount of heat transfer from a bare surface at a given temperature differential. Essentially, these values are calculated from the fundamental heat-transfer equation, Eq. (1), with the insulation thickness being zero. Listed below are three different sets of operating conditions for an 8-in (20-cm) pipe. To show the effectiveness of insulation, heat losses are shown for the bare pipe and for the pipe with 1 in of fiberglass insulation applied, even though a greater thickness would normally be used.

Operating temp., °F	Ambient temp., °F	Bare heat loss Btu/(h)(ft)	Heat loss with 1 in fiberglass, Btu/(h)(ft)	Reduction in heat loss, %
200	80	617	70	88.7
350	80	1882	188	90.1
500	80	3998	353	91.2

TABLE 5.60 Heat Loss from Bare Surfaces*

Nominal pipe size, inches	Temperature difference, °F															
	50	100	150	200	250	300	350	400	450	500	550	600	700	800	900	1000
½	22	47	79	117	162	215	279	355	442	541	650	772	1047	1364	1723	2123
¾	27	59	99	147	203	269	349	444	552	677	812	965	1309	1705	2153	2654
1	34	75	124	183	254	336	437	555	691	846	1016	1207	1637	2133	2694	3320
1¼	42	94	157	232	321	425	552	702	873	1070	1285	1527	2071	2697	3406	4198
1½	49	107	179	265	367	487	632	804	1000	1225	1471	1748	2371	3088	3899	4806
2	61	134	224	332	459	608	790	1004	1249	1530	1837	2183	2961	3856	4870	6002
2½	74	162	271	401	556	736	956	1215	1512	1852	2224	2643	3584	4669	5896	7267
3	89	197	330	489	677	897	1164	1480	1841	2256	2708	3219	4365	5685	7180	8849
3½	102	225	377	558	773	1024	1329	1690	2102	2576	3092	3675	4984	6491	8198	10100
4	115	254	424	628	869	1152	1496	1901	2365	2898	3479	4135	5607	7304	9224	11370
4½	128	282	471	698	965	1280	1662	2113	2628	3220	3866	4595	6231	8116	10250	12630
5	142	313	524	776	1074	1424	1848	2350	2923	3582	4300	5111	6931	9027	11400	14050
6	169	373	624	924	1279	1696	2201	2799	3481	4266	5121	6086	8254	10750	13580	16730
7	195	430	719	1064	1473	1952	2534	3222	4007	4910	5894	7006	9501	12380	15630	19260
8	220	486	813	1203	1665	2207	2865	3643	4531	5552	6666	7922	10740	13990	17670	21780
9	246	542	907	1343	1859	2464	3198	4066	5057	6197	7440	8842	11990	15620	19720	24310
10	275	606	1014	1502	2078	2755	3576	4547	5655	6930	8320	9888	13410	17470	22060	27180
11	300	661	1106	1638	2267	3005	3901	4960	6169	7560	9076	10790	14630	19050	24060	29660
12	326	718	1202	1779	2463	3265	4238	5338	6701	8212	9859	11720	15890	20700	26140	32210
14	357	783	1319	1952	2703	3582	4650	5912	7354	9011	10820	12860	17440	22710	28680	35350
16	408	901	1508	2232	3090	4096	5317	6759	8407	10300	12370	14700	19940	25970	32790	40410
18	460	1015	1698	2514	3480	4012	5987	7612	9467	11600	13930	16550	22450	29240	36930	45510
20	510	1127	1885	2790	3862	5120	6646	8449	10510	12880	15460	18380	24920	32460	40990	50520
24	613	1353	2263	3350	4638	6148	7980	10150	12620	15460	18570	22060	29920	38970	49220	60660
30	766	1690	2827	4186	5795	7681	9971	12680	15770	19320	23200	27570	37390	48700	61500	75790
Flat	98	215	360	533	738	978	1270	1614	2008	2460	2954	3510	4760	6200	7830	9650

* Losses given in Btu per hour per linear foot of bare pipe at various temperature differences and Btu per hour per square foot for flat surfaces.
Source: Reference 3.

PIPE INSULATION MATERIALS

Table 5.61 lists the principal insulations being used along with their important properties. To assure accurate information, current data sheets from the manufacturer should always be consulted before specifying a particular product. Table 5.61 also provides a brief description of each type of material and the benefits and drawbacks of each.

Calcium Silicate

Lime, silica, and reinforcing fibers are used as main raw materials in a chemical process to form rigid insulations. Since no organic binders are used, the products are noncombustible and maintain their physical integrity at very high temperatures. These materials are known for their exceptional durability and strength and are the high-quality standard in industrial plant environments, where physical abuse is always a problem. Calcium silicate insulations are more expensive than fibrous materials but are more thermally efficient at higher temperatures.

Cellular Glass

This product is also rigid and completely inorganic, composed of millions of completely sealed glass cells. It is unique among insulation materials in that it is made up of totally closed cells and will not absorb any liquids or vapors. Although its thermal conductivity is higher than most of the other insulations, the product is widely used in pipe systems operating below ambient temperatures and buried applications where moisture is often a problem. Similarly, lines carrying flammable liquids often use cellular glass around valves and fittings to minimize the hazard of a saturated insulation. The product is load-bearing, but it is also brittle and must be protected when installed on pipes where vibrations are likely to occur. The material is subject to thermal shock at temperatures above 400°F (204°C).

Expanded Perlite

A naturally occurring material, perlite is expanded at high temperature to form a structure of tiny air cells within a vitrified product. Organic and/or inorganic binders are used in conjunction with reinforcing fibers to hold the perlite structure together. At temperatures below 300°F (167°C), these products have low moisture absorption, but this increases after exposure to elevated temperatures. The products are rigid and load bearing, but are more brittle and less thermally efficient than the calcium silicate materials.

Mineral Rock or Slag Wool

Formed from molten rock or slag, these products have a higher recommended temperature limit than fiberglass, but they utilize similar organic binders for structural integrity. The products are not load bearing and contain varying amounts of unfiberized material typically called *shot*. The greatest drawbacks of the mineral rock or slag wool materials is the short fiber length and shot content, which allow vibration and physical abuse to cause severe damage, particularly after the organic binder is oxidized at around 400 to 500°F (204 to 260°C). However, the products have a lower first cost than most of the rigid insulations.

Glass Fiber

Fiberglass pipe insulations are formed from molten glass and bonded with organic resins. The principal products for commercial piping systems are the one-piece molded insulations, which install rapidly by hinging open and then closing around the pipe. Flexible wraparound prod-

TABLE 5.61 Properties of Various Industrial Pipe Insulation Materials

Insulation type	Temp. range, °F (°C)	Thermal conductivity (Btu)(in)/(h)/(ft²)/(°F) at T mean, °F(°C) (W·mm/m² · °C)			Compressive strength, lb/in² (kPa) at % deformation	Classification or flame-spread smoke developed*	Cell structure (permeability and moisture absorption*)
		75°F (42°C)	200°F (93°C)	500°F (260°C)			
Calcium silicate	1200 to 1500 (649 to 815)	(41.6) 0.37 (52)	(93.3) 0.41 (59)	(260) 0.53 (76)	100 to 250 @ 5% (689 to 1722)	Noncombustible	Open cell
Cellular glass	–450 to 900 (268 to 482)	0.38 (54)	0.45 (65)	0.72 (104)	100 @ 5% (689)	Noncombustible	Closed cell
Expanded perlite	1200 to 1500 (649 to 815)	—	0.46 (66)	0.63 (91)	90 @ 5% (620)	Noncombustible	Open cell
Mineral fiber	to 1900 (1038)	0.23 to 0.34 (33 to 49)	0.28 (40)	0.45 (65)	1 to 18 @ 10% (6.9 to 124)	Noncombustible to 25/50	Open cell
Glass fiber	to 850 (454)	0.23 (33)	0.30 (43)	0.62 (89)	0.02 to 3.5 @ 10% (0.13 to 24)	Noncombustible to 25/50	Open cell
Urethane foam	–100 to –450 to 225 (–73 to –268 to 107)	0.16 to 0.18 (23 to 26)	—	—	16 to 75 @ 10% (110 to 516)	25 to 75 140 to 400	95% closed cell
Isocyanurate foam	to 350 (177)	0.15 (22)	—	—	17 to 25 @ 10% (117 to 172)	25 55 to 100	93% closed cell
Phenolic foam	–40 to 250 (–40 to 121)	0.23 (33)	—	—	13 to 22 @ 10% (89 to 151)	25/50	Open cell
Elastomeric closed cell	–40 to 220 (–40 to 104)	0.25 to 0.27 (36 to 39)	—	—	40 @ 10% (275)	25 115 to 490 50 ≤ ¾ in 50–100 > ¾ in	Closed cell

* FHC from ASTM E 84 and UL 723 which indexes flame and smoke development to that generated by Red Oak which has 100/100 FHC.

ucts are also available for large-diameter pipes and vessels. Fiberglass products are also available for large-diameter pipes and vessels. Fiberglass products are very thermally efficient and easy to work with.

The only question about their use is for applications above the binder oxidation temperature of 400 to 500°F (205 to 260°C). Most manufacturers rate their products above the binder temperature and are confident of their performance due to the long glass fiber matrix. However, there are certain applications that combine severe vibration with high temperature. In such cases, fiberglass products, like mineral wools, may tend to sag on the pipe and lose some of their thermal efficiency.

Foams

Four products in the general foam category are now being used primarily for cold service and plumbing applications. Polyurethane- and isocyanurate-foamed plastics offer the lowest thermal conductivity since the cells are filled with fluorocarbon blowing agents that are heavier than air. However, the products still need to be sealed to prevent the migration of air and water vapor back into the cells. There have been problems with urethane regarding both dimensional stability and fire safety; the isocyanurates were developed, in part, to deal with these problems. However, a surface burning characteristic of 25/50 FHC (fire hazard classification) is still being sought for these materials.

Phenolic foams have the required level of fire safety but do not offer thermal efficiencies much better from that of fiberglass. They are limited in temperature range, but their rigid structure allows them to be used without special pipe saddle-supports on small lines.

Elastomeric closed-cell materials are the fourth type and are used primarily in plumbing and refrigeration work. These products are both flexible and closed-cell, permitting rapid installation without an additional vapor retarder lagging for moderate design conditions. These materials have a surface burning characteristic of 25/50 FHC only for thicknesses 0.75 in (1.9 cm) and below; this restricts their use in certain applications.

MATERIALS SELECTION AND APPLICATION

There are many factors involved in selecting the best insulation for specific application. First, the service requirements and locations must be analyzed to determine which products will, for example, meet the pipe temperature requirements and also be able to withstand the anticipated abuse. Also, special considerations such as fire protection, removability, and chemical resistance must all be reviewed in the selection process.

Finally, three insulation cost factors should be carefully analyzed. The *initial cost* of the material and installation is straightforward; competitive products can be readily compared. *Maintenance costs* are not as clear-cut and vary greatly from one material to another, with the nonrigid materials usually requiring more maintenance in industrial environments. The last cost element is that of the *heat lost* through the insulation system. If the same thickness is specified for several different materials, the product with the lowest thermal conductivity will provide the lowest lost-heat cost. Similarly, a material that requires less maintenance may well retain its original performance longer than a material that is easily damaged and degraded. All of these costs should be reviewed before choosing the lowest bid package, which only deals with initial costs.

The following paragraphs deal with insulation selection based on operating temperature, design considerations, and common industrial use.

Cryogenic

Cryogenic insulation systems [–455 to –150°F (–271 to 101°C)] are usually custom engineered due to the critical nature of such service. The two problems encountered are moisture migration and low insulation efficiency. Unequal vapor pressure serves to drive moisture to the cold pipe, which results ultimately in ice formation and the subsequent deterioration of the insulation. Multiple vapor barriers are used to prevent this, and cellular glass adds additional security since it is totally closed cell. The system must also be thermally efficient so as to keep the outside surface temperature above the ambient dew point.

This may require outlandish thicknesses of conventional insulations, although the more efficient plastic foams are frequently used for this reason. The most severe applications employ vacuum products with multiple layers of reflective foil or vacuum cavities with powder fill.

Low-Temperature

Refrigeration, plumbing, and HVAC systems operate in the range of −150 to 212°F (−101 to 100°C). Cellular glass, fiberglass, and foam products are the most frequently used products. Vapor-barrier requirements are still important for all applications below ambient temperatures, but the level of protection needed becomes less as ambient conditions are approached. Applications above ambient require little special attention with the exception that the plastic foams begin to reach their temperature limits around 200°F (93°C).

Much of this temperature service is in residential and commercial buildings. A variety of fire codes are in force for insulation—ranging from a requirement for noncombustibility to allowing smoke development up to 400. Many codes call for a flame spread of 25 or less, and products that carry a composite 25/50 FHC are suitable for virtually all applications. In general, foam products and cellular glass are used for most of the below-freezing applications while fiberglass is the primary product used for chilled water and above-ambient temperature work. Also, premolded PVC fitting covers with fiberglass inserts are frequently used as a quick and efficient way of insulating the wide variety of fittings encountered.

Intermediate-Temperature

Almost all steam and hot-process piping systems operate in the temperature range of 212 to 1000°F (100 to 538°C). The products most frequently used are calcium silicate, fiberglass, mineral wool, and expanded perlite. Choices are usually based on thermal conductivity and resistance to physical abuse where applicable. Fiberglass products are the most thermally efficient at lower temperatures, with calcium silicate being the best at higher temperatures. *It is important to use the insulation mean temperature when comparing thermal conductivities* in order to make the proper comparison.

Fiberglass pipe insulations reach their temperature limits within this range, usually at 850°F (454°C). However, they all have organic binders (as do the mineral wools) that begin to oxidize from 400 to 500°F (204 to 260°C). This should be recognized when applying the products to piping above this temperature, but should not be used as an arbitrary cutoff point for fibrous products unless the service conditions are severe enough to warrant it. Vibration and physical abuse conditions should be analyzed for each particular application.

Fire safety is another concern, in this instance from the standpoint of fire protection rather than fire hazard. Figure 5.168 shows the results of fire tests run with three high-temperature materials. The water of hydration in the calcium silicate prolonged the time required for the steel pipe to reach an unacceptable temperature.

Valves and fittings are most often insulated with field-mitered sections of the standard pipe insulation or shop-fabricated fitting covers. There is limited usage of mesh-enclosed blankets, which are laced around large valves. Also, cellular glass valve and flange covers are often used in conjunction with other insulations when the fluid being transported is volatile or has a low flash point.

Reviewing the entire range, fiberglass insulation is most often used in lower-temperature applications where physical abuse is not a problem because of its thermal efficiency and

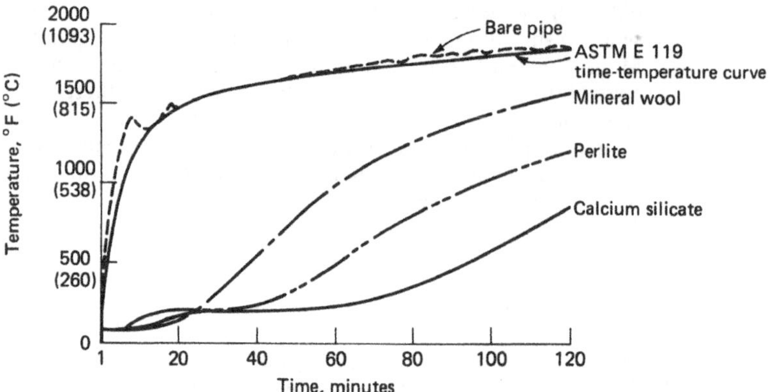

FIGURE 5.168 Fire-resistance test data for pipe insulation. *(Journal of Thermal Insulation.*[6]*)*

rapid installation. Calcium silicate is the standard type for higher-temperature work and in severe service. Expanded perlite provides the benefit of moisture resistance at lower temperatures, and mineral wool offers initial cost economies for higher-temperature work.

High-Temperature

Superheated steam, exhaust ducting, and some process operations are the few piping applications in the temperature range of 1000 to 1600°F (538 to 871°C). Again, calcium silicate, expanded perlite, and mineral wool are the usual materials employed. These products reach their temperature limits within this range, and higher temperature requirements are usually met by ceramic fiber blanket materials wrapped around the piping.

Installation Techniques and Specifications

The proper installation of pipe insulation is critical to the in-place performance of the product. Figure 5.169*a* illustrates installation methods for glass fiber pipe insulation with different jacket materials and pipe supports. Figure 5.169*b* illustrates insulation methods for a 90° elbow. References 7 and 8 provide detailed schematics and procedures for such work and should be consulted by plant maintenance workers involved in insulation application. Similarly, the proper selection of protective coatings and jackets will greatly influence the long-term product performance. These references, along with technical data from coating and jacketing manufacturers, provide appropriate information for proper selection.

INSULATION THICKNESS AND HEAT-LOSS CALCULATIONS

Terminology

The following symbols, definitions, and units will be used throughout the calculations.

t_a = ambient temperature, °F (°C)

t_s = surface temperature of insulation next to ambient, °F (°C)

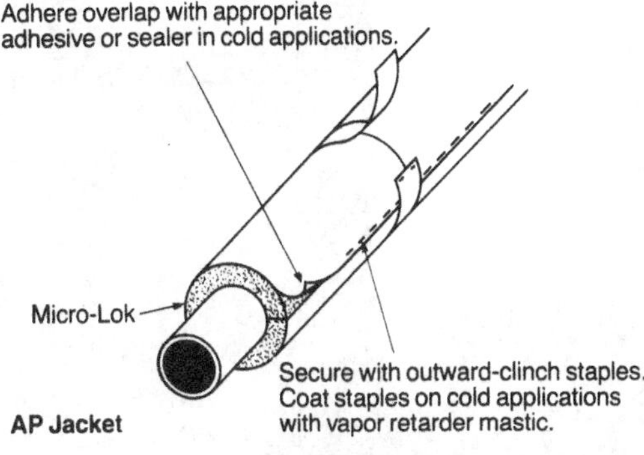

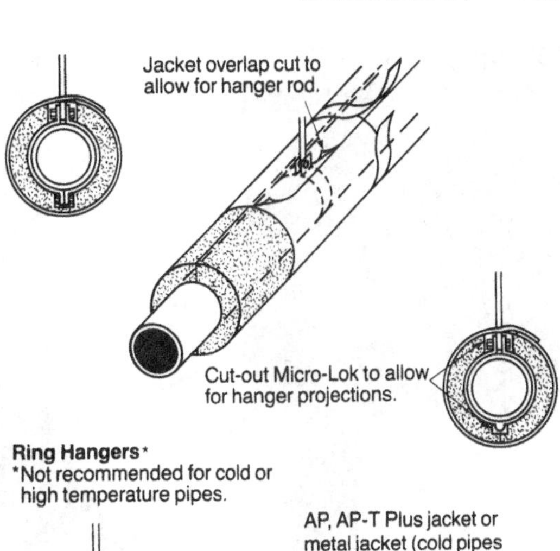

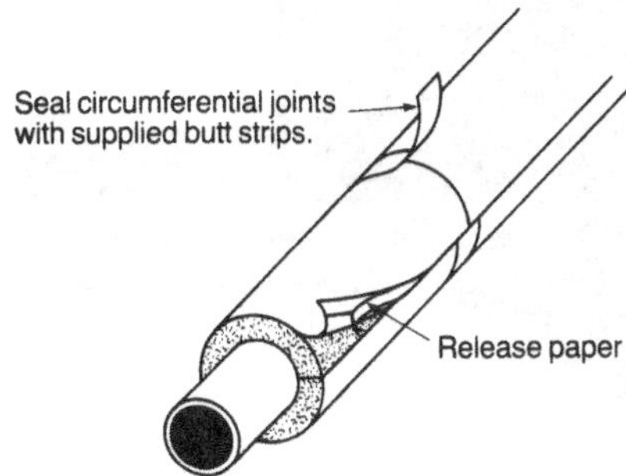

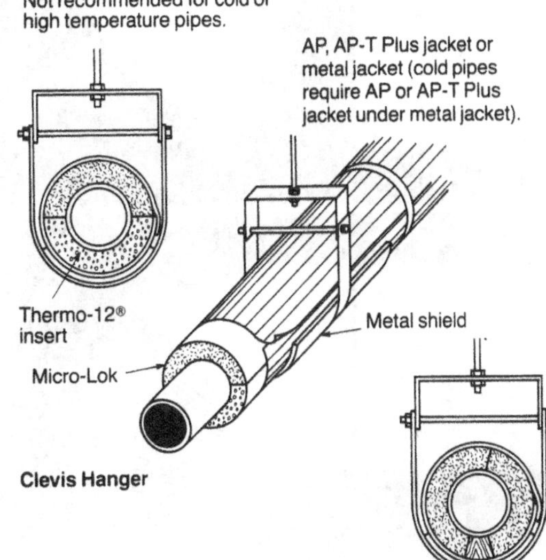

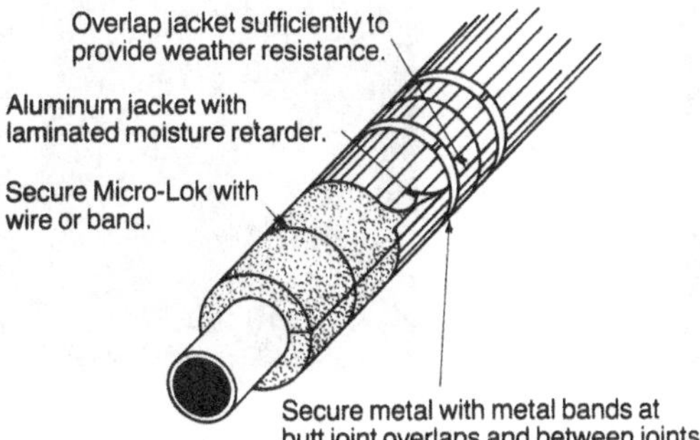

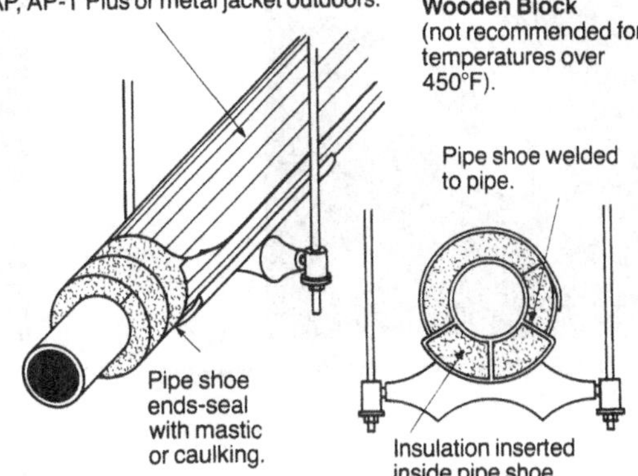

FIGURE 5.169*a* Installation methods.

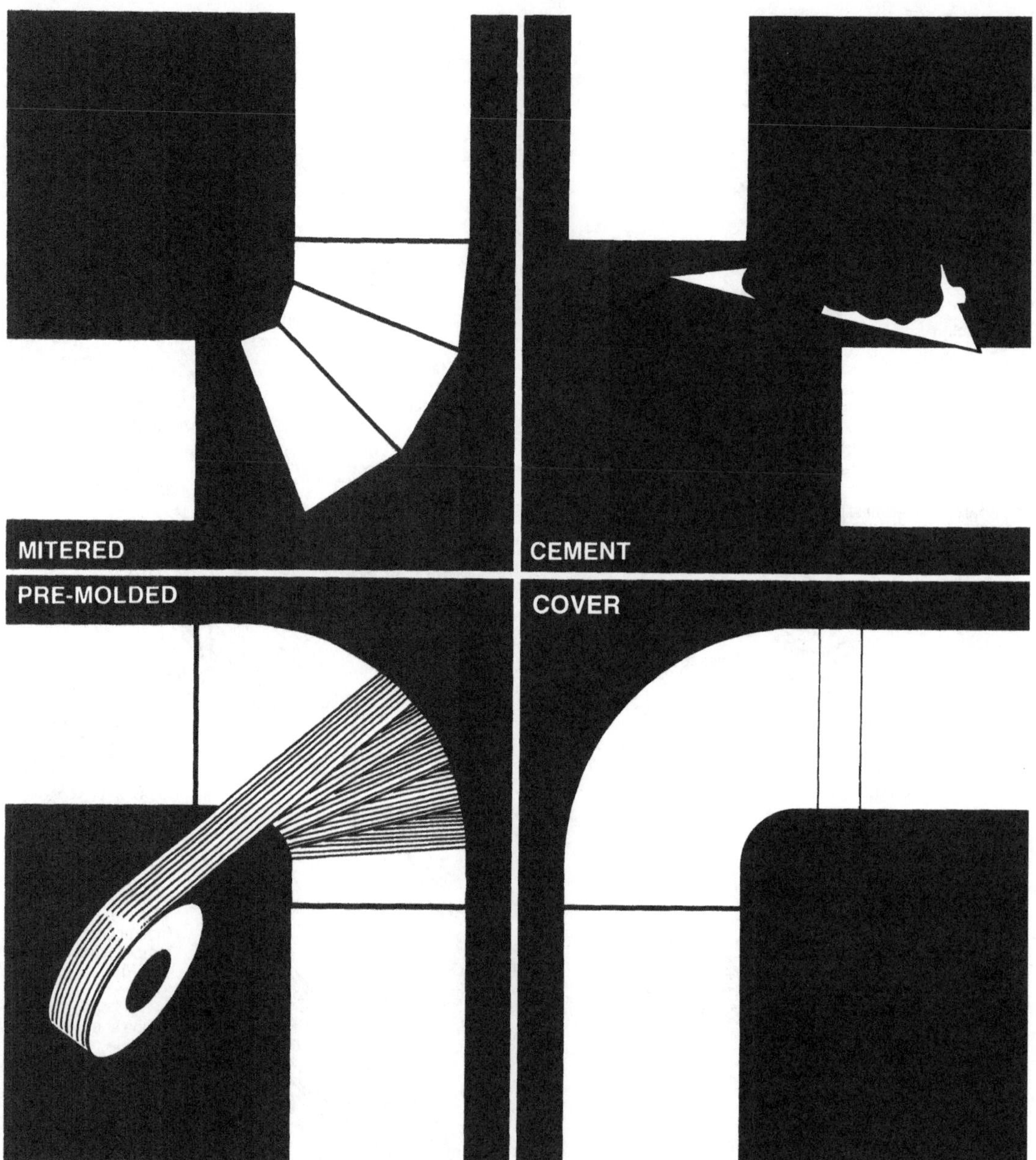

FIGURE 5.169*b* Insulating 90° elbows—four methods.

t_h = hot-surface temperature, normally the operating temperature (cold-surface temperature in cold applications), °F (°C)
k = thermal conductivity of insulation, always determined at the mean temperature, (Btu)(in)/(h)(ft²)(°F) [W·mm/m² · °C]
$t_m = (t_h + t_s)/2$, mean temperature of insulation, °F (°C)
tk = thickness of insulation, in (mm)
r_1 = actual outer radius of steel pipe or tubing, in (mm)
$r_2 = r_1 + tk$, radius to outside of insulation on piping, in (mm)
$eq\ tk = r_2 \ln (r_2/r_1)$, equivalent thickness of insulation on a pipe, in (mm)
f = surface air film coefficient, Btu/(h)(ft²)(°F) [W/m² · °C] $1/f$ = surface resistance, (h)(ft²)(°F)/Btu [m² · °C/W]
$R_1 = tk/k$, thermal resistance of insulation, (h)(ft²)(°F)/Btu [m² · °C/W]
Q_F = heat flux through a flat surface, Btu/(h)(ft²) [W/m²]
$Q_P = Q_F \times 2\pi r_2/12$, heat flux through a pipe, Btu/(h)(ft) (For SI, Q_P and $Q_F \times 2\pi r_2$, W/m)
Δ = difference by subtraction, unitless
RH = relative humidity, percent
DP = dew-point temperature, °F(°C)

SI Conversions from USCS Units

$$k_{\text{Eng}} \times 144.2279 = k_{SI}$$

$$R_{\text{Eng}} \times 0.17611 = R_{\text{SU}}$$

$$Q_{F,\ \text{Eng}} \times 3.155 = Q_{F,\ \text{SU}}$$

$$Q_{P,\ \text{Eng}} \times 0.962 = Q_{P,\ \text{SI}}$$

Specific SI units use kelvins rather than degrees Celsius, but in Δ calculations, there is no difference between the two. Degrees Celsius are used here for convenience.

Calculation Fundamentals

There are two concepts that must be understood before proceeding with the calculations. First, in a system in steady-state equilibrium, the heat transfer through each portion of the system is the same. Specifically, the heat transfer through the insulation is equal to the heat transfer from the insulation surface to the surrounding air.

Since the heat loss is proportional to both the temperature difference and thermal resistance of any portion of the system, the following equations result:

$$Q = \frac{t_h - t_a}{R_1 + R_s} = \frac{t_h - t_s}{R_I} = \frac{t_s - t_a}{R_s} \tag{2}$$

The second principle relates to the equivalent-thickness concept of pipe insulation. Since the outer insulation surface area is greater than the interior surface area, the heat flux "sees" a greater insulation thickness than the installed nominal thickness of the material. This equivalent thickness (*eq tk*) is used as the insulation thickness for the thermal calculation, but the nominal thickness is used for calculating the surface area per linear foot of insulated pipe. The equation for equivalent thickness is:

$$eq\ tk = r_2 \ln \frac{r_2}{r_1} \tag{3}$$

where r_I and r_2 represent the inner and outer radii of the insulation system, respectively. Figure 5.170 provides for easy conversion to equivalent thickness from any specific thickness and pipe size. Some materials are manufactured in true even thicknesses, whereas most others fol-

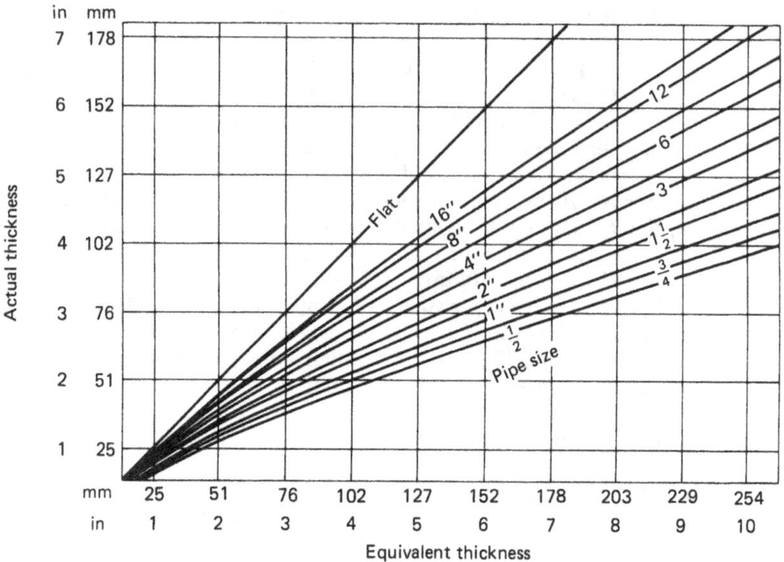

FIGURE 5.170 Equivalent thickness chart.[5]

low the schedule in ASTM C 585. In any event, the proper determination of insulation resistance for pipe insulation is

$$R_I = \frac{eq\ tk}{k}$$

Figures 5.171 and 5.172 provide typical thermal conductivity values for sample calculations.

After the material is selected, an appropriate thickness of the material as well as the heat loss or gain with that thickness must be determined. The proper amount of insulation to use depends on the *thermal design objective* of the system: What is the insulation supposed to accomplish? There are five broad categories that encompass most insulation objectives:

1. Personnel protection
2. Condensation control
3. Process control
4. Economics
5. Energy and environment conservation

Personnel Protection

To protect workers from getting burned on hot piping, the insulation surface temperature should be within the safe touch range of 130 to 150°F; 140°F(60°C) is the temperature most often specified. Both ambient temperature and surface resistance R are important in this calculation.

As noted in Table 5.59, the R_s values for aluminum are greater than for full surfaces and will result in higher surface temperatures for the aluminum over the same thickness of insulation.

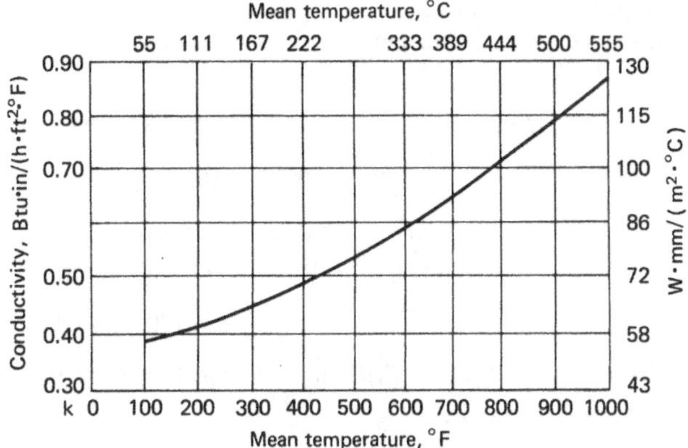

FIGURE 5.171 Thermal conductivity for typical calcium silicate pipe insulation.

To calculate the thickness required to achieve a specific surface temperature, the basic heat-loss equation is manipulated as follows:

$$Q = \frac{t_h - t_s}{R_I} = \frac{t_s - t_a}{R_s}$$

$$R_I = R_s \frac{t_h - t_s}{t_s - t_a} \qquad \text{and} \qquad R_I = \frac{eq\ tk}{k}$$

$$\therefore\ eq\ tk = kR_s \frac{t_h - t_s}{t_s - t_a}$$

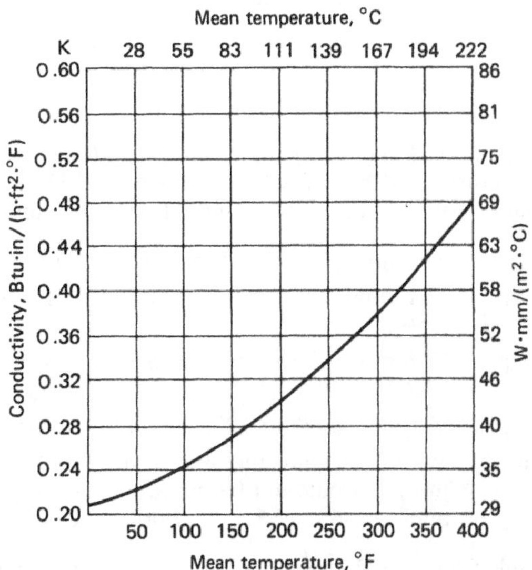

FIGURE 5.172 Thermal conductivity for typical fiberglass pipe insulation.

Example. An 8-in pipe is in operation at 700°F (371°C) ambient temperature with aluminum jacketing. Determine the thickness of calcium silicate required to keep the surfaces at 140°F (60°C).

Step 1. Determine k at $T_m = (740 + 140)/2 = 420$°F:

$$k = 0.49\ (\text{Btu})(\text{in})/(\text{h})(\text{Ft}^2)(\text{F})\ [71\ \text{W} \cdot \text{mm}/(\text{m}^2)(\text{C})]$$

from Fig. 5.170 for calcium silicate.

Step 2. Determine R_s from Table 5.59 for aluminum:

$$t_s - t_a = 140 - 85 = 55°\text{F}$$

So, $R_s = 0.85\ (\text{h})(\text{ft}^2)(°\text{F})/\text{Btu}\ [0.1497\ (\text{m}^2)(°\text{C})/\text{W}]$

Step 3. Calculate

$$eq\ tk = (0.49)(0.85)(700 - 140)/(140 - 85)$$

$$= 4.24 \text{ in}$$

$$\text{In SI units} = (71)(0.1497)(371.11 - 60)/(60 - 29.44)$$

$$= 10.8 \text{ mm}$$

Step 4. Determine from Fig. 5.170 that 4.24 in *eq tk* on an 8-in pipe can be accomplished with 3.25-in (83-mm) insulation. Always rounding off to the next half-inch increment, the recommendation would be 3.5-in (89-mm) calcium silicate.

Condensation Control

On cold systems, the insulation thickness must be sufficient to keep the insulation surface above the dew point of the ambient air. Table 5.62 gives the dew point for various temperature and relative-humidity combinations. The calculation is the same as for personnel protection with the surface temperature taking on the value of the dew-point temperature.

$$eq\ tk = kR_s \frac{(t_h - \text{DP})}{(\text{DP} - t_a)}$$

Example. A 4-in (10 cm) diam chilled water line is operating at 40°F (4.4°C) in an ambient temperature of 90°F (32.2°C) and 90 percent RH. Determine the amount of fiberglass pipe insulation with a kraft jacket required to prevent condensation.

Step 1. The dew point from Table 5.62 for 90°F, 90 percent RH is DP = 87°F (30.6°C)

Step 2. Determine k at $t_m = (40 + 87)/2 = 63.5$°F (17.5°C); $k = 0.23$ from Fig. 5.172 for fiberglass

Step 3. Determine R_s from Table 5.59 for fabric jacket and t_a – DP = 90 – 87 = 3°F (–16.1°C), $R_s = 0.54$

Step 4. Calculate $eq\ tk = (0.23)(0.54)\ (40 - 87)/(87 - 90) = 1.95$ in (49.5 mm)

Step 5. Actual thickness required from Fig. 5.170 for a 4-in (101-mm) pipe is 1.5 in (38.1 mm). It would not be unreasonable to specify 2-in (50.8-mm) thickness for this application since 1.5 in is borderline.

Table 5.63 gives the thickness of fiberglass pipe insulation required to prevent condensation for various operating temperatures and three different ambient conditions.

Process Control

There are many complex flow calculations required to determine the amount of insulation required to maintain process temperature or allow a specific temperature drop. The simple calculation below is for determining heat loss only and is basic to all the process control calculations. The heat loss from a pipe is

$$Q_p = Q_F \frac{2\pi r_2}{12} = \frac{t_h - t_a}{R_I + R_s} \frac{2\pi r_2}{12} \qquad \text{Btu/(h)(ft)} \qquad (4)$$

where

$$R_I = \frac{eq\ tk}{k}$$

In this calculation, a surface temperature t_s first must be estimated to arrive at a t_m and R_s. Then, when the t_s estimate is checked, the calculation can be redone with the new estimate if necessary.

Example. A 16-in diameter steam line operating at 850°F (454°C) in an 80°F (27°C) ambient temperature is insulated with 3.5 in (89 mm) of calcium silicate with an aluminum jacket. Determine the heat loss per linear foot and the surface temperature.

Step 1. Assume $t_s = 140°\text{F}$, $t_m = \frac{850 + 140}{2} = 495°\text{F}$; k from Fig. 5.171 for calcium silicate is 0.53.

Step 2. Determine R_s for aluminum from Table 5.59 for $t_s - t_a = 60°\text{F}$; $R_s = 0.85$.

Step 3. Determine *eq tk* from Fig. 5.170 for 3.5 in (9 cm) on a 16-in (40-cm) pipe, *eq tk* = 4.2 in:

$$R_I = \frac{eq\ tk}{k} = \frac{4.2}{0.53} = 7.92$$

Step 4. Calculate heat loss per square foot:

$$Q_F = \frac{850 - 80}{7.92 + 0.85} = 87.8\ \text{Btu/(h)(ft}^2\text{)}\ [(227\ \text{W/m}^2)]$$

Step 5. Check surface temperature assumption by

$$t_s = t_a + (R_s \times Q_F) = 80 + (0.85 \times 87.8)$$

$$= 155°\text{F}$$

This leads to a new $t_m = 502.5$, which is close enough to the assumed $t_m = 495$ that k will not change.

Step 6. Calculate heat loss per linear foot:

$$Q_p = Q_F \frac{2\pi r_2}{12} = 87.8 \frac{2\pi(8 + 3.5)}{12}$$

$$= 529\ \text{Btu/(h)ft} \qquad (508\ \text{W/m})$$

Freeze Protection

Water lines must either be heat-traced or have a controlled flow to prevent freezing. These calculations can be performed for various combinations of water temperature, line length, and ambient conditions. Table 5.64 shows the number of hours until freezing occurs along with the

TABLE 5.62 Dew-Point Temperature

Dry-bulb temp, °F	Percent relative humidity																		
	10	15	20	25	30	35	40	45	50	55	60	65	70	75	80	85	90	95	100
5	−35	−30	−25	−21	−17	−14	−12	−10	−8	−6	−5	−4	−2	−1	1	2	3	4	5
10	−31	−25	−20	−16	−13	−10	−7	−5	−3	−2	0	2	3	4	5	7	8	9	10
15	−28	−21	−16	−12	−8	−5	−3	−1	1	3	5	6	8	9	10	12	13	14	15
20	−24	−16	−11	−8	−4	−2	2	4	6	8	10	11	13	14	15	16	18	19	20
25	−20	−15	−8	−4	0	3	6	8	10	12	15	16	18	19	20	21	23	24	25
30	−15	−9	−3	2	5	8	11	13	15	17	20	22	23	24	25	27	28	29	30
35	−12	−5	1	5	9	12	15	18	20	22	24	26	27	28	30	32	33	34	35
40	−7	0	5	9	14	16	19	22	24	26	28	29	31	33	35	36	38	39	40
45	−4	3	9	13	17	20	23	25	28	30	32	34	36	38	39	41	43	44	45
50	−1	7	13	17	21	24	27	30	32	34	37	39	41	42	44	45	47	49	50
55	3	11	16	21	25	28	32	34	37	39	41	43	45	47	49	50	52	53	55
60	6	14	20	25	29	32	35	39	42	44	46	48	50	52	54	55	57	59	60
65	10	18	24	28	33	38	40	43	46	49	51	53	55	57	59	60	62	63	65
70	13	21	28	33	37	41	45	48	50	53	55	57	60	62	64	65	67	68	70
75	17	25	32	37	42	46	49	52	55	57	60	62	64	66	69	70	72	74	75
80	20	29	35	41	46	50	54	57	60	62	65	67	69	72	74	75	77	78	80
85	23	32	40	45	50	54	58	61	64	67	69	72	74	76	78	80	82	83	85
90	27	36	44	49	54	58	62	66	69	72	74	77	79	81	83	85	87	89	90
95	30	40	48	54	59	63	67	70	73	76	79	82	84	86	88	90	91	93	95
100	34	44	52	58	63	68	71	75	78	81	84	86	88	91	92	94	96	98	100
105	38	48	56	62	67	72	76	79	82	85	88	90	93	95	97	99	101	103	105
110	41	52	60	66	71	77	80	84	87	90	92	95	98	100	102	104	106	108	110
115	45	56	64	70	75	80	84	88	91	94	97	100	102	105	107	109	111	113	115
120	48	60	68	74	79	85	88	92	96	99	102	105	107	109	112	114	116	118	120

−15.0	−37.2	−34.4	−31.7	−29.4	−27.2	−25.6	−24.4	−23.3	−22.2	−21.1	−20.6	−20.0	−18.9	−18.3	−17.2	−16.7	−16.1	−15.6	−15.0
−12.2	−35.0	−31.7	−28.9	−26.7	−25.0	−23.3	−21.7	−20.6	−19.4	−18.9	−17.8	−16.7	−16.1	−15.6	−15.0	−13.9	−13.3	−12.8	−12.2
−9.4	−33.3	−29.4	−26.7	−24.4	−22.2	−20.6	−19.4	−18.3	−17.2	−16.1	−15.0	−14.4	−13.3	−12.8	−12.2	−11.1	−10.6	−10.0	−9.4
−6.7	−31.1	−26.7	−23.9	−22.2	−20.0	−18.9	−16.7	−15.6	−14.4	−13.3	−12.2	−11.7	−10.6	−10.0	−9.4	−8.9	−7.8	−7.2	−6.7
−3.9	−28.9	−26.1	−22.2	−20.0	−17.8	−16.1	−14.4	−13.3	−12.2	−11.1	−9.4	−8.9	−7.8	−7.2	−6.7	−6.1	−5.0	−4.4	−3.9
−1.1	−26.1	−22.8	−19.4	−16.7	−15.0	−13.3	−11.7	−10.6	−9.4	−8.3	−6.7	−5.6	−5.0	−4.4	−3.9	−2.8	−2.2	−1.7	−1.1
1.7	−24.4	−20.6	−17.2	3.0	−12.8	−11.1	−9.4	−7.8	−6.7	−5.6	−4.4	−3.3	−2.8	−2.2	−1.1	0.0	0.6	1.1	1.7
4.4	−21.7	−17.8	−15.0	−12.8	−10.0	−8.9	−7.2	−5.6	−4.4	−3.3	−2.2	−1.7	−0.6	0.6	1.7	2.2	3.3	3.9	4.4
7.2	−20.0	−16.1	−12.8	−10.6	−8.3	−6.7	−5.0	−3.9	−2.2	−1.1	0.0	1.1	2.2	3.3	3.9	5.0	6.1	6.7	7.2
10.0	−18.3	1.0	−10.6	−8.3	−6.1	−4.4	−2.8	−1.1	0.0	1.1	2.8	3.9	5.0	5.6	6.7	7.2	8.3	9.4	10.0
12.8	−16.1	−11.7	−8.9	−6.1	−3.9	−2.2	0.0	1.1	2.8	3.9	5.0	6.1	7.2	8.3	9.4	10.0	11.1	11.7	12.8
15.6	−14.4	−10.0	−6.7	−3.9	−1.7	0.0	1.7	3.9	5.6	6.7	7.8	8.9	10.0	11.1	12.2	12.8	13.9	15.0	15.6
18.3	−12.2	−7.8	−4.4	−2.2	0.6	3.3	4.4	6.1	7.8	9.4	10.6	11.7	12.8	13.9	15.0	15.6	16.7	17.2	18.3
21.1	−10.6	−6.1	−2.2	0.6	2.8	5.0	7.2	8.9	10.0	11.7	12.8	13.9	15.6	16.7	17.8	18.3	19.4	20.0	21.1
23.9	−8.3	−3.9	0.0	2.8	5.6	7.8	9.4	11.1	12.8	13.9	15.6	16.7	17.8	18.9	20.6	21.1	22.2	23.3	23.9
26.7	−6.7	−1.7	1.7	5.0	7.8	10.0	12.2	13.9	15.6	16.7	18.3	19.4	20.6	22.2	23.3	23.9	25.0	25.6	26.7
29.4	−5.0	0.0	4.4	7.2	10.0	12.2	14.4	16.1	17.8	19.4	20.6	22.2	23.3	24.4	25.6	26.7	27.8	28.3	29.4
32.2	−2.8	2.2	6.7	9.4	12.2	14.4	16.7	18.9	20.6	22.2	23.3	25.0	26.1	27.2	28.3	29.4	30.6	31.7	32.2
35.0	−1.1	4.4	8.9	12.2	15.0	17.2	19.4	21.1	22.8	24.4	26.1	27.8	28.9	30.0	31.1	32.2	32.8	33.9	35.0
37.8	1.1	6.7	11.1	14.4	17.2	20.0	21.7	23.9	25.6	27.2	28.9	30.0	31.1	32.8	33.3	34.4	35.6	36.7	37.8
40.6	3.3	8.9	13.3	16.7	19.4	22.2	24.4	26.1	27.8	29.4	31.1	32.2	33.9	35.0	36.1	37.2	38.3	39.4	40.6
43.3	5.0	11.1	15.6	18.9	21.7	25.0	26.7	28.9	30.6	32.2	33.3	35.0	36.7	37.8	38.9	40.0	41.1	42.2	43.3
46.1	7.2	13.3	17.8	21.1	23.9	26.7	28.9	31.1	32.8	34.4	36.1	37.8	38.9	40.6	41.7	42.8	43.9	45.0	46.1
48.9	8.9	15.6	20.0	23.3	26.1	29.4	31.1	33.3	35.6	37.2	38.9	40.6	41.7	42.8	44.4	45.6	46.7	47.8	48.9
51.7	11.1	17.2	22.2	25.6	28.9	31.7	33.9	36.1	37.8	40.0	41.7	42.8	43.9	45.6	47.2	48.3	49.4	50.6	51.7

TABLE 5.63 Minimum Insulation Thickness of Fiberglass Pipe Insulation Needed to Prevent Condensation (Based on Still Air and AP Jacket)

Operating pipe temperature	80°F (26.6°C) & 90% RH			80°F (26.6°C) & 70% RH			80°F (26.6°C) & 50% RH		
		Thickness			Thickness			Thickness	
	Pipe size, in	in	mm	Pipe size, in	in	mm	Pipe size, in	in	mm
0–34°F (−18–1°C)	Up to 1	2	51						
	1¼ to 2	2½	63	Up to 8	1	25	Up to 8	½	13
	2½ to 8	3	76	9 to 30	1½	38	9 to 30	1	25
	9 to 30	3½	89						
35–49°F (1–9°C)	Up to 1½	1½	38						
	2 to 8	2	51	Up to 4	½	13	Up to 30	½	13
	9 to 30	3	76	4½ to 30	1	25			
50–70°F (10–21°C)	Up to 3	1½	38						
	3½ to 20	2	51	Up to 30	½	13	Up to 30	½	13
	21 to 30	2½	63						

Source: Schuller International, Inc., Ref. 5.

gallons per minute flow required in a 100-ft (30.3-m) pipe run to prevent freezing. The calculations are based on fiberglass pipe insulation with $k = 0.23$, initial water temperature of 42°F (6°C), and an ambient air temperature of −10°F (−23°C). The flow rate for the 100-ft (30.3-m) length may be prorated for longer or shorter pipes.

Economics

With energy costs on the rise, insulating for economic reasons is becoming the standard for most insulation-thickness calculations. The first economic thickness equations were presented in 1926 by L.B. McMillan[8]. Since then, many monographs and computer programs have been

TABLE 5.64 Freeze Protection Data, Hours to Freeze, and Flow Rate (gal/min) Required to Prevent Freezing*

Nominal pipe size IPS	Insulation thickness					
	1 in		2 in		3 in	
	Hours to freeze	Flow rate, gal/min per 100 ft	Hours to freeze	Flow rate, gal/min per 100 ft	Hours to freeze	Flow rate, gal/min per 100 ft
½	0.30	0.087	0.42	0.282	0.50	0.053
¾	0.47	0.098	0.66	0.070	0.79	0.058
1	0.66	0.113	0.96	0.078	1.16	0.065
1½	0.90	0.144	1.35	0.096	1.67	0.078
2	1.72	0.169	2.64	0.110	3.31	0.088
2½	2.13	0.195	3.33	0.124	4.24	0.098
3	2.81	0.228	4.50	0.142	5.80	0.110
4	3.95	0.279	6.49	0.170	8.49	0.130
5	5.21	0.332	8.69	0.199	11.54	0.150
6	6.48	0.386	10.98	0.228	14.71	0.170
7	7.66	0.437	13.14	0.255	17.75	0.189
8	8.89	0.487	15.37	0.282	20.89	0.207

* Based upon 42°F (6°C) initial water temperature and −10°F (−23°C) ambient temperature.
Source: Schuller International, Inc., Ref. 5.

developed to perform these complex calculations.[10] Discounted cash-flow techniques, fuel escalation, maintenance and installed insulation costs, and after-tax effects are all involved in the computations.

However, the economic thickness of insulation (ETI) concept is still quite simple. Economic thickness is defined as that thickness of insulation at which the cost of the next increment is just offset by the energy savings due to that increment over the life of the project (see Fig. 5.173.).

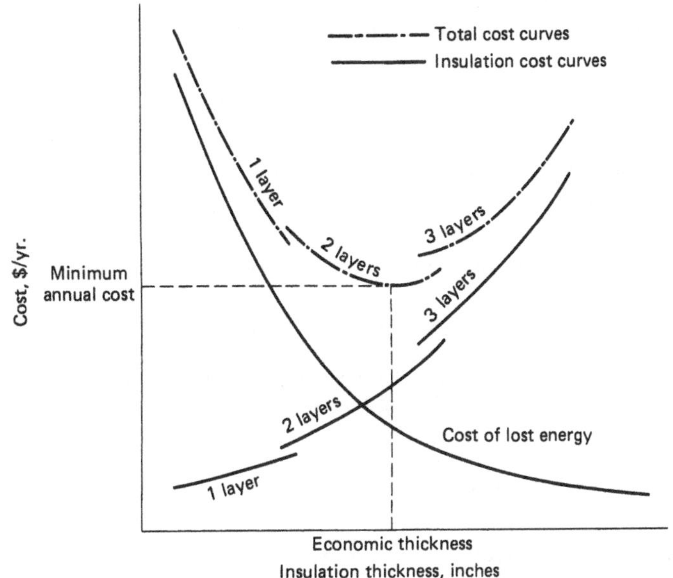

FIGURE 5.173 Economic thickness of insulation (ETI).

Energy and Environment Conservation

This era brings a new set of energy management challanges including a renewed focus on conservation measures and emisson reductions in the manufacturing sector. A national study completed in 1990 found that thermal insulation offers one of the most significant means to conserve energy and reduce CO_2 emissions. The study sponsored by the North American Insulation Manufacturers Association (NAIMA) and conducted by Dr. Harry L. Brown, Professor of Mechanical Engineering and Mechanics, Drexel University, found a potential reduction of 78 million tons of CO_2 emissions and equivalent fuel savings of up to 204 million barrels of oil a year by insulating bare pipe and equipment and upgrading existing insulation to the economic thickness of insulation (ETI). In order to determine the ETI insulation thickness, heat loss, surface temperature, annual payback energy cost, and annual savings, a computer program was developed with a published set of tables. The computer program entitled "3E-Plus for Energy * Environment * Economics" incorporates the economic parameters required to calculate the ETI insulation thickness, fuel savings, and reduction of CO_2 emissions. (See Ref. 12.)

Values used in these calculations are:

Annual hours of operation	8760
Fuel cost	\$2, \$6, \$10 per million Btu

Annual fuel inflation rate	5%
Interest rate or return on investment	10%
Effective income tax rate	30%
Physical plant depreciation period	15 yr
New insulation depreciation period	15 yr
Incremental equipment investment rate	2.00 $/million Btu/h
Percent of new insulation cost for annual insulation maintenance	5%
Percent of annual fuel bill for physical plant maintenance	5%
Ambient temperature	75°F
Emittance of outer jacketing (oxidized aluminum)	0.15
Wind speed	0 mi/h
Emittance of existing bare pipe surface	0.90
Reference thickness for payback calculations (bare pipe, no insulation)	0.0 in
Labor rate (means building construction cost data—1991)	$38.35

Below are the results calculated for a 6-in steam pipe operating at 450°F (232°C) using natural gas at $3.36 per MCF with a heating value of 1000 Btu per MCF. As seen by the results listed in Fig. 5.173, the 3E-Plus computer program is a powerful tool by which to begin conserving fuel and reducing emissions by including an insulation upgrade package in your energy management.

Economic thickness at 450°F	3 in
Heat loss	115 Btu/ft/h
Surface temperature	123°F
Payback	0.4 yr
Annual cost of owning insulation is	$8.41 ft/yr
Total annual savings	$89/ft/yr
CO_2 reduction	2131/lb/ln ft/yr
Total annual fuel savings for 1000 ft	$8900.00
Total CO_2 reduction	2,131,000 lb/yr

REFERENCES

1. McAdams, W. H.: *Heat Transmission,* McGraw-Hill, New York, 1954.
2. Sparrow, E. M., and R. D. Cess: *Radiation Heat Transfer,* McGraw-Hill, New York, 1978.
3. Greebler, P.: "Thermal Properties and Applications of High Temperature Aircraft Insulation," American Rocket Society, 1954. Reprinted in *Jet Propulsion,* November-December 1954.
4. American Society for Testing Materials: *Annual Book of ASTM Standards,* Vol. 04.06, Thermal Insulation; Environmental Acoustics.
5. Schuller International, Inc., Mechanical Insulations Division. Technical and Product Data Sheets, Denver, CO.
6. Kanakia, M., W. Herrera, and F. Hutto, Jr.: "Fire Resistance Tests for Thermal Insulation," *Journal of Thermal Insulation,* Vol. 1, No. 4, Technomic Publishing Company, Westport, CT, April 1978.

7. *Commercial & Industrial Insulation Standards,* Midwest Insulation Contractors Association, Omaha, NE, 1979.
8. Malloy, J. F.: *Thermal Insulation,* Reinhold, New York, 1969.
9. ASTM, Vol. 04.06, Standard C 585 (see Ref. 4).
10. McMillan, L. B.: "Heat Transfer through Insulation in the Moderate and High Temperature Fields: A Statement of Existing Data," No. 2034, The American Society of Mechanical Engineers, New York, 1934.
11. *Economic Thickness of Industrial Insulation,* Conservation Paper No. 46, Federal Energy Administration, Washington, DC, 1976. Available from Superintendent of Documents, U.S. Government Printing Office, Washington, DC 20402, Stock #041-018-00115-8.
12. NAIMA "3E-Plus" Computer Program, 44 Canal Center Plaza, #310, Alexandria, VA 22314.

CHAPTER 5.16
CORROSION CONTROL

Jose L. Villalobos
President
V & A Consulting Engineers
Oakland, California

INTRODUCTION

Over time, all materials deteriorate. It is the *rate of deterioration* that is important in the operation and maintenance of facilities. Facilities are designed with a specific design life. The rate of deterioration of the facility must be considered in estimating the life of the facility. Experience has shown that various materials will deteriorate at varying rates. The rate of deterioration is a dynamic function affected by environment, as well as type and length of service.

In the current regulatory climate, materials used for repair or maintenance of facilities must be evaluated in terms of their effects on both worker safety and the environment. By its nature, the deterioration process releases materials into the environment. Many existing facilities contain hazardous materials such as lead-based paints and asbestos. A project safety work plan must be written before any work involving these materials is undertaken. Also, material safety data sheets for all materials used in a repair must be read and precautions taken by the user to ensure a safe work environment.

Current plant engineering practice requires that facilities be monitored on a regular basis. This is imperative to maintain the useful life of a facility as designed and to extend the useful life of the facility if conditions change.

The total performance of a facility refers to the ability of both internal and external materials to fulfill their intended function over the useful life of the facility. As a general rule, the external components of a facility are subject to a harsher environment than internal components. For this reason, and because of recent experience with failures of new building materials, considerable care must be exercised when using new materials.

Atmospheric data that indicate the level of severity of the site in terms of deterioration or corrosion potential are generally available for a planned or existing facility. Data such as amount of rainfall; relative humidity; and levels of carbon dioxide, carbon monoxide, sulfur dioxide, sulfur trioxide, and nitrous oxide will greatly assist in the evaluation of the deterioration potential of a site. Some of the chemical components listed above can combine with moisture to form carbonic, sulfurous, sulfuric, or nitric acids. All of these acids will deteriorate the various facility components at varying rates.

Primarily due to new zoning requirements and design trends, various combinations of materials are being used in new residential, commercial, and industrial facilities. Concrete, paving brick, terrazzo, slate, bluestone, granite, glass, aluminum, copper, and plastics are being used in new construction and in repair of existing facilities.

This chapter will focus on the following materials, which are being used in the construction and repair of facilities today:

- Concrete
- Masonry
- Metals
- Wood
- Plastics

The primary factors that effect the deterioration of the materials listed are atmospheric exposure, exposure to water, and, to a lesser extent, exposure to soils.

Atmospheric exposure can result in corrosion, chemical attack, and deterioration of all five listed materials. Also, depending on location, ultraviolet light can have a negative impact on certain plastics. Various studies have shown that when the relative humidity rises over approximately 50 percent, corrosion rates for metals accelerate significantly.* Acid fog has been seen to cause localized pH as low as 2 in certain parts of the country. Many materials can become severely deteriorated by this type of exposure.

Water is considered to be the universal solvent. Given the right set of circumstances, water will over time dissolve or cause the deterioration of concrete, masonry, metals, wood, and plastic.

The single largest cause of failure of materials in facilities due to water-related damage is infiltration of water, which can be caused by events such as the failure of roof systems. The parapet is a common source of water infiltration caused by lack of proper flashing details, poor mortar joints, or failure of sealants. Differential expansion of a roof membrane can lead to splitting of the membrane, resulting in water infiltration. If improperly designed, vertical walls of a facility are also subject to the intrusion of water. High winds and rain also can cause water to infiltrate the facility and result in deterioration and corrosion of materials. Failure of glazing systems is another source of water infiltration.

* ASTM "Atmospheric Factors Affecting the Corrosion of Engineering Metals," STP 646, 1978.

If moisture is present, then corrosion of metals can result. To protect metals from corrosion, most uses of metal in facility construction today call for protective coatings specifically designed for the intended exposure. A very significant part of the design process is proper material selection. Certain metals perform their desired function quite well in certain environments, while others may fail in a matter of weeks. For example, 304 stainless steel is an excellent material for commercial kitchens because it does not corrode under the conditions generally found in this application. However, if 304 stainless is buried in a saline soil with a very low oxygen content, it will corrode. Most metals are subject to atmospheric exposure, and require some form of protective coating. If a metal is going to be immersed in water or some other liquid, it must be protected by a suitable coating system and possibly, depending upon the specific requirements, protected by cathodic protection. Cathodic protection is an electrical means of controlling corrosion of metals and is discussed later in this chapter.

Exposure to soil can have a negative impact on materials, depending on the chemical constituents of the soil. Some soils contain in excess of 2500 parts per million (ppm) of chlorides. These types of soils will cause significant corrosion of metals and reinforcing steel in concrete. If a soil groundwater has over 10,000 ppm of sulfate, then concrete in contact with this type of soil will be severely damaged. High-sulfate-containing soils are found in the southwestern United States. For new facilities, it is imperative that the soils at a proposed site be tested prior to design to allow the design engineer the opportunity to factor in the required protective measures for the specific facility.

CONCRETE

Causes for the Deterioration of Concrete

The visual symptoms of concrete deterioration are cracking, spalling, and disintegration. Each is visible and may occur individually or in combination. The American Concrete Institute* has identified six primary factors that positively affect the durability of concrete structures; these are:

1. Design of the structure to minimize exposure to moisture
2. Low water/cement ratio
3. Use of appropriate air entrainment
4. Quality of materials such as aggregates and mix water
5. Adequate curing before first freezing cycle
6. Special attention to construction practices

Deterioration generally results from the nonapplication of these listed items. Some of the more common forms of concrete deterioration are discussed in the following paragraphs.

Incomplete Design Details. Deterioration can often be traced to incomplete design details of joints between various concrete members; poorly sealed joints; inadequate drainage at foundations, horizontal surfaces (such as roofs and tops of walls), and incorrectly placed or incomplete weep holes; incompatibility of materials (such as aluminum in direct contact with concrete); use of chloride-containing admixtures (which can cause corrosion of the reinforcing steel); neglect of the cold-flow factor (in the deformation on concrete under stress); and incomplete provision for expansion joints, control joints, and contraction joints. During design *the designer must factor in the movement of water through concrete.* No matter how dense the concrete mix, unless there is a water or vapor barrier, moisture will migrate through concrete.

* ACI 201.2R-92.

For example, if high levels of chloride are present in the soils and groundwater around a concrete basement wall and no protection is provided for the concrete in the form of a water or moisture barrier, then eventually the chloride will migrate through the concrete until it comes into contact with the reinforcing steel. Once this happens, corrosion will begin, with the resultant spalling of the concrete.

Concrete is suitable for use in environments where the pH is greater than 5.5. If concrete is to be used and the environment has a pH of less than 5.5, then a coating or lining is required. Coatings must also be considered where chloride concentrations exceed 300 ppm in the soil or groundwater. Similarly, Type V cement is required where sulfate concentrations in the groundwater exceed 1500 ppm. Cathodic protection must be considered for concrete cylinder pipe where the soil resistivity is less than 2000 Ω-cm and the chloride content of the soil or ground water is greater than 300 ppm or the sulfate concentration is greater than 1000 ppm. Where prestressed concrete pipe is being considered, extreme care must be exercised in the protection of this type of pipe. Recent findings indicate that this type of pipe is subject to corrosion-related failures due to corrosion of the prestressing wire.

Concrete is typically a durable material in underground service. It is rarely affected by electrolytic corrosion as metals are. Unless the pH of the soil, groundwater, or process stream is less than 5.5, the chloride concentration is 300 ppm or more, or the sulfate concentration is greater than 1000 ppm, concrete is suitable for use in soil exposures. (For example, the San Diego, California, area may contain significant concentrations of chlorides and sulfates that are detrimental to concrete and reinforcing steel.) Attack on concrete and steel is likely when soils are acidic. When the pH of the soil is at or below 5.5, barrier coatings are required to protect the concrete surface. In areas of high sulfate concentrations in the groundwater (greater than 1500 ppm), modification of water/cement ratio, use of Type V cements (sulfate resistant), and barrier coatings must be considered for all buried concrete.

Concrete structures can be attacked by biologically formed sulfuric acid, which may occur in sewer pipelines. This acid formation is greatest where hydrogen sulfide is readily available and where aerobic conditions and moisture are present to oxidize the sulfides to sulfuric acid. These conditions exist where biological slimes are present on the pipe walls and aerobic conditions and moisture exist above the wastewater flow.

Construction Operations. The primary construction activity that can impact the degradation of concrete is improper placing of the concrete resulting in the segregation of materials followed by use before improper curing has occurred. Segregation can be minimized by using proper water/cement ratios. Proper compaction and vibration is mandatory.

Drying Shrinkage. Proper design should control damage caused by drying and shrinkage. Some of the factors that go into proper design are appropriate water/cement ratios, correct vibration, adequate reinforcement, and effective curing-membrane selection.

Temperature Stress. Variations in the atmospheric and internal temperatures can cause enough stress in concrete to cause cracking. Variations in temperature and the coefficient of expansion for concrete must be considered during design. This is particularly important during the curing of the concrete.

Moisture Absorption. Premature spalling can be caused by moisture absorption. The use of appropriate sealers and air entrainment can minimize this type of disintegration.

Corrosion of Reinforcing Steel. Corrosion of reinforcement can be attributed to two situations. First, certain types of cements, sands, aggregates, and water can cause corrosion of reinforcing steel. Also, additives containing high levels of chlorides can cause corrosion. Second, if a reinforced concrete structure is exposed to chloride-containing water, soil corrosion of the reinforcing steel may result. The use of deicing salts has caused a great deal of deterioration of bridge decks and parking garages because salt water is highly corrosive to concrete. Acids formed by various sources, such as pollution and acid rain, can attack concrete and result in corrosion of reinforcing steel. Good housekeeping can avoid some exposures of bare concrete to high chlorides. Spillage of salts from water-softening units on bare concrete can cause severe corrosion of reinforcing steel.

Wear. Erosion or abrasion are common factors that affect concrete, particularly in high traffic areas. Hydraulic structures, such as dams, are subject to cavitation erosion because of rapidly moving water. Flow velocities in excess of 10 f/s (3 m/s) should be avoided.

Impact. Floors in industrial facilities are particularly subject to damage by heavy live loading. Heavy reinforcement and high strength concrete are the usual methods for improving resistance to live-loading conditions.

Weathering. All concrete exposed to the atmosphere is subject to attack from the elements, including ultraviolet light and chemicals found in the atmosphere.

Carbonation. When concrete is exposed to carbon dioxide, a reaction producing carbonates takes place, which causes shrinkage of the concrete and cracking. Carbon dioxide may be introduced through the atmosphere or in liquids carrying carbon dioxide. The highest rates of carbonation in atmospheric exposures occur when the relative humidity is between 25 and 75 percent. Groundwater with high levels of carbon dioxide will also cause deterioration of concrete.

Exposure to Water. Concrete designed for immersion in water or other liquids must be protected. Recent observations at water and wastewater treatment facilities point to the need for protection of bare concrete. Certain types of water will leach components of concrete, leaving the aggregate behind. Eventually, enough of the cement material binding the aggregate together is removed to cause the concrete to fail. There are many products that can be used to provide a barrier between concrete and water in new and existing reinforced-concrete facilities.

Exposure to Soil. Soils with groundwater containing more than 1500 ppm of sulfate will cause deterioration of concrete. Once the concrete is damaged, the reinforcing steel is exposed and corrosion of the reinforcing steel can progress once the corrosion-inhibitive properties of the concrete are no longer present. Soils containing high levels of chloride can cause significant corrosion of reinforced concrete. When the chloride migrates through the concrete wall and comes in contact with the reinforcing steel, corrosion is initiated. Recently, floor slabs of homes built on soils with very high chloride levels have been experiencing corrosion failures, approximately 10 to 15 yr after construction.

Diagnosing Durability Problems

Deterioration of concrete may result from chemical attack, corrosion of the reinforcing steel, structural movement, and wear and tear. Various methods are available to identify possible causes for deterioration of concrete:

1. Review the original design to ensure that proper selection of materials has been completed. This should be completed by someone familiar with the original design intent.
2. Review the original mix design and the use of additives in the concrete. Also, look into the construction methods and records to determine if anything during construction could have caused the problem.
3. List the various exposures of the concrete and the seasonal variations that may exist. If the concrete is exposed to saline water, then the water quality should be investigated to determine actual exposures. Atmospheric properties of the project site should be documented. If a chemical that is part of the local atmosphere can have a negative impact on concrete, it may be part of the cause of the deterioration. One important factor to keep in mind is

that a chemical present in a small amount may cause significant deterioration of concrete, if the concrete is already under stress. The stress or loading of a concrete member is very important in making an assessment of the possible deterioration levels that can be expected at a facility.

4. List the types of failures and their possible causes. Conditions such as spalling, cracking, and corrosion of the reinforcing steel will assist in determining the cause of the concrete deterioration:
   - If the concrete is spalling, check for variation in temperature, chemical attack, corrosion of the reinforcing steel, and incomplete design details.
   - If the concrete is disintegrating, check for chemical attack, erosion, and general weathering.

Once the source and cause of the deterioration have been identified, corrective action can be taken. There are several accepted procedures and methods for repairing cracks in concrete structures, spalled concrete floors, and pavements. Table 5.65 outlines several types of materials for such repairs. The materials described also can be used for repair of masonry members.

Repairing Cracks

It has often been said that concrete is destined to crack. The purpose of proper design and the objective of the design engineer are to minimize cracking without the expectation of eliminating it. When cracking does develop, it is important to determine the basic causes, as well as its extent, before deciding on the method of repair.

The surface of concrete will often exhibit shrinkage cracks—these are not necessarily defects, but they are aesthetically undesirable. They are usually attributed to the use of a high-slump concrete that contains excessive water. Cracking occurs when excess moisture leaves the concrete too soon—before the concrete has sufficient tensile strength. Such cracks can be minimized by using a sheet membrane or curing compounds. Although this type of cracking may not necessarily lead to problems and it may often be ignored, applying a sealer based on a synthetic-rubber compound to protect the concrete from further damage is prudent.

Cracks can be described as active or dormant. An active crack will open and close with changes in temperature and with cyclic movement of the structure. Dormant cracks may not go through such movement, but they can still leak, collect dirt, interfere with traffic, and so on. A structural crack usually can be attributed to inadequate structural design, insufficient strength (material composition), or poorly designed joints. If joints are not provided in concrete slabs, the concrete will create its own joints by cracking, and these cracks will continue to develop until the concrete member comes to equilibrium.

Active cracks may be sealed with an elastomeric sealant. Dormant cracks may be sealed with a fluid epoxy sealer that can be pumped into the crack or allowed to flow in by gravity. An epoxy sealer of 100 percent solid content will seal the crack without shrinkage and will join the crack faces to re-form a monolithic structure. This seal will be strong enough to resist further cracking. However, should stress still occur, cracking will take place somewhere else in the structure.

Horizontal cracks may be filled by simply pouring in a liquid epoxy sealer until it overflows, indicating that the crack has been filled.

Vertical cracks may be filled with a liquid epoxy sealer. First, the face of the crack is sealed with a fast-setting epoxy compound, which is allowed to cure thoroughly. Small holes are then drilled into the crack through the epoxy seal, and nipples are installed in these holes and bonded with the same fast-setting epoxy. Low-viscosity epoxy sealer is then injected into the lowest nipple. Pressure is maintained until liquid begins to seep out of the next higher nipple. These two nipples are then plugged and the same procedure is resumed with the third nipple. This operation is continued until the entire crack is filled. After the sealer has cured, the nipples may be cut off flush with the concrete surface.

TABLE 5.65 Latex and Epoxy Adhesives and Bonding Agents for Concrete

	Latices			Epoxies				
						Epoxy-polyamide		
	Acrylic	Polyvinyl-acetate (nonre-emulsifiable)	Butadiene-styrene	Epoxy-polysulfide binder only	Binder with sand	Binder only	Binder with sand	Epoxy-coal tar binder only
Appearance	Milky white	Milky white	Milky white	Light straw to amber	Light straw to amber	Light straw to amber	Light straw to amber	Black
Solids content, %	45	55	48	100	100	95 to 100	95 to 100	100
Reference specifications	MIL-B-19235	MIL-B-19235	MIL-B-19235	MMM B-350A; AASHTO M-200	MMM G-650A; AASHTO M-200	AASHTO M-200	AASHTO M-200	AASHTO M-200
Chemical resistance								
Acids	Fair	Fair	Fair	Excellent	Excellent	Excellent	Excellent	Excellent
Alkalis	Very good	Very good	Very good	Excellent	Excellent	Excellent	Excellent	Excellent
Salts	Very good	Very good	Very good	Excellent	Excellent	Excellent	Excellent	Excellent
Solvents	Fair to good	Fair to good	Fair to good	Excellent	Excellent	Excellent	Excellent	Excellent
Compressive strength, lb/in^2 (2-in cubes; ASTM C 109)	3200 to 4100	3400 to 3600	3300 to 4000	8000 to 10,000	12,000 to 15,000	6000 to 8000	10,000 to 13,000	3000 to 4000
Tensile strength, lb/in^2 (1-in briquettes; ASTM C 190)	580 to 615	340 to 450	450 to 580	—	—	—	—	—
Tensile strength, lb/in^2 (ASTM D 638)	—	—	—	3000 to 3500	—	—	3500 to 4000	400 to 800
Tensile elongation, % (ASTM D 368)	—	—	—	2.5 to 15	—	6 to 25	—	35 to 40
Flexural strength, lb/in^2 (bar; ASTM C 348)	950 to 1400	1000 to 1250	1250 to 1650	—	—	—	—	—
Compressive double-shear strength, lb/in^2 (MMM G-650A)	—	—	—	900 to 1000	700 to 1000	400 to 500	500 to 650	300 to 400
Application notes	Suitable for indoor and outdoor exposure on concrete, steel, wood, thin section toppings; shotcrete, plaster bond within 45 to 60 min; not suitable for extreme chemical exposure			Suitable for filling cracks in concrete to bond both sides of crack into an integral member; preparation of epoxy mortars by adding sand	Suitable for bonding hardened concrete and other materials to hardened concrete; setting dowels; bonding plastic concrete to hardened concrete; bonding skid-resistant materials to hardened concrete	Suitable for filling cracks in concrete to bond both sides of crack into an integral member; preparation of epoxy mortars by adding sand	Suitable for bonding hardened concrete and other materials to hardened concrete; setting dowels; bonding plastic concrete to hardened concrete; bonding skid-resistant materials to hardened concrete	Suitable for preparation of epoxy mortars by adding sand; bonding skid-resistant materials to hardened concrete; membrane between asphalt and concrete
	Not for conditions of high hydrostatic head		Not for constant water immersion					
	Not for use with air entrainers	Can be used with accelerators, retarders, and water-reducing agents, but not with air entrainers	Not for use with air entrainers or accelerators	For maximum chemical and physical properties; highest cost; not for use on surfaces treated with rubber or resin-curing membranes, dirty surfaces, weak concrete, or bituminous surfaces		For maximum chemical and physical properties; not for use on surfaces treated with rubber or resin-curing membranes, dirty surfaces, weak concrete, or bituminous surfaces		For resistance to grease, oil, gasoline, and traffic; use on bituminous concrete; lower cost applications of nonskid membranes; not to be used for bonding new wet concrete to old or where black color will be undesirable

The tensile strength of a cracked concrete member may also be restored by stitching. U-shaped iron rods known as *stitching dogs* are inserted in drilled holes to transfer stress across the crack. Holes are drilled on both sides of a structural crack, far enough away not to cause additional breaks but not in parallel position, which would produce a plane of weakness. The legs of the stitching dogs are designed to be long enough to provide adequate pull strength. After the legs of the stitching dogs are inserted into the holes, the holes may be grouted with a nonshrinking grout. The crack itself should be sealed with an elastomeric or an epoxy sealer to prevent water from entering.

The elastomeric sealants should be based on polysulfide or urethane rubbers (preferably a two-component formulation in traffic grade) and comply with Federal Specification TT-S-00227E. The epoxy sealers may be epoxy-polysulfides (complying with Corps of Engineers Specification MMM B-350A) or epoxy-polyamides (complying with Specification M-200-65 of the American Association of State Highway and Transportation Officials). The epoxy compounds are described in Table 5.65.

In crack repair, certain procedures are detrimental to the successful repair of the crack:

- Filling cracks with new mortar will result in further cracking.
- Placing a topping over a crack to seal it, unless the topping is elastomeric, will inevitably result in the crack passing through the topping itself.
- Repairing a crack without relieving the restraints that caused it will cause cracking elsewhere.
- Repairing a crack with exposed reinforcing that has begun to corrode should not be done until the steel has been cleaned and protected with a rust-inhibitive paint.
- Burying a joint that has been repaired prevents frequent inspection to determine whether further failure has occurred.

Repairing Surfaces of Pavement and Slabs

A spalled concrete surface is the beginning of continued disintegration of the concrete. If the cost of replacement is at all comparable with the cost of repair, replacement is recommended. Replacement is also recommended when changing the level of the final surface is impractical. Otherwise, the concrete can be resurfaced with new concrete, epoxy topping, latex mortar, or iron topping.

Resurface with New Concrete. When there is no problem in changing the level of the surface, it may be resurfaced with new concrete. It is always recommended that new concrete be laid with a bonding agent. The old surface should be prepared by removing all loose material and contaminants.

A simple bonding can be made by scrubbing a neat cement slurry (a mixture of straight portland cement and water) into the surface. The new concrete may then be placed with the expectation of a good bond.

Concrete placement should follow recommended practice. The concrete mix should have a water/cement ratio of 0.50 or less and a slump of 3 to 4 in (8 to 10 cm). It is also recommended that reinforcement be embedded in the middle of the topping slab, rather than allowed to lie at the bottom of the slab.

Finishing procedures will depend on the required surface: use a wood float finish for a regular surface and steel troweling for a hard, smooth, sealed finish. The slab should be cured by covering it with sheet materials or liquid-curing membranes. Concrete overlays of this type should be at least 2½- to 3-in (6.5- to 8-cm) thick.

A more positive bond may be achieved by using an epoxy bonding agent, either a filled epoxy-polysulfide or an epoxy-polyamide bonding agent (see Table 5.65). The epoxy bonding agent may be a proprietary formulation or one meeting the specification of a governmental agency. The bonding agent must be thoroughly mixed and applied by brush, broom, or spray. Do not apply more bonding agent than can be covered with new concrete while the bonding agent is still tacky. If the film of bonding agent has set, apply fresh epoxy adhesive before applying new concrete.

Shearing of the new concrete course from the base slab at the bond line is unlikely if the bonding-agent film is still tacky when the topping is placed. However, should a crack develop in the base slab, there is a definite possibility that this crack will transfer through the epoxy bonding-agent glue line and through the new wearing course.

New urethane bonding agents, which are often used between slab waterproofing membranes, can inhibit crack transfer from the base slab to the topping. The urethane membrane is also a two-component system that is applied in the same way as an epoxy bonding agent. Being elastomeric, the urethane membrane can absorb more of the stresses that are set up as the concrete overlay cures and contracts. It is also flexible enough to stretch and bridge cracks that develop in the substrate as well as in the wearing course, thereby preventing transfer of cracks.

A relatively inexpensive bonding agent that may be used to ensure a positive bond of a new concrete overlay is a latex-reinforced cement slurry grout. Portland cement and sand (in a ratio of 1:3 by volume) are combined with a gauging liquid that is a mixture of equal volumes of water and an acrylic or polyvinyl acetate latex emulsion, as described in the table. This emulsion must be nonre-emulsifiable. The gauging liquid, based on the latex blend, is added to the cement-sand powder until a creamy paste is developed. This paste is scrubbed into the surface of the base slab, covering the surface thoroughly. New concrete is then placed on this bond line, which is approximately 1/16-in thick.

It is recommended that the latex slurry not be applied too far ahead of the placement of the concrete.

The latex bonding agent should not be used by itself because too much could be absorbed by a porous substrate, leaving a minimal thickness in the glue line. It may also dry too quickly to form an effective bond by the time new concrete is applied. It is best used in the form of a slurry grout.

Resurface with an Epoxy Topping. When the level of the surface must be kept close to the original elevation and a chemically resistant and tougher wearing surface must be provided, an epoxy topping may be the answer. After the base slab has been thoroughly cleaned by sandblasting, grinding, or acid etching, it is rinsed and allowed to dry. The epoxy mortar is supplied as a three-package proprietary system consisting of a base epoxy resin, a catalyst, and a measured quantity of dry, salt-free sand. The ingredients are mixed together to form a mortar which is applied to a thickness ranging from 1/8 to 3/8 in (3–10 mm), screeded, and troweled smooth. Most systems also include a primer, based on an epoxy system, to be applied before the epoxy mortar. There are many products available to complete this type of repair. Input from local vendors is recommended.

Resurface with a Latex Mortar. Although less resistant to chemicals than epoxies, latex mortar toppings also allow resurfacing without significantly changing surface elevation. The mortar may be made by blending 1 part (by volume) portland cement with 3 parts mason's sand. Latex is then added at the rate of 10 percent solids based on the cement content. Enough water is added to make a trowellable mixture. A typical formulation would be:

Portland cement (Type I)	1 bag [94 lb (43 kg)]
Mason's sand	3 bags [300 lb (135 kg)]
Latex emulsion (50 percent solids)	2 gal (11 L)
Water	4–5 gal (15–19 L)

Before the latex mortar is applied, a latex cement slurry should be applied as a bonding agent. The slurry may have the same latex as that used in the mortar topping. Latex mortar is best spread and leveled with a wood float rather than a steel trowel, because the latex tends to rise to the surface and cause a drag on the steel trowel.

After the latex mortar is finished, a sealer should be applied. The sealer, which functions as a curing membrane and protects the latex mortar against contamination from grease, oil, and deicing salts, may be based on a chlorinated rubber, a butadiene styrene rubber, or a methyl methacrylate resin. This type of sealer may be applied to all new concrete.

Resurface with an Iron Topping. Areas subject to very heavy traffic, particularly to steel-wheeled vehicles, require an extra-durable surface such as that produced by an iron topping. It is usually applied in thicknesses of 1 in (25 mm). Specially graded iron particles are substituted for a major percentage of the sand aggregate in a mortar or concrete mix. This topping may be bonded to a substrate, using a neat cement slurry, a latex cement slurry, or a urethane or epoxy bonding agent. This iron topping is usually available in a ready mixed form or can be blended at the jobsite following the recommendations of the supplier of the iron particles.

Repairing Disintegrated Concrete Members

Methods and materials for repairing columns, beams, piers, and precast concrete panels will generally be similar to those used in repairing concrete pavements. However, because such concrete members are either load bearing or an integral part of a structure, it is not always possible to remove and replace them. Therefore, repairs must be made to the existing structure using the best means available.

Before any repair or replacement of concrete sections is attempted, steel that is exposed by spalled or disintegrated concrete must be treated to prevent further corrosion, which may have been the initial cause of the disintegration. Rusty steel is best cleaned by sandblasting—a process that will also prepare the concrete surface by removing disintegrated and loose material. The steel should be treated with a rust-inhibitive chemical or coating, which may be a zinc-rich primer or another quality metal primer, and allowed to dry before new concrete is placed.

If the disintegration is deep, it may be necessary to use jackets or forms to hold the fresh concrete in place until it hardens. A bonding agent is recommended to ensure proper adhesion of the new concrete to the base substrate. A latex-cement slurry is the simplest and most economical bonding agent. However, for maximum adhesion, an epoxy bonding agent is recommended. This epoxy compound, painted on the reinforcing steel as well as on the surrounding concrete before placing the concrete, will also serve as a rust-inhibitive primer. A urethane bonding agent may not be suitable since this repair work is not on a deck over occupied areas requiring waterproofing properties.

Repair concrete should have a low water/cement ratio and a low slump. It should also be consolidated in the forms by direct vibration or by vibrating the form. After the concrete has been set for the minimum period of time, the forms may be removed and a liquid sealer may be applied. The sealer will continue to assist in the cure of the concrete and will protect the concrete against weathering.

If the disintegration is shallow, it may be possible to use a latex-fortified or an epoxy mortar to patch the area. The mixtures of latex and cement and sand described under "Repairing Surfaces of Pavement and Slabs" may be used. An epoxy mortar may also be used, but it may not blend readily with the surrounding concrete in color or appearance.

Proper preparation of the surface and proper priming are still necessary. If latex mortar is used, a sealer should be applied. It is not necessary to apply a sealer to an epoxy mortar since this compound is dense and resistant enough to require no additional protection.

Shotcrete or gunned mortar can be used on spalled areas by an experienced contractor who has the necessary equipment. The mortar or cement plaster is usually formulated to a 1:4.5 ratio of cement to sand, although richer 1:3 mixtures are often used. If, as in some cases, the shotcrete does not have the necessary adhesive qualities, a bonding agent (a latex-cement slurry or an epoxy compound) is used as a prime coat. Shotcrete applied in very thin layers may have to be reinforced with a latex emulsion admixture. When the necessary equipment can be obtained, this application can be made by plant maintenance people. Care should be taken in the application of shotcrete on vertical walls and ceilings. Equipment that produces vibration may cause the shotcrete to spall before it cures.

MASONRY*

Masonry structures, like all plant structures, are susceptible to deterioration caused by natural weathering and deleterious effects of the industrial environment. Steps can be taken to modify the rate of attack when the basic principles involved in weathering deterioration are understood.

Brick Masonry Construction

Although bricks may be made of many materials, the term *brick masonry* is normally applied only to that type of construction employing comparatively small units of burned clay or shale. Ordinary brick is economical, and, when hard burned and laid in good mortar, it is one of the most durable construction materials available for buildings.

Clay is produced naturally by the weathering of rocks. Shale, produced in much the same way but with compression and perhaps heating, is denser than clay and more difficult to mine. Various brick colors and textures result from different chemical compositions and methods of firing.

Clay is ground, mixed with water, and molded into bricks by several methods: (1) stiff-mud process in which stiff, plastic clay is pushed through a die and cut into desired lengths, (2) soft-mud process in which clay is pressed into forms, and (3) dry process in which relatively dry clay is put into molds and compressed at pressures from 550 to 1500 lb/in^2.

After some drying, green bricks are fired in large kilns. The total firing process takes between 75 and 100 h. The brick must be gradually brought to the vitrification point (the temperature at which clays begin to fuse) and then must be gradually cooled.

Common brick, also known as hard or kiln-run, is made from ordinary clay or shale and is fired in the usual manner. Overburned bricks, called "clinkers," are unusually hard and durable.

A standard brick is 8-in long, 2¼-in deep, and 3¾- to 3⅞-in wide and weighs approximately 4½ lb. It should be rough enough to assure good bonding with mortar and should not absorb more than 10 to 15 percent of its weight in water during a 24-h soaking.

Types of Brick. Common bricks are often classified according to their position in the kiln, as follows:

Arch and *clinker bricks* are close to the fire in the kiln and are overburned and extremely hard and durable. They are often irregular in shape and size.

Red, well-burned, and *straight-hard* are well-burned, hard, and durable bricks. Stretcher bricks come from these classifications and are selected for uniformity of hardness, size, and durability.

Rough-hard bricks are in the clinker class.

Soft and *salmon* bricks are farthest from the fire in the kiln and are underburned, soft, and less durable.

Special kinds of brick include the following:

Face brick is made from specially selected materials to control color, texture, hardness, uniformity, and strength. It is used in veneering and exterior tiers, chimneys, etc.

Pressed brick is made by the dry process and has regular smooth faces, sharp edges, and perfectly square corners. This type is generally used as face brick.

Glazed brick has the front surface glazed in white or other colors. It is used in dairies, hospitals, and other buildings where cleanliness and ease of cleaning are important.

Fire-brick is made from a special type of fire clay to resist high temperatures.

Imitation brick is usually made from portland cement and sand rather than from clay. It is not burned, but has qualities similar to good mortar.

---

* This material is adapted with permission from *Plant Engineering,* November 10, 1977, pp. 203–207.

Mortar. Mortar serves several functions in brick construction. It holds bricks together, compensates for brick irregularities, and distributes load or pressure among the units.

Properties of mortar depend, to a large extent, on the type and quantity of sand used. Good mortar is made from sharp, clean, and well-screened sand. When sand is too fine, the mortar has less "give" and water works out of it, making it stiff and difficult to trowel. It may also set before the bricks can be placed. Too much sand robs the mortar of its cohesive consistency and makes it difficult to work with.

Mortar may be classified into five general types, on the basis of composition: straight lime, straight cement, cement lime, masonry cement, or lime pozzolan.

Natural cement is an important constituent of masonry cement because of its gradual strength-gaining properties, high plasticity, excellent water retention, and good adherence to aggregates. Combining natural cement with portland cement, which has early-strength properties, provides a hydraulic cement ideally suited for masonry mortar. Cement-lime mortars are usually classified in accordance with the ratios of cement, lime, and sand, by volume, in Table 5.66.

Causes of Deterioration of Brick Masonry. Repeated natural destructive forces, mild though they may be, break down hard rock into clay. These same forces act on clay bricks, fired tile, and fired terra cotta to cause deterioration. Natural stones and the binding mortars of masonry construction are similarly affected. In addition, airborne chemicals in industrial atmospheres and pollutants from internal-combustion engines contribute greatly to rapid soiling and chemical destruction of these binding materials and masonry units.

Frost Damage. One of the more destructive agents of weathering is frost. Water expands 9 percent as it freezes. Under certain conditions, such expansion may produce stresses that disrupt the bricks and cause spalling.

To take up water, a brick must be porous. There are two measures of water absorption: one obtained after soaking the brick for 24 h in cold water and another, larger one obtained after boiling the brick for 5 h. The difference between the two values represents the so-called sealed pores (pores that are not accessible to water under normal conditions, such as wetting by rain). If all open pores are filled with water, the unfilled sealed pores provide space into which water can expand on freezing with little or no development of stress.

Efflorescence. Pitting and spalling of clay products and natural stones is associated with efflorescence, a phenomenon in which salts percolate through the member and crystallize on the surface of a brick, stone, or mortar joint. These salts may be the sulfates of calcium, magnesium, sodium, potassium, and, in some cases, iron. Soluble salts may be present in the clay, or they may be formed by the firing process (oxidation of pyrites, or reaction of sulfurous fuel gases with carbonates in the clay).

Other sources of soluble salts are portland cement and hydraulic lime mortars, which contain soluble sulfates and carbonates of sodium and potassium; mortar containing magnesium

TABLE 5.66 Cement Lime Mortars for Masonry Construction

Mortar	Use	Proportions*			Minimum compressive strength, lb/in²	(k bar)
		Portland cement	Hydrated lime	Damp, loose sand†		
Type M	Maximum compressive strength	1	¼	2⅞–3¾	2500	17.3
Type S	Maximum bond	1	½	3⅜–4½	1800	12.4
Type N	General purpose	1	1	4½–5	750	51.8
Type O	Non-load-bearing interior construction	1	2	6¾–9	350	24.1

* Parts by volume.
† Sand quantity is 2¼ to 3 times combined volume of cement and lime.

lime, which can produce destructive magnesium sulfate; and gypsum plaster or dry wall. Salts may also be drawn by capillary action from the ground soil and limestone or concrete copings.

For efflorescence to occur, salts must be present, water must be available to take them into solution, and a drying surface on which evaporation can proceed to deposit crystals at the surface must exist. Water is always available and drying is periodic. The potential for efflorescence can be reduced by specifying well-fired brick that, generally, contains less soluble salts, or by specifying special-quality brick formulated with minimum soluble salts. The place where efflorescence appears is no indication of its source because solutions of salts may migrate considerable distances. Efflorescence on a particular surface merely indicates that it provided a convenient drying area. When a very dense impermeable mortar touches a more permeable brick, efflorescent salts will often appear on the brick, although the salts may have migrated from the mortar.

Glazed brick veneers do not always eliminate efflorescence. Although the glazing is impermeable to water entry from the exterior, moisture can still move slowly from the interior of the building toward the exterior face. Salts transferred to the exterior glazing create enough pressure to produce shaling of the glazed face. This phenomenon is common, especially when the brick has been inadequately fired. The result is unattractive, and the brick is exposed to further degradation.

Dimensional Changes. Expansion or contraction of building units may not in itself be harmful. However, the continuation of differential movements of dissimilar materials may give rise to difficulties. Live-load changes and foundation settlement can cause whole buildings to move. Building joints must be properly designed and located to accommodate these movements. Lime-sand mortar, used in older structures, is able to accommodate large movements without distress. Modern, higher-strength cement mortars, on the other hand, are less flexible and may shrink excessively on setting to cause cracking.

The movement of moisture in porous building materials can cause expansion and contraction. Expansion generally takes place with wetting; shrinkage occurs with drying. Water taken up by new brick when it is laid in fresh mortar causes the unit to expand Meanwhile, the mortar is shrinking. Such action can be minimized by wetting kiln-fresh bricks before they are used, avoiding excessively rigid mortars, and providing adequate joints. Reinforced concrete frames may also lead to failure of brick cladding when there is vertical shrinkage of the frame and when no movement joints are provided.

Steel columns clad with brick, common in structures built two decades or more ago, also can cause brick failure. As water enters this cladding, through either the brick or the mortar, the steel column rusts. Expansion of the rust pushes the brick cladding away from the column.

Mortar Deterioration. Mortar may decay from the formation of calcium sulfoaluminate (which causes expansion and loss of mortar strength) and by the attack of pollutants in the atmosphere. Portland cement contains tricalcium aluminate, which reacts with sulfates in solution to form calcium sulfoaluminate. Exhaust gases from automobiles contain sulfur dioxide, sulfur trioxide, and nitrous oxides. These oxides react with moisture in the atmosphere to form sulfurous acid, sulfuric acid, and nitric acid, which are the attacking agents. As attack continues over the years, the mortar joints may crack, the surface of the joint may spall off, and the mortar may become softer and more crumbly.

Sulfate-resistant cements used in mortar will inhibit this disintegration.

Repointing Mortar Joints

Repointing, or tuckpointing, is the process of removing deteriorated mortar from masonry joints and replacing it with new mortar to correct some perceptible problem, such as falling mortar, loose bricks, or damp walls. All contributing factors to the problem should be thoroughly investigated and corrected before repointing because the great amount of hand work and special materials required make repointing expensive and time consuming. Matching bricks may have to be obtained or specially made. Existing mortar may have to be analyzed before a repointing mortar can be formulated to match its color, texture, and physical properties.

It is a common error to assume that hardness or high strength is a measure of durability. A mortar that is stronger and harder than the masonry units will not "give," causing stress concentrations in the masonry units. These stresses are usually relieved by cracking and spalling. Mortar should contain as much sand as possible (consistent with workability) to help reduce shrinkage while drying. It should have good cohesive and adhesive qualities, be easy to handle on the pointing tool, and have good water retention (to resist rapid loss of water through absorption by the brick). It should not be sticky.

There is some controversy as to whether high-lime mortar is preferable to portland cement mortar for tuck pointing. High-lime mortar is suggested for use on old buildings because it is soft and porous, has low volume change, and is slightly soluble in water.

A slight amount of high-lime mortar will dissolve in rain and precipitate in small cracks; during drying, these small cracks and voids will seal. A small amount of white portland cement will accelerate setting of this normally slow-setting mortar. Even if the building was originally constructed with cement mortar, high-lime mortar may be recommended to reduce shrinkage and potential stresses at the edges of the masonry.

Mortar Mixes. The mixes outlined in Table 5.67 provide a starting point for developing a visually and physically acceptable mortar.

TABLE 5.67 Trial Mortar Mixes for Repointing Masonry

Mortar	Portland cement	Hydrated lime	Sand	Acrylic latex (50% solids)
Formula A	¼ bag	1 bag	3 ft^3	—
Formula B	1 bag	1–1½ bags	5–6½ ft^3	—
Formula C	1 bag	—	3 ft^3	2 gal

Sometimes, small amounts (5 to 10 percent) of finely divided iron are added to the repointing mixes to provide slight expansion rather than normal shrinkage. Too much iron may produce excessive expansion and may cause iron-rust stain. Repointing mortars can be further modified by adding (1) water-reducing agents to keep water content or water-cement ratio low, (2) waterproofing admixtures such as stearate soaps to minimize the absorption of water, (3) air-entraining agents to increase resistance to freeze-thaw weathering in areas of extreme exposure, and (4) mineral oxide colors to blend the mortar color with the brick.

Requirements for Mortar Materials. Materials used in preparing pointing mortars should comply with the following specifications and requirements:

Lime. "Standard Specification for Hydrated Lime for Masonry Purposes (ASTM C 207)," Type S; Federal Specification SS-L-351B.

Cement. "Standard Specification for Portland Cement (ASTM C 150)," Type I or II; Federal Specification SS-C-192G(3). The cement should not have more than 0.60 percent alkali (sodium oxide), or not more than 0.15 percent water-soluble alkali by weight.

Sand. "Standard Specification for Aggregate for Masonry Mortar (ASTM C 144)," Federal Specification SS-A-281B(1), paragraph 3.1.

Water. Potable water free from acids, alkalis, and organic materials.

Execution of the Work. Generally, old mortar should be cut out to a depth of ¾ to 1 in (2–2.5 cm) to ensure an adequate bond between old and new mortar and to prevent popouts. For joints that are less than ⅜-in (1-cm) wide, cutting back ½ in (1.3 cm) is usually sufficient. Using power tools is risky, unless they are handled by a skilled mason, because bricks can be easily damaged by such equipment. The use of hand chisels is still the best procedure.

Dry mortar ingredients should be mixed first. They should be prehydrated with only enough water to make a damp, stiff mortar to help prevent drying shrinkage. After an hour or two, the mortar is mixed with additional water to provide trowelability.

The joints should be thoroughly cleaned and moistened before the mortar is placed. A chemical bonding agent may be used; however, care must be exercised in painting it into the joint so that it is not smeared on the brick face. Mortar must not be placed before the bonding agent has set.

The mortar is best placed in ¼-in (6-mm) layers and then packed until the void is filled. When the final layer of mortar is thumbprint hard, the joint should be tooled to the desired shape with the correct size of pointing tool. If old bricks have worn, rounded corners, the final mortar surface should be slightly recessed to avoid leaving uneven joints that may be damaged easily.

The small amount of excess mortar left on the wall from a careful repointing job can be removed with a bristle brush before it hardens. Hardened mortar can be removed with a wooden paddle or chisel. Care should be exercised in using any chemicals, especially acids.

Grouting techniques can be used to repoint joints that show only minor defects such as very shallow deterioration, hairline cracks, or slight loss of adhesion to the brick. One method, often referred to as a bagging operation, can be used on glazed brick. A grouting mortar such as formula C in Table 5.67 is mixed to the consistency of a creamy paste and brushed onto the wall surface. After a short period (15 to 30 min, depending on temperature and degree of set) burlap bags or rags are used to bag or wipe the grout from the surface of the nonabsorptive glazed brick; grout is left only in the mortar joint. After an entire section is done, the area may be washed to remove any grout remaining on the glazed brick faces.

Another method involves the same bagging or grouting operation, but it may be done on unglazed or any standard face brick. This method involves masking each brick with tape cut to the same dimensions as the brick. The same type of latex grout is applied by brush over a wider area. The masking tape is removed after the grout sets, leaving neat, clean, and sharp mortar joints. This method is commonly called mask and grout.

Some mortar joints, especially very narrow ones, may be sealed with an elastomeric sealant applied by a caulking gun. Cracked bricks may also be repaired by widening the crack and sealing it with an elastomeric sealant.

Cleaning Plant Buildings

There are a number of valid reasons for cleaning building exteriors. Before building units are repaired or replaced, the original colors and textures of these units must be known so they can be properly matched. And, when new materials are installed, uniform appearance and weathering of the entire structure are desirable.

Preventive maintenance is an often-overlooked reason for cleaning. Dirt provides a much greater surface area than clean building materials; and, the more surface area that is exposed to atmospheric pollutants, the greater are the possibilities for destructive chemical reactions to be started. Dirty areas remain wet longer, resulting in more severe freeze-thaw cycling. And wet, dirty areas can support microorganisms that can cause disintegration, dissolution, and staining.

Selecting an appropriate cleaning method can be challenging, because dirt composition is so complex. Dirt is a surface deposit of finely divided solids held together by various organic materials. The solids are primarily carbon soot, siliceous dust, and inorganic sulfates. The organic binders consist largely of hydrocarbons from incomplete combustion products of various fuels. A combination of adsorption and electrostatic attractive forces hold the dirt to the masonry. Other adherent factors include efflorescent salts, leached cementitious materials, and recrystalized carbonates.

Acidic cleaners can be very damaging, particularly to marble and limestone, and alkaline cleaners can also be harmful. It is recommended that cleaners be tested on small areas to determine their effectiveness and reaction to the substrates.

One of the most versatile techniques for cleaning building exteriors is water washing. Although it requires minimal expenditure for materials and equipment, it can be time consuming, particularly if hand scrubbing is involved. There are three types of water-washing procedures: low pressure, high pressure, and steam cleaning.

The low-pressure wash is carried out over an extended period. The prolonged spraying loosens heavy dirt deposits; then, moderate pressure [200 to 600 lb/in^2 (1380 to 4140 bars)] can be used to flush away loosened dirt.

High-pressure water cleaning involves equipment capable of supplying water to a special high-pressure gun that jets the water at pressures of 1000 to 1800 lb/in^2 (6.9 to 12.4 kb). The gun can deliver up to 1700 gal of water per hour. A special nozzle can aerate the water, thereby minimizing physical damage to the masonry.

Steam cleaning involves the use of low-pressure [10 to 30 lb/in^2 (6.90 to 2100 kb)], large-diameter [½ in (12 mm)] nozzles; steam is generated from a flash boiler. The equipment is relatively more expensive and presents some safety hazards to the operators. It is also possible to add detergents or surfactants to the water in the flash boiler. Adequate dirt removal requires an average working time of 1 min/ft^2 (9 min/m^2).

Chemical cleaners can be acidic (low pH) or alkaline (high pH). Both types are used with surfactants (1 to 2 percent) to promote detergency and wetting. Acidic cleaners are often based on hydrofluoric acid or phosphoric acid in concentrations of less than 5 percent in water. Alkaline cleaners are often based on sodium hydroxide, ammonium hydroxide, or ammonia. Sometimes, all purpose cleaners, such as ammonium bifluoride, are used.

Abrasive blasting, both dry and wet processes, are effective on all substrates, but they require experienced mechanics to minimize damage to surfaces. Pressures used are usually between 20 and 110 lb/in^2 (140 and 760 bars), and working distances are from 3 to 12 in (7.6 to 30 cm). The abrasives are usually silica sand, but crushed slags and coal wastes are often used. The mesh sizes are very fine, either 0 or 00. Round particles are less abrasive and damaging than crushed grains.

Cleaning may precede or follow replacement of masonry units and repointing, depending on conditions. Cleaning before repair helps reveal original colors and textures and prepares substrates for receiving new mortars and bonding agents. Cleaning after repairs helps remove any excess droppings, splashings, or other accidental contamination. Whether done before or after, cleaning is a very important aspect of masonry and concrete restoration.

Sealing the Surface

A final step in the entire restoration process involves sealing all porous surfaces on the exterior facade. This procedure waterproofs the masonry to minimize the ingress of water, protects against attack by pollutants and other chemicals, minimizes the collection of dirt, and protects against graffiti damage.

One of the best sealers is a methyl methacrylate in organic solvent, containing 15 to 20 percent solids and, preferably, having a matte finish. The sealer may be applied in one or two coats, depending on the porosity of the substrate, by brush, spray, or roller. The sealer is water white, will not yellow or embrittle, and may be effective for 5 to 10 yr or longer.

Silicone compounds are not particularly recommended, primarily because they are effective only for relatively short periods and are water-repellent rather than waterproofing.

METALS

Principles of Corrosion Control

When people first began mining ores and reducing iron from its natural state, they reversed a fundamental process of nature. This process is the breakdown and blending of all unnatural materials into a neutral state. Iron is not natural or stable in its refined state and, as occurs with many materials, it tends to return to its natural state. The primary destructive forces that try to return steel, iron, and other materials to their natural states are known collectively as corrosion.

Corrosion damage is responsible for a greater loss in material and economic investment than any other factor in our modern society. Billions of dollars are spent annually replacing or repairing plant equipment and structures damaged by corrosion. Although minor maintenance of equipment and structures will always be necessary, material and economic losses can be minimized by reducing corrosion activity. Some of the methods that may be used to reduce corrosion include materials selection, proper design, coatings, corrosion monitoring, chemical treatment, and cathodic protection.

Corrosion deterioration can have a serious impact on a variety of facilities and equipment. The types of facilities and structures may include industrial plants, water and wastewater plants, pipelines, tanks, and the like. Hazards due to leakage, such as chemical spills, may occur. Volatile liquids such as fuel are more likely to ignite and explode when leaks have developed in the storage facility. Corrosion damage may also cause system downtime for facilities such as water-treatment plants, liquid-storage tanks, pipelines, metal-cased water wells, bridges, drydock structures, and many others. Funds allocated for capital improvements or new equipment may have to be diverted to rebuild existing structures and utilities damaged by corrosion.

Department of Transportation (DOT) regulations prescribe minimum requirements for the protection of metallic pipelines from internal, external, and atmospheric corrosion. These regulations are divided into the following three parts:

1. New construction
2. Existing pipelines and tanks
3. Monitoring of corrosion control and record keeping

These regulations and the need to ensure continued operation of facilities, structures, and pipelines make it necessary to establish a continuing corrosion-control program. Corrosion control can be accomplished by selection of appropriate materials, design practices, coatings, and proactive corrosion-control systems, such as cathodic protection. The program must be planned and budgeted to meet the requirements for new and existing construction and to establish continual monitoring of corrosion protection with the requisite records.

Even though the bulk of facilities and structures that are part of the U.S. infrastructure are currently not governed by DOT regulations, consideration of corrosion and corrosion-control measures should be incorporated into all facilities and structures during the planning, design, construction, and postconstruction phases. By taking this level of care in the planning and implementation of projects, regardless of their specific purpose or location, the long-term cost of the project will be greatly reduced. For example, a pipeline or plant facility that is well planned and has a corrosion control system that is maintained can last well over 50 yr without any major leaks, material failures, or need for repair. Whereas, without corrosion-control measures, the pipeline or plant facility may only last 7 to 10 yr before it must be repaired or replaced.

Corrosion protection systems should not be forgotten or ignored after they are installed. A comprehensive maintenance policy should be implemented to ensure that proper operation of the system is preserved. The policy should include regular testing, inspection, and record keeping.

In summary, the following key features should be incorporated into any project involving a facility or structure:

1. Planning to eliminate potential corrosion cells
2. Designing to eliminate corrosion cells which may include:
    a. Modifying the environment to eliminate the potential corrosion cell
    b. Selecting materials that are inert or corrosion resistant for the planned service

c. Coating the structure or facility with a suitable coating system
d. Providing a cathodic protection system for metal-based facilities or structures
e. Providing a plan to operate and maintain the corrosion system after construction

Causes of Corrosion

Corrosion is defined as the deterioration of a material by chemical or electrochemical action as a result of a reaction with the environment into which it is placed. Chemical attack of a metal is the simple dissolution of a metal as it reacts to a particular chemical. Electrochemical attack requires four components, an electrolyte, an anode, a cathode, and a metallic connection between the anode and cathode (see Fig. 5.174).

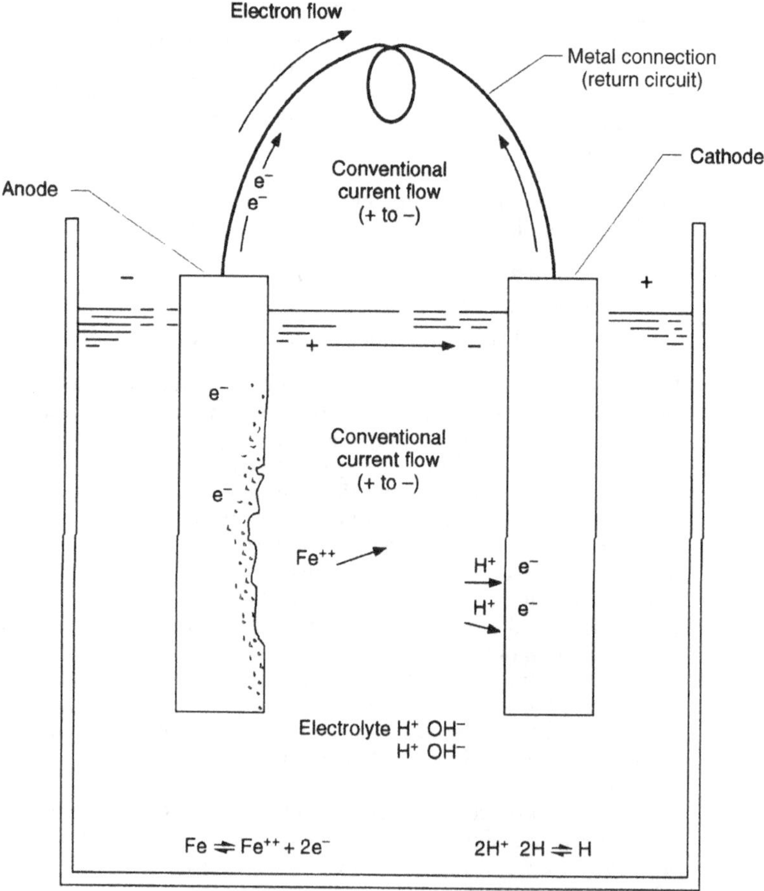

FIGURE 5.174 Four essential elements of the basic corrosion cell are electrolyte, anode, cathode, and return circuit.

One of the most important factors is the rate of corrosion. In certain cases the rate of corrosion is not critical to the operation of a facility and in other cases it can be a very critical factor. The factors that tend to accelerate corrosion rates are:

1. Temperature
2. Low or high pH of environment
3. High mechanical stress
4. Soils high in concentrations of aggressive chemicals
5. Metallic connections between dissimilar metals
6. Small anodic (corroding) areas in contact with large cathodic (protected) areas

The deterioration of metals, commonly referred to as corrosion, is a critical factor affecting the useful life of metals in facilities. Most importantly, the rate of corrosion affects how long a particular metal component will function in its intended use. Some metals corrode at a very slow rate, which makes them good candidates for certain applications. For example, aluminum has been found to be an excellent material for hatch covers in atmospheric exposures. On the other hand, the use of 304 stainless steel for pipe hangers in piers over salt water has resulted in failures within a year. The application, the environment, and the intended service are critical to proper material selection.

Atmospheric Exposure

Facilities and structures exposed to the atmosphere are subject to corrosion primarily from the chemical contaminants of the local atmosphere. For example, a galvanized metal roof will last 30 to 50 yr in a rural environment but only 5 to 10 yr in a heavily polluted, industrial environment. If structures and facilities are near the ocean, the impact of corrosion due primarily to the chlorides in the atmosphere will be significant if no protective measures are taken. For atmospheric corrosion to occur there must be a metallic substrate, moisture, and a chemical agent.

Submerged Exposure

Facilities and structures exposed to immersion in service are subject to corrosion due to exposure to a liquid (electrolyte). The most common example of this type of corrosion cell is an anchor bolt attached to the wall of a concrete tank which is holding water or some other electrolyte. The anchor bolts are usually tied to the reinforcing steel in the concrete wall of the tank. In this situation all of the conditions for a corrosion cell exist as well as several factors that affect the rate of corrosion. They are:

1. The anchor bolt is a small anode tied to a very large cathode (all of the reinforcing steel in the tank walls).
2. The anchor bolt is usually under stress.
3. There is a metallic path between the anodic and cathodic sites.
4. There is a large potential difference between the anchor bolt and the reinforcing steel in the tank.
5. There is an electrolyte present to facilitate the corrosion process.

In summary, for submerged exposures, most metals require some form of protection from corrosion. The key factors that affect the corrosion rate in this type of exposure are:

- The properties of the electrolyte (for example, seawater is more corrosive than fresh water)
- The type of immersed material (for example, monel is more resistant than mild steel when immersed in water)

Corrosion control measures for immersion-type applications include:

- Material selection
- Coating systems
- Cathodic protection

Buried Exposure

In the case of buried exposure, facilities and structures are subject to corrosion primarily due to the interaction between the facility or structure and the soil environment. The properties of the soil are therefore very important when planning a pipeline or other buried facility or structure. Based on the experience of corrosion engineers, it appears that soil resistivity is the one soil property that is the best indicator of corrosion potential for buried metal-based structures and facilities. The lower the soil resistivity, the greater is the tendency for corrosion. Another key factor in buried exposure is the amount of chloride present in the soil. Levels of 500 ppm or more in a soil will cause corrosion of metal-based systems.

The presence of ground water and the properties of that ground water will greatly affect the corrosivity of a soil. The more brackish the ground water the more corrosive it will be to metal-based systems.

The various corrosion-control measures listed previously cannot all be used in buried exposures. From a practical point of view there is not much that can be done to modify the environment to make it less corrosive, and material selection options are often limited by engineering requirements. Although clean, uniform, well-draining backfill materials will reduce corrosion activity, chemicals present in the soil will tend to leach into the backfill due to movement of water through the soil.

Other types of corrosion cells can be caused by manmade conditions. Three humanmade corrosion cells—dissimilar metals, damaged coating, and stray current—are illustrated in Fig. 5.175.

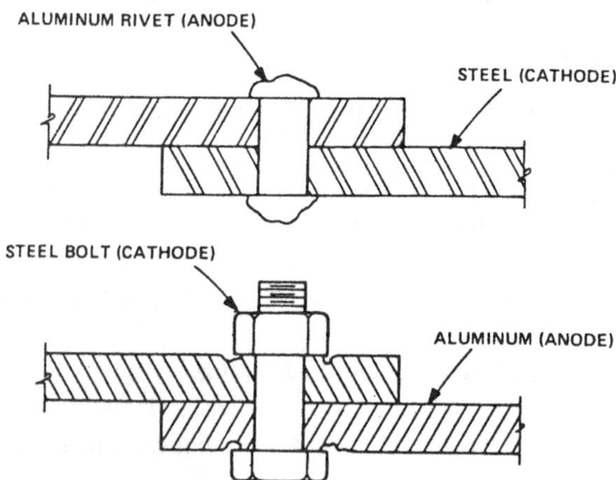

FIGURE 5.175 In a galvanic couple of aluminum and steel, anodic aluminum corrodes while cathodic steel is relatively unaffected. It is important that fasteners be compatible with building materials.

When two dissimilar metals are connected together and buried or submerged, an electrolytic corrosion reaction will occur. This is due to the difference in electrical potential between the two metals. This difference in potential produces a driving force, causing corrosion activity on the least noble of the metals.

Galvanic series of metals and alloys in seawater are listed in Fig. 5.176. These values are typical values which are observed in the field.

FIGURE 5.176 Galvanic series of metals and alloys in seawater. Metals and alloys are listed in the order of their corrosion potential in seawater. When two metals are coupled, those close together in the list are less susceptible to galvanic corrosion than those widely separated in list.

If an anode and a cathode are placed in a jar containing a salt or acid electrolyte, and connected together by a piece of copper wire as shown in Fig. 5.174, an electric current will flow through the wire. Except for differences in the details and materials of construction, this cell is the same as a flashlight battery.

The current through the metal connection flows from the cathode to the anode. The current flows from the anode into the electrolyte. This current flows through the electrolyte and is picked up at the cathode. The electrical circuit is then completed by current flow through the metallic connection between the anode and the cathode.

If the cell is permitted to operate long enough, the anode will deteriorate, but the cathode will not. This is because anode ions are going into solution in the electrolyte, forming an oxide or similar compounds. This process releases hydrogen ions from the solution which migrate to the cathode and deposit electrons on it. If the anode were steel with a copper cathode, the steel would corrode and provide protection to the copper. This is the basic mechanism of a corrosion cell.

Material Selection

In selecting materials for use at a plant or other facility, two main factors must be considered. The materials must perform the desired function in a safe and economical manner and operate satisfactorily over the design life of the facility. As corrosion-caused deterioration of materials is a likely mode of failure for buried or submerged components, it is important to select materials able to withstand the environment in which they will function. The following discussion will focus on the implications of corrosion on material selection for a variety of environments.

Corrosion of metals can occur in a number of situations. Corrosion activity may be initiated at internal or external structure surfaces. External surfaces may be buried, submerged, exposed to the atmosphere, or some combination of the above. Corrosion of internal surfaces may be due to the environment created by storage or flow of process streams.

Another corrosion mechanism, galvanic corrosion, is possible when two dissimilar metals such as copper and steel are connected together, whether buried or submerged. An electrolytic corrosion reaction will occur due to the difference in electrical potential between the two metals in contact with each other. This difference in potential produces a driving force, causing corrosion activity on the less noble of the two metals. Such galvanic couples must be avoided or isolated so that the possibility of corrosion damage is minimized.

As discussed previously, some soil environments and ground waters can be corrosive to buried metal structures. Resistivity measurements, which can be made both in the field and in the laboratory, indicate how corrosion currents will flow through soils or ground waters. High concentrations of chlorides and sulfates contribute to a reduction in resistivity and an increase in the corrosion activity of a material. In the presence of oxygen, chloride ions can be extremely corrosive to steel. Similarly, high levels of sulfates can cause a reduction in soil or groundwater resistivity and corrosion of steel and concrete.

Surfaces exposed to fluids include the interiors of pipelines, tanks and other basins. For example, metals in contact with wastewater streams are subject to corrosion, particularly when exposed in splash zones and where the surface is subject to exposure to hydrogen sulfide and its by-products.

Atmospheric exposures can include a variety of pollutants as well as airborne salts and locations in which structures may be exposed to direct salt spray. Atmospheric exposure may, therefore, be severe. The combination of wet/dry cycling in the presence of chlorides found in salt spray creates the potential for significant corrosion activity on all exposed metal surfaces. In addition to marine exposures, plant processes and sanitary sewers are likely to generate airborne substances such as hydrogen sulfide which also is deleterious to metallic and concrete structures.

The following describes the performance that can be expected from various materials.

Steel

Coatings must be considered for all applications of steel. Cathodic protection should be considered for steel pipe where soil or groundwater resistivity is less than 10,000 Ω-cm, and where steel will be in contact with process streams. Cathodic protection of steel is strongly recommended where resistivity is less than 5000 Ω-cm. For all exposures, steel should be electrically isolated from dissimilar metals to prevent the formation of unfavorable galvanic corrosion cells. In areas where abrasive materials are likely to damage coatings, cathodic protection by impressed current or galvanic anodes may be desirable.

Low resistivities and high chloride concentrations in the soil may lead to corrosion of buried steel pipelines or structures. Cathodic protection should be considered for all buried steel pipelines or structures. Where cathodic protection is not provided, corrosion monitoring equipment should be incorporated into the design to allow the operating staff to monitor the condition of the pipelines or structures. Nonwelded joints should be bonded for electrical continuity. In addition, coatings should also be considered. Coatings may be used alone or in conjunction with cathodic protection.

Bare or galvanized steel is subject to corrosion when exposed to aggressive fluids. Corrosion is most severe in the splash zone, where readily available oxygen hastens the corrosion process. Submerged steel should be coated with a material suitable for use in the anticipated exposure. Where there are concerns regarding the corrosion of steel in contact with process streams, cathodic protection should be provided for steel structures considered to be in a corrosive exposure. This type of corrosion control should be incorporated along with suitable coatings.

Corrosion of steel structures at facilities generally occurs as a result of exposure to atmospheric chlorides (near marine environments), chlorine process streams, hydrogen sulfide, or other air pollutants. Steel structures of particular concern should be coated. Even in those areas where temperatures are maintained above the dew point and sulfide concentrations are expected to be relatively low, steel should receive a protective coating.

Hydrogen sulfide attacks steel rapidly and it can be assumed that bare or galvanized steel will not endure in outdoor ambient conditions where hydrogen sulfide is present. The use of galvanized or bare steel should therefore be limited. Steel structural materials or piping exposed to the atmosphere should be coated with suitable coatings for nearly all exposures.

Aluminum

Aluminum is suitable for atmospheric exposure. It may not be used in direct contact with concrete; there must be physical isolation, generally by coating the aluminum or placing it in PVC cast in the concrete.

Aluminum is not recommended for use when in contact with soil or process streams with high or low pH. In very broad terms, dry, sandy, and well-aerated soils are not corrosive to aluminum. However, as moisture and dissolved salts increase, the soil becomes more aggressive to aluminum. Where levels of chlorides, sulfates, or pH are high, contact with the soil may be detrimental to buried aluminum. Its use, therefore, is not generally recommended for underground applications.

Aluminum has excellent resistance to atmospheric sulfides and other pollutants. Its resistance to atmospheric corrosion is due to a tightly adherent oxide film, and destruction of this film by either mechanical or chemical means exposes a very reactive surface. If the protective oxide film is disturbed, the presence of salts, including chlorides, can cause rapid pitting of aluminum. Further, electrical coupling to iron, stainless steel, or copper will accelerate this deterioration.

Because aluminum is an electrically active metal (standard potential of −1.66 V versus standard hydrogen electrode), its use in water is limited. If coupled to any of the common engineering alloys (steel, iron, stainless steel, and copper and its alloys), aluminum will become the anode and galvanically corrode. For this reason, aluminum must be electrically isolated from dissimilar materials.

Severe pitting of aluminum can occur where iron or copper ions are in a solution in contact with aluminum, creating a galvanic couple in which aluminum is attacked at localized areas. Chloride ions can also lead to pitting in aluminum. This is most likely to occur in crevices and where stagnant water collects. The introduction of chemicals like ferric chloride or chlorine into process streams would cause severe degradation to aluminum, leading to rapid failure.

When exposed to fluids, aluminum is stable only in a small band of pH values from 4 to 8.5. Because it is an amphoteric material, it is attacked by both acids and bases. Contact with solutions with pH greater than 8.5 or less than 4 will cause corrosion of the aluminum. For example, aluminum handrails to be installed in concrete (pH 12 to 13) should be placed in plastic shields cast into the concrete, and a sealant should be placed between the plastic shield and the aluminum. Under these conditions, aluminum will provide acceptable performance.

Alloys typically used include 5052 and 6061, but expect that most alloys will show similar atmospheric corrosion characteristics. The aluminum-copper precipitation hardening grades (2000 series) would generally show somewhat greater corrosion, and the pure aluminum and Al-clad varieties (1000 series) would exhibit somewhat less corrosion. Aluminum is not recommended in areas where spillage of chlorine, sodium hydroxide, or other strong acids or bases may occur. Anodized aluminum has shown excellent performance in certain applications. Although it would most likely be dinged and abraded in uses like handrails, anodized aluminum can be used in a number of other areas, such as electrical switchgear enclosures.

Copper and Brass

Consideration must be given to coating and cathodic protection of copper and brass used in contact with soil. Care must be taken to prevent direct electrical connection of piping operated at different temperatures. Copper and brass must not be used in contact with process streams containing chlorine in excess of 2 ppm or in environments with pH less than 5.5.

Copper and brass typically perform quite well in underground applications where the pH is neutral to alkaline and the concentration of aggressive ions, such as chlorides and sulfates, is low. They are often used for potable water lines and fittings. Copper is a corrosion-resistant material that is not dependent on the formation of an oxide or other surface film to be protected from corrosion. Because it is cathodic to iron and aluminum, it will hasten their corrosion when coupled to them. Isolation of copper from most materials commonly used in plant construction is therefore required in buried service. Because copper is an excellent electrical conductor, and maintains a low resistivity interface with the soil, bare copper cable is often used for grounding systems. Unfortunately, this can lead to galvanic corrosion of steel and iron, as described previously, and also to very high current requirements for cathodic protection systems. Various solutions have been proposed, such as grounding cells, which are essentially dielectric until large potential differences (ground faults) occur. When ground faults do occur, the cell short circuits and dissipates the charge.

Copper is subject to changes in corrosion resistance with changes in temperature, so electrolytic corrosion can occur in hot and cold water lines buried in a common trench. To prevent this, the two lines should be isolated from each other at points of electrical contact. This can be accomplished with the use of insulating-type couplings. However, where high concentrations of chlorides (300 ppm or more) and sulfates (1000 ppm or more) and low pH values (5.5 or less) are found, copper or brass piping should not be used without a tape wrap coating and cathodic protection. Furthermore, when copper or brass is used in an aggressive environment, it should be electrically isolated from other structures. If copper piping is used for connection of copper service lines to plastic mains, this should be accomplished by using brass tapping saddles.

Although copper and brass typically have good corrosion resistance in aqueous solutions, they may be subject to corrosion in plant environments, depending upon the process stream. The presence of sulfides and ammonia compounds in wastewater can lead to dissolution of cuprous compounds. Further, if copper is coupled to a less noble metal like steel or aluminum, galvanic corrosion of the less noble metal may result. Because copper is a fairly soft material, it is also subject to erosion. This type of corrosion is accelerated by high fluid velocities, high temperatures, and abrasive particulate matter.

Copper and brass should not be used in any process flow stream that will allow exposure of the metal to solutions carrying chlorine (2 ppm or more). This is especially critical in reclaimed water systems, since chlorine can cause severe corrosion of copper and brass. Brasses containing over 15 percent zinc may suffer dezincification. This form of corrosion is especially prevalent in stagnant, acidic solutions.

Although copper and brass typically have excellent atmospheric corrosion resistance, they are actively attacked by hydrogen sulfide. Due to the probability of the presence of hydrogen sulfide in the atmosphere at a wastewater treatment plant and the possible implications in the operation of electrical systems, copper, brass, and bronze should not be used without protective coatings. Since it is impossible in many cases to eliminate the use of cuprous alloys, particularly in electrical systems such as switches and fuses, extreme care should be taken to prevent rapid deterioration. One means to reduce corrosion in electrical applications is to use well-sealed junction boxes (typically NEMA 4X). These boxes should be constructed of either stainless steel or a noncorroding, nonmetallic material (e.g., glass reinforced plastic, polycarbonate, or polyester), and completely sealed. Conduit entrances should be sealed tightly to the boxes and conduits should be internally sealed to prevent transmission of corrosive gases between boxes. Vapor corrosion inhibitors may also be required in certain applications. Further protection can be afforded by the use of spray- or dip-type corrosion-inhibiting coatings. A limited number of these products have been tested, and have in some cases yielded better results than the vapor-phase corrosion inhibitors.

Stainless Steel

Coating and cathodic protection should be considered for stainless steel in soil and water. Stainless steel may be used in most atmospheric exposures and may also be used as hardware for connection to steel. Stainless steel should not be used for complex structures with overlapping bolted connections in soil or fluid exposures. Bolted connections of this type in soil or fluid exposures can experience very rapid crevice corrosion.

In soil, stainless steel is fairly resistant to uniform corrosion, which occurs over the entire surface; however, it may be subject to pitting corrosion. Stainless steel is most often used in situations where contamination of the material carried in the pipe is the prime concern. However, as pitting of the buried structures might occur, where soil conditions surrounding the pipe vary, it would be prudent to install stainless steel pipe with a uniform, well-installed backfill where differential oxygen corrosion cells will not occur. Coatings and cathodic protection of buried pipelines in corrosive soils should be considered. In noncorrosive soils, coatings for stainless steel are recommended.

Stainless steel is typically resistant to corrosion in flowing waters. Of the various types of stainless steel, the austenitic grades (300 series) have shown the best performance. In stagnant waters, however, pitting of stainless steel may occur. Oxidizing metal salts such as ferric chloride may also attack stainless steel. Types 304 and 316 alloys are more resistant to chlorine, hypochlorous acid (HOCl), and hypochlorite ions than other alloys that might be used in the process streams. Stainless alloys are also resistant to hydrogen sulfide and other organic materials likely to be found in plant or wastewater environments.

Cathodic protection of stainless steel is an option for preventing pitting. Although stainless steel is essentially immune to uniform corrosion, pitting has been encountered in many aqueous environments and can be prevented by the use of cathodic protection. By electrically coupling stainless steel to a large immersed structure made of steel, zinc, or other metal that is more anodic to stainless steel, pitting can be reduced or eliminated. It should be recognized, however, that where galvanic couples exist, stainless steel will increase the magnitude of corrosion deterioration of the structure to which it is bonded. It is therefore recommended that electrical contact be eliminated where the anode/cathode ratio is not favorable (i.e., small anode to large cathode).

Stainless steel has been used with much success in both outdoor and indoor applications. Of the various types of stainless steel, the austenitic grades (typically 302, 304, and 316) generally have the best corrosion resistance. Of the three alloys listed above, 316, although more expensive than the others, is the most resistant to pitting.

The austenitic alloys are resistant to hydrogen sulfide, chlorides, and moisture. There are also advantages to using stainless steel in combination with other metals. This is true where the more anodic material has a much larger surface area than the cathodic material. For example, galvanic corrosion has not been a problem where stainless steel fasteners are used to hold down aluminum deck plates. This is because the amount of stainless steel (cathodic material) used to hold down the aluminum (anodic material) is quite small when comparisons of surface area ratios are made. Overall, stainless steel has been demonstrated to provide excellent corrosion resistance in severe atmospheric environments.

Zinc

Although zinc can be used in the form of rolled sheet and strip, its widest use in building is for protective plating. When exposed to weather, zinc on the surface of galvanized steel develops a passive film that protects the underlying zinc and, thus, the steel. When the surface is scratched or scored, the anodic zinc corrodes, protecting the cathodic steel from attack.

Zinc can be applied to metal surfaces with hot-dip galvanizing, electrolytic processes, metal spraying, sherardizing, and zinc-rich paints. Hot-dip galvanizing provides excellent coverage and bonding leaving a film 3- to 5- (75- to 125-μm) mil thick. The interface is actually a zinc-iron alloy. Electrolytic processes provide a thin coating up to 1-mil (25-μm) thick. Metal spraying provides a 4- to 20-mil (100- to 500-μm) thick zinc film. Sherardizing, the heating of small components in a container of zinc dust, produces films of 0.5- to 1.5-mils (12- to 36-μm) thick. The dry film of zinc-rich paint should be at least 90 percent zinc to provide adequate galvanic protection to a steel substrate.

Atmospheric exposure gives the zinc surface a coating of zinc carbonate that reacts with sulfurous and sulfuric acids in the atmosphere to form zinc sulfate, which is water-soluble and can be washed off by rain.

Zinc sheeting is susceptible to attack from condensation on its underside. Proper ventilation of roof spaces and use of vapor barriers and suitable underlays can minimize such conditions. Zinc surfaces should be designed to allow rapid drying.

Local air pollution should be considered when designing zinc roofs, and proper fasteners should be specified. Zinc should be prevented from contacting copper and the chlorides and sulfates in concrete and mortar. Acids from certain timbers and some wood preservatives can also attack zinc. Bituminous coatings are recommended for protection.

Weathered galvanized members can be painted after being properly cleaned. Freshly galvanized members can also be painted. The Zinc Institute now recommends solvent cleaning or detergent washing, rather than acid etching, before painting. Suitable primers include zinc dust, zinc oxide primers, vinyl wash primers, and latex emulsion primers.

*WOOD AND PLASTIC**

Wood and plastic building materials resist many of the natural and industrial environmental factors that affect concrete, masonry, and metals. However, wood and plastic are susceptible to other types of attack—fungi and insects can destroy wood, and ultraviolet light and temperature can affect some plastics.

Wood

When exposed to the elements, wood is affected by water and light. Pollutants have little effect on wood except to dirty it. It has good resistance to chemical corrosion, and its thermal expansion is slight.

* This material adapted with permission from *Plant Engineering*, February 16, 1978, pp. 149–152.

Wood normally contains 12 to 18 percent water, but it can absorb moisture and swell up to 5 percent. Ultraviolet light can cause chemical and color change (accelerated by the extraction of water-soluble materials). The stresses set up by fluctuating moisture content and ultraviolet light cause splitting and checking (defects in the wood surface that allow more water to enter to produce gross dimensional changes). Weathering effects of moisture and sunlight can be reduced by applying a film-forming or penetrating finish.

Biological Attack. Wood is subject to decay caused by certain types of fungi. In extreme cases, such as in tropical climates, unprotected timber may be destroyed in a few months. Keeping wood adequately dry (less than 20 percent moisture) minimizes fungal damage. Structural timber should be protected from the weather by a well-ventilated shelter, and wood members should not be framed close to the ground.

Deprivation of air causes fungi to become dormant and, finally, die. Encasement in concrete or heavy bituminous coatings can shut off the air supply; ordinary painting will not. Embedment of timbers in soil can also shut off access to air, but it can cause other problems.

Temperatures between 50 and 90°F (10 and 32°C) are optimum for the growth of fungi. They become dormant at temperatures over 110°F (93°C) and near 32°F (0°C), but can reactivate when temperatures moderate.

Protection against fungal decay can be provided by various chemical treatments.

In wood exposed to marine atmospheres, attack by marine borers may lead to destruction. Aggressive organisms such as teredo or shipworms, various mollusks, and wood lice attack wood by burrowing or boring and gnawing.

Their terrestrial counterparts—termites, beetles, caterpillars, bees, and ants—eat the wood. Such attack reduces the strength of the timbers and weakens and ultimately destroys the structure.

Termites are one of the most destructive organisms to timber. They are found in almost all areas of the country, with greater concentrations in warmer and more humid regions. Their diet is based on cellulose. There are wood-dwelling and earth-dwelling types. Visible symptoms of termite attack are subtle and difficult to recognize.

Protection of most wood against termites is provided by chemical treatments or by poisoning the ground with chemicals such as copper sulfate, sodium fluosilicate, borax, paradichlorobenzene, and various commercial poisons.

Chemical Preservatives. A number of species of wood have natural decay resistance. The heartwood of several species native to North America—especially redwood, cedar, cypress, and juniper—has good decay resistance; however, the sapwood of substantially all common species has poor resistance. Regardless of the species, chemical protection of woods is recommended. The preservative treatment makes the wood poisonous to fungi, insects, and marine borers. There are three classes of preservatives: waterborne, oilborne, and creosote.

Waterborne Salts. Waterborne salts for wood preservation include zinc chloride, chromated zinc chloride, copperized chromated zinc chloride, zinc meta arsenite, chromated zinc arsenate, chromated copper arsenate, ammoniacal copper arsenite, acid copper chromate, and fluor chrome arsenate phenol. These chemicals leave little odor and have little effect on the appearance of the wood, which may also be painted. The zinc chloride salts, at high penetrations, also provide fire retardance. Wood should be reseasoned before use because this type of chemical treatment injects a large amount of water into the wood.

Oilborne Preservatives. These include penta (pentachlorophenol) and copper naphthenate. Penta does not change the color of the wood, but copper naphthenate gives wood a green shade. Paintability after treatment depends on the type of oil or solvent used as a vehicle.

Creosote. Creosote is excellent for preventing decay, especially in exterior use or in contact with water. Its advantages include relative insolubility in water (giving it a high degree of permanence), good penetration, good availability, and low cost; in addition, it causes little dimensional change in the wood. Disadvantages include its potential fire hazard; a distinctive, sometimes unpleasant odor that can affect foods; volatile vapors that can affect plant life; black color; and staining of adjacent woods or porous materials. Treated wood cannot be painted, and on hot days it may sweat and become wet and tacky.

Treatment Processes. Methods for treating timber with chemicals include pressure, coating, dipping and steeping, thermal, and diffusion.

Pressure Process. These processes produce relatively deep penetration of the preservative into the wood. Although various processes differ in detail, the basic principle of all is the same—wood is placed in a pressure vessel that is filled with preservative. Pressurization then drives preservative into the wood to meet penetration and retention specifications.

Coating. Coating by brush or spray may be done at the site. At least two, preferably three, applications are made; each coat is applied after the previous one has been absorbed. Brush and spray treatments are generally used only when more effective treatments are not practical.

Dipping and Steeping. This involves immersing the wooden member in the preservative liquid. From a few minutes to as long as several days of immersion may be required. It provides greater penetration of the preservative than brush and spray treatments, but it generally less effective than pressure processes.

Thermal Treatment. Thermal treatment, or hot and cold dipping, is similar to dipping, except the member is first heated in the preservative in an open tank and is then submerged in cold preservative. This procedure provides deeper penetration of preservative than dipping and steeping.

Diffusion Processes. Diffusion processes can be used while the timber is in place. Water-soluble preservatives, carried in bandages or pastes or in retaining rings applied to the member, diffuse into the water present in the wood.

All of these methods should meet the standards and specifications of the American Wood-Preservers' Association (AWPA); see Tables 5.68 to 5.70. Any use of treated wood should be checked for conformance to current environmental regulations.

Repairing Termite-Damaged Wood. Wood that has been weakened or damaged by termite attack may be repaired without removing the wooden member from the structure. The procedure is similar to pressure-grouting cracks in concrete. For termite-damaged wood, a form sleeve must be built around the damaged wooden member; then, a low-viscosity epoxy-resin

TABLE 5.68 Selected AWPA Standards for Preservatives

Standard numbers	Preservative	Symbol	Trade names†
P1 or P13	Coal tar creosote	—	—
P2 or P12	Creosote-coal tar solution	—	—
P5	Acid copper chromate	ACC	Celcure*
	Ammoniacal copper arsenite	ACA	Chemonite*
	Chromated copper arsenate	CCA	Type A: Greensalt* Langwood* Type B: Boliden* CCA Koppers CCA-B‡ Osmose K-33* Type C: Chrom-Ar-Cu(CAC)* Osmose K-33C* Wolman* CCA Wolmanac* CCA
	Chromated zinc chloride	CZC	—
	Fluor chromate arsenate phenol	FCAP	Osmosalts* (Osmosar*) Tanalith* Wolman* FCAP Wolman* FMP
P8 and P9	Pentachlorophenol	Penta	—

* Reg. U.S. Pat. Off.
† *Source:* American Wood-Preservers' Association.
‡ Koppers is now known as Kop-coat.

TABLE 5.69 Selected AWPA Standards for Pressure-Treatment Processes

AWPA* standard	Product
C1	All timber products (general)
C2	Lumber, timber, bridge ties, and mine ties
C3	Piles
C4	Poles
C9	Plywood
C11	Wood blocks for floors and platforms
C14	Wood for highway construction
C18	Piles and timbers for marine construction
C23	Round poles and posts used in building construction
C29	Lumber to be used for the harvesting, storage, and transportation of foodstuffs

* American Wood-Preservers' Association.

TABLE 5.70 Preservatives and Minimum Retentions for Various Wood Products[a]

		Recommended minimum net retention, lb/ft^{3c}						
	AWPA product[b] standard	Waterborne preservatives[d]					Oilborne[d,e]	
Product and service condition		CCA	ACA	ACC	CZC	FCAP	Penta[f]	Creosote
LUMBER and TIMBER								
Above ground	C2	0.23	0.23	0.25	0.46	0.22	0.40	8
Soil or water contact								
Nonstructural	C2	0.40	0.40	0.50	NR	NR	0.50	10
Structural	C14	0.60	0.60	NR	NR	NR	NR	12
In salt water	C14	2.5	2.5	NR	NR	NR	NR	25
PLYWOOD								
Above ground	C9	0.23	0.23	0.25	0.46	0.22	0.40	8
Soil or water contact	C9	0.40	0.40	0.50	NR	NR	0.50	10
PILING								
Soil or fresh water	C3	0.80	0.80	NR	NR	NR	0.60	12
In salt water								
Severe borer hazard (Limnoria)	C18	2.5 & 1.5[g]	2.5 & 1.5[g]	NR	NR	NR	NR	NR
Moderate borer hazard (Pholads)	C18	NR	NR	NR	NR	NR	NR	20
Dual treatment (Limnoria and pholads)								
First treatment	C18	1.0	1.0	NR	NR	NR	NR	—
Second treatment	C18	—	—	NR	NR	NR	NR	20
POLES								
Utility in normal service	C4	0.60	0.60	NR	NR	NR	0.38	7.5
Utility in severe decay & termite areas	C4	0.60	0.60	NR	NR	NR	0.45	9.0
Building poles (structural)	C23	0.60	0.60	NR	NR	NR	0.45	9.0
POSTS								
Fence								
Round, half-round, quarter-round	C14	0.40	0.40	0.50	NR	NR	0.40	8
Sawn four sides	C14	0.50	0.50	0.62	NR	NR	0.50	10
Guardrail and sign								
Round	C14	0.50	0.50	NR	NR	NR	0.50	10
Sawn four sides	C14	0.60	0.60	NR	NR	NR	0.60	12

[a] Key to symbols: ACA, ammoniacal copper arsenate; ACC, Acid copper chromate; CCA, chromated copper arsenate; CZC, chromated zinc chloride; FCAP, Fluor chrome arsenate phenol; NR, not recommended; Penta, pentachlorophenol.
[b] See Table 5.69.
[c] Minimum net retentions conforming to AWPA standards for softwood lumber and plywood. Retentions for piles, poles, and posts are for southern pine. AWPA Standard C1 applies to all processes.
[d] See Table 5.68.
[e] Creosote, creosote-coal tar solution, and oilborne penta are not recommended for applications that require clean, paintable, or odor-free wood.
[f] Penta can be applied in liquid petroleum gas or light solvents to provide a clean, paintable surface
[g] Two assay zones: 0 to 0.5 in and 0.5 to 2.0 in.

compound is injected until all voids have been filled. The form sleeve may be treated with oil or wax to prevent the epoxy resin from sticking to the form. After the epoxy resin has hardened, the wood member's structural strength may be even greater than originally. If, as so often happens, the termite damage is inside the wooden member and the outside shell is intact, the epoxy-resin sealer may be invisible. However, should it be exposed, its light amber color may blend with the wood's color.

Sometimes, epoxy-resin compounds used for repairing wood may be manufactured with protective chemicals such as pentachlorophenol to help prevent further attack by termites. Termites cannot damage the epoxy compound. These special epoxy-resin compounds are generally available in boat yards or marinas, where they are widely used to repair wood damaged by rot or marine borers.

Plastics

Use of reinforced plastics in all construction is growing by millions of pounds yearly. These materials are finding extensive use in ceiling and floor systems, piping, skylights, translucent panels, structural shapes, grating, and the like.

Plastics are usually strong, durable, lightweight, and resilient; they are easy to manufacture and install and have low maintenance costs. Color can be built in over a wide range, enhancing their use for exterior decoration.

Plastics, a term used for synthetic or modified natural polymers, are classified into two major groups: thermoplastics, which can be softened by heat and lend themselves readily to molding, and thermosetting plastics, which do not soften under heat once they are cured. Plastics used for exterior applications in the construction industry can be formulated to almost any set of properties, including light stability, rigidity, opacity, water absorption, and abrasion and fire resistance.

Weathering of plastics involves ultraviolet radiation, infrared radiation, water, temperature, microorganisms, industrial gases, and stresses from wind and snow loadings. Some plastics discolor after weathering, but discoloration can usually be minimized by selecting materials containing ultraviolet absorbers. Certain pigments can also increase stability for exterior exposure.

Thermoplastic polymers should not be used in areas where high temperatures are present to cause distortion. Distortion temperatures of commercial plastic materials should be a guide in this respect. Low temperatures can cause embrittlement. Oxidation can cause changes in the molecular structure of a plastic similar to those in paint films. In most cases, plastics are reasonably resistant to industrial pollution and microorganisms, but they will discolor as dirt collects. However, frequent simple washings can maintain plastics adequately.

The most important generic types of plastics for exterior application include polyvinyl chloride, glass-reinforced polyester, acrylics, phenolics, and amino resins.

Polyvinyl Chloride (PVC). One of the most extensively used plastics in construction, PVC is often used for roofing panels, gutters and downspouts, pipes, cladding, wall and floor coverings, and window frames. PVC film has also been used extensively on metal sheeting as a protective and decorative finish. Heavy use is made of PVC in hidden items, such as water stops, vapor barriers, and waterproofing membranes.

PVC is a thermoplastic material with a wide range of properties determined by stabilizers, plasticizers, ultraviolet absorbers, lubricants, and other additives. For fire-resistant properties, antimony trioxide is often used. Methods of manufacture include extrusion, injection molding, blow molding, and calendering.

When properly formulated and fabricated under controlled conditions, a PVC product can have a life of 20 to 30 yr. Certainly, a PVC waterstop, buried in a concrete foundation wall, must last for the life of the structure.

Translucent or transparent PVC sheet does not weather as well as the opaque form because stabilizers and ultraviolet absorbers affect light-transfer properties.

If weathering causes color change or differential fading, the PVC member can be painted after proper preparation of the surface by washing and light sanding.

Glass-Reinforced Polyester. Laminates of glass-reinforced polyester find wide use in automobile bodies, aircraft, boats, swimming pools, tanks, prefabricated housing systems, curtain walls, and lightweight building panels.

Polyesters are thermosetting resins produced by the reaction of mixtures of glycols and dibasic acids. The compound is comparable to an alkyd resin used in paints. However, it is further modified by dissolving it in styrene, and it is cured into a thermosetting plastic by the addition of catalysts and accelerators. The plastic can be modified with additives similar to those used in the PVC compounds.

The polyester resin, by itself, is not strong enough for industrial use, so it must be reinforced with glass fibers. A laminate is manufactured by spreading the catalyzed polyester resin onto a form or mold. While it is still uncured, woven glass cloth, swirled mat, or chopped strands are laid up into the film. Then, more coats of catalyzed polyester and glass fibers are added until the necessary number of layers is installed. The final coat of polyester is normally heavy enough to cover the glass fibers thoroughly. The last layer of resin will be exposed to weathering and can have color and all the necessary additives built in.

The polyester resin may be modified with methyl methacrylate resins for better clarity and transparency. Recent developments involve the use of acrylates or polyvinyl fluoride as surface coatings bonded to the laminate for longer gloss retention. The product is cured at ambient temperatures, although heat will accelerate the cure and ensure a more satisfactory laminate. The same procedures are also followed in producing epoxy-resin laminates, although these are not as widely used in the construction industry.

A glass-reinforced polyester laminate may fail if the surface resin is not thick enough to cover the glass strands, or if weathering wears away the surface color. Change of color, fading, and loss of gloss may also develop. Good maintenance depends on regular and thorough washing to remove dirt. When the surface must be refurbished, steel wool may be used to remove dirt and loose resin and fibers. A surface layer of polyester resin or an acrylic sealer may be applied. Frequent washing to remove dirt is usually the best method of maintenance.

Acrylics. Methyl methacrylate is the basic monomer for acrylic resins, materials that found their first extensive uses in cockpit covers of airplanes. Today, methyl methacrylate and its modifications are used in making window panes, fascia panels, skylights, sunshades, bath and shower enclosures, roof lights, and the like. These thermoplastic materials are water-white and have almost the same light transmission properties as glass. They have good impact resistance, can be easily formed and machined, are easy to handle and install, and possess outstanding weathering resistance and durability. However, acrylics have low abrasion resistance and a very high coefficient of thermal expansion.

Periodic cleaning and washing of acrylic members is recommended as the best method of maintenance. However, they are easily scratched. Polishing with a soft rouge can remove scratches without affecting transparency. Colored acrylic members may be produced, and there is little or no fading or change in color. Painting is seldom required.

Phenolics and Amino Resins. Reacting aldehydes with phenol and amino compounds (such as urea and melamine) can produce thermosetting plastics with good chemical resistance. One of the first synthetic resins of this type was Bakelite, a phenolic resin. This group of plastics is used in making laminates, usually with reinforcement, for curtain wall paneling, wall linings, corrugated roofing, etc. Melamine formaldehyde is used in making the laminate known as Formica. These plastics, particularly the phenolics, can change color and weathering and develop very fine crazing patterns with aging. Painting can restore surface appearance and color after proper washing and light sanding. Frequent washing is recommended for maintenance.

Other plastics are used in plant structures, although less extensively. Among these are the acrylonitrile butadiene styrene (ABS) resins, the polyvinyl fluoride resins, the polycarbonate resins, and the polyurethanes. The epoxy resins have been used extensively in structural applications (such as flooring) and adhesives. Specialized applications include use in chemically resistant coatings and in plasters for exposed aggregate wall finishes.

As research and development proceed, improved formulations of available plastics and completely new resins will find their way into plant structures.

BIBLIOGRAPHY

AMERICAN CONCRETE INSTITUTE:

201.2R-92	GUIDE TO DURABLE CONCRETE
212.3R	CHEMICAL ADMIXTURES FOR CONCRETE
222R	CORROSION OF METALS IN CONCRETE
224R	CONTROLLING CRACKING IN CONCRETE STRUCTURES
503R	USE OF EPOXY COMPOUNDS WITH CONCRETE
503.2	STANDARD SPECIFICATION FOR BONDING PLASTIC CONCRETE TO HARDENED CONCRETE WITH A MULTI-COMPONENT EPOXY ADHESIVE
506R	GUIDE TO SHOTCRETE

ASTM:

C 881	SPECIFICATION FOR EPOXY-RESIN-BASE BONDING SYSTEM FOR CONCRETE
C 1059	SPECIFICATION FOR LATEX AGENTS FOR BONDING FRESH TO HARDENED CONCRETE

GLOSSARY OF CORROSION-RELATED TERMS

Active The negative direction of electrode potential. Also used to describe a metal that is corroding without significant influence of reaction product.

Aeration cell An electrolytic cell whose electromotive force is due to electrodes of the same material located in different concentrations of dissolved air.

Anion A negatively charged ion of an electrolyte that migrates toward the anode under the influence of a potential gradient.

Anode The electrode of an electrolytic cell at which oxidation occurs.

Cathode The electrode of an electrolytic cell at which reduction occurs.

Cathodic protection A means of applying an external electric current to reduce corrosion virtually to zero. A metal surface can be maintained in a corrosive environment without deterioration as long as the cathodic protection system provides the external current. Cathodic protection can be *galvanic*, such as that on galvanized pipe, or it can be *impressed current*, such as that usually found on large potable water tanks and pipelines.

Cation A positively charged ion of an electrolyte that migrates toward the cathode under the influence of a potential gradient.

Corrosion cell Electrochemical system consisting of an anode and cathode immersed in an electrolyte. The anode and a cathode may be separate metals or dissimilar areas on the same metal.

Corrosion The deterioration of material, usually a metal, by reaction with its environment.

Corrosion fatigue Damage to a metal from a combination of corrosion and fatigue (cyclic stresses).

Corrosion potential The potential of a corroding surface in an electrolyte relative to the reference electrode. Also called *rest potential, open-circuit potential, freely corroding potential.*

Bimetallic corrosion or **couple** See Galvanic corrosion.

Electrode potential The potential of an electrode measured against a reference electrode.

Electrolysis The chemical change in an electrolyte resulting from the passage of electricity.

Electrolyte A chemical substance or mixture, usually liquid, containing ions which migrate in an electric field.

Electromotive force series (EMF Series) A list of elements arranged according to their standard electrode potentials, the sign being positive for elements whose potentials are cathodic to hydrogen and negative for those anodic to hydrogen.

Environment The surroundings or conditions (physical, chemical, or mechanical) in which a material exists.

Erosion Deterioration of a surface by abrasive action of moving fluids.

Erosion corrosion Combined effects of erosion and corrosion on a metal surface.

Galvanic corrosion Corrosion associated with the current resulting from the electrical coupling of dissimilar electrodes in an electrolyte.

Galvanic series A list of alloys arranged according to their corrosion potentials in a given environment.

General corrosion A form of deterioration that is distributed more or less uniformly over a surface.

Holiday Any discontinuity or bare spot in a coated surface.

Ion An electrically charged atom or group of atoms.

Local cell Galvanic cell produced by differences in the composition of the metal or electrolyte.

Noble metal A metal or alloy such as silver, gold, or platinum having high resistance to corrosion or oxidation.

Open-circuit potential See Corrosion potential.

Oxidation Loss of electrons by a constituent of a chemical reaction.

Passivation A reduction of the anodic reaction rate of an electrode involved in corrosion.

Passive Metal corroding under the control of a surface reaction product.

Patina A green coating that slowly forms on copper and copper alloys exposed to the atmosphere. A patina contains mainly copper sulfates, carbonates, and chlorides.

Pits Localized corrosion of a metal surface, confined to a small area, which takes the form of cavities.

Potential The reversible work required to move a unit charge from the electrode surface through the solution to the reference electrode.

Reduction Gain of electrons by a constituent of a chemical reaction.

Reference electrode A reversible electrode used for measuring the potentials of other electrodes.

Rust A reddish brown corrosion product of iron and ferrous alloys. It is primarily hydrated ferric oxide.

Sacrificial protection Reduction of corrosion of a metal in an electrolyte by galvanically coupling it to a more anodic metal. A form of cathodic protection.

Tuberculation The formation of localized corrosion products scattered over the surface in the form of knoblike mounds (tubercules).

Tubercules See Tuberculation.

Voids A term generally applied to paints to describe holidays, holes, and skips in the film. Also used to describe shrinkage in castings or welds.

Working electrode cell The test or specimen electrode in an electrochemical cell.

Sources of Glossary Terms

1. NACE, *Materials Protection,* Vol. 7, No. 10, pp. 68–71 (1968).
2. NACE, *Materials Protection,* Vol. 4, No. 1, pp. 73–80 (1965).
3. NACE, Standard MR-01-75 (1978 Revision).
4. NACE, H.H. Uhlig, *Corrosion Handbook,* Wiley (1948).
5. ASM, *Metals Handbook,* Vol. 1, 8th Edition (1961).

CHAPTER 5.17

VALVES*

Dresser Valve Division
Stafford, Texas

INTRODUCTION

A valve may be defined as a mechanical device by which the flow of liquid or gas may be started, stopped, or regulated by a movable part that opens, shuts, or partially obstructs one or more ports or passageways.

Plant engineers describe a valve as one of the most essential control instruments used in industry.

By the nature of their design and material, valves can open and close, turn on and turn off, regulate, modulate, or isolate an extremely large array of liquids and gases, from the most basic to the most corrosive or toxic. They range in size from a fraction of an inch to 30 ft (9 m) in diameter. They can handle pressures ranging from vacuum to more than 20,000 lb/in^2 (140 MPa) and temperatures from the cryogenic region to 1500°F (815°C). Some applications require absolute sealing; in others leakage is not a factor.

CATEGORIES OF VALVES

Because of all these variables, there can be no universal valve; therefore, to meet the changing requirements of industry, innumerable designs and variations have evolved over the years as new materials have been developed. All these designs fall into 10 major categories: gate valves, plug valves, globe valves, ball valves, butterfly valves, diaphragm valves, pinch valves, check valves, relief valves, and control valves. While control valves overlap with other categories, features essential for flow control distinguish them. Special positioners and actuators required for operation are included in the discussion of control valves.

* Updated from the second edition chapter written by the Valve Manufacturers Association.

These basic categories are described in the following paragraphs. It would be impossible to mention every feature of every valve manufactured, and we have not attempted to so this. Instead, a general overview of each type is presented in outline format, giving service recommendations, applications, advantages, disadvantages, and other information helpful to the reader. In many cases, a disadvantage inherent in a type of valve has been overcome or corrected by a particular manufacturer. Therefore, for specific applications, manufacturers' recommendations should be sought.

Gate Valves

A gate valve is a multiturn valve in which the port is closed by a flat-faced vertical disk that slides at right angles over the seat (Fig. 5.177).

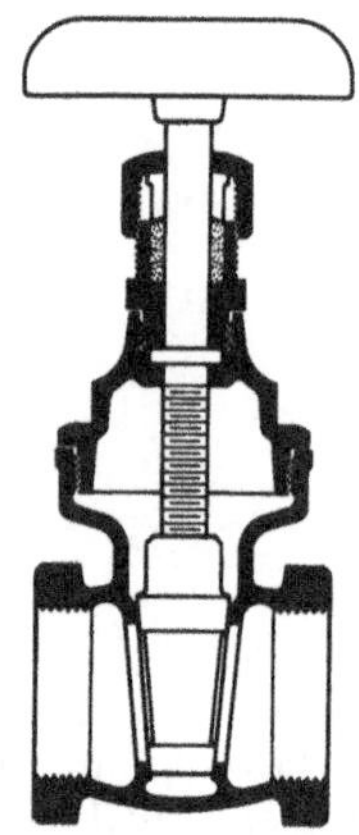

FIGURE 5.177 Gate valve.

Recommended

- For fully open or fully closed, nonthrottling service
- For infrequent operation
- For minimum resistance to flow
- For minimum amounts of fluid trapped in line

Applications

- General service, oil, gas, air, slurries, heavy liquids, steam, noncondensing gases and liquids, corrosive liquids

Advantages

- High capacity
- Tight shutoff
- Low cost
- Simple design and operation
- Little resistance to flow

Disadvantages

- Poor flow control
- High operating force
- Cavitates at low pressure drop
- Must be kept in fully open or fully closed position
- Throttling position will erode seat and disk
- Body cavity pressure lock on certain designs

Variations

- Gate: solid wedge, flexible wedge, split wedge, double disk, slab, expanding slab, through conduit
- Stem: standard, packing, outside screw and yoke

Materials

- Body: bronze, iron, ductile iron, carbon steel, low-alloy steel, stainless steel, Monel, nickel alloys, PVC plastic
- Trim: various

Special Installation and Maintenance Instructions

- Lubricate on regular schedule.
- Correct packing leaks immediately.
- Always cool system when closing down a hot line and checking closed valves.
- Never force valves closed with wrench or pry.
- Open valves slowly to prevent hydraulic shock in line.
- Close valves slowly to help flush trapped sediment and dirt.

Ordering Specifications

- Type of end connections
- Type of wedge or gate
- Type of seat
- Type of stem assembly
- Type of bonnet assembly
- Type of stem packing
- Pressure rating: operating and design
- Temperature rating: operating and design

Plug Valves

A plug valve is a quarter-turn valve that controls flow by means of a cylindrical or tapered plug with a hole through the center, which can be positioned from open to closed by a 90° turn (Fig. 5.178).

Recommended

- For fully open or fully closed service
- For frequent operation
- For low pressure drop across the valve
- For minimum resistance to flow
- For minimum amount of fluid trapped in line

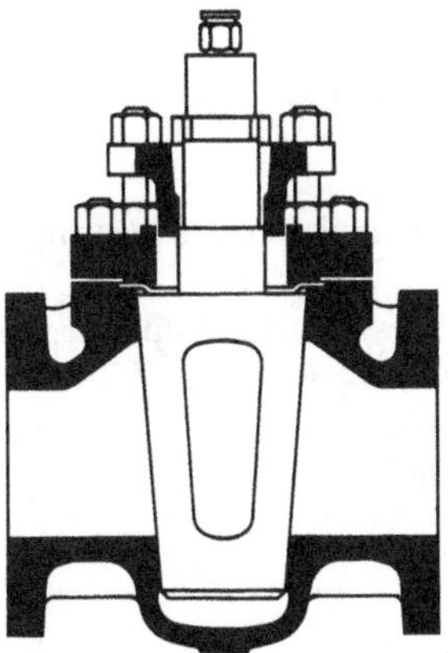

FIGURE 5.178 Plug valve.

Applications

- General service, blow-down service, liquids, gases, steam, corrosives, abrasive media, slurries

Advantages

- High capacity
- Low cost
- Tight shutoff
- Quick operation

Disadvantages

- High torque for actuation
- Seat wear
- Cavitation at low pressure drop

Variations

- Lubricated, nonlubricated, multiport

Materials

- Bronze, iron, ductile iron, carbon steel, low-alloy steel, stainless steel, Monel, nickel alloys, PVC plastic, plastic-lined

Special Installation and Maintenance Instructions

- Allow space for operation of handle on wrench-operated valves.
- For lubricated plug valves, lubricate before putting into service.
- For lubricated plug valves, lubricate on regular schedule.

Ordering Specifications

- Body material
- Plug material
- Temperature rating
- Pressure rating
- Port arrangement, if multiport valve
- Lubricant, if lubricated valve

Globe Valves

A globe valve is a multiturn valve in which closure is achieved by means of a disk or plug that seals or stops the fluid on a seat generally parallel to the line flow (Fig. 5.179).

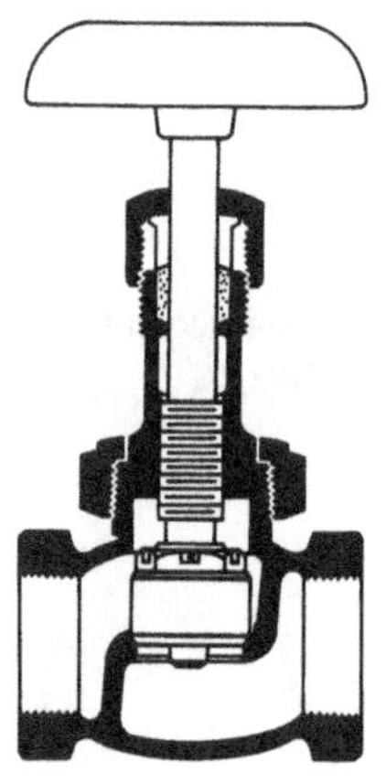

FIGURE 5.179 Globe valve.

Recommended

- For throttling service or flow regulation
- For frequent operations
- For positive shutoff of gases or air
- Where some resistance to flow is acceptable

Applications

- General service, liquids, vapors, gases, corrosives, slurries

Advantages

- Efficient throttling with minimum wire drawing or disk or seat erosion
- Short disk travel and fewer turns to operate, saving time and wear on stem and bonnet
- Accurate flow control
- Available in multiports

Disadvantages

- High pressure drop
- Relatively high cost

Variations

- Standard, Y pattern, angle, three-way

Materials

- Body: bronze, all iron, cast iron, forged steel, Monel, cast steel, stainless steel, plastics
- Trim: various

Special Installation and Maintenance Instructions

- Install so pressure is under disk, except in high-temperature steam service.
- Lubricate on strict schedule.
- Flush foreign matter off seat by opening valve slightly.
- Correct packing leaks immediately by tightening the packing nut.

Ordering Specifications

- Type of end connection
- Type of disk
- Type of seat
- Type of stem assembly
- Type of bonnet assembly
- Pressure rating
- Temperature rating

Ball Valves

A ball valve is a quarter-turn valve in which a drilled ball rotates between seats, allowing straight-through flow in the open position and shutting off flow when the ball is rotated 90° and blocks the flow passage (Fig. 5.180).

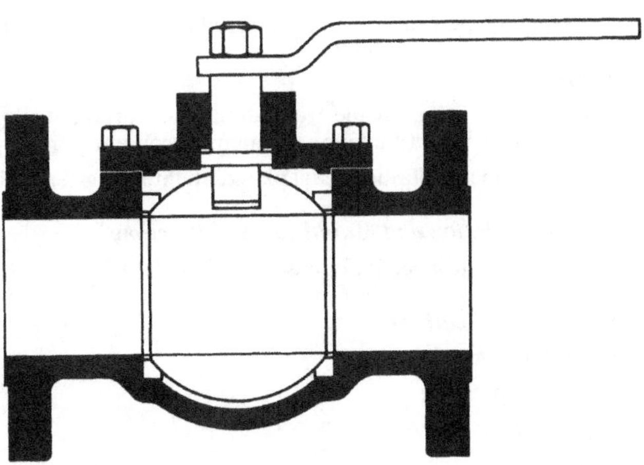

FIGURE 5.180 Ball valve.

Recommended

- For on-off, nonthrottling service
- Throttling service with special trim
- Where quick opening is required
- Where minimum resistance to flow is needed

Applications

- General service, high temperatures, cryogenic service, slurries

Advantages

- Low cost
- High capacity
- Bidirectional shutoff
- Straight-through pattern
- Low leakage
- Self-cleaning
- Low maintenance
- No lubrication requirement
- Compact
- Tight sealing with low torque
- Good throttling characteristics with special trim

Disadvantages

- Poor throttling characteristics with standard ball designs
- Susceptible to seal wear with soft seats
- Prone to cavitation with standard ball designs

Variations

- Construction: top entry, split body, end entry, or welded
- Style: floating ball, trunnion mounted, rising stem, or throttling
- Ports: three-way, venturi, full ported, reduced port, or venturi

Materials

- Body: bronze, iron, ductile iron, aluminum, carbon steel, low-alloy steel, stainless steel, Monel, nickel alloys, titanium, tantalum, zirconium, polypropylene, PVC plastics
- Seat: elastomeric, plastic, nickel plated, stellite, tungsten carbide, or ceramic

Special Installation and Maintenance Instructions

- Allow sufficient space for operation of handle.

Ordering Specifications

- Operating temperature
- Operating pressure
- Type of construction
- Style

- Type of port in ball
- Body material
- Seat material
- Type of actuation

Butterfly Valves

A butterfly valve is a quarter-turn valve that controls flow by means of a circular disk pivoting on its central axis (Fig. 5.181).

FIGURE 5.181 Butterfly valve.

Recommended

- For fully open or fully closed service
- For throttling service
- For frequent operation
- Where positive shutoff is required for gases or liquids
- Where minimum amount of fluid trapped in line is allowed
- For low pressure drop across valve

Applications

- General service, cryogenic service, high-temperature service, liquids, gases, slurries, liquids with suspended solids

Advantages

- Compact, lightweight, low-cost
- Low maintenance
- Minimum number of moving parts
- No pockets
- High capacity
- Straight-through flow
- Self-cleaning

Disadvantages

- High torque for actuation
- Limited pressure-drop capability
- Prone to cavitation

Variations

- Construction: wafer, lug wafer, flanged, screwed, fully lined
- Style: single offset, double offset, triple offset

Materials

- Body: bronze, iron, ductile iron, carbon steel, low-alloy steel, stainless steel, Monel, nickel alloys, PVC plastic, plastic/elastomer lined
- Disk: all metals, elastomer coatings such as TFE, Kynar, Buna-N, neoprene, Hypalon
- Seat: elastomeric, plastic, nickel plated, stellite, tungsten carbide, ceramic

Special Installation and Maintenance Instructions

- May be operated by lever, handwheel, or chainwheel.
- Allow sufficient space for operation of handle if lever-operated.
- Valves should remain in closed position during all handling and installation operations.

Ordering Specifications

- Type of construction
- Style
- Type of seat
- Body material
- Disk material
- Seat material
- Type of actuation
- Operating pressure
- Operating temperature

Diaphragm Valves

A diaphragm valve is a multiturn valve that effects closure by means of a flexible diaphragm attached to a compressor. When the compressor is lowered by the valve stem, the diaphragm seals and cuts off flow (Fig. 5.182).

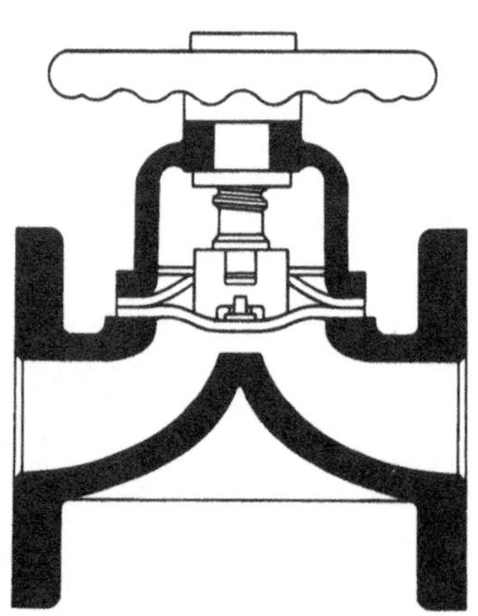

FIGURE 5.182 Diaphragm valve.

Recommended

- For fully open or fully closed service
- For throttling service
- For service with low operating pressures

Applications

- Corrosive fluids, sticky and/or viscous materials, fibrous slurries, sludges, foods, pharmaceuticals

Advantages

- Low cost
- No packing glands
- No possibility of stem leakage
- Resistant to problems of clogging, corroding, or gumming of media

Disadvantages

- Diaphragm subject to wear
- High torque under live-line closure

Variations

- Weir type and straight-through type

Materials

- Metallic, solid plastic, lined—wide variety of each

Special Installation and Maintenance Instructions

- Lubricate on a regular schedule.
- Do not use bars, wrenches, or cheaters to close.

Ordering Specifications

- Body material
- Diaphragm material
- End connections
- Type of stem assembly
- Type of bonnet assembly
- Type of operation
- Operating pressure
- Operating temperature

Pinch Valves

A pinch valve is a multiturn valve that effects closure by means of one or more flexible elements, such as diaphragms or rubber tubes, that can be pressed together to cut off flow (Fig. 5.183).

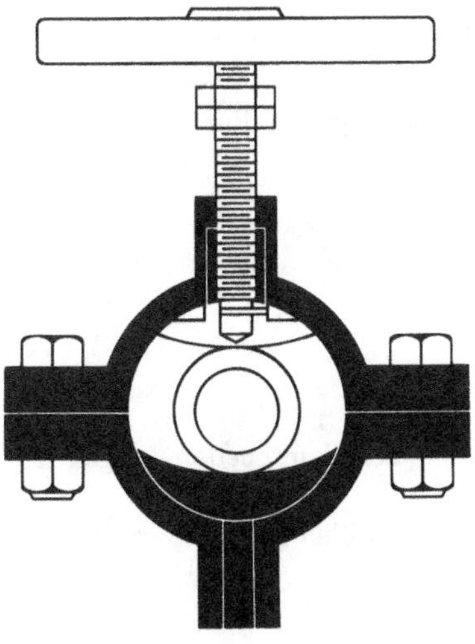

FIGURE 5.183 Pinch valve.

Recommended

- For on-off service
- For throttling service
- For moderate temperatures
- Where pressure drop through valve is low
- For services requiring low maintenance

Applications

- Slurries, mining slurries, liquids with large amounts of suspended solids, systems that convey solids pneumatically, food service

Advantages

- Low cost
- Low maintenance
- No internal obstruction or pockets to cause clogging
- Simple design
- Noncorrosive and abrasion-resistant

Disadvantages

- Limited vacuum application
- Difficult to size

Variations

- Exposed sleeve or body, encased metallic sleeve or body

Materials

- Rubber, white rubber, Hypalon, polyurethane, neoprene, white neoprene, Buna-N, Buna-S, Viton-A, butyl rubber, silcone, TFE

Special Installation and Maintenance Instructions

- Large sizes may require supports above or below the line if pipe supports are inadequate.

Ordering Specifications

- Operating pressure
- Operating temperature
- Sleeve material
- Exposed or encased sleeve

Check Valves

A check valve is self-actuating and is designed to check reversal of flow. Fluid flow in the desired direction opens the valve; reversal of flow closes it. There are three basic styles of check valves: (1) swing check, (2) lift check, and (3) butterfly check.

Swing Check Valves. A swing check valve has a hinged disk designed to open with flow and close when backflow begins (Fig. 5.184).

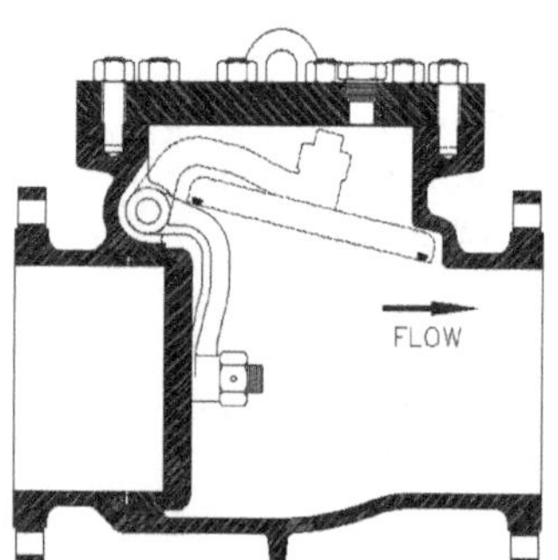

FIGURE 5.184 Swing-style check valve.

Recommended

- Where minimum resistance to flow is needed
- Where there is infrequent change of direction in the line
- For vertical lines having upward flow

Applications

- General service, high-temperature service, cryogenic service, steam

Advantages

- Self-actuating.
- Unobstructed flow path.
- Allows passage of pipeline inspection gauges, if a full-opening swing.
- Turbulence and pressure within valve are very low.
- Resistant to debris.

Disadvantages

- Flow conditions can cause valve to slam.

Variations

- Tilting-disk check valve, turbine extraction

Materials

- Body: bronze, iron, ductile iron, aluminum, carbon steel, low-alloy steel, stainless steel, Monel, nickel alloys, titanium, tantalum, zirconium, polypropylene, PVC plastics
- Trim: various

Special Installation and Maintenance Instructions

- In vertical lines, pressure should always be under seat.
- If valve fails to seal, check seating surfaces.
- If seat is damaged or scored, regrind or replace.
- Before reassembling, clean internal portions thoroughly.

Lift Check Valve. A lift check valve is similar in design to a globe valve except that the disk is lifted by the forward line pressure and closed by gravity and backflow (Fig. 5.185).

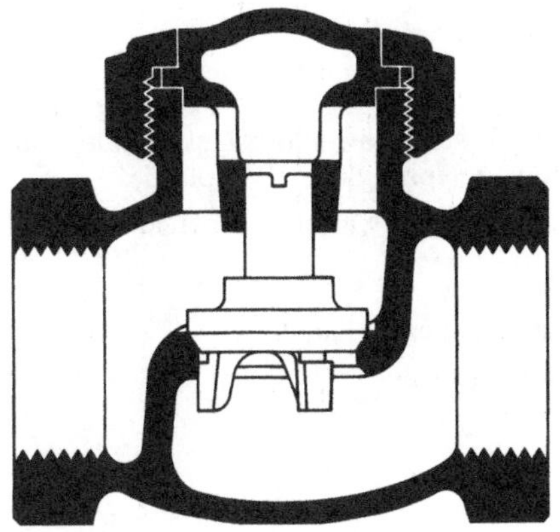

FIGURE 5.185 Lift-style check valve.

Recommended

- Where there are frequent changes of direction in the line
- Pulsating flow when lift check has internal dampening
- Nozzle variant for low pressure drops

Applications

- Steam, air, gas, water, and vapor lines with high flow velocities

Advantages

- Minimum travel of disk for fully open position
- Quick-acting

Disadvantages

- Nonpiggable
- Pressure drop in some designs
- Restricted flow path

Variations

- Three body patterns: horizontal, angle, vertical
- Ball check, piston check, nozzle check, spring-loaded check, stop check

Materials

- Body: bronze, all iron, cast iron, forged steel, Monel, stainless steel, PVC, Penton, impervious graphite, TFE-lined
- Trim: various

Special Installation and Maintenance Instructions

- Line pressure should be under seat.
- Vertical flow up or down must be specified when ordering.
- If backflow leaks, check disk and seat.

Butterfly Check Valve. A butterfly check valve has a split disk. Each half of the disk is hinged on a shaft in the center of the valve. The half disks open like butterfly wings to allow flow through the valve. Backward flow automatically closes the half disks.

Recommended

- Where low resistance to flow in the line is needed
- Where there is frequent change of direction in the line

Applications

- General service, high-temperature service, cryogenic service, steam

Advantages

- Less expensive
- Simplicity of design permits construction in large diameters
- May be installed in virtually any position

Disadvantages

- Typically wafer style
- High wear in pulsating service

Variations

- Fully lined, soft seated, flexible or rigid sealing members

Materials

- Body: bronze, iron, ductile iron, aluminum, carbon steel, low-alloy steel, stainless steel, Monel, nickel alloys, titanium, tantalum, zirconium, polypropylene, PVC plastics
- Sealing members: elastomeric or plastic for flexible members, as body for rigid members

Special Installation and Maintenance Instructions

- On line valves, liner should be protected from damage during handling.
- Make sure valve is installed so that forward flow opens valve.

Relief Valves

Pressure-relief valves protect life and property. They are installed on pressurized systems in almost all types of industries such as refineries, oil and gas production facilities, pipelines, electric power generating plants, paper mills, chemical plants, and other industrial facilities worldwide. Pressure-relief designs are available for both compressible and noncompressible fluid applications in a wide range of sizes, material selections, pressure classes and temperature ranges.

Sizing and selection of relief valves must comply with criteria specified in applicable codes and standards as well as the manufacturer's data concerning the flow characteristics for each design type. Pressure-relief valve manufacturers provide catalogs and computer sizing programs to aid in the sizing and selection of their specific products (Fig. 5.186).

FIGURE 5.186 Relief valves.

Recommended

- Overpressure protection of pressurized systems and vessels

Applications

- Compressible and noncompressible fluids (steam, liquids, gases, vapor)

Advantages

- Prevents catastrophic failures of pressurized systems and vessels.
- Standard designs are self-actuating, requiring no external power source.
- Designs are fail safe. Any failure will cause the relief valve to assume the open and flowing position.
- Compliant with applicable ASME boiler and pressure vessel codes and API standards.

Variations. The following definitions are excerpts from the ASME PTC25 performance test codes, accompanied by typical illustrations and additional commentary on the valve types described.

Pressure-Relief Device. A pressure-relief device is a device designed to prevent internal fluid pressure from rising above a predetermined maximum pressure in a pressure vessel exposed to emergency or abnormal conditions (Fig. 5.187).

FIGURE 5.187 Pressure-relief device.

Reclosing Pressure-Relieving Devices. A pressure-relief valve is a spring-loaded pressure-relief device, which is designed to open to relieve excess pressure and to reclose and prevent the further flow of fluid after normal conditions have been restored. It is characterized by rapid-opening pop action or by opening at a rate generally proportional to the increase in pressure over the opening pressure. It may be used for either compressible or incompressible fluids, depending on design, adjustment or application.

Safety Valve. A safety valve is a pressure-relief valve actuated by inlet static pressure and characterized by rapid opening or pop action. It is normally used for steam and air services.

Comment: A safety valve is an automatic pressure-relieving device actuated by the static pressure upstream of the valve and characterized by full-opening pop action. These valves are designed for boiler applications in accordance with the requirements of ASME Boiler and Pressure Vessel Code Section I for nonnuclear use or Section III for nuclear applications. Section I valves may also be used for Section VIII applications for both air and steam service. Most commonly used in electric power plants, auxiliary steam systems, and industrial boilers. Also, there are safety valve designs for heating boiler applications. These valves comply with ASME Code Section IV (Fig. 5.188).

Relief Valve. A relief valve is a pressure-relief device actuated by inlet static pressure having a gradual lift generally proportional to the increase in pressure over opening pressure. It may be provided with an enclosed spring housing suitable for closed-discharge system application and is primarily used for liquid service.

Comment: There are two categories of liquid valves, those that are designed for ASME Boiler and Pressure Vessel Code Section VIII or Section III for nuclear applications and designs for noncode applications where liquid relief valves are used (Fig. 5.189).

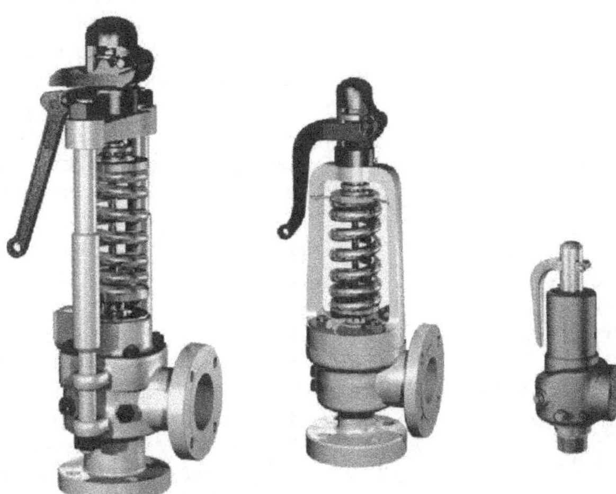

FIGURE 5.188 Reclosing-style relief valves.

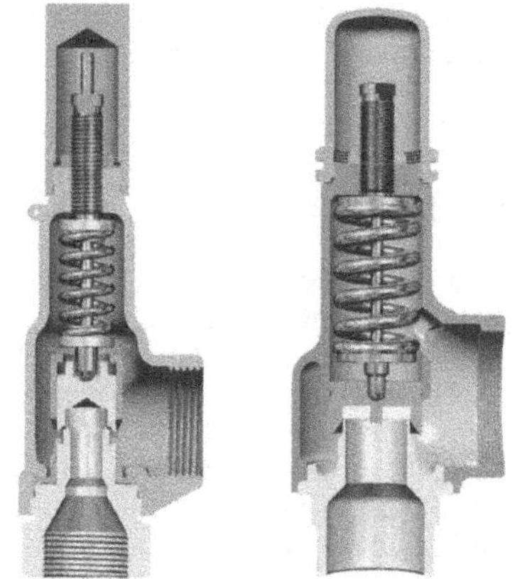

FIGURE 5.189 Liquid service relief valves: (*a*) ASME Code Section VIII, and (*b*) noncode.

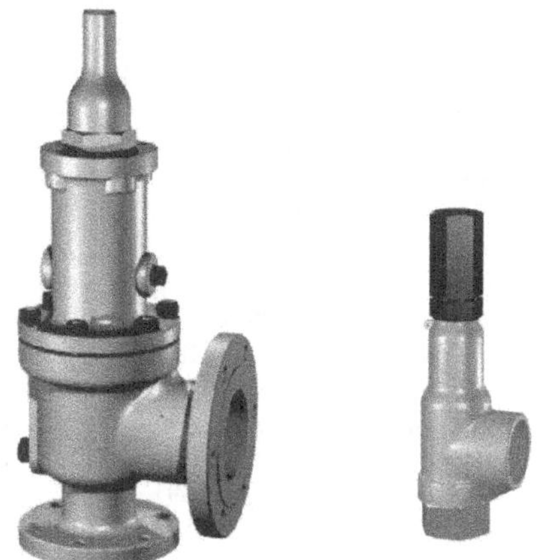

FIGURE 5.190 Safety relief valves.

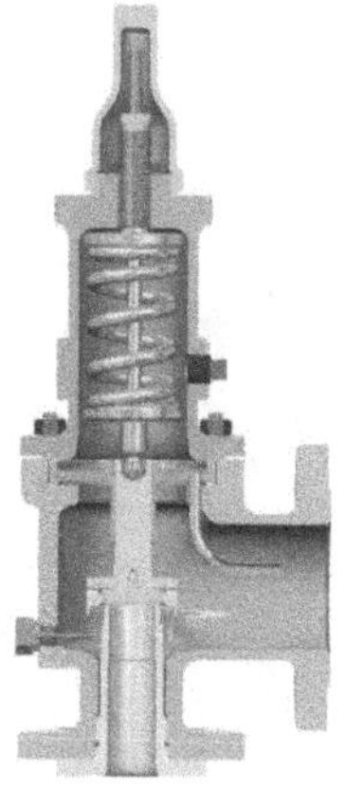

FIGURE 5.191 Conventional safety relief valve.

Safety Relief Valve. A safety relief valve is a pressure-relief valve characterized by rapid opening or pop action, or by opening in proportion to the increase in pressure over the opening pressure, depending on the application. It may be used on liquid or compressible fluid.

Comment: A safety relief valve is an automatic pressure-actuated relieving device suitable for use either as a safety valve or a relief valve, depending on the application. Safety relief valves are designed in accordance with ASME Boiler and Pressure Vessel Code Section VIII or Section III for nuclear applications. These valves are general-purpose relieving devices suitable for varied applications on compressible and noncompressible fluids. These valves are most commonly used in refineries, oil and gas production facilities, pipelines, paper mills, chemical plants, and other industrial facilities (Fig. 5.190). The following descriptions are for design variations of safety relief valves for specific applications.

Conventional Safety Relief Valve. A conventional safety relief valve is a pressure-relief valve which has its spring housing vented to the discharge side of the valve. The operational characteristics (opening pressure, closing pressure and relieving capacity) are directly affected by changes in the backpressure on the valve.

Comment: Conventional safety relief valves are for applications where excessive variable or built-up backpressure is not present in the system into which the valve discharges (Fig. 5.191).

Balanced Safety Relief Valve. A balanced safety relief valve is a pressure-relief valve which incorporates means of minimizing the effect of backpressure on the operational characteristics (opening pressure, closing pressure and relieving capacity).

Comment: These valves are typically equipped with a bellows which balances or eliminates the effect of variable or built-up backpressure that may exist in the system into which the safety relief valve discharges (Fig. 5.192).

Pilot-Operated Pressure-Relief Valve. A pilot-operated pressure-relief valve is a pressure-relief valve in which the major relieving device is combined with and is controlled by a self-actuated auxiliary pressure-relief valve.

Comment: Pilot-operated relief valves are available in both pop-action and modulating-action designs. These valves are suitable for applications where it is desired to maintain system operating pressure very close to the valve's opening pressure (Fig. 5.193).

Power-Actuated Pressure-Relief Valve. A power-actuated pressure-relief valve is a pressure-relief valve in which the major relieving device is combined with and controlled by a device requiring an external source of energy.

Comment: There are a number of designs that employ pneumatic, hydraulic, or electrical systems to actuate relief valves. Some designs attach directly to the spring-loaded relief valves and others may use a ball valve combined with a motor-driven operator, such as the one shown in Fig. 5.194.

Internal-Spring Safety Relief Valve. An internal-spring safety relief valve is a pressure-relief valve characterized by containing the spring and all or part of the operating mechanism within the pressure vessel.

Temperature-Actuated Pressure-Relief Valve. A temperature-actuated pressure-relief valve is a pressure-relief valve that may be actuated by external or internal temperature or by pressure on the inlet side.

Vacuum Relief Valve. A vacuum relief valve is a pressure-relief device designed to admit fluid to prevent an excessive internal vacuum; it is designed to reclose and prevent further flow of fluid after normal conditions have been restored.

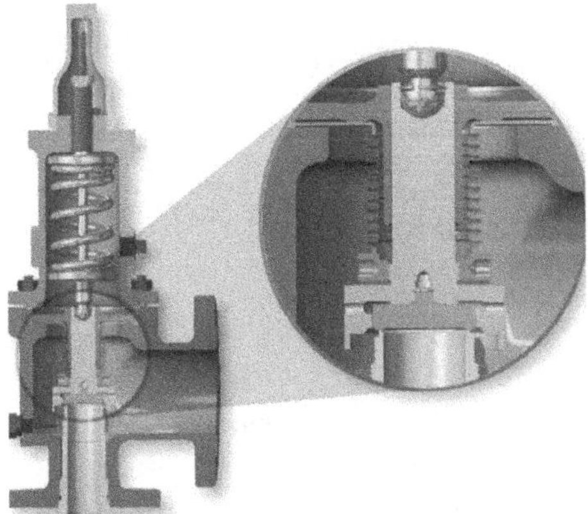

FIGURE 5.192 Balanced safety relief valve.

Materials for Body and Trim

- Pressure-relief valves are offered in many material combinations. Manufacturers recommend materials for specific applications. Proper selection of materials requires that service conditions be specified: temperature, pressure, media contained within the system, and any external environmental conditions that need consideration.
- Typical materials available for pressure-relief valves include bronze, cast iron, carbon and alloy steels, stainless steels, Monel, Hastelloy, and Inconel.
- Materials for severe applications are also available from most manufacturers. Special nonstandard valves can be supplied in such exotic materials as titanium and duplex stainless steels.

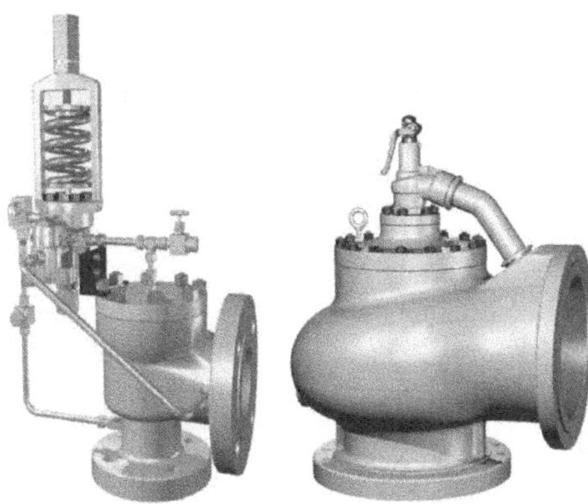

FIGURE 5.193 Pilot-operated relief valves.

Special Installation and Maintenance Instructions

- Installation of pressure-relief valves should be in accordance with the appropriate codes and standards and jurisdictional authorities that apply to the specific plant location.
- Manufacturers provide installation and maintenance manuals with recommendations and instructions for their specific valve types.

Control Valves

Globe-Style Control Valves. Use stem-, post-, or cage-guided valve plugs to control the flow of process media in a pipeline. The plug is throttled in a linear motion to control the flow, and closes on a seat parallel to the pipeline to provide shutoff. Various trim configurations can be utilized for differing degrees of cavitation and flashing (Fig. 5.195).

Recommended

- For throttling service and flow regulation
- For applications requiring rugged designs due to high cycling requirements
- For applications requiring precise control
- For applications requiring cavitation containment and elimination
- For applications requiring noise reduction

FIGURE 5.194 Power-actuated relief valve.

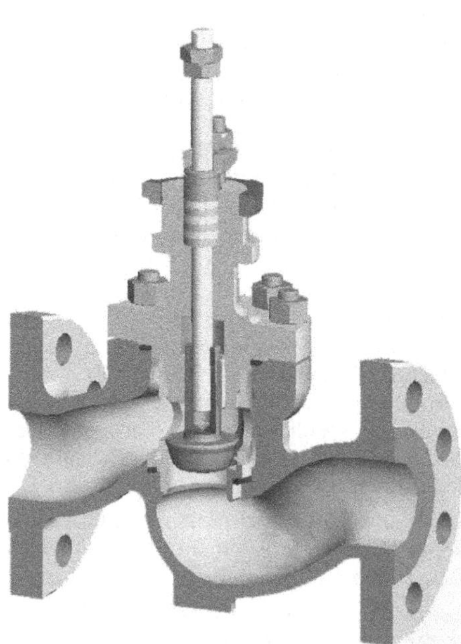

FIGURE 5.195 Globe-style control valve.

Applications

- From general service (stem-guided) to severe service (cage-guided)
- Liquids, vapor, gases, steam
- High pressure drops
- High or low temperatures
- High-pressure shutoff requirements

Advantages

- Precise flow control
- Versatility—cover a wide range of applications
- In-line repairable—top-entry designs
- Long-term performance

Disadvantages

- Less capacity per size
- Higher costs per size
- Heavier weight

Variations

- Body: standard straight-through angle patterns, Y patterns, three-way valves
- Trims: low noise, anticavitation, linear, equal percentage, quick opening

Materials

- Body: bronze, cast iron, carbon steel, chrome molybdenum, stainless steel, exotic alloys, Monel, Hastelloy
- Trim: typically stainless steel, hardened materials

Special Installation and Maintenance Instructions

- Remove trim internals if welding valve body into pipeline.
- Remove trim internals to flush system to prevent damage to plug and seats.

Ordering Specifications

- Size
- End connections
- Body and bonnet material
- ANSI pressure rating
- Temperature of service
- Inlet/outlet pressure
- Flow rates of service

Rotary Control Valves. A rotary control valve utilizes an eccentric plug that rotates nominally 50° to an open position or to intermediate positions for throttling control. The eccentric action of the plug ensures that rubbing contact with the valve seat occurs only within a few degrees of the closed position, minimizing friction and optimizing controllability (Fig. 5.196).

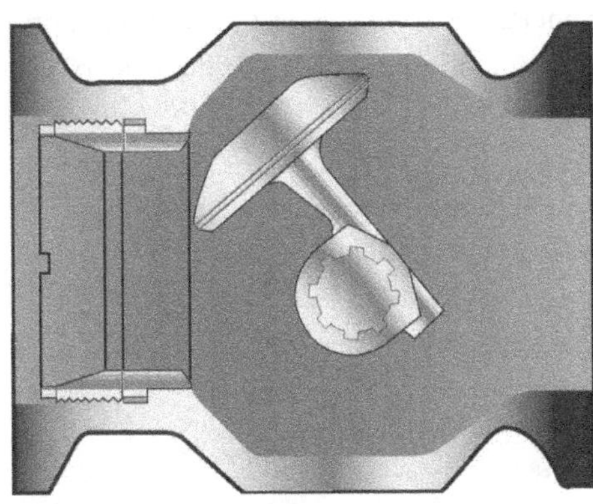
FIGURE 5.196 Rotary-style control valve.

Recommended

- For accurate throttling control
- For tight shutoff
- Where fugitive emissions must be kept to a minimum
- For a wide range of fluids and gases and service conditions

Applications

- General service, liquids, gases, steam, corrosives, abrasive media, slurries

Advantages

- Simple, lightweight construction
- Reliable
- Precision of throttling control
- Ample flow capacity yet good resistance to cavitation and noise

Disadvantages

- Limited pressure drop
- No in-line trim maintenance

Variations

- Flangeless, flanged, and threaded ends
- Metal or soft seat

Materials

- Body: carbon steel, stainless steel, Hastelloy, Monel, Alloy 20, other alloys
- Trim: stainless, ceramic

Special Installation and Maintenance Instructions

- Ensure that the plug and seat are centered before tightening the seat ring retainer.
- Carefully handle valve bushing O-rings to prevent tears and to ensure shaft seal integrity.

Ordering Specifications

- Body material
- Seat material
- ANSI rating
- End connection
- Full or reduced trim

Valve Positioners. A valve positioner is a device that precisely positions the valve plug or stem. Valve positioners are used on many valves to overcome frictional forces inherent to the valve and actuator assembly. The positioner's primary function is to accurately move the valve to the desired position in response to a signal from a controller. The position command signal can be pneumatic, analog, or digital (Fig. 5.197).

Recommended

- For throttling service

FIGURE 5.197 Smart valve-interface valve positioner.

Applications

- Required for pneumatic springless piston-type actuators
- Any application where improved throttling control performance is needed for better process control

Advantages

- Improved throttling control accuracy
- Reduced hysteresis and deadband
- Better linearity
- Quickly changeable valve characterization (linear or equal percentage)
- Decreased process variability
- Increased process yields

Disadvantages

- Increased cost for valve package

Variations

- Single-acting (for pneumatic spring-diaphragm actuators) or double-acting (for pneumatic springless piston-type actuators)
- Pneumatic (3 to 15 or 6 to 30 psi), electropneumatic (4 to 20 mA, Hart Communications Foundation Protocol), Fieldbus protocol

Materials

- Various

Special Installation and Maintenance Instructions

- Use clean instrument supply air for prolonged life.

Ordering Specifications

- Type of input signal (pneumatic, analog, or digital)
- Single- or double-acting
- Agency approvals [e.g., Factory Mutual (FM), Canadian Standards Association (CSA), European Committee for Electromechanical Standardization (CENELEC)]

Actuators. An actuator is a device that operates a valve and moves the flow-control element to the desired position—opened, closed, or throttling. The actuator must overcome valve unbalance and frictional forces, and provide enough loading to shut off the control element against the seat. The actuator is typically powered by pressurized air.

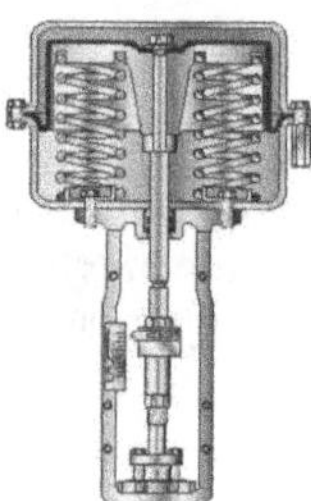

FIGURE 5.198 Spring-diaphragm-style actuator.

Main Types

- Pneumatic spring diaphragm (Fig. 5.198)
- Pneumatic cylinders and pistons (Fig. 5.199)

PNEUMATIC SPRING DIAPHRAGM

Advantages

- Reliable, low-cost performance
- Low cost of ownership
- Low-friction drive

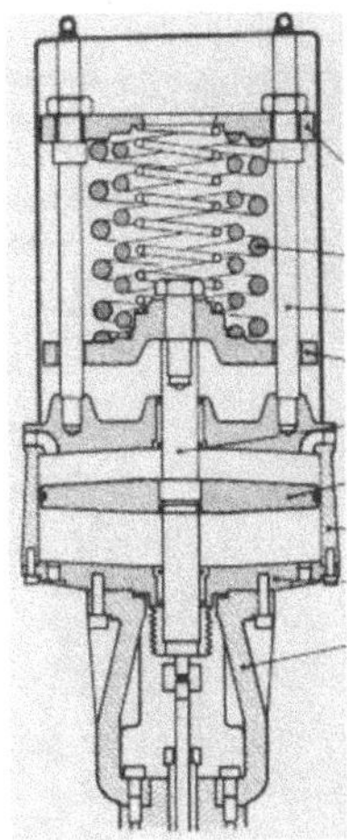

FIGURE 5.199 Piston-style actuator.

Disadvantages

- Limited stem travel
- Limited supply pressures

PNEUMATIC CYLINDER AND PISTON

Advantages

- Constant effective area
- Higher supply pressure
- Longer stem travel

Disadvantages

- Higher frictional load
- Expensive
- Higher supply pressure required

Variations

- Linear
- Rotary

Ordering Specifications

- Stroking speed
- Maximum supply pressure available
- Fail position
- Stroke length

CODES AND STANDARDS

Various professional and industrial organizations have created codes and standards that are applicable to the design, selection, and use of industrial products. Certain of these standards are peculiar to specific industries, such as those set up by the American Petroleum Institute (API), American Water Works Association (AWWA), American Gas Association (AGA), American National Standards Institute (ANSI), American Society of Mechanical Engineers (ASME), and Manufacturers Standardization Society (MSS) of the Valve and Fittings Industry. These are but a few of the existing organizations.

The valve specified should be aware of those codes that apply to a specific industry and the application in question. For further information on the codes and standards available, these organizations may be contacted at the addresses listed here:

AWWA, 6666 W. Quincy Avenue, Denver, CO 80235

ANSI, 1430 Broadway, New York, NY 10018

AGA, 1515 Wilson Boulevard, Arlington, VA 22209

API, 1220 L Street, NW, Washington, DC 20005

ASME, 345 East 47th Street, New York, NY 10017

MSS, 127 Park Street, NE, Vienna, VA 22180

BIBLIOGRAPHY

Skouson, P.: *Valve Handbook,* McGraw-Hill, New York, 1998.

CHAPTER 5.18
LUBRICANTS: GENERAL THEORY AND PRACTICE*

GENERAL THEORY

Functions of Lubricants

Lubricants perform a variety of functions. The primary, and most obvious, function is to reduce friction and wear in moving machinery. In addition, lubricants can:

- Protect metal surfaces against rust and corrosion
- Control temperature and act as heat-transfer agents
- Flush out dirt and wear-debris contaminants
- Transmit power
- Absorb or damp shocks
- Form seals

Because reducing friction is such an important function of lubricants, it is necessary to understand how they perform.

* Updated for this third edition by William H. Conner, Elf Lubricants, Linden, New Jersey.

Friction

Friction is the resistance to movement between parts of a machine. Two categories of friction can be identified: solid (or dry) friction and fluid friction.

Solid Friction. Solid friction occurs when there is physical contact between two solid bodies moving relative to each other. The type of motion divides solid friction into two categories, *sliding* and *rolling* friction.

Sliding Friction. This is the resistance to movement as one part slides past another. Solid surfaces, which appear smooth to the eye, do in fact consist of many peaks and valleys. The resistance to motion is due primarily to the interlocking of these peaks or asperities. When asperities come in contact without a lubricating film, enough heat is generated to melt the metal at that microscopic point. These tiny points can weld and create larger, harder points of metal. The process can continue until visible metal deformation or galling can cause seizure of the bearing or gear. Industry generally refers to exceptionally high loading as *extreme pressure* conditions, but without a lubricant, even light loads on a bearing can cause galling.

Rolling Friction. This is the resistance to motion as one solid body rolls over another. It is caused primarily by the deformation of the rolling elements and support surfaces under load. For a given load, rolling friction is significantly less than sliding friction.

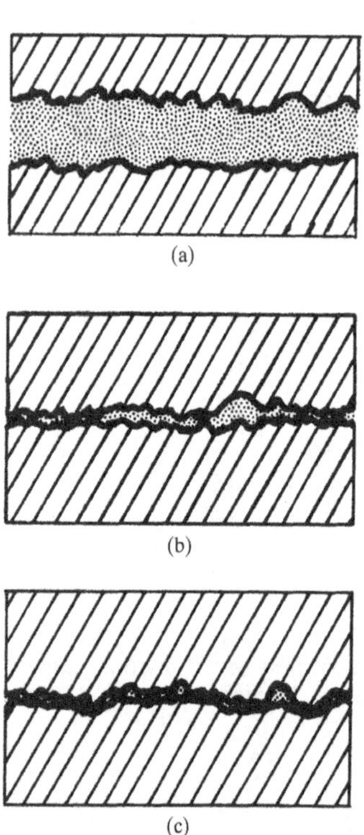

FIGURE 5.200 (*a*) Full-film lubrication. (*b*) Mixed-film lubrication. (*c*) Boundary lubrication.

Fluid Friction. Fluid friction is the resistance to motion between the molecules in a fluid. For a given load, fluid friction is usually significantly less than solid friction. The film thickness, relative to the height of the surface asperities, distinguishes three types of lubrication:

- Full or thick-film lubrication
- Mixed-film lubrication
- Boundary lubrication

Full or Thick-Film Lubrication. This is the state in which the lubricant film between two surfaces is of sufficient thickness to completely separate the asperities on the two surfaces. In this case, only fluid friction exists between the moving surfaces and no metal-to-metal contact will occur (Fig. 5.200*a*).

Mixed-Film Lubrication. This is the state in which the lubricant film between the two surfaces is of sufficient thickness to separate them, but some asperities may rupture the film and make contact (Fig. 5.200*b*).

Boundary Lubrication. This is the state in which lubricant may be present, but not enough to cause separation of the bearing surfaces (Fig. 5.200*c*).

Formation of the Lubricant Film

The lubricant film may be formed and maintained in one of two ways:

- Hydrostatically
- Hydrodynamically

Hydrostatic Lubrication. Hydrostatic conditions occur when fluid is pumped between the bearing surfaces with enough pressure to separate them. The surfaces may or may not be moving with respect to each other.

Hydrodynamic Lubrication. Hydrodynamic lubrication depends on motion between the two solid surfaces to generate and maintain the lubricating film. In a plain bearing that is not rotating, the shaft will rest on the bottom of the bearing and will tend to squeeze any lubricant out from between the surfaces. When the shaft begins to rotate, a very thin film of lubricant will tend to adhere to the shaft surface and will be drawn between the shaft and the bearing in the loaded region. A self-acting film that will ultimately separate the load-bearing surfaces is established. A lubricant film generated in this manner is called a *hydrodynamic film.*

Five requirements must be met for hydrodynamic lubrication to operate successfully:

1. There are converging surfaces leading to the load zone (or *wedge area*).
2. The bearing is of adequate size.
3. There is an adequate supply of lubricant.
4. The viscosity of the lubricant is adequate.
5. There is sufficient shaft speed.

The thickness of the hydrodynamic oil film developed in a properly designed plain bearing is dependent on the oil viscosity, the bearing load, speed, metallurgy, and the quality of the bearing surfaces. The dimensionless bearing parameter ZN/P conveniently describes the combined effect of viscosity Z, speed N, and load P.

The thickness of the hydrodynamic film and the amount of friction developed in the bearing can be predicted by means of the bearing parameter ZN/P. Plotting the bearing coefficient of friction versus the bearing parameter for a particular bearing and lubricant gives a characteristic curve similar to the one in Fig. 5.201. Experience has shown that the thickness of the

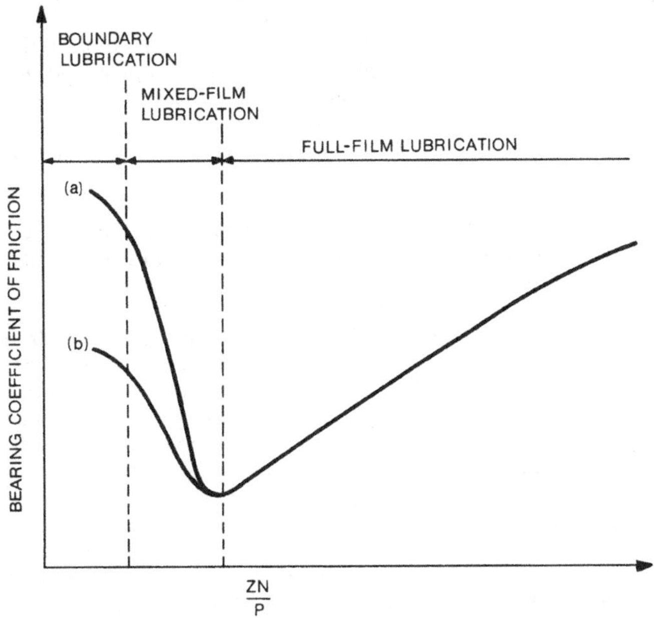

FIGURE 5.201 Typical ZN/P versus coefficient of friction curve.

lubricant film developed in a bearing can be determined by estimating where on the curve a bearing is operating (see Fig. 5.201).

Changes in the quality of the bearing's metallurgy or surface finish, or the lubricant's "oiliness" or film strength, will cause shifts in coefficient of friction under boundary or mixed-film conditions. For example, when holding everything else constant, adding an oiliness additive to the lubricant will shift the bearing performance curve from curve *a* to curve *b*. As this indicates, it is possible to reduce the amount of friction generated in a particular bearing under boundary or mixed-film lubrication conditions through the use of certain additives.

TYPES OF LUBRICANTS

There are three major categories of lubricants:

- Liquid lubricants
- Greases (semisolid lubricants)
- Solid lubricants

Each lubricant has its own physical properties that affect its performance in different applications. A knowledge of the various types of lubricants on the market today and a basic understanding of their advantages and limitations is most helpful in the selection of the best lubricant for a particular application.

Liquid Lubricants

Liquid lubricants are the most widely used. The most common are petroleum oils, synthetic fluids, and animal or vegetable oils. Many other fluids can fulfill a lubrication function under special conditions when the use of oils may be precluded.

Petroleum or Mineral Oils. Petroleum or mineral oils refined from petroleum crude have many characteristics that make them the most popular choice worldwide for many lubricating applications: they are abundant in nature, easily refined into many products of both high and low quality, fairly inexpensive, compatible with most machine construction materials, able to accept additives for performance enhancement, and essentially nontoxic, with many viscosity grades available.

Synthetic Fluids. Synthetic fluids include all human-made fluids used for lubricating purposes. Included in this category are synthesized hydrocarbons, esters, silicones, polyglycols, and phosphate esters.

Animal and Vegetable Oils. Animal and vegetable oils, as the terms indicate, are oils derived from those sources. They can be used where food contact is likely to occur and the lubricant must be edible. Their main disadvantage is that most of them tend to deteriorate rapidly in the presence of heat. Petroleum oils, on the other hand, can be refined to such a degree that all undesirable components can be removed and thus be labeled *USP* or *white oil,* suitable for medicinal, cosmetic, or incidental food-contact applications.

Animal and vegetable oils have polar molecules which give them a good affinity for metal surfaces. This gives rise to the term *oiliness,* meaning an oil with a more tenacious film strength.

In the past, oils made from animal fats, such as sperm-whale oil and lard oil, were frequently used for their oiliness properties. Today, however, these are frequently being replaced with synthesized fatty oils which perform the same function.

Greases

Greases are liquid lubricants with dispersed thickeners to give them a solid or semisolid consistency. The liquid content of a grease performs the actual lubricating function. The thickener acts to hold the lubricant in place and to prevent leakage. Grease can help block the entrance of contaminants into a bearing or gear housing.

Many types of thickeners are used in the manufacture of modern greases. Each type imparts certain properties to the finished product. Table 5.71 describes some of the typical properties and applications of greases manufactured with certain common thickeners.

Solid Lubricants

Solid lubricants, such as graphite, molybdenum disulfide (moly) and polytetrafluoroethylene (PTFE), are not only used by themselves but are also frequently added to oils and greases to improve their performance under boundary lubrication conditions.

IMPORTANT LUBRICANT CHARACTERISTICS

Various physical and chemical properties of lubricants are measured and used to determine a lubricant's suitability for different applications.

Oil Properties

Viscosity. Of the various lubricant properties and specifications, viscosity (also referred to as the *body* or *weight*) normally is considered the most important. It is a measure of the force required to overcome fluid friction and allow an oil to flow—that is, the resistance to flow.

Industry uses several different systems to express the viscosity of an oil. Figure 5.202 gives a comparison of some of the most common. Lubricant specifications usually express viscosity in Saybolt universal seconds (SUS or SSU) at 100 and 210°F (37.8 and 98.9°C) and/or in centistokes (cSt) at 40 and 100°C (104 and 212°F). Viscosity expressed in centistokes is called the *kinematic viscosity.*

With the general move toward metrication and the establishment of the International Organization for Standardization (ISO) viscosity grade identification system, the centistoke has become the preferred unit of measure. The ISO viscosity grade system contains 18 grades covering a viscosity range from 2 to 1500 cSt at 40°C. Each grade is approximately 50 percent more viscous than the previous one.

Laboratories determine oil viscosity experimentally using a viscometer (Fig. 5.203). The viscometer measures an oil's kinematic viscosity by the time (in seconds) it takes a specific volume of lubricant to pass through a capillary of a specified size, at a specified temperature. The kinematic viscosity is then derived by calculations based on constants for the viscometer and the time it took the sample to pass through the instrument.

Viscosity Index. The viscosity index (VI) is an empirical measure of an oil's change in viscosity with temperature. The greater the value of the viscosity index, the less the oil viscosity will change with temperature. Originally ranging from 0 to 100, viscosity indexes greater than 100 are now achieved through better techniques of refining petroleum, through the use of certain synthetic oils, and through the use of additives. The higher the number, the smaller the relative change in viscosity with temperature. Paraffinic lubricants have higher viscosity indexes than naphthenic lubricants.

TABLE 5.71 Grease Application Guide

Thickeners	Lithium	Lithium complex	Calcium	Calcium complex	Aluminum complex	Sodium	Polyurea	Clay
Properties								
Dropping point, °F	350–375	500+	200–225	500+	500+	325–350	550	None
Average max usable temp, °F	275	300	175	325	300	250	350	275
High temperature characteristics	Good	Good	Poor	Good	Good	Fair–good	Excellent	Good
Thermal stability	Good	Good	Poor	Good	Good	Fair–good	Excellent	Good
Low temperature characteristics	Good	Good	Fair	Fair–good	Fair–good	Fair	Good	Good
Pumpability	Excellent	Excellent	Fair–good	Fair	Fair–good	Poor	Good–excellent	Good
Mechanical stability	Excellent	Excellent	Good	Fair–good	Excellent	Poor	Excellent	Good
Oil separation	Good	Excellent	Poor	Excellent	Good	Fair–good	Excellent	Excellent
Water resistance	Good	Excellent	Excellent	Good	Excellent	Poor	Excellent	Good
Texture	Smooth & buttery	Smooth & buttery	Smooth & buttery	Smooth & buttery	Smooth & buttery	Buttery to fibrous	Smooth & buttery	Smooth & buttery
Rust protection	Fair to good	Good	Poor	Excellent	Good	Excellent	Excellent	Poor
Oxidation stability	Fair to good	Good	Poor	Good	Good	Poor	Excellent	Good
Other properties				Good inherent EP properties		Good adhesive and cohesive properties		
Applications								
	Multipurpose All applications except extra high temperatures	Multipurpose Many uses	Where water is dominant factor Wet, moderately low temperature conditions Plain and roller bearings, water pumps, slides	High temperatures Corrosive conditions Do not use in centralized lubrication systems	Multipurpose Moderately high temperatures	Antifriction and plain bearings Electric motors, fans Must be used in dry conditions	Multipurpose High temperatures Antifriction and plain bearings Electric motors, fans Wet conditions Corrosive conditions	Multipurpose High temperatures

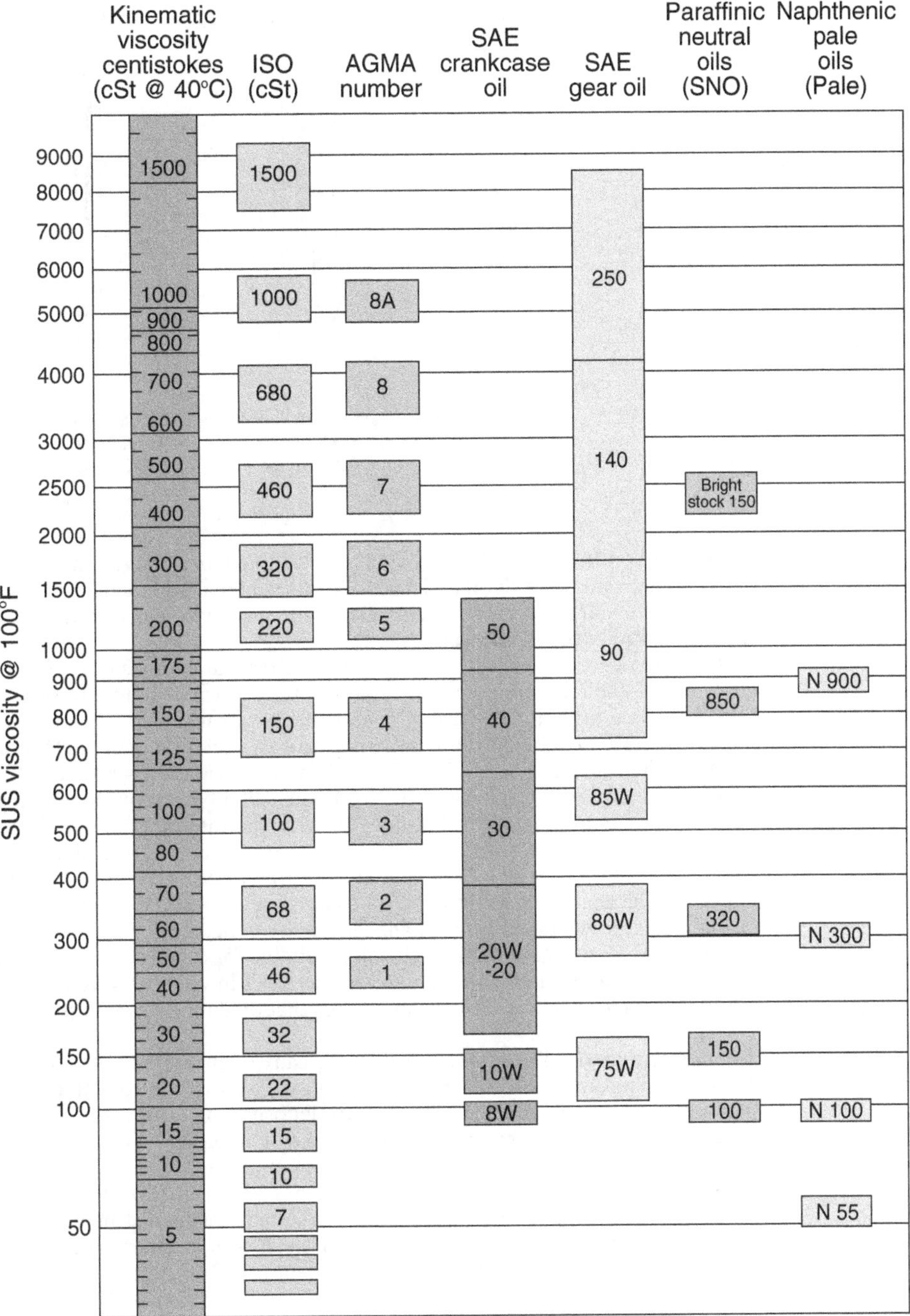

FIGURE 5.202 Base oil and finished product viscosity grade comparisons. *(Texaco Lubricants Company.)*

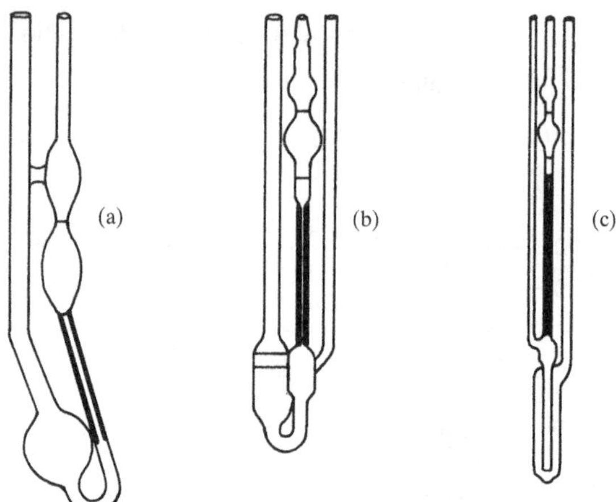

FIGURE 5.203 Capillary tube viscosimeters used to measure kinematic viscosity: (*a*) modified Ostwald, (*b*) Ubbelohde, (*c*) Fitz Simmons. *(Gulf Oil Corporation.)*

Oxidation Stability. Oxidation stability refers to a lubricant's ability to resist the tendency of oxygen to combine with oil molecules. Products of oxidation include sludge, varnish, resins, and corrosive and noncorrosive acids. Oxidation usually brings with it an increase in the viscosity and acidity of the lubricant.

The rate of oxidation is dependent on the chemical composition of the oil, the ambient temperature, the amount of surface area exposed to air, the length of time the lubricant has been in service, and the presence of contaminants which can act as catalysts to the oxidation reaction. Paraffinic lubricants are more resistant to oxidation than naphthenic types.

Depending on the intended end use of the oil, oxidation stability will be measured or expressed in different ways. All of the oxidation stability tests are based on placing a sample of oil under conditions that will greatly increase the rate of oxidation. Buildups of reaction products are then measured. The American Society for Testing and Materials (ASTM) D-943 test is the most widely used. Conducted under prescribed conditions, it measures the time (in hours) for the acidity of a sample of oil to increase a specified amount. The more stable the oil, the longer it will take for the change in acidity to occur.

Analysis to determine whether the oil is suitable for further service is based on a comparison between the used oil and the new oil. Increases in viscosity, acidity, and development of insoluble contaminants are usually indicators that oxidation has occurred.

Thermal Stability. Thermal stability is a measure of an oil's ability to resist chemical change due to temperature in the absence of oxygen. High temperatures can cause oil molecules to split and produce sludge and carbonaceous deposits. The term *thermal stability* is frequently misused in reference to the oxidation resistance of an oil.

Chemical Stability. Chemical stability defines an oil's ability to resist chemical change. Usually it, too, is used to refer to the oxidation stability of an oil. Chemical stability, other than resistance to oxidation, sometimes can refer to an oil's inertness in the presence of various metals and outside contaminants.

Carbon Residue. The carbon-forming tendencies of an oil can be determined with a test in which the weight percentage of the carbon residue of a sample is measured after evaporation and pyrolysis.

Neutralization Number. The neutralization number (neut. no.) is a measure of the acidity or alkalinity of an oil. Usually reported as the total acid number (TAN) or total base number (TBN), it is expressed as the equivalent milligrams of potassium hydroxide required to neutralize the acidic or basic content of a 1-g sample of oil. Increases in the TAN or decreases in the TBN are usually indicators that oxidation has occurred.

Lubricity. *Lubricity* is the term used to describe an oil's "oiliness" or "slipperiness." If two oils of the same viscosity are used in the same applications and one causes greater reduction in friction and wear than the other, it is said to have better lubricity than the first. This is strictly a descriptive term.

Saponification Number. The saponification number (SAP no.) is an indicator of the amount of fatty material present in an oil. The SAP number will vary from 0 (for an oil containing no fatty material) to 200 (for 100 percent fatty material).

Demulsibility. *Demulsibility* is the term used to describe an oil's ability to shed water. The better the oil's demulsibility, the more rapidly the oil will separate from water after the two have been mixed together.

API Gravity. API gravity is a relative measure of the unit weight of a petroleum product. It is related to the specific gravity in the following manner:

$$\text{API gravity @ 60°F (15°C)} = \frac{141.5}{\text{specific gravity @ 60°F (15°C)}} - 131.5$$

Pour Point. Pour point is the lowest temperature at which an oil will flow in a certain test procedure. It is usually not advisable to use an oil at temperatures lower than 15°F (8°C) above its pour point.

Flash Point. Flash point is the oil temperature at which vapors from the oil ignite briefly when an open flame is passed over a test sample.

Fire Point. Fire point is the oil temperature at which vapors from the oil will ignite when an open flame is passed over a test sample and continue to burn without the assistance or presence of that external flame source. The fire point is usually about 15 to 20°F (8 to 11°C) above the flash point.

Grease Properties

Penetration. Penetration is an indicator of a grease's relative hardness or softness and not a criterion of quality. Measured on a penetrometer at 77°F (25°C), it is the depth of penetration (in tenths of millimeters) into the grease of a standard 150-g cone. The softer the grease, the greater the penetration number will be.

If the penetration test is performed on an undisturbed sample, the results are reported as *unworked penetration.* If the sample has been subjected to extrusion by a reciprocating perforated piston for a number of strokes (most commonly 60 strokes) prior to the penetration test, the results are reported as *worked penetration.* It is desirable to have as little difference between the worked and the unworked penetration as possible.

NLGI Consistency Numbers. The National Lubricating Grease Institute (NLGI) has developed a number system ranging from 000 (triple zero) to 6 to identify various grease consistencies. This system is used by most of industry. Table 5.72 gives the NLGI numbers, their corresponding worked penetration ranges, and their descriptions (their corresponding consistencies). Most multipurpose greases are of either a number 1 or 2 consistency.

TABLE 5.72 NLGI Classification of Greases

NLGI consistency grade	Worked penetration ASTM D-217-60T	Description
000	445–475	Very fluid
00	400–430	Fluid
0	355–385	Semifluid
1	310–340	Very soft
2	265–295	Soft
3	220–250	Semistiff
4	175–205	Stiff
5	130–160	Very stiff
6	85–115	Hard

Dropping Point. Dropping point is the temperature at which the first drop of oil falls from a sample of grease under test conditions. Generally it is not advisable to use a grease at temperatures higher than 125 to 150°F (69 to 85°C) below its dropping point.

Thickeners. Many greases use metallic soaps as thickeners. Soap consists of a metallic polar head with a long hydrocarbon tail. Other thickeners are organic compounds that do not use metal in the molecule. It has been found that metals can actually contribute to the oxidation of the oil in greases by catalytic action. Also, some metals have now been listed as hazardous materials, which makes the disposal of grease and grease containers more dangerous and costly. Table 5.71 shows a comparison of some of the key properties of greases manufactured with different soaps and their typical applications.

Base Oils. Lubricating oil makes up the bulk of a grease, 50 to 98 percent. Greases can be made with base oil of almost any viscosity, from very light to very heavy. The NLGI number or consistency is independent of the base-oil viscosity. A grease with NLGI grade 2 can be made with an ISO VG 22 oil as well as with an ISO VG 460 oil. The base oil should correspond to the same viscosity as the application would demand if a liquid oil were being employed. Unfortunately, manufacturers rarely include the base-oil viscosity on the package. One must look at the manufacturer's product technical data sheet to find the viscosity, and even there it is sometimes omitted.

ADDITIVES USED IN LUBRICATING OILS

It is possible, through the use of chemical additives, to accomplish the following:

- Limit the chemical change or deterioration of the lubricant.
- Protect the machine from the effects of outside contaminants.
- Improve the physical properties of the lubricant.
- Reduce the wear on the machine.

Since industrial lubricating oils are frequently described by the additives they contain, it is helpful to understand the functions of the major types of additives. Following are general definitions of some of the most common, listed in alphabetical order:

Antifoam agents. Promote the rapid breakup of foam bubbles and release of entrapped air.

Antiseptic agents or bactericides. Prevent the growth of microorganisms and bacteria. These are found primarily in water-soluble coolants.

Antiwear agents. Decrease the coefficient of friction and reduce wear under boundary or mixed-film lubrication conditions.

Demulsifiers. Assist the natural ability of an oil to separate rapidly from water. These agents can be helpful in preventing rust since they help to keep water out of the oil and thus away from the metal surfaces.

Detergent-dispersant agents. Detergents keep machinery parts clean, while dispersants keep sludge from forming, settling, and turning into varnish. They are most commonly found in engine oils.

Emulsifiers. Permit the mixing of oil and water to form stable emulsions. They are used primarily in the manufacture of water-soluble oils.

Extreme-pressure agents. Improve the load-carrying ability of the lubricant by preventing welding and galling. They can reduce operating temperature, which reduces thermal stress on the oil. The majority of the extreme-pressure oils on the market today are of the sulfur-phosphorus type and are noncorrosive to most metals, including brass. This was not true of some of the earlier formulations, and many misconceptions still exist in this regard.

Fatty compounds ("oiliness"). Improve the lubricity or slipperiness of an oil. Friction is reduced by formation of an adsorbed film.

Oxidation inhibitors. Prevent or retard the oxidation of a lubricant, thereby reducing the formation of deposits and acids.

Pour-point depressants. Lower the pour point of paraffinic or mixed-base petroleum oils.

Rust and corrosion inhibitors. Improve an oil's ability to protect metal surfaces from rust and corrosion.

Tackiness agents. Improve the adhesive qualities of an oil.

Viscosity index improvers. Help keep the oil from thinning as fast with temperature rise, thus allowing a wider range of serviceability. These additives are long-chain polymers, most widely used in motor oils to create multigrade oils.

LUBRICANT SELECTION

Proper lubricant selection depends on the system needs and cost considerations. The final choice depends on the equipment design, operating conditions, and method of application, as well as health, safety, and environmental considerations. Whenever possible, these recommendations should be followed. In addition, most reputable oil suppliers keep in close contact with major equipment builders and are available to consult with users on lubricant selection. In addition, many PC software programs have recently become available that guide the user in making intelligent selections for specific equipment.

The recommendations included in this chapter are based on standard practices and are intended solely as guidelines.

General Selection Guidelines

The design of the equipment and the expected operating conditions will determine which functions the lubricant is expected to perform and will dictate the type of lubricant and additives that will be best suited.

The oil of proper viscosity for an application is a function of speed, load, and temperature (both ambient and operating). Conditions of high loads and slow speeds will require high-viscosity oil (ISO VG 460 or greater). Similarly, lower-viscosity oil is best suited to conditions of low loads and high speeds. Ideally, one would select the oil of lowest possible viscosity that is capable of maintaining a lubricant film between the moving surfaces. Selection of higher-viscosity oil than is needed can result in some power losses and temperature rise due to the higher internal fluid friction of the lubricant. This higher temperature puts thermal stress on the fluid, which may shorten its useful life. When lubricants are chosen for plantwide use, there will be compromises, but a higher than required viscosity for a particular application is also known to provide a better cushion of protection and even longer component life.

The effect of ambient and operating temperature on the selection of the lubricant should not be overlooked. Since oil viscosity decreases as temperature increases, it is necessary to select a lubricant that will provide an adequate oil film under the stabilized operating conditions. Fluids with high viscosity indexes can offer a wider operating temperature range. Generally, fluids which derive their higher VI from additives are susceptible to both temporary and permanent shearing and are therefore not usually chosen for gears and bearings. Synthetic fluids may have a naturally high VI and not be susceptible to shearing. The drawback to synthetics is their cost, and this must be weighed in the decision.

Operating Limits of Petroleum Oils

As a result of additive technology, a suitable petroleum-based lubricating oil can be found for most applications. Exceptions can exist where fire-resistant fluids are required or extreme temperature conditions exist. There are a variety of fire-resistant fluids available which may replace petroleum oils in hazardous situations. These include the following:

- Water/glycol base
- Oil in water emulsions
- High water base (95/5)
- Phosphate esters
- Polyol esters

The water-containing fluids are suitable where operating temperatures do not exceed 140°F (60°C). It is important that the hydraulic system or other application be suitable for the prospective fluid. System components such as seals, paints, and plastics, and pump pressures and other criteria, must be compatible with the candidate fluid. It is recommended that the manufacturer of the equipment be consulted before final selection is made.

Synthetic lubricants should be considered under three types of extreme temperature conditions:

- Excessively high temperatures
- Excessively low temperatures
- Wide temperature variations

There is no magic temperature that can be called the upper operating limit for petroleum oil. Oxidation is the life-ending factor for petroleum lubricants, and it is directly affected by time and temperature conditions. The ideal temperature for gears, bearings, and circulating systems ranges from 110 to 140°F (45 to 60°C). In that range, minor water or condensate contamination is driven off, while oxidation is not a great factor. There is a rule of thumb that the rate of oxidation doubles for every 20°F (11°C) rise in temperature above 140°F (60°C). With good-quality Group I base stocks, fortified with oxidation inhibitors, petroleum oil can serve for 5 yr at 140°F (60°C). There are few systems that remain intact for 5 yr without top-up or replacement. Factors such as broken lines, leaking seals, and outside contaminants come into play in the real industrial world during any 5-yr span.

So what is the upper temperature limit for Group I petroleum oil with a good oxidation inhibitor? A circulating system in which the oil leaves the bearings at 300°F (150°C) but is cooled to 140°F (60°C) and filtered shortly thereafter could be a stable arrangement. But a steady state of 250°F (120°C) would be very severe on petroleum oil, and the oil would need changing in a few weeks. Indeed, following the rule of thumb, oil maintained at 200°F (93°C) would be ready for changing in about 8 months. The newer hydrocracked Group II and Group III basestocks and synthetic fluids will serve successfully at higher temperatures.

For low-temperature service conditions, petroleum oils should not be used at less than 10 to 15°F (5 to 8°C) above their pour points.

For applications where there are wide temperature variations, oils with high viscosity indexes will help by providing a wider range of acceptable viscosity. But the correct oil viscosity at the operating temperature is the deciding factor.

Plain-Bearing Lubrication

Plain bearings, also called *journal* or *sleeve bearings,* comprise one of the simplest machine components. The type of motion between the bearing and the shaft is pure sliding.

In plain bearings, the lubricant must reduce sliding friction and prevent rust and corrosion. Grease in a bearing does not carry away much heat, but it can help prevent the entry of contaminants.

Barring any unusual operating conditions, plain bearings will operate satisfactorily with a good rust- and oxidation-inhibited lubricant of the correct viscosity. Special operating conditions may require the use of oils containing other additives. Antiwear or extreme-pressure oils may be desirable for plain bearings operating under high loads.

Most plain bearings are designed to operate under full-film hydrodynamic lubrication. Referring to Fig. 5.201 and assuming the bearing load and oil viscosity to be constant, the lubricant film development would be expected to follow the *ZN/P* curve as the shaft speed increases. If oil of the proper viscosity is selected for the load and speed conditions, full-film hydrodynamic lubrication will prevail during continuous operation.

Table 5.73 offers a general guide to viscosity selection for plain bearings subjected to average loading.

TABLE 5.73 Oil Viscosity Selection for Plain Bearings

ft/min	m/min	cSt	SUS
Below 100	30	150–325	600–1500
100–500	30–150	46–150	300–600
500–1000	150–300	22–68	150–300
1000–2500	300–800	15–32	75–150
over 2500	800	5–15	40–75

Speed factor = FPM
FPM = RPM × shaft diam (ft) × 3.14

Plain bearings may be grease lubricated if contact surface operating speed does not exceed approximately 30 ft/min (9 m/min). At higher speeds, excessive temperature buildup and grease breakdown will occur.

In general, NLGI Grade 0 or 1 greases are used for centralized systems and harder greases (NLGI Grade 2 or 3) are used for compression cups and open journals. Each application should be considered on its own merits, taking into consideration the operating conditions. Temperature and water contamination require particular attention.

Plain bearings are frequently grooved (Fig. 5.204) to improve the distribution and flow of the lubricant. Normally, two important rules should be followed when grooving a plain bearing:

- Grooves should not extend into the load-carrying area of the bearing because this would increase unit pressure.
- Groove edges should be rounded to prevent scraping the lubricant off the journal.

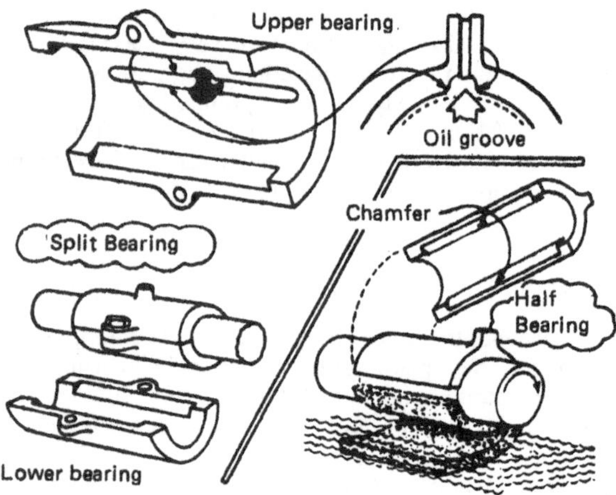

FIGURE 5.204 Oil groove. (*Association of Iron and Steel Engineers; reprinted with permission from* The Lubrication Engineers Manual, *copyright 1971.*)

Antifriction-Bearing Lubrication*

Antifriction or rolling-element bearings use balls or rollers to substitute rolling friction for sliding friction. This type of bearing has closer tolerances than do plain bearings and is used where precision, high speeds, and heavy loads are encountered.

In antifriction bearings, a lubricant facilitates easy rolling; reduces the friction generated between the rolling elements and the cages or retainers; prevents rust and corrosion; and, in the case of grease, serves as a seal to prevent the entry of foreign material.

High-quality rust- and oxidation-inhibited (R & O) oils are generally recommended for bearings. Extreme-pressure and antiwear additives may also be desirable under conditions of heavy or high-shock loads. Oils with no additives can easily oxidize and turn gummy. Only once-through drip applicators can justify the use of straight nonadditivized oils. In fact, R & O oils are probably just as inexpensive today.

When temperature control and cooling are a consideration, oil-circulating systems are the choice. A bearing lubricated with either oil or grease does not carry away heat on its own. Table 5.74 gives general guidelines for the selection of proper viscosity oils for antifriction bearings.

Most antifriction bearings are grease lubricated because of the economics of simple seal and housing designs. Greased bearings offer adequate protection from dirt and water, and require infrequent attention.

The selection of the proper type and grade of grease depends on the operating conditions and the method of application. Generally, soft greases (e.g., NLGI 1) are preferred for use at low temperatures and in central systems. Harder greases (e.g., NLGI 2 or 3) perform better at high speeds. Ball bearings do best with NLGI 2 or 3 grease, while spherical, cylindrical, needle, or tapered roller bearings with broad face line contact design require NLGI 2 or less.

Care should be taken not to overgrease antifriction bearings. Generally, the bearing housing should be one-third to one-half full. Overfilling can lead to several problems: ruptured seals, excessive temperature buildup, and eventual failure due to starvation of the bearing for lubricant. With repeated greasing and continued high temperature, the oil in the grease may

* See also Chap. 5.5.

TABLE 5.74 Oil Viscosity Selection for Antifriction Bearings

Speed factor bearing bore, mm × r/min	Operating temperatures		Viscosity	
	°F	°C	ISO viscosity grade	SUS at 100°F
Up to 75,000	−40–32	−40–0	15–32	70–150
	32–150	0–65	32–100	150–800
	150–200	65–93	100–220	800–1200
	200–250	93–21	220–680	1100–3000
75,000–200,000	−40–32	−40–0	7–22	50–100
	32–150	0–65	22–68	100–300
	150–200	65–93	68–100	300–800
	200–250	95–121	150–320	700–2100
200,000–400,000	−40–32	−40–0	7–15	50–70
	32–150	0–65	15–46	70–200
	150–200	65–93	32–68	150–300
	200–250	93–121	68–150	400–900
Above 400,000	−40–32	−40–0	5–10	40–80
	32–150	0–65	10–32	80–150
	150–200	65–93	22–46	100–200
	200–250	93–121	68–100	300–800

be driven off, leaving the soap in the bearing. Soap makes up about 8 to 10 percent of grease. It is easy to see that if the oil portion of the grease repeatedly is eliminated, the soap can rapidly fill the voids in the bearing. Eventually the bearing will accept no more grease because it becomes filled with soap, a nonlubricant. The bearing then soon fails, and the unlucky grease sales representative is told that the grease is poor quality.

Many PC software programs are available that can aid in the selection of lubricants in the initial design and in failure analysis. These can be obtained from many lubricant manufacturers.

Gear Lubrication*

The motion between gear teeth as they go through mesh is a combination of sliding and rolling. The type of gear, the operating load, speed, temperature, method of application of the lubricant, and metallurgy of the gears are all important considerations in the selection of a lubricant.

Industrial gearing may be *enclosed,* where the gears and the bearings that support them are operated off the same lubricant system; or *open,* where the bearings are lubricated separately from the gears themselves.

Due to the high sliding contact encountered in enclosed worm and hypoid gears, lubricant selection for these should be considered separately from lubrication of other types of enclosed gears.

As with all equipment, the first rule in selecting a gear lubricant is to *follow the manufacturer's recommendation,* if at all possible. In general, one of the following types of oils is used:

Rust- and Oxidation-Inhibited (R & O) Oils. R & O oils are high-quality petroleum-based oils containing rust and oxidation inhibitors. These oils provide satisfactory protection for most lightly to moderately loaded enclosed gears.

* See also Chap. 5.4.

Extreme-Pressure (EP) Oils. EP oils are usually high-quality petroleum-based oils containing sulfur- and phosphorus-based extreme-pressure additives. These products are especially helpful when high-load conditions exist and are a must in the lubrication of enclosed hypoid gears.

Compounded Oils. These are usually petroleum-based oils containing 3 to 5 percent fatty or synthetic fatty oils (usually animal fat or acidless tallow). They are usually used for worm-gear lubrication, where the fatty content helps reduce the friction generated under high-sliding conditions.

Heavy Open-Gear Compounds. These are very heavy bodied tarlike substances designed to stick tenaciously to the metal surfaces. Some are so thick they must be heated or diluted with a solvent to soften them for application. These products are used in cases where the lubricant application is intermittent.

A number of gear lubrication models and viscosity selection guides exist. In the United States, the most widely used selection method employs the American Gear Manufacturers Association (AGMA) standards. Under its specifications for enclosed industrial gear drives, the AGMA has defined lubricant numbers, which designate viscosity grades for gear oils. Table 5.75 identifies the AGMA viscosity numbers with their corresponding ISO viscosity grades.

As a rule, low speeds and high pressures require high-viscosity oils. Intermediate speeds and pressure require medium-viscosity oils, and high speeds and low pressures require low-viscosity oils. Table 5.76 gives some very general guidelines for viscosity and type of lubricant for industrial gearing.

Open gears operate under conditions of boundary lubrication. The lubricant can be applied by hand or via drip-feed cups, mechanical force-feed lubricators, or sprays. Heavy-bodied residual oils with good adhesive and film-strength properties are required to survive the relatively long, slow, heavy tooth pressure while maintaining some film between applications of lubricant.

Several PC software programs exist to aid in lubricant selection to reduce wear, scuffing, and pitting of gear-tooth surfaces. See "Specifications and Standards" at the end of this chapter.

Compressor Lubrication

The compressor model and type, the loading, the gas being compressed, and other environmental conditions dictate the type and viscosity of the oil to be used. Most compressors are lubricated with petroleum oils; however, there has been considerable interest in synthetic lubricants for compressor lubrication in recent years.

Compressing gases other than air creates problems that require special lubrication consideration, because of possible chemical reactions between the gas being compressed and the lubricant. Since no two cases are alike, it is recommended that the compressor manufacturer and lubricant supplier be consulted for recommendations for a particular operation.

Oils for use in compressors should have the following characteristics:

Proper Viscosity. The operator's manual should be consulted for the manufacturer's viscosity recommendations for the prevailing operating temperatures and conditions.

Good Oxidation Stability. A good compressor oil must have high oxidation stability to minimize the formation of gum and carbon deposits. Such deposits can cause valve sticking. This can lead to very high temperature conditions, compressor malfunction, and fire or explosion of reciprocating compressors.

Good Demulsibility. A good compressor oil must be able to shed water readily to prevent formation of emulsions which could interfere with proper lubrication.

TABLE 5.75 Viscosity Ranges for AGMA Lubricants

Rust- and oxidation-inhibited gear oils, AGMA lubricant no.	Viscosity range,[a] mm^2/s (cSt) at 40°C	Equivalent ISO grade[a]	Extreme-pressure gear lubricants,[b] AGMA lubricant no.	Synthetic gear oils,[c] AGMA lubricant no.
0	28.8–35.2	32	—	0 S
1	41.4–50.6	46	—	1 S
2	61.2–74.8	68	2 EP	2 S
3	90–110	100	3 EP	3 S
4	135–165	150	4 EP	4 S
5	198–242	220	5 EP	5 S
6	288–352	320	6 EP	6 S
7, 7 comp[d]	414–506	460	7 EP	7 S
8, 8 comp[d]	612–748	680	8 EP	8 S
8A comp[d]	900–1100	1000	8A EP	—
9	1350–1650	1500	9 EP	9 S
10	2880–3520	—	10 EP	10 S
11	4140–5060	—	11 EP	11 S
12	6120–7480	—	12 EP	12 S
13	190–220 cSt at 100°C (212°F)[e]	—	13 EP	13 S

Residual compounds,[f] AGMA lubricant no.	Viscosity ranges,[e] cSt at 100°C (212°F)
14R	428.5–857.0
15R	857.0–1714.0

[a] Per ISO 3448, *Industrial Liquid Lubricants—ISO Viscosity Classification;* also ASTM D 2422 and British Standards Institution B.S. 4231.

[b] Extreme-pressure lubricants should be used only when recommended by the gear manufacturer.

[c] Synthetic gear oils 9S to 13S are available but not yet in wide use.

[d] Oils marked *comp* are compounded with 3 to 10 percent fatty or synthetic fatty oils.

[e] Viscosities of AGMA lubricant number 13 and above are specified at 100°C (210°F) as measurement of viscosities of these heavy lubricants at 40°C (100°F) would not be practical.

[f] Residual compounds—diluent type, commonly known as *solvent cutbacks,* are heavy oils containing a volatile, nonflammable diluent for ease of application. The diluent evaporates, leaving a thick film of lubricant on the gear teeth. Viscosities listed are for the base compound without diluent.

Caution: These lubricants may require special handling and storage procedures. Diluent can be toxic or irritating to the skin. Do not use these lubricants without proper ventilation. Consult lubricant supplier's instructions.

Source: ANSI/AGMA 9005-D94, *Industrial Gear Lubrication.* Reprinted with permission of the American Gear Manufacturers Association, Alexandria, Va.

TABLE 5.76 Oil Selection for Enclosed Gear Drives

Service	ISO viscosity grade	Oil type*
Helical, herringbone, straight-bevel, spiral-bevel, and spur-gear drives		
Operating at normal speeds and loads	220	EP or R & O
Operating at normal speeds and high loads	220	EP
Operating at high speeds (above 3800 r/min)	68	EP or R & O
Wormgear drives	680	Compounded or EP
Hypoid-gear drives		
Normal speeds (1200–2000 r/min)	220	EP
High speeds (above 2000 r/min)	180	EP
Low speeds (below 1200 r/min)	460	EP

* EP = extreme pressure; R & O = rust and oxidation inhibited.

Anticorrosion and Antirust Properties. A compressor lubricant must protect the valves, pistons, rings, and bearings against rust and corrosion. This is especially important in a humid atmosphere or in compressors that operate intermittently.

Nonfoaming Properties. This requirement is universal.

Low Pour Point. This property is necessary for low-temperature start-up. Usually it is a factor only in portable air compressors that will be used outdoors.

SPECIFICATIONS AND STANDARDS

"Industrial Gear Lubrication," ANSI/ AGMA 9005-D94, American Gear Manufacturers Association, Alexandria, Va., August 1, 1994.

SECTION 6

IN-PLANT PRIME POWER GENERATION AND COGENERATION

blank page 6.2

CHAPTER 6.1
BOILERS

Russell N. Mosher
Assistant Executive Director
American Boiler Manufacturers Association
Arlington, Virginia

TERMINOLOGY

Since the term *boiler* alone does not adequately describe the complete system for providing the motive force or energy, it is necessary to know and understand the component functions that go into making the boiler a complete unit. Comprehending the definitions in the terminology used by the industry is the first step in sorting out the perceived mysteries of understanding.

A *boiler* is a closed vessel in which water is heated, steam is generated, or steam is superheated (or any combination of these) under pressure or vacuum by the application of heat from combustible fuels, electricity, or nuclear energy. Boilers are generally subdivided into four classic types—residential, commercial, industrial, and utility.

Residential boilers produce low-pressure steam or hot water primarily for heating applications in private residences.

Commercial boilers produce steam or hot water primarily for heating applications in commercial use, with incidental use in process operations.

Industrial boilers produce steam or hot water primarily for process applications, with incidental use as heating.

Utility boilers produce steam primarily for the production of electricity.

Within these four generic types of boilers, specific types of boilers emerge, with their classification based on their use. An example would be a *heat-recovery boiler* (Fig. 6.1) which recovers normally unused energy and converts it into usable heat. Likewise, a *fluidized-bed boiler* is one which utilizes a fluidized-bed combustion process. In this type of process, fuel is burned in a bed of granulated particles which are maintained in a mobile suspension by an upward flow of air and combustion products. This technology, in use for over 40 years, is currently being adapted as a combustion process to be installed within a boiler.

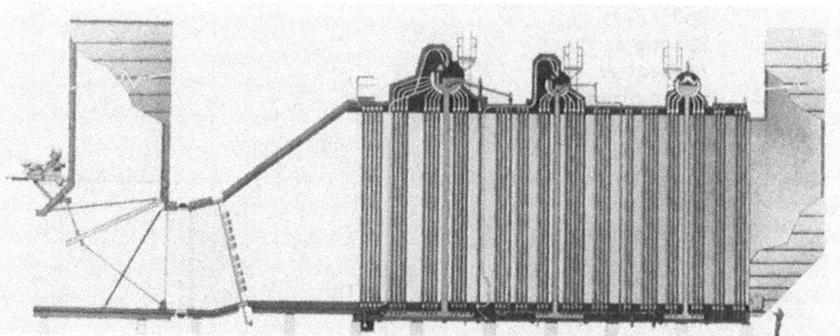

FIGURE 6.1 Heat-recovery boiler.

Generally, boilers are furnished either *packaged* or *field-erected.* A *packaged boiler* is one which is equipped and shipped complete with fuel-burning equipment, mechanical-draft equipment, automatic controls, and accessories. It is usually shipped in one or more major sections. *Field-erected boilers* are those that are shipped from the factory as tubes, casings, drums, fittings, etc., and completely assembled in the field.

The generally accepted reference for the *capacity* of a boiler is the manufacturer's stated output rate for which the boiler is designed to operate over a period of time. The *maximum continuous rating* is the maximum load in pounds (kilograms) of steam per hour for a specific period of time for which the boiler is designed. Likewise, then, the *capacity factor* is the total output over a period of time divided by the product of the boiler capacity and the time period.

INTRODUCTION

Modern boilers provide most of the motive force in the world and are probably the least understood of all mechanical pieces of equipment. They are subjects for engineering, congressional legislation, agency legislation, and physical laws, all of which shape their destiny. Despite all this attention this age-old energy source is still shrouded in mystery.

This chapter explains the theoretical and practical aspects of modern boiler equipment. The objective is to enlighten the reader on the principles of boiler design, to characterize the available equipment, and to explain the need to achieve maintenance for maximum and sustained utilization of equipment.

BOILER APPLICATION

Modern boilers range in size from those required to provide steam or hot water to heat homes, through midsized units which provide energy to drive presses, to very large units used as the primary motive force in producing electric power. Boilers can be arranged for firing almost any type of fuel available, provided the designer is cognizant of the fuel to be employed prior to making the initial calculations for sizing.

The primary purpose of a boiler is to generate steam or hot water at pressures and/or temperatures above that of the atmosphere. Steam or hot water is produced by the transfer of heat from the combustion process taking place within the boiler, thereby elevating its pressure and temperature.

With this higher pressure and temperature, it follows that the containment vessel or pressure vessel must be designed in such a way as to encompass the desired design limits with a reasonable factor of safety. For the sake of economy, the capacity of the unit must be generated and delivered with minimum losses.

In smaller boilers used in home heating applications, the maximum operating pressure for steam is usually 15 psig (104,000 N/m^2). In the case of hot water, this is equal to 450°F (232°C).

Larger boilers are designed for various pressures and temperatures, depending upon the application within the heat cycle for which the unit is being designed. A boiler designed to heat a large college campus may require a certain capacity at an elevated pressure and a superheated temperature which provides the force to transmit the steam to its final use point. In other cases, very high pressures and temperatures are required in order to implement chemical reactions, to provide drying steam in a paper cycle, or to provide the needed energy to drive a large piece of mechanical equipment.

The dependability and safety record displayed by today's modern boilers is the product of almost 100 years of design experience, control fabrication, and monitored operation of boilers. The properties of steam and water have been accurately graphed for use by the engineer.

With the use of computers, the boiler design engineer has gained a new understanding of boiler thermal dynamics and heat transfer and has expanded the understanding of the burnability of fuels in a safe and efficient manner, thus developing units to produce the large amounts of steam required today. Advances in metallurgical fields have yielded better-quality steels and alloys, which allow use of the high pressures and temperatures required.

A large central-station boiler is designed on the basis of the output cost of electricity produced. Its operation is under close control, and the load cycle follows a very well defined and predictable pattern based on area power requirements.

The industrial boiler, however, is usually a single unit installed primarily as an important step in the production of a product. Its use is merely a means whereby the product can be fabricated in the shortest amount of time, with the lowest materials cost, and shipped. Hence, it is frequently called upon to perform a difficult task, often under unfavorable conditions of steam load, water, and fuel. Plant load cycles are usually highly unpredictable, and the boiler must be ready at any time to achieve the required capacity in the shortest amount of time without hesitation.

Many factors must be taken into consideration when one is designing a boiler. After the decision has been made as to what fuel must be burned, it is necessary to determine how much input steam is necessary to satisfy the requirements or demands upon the boiler. Operating parameters include minimum, maximum, and normal load range; length of time in a cycle; and type of load, whether constant or fluctuating. All of these parameters must be analyzed for proper size selection.

The basic components of a boiler are the *furnace* and the *convection* sections. In the *furnace* section (Fig. 6.2), the products of combustion are consumed and heat is released and transferred into the water, thereby producing steam or heating the water. This space must be designed for the three "T's" of combustion—time, turbulence, and temperature. *For complete combustion, it is necessary for the fuel to have sufficient time to be completely consumed;* there must be sufficient turbulence for complete mixing of fuel and air for efficient burning; and there must be a high enough temperature to allow the products to be ignited.

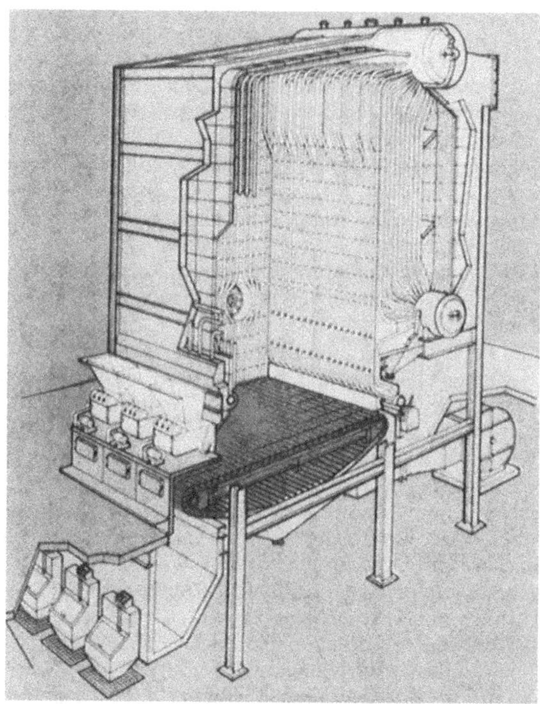

FIGURE 6.2 Furnace section. (*Babcock & Wilcox Company.*)

The shape of the furnace is controlled by the type of fuel and burning method. Adequate provisions must be made for instituting and maintaining ignition and combustion of the fuel. For those units with solid-fuel firing, adequate provision must be made to allow for the removal of unburned combustibles and/or ash.

The *convection* section (Fig. 6.3) of a boiler is that portion in which heat contained in the combustion gases is transferred to the water for the production of steam. The selection of heating surface and tube spacing depends completely upon the type of fuel producing the flue gas with its entrained particles. Adequate provision must be made in this section to allow any unburned particles to pass through and be collected in downstream separators. Pressure drop and volumetric flow are influencing factors which dictate the overall design of the convection section. Generally, a higher gas volume through a fixed flow area produces a greater pressure drop. A greater pressure drop enhances the transfer to a point where economic decisions must be made on inputs of auxiliary equipment to produce the necessary flow of these gases.

The steam and water circulation rate within the pressure vessel decides the effectiveness of the heat-transfer surface. When new feedwater is added to the system, precipitates fall out which might be removed as blowoff. Provisions are usually made in the lower portion of the convection section whereby a boiler operator may remove these particular precipitates by opening boiler blowoff valves.

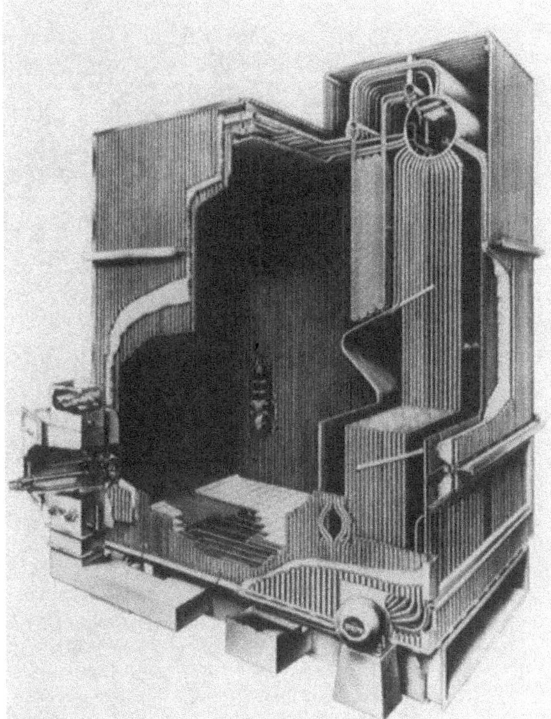

FIGURE 6.3 Convection section. (*Combustion Engineering, Inc.*)

Many applications utilizing boiler equipment require steam at a very high rate of purity. Boiler designers therefore install steam-separating equipment in the boiler drum to remove entrained moisture and solids before the steam is taken from the boiler to the system. These steam separators come in a variety of types including cyclones, mist deflectors, and baffle plates.

In some applications, the heating surface or tubes are of the bare-tube type. In other cases, the heating surface will be of the extended surface or fin-tube type. Utilizing fin tubes allows greater tube surface within the convection section. Greater heat-transfer levels can be achieved with the use of this type of tubing.

BOILER CLASSIFICATIONS

Characteristically, boiler types are generally classified as either *firetube* or *watertube.*

Firetube Boilers

In the *firetube* boiler (Fig. 6.4), the flame and products of combustion pass through the tubes. The heated water or other medium surrounds the internal furnace and the tube bundles.

Various types of furnaces are used in conjunction with firetube boilers. Some are long, cylindrical tubes while others are firebox (Fig. 6.5) arrangements allowing the burning of solid fuels. In most cases, the firetube boiler includes a shell to contain the water and steam space. Within the shell will be tube sheets and tubes which are portions of the pressure vessel containment. The furnace or firebox provides space for the combustion process from the heat source.

FIGURE 6.4 Firetube boiler. (*Cleaver-Brooks Division.*)

FIGURE 6.5 Firebox boiler. (*Kewanee Boiler Corporation.*)

Many types of firetube boilers are being supplied to industry. One type is the *horizontal return tubular boiler* (Fig. 6.6). In this unit, the products of combustion travel across the shell and back through the tubes within the pressure vessel. These units are usually horizontally brick-set.

FIGURE 6.6 Horizontal-return tubular boiler. (*Zurn Industries, Inc.*)

Another type of firetube unit is the *Scotch marine boiler.* This design was developed originally for shipboard installation. This type of boiler can be fired with either solid, liquid, or gaseous fuels.

Because the Scotch marine boiler is a very compact type of unit, it has become readily adaptable for stationary service. When modifications to the basic type are made in adapting it to process and heating use, it is called a *modified Scotch marine boiler.*

Another type of firetube unit currently being marketed is the *vertical-type boiler.* In this particular unit the fuel or heat source is in the bottom, and the products of combustion rise up through tubes and are emitted from the top of the unit.

Watertube Boilers

Watertube boilers come in a variety of arrangements and designs. In this type of unit, the products of combustion usually surround the tubes (Fig. 6.7), and the water is inside the tubes which are inclined upward toward a vessel or drum at the highest point of the boiler. The configuration of these tubes generally describes the type of boiler. Some manufacturers offer straight-tube units while others offer units with bent tubes (Fig. 6.8). Other configurations of watertube boilers describe the various types in terms of the variations of pressure vessel arrangement.

In a *box-header watertube* boiler, the watertubes are connected to rectangular headers which are arranged so that the circulating water and steam mixture will rise toward a collection drum. The box headers are usually on either end of the tube bundles, and the products of combustion pass between the headers and around tube bundles.

FIGURE 6.7 Commercial watertube boiler. (*Bryan Steam Boiler Company.*)

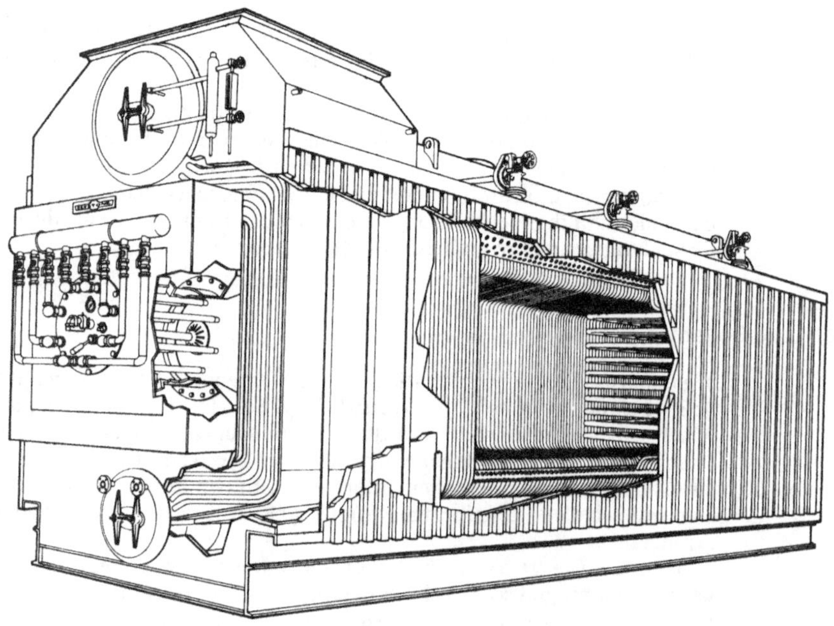

FIGURE 6.8 Bent-tube boiler.

Some boilers are of the *long-drum* type; that is, when viewed from the front of the boiler (Fig. 6.9), the drum is the length of the boiler. Its corollary is the *cross-drum* boiler. When viewed from the front of the unit, the drums are installed perpendicular to the long centerline or across the boiler.

FIGURE 6.9 Long-drum boiler. (*Zurn Industries, Inc.*)

Firetube units are generally furnished in applications up to approximately 30,000 lb (13,500 kg) of steam per hour. They are furnished for low-pressure operation [15 psig (104 kN/m^2) and under] and as power boilers [up to approximately 300 psig (2100 kN/m^2) of steam pressure]. Watertube boilers for use in industrial applications are furnished in capacities up to almost 1 million lb (450,000 kg) of steam per hour. Design pressures vary from 100 psig (700 kN/m^2) up through 1200 or 1400 psig (8.3 or 9.6 MN/m^2) with steam temperatures ranging from saturated to 1000°F (540°C).

Packaged Boilers

Many manufacturers supply both watertube and firetube units already packaged, or shop-assembled. A *packaged* steam or hot-water boiler is one which is generally shop assembled and includes all major components: burner, draft equipment, pressure vessel, trim, and controls. The main limitation to packaging boilers is the capability of their being handled by trucking or railroad equipment. Most manufacturers have designed complete lines of firetube and watertube boilers which can be shipped by either truck or railroad. Some manufacturers have utilized shop-assembling procedures to fabricate a boiler which can be assembled easily in the field. These components may be a single portion of the vessel or multiple major parts which can be brought together in the field.

BOILER COMPONENTS

To understand the operation of a boiler, it is necessary to observe what happens from input to output of the unit. Several cycles are involved in the complete operation of the unit. The heat cycle, the water and steam cycle, and the boiler-water circulation cycle all interact to produce the output of the boiler. Fuel and water are brought to the unit; water is heated to its final predesignated condition (water and/or steam) and transported to the point of its end use. When the heat has been taken out of the water, the remaining steam and water mixture (or condensate), if usable, is returned to the unit and recycled.

Furnace*

In the fuel cycle, the solid, liquid, or gaseous fuel is delivered to the boiler where it is mixed with air and burned. This liberation of heat is usually achieved in the *furnace* portion of the boiler (Fig. 6.1). Furnaces can be of either the *refractory* or the *water-cooled* type.

In the *refractory-type* furnace, refractory brick forms the envelope of the furnace. These refractory furnaces are usually backed with insulation and a casing material. For the *water-cooled* wall-type furnace, the envelope consists of tubes placed close to each other, which thereby absorb heat and help in the production of steam. These water-cooled furnaces can have either tube and tile, tangent-tube, or welded-membrane walls.

The basic function of the furnace is to allow for the combustion of the fuel. It is necessary that the furnace size be sufficient to allow for adequate combustion of the fuel, time for its combustion, and for enough turbulence to permit efficient combustion.

Boiler Section

This is usually referred to as the *boiler* or *convection section* of the unit. Closely spaced tubes are arranged to allow passing of the products of combustion around the tubes or through the tubes, depending on the type of unit. Most of the steam is generated in the boiler portion of the unit. In watertube units, if additional steam temperature is required by the process, the steam is then routed to a superheater.

Superheater

In a superheater unit (Fig. 6.10), the steam is directed back through the products of combustion to take on additional heat. This additional heat results in considerable energy gain by the steam which will be liberated in end use. This end use can be a steam turbine or other type of equipment requiring considerable energy release for its operation.

Superheaters are either of the radiant or convection type. In a radiant superheater, the tubes are usually located in the furnace section of the boiler. Convection-type superheaters are usually located behind the screenwall of the convection section. Radiant-type superheaters receive their heat by direct radiation from the flame, while convection superheaters receive their heat primarily from the passage of the products of combustion around the tubes.

---

* See also Chap. 6.3.

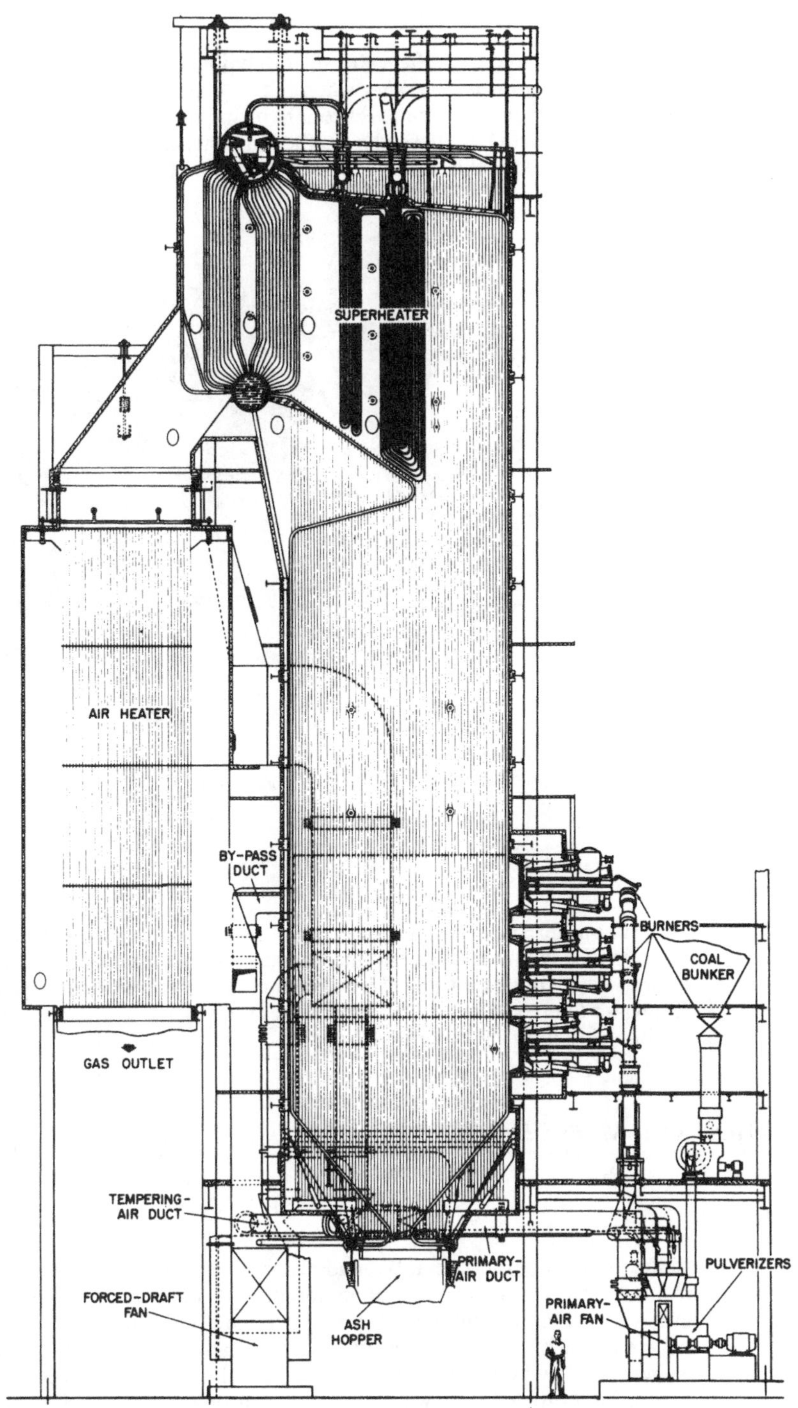

FIGURE 6.10 Boiler with superheater unit. (*Riley Stroker Corporation.*)

Air Heaters

It is often desirable to preheat the air for combustion prior to bringing it in contact with the fuel (Fig. 6.11). This is necessary when burning fuels of very high moisture content. In an air heater, the ambient air volume is brought in and preheated by utilizing sensible heat from the boiler flue gas being discharged from the unit. This increases overall efficiency, eliminating the use of extra fuel for this purpose. This is one type of *heat recovery unit.*

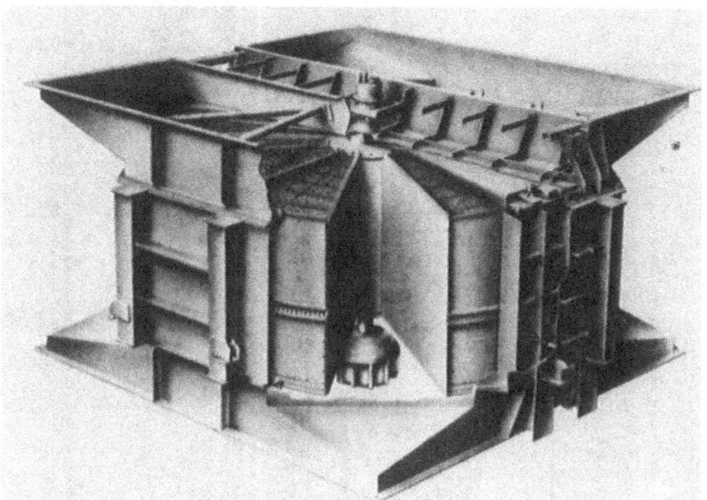

FIGURE 6.11 Air heater. (*C-E Air Preheater.*)

Economizers

An economizer (another type of heat recovery unit) is a boiler component which preheats incoming feedwater from its supplied temperature, utilizing sensible heat from the boiler outlet flue gas being exhausted from the unit. As in the air-heater principle, raising this inlet feedwater temperature (Fig. 6.12) increases the efficiency of the unit by eliminating the use of additional fuel for this operation.

OPERATION AND MAINTENANCE

Start-Up

Before any preparation can be made to start up a boiler, new or otherwise, the operator's manual furnished by the boiler manufacturer for the particular make and model unit must be available. It is important that operating personnel carefully follow the procedures in the manual, particularly the safety precautions, before attempting to activate the equipment.

When a new boiler is prepared for its initial operation, procedures should be followed to ensure high efficiency at which the unit can operate a long life, and the economies to be expected from an engineered piece of equipment. Even though the unit has been checked by the manufacturer, the following precautions should be observed:

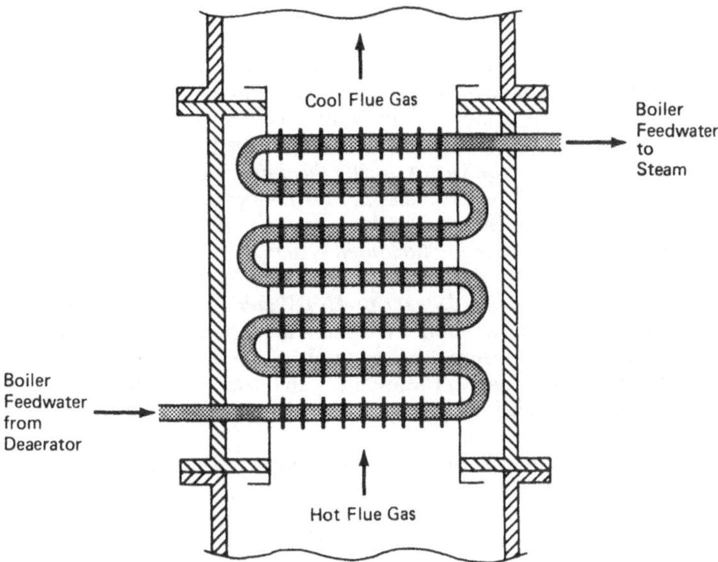

FIGURE 6.12 Principle of economizer. (*Enerex, Inc.*)

1. The unit should be thoroughly examined on both the water side and the fire side to make sure that no foreign material is present.
2. All piping such as blowdown piping, steam piping, and feedwater piping should be checked to ensure that the piping has been properly installed so that there will be no danger to any individuals. Items such as the gauge, gauge glasses, and controls should be checked for any evidence of damage or breakage either incurred during transportation or caused by installation personnel working around the equipment after it was placed.
3. Electrical equipment such as motors, pumps, blowers, and compressors should be operated whenever possible to assure proper rotation. Items such as control valves, interlocks, motorized valves, and limit switches should be checked wherever possible to ensure their proper operation.
4. All fuel lines should be checked per the installation instructions in the manual.

After a thorough inspection of the unit has been completed, the next steps to start-up follow.

Caution: *All manufacturers of boilers usually supply operating instruction manuals with their equipment. Before starting the unit, be sure that the manual of instructions has been thoroughly read and understood.*

Drying Out and Boiling Out

Since the refractories and insulation of the unit may contain absorbed moisture and since (at initial start-up) the boiler is filled with water that is at the supplied temperature (which is colder than normal), the initial firing of the unit should be maintained at as low a level as possible. The boiling-out period should be continued at approximately 50 percent of the unit operating pressure for a long enough time to ensure that all oils and materials to be removed by the boiling-out process have been dislodged. Experience has indicated that a 12-h minimum boil-out period is generally sufficient to complete the cleaning. However, factors such as chemical concentration, amount of material to be removed, and pressure may modify boil-out time.

Cleaning

The system must be carefully cleaned before the boiler is connected into the system. Many clean boilers have been ruined with system contaminants such as pipe dope, cutting oil, and metal shavings or chips. Many contractors will use a new boiler for heating and curing a building under construction. Special care must be taken to ensure that *adequate water treatment is provided by the contractor during this initial use of the boiler.* Succeeding owners can receive a badly scaled or damaged boiler through contractor misuse. Moreover, as new zones are cut into a system, *cleaning of these zones is required to prevent damage to the boiler. Only one boiler should be used to boil out a system.*

Cleaning Improves a Steam or Hot-Water Heating System. One important phase in completing boiler installation is too often neglected in the specifications. *No provision usually has been made for cleaning the system.* It is sometimes drained for changes and adjustments but never actually cleaned. The architect, engineer, or contractor selects boilers for applicable installations. The selection may represent the best system; but it will be better if it is a clean system.

How to Tell If a System Needs Cleaning. There are definite symptoms of an unclean system. A typical checklist follows. If any of the items are positive, the system needs cleaning.

1. Obviously discolored, murky, dirty water
2. Gases vented at high points in the heating area that ignite and burn with an almost invisible bluish flame
3. A pH alkalinity test that gives a pH test reading below 7 (A pH lower than 7 indicates the water in the system is acid.)

No matter how carefully a system is installed, certain extraneous materials do find their way accidentally into the system during construction. Pipe dope, thread-cutting oil, soldering flux, rust preventives or slushing compounds, coarse sand, welding slag, and dirt, sand, or clays from the jobsite are usually found. Fortunately the amounts of these are usually small and do not cause trouble. However, in some instances there may be sufficient quantities to break down chemically during the operation of the system, causing gas formation and acid in the water system. Hot-water systems, in most cases, naturally operate with a pH of 7 or higher. The condition of the water can be quickly tested with Hydrion paper, which is used in the same manner as litmus paper except that it gives specific pH readings. A color chart on the side of the small Hydrion dispenser gives the readings in pH units. Hydrion paper is inexpensive and readily obtainable through appropriate wholesale and retail channels.

A system that tests acid (below 7 on the scale, sometimes as low as 4) will usually have the following symptoms:

1. Gas formation (air trouble)
2. Pump seal and gland problems
3. Air vent sticking and leaking
4. Frequent operation of relief valves
5. Piping leaks at joints

Once this condition exists, the symptoms continue until the situation is corrected by cleaning the system. Many times, because of gas formation, automatic air vents are added throughout the system to attempt a cure. The excessive use of automatic air vents can defeat the function of the air-elimination system since the small quantities of entering air must be returned to the expansion tank to maintain the balance between the air cushion and the water volume.

If a system is permitted to deteriorate with resultant leaks and increased water losses, serious boiler damage can occur. Therefore, the chief consideration is to maintain a closed system that is clean, neutral, and watertight.

How to Clean a Heating System. Cleaning a system (either steel or copper piping) is neither difficult nor expensive. The materials for cleaning are readily available. Trisodium phosphate, sodium carbonate, and sodium hydroxide (lye) are the materials most commonly used for cleaning. They are available at paint and hardware stores.

The preference is in the order named, and the substances should be used in the following proportions; use a solution of *only one type* in the system.

Trisodium phosphate, 1 lb for each 50 gal (1 kg for 420 L) in the system

Sodium carbonate, 1 lb for each 30 gal (1 kg for 240 L) in the system

Sodium hydroxide (lye), 1 lb for each 50 gal (1 kg for 420 L) in the system

Fill and vent the system and circulate the cleaning solution throughout, allowing the system to reach design or operating temperatures if possible. After the solution has been circulated for a few hours, the system should be drained completely and refilled with fresh water. Usually, enough of the cleaner will adhere to the piping to give an alkaline solution satisfactory for operation. A pH reading between 7 and 8 is preferred, and a small amount of cleaner can be added if necessary.

A clean, neutral system should *never* be drained except for an emergency or for such servicing of equipment as may be necessary after years of operation. Antifreeze solution in the system should be tested from year to year as recommended by the manufacturers of the antifreeze used. Without a doubt, the clean system is the better system.

Arrangements for Cleaning Heating Systems. Much of the dirt and contamination in a new system can be flushed out prior to boil-out of the system. This is accomplished by first flushing the system to waste with clear water and then using a chemical wash.

The boiler and circulating pump are isolated with valves, and city water is flushed through the successive zones of the system, carrying chips, dirt, pipe joint compound, etc., to waste with it. This is followed by a chemical flush. Removal of pipe chips and other debris before operating the isolation valves of the boiler and pump will help to protect this equipment from damage by such debris. After this flushing process is complete, the usual boil-out procedure is accomplished.

Caution: *If one zone is flushed and boiled out before other zones are completed or connected, this flushing process should be repeated on completion of additional zones, loops, or sections of the piping.*

When a boiler is fired for the first time (or started again after repairs or inspection), vapor and water may be observed as a white plume in stack discharge or as condensate on the boiler fire sides and services. Generally, this condition is temporary and it will disappear after the unit reaches operating temperature. This condensation should not be confused with the stack plume that occurs when the boiler is operating during extremely cold weather.

When cool-down of a boiler is required, the unit should be permitted to cool over a period of 12 h, losing its heat to the atmosphere. Forced cooling is not recommended; it will possibly loosen tubes in the tube sheets or cause other damage to the pressure parts.

Water Treatment*

Water treatment is required for satisfactory operation of a boiler at the initial start-up to prevent any deposition of scale and to prevent any corrosion from acids, oxygen, and other harmful substances that may be in the water supply. A qualified water-treatment specialist should be consulted and the water should be appropriately treated.

The basic aims and objectives of boiler-water conditioning are to:

---

* See also Chap. 3.3.

1. Prevent the accumulation of scale and deposits in the boiler
2. Remove dissolved gases from the water
3. Protect the boiler against corrosion
4. Eliminate carryover and/or timing (steam)
5. Maintain the highest possible boiler efficiency
6. Decrease the amount of boiler downtime for cleaning

Water treatment should be checked and maintained whenever the boiler is fired.

Caution: *The purchaser should be sure that the boiler is not operating for approval tests or any other operation of firing without water treatment.*

It should also be noted that water boilers may well need chemical treatment for the first filling of water as well as additional periodic chemical treatment depending on the system's losses and the makeup requirements. Water treatment may vary from season to season or over a period of time and, therefore, there should be a requirement that the water-treatment procedure be checked no fewer than four times a year and possibly more frequently if the local water conditions require it. When the system is drained and then refilled, chemical treatment is required inasmuch as raw water has been put into the boiler system.

There are two major methods of boiler feedwater treatment, external and internal.

External Feedwater Treatment. This type of treatment is performed in separate tanks, containers, or other necessary devices for the removal of oxygen and other detrimental gases, and the removal of magnesium carbonate, calcium carbonate, silica, iron, etc. There are also filters available for the removal of foreign matter. A common method of removing gas from the boiler water is to use deaerating feedwater heaters. One method of removing the magnesium and calcium carbonates is to use sodium zeolite softeners. There are filters and other equipment presently manufactured that will cover virtually every requirement for water treatment.

Internal Feedwater Treatment. Internal feedwater treatment is generally nothing more than the addition of the proper chemicals to prevent the deposition of scaling materials on the hot surfaces of the boiler. A sludge formed by the chemicals with calcium and/or magnesium carbonates drops to the bottom of the boiler or remains in suspension. In steam boilers, this sludge can be removed by proper blowdown procedures. The chemicals that are to be added to the boiler water, the blowdown procedure, and the analysis and maintenance of the feedwater conditioning should be handled by a water-treatment consultant.

Care of Idle Boilers

Boilers that are used on a seasonal basis and will be idle for a long period of time (more than 30 days) should be laid up by using either a dry or a wet method of protection during the periods of inactivity.

Boilers Laid Up Dry. If a boiler is subject to freezing temperatures or if it is to be idle for an excessive period of time, the following procedures should be carried out so that the boiler is not damaged during its period of inactivity:

1. Drain and clean the boiler thoroughly (both fire and water sides) and dry the boiler out.
2. Place lime or another water-absorbing substance in open trays inside the boiler and close the unit tightly to exclude all moisture and air.
3. All allied equipment such as tanks, pumps, etc. should be thoroughly drained.

Boilers Laid Up Wet. In order to protect the boiler during short periods of idleness, it should be laid up wet and in the following manner:

1. Fill the boiler to overflowing with hot water. The water should be at approximately 120°F (45°C) to help drive out the free oxygen. Add enough caustic soda to the hot water to maintain approximately 350 parts per million (ppm) of alkalinity and also add enough sodium sulfite to produce a residue of 50 to 60 ppm of this chemical.
2. Check all boiler connections for leaks and take a weekly water sample to make sure that alkalinity and sulfite content are stable.

Restarting Boilers. Upon restarting a boiler that has been laid up dry, laid up wet, or has been cooled down for repairs, be sure to follow the recommended start-up procedure as defined in the operating manual provided by the manufacturer.

Preparation for Lay-Up. When a boiler is being cleaned in preparation for lay-up, the water side of the unit should be cleaned and then the unit should be fired to drive off gases. The fire side should then be cleaned. An oil coating on the fire-side metal surfaces is beneficial when the boiler is not used for extended periods of time. Another helpful treatment would consist of completely filling the boiler with an inert gas and sealing it tightly to prevent any leakage of the inert gas. This will help prevent oxidation of the metal. Fuel-oil lines should be drained and flushed of residual oil and refilled with distillate fuel. If all boilers are to be laid up, care of oil tanks, lines, pumps, and heaters is similarly required.

Burner Care

A planned preventive maintenance program is a direct route to safe, dependable boiler unit operation. Boilers are supplied with engineering fuel-burning equipment that must be maintained through a regular, conscientious maintenance program to keep it in satisfactory operating condition. Oil nozzles, igniters, electrodes, and internal burner parts should be checked as part of a regular monthly maintenance program. The settings of spark gaps and nozzle openings as well as their general dimensions should be checked for both wear and cleanliness. Specific instructions as to the method of cleaning, methods of adjustment, and particular dimensions are contained in the instruction manual furnished by the boiler manufacturer.

The best method of keeping a planned preventive maintenance program in effect is to keep a daily log of pressure, temperature, and other gauge data as well as of water-treatment data. In the event that a variation appears from the normal readings, the trouble can be quickly analyzed and corrected to avoid serious problems. For example, an oil-fired unit showing a drop in oil pressure can indicate a faulty regulating valve, a plug strainer, an air leak in the suction line, or a change in the operation of some other piece of equipment in the oil line. A decrease in oil temperature can indicate malfunction of the temperature controls or a malfunction or fouling of the heating element.

For example, in a gas-fired unit, a decrease in the gas pressure can indicate a malfunction of the regulator, a drop in the gas supply pressure, or some restriction in the gas flow possibly caused by one or more controls or valves not operating properly.

Items to Be Checked Periodically. Such items as linkages and other mechanical fastenings and stops should be periodically checked for tightness and visually checked for any movements or vibration. Any items that are loose or that have changed in position should be thoroughly checked and readjusted as necessary.

Stacks should be checked daily for haze or smoke conditions. A cloudy, hazy, or smoky stack indicates a possible need for burner adjustment. The fire may not be receiving enough air; there may be improper control of air/fuel ratios; there could be a change of fuel delivered, etc.

Stack temperatures should be checked and noted on the log; however, one must bear in mind that a rise in stack temperature does not always mean poor combustion or a fouled water site or fire site. Stack temperatures will vary proportionally as the low point changes. Stack temperatures should be observed in relation to the firing rate and a comparison should be made with previous records of the same firing rates. Stack temperatures can vary as much as 100°F (55°C) from high fire levels to low fire levels, and therefore caution must be exercised before interpreting the stack temperature reading.

Fire-Side Maintenance. Periodic inspection and fire-side cleaning should be performed when the exit fuel gas temperature is more than 100 to 175°F (55 to 97°C) above normal operating temperature. Cleaning should be performed immediately upon shutdown. The boiler manufacturer's recommended procedures for all fire-side maintenance should be followed.

Burner, access, or head gaskets should be inspected and replaced as required. All refractories should be inspected for excessive cracking, chipping, erosion, or loose sections. This inspection should be carried out when the boiler is open for cleaning, or at least once a year.

Fuel solenoid valves and motorized valves should be visually checked by observing the fire when the unit shuts down. If the fire does not cut off immediately, the valve could be fouling or showing wear. If this occurs, the valve should be repaired or replaced immediately to avoid any serious problems.

All switches, controls, safety devices, and other equipment associated with the boiler should be periodically checked. Do not assume that all safety devices, switches, controls, etc. are operating properly. They should all be checked periodically on a planned maintenance schedule and any malfunctions noted and repaired.

Spare Parts

Be sure spare parts for your equipment are readily available. Consult the boiler manufacturer for a suggested parts list.

Manual

The operator's manual supplied with your boiler is an excellent guide to the control functions, control care, and control adjustments. It is important to know and use the instruction manual. The manual should be kept in a place where it will be readily available for the operator's use.

Water-Level Controls

The purpose of water-level controls is to maintain the water inside the boiler at the proper operating level. All water-level controls have a range of operations—not one set point. Water-level controls with gauge glasses should be so set that the water level is never out of sight, either low or high.

Water columns, gauge glasses, and low-water cutoffs on a steam boiler should be flushed at least once every shift. The purpose of this flushing is to prevent any accumulation of sludge or dirt that could possibly cause a control failure. When flushing the water column, it is advisable to test the operation of the low-water cutoff.

The water-level control that is most frequently found on boilers is a combination level indicator and low-water cutoff that is often incorporated in a water-column arrangement. This combination control allows for visual inspection of the water level, and in addition it functions to interrupt the electric current to a burner circuit in the event that an unsafe water level should develop.

Local water conditions and the introduction of treatment chemicals to a boiler will vary the amount of sediment accumulation in a control float bowl or a water column. For heating boilers and power boilers it is recommended that the boiler safety control be blown down regularly at least once a week when the boiler is in operation; however, power boilers may require a more frequent blowdown depending on operating and water conditions. When blowing down a control, it is advisable to check the operation of the low-water cutoff at a low-fire burner setting.

Monthly, the low-water cutoff should be tested under actual operating conditions. With the burner operating and the boiler steaming at the proper water level, close all the valves in the feedwater and condensate return lines for the duration of the test and shut off the feedwater pump, if required, so that the boiler will not receive any placement water. Then carefully observe the waterline to determine where the cutoff switch stops the burner in relation to the lowest permissible waterline established by the boiler manufacturers. The boiler water level should never be allowed to drop below the lowest visible part of the water-gauge glass. If the cutoff does not function during this test, immediately stop the operation of the burner. Then determine the cause of the failure and remedy it. The slow steaming evaporation test should then be repeated to verify that the control does function correctly.

If the burner cutoff level is not at, or slightly above, the lowest permissible water level, the low-water cutoff should be moved to the proper elevation, or should be serviced, repaired, or even replaced if necessary.

The low-water cutoff should be dismantled and checked at yearly intervals by a qualified technician to the extent necessary to ensure freedom from obstructions and proper functioning of the working parts. Inspect connecting lines to the boiler for any accumulation of sediment or scale, and clean as required. Examine all visible wiring for brittle or worn insulation, make sure electric contacts are clean, and where applicable, check mercury switches for any discoloration or mercury separation. Normally, operating mechanisms should not be repaired in the field. Replacement parts and complete replacement mechanisms, including necessary gaskets and installation instructions, are available from the manufacturer.

On boilers, the low-water cutoff may be checked periodically by manually tripping the control. Instructions on the method of tripping the specific control are found in the operator's manual supplied by the manufacturer.

Allied Boiler Equipment

It is recommended that a thorough check be made of all grease fittings, oil fittings, and other lubrication points and that a maintenance program be instituted to ensure proper lubrication of all moving parts as required by the manufacturer. Such items as air compressors, blower bearings, motors, and other mechanically operated equipment do require lubrication from time to time. Consult the instruction manual for the detailed points and instructions; do not fail to establish a continuing maintenance program.

Depending upon the particular conditions of installation, such items as oil strainers, air filters, screens, etc. will accumulate foreign matter. All filters, screens, and strainers should be periodically cleaned to avoid any obstruction or malfunction. A smoke condition in the burner can be caused by an obstructed air inlet to the blower or by an obstructed compressor air filter just as well as by excess fuel input. A reduction in oil flow can be caused by a dirty oil strainer. A regular maintenance program should be set up according to the particular installation conditions to prevent any accumulation of foreign material.

Fuel-Oil System

Tanks should be checked annually for the presence of water and sludge. Fill should be checked for tightness of covers and proper gaskets after each delivery. Make certain the fill box is above grade to prevent water seepage into the tanks. Check vent pipes for obstructions.

Heaters

Heaters should be checked annually for the presence of water or sludge and for "coking" of heat-exchanger surfaces.

Pumps

Pumps should be checked at least monthly for leaky shaft sills and worn or loose drive mechanisms.

Piping

Oil lines should be checked at least monthly for external leaks and damage.

General Precautions. The oil-line vacuum gauge should be checked daily. Gauge readings that increase or are erratic indicate potential trouble. In this case, strainers should be checked and cleaned, and oil lines checked for obstructions or internal sludge buildup.

When taking an oil circulation and heating system out of service, precautions should be taken so that start-up will not be obstructed by congealed fuel oil. This may require flushing the lines with a lighter-grade oil and/or shutting down the system on the lighter-grade fuel.

Electric Contacts

Electric contacts on such items as starters, contactors, and controls should be periodically checked for cleanliness and arc burns. Dust should not be permitted to accumulate on any contact. Covers on controls must be in place and control cabinet doors should be closed except during the period when access is actually required for service.

Caution: *Power must be shut off before any control cover or cabinet is opened.*

Steam Systems

It is recommended that a periodic check of the steam system be conducted to prevent boiler malfunction due to external problems.

- Check to ensure that cold makeup water is not being fed into a hot boiler.
- Check the feedwater-treatment equipment to make sure that it is operating properly.
- Check items such as feed pumps, valves, and other equipment to assure proper performance.
- Check safety valves in accordance with instructions by ASME Boiler & Pressure Vessel Code, Sec. VI, "Recommended Rules for Care and Operation of Heating Boilers."

Hot-Water Systems

The following checks should be made periodically to prevent boiler problems:

- Expansion tanks should be checked to ensure proper performance.
- Water circulators should be checked to make sure that circulation is maintained in all operating conditions.
- Extreme fluctuations of pressure gauges can be indicators of system problems. Check the system for a possible lock in the expansion tank. The air-removal device or connection on top of the boiler should be checked to ensure that it is functioning properly.
- Piping should be checked to make sure that there are no stoppages due to the piping being air-bound.
- Air vents at the high points of the systems should be checked to ensure that they are not bleeding air into the system during pump starts.

- Weeping safety valves not only cause undesirable watermarks and steam losses but may indicate valve malfunction. A check for excessive amounts of system makeup water should be made by means of a water meter on the inlet line.
- A check should be maintained whenever the equipment is switching from a cooling to a heating cycle to ensure that the boiler is not shocked by extreme system changes. When a boiler is used with a cooling/heating system, switching to the chilling cycle should be handled with care. Proper controls should be installed in the system to prevent chilled water from entering the boiler.

TABLE 6.1 Recommended Periodic Testing and Verification Checklist

Item	Frequency	Accomplished by	Remarks
Gauges, monitors, & indicators	Daily	Operator	Make visual inspection and record readings in log
Instrument & equipment settings	Daily	Operator	Make visual check against factory-recommended specifications
Firing rate control	Weekly	Operator	Visual inspection
	Semiannually	Service technician	Verify factory settings; check with combustion test instruments
Fuel valves			
Pilot valves, Main gas valves, Main oil valves	Weekly	Operator	Open limit switch; make audible and visual check; check valve position indicators; check fuel meters
	Annually	Service technician	Perform leakage tests; refer to manufacturer's instructions
Combustion safety controls			
Flame failure	Weekly	Operator	Close manual fuel supply for (1) pilot, (2) main fuel cock and/or valve(s); check safety shutdown timing; log
Flame signal strength	Weekly	Operator	If flame signal meter has been installed, read and log; for both pilot and main flames, notify service organization if readings are very high, very low, or fluctuating. Refer to manufacturer's instructions
Pilot turn down tests	As required/ annually	Service technician	Required after any adjustments to flame scanner mount or pilot burner; verify annually
Refractory hold in	As required/ annually	Service technician	See pilot turn down test
Low-water cutoff	Monthly	Operator	
High limit safety control	Annually	Service technician	Refer to manufacturer's instructions
Operating control	Annually	Service technician	Refer to manufacturer's instructions
Low draft interlock	Annually	Service technician	Refer to manufacturer's instructions
Atomizing air steam interlock	Annually	Service technician	Refer to manufacturer's instructions
High- & low-gas-pressure interlock	Annually	Service technician	Refer to manufacturer's instructions
High- & low-oil-pressure interlock	Annually	Service technician	Refer to manufacturer's instructions
High- & low-oil-temperature interlock	Annually	Service technician	Refer to manufacturer's instructions
Fuel valve interlock switch	Annually	Service technician	Refer to manufacturer's instructions
Purge switch	Annually	Service technician	Refer to manufacturer's instructions
Burner position interlock	Annually	Service technician	Refer to manufacturer's instructions
Rotary cup interlock	Annually	Service technician	Refer to manufacturer's instructions
Low fire start interlock	Annually	Service technician	Refer to manufacturer's instructions
Automatic changeover control (dual fuel)	At least annually	Service technician	Under supervision of gas utility
Safety valves	As required	Operator	In accordance with procedure in ASME boiler code

When a hot-water system is in use, it is recommended that circulation be maintained. If it is necessary to shut down the circulators, the system will cool down to ambient temperature and can then cause damage at start-up. This is referred to as *thermal shock.* Particular instructions and recommendations made in the instruction manual should be observed to assure the long life of the boiler.

When a boiler is being drained for inspection, it is recommended that a flow of water be maintained in the boiler with a high-pressure hose to keep any sediment thoroughly agitated and in suspension to prevent caking of the sludge that can be extremely difficult to remove at a later date.

Recommended Periodic Testing and Verification

Once equipment has been placed in service, it becomes the owner's responsibility to maintain it. Maintenance should include periodic testing and verification of controls and safety devices. Records or logs of such maintenance should be kept by the owner and/or boiler operator.

The maintenance and testing is in addition to those inspections required by the various governmental agencies or insurance companies. A list showing the recommended frequency of periodic testing and verification is found in Table 6.1.

BIBLIOGRAPHY

Boiler & Pressure Vessel Code, American Society of Mechanical Engineers (ASME), New York, 1980.

"Combustion Engineering," Combustion Engineering, Inc., Stamford, Conn., 1967.

D. Gunn and R. Horton: *Industrial Boilers,* John Wiley & Sons, New York, 1989.

A. Kohan: *Boiler Operator's Guide,* 3d ed., McGraw-Hill, New York, 1993.

D. Lindsley: *Boiler Control Systems,* McGraw-Hill, New York, 1991.

Packaged Firetube Engineering Manual, American Boiler Manufacturers Association, 1971.

Carl D. Shields: *Boilers,* McGraw-Hill, New York, 1961.

"Steam/Its Generation and Use," Babcock & Wilcox Co., Wilmington, N.C., 1980.

CHAPTER 6.2
FUELS AND COMBUSTION EQUIPMENT*

P. Eric Ralston
Vice President and General Manager
Environmental Equipment Division
Babcock and Wilcox
Barberton, Ohio

GLOSSARY

Grindability A term used to measure the ease of pulverizing a coal in comparison with a standard coal chosen as 100 grindability.

Gross (higher) heating value The heat released from the combustion of a unit of fuel quantity (mass) with the products in the form of ash, gaseous CO_2 (carbon dioxide), SO_2 (sulfur dioxide), N_2 (nitrogen), and liquid water exclusive of any water added directly as vapor.

Net (lower) heating value Calculated from the gross heating value as the heat produced by a unit quantity of fuel when all water in the products remains as vapor. This calculation (ASTM Standard D 407) is made by deducting 1030 lb (470 kg) of water derived from the fuel, including both the water originally present as moisture and that formed by combustion.

* A portion of this material was adapted by permission from *Steam: Its Generation and Use,* published by Babcock & Wilcox Co. in 1992.

Proximate analysis The determination of moisture, volatile matter, and ash and the calculation of fixed carbon by difference.

Ultimate analysis The determination (using a dried sample) of carbon, hydrogen, sulfur, nitrogen, and ash and the estimation of oxygen by difference.

SOLID FUELS

Characteristics

Coal

Coal Analysis. Customary practice in reporting the components of a coal is to use proximate and ultimate analyses (see Glossary).

The scope of each is indicated in the analyses of a West Virginia coal (Table 6.2) in which the ultimate analysis has been converted to the as-received basis. The analysis on the as-received basis includes the total moisture content of the coal received at the plant. Similarly, the as-fired basis includes the total moisture content of the coal as it enters the boiler furnace or pulverizer. Standard laboratory procedures for making these analyses appear in ASTM D 271, "Sampling & Analysis, Laboratory, Coal & Coke."

TABLE 6.2 Coal Analyses on As-Received Basis (Pittsburgh Seam Coal, West Virginia)

Proximate analysis		Ultimate analysis	
Component	% by wt	Component	% by wt
Moisture	2.5	Moisture	2.5
Volatile matter	37.6	Carbon	75.0
Fixed carbon	52.9	Hydrogen	5.0
Ash	7.0	Sulfur	2.3
Total	100.0	Nitrogen	1.5
		Oxygen	6.7
Heating value,		Ash	7.0
Btu/lb	13,000	Total	100.0
(kJ/kg)	(30,238)		

A list of other testing standards is given in Table 6.3.

ASTM Classification by Rank. Coals are classified in order to identify their end use and also to provide data useful in specifying and selecting burning and handling equipment and in the design and arrangement of heat-transfer surfaces.

One classification of coal is by rank, i.e., according to the degree of metamorphism, or progressive alteration, in the natural series from lignite to anthracite. In the ASTM classification, the basic criteria are the fixed-carbon content and the calorific values (in British thermal units) calculated on a mineral-matter-free basis.

In establishing the rank of coals, it is necessary to use information showing an appreciable and systematic variation with age. For the older coals, a good criterion is the "dry, mineral-matter-free fixed carbon or volatile [matter]." However, this value is not suitable for designating the rank of the younger coals. A dependable means of classifying the latter is the "moist, mineral-matter-free Btu," or calorific value, which varies little for the older coals but appreciably and systematically for younger coals. Table 6.4, ASTM D 388, is used for classification according to rank or age.

TABLE 6.3 ASTM Standards for Testing Coal, Specifications, and Definitions of Terms

ASTM Standards for Testing Coal	
*D 1756	Carbon Dioxide in Coal
*D 2361	Chlorine in Coal
*D 291	Cubic Foot Weight of Crushed Bituminous Coal
*D 440	Drop Shatter Test for Coal
*D 547	Dustiness, Index of, of Coal and Coke
*D 1857	Fusibility of Coal Ash
*D 1412	Equilibrium Moisture of Coal at 96 to 97% Relative Humidity and 30°C
*D 2014	Expansion or Contraction of Coal by the Sole-Heated Oven
*D 720	Free-Swelling Index of Coal
D 409	Grindability of Coal by the Hardgrove-Machine Method
*D 2015	Gross Calorific Value of Solid Fuel by the Adiabatic Bomb Calorimeter
D 1812	Plastic Properties of Coal by the Gieseler Plastometer
D 2639	Plastic Properties of Coal by the Automatic Gieseler Plastometer
D 197	Sampling and Fineness Test of Powdered Coal
*D 271	Sampling and Analysis, Laboratory, of Coal and Coke
D 492	Sampling Coals Classified According to Ash Content
*D 2234	Sampling, Mechanical, of Coal
*D 2013	Samples, Coal, Preparing of Analysis
*D 410	Screen, Analysis of Coal
D 311	Sieve Analysis of Crushed Bituminous Coal
D 310	Size of Anthracite
*D 431	Size of Coal, Designating from its Screen Analysis
*D 1757	Sulfur in Coal Ash
*D 2492	Sulfur, Forms of, in Coal
*D 441	Tumbler Test for Coal
Specifications	
*D 388	Classification of Coals by Rank
*E 11	Wire-Cloth Sieves for Testing Purposes
E 323	Perforated-Plate Sieves for Testing Purposes
Definitions of Terms	
*D 121	Coal and Coke
D 2796	Lithological Classes and Physical Components of Coal
*D 407	Gross Calorific Value and Net Calorific Value of Solid and Liquid Fuels

* Approved as American National Standard by the American National Standards Institute.

The basis for the two ASTM criteria (the fixed-carbon content and the calorific value calculated on a moist, mineral-matter-free basis) are shown in Fig.6.13 for over 300 typical coals of the United States. The classes and groups of Table 6.4 are indicated in Fig. 6.13. For the anthracitic and low- and medium-volatile bituminous coals, the moist, mineral-matter-free calorific value changes very little; hence the fixed-carbon criterion is used. Conversely, in the case of the high-volatile bituminous, subbituminous, and lignitic coals, the moist, mineral-matter-free calorific value is used, since the fixed-carbon value is almost the same for all classifications.

Other Classifications of Coal by Rank. There are other classifications of coal by rank (or type) which are currently in limited use on the European continent. These are the International Classification of Hard Coals by Type, and the International Classification of Brown Coals. Other criteria for the classification of coal by rank have been proposed by various authorities.

Volatile Matter and Heating Value. The composition of the fixed carbon in all types of coal is substantially the same. The variable constituents of coals can, therefore, be considered as concentrated in the volatile matter. One index of the quality of the volatile matter, its heat-

TABLE 6.4 Classification of Coals by Rank[a] (ASTM D 388)

Class	Group	Fixed carbon limits, % (dry, mineral-matter-free basis) Equal or greater than	Fixed carbon Less than	Volatile matter limits, & (dry, mineral-matter-free basis) Greater than	Volatile matter Equal or less than	Calorific value limits, Btu/lb (moist,[b] mineral-matter-free basis) Equal or greater than	Calorific value Less than	Agglomerating character
I. Anthracitic	1. Meta-anthracite	98	—	—	2	—	—	Nonagglomerating
	2. Anthracite	92	98	2	8	—	—	
	3. Semianthracite[c]	86	92	8	14	—	—	
II. Bituminous	1. Low volatile bituminous coal	78	86	14	22	—	—	Commonly agglomerating[e]
	2. Medium volatile bituminous coal	69	78	22	31	—	—	
	3. High volatile A bituminous coal	—	69	31	—	14,000[d]	—	
	4. High volatile B bituminous coal	—	—	—	—	13,000[d]	14,000	
	5. High volatile C bituminous coal	—	—	—	—	11,500	13,000	
						10,500[e]	11,500	Agglomerating
III. Subbituminous	1. Subbituminous A coal	—	—	—	—	10,500	11,500	Nonagglomerating
	2. Subbituminous B coal	—	—	—	—	9,500	10,500	
	3. Subbituminous C coal	—	—	—	—	8,300	9,500	
IV. Lignitic	1. Lignite A	—	—	—	—	6,300	8,300	
	2. Lignite B	—	—	—	—	—	6,300	

[a] This classification does not include a few coals, principally nonbanded varieties, which have unusual physical and chemical properties and which come within the limits of fixed carbon or calorific value of the high volatile bituminous and subbituminous ranks. All of these coals either contain less than 48% dry, mineral-matter-free fixed carbon or have more than 15,500 moist, mineral-matter-free British thermal units per pound.

[b] Moist refers to coal containing its natural inherent moisture but not including visible water on the surface of the coal.

[c] If agglomerating, classify in low volatile group of the bituminous class.

[d] Coals having 69% or more fixed carbon on the dry, mineral-matter-free basis shall be classified according to fixed carbon, regardless of calorific value.

[e] It is recognized that there may be nonagglomerating varieties in these groups of the bituminous class, and there are notable exceptions in high volatile C bituminous group.

ing value, is perhaps the most important property as far as combustion is concerned, and this bears a direct relation to the properties of the pure coals (dry, mineral-matter-free). The volatile matter in coals of lower rank is relatively high in water and CO_2 and consequently low in heating value. The volatile matter in coals of higher rank is relatively high in hydrocarbons, such as methane (CH_4), and consequently is relatively high in heating value.

The relationship of the heating value of the volatile matter to the heating value of the pure coal is shown in Fig. 6.14 for a large number of coals.

Commercial Sizes of Coal

BITUMINOUS. Sizes of bituminous coal are not well standardized, but the following sizings are common:

Run of Mine. This is coal shipped as it comes from the mine without screening.

Run of Mine [8 in (20 cm)]. This is run of mine with oversized lumps broken up.

Lump [5 in (12.5 cm)]. This size will not go through a 5-in (12.5-cm) round hole.

Egg [5 × 2 in (12.5 × 5.1 cm)]. This size goes through 5-in (12.5-cm) holes and is retained on 2-in (5.1-cm) round-hole screens.

Nut [2 × 1¼ in (5.1 × 3.2)]. This size is used for small industrial stokers and for hand firing.

Stoker Coal [1¼ × ¾ in (3.2 × 1.9 cm)]. This is used largely for small industrial stokers and for domestic firing.

Slack [¾ in (1.9 cm) and under]. This is used for pulverizers and industrial stokers.

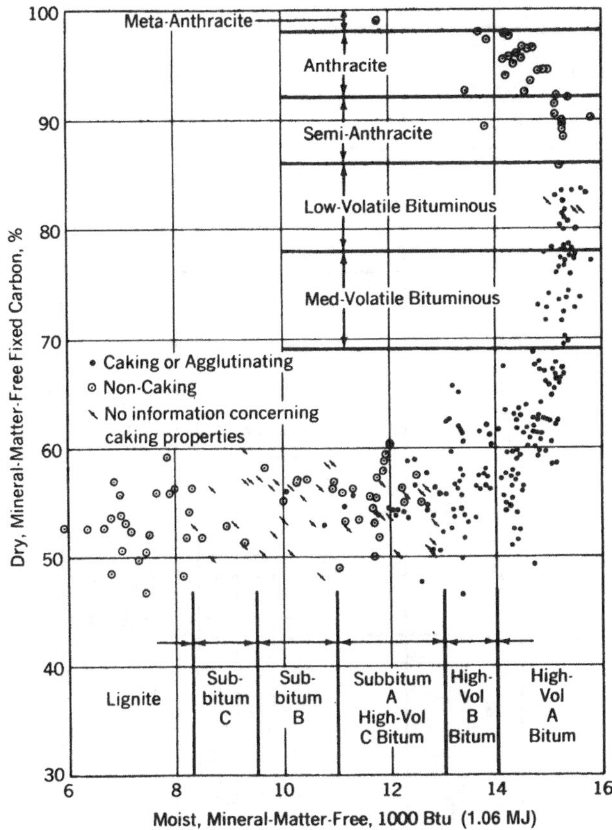

FIGURE 6.13 Distribution plot for over 300 coals of the United States, illustrating ASTM classification by rank as defined in Table 6.4.

ANTHRACITE. Definite sizes of anthracite are standardized in Table 6.5.

Moisture Determination. Moisture in coal is determined quantitatively by definite prescribed methods. It is preferable to determine the moisture in two steps: (1) prescribed air drying to equilibrium at 10 to 15°C above room temperature, and (2) prescribed oven drying for 1 h at 104 to 110°C, after pulverizing (see ASTM Standard D 271).

Since it is the surface moisture that must be evaporated prior to efficient pulverizing of coal, the air-dried component of the total moisture value should be reported separately.

ASTM Standard D 1412 provides a means of estimating the bed moisture of either wet coal showing visible surface moisture or coal that may have lost some moisture. It may be used for estimating the surface moisture of wet coal, i.e., the difference between total moisture, as determined by ASTM Standard D 271, and equilibrium moisture.

Grindability Index. A coal is harder or easier to grind if its grindability index is less or greater, respectively, than 100. The capacity of a pulverizer is related to the grindability index of the coal.

Coal Ash. The presence of ash is accounted for by minerals associated with initial vegetal growth or by those which entered the coal seam from external sources during or after the period of coal formation.

The composition of the coal ash is customarily determined by chemical analysis of the residue produced by burning a sample of coal at a slow rate and at moderate temperature

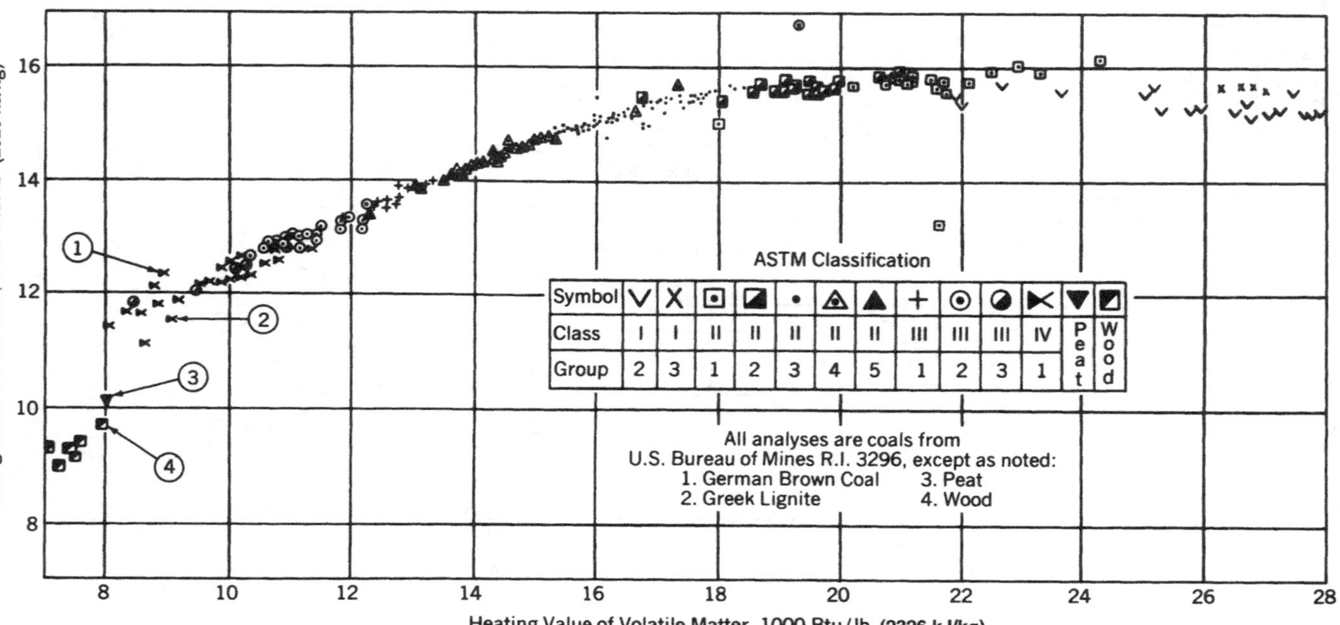

FIGURE 6.14 To illustrate a suggested coal classification using the relationship of the respective heating values of "pure coal" and the volatile matter.

TABLE 6.5 Commercial Sizes of Anthracite (ASTM D 310)
(Graded on round-hole screens)

	Diameter of holes, in*	
Trade name	Through	Retained on
Broken	4⅛	3¼ to 3
Egg	3¼ to 3	2 7/16
Stove	2 7/16	1⅝
Nut	1⅝	13/16
Pea	13/16	9/16
Buckwheat	9/16	5/16
Rice	5/16	3/16

* Metric standards will vary in names and sizes.

(1350°F) under oxidizing conditions in a laboratory furnace. It is thus found to be composed chiefly of compounds of silicon, aluminum, iron, and calcium, with smaller amounts of magnesium, titanium, sodium, and potassium. The analyses of coal ash in Table 6.6 indicate what may be expected of selected coals from various areas of the United States.

Coals may be classified into two groups on the basis of the nature of their ash constituents. One is the bituminous-type ash and the other is the lignite-type ash. *Lignite-type ash* is ash having more CaO plus MgO than Fe_2O_3. By contrast, *bituminous-type ash* has more Fe_2O_3 than CaO plus MgO.

FREE-SWELLING INDEX. The ASTM Standard method D 720 is used for obtaining information regarding the free-swelling property of a coal. Since it is a measure of the behavior of rapidly heated coal, it may be used as an indication of the caking characteristics of coal burned as a fuel.

ASH FUSIBILITY. The determination of ash fusion temperatures (initial deformation, softening, hemispherical, and fluid) is a laboratory procedure, developed in standardized form (ASTM Standard D 1857).

Ash melts when heated to a sufficiently high temperature. If insufficiently cooled, the ash particles remain molten or sticky and tend to coalesce into large masses in the boiler furnace or other heat-absorption surfaces. This problem is dealt with by adequate design of burners and furnace arrangement, knowing the ash fusion temperatures for the fuels to be burned.

VISCOSITY OF COAL-ASH SLAG. Measurement of viscosity of coal-ash slags provides reliable data that can be used for determining suitability of coals for use in slag-tap-type boilers.

Other Solid Fuels

Coke. When coal is heated with a large deficiency of air, the lighter constituents are volatilized and the heavier hydrocarbons crack, liberating hydrogen and leaving a residue of carbon. The carbonaceous residue containing the ash and a part of the sulfur of the original coal is called *coke.*

Undersized coke, called *coke breeze,* usually passing a ⅝-in (17-mm) screen, is unsuited for charging blast furnaces and is available for steam generation. A typical analysis of coke breeze appears in Table 6.7.

Coke from Petroleum. The heavy residuals from the various petroleum cracking processes are presently utilized to produce a higher yield of lighter hydrocarbons and a solid residue suitable for fuel. Solid fuels from oil include delayed coke, fluid coke, and petroleum pitch. Some selected analyses are given in Table 6.8.

Some of these cokes are easy to pulverize and burn, while others are quite difficult. The low-melting-point pitches may be heated and burned like heavy oil, while those with higher melting points may be pulverized and burned.

TABLE 6.6 Ash Content and Ash Fusion Temperatures of Some U.S. Coals and Lignite

	Rank						
	Low-volatile bituminous	High-volatile bituminous				Subbituminous	Lignite
Seam location	Pocahontas No. 3 West Virginia	No. 9 Ohio	Pittsburgh West Virginia	No. 6 Illinois	Utah	Wyoming	Texas
Ash, dry basis, %	12.3	14.10	10.87	17.36	6.6	6.6	12.8
Sulfur, dry basis, %	0.7	3.30	3.53	4.17	0.5	1.0	1.1
Analysis of ash, % by wt							
SiO_2	60.0	47.27	37.64	47.52	48.0	24.0	41.8
Al_2O_3	30.0	22.96	20.11	17.87	11.5	20.0	13.6
TiO_2	1.6	1.00	0.81	0.78	0.6	0.7	1.5
Fe_2O_3	4.0	22.81	29.28	20.13	7.0	11.0	6.6
CaO	0.6	1.30	4.25	5.75	25.0	26.0	17.6
MgO	0.6	0.85	1.25	1.02	4.0	4.0	2.5
Na_2O	0.5	0.28	0.80	0.36	1.2	0.2	0.6
K_2O	1.5	1.97	1.60	1.77	0.2	0.5	0.1
Total	98.8	98.44	95.74	95.20	97.5	86.4	84.3
Ash fusibility							
Initial deformation temperature, °F*							
Reducing	2900+	2030	2030	2000	2060	1990	1975
Oxidizing	2900+	2420	2265	2300	2120	2190	2070
Softening temperature, °F*							
Reducing		2450	2175	2160		2180	2130
Oxidizing		2605	2385	2430		2220	2190
Hemispherical temperature, °F*							
Reducing		2480	2225	2180	2140	2250	2150
Oxidizing		2620	2450	2450	2220	2240	2210
Fluid temperature, °F*							
Reducing		2620	2370	2310	2250	2290	2240
Oxidizing		2670	2540	2610	2460	2300	2290

* °C = (°F − 32) × 5/9.

TABLE 6.7 Analyses—Bagasse and Coke Breeze

Analyses (as-fired), % by wt	Bagasse	Coke breeze
Proximate		
Moisture	52.0	7.3
Volatile matter	40.2	2.3
Fixed carbon	6.1	79.4
Ash	1.7	11.0
Ultimate		
Hydrogen, H_2	2.8	0.3
Carbon, C	23.4	80.0
Sulfur, S	Trace	0.6
Nitrogen, N_2	0.1	0.3
Oxygen, O_2	20.0	0.5
Moisture, H_2O	52.0	7.3
Ash, A	1.7	11.0
Heating value, Btu/lb	4000	11,670
(kJ/kg)	(9304)	(27,144)

TABLE 6.8 Selected Analyses of Solid Fuels Derived from Oil

Analyses (dry basis), % by wt	Delayed coke		Fluid coke	
Proximate:				
VM	10.8	9.0	6.0	6.7
FC	88.5	90.0	93.7	93.2
Ash	0.7	0.1	0.3	0.1
Ultimate:				
Sulfur	9.9	1.5	4.7	5.7
Heating value,				
Btu/lb	14,700	15,700	14,160	14,290
(kJ/kg)	(34,192)	(36,518)	(32,936)	(33,239)

Wood. Analyses of wood ash and selected analyses and heating values of several types of wood are given in Table 6.9.

Wood or bark with a moisture content of 50 percent or less burns quite well; however, as the moisture content increases above this level, combustion becomes more difficult. With moisture content above 65 percent, a large part of the heat in the wood is required to evaporate the moisture, and little remains for steam generation.

Bagasse. Bagasse is the refuse from the milling of sugar cane. It consists of matted cellulose fibers and fine particles. Mills grinding sugar cane commonly use bagasse for steam production. A selected analysis of bagasse is given in Table 6.7.

Other Vegetable Wastes. Food and related industries produce numerous vegetable wastes that are usable as fuels. They include such materials as grain hulls, the residue from the production of furfural from corncobs and grain hulls, coffee grounds from the production of instant coffee, and tobacco stems.

Solvent-Refined Coal. Solvent-refined coal is made by dissolving the organic material in coal in a coal-derived solvent. The finished product is a solid with a melting point of 284 to 293°F (140 to 145°C). It can be burned as an oil or a solid.

Municipal Solid Waste. Municipal solid waste (MSW) can be burned as received or by preparing a refuse-derived fuel (RDF) by shredding the MSW and removing ferrous and nonferrous inorganic materials. The heating value of MSW can range from 3500 to 6500 Btu/lb (8.1 to 15.2 MJ/kg), and the moisture and ash can vary just as widely in similar or opposite directions.

TABLE 6.9 Analyses of Wood and Wood Ash

Wood analyses (dry basis), % by wt	Pine bark	Oak bark	Spruce bark*	Redwood bark*
Proximate analysis, %				
Volatile matter	72.9	76.0	69.6	72.6
Fixed carbon	24.2	18.7	26.6	27.0
Ash	2.9	5.3	3.8	0.4
Ultimate analysis, %				
Hydrogen	5.6	5.4	5.7	5.1
Carbon	53.4	49.7	51.8	51.9
Sulfur	0.1	0.1	0.1	0.1
Nitrogen	0.1	0.2	0.2	0.1
Oxygen	37.9	39.3	38.4	42.4
Ash	2.9	5.3	3.8	0.4
Heating value, Btu/lb	9,030	8,370	8,740	8,350
(kJ/kg)	(21,004)	(19,469)	(20,329)	(19,442)
Ash analysis, % by wt				
SiO_2	39.0	11.1	32.0	14.3
Fe_2O_3	3.0	3.3	6.4	3.5
TiO_2	0.2	0.1	0.8	0.3
Al_2O_3	14.0	0.1	11.0	4.0
Mn_3O_4	Trace	Trace	1.5	0.1
CaO	25.5	64.5	25.3	6.0
MgO	6.5	1.2	4.1	6.6
Na_2O	1.3	8.9	8.0	18.0
K_2O	6.0	0.2	2.4	10.6
SO_3	0.3	2.0	2.1	7.4
Cl	Trace	Trace	Trace	18.4
Ash fusibility temp, °F				
Reducing				
Initial deformation	2,180	2,690		
Softening	2,240	2,720		
Fluid	2,310	2,740		
Oxidizing				
Initial deformation	2,210	2,680		
Softening	2,280	2,730		
Fluid	2,350	2,750		

* Saltwater stored.

Coal Processing and Transporting

The purpose of coal preparation is to improve the quality of the coal or to make it suitable for a specific purpose by (1) cleaning to remove inorganic impurities, (2) special treatment (such as dedusting), or (3) sizing by crushing and screening.

Economic Factors. All coals have certain properties which place limitations on their most advantageous use.

To define the limitations of various types of coal-burning equipment in service, specifications covering several important properties of coal are necessary. For certain types of stokers, for instance, a minimum ash-softening temperature, maximum ash content, and maximum sulfur content of the coal are specified to prevent excessive clinkering. For pulverized-coal firing, it is necessary to specify ash-slagging and ash-fouling parameters for a dry-ash installation. Within these and other equipment limitations there is usually a wide range of coals that can be satisfactorily burned in a specific steam boiler, and the choice depends primarily on economics—which coal will produce steam at the lowest overall cost, including cost at the mine, shipment, storage, handling, operating, and maintenance costs. To burn a wide range of coals often requires a larger initial investment than would otherwise be necessary. However, since fuel cost represents a large part of the overall operating cost the investment may be more than offset by fuel savings.

Nature of Impurities in Coal. Impurities can be divided into two general classifications—inherent and removable. The inherent impurities are inseparably combined with the coal. The removable impurities are segregated and can be eliminated, by available cleaning methods, to the extent that is economically justified.

Mineral matter is always present in raw coal and forms ash when the coal is burned. The ash-forming mineral matter is usually classified as either inherent or extraneous. Ash-forming material organically combined with the coal is considered as inherent mineral matter. Generally the inherent mineral matter contained in coal is about 2 percent or less of the total ash. Extraneous mineral matter is ash-forming material that is foreign to the plant material from which the coal was formed. It consists usually of slate, shale, sandstone, or limestone.

Sulfur is always present in raw coal in amounts ranging from traces to as high as 8 percent or more. This results in the emission of sulfur oxides in the stack gases when the coal is burned.

Moisture, which is inherent in coal, may be considered to be an impurity. It varies with different ranks of coal, increasing for lower-rank coals from 1 to 2 percent in anthracite to 50 percent or more in lignite and brown coals. Moisture that collects on the exposed surfaces of coal is commonly called *surface moisture* and is removable.

Cleaning Methods. The principal benefit of cleaning is the reduction in ash content. With ash content reduced, shipping costs and the requirements for storage and handling decrease because of the smaller quantity of coal necessary per unit of heating value.

Cleaning at Mine Face. Efforts to reduce the impurities loaded in the mine usually result in increased mining cost; hence economics plays an important part in determining the amount of cleaning done.

Mechanical Picking. Many mechanical picking devices employ differences in physical picking dimensions of coal and impurities as a means of separation. Bituminous coal fractures into rough cubes, whereas slate and shale normally fracture as thin slabs. A slot-type flat picker can be installed on shaker screens, shaking conveyors, and chutes.

Froth Flotation. The incoming coal feed may be agitated in a controlled amount of water, air, and reagents that cause a surface froth to form, the bubbles of which selectively attach themselves to coal particles and keep them buoyant while the heavier particles of pyrite, slate, and shale remain dispersed in the water. This method can be used in cleaning smaller particles, $\frac{1}{10}$ in by 0 (no minimum), particularly particles smaller than 48 mesh.

Gravity Concentration. Removal of segregated impurities in coal by gravity concentration is based on the principle that heavier particles separate from the lighter ones when settling in a fluid. This principle is applicable because most common solid impurities are heavier than coal (Table 6.10).

The commercial processes used in gravity concentration can be divided into two main classifications—wet and dry (or pneumatic).

Cleaning by Gravity Concentration

Wet Processes. Most cleaning of bituminous coal and lignite is accomplished by wet processes. Table 6.11 lists the methods and types of equipment generally used and the extent of application of each method, expressed in percent.

Dry Processes. The principal pneumatic cleaning process uses tables equipped with riffles, with air as the separating medium. Admitted through holes in the tables, the air is blown up through the bed of coal. The motion of the tables plus the airflow segregates the coal and impurities.

TABLE 6.10 Typical Specific Gravities of Coal and Related Impurities

Material	Specific gravity
Bituminous coal	1.10–1.35
Bone coal	1.35–1.70
Carbonaceous shale	1.60–2.20
Shale	2.00–2.60
Clay	1.80–2.20
Pyrite	4.80–5.20

TABLE 6.11 Types of Wet Processes Used for Cleaning of Bituminous Coals and Lignite

Equipment or method used	Percent of cleaning done by wet processes
Jigs	46.7
Dense-media method	29.2
Concentrating tables	13.9
Flotation method	2.6
Classifiers	1.4
Launderers	1.3
Total	95.1

Washability Characteristics of Coal. There is a general correlation between specific gravity and ash content, although the relation differs for various coals. Ash contents corresponding to various specific gravities for bituminous coal are shown in Table 6.12.

TABLE 6.12 Typical Ash Contents of Various Bituminous Coal Specific Gravity Fractions

Specific gravity fraction	Ash content, % by wt
1.3–1.4	1–5
1.4–1.5	5–10
1.5–1.6	10–35
1.6–1.8	35–60
1.8–1.9	60–75
Above 1.9	75–90

Float-and-sink tests run in a laboratory provide data useful for rating and controlling cleaning equipment and evaluating efficiencies obtained, both quantitative and qualitative.

Special Treatment

Dedusting. Dedusting is a type of air separation with classification according to size. Air passed through the coal entrains a large percentage of the "fines." The fine coal is recovered from the air with cyclone separators and bag filters. This process is often employed to remove fines prior to wet cleaning.

Mechanical Dewatering and Heat Drying. The larger sizes of coal can be easily dewatered, and natural drainage is sometimes provided by special hoppers (or bins), screen conveyors, perforated-bucket elevators, and fixed screens. However, when the fine sizes must be dried or when lower moisture content is required for the large coal, mechanical dewatering or thermal drying is necessary.

Mechanical dewatering devices may be shaker screens, vibrating screens, filters, centrifuges, and thickening or desliming equipment. Thermal drying is used to obtain low-moisture-content coal, especially for the finer sizes. Various types of thermal dryers are rotary, cascade, reciprocating-screen, conveyor, suspension (including flash and venturi dryers), and fluidized-bed.

Dustproofing. Oil is commonly used for dustproofing coal. When coal is sprayed with oil, the film causes some dust particles to adhere to the larger pieces of coal and others to agglomerate into larger lumps not easily airborne. Calcium chloride absorbs moisture from the air, providing a wet surface to which the dust adheres.

Freezeproofing. To prevent freezing of coal during transit or storage, spray oil may be used, and it is applied to the coal in the same manner as dustproofing. As an alternative and less costly method, the car hoppers may be heavily sprayed with oil. Another freezeproofing method is thermal drying of the fine coal.

Sizing by Crushing and Screening

Sizing Requirements. Generally acceptable coal sizings for various types of fuel-burning equipment are given in the section covering the equipment. The degradation in coal sizing resulting from handling is an important consideration in establishing sizing specifications for steam plants, where the maximum permissible quantity of fines in the coal is set by firing-equipment limitations.

Crushers and Breakers. Many types of crushing and screening equipment are used commercially. Representative types are the Bradford breaker, single-roll crusher, double-roll crusher, and hammer mill.

Screens. Many types of screens are used for sizing, including the gravity bar screen or grizzly, revolving screen, shaker screen, and vibrating screen.

Transportation of Coal. Transportation costs may represent a large portion of the delivered cost of coal. Transportation may affect the condition of the received coal if freezing occurs in transit or if there is a change in moisture content or a degradation of size.

When freezing in transit is anticipated, its effect may be ameliorated by special treatment of the coal at the point of loading.

The moisture content of the coal will depend on sizing, condition at time of loading, and weather conditions during transit.

Size degradation depends in a large measure on the friability of the coal but it is also affected by the amount of handling and shaking in transit.

Handling and Storage at Consumer's Plant

Unloading. When coal is received by rail, if the surface moisture of the coal is high, it may be necessary to rap the car sides with a sledge or to use a slice bar from above in order to start the coal flowing. If this high-moisture coal has been in transit several days at freezing temperatures, it may be frozen into a solid mass, and unloading is a serious problem. In hot, dry weather the coal, on arrival, may be so dry that a high wind can blow the fines away as dust.

Frozen Coal. Successful equipment for unloading frozen coal includes steam-heated thawing sheds, oil-fired thawing pits, and the radiant-electric railroad-car thawing system. Another device is a car shakeout, in which a motor-driven eccentric shaker clamped to the top flanges of the car transmits a vibratory motion to the car body.

When expensive equipment is not economically justified, the methods used depend primarily on manual labor and slice bars, sledges, portable oil torches, and steam or hot water.

Mechanical Handling. Extensive equipment, covering the range of plant requirements from a few tons per day to the largest, is available for transferring coal from the railroad cars either to outside storage or to inside bunkers or hoppers. When pulverizers are used, it is desirable to include a magnetic separator somewhere in the system. Frequently a coal crusher will make it economically possible to use coal that is not screened or sized.

Outside Storage at or near Plant. Outside storage of coal at or near the plant site is often necessary, and in some cases required, to ensure continuous plant operation.

The changes that may affect the value of stored coal are loss in heating value, reduction of coking power, reduction of average particle size (weathering or slacking), and, most important, losses from self-ignition or spontaneous combustion. Direct loss of stored coal by wind or water erosion is possible and may become a serious nuisance.

Oxidation of Coal. Many constituents of pure coal begin to oxidize when exposed to air. Oxidation of coal is affected by length of time exposed to air, coal rank, size or surface area, temperature, amount and size of pyrites particles, moisture, and amount of mineral matter or ash.

All coals may be safely stored, provided proper procedures are used. Coal can be successfully stored by the following methods outlined by the Bureau of Mines.

Selection of Storage Site. The site should be level solid ground, free of loose fill, properly graded for drainage, and compacted by a bulldozer. Consideration should be given to access and provision for coal delivery as well as to protection from prevailing winds, tides, flooded rivers, and spray from salt water.

Storage in Small Amounts. No special precautions are required for anthracite. Sized (double-screened) bituminous coals may be stored in small conical piles from 5 to 15 ft (1.5 to 4.6 m) high. Slack sizes of bituminous coal may be compacted with a bulldozer. Subbituminous coal and lignite should be stored in small piles and thoroughly packed.

Storage in Large Amounts. Anthracite coal is very safe to store in large high piles that permit drainage.

Bituminous coal should be stockpiled in multiple horizontal layers. Each layer should be spread out to vary from 1 to 2 ft (22 to 44 cm) in thickness and thoroughly packed to eliminate air spaces. The top should be slightly crowned and symmetrical to permit even runoff of water. All sides and the top should be covered with a 1-ft (22-cm) compacted layer of fines and then capped with a 1-ft (22-cm) layer of sized lump coal.

Subbituminous coal and lignite should be stockpiled by using the same system of layering and compacting, but the coal layers should be thinner, not more than 1 ft (22 cm) thick, to ensure good compaction.

Inside Storage and Handling. Dry anthracite coal flows easily; however, other fine, wet coals feeding from a bottom-outlet bunker may have a tendency to "rat-hole" or "pipe" all the way to the top surface. When this occurs, coal flow to the feeders is intermittent and spasmodic. From studies of actual plant operations it has been found that hang-ups of coal may begin when there is a surface-moisture level of 5 to 6 percent.

Bunker Design. The purpose of the bunker is not only to store a given capacity of coal but also to function efficiently as part of the system in maintaining a continuous supply of fuel to the pulverizer or stoker. In the design of bunkers, careful consideration should be given to capacity, shape, material, and location.

Bunker Fires. A fire in the coal bunker should be recognized at once as a serious danger to personnel and equipment. Steam or carbon dioxide may be piped in directly to covered bunkers and to the affected areas in open-top bunkers. If the hot coal is run through the fuel-burning equipment, special care should be taken to feed uniformly, without interruption.

Stokers

Mechanical stokers are designed to feed fuel onto a grate within the furnace and to remove the ash residue. The grate area required for a stoker of given type and capacity is determined from allowable rates established by experience. Table 6.13 lists typical fuel-burning rates for various types of stokers, based on using coals suited to the stoker type in each case. The prac-

TABLE 6.13 Stoker/Grate System Overview

Stoker type	Grate type	Fuel	Typical release rate, 1000 Btu/h ft^2 (MW/m^2)	Steam capacity, 1000 lb/h (kg/s)
Underfeed:				
Single retort	—	Coal	425 (1.34)	25 (3.15)
Double retort	—	Coal	425 (1.34)	30 (3.78)
Multiple retort	—	Coal	600 (1.89)	500 (63.0)
Overfeed:				
Mass	Vibrating: water-cooled	Coal	400 (1.26)	125 (15.8)
	Traveling chain	Coal	500 (1.58)	80 (10.1)
	Reciprocating	Refuse	300 (0.95)	350 (44.1)
Spreader	Vibrating:			
	air-cooled	Coal	650 (2.05)	150 (18.9)
		Wood	1100 (3.47)	700 (88.2)
	water-cooled	Wood	1100 (3.47)	700 (88.2)
	Traveling	Coal	750 (2.37)	390 (49.1)
		Wood	1100 (3.47)	550 (69.3)
		RDF*	750 (2.37)	400 (50.4)

* Refuse-derived fuel.

tical steam-output limit of boilers equipped with mechanical stokers is about 500,000 lb/h (63 kg/s).

Almost any coal can be burned successfully on some type of stoker. In addition, many by-products and waste fuels, such as coke breeze, wood wastes, pulpwood bark, and bagasse, can be used either as a base or auxiliary fuel.

Mechanical stokers can be classified into four main groups, based on the method of introducing the fuel to the furnace:

1. Spreader stokers
2. Underfeed stokers
3. Water-cooled vibrating-grate stokers
4. Chain-grate and traveling-grate stokers

Spreader Stokers. The spreader stoker is the one most commonly used in the capacity range up to 500,000 lb/h (63 kg/s) because it responds rapidly to load swings and can burn a variety of fuels. The spreader stoker is capable of burning a wide range of coals, from high-rank eastern bituminous to lignite or brown coal, as well as a variety of by-product waste fuels.

As the name implies, the spreader stoker projects fuel into the furnace over the fire with a uniform spreading action, permitting suspension burning of the fine fuel particles (Fig. 6.15). The heavier pieces, which cannot be supported in the gas flow, fall to the grate for combustion in a thin, fast-burning bed.

Grates for Spreader Stokers. Spreader-stoker firing is old in principle, and the first grate design was a stationary type, with the ash removed manually through the front doors.

Stationary grates were soon followed by dumping-grate designs, in which grate sections are provided for each feeder and the undergrate air plenum chambers are correspondingly divided. This permits the temporary discontinuance of the fuel and air supply to a grate section for ash removal without affecting other sections of the stoker.

The continuous-ash-discharge traveling grate has no interruptions for removing ashes and because of the thin, fast-burning fuel bed, this design increased average burning rates approximately 70 percent over the stationary and dumping-grate types. Continuous-cleaning grates of reciprocating and vibrating designs have also been developed.

The normal practice of all continuous-ash-discharge spreader stokers is to remove the ashes at the front or feeding end of the stoker. This permits the most satisfactory fuel-

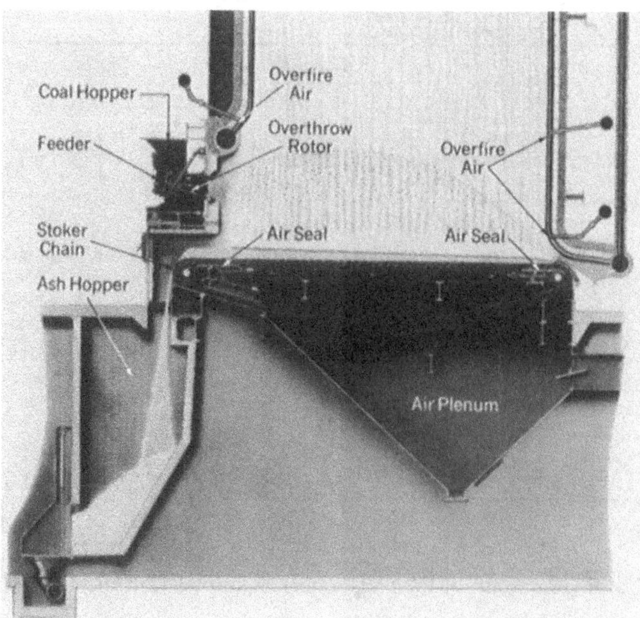

FIGURE 6.15 Traveling-grate spreader stoker with front-ash discharge.

distribution pattern and provides maximum residence time on the grates for complete combustion of the fuel.

Overfire Air. An overfire air system, with pressures from 15 to 30 in (3.7 to 7.5 kPa) of water, is essential to successful suspension burning. It is customary to provide at least two rows of evenly spaced high-pressure-air jets in the rear wall of the furnace and one in the front wall. This air mixes with the furnace gases and creates the turbulence required for complete combustion.

Carbon Reinjection. Partial suspension burning results in a greater carryover of particulate matter in the flue gas than occurs with other types of stokers. In general the arrangement of the collection equipment is such that the coarse carbon-bearing particles can be returned to the furnace for further burning and the fine material discharged to the ash removal system.

Reintroducing the fly carbon into the furnace results in an increase in boiler efficiency of 2 to 3 percent.

Fuels and Fuel Bed. All spreader stokers, and in particular the traveling-grate spreader type, have an extraordinary ability to burn fuels with a wide range of burning characteristics, including coals with caking tendencies. High-moisture, free-burning bituminous and lignite coals are commonly used, and some low-volatile fuels, such as coke breeze, have been burned in a mixture with higher-volatile coal. Anthracite coal, however, is not a satisfactory fuel for spreader-stoker firing.

Coal size segregation is a problem with any type of stoker, but the spreader stoker can tolerate a small amount of segregation because the feeding rate of the individual feeder-distributors can be varied. Size segregation, where fine and coarse coal are not distributed evenly over the grate, produces a ragged fire and poor efficiency.

Firing of By-Product Waste Fuels. By-product wastes having considerable calorific value can be used as a base fuel or as a supplementary fuel in the generation of steam for power, heating, or industrial processes. Bark from wood-pulping operations and bagasse from sugar refineries are good examples. Others include coffee-ground residue from instant-coffee manufacture, corncobs, coconut and peanut hulls from furfural manufacture, bark and sawdust from woodworking plants, and municipal solid waste. Spreader-stoker firing provides an excellent way to burn these wastes.

Waste fuels with high moisture content may present problems in maintaining combustion unless there is enough auxiliary fuel to maintain the average moisture of the total fuel input at a maximum of about 50 percent. Preheated air, at temperatures dependent upon the fuel moisture content, aids in drying and igniting the fuel as it is fed into the furnace. Air temperatures up to 450°F (240°C) are common in bark-fired units. High air temperatures may require the use of alloy grate materials in order to reduce maintenance.

Underfeed Stokers. Underfeed stokers are used principally for heating and for small industrial units with a capacity of less than 30,000 lb/h (3780 kg/s). Underfeed stokers are generally of two types: the horizontal-feed, side-ash-discharge type, Fig. 6.16; and the gravity-feed, rear-ash-discharge type, Fig. 6.17.

In the side-ash-discharge underfeed stoker (Fig. 6.16), fuel is fed from the hopper by means of a reciprocating ram to a central trough called the *retort.* On very small heating stokers, a screw conveys the coal from the hopper to the retort. A series of small auxiliary pushers in the bottom of the retort assist in moving the fuel rearward, and as the retort is filled, the fuel is moved upward to spread to each side over the air-admitting tuyères and side grates. The fuel rises in the retort and burns, and the ash is intermittently discharged to shallow pits, quenched, and removed through doors at the front of the stoker.

The single-retort and double-retort, horizontal-type stokers are generally limited to 25,000 to 30,000 lb of steam per hour with burning rates of 425,000 Btu/(ft^2)(h) (1.34 MW_T/m^2) in furnaces with water-cooled walls. The multiple-retort, rear-end-cleaning type (Fig. 6.17), has a retort and grate inclination of 20 to 25°. This type of stoker can be designed for boiler units generating up to 500,000 lb/h (63,000 kg/s). Burning rates up to 600,000 Btu/(ft^2)(h) (1.89 MW_T/m^2) are practicable.

The burning rates for underfeed stokers are directly related to the ash-softening temperature. For coals with ash-softening temperature below 2400°F (1316°C), the burning rates are progressively reduced.

With multiple-retort stokers, overfire-air systems are also frequently provided.

The size of the coal has a marked effect on the relative capacity and efficiency of an underfeed stoker. The most desirable size consists of 1 in (25.4 mm) top size × 0.25 in (6.4 mm) with a maximum 20 percent through 0.25-in (6.4-mm) round screen. A reduction in the percentage of fines helps to keep the fuel bed more porous and extends the range of coals with a high coking index.

In general, underfeed stokers are able to burn caking coals. The range of agitation imparted to the fuel bed in different stoker designs permits the use of coals with varying degrees of caking properties. The ash-fusion temperature is an important factor in the selection of the coals. Usually, the lower the ash-fusion temperature, the greater the possibility of clinker trouble.

Water-Cooled Vibrating-Grate Stokers. An entirely different design of stoker is the water-cooled vibrating-grate hopper-feed type, Fig. 6.18. This stoker consists of a tuyère grate surface mounted on, and in intimate contact with, a grid of water tubes interconnected with the boiler circulation system for positive cooling. The entire structure is supported by a number of flexing plates, allowing the grid and its grate to move freely in a vibrating action that conveys coal from the feeding hopper onto the grate and gradually to the rear of the stoker. Ashes are automatically discharged to an ash pit.

Vibration of the grates is intermittent, and the frequency of the vibration periods is regulated by a timing device to conform to load variations, synchronizing the fuel feeding rate with the air supply.

Water cooling of the grates makes this stoker especially adaptable to multiple-fuel firing since a shift to oil or gas does not require special provision for protection of the grates. A normal bed of ash left as a cover gives adequate protection from furnace radiation.

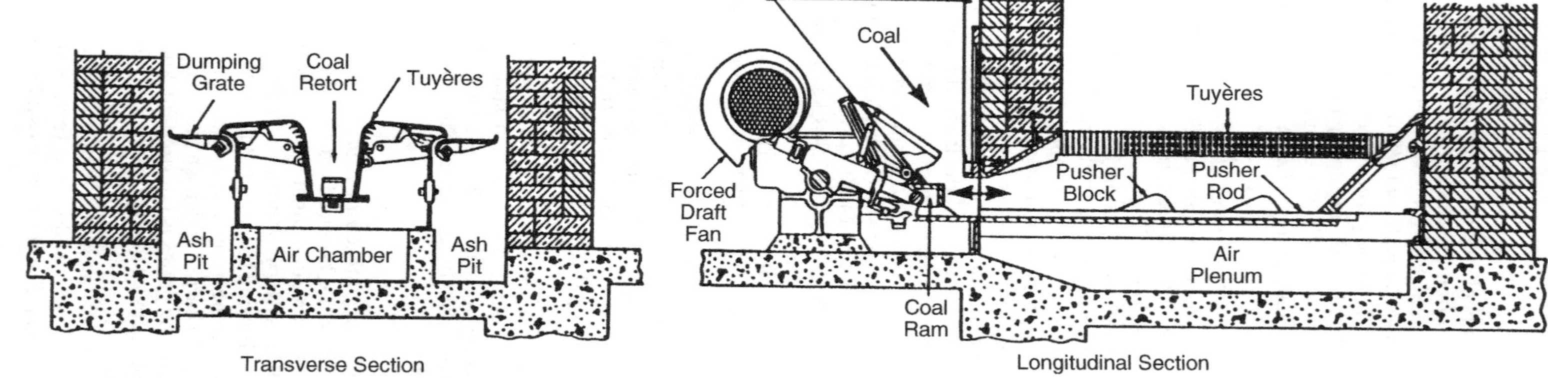

FIGURE 6.16 Single-retort underfire stoker with horizontal feed, side ash discharge.

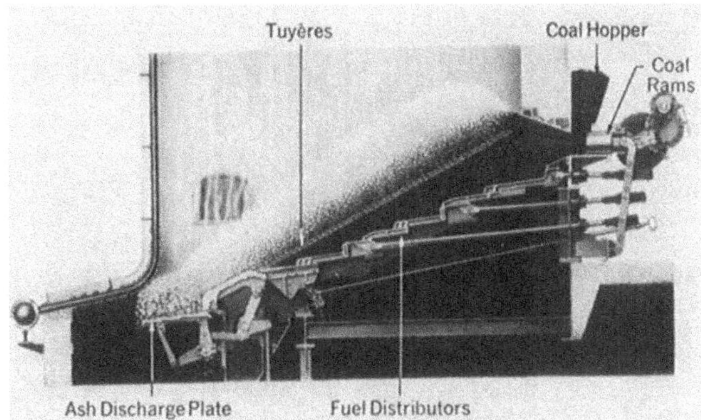

FIGURE 6.17 Multiple-retort gravity-feed type, rear-ash-discharge underfeed stoker.

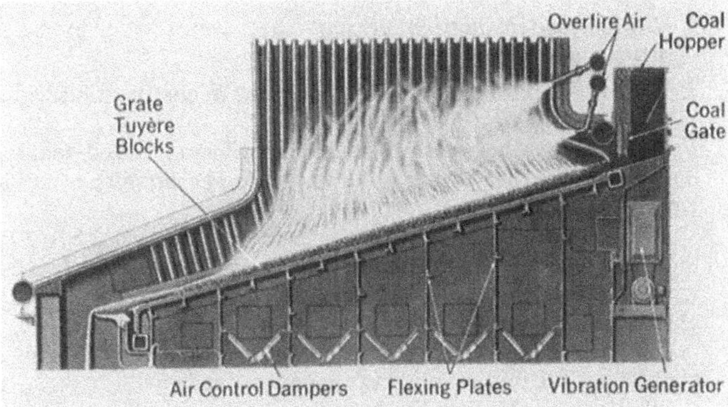

FIGURE 6.18 Water-cooled vibrating-grate stoker.

Chain-Grate and Traveling-Grate Stokers. Traveling-grate stokers, including the specific type known as the chain-grate stoker (Fig. 6.19), have assembled links, grates, or keys joined together in endless belt arrangements that pass over the sprockets or return bends located at the front and the rear of the furnace. Coal enters the furnace after passing under an adjustable gate to regulate the thickness of the fuel bed, is heated by radiation from the furnace gases, and is ignited. The fuel bed continues to burn as it moves along, and ash is discharged from the end of the grate into the ash pit. Generally these stokers use furnace arches (front and/or rear) to improve combustion by reflecting heat into the fuel bed.

Chain-and traveling-grate stokers can burn a wide variety of fuels.

Preparation and Utilization of Pulverized Coal

The capacity limitations imposed by stokers are overcome by the pulverized-coal system. This method of burning coal also provides:

1. Ability to use coal from fines up to 2-in (5-cm) maximum size
2. Improved response to load changes

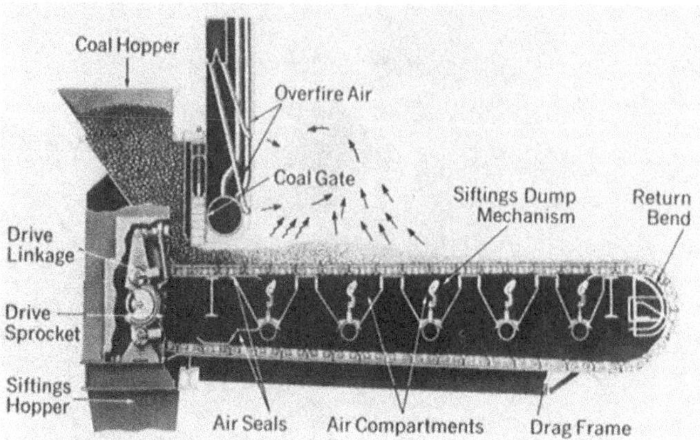

FIGURE 6.19 Chain-grate stoker. (*Courtesy of Detroit Stoker Company.*)

3. An increase in thermal efficiency because of lower excess air used for combustion and lower carbon loss than with stoker firing
4. Improved ability to efficiently burn coal in combination with oil and gas

Pulverized-Coal Systems. The function of a pulverized-coal system is to pulverize the coal, deliver it to the fuel-burning equipment, and accomplish complete combustion in the furnace with a minimum of excess air.

A small portion of the air required for combustion (15 to 20 percent) is used to transport the coal to the burner. This is known as *primary air* and is also used to dry the coal in the pulverizer. The remainder of the combustion air (80 to 85 percent) is introduced at the burner and is known as *secondary air.*

Two principal systems—the bin system and the direct-firing system—have been used for processing, distributing, and burning pulverized coal. The direct-firing system is almost exclusively the one being installed today.

Bin System. The bin system is primarily of historical interest. In this system the coal is processed at a location apart from the furnace, and the end product is pneumatically conveyed to cyclone collectors, which recover the fines and clean the moisture-laden air before returning it to the atmosphere. The pulverized coal is discharged into storage bins and later conveyed by pneumatic transport through pipelines to utilization bins and from there to the furnace.

Direct-Firing System. The pulverizing equipment developed for the direct-firing system permits continuous utilization of raw coal directly from the bunkers. This is accomplished by feeding coal of a maximum top size directly into the pulverizer, where it is dried as well as pulverized, and then delivering it to the burners in a single continuous operation. Components of the direct-firing system are illustrated in Fig.6.20.

There are two direct-firing methods in use—the pressure type and the suction type.

PRESSURE FIRING. In the pressure method, the primary-air fan, located on the inlet side of the pulverizer, forces the hot primary air through the pulverizer, where it picks up the pulverized coal and delivers the proper coal-air mixture to the burners. On large installations, primary-air fans operating on cold air force the air through the air heater first and then through the pulverizer.

SUCTION FIRING. In the suction method, the air and entrained coal are drawn through the pulverizer under negative pressure by an exhauster located on the outlet side of the pulverizer. With this arrangement the fan handles a mixture of coal and air, and distribution of the mixture to more than one burner is attained by a distributor beyond the fan discharge.

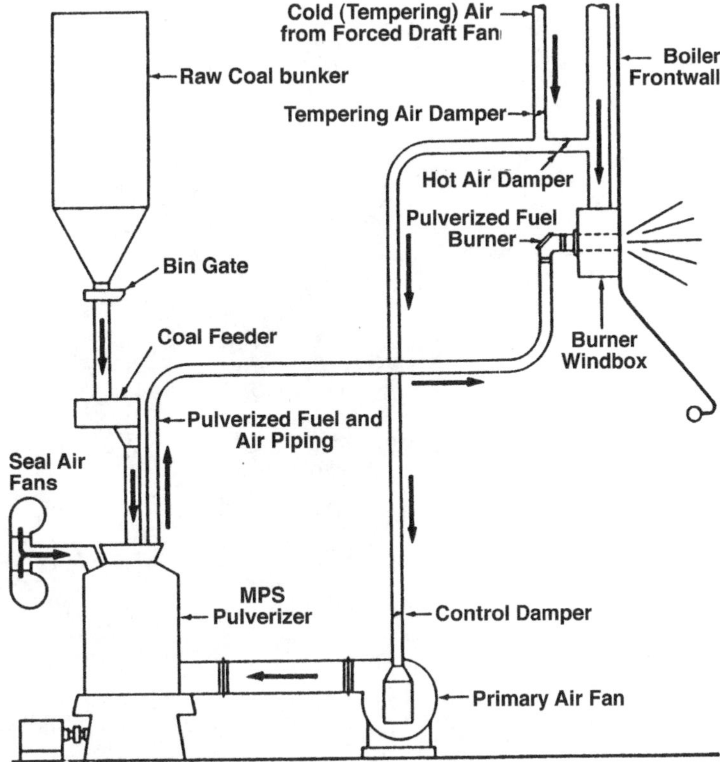

FIGURE 6.20 Direct-fired, hot-fan system for pulverized coal.

In the direct-firing system the operating range of a pulverizer is usually not more than 3 to 1 (without change in the number of burners in service) because the air velocities in lines and other parts of the system must be maintained above the minimum levels to keep the coals in suspension. Most boiler units are provided with more than one pulverizer, each feeding multiple burners. Load variations beyond 3 to 1 are accommodated by shutting down (or starting up) a pulverizer and the burners it supplies.

Types of Pulverizers. All pulverizing machinery operates to grind by impact, attrition, compression, or a combination of two or more of these.

Medium-Speed Pulverizers. There are two groups of medium-speed (75 to 225 r/min) pulverizers, classified as the ball-and-race and roller types. The principle of pulverizing by a combination of crushing under pressure, impact, and attrition between grinding surfaces and material is used in each group, but the method is different.

MEDIUM-SPEED BALL-AND-RACE PULVERIZERS. The ball-and-race pulverizer works on the ball-bearing principle. The Type EL pulverizer, illustrated in Fig. 6.21, has one stationary top ring, one rotating bottom ring, and one set of balls that make up the grinding elements. The pressure required for efficient grinding is obtained from externally adjustable dual-purpose springs. The bottom ring is driven by a yoke which is attached to the vertical main shaft of the pulverizer. The top ring is held stationary by the dual-purpose springs.

MEDIUM-SPEED ROLLER-TYPE PULVERIZERS. This type of pulverizer can be of a design in which the ring is stationary and the rolls rotate or one in which the rolls are mounted off the mill housing and the ring rotates. In the first type, grinding elements consisting of three or more cylindrical rolls, suspended from driving arms, revolve in a horizontally positioned

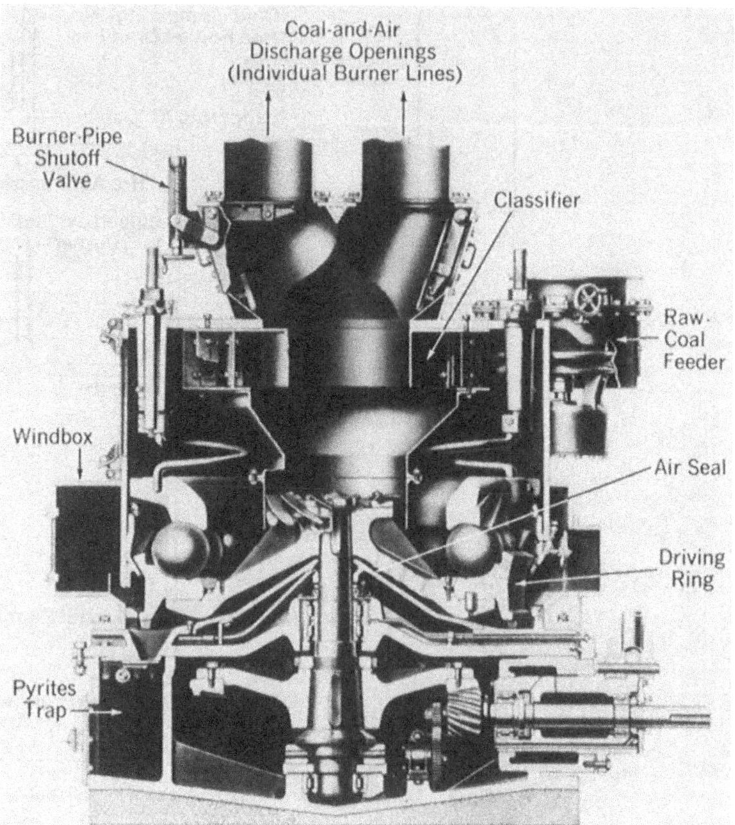

FIGURE 6.21 Babcock & Wilcox Type EL single-row ball-and-race pulverizer.

replaceable race. The principal components of the second type, the bowl mill, are a rotating bowl equipped with a replaceable grinding ring, two or more tapered rolls in stationary journals, a classifier, and a main drive.

TUBE MILL. One of the oldest practical pulverizers is the tube mill, in which a charge of mixed-size forged-steel balls in a horizontally supported grinding cylinder is activated by gravity as the cylinder is rotated. The coal is pulverized by attrition and impact as the ball charge ascends and falls within the coal.

High-Speed Pulverizers. High-speed pulverizers use impact as a primary means of grinding through the use of hammerlike beaters, wear-resistant pegs rotating within a cage, and fan blades integral with the pulverizer shaft.

Selecting Pulverizer Equipment. A number of factors must be considered in the selection of pulverizer equipment. If selection anticipates the use of a variety of coals, the pulverizer should be sized for the coal that gives the highest base capacity. *Base capacity* is the desired capacity divided by the capacity factor. The latter is a function of the grindability of the coal and the fineness required (see Fig. 6.22).

The percentage of volatile matter in the fuel has a direct bearing on the recommended temperature for combustion of the mixture of primary air and fuel. The generally accepted safe values for pulverizer exit temperatures for fuel-air mixtures are given in Table 6.14. The temperature of the primary air entering the pulverizer may run 650°F (345°C) or higher, depending on the amount of moisture in the coal.

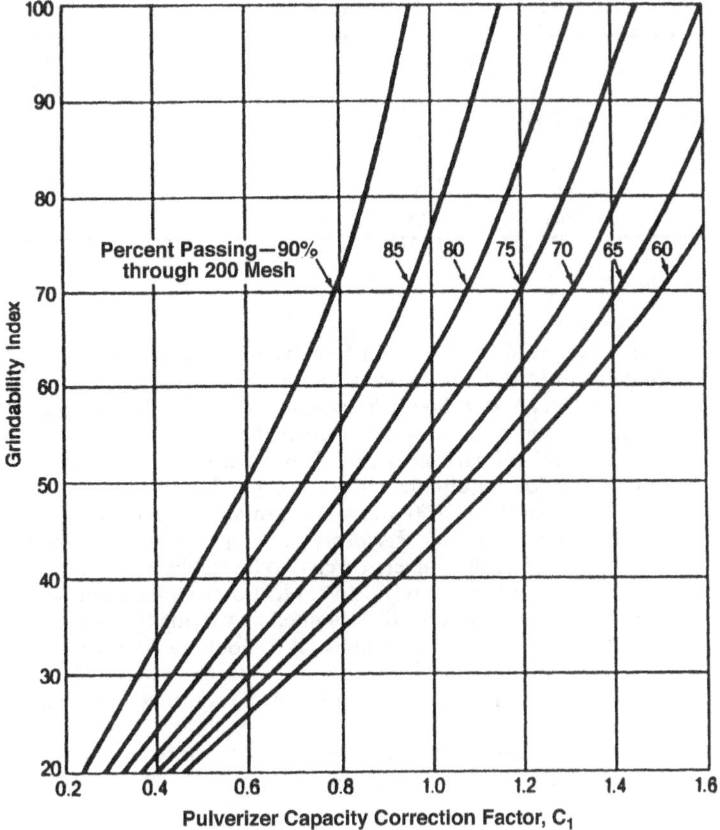

FIGURE 6.22 Capacity correction factor.

Fine grinding of coal is necessary to assure complete combustion of the carbon for maximum efficiency. The required pulverized fuel fineness is expressed as the percentage of the product passing through various sizes of sieves. Coal classification by rank and end use of product determine the fineness to which coal should be ground (Table 6.15).

Burning Equipment for Pulverized Coal. The burner is the principal component of equipment for firing pulverized coal. Coal must be pulverized to the point where particles are small enough to assure proper combustion (Table 6.15). In the direct-firing system the coal is dried

TABLE 6.14 Typical Pulverizer Outlet Temperature

Fuel type	Volatile content, %*	Exit temperature, °F (°C)†
Lignite and subbituminous	—	125–140 (52–60)
High volatile bituminous	30	150 (66)
Low volatile bituminous	14–22	150–180 (66–82)
Anthracite, coal waste	14	200–210 (93–99)
Petroleum coke	0–8	200–250 (93–121)

* Volatile content is on a dry, mineral-matter-free basis.

† The capacity of pulverizers is adversely affected with exit temperatures below 125°F (52°C) when grinding high-moisture lignites.

TABLE 6.15 Typical Pulverized Coal Fineness Requirements—Percent Passing 200 U.S. Standard Sieve

	High rank* fixed carbon, %			Low rank (fixed carbon <69%) heating value, Btu/lb†		
Furnace or process:	97.9 to 86.0	85.9 to 78.0	77.9 to 69.0	Above 13,000	13,000 to 11,000	Below 11,000
Marine boiler	—	85	80	80	75	—
Water-cooled	80	75	70	70	65	60
Cement kiln	90	85	80	80	80	—
Blast furnace	N/A	N/A	N/A	80	80	N/A

* ASTM classification.
† Btu/lb × 2.326 = kJ/kg.

and delivered to the burner in suspension in the primary air, and this mixture must be adequately mixed with the secondary air at the burner.

Piping and Nozzle Sizing Requirements. Size selection of nozzles for pulverized-coal piping and burners requires flow velocities that are high enough to keep the coal particles in suspension in the primary air stream. This generally requires 40 to 70 percent of the pulverizer's full-load airflow requirement at zero output. Horizontally arranged burner nozzles should be sized for no less than 3000 ft/min at the minimum pulverizer capacity.

Standards of Burner Performance. Operators of pulverized-coal equipment should expect ignition of the pulverized coal to be stable, without the use of supporting fuel, over a load range of approximately 3 to 1. Most boilers are equipped with many burners so that a wider capacity range is readily obtained by varying the number of burners and pulverizers in use. The loss of unburned combustibles should be less than 2 percent with excess air in the range of 15 to 22 percent, measured at the furnace outlet. The design should avoid the formation of deposits that may interfere with the continued efficient and reliable performance of the burner.

BURNERS. The most frequently used burners are of the circular single-register type designed for firing pulverized coal only (Fig. 6.23). This type can be equipped to fire any com-

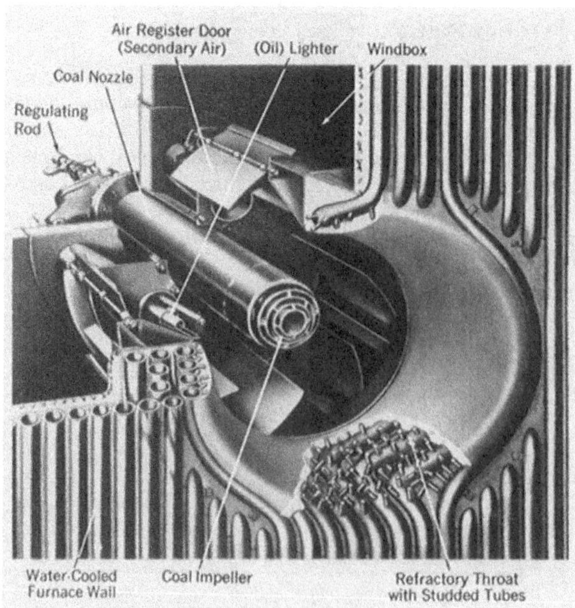

FIGURE 6.23 Circular register burner for pulverized-coal firing.

bination of the three principal fuels. However, combination firing of pulverized-coal with oil in the same burner should be restricted to short emergency periods because of the possibility of coke formation on the pulverized-coal element.

LIGHTERS (IGNITERS) AND PILOTS. In starting up the burner on pulverized coal, it is necessary to keep the igniters in operation until the temperature in the combustion zone becomes high enough to assure self-sustaining ignition of the main fuel. The self-igniting characteristics of pulverized coal vary from one fuel to another. In some instances completely reliable ignition is obtained down to a quarter load. When firing pulverized coal with a volatile matter content less than 25 percent, it may be necessary to activate the igniters even at high loads.

Excess Air Pulverized coal requires more excess air for satisfactory combustion than either oil or natural gas. An acceptable quantity of unburned combustible coal is usually obtained with 15 percent excess air at high loads. This allows for the normal maldistribution of primary-air, coal, and secondary air.

LIQUID FUELS

Characteristics

Fuel Oil. It is common practice in refining petroleum to produce fuel oils complying with several specifications prepared by the ASTM and adopted as a commercial standard by the National Bureau of Standards (Table 6.16).

Fuel oils are graded according to gravity and viscosity, the lightest being No. 1 and the heaviest No. 6. Grades 5 and 6 generally require heating for satisfactory pumping and burning. The range of analyses and heating values of the several grades of fuel oils are given in Table 6.17.

The gross heating value, density, and specific gravity of various fuel oils for a range of API gravities are shown in Fig. 6.24. The abscissa on this figure is the API (American Petroleum Institute) gravity and sp gr at 60–60°F (15°C) represents the ratio of oil density at 60°F (15°C) to water density also at 60°F (15°C).

Fuel oils are generally sold on a volume basis, with 60°F as the base temperature. Correction factors are given in Fig. 6.25 for converting known volumes at other temperatures to the 60°F (15°C) standard base. This correction is also dependent on the API gravity range as illustrated by the three parametric curves of Fig. 6.24.

Since equipment for handling and, especially, burning of fuel oil is usually designed for a maximum oil viscosity, it is necessary to know the viscosity characteristics of the fuel oil to be used. If the viscosities of heavy oils are known at two temperatures, viscosities at other temperatures can be closely predicted by a linear interpolation between these two values located on the standard ASTM chart of Fig. 6.26. Viscosities of light oils at various temperatures within the region designated as No. 2 fuel oil can be found by drawing a line parallel to the No. 2 boundary lines through the point of only one known viscosity and temperature.

Shale Oil Oil shale is not actually a shale nor does it contain oil. It is generally defined as a fine-grained, compact, sedimentary rock containing an organic material called *kerogen.* Heating the oil shale to about 875°F (468°C) decomposes this material to produce shale oil.

Pitch and Tar The liquid and semiliquid residues from the distillation of petroleum and coal are known as pitch and tar. Most of these residues are suitable for use as boiler fuels. Some handle as easily and burn as readily as does kerosene, whereas others give considerable trouble.

Coal Oil Mixture (COM) Pulverized coal can be mixed with oil and kept in suspension by agitation, recirculation, or by use of additives. Transportation by pipeline is possible, but truck, train, or barge shipment may be more economical. COM is burned with equipment similar to that used for oil firing.

TABLE 6.16 ASTM Standard Specifications for Fuel Oils[a]

Grade of fuel oil[b]	Flash point, °F (°C)	Pour point, °F (°C)	Water and sediment, % by vol	Carbon residue on 10% bottoms, %	Ash, % by wt	Distillation temperatures, °F (°C)			Saybolt viscosity, s				Kinematic viscosity, centistokes				Gravity, deg API	Copper strip corrosion
						10% point	90% point		Universal at 100°F (38°C)		Furol at 122°F (50°C)		At 100°F (38°C)		At 122°F (50°C)			
	Min	Max	Max	Max	Max	Max	Min	Max	Min	Max	Min	Max	Min	Max	Min	Max	Min	Max
No. 1	100 or legal (38)	0	trace	0.15	—	420 (215)	—	550 (288)	—	—	—	—	1.4	2.2	—	—	35	No. 3
No. 2	100 or legal (38)	20[c] (−7)	0.10	0.35	—	[d]	540[c] (282)	640 (338)	(32.6)[f]	(37.93)	—	—	2.0[e]	3.6	—	—	30	—
No. 4	130 or legal (55)	20 (−7)	0.50	—	0.10	—	—	—	45	125	—	—	(5.8)	(26.4)	—	—	—	—
No. 5 (Light)	130 or legal (55)	—	1.00	—	0.10	—	—	—	150	300	—	—	(32)	(65)	—	—	—	—
No. 5 (Heavy)	130 or legal (55)	—	1.00	—	0.10	—	—	—	350	750	(23)	(40)	(75)	(162)	(42)	(81)	—	—
No. 6	150 (65)	—	2.00[g]	—	—	—	—	—	(900)	(9000)	45	300	—	—	(92)	(638)	—	—

No. 1: A distillate oil intended for vaporizing pot-type burners and other burners requiring this grade of fuel.

No. 2: A distillate oil for general purpose domestic heating for use in burners not requiring No. 1 fuel oil.

No. 4: Preheating not usually required for handling or burning.

No. 5 (Light): Preheating may be required depending on climate and equipment.

No. 5 (Heavy): Preheating may be required for burning and, in cold climates, may be required for handling.

No. 6: Preheating required for burning and handling.

[a] Recognizing the necessity for low-sulfur fuel oils used in connection with heat treatment, nonferrous metal, glass, and ceramic furnaces and other special uses, a sulfur requirement may be specified in accordance with the following table:

Grade of Fuel Oil	Sulfur, Max, %
No. 1	0.5
No. 2	0.7
No. 4	No limit
No. 5	No limit
No. 6	No limit

Other sulfur limits may be specified only by mutual agreement between the purchaser and the seller.

[b] It is the intent of these classifications that failure to meet any requirement of a given grade does not automatically place an oil in the next lower grade unless in fact it meets all requirements of the lower grade.

[c] Lower or higher pour points may be specified whenever required by conditions of storage or use.

[d] The 10% distillation temperature point may be specified at 440°F (226°C) maximum for use in other than atomizing burners.

[e] When pour point less than 0°F is specified, the minimum viscosity shall be 1.8 cs (32.0 s, Saybolt Universal) and the minimum 90% point shall be waived.

[f] Viscosity values in parentheses are for information only and not necessarily limiting.

[g] The amount of water by distillation plus the sediment by extraction shall not exceed 2.00%. The amount of sediment by extraction shall not exceed 0.50%. A deduction in quantity shall be made for all water and sediment in excess of 1.0%.

Source: ASTM D 396.

TABLE 6.17 Range of Analyses of Fuel Oils

Characteristic	Grade of fuel oil				
	No. 1	No. 2	No. 4	No. 5	No. 6
Weight, %					
Sulfur	0.01–0.5	0.05–1.0	0.2–2.0	0.5–3.0	0.7–3.5
Hydrogen	13.3–14.1	11.8–13.9	(10.6–13.0)*	(10.5–12.0)*	(9.5–12.0)*
Carbon	85.9–86.7	86.1–88.2	(86.5–89.2)*	(86.5–89.2)*	(86.5–90.2)*
Nitrogen	Nil–0.1	Nil–0.1	—	—	—
Oxygen	—	—	—	—	—
Ash	—	—	0–0.1	0–0.1	0.01–0.05
Gravity					
Deg API	40–44	28–40	15–30	14–22	7–22
Specific,	0.825–0.806	0.887–0.825	0.966–0.876	0.972–0.922	1.022–0.922
lb/gal	6.87–6.71	7.39–6.87	8.04–7.30	8.10–7.68	8.51–7.68
Pour point, °F	0 to –50	0 to –40	–10 to +50	–10 to +80	+15 to +85
Viscosity					
Centistokes @ 100°F	1.4–2.2	1.9–3.0	10.5–65	65–200	260–750
SSU @ 100°F	—	32–38	60–300	—	—
SSF @ 122°F	—	—	—	20–40	45–300
Water & sediment, vol %	—	0–0.1	Tr to 1.0	0.05–1.0	0.05–2.0
Heating value, Btu/lb, gross (calculated)	19,670–19,860	19,170–19,750	18,280–19,400	18,100–19,020	17,410–18,990

* Estimated.

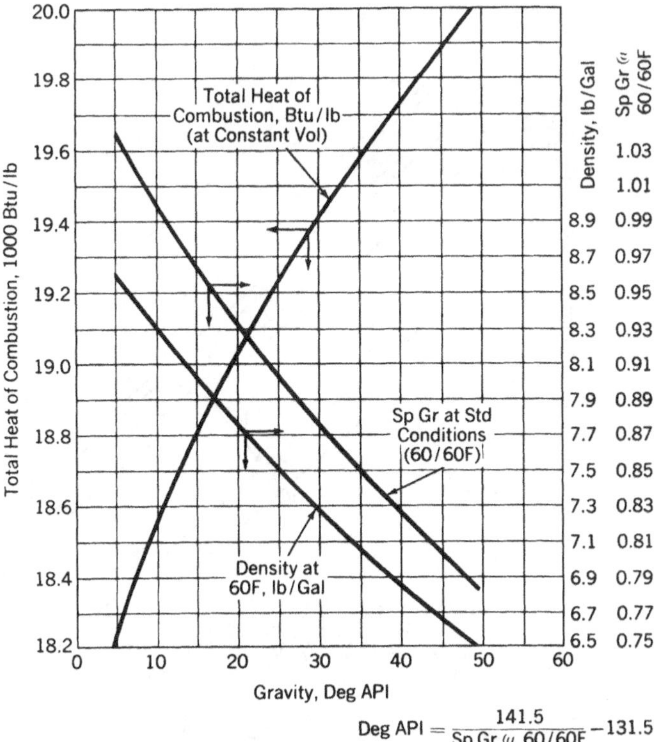

$$\text{Deg API} = \frac{141.5}{\text{Sp Gr @ 60/60F}} - 131.5$$

FIGURE 6.24 Heating value, weight (pounds per gallon), and specific gravity of fuel oil for a range of API gravities.

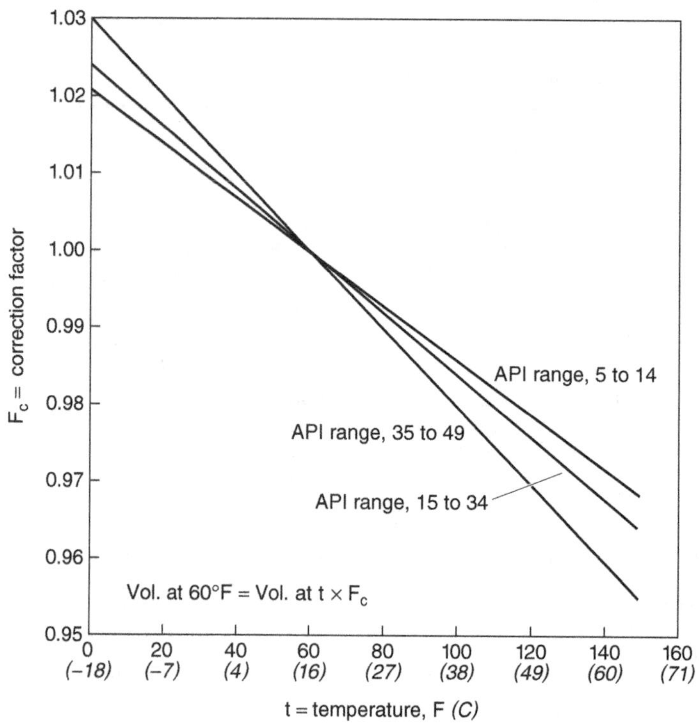

FIGURE 6.25 Oil volume-temperature correction factors.

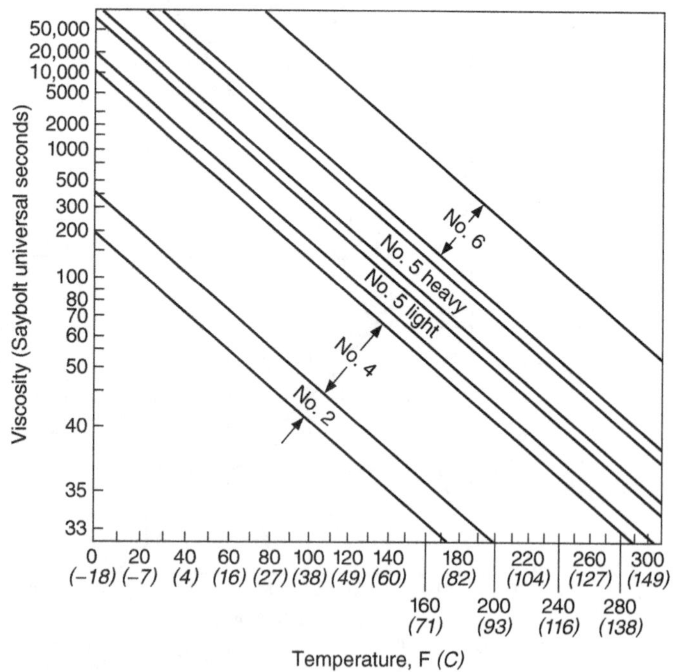

FIGURE 6.26 Approximate viscosity of fuel oil at various temperatures.

Preparation and Utilization of Oil

Preparation. Most petroleum is refined to some extent before use, although small amounts are burned without processing. The refining of crude oil yields a number of products having many different applications. Those used as fuel include gasoline, distillate fuel, residual fuel, jet fuels, still gas, liquefied gases, kerosene, and petroleum coke.

Transportation, Handling, and Storage. A worldwide system for distributing petroleum (and its products) has been developed because petroleum has a high calorific value per unit volume, is in easily handled liquid form, and has varied applications.

The serious hazard inherent in possible oil-storage-tank failure is overcome by storing oil in underground tanks or by protecting surface tanks by surrounding them with cofferdams of sufficient capacity to hold the entire contents of any tank so protected. The National Fire Protection Association has prepared a standard set of rules for the storage and handling of oils.

To facilitate pumping heavy fuel oil, heating equipment is usually provided in storage and transportation facilities. Storage tanks, piping, and heaters for heavy oils must be cleaned at intervals because of fouling or sludge formation. Various commercial compounds are helpful in reducing sludge.

Oil-Burning Equipment

The burner is the principal component of equipment for firing oil. Burners are normally located in the vertical walls of the furnaces.

Oil Burners. The most frequently used burners are the circular type. Figure 6.27 shows a single circular-register burner for gas and oil firing.

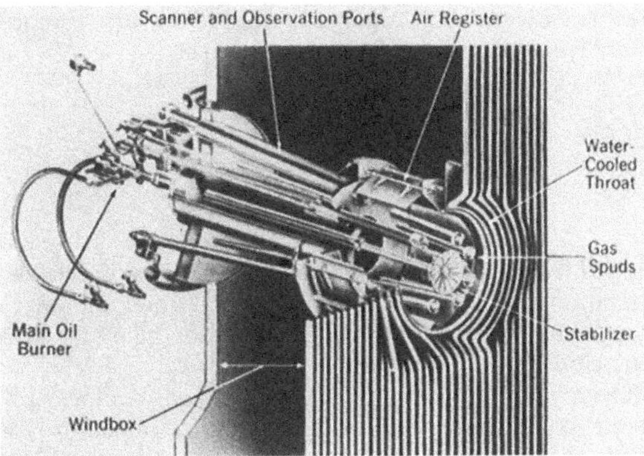

FIGURE 6.27 Circular register burner with water-cooled throat for oil and gas firing.

The maximum capacity of the individual circular burner ranges up to 300×10^6 Btu/h (316×10^4 kJ/h). In circular burners the tangential "doors" built into the air register provide the turbulence necessary to mix the fuel and air and control flame shape. Although the fuel mixture as introduced to the burner is fairly dense in the center, the direction and velocity of the air, plus dispersion of the fuel, completely and thoroughly mix it with the combustion air.

In order to burn fuel oil at the high rates demanded of modern boiler units it is necessary that the oil be *atomized,* i.e., dispersed into the furnace as a fine mist, to expose a large amount of oil particle surface to the air and ensure prompt ignition and rapid combustion.

For proper atomization, oil of grades heavier than No. 2 must be heated to reduce viscosity to 135 to 150 SSU (Saybolt seconds universal). Steam or electric heaters are required to raise the oil temperature to the required level, i.e., approximately 135°F (57°C) for No. 4 oil, 185°F (74°C) for No. 5 oil, and 200 to 220°F (93 to 104°C) for No. 6 oil.

Steam or Air Atomizers. Steam atomizers are the most widely used. In general they operate on the principle of producing a steam-fuel emulsion which, when released into the furnace, atomizes the oil through the rapid expansion of the steam. The atomizing steam must be dry because entrained moisture causes pulsations which can lead to loss of ignition. Where steam is not available, moisture-free compressed air can be substituted.

Steam atomizers are available in sizes up to 300×10^6 Btu/h input—about 16,500 lb (7500 kg) of oil per hour. Oil pressure is much lower than that required for mechanical atomizers. Maximum oil pressure can be as much as 300 lb/in^2 (2040 kPa) and maximum steam pressure 150 lb/in^2 (1020 kPa). The steam atomizer performs more efficiently over a wider load range than other types. It normally atomizes the fuel properly down to 20 percent of rated capacity.

A disadvantage of the steam atomizer is its consumption of steam. A good steam atomizer can operate with a steam consumption as low as 0.02 lb of steam per pound of fuel oil at maximum atomizer capacity.

Mechanical Atomizers. In mechanical atomizers the pressure of the fuel itself is used as the means for atomization.

The return-flow atomizer is used in many units where the use of atomizing steam is objectionable or impractical. The oil pressure required at the atomizer for maximum capacity ranges from 600 to 1000 lb/in^2 (4080 to 6700 kPa), depending on capacity, load range, and fuel. Mechanical atomizers are available in sizes up to 180×10^6 Btu/h (190×10^6 kJ) input—about 10,000 lb (4500 kg) of oil per hour.

Excess Air. It is necessary to supply more than the theoretical quantity of air to ensure complete combustion of the fuel in the furnace. The amount of excess air provided should be just enough to burn the fuel completely in order to minimize the sensible heat loss in the stack gases. The excess air normally required for oil firing, expressed as percent of theoretical air, is generally between 5 and 7 percent.

GASEOUS FUELS

Characteristics

Natural Gas. Of all chemical fuels, natural gas is considered the least troublesome for steam generation. It is piped directly to the consumer, eliminating the need for storage at the consumer's plant. It is substantially free of ash and mixes intimately with air to provide complete combustion at low excess air without smoke.

The high hydrogen content of natural gas compared with that of oil or coal results in the production of more water vapor in the combustion gases, thus causing a correspondingly lower efficiency of the steam-generating equipment. This can be taken into account in the design of the equipment and evaluated in comparing the cost of gas with other fuels.

Analyses of natural gas from several U.S. fields are given in Table 6.18.

Gaseous Fuels from Coal. A number of gaseous fuels are derived from coal either as by-products or from coal gasification processes. Table 6.19 lists selected analyses of these gases according to the various types described in the following paragraphs.

Coke-Oven Gas. A considerable portion of coal is converted to gases or vapors in the production of coke. The noncondensable portion is called *coke-oven gas.* Constituents depend on the nature of the coal and coking process used (Table 6.19).

Blast-Furnace Gas. The gas discharged from steel-mill blast furnaces is used at the mills in heating furnaces, in gas engines, and for steam generation. This gas is quite variable in quality but generally has a high carbon monoxide content and low heating value (Table 6.19).

TABLE 6.18 Selected Samples of Natural Gas from U.S. Fields

Sample no.	1	2	3	4	5
Source:	Pa.	S.C.	Ohio	La.	Ok.
Analyses:					
Constituents, % by vol					
Hydrogen, H_2	—	—	1.82	—	—
Methane, CH_4	83.40	84.00	93.33	90.00	84.10
Ethylene, C_2H_4	—	0.25	—	—	—
Ethane, C_2H_6	15.80	14.80	—	5.00	6.70
Carbon monoxide, CO	—	—	0.45	—	—
Carbon dioxide, CO_2	—	0.70	0.22	—	0.80
Nitrogen, N_2	0.80	0.50	3.40	5.00	8.40
Oxygen, O_2	—	—	0.35	—	—
Hydrogen sulfide, H_2S	—	—	0.18	—	—
Ultimate, % by wt					
Sulfur, S	—	—	0.34	—	—
Hydrogen, H_2	23.53	23.30	23.20	22.68	20.85
Carbon, C	75.25	74.72	69.12	69.26	64.84
Nitrogen, N_2	1.22	0.76	5.76	8.06	12.90
Oxygen, O_2	—	1.22	1.58	—	1.41
Specific gravity (rel to air)	0.636	0.636	0.567	0.600	0.630
HHV					
Btu/ft^3 at 60°F and 30 in Hg	1,129	1,116	964	1,022	974
(kJ/m^3 at 16°C and 102 kPa)	(42,065)	(41,581)	(35,918)	(38,079)	(36,290)
Btu/lb (kJ/kg) of fuel	23,170 (53,893)	22,904 (53,275)	22,077 (51,351)	21,824 (50,763)	20,160 (46,892)

TABLE 6.19 Selected Analyses of Gaseous Fuels Derived from Coal

Analysis no.	Coke oven gas 1	Blast furnace gas 2	Carbureted water gas 3	Producer gas 4
Analyses, % by volume				
Hydrogen, H_2	47.9	2.4	34.0	14.0
Methane, CH_4	33.9	0.1	15.5	3.0
Ethylene, C_2H_4	5.2	—	4.7	—
Carbon monoxide, CO	6.1	23.3	32.0	27.0
Carbon dioxide, CO_2	2.6	14.4	4.3	4.5
Nitrogen, N_2	3.7	56.4	6.5	50.9
Oxygen, O_2	0.6	—	0.7	0.6
Benzene, C_6H_6	—	—	2.3	—
Water, H_2O	—	3.4	—	—
Specific gravity (relative to air)	0.413	1.015	0.666	0.857
HHV—Btu/ft^3 (kJ/m^3)				
at 60°F (16°C) and	590	—	534	163
30 in Hg (102 kPa)	(21,983)	—	(19,896)	(6,073)
at 80°F (27°C) and	—	83.8	—	—
30 in Hg (102 kPa)	(3,122)	(3,122)		

Water Gas. The gas produced by passing steam through a bed of hot coke is known as *water gas.* Carbon in the coke combines with the steam to form hydrogen and carbon monoxide.

Water gas is often enriched by passing the gas through a checkerwork of hot bricks sprayed with oil, which in turn is cracked to a gas by the heat. Such enriched water gas is called *carburetted water gas* (Table 6.19).

Producer Gas. When coal or coke is burned with a deficiency of air and a controlled amount of moisture (steam), a product known as *producer gas* is obtained. This gas, after removal of entrained ash and sulfur compounds, is used near its source because of its low heating value (Table 6.19).

"Synthetic Natural Gas." "Synthetic natural gas" is made from coal by one of the many gasification processes resulting in a low-Btu gas which is then methanated.

Carbon Monoxide. In the petroleum industry, the efficient operation of a fluid-catalytic-cracking unit produces gases rich in carbon monoxide. To reclaim the thermal energy represented by these gases, the fluid-catalytic-cracking unit can be designed to include a CO boiler that uses the CO as fuel to generate steam for use in the process.

Processing and Utilization of Natural Gas

Processing. Propane and butane are often separated from the lighter gases and are widely used as bottled gas. They are distributed and stored liquefied under pressure. Natural gas containing excessive amounts of hydrogen sulfide is commonly known as "sour" gas. The sulfur is removed before distribution.

Transportation, Handling, and Storage. A pipeline is an economical means for transporting natural gas overland. Tankers are employed for overseas transportation of natural gas. The gas is liquefied under pressure (liquefied natural gas, LNG) for ease of transportation.

Gas-Burning Equipment

Natural-Gas Burners. An example is the variable-mix multispud gas element (Fig. 2.27) for use with circular-type burners. Simultaneous firing of natural gas and oil in the same burner is possible.

To provide safe operation, ignition of a gas burner should remain close to the burner wall throughout the full range of allowable gas pressures, not only with normal airflows, but also with much more airflow through the burner than is theoretically required.

Burners for Other Gases. Many industrial applications utilize coke-oven gas, blast-furnace gas, refinery gas, or other industrial by-product gases. With these gases, the heat release per unit volume of fuel gas may be very different from that of natural gas. Hence, gas elements must be designed to accommodate the particular characteristics of the gas to be burned. Other special problems may be introduced by the presence of impurities in industrial gases, such as sulfur in coke-oven gas and entrained dust in blast-furnace gas.

Lighters (Igniters) and Pilots. Usually the ignition device is a spark device energized only long enough to ensure that the main flame is self-sustaining. Although ignition should be self-sustaining within 1 or 2 s after the fuel reaches the combustion air, in a fully automated burner it is customary to allow 10 to 15 s "trial for ignition" so that the fuel can reach the burner after the fuel shutoff valve on the burner is opened. There are applications where a continuously burning lighter or pilot is needed. This is particularly true in the use of a by-product fuel, such as gas from a chemical process.

Excess Air. It is possible to operate most units with as little as 5 to 7 percent excess air at the furnace outlet at full load, and some boilers have operated with less than 2.5 percent excess air without excessive loss of unburned combustibles.

At partial loads on all units, regardless of the fuel fired, it is necessary to increase the excess air as the load is reduced. Burner dampers are designed not to close tightly in order to permit the air to protect the idle burner(s) from overheating by radiant heat from nearby operating burners.

EFFECT OF COAL AND MULTIFUEL FIRING ON INDUSTRIAL BOILER DESIGN*

Coal is a complex fuel, and it is necessary to establish its source as well as its physical and chemical characteristics before considering boiler size or selecting equipment. Adding wood, oil, or gas to obtain combined fuel firing further affects the design criteria employed to properly design the boiler furnace.

A properly designed furnace performs two functions: (1) burning the fuel completely and (2) cooling the products of combustion sufficiently so that the convection passes of the boiler unit are maintained in a satisfactory condition of cleanliness with a reasonable amount of soot blowing. When the average temperature of gas leaving a coal-fired furnace is too high, the ash particles are molten or sticky, a condition which leads to an excessive need for cleaning the ash deposits from the upper furnace and the high-temperature zones of the convection passes.

The fouling and slagging classification of the coal, as characterized by the ash analysis, establishes furnace sizing, spacing and arrangement of tubes in convection passes, and placement of soot blowers for the furnace walls and convection passes. The slagging characteristic governs burner clearances, heat input per unit of furnace cross-sectional area, and the number of furnace-wall soot blowers provided to control buildup of slag on the furnace walls. The fouling characteristic sets a relationship of gas temperature entering the tube bank for a given side spacing of tubes and tube bank depth according to soot-blower cleaning radius.

Effect of Design Parameters on Boiler Sizing

The substantial effect on boiler sizing of the coal-ash classification is made clear by comparison of units sized for three types of coal. In general, an increased fouling tendency requires an increase in the furnace surface to lower the gas temperature entering the superheater, and an increased slagging potential of a coal causes furnaces designed for western coals to be substantially more conservative than those designed to fire eastern bituminous coals.

Figure 6.28 shows three boiler sizes. Boiler A is designed to fire a West Virginia bituminous coal classified as having a low slagging and low fouling potential. The slightly larger boiler B is designed to fire an Alabama bituminous coal classified as having a medium slagging and a medium fouling potential. Boiler C is the largest, designed to fire a Wyoming subbituminous lignitic ash coal classified as having a severe slagging and medium fouling potential.

The difference in the size of the unit for the West Virginia coal and the unit for the Alabama coal can be attributed primarily to the difference in fouling potential. The furnace height has been increased substantially to reduce the temperature of the gas entering the superheater, thereby reducing the fouling problem. The increased furnace depth and height of the Wyoming coal unit over the dimensions of the unit sized for the Alabama coal are primarily attributed to the difference in the slagging potential of the ash. The lignitic ash results in

* A portion of the material in this section has been adapted with permission from the paper of the same title by J. D. Blue, J. L. Clement, and V. L. Smith that was presented at the TAPPI Engineering Meeting, October, 1974.

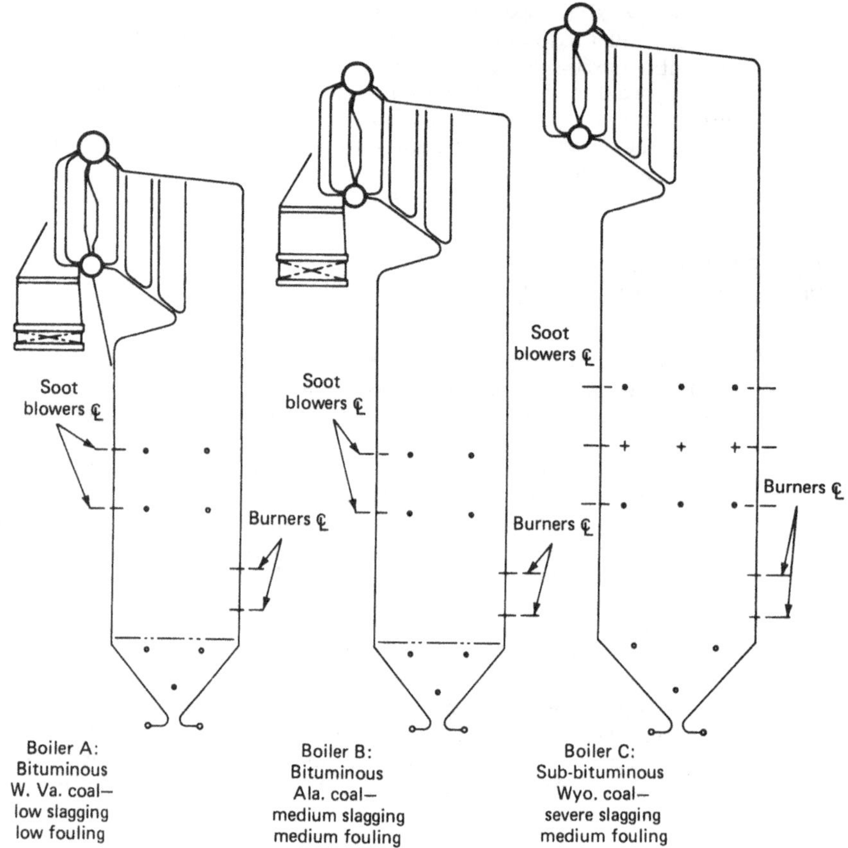

FIGURE 6.28 Influence of ash characteristics on furnace size.

considerably more slag buildups on the furnace walls, thereby decreasing heat transfer and requiring considerably more furnace surface. The furnace depth has been increased to control slagging by reducing the input per plan area.

Oil/Gas Capability

A boiler designed to use coal and/or wood as the primary fuel is frequently required to provide steam-generating capability when fired with oil and/or natural gas.

The design of a boiler for alternative fuels generally requires a compromise. For example, the steam temperature resulting from the effects of a fixed furnace and superheater surface on various fuels is illustrated in Fig. 6.29. The superheater surface in this illustration is designed to provide full-rated steam temperature on oil. The operation with gas or bark as a fuel produces the desired steam temperature at 45 and 55 percent of full load rating, respectively. The steam temperatures on these fuels would have to be moderated at higher loads to control the terminal temperature. The extent of alloy tubing used in fabrication of the superheater is governed by gas firing where the highest potential steam temperatures exist.

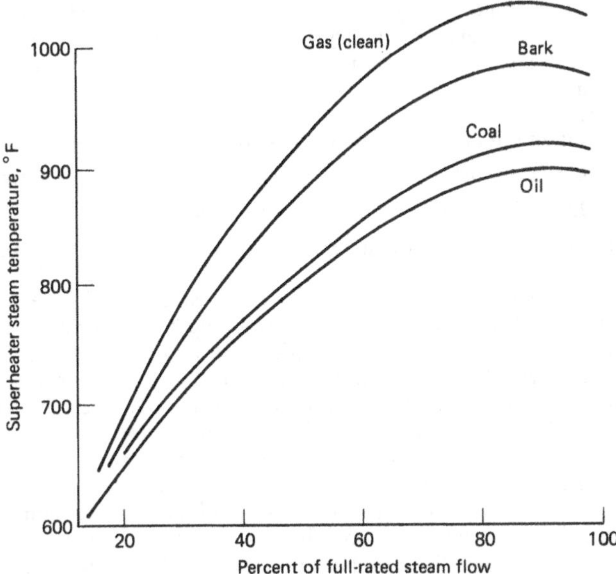

FIGURE 6.29 Effect of fuel on uncontrolled superheater steam temperature.

Stoker or Pulverized Coal

A decision to design for stoker or pulverized-coal firing is generally one of economics. A comparison of boilers designed to generate equal steam rates is characterized as follows:

1. The furnace volume is less for stoker firing because burning lower in the furnace gives more effective use of total furnace surface.
2. The furnace width is less for pulverized-coal firing.
3. Pulverized-coal-fired boilers are more responsive to load change.
4. Stoker firing requires a lower operating horsepower.
5. Thermal efficiency is higher for pulverized coal (PC) by virtue of the lower unburned-carbon loss (UCL).

Furnace Maintenance

Preventive maintenance includes a policy of operating equipment properly and within its range of capability and keeping the equipment clean and in prime operating condition. This is verified by instrumentation and in-service observations. Preventive maintenance also includes regularly scheduled outages to make those inspections which cannot be made during operation and to perform necessary repairs.

In-Service Maintenance

Safety. Primary emphasis should be on safe operation, the avoidance of conditions that could result in explosive mixtures of fuel in the furnace or other parts of the setting, and the protection of furnace pressure parts to prevent excessive thermal stresses or overheating that would result in failure.

Furnace and Setting. The prevention of furnace explosions deserves top priority because of potential personnel injury, the high cost of repairs, and the effect of outage time on the industrial processes. Four items are of major importance in the prevention of furnace explosions:

1. Optimum operating procedures and operator training
2. Optimum burner observation with prompt detection of flame failure
3. Detection of unburned combustibles in the flue gas
4. Positive, immediate indication of fuel-air relation at the burners

The majority of furnace explosions result from failure to detect a loss of ignition, even though other indications such as dropping boiler pressure, steam temperature, and exit-gas temperature show that fuel is either not being burned or is being burned incompletely. This emphasizes the fact that *nothing takes the place of seeing the fire.* Therefore, a reliable remote indication of positive ignition on all burners must be relayed to the operator at the central control point. A *combustibles alarm* to indicate the presence of unburned combustibles in the flue gas is considered a good backup for flame observation.

The measurement of furnace pressure on both pressurized and suction-type units is required to assure that design pressures of the casing or containment are not exceeded. Some form of differential gas- or air-pressure measurement, in conjunction with time lapse, is required as an indication that the setting has been adequately purged prior to firing. This is a necessary operating guide to assure a proper fuel-air ratio at the burners.

Variables in Efficiency Losses. There are two variables in dry gas loss: stack temperature of the gas and weight of the gas leaving the unit. Stack temperature varies with the degree of deposit on the heat-absorbing surfaces throughout the unit, and it varies with the amount of excess combustion air. The effect of excess air is twofold: (1) It increases the gas weight, and (2) it raises the exit-gas temperature. Both effects increase dry gas loss and thereby reduce efficiency. A rough rule of thumb gives an approximate 1 percent reduction in efficiency for about a 40°F (22°C) increase in stack-gas temperature on coal-fired installations.

In coal-fired units, unburned-combustible loss includes the unburned constituents in the ash-pit refuse and in the flue dust.

Monitoring Efficiency. Continuous monitoring of flue-gas temperatures and flue-gas oxygen content by a regularly calibrated recorder or indicator and by periodic checks on combustibles in the refuse will indicate if original efficiencies are being maintained. If conditions vary from the established performance base, corrective adjustments or maintenance steps should be taken.

High exit-gas temperatures and high draft losses with normal excess air indicate dirty heat-absorbing surfaces and the need for soot blowing.

High excess air normally increases exit-gas temperatures and draft losses and indicates the need for an adjustment to the fuel/air ratio. The high excess air may, however, be caused by excessive casing leaks or by cooling air, sealing air, or air-heater leaks.

High combustibles in the refuse indicates a need for adjustments or maintenance of fuel-preparation and -burning equipment.

Water-Side Cleanliness. One of the best preventive steps that can be taken to assure safe, dependable operation is the maintenance of boiler-water conditions that will ensure against any internal tube deposits that could cause overheating and failure of furnace tubes.

Outage Maintenance. Outages for preventive maintenance should be scheduled as required to prevent equipment failures.

Water-Side Cleaning. Chemical or acid cleaning is the quickest and most satisfactory method for the removal of water-side deposits. It is, however, extremely important to use a procedure of known reliability under careful control.

Gas-Side Inspection. During outage periods, the furnace should be thoroughly inspected; the objectives should be:

1. To detect any possible signs of overheating of tubes; furnace wall tubes should be examined for swelling, blistering, or warping
2. To discover any possible signs of erosion or corrosion
3. To detect any misalignment of tubes from warpage
4. To locate any deposits of ash or slag, not removed by sootblowing, that interfere with heat transfer or free gas flow in the furnace
5. To determine the condition of fuel-preparation and -burning equipment, particularly if routine sampling of ash has indicated the presence of increasing amounts of unburned combustible material
6. To determine the condition of any refractory exposed to furnace gases, such as burner throats or furnace walls

Gas-Side Cleaning. When ash deposits contain an appreciable amount of sulfur, they should be removed prior to any extended outage, since these deposits can absorb ambient moisture to form sulfuric acid that will corrode furnace pressure parts.

CHAPTER 6.3
ELECTRIC GENERATORS

Clemens M. Thoennes
Sales Programs Development
General Electric Company
Schenectady, New York

GLOSSARY

Armature The member of a rotating electric machine in which an alternating voltage is generated by virtue of relative motion with respect to a magnetic-flux field.

Class F ANSI and IEC temperature standard of 155°C.

Core An element of magnetic material, serving as part of a path for magnetic flux. In rotating machines, this is frequently part of the stator, a hollow cylinder of laminated magnetic steel, slotted on the inner surface for the purpose of containing the armature windings.

Exciter The source of all or part of the dc field current for the excitation of an electric machine.

Field An insulated winding in rotating synchronous electric machinery whose purpose is the production of the main electromagnetic field of the machine.

Permeability A property of materials which expresses the relationship between magnetic induction and magnetizing force. An indication of the relative ability to conduct magnetic flux.

Power factor The ratio of the active power in watts to the apparent power in voltamperes. It is also the cosine of the phase angle between the voltage and current.

Reactance The imaginary part of impedance.

Reactive power The imaginary power required to magnetize the air gaps in motors and transformers in ac electric circuits.

Short-circuit ratio In synchronous electric machines, the ratio of the field current for rated open-circuit armature voltage and rated frequency to the field current for rated armature current on a sustained symmetrical short circuit at rated frequency.

Slip Difference between synchronous and actual speed, usually expressed as a percentage of synchronous speed.

Synchronous speed Speed in rpm (revolutions per minute) relating to ac frequency. It equals 120 times the frequency in hertz per number of poles.

INTRODUCTION

There are three basic types of rotating electric generators: synchronous ac, induction ac, and rotating dc.

Virtually all of the power generated by electric utilities and industrial turbine generators is supplied by synchronous ac generators. This type of generator includes an excitation system which is used to regulate the output voltage and power factor. The emphasis in this chapter will therefore be on synchronous ac generators.

Induction generators are squirrel-cage induction motors which are driven above synchronous speed. They do not have an excitation system and hence cannot control voltage or power factor. The system must supply the excitation. Induction generators are generally applied where relatively small waste energy or hydro potential exists; they are driven by a steam turbine, a gas expander, or a hydraulic turbine to recover the power in the energy stream. In these cases it is economical to adjust power factor and voltage on other larger synchronous generators in the system.

Rotating dc generators have been replaced almost entirely by static silicon rectifiers. The demand for rotating dc generators is limited to a few very special applications such as elevators and large excavators. No practical method has been developed for reducing the high maintenance associated with the commutators and brushes of dc generators.

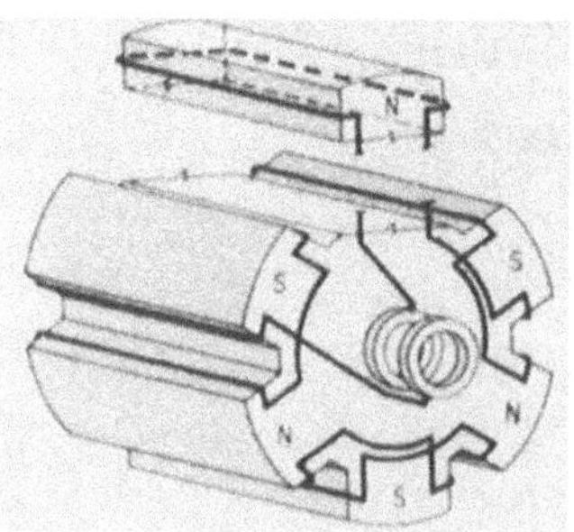

FIGURE 6.30 Simplified six-pole generator field.

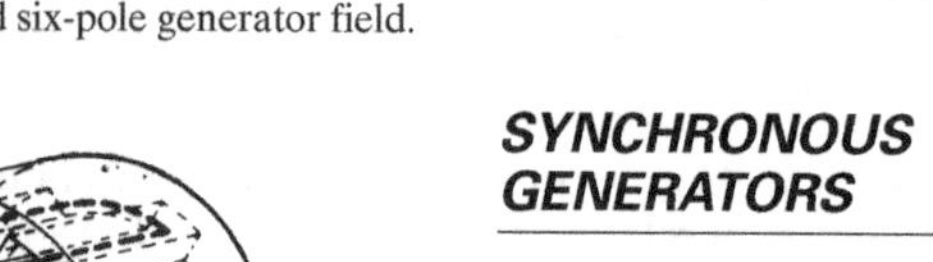

SYNCHRONOUS GENERATORS

The fundamental principle of operation of synchronous ac generators is that relative motion between a conductor and a magnetic field induces a voltage in the conductor. The magnitude of the voltage is proportional to the rate at which the conductor cuts lines of flux. The most common arrangement is with a cylindrical electromagnet rotating inside a stationary conductor assembly. The electromagnet is called the field and it is shown in simplified schematic form in Fig. 6.30. The conductors constitute the armature and are illustrated in Fig. 6.31. An external source of dc power is applied through the collector rings on the rotor. The flux strength and hence the induced voltage in the armature are regulated by the dc current and voltage supplied to the field. Alternating current is produced in the armature by the reversal of the magnetic field as north and south poles pass the individual conductors.

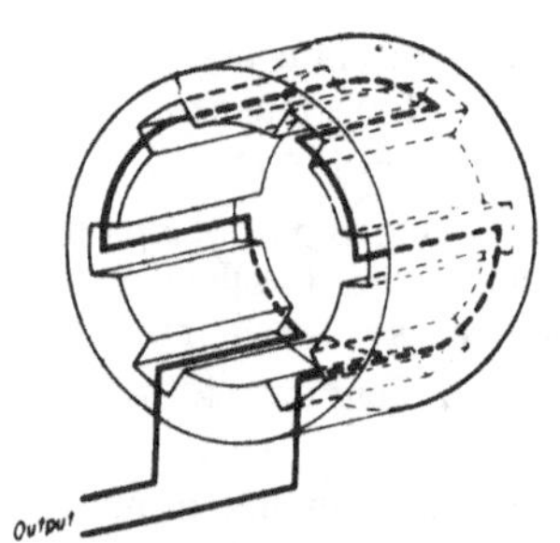

FIGURE 6.31 Simplified generator armature.

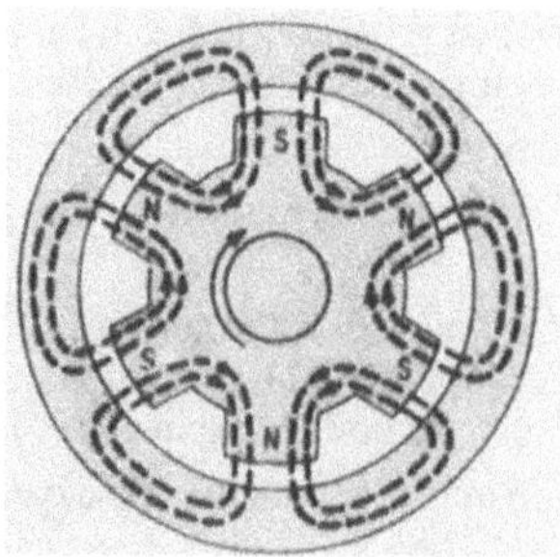

FIGURE 6.32 Generator magnetic circuit.

Lines of magnetic flux always form a closed circuit as shown in Fig. 6.32. Confining the flux field in materials with high permeability (low resistance to magnetic flux) intensifies the flux density. The permeability of certain steels is thousands of times greater than air. The flux density at the pole faces is proportional to the ampere turns on the poles and the combined permeability of all the materials in the circuit including the rotor core, the stator core, and the air gap.

The stator core is built up with steel laminations to provide both the high-permeability magnetic path and a high-resistance electric path to minimize induced voltage and inherent heat generation.

The simplified drawing of an armature winding in Fig. 6.31 shows a single phase only. All generators except those with very small ratings have three phases, each phase consisting of several conductors.

There are two parameters that limit the output of a generator:

1. *Flux density saturation.* As field-exciting current is increased, a point is reached where the flux density no longer increases because of iron saturation in the core. Normally the generator rating in kilovoltamperes (kVA) is near this flux saturation point.
2. *Temperature rise in the windings and insulation due to losses.* This includes losses due to excitation current in the field windings, ac current in the armature windings, the magnetic circuit, and any stray currents or magnetic fields that are generated.

Synchronous ac generators are classified by their method of cooling and excitation system. The design chosen is determined by the type of prime mover driving the generator, the power required, and the operating duty (e.g., continuous versus intermittent operation, clean versus dirty environment).

AIR-COOLED GENERATORS

Air-cooled generators are produced in two basic configurations: open ventilated (OV) and totally enclosed water-to-air-cooled (TEWAC). In the OV design, outside air is drawn directly from outside the unit through filters, passes through the generator, and is discharged outside the generator. In the TEWAC design, air is circulated within the generator and passes through frame-mounted, air-to-water heat exchangers.

The stator frame for an air-cooled generator is divided into an inner and an outer section, both of which mount on a single-base fabrication. The inner frame is a very simple structure, designed to support the stator core and winding, while providing some guidance to the air flow in the machine. The stator core, made from grain-oriented silicon steel for low loss and high permeability, is mounted rigidly in the inner frame. Isolation of the core vibration from the remainder of the structure is accomplished through the use of flexible pads between the feet on the inner frame and the base structure.

The outer frame is a simple fabricated enclosure that supports either the air inlets and silencers if the unit is open-ventilated (Fig. 6.33) or the roof and cooler enclosure if the unit is totally enclosed water-to-air-cooled. The outer frame further acts as an air guide to complete the ventilation paths, and as a soundproof enclosure to keep noise levels low. Since the rotor is pedestal mounted, the end shields are very simple structures. As with the inner frame, the outer frame is designed to be free of resonances in the range of operating frequencies.

The entire generator is mounted on a single fabricated base that supports the pedestals, the inner and outer frames, and the brush rigging or the exciter. The base contains piping for oil supplies, conduit for wiring, and a number of components associated with the main leads, such as lightning arresters and surge capacitors. The structural vibration of the base must be well away from any frequency of concern.

The stator winding is a conventional lap-wound design. The materials of modern generators are all designed and tested to provide reliable performance at Class F temperatures for the life of the machine. The stator bar copper is stranded and insulated with Class F

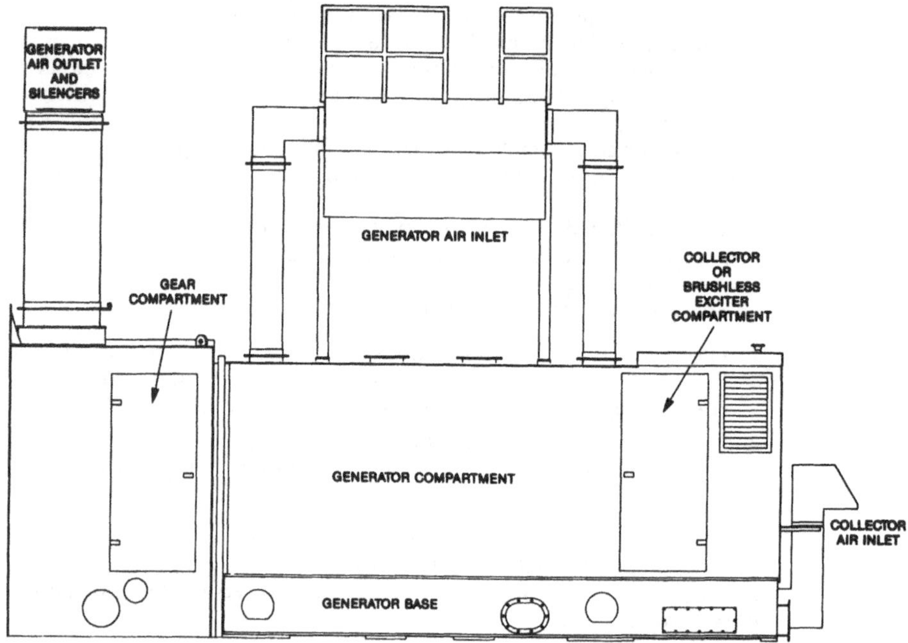

FIGURE 6.33 Generator packaging.

materials and is Roebelled for minimum losses. The exterior of the bar is taped with a conducting armor in the slot section, and a semiconducting grading system is applied to the end arms. In this way the bar is fully protected from the effects of high electrical voltage gradients.

The bars are secured in the slots with fillers and top-ripple springs to restrain the bars radially and with side-ripple springs to increase friction between the bar and the slot wall. The side-ripple springs are also conducting to ensure proper grounding of the bar surface.

The end-winding support system utilizes resin-impregnated glass roving ties (Fig. 6.34). All the strands are brazed together in a solid block and then the top and bottom bars are brazed together with solid copper plates. This provides a solid electrical connection and a rugged mechanical joint.

The rotor (Fig. 6.35) is a simple single-piece forging, pedestal mounted, with tilting-pad bearings for smooth operation. On smaller units, the rotor is sufficiently short so that the second critical speed is above running speed, thus simplifying balance. The retaining ring is nonmagnetic stainless steel with low losses and good stress-corrosion resistance. The rings are shrunk onto the rotor body, thus eliminating any risk of top burn breakage. The retaining ring is secured to the rotor body with a snap ring, a design that minimizes the stresses in the tip of the retaining ring.

High-efficiency, radial-flow fans are mounted on the centering ring at each end of the rotor. The fans provide cooling air for the stator winding and core. The rotor winding, which is a directly cooled radial flow design, is self-pumping and does not rely on the fan for air flow. The overall ventilation pattern is shown in Fig. 6.36.

The rotor winding fits in a rectangular slot and is retained by a full-length wedge on the shorter machines. When cross slots are required on longer rotors, several wedges are used in each slot. The rotor slot insulation, turn insulation, and other materials in contact with the winding are full Class F materials.

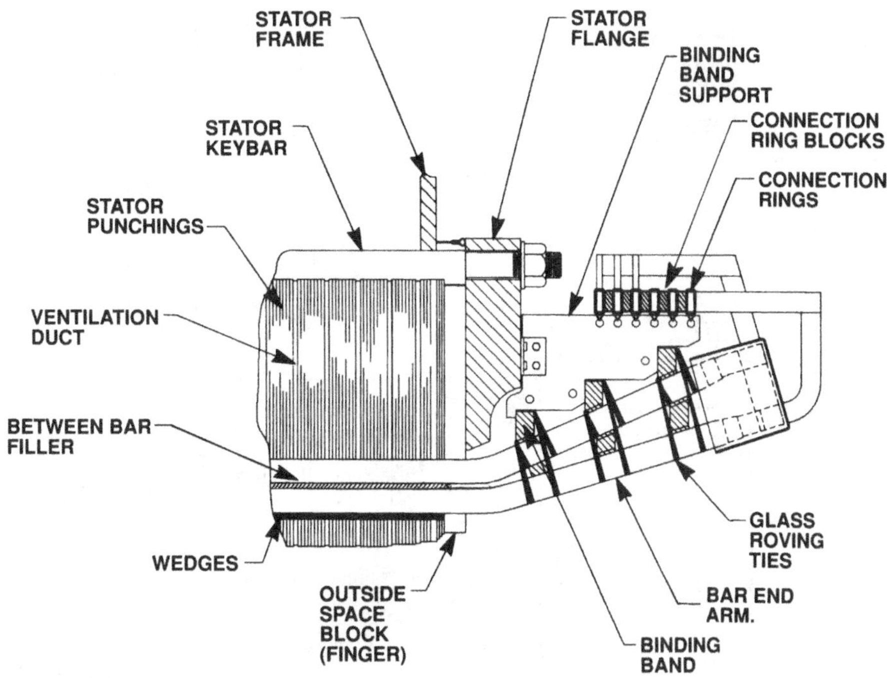

FIGURE 6.34 Stator end-winding section.

HYDROGEN-COOLED GENERATORS

To keep the size, weight, ability to ship, and cost of a larger generator within reason, a more optimal cooling medium needs to be used. Hence the development of hydrogen cooling.

How well the armature winding of a generator is cooled has a significant influence on the overall size of a synchronous generator. The cooling of the armature winding is dependent on a number of factors: cooling medium (air, hydrogen, water), insulation thickness, and overall electrical losses (I^2R + load loss). As Table 6.20 shows, relative heat removal capability improves from air to hydrogen, with increased hydrogen pressure, and even more significantly with the use of water cooling. Conventional hydrogen cooling can be utilized on generators rated approximately 300 MVA and below, while direct water cooling of armature windings is applied to units above 250 MVA. This division results from design optimization. While it is

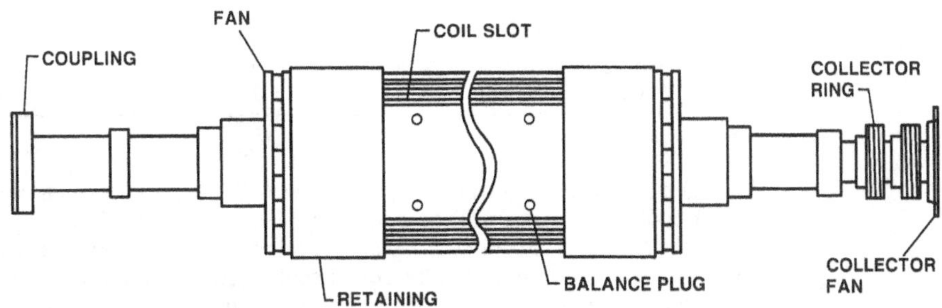

FIGURE 6.35 Generator field.

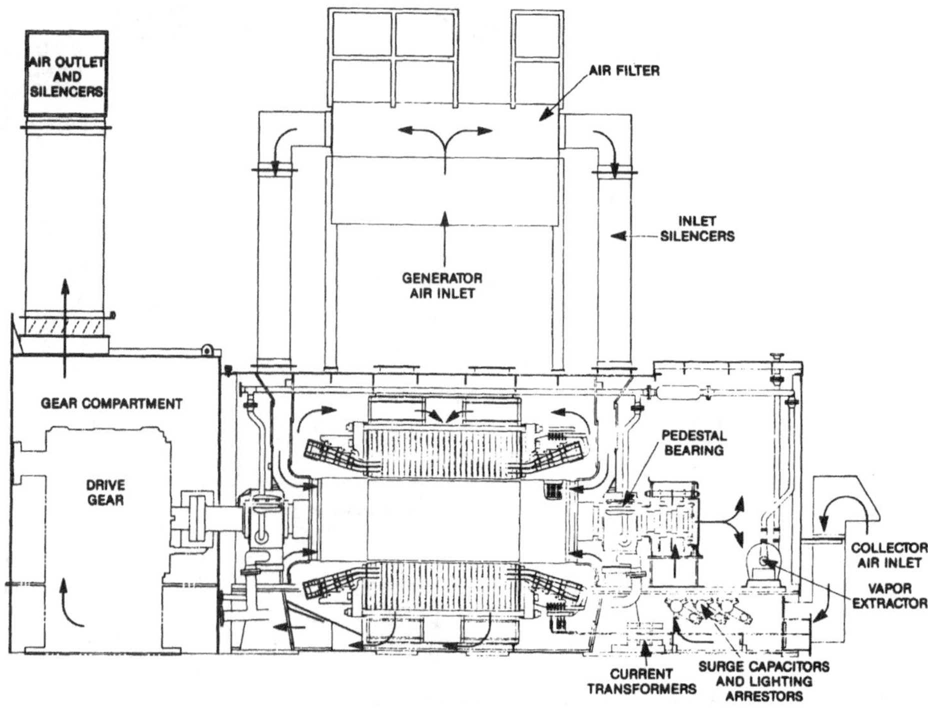

FIGURE 6.36 Generator cross section and ventilation paths.

possible to apply water cooling on machines rated below 250 MVA, the cost/performance benefit suffers. Water cooling adds manufacturing complexity, as well as requires the need for an auxiliary water-cooling and deionizing skid, plus associated piping, control, and protection features. At higher ratings, the cost of this complexity is offset by the advantage of producing a generator of significantly smaller size than a comparable conventionally cooled generator.

TABLE 6.20 Air, Hydrogen, Water Heat Removal Comparison

Fluid	Relative specific heat	Relative density	Relative practical vol. flow	Approx. rel. heat removal ability
Air	1.0	1.0	1.0	1.0
Hydrogen, 30 lb	14.36	0.21	1.0	3.0
Hydrogen, 45 lb	14.36	0.26	1.0	4.0
Water	4.16	1000.0	0.012	50.0

Hydrogen-cooled generator construction (Fig. 6.37), except for the frame, is very similar to that of air-cooled generators. Because of the need to contain 30 to 45 psig (210–310 bars) hydrogen, the stator frame uses thick plate cylindrical construction. End shields are appropriately more rugged, and contain a hydrogen seal system to minimize leakage. Conventional hydrogen cooling, while available for generators rated below 100 MVA, is most often applied to steam-turbine-driven units above 100 MVA, as well as gas turbines above 100 MVA.

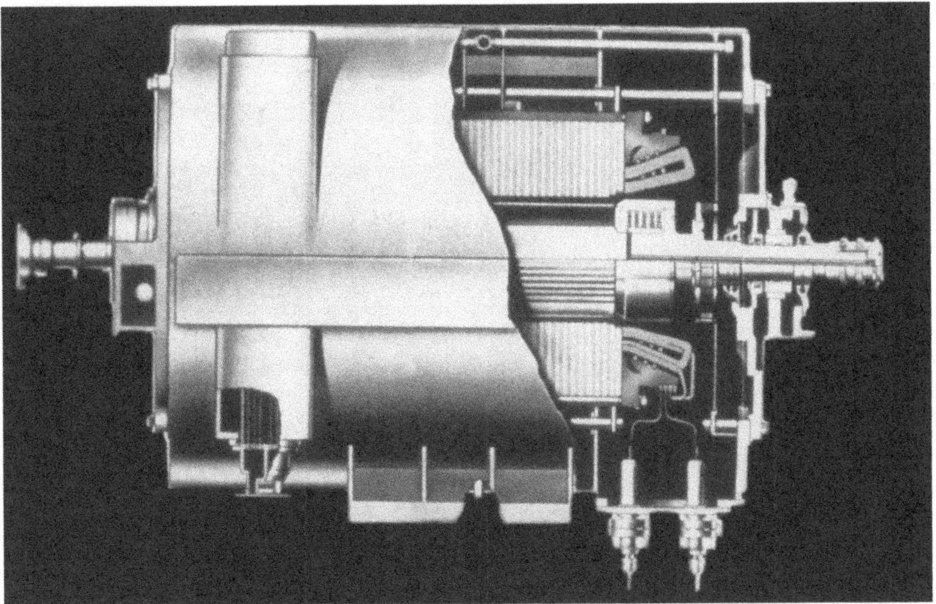

FIGURE 6.37 Hydrogen-cooled generator.

HYDROGEN/WATER-COOLED GENERATORS

Even more compact generator designs can be achieved through the use of direct water cooling of the generator armature winding (Fig. 6.38). These designs employ hollow copper strands (Fig. 6.39) through which deionized water flows. The cooling water is supplied via a closed-loop auxiliary-base-mounted skid. The cool water enters the winding through a distribution header on the connection end of the generator and the warm water is discharged in a similar manner on the turbine end of the generator (Fig. 6.40).

The armature voltage and current of hydrogen/water-cooled generators are significantly higher than those of air- or hydrogen-cooled units. As a result, the insulation voltage stress and forces on the armature windings can be several orders of magnitude larger than those experienced on lower-rated units. These present unique design requirements which must be addressed with specially configured epoxy-mica-based material systems.

Excitation System

The excitation system provides magnetizing power (about 1 percent of generator output power) for the rotating field winding and accurately controls the amount of magnetizing power to maintain close regulation of the generator output voltage and power factor.

Several excitation systems presently exist; these are classified according to the exciter power source:

- DC generator with commutator
- AC generator and stationary rectifiers
- AC generator and rotating rectifiers (brushless)
- Transformers on the main generator and rectifiers (static excitation)

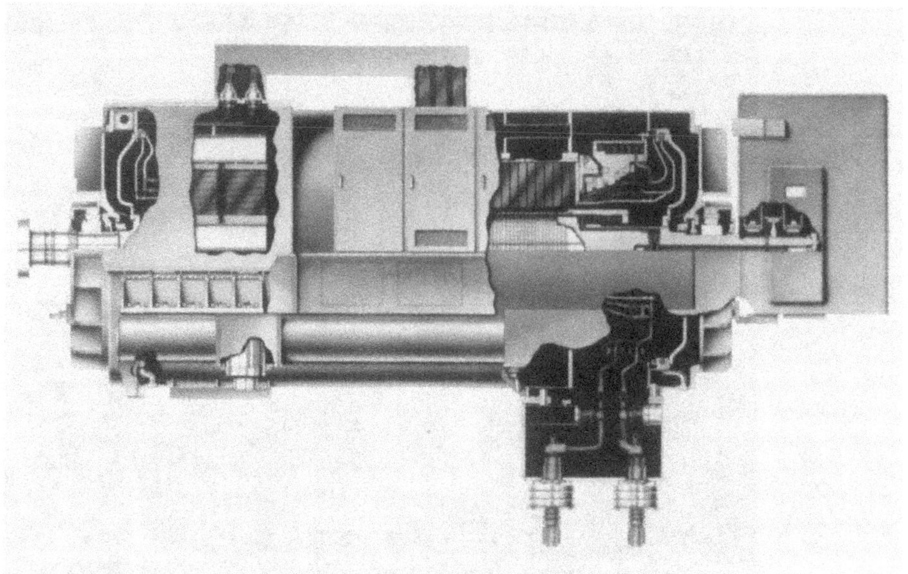

FIGURE 6.38 Water-cooled generator.

A schematic diagram of the dc generator with commutator connected to the main shaft is shown in Fig. 6.41. (They can be driven by separate motors or steam turbines.) The excitation power is taken from the commutator on the dc-generator rotor and applied to the main-generator rotating field through collector rings. The main-generator output voltage is controlled by using a voltage regulator to vary the excitation of the dc-generator stator.

Since commutator-type dc-generator excitation systems are inherently high in maintenance due to the commutator and brushes, the invention of solid-state rectifiers has reduced the use of this equipment in favor of ac generators rectified to dc using silicon diode rectifiers. The two methods of implementing these systems are based on either stationary rectifiers or rotating rectifiers.

FIGURE 6.39 Hollow copper strand construction.

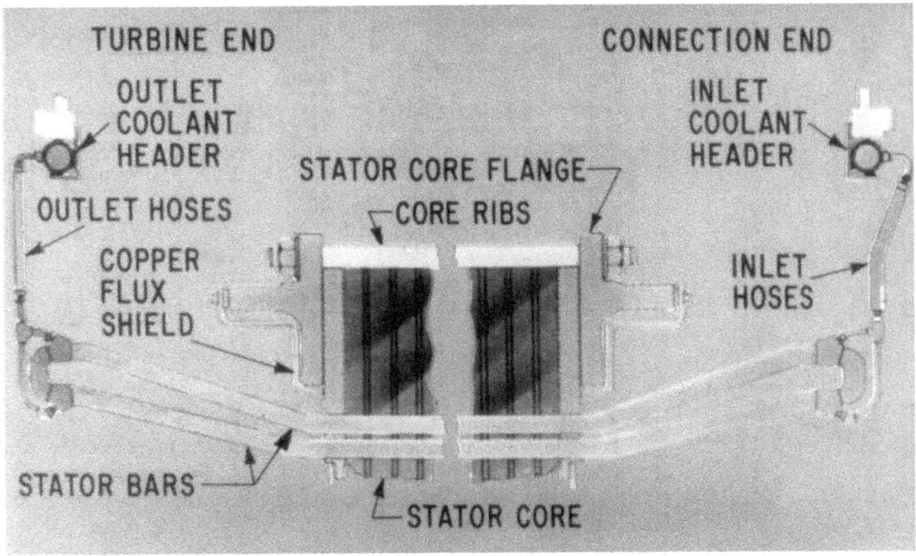

FIGURE 6.40 Water-cooled stator winding.

The stationary rectifier system is shown schematically in Fig. 6.42. The ac exciter has a rotating field, as does the main generator. The exciter output is taken from its stationary armature windings, converted to dc by silicon diode rectifiers, and applied to the main-generator rotating field through collector rings. The control system is similar to the dc-generator system, except that the excitation to the exciter rotating field is transferred by collector rings. This type of system is used for generators larger than 400,000 kVA where the excitation power can be as high as 7000 kW.

An alternative ac exciter system known as the *brushless* or *rotating rectifier* is shown in Fig. 6.43. It reverses the exciter field and armature and thereby eliminates both sets of collector rings. The main generator output voltage is controlled through the exciter field in the stator. The exciter armature and silicon diode rectifiers are on the main shaft, directly connected to the main-generator field, and generator control is effected through the air gap of the exciter by varying the stationary exciter field current. This system eliminates all collector rings, hence the name *brushless exciter.*

A static excitation system (see Fig. 6.44) eliminates the need for a separate generator for excitation. The excitation power is provided by the main-generator terminals through excita-

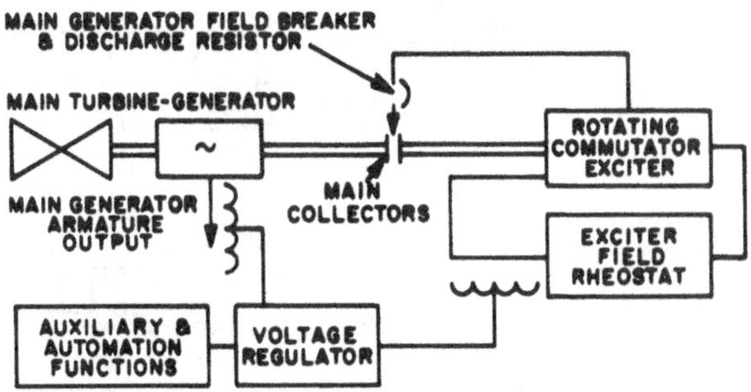

FIGURE 6.41 DC commutator excitation system.

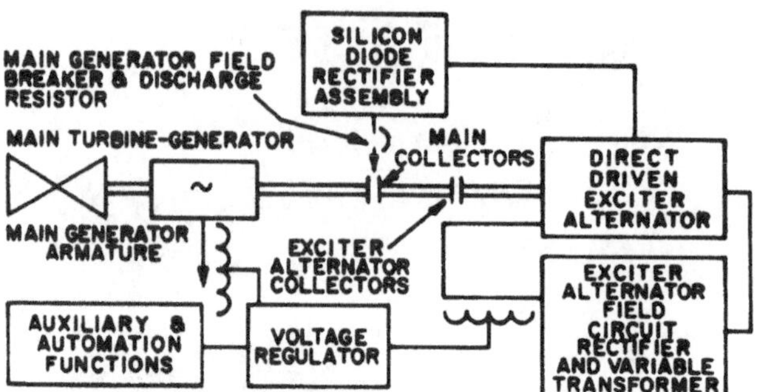

FIGURE 6.42 AC generator with stationary rectifier excitation system.

tion transformers. The controlled ac output of the transformers is converted to dc by silicon diode rectifers and applied to the main-generator field through collector rings.

When comparing excitation systems, each has its advantages. While the brushless system eliminates collector rings, a failure of the rotating rectifier can cause a shutdown. In contrast, the static excitation system normally provides parallel sets of stationary rectifiers, so a full load can be carried with one bank out of service. However, this system requires periodic brush maintenance, which can be done while operating.

For applications where more than a base-level excitation system performance is required, static compound power source exciters can be utilized.

In a compound system, excitation power is derived from both a voltage source and a current source. The voltage source supports operation during no-load conditions, when the generator is supplying load current, a portion of the field excitation is derived from the generator load current. Combining the potential and current sources enables full excitation power to be supplied through system disturbances with severely depressed generator line voltage. This performance feature can be valuable in certain power system applications.

Figure 6.45 illustrates a simplified block diagram of a typical system. The rectifier bridge is a shunt-thyristor type, meaning that the bridge component arrangement is such that the firing point of the thyristors can be used to control or "shunt" the amount of excitation current that reaches the generator field.

The principal function of the excitation system is to furnish power in the form of direct current and voltage to the generator field, creating the magnetic field. The excitation system also comprises control and protective equipment which regulates the generator electrical output. In today's complex power system transmission design, the performance and protection fea-

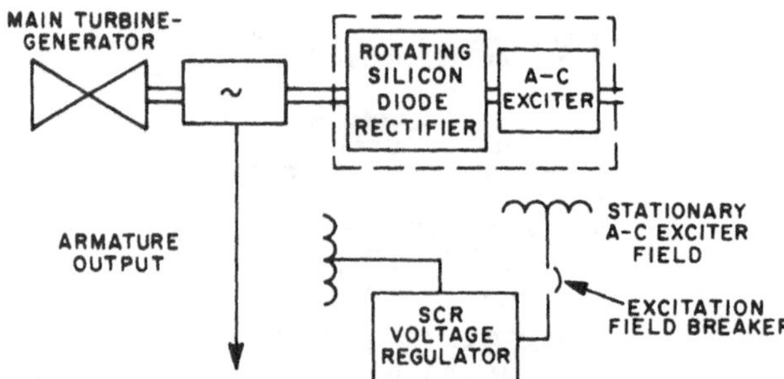

FIGURE 6.43 AC generator with rotating rectifier (brushless) excitation system.

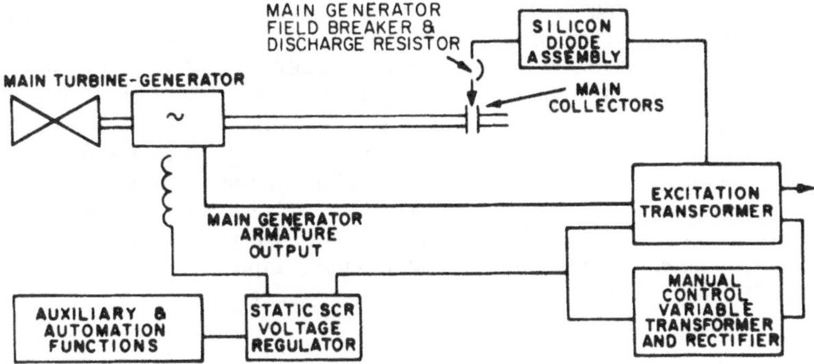

FIGURE 6.44 Static excitation system.

tures of an excitation system should be evaluated just as carefully as hardware design characteristics.

Excitation voltage is a key factor in controlling generator output. One desirable characteristic of an excitation system is its ability to produce high levels of excitation voltage (ceiling) very rapidly following a change in terminal voltage. A *high initial response* (HIR) excitation system is defined by IEEE as one that reaches 95 percent of the specified ceiling voltage in 0.1 or less. For units tied into a power system grid, such quick action to restore power system conditions reduces the tendency for loss of synchronization.

A second desirable performance feature of an excitation system is the level or amount of ceiling voltage it can achieve. *Response ratio* (RR) (or *nominal response*) is a useful term for quantifying the forcing or ceiling voltage available from the exciter. Response ratio is the average rate of rise in exciter voltage for the first half second after change initiation divided by the rated generator field voltage. Thus, it is expressed in terms of per unit (pu) of rated field voltage.

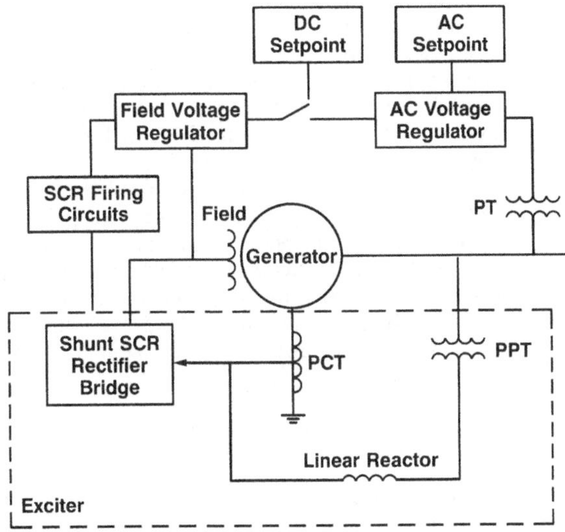

FIGURE 6.45 Shunt-SCR simplified block diagram.

In general, conventional rotating exciters, such as rotating rectifiers, have slower response time due to the time constants of the rotating magnetic components. In fast-acting static exciters, maximum exciter output is available almost instantaneously by signaling the controlling thyristors to provide full forcing. Figure 6.46 illustrates an important distinction between HIR exciters and conventional exciters. Since RR is proportional to the area under each curve, the HIR exciter, having achieved the specified 150 percent ceiling almost instantaneously, will exhibit a higher RR. Following this reasoning, the compound exciters can provide the highest overall performance levels since they employ an HIR thyristor-controlled rectifier bridge and, in addition, can capitalize on the fault current itself to drive the magnetic components of the exciter to high ceiling output voltages.

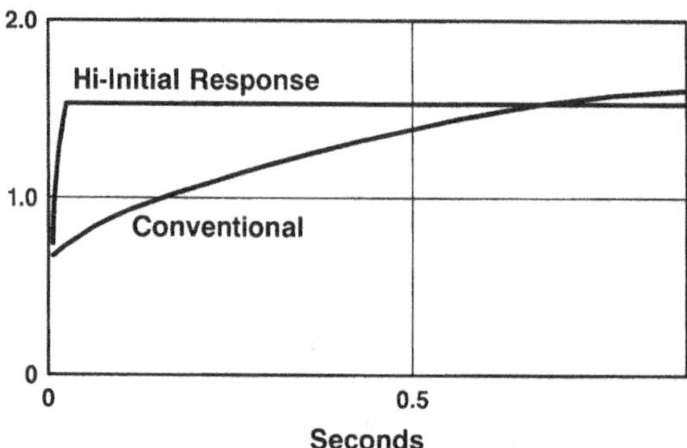

FIGURE 6.46 Excitation system performance from unity power factor.

Operation

When a generator is operating alone with an isolated load, the power factor and reactive power are determined by the load. When it operates as part of a system of generators and loads (i.e., power grid), however, the reactive power supplied by each generator can vary depending on the level of excitation of each machine. Therefore, changing the excitation of one generator in a system will change the power factor of that unit, while the voltage and frequency of the power grid will remain constant by the collective action of the other generators.

The effect of varying excitation or field current on generator performance is typically shown in excitation *V* curves and reactive capability curves. Typical curves of this type are shown as Figs. 6.47 and 6.48, respectively, for a hydrogen-cooled generator. The excitation *V* curve shows the amount of excitation required to produce a desired power factor at any power output of the generator, while the reactive capability curve defines the limits of operation. Using these curves, an operator can cause a generator to produce reactive power by increasing the field current (excitation) or absorb reactive power by decreasing the field current. This would correspond to moving along the vertical axis on the reactive capability curve. To move horizontally along this curve, the power supplied to the generator by the driving machinery would have to be increased. This highlights two important facts of generator operation in a power system:

1. The reactive power (power factor) of the power delivered by the generator is controlled by the generator field current (excitation).
2. The amount of power delivered by the generator is controlled by the machinery driving the generator.

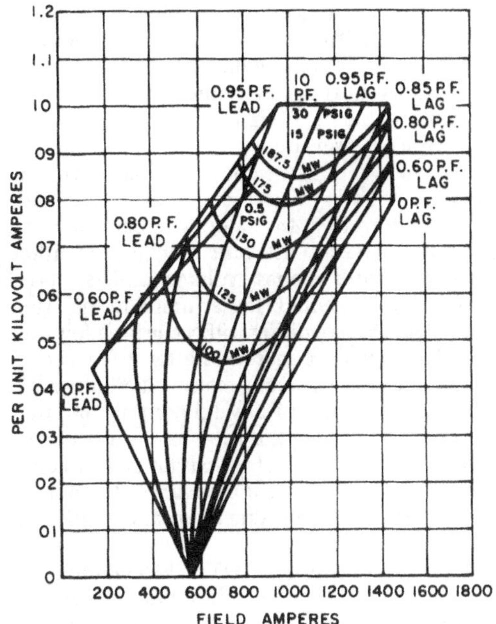

FIGURE 6.47 Generator excitation *V* curves.

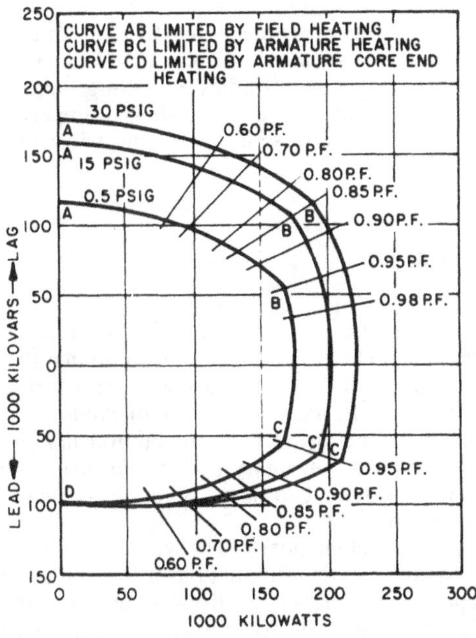

FIGURE 6.48 Reactive capability curve.

The stability of a generator is a function of its short-circuit ratio and the response time of its excitation system. The short-circuit ratio is the ratio of field current for open-circuited armature voltage at rated frequency to the field current for rated armature current on a sustained symmetrical short circuit at rated frequency. The higher the short-circuit ratio of a generator, the more stable the design.

A typical type of upset condition examined when determining generator stability is a short circuit. In examining the transient conditions during upsets, the term *reactance* is used. It is defined here as the ratio of open-circuit armature voltage at some field current to the short-circuit armature current at the same field current and is expressed in ohms (Ω). Reactances are usually given as a ratio to a base reactance in either percent or per unit. The base is generally taken as the ratio of phase voltage to phase current. Following a short circuit, the reactance can decrease to about 0.15 per unit and the armature current can increase to 10 or 15 times rated value. This condition will occur for a few cycles and is called the subtransient reactance (X''_d). The reactance increases after a few cycles to the transient reactance (X'_d) and continues to increase to a steady-state value known as the synchronous reactance (X_d). The synchronous reactance is typically 1.5 per unit, and the armature current will be about twice the rated current, assuming a three-phase short circuit. Since the synchronous reactance is inversely proportional to the short-circuit ratio when the saturation effects of the magnetic circuit are neglected, and since it is desirable to minimize the changes in generator operation (changes in armature current) due to upset conditions, a short-circuit ratio is a good measure of generator stability.

When a generator is started, it is first brought to rated speed by the driving equipment, and the field current is adjusted to produce rated generator voltage on an open circuit. The value of field current is shown as the extreme lower end of the excitation *V* curve. The frequency, voltage, and phase angle of the generator output are checked to ensure they are consistent with the power system, and the circuit breakers are closed to connect the generator to the system.

During operation, a number of protective devices are used to detect abnormal conditions and promptly isolate the troubled area to prevent damage to the generator. Relays can be provided to trip the generator upon the occurrence of any of the following conditions: unbalanced differential phase currents, phase-to-phase or ground faults, external phase-to-phase or ground faults, loss of excitation, out-of-step reverse power flow, unbalanced loading, lightning, and overvoltage. Alarms also indicate a field ground fault or excessive stator winding temperatures, which are measured by resistance temperature detectors (RTDs). The latter two conditions may not require an instantaneous trip, so the operator can plan the maintenance shutdown.

INDUCTION GENERATORS

The stator of an induction generator is similar to that of a synchronous generator. The rotor differs from the synchronous generator rotor in that there is no excitation and the conductors are shorted together at the rotor ends by an annular ring. This arrangement resembles a squirrel cage, which lends its name to the type of winding.

The induction generator supplies real power in *kilowatts,* which displace high-cost energy from the system. The imaginary power, *kilovars,* is drawn by the induction generator; it requires installed capability by some other device on the system, but consumes only a negligible amount of energy.

An induction machine operates at synchronous speed at zero load. The rotor turns at the same speed as the rotating flux field in the stator, and no lines of flux are cut. When a load torque is applied, the rotor speed drops off or "slips" until full torque is reached at 2 to 5 percent slip. As a generator, the driver must overspeed the generator by 2 to 5 percent to achieve full electric output.

Induction generators cannot operate independently in an isolated system. They can only function in parallel with synchronous generators that regulate voltage and supply the kilovars necessary to overcome the lagging power of the induction generation.

Induction generators are simpler and lower in initial cost than synchronous generators. They have been applied to recover power by expanding waste-gas streams and low-pressure steam. In some applications an energy-recovery turbine or expander drives an induction generator-motor and another pump or compressor on the same shaft. The generator-motor can either supply or absorb torque when the power of the other two devices is out of balance.

DC GENERATORS

The operating principle of the dc generator is very similar to that of the ac generator. In the dc generator the field is located in the stator while the armature rotates, generating alternating current in the armature windings. The commutator and brushes provide a means of transferring the output from the rotor to the stator, as well as of mechanically rectifying the alternating current. Figure 6.49 illustrates a dc-generator brush rigging and commutator. The commutator is a wearing surface for the brushes. It consists of individual copper segments insulated from each other by mica and connected to the armature windings. The armature-winding connections to the commutator and the brush spacing have to be carefully arranged so that brushes of opposite polarity contact windings which are 180 electrical degrees out of phase.

Many dc generators driven by motors have been installed in industrial plants, such as steel mills, to provide power for variable-speed drives. However, the advances in static silicon-rectifier dc power sources have reduced the market for dc generators primarily to replacement and repair parts, with very few new installations.

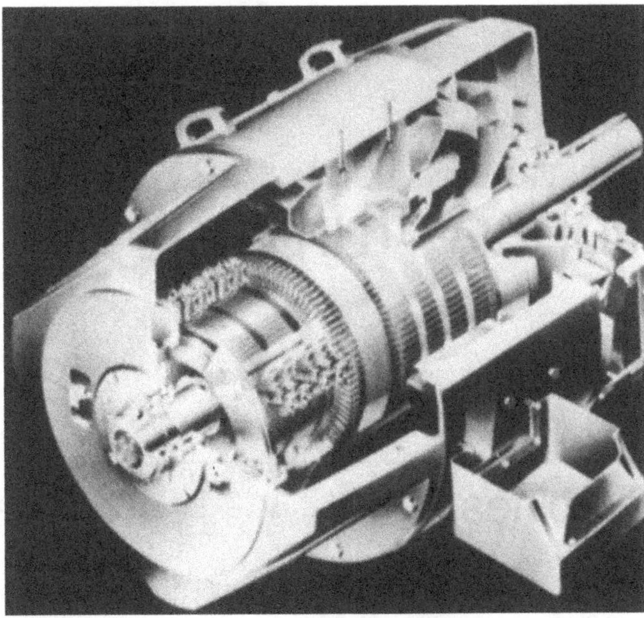

FIGURE 6.49 DC generator. (*General Electric Co.*)

ACKNOWLEDGMENT

The author acknowledges the support and contribution provided by the Engineering Staff of GE's Power Generation Business.

BIBLIOGRAPHY

"American Standard Requirements for Cylindrical Rotor Synchronous Generators," C50.13-1965, American Standards Association (now ANSI) New York, 1965.

"Electric Generators," special report, *Power,* McGraw-Hill, New York, March 1966.

Fenton, R. E. and J. J. Gibney: *GE Generators–An Overview,* Turbine State-of-the-Art Conference, GER-3688, Asheville, North Carolina, August 1991.

Fink, Donald G. (ed.): *Standard Handbook for Electrical Engineers,* 11th ed., McGraw-Hill, New York, 1978.

"Guide for Operation and Maintenance of Turbine-Generators," IEEE Standard 67-1972 (ANSI C50.30-1972), The Institute of Electrical and Electronics Engineers, Inc., New York, 1972.

Jay, Frank (ed.): *IEEE Standard Dictionary of Electrical and Electronics Terms,* 2d ed., The Institute of Electrical and Electronics Engineers, Inc., New York, 1977. Distributed in cooperation with Wiley-Interscience, New York.

O'Brien, K. G., et al.: *Generator Excitation Systems,* Turbine State-of-the-Art Conference, GER-3581B, Asheville, North Carolina, August 1991.

CHAPTER 6.4
GAS TURBINES

Clemens M. Thoennes
Sales Programs Development
General Electric Company
Schenectady, New York

GLOSSARY

Aircraft-derivative gas turbine An aircraft jet engine modified for ground applications to produce shaft power instead of thrust.

Base rating The designed rating point of a gas turbine at which it is suitable for continuous operation. Referenced to standard ISO conditions. (See "ISO rating.")

Cogeneration The sequential production of heat and power or recovery of low-level energy for power production.

Combined cycle A combined steam and gas turbine arrangement in which the gas turbine exhaust is ducted to a heat-recovery steam generator which supplies steam to the steam turbine.

Compression ratio The ratio of the compressor discharge pressure to the suction pressure.

Firing temperature The mass-flow mean total temperature of the working fluid measured in a plane immediately upstream of the first-stage turbine buckets.

Fuel consumption The input fuel heating value per unit of time to a gas turbine, generally measured in Btu/h (kJ/h), also called *heat consumption.* It is generally stated in terms of the lower heating value (LHV) of the fuel by gas turbine manufacturers, but can also be in terms of pounds per hour, where typical heating values are 18,500 Btu/lb (42,940 kj/kg) for liquid fuels and 21,500 Btu/lb (49,902 kj/kg) LHV for natural gas. (Also see "Specific Fuel Consumption.")

Heat rate The fuel consumption of a gas turbine divided by the output. For mechanical-drive gas turbines this is the net output including the on-base auxiliary power losses. For generator-drive gas turbines this includes these auxiliaries plus the generator losses. It does not include the power requirements for off-base lubrication oil cooling or heavy fuel treatment, unless specified as in certain totally packaged designs. Expressed as British thermal units or kilojoules per kilowatthour or horsepower hour. It is usually expressed in terms of the LHV of the fuel.

Heavy-duty industrial gas turbine A type of gas turbine designed specifically for ground applications using a design philosophy similar to that of the steam-turbine industry. Casings are split on the horizontal centerline, with on-site maintenance planned after long periods of operation.

Heavy fuel Liquid petroleum fuels that are ash bearing and not true distillates. These can be crude oil or residuals (No. 5 or 6), or a blend of a distillate and residuals.

High heating value The gross heating value of the fuel. This includes the latent heat required to vaporize the water in the products of combustion, which is not truly available to a combustion device having an exhaust temperature higher than 212°F (100°C).

Hot gas path The path of the working fluid of a gas turbine during and after combustion. It includes the fuel nozzles, combustion chamber and liner (if required), transition pieces to the turbine, stationary and rotating airfoils (nozzles and buckets), and exhaust plenum and ducting.

ISO rating The rated output of a gas turbine at standard site conditions as specified by the International Standards Organization: sea-level altitude, standard atmospheric pressure of 14.7 psia (101.4 kPa) at the turbine inlet and exhaust, 59°F (15°C) ambient temperature, and 60 percent relative humidity.

NO_2 Oxides of nitrogen include both NO and NO_2. Emission limits are generally based on parts per million by volume (ppmv) of the total of NO and NO_2 emitted by combustion devices.

Peak rating The designed rating point for gas turbines for peak-load service generally operated at less than 1000/yr. Not generally used for industrial applications which are base-loaded.

Regenerative cycle A gas turbine that includes a gas-to-gas heat exchanger which transfers heat from the exhaust to the compressor discharge air to reduce fuel consumption.

Simple cycle A gas turbine that exhausts to the atmosphere without heat recovery.

Specific fuel consumption (SFC) The gas turbine fuel consumption per unit of output. SFC is usually stated in terms of pounds per kilowatthour or pounds per horsepower hour using LHV.

Specific work The output of a gas turbine per unit of air flow. It can be horsepower seconds per pound or British thermal units per pound (kilojoules per kilogram).

Thermodynamic efficiency The net output of a gas turbine divided by the input. It is the reciprocal of heat rate after normalizing units. For example:

$$\text{Thermal efficiency} = \frac{1}{\text{heat rate (Btu/kWh)} \times 3412\ \text{Btu/kWh}}$$

PRINCIPLES

Thermodynamic Fundamentals

A schematic diagram for a simple-cycle, single-shaft gas turbine is shown in Fig. 6.50. Air enters the axial flow compressor at point 1 of the schematic at ambient conditions. Since these conditions vary from day to day and from location to location, it is convenient to consider some standard conditions for comparative purposes. The standard conditions used by the gas turbine industry are 59°F (15°C), 14.7 psia (1.013 bar), and 60 percent relative humidity, which were established by the International Organization for Standardization (ISO). These conditions are frequently referred to as ISO conditions.

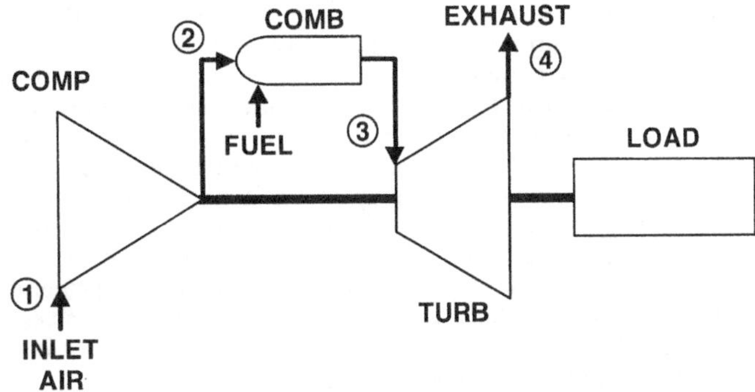

FIGURE 6.50 Simple-cycle, single-shaft gas turbine.

Air entering the compressor at point 1 is compressed to some higher pressure. No heat is added; however, the temperature of the air rises due to compression, so that the air at the discharge of the compressor is at a higher temperature and pressure.

Upon leaving the compressor, air enters the combustion system at point 2, where fuel is injected and combustion takes place. The combustion process occurs at essentially constant pressure. Although very high local temperatures are reached within the primary combustion zone (approaching stoichiometric conditions), the combustion system is designed to provide mixing, dilution, and cooling. Thus, by the time the combustion mixture leaves the combustion system and enters the turbine at point 3, it is at a mixed average temperature.

In the turbine section of the gas turbine, the energy of the hot gases is converted into work. This conversion actually takes place in two steps. In the nozzle section of the turbine, the hot gases are expanded and a portion of the thermal energy is converted into kinetic energy. In the subsequent bucket section of the turbine, a portion of the kinetic energy is transferred to the rotating buckets and converted to work.

Some of the work developed by the turbine is used to drive the compressor, and the remainder is available for useful work at the output flange of the gas turbine. Typically, more than 50 percent of the work developed by the turbine sections is used to power the axial flow compressor.

The thermodynamic cycle upon which all gas turbines operate is called the *Brayton cycle.* Figure 6.51 shows the classical pressure-volume (PV) and temperature-entropy (TS) diagrams for this cycle. The numbers on this diagram correspond to the numbers also used in Fig. 6.50. Every Brayton cycle can be characterized by two significant parameters: pressure ratio

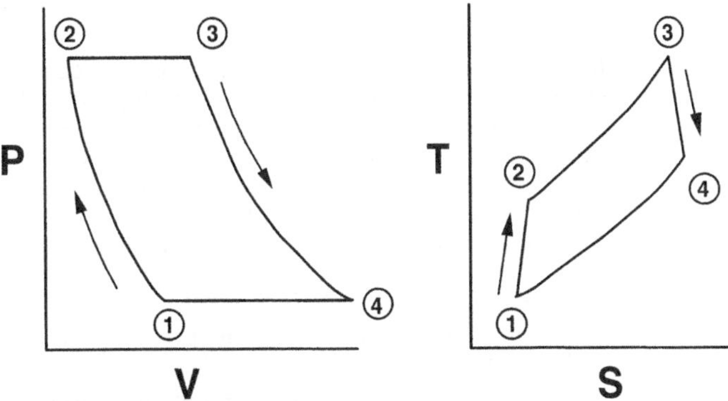

FIGURE 6.51 Brayton cycle.

and firing temperature. The pressure ratio of the cycle is the pressure at point 2 (compressor discharge pressure) divided by the pressure at point 1 (compressor inlet pressure). In an ideal cycle, this pressure ratio is also equal to the pressure at point 3 divided by the pressure at point 4. However, in an actual cycle, there is some slight pressure loss in the combustion system and, hence, the pressure at point 3 is slightly less than at point 2. The other significant parameter is the *firing temperature,* which is the highest temperature reached in the cycle. The most accepted definition of firing temperature is the mass-flow mean total temperature at the first-stage nozzle trailing edge plane. In a gas turbine without first-stage turbine nozzle cooling (in which air enters the hot gas stream after cooling down the nozzle), the total temperature immediately downstream of the nozzle is equal to the temperature immediately upstream of the nozzle. With turbine nozzle cooling, this cooling air mixes with the hot gases expanding through the nozzle. This definition utilizes a temperature that is indicative of the cycle temperature represented by point 3 of Fig. 6.51.

An alternate method of determining firing temperature is defined in ISO document 2314, "Gas Turbines-Acceptance Tests." The firing temperature here is really a reference turbine inlet temperature, and is not generally a temperature that exists in a gas turbine cycle. It is calculated from a heat balance on the combustion system, using parameters obtained in a field test. This ISO reference temperature will always be less than the true firing temperature, in many cases by 100°F (37°C) or more for machines using air extracted from the compressor for internal cooling. Figure 6.52 shows how these various temperatures are defined.

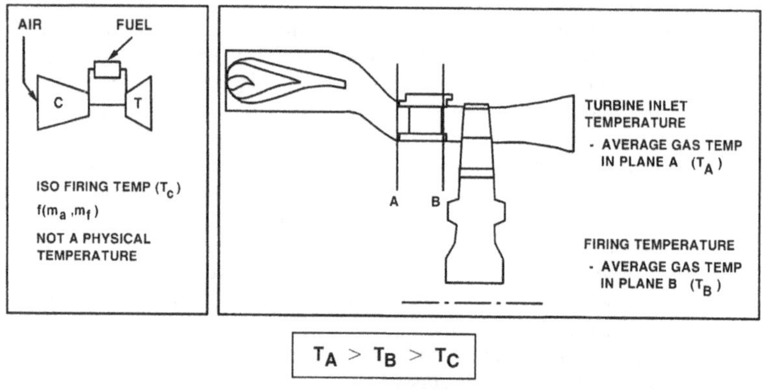

FIGURE 6.52 Definition of firing temperature.

The thermal efficiency of the Brayton cycle can be calculated using classical thermodynamic analysis. The compression ratio of the working fluid and the temperatures of heat addition and heat rejection are very important parameters. The results of such an analysis are shown in Fig. 6.53. These calculations are based on an ambient temperature of 59°F (15°C), actual component efficiencies, and real gas relationships. The results are plotted as thermal efficiency versus specific work for two different firing temperatures.

The observations that can be made from these curves are:

- Thermal efficiency increases as heat is added.
- For a given firing temperature there is an optimum pressure ratio for achieving maximum thermal efficiency.
- For a given firing temperature there is an optimum pressure ratio for achieving the maximum specific work which is different from the optimum thermal-efficiency pressure ratio.

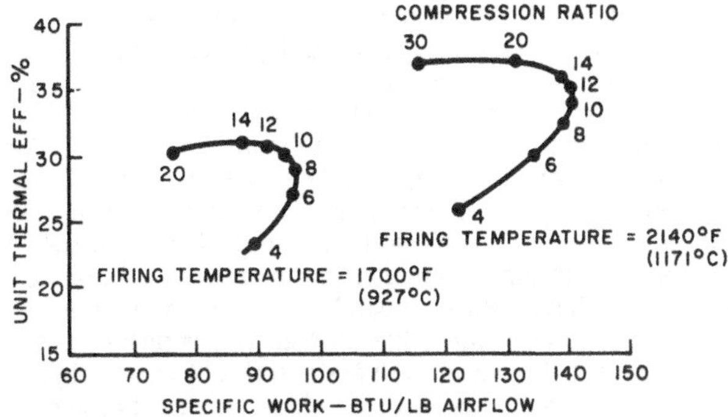

FIGURE 6.53 Efficiency vs. specific work of gas turbine cycles.

Design Features

There are many different design features among the gas turbines available for industrial plant applications. Some of the more important characteristics are:

- One or more shafts
- Heavy-duty industrial type or aircraft derivative
- Combustion-chamber design

As shown in Fig. 6.50, single-shaft gas turbines are configured in one continuous shaft and, therefore, all stages operate at the same speed. These units are typically used for generator-drive applications where significant speed variation is not required.

A schematic diagram for a simple-cycle, two-shaft gas turbine is shown in Fig. 6.54. The low-pressure or power turbine rotor is mechanically separate from the high-pressure turbine and compressor rotor. This unique feature allows the power turbine to be operated at a wide range of speeds, and makes two-shaft gas turbines ideally suited for variable-speed applications.

All of the work developed by the power turbine is available to drive the load equipment since the work developed by the high-pressure turbine supplies all the necessary energy to drive the compressor. Further, the starting requirements for the gas turbine load train are reduced since the load equipment is mechanically separate from the high-pressure turbine.

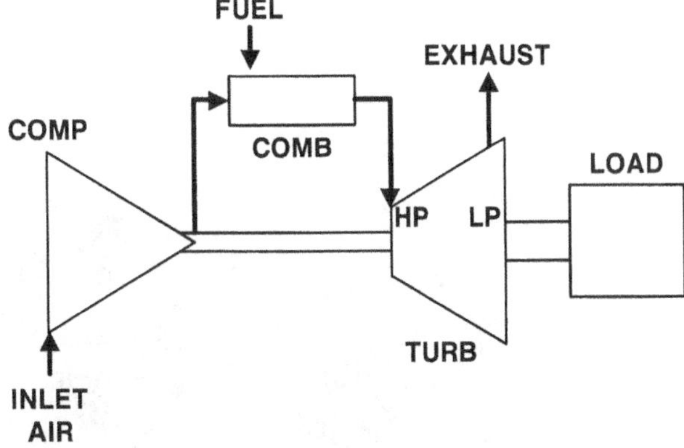

FIGURE 6.54 Simple-cycle, two-shaft gas turbine.

The designs of gas turbines have evolved from two distinct philosophies. Industrial-type units have been based on the technology developed in the steam-turbine industry for large central stations. Of robust construction, with casings split along the horizontal centerline, these units are designed for long periods of continuous operation, generally have the capability to burn a variety of fuels, and are maintained on-site. Aircraft-derivative gas turbines are jet engines modified to produce shaft power instead of thrust. Of lightweight construction, aircraft-derivative gas turbines are generally derated from flight-takeoff firing temperatures to allow long periods of continuous operation; they can usually be maintained on site, or are suitable for quick change-out and replacement with a spare engine. Aircraft-derivative units generally do not have the fuel flexibility of heavy-duty units.

Another distinguishing characteristic of gas turbine designs is the type of combustion section. There are three general types: a series of small cylindrical chambers or cans, an annular chamber surrounding the shaft, and large single off-base combustors. The series of small cylindrical combustors is best suited to full-scale combustion development testing, a key factor in successfully introducing a new model. New materials and designs can be developed without going to the expense of prototype testing. Also, investigations of unusual fuels and methods of reducing objectionable emissions such as NO_x can be easily made. The annular combustion chamber has minimum ducting, weight, and length and is therefore best suited to aircraft-type units.

Figures 6.55 and 6.56 illustrate many of the characteristics of the different gas turbine designs mentioned previously. Figure 6.55 is the General Electric Model Series 7001FA single-shaft heavy-duty gas turbine, and Fig. 6.56 is the General Electric LM6000 aircraft-derivative unit.

FIGURE 6.55 Cross section of an MS7001FA single-shaft gas turbine. (*General Electric Co.*)

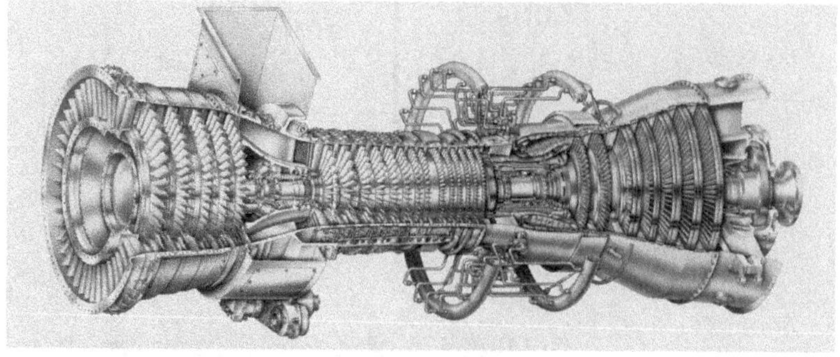

FIGURE 6.56 Cross section of an LM6000 two-shaft gas turbine. (*General Electric Co.*)

The compressor for the MS7001FA is an axial-flow, 18-stage compressor with extraction provisions at stages 9 and 13. Stages 0 and 1 have been designed for operation in transonic flow using design practices applied by aircraft gas turbine designers for high-bypass-ratio aircraft engines. Compressor surge control is accomplished through variable inlet guide vanes (VIGV) and selective bleed. At 100 percent speed, the VIGV are fully open for simple-cycle applications. For combined-cycle applications the VIGV are positioned at an intermediate setting and opened as a function of load and exhaust temperature to maintain maximum thermal efficiency. The 9th and 13th stage bleed valves close during start-up when the generator breaker closes.

The low stage loading has resulted in a very rugged compressor with a high level of compressor efficiency.

The MS7001FA combustion system consists of 14 combustion chambers with 14-in (36-cm) nominal-diameter combustion liners. Transition pieces conduct the combustion gases to the first-stage nozzle.

The MS7001FA turbine is a three-stage design, with the first-stage blade unshrouded and the second- and third-stage blades equipped with integral *Z*-form tip shrouds.

Each of the three rotor stages consists of 92 investment-cast blades. The first- and second-stage blades and all three nozzle stages are air cooled. The first-stage blade is made of directionally solidified construction and is convectively cooled via serpentine passages, with turbulence promoters formed by coring techniques during the casting process (Fig. 6.57). The cooling air leaves the blade through holes in the tip as well as in the trailing edge.

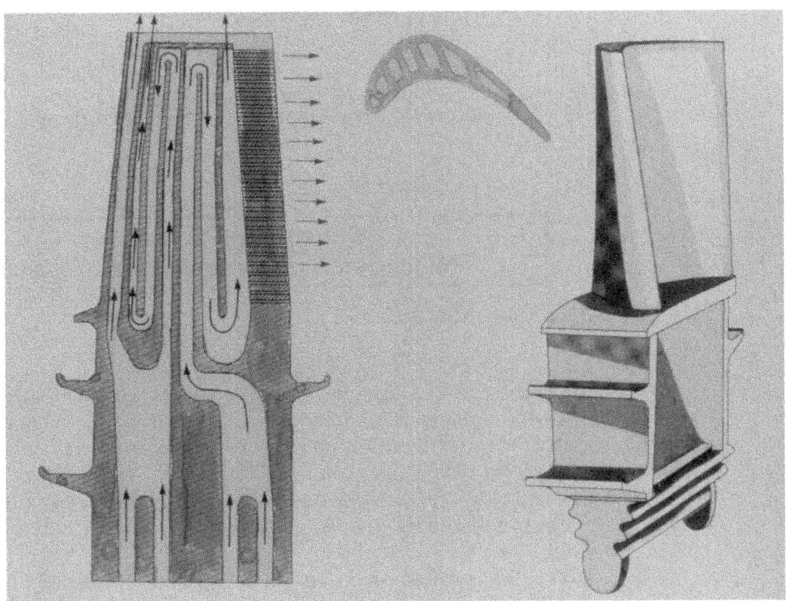

FIGURE 6.57 First-stage bucket cooling passages.

The second-stage blade is cooled by convective heat transfer using STEM (shaped tube electrode machining)–drilled radial holes, with all cooling air exiting through the tip.

The first-stage nozzle contains a forward and aft cavity in the vane, and is cooled by a combination of film, impingement, and convection techniques in both the vane and sidewall regions. There are a total of 575 holes in each of the 24 segments.

The second-stage nozzle is cooled by a combination of impingement and convection techniques, while the third-stage nozzle is cooled by convection only.

The efficient use of cooling air made possible by these advanced cooling methods is further enhanced by the reduced vane-surface area of the first-stage nozzle, which is achieved by low solidity.

The LM6000 is a dual-rotor "direct-drive" gas turbine. The LM6000 takes advantage of the fact that the low-pressure rotor normal operating speed of its parent turbofan aircraft engine is approximately 3600 rpm. The LM6000 gas turbine concept provides for direct coupling of the gas turbine low-pressure system to the load and maintains an extraordinarily high degree of commonality with the aircraft engine, as illustrated in Fig. 6.58. This is unlike the traditional aeroderivative approach, also shown in Fig. 6.58, which maintains a high degree of commonality with the aircraft engine in the gas generator only, and adds a unique power turbine. By maintaining high commonality, the LM6000 offers reduced parts cost benefits and demonstrated reliability.

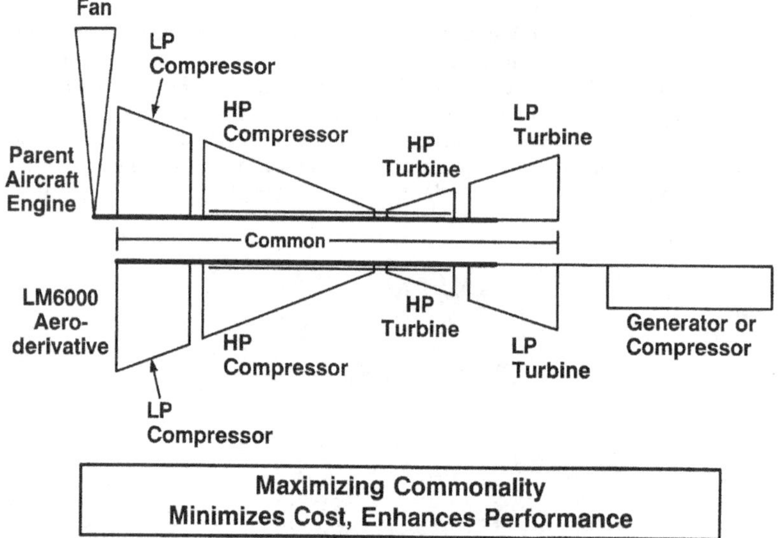

FIGURE 6.58 LM6000 concept.

The LM6000 consists of a low-pressure rotor made up of a five-stage low-pressure compressor (LPC) with variable-inlet guide vanes, driven by a five-stage low-pressure turbine via a concentric shaft through the high-pressure rotor. This low-pressure rotor is also the driven equipment driver, providing the option for either cold-end or hot-end drive arrangements. The high-pressure rotor consists of a 14-stage high-pressure compressor with six stages of variable guide vanes driven by a two-stage air-cooled high-pressure turbine. The overall compression ratio is 29:1.

The LM6000 utilizes an annular combustor with 30 individually replaceable fuel nozzles, and is equipped with an engine-mounted accessory drive gearbox for starting the unit and driving critical accessories.

PERFORMANCE CHARACTERISTICS

Gas Turbine Ratings

Since the introduction of the first industrial gas turbines in the 1950s there has been a continuous growth in performance. During this period there have been significant developments in

TABLE 6.21 GE Gas Turbine Performance Characteristics: Generator-Drive Gas Turbine Ratings

Model no.	Fuel gas/ distillate	Output, kW	Heat rate, Btu/kWh (LHV)	Exhaust flow, lb/h	Exhaust temp., deg. F	Freq., Hz
PG5271(RA)	G	20,260	12,820	781,000	969	50 and 60
	D	19,940	12,920	783,000	970	
PG5371(PA)	G	26,300	11,990	985,000	909	50 and 60
	D	25,800	12,070	988,000	910	
PG6541(B)	G	38,340	10,860	1,104,000	1,002	50 and 60
	D	37,520	10,970	1,107,000	1,003	
PG7111(EA)	G	83,500	10,480	2,351,000	986	60
	D	82,100	10,560	2,358,000	986	
PG7171(E/F)	G	125,000	10,030	3,309,000	991	60
	D	122,410	10,130	3,318,000	992	
PG7221(FA)	G	159,000	9,500	3,387,000	1,093	60
	D	144,800	9,580	3,397,000	1,095	
PG9171(E)	G	123,400	10,100	3,256,000	1,001	50
	D	121,300	10,170	3,265,000	1,002	
PG9301(F)	G	212,200	9,995	4,860,000	1,081	50
	D	208,000	10,080	4,875,000	1,082	
PG9311(FA)	G	226,500	9,570	4,877,000	1,093	50
	D	222,000	9,650	4,892,000	1,095	
LM6000(PA)	G	39,970	8,790	982,300	840	60
	D	39,920	8,850	982,100	856	
LM6000(PA)	G	39,170	8,960	982,300	840	50
	D	39,120	9,030	982,100	856	

the metallurgy of hot-gas path parts and in coatings, cooling techniques, instruments, control systems, and component efficiencies. Ratings for specific frame sizes have grown threefold. Indications are that ratings and efficiency values will continue to increase as new techniques are developed for increased air and water cooling of hot-gas parts.

Therefore, any table of specifications for gas turbines can only represent a "snapshot" in time of what is a dynamic, ever-changing picture. Nevertheless, Tables 6.21 and 6.22 are offered to represent the state of the art of gas turbine technology in the early 1990s.

In Table 6.21 the models are heavy-duty gas turbine-generator sets. The PG7221(FA) incorporates the MS7001FA gas turbine, the technological state of the art in the early 1990s. Aircraft-derivative state of the art is embodied in the LM6000(PA) unit.

TABLE 6.22 Mechanical-Drive Gas Turbine Ratings

Model no.	Cycle (SC-simple) (RC-regenerative)	Fuel (G-gas) (D-dist.)	Output, hp	Heat rate, Btu/hph (LHV)	Output shaft speed, rpm	Exhaust temp., F	Exhaust flow, lb/h
M3142(J)	SC	G	14,600	9,530	6,500	979	415.0
		D	14,250	9,680	6,500	979	415.0
M3142R(J)	RC	G	14,000	7,410	6,500	668	415.0
		D	13,650	7,500	6,500	668	415.0
M5261(RA)	SC	G	26,400	9,380	4,860	988	740.4
M5352(B)	SC	G	35,000	8,830	4,670	915	977.9
M5382(C)	SC	G	38,000	8,700	4,670	960	993.4
M5352R(C)	RC	G	35,600	6,990	4,670	970/693	956.2
M6501(B)	SC	G	50,010	7,930	4,860	1,022	1,039.6
M7111(EA)	SC	G	108,200	7,790	3,460	1,001	2,224.7
LM6000(PA)	SC	G	56,130	6,370	3,600	836	999.5

Factors Affecting Gas Turbine Performance

Since the gas turbine is an ambient-air-breathing engine, its performance will be changed by anything affecting the mass flow of the air intake to the compressor, most obviously changes from the reference conditions of 59°F (15°C) and 14.7 psia (101.4 kPa). Figure 6.59 shows how ambient temperature affects output, heat rate, heat consumption and exhaust flow for a single-shaft MS7001EA. Each turbine model has its own temperature-effect curve, since it depends on the cycle parameters and component efficiencies as well as air mass flow.

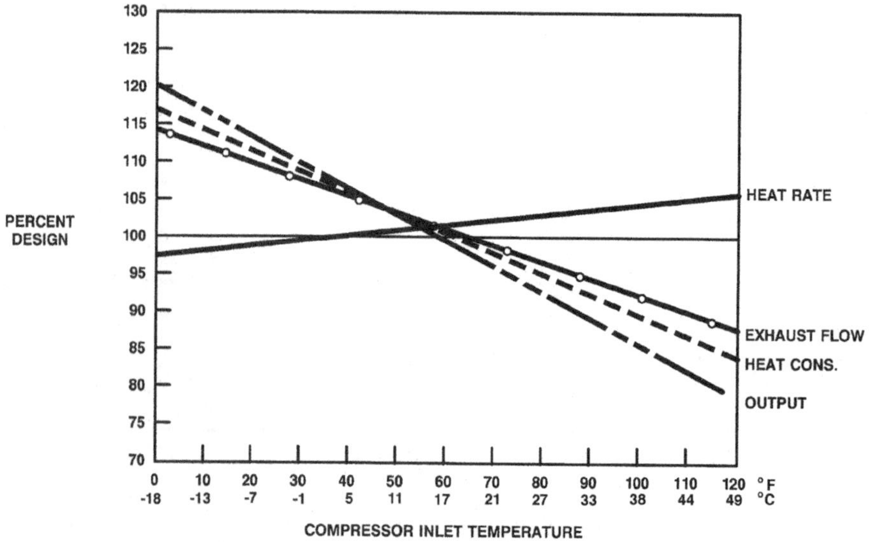

FIGURE 6.59 Effect of ambient temperature on MS7001EA.

Correction for altitude or barometric pressure is simpler. The less-dense air reduces the airflow and output proportionately; heat rate and other cycle parameters are not affected. A standard altitude correction curve is presented in Fig. 6.60.

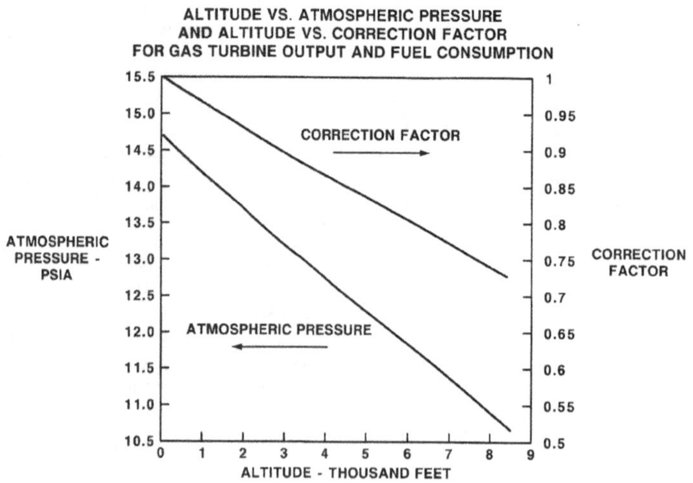

FIGURE 6.60 Altitude correction curve.

Similarly, humid air, being less dense than dry air, will also have an effect on output and heat rate as shown in Fig. 6.61. In the past, this effect was thought to be too small to be considered. However, with the increasing size of gas turbines and the utilization of humidity to bias water and steam injection for NO_x control, this effect has greater significance.

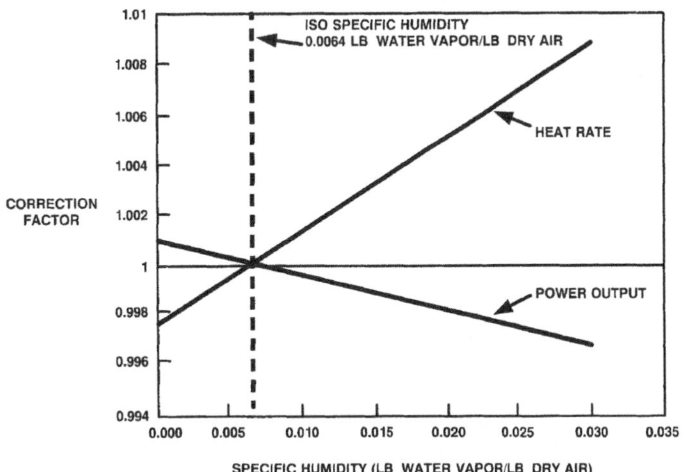

FIGURE 6.61 Humidity effect curve.

Inserting air filtration, silencing, evaporative coolers, chillers, and exhaust heat recovery devices in the inlet and exhaust systems causes pressure drops in the system. The effects of these pressure drops are unique to each design. Shown in Fig. 6.62 are the effects on the MS7001EA.

4 INCHES H_2O INLET DROP PRODUCES:

1.42% POWER OUTPUT LOSS
0.45% HEAT RATE INCREASE
1.9°F EXHAUST TEMPERATURE INCREASE

4 INCHES H_2O EXHAUST DROP PRODUCES:

0.42% POWER OUTPUT LOSS
0.42% HEAT RATE INCREASE
1.9°F EXHAUST TEMPERATURE INCREASE

FIGURE 6.62 MS7001EA pressure drop effects.

Fuel type will also impact performance. Tables 6.21 and 6.22 show that natural gas produces nearly 2 percent more output than does distillate oil. This is due to the higher specific heat in the combustion products of natural gas, resulting from the higher water vapor content produced by the higher hydrogen/carbon ratio of methane.

OPERATION AND MAINTENANCE

Starting Procedures

In order to start up a gas turbine, another small prime mover is required to accelerate the unit to a preselected speed until firing occurs and the unit becomes self-sustaining. The starting device is then uncoupled from the gas turbine by a clutch. Starting devices can be:

- Motors
- Diesel engines
- Expansion turbines
- Steam turbines
- Generators (via frequency conversion)

Normal Operation

Most gas turbine-generators normally operate at synchronous speed at full capability. Fuel flow is governed to maintain the firing temperature at its design limit. Therefore, the output of the unit will vary with ambient temperature. Units that are synchronized in a grid can also operate at part load using a droop or speed/load control characteristic. This is illustrated in Fig. 6.63. Because the unit is synchronized to the system, the speed is essentially constant. Therefore, varying the speed set-point effectively varies the load. The family of diagonal lines represents different settings of the speed/load control knob. Isolated gas turbine-generators can be furnished with an isochronous control mode. Load changes result in transient speed excursions which are instantaneously corrected by modulating fuel flow. Whether a gas turbine is on droop or isochronous control, the maximum firing temperature control will always provide an upper limit to prevent overfiring. Many other backup and protective controls and alarms are also provided.

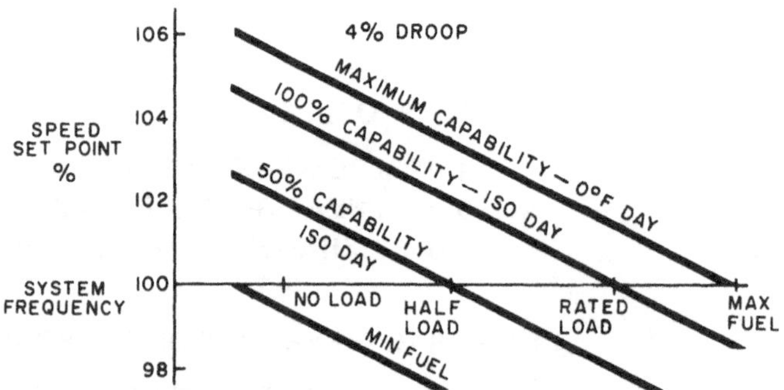

FIGURE 6.63 Typical droop speed control characteristics.

Mechanical-drive gas turbines normally operate on speed/load control with the set-point provided by the process control system. Figure 6.64 depicts a typical performance curve for a two-shaft mechanical-drive gas turbine, with the load characteristic of a process compressor system superimposed. A process controller might receive the suction or discharge pressure signal of the driven compressor and generate the appropriate speed/load set-point of the gas turbine. Again the fuel flow is still limited by the maximum-firing-temperature control.

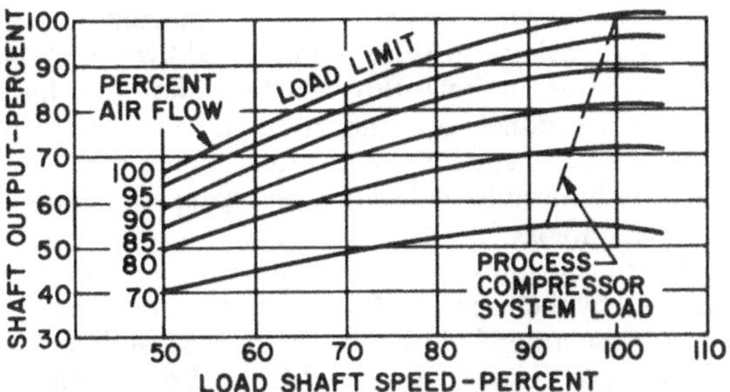

FIGURE 6.64 Two-shaft mechanical-drive gas turbine performance curve with a process compressor system load curve.

Gas Turbine Emissions

The gas turbine is one of many types of combustion devices that have been subject to strict environmental codes in recent years. Limits have been placed on the following types of objectionable emissions:

- Oxides of nitrogen, NO_x
- Oxides of sulfur, SO_x
- Particulates
- Unburned hydrocarbons
- Carbon monoxide, CO

The development of combustion systems has progressed to the point where typical gas turbine emissions of particulates, unburned hydrocarbons, and carbon monoxide fall well below the environmental limits. Figure 6.65 shows the reverse-flow cannular combustion system that has been the object of some of the most intensive development programs. Liquid fuels are

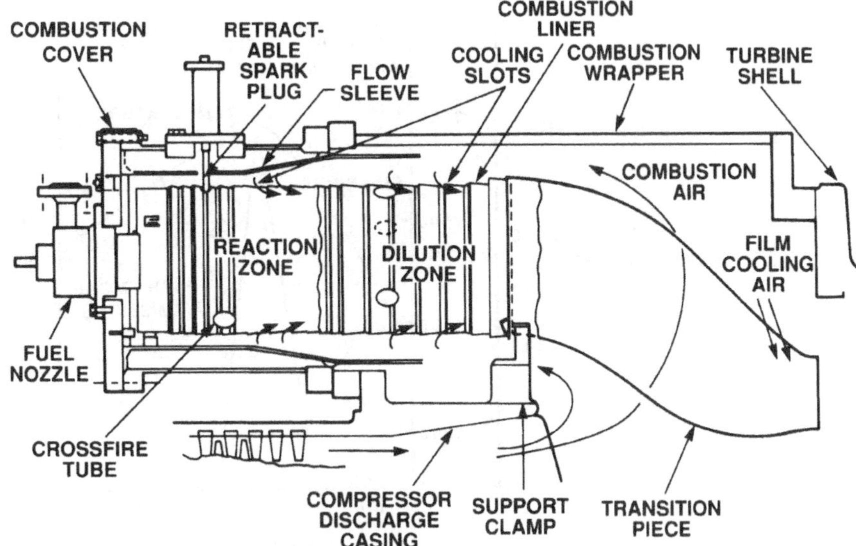

FIGURE 6.65 Reverse-flow combustion system.

atomized by high-pressure air as they are injected through the fuel nozzle. This has been very effective in reducing emissions of unburned hydrocarbons and particulates. Most of the compressor discharge air flows through the combustion liner downstream of the reaction zone. This cools the products of combustion and reduces the formation of NO_x. Injection of water or steam into the reaction zone further reduces NO_x formation. In response to continuing reduction of allowable NO_x emissions, water or steam injection is being supplanted by dry low-NO_x technology.

There are two sources of NO_x emissions in the exhaust of a gas turbine. Most of the NO_x is generated by the fixation of atmospheric nitrogen in the flame. This is called *thermal* NO_x. NO_x is also generated by the conversion of a fraction of any nitrogen chemically bound in the fuel [called *fuel-bound nitrogen* (FBN)]. Lower-quality distillates and low-Btu coal gases from gasifiers with hot-gas cleanup carry varying amounts of bound nitrogen, which must be taken into account when emissions calculations are made. The methods described below to control thermal emissions are ineffective in controlling the conversion of FBN to NO_x. In fact, these methods result in the conversion of a greater fraction of FBN to NO_x. If fuels with high bound nitrogen levels become common, other control techniques will have to be used.

Thermal NO_x is generally regarded as being generated by a chemical reaction sequence called the *Zeldovich mechanism.* This set of well-verified chemical reactions postulates that the rate of generation of thermal NO_x is an exponential function of the temperature of the flame. It, therefore, follows that the amount of NO_x generated is a function not only of the temperature but also of the time the hot gas mixture is at flame temperature. It turns out to be a linear function of time. Thus, temperature and residence time determine NO_x emission levels and are the principal variables that a gas turbine designer can alter to control emission levels.

Since flame temperature of a given fuel is a unique function of the equivalence ratio, the rate of NO_x generation in a flame can be cast as a function of the equivalence ratio. This is illustrated in Fig. 6.66, which shows that the highest rate of NO_x production occurs at an equivalence ratio of 1 when the temperature is equal to the stoichiometric, adiabatic flame temperature.

To the left of the maximum temperature point, there is more oxygen available than there is fuel and the flame temperature is lower. This is called *fuel lean operation.* In this case, the equivalence ratio is less than unity.

Since the rate of NO_x formation is a function of temperature and time, it follows that some difference in NO_x emissions can be expected when different fuels are burned in a given com-

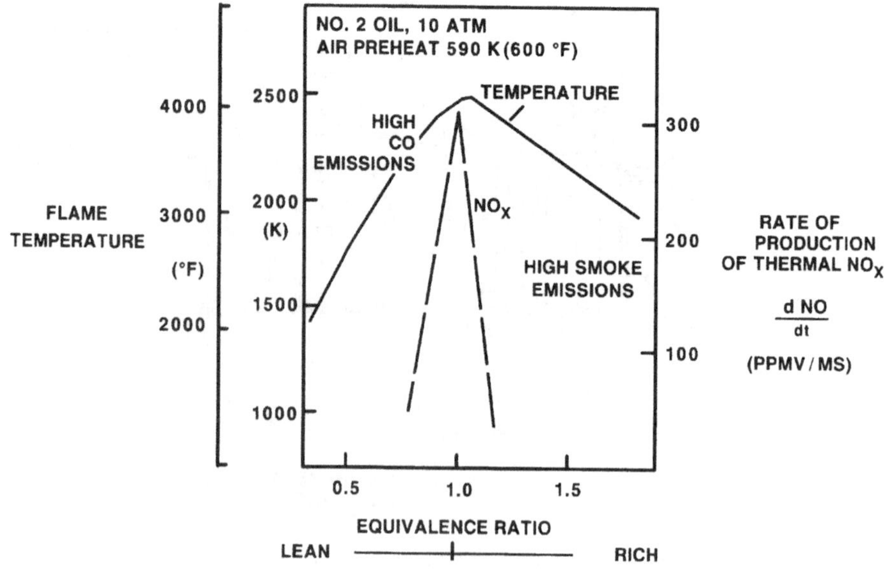

FIGURE 6.66 Rate of NO_x production.

bustion system. Since distillate oil and natural gas have approximately a 100°F (37°C) flame temperature difference, a significant difference in NO_x emissions can be expected, all other things (reaction zone equivalence ratio, water injection rate, etc.) being equal.

As can be seen from Fig. 6.66, the rate of NO_x production falls drastically as temperature decreases (i.e., the flame becomes fuel lean). This is because of the exponential effect of temperature in the Zeldovich mechanism and is the reason why diluent injection (usually water or steam) into a gas turbine combustor flame zone reduces NO_x emissions. For the same reason, very lean combustors can be used to control emissions. This is desirable for reaching the lower NO_x levels now being required in many applications. There are, however, two design challenges with very lean combustors: First, care must be taken to ensure stability at the design operating point; second, it is necessary to have turndown capability, as a gas turbine must ignite, accelerate, and operate over the load range. At lower loads, the flame could be fuel lean and not burn well or it could become unstable and blow out.

In response, designers use staged combustors so that only a portion of the flame-zone air is brought into contact with the fuel at lower load or during start-up. Staged combustors can be of two basic types: *fuel staged* or *air staged.* In the simplest and most common configuration, a fuel-staged combustor has two flame zones, each receiving a constant fraction of the combustor air flow. Fuel flow is divided between the two zones so that at each machine the amount of fuel fed to a stage is matched to the amount of air available. An air-staged combustor has a mechanism for diverting a fraction of the air flow from the flame zone to the dilution zone at low load to increase turndown. These methods can be combined.

Maintenance

Periodic inspection, repair, and replacement of parts are required to maintain gas turbines. The frequency of maintenance is heavily dependent on the type of fuel, the start-up frequency, and the environment. Although control systems carefully sequence start-up, there is an inherent thermal cycle which reduces parts life if frequently repeated. The parts life of peaking gas turbines that run for 4 h/day is lower than that of continuous-duty units. However, most industrial plants operate for many more than 100 fired hours per start and therefore do not have this problem.

The general environment can also affect the parts life of gas turbines. Many plants are located in areas with corrosive or abrasive matter in the atmosphere. Desert sandstorms, saltwater mist, chemical fumes, and airborne fertilizers are examples. However, the effects of these types of environments can be minimized by multistage high-efficiency inlet-air filters and mist eliminators as well as the presence of correct materials and protective coatings in the compressor and turbine.

The most important factor in gas turbine maintenance is the type of fuel burned. Natural gas is the cleanest fuel and incurs minimum maintenance costs and downtime. It is common for gas turbines in base-load industrial service to operate at full load on maximum-exhaust-temperature control continuously for 3 yr. Not many industrial plants or processes can operate for such long periods; hence, gas turbines are generally maintained at shorter intervals during process outages.

Normally, No. 2 distillate oil contains very little contamination, but it does burn with greater radiation, or luminosity, than natural gas. This decreases the life of hot-gas-path parts. The low lubricity of distillate oil decreases the life of parts of fuel forwarding and metering systems as well. Heavy fuel oils, both crude and residual, generally burn with additional radiation and have contaminants which accelerate corrosion and deposition of the hot-gas-path parts. Sodium and potassium must be removed from these fuels to prevent hot corrosion, and vanadium must be inhibited by the use of magnesium additives.

Preventive maintenance practices generally consist of several different types of maintenance procedures:

- Running inspection
- Combustion inspection
- Hot-gas-path inspection
- Major inspection

TABLE 6.23 Typical Achieved Crew Size and Skills—Combustion, Hot-Gas-Path, and Major Inspection
Baseload duty (gas fuel) MS7001E gas turbine

	Worker-hours		
Trade skill	Combustion	Hot-gas-path	Major
Millwright	136	2340	4730
Crane operator	3	68	112
Instr/elec/NDT tech	13	36	78
Carpenter	0	48	60
Welder	0	4	12
Pickup/driver	0	200	300
	152	2696	5292
Elapsed times	(3 × 10 h shifts)	17 days (2 × 10 h shift/day)	22 days (2 × 10 h shift/day)

Running inspections include load versus exhaust temperature measurements, vibration monitoring, and fuel-flow and fuel-pressure measurements. Sophisticated electronic equipment is planned to enhance trend monitoring and on-line diagnostics.

In a combustion inspection the unit is shut down and some disassembly is required to repair or replace combustion parts such as fuel nozzles and liners. Visual or boroscope inspections can also be made of turbine nozzles and buckets during these inspections.

A hot-gas-path inspection includes disassembly of the turbine casing. A major inspection includes a disassembly of the compressor casing as well as the turbine casing. A major inspection essentially returns the gas turbine to its new, or *zero time,* condition. For an MS7000 operating on natural gas or distillate, combustion, hot-gas-path, and major inspections occur at 8,000-, 24,000-, and 48,000-fired-hour intervals, respectively.

Many gas turbine parts are fabricated from expensive superalloys. Minimum maintenance costs can be achieved by repairing these parts during an inspection to extend their life. Spare sets of parts can be used as replacements to minimize downtime. In some critical continuous-process plants, it is more economical to maintain production without outages rather than extend parts life by repairs.

Typical crew sizes and trade skills needed to perform combustion, hot-gas-path, and major inspections on an MS7001E unit are shown in Table 6.23. Furthermore, as an indication of typical maintenance worker-hour requirements which may be used in initial planning phases, Table 6.23 also presents average worker-hours per downtime (calendar) hour for some of the more prevalent types of inspection activity that occur during the life of a gas turbine.

APPLICATIONS IN PLANTS

General Discussion

There are several different application categories for stationary gas turbines. These include:

- Pipeline pumping stations
- Offshore platforms
- Electric utility stations, including:
  Base-load
  Midrange (1500 to 3000 h/yr)
  Peaking duty
- Industrial plants

Pipeline pumping stations are generally base-loaded around the year, or through all except the summer months. There are many applications of simple-cycle gas turbines in remote areas.

TABLE 6.24 Typical Performance of a Combined Cycle, Based on 59°F (15°C) Sea-Level Site, with Natural Gas Fuel

Plant designation	Output, kW	Heat rate (LHV), Btu/kWh (kJ/kWh)	Gas turbine configuration
STAG* 107EA	124,100	7055 (7440)	One MS7001EA
STAG* 207EA	249,400	7020 (7410)	Two MS7001EA

* Registered trademark of General Electric Co.

There have also been a very small number of combined steam and gas turbine cycles in this category.

Most offshore-platform applications have been simple cycles due to weight and "footprint" constraints, with wide application of aeroderivative units.

In electric utility service, thousands of gas turbines around the world have been applied to serve peak loads (up to 1500 h/yr) in the simple-cycle mode. Because of limited operation, fuel consumption is not as significant a factor as are capital costs, operating labor, and maintenance. Most gas turbines that are applied in midrange or base-load electric utility service combine steam and gas turbine cycles, but a small number also have used regenerative cycles.

Many of the gas turbines applied in electric utility combined-cycle service are supplied as part of a complete package by the gas turbine manufacturer. The manufacturer supplies or specifies all the major equipment, such as heat-recovery steam generators (HRSGs), steam turbines, and plant controls, to optimize plant performance through an integrated approach. Table 6.24 lists typical performance specifications for three versions of combined cycles based on the MS7001E gas turbine.

Most gas turbines applied in industrial plants are in base-load service. There are many simple-cycle gas turbines applied throughout the world in industrial plants where fuel supplies are abundant. However, generally all gas turbines applied in industrial plants are equipped with some type of heat recovery to improve overall energy efficiency. Figure 6.67 illustrates some of the ways in which the high-temperature exhaust of gas turbines has been recovered in industrial plants. In Fig. 6.67 the exhaust gases are used to generate low-pressure process steam. The HRSGs can be unfired or have supplementary firing to increase steam output. In Fig. 6.67 higher-pressure steam is generated for a steam turbine. Typical upper limits for steam conditions of unfired HRSGs are 850 psig, 825°F (5964 kPa, 441°C). Fired HRSGs have been applied with steam conditions as high as 1450 psig, 950°F (10,100 kPa, 510°C). In Fig. 6.67 a two-pressure HRSG is shown. When high-pressure turbine inlet steam is generated in an unfired HRSG, typical stack temperatures are a relatively high 400 to 450°F (204 to 232°C). Additional heat can be recovered when a 25- to 150-psig (276- to 1138-kPa) saturated steam-generation section is included.

In Fig. 6.67 a regenerative-cycle gas turbine is followed by a low-pressure process steam generator. One of the consequences of the low fuel consumption of the regenerative-cycle gas turbine is a reduction of the regenerator exhaust gas temperature to approximately 600°F (316°C). This arrangement should be selected when only a relatively small amount of process steam is required.

Finally, in Fig. 6.67, the heat in the exhaust gas is used directly in the process or as preheated combustion air for a fired process heater.

In all these cycles the process is known as *cogeneration,* and the fuel utilization effectiveness is improved by recovering heat from the gas turbine exhaust. A parameter used to define the thermal performance of a cogeneration system is *fuel chargeable to power* (FCP). The FCP is the incremental fuel-power ratio for the cogeneration system relative to the case with which it is being compared (usually a noncogeneration alternative). For a plant generating electric power only (an industrial or a utility), the fuel chargeable to power and net plant heat rate are interchangeable terms. Net plant heat rate in Btu/kWh is the more commonly used term for plants generating electric power only.

The FCP concept is illustrated in Fig. 6.68. Stated in simple terms, the FCP is the total fuel burned in the cogeneration system minus the fuel which would have been required if all

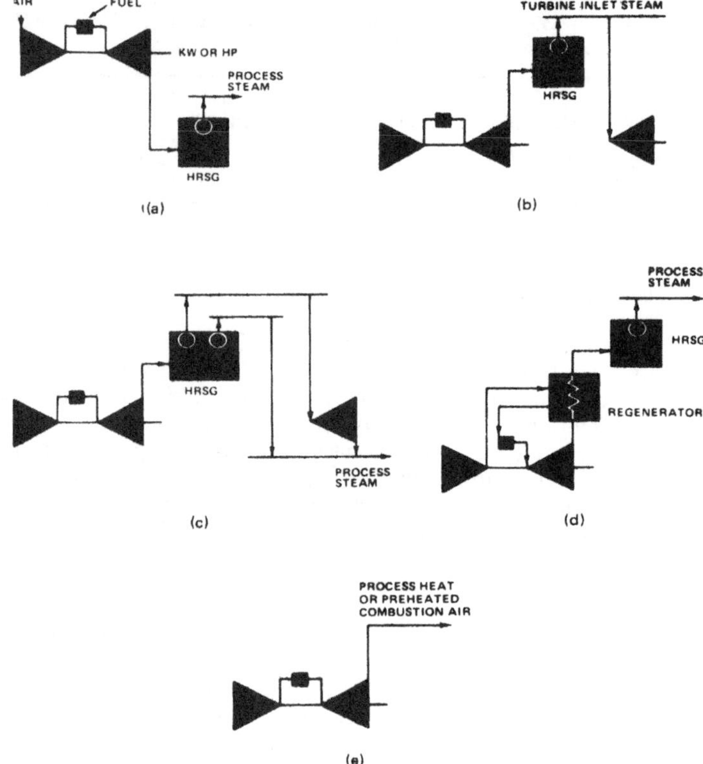

FIGURE 6.67 Industrial gas turbine heat-recovery cycles.

power were purchased (process fuel credit) divided by the gross power generated minus the difference in powerhouse auxiliaries.

The heat recovery capability and fuel chargeable-to-power for typical gas turbines is shown in Table 6.25.

Steam turbines are often used in cogeneration systems that produce heat for industrial processes as well as power. A typical application is shown in Fig. 6.69. In this case an

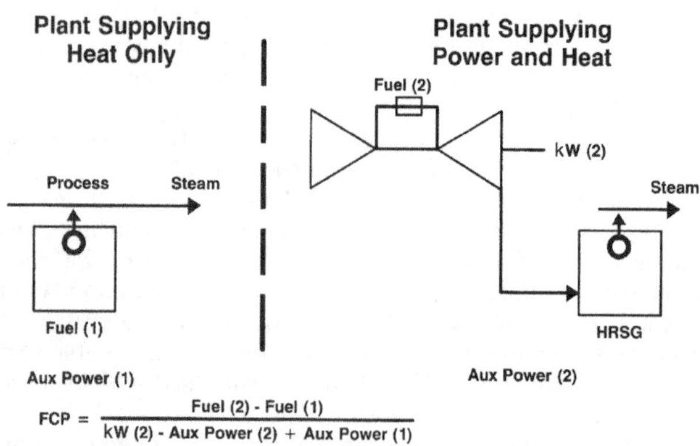

FIGURE 6.68 Fuel chargeable to power.

TABLE 6.25 Steam Generation and Fuel Chargeable-to-Power with Gas Turbine and Exhaust Heat Boilers

Generator drives—natural gas fuel

Gas turbine type	MS5001 (PA)	MS6001 (B)	MS7001 (EA)	MS7001 (F)	LM2500-PE	LM5000-PC	LM6000-PA
Gas turbine model	PG5371 (PA)	PG6541 (B)	PG7111 (EA)	PG7221 (F)	PGLM2500-PE	PGLM5000-PC	PGLM6000-PA
ISO base rating, kW	26,300	38,340	83,500	159,000	21,790	33,630	39,970
Performance at 59°F, sea level, natural gas fuel output, kW							
Unfired	25,890	38,000	82,680	156,500	21,540	33,190	39,700
Supp fired	25,710	37,820	82,330	155,700	21,400	32,970	39,550
Fully fired	25,430	37,530	81,780	154,400	21,220	32,620	39,320
Speed, rpm	5,100	5,100	3,600	3,600	3,600	3,600	3,600
Fuel, MBtu/h (HHV)	344.6	462.3	967.8	1,678.1	236.5	355.4	390.0
Exhaust flow, lb/h	971,400	1,083,000	2,343,000	3,387,000	535,000	950,800	982,300
Exhaust temp., °F							
Unfired	906	1,007	968	1,103	993	823	844
Supp fired	909	1,010	990	1,106	996	826	846
Fully fired	913	1,014	993	1,111	1,001	830	849
HRSG performance fuel, MBtu/h (HHV)							
Supp fired	221.5	214.2	475.7	564.5	107.9	242.1	244.4
Fully fired	878.7	904.4	1,989.6	2,592.3	438.4	845.5	850.0

Steam conditions, psig/°F	MS5001 (PA) HRSG Steam K lb/h	MS5001 (PA) FCP GT Btu/kWh	MS6001 (B) HRSG Steam K lb/h	MS6001 (B) FCP GT Btu/kWh	MS7001 (EA) HRSG Steam K lb/h	MS7001 (EA) FCP GT Btu/kWh	MS7001 (F) HRSG Steam K lb/h	MS7001 (F) FCP GT Btu/kWh	LM2500-PE HRSG Steam K lb/h	LM2500-PE FCP GT Btu/kWh	LM5000-PC HRSG Steam K lb/h	LM5000-PC FCP GT Btu/kWh	LM6000-PA HRSG Steam K lb/h	LM6000-PA FCP GT Btu/kWh
Unfired														
160/371	143.4	6650	193.2	6060	403.5	5840	704.0	5320	93.4	5770	117.8	6440	127.8	5960
420/655	114.4	7240	159.0	6420	331.5	6200	592.9	5530	76.5	6110	90.9	6950	99.8	6370
630/755	104.4	7560	148.0	6610	307.0	6410	559.2	5630	71.0	6280	81.0	7230	89.5	6610
895/830	96.2	7870	139.4	6800	288.5	6600	533.0	5740	66.8	6440	—	—	—	—
895/830	96.2	6360	139.4	5920	288.5	5680	533.0	5270	66.8	5640	—	—	—	—
160/371	32.3	—	27.6	—	63.0	—	61.4	—	14.3	—	—	—	—	—
1315/905	—	—	130.8	5920	269.5	5680	510.0	5270	—	—	—	—	—	—
160/371	—	—	34.3	—	78.7	—	74.9	—	—	—	—	—	—	—
1525/955	—	—	125.8	5920	258.5	5870	495.0	5270	—	—	—	—	—	—
160/371	—	—	37.8	—	86.9	—	82.0	—	—	—	—	—	—	—
Supp fired														
420/655	301.0	5960	338.0	5630	730.0	5380	1059.0	5080	167.2	5380	297.5	5750	307.5	5380
630/755	289.5	5980	324.5	5670	701.0	5410	1017.0	5100	160.6	5410	285.5	5790	295.5	5400
895/830	281.0	6030	315.0	5710	681.0	5440	988.0	5130	156.8	5380	277.5	5810	287.0	5430
1315/905	273.5	6100	306.5	5770	663.0	5490	962.0	5170	151.8	5490	270.0	5870	279.5	5470
1525/955	269.0	6080	301.5	5740	652.0	5470	946.0	5150	149.4	5460	265.5	5860	274.5	5470
Fully fired														
630/755	777.0	4610	836.0	4710	1826.0	4390	2526.0	4370	406.5	4540	734.0	4780	745.0	4570
895/830	757.0	4610	815.0	4690	1779.0	4390	2460.0	4380	396.0	4540	715.0	4790	726.0	4560
1315/905	740.0	4610	796.0	4710	1739.0	4390	2405.0	4380	387.0	4550	699.0	4780	710.0	4550
1525/955	726.0	4650	782.0	4700	1708.0	4380	2362.0	4380	380.0	4550	686.0	4800	697.0	4560

* Gas turbines and boilers fueled with natural gas and all fuel data based on higher heating value (HHV).
Unfired single-pressure boilers 92% effectiveness for SH and evaporator; supplementary fired to 1600°F, 86.8% to 90.5% effectiveness; fully fired to 10% excess air with 300°F stack temperature.
For two-pressure boilers, criterion of minimum 300°F stack temperature may require less than 92% low-pressure boiler effectiveness.
Assumes 0% exhaust bypass stack damper leakage, 3% blowdown, 1½% radiation and unaccounted losses, and 228°F feedwater for all cases.
Standard gas turbine inlet losses; exhaust 10° H_2O for unfired, 14° H_2O for supplementary fired, and 20° H_2O for fully fired.
LM2500, LM5000, and LM6000 values based on guarantee, not average engine performance.
Fuel chargeable to gas turbine power assumes GT credit with PH auxiliaries and equivalent 84% boiler fuel required to generate steam.
Lower heating value (LHV)—21,515 Btu/lb; HHV = LHV × 1.11.

automatic-extraction noncondensing unit supplies steam at two different pressure levels to the process. A typical value of fuel chargeable to power for noncondensing steam turbine cycles is 4200 Btu/kWh (4431 kJ/kWh) HHV. This is an equivalent thermal efficiency of 80 percent, which is far higher than that of most other types of prime movers. The high efficiency of the noncondensing steam turbine cycle is due to the fact that heat losses to the surroundings are minimized. The only losses are the boiler inefficiency (stack losses), generator, seals, bearing friction, radiation, and additional auxiliary power requirements.

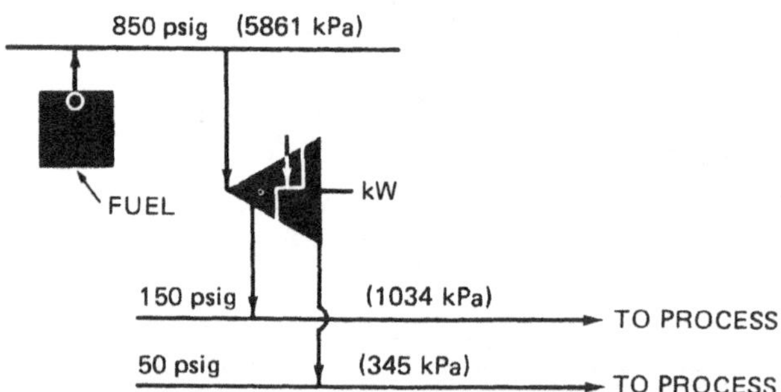

FIGURE 6.69 Typical noncondensing steam turbine application.

One method of displaying the many options available by using a gas turbine in a cogeneration application is shown in Fig. 6.70. This diagram has been developed for the GE MS7001EA gas turbine-generator.

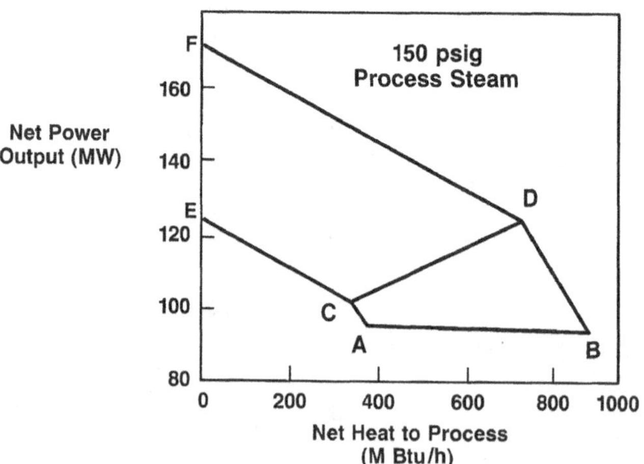

FIGURE 6.70 Performance envelope for gas turbine cogeneration system.

Point A represents the MS7001EA gas turbine-generator exhausting into an unfired low-pressure HRSG. Point C is a combined-cycle configuration based on use of a two-pressure-level unfired HRSG. The steam turbine in the C cycle is a noncondensing unit expanding the HP HRSG steam to the 150-psig (1034 kPa) process steam header.

Points B and D in Fig. 6.70 represent operation of the HRSG with supplementary firing to a 1600°F (878°C) average exhaust-gas temperature entering the heat-transfer surface. The

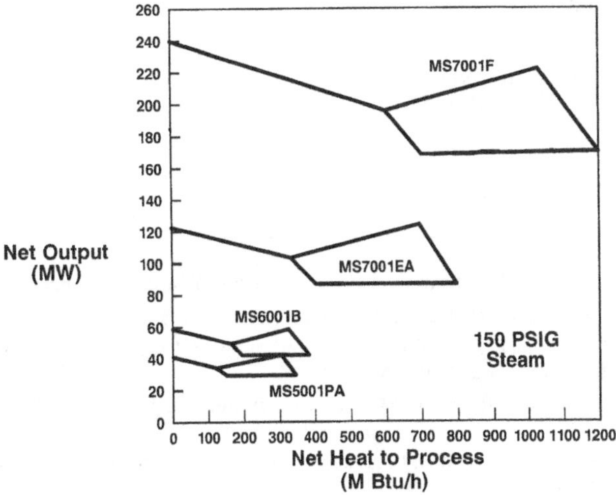

FIGURE 6.71 Gas turbine cogeneration systems MS options, 60 Hz.

temperature used for the HRSG firing in Fig. 6.70 has been arbitrarily limited to 1600°F (878°C) even though higher firing temperatures (and thus steam production rates) are possible in the exhaust of this unit.

The envelope defined by A, B, C, and D in Fig. 6.70 represents the most thermally optimized use of a gas turbine in a cogeneration application (i.e., provides the lowest FCP). Operation along the line CE, DF, or any intermediate point to the left of line CD represents the use of condensing steam turbine power generation with the E and F points applicable for combined-cycle operation without any heat supplied to process. Thus, the cycles along line EF are combined cycles providing power alone.

Performance envelopes for many of the gas turbines included in Table 6.25 are presented in Figs. 6.71 and 6.72. These data are on the same basis as Fig. 6.70 except for point C. Point C for all units except the various MS7001 models is based on 850 psig (5464 kPa), 825°F (441°C) initial steam temperature to the noncondensing steam turbine. Furthermore, the only condensing power illustrated is based on unfired, two-pressure-level HRSG designs.

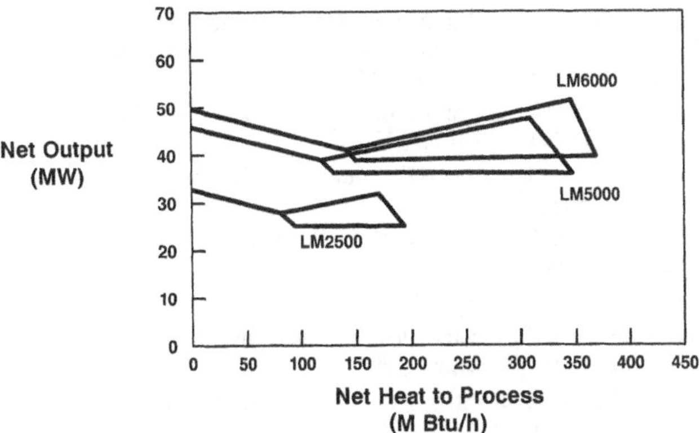

FIGURE 6.72 Gas turbine cogeneration systems LM options, 60 Hz.

ACKNOWLEDGMENT

The author acknowledges the support and contribution provided by the Engineering Staff of GE's Power Generation Business.

BIBLIOGRAPHY

1. "Performance Specifications 1980," *Gas Turbine World*, December 1979.
2. In *Power Engineering,* a series of seven articles:
   (a) Carlstrom, L. A., H. F. Heissenbuttel, and A. H. Perugi: "Gas Turbine Combustion System: Key to Improved Availability," May 1978.
   (b) Patterson, J. R., and C. M. Grant "Operating Gas Turbines for Extended Component Life," June 1978.
   (c) DuBois, M. R., and R. J. Fresneda: "Inspection and Maintenance of Gas Turbine Nozzles, Buckets and Rotors," July 1978.
   (d) Scheper, G. W., A. J. Mayoral, and E. J. Hipp: "Maintaining Gas Turbine Compressors for High Efficiency," August 1978.
   (e) Bingham, P. J., P. H. Huhtanen, and H. G. Starnes: "Maintenance of Gas Turbine Accessory Equipment," Sept. 1978.
   (f) Kiernan, J. G., A. D. Foster, and D. T. Harden: "Gas Turbine Fuels and Fuel Systems," October 1978.
   (g) Stretch, R. H., J. N. Shinn, and D. B. Brudos: "Calibration and Troubleshooting of Gas Turbine Control," November 1978.
3. Doherty, M. C., and D. R. Wright: "Application of Aircraft Derivative and Heavy Duty Gas Turbines in the Process Industries," ASME Paper 79-GT-12, *ASME Gas Turbine Conference,* San Diego, CA, March 1979.
4. Hilt, M. B., and R. H. Johnson: "Nitric Oxide Abatement in Heavy-Duty Gas Turbine Combustors by Means of Aerodynamics and Water Injection," ASME Paper No. 72-GT-53, March 1972.
5. Hilt, M. B., and J. Waslo: "Evolution of NO_x Abatement Techniques Through Combustor Design for Heavy-Duty Gas Turbines," *Journal of Engineering for Gas Turbines and Power,* Vol. 106, October 1984, pp 825–83.
6. Dibelius, N. R., M. B. Hilt, and R. H. Johnson: "Reduction of Nitrogen Oxides from Gas Turbines by Steam Injection," ASME Paper No. 71-GT-58, December 1970.
7. Miller, H. E.: "Development of the Quiet Combustor and Other Design Changes to Benefit Air Quality," American Cogeneration Association, San Francisco, March 1988.
8. Cutrone, M. B., M. B. Hilt, A. Goyal, E. E. Ekstedt, and J. Notardonato: "Evaluation of Advanced Combustor for Dry NO_x Suppression with Nitrogen Bearing Fuels in Utility and Industrial Gas Turbines," ASME Paper 81-GT-125, March 1981.
9. Zeldovich, J.: "The Oxidation of Nitrogen in Combustion and Explosions," *Acta Physicochimica USSR,* Vol. 21, No. 4, 1946, pp 577–628.
10. Washam, R. M.: "Dry Low NO_x Combustion System for Utility Gas Turbine," ASME Paper No. 83-JPGC-GT-13, September 1983.
11. Davis, L. B., and R. M. Washam: "Development of a Dry Low NO_x Combustor," ASME Paper No. 89-GT-255, June 1989.
12. Wilson, D. G.: *The Design of High Efficiency Turbomachinery and Gas Turbines,* M.I.T. Press, 1988.
13. Brandt, D. E.: "Gas Turbine Design Philosophy," 1991 Turbine State of the Art Conference: GER-3434A: Asheville, NC: August, 1991.
14. Fogg, H. E.: "Aeroderivative Gas Turbines," 1991 Turbine State of the Art Conference, GER-3695, Asheville, NC, August 1991.

15. Hopkins, J. E.: "Gas Turbine Performance Characteristics," 1991 Turbine State of the Art Conference, GER-3567B, Asheville, NC, August 1991.

16. Walsh, E. J., and M. A. Freeman: "Gas Turbine Operation and Maintenance Considerations," 1991 Turbine State of the Art Conference, GER-3620A: Asheville, NC, August 1991.

17. Kovacik, J. M.: "Cogeneration Application Considerations," 1991 Turbine State of the Art Conference, GER-3430B, Asheville, NC, August 1991.

18. Davis, L. B.: "Low NO_x Combustion for Gas Turbines," 1991 Turbine State of the Art Conference, GER-3568B, Asheville, NC, August 1991.

CHAPTER 6.5
STEAM TURBINES

Clemens M. Thoennes
Sales Programs Development
General Electric Company
Schenectady, New York

GLOSSARY

Automatic-admission turbine A steam turbine with the capacity to admit steam at two or more pressures. Valve gear at the low-pressure opening can automatically control the pressure in that header by internally varying the downstream turbine flow-passing capability.

Automatic-extraction turbine A steam turbine with the capacity to extract steam. The pressure of the extracted steam is controlled by a valve gear at that opening, as with an automatic-admission turbine. *Note:* Steam turbines can be furnished with automatic extraction and admission capability at the same opening.

Available energy The difference in enthalpy between an inlet steam condition (a specific pressure and temperature) and an exhaust pressure along a path of constant entropy.

Back-pressure turbine A steam turbine that exhausts at a pressure equal to or greater than atmospheric pressure. It is also referred to as a *noncondensing* turbine.

Bottoming cycle An energy recovery cycle that uses waste heat from another source to generate power. A steam turbine bottoming cycle uses steam generated by a waste-heat exhaust stream.

Bucket A blade located on the rotor of a steam turbine which transfers energy from the steam to the rotating shaft.

By-product power Power generated coincidentally when supplying useful heat. (See also "Noncondensing power" and "Cogeneration.")

Cogeneration The simultaneous production of power and other forms of useful energy—such as heat or process steam. U.S. government agencies restrict the definition of cogeneration to electric power generation only, while various industrial associations extend it to include mechanical-drive power.

Feedwater heater A steam-to-water heat exchanger that heats the boiler feedwater, generally with steam extracted from a steam turbine. In a closed feedwater heater the two fluids are separated by the use of shell and tube construction. In an open feedwater heater the fluids are mixed. A *deaerator* is an open feedwater heater that separates entrained gases from the feedwater by vigorous agitation with steam.

Fuel chargeable to power (FCP) See Chap. 6.4.

Impulse design A stage design approach characterized by a large pressure drop occurring in the nozzles and relatively little pressure drop across the buckets.

Mechanical-drive steam turbine A turbine used to drive devices other than electric generators, such as pumps or compressors. It is generally designed to operate over a wide speed range.

Mollier diagram A plot of enthalpy versus entropy of a fluid, which includes lines of constant pressure and temperature.

Noncondensing power Power generated by steam that is expanded through a turbine and either exhausts or extracts at a pressure equal to or greater than atmospheric pressure.

Nozzle Stationary blade in steam path generally utilized to turn, accelerate, and direct steam flow for efficient energy transfer to the rotating blades, or *buckets.*

Pressure control The ability of steam turbine governor systems to maintain constant pressure in a steam line by the action of a valve gear, as the turbine supplies steam to the header or draws steam from it.

Pressure rise point In an automatic-extraction steam turbine the maximum exhaust flow with zero extraction flow.

Reaction design A stage design approach characterized by an approximately equal pressure drop in the nozzle and associated bucket.

Steam rate The mass flow rate of steam required to produce a unit of output, in pounds per kilowatthour (kilograms per kilowatthour) or pounds per horsepower hour. The *theoretical steam rate* (TSR) assumes a perfect expansion process between two conditions. An *actual steam rate* is based on the actual expansion, including the inefficiency of the turbine and generator.

Uncontrolled extraction An opening in a steam turbine casing between two stages. The pressure of the extraction steam available is uncontrolled and is a function of the steam flow to the following stage.

TYPES OF TURBINES

There are many different types of steam turbines in industrial-plant service. These can be broadly classified as either *condensing* or *noncondensing.* Condensing steam turbines have a subatmospheric exhaust pressure, while noncondensing units exhaust at atmospheric or higher pressure. When steam is expanded to subatmospheric pressure in a condensing steam turbine, its temperature is generally reduced to less than 130°F (54°C). This low-temperature energy is usually not useful and is generally classified as waste heat. On the other hand, steam exhausted from noncondensing steam turbines is much higher in temperature and pressure and is useful in many industrial processes or heating applications.

Figure 6.73 shows schematic diagrams of three different configurations of noncondensing steam turbines. All three exhaust into a low-pressure header. The single automatic-extraction noncondensing (SAXNC) unit also extracts steam at another higher-pressure header. The double automatic-extraction noncondensing (DAXNC) unit extracts steam at two additional pressure levels. The term *automatic extraction* implies that the steam flow is automatically controlled (governed) internally to maintain constant pressure in the header independent of the extraction flow. Figure 6.74 is a cross-sectional view of a typical single automatic-extraction noncondensing steam turbine. The high-pressure section consists of the turbine inlet valve gear, a high-pressure turbine control stage, five additional turbine stages, and the extraction opening. The exhaust section includes the extraction valve gear, three additional turbine stages, and the exhaust opening. The turbine governor operates both sets of valve gears in concert to control the pressure in the extraction steam line and one other variable, such as exhaust pressure or the power output. Figure 6.75 is an axial cross-sectional view of the upper-inlet valve gear. It consists of six poppets which are individually operated through cam action lifts. This arrangement minimizes throttling losses for high-efficiency operation

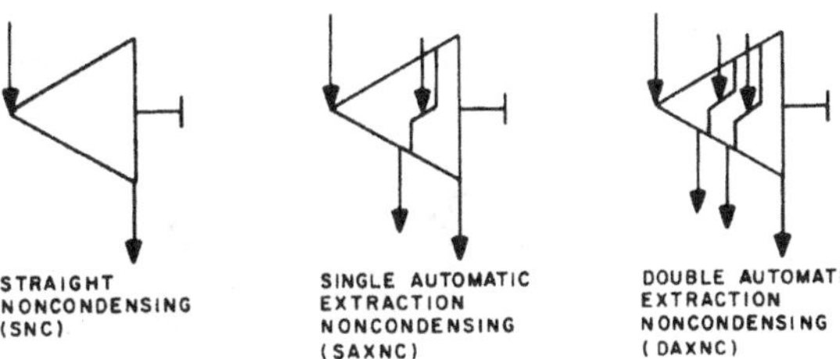

FIGURE 6.73 Schematic diagram of types of noncondensing steam turbines.

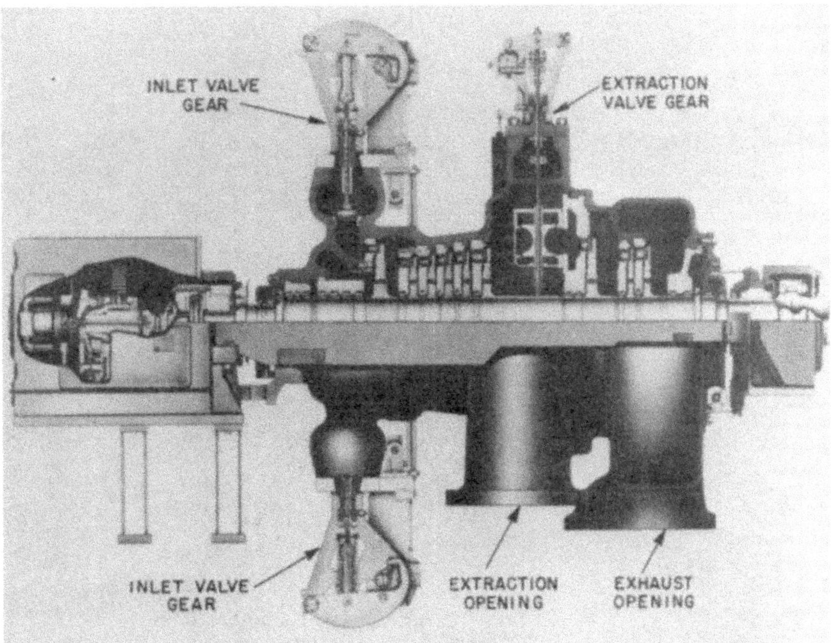

FIGURE 6.74 Cross-sectional view of a single automatic-extraction noncondensing steam turbine. (*General Electric Co.*)

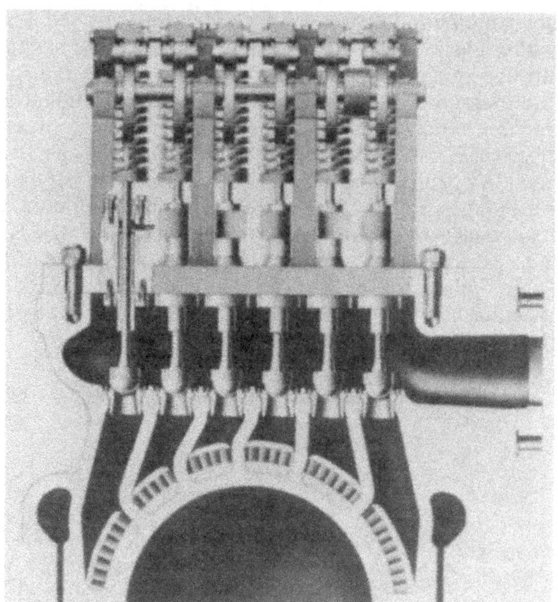

FIGURE 6.75 Axial cross-sectional view of a steam turbine inlet valve gear. (*General Electric Co.*)

FIGURE 6.76 Axial cross-sectional view of a spool-type extraction valve. (*General Electric Co.*)

over a wide range of steam flow. Units can be furnished with the upper-inlet gear only or an upper- and lower-inlet valve gear for increased flow capacity. Figure 6.76 is an axial cross-sectional view of a spool-type extraction. It is a variable restriction which is mounted downstream of the extraction opening. Notice that it is the steam flow to the exhaust, and not the extracted steam, which actually passes through the extraction valve.

Condensing steam turbines in industrial plants can also include automatic-extraction capability. Figure 6.77 is a schematic illustration of condensing steam turbine types; straight condensing, single automatic-extraction condensing (SAXC), and double automatic-extraction condensing (DAXC). A cross-sectional view of a typical DAXC steam turbine is shown in Fig. 6.78. Notice the relatively large exhaust casing required to pass the low-density subatmospheric exhaust steam flow. A DAXC steam turbine has the capability to control two extraction-pressure levels and also independently control the amount of power produced by varying steam flow to the condensing exhaust.

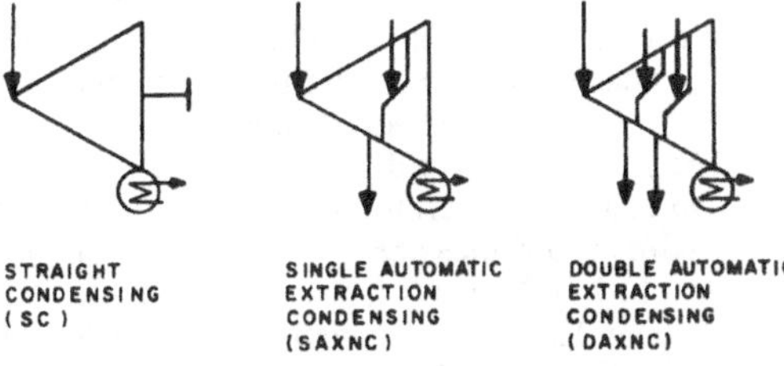

FIGURE 6.77 Schematic diagram of types of condensing steam turbines.

FIGURE 6.78 Cross-sectional view of a double automatic-extraction condensing steam turbine. (*General Electric Co.*)

Automatic admission capability is another feature that can be specified for industrial steam turbines. Some industrial plants have an excess of low-pressure steam which can be admitted through an extraction/admission opening and then expanded through the low-pressure section. In some cases, the turbines extract steam during normal operation and only admit steam during process upsets or outages.

The control stage and additional stages between each of the turbine pressure levels constitute a turbine section. For example, on a DAXC or DAXNC unit, there are three such sections, referred to as HP, IP and LP for high, intermediate, and low pressure. Each section is designed to pass a specified maximum flow. Turbine section efficiency falls off as the actual section flow (as a fraction of design flow) is reduced. The proper sizing of each turbine section is a key part of turbine design optimization and should take into account the expected duration of operation at different flows. Large, but infrequent or upset condition flows are often handled better by the use of external pressure-reducing stations to bypass one or more turbine sections.

Steam turbines can also be furnished with uncontrolled extraction openings. However, the pressure at an uncontrolled extraction varies with steam flow. The variation from normal conditions in absolute pressure is approximately proportional to the variation in flow through the following stage. Therefore, uncontrolled extractions are not generally suitable to supply process steam headers. On the other hand, uncontrolled extractions are suitable to provide feedwater heating in many cases. When steam is extracted from a turbine to heat only the feedwater for that turbine, then the requirements for extraction steam will also be proportional to steam flow. Large steam turbines in central stations can have as many as six uncontrolled extractions to supply different stages of feedwater heaters.

The steam turbines illustrated in Figs. 6.74 to 6.78 are of multivalve, multistage construction. Smaller steam turbines are available with a single throttling valve on the inlet and with one (or more) turbine stage(s). These units are classified as single-valve/multistage or single-valve/single-stage. This type of construction reduces the initial cost of the steam turbine, but at a penalty in efficiency, which can be less than half that of the multistage/multivalve type. Single-valve/single-stage units are generally applied in small [fewer than 1000 hp (746 W)], mechanical-drive service.

TURBINE DESIGN CHARACTERISTICS

Turbine Stage Design and Construction

Steam turbine stage designs are generally either of an *impulse* or *reaction* type.

In the impulse design, most of the stage pressure drop occurs in the stationary nozzles, with relatively little pressure drop in the buckets (or moving blades). Because there is little pressure drop across the buckets, it is possible to mount them on the periphery of a wheel without generating a large axial thrust on the rotor. This *wheel and diaphragm* construction, typical of the impulse design, is shown in Fig. 6.79.

In the reaction design, approximately equal pressure drop occurs in the nozzles and buckets. To reduce the associated axial thrust on the rotor, a *drum* construction is used with the reaction design, as shown in Fig. 6.80. Even with a drum rotor, it is usually necessary to build a *balance piston* into the rotor to balance the stage thrust, unless the turbine section is double-flowed.

Peak efficiency is obtained in an impulse stage with more work per stage than in a reaction stage for a given stage diameter. It is normal, therefore, for an impulse turbine section to require either fewer stages on the same diameter or the same number of stages on a smaller diameter.

With less pressure drop across the bucket, the bucket tip leakage of an impulse design is much less than that of a reaction design. There is more pressure drop across the diaphragm of an impulse design than that of a reaction design; however, the leakage diameter is considerably less. In addition, there is sufficient room in the inner web of the diaphragm to mount spring-backed packings with a generous amount of radial movement and a large number of labyrinth packing teeth. In total, then, the shaft leakage around the nozzles is lower for an

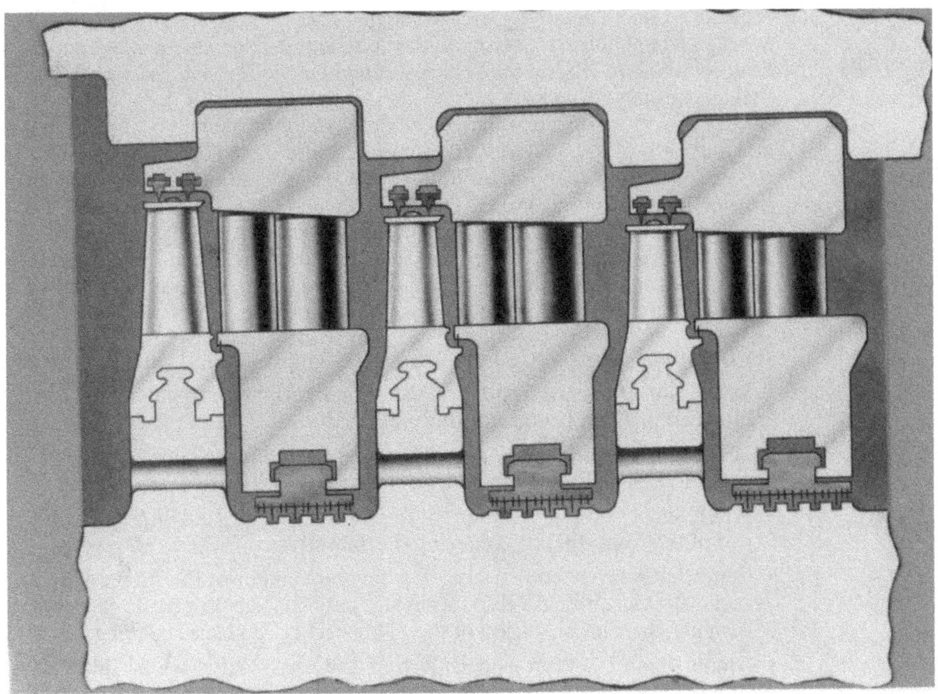

FIGURE 6.79 Typical impulse stages, wheel and diaphragm construction.

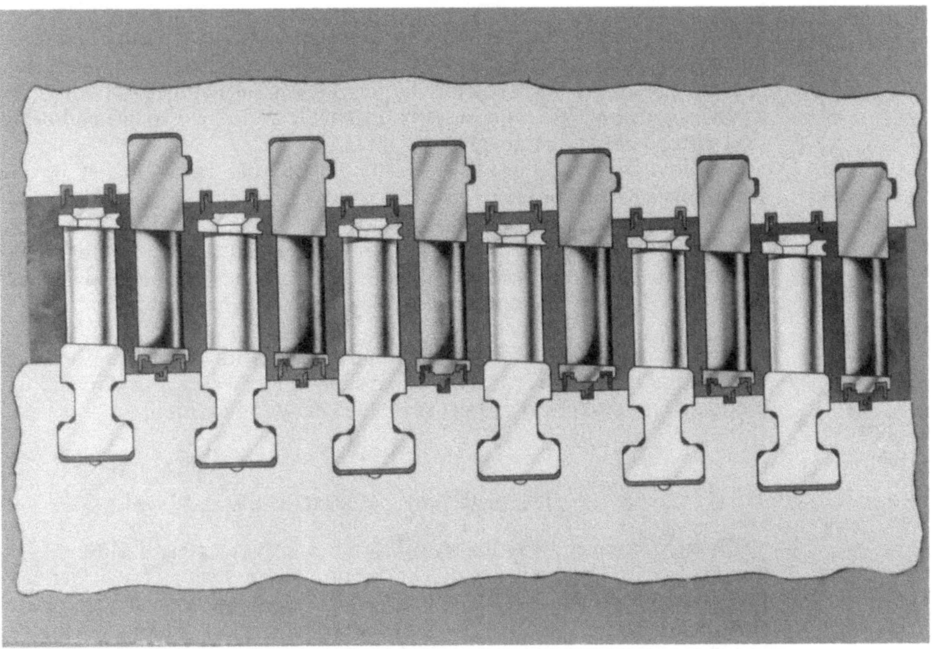

FIGURE 6.80 Typical reaction stages, drum rotor construction.

impulse stage than for a reaction stage. With the impulse stage, this leakage is diverted through wheel holes. However, with the reaction stage, this leakage reenters the steam path between the nozzles and buckets, resulting in an additional loss, due to the disturbance of the main steam flow.

Leakage flows in a turbine stage can easily be calculated and the results show that for stages typical of today's industrial steam turbines tip leakage flows are 2 to 4 times greater for a reaction design than for an impulse design, and shaft packing flows are 1.2 to 2.4 times greater for a reaction design. The importance of these leakages becomes greater with the smaller volume flows typical of industrial applications.

In addition to lower leakage flows, the impulse wheel-and-diaphragm construction has significant advantages in the areas of thermal stress and, therefore, low-cycle fatigue resistance. Specific characteristics impacting these areas include:

- Rotor body diameters are small, thereby significantly reducing thermal stress.
- Bucket dovetails are located on the wheel peripheries away from the rotor surface, separating centrifugal from rotor thermal stress.
- Fewer stages and smaller stage diameters permit more compact designs with smaller components, especially shells.
- Wheel shapes on the periphery of the rotor act as fins and improve heat transfer to the interior of the rotor, thereby reducing thermal stress.
- Centerline-supported diaphragms allow movement of the diaphragms with respect to the shell without rubbing; if the stationary blading is held in a blade carrier (as in reaction/drum designs), thermal distortion of the blade carrier can cause rubbing.
- Ample room between wheels permits the use of large wheel fillets to minimize thermal stress.

Thermodynamic Considerations

The properties of steam are well documented in steam tables.[1] These tables, or associated numerical formulations, are useful in the calculation of detailed steam turbine cycle performance. A familiar graphical representation of steam properties is the *Mollier diagram,* which is a plot of enthalpy (h) versus entropy (s). An ideal turbine expansion is represented by a vertical path on the Mollier diagram.

Several terms and concepts are commonly used in the discussion of actual steam turbine thermodynamic performance characteristics.

Available Energy. The available energy is the difference in enthalpy between an initial thermodynamic state (pressure, temperature) and a final thermodynamic state at a lower pressure and at constant entropy, associated with an ideal turbine expansion.

Theoretical Steam Rate. The theoretical steam rate is the ratio between mass flow and work output for an ideal turbine expansion, in lb/kWh or kg/kWh. The work output is the product of the mass flow and the available energy. Therefore,

$$\text{TSR (lb/kWh)} = 3412.14\ \text{(Btu/kWh)/AE (Btu/lb)}$$

$$\text{TSR (kg/kWh)} = 3600.00\ \text{(kJ/kWh)/AE (kJ/kg)}$$

The performance of actual steam turbines includes several different types of losses and irreversibilities. The efficiency of a steam turbine-generator (TG) is defined as the output at the generator terminals divided by the available energy. Typical industrial steam turbine efficiencies are 70 to 80 percent. The *actual steam rate (ASR)* is defined as the TSR divided by the turbine-generator efficiency.

Example. Consider a turbine operating with inlet steam conditions of 1250 psig, 900°F (8720 kPa, 482°C). For an isentropic expansion to 4.0 in HgA (inches of mercury absolute) (13.55 kPa), the available energy determined from the steam tables is:

$$AE = 521 \text{ Btu/lb } (1212 \text{ kJ/kg})$$

The theoretical steam rate is:

$$TSR = 3412.14/521 = 6.541 \text{ lb/kWh}$$

$$= 3600.00/1212 = 2.97 \text{ kg/kWh}$$

For a steam turbine efficiency of 0.75, the actual steam rate would be:

$$ASR = 6.541/0.75 = 8.72 \text{ lb/kWh}$$

$$= 2.97/0.75 = 3.96 \text{ kg/kWh}$$

The actual steam rate can be used to determine the required steam flow to produce a given output. To produce 20,000 kW, the required steam flow would be:

$$\text{Steam flow} = ASR \times \text{output} = 8.72 \times 20{,}000 = 174{,}400 \text{ lb/h}$$

$$= 3.96 \times 20{,}000 = 79{,}200 \text{ kg/h}$$

When steam turbines exhaust or extract into a process header, the value of enthalpy at that point is also of interest. In order to determine exhaust or extraction enthalpy, actual steam rates or turbine efficiencies are commonly defined for the various sections of the turbine.

Performance characteristics of industrial steam turbine-generators are generally plotted as curves relating throttle (and extraction) flows to generator output. Figure 6.81 shows a typical performance curve of a straight condensing or noncondensing steam turbine. The plot of throttle flow versus generator output approximates a straight line. The plot is known as the *Willans line.* The intercept of this line at zero load represents the steam flow required to supply the no-load losses of the set.

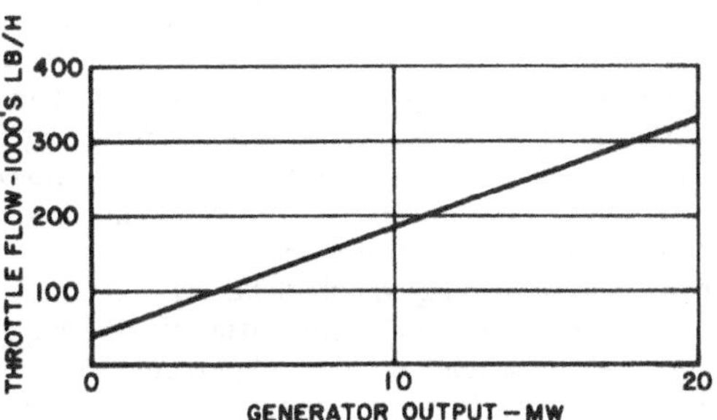

FIGURE 6.81 Performance curve of a straight condensing or straight noncondensing steam turbine.

The performance curve of a typical single automatic-extraction turbine is shown in Fig. 6.82. The zero extraction line is similar to the performance of a straight condensing or noncondensing unit. The family of parallel lines indicates the performance for various values of extraction flow. A certain minimum amount of flow is required at the exhaust end to cool the turbine buckets. This limit is indicated on the left-hand side of the plot. Also, the geometry of

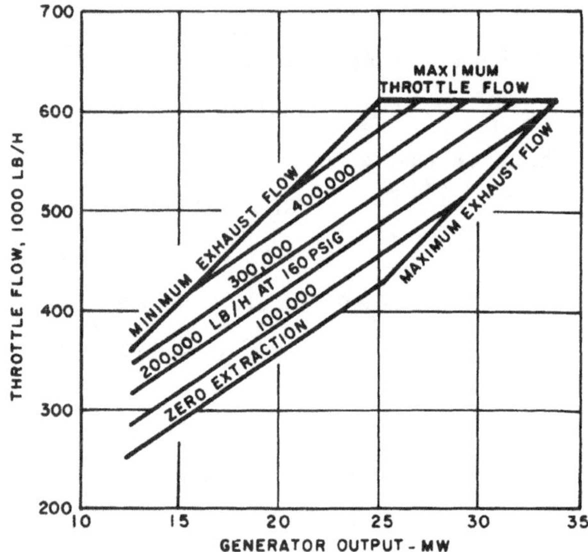

FIGURE 6.82 Performance curve of a single automatic-extraction steam turbine.

the section will limit it to a maximum flow capability, which is shown on the right-hand side. The intercept of the maximum exhaust flow line and the zero extraction line is called the *pressure rise* point. Additional flow can be put through the exhaust end, but only if the pressure ahead of the extraction valve gear increases. This also causes an increase in the extraction pressure. This is not a normal mode of operation.

Performance of Mechanical-Drive Steam Turbines

Steam turbines are also selected as drivers for pumps and compressors that generally require variable speed. Process compressors as large as 60,000 hp (44,800 kW) are in service with variable-speed steam turbine drives. Special robust buckets are required for these turbines to withstand the vibratory stresses inherent with continuous operation over a wide speed range. In other respects, large multistage/multivalve mechanical-drive steam turbines are similar to corresponding turbine-generator sets.

Single-valve/single-stage mechanical-drive steam turbines are very common in small sizes [less than 1000 hp (746 kW)] in industrial plants. This type of construction has a low cost, but efficiency drops to a range of 40 to 50 percent. These small units are often used in parallel or as a backup to motor drives to provide added reliability against a loss of either electric power or steam supply. Small mechanical-drive steam turbines also can be selected as drives for hazardous locations. However, the low efficiency of these types of turbines is an increasingly important disadvantage with high energy costs.

Control Systems

The philosophy applied to steam turbine control systems has developed over time, and is summarized in Table 6.26.

TABLE 6.26 Steam Turbine Control Philosophy

1. Clear separation between control and protection shall be provided.
2. Controls use two out of three (3) redundancy from sensor to actuator for all vital and important functions.
3. A single failure in the controls will not cause a shutdown. It will cause a diagnostic alarm, and it is repairable on-line.
4. Controls comply with IEEE 122 std. (e.g., can reject rated load without causing a turbine trip).
5. A protection system backup is provided for all control functions.
6. A double set of steam valves is provided for all major admissions. One set for controls and one set for protection.
7. Protection (trips) are classified according to criticality: vital have conceptual redundancy.

The main functions of a modern steam turbine control system are:

- Speed and acceleration control during start-up
- Initialization of generator excitation
- Synchronization and application of load in response to local or area generation dispatch commands
- Pressure control of various forms: inlet, extraction, back pressure, etc.
- Unloading and securing of the turbine
- Sequencing of the above functions under constraint of thermal stress
- Overspeed protection during load rejection and emergencies
- Protection against serious hazards, e.g., loss of lubrication oil pressure, high exhaust temperature, high bearing vibration
- Testing of steam valves and other important protective functions

Additional control and monitoring functions are also required in most applications, such as:

- Monitoring and supervision of a large number of pressures, temperatures, etc., to provide guidance and alarms for operators
- Start-up and monitoring of turbine-generator auxiliaries such as lubrication oil, hydraulic, and steam seal systems
- Display, alarm, and recording of the above functions and data
- Diagnosis of turbine or generator problems
- Health check and diagnostics of the electronic system itself

The first group of functions must be performed with high control bandwidth or very high reliability, or both, to ensure long-term reliable operation and service of the turbine. It is for these reasons that turbine controls and protection are closely coupled with the turbine detailed design. The first group of functions, together with the input and output (I/O) devices required, are included in the turbine unit control system, which is an integral part of the steam turbine hardware.

Steam turbine unit control systems have many capabilities to control steam pressures. The most common feature is steam pressure control of straight noncondensing or automatic-extraction noncondensing units. The synchronized generator is "locked into" the grid frequency, and the turbine inlet and extraction control valves maintain flow through the sections of the turbine in response to extraction and exhaust steam pressure signals. This type of system follows heat demands. The electricity that is generated in this manner is called *by-product power.*

When automatic-extraction steam turbines have a condensing exhaust, the control system can control pressure as well as power generation by varying flow to the condenser. The amount of power called for by the steam turbine unit control system can be continuously modulated by the plant control system.

Another control mode is initial pressure control. When a steam turbine is supplied by a heat recovery boiler or a by-product fuel boiler, the amount of steam generated can vary independently. Initial pressure control allows the steam turbine to draw all the available steam out of the header while maintaining constant pressure.

A characteristic of the unit control system is that all essential turbine control and protection functions are included to allow a unit to operate safely even if other supporting systems should fail. Another characteristic is that the control point interface (i.e., the interface between the turbine and the control system) remains in the turbine supplier's scope, while interface to plant controls can be made at data-point level, which does not include critical and rapidly varying commands and feedback signals and, therefore, is a more suitable point of interface to plant controls. Yet another characteristic of unit control functions is that they must be performed either continuously or very frequently to provide satisfactory control.

The second group of functions can be performed less frequently (i.e., every few seconds or more), and turbine operation may be continued in most cases during short-term interruptions in the monitoring functions as long as the unit control is performing correctly.

The second group of functions is called TGM, for turbine generator monitoring. The TGM functions can be included in the unit control system or in the plant control system.

OPERATION AND MAINTENANCE

Starting Procedures

Before a steam turbine-generator can be started, proper operation of a number of auxiliary systems must be assured, including:

- Bearing oil pump and seal oil pump, where applicable
- Hydraulic control fluid pump
- Condenser vacuum pump, where applicable
- Generator cooling system (air or hydrogen pressure and circulating water)
- Emergency trip system

Drains in all steam lines must be opened to prevent slugs of water from entering the turbine. Units of 10 MW or larger are generally furnished by a turning gear. This is then engaged to rotate the unit slowly. The steam lines are preheated and the gland seal exhaust system is put into operation. Steam is very gradually admitted to the turbine, and the turbine rolls off the turning gear and is gradually accelerated. The rate of acceleration selected will depend on the casing metal temperature, which depends on the duration of the previous outage. Acceleration time can vary from 10 to 30 min. During this period vibration and turbine shell temperature are monitored.

As the unit approaches synchronous speed, the generator breaker is in the open position and the excitation system regulates voltage to match the bus voltage. The operator then matches the frequency and phase angle before the generator breaker is closed. The unit then picks up load at a rate again depending on the temperature of the unit before start-up. Typical loading times are 30 to 60 min, depending on whether the unit was hot or cold before starting. Many of these starting functions can be handled automatically by modern turbine unit control systems, as discussed previously.

Normal Operation

During periods of normal operation several parameters are monitored, including turbine shell temperature and pressure, exhaust hood temperature, bearing oil temperature and pressure, condenser vacuum, shaft vibration, hydraulic oil pressure, and generator gas and winding temperatures.

Control systems generally contain many automatic protection and alarm systems. A typical protection system includes a main stop valve which is spring-loaded to close and is mounted ahead of the turbine inlet valve. Hydraulic oil keeps this valve open if several protection devices all indicate safe operation. These include an overspeed governor, bearing oil pressure relay, a manual trip relay, and where applicable, a low-vacuum relay and nonreturn valve relays.

Maintenance

A limited amount of running maintenance is required in addition to data logging. This includes periodic lubrication of valve gear (monthly or quarterly) and H_2 replenishment for hydrogen-cooled generators (three times a month).

The turbine shell is generally removed for a warranty inspection after the first year's operation. Shutdowns thereafter can be at intervals of 3, 4, or 5 yr. Long periods of operation with minimal maintenance require steam of high purity. Carryover of certain contaminants in the steam can cause deposits, erosion, and stress cracking. A list of common deposit- and corrosion-causing contaminants is given in Table 6.27.

TABLE 6.27 Common Water Contaminants

Deposit-forming	Corrosion-causing
Calcium salts	Ammonia
Magnesium salts	Oxygen
Silica	Chlorides
Iron salts	Sulfides
Organic matter	Carbonates
Copper salts	Bicarbonates
Sulfates	
Nitrates	

These contaminants can enter the steam supply system with the makeup water or in process heat exchange equipment. They can exist in the boiler drum in relatively high concentrations without causing problems. It is only when they are carried over into the exiting steam that they enter the turbine. Efficient boiler drum separators can limit total dissolved solids to as little as 0.5 to 1.0 ppm. Silica is difficult to separate from the steam and must be controlled in the boiler feedwater. Care must also be taken with the fluid used to attemperate the steam exiting the superheater. Contaminated process returns can bypass the steam separation in the drum as attemperator fluid. This should be maintained at less than 1.0 ppm total dissolved solids.

Primary water treatment conditions the makeup water before it enters the cycle. Secondary water treatment is the addition of chemicals to the cycle to "polish" the feedwater. Boiler blowdown is the removal of solids from the cycle to prevent high concentrations in the drum. Monitoring purity is a very important part of steam turbine maintenance.

The effect of steam impurity may be observed by visually inspecting steam path components (turbine buckets and nozzle partitions) for damage or deposits. These conditions can significantly affect availability if they are not identified until the unit is opened for a regularly scheduled outage. This is particularly true if new parts are needed but have not been ordered. One feature available to help in planning for maintenance outages is the borescope-access port. These ports permit the insertion of optical devices (i.e., borescopes) for the visual inspection of buckets and nozzle partitions without having to disassemble the major turbine components.

Many types of deposits can be removed by washing the turbine at reduced speed with wet steam. Persistent deposits are removed by opening the turbine. Water washing and rinsing remove most soluble deposits. Steam cleaning and/or blasting with a fine-grade abrasive may be required for hard deposits like silica or iron oxide.

APPLICATIONS IN PLANTS

Cogeneration

Cogeneration is the simultaneous generation of power and useful heat. Steam turbines are the most common prime movers used in cogeneration.

In *thermally optimized* steam turbine cogeneration cycles, steam is expanded in noncondensing or automatic-extraction noncondensing steam turbine-generators which extract and/or exhaust into the process steam header(s). The fuel chargeable to power (FCP) for these systems is typically in the 4000- to 4500-Btu/kWh HHV range. The influence of initial steam conditions and process steam pressure on the amount of cogenerated power per 100 million Btu/h net heat to process (NHP) is shown in Fig. 6.83. The increase in cogenerated power through the use of higher initial steam conditions, as well as lower process pressures, is readily apparent.

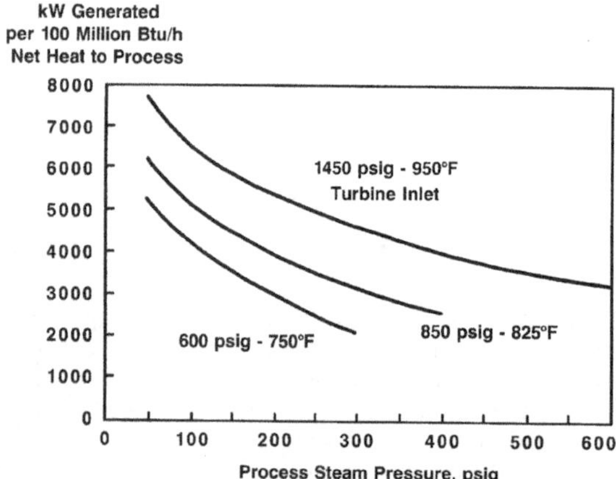

FIGURE 6.83 Cogeneration power with steam turbines.

Studies have shown that the higher steam conditions can be economically justified more easily in industrial plants having relatively large process steam demands. Data given in Fig. 6.84 provide guidance with regard to the initial steam conditions normally considered for industrial cogeneration applications. Higher energy costs experienced since the mid-1970s are favoring the upper portion of the bands shown in Fig. 6.84. However, particularly above 1800 psig (12,411 kPa), the capital costs should be carefully understood. Material changes in the turbine and steam generator can significantly increase first cost.

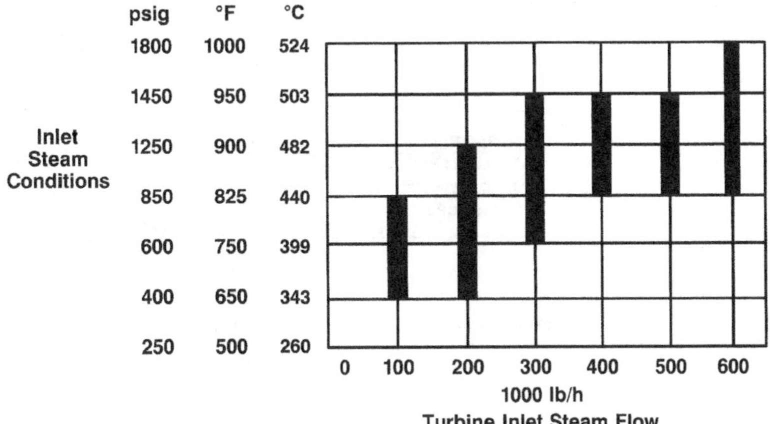

FIGURE 6.84 Range of initial steam conditions normally selected for industrial steam turbines.

Even through the use of most effective steam turbine cogeneration systems, the amount of power that can be cogenerated per unit of heat energy delivered to process will usually not exceed about 85 kW per million Btu net heat supplied. This is generally less power than is required to satisfy most industrial plant electrical energy needs. Thus, with thermally optimized steam turbine cogeneration systems, a purchased power tie or condensing power generation is necessary to provide the balance of the plant power needs.

Condensing steam turbine power generation, although not necessarily energy-efficient, has proven economic in many industrial applications. Favorable economics are often associated with systems where:

- Condensing power is used to control purchased power demand.
- Low-cost fuels or process by-product fuels are available.
- Adequate low-level process energy is available for a bottoming cogeneration system.
- Condensing provides the continuity of service in critical plant operations where loss of the electric power can cause a major disruption in process operations and/or plant safety.
- Utility-specific situations favor power sales, particularly if low-cost fuels are available.

Steam turbine efficiency is another important parameter in cogeneration plants. In noncondensing steam turbine applications, the turbine efficiency has only a minor effect on FCP. However, the amount of high-efficiency by-product power generated is directly proportional to the turbine efficiency. Converting fuel energy into low-temperature heat is a simple process, and high efficiencies of 85 to 90 percent can be achieved. However, converting fuel energy into power is much more difficult, and efficiencies are only 33 to 35 percent in the largest, most modern central stations. Therefore, high-efficiency multistage/multivalve noncondensing steam turbines should be preferred over inefficient single-valve/single-stage designs because more of the fuel energy is converted to a more valuable quantity—power.

Feedwater Heating

Feedwater heating is another method of enhancing the amount of by-product power generated for a given process heat load. Figure 6.85 illustrates a single automatic-extraction noncondensing steam turbine with a single heater and a second higher-pressure feedwater heater. The second heater increases the amount of by-product power by 2550 kW.

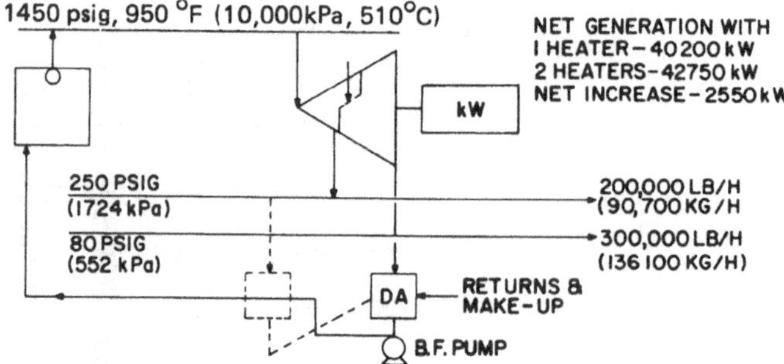

FIGURE 6.85 Effects of feedwater heating on by-product power generation.

As was the case with gas turbines, or any other type of heat engine applied to a cogeneration system, FCP or the quantity of by-product power is not the only factor to be considered. The investment in this equipment must be justified in terms of rate of return or a similar criterion.

In the case of large variable-speed drivers, mechanical-drive steam turbines often are the only viable solution. For example, a schematic diagram of a large ethylene plant driver is shown in Fig. 6.86. The charge gas compressor has a speed and power beyond that of suitable gas turbines or direct-drive motors. Process heat recovery produces steam at 1500 psig, 950°F (10,446 kPa, 510°C) and there is a large process heat demand at 300 psig (2070 kPa). A single automatic-extraction condensing steam turbine is ideally suited as a driver in these types of plants.

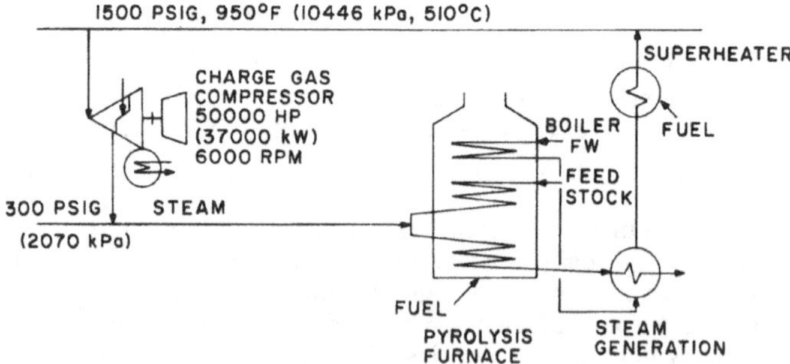

FIGURE 6.86 Schematic diagram of a large mechanical-drive steam turbine in an ethylene plant.

ACKNOWLEDGMENT

The author acknowledges the support and contribution provided by the Engineering Staff of GE's Power Generation Business.

REFERENCES

1. *ASME Steam Tables,* The American Society of Mechanical Engineers, New York, 1967.
2. *Theoretical Steam Rate Tables,* The American Society of Mechanical Engineers, New York, 1969.
3. Salisbury, J. K.: *Steam Turbines and Their Cycles,* Wiley, New York, 1950.
4. Newman, L. E.: *Modern Turbines,* Wiley, New York, 1944.
5. Kovacik, J. M., and W. B. Wilson: "Turbine Systems to Reduce Petroleum Refining and Petrochemical Plant Energy Costs," *Petroleum Division Conference,* ASME Paper 76-PET-62, Mexico City, Mexico, September 1976.
6. Kovacik, J.: "Cogeneration Application Considerations," 1991 Turbine State of the Art Conference, GER-3430B, Asheville, NC, August 1991.
7. Couchman, R. S., et al.: "Steam Turbine Design Philosophy and Technology Programs," 1991 Turbine State of the Art Conference, GER-3705, Asheville, NC, August 1991.
8. Kure-Jensen, J., et al.: "Steam Turbine Control System," 1991 Turbine State of the Art Conference, GER-3687, Asheville, NC, August 1991.

CHAPTER 6.6

DIESEL AND NATURAL GAS ENGINES

Waukesha Engine Division
Dresser Industries, Inc.
Waukesha, Wisconsin

GLOSSARY

Brake horsepower (bhp) The horsepower delivered by the engine shaft at the output end.[1] The name is derived from the fact that it originally was determined by a braking device on the engine flywheel.

Brake mean effective pressure (BMEP) The average cylinder pressure to give a resultant torque at the flywheel.[2]

$$\text{BMEP (lb/in}^2\text{)} = \frac{792{,}000 \times \text{bhp}}{(\text{r/min}) \times \text{displacement (CID)}} \quad \text{(four-stroke cycle)}$$

$$\text{BMEP (lb/in}^2\text{)} = \frac{396{,}000 \times \text{bhp}}{(\text{r/min}) \times \text{displacement (CID)}} \quad \text{(two-stroke cycle)}$$

$$\text{BMEP (kPa)} = \frac{600{,}000 \times \text{kW}}{(\text{r/min}) \times \text{displacement (L)}} \quad \text{(two-stroke cycle)}$$

$$\text{BMEP (kPa)} = \frac{1{,}200{,}000 \times \text{kW}}{(\text{r/min}) \times \text{displacement (L)}} \quad \text{(four-stroke cycle)}$$

Compression ratio The ratio of the volume of cylinder space above the top piston ring with the piston at *bottom* dead center to the volume in the cylinder above the top ring when the piston is at *top* dead center.

Cycle The complete series of events in each cylinder, including introduction and compression of air, burning of fuel, expansion and expulsion of the working medium in the engine.[1]

Diesel engine An engine in which the fuel is ignited entirely by the heat resulting from the compression of the air supplied for combustion.[1]

Displacement The volume of an engine's cylinder swept by the piston. It is equal to the area of each piston multiplied by the stroke multiplied by the number of cylinders and is expressed in cubic inches displacement (CID) or liters.

Duty cycle A term used to describe the load pattern imposed on the engine. Continuous, or heavy-duty, service is generally considered to be 24 h/day with little variation in load or speed.[2] Intermittent, or standby, service is classed as duty where an engine is called upon to operate only in emergencies or at infrequent intervals.

Four-stroke cycle engine An engine completing one cycle in four strokes of the piston or two shaft revolutions. The cyclic events are designated by the following strokes: (1) induction or suction stroke, (2) compression stroke, (3) power or expansion stroke, (4) exhaust stroke.[1]

Gas engine An engine in which the fuel in its natural state is a gas and the air-fuel mixture is ignited by a spark within the combustion chamber or the prechamber.

Intercooler A device used to cool the intake air after it leaves the turbocharger compressor but before it enters the engine cylinders. This device is sometimes called an *aftercooler* or *charge air cooler.*

Naturally aspirated A term used to describe an engine which does not use a device to increase the pressure on the intake air of the engine before it enters the cylinder.

Opposed piston engine One that uses the working medium simultaneously between two pistons in the same cylinder.[1]

Power The rate of doing work. Engine power output is expressed in units of horsepower, equivalent to 550 ft·lb/s, or kilowatts, equivalent to 1000 J/s. Here are some conversion formulas:

Useful Equivalents

$$1 \text{ hp (U.S.)} = 550 \text{ ft} \cdot \text{lb/s}$$

$$1 \text{ hp (metric)} = 0.9864 \text{ hp (U.S.)}$$

$$1 \text{ hp (U.S.)} = 1.0138 \text{ hp (metric)}$$

$$1 \text{ kW} = 1000 \text{ J/s}$$

$$= 1000 \text{ N} \cdot \text{m/s}$$

$$1 \text{ kW} = 1.341 \text{ hp (U.S.)}$$

Engine power expressed by various engine manufacturers throughout the world should be specified as to the conditions under which the engine was run or the conditions to which the horsepower is corrected. Rating societies such as the Diesel Engine Manufacturers Association (DEMA), the Society of Automotive Engineers (SAE), German Industry Standards (DIN), British Standards (BS), and International Standards Association (ISO) all have their ratings for temperature and barometric conditions to which engine manufacturers relate their data. Manufacturers usually specify their engine performance and rating standards as well as methods of correcting the power available to expected site conditions of temperature and barometric pressure.

Speed regulation The incremental difference between no-load (NL) speed and full-load (FL) speed divided by the full-load speed. This is sometimes referred to as *speed droop*.

$$\frac{(\text{NL speed}) - (\text{FL speed})}{(\text{FL speed})} \times 100 = \text{percent speed regulation}$$

Torque Twisting effort of an engine described in pound · feet^2:

$$T \text{ (U.S.)} = 5252 \times \frac{\text{bhp}}{\text{r/min}}$$

Turbocharger A rotary air compressor driven by a turbine using exhaust gas as the driving fluid and compressing intake air into the engine.

Two-stroke cycle engine An engine completing one cycle in two strokes of the piston or one shaft revolution. The cyclic events are designated by the following strokes: (1) induction and compression stroke, (2) expansion and exhaust stroke.[1]

INTRODUCTION

Stationary internal combustion engines are employed in plant applications primarily for emergency and prime electric power generation as well as for compressors, blowers, chillers for air conditioning, and refrigerating and pumping equipment. Diesel engines are widely used for emergency fire and flood pumps as well as for in-plant engine-generator sets and they have found acceptance in some areas as prime movers for compressor equipment used for air conditioning or refrigeration. In sewage treatment plants, where gas is produced as a by-product and captured for use as engine fuel, gas engines have become popular for driving blowers required for the aeration process, for pumping effluent, and for prime power generation. Landfill gas, a by-product of the urban sanitary landfill, is one of the latest alternative gaseous fuels used to generate electricity for sale to the utility power grid.

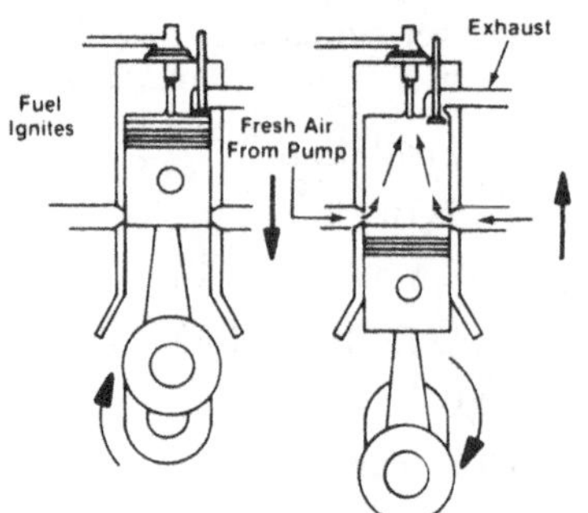

FIGURE 6.87 Two-stroke diesel engine. (*Waukesha Engine Division, Dresser Industries, Inc.*)

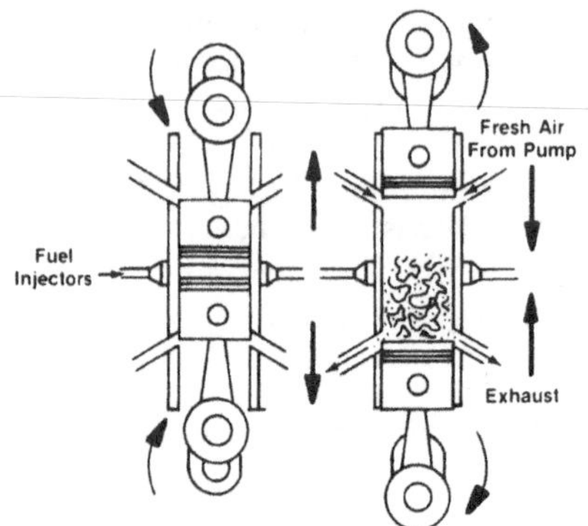

FIGURE 6.88 Opposed-piston diesel engine. (*Waukesha Engine Division, Dresser Industries, Inc.*)

DIESEL ENGINES

Principles

The diesel engine is an engine in which the cylinder is charged with air, which is then compressed until it is hot enough to ignite fuel injected into the combustion space. The fuel is ignited by the hot air, and the expanding gases drive the piston down on the power stroke. The compression ratios of diesel engines cover an approximate range from 12:1 to as high as 23:1.

Engine types fall into two categories: two-stroke cycle and four-stroke cycle. Typical two-stroke-cycle engine arrangements are shown in Figs. 6.87 and 6.88. A typical four-stroke-cycle engine arrangement is shown in Fig. 6.89.

There are many types of combustion chamber designs which manufacturers use to mix fuel and air. There are claimed advantages for each type. These may be fuel economy, ability to make use of simpler fuel systems, wide speed-range coverage, or firing pressure minimization. These systems generally fall into an *open-chamber* (Fig. 6.90) or *divided-chamber* (Fig. 6.91) category.

The open chamber design has the advantage of easier starting and has less heat rejected to the cooling system, but has greater demands on the injection system. The divided-chamber engines can operate with less sophisticated injection equipment since the air and fuel mixing is aided by more rapid

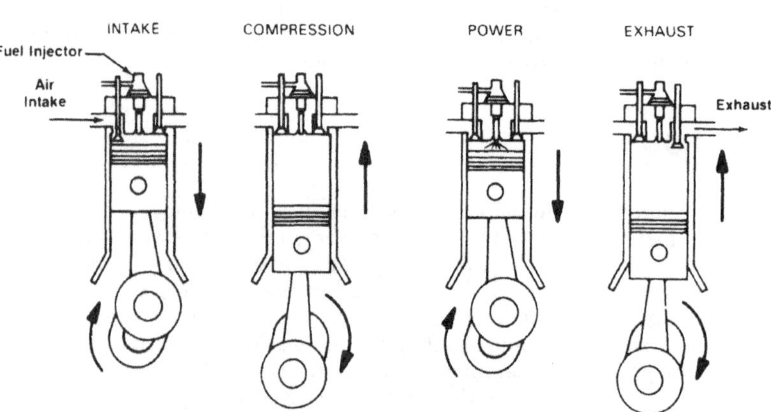

FIGURE 6.89 Four-stroke diesel engine. (*Waukesha Engine Division, Dresser Industries, Inc.*)

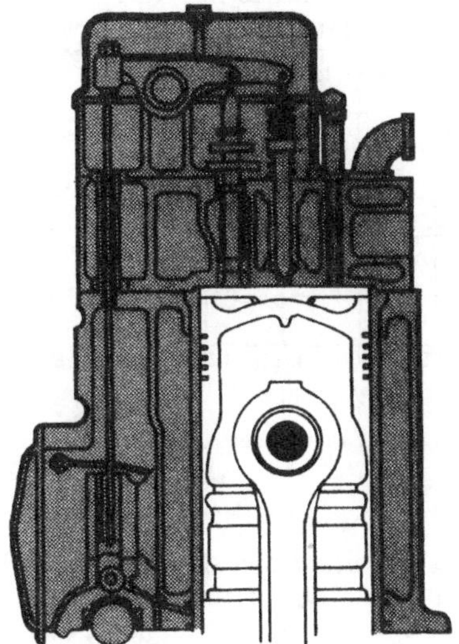

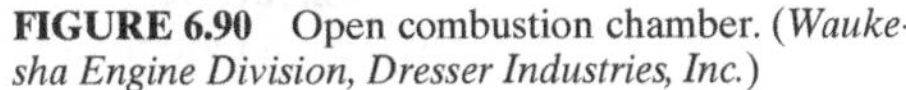

FIGURE 6.90 Open combustion chamber. (*Waukesha Engine Division, Dresser Industries, Inc.*)

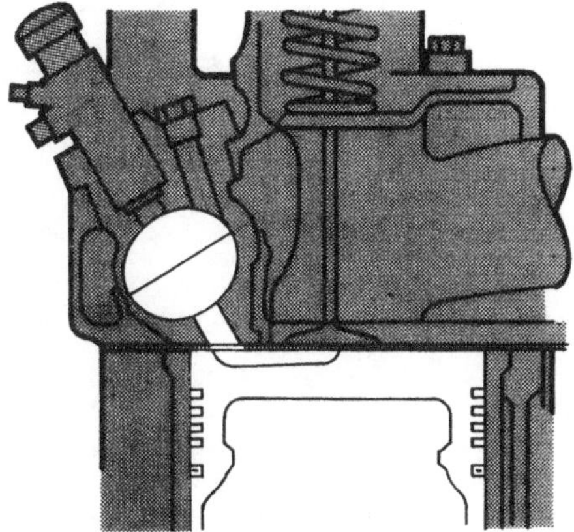

FIGURE 6.91 Divided combustion chamber. (*Waukesha Engine Division, Dresser Industries, Inc.*)

air motion. Better fuel economy can be attained with an open-chamber design, while a divided chamber permits lower emissions.

Engines are frequently *turbocharged* to increase the horsepower taken from a given displacement engine. Turbocharging utilizes some of the waste heat energy and velocity of engine exhaust gas to drive a turbine connected to a high-speed centrifugal compressor. The power from a given package size can be more than doubled, provided the engine components are strong enough to withstand the higher cylinder pressures. The highly turbocharged engines usually have a means of reducing the air temperature after the air leaves the compressor by means of an air-to-air cooler or an air-to-water cooler. These devices are known as *charge air coolers*, *aftercoolers*, or *intercoolers*.

Performance Characteristics

The performance of a diesel engine is affected by air temperature, air pressure, and humidity. Correction factors are employed to ensure that the power specified takes into account the losses that are expected with altitude or temperature conditions at the site. These correction factors are usually specified by the Society of Automotive Engineers (SAE), German Industry Standards (DIN), British Standards (BS), or the International Organization for Standardization (ISO).

Fuels

Introduction. Diesel engines can be designed to use a wide variety of fuels; however, if fuel other than diesel grade 1 or 2 is used, the manufacturer should be consulted. Jet A fuel can be used in diesel engines if it has a 40 octane minimum.

Fuel consumption rates on diesel engines are generally specified as brake-specific fuel consumption and its units are in pounds per brake horsepower hour, grams per brake horsepower hour, or grams per kilowatthour. (See Fig. 6.92.)

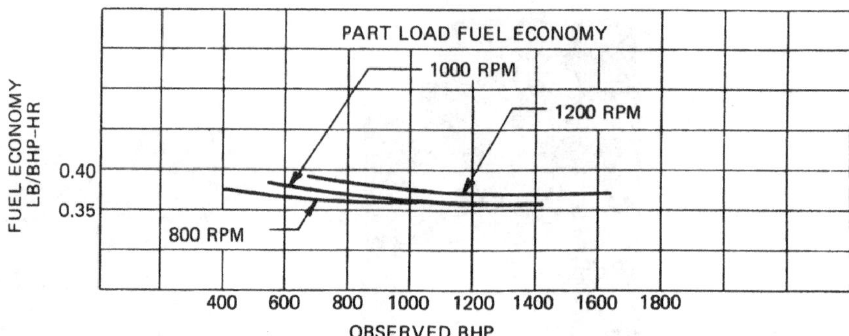

FIGURE 6.92 Fuel consumption rates for the diesel engine. (*Waukesha Engine Division, Dresser Industries, Inc.*)

Diesel Fuel-Oil Specifications. It is important that the fuel oil purchased for use in an engine be as clean and water-free as possible. Dirt in the fuel can clog injector nozzles and ruin the finely machined precision parts in the system. Water in the fuel will accelerate corrosion of these parts. Reputable fuel suppliers deliver clean, moisture-free fuel. Most of the dirt and water in the fuel is introduced through careless handling, inadequate filtration, dirty storage tanks or lines, and poorly fitted tank covers.

There are fuel composition requirements that must be met when purchasing diesel fuel. Table 6.28 lists fuel properties for No. 2 diesel fuel and their limits. Definitions of the more critical properties follow.

TABLE 6.28 Fuel-Oil Recommendations Chart

Fuel-oil physical properties	Limits	ASTM test method
Fuel grade	Diesel no. 2	
API gravity	30 min	D 287
Cetane number	40 min*	D 613
Sulfur, %	0.7 max	D 129
SSU viscosity, at 100°F (37.7°C)	30–50	D 88
Water and sediment, %	0.1	D 96
Pour point, min	10°F (5.5°C) below ambient air	D 97
Carbon residue	0.25%	D 189
Ash, % max.	0.02	D 482
Aklali or mineral acid	Neutral (pH 7)	D 974
Distillation point		D 158
10% min	450°F (232°C)	
50%	475–550°F (246–288°C)	
90% max	675°F (357°C)	
End point max	725°F (385°C)	
Cloud point	†	D 97

* For automatic starting units, a fuel with 50 cetane minimum is recommended.
† Cloud point should not be more than 10°F (5.5°C) above pour point.
Source: Waukesha Engine Division installation manual.

Diesel Fuel Properties

API Gravity. The specific gravity, or density in pounds per gallon.

Ash. The mineral residue in fuel. High ash content leads to excessive oxide buildup in the cylinder and/or injector.

Cetane Number. Ignitability of the fuel. The lower the cetane number, the harder it is to start and run the engine. Low-cetane fuels ignite later and burn more slowly. Explosive detonation could be caused by having excessive fuel in the chamber at the time of ignition.

Cloud and Pour Points. The pour point is the temperature at which the fuel will not flow. The cloud point is the temperature at which the wax crystals separate from the fuel. The cloud point must be no more than 10°F (5.5°C) above the pour point so the wax crystals will not settle out of the fuel and plug the filtration system. The cloud point should also be at least 10°F (5.5°C) below the ambient temperature to allow the fuel to move through the lines.

Distillation Point. Temperature at which certain portions of the fuel will evaporate. The distillation point will vary with the grade of fuel used.

Sulfur. Amount of sulfur residue in the fuel. The sulfur combines both with any moisture in the fuel and the water vapor formed during combustion to form sulfuric acid. This acid can quickly corrode engine parts. The lower the sulfur content of the fuel, the better.

Viscosity. Influences the size of the atomized droplets during injection. Improper viscosity will lead to detonation, power loss, excessive smoke, and unnecessary wear on the injector system.

Fuel Systems

Installation of a diesel engine requires special planning to handle fuel delivery, storage, and piping (Fig. 6.93).

Fuel Tanks. Most diesel engine installations utilize a two-tank system, with both a main storage tank and a day tank.

Day Tanks. A day tank is designed to keep a clean supply of fuel oil close to the engine and to provide an immediate supply of fuel when the engine is started. By locating the day tank close to the engine, the fuel-transfer pump will be able to draw fuel more easily, without having to develop high suction pressures.

The day tank is generally sized to hold enough fuel for several hours of operation or whatever local fire codes allow. It is often a 275-gal (1000-L) standard commercial tank installed at or above floor level.

If the day tank is positioned above the level of the engine fuel-transfer pump, the flow of the fuel will maintain a constant pressure at the fuel-transfer pump inlet.

A positive shutoff should be added just beyond the day tank whenever the tank is above the level of the fuel-injection pump. If the day tank is mounted on a structure that is subject to vibration, a flexible connector should be added between the tank and the fuel piping. (Actually, flexible connections are recommended between an engine and any support system, whether it involves fuel, air, water, or oil.) The weight of both the tank and the fuel must be considered when designing and mounting the tank.

Main Storage Tanks. Spherical or cylindrical tanks should be used for added strength. Avoid square tanks. Main storage tanks are generally large enough to hold a 10-day fuel supply. (Local codes may dictate tank sizes.) If fuel delivery is uncertain due to weather, traffic, or any other reason, the tank size should be increased.

The location of the main storage tank is influenced by the method of delivery. If the fuel oil will be delivered by railroad tank cars, the tank and filler opening should be near the railroad siding. If a truck will be delivering the fuel, the tank and filler opening should be close to the road. Tanks should always be located so as to minimize the length of the fuel lines.

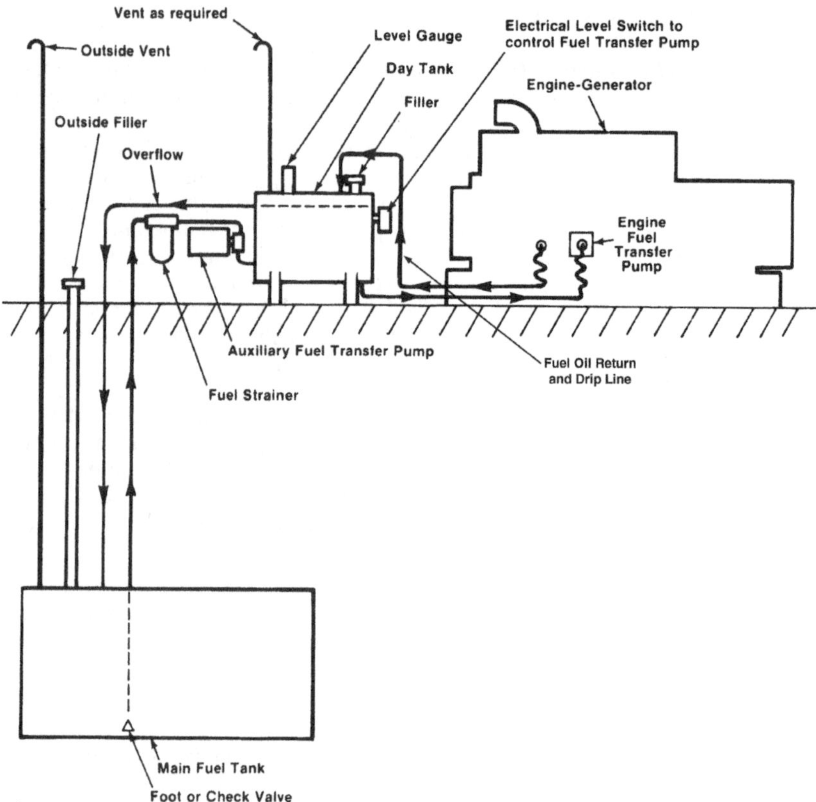

FIGURE 6.93 Typical diesel fuel-oil system. (*Waukesha Engine Division, Dresser Industries, Inc.*)

Underground fuel-oil tank storage is no longer permitted. Above-ground storage does require containment around the tank, and heating may be required in cold climates. As with day tanks, the fuel pickup should never draw from the bottom of the main storage tank. The tank should be installed lower at one end to allow the dirt and water that settle out of the fuel to be drained off or pumped out. Underground tanks should be bottom pumped at least twice each year to remove all accumulated water and sludge. Above-ground tanks are more subject to condensation, so they should be bottom pumped more frequently.

Caution: *Diesel fuel tanks, fittings, and lines should never be made of galvanized steel nor should they be of a zinc alloy material.* The sulfur in the fuel will corrode these metals, gumming up the injection pump and injectors.

Heating lines can be added to warm the fuel and keep it at a temperature at which it can be easily pumped to the engines. The manufacturers of such components should be consulted for further information.

The vent pipe to the outside must have at least a 1-in (2.5-cm) diameter, and an approved flame arrester must be incorporated into the vent.

Strainers and Filters. Strainers and filters are an important part of any diesel fuel system. Without the cleaning action of these components, the dirt and grime in the fuel would destroy the finely machined parts in the injectors and the injection pump.

A filter should be placed just before each meter and each pump. To ensure proper maintenance, it should be located in an easy-to-reach position. Also, try to leave enough room under each filter for a catch basin to avoid messy, dangerous fuel-oil spills.

Separators should be added to a system, particularly at the main storage tank outlet, to remove sediment.

Shutoff Valves. A shutoff valve should be incorporated in the fuel system at the fuel tank outlet and at the point where the fuel line enters the building or engine room and wherever applicable local codes so dictate.

Fuel-Transfer Pumps. Transfer pumps are used to supply fuel to the injection pump, or to raise fuel to a tank or engine at a higher level. Centrifugal pumps cannot be used as transfer pumps because they are not self-priming. Positive-displacement pumps must be used.

Fuel Return Lines. Fuel return lines take the hot excess fuel not used in the engine cycle away from the injector and back to either the fuel storage tank or the day tank. The heat from the excess fuel is dissipated in the tank. ***CAUTION:*** *Never run a fuel return line directly back to the engine fuel supply lines. The fuel will overheat and break down.*

The fuel return lines should always enter the storage or day tank above the highest fuel level expected. This will prevent fuel in the storage tank from running back into the fuel return line.

NATURAL GAS ENGINES

Principles

Gas engines can be two-cycle or four-cycle and either naturally aspirated, turbocharged or intercooled (aftercooled). Gas engines may be stoichiometric (air fuel ratio of approximately 16:1, or rich burning) or lean combustion (air fuel ratio of approximately 30:1, or lean burning). The fuel can be introduced into the intake air by a carburetor or injected into the intake port just ahead of the inlet valve or directly into the combustion chamber.

The combustion chamber can be either open type or divided type (see section on diesel engines).

Performance Characteristics

Like diesel performance, the performance of a gas engine is affected by air temperature, air pressure, and humidity. Correction factors are used to ensure that the power specified will take into account the losses expected with the altitude and temperature conditions anticipated at the site. The rating societies mentioned previously, in the discussion of diesel engine performance, supply methods of correcting power on spark-ignited engines.

The fuel consumption of a gas engine usually is expressed in British thermal units per brake horsepower (kilocalories per metric horsepower) (see Fig. 6.94).

Fuels

A gas engine can be adjusted to accept a wide variety of fuels. Among these are:

1. Pipeline-quality natural gas
2. Digester gas from sewage treatment plants
3. Methane from sanitary landfills (landfill gas)

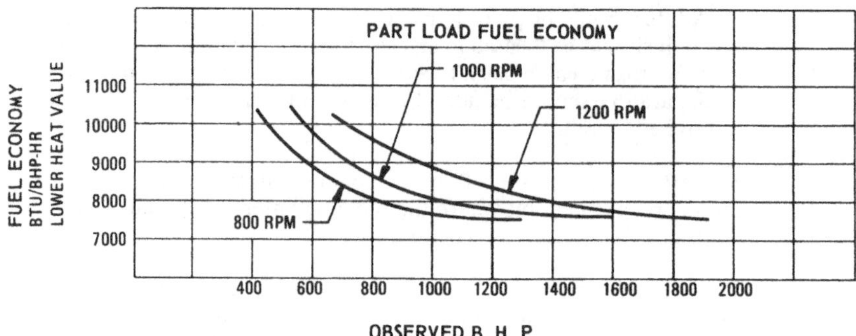

FIGURE 6.94 Fuel consumption rates for the natural gas engine. (*Waukesha Engine Division, Dresser Industries, Inc.*)

4. HD-5 propane and LP gas
5. Pyrolytic gas from hydrocarbon sources
6. Field gas (wellhead gas)

The octane number for various gaseous fuels available for gas engines can be calculated from their known constituents. The heating value of engine fuel can also be obtained with a calorimeter and is nearly always expressed as lower heating value (LHV).

Fuel System. The gas engine can be *naturally aspirated* or *turbocharged*. Since the manifold pressure of the turbocharged engine can be as high as 23 psig (160 kPa), the pressure of the natural gas supply should be in the 30- to 50-psig (200- to 350-kPa) range in order to have the engine operate properly. Conversely, the gas supply to the naturally aspirated engine can be at a pressure as low as 0.5 (3.5 kPa) psig [but is normally preferred at a 5- to 10-psig (3.5- to 7-kPa) span] since the engine manifold operates at atmospheric pressure or less. Turbocharged engines of the draw-through carburetor type are now available for operation on gas pressures as low as 0.5 psig (3.5 kPa). They are designed with special carburetors located on the inlet side of the turbocharger compressor.

The gas pressure regulators to reduce the supply pressures to the required engine carburetor inlet pressures are usually mounted on the engine. Experience has proved that it is best to supply one regulator per carburetor (on dual-carburetor engines), mounted as close to the carburetor as possible. This minimizes line drops and governing instability.

The type and number of gas shutoff devices and pressure-sensing switches are determined by applicable safety codes and control circuit requirements. A low-gas-pressure switch is commonly specified.

SYSTEMS COMMON TO BOTH DIESEL AND NATURAL GAS ENGINES

Lubrication System

Although the lubrication system is one of the simplest systems of the engine, its importance should not be underestimated. It is the most important support system of the engine and cannot be neglected except at the expense of premature engine failure.

The lubrication system of an industrial engine is almost completely assembled before it leaves the factory. On large-displacement engines, the free-standing lube-oil filter (and in some cases the oil cooler) is the only major lubrication component usually shipped free of the engine. (Smaller engines have the oil filter mounted directly on the engine.)

Lubricating Oil Filter Installation. Position the lube-oil filter as close to the engine as possible.

Caution: *Do not put the filter near the exhaust outlet or other places where the temperature could become excessively warm. Excessive heat will speed oil deterioration and will also create a fire hazard in the event of an oil spill or line rupture.*

It is important to use pipe of adequate size between the engine and lube-oil filter in order to maintain the proper oil pressure to the engine. Consult the engine supplier for recommendations. If they are not galvanized or zinc-plated, black iron or steel pipes should be used to carry oil. ***Caution:*** *After welding, flush pipes with muriatic acid to remove all welding scale, and rinse thoroughly to neutralize the acid and ensure clean piping.*

Flexible Connections. Flexible connections designed to handle hot lubricating oil at pressures up to 100 lb/in^2 (690 kPa) should be used between the engine and the free-standing oil filter. Position the connections as close to the engine as possible. Supports should be added under the oil filter lines to minimize vibration and prevent breakage.

Lubricating Oil Recommendations. Lube-oil selection is the responsibility of the engine operator and the oil supplier. The refiner is responsible for the performance of the lubricant.

Most engine warranties are limited to the repair or replacement of parts that fail due to defective material or workmanship during the warranty period. These warranties do not include satisfactory performance of lube oil. That is considered the responsibility of the oil supplier.

Assistance in lubricant selection can be obtained from a publication of the Engine Manufacturer's Association, One Illinois Center, 111 E. Wacker Drive, Chicago, IL 60601. This book, *EMA Lubrication Oils Data Book for Heavy-Duty Automotive and Industrial Engines,* has a table of lubricants and their performance grades. Consult the engine manufacturers concerning oil recommendations for their various engines. It is common to find that different oils are used for gas and diesel engines, as well as for different models of each type.

Accessories available for lube-oil systems are such items as flowmeters and oil-level regulators. These can be unit-mounted and automatically add oil as it is consumed as well as measure and record the quantity of oil consumed. It is also possible to obtain engine-mounted switches for low and high engine lube-oil levels to signal a warning and/or cause engine shutdown.

On larger engines, use an air-motor- or electric-motor-driven prelube pump to fill and pressurize the lube-oil system before cranking and operating the engine. In addition, use a lube-oil heater to keep the oil warm and in condition for positive lubrication of vital areas on start-up wherever low ambient temperature may be encountered.

Cooling Systems

Cooling systems in liquid-cooled engines are affected by the mineral content and corrosiveness of water put into the system—be it cooled by radiator, heat exchanger, or standpipe. In all cases, recommended practice is to use treated water so as to minimize long-term effects on the engine.

Radiator Cooling. The most common cooling arrangement is that of the stationary engine with a unit-mounted radiator. In this case, the cooling fan is belt-driven from the front of the engine. Usually a pusher fan is used to prevent hot air from being drawn over the set and its operator. The radiator has to be sized for the maximum expected ambient operating temperature (including room heat sources such as generator losses), the maximum rated engine horsepower or kilowatt load, and the type of coolant used (ethylene glycol). The radiator has to cool the heat rejected to the engine-jacket water, the heat rejected from the engine lube oil, and (if applicable) the heat rejected from the intercooler circuit. Some radiators are dual-core units, with one core for the jacket-water circuit and a second core for the intercooler oil-cooler circuit. This is required on natural gas engines to keep the water entering the intercooler at 130°F (54°C) maximum.

A radiator-cooled unit (Fig. 6.95) is self-contained and, therefore, adaptable to mobile installations and/or relocation. Since the radiator is sealed, it does not waste water and does not contaminate drain water. However, the horsepower required to drive the fan is lost for other uses, and the relatively large radiator airflow both in and out of the engine room has to be handled judiciously. Other installation items are louvers (fixed or variable), vertical discharge air scoops, and air discharge duct adaptors.

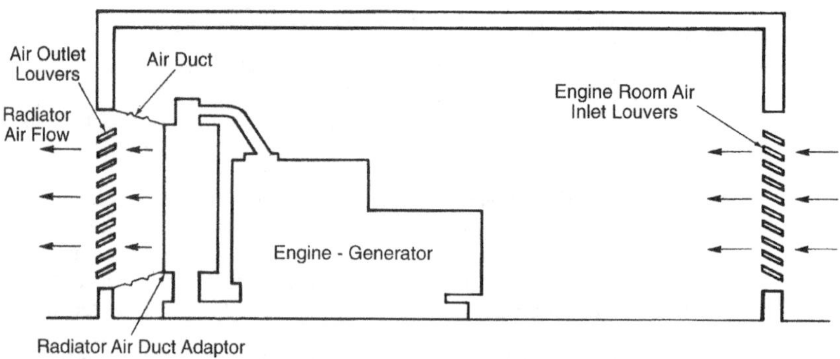

FIGURE 6.95 Radiator cooling. (*Waukesha Engine Division, Dresser Industries, Inc.*)

Cooling by use of a remote-mounted radiator is also common. This system is quite flexible since the radiator can be mounted outside, which reduces engine-room airflow. Also, the core can be horizontal, which makes the radiator insensitive to wind direction. Roof mounting of the radiator is quite common; if the roof height causes an increase over the recommended jacket-water pressure, a hot well (Fig. 6.96) and auxiliary pump can be added. Other installa-

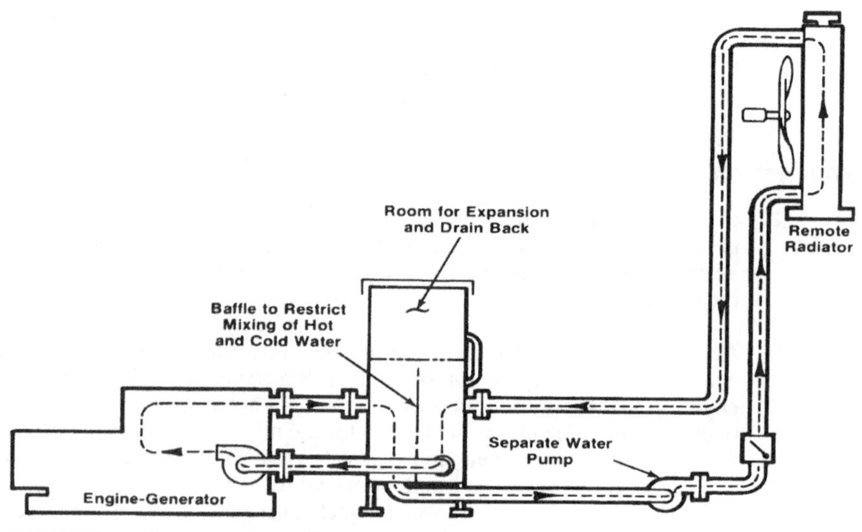

FIGURE 6.96 Hot well cooling. (*Waukesha Engine Division, Dresser Industries, Inc.*)

tion considerations are automatic louvers, vertical versus horizontal core, remote surge tank, piping sizes to control line-pressure drops, and sound-limit requirements.

Heat-Exchanger Cooling. In applications where cool water is plentiful, or where it is desirable to preheat a water supply for other processes, engines using heat-exchanger (Fig. 6.97) cooling are used. If conservation of raw water is important, raw water can be controlled thermostatically to limit its flow to the minimum required.

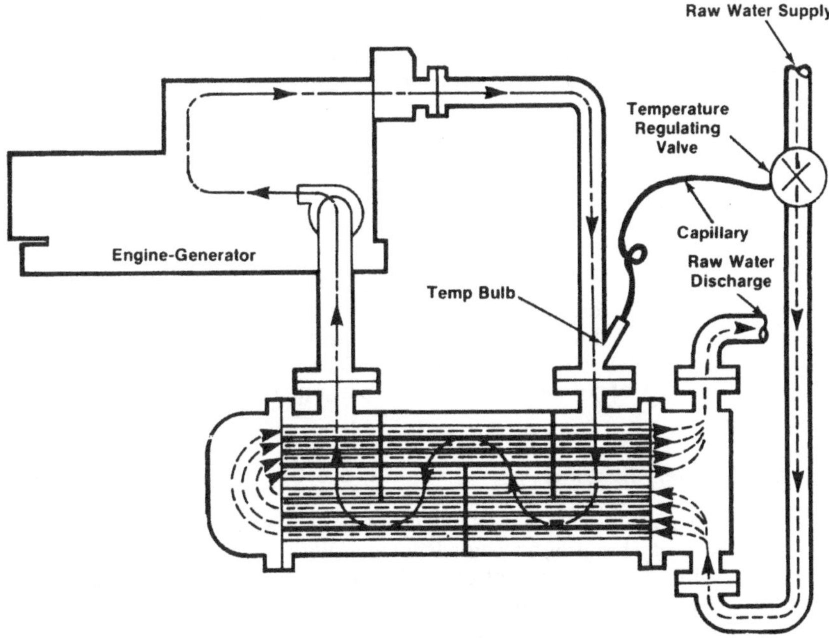

FIGURE 6.97 Heat-exchanger cooling. (*Waukesha Engine Division, Dresser Industries, Inc.*)

The raw-water pressures are not imposed on the engine jacket; engine cooling is not greatly affected by ambient temperatures; engine-room airflow requirements are much lower than unit-mounted radiator applications; and engine-jacket-water heat can be used constructively.

City-Water Cooling or Standpipe Cooling. Occasionally used with standby-service engine-generator systems, city-water cooling is simply the blending of cold raw water into the jacket water during operation through the use of water-pressure-regulating and thermostatic-control valves and diverting the excess to waste (Fig. 6.98). ***Caution:*** *Although inexpensive to install, this method introduces minerals and corrosive elements into the engine water jacket and is dependent on a municipal water supply which could become inoperative in time of disaster.)*

Ebullition (Ebullient) Cooling. Engines and engine-generator systems, intended for applications where recovery of the heat in jacket water (for utility heating, air conditioning through absorption chillers, or processing needs) may be of genuine economic value, are sometimes equipped with ebullition cooling (Fig. 6.99).

Ebullient cooling is a process whereby jacket water is circulated through the engine at near-boiling temperature, vaporized at the top of the engine, and then condensed and recirculated. Cooling is accomplished by capturing the heat of vaporization for process or other uses.

By omitting the jacket-water pump and taking advantage of the natural circulation of water with temperature differences, the engine may be operated at jacket-water temperatures

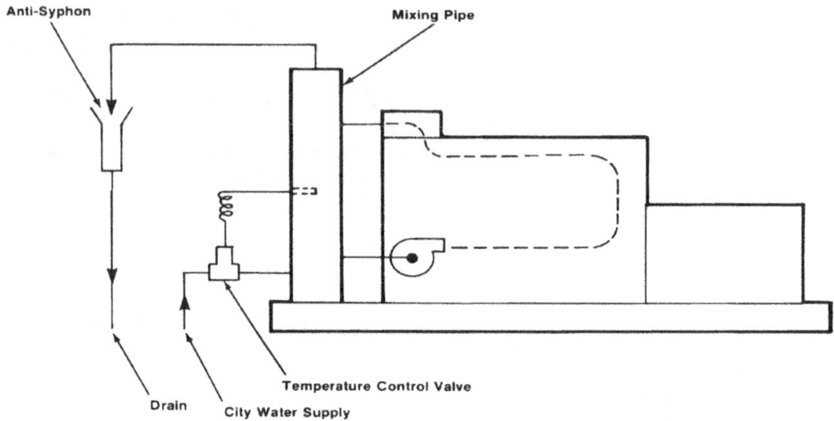

FIGURE 6.98 City-water cooling. (*Waukesha Engine Division, Dresser Industries, Inc.*)

well into the boiling range [15 psig, 250°F max (103 kPag, 127°C)]. A considerable amount of heat in the form of low-pressure steam may be recovered. When ebullition cooling is being considered, however, the engine manufacturer should be consulted since some engine designs are not offered with this option. In general, the two big advantages of ebullient cooling are the recovery of normally wasted heat energy and optimum use of water.

Waste-Heat Recovery Systems

Where the waste heat of an engine can be used, electric power generation with waste-heat recovery has pronounced advantages. With a conventional engine generator, something on the order of 32 percent of the fuel energy is converted to useful electric power, while the remaining 68 percent is wasted to the atmosphere. With an engine jacket-water and exhaust-heat recovery system, the recovered heat energy is about 50 percent of the fuel burned, reducing waste heat by about 18 percent.

A typical heat balance on engines of identical size for both gas and diesel engine-generators, is shown in Table 6.29.

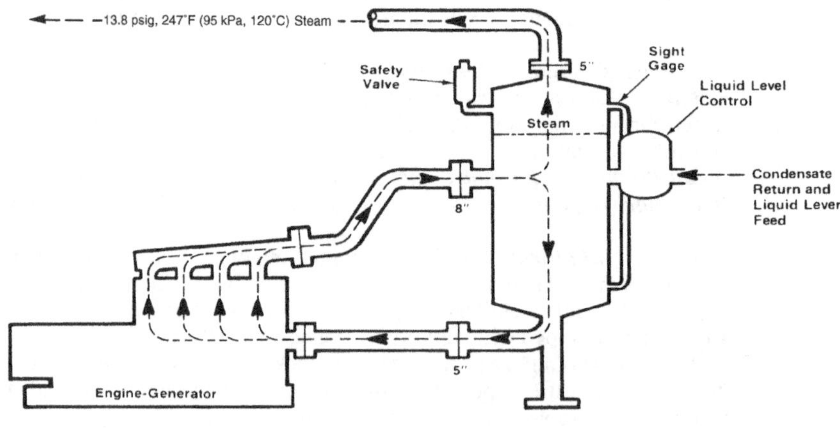

FIGURE 6.99 Ebullition cooling. (*Waukesha Engine Division, Dresser Industries, Inc.*)

TABLE 6.29 Typical Heat Balance

	Conventional cooling system		Cooling system with engine jacket and exhaust heat recovery		
500-kW natural gas engine generator*					
Electric power	30%			30%	
Jacket-water heat	38%	70% wasted		38%	54% recoverable
Exhaust heat	24%		Exh recoverable	16%	
			Exh lost	8%	16% wasted
Radiated heat lost to atmosphere	8%			8%	
	100%			100%	
500-kW diesel engine generator†					
Electric power	35%			35%	
Jacket water	32%	65% wasted		32%	48% recoverable
Exhaust heat	24%		Exh recoverable	16%	
			Exh lost	8%	17% wasted
Radiated heat lost to atmosphere	9%			9%	
	100%			100%	

* Based on 7900 Btu (bhp)(h) at rated load (LHV) and 95 percent generator efficiency.

† Based on 0.380 lb/(bhp)(h) at rated load [18,200 Btu/(lb)(LHV)] and 95 percent generator efficiency.

Figure 6.100 shows a simple representative arrangement for engine heat recovery. Engine jacket water and exhaust are passed through a first heat exchanger, where the water temperature is increased as the exhaust gas is cooled. It then passes through a second heat exchanger, where the building or process water is heated and the jacket water is returned to the engine for cooling.

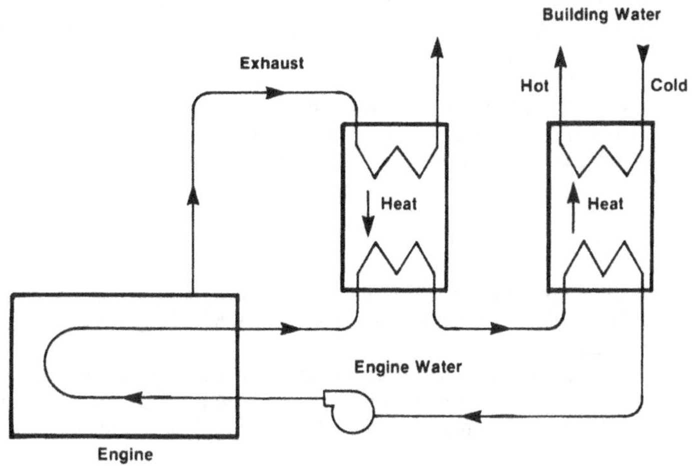

FIGURE 6.100 Representative heat-recovery arrangement. (*Waukesha Engine Division, Dresser Industries, Inc.*)

The amount of waste heat readily recoverable in a representative system of this kind is approximately 71 Btu/(bhp)(min) for a gas engine and 56 Btu/(bhp)(min) for a diesel engine.

Exhaust System

Every engine manufacturer specifies a maximum exhaust backpressure limit at the exhaust outlet of the engine (Fig. 6.101). Typical values can range from 12 to 20 in (30 to 52 cm) H_2O depending on the engine models. To exceed these values could result in engine damage, interfere with good long-life engine operation, or detract from good performance.

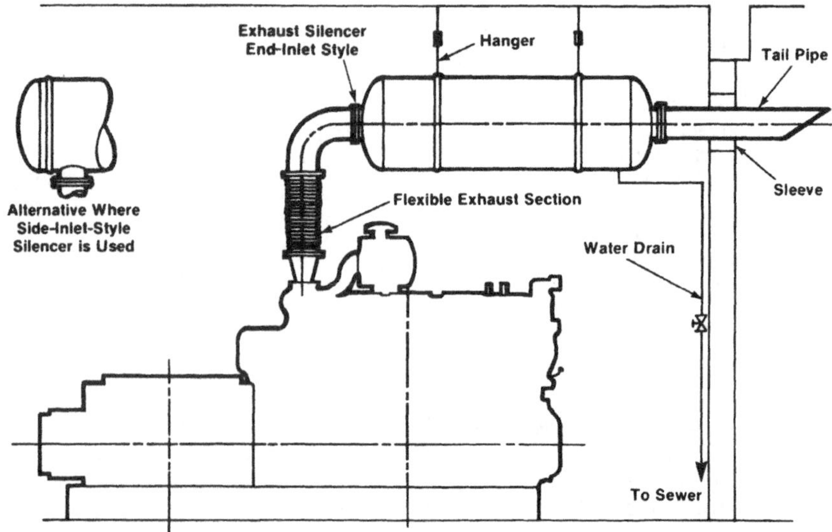

FIGURE 6.101 Horizontal exhaust schematic. (*Waukesha Engine Division, Dresser Industries, Inc.*)

Since sound levels are becoming more critical daily, silencers are selected to give adequate noise attenuation. Unfortunately, this usually increases the exhaust flow pressure or drop through the silencer.

Silencers and exhaust pipes in engine rooms are usually insulated to reduce heat radiation and noise. A flexible exhaust connection is mounted at the engine to isolate engine vibration; also, the exhaust line is sloped away from the engine with a drain to avoid the accumulation of condensation. Additional flexible connectors are necessary to compensate for thermal growth.

Emissions

Exhaust products of either diesel or natural gas engines contain oxides of nitrogen (NO and NO_2), carbon monoxide (CO), hydrocarbons (HC), and sulfur oxides (SO_2 and SO_3) if the fuel contains sulfur.

Emissions of these products are subject to regulation under federal, state, and local laws. The plant engineer should be aware of the regulations covering a specific site and, if necessary, consult the engine manufacturer for emission information.

Starting Systems

The most frequently used starter systems are electric motors and air starters. The electric starters are usually 12- or 24-V dc cranking motors. The higher voltage is preferred on larger engines because of the high power required for annunciator controls and/or power circuit-breaker tripping. Adequately sized and maintained lead-acid batteries are the most practical and economical solution for the battery source today. However, nickel-cadmium and other types of battery requiring less maintenance are often specified.

All the batteries require an adequately sized battery charger to keep the battery fully charged both with the engine operating and when down. In low ambient temperatures, both batteries and engines must be kept warm to ensure good starting and operation.

Many starting systems use one or more air starters per engine. The common air motor will crank an engine with an applied air pressure of between 90 and 150 psig. Starters can be controlled through a hand or a solenoid valve both manually or automatically. They also have an inline oiling system to ensure motor life.

The air starting system usually includes a remote high-pressure air receiver [approximately 250 psig (1.7 mPa)] with an air regulator to drop pressure to the cranking motor. Again the pipe losses and line sizes have to be considered to ensure good cranking. The compressors to pressurize the receivers are normally driven by electric motors and are automatically started and stopped by a pressure-switch sensor at the air receiver.

Engine-Speed Governing System

The engine governor is the most important device with respect to engine performance. On smaller units, engine-mounted mechanical governors provide adequate speed and frequency control with approximately 3 to 10 percent steady-state speed regulation from no load to full load.

Hydraulic isochronous governors are also available for medium to large engines, and these provide excellent speed control with adjustable speed regulation of 0 to 5 percent. In engine-generator applications, this permits running isochronously as a single unit or with adjustable regulation for multiunit manual parallel operation and load sharing.

More precise governing is available on both small and large engines with electric governors. This type depends on engine speed with a magnetic pulse pickup on the flywheel ring gear teeth, and provides speed and load control with an electric or electrohydraulic actuator connected to the engine throttle or fuel injection pump.

The electronic control of the governor is solid-state and is usually mounted in the control panel or generator switch gear.

Electric governing permits operation of multiengine generator units in parallel isochronously, that is, at constant frequency under steady-state operation regardless of load, with each engine taking an equal share of the overall load.

In addition, the electric governor is extremely fast with respect to transient load response and will handle up to 50 percent sudden load changes with a speed deviation of approximately 3 percent and recovery within approximately 2 s for diesel engines, and with a speed deviation of approximately 4 to 5 percent and recovery within approximately 3 s for carburetted gas engines. Response varies by engine type: naturally aspirated, turbocharged-intercooled stoichiometric, or lean combustion.

Where automatic operation includes coming on-line in parallel with other engine generators or the utility, the electric governor with isochronous load-sharing control is desirable. In conjunction with the proper switchboard relaying, units can be added or removed on a load-sharing basis or on load control basis as the situation may demand. A good 24-V dc battery system is required as a power supply to the electric governor.

INSTALLATION AND MAINTENANCE OF DIESEL AND NATURAL GAS ENGINES

The engine installation should be designed with maintenance requirements in mind. Serviceable components such as filters, fittings, and connections should be readily accessible to the engine operator. Routine engine maintenance will not be neglected if the operator has easy access to the engine.

Sufficient service space must be present on all sides of the engine to allow for removal of even the largest engine components. An overhead crane should be incorporated into the engine-room design to assist the mechanic-operator in removing heavy assemblies. Air-line connections will be necessary for air power tools, as will scaffolding for servicing the engine.

Ventilation

In engine-generator installations sufficient airflow must be provided into the engine room for ventilation and combustion air. It is also good practice to calculate the amount of heat transferred to the room air (i.e., engine and generator radiator heat, plus any other heat sources) to determine the temperature rise of the engine-room air. In many cases it is necessary to increase the engine-room airflow to maintain reasonable operating temperatures.

The following are *general rule of thumb values* that assume the only radiating heat source in the engine room is the engine-generator set. For greater accuracy, an independent engineering study should be made covering the following points:

Cubic feet per minute of air required to limit air room temperature rise to 18°F (10°C), over normal ambient = 45 × kilowatt rating

Cubic feet per minute of combustion air required = 3.5 × kilowatt rating for diesel engines

Cubic feet per minute of combustion air required = 2.4 × kilowatt rating for gas engines

The total air requirement equals the sum of the cubic feet per minute of combustion air plus the cubic feet per minute required to limit the room temperature rise.

Other ventilation considerations are filters for sandy or dusty areas and louvered openings at both inlet and outlet air openings. The louvers can be motor-operated and temperature-controlled.

Cooling System

Potential problems with the engine cooling system can be avoided if the following considerations are incorporated into the design and installation of the cooling system.

Excessive fittings, elbows, and connectors in the system piping will impede coolant flow. Use of fittings should be kept to a minimum.

An expansion-tank balance line should be incorporated into the cooling system, running to the suction side of the water pump. This balance line will maintain a net positive suction at the inlet of the pump and reduce the possibility of air locks and cavitational erosion.

All filters, fill points, and bleed cocks should be installed in an easy-to-reach location.

Place the radiator away from a wall or any other obstruction that causes air recirculation or restricts airflow. These obstructions would also include any dirt source, vehicle travel path, air-conditioning units, or exhaust stacks and chimneys. Remember that the radiator must be in a location where it can be cleaned and serviced.

In installations where gaseous or LP fuels are used, keep all floor drains and service trenches out of the engine enclosure. LP and some constituents of natural gas can be heavier than air and will quickly flow into such low spots, creating a fire hazard.

Exhaust System

Plan the exhaust system so that the gases are expelled to a safe outside area, consistent with all local building and environmental codes. Do not discharge gases near windows, ventilation shafts, or air inlets. The exhaust outlet must be designed to keep out water, dust, and dirt.

To avoid metal stress and turbocharger damage, support the exhaust system independently, keeping the weight of the piping off the engine. Roller-type supports and flexible exhaust connections should be used to absorb thermal expansion. (If overhead cranes and hoists are used in the engine room, the exhaust-system piping may have to be supported from below.)

A condensate trap and drain should be designed into the exhaust system. The drain should be in an easy-to-reach location.

If the exhaust systems of more than one engine are to be connected to a common exhaust, the engine manufacturer should be consulted beforehand. Exhaust-system backflow (common in such connections) could result in an engine that is not running.

Exhaust-system backpressure should be checked periodically. The backpressure must fall within the limits established by the engine manufacturer.

Air Induction

As with other engine systems, accessibility is the key to air-induction system maintenance. The filter element should be positioned so that it can be easily removed and replaced. The filter should always be positioned at the entrance to the air induction system; when combustion air is ducted in from outside the engine room, the filter should be at the opening to the piping. All systems should be equipped with a restriction indicator to show excess pressure drop due to filter-element plugging.

Always locate the air inlet away from concentrations of dirt, exhaust stacks, fuel tanks, tank vents, and stockpiles of chemicals and industrial wastes. Try to duct air to the engine from a cool, dry, dirt-free area. The ambient temperature at the air inlet location should ideally be 60 to 90°F (15 to 32°C).

Run all air ducts away from engine exhaust pipes, heating lines, or other hot areas. Remember to allow clearance for overhead lifts and cranes when air ducts run through the engine room.

Air ducts should be thoroughly sealed to avoid drawing dirty air in behind the filter. The ducting must be checked periodically for leaks.

Air-system ducting should be seamless, welded-seam, or PVC piping. Flanged fittings with gaskets, not threaded connections, should be used between pipe sections to avoid restrictions in the system. The best ducting system is as short and straight as possible, using long-radius bends and low-restriction fittings. Never allow air-duct restriction to exceed 2 in (50.8 mm) of water column. Air-ducting systems must be leak-free under vacuum conditions.

Engine Alignment

The alignment of the engine mount and the alignment between the engine and the driven equipment is critical to long engine life. Alignment should be checked periodically according to the manufacturer's recommendations.

SELECTION OF ENGINES

There are several factors to consider in the selection of an engine.

Type of Fuel

The type of fuel chosen will depend to a large degree on availability, price, local building codes, and pollution restrictions. For instance, in some communities, natural gas is not available for industrial use, while in others, local building codes will prohibit storage of diesel fuel. The price of one fuel may preclude its use in comparison with another type of fuel. Moreover, emissions restrictions may have a bearing on what type of fuel may be used. All of these are factors in choosing an engine.

Horsepower Load and Speed

The load and speed of the equipment to which the engine will be coupled are key considerations in engine selection. The aim is to match the prime mover to the power required at a speed which will be compatible with the equipment to be driven, while maintaining optimum efficiency.

Duty Cycle

The duty cycle should be examined to determine whether continuous or intermittent operation of the equipment is required, because this affects engine selection.

Brake Mean Effective Pressure (BMEP)

This is important because the higher the figure, the greater the chance for higher stress on working parts of the engine. This could mean higher maintenance costs and earlier need for rebuilding.

Fuel and Oil Consumption

With the cost of petroleum or fossil-based fuels tending to rise over time, this is becoming an increasingly significant factor in the selection of an engine.

Torsional Compatibility

A torsional analysis should be conducted to determine whether the engine and driven equipment are compatible with respect to operating stress in the shaft system.

SELECTION OF ENGINE GENERATORS

Both diesel and gas engines perform very satisfactorily for in-plant power generation when properly applied and installed. The varied nature of these applications depends on factors such as:

1. The specific characteristics of the plant's product and its production process as well as its sensitivity to power failure
2. The plant power needs in terms of availability of purchased power and the reliability of normal power sources
3. Economic considerations in terms of location, cost of purchased power, demand charges, and equipment
4. Economic considerations of in-plant generated power in terms of capital outlay, operation, maintenance, and fuel cost

Once the decision has been made that in-plant engine-driven power generation will be used to provide electric power, whether on the basis of prime, cogeneration, peaking service, emergency standby, or a combination of any of these, there are a number of considerations to be investigated.

Purpose

The purpose of the installation should be stated and careful thought given to definition of the electrical load requirements.

- Will the power generators supply all of the plant load or only a part of the total?
- What is the power factor characteristic of the load? (Engine generators are normally rated at 0.8 power factor lagging.)
- Will there be any attempt to control or improve power factor?
- What is the largest block kilowatt loading anticipated and what large motors are to be started and on what basis—across the line, reduced voltage?
- Are computer loads, SCRs, inverters, x-ray, or heavy welding equipment involved?
- What are the code requirements? NFDA? CSA? Others?

Amount of Equipment

Selection of the size and number of units to handle the load demands thoughtful consideration.

Emergency standby protection (see also Chap. 5.12) may be provided by a single unit rated to protect a known segment of load that can be isolated with a two-way transfer (normal emergency) switch (Fig. 6.102). The size of this type of unit may range from 50 kW or under to approximately 1500 kW (Fig. 6.103).

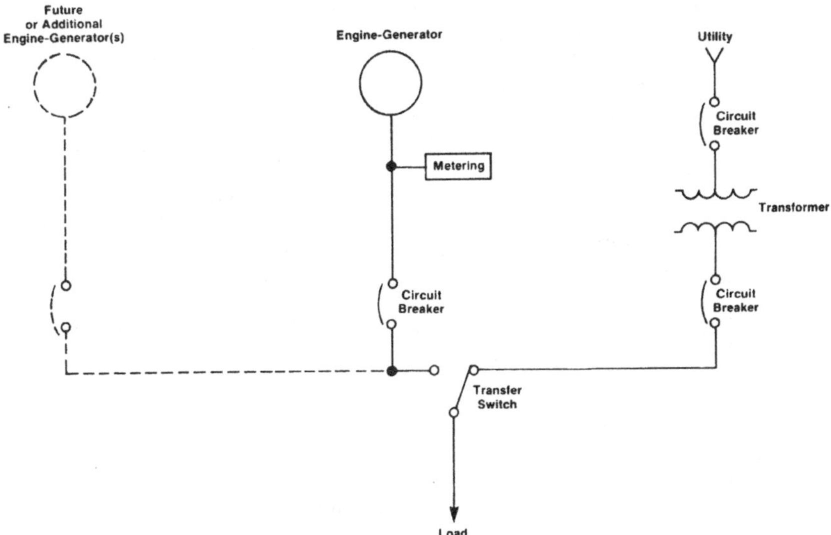

FIGURE 6.102 Normal emergency standby without utility parallel. (*Waukesha Engine Division, Dresser Industries, Inc.*)

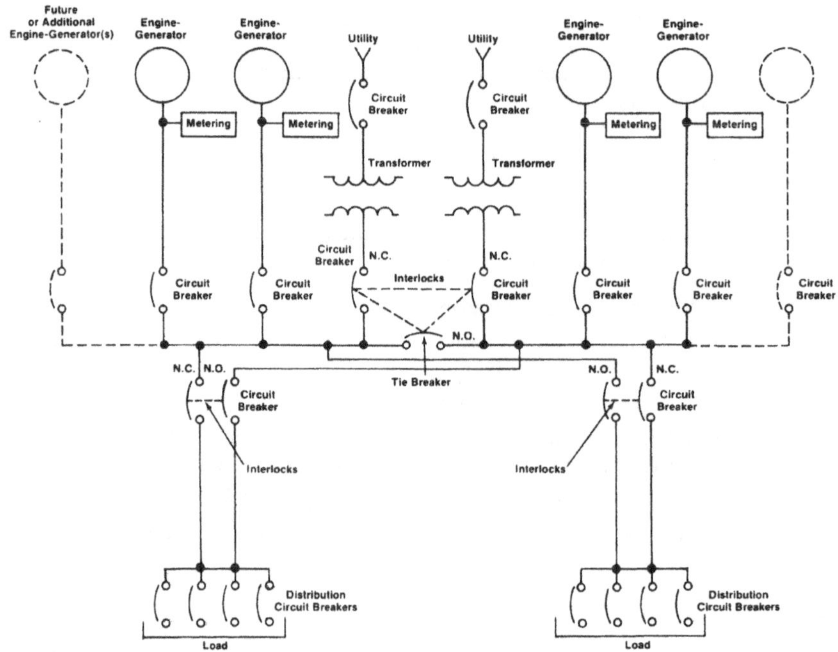

FIGURE 6.103 Normal emergency standby with peaking capability or selection of critical loads for limited kilowatt generation (without utility parallel). (*Waukesha Engine Division, Dresser Industries, Inc.*)

For loads well over the 1000-kW range, it is customary to provide multiple engine-generator units, sometimes as many as eight units operating in parallel on a common or split bus for total load capability to 10,000 kW and beyond.

Generally available emergency standby units are rated at 1800 r/min synchronous speed for 60-Hz service up to approximately 1000 kW and at 1200 r/min up to 2000 kW, while prime power units are available in both 1200 and 900 r/min synchronous speed for 60-Hz service up to 3000 kW rating per unit.

The same ratings are available for 50-Hz service at synchronous speeds of 1500, 1000, and 750 r/min. Much larger units and units with slower speeds are also available.

When selecting size and number of units, it is most helpful to study the plant electrical load profile over the course of a year. If this is not available, a profile based on connected loads with anticipated load and diversity factors can be developed. Single-unit emergency standby generators are often applied with load factors of only 50 to 75 percent of rating. Multiunit prime power installations favor load factors in the 75 to 90 percent range, as this selection results in optimum fuel economy and overall operating efficiencies. Future growth of plant electrical loads must be anticipated.

Fuel Selection

As with engines in direct-drive applications, this may be determined by availability and cost.

Natural Gas. This is an excellent engine fuel and is often available on an uninterrupted basis. Where necessary, LPG or propane can be stored as backup or secondary fuel supply.

Sewage treatment plants are increasingly turning to power generation, utilizing the process-waste gas from digesters to fuel the engine. Coal gasification, wood-chip processing, and numerous other process-waste gases may be suitable for burning in a gas engine but should never be used without consulting the engine manufacturer for specific approval.

No. 2D Diesel Fuel Oil. This is commonly used in small and medium-size engines. Large diesel engines have the capability of burning heavier fuels such as No. 4D and heavier residual fuels; these may require special handling, heating, centrifuging, and filtering. Diesel fuel has the advantage of large-volume on-site storage capability.

Location of Equipment

Locating the engine generator(s) in the plant may involve many practical considerations. Often it is advantageous to locate them near other heavy plant equipment such as boilers, large air conditioners, or compressors. Standby power generation equipment is normally automatically controlled, and by locating it close to other equipment requiring periodic operator attention, it may also get the attention it deserves.

Figure 6.104 shows typical envelope dimensions for both gas and diesel generator sets. Minimum clearance on all sides is also charted and can be used for preliminary space esti-

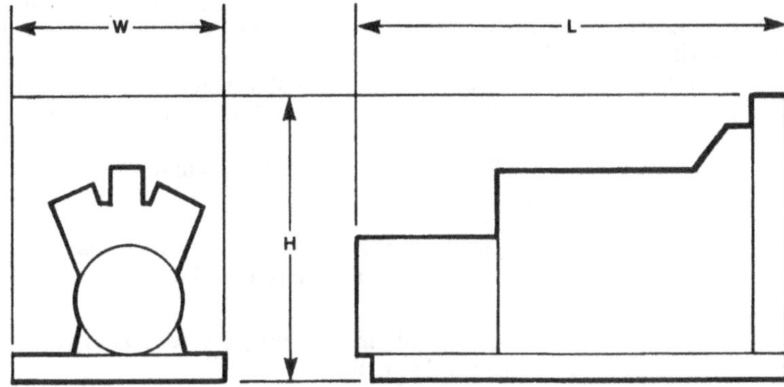

(Length, Width, and Height Are In Inches)

kW	RPM	L	W	H	WT, LB
50	1800	88	30	50	2,300
100	1800	110	36	55	4,200
150	1800	114	36	60	4,500
200	1800	114	42	60	6,000
250	1800	114	42	60	7,000
300	1800	140	56	90	13,000
400	1800	160	69	90	16,000
500	1800	180	72	90	19,000
600	1200	210	74	121	32,000
700	1200	216	84	124	32,500
800	1200	222	84	124	33,000
900	1200	222	96	124	35,000
1000	1200	222	96	127	36,000
*1200	1200	186	80	98	38,000
*1500	1200	262	78	113	42,000
*2000	900	330	72	139	45,000
*2500	900	363	72	139	52,000
*3000	900	363	72	139	52,000

* - Indicates less cooling radiator

FIGURE 6.104 Approximate space envelope for gas and diesel engine generators. (*Waukesha Engine Division, Dresser Industries, Inc.*)

mates. For final determinations, always work from the equipment manufacturer's specific space requirements for optimum clearances both around the unit and overhead for efficient maintenance, major overhaul, and repair.

Cooling

Adequate cooling of both gas and diesel engines is essential to good performance and equipment life.

On emergency standby units, it is common to waste the recoverable heat. On prime power applications, there is increasing attention given to heat recovery from the engine jacket water and exhaust for process use in the plant.

The Electric Control System

This system, together with its attendant complexities, requires determination of the power-generation voltage and control voltages.

Decisions as to the mode of operation, such as manual, semiautomatic, automatic, or unattended operation, must be made. Adequate metering, monitoring, readout, and display decisions must be made. Independent engine generation on isolated plant loads or multiunit engine generation in parallel operation against plant loads must be determined. Will the engine generators always operate isolated from the utility or will there be occasions to parallel with the utility's power?

Both peaking service and cogeneration may call for parallel operation with the utility, and this requires review with the power company at the planning stage.

Electric Controls. After the kilowatt size of each unit is selected, the generated output voltage should be considered.

1. Voltages of 208/120 wye or 240/139 wye can be economically utilized in the lower kilowatt ranges since no transformers should be required in plants (facilities) using only these voltages.
2. Voltage of 480/277 wye is the most commonly generated voltage with minimum capital cost in both generator and switchgear.
3. Voltage of 4160/2400 wye is the common generated voltage for larger plants and more extensive power distribution systems. This voltage level increases the cost of the engine-generator unit and its associated generator switchgear, but may effect savings in the plant distribution system.

A further consideration is the engine generator and panel control voltage. Generally, this is supplied from a dc battery source which allows control and operation of the engine and breakers without ac voltage from either utility or generators. Engine-generator sets in the 150- to 1500-kW range require a 24-V dc battery source to start and run. An alternate is 100 to 150 psig air pressure to start and 24, 48, or 125 V dc to run and control. For large, multiunit engine-generator sets and their associated switchgear, 48 or 125 V dc is recommended for the most reliable circuit-breaker operations.

Metering. When 24-V dc engine-starting batteries are used for control purposes, the control should be run from the set of batteries with the highest voltage to minimize the effects of voltage dip when starting engines or tripping circuit breakers.

Single-unit metering would include at least a frequency meter, voltmeter, ammeter, and wattmeter; for emergency service, 3½-in (9-cm) ± 2 percent accuracy panel meters are both generally adequate. Prime power service with long-term continuous operation, often with multiple units in parallel, demands 4½-in (11.5-cm) switchboard meters of 1 percent accuracy. These sets should also include an indicating wattmeter for each engine generator. Additional meters that may be considered are those reading kilowatthours, kilovoltamperes reactive (kvar), and power factor. Of increasing interest on prime power multiunit applications are recording meters on the output bus for voltage, frequency, and kilowatts and/or kilowatthours. More recently available are digital readout meters providing greater accuracy, but with limitations with respect to indication of transients, which is an important characteristic of engine-generator sets.

Protection. The electrical protection of the generator system can vary from simple molded-case circuit breakers to insulated-case breakers and to metal-frame air circuit breakers. Both insulated-case and air breakers are available with either fixed mounting or drawout provisions. For optimum flexibility in long-term use, the drawout breaker is superior and recommended. Since engine generators have a limited short-circuit capability, care should be taken in breaker selection for both generator and distribution to achieve proper selective trip coordination. Where selective trip coordination is required, the generator must be supplied with three per unit short-circuit sustain capability for 10 s.

Additional protection to consider is generator differential and ground fault. Usually, generator-differential protection is used on 4160-V systems where generator internal damage requires early detection to protect capital investment.

Ground-fault protection also protects equipment and personnel and is less expensive to install. Most circuit-breaker manufacturers provide optional ground-fault tripping integral with the breaker. Many different types of ground-fault protection are available, and consideration must be given to the selection and specification of protection consistent with the existing plant grounding system and specific application. More sophisticated methods of fault detection are available, but are usually associated with megawatt generator sizes since protection cost is very high.

When engine generators are operated in parallel with the utility, certain cautions must be observed. The utility, being an "infinite" source, can cause severe damage to the engine generator should the utility's protection system open and then reclose (out of phase). Conversely, should the upstream utility protection system open and then not reclose, the generators can feed a fault from the reverse direction, which can be hazardous to both personnel and equipment. Whenever parallel operation with the utility is being considered, review the relaying used with the local utility company to protect against the occurrences mentioned.

Many breaker and/or transfer switch configurations are possible, each with its own advantages and disadvantages dependent upon application (Figs. 6.105 to 6.107).

Engine Control. After the power distribution format is known, the engine-control (starting and stopping) mode(s) must be considered. For single emergency standby units the unit should start, come up to rated frequency and voltage, and provide power whenever the utility source fails. Normal engine protection would include shutdown on engine low oil pressure, high water temperature, overspeed, and failure to start. Additional features available and actually included for hospital duty are warnings before shutdown on low oil pressure and high water temperature. On prime power units, additional considerations are warning and/or shutdown on high oil temperature, low oil level, high vibration, and engine overload.

Some extremely critical applications may call for the generator set not to stop for any reason, as nuisance shutdown cannot be tolerated and the possibility of generator set failure or destruction may be less costly than the process it controls.

Engine control is only a part of the overall system control. System control is offered in varying degrees of complexity. Single standby units have very simple system controls, while prime power multiunit applications must determine the quantity of units required to satisfy power demand while also controlling plant loads to prevent high peak loading or overloading. When operating in parallel with the utility, controls can be programmed to accept: (1) fixed utility supply with load variations picked up by the engine-generator sets, (2) variable utility supply with peak demands picked up by the generator sets, or (3) fixed generator set output with load variations being picked up by the utility (this mode includes capability of supplying power to the utility over low-plant-load conditions).

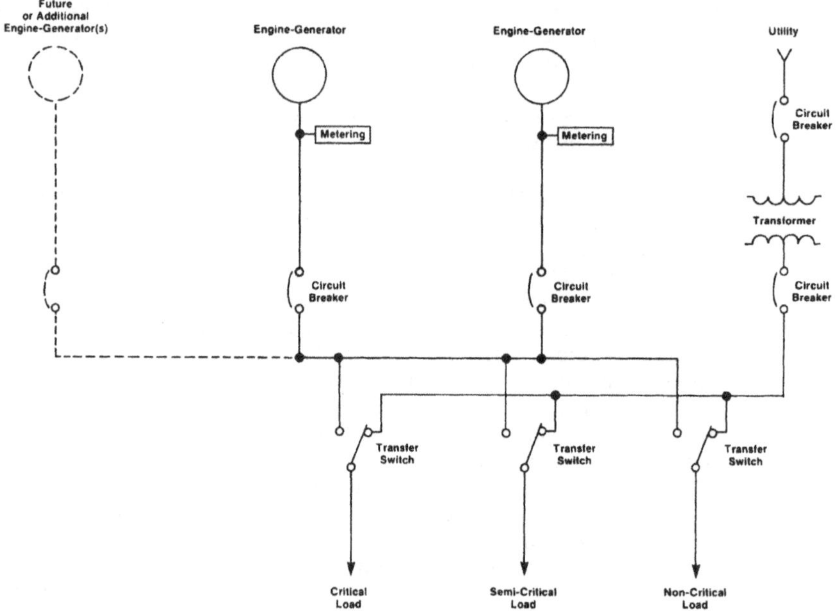

FIGURE 6.105 Multiunit standby or cogeneration in parallel with utility or in independent operation. (*Waukesha Engine Division, Dresser Industries, Inc.*)

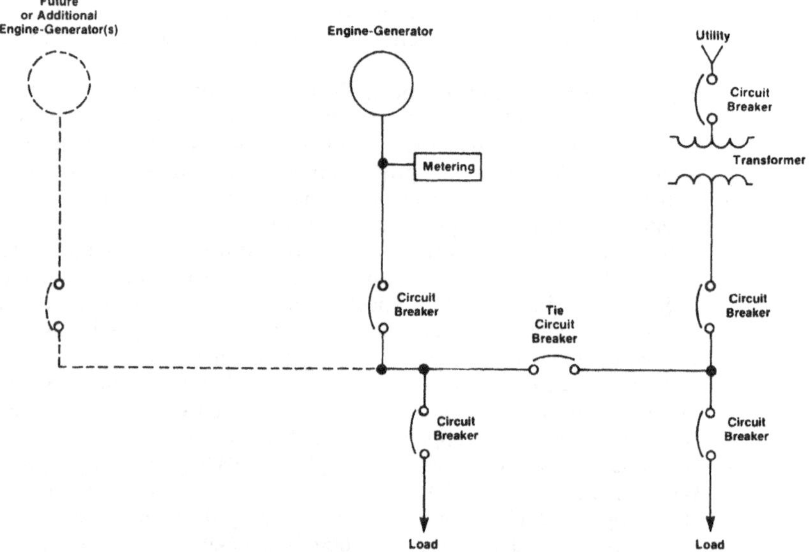

FIGURE 6.106 Isolated loading for peak shaving without parallel utility (breakers interlocked) or peak shaving and parallel operation with utility. (*Waukesha Engine Division, Dresser Industries, Inc.*)

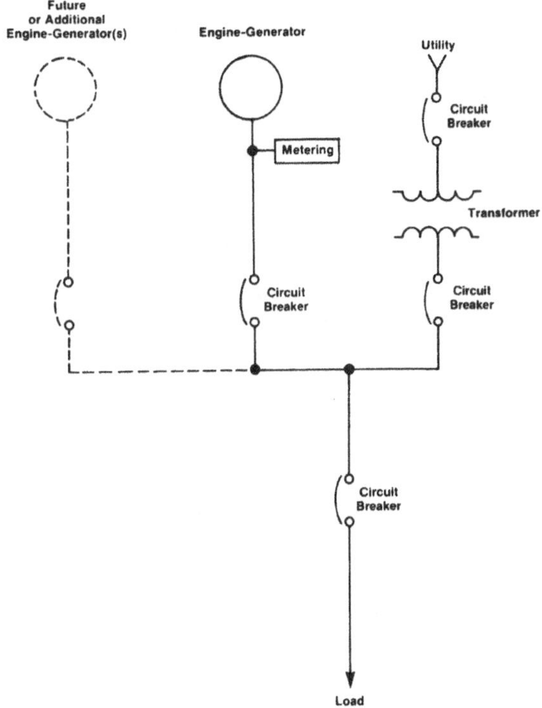

FIGURE 6.107 Isolated loading for standby (breakers interlocked) or peak shaving and cogeneration with parallel operation with utility. (*Waukesha Engine Division, Dresser Industries, Inc.*)

Generally speaking, the control logic can do almost anything the engineer may require in a given situation; however, simplicity is the key to both cost and long-term performance. Simpler systems are easier to maintain, whereas complex systems may require a wider variety of engineering skills to maintain and troubleshoot problems.

The system control can start and stop generator sets on the basis of actual kilowatt demand; an alternative simple approach would be to know the load profile and manually operate or utilize real-time devices to start and stop units.

Today solid-state programmable controllers, which can be used in place of the more conventional relay logic in control circuits, are available. Some of these controllers are versatile and allow control logic to be changed simply. Many allow computer input-output links for record-keeping purposes.

These versatile programmable controllers provide flexibility and a ready means of tailoring sequence, time, and load functions to a particular plant's needs. Selected performance indicators can be linked to provide readouts and printouts as may be required for efficient operation and recording purposes.

REFERENCES

1. *Standard Practices for Law and Medium Speed Stationary Diesel and Gas Engines,* Diesel Engine Manufacturers Association, Cleveland, Ohio, 1972.
2. Gunther, F. J.: "Engines," in *Pump Handbook,* Igor J., Karassik, William C. Krutzsch, Warren H. Fraser, and Joseph P. Messina (eds.), sec. 6.1.3, McGraw-Hill, New York, 1986.

APPENDIX
METRIC CONVERSION TABLE

Editor's note: Metric conversions, in terms of SI (International System) units, are given in most of the text, graphs, and tables in this handbook. However, practical considerations precluded direct conversion in every instance.

This appendix presents a table of convenient conversion factors into SI units from U.S. Customary and non-SI metric units for frequently used physical quantities involved in plant engineering.

CONVERSION TO SI UNITS: LISTING BY PHYSICAL QUANTITY

The first two digits of each numerical entry represent a power of 10. For example, the entry "–02 2.54" expresses the fact that 1 inch = 2.54×10^{-2} meter.

To convert from	to	Multiply by
	Acceleration	
foot/second^2 (ft/s^2)	meter/second^2 (m/s^2)	–01 3.048
inch/second^2 (in/s^2)	m/s^2	–02 2.54
	Area	
acre	meter^2 (m^2)	+03 4.046
circular mil (cmil)	m^2	–10 5.067
foot^2 (ft^2)	m^2	–02 9.290
hectare (ha)	m^2	+04 1.00
inch^2 (in^2)	m^2	–04 6.4512
mile^2, U.S. statute (mi^2)	m^2	+06 2.589
yard^2 (yd^2)	m^2	–01 8.361
	Density	
gram/centimeter^3 (g/cm^3)	kilogram/meter^3 (kg/m^3)	+03 1.00
lbm/inch^3 (lb/in^3)	kg/m^3	+04 2.768
lbm/foot^3 (lb/ft^3)	kg/m^3	+01 1.602
slug/foot^3 (slug/ft^3)	kg/m^3	+02 5.154

To convert from	to	Multiply by
Energy		
British thermal unit, ISO/TC 12 (Btu)	joule (J)	+03 1.055
Btu, International Steam Table	J	+03 1.055
Btu, mean	J	+03 1.056
Btu, thermochemical	J	+03 1.054
Btu, 39°F	J	+03 1.060
Btu, 60°F	J	+03 1.055
calorie, International Steam Table (cal)	J	+00 4.187
calorie, mean (cal)	J	+00 4.190
calorie, thermochemical (cal)	J	+00 4.184
calorie, 15°C (cal)	J	+00 4.186
calorie, 20°C (cal)	J	+00 4.182
calorie, kilogram, International Steam Table (cal)	J	+03 4.187
calorie, kilogram, mean (cal)	J	+03 4.190
calorie, kilogram, thermochemical (cal)	J	+03 4.184
foot-lbf (ft · lb)	J	+00 1.356
kilowatthour (kWh)	J	+06 3.60
ton (nuclear equivalent of TNT)	J	+09 4.20
watthour (Wh)	J	+03 3.60
Energy/area/time		
Btu, thermochemical/foot^2 second (Btu/ft^2 · s)	watt/meter^2 (W/m^2)	+04 1.135
Btu, thermochemical/foot^2 minute (Btu/ft^2 · min)	W/m^2	+02 1.891
Btu, thermochemical/foot^2 hour (Btu/ft^2 · h)	W/m^2	+00 3.152
Btu, thermochemical/inch^2 second (Btu/in^2 · s)	W/m^2	+06 1.634
calorie, thermochemical/cm^2 minute (cal/cm^2 · min)	W/m^2	+02 6.973
watt/centimeter^2 (W/cm^2)	W/m^2	+04 1.00
Force		
dyne (dyn)	newton (N)	−05 1.00
kilogram force (kgf)	N	+00 9.307
pound force, avoirdupois (lbf)	N	+00 4.448
ounce force, avoirdupois (ozf)	N	−01 2.730
Length		
caliber	meter (m)	−04 2.54
chain, surveyor's or gunter	m	+01 2.012
chain, engineer's or ramsden	m	+01 3.048
cubit	m	−01 4.572
fathom	m	+00 1.829
foot (ft)	m	−01 3.048
foot, U.S. survey (ft)	m	−01 3.048
furlong	m	+02 2.012
inch (in)	m	−02 2.54
link, engineer's or ramsden	m	−01 3.048
link, surveyor's or gunter	m	−01 2.012
micrometer (μm)	m	−06 1.00
mil	m	−05 2.54
mile, U.S. statute (mi)	m	+03 1.609
mile, international nautical (mi)	m	+03 1.852
mile, U.S. nautical (mi)	m	+03 1.852
yard (yd)	m	−01 9.144

To convert from	to	Multiply by
Mass		
carat, metric	kilogram (kg)	–04 2.00
ounce mass, avoirdupois (oz)	kg	–02 2.835
ounce mass, troy (oz t) or apothecary (oz ap)	kg	–02 3.110
pound mass, avoirdupois (lb)	kg	–01 4.536
pound mass, troy (lb t) or apothecary (lb ap)	kg	–01 3.782
ton, long	kg	+03 1.016
ton, metric	kg	+03 1.00
ton, short, 2000 lb	kg	+02 9.027
Power		
Btu, thermochemical/second (Btu/s)	watt (W)	+03 1.054
Btu, thermochemical/minute (Btu/min)	W	+01 1.757
calorie, thermochemical/second (cal/s)	W	+00 4.184
calorie, thermochemical/minute (cal/min)	W	–02 6.973
foot-lbf/hour (ft · lb/h)	W	–04 3.766
foot-lbf/minute (ft · lb/min)	W	–02 2.260
foot-lbf/second (ft · lb/s)	W	+00 1.356
horsepower (hp; 550 ft · lb/s)	W	+02 7.457
horsepower, boiler (hp)	W	+03 9.809
horsepower, electric (hp)	W	+02 7.46
horsepower, metric (hp)	W	+02 7.855
horsepower, U.K. (hp)	W	+02 7.457
horsepower, water (hp)	W	+02 7.460
kilocalorie, thermochemical/minute (kcal/min)	W	+01 6.973
kilocalorie, thermochemical/second (kcal/s)	W	+03 4.184
Pressure*		
atmosphere (atm)	newton/meter^2 (N/m^2)	+05 1.013
bar	N/m^2	+05 1.00
centimeter of mercury (cmHg), 0°C	N/m^2	+03 1.838
centimeter of water, 4°C	N/m^2	+01 9.806
dyne/centimeter^2 (dyn/cm^2)	N/m^2	–01 1.00
foot of water, 39.2°F	N/m^2	+03 2.989
inch of mercury (inHg), 32°F	N/m^2	+03 3.386
inch of mercury (inHg), 60°F	N/m^2	+03 3.377
inch of water, 39.2°F	N/m^2	+02 2.490
inch of water, 60°F	N/m^2	+02 2.488
kgf/centimeter^2 (kg/cm^2)	N/m^2	+04 9.807
kgf/meter^2 (kg/m^2)	N/m^2	+00 9.807
lbf/foot^2 (lb/ft^2)	N/m^2	+01 4.788
lbf/inch^2 (lb/in^2; psi)	N/m^2	+03 6.895
millibar	N/m^2	+02 1.00
millimeter of mercury (mmHg), 0°C	N/m^2	+02 1.333
pascal (Pa)	N/m^2	+00 1.00
psi (lb/in^2)	N/m^2	+03 6.895
Speed		
foot/hour (ft/h)	meter/second (m/s)	–05 8.467
foot/minute (ft/min)	m/s	–03 5.08
foot/second (ft/s; fps)	m/s	–01 3.049
inch/second (in/s)	m/s	–02 2.54

*The pascal is preferred here for most purposes. It is less cumbersome to use.

To convert from	to	Multiply by
	Speed (*continued*)	
kilometer/hour (km/h; kph)	m/s	–01 2.778
knot, international	m/s	–01 5.144
mile/hour, U.S. statute (mi/h; mph)	m/s	–01 4.470
mile/minute, U.S. statute (mi/min)	m/s	+01 2.682
mile/second, U.S. statute (mi/s)	m/s	+03 1.609
	Temperature	
degree Celsius (°C)	kelvin (K)	K = °C + 273.15
degree Fahrenheit (°F) 459.67)	K	K = 5/9 (°F
degree Fahrenheit (°F)	degree Celsius (°C)	°C = 5/9 (°F – 32)
degree Rankine (°R)	K	K = 5/9 (°R)
	Viscosity	
centistoke (cSt)	meter²/second (m²/s)	–05 1.00
stoke (St)	m²/s	–04 1.00
foot²/second (ft²/s)	m²/s	–02 9.290
centipoise (cP)	newton second/meter² (N · s/m²)	–03 1.00
lbm/foot second (lb/ft · s)	N · s/m²	+00 1.488
lbf second/foot² (lb · s/ft²)	N · s/m²	+01 4.788
poise (P)	N · s/m²	–01 1.00
poundal second/foot²	N · s/m²	+00 1.488
slug/foot second (slug/ft · s)	N · s/m²	+01 4.788
	Volume	
acre foot (ac · ft)	meter³ (m³)	+03 1.283
barrel, petroleum (bbl; 42 gal)	m³	–01 1.590
board foot	m³	–03 2.360
cord	m³	+00 3.625
dram, U.S. fluid (fl dr)	m³	–06 3.697
fluid ounce, U.S. (fl oz)	m³	–05 2.957
foot³	m³	–02 2.832
gallon, U.S. dry (gal)	m³	–03 4.405
gallon, U.S, liquid (gal)	m³	–03 3.785
inch³ (in³)	m³	–05 1.639
liter (L)	m³	–03 1.00
ounce, U.S. fluid (fl oz)	m³	–05 2.957
pint, U.S. dry (pt)	m³	–04 5.506
pint, U.S. liquid (pt)	m³	–04 4.732
quart, U.S. dry (qt)	m³	–03 1.101
quart, U.S. liquid (qt)	m³	–04 9.466
stere	m³	+00 1.00
yard³ (yd³)	m³	–01 7.646

INDEX

C

D

G

H

N

O

P

T

U

ABOUT THE EDITOR-IN-CHIEF

Robert C. Rosaler, P.E., is an engineering and management consultant with more than 40 years of experience as an engineering executive in both large and small corporations. He is also editor-in-chief of *HVAC Systems and Components Handbook* and *HVAC Maintenance and Operations Handbook,* both published by McGraw-Hill.

Environmental Management

Blackwell Readers on the Natural Environment

Published
The Human Impact Reader
Edited by Andrew Goudie

The Environmental Management Reader
Edited by Lewis A. Owen and Tim Unwin